# HEAT EXCHANGERS

# HEAT EXCHANGERS

## Thermal-Hydraulic Fundamentals and Design

ADVANCED STUDY INSTITUTE BOOK

*Edited by*

### S. Kakaç
Middle East Technical University
Ankara, Turkey

### A. E. Bergles
Iowa State University
Ames, Iowa, U.S.A.

### F. Mayinger
Institute of Thermodynamics (A)
Technical University of Munich
Munich, F.R. Germany

**HEMISPHERE PUBLISHING CORPORATION**
Washington   New York   London

**McGRAW-HILL BOOK COMPANY**

New York   St. Louis   San Francisco   Auckland   Bogotá
Hamburg   Johannesburg   London   Madrid   Mexico   Montreal
New Delhi   Panama   Paris   São Paulo   Singapore   Sydney
Tokyo   Toronto

**HEAT EXCHANGERS: Thermal-Hydraulic Fundamentals and Design**

Copyright © 1981 by Hemisphere Publishing Corporation. All rights reserved. Printed in the United States of America. No part of this publication may be reproduced, stored in a retrieval system, or transmitted, in any form or by any means, electronic, mechanical, photocopying, recording, or otherwise, without the prior written permission of the publisher.

1 2 3 4 5 6 7 8 9 0   B C B C   8 9 8 7 6 5 4 3 2 1

**Library of Congress Cataloging in Publication Data**

Main entry under title:

Heat exchangers.

"Advanced Study Institute book."
Papers from the NATO Advanced Study Institute on "Heat Exchangers—Thermal-Hydraulic Fundamentals and Design," held in Istanbul, Turkey, Aug. 4-15, 1980.
  Bibliography: p.
  Includes index.
  1. Heat exchangers—Congresses. I. Kakaç, S. (Sadik)  II. Bergles, A. E., date.  Mayinger, F., date.  IV. NATO Advanced Study Institute.  V. NATO Advanced Study Institute on "Heat Exchangers—Thermal-Hydraulic Fundamentals and Design" (1980 : Istanbul, Turkey)
TJ263.H42          621.402'2          81-4106
ISBN 0-07-033284-3                    AACR2

# Contents

Preface    ix

## INTRODUCTION

Introduction to Heat Exchangers: Thermal and Hydraulic Fundamentals and Design—Preview of the Contributions
   S. Kakaç    3

Classification of Heat Exchangers
   R. K. Shah    9

## THERMAL-HYDRAULIC FUNDAMENTALS: SINGLE-PHASE

Air-cooled Heat Exchangers
   A. Žukauskas    49

Total and Local Heat Transfer and Pressure Drop of Staggered and In-Line Tube Bundles
   E. Achenbach    85

Thermal Hydraulic Investigation of Yawed Tube Bundle Heat Exchangers
   H.-G. Groehn    97

Compact Heat Exchangers
   R. K. Shah    111

Heat Transfer in Concurrent Flow Double Pipe Heat Exchangers
   M. D. Mikhailov and B. K. Shishedjiev    153

Plate Heat Exchangers
   K. J. Bell    165

Heat Transfer between Packed, Agitated and Fluidized Beds and Submerged Surfaces
   E. U. Schlünder    177

Fluidized Bed Heat Exchangers: Theory and Practice
   B. I. Kılkış    209

## THERMAL-HYDRAULIC FUNDAMENTALS: TWO-PHASE

Boiling and Evaporation-I
   J. G. Collier    235

Boiling and Evaporation-II
   J. G. Collier    261

Condensers: Basic Heat Transfer and Fluid Flow
  D. Butterworth   **289**

Heat Transfer with Natural Convection Boiling in Multicomponent Mixtures
  K. Stephan   **315**

Heat Transfer with Condensation in Multicomponent Mixtures
  K. Stephan   **337**

Film Evaporation and Condensation in Desalination
  S. Sideman   **357**

Experimental and Theoretical Investigation of Deposition Motion of Liquid Droplets in Two-Phase Flow through a Vertical Tube
  E. N. Ganić and K. Mastanaiah   **377**

# RADIATIVE HEAT TRANSFER IN HEAT EXCHANGERS

Radiative Heat Transfer in Heat Exchangers
  M. N. Özışık   **399**

# HEAT EXCHANGER DESIGN: RATING, SIZING, AND OPTIMIZATION

Heat Exchangers—Basic Methods
  W. M. Rohsenow   **429**

Heat Exchanger Design Methodology—An Overview
  R. K. Shah   **455**

Finite Element Analysis of Heat Exchangers
  M. D. Mikhailov and M. N. Özışık   **461**

Exact Explicit Equations for Some Two- and Three-Pass Cross-Flow Heat Exchangers Effectiveness
  B. S. Bačlić and D. D. Gvozdenac   **481**

Compact Heat Exchanger Design Procedures
  R. K. Shah   **495**

Construction Features of Shell and Tube Heat Exchangers
  K. J. Bell   **537**

Preliminary Design of Shell and Tube Heat Exchangers
  K. J. Bell   **559**

Delaware Method for Shell Side Design
  K. J. Bell   **581**

The Design of Boilers
  J. G. Collier   **619**

Condensers: Thermohydraulic Design
    D. Butterworth    **647**

Some Design Aspects of Thin Film Evaporators/Condensers in Water Desalinization
    S. Sideman    **681**

On the Heat Transfer and Fluid Flow in Falling Film Shell-and-Tube Evaporators
    E. N. Ganić    **705**

Thermal Design Theory for Regenerators
    R. K. Shah    **721**

Analytical and Experimental Thermal-Hydraulic Optimization of Finned Tube Bundle Heat Exchangers for Dry Cooling Towers
    O. Fischer and K. H. Bucher    **765**

Offshore Heat Exchanger Design Practice throughout a Project
    M. A. Taylor    **779**

Problems Facing the Designer of Offshore Heat Exchangers
    M. A. Taylor    **789**

A Survey of Dry Cooling Tower Technology for Power Generation Application
    A. Montakhab    **799**

## ADVANCED SURFACE SELECTION AND PERFORMANCE

Principles of Heat Transfer Augmentation. I: Single-Phase Heat Transfer
    A. E. Bergles    **819**

The Development of a High Performance Heat Transfer Surface
    C. M. B. Russell    **843**

Principles of Heat Transfer Augmentation. II: Two-Phase Heat Transfer
    A. E. Bergles    **857**

Applications of Heat Transfer Augmentation
    A. E. Bergles    **883**

## OPERATIONAL CONSIDERATIONS

The Transient Response of Heat Exchangers
    R. K. Shah    **915**

Dynamic Behaviour of Double-Phase-Change Heat Exchangers
    F. Mayinger and M. Schult    **955**

Vibration in Heat Exchangers
    F. Mayinger and H. G. Gross    981

Heat Exchanger Fouling and Corrosion
    J. G. Collier    999

Fouling of Heat Transfer Surfaces
    J. D. R. S. Pinheiro    1013

Fouling of Heat Transfer Equipment: Summary Review
    M. G. O'Callaghan    1037

Dynamic Response of Tubes in Cross Flow Subjected to Axial Forces near the Buckling Load
    E. P. Heinecke    1049

Why Laminar Flow Heat Exchangers Can Perform Poorly
    W. M. Rohsenow    1057

Operational Problems Encountered in Heat Exchangers Used for Geothermal Energy Utilization
    O. Yeşin and Z. Kutan    1073

## PROBLEMS AND PROSPECTS FOR THE FUTURE

Unresolved Problems in Heat Exchanger Design
    D. Butterworth    1087

Suggestions for Further Research and Development on Heat Exchangers
    E. Achenbach, K. J. Bell, A. E. Bergles, D. Butterworth, J. G. Collier,
    E. N. Ganić, F. Mayinger, M. D. Mikhailov, M. N. Özışık,
    C. M. B. Russell, R. K. Shah, K. Stephan, G. Yadigaroglu,
    and A. A. Žukauskas    1107

Index    1113

# Preface

This volume represents the archival record of the NATO Advanced Study Institute on "Heat Exchangers: Thermal-Hydraulic Fundamentals and Design," held in Istanbul, Turkey, August 4–15, 1980. This is the third NATO ASI on the subject of heat transfer. The previous ASI's in 1978 and 1976 are recorded in volumes published in 1979 and 1977, respectively, by Hemisphere Publishing Corporation, Washington, D.C.

The NATO ASI's are intended to be high level teaching activities in scientific and technical areas of current concern. Certainly, the subject of heat exchangers needs no justification in this regard. The worldwide demand for efficient, reliable, and economical heat exchange equipment is accelerating rapidly, particularly in large-scale power and process industry facilities. Advanced analytical and experimental techniques are being directed toward heat exchanger design and development. In spite of this increasing sophistication, the final product requires a liberal dose of old-fashioned engineering. It is only natural, then, that the lectures and research reports presented in Istanbul address both the science and the art of heat exchangers.

The invited lecturers reviewed the current state-of-knowledge of thermal-hydraulic fundamentals of heat exchangers. The invited lecturers were: K. J. Bell, Oklahoma State University, Stillwater, Oklahoma, U.S.A.; A. E. Bergles, Iowa State University, Ames, Iowa, U.S.A.; D. Butterworth, U.K. Atomic Energy Authority, Harwell, U.K.; J. G. Collier, U.K. Atomic Energy Authroity, Harwell, U.K.: S. Kakaç, Middle East Technical University, Ankara, Turkey; F. Mayinger, University of Hannover, Hannover, F.R. Germany; M.D. Mikhailov, Applied Mathematics Center, Sofia, Bulgaria; M. N. Özişik, North Carolina State University, Raleigh, North Carolina, U.S.A.; W. M. Rohsenow, Massachusetts Institute of Technology, Cambridge, Massachusetts, U.S.A.; C. M. B. Russell, The Lummus Co., London, U.K.; E. U. Schlünder, University of Karlsruhe, Karlsruhe, F.R. Germany; R. K. Shah, General Motors Corp., Harrison Radiator Division, Lockport, New York, U.S.A.; S. Sideman, Technion, Haifa, Israel; R. I. Soloukhin, B.S.S.R. Academy of Sciences, Minsk, U.S.S.R.; K. Stephan, University of Stuttgart, Stuttgart, F.R. Germany; M. A. Taylor, British National Oil Co., London, U.K.; A. A. Žukauskas, Lithuanian Academy of Sciences, Vilnius, U.S.S.R. Specially edited versions of their lectures will be found in the volume. In addition, thirteen research reports were contributed from different countries.

Initiators of any institute scheduled for a particular place and time suffer constraints that do not apply to compliers of a comprehensive handbook. In spite of these inevitable limitations, the assembly of material here has exceeded, in breadth and coherence, the highest hopes of the institute organizers. This is due in no small part to the unusual willingness of the lecturers and contributors to adapt their work to the aims set by the organizers.

The material in the volume is arranged in eight parts. Only a brief summary of the contents is required as the first lecture gives a detailed preview of the lectures and contributed papers in this volume. After a comprehensive classification of heat exchangers, the thermal-hydraulic fundamentals—both single-phase and two-phase—are reviewed. The emphasis is on flow and geometrical situations found in the major types of heat exchangers. Heat exchanger design is then considered from the points of view of rating, sizing, and optimization. The current status regarding advanced surfaces and other heat transfer augmentation techniques is reviewed and applications are discussed. Important operating considerations, such as transient behavior, vibration, and fouling, are discussed. Finally, problem areas and priority research topics are suggested.

We hope that this volume will continue the major objective of the institute—to provide an international forum for the dissemination of information on the thermal-hydraulic fundamentals and design of heat exchangers—by serving as a reference for engineers and for specialized courses on the subject. Perhaps the volume will stimulate development of new advanced undergraduate, graduate, and continuing education courses in this area. We appreciate the cooperation of Hemisphere Publishing Corporation in making the proceedings widely available.

The sponsorship of the NATO Scientific Affairs Division and the additional support of the Scientific and Technical Research Council of Turkey and the Middle East Technical University are gratefully acknowledged. Thanks are also due to the Nuclear Engineering Department and staff of Bogazici University for their cooperation in hosting the Institute.

Mr. J. G. Collier's collaboration on the organizing committee is gratefully acknowledged. The editors wish to extend a word of appreciation to the Turkish industrial firms, Gamma Industrial Plants Inc., Alarko Holding, Selnikel, Üntes, Teba, Sungurlar and Profilo Electrical Equipment Inc, and to the Turkish Society of Thermal Technology and Sciences for their support.

Our special gratitude goes to Drs. F. Arınç, F. Borak, B. Kılkış, Y. Yener, A. O. Yeşin, H. Yüncü and Messrs. E. Paykoç and I. Yegen for their invaluable efforts in making the Institute a success and to Mrs. A. Emrin Çiçek for her efficient secretarial work.

A world of appreciation is also due the members of the Programme Committee, Session Chairmen, and Co-Chairmen for their efforts in programming, organizing, and in expediting the technical sessions.

Much of the planning of this NATO ASI was carried out at the Institut für Verfahrenstechnik der Universität Hannover. A. E. Bergles wishes to acknowledge the support for ASI planning and preparing of lectures given by the Alexander von Humboldt Foundation during 1979-1980.

Finally, our heartfelt thanks to all invited lecturers and authors, who provided the substance of the institute, and the participants for their attendance, questions, and comments.

*S. Kakaç*
*A. E. Bergles*
*F. Mayinger*

# INTRODUCTION

# Introduction to Heat Exchangers: Thermal and Hydraulic Fundamentals and Design—Preview of the Contributions

**S. KAKAÇ**
Middle East Technical University
Ankara, Turkey

A heat exchanger is a device which provides for transfer of thermal energy between fluids at different temperatures. Heat exchanger applications are important in an extremely wide range of industrial plants. Heat exchangers are used in the process, power, automative, air conditioning, refrigeration, cryogenics, heat recovery, manufacturing industries; and they are key components of many products available in the marketplace.

During the Institute almost every facet of the subject of heat exchangers was considered. This brief lecture introduces some of the topics which will be considered in detail in the following lectures.

Starting with a definition, heat exchangers are classified according to the transfer processes, surface compactness, flow arrangements, heat transfer mechanisms, number of fluids and construction features.

Air coolers which use atmospheric air as the coolant are widely used in industry. Lectures on air-cooled heat exchangers review the various schemes and types of air-cooled heat exchangers. Heat transfer and fluid flow friction characteristics of different configurations of smooth, rough and finned banks of tubes in cross-flow of air are examined in detail. The study covers the range of Reynolds numbers from 1 to $4 \times 10^6$, which characterize different flow regimes. Analytical expressions and monograms on heat transfer and fluid flow friction of various banks of tubes in cross-flow are given.

Cross-flow heat exchangers applied to nuclear gas-cooled reactors operate at high Reynolds numbers. Similar to a single circular cylinder in cross-flow there exists a critical Reynolds number beyond which the heat transfer and pressure drop change their dependence upon the Reynolds number. The magnitude of the critical Reynolds number is influenced by the surface roughness of the tubes. Total and local heat transfer and pressure drop of staggered and in-line tubes bundles with highly smoothed surfaces are presented.

Thermal hydraulic investigation of a large number of yawed in-line and staggered tube arrangements with different pitches is also presented.

Compact heat exchangers are presented. The surface basic characteristics of compact heat exchanger surfaces for single phase convection are discussed. Starting with the definitions of important dimensional and dimensionless groups, experimental methods are described to obtain heat transfer and fluid flow friction characteristics simply referred to as the surface basic characteristics. Theoretical solutions and correlations for simple geometries are presented.

Next, plate heat exchangers are presented. The three types of plate heat exchangers are briefly surveyed and their major construction features described. The plate and frame type is discussed and several applications are presented.

In another lecture, a new algorithm is given for the calculation of fully developed heat transfer coefficients in concurrent flow double pipe heat exchangers if only the velocity distribution is known.

Heat transfer between packed, agitated and fluidized beds and submerged surfaces is treated by a common theoretical concept. It is shown that the penetration model, which is well known from gas liquid systems, also applies to gas-solid systems. The results of the theoretical approach are compared with numerous experimental data obtained from the literature.

The recent developments on transport phenomena in fluidized bed heat exchangers are presented; and theory and practice of the fluidized bed heat exchangers are discussed.

Many speakers discuss the thermal and hydraulic fundamentals of two-phase heat exchangers: The basic process involved in boiling and evaporation are described including the thermodynamics of vapor formation, evaporation at planar interfaces, homogeneous and heterogeneous nucleation, sizing of active nucleation sites, bubble growth and bubble detachment and frequency. The natural convection pool boiling, boiling outside tubes and tube bundles are discussed.

The various heat transfer regimes in natural convection pool boiling are identified and practical correlations for each regime are given. Forced convection boiling of single component liquids in vertical and horizontal tubes is described. The various heat transfer regimes which occur are identified and described with the aid of the three-dimensional representation - the "boiling surface". Appropriate correlations for each heat transfer regime are given so that this surface may be constructed for any particular fluid or geometry.

The various modes of condensation (dropwise, filmwise, etc.) are described and the resistances to heat transfer coefficient for condensing inside and outside tubes at various orientations and with low and high vapour velocities are presented. The prediction of heat and mass transfer in multicomponent mixtures is discussed. Pressure drop calculation methods for both the shell side and tube side are given.

Heat transfer in natural convection boiling of multicomponent mixtures is usually lower than that of the pure components. Most of the hypotheses to explain this effect originate from the different equilibrium compositions of the vapour and the liquid phases. Based on this effect heat transfer coefficients shown that heat transfer to boiling multicomponent mixtures may be predicted to a good approximation from the data of pure components and the binary systems. For this reason, the nearly 5000 existing experimental data points for natural convection boiling heat transfer of pure substances are analysed and an additional term in this correlation is introduced to describe the influence of mass transport on heat transfer in multicomponent systems.

Heat transfer in condensation of multicomponent mixtures depends on how the liquid phases are formed near a cooled wall. They can either be homogeneous or consist of several immiscible phases. Condensation with a homogeneous liquid phase is most often met in practical applications. In this case, the condenser design usually is based on the assumption that the temperature of the vapor-liquid interface is constant and given by the boiling temperature. The temperature of the vapour-liquid interface, however, may change considerably according to the rate of condensation, an effect which is mostly neglected. These and other cases are discussed fully. Approximate calculation of heat trasnfer coefficients for the cases where the liquid consists of several phases is presented. When the vapor condenses at an immiscible liquid interface an additional heat resistance must be taken into account.

Horizontal-tube evaporator-condenser (desalination) units are characterized by short circumferential thin films inside and outside the tubes. The external film is sustained by liquids draining upon it from the tubes above it. The experimental and theoretical studies of these films, either laminar or turbulent, flowing over smooth or grooved conduits of different cross sections are presented.

Because of the fundamental importance in two-phase flow, a lecture on the theoretical and experimental study of the deposition motion of droplets from a turbulent gas stream in a vertical tube has also been included.

Clearly, an essential part of research in two-phase flow is the need for sophisticated experimental techniques and instrumentation to cover every physical variable associated with the fine structure of the two-phase flow. In this connection, an interesting research work on laser Doppler diagnostics of fast two-phase flow is introduced which is important in heat exchangers with two-phase flow. Basic principles of operation and technical realization of laser Doppler velocimeters with direct spectrum analyses, based on the high resolution interference spectroscopy methods are reported.

Fundamentals of radiative heat transfer pertinent to application in heat exchangers are presented with particular emphasis to: heat transfer from extended surfaces with radiation; effects of scattering of radiation within the fluid on heat transfer; forced convection to non-participating fluid flowing through conduits with radiation boundary conditions; forced convection to absorbing and emitting fluid flowing through conduits.

Many lecturers will discuss the design of heat exchangers: basic design methods are reviewed first. The discussion of individual resistances in the overall heat transfer coefficient and their magnitude, temperature distribution in heat exchangers, mean temperature difference, effectiveness-NTU method are discussed.

An overview is presented on various quantitative and qualitative steps involved in arriving at the optimum heat exchanger design.

A new method of analysis is presented for heat transfer through an array of extended surfaces containing any number of fins. The method has numerous applications in compact heat exchangers.

Exact explicit equations for some two and three pass cross-flow heat exchangers effectiveness are given. The results of the analytical solutions for all possible combinations of these types of cross-flow heat exchangers with one fluid unmixed throughout and the other mixed between passes and unmixed in each pass are presented.

The most common heat exchanger design problems are the rating and sizing problems. A detailed procedure for rating and sizing problems for compact heat exchangers is outlined with a specific example to obtain solutions to these problems for a direct transfer type two-fluid compact heat exchanger. A general methodology is then discussed for optimization of heat exchanger.

The next, the major construction features of shell and tube exchangers are described. The flexibility of the basic design and the wide range of mechanical design options available to the designer are emphasized. Selection of features to meet a variety of service situations is illustrated. Preliminary design of shell and tube heat exchangers is discussed and Delaware method for shell side design is presented.

The application of two-phase flow and heat transfer fundamentals involves the design of boilers and condensers. The design of both fossil-fired and waste heat-boilers is introduced. The various types of equipment used in central station generating plant and in process industry are reviewed first. Specific two-phase flow and boiling processes which are relevant to boiler design are considered and problems may arise are discussed.

Moving on to the condensers, it is shown how the methods described in previous lectures on condensers are applied to the thermal and hydraulic design of shell-and-tube condensers. An important step in this is the calculation of local and mean temperature differences for multipass units and for fluid streams with non-linear temperature-enthalpy curves. This problem is discussed in some detail. Spray condensers are described and simple design methods for them outlined.

The advantages of film evaporator/condensers are detailed and evaporation and condensation of laminar and turbulent films, as manifested in various water desalination schemes, are reviewed with special emphasis on the more recent and promising horizontal-evaporator-condenser units. The importance of the interacting effects of the phase change phenomena occurring across the metal wall is noted and the importance of the thermal conductivity of the wall for operation with intermittent nucleate boiling and enhanced heat transfer areas are discussed. Recent work on the tube shapes yielding highest heat transfer rates is presented. Numerous studies dealing independently with either film condensation or film evaporation, inside or outside the tube, are reviewed. Overall heat transfer coefficients in the horizontal tube evaporator/condenser are discussed.

A survey of the state-of-the-art of thin film heat transfer, liquid/vapour interaction and entrainment in shell-and-tube heat exchangers is presented. One of the main difficulties encountered in designing falling-film heat exchangers is maintaining the complete wettability of the tubes. Experimental data for the liquid film breakdown and heat transfer are summarized and their application for design illustrated. The problem of vapour/liquid interaction and entrainment in falling film evaporators is anlaysed and the heat exchanger design application is demonstrated.

Starting with an introduction on the similarities and differences between rotary and fixed-matrix regenerators, a detailed analysis of heat transfer is given for thermal design of regenerators and the influence of wall thermal resistance, rotation, bypass, carryover leakages and influence of longitudinal wall heat conduction on the regenerator effectiveness are presented.

Offshore heat exchanger design practice throughout a project is introduced. The problems facing the designer of offshore heat exchangers and solutions to overcome these difficulties are indicated.

During the Institute, thermal-hydraulic problems of dry-cooling towers are also discussed. The optimization of the heat exchanger design for the needs of cooling towers in modern power plants is an important problem. In this respect, development of analytical models for the detailed thermal-hydraulic calculation of finned tube bundle heat exchangers, analytical thermal-hydraulic optimization of heat exchanger geometry and cooling tower systems, thermal-hydrualic and performance tests on existing and optimized heat exchangers are presented.

Design considerations, economic and technical aspects of using dry and wet/

dry cooling systems for power generation are also summarized.

Energy and materials saving consideration, as well as economic incentives have led to recent expansion of efforts to produce more efficient heat exchangers. There are circumstances when, for thermodynamic reasons, it is essential to transfer a given heat load at the lowest possible temperature driving force. In general, this means an increase in heat transfer coefficient. Various ways of augmenting the heat transfer rates in single phase, and two-phase systems are discussed. Performance evaluation criteria, in terms of thermal-hydraulic goals or cost are proposed. These criteria are applied to several of the single phase and two-phase surfaces. The various applications of advanced surfaces are reviewed and discussed.

The development of a high performance heat transfer surface with vortex generators is also introduced, and vorticity and turbulence compared.

Increasingly, attention is being turned to the dynamic and transient aspects of heat exchangers. The transient behavior of heat exchangers is important for the precise control of the system. The transient response analysis is presented for direct transfer type and regenerative exchangers. This includes the formulation, exchanger variable and specific solutions for counter flow, cross-flow, and periodic-flow exchangers.

The literature on the experimental and theoretical studies on the dynamic and transient behaviour of heat exchangers with double-phase-change are very limited. This type of heat exchangers is used in the process industry steam reformers where by condensing vapour on one side of the heat transport system, liquid of the same substance at lower pressure or of different substance is evaporated on the other side. Theoretical model and a computer program have been developed to describe and predict the dynamic behaviour of such heat exchangers.

Vibration phenomena are frequently the cause of serious damage in the tube assemblies of heat exchnagers. In experimental and theoretical studies, different inducing and exciting mechanisms for vibrations in cross-flow heat exchangers are investigated and are discussed in one of the lectures.

Most of the heat transfer processes result in depostion of undesirable scale commonly referred to as fouling. Fouling of heat transfer surfaces introduces perhaps the major uncertainty into the design and operation of heat exchange equipment, leading very often to extra capital and running costs and/or reduces efficiency of the heat exchange equipment. Over the last decade increasing efforts have been directed towards a better understanding of fouling and several models were proposed to predict the fouling in design and in the operation of heat exchangers. The various types of fouling are described and current theories of fouling are reviewed. Fouling in equipment involving boiling and evaporation is often more severe than in single phase heat exchnagers; heat transfer and pressure drop characteristics by fouling layers should be modified; these are identified and illustrated. A critical review of the problem of fouling is undertaken, with special emphasis being placed on the mechanisms and models more closely associated with solubility, particulate and reaction fouling.

In heat exchangers, temperature differences in the primary circuit are inevitable in most cases. These temperature differences may ensue in the non-uniform thermal expansion of the total array leads to the buckling of a part of the tubes. This problem is presented for in-line and staggered tube arrangements in cross-flow subjected to axial force near the buckling load.

Heat exchangers with non-interconnecting passages (parallel tubes or plates) are usually sized assuming uniform flow in each of the passages. Particularly in laminar flow, their performance can depart from these predictions due to non-uniform flow distribution resulting from (a) superimposed gross natural convection in horizontal orientation, (b) the effect of the viscosity temperature relation permitting two different flow rates for the same pressure drop when liquids are cooled, (c) having non-uniform size passages resulting from large passages can reduce performance. These are introduced under the title of "Why laminar slow heat exchangers can perform poorly".

Geothermal energy is an alternative source of energy which has been receiving increased attention in a number of countries, including Turkey, in recent years. Because of the high temperatures and the presence of impurities, geothermal fluid is seldom used as it comes from the earth, but it is usually passed through a surface-mounted heat exchanger. This presents a considerable amount of operational problems in heat exchangers used for geothermal energy utilization; these operational problems are fully discussed.

Five practically important but unresolved problems in heat exchanger design are discussed: viz. flow-induced tube vibration, fouling, mixture boiling, flow distribution in two-phase flow and detailed turbulence flow modelling. For each of these topics, the current state of knowledge is reviewed and the outstanding problems highlighted.

Finally, suggestions for further research on heat exchangers are presented by invited lecturers.

# Classification of Heat Exchangers

**RAMESH K. SHAH**
Harrison Radiator Division
General Motors Corporation
Lockport, New York 14094 USA

ABSTRACT

Starting with a definition, heat exchangers are classified according to the transfer processes, degree of surface compactness, construction features, flow arrangements, number of fluids, and heat transfer mechanisms. With a detailed classification in each category, the terminology associated with a variety of these exchangers is introduced and practical applications are outlined. A brief description is also provided on the differences in the design procedure for the different types of exchangers.

INTRODUCTION

A heat exchanger is a device which provides for transfer of internal thermal energy between two or more fluids at differing temperatures. Heat transfer between the fluids take place through a separating wall. Since the fluids are separated by a heat transfer surface, they do not mix. Common examples of such heat exchangers are the shell-and-tube exchangers, automobile radiators, condensers, evaporators, air preheaters, and "dry" cooling towers. If no phase change occurs in any of the fluids in the exchanger, it is sometimes referred to as a sensible heat exchanger. There are no internal thermal energy sources in a heat exchanger, ruling out fired heaters, electric heaters, and nuclear fuel elements. If the fluids are immiscible, the separating wall may be eliminated, and the interface between the fluids serves as a heat transfer surface as in a direct contact heat exchanger.

A heat exchanger consists of the active heat exchanging elements such as a core or a matrix containing the heat transfer surface, and passive fluid distribution elements such as headers, manifolds, tanks, inlet and outlet nozzles or pipes, or seals. Usually there are no moving parts in a heat exchanger; however, there are exceptions such as a rotary regenerative exchanger, in which the matrix is mechanically driven to rotate at some design speed.

The heat transfer surface is the surface of the exchanger core which is in direct contact with fluids and through which heat is transferred by conduction. That portion of the surface which also separates the fluids is referred to as primary or direct surface. To increase heat transfer area, appendages known as fins may be intimately connected to the primary surface to provide extended, secondary, or indirect surface. Fins may form flow passages for the individual fluids but do not separate the fluids. These secondary surfaces or fins may also be introduced primarily for structural strength purposes, or to provide thorough mixing of a highly viscous liquid.

# CLASSIFICATION

Heat exchangers are used in the process, power, automotive, air conditioning, refrigeration, cryogenics, heat recovery, alternate fuels, and manufacturing industries, as well as key components of many products available in the marketplace. These heat exchangers may be classified according to the transfer processes, degree of surface compactness, construction features, flow arrangements, number of fluids, and fluid phase changes or process function. These classifications are summarized in Fig. 1. A brief description follows.

## 1. CLASSIFICATION ACCORDING TO TRANSFER PROCESSES

In this category, heat exchangers are classified into direct contact type and indirect contact type.

In a direct contact type, heat is transferred through direct contact between the hot and cold <u>immiscible</u> fluids. Generally, one of the fluids is a gas and the other a very low vapor pressure liquid, and are readily separable after the energy exchange. A water cooling tower with forced or natural draft air flow is the most common application of direct contact exchangers.† Other applications are the jet condenser for water vapor and other vapors using a water spray.

In an indirect contact type heat exchanger, heat is transferred first from the hot fluid to an impervious surface and then to the cold fluid. This type of heat exchanger, also referred to as a <u>surface heat exchanger</u>, can be further classified into the direct transfer type, storage type, and fluidized bed exchangers.

In a direct transfer type heat exchanger, two fluids are separated by a thin wall (parting plates or tube walls) through which heat flows. Although simultaneous flow of both fluids is required in the exchanger, there is no mixing of two fluids. There are no moving parts in the exchanger. This type of exchanger is designated as a <u>recuperative heat exchanger</u>, or simply as a <u>recuperator</u>.‡ Some examples of direct transfer type heat exchangers are: tubular, plate and extended surface exchangers. The $\varepsilon$-$N_{tu}$ and LMTD (log-mean temperature difference) methods are used for heat transfer analysis of this type of exchanger [1].

In contrast, for a storage type heat exchanger, the same flow passages are alternately occupied by one of the two fluids. The heat transfer surface is of cellular structure usually referred to as a matrix. During the hot gas flow through a passage, thermal energy is stored in the matrix wall. During the cold gas flow through the same passage later, the matrix wall delivers thermal energy to the cold fluid. Thus heat is not transferred through the wall as in a direct transfer type exchanger, but is alternately stored and rejected by the matrix wall. This storage type heat exchanger is also referred to as a <u>regenerative</u>

---

†Generally in a water cooling tower, more than 90% of the energy transfer is by virtue of mass transfer, and heat transfer as such is a minor mechanism.

‡In vehicular gas turbines, a stationary heat exchanger is usually referred to as a recuperator and a rotating heat exchanger as a regenerator. However, in industrial gas trubines, by long tradition and from a thermodynamic sense, a stationary heat exchanger is generally referred to as a regenerator.

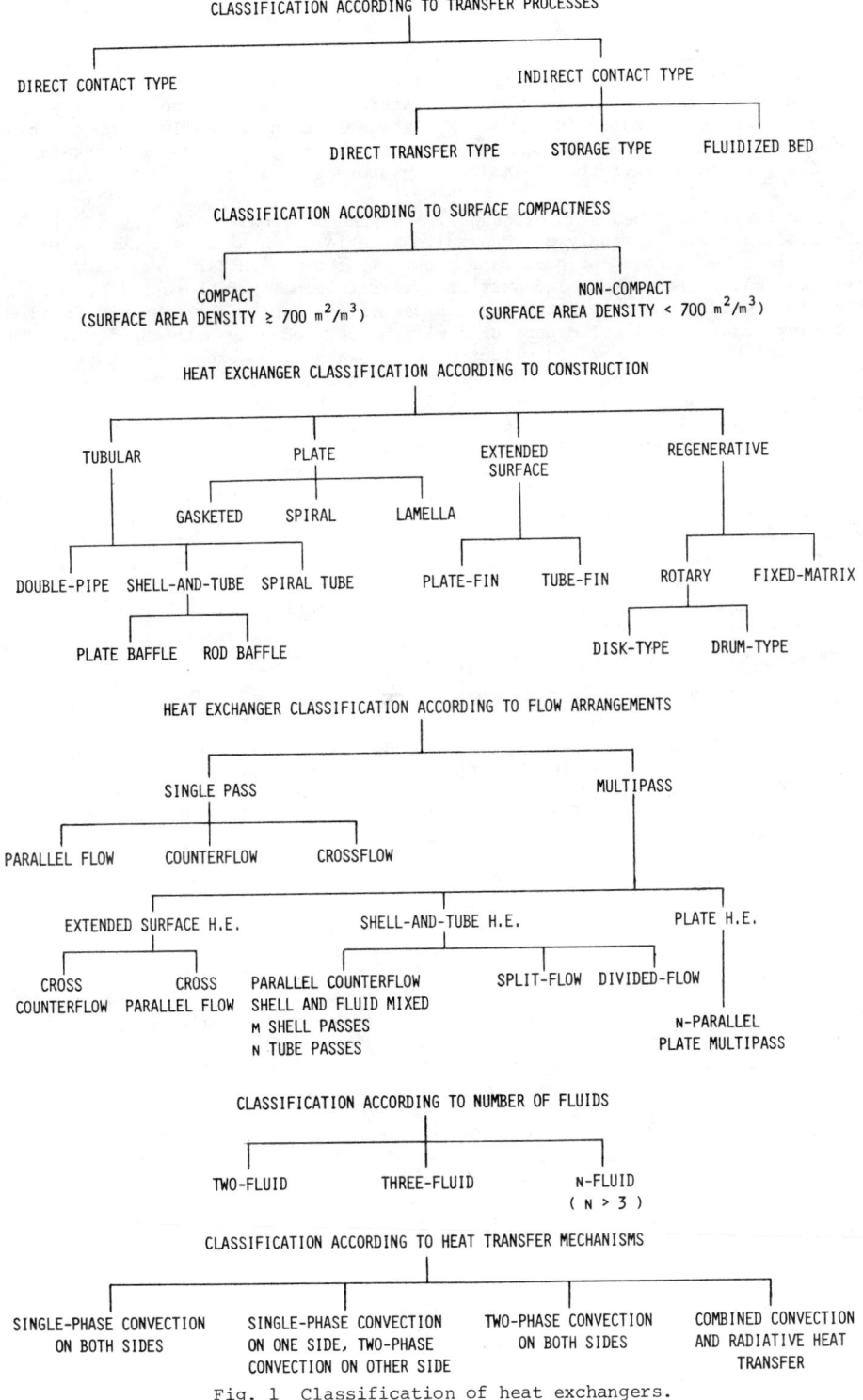

Fig. 1 Classification of heat exchangers.

heat exchanger, or simply as a regenerator. Some examples of storage type heat exchangers are: vehicular gas turbine rotary regenerator, Ljungstrom rotary air preheater, and fixed-matrix air preheaters for blast furnace stoves, glass furnaces, and open hearth furnaces. For the heat transfer analysis of the regenerators, the $\varepsilon$-$N_{tu}$ method of the direct transfer type exchanger needs to be modified to account for the thermal energy storage capacity of the matrix.

In a fluidized bed heat exchanger, one side of a two-fluid exchanger is immersed in a bed of finely divided solid material such as a tube bundle immersed in a bed of sand or coal particles. If the upward fluid velocity on the bed side is low, the solid particles will remain fixed in position, and the fluid will flow through the interstices of the bed. If the upward fluid velocity is high, the solid particles will be carried away with the fluid. At a "proper" value of the fluid velocity, the upward drag force is slightly higher than the solid weight. As a result, the solid particles will float with a slight increase in the bed volume, and the bed behaves as a liquid. This characteristic of the bed is referred to as fluidized condition. Then the fluid pressure drop through the bed remains almost constant, independent of the flow rate. A strong mixing of the particles occurs. This results into an isothermal temperature for the total bed (gas and particles) with an apparent thermal conductivity of the solid particles as infinity. Very high heat transfer coefficients are achieved on the fluidized side. The common applications of the fluidized bed heat exchanger are drying, mixing, adsorption, reactor engineering and waste heat recovery. Since the initial temperature difference ($t_{h,i}-t_{c,i}$) is reduced due to fluidization, the exchanger effectiveness is lower and should be properly evaluated [2].

## 2. CLASSIFICATION ACCORDING TO THE SURFACE COMPACTNESS

Loosely defined, a compact heat exchanger is one which incorporates a heat transfer surface having a high "area density." That is a high ratio of heat transfer surface to volume. The heat exchanger is not necessarily of small bulk and mass. However, if it did not incorporate a surface of high area density, it would be much more bulky and massive.

Somewhat arbitrarily, we will specify that a compact surface has an area density $\beta$ greater than 700 $m^2/m^3$ (213 $ft^2/ft^3$). A heat exchanger of any structural construction of the next section is considered as compact if it employs a compact surface on either one or more sides of a two-fluid or a multi-fluid heat exchanger.

A spectrum of surface area density of heat exchanger surfaces is shown in Fig. 2. On the bottom of the figure, two scales are shown: the hydraulic diameter $D_h$ in mm and equivalent heat transfer surface area density $\beta$ $m^2/m^3$. Different heat exchanger surfaces are shown in the rectangles. The short vertical sides of a rectangle when projected on the $\beta$ (or $D_h$) scale indicate the range of surface area density (or hydraulic diameter) for the particular surface in question. What is referred to as $\beta$ in this figure is either $\beta_1$, $\beta_2$ or $\beta_3$ as defined below.

Shell-and-Tube H.E.: $\quad \beta_1 = \dfrac{A_h + A_c}{V_{total}} \quad\quad q \simeq U(\beta_1/2)V_{total}\Delta t_m \quad\quad$ (1)

Plate and Extended Surface Heat Exchangers: $\quad \beta_2 = \dfrac{A_h}{V_h} \text{ or } \dfrac{A_c}{V_c} \quad\quad \begin{array}{l} q = U(\beta_2 V_h)\Delta t_m \text{ or} \\ U(\beta_2 V_c)\Delta t_m \end{array} \quad\quad$ (2)

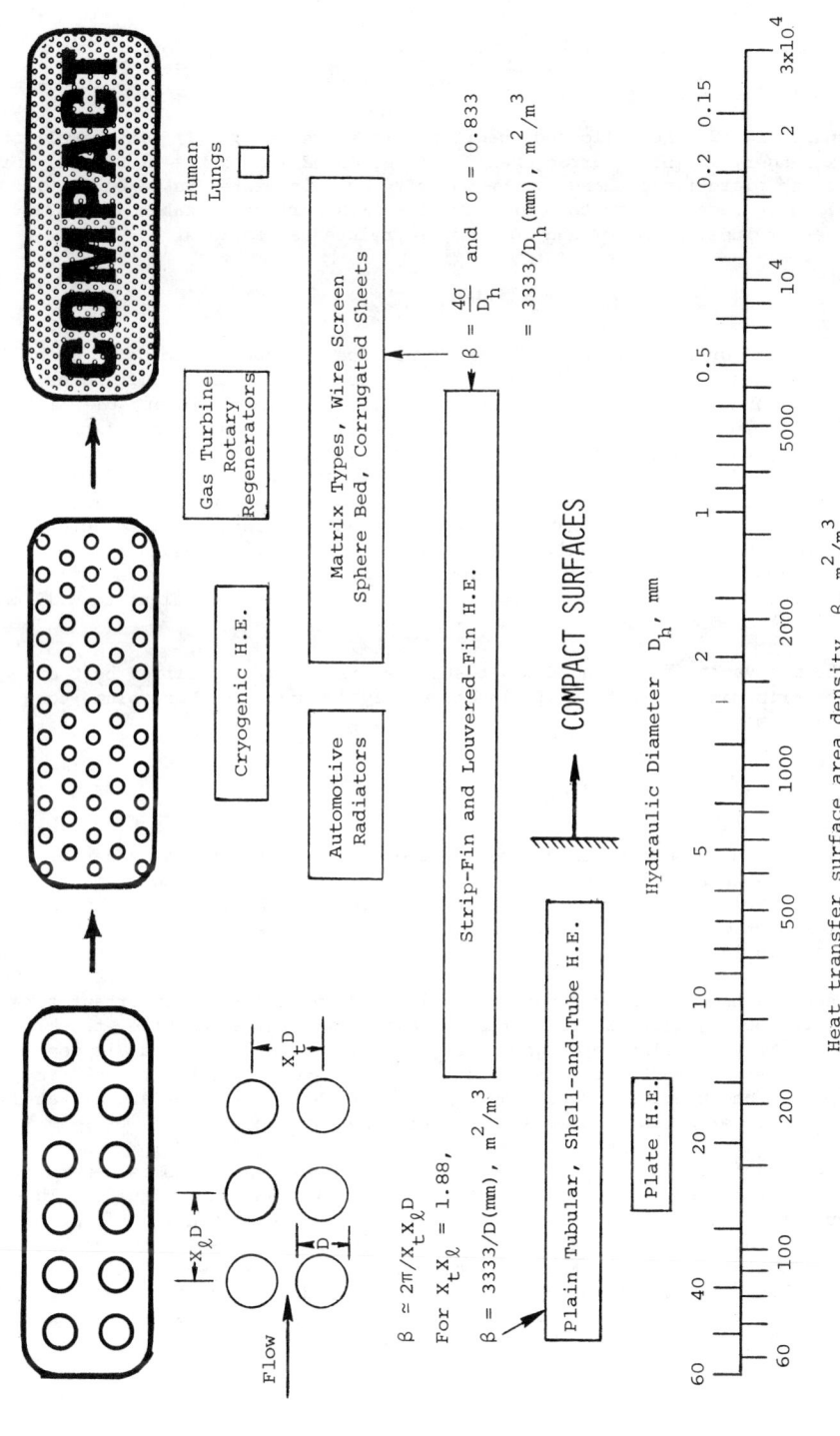

Fig. 2 Heat transfer surface area density spectrum of exchanger surfaces.

Regenerators:
$$\beta_3 = \frac{A_h}{V_{total}} \text{ or } \frac{A_c}{V_{total}} \qquad q = U(\beta_3 V)\Delta t_m \qquad (3)$$

To prepare Fig. 2, the following specific values are used: (1) for a shell-and-tube exchanger, an inline arrangement[†] is considered with $X_t X_\ell = 1.88$; (2) for a plate and plate-fin exchanger, the porosity between plates is taken as 0.8333; and (3) for a regenerator, the porosity of matrix surface is taken as 0.8333. With these values, $\beta$ m²/m³ and $D_h$ mm are related as follows.

$$\beta = 3333/D_h \qquad (4)$$

Based on our definition of a compact surface, a tube bundle using 5 mm (0.2 in.) tubes in a shell-and-tube exchanger comes close to qualifying as a compact exchanger. As $\beta$ varies inversely with the tube diameter, the 25.4 mm (1 in.) tubes used in a power plant condenser result in a non-compact exchanger. In contrast, the modern automobile "radiator" (5.5 fins/cm, 14 fins/in.) has an area density of the order of 1100 m²/m³ (335 ft²/ft³) equivalent to 3 mm (0.12 in.) tubes. The regenerators in some vehicular gas turbine engines currently under development have matrices with an area density of the order of 6600 m²/m³ (2000 ft²/ft³) equivalent to 0.5 mm (0.02 in.) tubes in a bundle. The human lungs are one of the most compact heat-and-mass exchangers having a surface area density of about 17500 m²/m³ (5330 ft²/ft³) equivalent to 0.19 mm (0.0075 in.) diameter tubes.

The motivation for using compact surfaces is to gain specified heat exchanger performance, $q/\Delta t_m$, within acceptably low mass and box volume constraints. As

$$\frac{q}{\Delta t_m} = U \beta V \qquad (5)$$

clearly, a high $\beta$ minimizes volume. Moreover, compact surfaces generally result in a higher overall conductance U; again a contribution to a smaller volume. As compact surfaces can achieve structural stability and strength with a thinner section, the gain in lower exchanger mass is even more pronounced than the gain in lower volume.

These gains of exchanger mass and volume are particularly important for the propulsive power plants of transportation systems -- the automobile, truck, and airplane. They are also important in cryogenic, refrigeration and air conditioning systems, and in the dry cooling towers used as the heat "sink" in power plants and chemical processing complexes. These last uses are dictated by the much lower costs as well as manufacturing labor costs.

The convective heat transfer coefficient for gaseous fluids is generally one or two orders of magnitude lower than water, oil and other liquids. Thus to reduce the size and weight of a gas-to-liquid heat exchanger, the heat transfer surface on the gas side needs to be much more compact than can be practically realized with circular tubes. Hence, for a somewhat "balanced" design, a compact surface is employed on the gas side. Thus, major applications of compact heat exchangers are gas-to-gas, gas-to-liquid and gas-to-condensing or evaporating fluid heat exchangers.

---

[†] The tube array is idealized as infinite with thin walled circular tubes.

Various techniques are employed to make heat transfer surfaces compact. Such compact surfaces have fins between plates, finned circular tubes, or densely packed continuous or interrupted cylindrical flow passages of various shapes.

The uniquenesses of compact heat exchangers are: (1) many surfaces available having different orders of magnitude of surface area density; (2) flexibility in distributing the area on the hot and cold sides as desired by design considerations; and (3) generally substantial cost, weight or volume savings.

Constraints on the application of compact surfaces are: (1) usually at least one of the fluids should be a gas; (2) fluids must be clean and relatively non-corrosive; (3) allowed pressured drop is generally small and operating pressures and temperatures are somewhat limited; (4) the market potential must be large enough to warrant the sizeable manufacturing, research and development costs.

A modular design approach, where core modules are assembled in series and in parallel to achieve the desired performance, can be effective in reducing costs.

## 3. CLASSIFICATION ACCORDING TO CONSTRUCTION FEATURES

Heat exchangers are frequently characterized by construction features. Examples described below are the major construction types: tubular, plate, extended surface, and regenerative exchangers. Although the $\varepsilon$-$N_{tu}$ or LMTD method is identical for tubular, plate and extended surface exchangers, the influence of the following factors must be accounted for in the exchanger design: the corrections due to leakage and bypass streams in a shell-and-tube exchanger, end effects due to few plates in a plate exchanger, and the temperature ineffectiveness of the fin in an extended surface exchanger. Similarly the $\varepsilon$-$N_{tu}$ method must be modified to account for the thermal capacity of the regenerator matrix. Thus the design theory differs in detail for each construction type and will be discussed in detail later in this lecture series.

### 3.1 Tubular Heat Exchangers

These exchangers are generally built of circular tubes.† There is a considerable flexibility in the design because the core geometry can be varied easily by changing the tube diameter, length and arrangement. Tubular exchangers can be designed for high pressures relative to the environment and high pressure differences between the fluids. These exchangers are further classified as follows.

<u>Shell-and-Tube Heat Exchanger</u>. This exchanger, as shown in Fig. 3, is built of round tubes mounted in a cylindrical shell with the tube axis parallel to that of the shell. One fluid flows inside the tubes, the other flows across and along the tubes. The major components of this exchanger are tubes (or tube bundle), shell, front end head, rear end head, baffles, and tubesheets.

Various front and rear head types and shell types have been standardized by TEMA [3].‡ They are identified by an alphabetic character as shown in Fig. 4.

---

†Heat exchangers having elliptical or rectangular tubes have been built for some applications.

‡TEMA is the abbreviation for Tubular Exchanger Manufacturers Association.

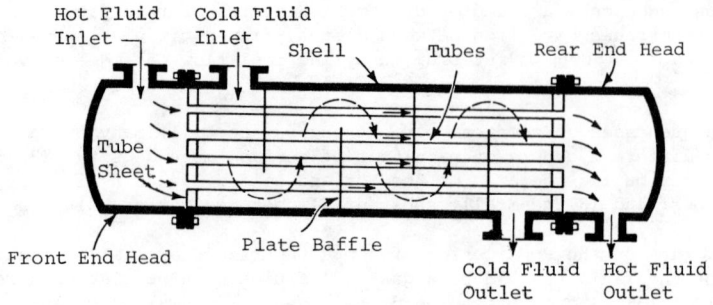

Fig. 3  A shell-and-tube exchanger with one-shell pass and one-tube pass.

Variety of combinations of head types and shell types are possible. A brief description on the seven types of shells follows.

The "E" shell is the most common due to its cheapness and simplicity. In this shell, the shell fluid enters at one end of the shell and leaves at the other end, i.e., there is one pass on the shell side. The tubes may have a single or multiple passes and are supported by transverse baffles. This shell is the most common for single-phase shell fluid applications. To increase the effective temperature difference and hence exchanger effectiveness, a pure counterflow arrangement is desirable for a two tube-pass exchanger. This is achieved by the use of "F" shell having a longitudinal baffle and resulting in two shell passes.

The split and divided flow shells, such as "G", "H", and "J" are used for specific applications. The split flow "G" shell has horizontal baffles with ends removed; the shell nozzles are 180° apart at the midpoint of tubes. The "G" shell has the same pressure drop as that for the "E" shell, but the LMTD factor F and hence the exchanger effectiveness is higher for the same surface area and flow rates. The "G" shell can be used for single-phase flows, but is very often used as horizontal thermosiphon reboiler. In this case, the longitudinal baffle serves to prevent flashing out of the lighter components of the shell fluids and provides increased mixing. The double split flow "H" shell is similar to the "G" shell, but with two inlet and two outlet nozzles and two horizontal baffles. The divided flow "J" shell has two inlets and one outlet or one inlet and two outlet nozzles, single nozzle at the midpoint of tubes and two nozzles near the tube ends. The "J" shell has approximately 1/8 the pressure drop of a comparable "E" shell and is therefore used for low pressure drop applications such as a condenser in vacuum. For a condensing shell fluid, the "J" shell is used with two inlets for the gas phase and one central outlet for the condensates and leftover gases.

The "K" shell is a kettle reboiler with the tube bundle in the bottom of the shell covering about 60% of the shell diameter. This shell is used for pool boiling applications. The liquid covers the tube bundle, and the vapor occupies the upper space without tubes. The vertical baffle acts as a weir; the excess liquid overflows and is drained.

The two fluids in the "X" shell are in a crossflow arrangement. No baffles are used in the "X" shell, however, support plates are used to suppress flow-induced vibrations. The "X" shell provides very low pressure drop for the shell fluid. It is most frequently used for gas cooling with finned tubes and/or condensation.

# CLASSIFICATION OF HEAT EXCHANGERS

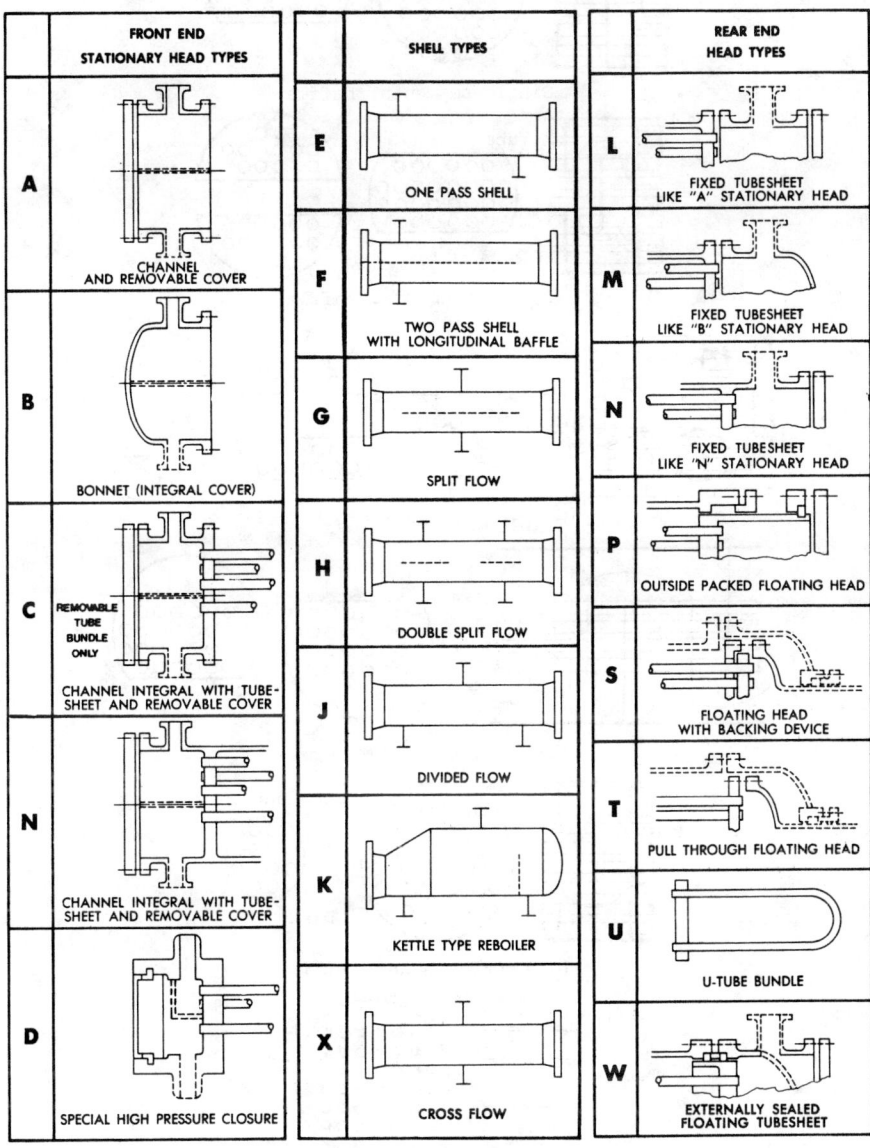

Fig. 4  Standard shell types and front end and rear end head types, from TEMA [3].

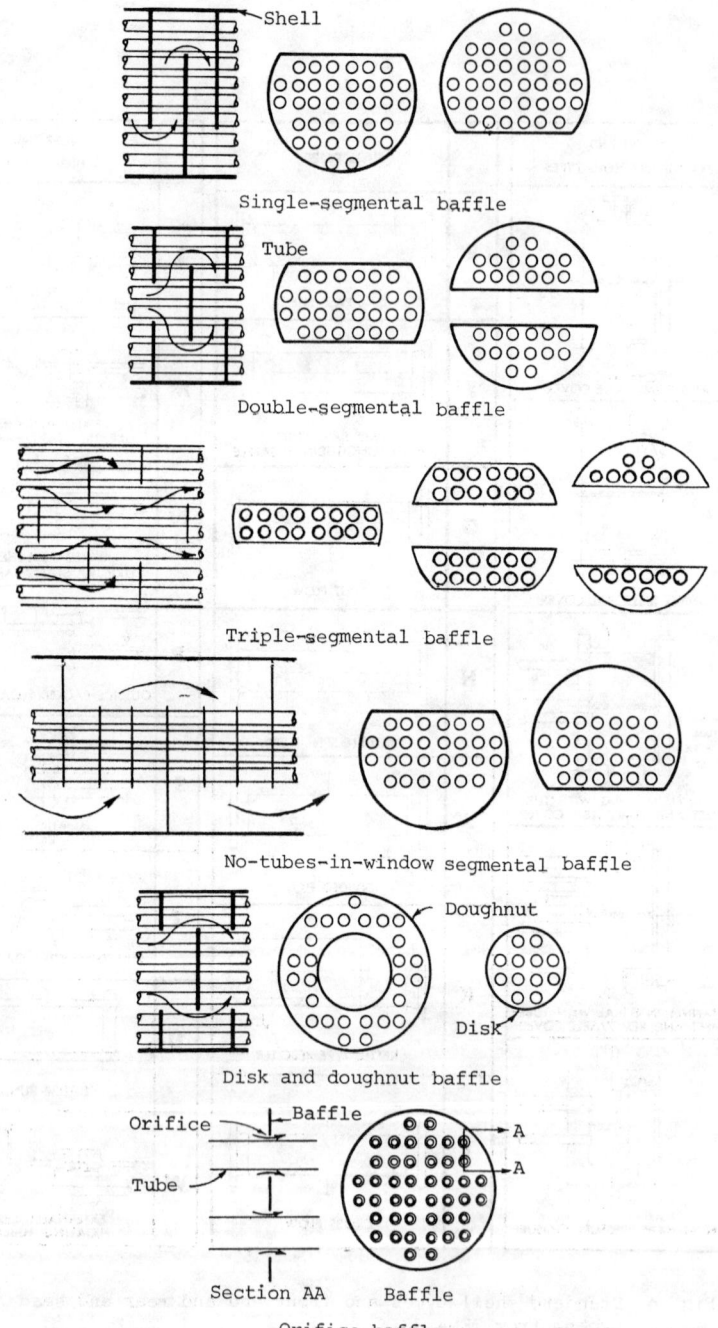

Fig. 5 Plate baffle types.

# CLASSIFICATION OF HEAT EXCHANGERS

Baffles may be classified as transverse and longitudinal types. The purpose of longitudinal baffles is to control the flow direction of the shell fluid[†] such that an overall arrangement of the two fluid streams is achieved. For example, the two-pass "F" shell has a longitudinal baffle (see Fig. 4). The transverse baffles may be classified as plate baffles and rod (or bar) baffles. The plate baffles are used to support the tubes, to direct the fluid in the tube bundle approximately at right angles to the tubes, and to increase the turbulence of the shell fluid. Shown in Fig. 5 are single and multi-segmental, disk and doughnut, and orifice baffles. The single and double segmental baffles are most frequently used. The triple and no-tubes-in-window segmental baffles are used for low pressure drop applications. The choice of baffle type, spacing and cut are largely determined by flow rate, allowable pressure drop, tube support, and flow-induced vibrations. The disk and doughnut and orifice baffles are rarely used. The rod (or bar) baffles are used to support the tubes and to increase the turbulence of the shell fluid, and are shown in Fig. 6. The flow in a rod baffled heat exchanger is parallel to the tubes, and flow-induced vibrations are virtually eliminated by the baffle support of the tubes.

---

[†] The fluid flowing in the tubes is referred to as the tube fluid, the fluid flowing outside the tubes is referred to as the shell fluid.

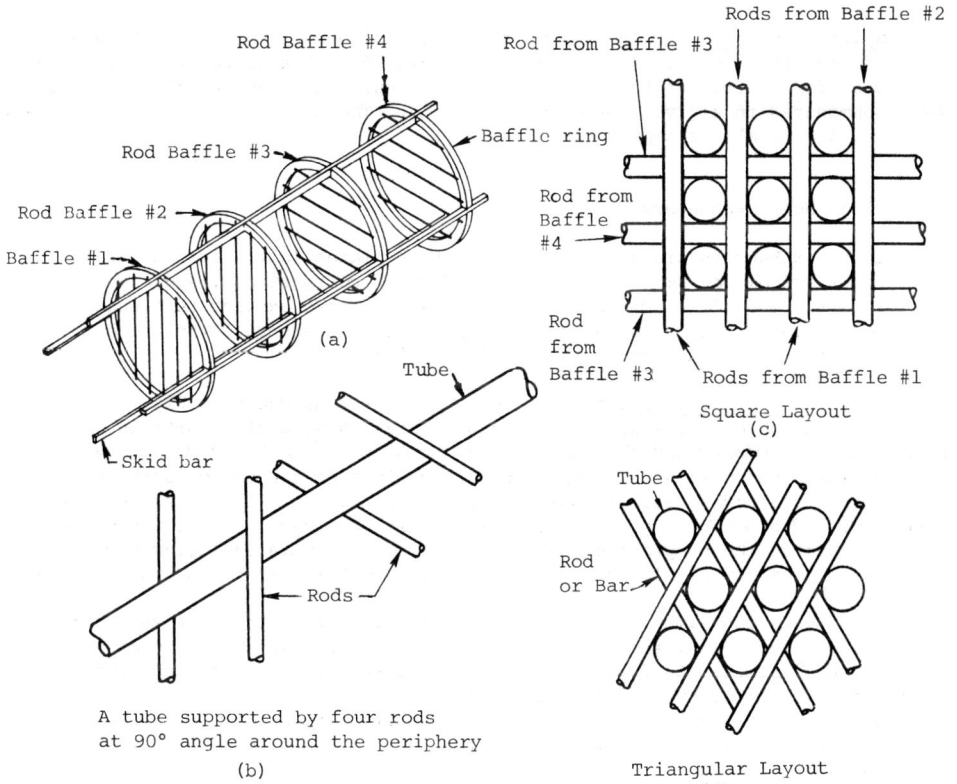

Fig. 6 (a) Four rod baffles held by skid bars (no tubes shown), (b) a tube supported by four rods, (c) a square layout of tubes with rods, and (d) a triangular layout of tubes with rods.

life is sometimes limited. The pin-hole leaks are hard to detect. Some of the largest units have about 1500 $m^2$ (16000 $ft^2$) of surface.

Plate exchangers are widely used in the dairy, beverage, food and pharmaceutical industries, synthetic rubber industry, and paper mills. They have also been used in chemical and petroleum-refining industries in competition with shell-and-tube exchangers.

Spiral plate heat exchanger. This consists of two relatively long strips of sheet metal, provided with spacer studs, and wrapped helically to form a pair of spiral channels, as shown in Fig. 8. The basic spiral element is sealed by either welded at each side of the channel or by providing gasket at each end cover to obtain the following arrangements of the two fluids: (1) both fluids in spiral counterflow; (2) one fluid in spiral flow, other in crossflow across the spiral; and (3) one fluid in spiral flow, and the other in a combination of crossflow and spiral flow.

The spiral plate exchanger has relatively a large diameter because of the spiral turns. This is a non-compact heat exchanger with a maximum of about 185 $m^2$ (2000 $ft^2$) surface area for a 1.47 m (58 in.) diameter maximum shell diameter. The heat transfer coefficients are not as high as in a plate exchanger if the plates are not corrugated.

The advantages of this exchanger are: It can handle viscous and fouling liquids more readily because of a single passage; the fouling rate is found to be very low compared to the shell-and-tube unit. It is more amenable to chemical, flush and reversing fluids cleaning techniques because of a single passage. From the cost viewpoints, steel or alloy spiral exchangers are either competitive or less expensive than shell-and-tube exchangers.

The disadvantage of this exchanger are: The maximum size is limited. The maximum operating pressure is limited to 1000 kPa (150 psi) for large units. The maximum operating temperature is limited to 500°C (930°F) with compressed asbestos gaskets. Field repair is also difficult due to construction features.

This exchanger is used in cellulose industry for cleaning relief vapors in sulfate and sulfite mills, and is also used as thermosiphon and kettle reboilers.

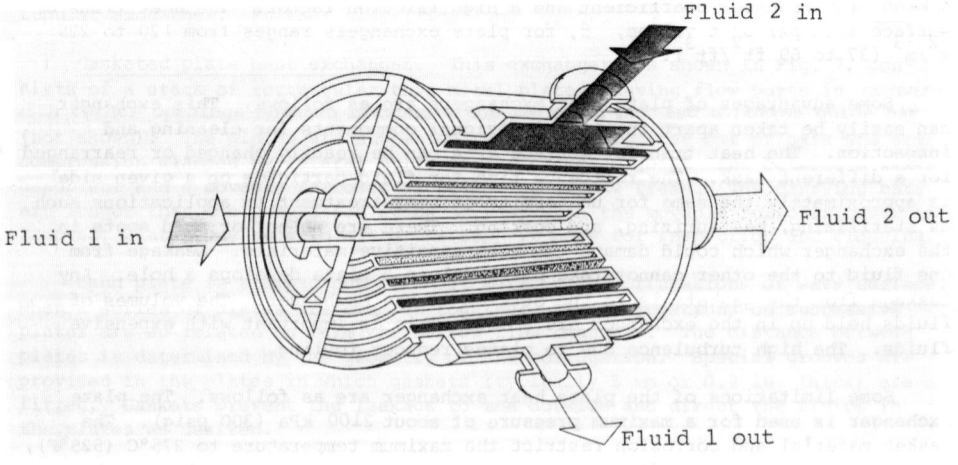

Lamella heat exchanger. It consists of an outer tubular shell surrounding an inside bundle of heat transfer elements. These elements, referred to as lamellae, are flat tubes (high aspect ratio rectangular tubes with rounded corners) stacked close to each other to form narrow channels on the "shell" side. The tubes are of increasing width (aspect ratio) from either end to the center of the shell to fully utilize the available space. Thus the tube bundle (having an overall circular shape) represents a single tube row. There are no baffles. One fluid flows inside the lamella and the other in the spaces between them. The exchanger thus has a single pass and the flow arrangement is generally counterflow. The tube walls are either plain or have dimples. High heat transfer coefficients are usually obtained because of small hydraulic diameters. This design is capable of pressures up to 2000 kPa (300 psig) and temperature limits of 200°C (430°F) for teflon gaskets and 500°C (930°F) for asbestos gaskets. This exchanger is used in pulp and paper industry, chemical process industry and also for other applications in competition with the shell-and-tube exchanger.

## 3.3 Extended Surface Heat Exchangers

The tubular and plate exchangers described previously are all prime surface heat exchangers. The design thermal effectiveness is usually 60% and below, and the heat transfer surface area density is usually less than 300 $m^2/m^3$ (91 $ft^2/ft^3$). In many applications, a much higher (up to about 98%) exchanger effectiveness is essential, and the box volume and mass are limited so that a much more compact surface is mandated. Usually either a gas or a liquid having a low heat transfer coefficient is the fluid on one or both sides. This results in a large heat transfer surface area requirements. For low density fluids (gases), pressure drop constraints tend to require a large flow area. So a question arises how can we increase both the surface area and flow area together in a reasonably shaped configuration.

As mentioned initially, appendages or fins on the primary surfaces increase the surface area density. Flow area is increased by the use of thin gauge material and sizing the core properly. The heat transfer coefficient on the extended surfaces may be higher or lower than that on the unfinned surfaces. For example, the interrupted (strip, louver, etc.) fins provide both an increased area and increased heat transfer coefficient; while the internal fins in a tube may result into a slight reduction in the heat transfer coefficient depending upon the fin spacing. Plate-fin and tube-fin geometries are the two most common types of extended surface heat exchangers.

Plate-fin heat exchanger. This type has fins or spacers sandwiched between parallel plates (referred to as parting plates or parting sheets) or formed tubes as shown in Fig. 9. While the plates separate the two fluid streams, the fins form the individual flow passages. Alternate fluid passages are connected in parallel by suitable headers to form the two or more sides of the exchanger. Fins are attached to the plates by a mechanical fit, gluing, soldering, brazing, welding, or extrusion. Fins are used on both sides in a gas-to-gas heat exchanger application. In a gas-to-liquid application, fins are usually used only on the gas side; if employed on the liquid side, they are primarily for structural strength and flow mixing purposes. Fins are also sometimes used for pressure containment and rigidity. The plate-fin exchanger is referred to as a matrix heat exchanger in Europe. Since bars are used to seal the fluid passages at the ends in some exchangers, those are also referred to as bar and plate exchangers. In the automotive industry, fins in the plate-fin unit is referred to as centers in order to distinguish them from fins outside of the tubes in a tube-fin exchanger. The latter are simply referred to as fins.

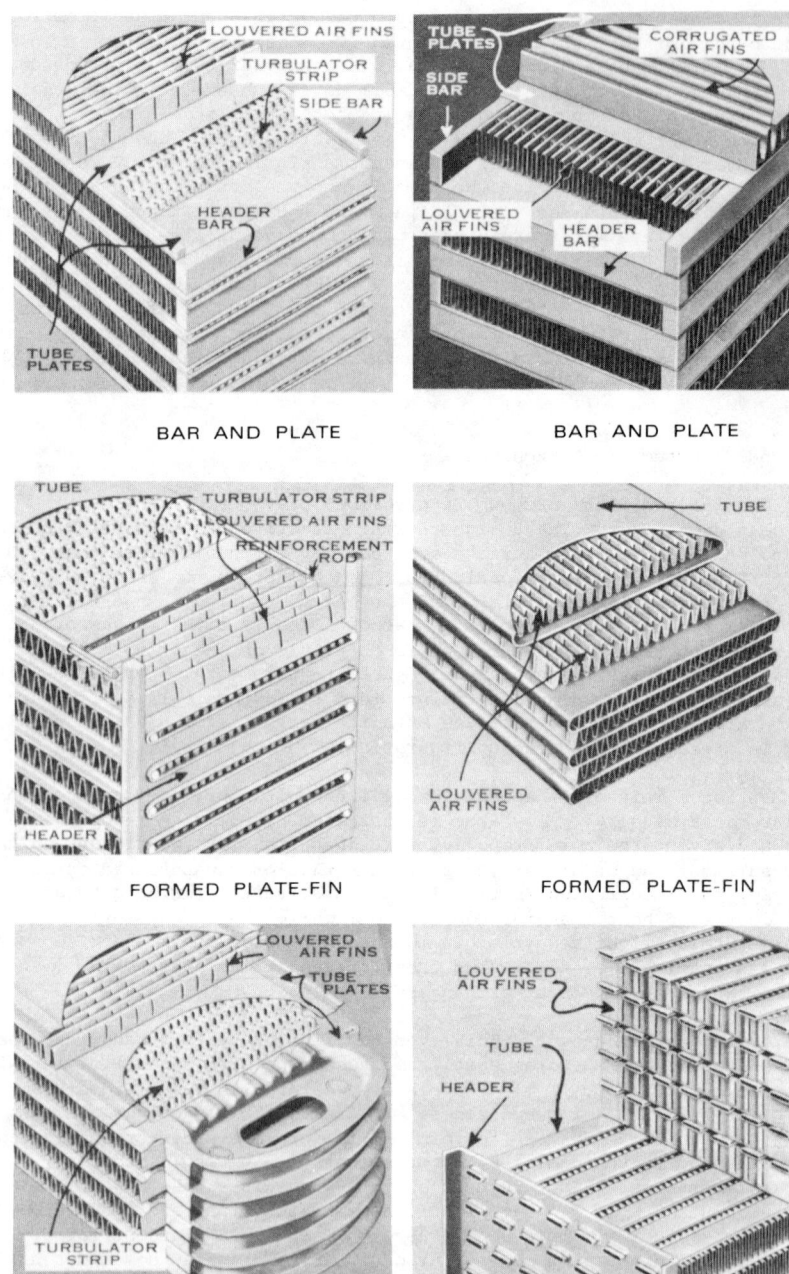

Fig. 9 Plate-fin heat exchangers (courtesy of Harrison Radiator Div., GMC, Lockport, NY).

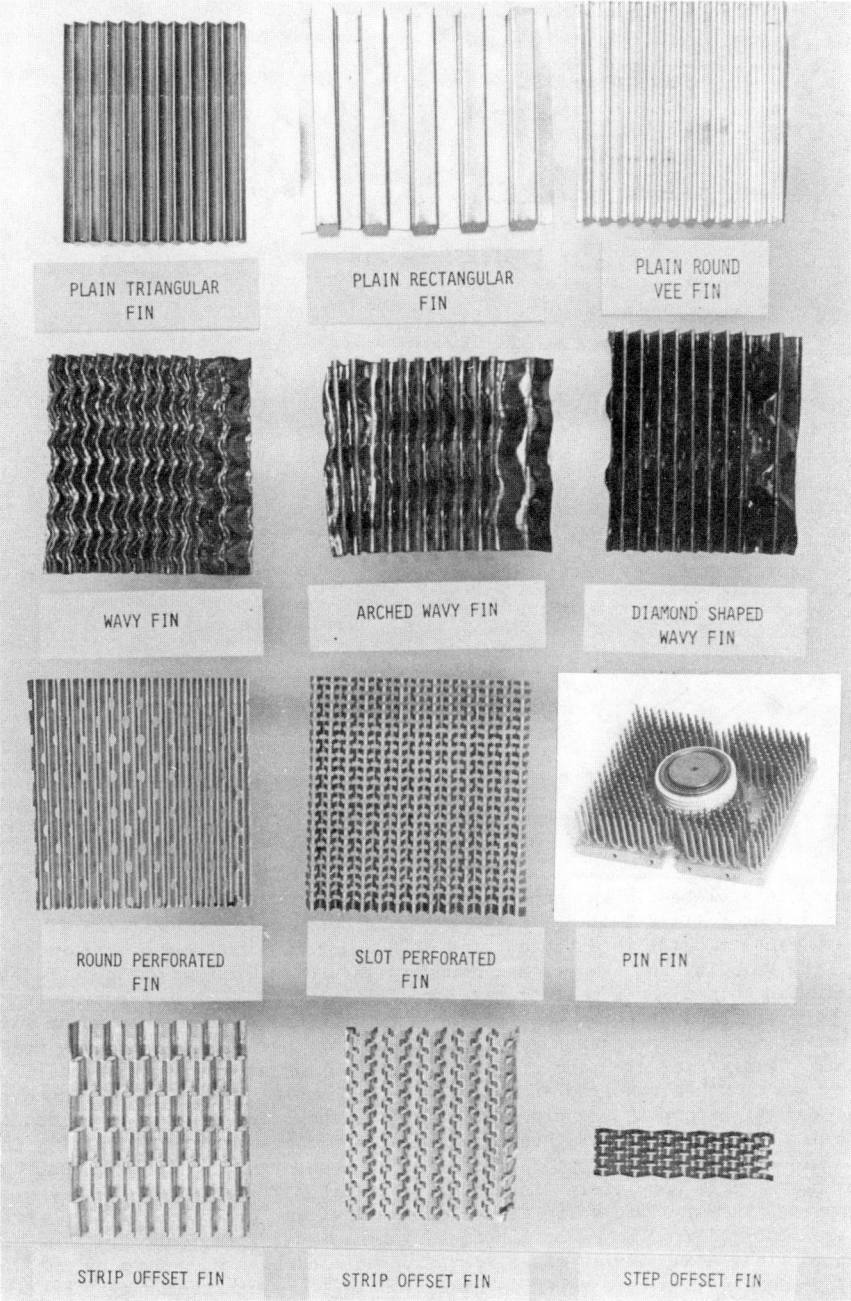

Fig. 10 Plain, wavy and interrupted fin geometries for plate-fin exchangers (Pin fin geometry, courtesy of PinFin, Inc., Little Compton, RI; all other geometries courtesy of Harrison Radiator Div., GMC, Lockport, NY)

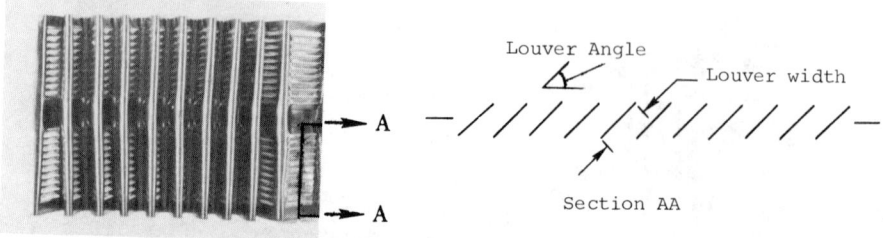

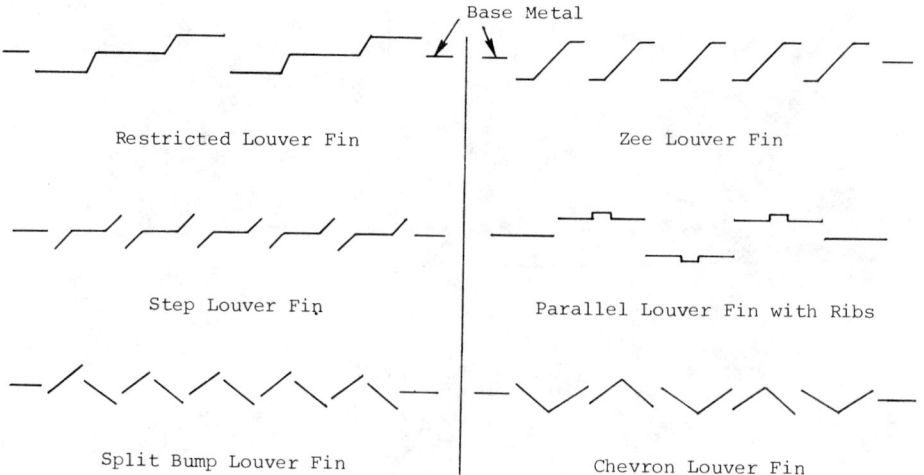

Fig. 11  Louver fin geometries for plate-fin exchangers.

Fins used in a plate-fin exchanger are shown in Figs. 10 and 11 and they may be categorized as follows: (1) Plain (uncut surfaces) and straight fins; (2) plain but wavy fins; and (3) interrupted fins such as strip, louver and perforated. fins. The velocity and temperature boundary layers thicken on plain surfaces resulting in both a lower heat transfer coefficient and a lower friction factor. Plain fins are used when the pressure drop is critical and interrupted or wavy fins cannot meet the pressure drop requirement together with a flow area constraints. Plain fins are made such that the flow passages have triangular, rectangular or other noncircular shapes. When the plain fin is formed such that it has a wavy surface in the flow direction (as shown in Fig. 10), the boundary layers are either thinned or interrupted when the flow is turned resulting in both a higher heat transfer coefficient and a higher friction factor. Boundary layers can be more completely disrupted if the fin surface is made highly discontinuous. Examples are strip fins, louver fins, and perforated fins. Strip fins are also referred to as offset fins, lance-offset fins, serrated fins or segmented fins. Many variations of interrupted fins have been used by the industry; some of them are shown in Figs. 10 and 11. Interrupted fins employ the materials of construction more efficient than plain fins and are used therefore when allowed by the design constraints.

Plate-fin exchangers are generally designed for moderate pressures (less than about 700 kPa or 100 psig).† The temperature limitation depends upon the method of bonding and the material employed. Such exchangers have been used for temperatures of about 800°C (1500°F). A variety of materials are used for plates and fins. Plate-fin exchangers have been built with a surface area density of up to 5900 $m^2/m^3$ (1800 $ft^2/ft^3$). Plate-fin exchangers are widely used in electric power plants (gas turbine, steam, nuclear, fuel cell, etc.), in propulsive power plants (automobile, truck, airplane, etc.), in thermodynamic cycles (heat pump, refrigeration, etc.), and in electronics, cryogenics, air conditioning, and waste heat recovery systems.

Tube-fin heat exchanger. In a gas-to-liquid exchanger, the heat transfer coefficient on the liquid side is generally high and no fins are required on that side. In addition, if the pressure is high for one fluid, it is generally economical to employ tubes. In a tube-fin exchanger, tubes of round, rectangular or elliptical shapes are generally used. Fins are generally used on the outside and also used inside the tubes in some applications. They are attached to the tubes by a tight mechanical fit, tension wound, gluing, soldering, brazing, welding or extrusion. Two tube-fin exchangers are shown in Fig. 12. This exchanger is also sometimes referred to as the finned tube exchanger. More commonly used geometries for fins outside the tubes are shown in Figs. 13-15.

Fins outside the tubes may be categorized as follows: (1) normal fins on individual tubes as shown in Fig. 13; (2) longitudinal fins on individual tubes as shown in Fig. 14; and (3) continuous (plain or interrupted) fins on an array of tubes, as shown in Fig. 15.‡ The first two types of fins are probably more rugged and practical in large tube-fin exchangers. The exchanger with continuous fins is usually cheaper on a unit heat transfer surface area basis because of its simple and mass production type construction features.

Fins inside the tubes are of two types: integral fins, as in internally finned tubes, and attached fins. Internally finned tubes are shown in Fig. 16.

Tube-fin exchangers can withstand high pressures on the tube side. The highest temperature is again limited to the type of bonding and the material

---

†Some cryogenics plate-fin exchangers are exceptions; they are designed for operating pressures of about 8300 kPa (1200 psig).

‡An exchanger having continuous fins on tubes is also referred to as a plate-fin and tube exchanger. To avoid confusion, we will refer to it as a tube-fin exchanger having plain, wavy, louver, etc. continuous fins.

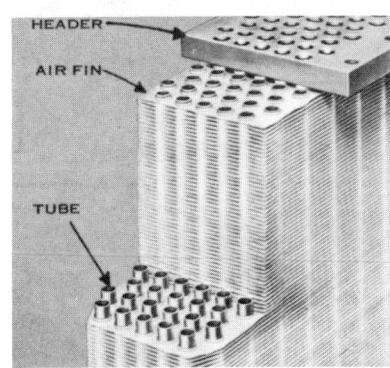
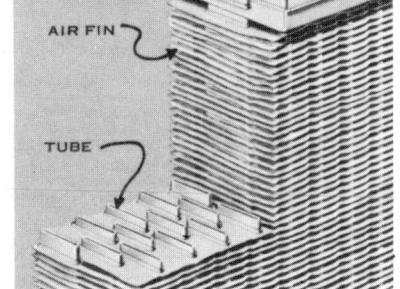

(a)  Round Tube and Plain Fin      (b)  Flat Tube and Plain Fin

Fig. 12  Tube-fin heat exchangers (courtesy of Harrison Radiator Div., GMC, Lockport, NY).

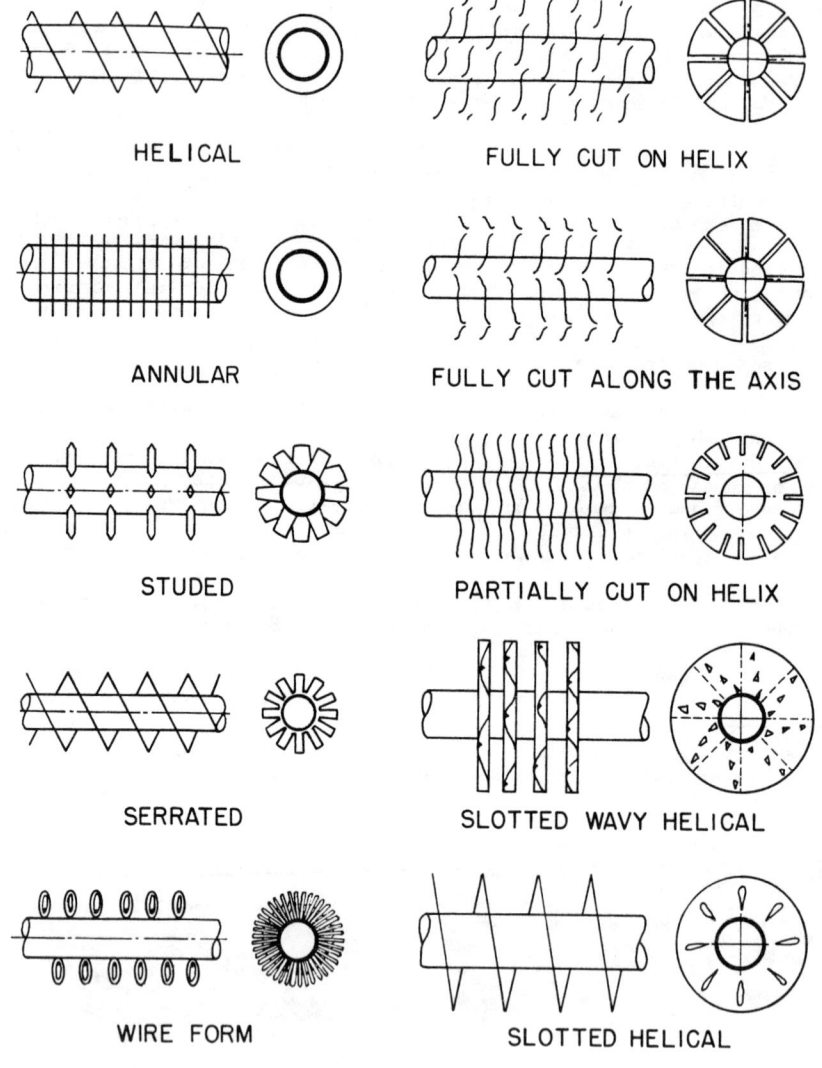

Fig. 13  Normal fins on individual tubes.

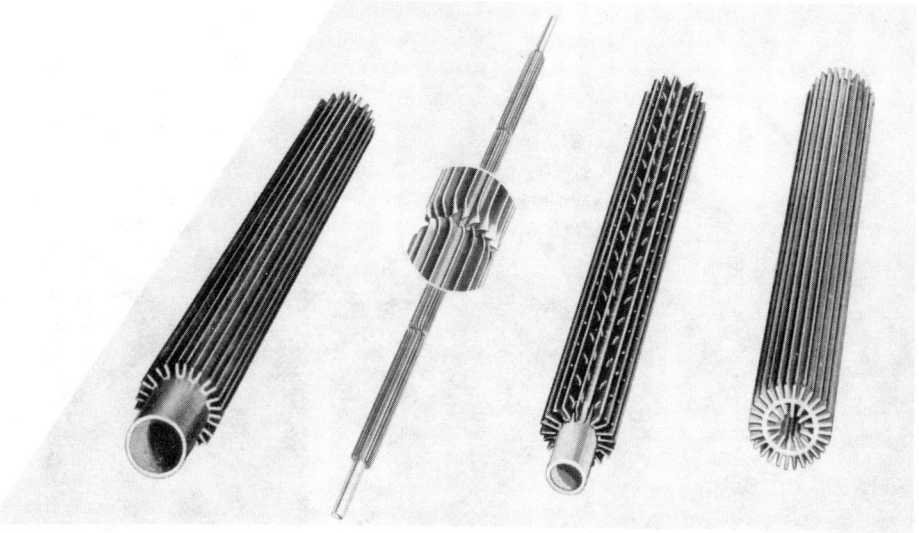

(a)　　　　　　　(b)　　　　　　　(c)　　　　　　　(d)

Fig. 14　Longitudinal fins on individual **tubes**: (a) continuous plain, (b) cut and twisted, (c) perforated, and (d) internal and external longitudinal fins (courtesy of Brown Fintube Company, Tulsa, OK).

employed. Tube-fin exchangers have usually lower compactness compared to the plate-fin unit. Tube-fin exchangers with an area density of about 3300 $m^2/m^3$ (1000 $ft^2/ft^3$) have been built. These are also used for the specific applications for power plants, propulsive systems, and air conditioning and refrigeration industries.

An air-cooled exchanger† is a tube-fin exchanger in which hot process fluids, usually liquids or condensing fluids, flow inside the tubes, and atmospheric air is circulated outside by forced or induced draft over the extended surface. Characteristics of this type of exchangers are shallow tube bundles (short air flow length) and large face area due to the design constraint on the fan power and low density of air.

3.4　Regenerative Heat Exchangers

These are the storage type heat exchangers described earlier and also referred to simply as <u>regenerators</u>. The heat transfer in a regenerator is generally of cellular structure referred to as a matrix. In order to have continuous operation in a regenerator, either the matrix must be moved periodically in and out of the fixed streams of gases, as in a <u>rotary</u> regenerator, or the gas flows must be diverted to and from the fixed matrices as in a <u>fixed-matrix</u> regenerator. The latter regenerator is also sometimes referred to as a <u>periodic-flow</u> regenerator‡ or reversible heat accumulator.

---

†If the process fluid is water, it is also referred to as a <u>dry cooling tower</u>.

‡Both the rotary matrix and fixed matrix regenerators have been designated as periodic-flow heat exchangers by Kays and London [1], because from the viewpoint of an observer riding on the matrix, identical periodic conditions are experienced.

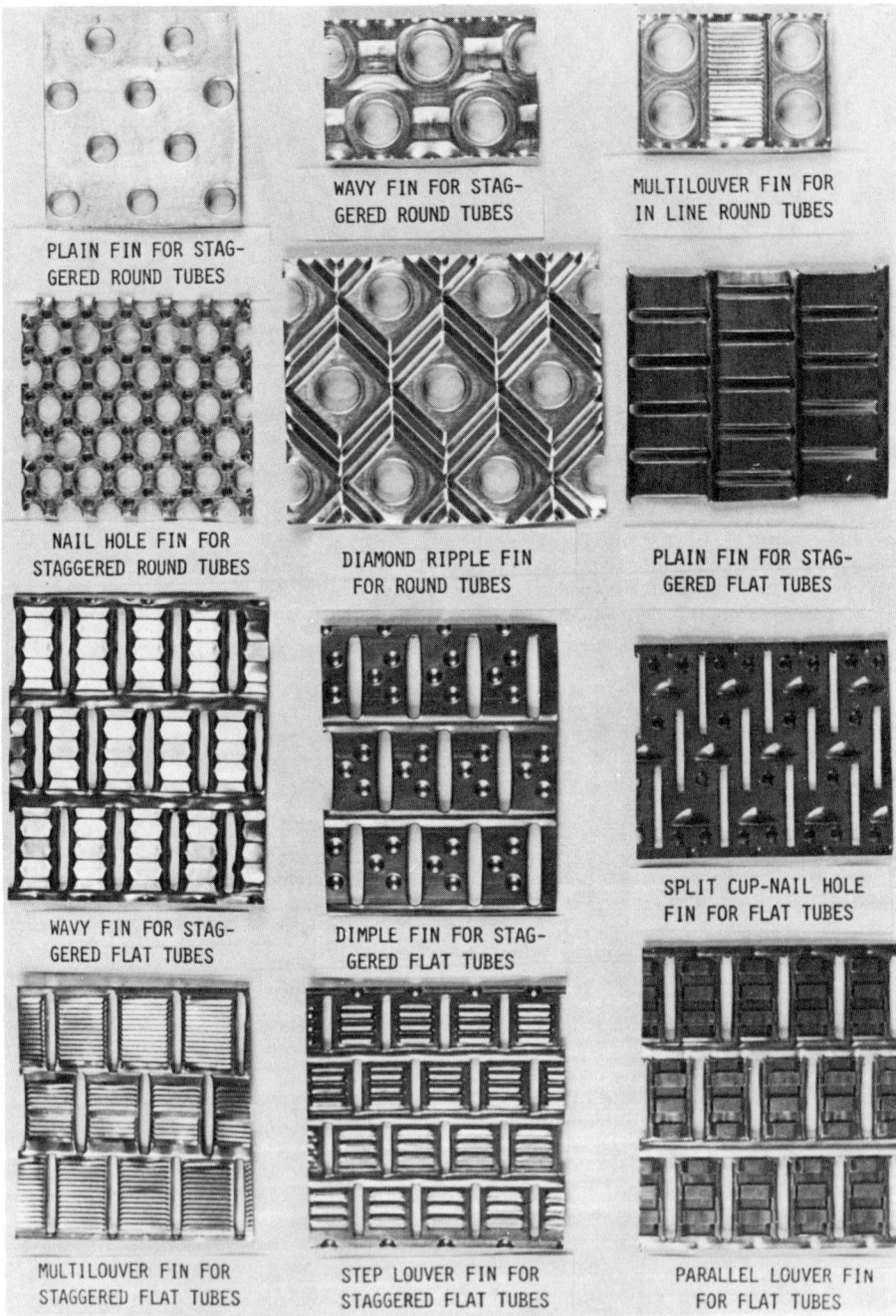

Fig. 15 Continuous fins on an array of tubes (courtesy of Harrison Radiator Div., GMC, Lockport, NY).

Fig. 16  Internally finned tubes (courtesy of Forged-Fin Division, Noranda Metal Industries, Inc., Newton, CT)

Thus, in a rotary regenerator, the matrix continuously rotates with a constant fraction of the core continuously in the hot-fluid stream and the remaining fraction in the cold-fluid stream; the outlet fluid temperature varies across the flow area. In a fixed matrix regenerator, the hot and cold fluids are ducted through the use of valves to the different parts of the regenerator in alternate operating periods; the outlet fluid temperature varies with time.

The thermodynamically superior counterflow arrangement is usually employed for storage type heat exchangers. When the rotational speed or the frequency of switching hot and cold fluids through such a regenerator is increased, its thermal performance approaches that of a pure counterflow heat exchanger. For some applications, a parallel flow arrangement may be used, but there is no counterpart of the single-pass or multipass crossflow arrangements so common in recuperators. Note that for a counterflow rotary regenerator, there are no complexities in the header design, but the design of seals to prevent leakages of hot and cold fluids becomes a difficult task especially if the two fluids are at significantly differing pressures.

Major advantages of the regenerators are: A much more compact surface may be employed compared to a recuperator. The cost of the regenerator surface per unit of transfer area is generally substantially lower than the equivalent recuperator. The matrix surface has self-cleaning characteristics because of periodic flow reversals. Because of compact surface area density and the counterflow arrangement, the regenerator is ideally suited for the gas-to-gas heat exchanger applications requiring high exchanger effectiveness generally exceeding 85%. A major disadvantage of the regenerator is the unavoidable carryover of a small fraction of the flow trapped in the passage at the moment of periodic flow switching. Where fluid contamination is prohibited, regenerators cannot be used. Other disadvantages are listed separately next for rotary and stationary regenerators.

Rotary regenerators. These are of two types: (1) a disk type in which the matrix (heat transfer surface) is in a disk form and fluids flow axially as in Fig. 17a; (2) a drum type in which the matrix is in a hollow drum form and fluids flow radially as in Fig. 17b. Two disk-type regenerators are shown in Figs. 18 and 19.

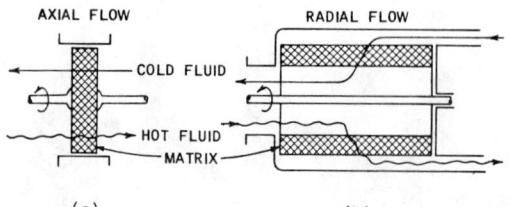

Fig. 17 Rotary regenerators: (a) disk-type, (b) drum-type [1].

The drum-type regenerator offers some important advantages in a vehicular gas turbine power plant application: The drum is wrapped around the turbomachinery thus completely eliminating interconnecting ducts, providing uniform flow distribution within the regenerator, and providing an insulating and noise blanket around the engine. However, it requires handling and putting together many thousands of flow passages of the drum width, and also requires expensive clearance type seals maintained by a roller assembly. Thus the cost of fabricating a drum-type regenerator is significantly higher than that for a disk-type regenerator and hence the former is not used in any applications. The disk-type regenerator, simply referred to as a rotary regenerator, will be discussed next.

In this exchanger, any of the plate-fin surface geometries could be used in the matrix. However, generally interrupted passage surfaces (such as strip fins, louver fins) are not used because a transverse (to the main flow direction) flow leakage will be present, if the two fluids are at different pressures. This leak mixes the two fluids (contaminates the lower pressure fluid) and reduces the exchanger effectiveness. Hence, the matrix generally has continuous (uninterrupted) flow passages. Flat or wavy spacers are used to stack the "fins"[†] (see Fig. 20c). The fluid is unmixed at any cross section for these surfaces. Some examples are shown in Fig. 20. The herringbone or skewed passage matrix does not require spacers for stacking the "fins". The design Reynolds number range for the rotary regenerator is 100-1000.

The matrix in the regenerator is rotated by a hub shaft or a peripheral ring gear drive. Every matrix element is periodically passed from the hot to the cold stream and back again. The time required for a complete rotation of the matrix is equivalent to the total period of a fixed-matrix regenerator. In a rotary regenerator, the stationary seal locations control the desired frontal areas for each fluid and also serve to minimize the primary leakage from the high pressure fluid to the low pressure fluid. The design flexibility of selecting different frontal area is not possible for a fixed-matrix regenerator but instead, different hot and cold flow periods are selected. The seal leakage is obviously nonexistent in a fixed-matrix regenerator.

A temperature distribution for a regenerator is shown in Fig. 21. It is interesting to note that the wall temperature periodically fluctuates between the solid line limits shown.

---

[†] It should be emphasized that in a regenerator matrix, however, all the surface acts as a direct heat absorbing and rejecting surface (a primary surface); there is no secondary surface.

CLASSIFICATION OF HEAT EXCHANGERS 33

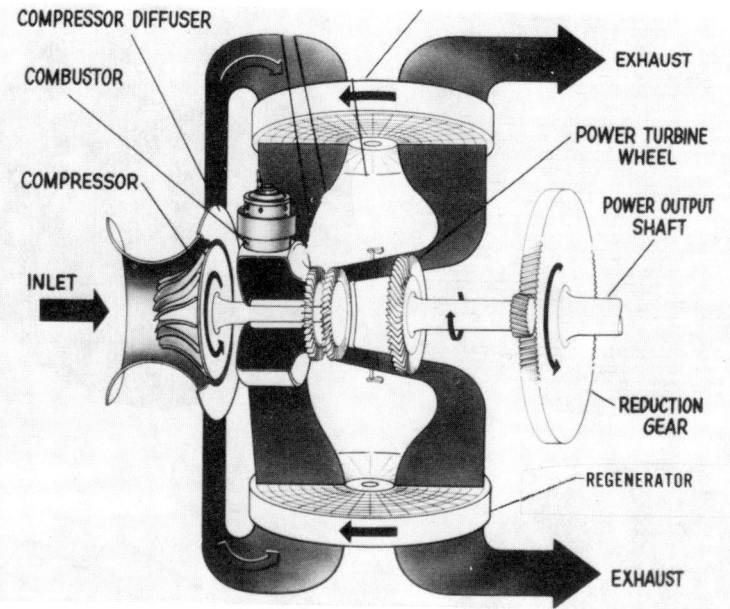

Fig. 18 Two disk-type rotary regenerators for a vehicular gas turbine engine.

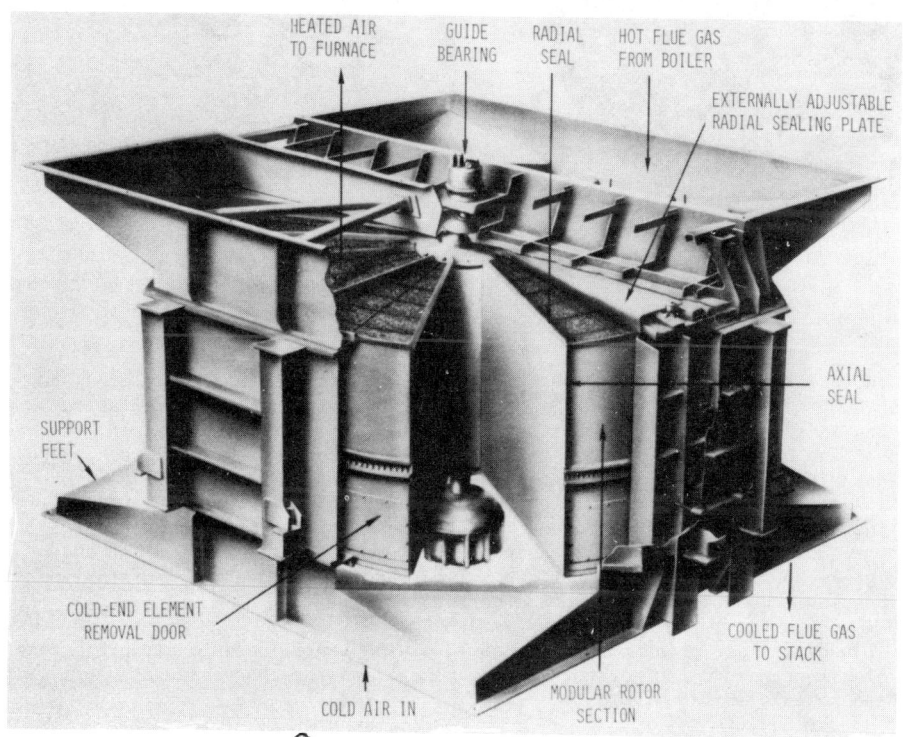

Fig. 19 A Ljungstrom® air preheater (courtesy of C-E Air Preheater, Combustion Engineering, Inc., Wellsville, NY).

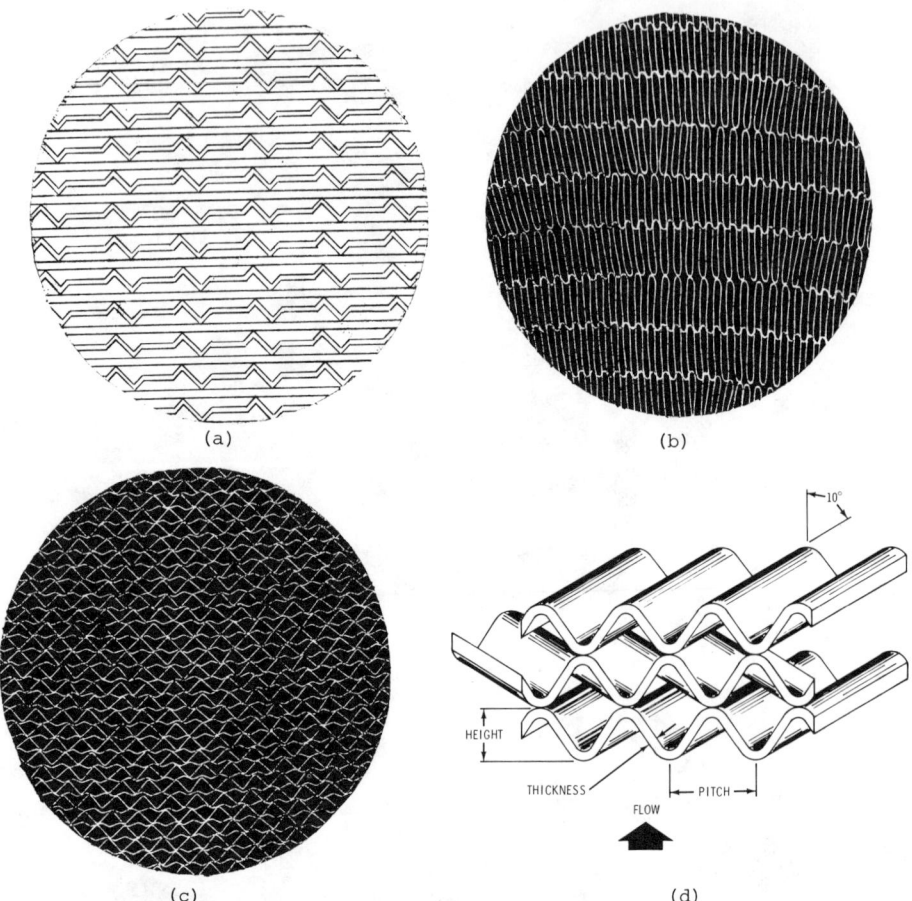

Fig. 20 Continuous passage matrices for a rotary regenerator: (a) notched plate, (b) deepfold rectangular, (c) triangular, and (d) herringbone matrices.

Rotary regenerators have been designed for surface area density of up to about 6600 $m^2/m^3$ (2000 $ft^2/ft^3$). They can employ thinner stock material resulting in the lowest amount of material for the same effectiveness and pressure drop of any heat exchanger known today. The metal rotary regenerators have been designed for temperatures up to about 870°C (1600°F). For higher temperature applications, ceramic matrices are used. Because of periodic flow reversal, fouling is generally not a problem. This regenerator cannot withstand large pressure differences (greater than about 400 kPa or 60 psi) between hot and cold gases, because the design of seals (wear-and-tear and thermal distortion) is the single most difficult problem to resolve.

Ljungstrom air preheaters for thermal power plants, ceramic air preheaters for high temperature incinerators, and regenerators for the vehicular gas turbine power plant are typical applications of rotary regenerators.

# CLASSIFICATION OF HEAT EXCHANGERS

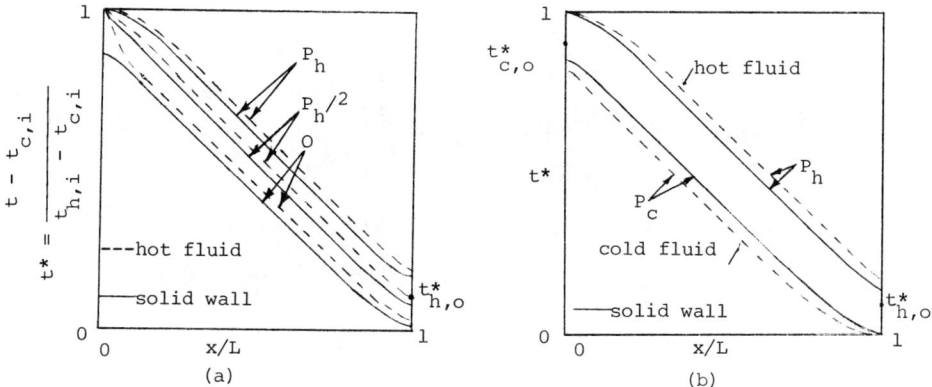

Fig. 21 (a) Hot side solid fluid temperature excursion, and (b) balanced regenerator temperature distributions at switching instant.

<u>Fixed-Matrix Regenerator</u>. This type is also referred to as a <u>periodic-flow</u> or <u>valved</u> or <u>stationary</u> regenerator. For continuous operation, this exchanger has at least two identical matrices operated in parallel, but usually three or four as shown later in Figs. 23 and 24.

Fixed-matrix regenerators are of two major categories: (1) non-compact regenerators used for high temperature applications (925-1500°C or 1700-2700°F) with corrosive gases, such as a Cowper stove (Fig. 22) for a blast furnace for steel industries and air preheaters for coke manufacture and glass melting tanks.

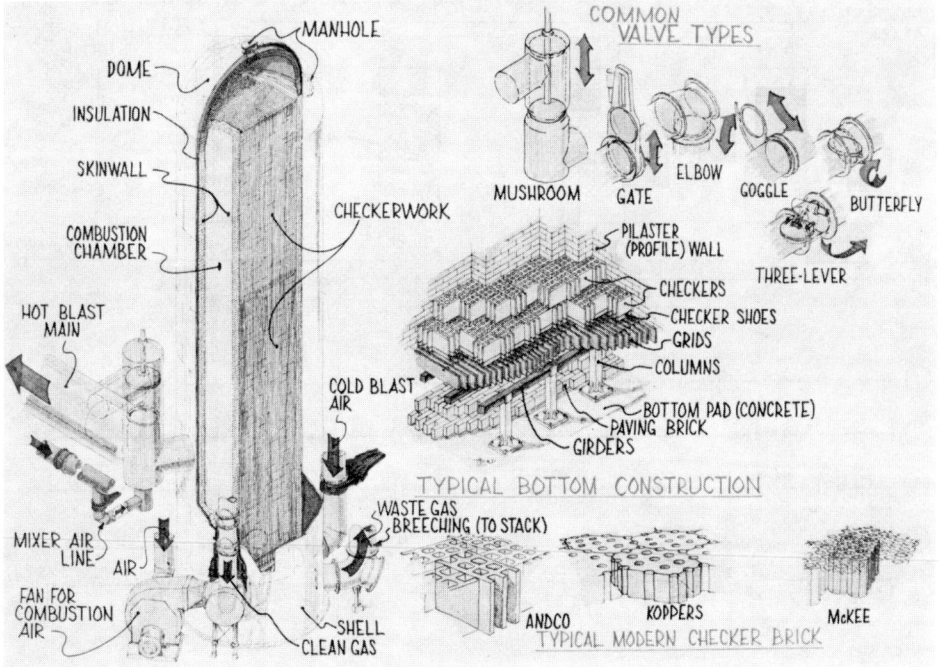

Fig. 22 A Cowper stove (courtesy of Andco Industries, Inc., Buffalo, NY).

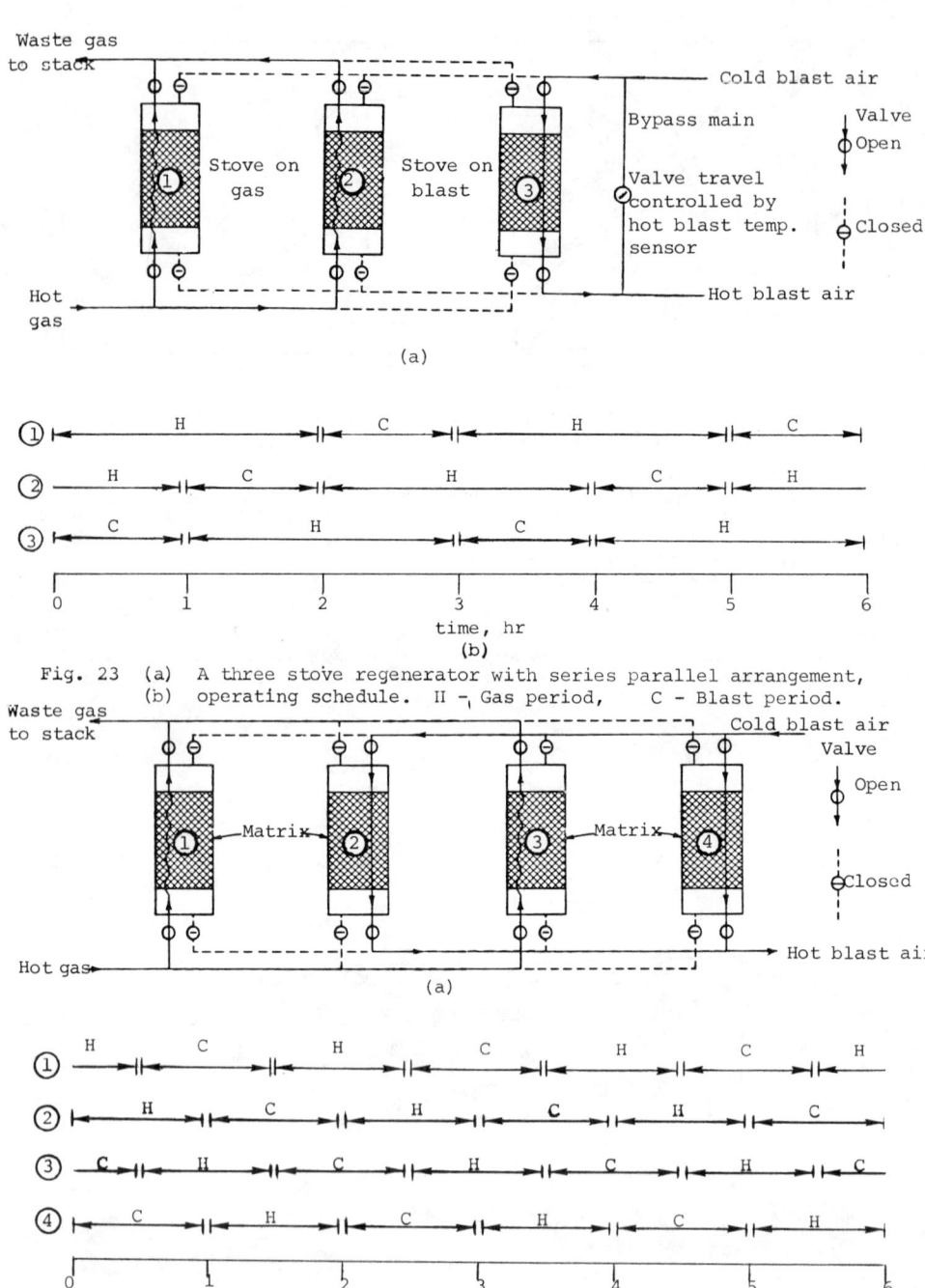

Fig. 23 (a) A three stove regenerator with series parallel arrangement, (b) operating schedule. H - Gas period, C - Blast period.

Fig. 24 (a) A four stove regenerator with staggered parallel arrangement, (b) Operating schedule. H - Gas period, C - Blast period.

(2) highly compact regenerators from low to high temperature applications such as refrigeration, cryogenic process for air separation, and Stirling engine. The regenerator, a key thermodynamic element in the Stirling engine cycle, has only one matrix, and hence it does not have continuous fluid flows as in other heat exchangers.

In a Cowper stove, it is highly desirable to have the temperature of the outlet heated air approximately constant with time. The difference between the outlet temperatures of the heated air at the beginning and end of a period is referred to as a <u>temperature swing</u>. To minimize the temperature swing, three or four stove regenerators, as shown in Figs. 23 and 24, are employed. In the "series parallel" arrangement of Fig. 23, part of the cold air (blast) flow is bypassed around the stove and mixed with the heated air (hot blast) leaving the stove. Since the stove cools as the blast is blown through it, it is necessary to constantly decrease the amount of blast bypassed while increasing the blast through the stove by a corresponding amount. In the "staggered parallel" arrangement of Fig. 24, two stoves on air are maintained out of phase by one-half period. In this arrangement, cold blast is routed through a "hot" stove and a "cool" stove (i.e. through which cold blast has blown for one-half period) rather than being bypassed. The amount of blast through the hot stove is constantly increased while that through the cool stove is decreased by the same amount. At the end of one-half period, the hot stove's inlet valve is fully open, the cool stove's inlet valve is fully closed. At this point, the cool stove is put "on gas", the "hot stove" becomes the "cool stove", and a new "hot" stove is switched in.

The heat transfer surface used in the aforementioned high temperature regenerator is made of refractory bricks simply referred to as <u>checkers</u>. The commonly used checker shapes and their surface area density are shown in Fig. 25. The checker flow passage (referred to as flue) size is relatively large primarily due to the fouling problem.

The surface geometries used for the compact fixed-matrix regenerator are similar to those used for rotary regenerators, but in addition used are the quartz pebbles, steel or copper or lead shots, copper wool, randomly packed woven screens and crossed rods.

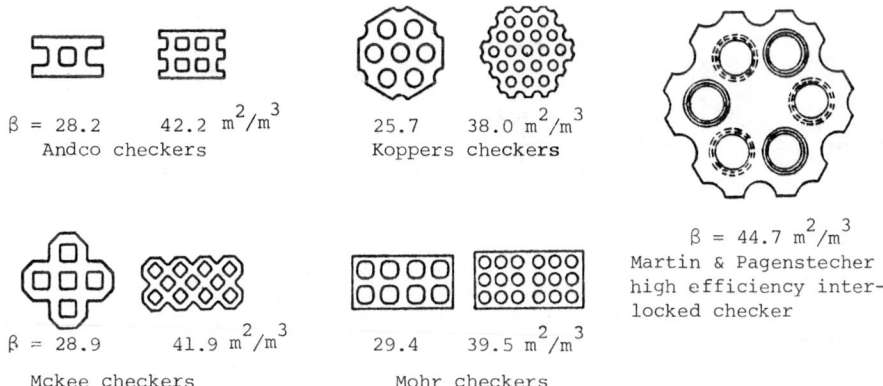

Fig. 25 Checkers used for blast furnace regenerators.

## 4. CLASSIFICATION ACCORDING TO FLOW ARRANGEMENTS

Common flow arrangements of the fluids in a heat exchanger are classified in Fig. 1. The choice of a particular flow arrangements is dependent upon the required exchanger effectiveness, fluid flow paths, packaging envelope, allowable thermal stresses, temperature levels and other design criteria. Some of the basic flow arrangements for a two-fluid heat exchanger are described below for single-pass and multipass heat exchangers.

A <u>fluid</u> is considered to have made <u>one</u> pass if it flows through a section of the heat exchanger through its full length. If the fluid is reversed and flows through an equal or different section, it is considered to have made a second pass of equal or different size. A <u>heat exchanger</u> is considered as being multipass, if it does not represent any of the single-pass flow arrangements when "unfolded."[†] A two-pass and a single-pass crossflow heat exchangers are shown in Fig. 26. Fluid 2 in both these exchangers makes two passes.

### 4.1 Single-Pass Heat Exchangers

<u>Counterflow[‡] heat exchanger</u>. In this type, shown in Fig. 27, the fluids flow in directions opposite to each other. Temperature variation of the two fluids in such an exchanger may be idealized as one-dimensional as shown in Fig. 28. As will be shown later, the counterflow arrangement is thermodynamically superior to any other flow arrangement. Moreover, the structural temperature differences that produce thermal stresses are minimized. However, with very compact heat exchanger surfaces, there are manufacturing difficulties associated with the true counterflow arrangement because it is necessary to

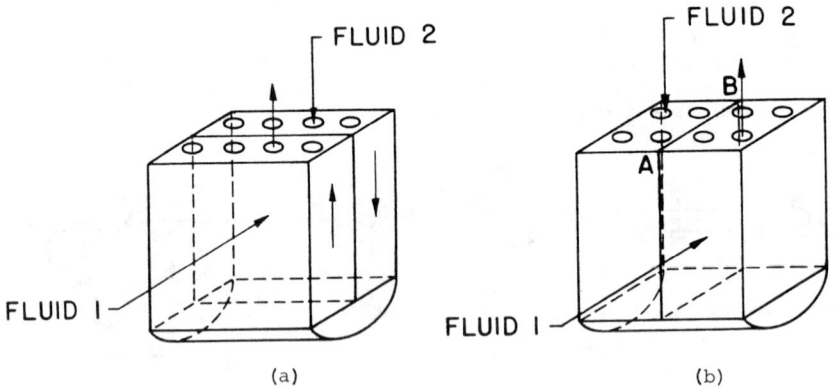

Fig. 26 (a) A two-pass cross-counterflow heat exchanger; (b) a single-pass crossflow heat exchanger; the plane through AB may be idealized as adiabatic.

---

[†] Folding is resorted to for the purpose of controlling envelope size. For instance in Fig. 26(a), the option of two passes for Fluid 2 and one pass for Fluid 1 is elected. Another option, achieved by unfolding the exchanger of Fig. 26(a), would be one pass for Fluid 2 and two passes for Fluid 1.

[‡] Counterflow is also referred to as counter-current flow.

# CLASSIFICATION OF HEAT EXCHANGERS

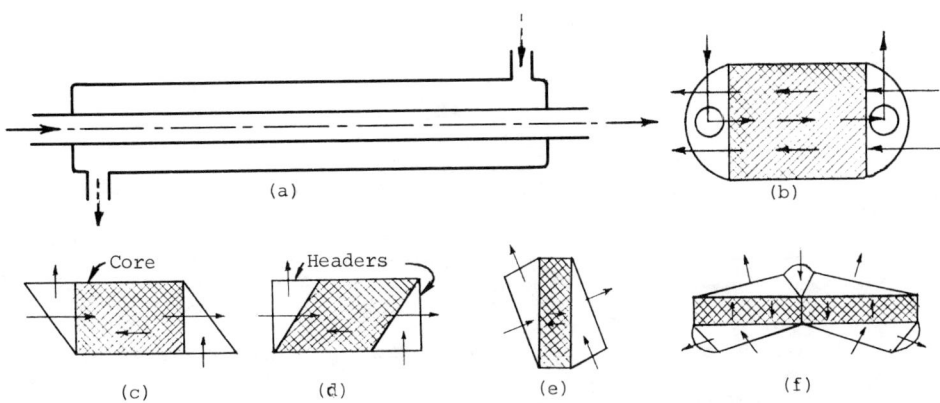

Fig. 27  (a) A double-pipe heat exchanger with pure counterflow; (b)-(f) Plate-fin exchangers with counterflow core and crossflow headers.

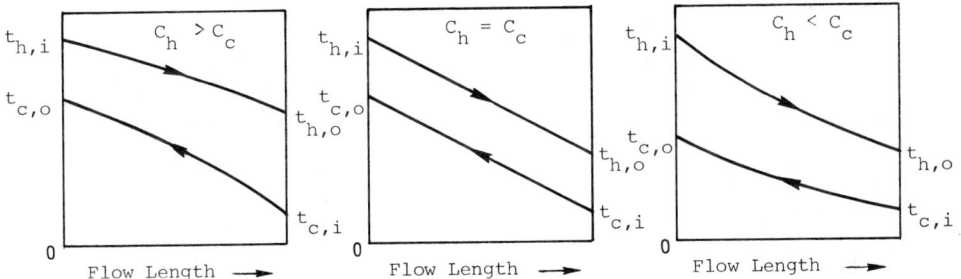

Fig. 28  Temperature distributions in a counterflow heat exchanger.

separate the fluids, and the headering problem is complex and difficult. Some header arrangements are shown in Fig. 27 (b)-(f).

Parallel flow[†] heat exchanger. In this type, the fluid streams enter together at one end, flow through in the same direction and leave together at the other end. Figure 27(a) with the dashed arrows reversed would then describe parallel flow. Fluid temperature variations, idealized as one-dimensional, are shown in Fig. 29. Thermodynamically, this is one of the poorest flow arrangements. Moreover, large temperature differences exist at the inlet side which may induce high thermal stresses. This flow arrangement may be used in the following applications: (1) it often produces more uniform tubewall temperature and not as high tubewall temperature as in a counterflow arrangement. For this reason, it is sometimes used with temperature sensitive materials; (2) the desired exchanger effectiveness is low and is to be maintained approximately constant over a large flow rate range; (3) the application allows piping only suited to parallel flow; and (4) it provides early initiation of nucleate boiling for boiling applications.

Crossflow heat exchanger. In this type, shown in Fig. 30, the fluid flows

---

[†]Parallel flow is also referred to as co-current flow.

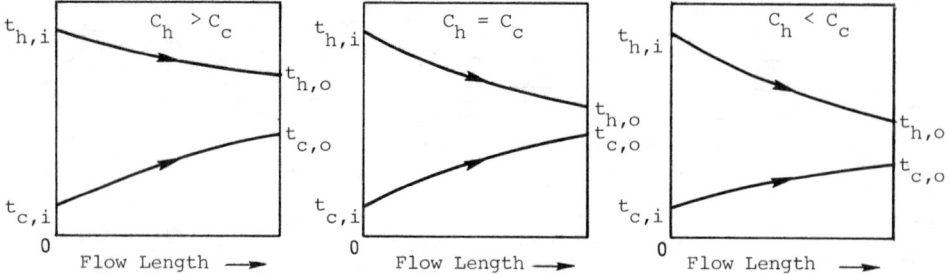

Fig. 29  Temperature distributions in a parallel flow heat exchanger.

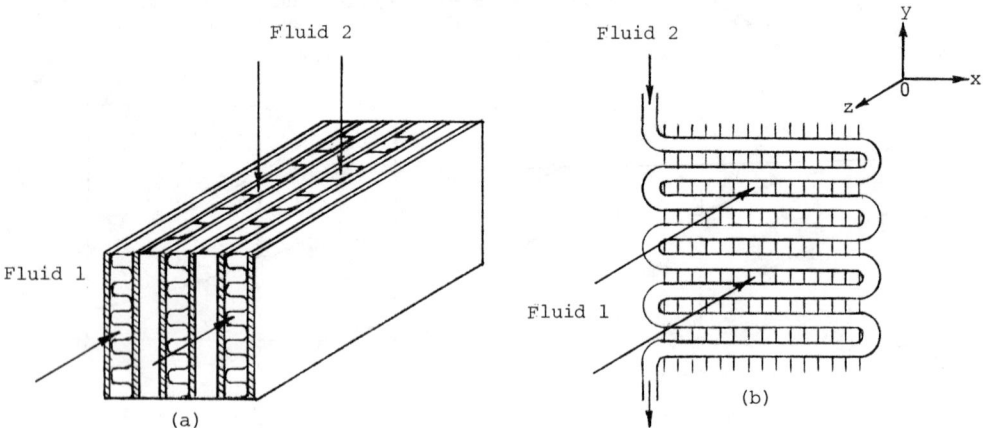

Fig. 30  (a) A plate-fin unmixed-unmixed crossflow heat exchanger; (b) a serpentine tube-fin unmixed-mixed crossflow heat exchanger.

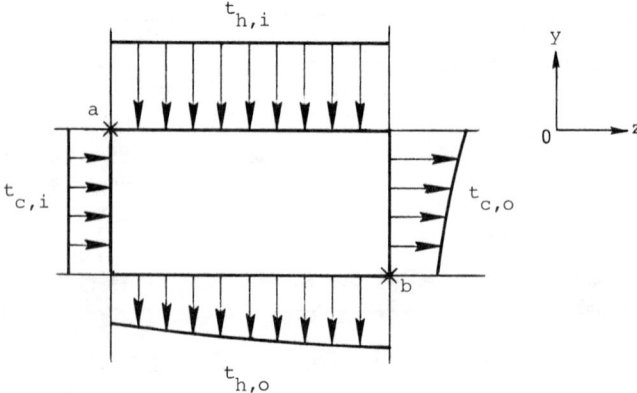

Fig. 31  Temperature distribution in an unmixed-unmixed crossflow heat exchanger.

# CLASSIFICATION OF HEAT EXCHANGERS

are normal to each other. Typical fluid temperature variations are idealized as two-dimensional and are shown in Fig. 31 only for the inlet and outlet sections. Thermodynamically, the effectiveness for the crossflow exchanger falls in between that for the counterflow and parallel flow arrangements. The largest structural temperature differences exist at the "corners" of the entering and leaving hot and cold fluids, such as points a and b in Fig. 31. This is one of the most common flow arrangements used for compact heat exchangers, because it greatly simplifies the header design at the entrance and exit of each fluid. If the desired heat exchanger effectiveness is high (say greater than 80%), the size penalty for crossflow may become excessive. In such a case, a counterflow unit is preferred.

In a crossflow arrangement, "mixing" of either fluid stream may or may not occur depending upon the design. A fluid stream is considered "unmixed" when it passes through individual flow channels or tubes with no fluid mixing between adjacent flow channels. Fluids 1 and 2 in Fig. 30a are "unmixed." Fluid 1 in Fig. 30b is unmixed, while Fluid 2 is considered "mixed" because of only one flow channel. The temperature of an unmixed fluid, such as in Fig. 30, is a function of two coordinates y and z, and it cannot be treated as constant across a cross section perpendicular to the general flow direction. Typical temperature distributions of the unmixed fluids at exchanger outlet sections are shown in Fig. 31. The temperature of a mixed fluid (Fluid 2 in Fig. 30b) is a function of only one coordinate y. The temperature change per pass (in the x direction) of Fluid 2 in Fig. 30b is small compared to the total. In a plate-baffled shell-and-tube exchanger, the shell fluid is considered mixed and the tube fluid unmixed. Thus, three idealized flow arrangement combinations for a crossflow exchanger are: (1) both fluids unmixed; (2) one fluid mixed, the other unmixed; and (3) both fluids mixed (practically a less important case). For equal $N_{tu}$ and $C^*$, the exchanger effectiveness is in decreasing order for these three flow arrangements. The higher the mixing, the lower is the exchanger effectiveness having all other parameters constant.

## 4.2 Multipass Heat Exchangers

Heat exchangers of any of the foregoing three basic flow arrangements can be put into series to make a multipass unit. One of the major advantages of multipassing is to increase the exchanger overall effectiveness over individual pass effectivenesses. If the overall direction of the two fluids is chosen as counterflow (see Figs. 32 and 35), the exchanger overall effectiveness approaches that of a pure counterflow exchanger as the number of passes increases. The multipass arrangements are classified according to the type of construction, for example, extended surface, shell-and-tube, or plate exchangers (Fig. 1).

Overall cross-counterflow arrangement.[†] In this arrangement, two or more passes are put in series with each pass usually having crossflow, although any one of the foregoing three basic flow arrangements could be employed. The fluid on one side usually goes straight through, while the fluid on the other side turns to flow from pass to pass. Usually the flow direction is chosen such that an overall counterflow is obtained, as shown in Fig. 32.

The exchanger effectiveness for this arrangement depends upon whether the fluids are mixed or unmixed between passes on each side. If guide vanes for a plate-fin or hairpin/U tubes for a tubular exchanger are used in the header, the fluid is considered unmixed in the header.

---

[†]This arrangement is most common for extended surface exchangers.

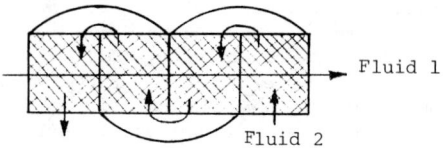

Fig. 32  A four-pass cross-counterflow heat exchanger.

Multipassing retains the header and ducting advantages of the simple crossflow heat exchanger, while it is possible to approach the thermal performance of a true counterflow heat exchanger for an overall cross-counterflow arrangement. The structural temperature differences are reduced by a factor n of an n equal pass exchanger relative to a single-pass crossflow design for the same terminal temperatures. For high temperature applications, different materials may be used in different passes to reduce material cost.

Overall cross-parallel flow arrangement. This arrangement is similar to the preceding one except that the overall direction of both fluids is the same. Thus if the direction of the Fluid 2 in Fig. 32 is reversed, overall parallel flow would be achieved. Sometimes this arrangement is used to prevent the freezing of the hot fluid in the core near the inlet of the cold fluid (for example, air). The exchanger overall effectiveness for this arrangement approaches that of a pure parallel flow exchanger as the number of passes increases.

Parallel counterflow, shell fluid mixed. This is one of the most common flow arrangements used in shell-and-tube heat exchangers. One of the simplest flow arrangements is one shell-pass two tube-passes,[†] as shown in Fig. 33. The heat exchanger with this arrangement is also referred to as a 1-2 heat exchanger.

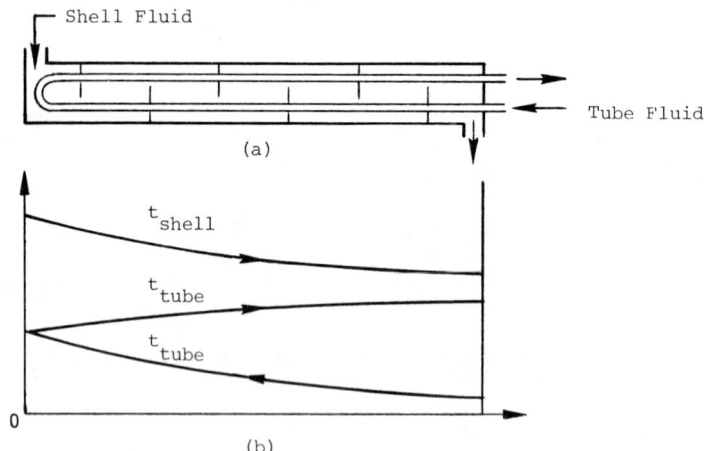

Fig. 33  (a) A 1-2 heat exchanger (one shell pass and two tube passes); (b) corresponding temperature distribution.

---

[†]Baffles serve to "mix" the shell fluid in the sense previously considered, and maintain the direction of the shell fluid normal to the tubes. The heat transfer coefficient for flow normal to tubes is considerably higher than that for flow parallel to the tubes.

# CLASSIFICATION OF HEAT EXCHANGERS 43

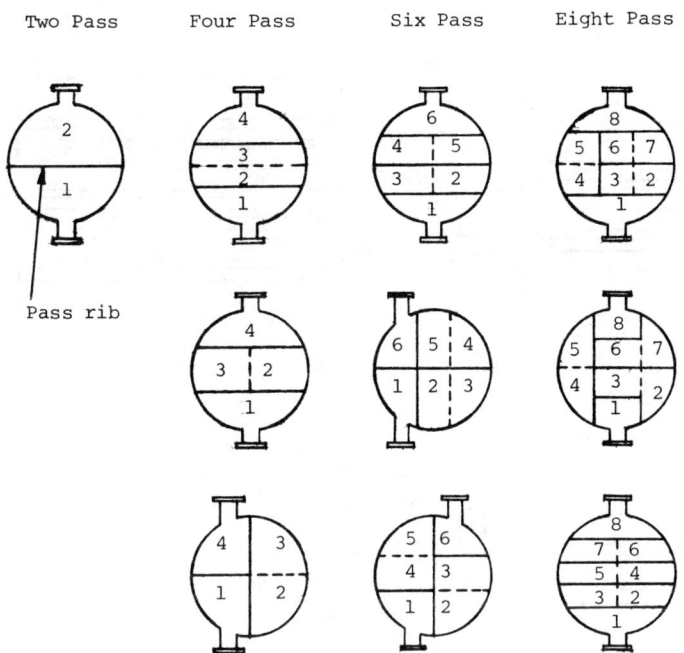

Fig. 34 Common tubeside multipass arrangements in shell-and-tube exchangers (tubes are not shown in order to simplify the sketches).

As the tubes are rigidly mounted only at one end, thermal expansion is readily accommodated. As the shell fluid is idealized as well mixed, its temperature is constant at any cross section. Hence, reversing the tube fluid flow direction will not change the idealized temperature distribution of Fig. 33b and the exchanger effectiveness.

Increasing the even number of tube passes[†] from two to four, six, etc. decreases the exchanger effectiveness slightly; and in the limit when the number of tube passes approaches infinity with one shell pass, the exchanger effectiveness approaches that for a single-pass crossflow with both fluids mixed. Common tubeside multipass arrangements are shown in Fig. 34.

The odd number of tubes per shell has slightly better effectiveness when the shell fluid flows countercurrent to the tube fluid for more than half the tube passes. However, this is an uncommon design and may result in structural and thermal problems in manufacturing and design.

Since the 1-n exchanger has lower effectiveness, multipassing of this basic arrangement may be employed to approach the counterflow effectiveness. The heat exchanger with the most general flow arrangement would have m shell passes and n tube passes. Figure 35 represents two such exchangers.

<u>Divided-flow, shell fluid mixed</u>. In this arrangement, the shell fluid enters at the center, divides into two equal streams and leaves at both ends as shown in Fig. 36. TEMA J-type shell has this type of flow arrangement. Kays and London [1] call this arrangement as the split-flow, shell fluid mixed.

---
[†]This exchanger is referred to as the 1-n exchanger.

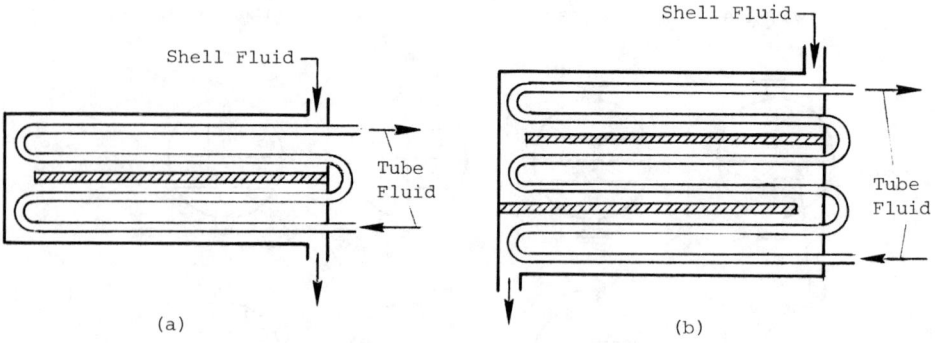

Fig. 35 (a) A 2 shell-pass 4 tube-pass exchanger; (b) 3 shell-pass 6 tube-pass exchanger.

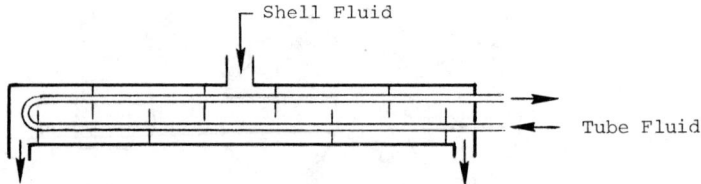

Fig. 36 Divided-flow, shell fluid mixed.

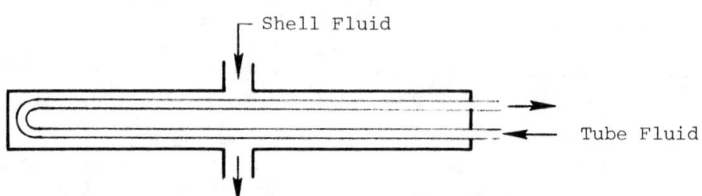

Fig. 37 Split-flow, shell fluid mixed.

<u>Split-Flow, shell fluid mixed</u>. This arrangement differs from the preceding one in two ways: a longitudinal baffle is used, and only one exit nozzle is employed. This arrangement, as shown in Fig. 37, is a variant of the 1-2 exchanger. TEMA G-type shell has this type of flow arrangement. As the "mixing" is less severe than for the 1-2 exchanger of Fig. 33, the effectiveness is higher, particularly at high $N_{tu}$ for this arrangement.

<u>n Parallel plates multipass arrangements</u>. In a plate exchanger, there exists a large number of feasible multipass flow arrangements. Some of them are as shown in Fig. 38. Essentially, these are combinations of parallel flow and counterflow arrangements with heat transfer taking place in adjacent channels. These arrangements can be obtained simply by properly gasketing around the ports in the plates.

# CLASSIFICATION OF HEAT EXCHANGERS

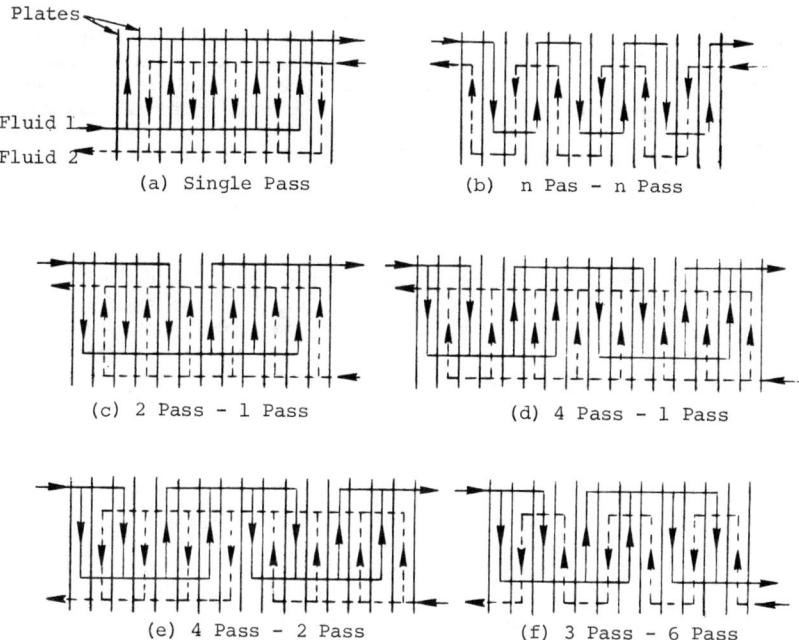

Fig. 38  n-parallel plates single and multipass arrangements.

## 5. CLASSIFICATION ACCORDING TO NUMBER OF DIFFERENT FLUIDS

Most processes of thermal energy recovery or dissipation involve transfer of thermal energy between two fluids. Hence, two-fluid heat exchangers are the most common. Three-fluid heat exchangers are widely used in cryogenics and some chemical processes, e.g., air separation systems, helium-air separation unit, purification and liquefaction of hydrogen, ammonia gas synthesis, etc. The design theory of three-fluid and multifluid heat exchangers is algebraically very complex.

## 6. CLASSIFICATION ACCORDING TO HEAT TRANSFER MECHANISMS

The basic heat transfer mechanisms employed for transfer of thermal energy from fluids on one side of the exchanger to the wall (separating the fluids on the other side) are: single-phase convection (forced or free), two-phase convection (condensation or evaporation, by forced or free convection) and combined convection and radiation heat transfer. Any of these mechanisms individually or any combinations of them could be active on each side of the exchanger. Such a classification is provided in Fig. 1.

Single-phase convection occurs on both sides of the following two-fluid exchangers: automotive radiators and passenger space heaters, regenerators, intercoolers, economizers, etc. Single-phase convection on one side and two-phase convection on the other side (with or without desuperheating and sub-cooling, and with or without noncondensables) occur in the following two-fluid exchangers: steam power plant condensers, automotive and air-cooled condensers, evaporators, steam generators, etc. Two-phase convection could occur on each side of a two-fluid heat exchanger such as condensing on one side, evaporating on the other side. Multi-component two-phase convection occurs in condensation of mixed vapors in distillation of hydrocarbons. Radiant heat transfer combined with convective heat transfer plays a role in liquid metal heat exchangers and high temperature fixed-matrix regenerators.

SUMMARY

Heat exchangers have been classified according to transfer processes, degree of surface compactness, construction features, flow arrangements, number of fluids, and heat transfer mechanisms. A summary is provided in Fig. 1. Common terminology associated with the major components of these exchangers is also introduced. Major applications of many heat exchangers are mentioned.

ACKNOWLEDGEMENTS

The author is grateful to Prof. A.L. London of Stanford University and Dr. J. Taborek of Heat Transfer Research, Inc. for providing a critical review on this article.

NOMENCLATURE

| | |
|---|---|
| $A$ | total heat transfer surface area (both primary and secondary, if any) on one side of a heat exchanger, $m^2$ |
| $D$ | tube outside diameter, m |
| $D_h$ | hydraulic diameter of flow passages, m |
| $q$ | heat transfer rate in the exchanger, W |
| $\Delta t_m$ | true mean temperature difference, °C |
| $V$ | heat exchanger volume, $m^3$ |

Subscripts

| | |
|---|---|
| c | cold fluid side |
| h | hot fluid side |

REFERENCES

1. W.M. Kays and A.L. London, <u>Compact Heat Exchangers</u>, Second Edition, McGraw-Hill, New York (1964).

2. M. Suo, Calculation methods for performance of heat exchangers enhanced with fluidized beds, <u>Letters in Heat and Mass Transfer</u>, Vol. 3, 555-564 (1976).

3. Tubular Exchanger Manufacturers Association, <u>Standards of TEMA</u>, Sixth Edition, New York (1978).

4. A.C. Mueller, Heat exchangers, in <u>Handbook of Heat Transfer</u>, edited by W.M. Rohsenow and J.P. Hartnett, Chapter 18, pp. 1-113, McGraw-Hill (1973).

# THERMAL-HYDRAULIC FUNDAMENTALS: SINGLE-PHASE

# Air-Cooled Heat Exchangers

**A. ŽUKAUSKAS**
Academy of Sciences of the Lithuanian SSR
Vilnius, USSR

ABSTRACT

The lecture is introduced with a review of various schemes and types of air-cooled heat exchangers. The heat transfer and pressure drop of different configuration of smooth, rough and finned banks of tubes in crossflow of air are examined in detail. The study covers the range of Reynolds numbers from 1 to $10^7$, which gives an opportunity to characterize in detail the process of heat transfer in subcritical, critical and supercritical flow regimes. Analytical expressions and nomograms on heat transfer and pressure drop of various banks of tubes in crossflow are given.

NOMENCLATURE

a Relative transverse pitch, $s_1/D$

A Plain tube surface, $m^2$

$A_{tot}$ Finned tube full surface, $m^2$

b Relative longitudinal pitch, $s_2/D$

c Constant

$c_p$ Specific heat, J/kg.K

$c_z, c_z'$ Factor for the number of rows in a bank

d Tube diameter at fin base, m

D Outside diameter of tube, m

E Efficiency of fin

h Height of fin, m

H Height of channel, m

k Height of surface elements, m

$k^+$ Dimensionless roughness height, $ku_*/\nu$

| | |
|---|---|
| $k_q$ | Blocking ratio, $D/H$ |
| $K$ | Complex dimensionless terms, $Nu_f Pr_f^{-n}(P_f/Pr_w)^{-p}$ |
| $L$ | Length, m |
| $m$ | Power index of Re |
| $n$ | Power index of Pr |
| $p$ | Power index of $Pr_f/Pr_w$ |
| $\bar{p}$ | Pressure coefficient, $1-2(P_{\varphi=0} - P_\varphi)/\rho \bar{u}^2$ |
| $p_b$ | Main flow pressure, $N/m^2$ |
| $P_\varphi$ | Local pressure on the tube wall, $N/m^2$ |
| $\Delta p$ | Pressure drop, $N/m^2$ |
| $q_w$ | Specific heat flux, $W/m^2$ |
| $s$ | Fin spacing, m |
| $s_1$ | Transverse pitch of bank of tubes, m |
| $s_2$ | Longitudinal pitch of bank of tubes, m |
| $s_2'$ | Diagonal pitch of staggered bank, m |
| $t$ | Fin thickness, m |
| $Tu$ | Turbulence intensity, % |
| $u$ | Local fluid flow velocity, m/s |
| $\bar{u}$ | Mean velocity in the minimum inter-tube space, m/s |
| $u_b$ | Main flow velocity, m/s |
| $u_\varphi$ | Local velocity in the outer boundary layer, m/s |
| $u_*$ | Friction velocity, m/s |
| $u^+$ | Dimensionless velocity, $u/u_*$ |
| $y$ | Distance measured normal to the wall, m |
| $y^+$ | Dimensionless distance, $yu_*/\nu$ |
| $z$ | Number of tube rows in a bank |
| $\alpha, \bar{\alpha}$ | Local and average heat transfer coefficient, respectively, $W/m^2K$ |
| $\alpha_c, \bar{\alpha}_T$ | Reduced and average heat transfer coefficient of the finned tube, respectively, $W/m^2K$ |
| $\bar{\alpha}_F$ | Average heat transfer coefficient of fin height and circle, $W/m^2K$ |

# AIR-COOLED HEAT EXCHANGERS

$\bar{\alpha}_h$    Average heat transfer coefficient of fin height, $W/m^2K$

$\delta$    Thickness of the hydrodynamic boundary layer, m

$\varepsilon$    Finning factor of tube, $A_{tot}/A$

$\varepsilon_q$    Eddy diffusivity for heat transfer, $m^2/s$

$\varepsilon_\tau$    Eddy diffusivity for momentum transfer, $m^2/s$

$\upsilon$    Temperature, calculated from the surface, $^{\circ}C$

$\upsilon_*$    Friction temperature, $q_w/\rho c_p u_*$, $^{\circ}C$

$\upsilon^+$    Dimensionless temperature, $\upsilon/\upsilon_*$

$\theta$    Momentum thickness, m

$\lambda$    Thermal conductivity, W/mK

$\mu$    Dynamic viscosity, $m^2/s$

$\nu$    Kinematic viscosity, $m^2/s$

$\rho$    Density, $kg/m^3$

$\tau$    Shear stress, $N/m^2$

$\varphi$    Angle measured from the front stagnation point, deg

$\psi$    Factor for the non-uniform heat transfer of a fin

Eu    Euler number, $2\Delta p/\rho \bar{u}^2 z$

Nu    Nusselt number, $\alpha D/\lambda$

Pr    Prandtl number, $c_p \mu/\lambda$

$Pr_t$    Turbulent Prandtl number, $\varepsilon_t/\varepsilon_q$

Re    Reynolds number, $\bar{u}D/\nu$

St    Stanton number, $\alpha/\rho c_p \bar{u}$

Subscripts

f,b    Conditions of the main flow

w    Conditions on the wall

x, $\varphi$    Local conditions

## 1. INTRODUCTION

Coolers which employ atmospheric air as the coolant are quite common in industry. As to the advantages of air as the coolant against water, one might enumerate several separate aspects.

1. The large amount of air in the Earth's atmosphere is a convenient alternative to the limited resources of fresh water and to its progressively increasing global consumption.

2. Because of lower corrosion rates in air, the expensive alloy steels can be replaced by carbon steels or by light alloys.

3. In many implementations, the replacement of water by air is at the same time a way of decreasing the weight of the apparatus.

4. With air, closed circulation loops are not necessary, and visual observation of the internal flows is much simpler.

5. Any problems of scale formation are excluded.

6. Many of the air-cooled heat exchangers are portable and independent of water supply.

7. The pressure of air being lower, it is immiscible with the heat-carrier fluid.

Of the disadvantages of air as a coolant, the most prominent one lies in its heat transfer parameters, which are considerably lower, than in water. It is usually overcome by increasing the heat transfer surface.

Employing atmospheric air as a coolant is reasonable, when the heated surface temperature is at least 50 °C higher, than that of the surrounding air, and also when cooling water is not available in the necessary amounts, or its freezing is a real hazard. In some technologies water is excluded for the production reasons, and in others certain limitations of either the size or the weight of the heat exchanger are envisaged.

Air coolers are employed in thermal and nuclear power generation, in refrigeration, in traffic, in systems of air conditioning and room heating, domestic and industrial. Of the high variety of constructions which employ air as one of the fluids, we separate radiators, air heaters, condensers, cooling towers, and special heat exchangers.

Radiators are compact heat exchangers which are employed to emit the heat from separate functional sets of vehical motors directly into the surrounding air, or for room heating. They employ liquid carriers in finned or membranne surfaces, to transfer the heat into air.

Air heaters are employed in thermal power stations and in gas turbines. There the heat is transferred by air from a gas flow, and transported to the combustion chamber, to intensify combustion and to increase thermal loads along the gas flow, so that the pressurized heat transfer surfaces can be reduced. In the opposite sides of the heat transfer surface, an up to 3 to 4 times difference of the heat transfer coefficient is maintained. Air heaters operate on tube or plate surfaces.

Condensers are used to condense vapour and to transfer the latent heat separated in the process. Air-cooled finned-tube condensers are widely used in low-power refrigerators.

In cooling towers, the waste heat of industrial plants is transferred to the surrounding air. They are also used in refrigeration, in air conditioning and whenever they can reasonably replace heat transfer into the air or to a cooling pond through a solid wall.

# AIR-COOLED HEAT EXCHANGERS

Special heat exchangers with air are constructed for nuclear power plants and for other modern technologies.

From the point of view of their performance, air coolers are recuperators, regenerators and mixers. In recuperator air coolers, two fluid flows of different temperatures are separated by a solid wall.

We shall limit our consideration on recuperator coolers. To begin with, we observe, that these can be of very different constructions depending on the fields of their application. This can be found in publications [1 through 3], as well as in the schematic representation in Fig. 1.

It was noted, that the heat transfer coefficient and specific heat of air are considerably lower, than of water. To transfer even a small amount of heat from a liquid, the air-side surface and flow rate must be higher. Thus in shell-and-tube heat exchangers, the liquid is circulated in the tubes, and the air-in the shell. They operate by forced convection. Various ways of augmenting convective heat transfer constitute at present a highly actual and a complex problem both in the theory of heat-and mass transfer, and in the construction optimization of the apparatus.

We consider now the heat transfer and the pressure drop in tube heat exchangers of plain, rough and finned tubes.

## 2. BANKS OF PLAIN TUBES

Heat exchangers are mostly made of tubes due to the simplicity of their manufacture and exploitation. Therefore, studies of fluid mechanics and heat transfer in many cases concentrate on these processes in flows past single tubes and banks of tubes.

The replacement of a single tube by a set of tubes or a bank of certain arrangement introduces a high degree of complexity in the process of heat transfer. The pattern of flow around a tube in a bank is greatly influenced by the surrounding tubes. The changes of the pressure gradient are even more pronounced in a contraction between adjacent tubes of a transverse row. This leads to corresponding changes of the velocity distribution in the boundary layer, and of the flow pattern in the wake.

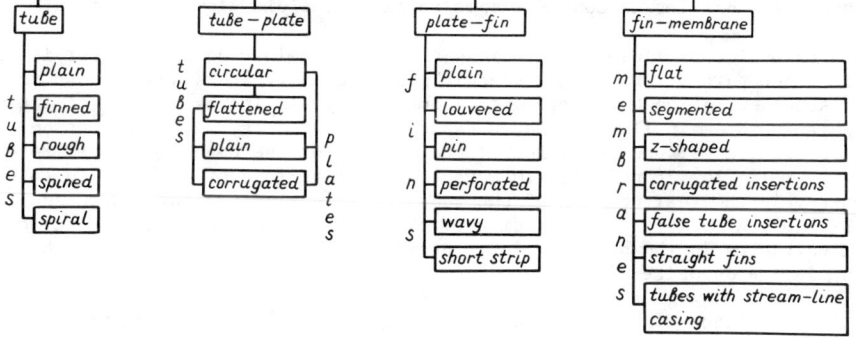

Fig.1. Schematic Representation of Constructions Types

The pattern of flow dynamics around a tube in a bank is determined by the arrangement and by other geometrical parameters of the bank. The two most common bank arrangements are the staggered and in-line ones, Fig. 2, and the common way of definition for them is by their relative transferse $a = s_1/D$ and longitudinal $b = s_2/D$ pitches.

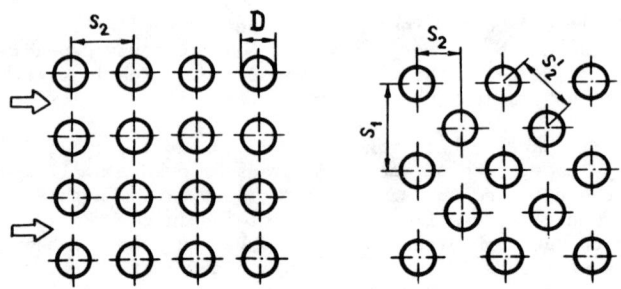

Fig.2. Arrangement of Tubes in Banks

Developments in the field of steam boilers have led to investigations of heat transfer from single tubes and banks of tubes in gas flows. A number of investigations has been devoted to this problem in air flows, mostly in the ranges of Reynolds number only from $1 \times 10^3$ to $1 \times 10^5$.

In this paper I shall be mainly concerned with the heat transfer of single tubes and banks of tubes in crossflow in a wide range of Reynolds numbers, and particularly in the critical and the supercritical flow regimes. As to the lower range of Reynolds number, I shall limit myself to the results of some recent experiments. It will be in fact a general review of investigations performed at the Institute of Physical and Technical Problems of Energetics of the Academy of Sciences of the Lithuanian SSR [4 through 9]. They refer to the heat transfer and pressure drop of banks of tubes of 100 different arrangements, in crossflows of gases and liquids in the range of Reynolds number from 1 to $2 \times 10^6$.

2.1. General Flow Pattern and Local Heat Transfer

The flow regime - laminar or turbulent-in the boundary layer has a considerable effect on the heat transfer. Therefore, before passing to the problems on banks of tubes let us consider some details of fluid flow past a single tube.

When a viscous fluid flows across a tube, a laminar boundary layer is formed on the front part of the tube. Its thickness gradually increases. Due to the shear stresses in the boundary layer, some energy dissipates. At the same time, pressureoin the main flow decreases ($dp/dx < 0$), velocity in the direction of the flow increases, and particles of the fluid from the boundary layer are drawn into the main flow, continuing to move along the surface of the tube in spite of friction. In the rear portion of the tube the main flow pressure increases ($dp/dx > 0$). As velocity in the direction of flow decreases, the velocity of the fluid particles in the boundary layer decreases to zero and eventually starts moving in the opposite direction.

The existence of layers in which fluid flows in opposite directions gives rise to a vortex. As soon as a vortex separates from the tube, another vortex is formed. In this way the rear portion of the tube is in the region of a mixed

# AIR-COOLED HEAT EXCHANGERS

vortical flow.

The laminar boundary layer separates from the sides of the tube at the angle $\varphi = 82°$.

So the pattern of flow on a tube is governed by the distribution of pressure and velocity on its surface. Outside the boundary layer, a relation between the two distributions is clearly observed:

$$u_\varphi = u_b \sqrt{1 + \frac{2(p_b - p_\varphi)}{u_b^2}} \qquad (1)$$

In curves Fig. 3 for the pressure distribution on a tube the different flow regimes are reflected. Curve 1 stands for the analytical expression of the pressure distribution, when both friction and viscosity are ignored. It shows a symmetric distribution of pressure with maxima in the front and the rear stagnation points, and a minimum in the middle cross section.

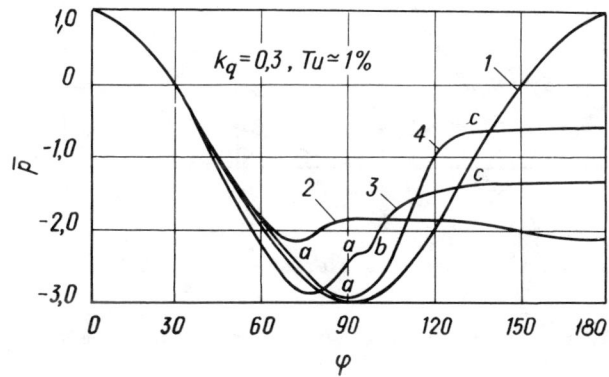

Fig.3. Distribution of Pressure Coefficient on the Surface of a Single Tube

Curve 2 represents the distribution of pressure on a tube surface in the subcritical flow regime. Under the action of the viscous forces, a laminar boundary layer is formed on the tube, its thickness increasing downstream. The kink in curve 2 at $\varphi = 82°$ and at point a corresponds to the separation of the boundary layer and to the beginning of a complex vortical flow in the rear part of the tube.

With an increase of the Reynolds number ($Re > 1.5 \times 10^5$), the critical flow regime is established. On the front part of the tube, separation of the boundary layer and the separation bubble are observed. These are followed by the reattachment of the boundary layer and the development of a turbulent boundary layer on the rear part of the tube. The turbulent boundary layer can resist the effect of the increased velocity gradient in the main flow (dp/dx > 0). It is prolonged downstream and separates at $\varphi \cong 140°$.

Curve 3 in Figure 3 represents the distribution of pressure on the surface of a tube at $Re = 4.5 \times 10^5$. Here point a corresponds to the separation of the laminar boundary layer, followed by the separation bubble, b - to the reattachment and c - to the separation of the turbulent boundary layer. A further increase of Re leads to the supercritical flow regime, noted for the absence of the separation bubble. The laminar-turbulent transition occurs inside the boundary layer, which is still more prolonged downstream and separates at $\varphi=140°$.

In curve 4 for the pressure distribution in air at Re = $10^6$ a is the laminar-turbulent transition in the boundary layer, c - separation of the turbulent boundary layer. With a still further increase of Re, combined with the increase of turbulence, the laminar-turbulent transition in the boundary layer is removed upstream, and the point of the turbulent boundary layer separation is practically stable.

In Fig. 4 we see the distribution of the local heat transfer along the perimeter of the tube as a function of Reynolds number.

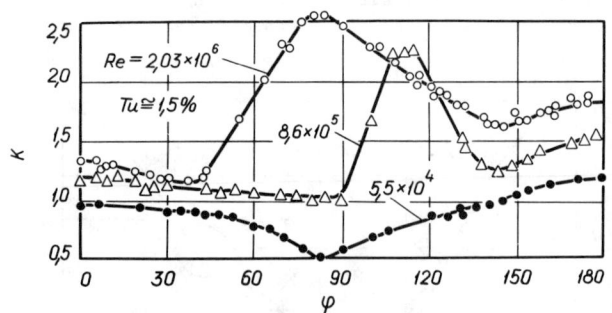

Fig.4. Local Heat Transfer of a Single Tube

In any flow regime, a laminar boundary layer is formed on the front portion of a tube. With its development around a tube, a rapid decrease in the coefficient of heat transfer is observed. At Re = 5.5 x $10^4$ the minimum at $\varphi$ = 82° denotes the separation of the boundary layer.

Reynolds number increasing over 1.5 x $10^5$, a laminar-turbulent transition occurs in the boundary layer, and at Re = 8.6 x $10^5$ the curve has two minima, the first one at about $\varphi$ = 90° is relevant to the laminar-turbulent transition, and the second at $\varphi$ =140° - to the separation of the boundary layer.

With a further increase of Re to 2 x $10^6$, the initial point of transition moves sharply upstream to $\varphi$ = 35°.

In banks of various arrangements, flow around a tube in the first row is similar to that on a single tube, but it is different on tubes in the inner rows.

Let us consider now some figures for the local heat transfer of a single tube and a tube in a bank in the subcritical range of Re. In Fig. 5 curve 1 stands for a single tube, curve 2 - for an in-line bank, curve 3 - for a staggered bank. In both banks higher intensities of heat transfer from both the front and the rear portions of the tube are evidently the results of a higher turbulence of the flow in the inner rows. A pronounced maximum of the heat transfer in the case of an in-line bank is noted at $\varphi$ =50°, which is the impact point of the stream on the tube surface.

# AIR-COOLED HEAT EXCHANGERS

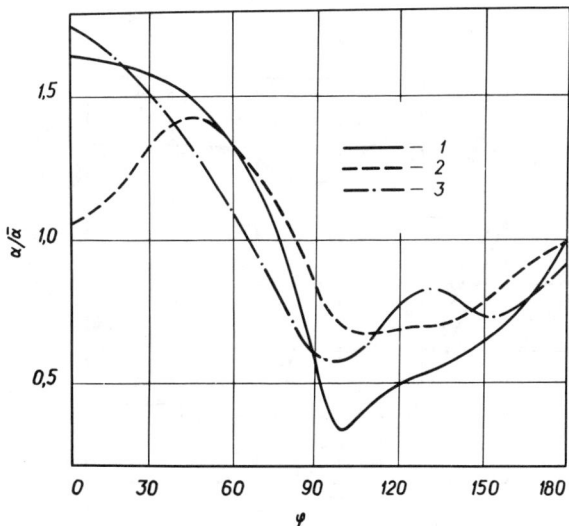

Fig.5. Variation of Local Heat Transfer of a Single Tube and a Tube in a Bank. Curve 1 - Single Tube; Curve 2 - in an In-Line Bank; Curve 3 - in a Staggered Bank

In the inner rows of staggered banks, some increase in the heat transfer is observed at $\varphi = 120°$. In closely-spaced staggered banks it can be attributed to large pressure gradients and consequently increased velocity rates.

For in-line banks, the location of the impact point is dependent on the longitudinal pitch and on the Reynolds number. The investigations in the wake of a tube at $Re = 1.3 \times 10^4$ may be of interest in this respect. A set of two tubes was studied with a distance L/D varied from 1.6 to 9. Fig. 6 suggests a qualitative change of the pressure distribution along the perimeter of the second tube at $L/D \leqslant 3$. A substantial difference may be noted, if compared to a single tube in an infinite flow. With L/D=6 the point of attack and the point of impact coincide at $\varphi = 0$, and with L/D=1.6, the point of impact is shifted to $\varphi = 75°$.

Here the coefficient of pressure is described by

$$\bar{p} = 1 - (p_{\varphi=0} - p_{\varphi})/(\rho \, \bar{u}^2/2) \qquad (2)$$

The general pattern of a flow based on the results of [10] is represented in Fig. 7.

Fig. 8 illustrates the dependence of the local heat transfer on Re in an inner row of a staggered bank. A distinct change in the character of the local heat transfer is observed with an increase of Re from $1.5 \times 10^5$ (curve 2) to $9.3 \times 10^5$ (curve 3).

With an increase of Re to $9.3 \times 10^5$ (curve 3), the laminar-turbulent transition in the boundary layer is initiated at $\varphi = 30°$, instead of at $\varphi = 90°$ for $Re = 8 \times 10^4$ (curve 1).

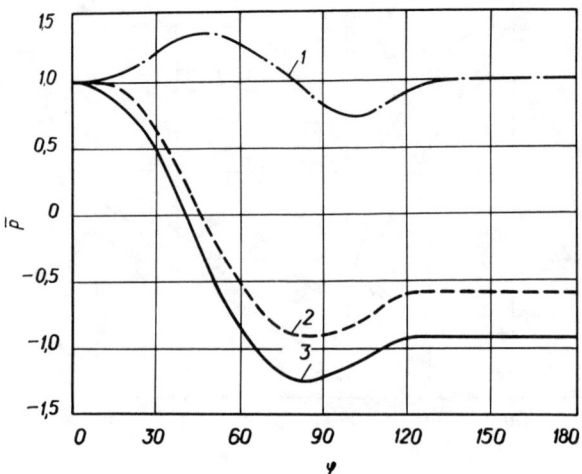

Fig.6. Variation of Pressure Coefficient as a Function of Longitudinal Pitch. Curve 1 - L/D = 3; Curve 2 - L/D = 6; Curve 3 - L/D = 9

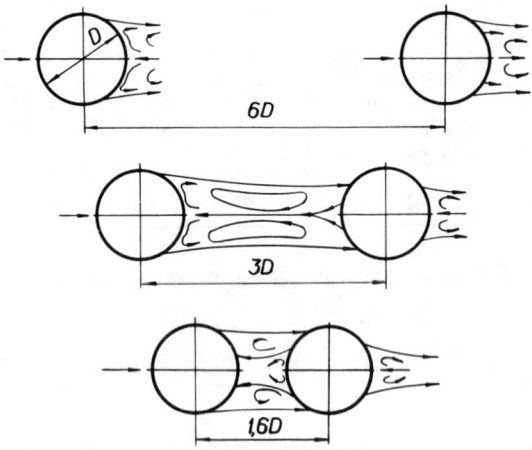

Fig.7. Pattern of Flow Past a Tube in a Longitudinal Row

# AIR-COOLED HEAT EXCHANGERS

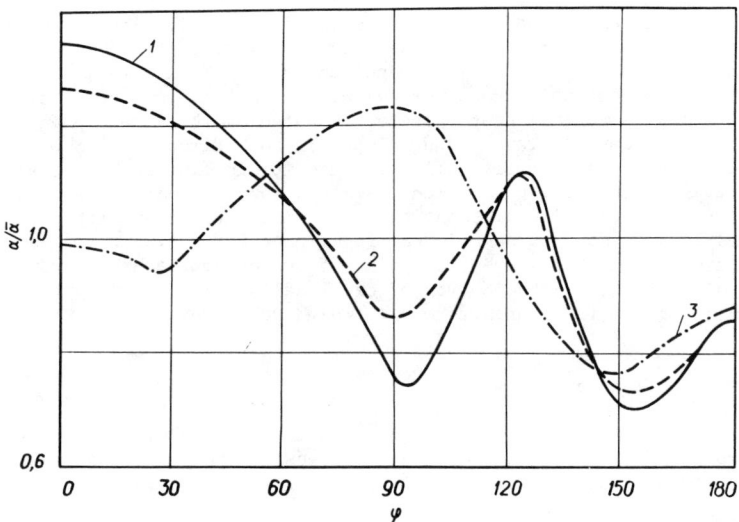

Fig.8. Local Heat Transfer of a Tube in a Staggered Bank. Curve 1 – Re=8 x $10^4$; Curve 2 – Re=1.5 x $10^5$; Curve 3 – Re=9.3 x $10^5$

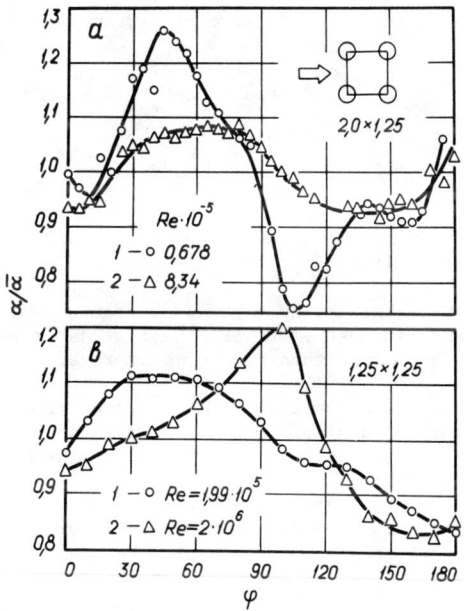

Fig.9. Local Heat Transfer of a Tube in an In-Line Bank: a – 2 x 1.25, b – 1.25 x 1.25

Fig. 9 presents analogical measurements in air flows for in-line banks. Here the upper and the lower curves present the local heat transfer of a tube in the fifth row of a 2 x 1.25 in-line bank, and of a 1.25 x 1.25 in-line bank, respectively. The impact point does not coincide with the front stagnation point in the former bank. A maximum heat transfer is noted at the impact point at $\varphi = 50°$. A laminar boundary layer is developed in both directions from the impact point, a very characteristic flow feature for in-line banks. At about $\varphi \cong 100°$, stability in the laminar boundary layer is disturbed and a turbulent transition follows. The point of transition is shifted upstream with an increase of Re. This is more evident in the lower plot for a 1.25 x 1.25 bank. At the same time for $Re = 2 \times 10^6$, the point of impact approaches the front stagnation point $\varphi = 0$, and the curves of the local heat transfer approach in their tendencies those for staggered banks at high values of Re.

## 2.2. Average Heat Transfer

The heat transfer of a tube in a bank is determined mainly by flow velocity, physical properties of the fluid, heat flux intensity and the arrangement of the tubes. To generalize the experimental data, we use the following equation

$$Nu = c\, Re^m \cdot Pr^n \quad (3)$$

For gases, for which the Prandtl numbers are equal and constant, Eq.(3) becomes

$$Nu = c\, Re^m \quad (4)$$

For practical purposes, the power index of the Prandtl number $n = 0.36$ is sufficiently accurate for all sorts of banks.

The final results for the average heat transfer in the form of

$$K = Nu \cdot Pr^{-0.36} = f(Re) \quad (5)$$

or

$$Nu = f(Re) \quad (6)$$

are presented in the following figures. The results are referred to the tube diameter and to the flow velocity in the minimum inter-tube space.

The experiments suggest that the heat transfer from a tube is determined by its position in the bank. In most cases, the heat transfer from tubes in the first row is considerably lower than in the inner rows.

In the range of low Reynolds numbers, the heat transfer from a tube in the first row is similar to that of a single tube or a tube in an inner row. The increase of flow turbulence inside a bank at higher Reynolds numbers leads to an increase of the heat transfer of the inner tubes, as compared to the first row. The rows tubes in a bank in fact act as a turbulence grid. In most banks, the heat transfer becomes stable from the third or the fourth row in the subcritical flow regime. A comparison of heat transfer between the first row and the inner rows in a fully developed flow reveals the influence of turbulence on the intensity of heat transfer.

The heat transfer of inner tubes generally increases as the longitudinal pitch decreases. This correlates well with known investigations of the heat transfer of a tube placed at various distances from the turbulence grid.

As a result of turbulence, the heat transfer in inner tubes exceeds by 30 to

# AIR-COOLED HEAT EXCHANGERS

100 % that of the first row, the difference depending on the longitudinal pitch. Thus the heat transfer in inner rows is mainly determined by turbulence intensity, which increases with a decrease in the distance from the turbulizer, i.e., from the preceding row.

The heat transfer of a tube in the second row is in most cases 10 to 30 % lower than that of inner tubes.

The following discussion deals mainly with the heat transfer of a tube in an inner row of a bank.

## 2.3. Heat Transfer at Low Reynolds Numbers

Fig. 10 presents the heat transfer of the two arrangements in different pitches: curves 1 and 3 for 1.5 x 1.5, curves 2 and 4 for 2 x 2.

At $Re < 10^3$ in in-line banks, the first-row tubes give the highest heat transfer. The inner rows are in the wake of the preceding ones, and their heat transfer is lowered by lower flow velocities in the recirculation regions. With decreasing longitudinal pitches, heat transfer in in-line banks decreases.

On the contrary, in banks of staggered arrangements, the first-row tubes exhibit lower heat transfer intensities than the inner ones. This is an evidence for different flow patterns in the two bank arrangements.

A comparison of the data for the average heat transfer reveals also higher values for staggered banks in the whole range of Re covered. The power index for Re in the range under $10^3$ is from 0.63 to 0.4 or 0.5 for in-line banks, and from 0.6 to 0.4 or 0.5 for staggered banks. At $Re < 10^2$ some of the banks have as high values of m as 0.33 and 0.37.

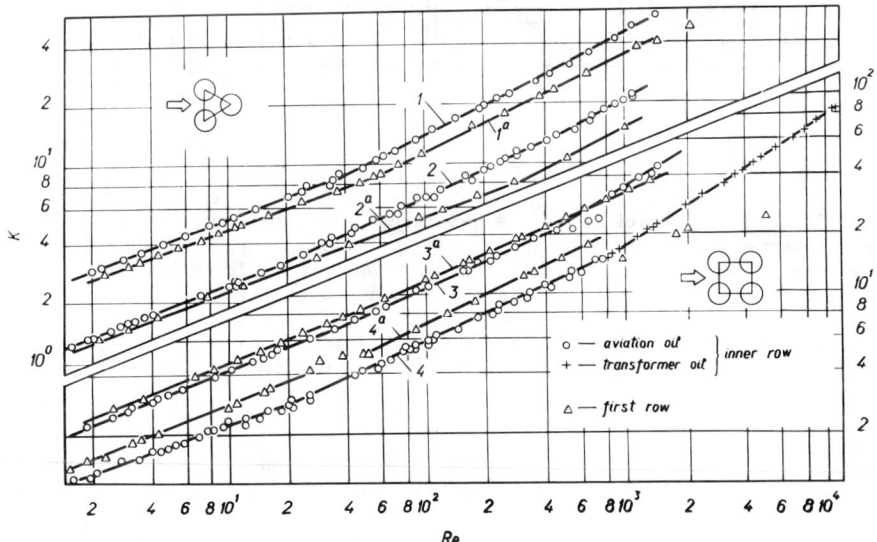

Fig.10. Heat Transfer of Tubes of In-Line and Staggered Banks. Curves 1 and 3 - a Bank 1.5 x 1.5; Curves 2 and 4 - a Bank 2 x 2

## 2.4. Heat Transfer in the Subcritical Flow Regime

This regime covers Re from $10^3$ to $2 \times 10^5$. The transition of the predominantly laminar to the mixed flow takes place at different Reynolds numbers, depending on the tube arrangements. For $Re > 10^3$ a laminar boundary layer forms on the front of an inner tube, but the main portion of it is influenced by a vortical flow. The character of the heat transfer is determined by the flow regime in the boundary layer. Thus the power index m of the Reynolds number varies from 0.55 to 0.7 for banks of different arrangements.

Figures from 11 to 13 give our results of the heat transfer from various banks of tubes in crossflow of liquids and air.

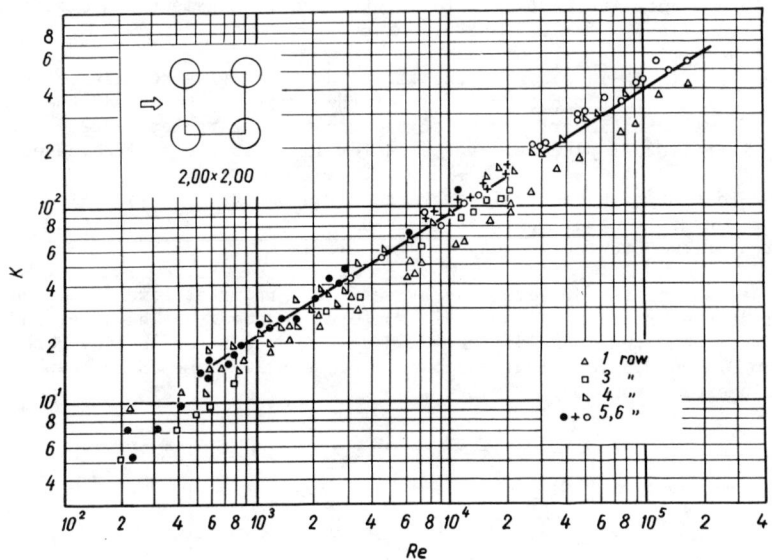

Fig.11. Heat Transfer of In-Line Bank 2 x 2

Fig. 14 shows a comparison of the heat transfer from 15 various banks of in-line arrangements. It suggests an increase of m with a constant longitudinal and a decreasing transverse pitch. In fact, the value of m is influenced by changes in the ratio of longitudinal and transverse pitches.

In this flow regime, m = 0.63 is acceptable for most banks of in-line arrangement. The average heat transfer of an inner tube is calculated from

$$Nu = 0.24 \, Re^{0.63} \qquad (7)$$

In banks with a/b < 0.7, experimentally measured heat transfer is much lower than calculated by Eq. (7). Banks of this type are considered inefficient as heat exchangers.

Final experimental results of the heat transfer of different staggered banks in flows of viscous fluids are presented in Fig. 15. The power index of Re is equal to 0.60 for all banks.

# AIR-COOLED HEAT EXCHANGERS

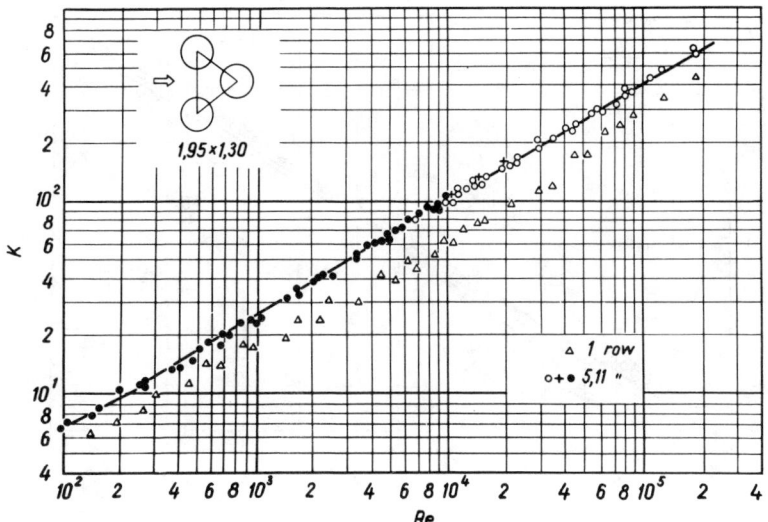

Fig.12. Heat Transfer of Staggered Bank 1.95 x 1.30

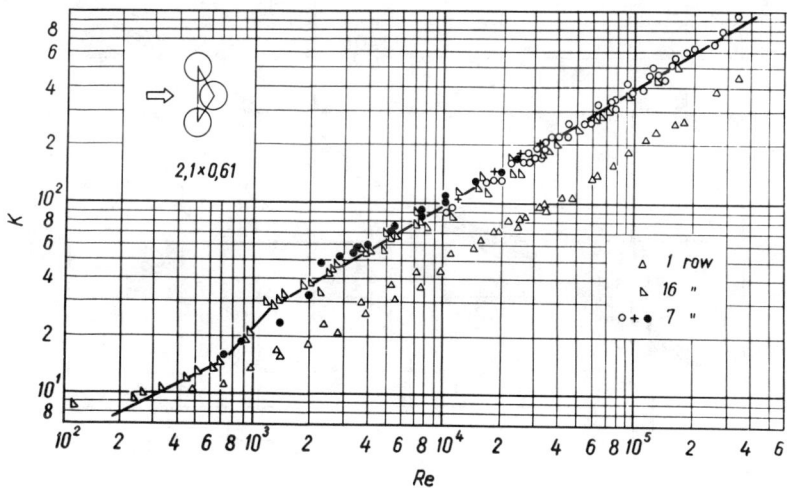

Fig.13. Heat Transfer of Staggered Bank 2.1 x 0.61

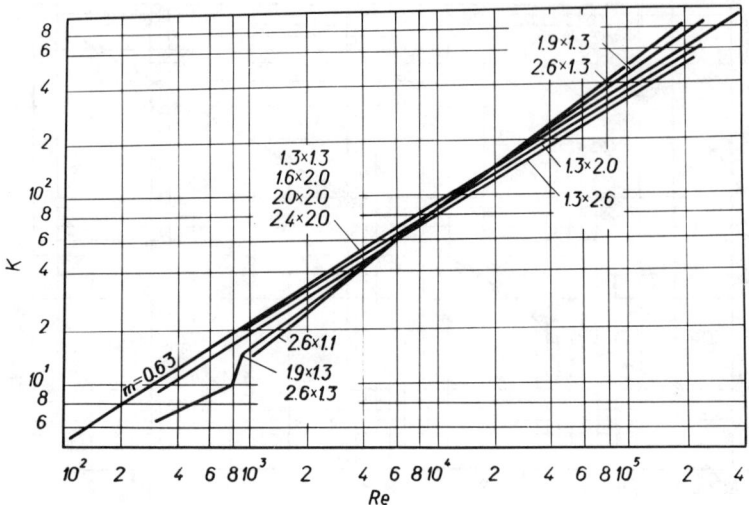

Fig.14. Comparison of Heat Transfer of Various In-Line Banks

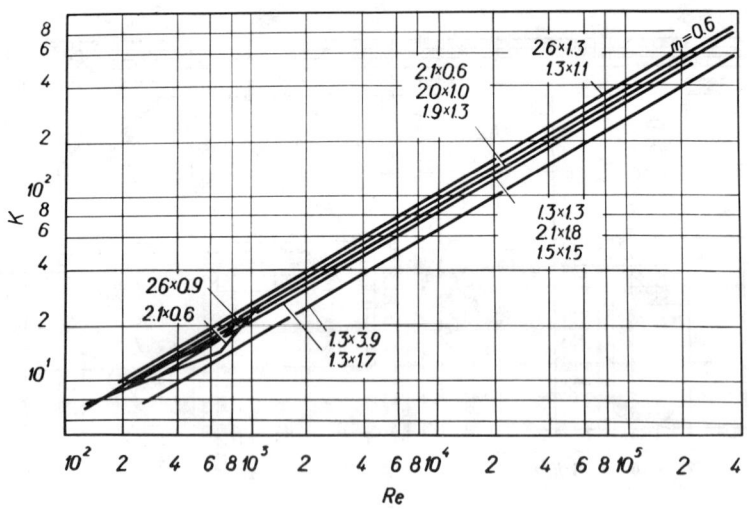

Fig.15. Comparison of Heat Transfer of Various Staggered Banks

The effect of the pitch is clear. Heat transfer increases with a decrease in the longitudinal pitch and, to a lesser extent, with an increase of the transverse pitch. The variation of c may be evaluated by the geometrical parameter a/b to the power 0.2 for a/b < 2. For a/b > 2, c = 0.35. In such banks the minimum inter-tube space is diagonal with respect to the main flow. Thus changes in c involve certain changes in the conditions of flow through a bank. The general formulas of the heat transfer of inner tubes in various staggered banks are

for a/b < 2

$$Nu = 0.3(a/b)^{0.2} \cdot Re^{0.6} \qquad (8)$$

fond for a/b > 2

$$Nu = 0.35 \, Re^{0.6} \qquad (9)$$

2.5. Heat Transfer in the Critical Flow Regime

In Fig.16 the mean heat transfer in presented from our experiments on in-line banks of pitches 1.5 x 1.25 and 2 x 1.25 for air flows in the first and inner rows. In the range of Re < 2 x $10^5$, the power index of Re is m = 0.6 to 0.65 in the relation of heat transfer. But for Re above 2 x $10^5$, it must be increased to m = 0.76 to 0.8, marking an establishment of a new law of the heat transfer.

Fig. 17 presents the experiments of the heat transfer with staggered banks of pitches 1.25 x 1.25, 1.5 x 1.5, 2 x 1.25. Here at Re < 2 x $10^5$ the power index of Re is m = 0.6, and for Re > 2 x $10^5$ it is increased to 0.8

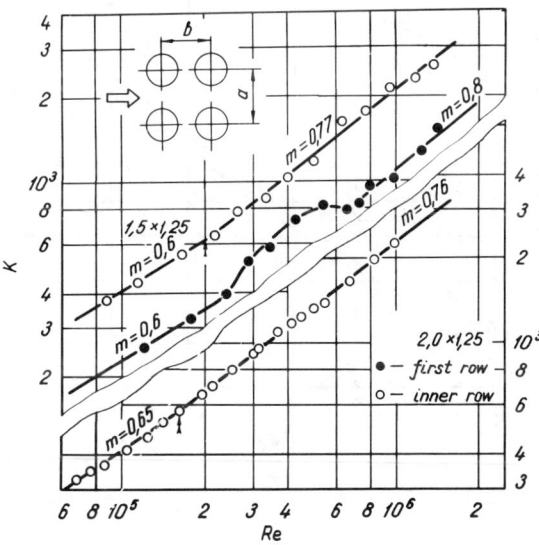

Fig.16. Heat Tranfser of In-Line Banks at High Reynolds Numbers

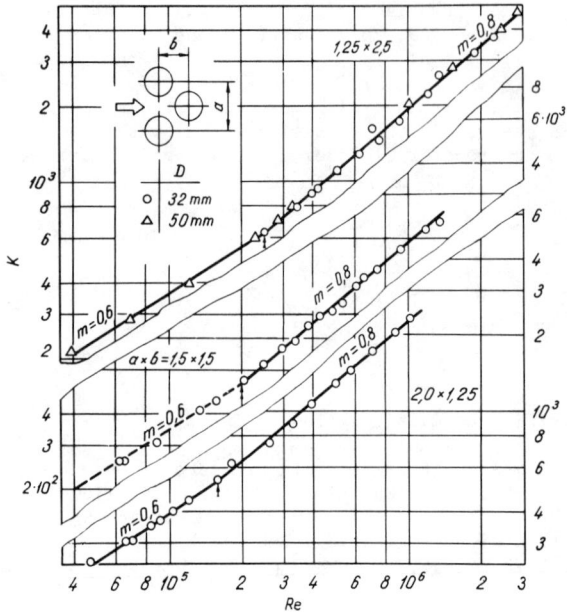

Fig.17. Heat Transfer of Staggered Banks at High Reynolds Numbers

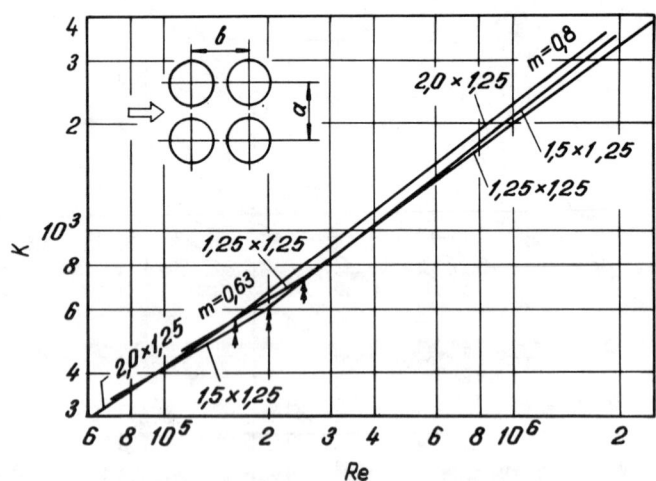

Fig.18. Comparison of Heat Transfer of Various In-Line Banks at High Reynolds Numbers

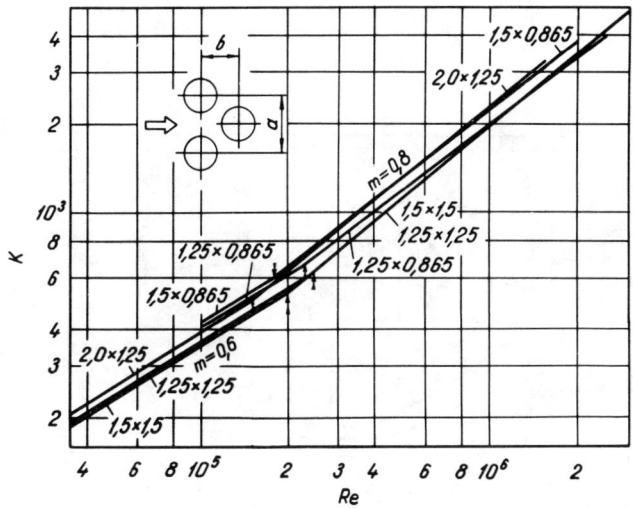

Fig.19. Comparison of Heat Transfer of Various Staggered Banks at High Reynolds Numbers

Figures 18 and 19 present the heat transfer in inner rows of in-line and staggered banks in crossflow of air. In the subcritical regime, the results correlate well with the results of heat transfer in banks of tubes considered above.

The influence of pitch on heat transfer in the critical regime is similar to that in the subcritical regime. The heat transfer of all staggered banks with widely spaced tubes is more intensive, but never differs from others by more than 25 %.

Similar results are obtained for in-line banks. Heat transfer is most intensive in banks with large transverse and small longitudinal pitches. The heat transfer intensity of most in-line banks can be calculated from

$$Nu = 0.029 \cdot Re^{0.8} \qquad (10)$$

The following relation are recommended for calculating heat transfered from tubes in inner rows of staggered banks:

$$Nu = 0.027 \cdot Re^{0.8} (a/b)^{0.2} \qquad (11)$$

2.6. Pressure Drop

As suggested by the analysis of experimental results for the simplicity of calculation, a graphical interpretation of the data is most convenient. General graphs have been compiled from our results described previously, including the results of other authors on the pressure drop across banks in flows of gases and liquids. A satisfactory correlation has been achieved.

The graphs of the pressure drop coefficient as a function of Re for in-line and staggered banks are presented in Fig. 20 and 21, calculated for one row of a bank.

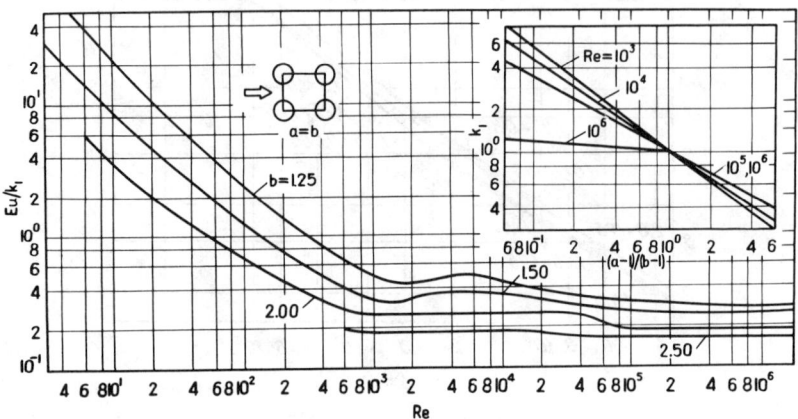

Fig.20. Pressure Drop of In-Line Banks, as Referred to the Relative Longitudinal Pitch b

The general graphs of in-line banks are based on the resistance of banks of square arrangements, with the longitudinal pitch as reference. Graphic correlations have been introduced for other banks to account for different pitches and Reynolds numbers.

The graphs of staggered banks are based on the equidistant arrangement with corrections for the evaluation of other arrangements and of Re.

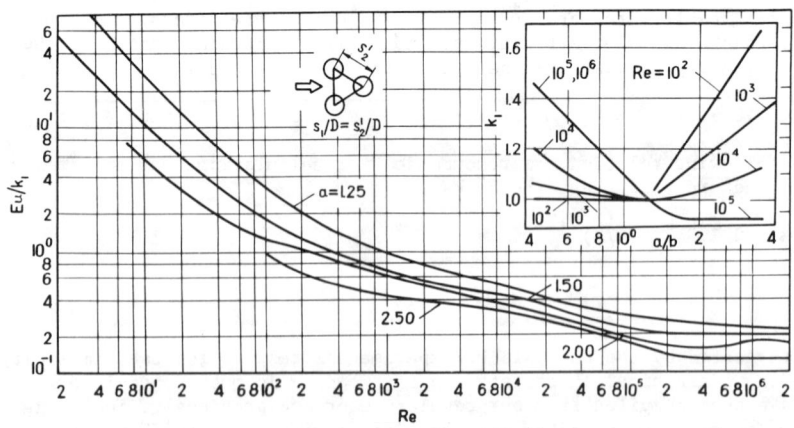

Fig.21. Pressure Drop of Staggered Banks as Referred to the Relative Transverse Pitch a

# AIR-COOLED HEAT EXCHANGERS

With an increasing Reynolds number, the flow in a bank becomes increasingly turbulent, and the pressure drop across in-line and staggered banks becomes equal.

## 3. BANKS OF ROUGH-SURFACE TUBES

Introduction of a surface roughness is one of the methods of the heat transfer augmentation. The effect is due to a change in the fluid dynamics of the boundary layer, so that for similar conditions, the laminar-turbulent transition in it occurs at a lower value of the Reynolds number. We may differ "open" roughness of a surface, in which reattachment is observed downstream of each separate surface element, and compact roughness with $s/d < 5$, where s - is the distance between two adjacent surface elements. Now we consider tubes with compact roughness elements, as in Fig. 22.

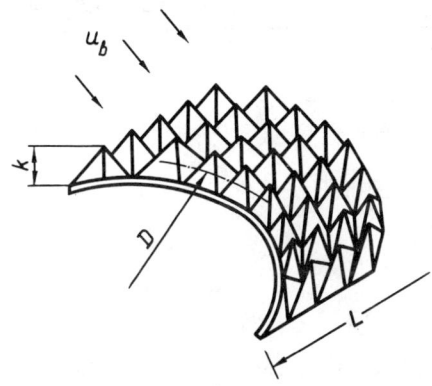

Fig.22. The Type of Surface Roughness Studied

On a smooth tube in an in-line bank, the maximum heat transfer is observed in the vicinity of the front stagnation point, Fig. 23. The heat transfer decreases with the angular distance, because the boundary layer thickness increases up to its transition into turbulent layer or separation. The presence of surface roughness causes supplementary turbulization of the boundary layer. This effect is determined by the ratio of k, the height of the surface elements, to $\delta$, the boundary layer thickness. With k considerably less than $\delta$, the velocity fluctuation it causes does not effect the heat transfer. Thus, with a suitably chosen ratio $k/\delta$, the heat transfer is augmented through a combined effect of the velocity fluctuation in the boundary layer, and of the main flow turbulence. The boundary layer is partially destroyed, and its laminar-turbulent transition is shifted upstream, from $\varphi \cong 90°$ to $60°$ at $Re = 2 \times 10^5$ (Fig. 23).

In a turbulent boundary layer, the surface elements cause an increase of the shear stress, provided their height exceeds the thickness of the viscous sublayer. The condition may be described by the dimensionless roughness parameter $k^+ \geq 5$.

The well-known versatile character of the velocity distribution is no longer valid. The velocity - angular distance curve becomes logarithmic and goes parallel to Curve 2, Fig. 24. In the presence of a pressure gradient, the velocity distribution in the viscous sub-layer (curve 1) is described by

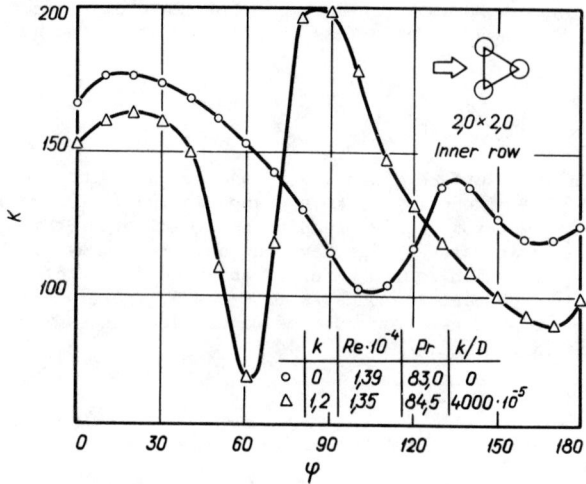

Fig.23. The Effect of Surface Roughness on the Local Heat Transfer of Tube in Bank

$$u^+ = y^+ + \frac{1}{2} \left( \frac{\nu}{\rho u_*^3} \cdot \frac{dp}{dx} \right) y^{+2} \tag{12}$$

if

$$dp/dx \quad 0 \ ; \quad u^+ = y^+$$

Velocity distribution outside the viscous sub-layer is

$$\frac{u}{u_\psi} = 1 - \frac{u_*}{u_\psi} \int_\eta^1 [\tau(\eta,\Phi)]^{1/2} \bar{\ell}^{-1}(\eta)] d\eta \tag{13}$$

where $\Phi = \frac{\delta}{\tau_w} \cdot \frac{dp}{dx}$ parameter of the pressure gradient.

Both $\tau = \tau/\tau_w$ and $\bar{\ell} = \ell/\delta$ in the height of the boundary layer are determined by the pressure gradient $\Phi$ and by the height of the surface elements k.

$$\left. \begin{array}{c} \tau/\tau_w \\ \ell/\delta \end{array} \right\} = f(\Phi, k^+, \eta) \tag{14}$$

where $\eta = y/\delta$, and $\tau/\tau_w$ and $\ell/\delta$ are approximated by a polynomial $\tau/\tau_w = \sum_{i=0}^{n} a_i \eta^i$.

In the heat transfer studies of rough-surface tubes, a good deal of information may be gained from the temperature distribution. It is different on the top of the surface elements, and on their inter-spaces. To coordinate the data, we introduce $\upsilon^+ = f(y^+)$.

For the temperature distribution, we use

$$\upsilon^+ = Pr \ y^+ \tag{15}$$

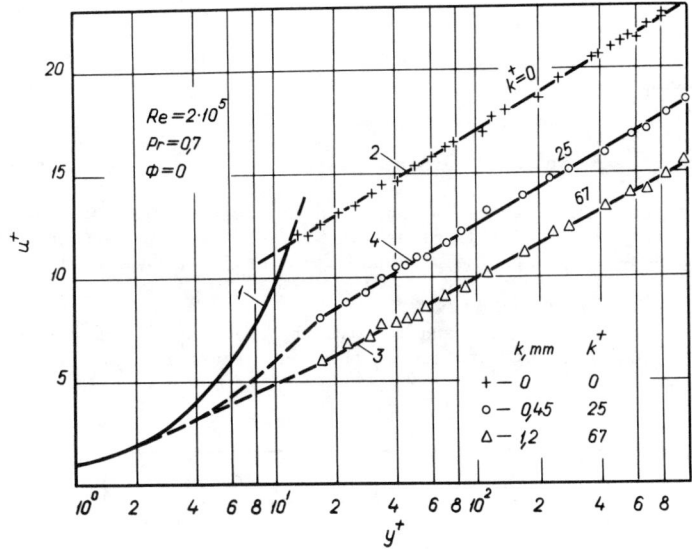

Fig.24. Stratification of the Velocity Distribution on Rough-Surface Tubes in Air

in the viscous sub-layer (Fig. 25 Curve 1), and

$$\upsilon^+ = \frac{Pr_t}{\chi} \ln y^+ + C(Pr, k^+) \tag{16}$$

outside it (Fig. 25 Curve 2). Here $Pr_t$ is the turbulent Prandtl number, $\chi = 0.4$ and $C(Pr, k^+)$ is a function of the physical properties, of the shape and the size of the surface elements.

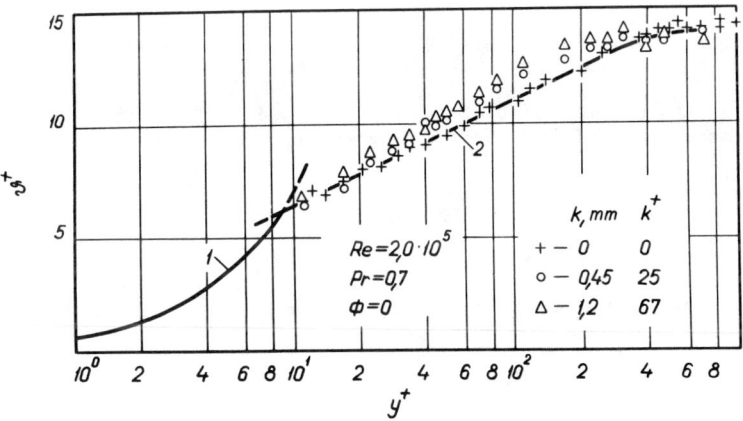

Fig.25. Temperature Distribution of Rough-Surface Tubes in Air

For the local heat transfer take

$$St_x = \frac{1}{u_\varphi^+ \upsilon_\varphi^+} \tag{17}$$

where $u_\varphi^+$ and $\upsilon_\varphi^+$ refer to the external part of the boundary layer.

The effect of the surface roughness on the heat transfer is largely determined by Pr. The relation is closer at higher Pr, when the thermal resistance concentrates near the wall. The viscous sub-layer is turbulent on rough surfaces.

From Fig. 26, at Pr = 0.7 and k = 0.2 mm, the heat transfer is close to that of a smooth tube. Augmentation of the heat transfer commences at higher values of k.

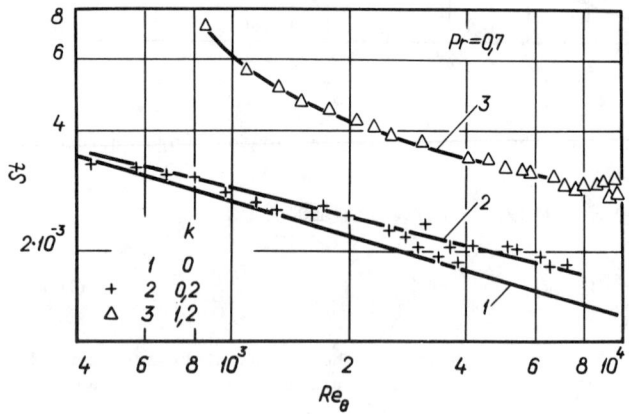

Fig.26. Heat Transfer of Smooth and Rough-Surface Tubes in Air

We performed a comparison of the average heat transfer and the pressure drop of 7 different banks of rough-surface tubes, Fig. 27 [4, 11].

For air, at the higher values of Re, the average heat transfer of both smooth and rough-surface tubes fall on a single line, and separate only at Re = $7 \times 10^4$, depending on (k/D) (Fig. 27 Curve 1). The rough-surface tube banks, which exceed the critical condition, give considerable augmentations of the heat transfer, which amount to 1.8 times at Re = $10^6$ and k/D = $450 \times 10^{-5}$ in air.

The pressure drop increases with Re, until it acquires a constant value.

For the average heat transfer in air at Re from $7 \times 10^4$ to $7 \times 10^5$ and k/D from $450 \times 10^{-5}$ to $800 \times 10^{-5}$ we suggest

$$Nu = 0.0091 \, Re^{0.85} \, (k/D)^{-0.15} \, (a/b)^{0.2} \tag{18}$$

Here the values refer to the tube diameter and to the velocity in the minimum inter-tube space.

# AIR-COOLED HEAT EXCHANGERS

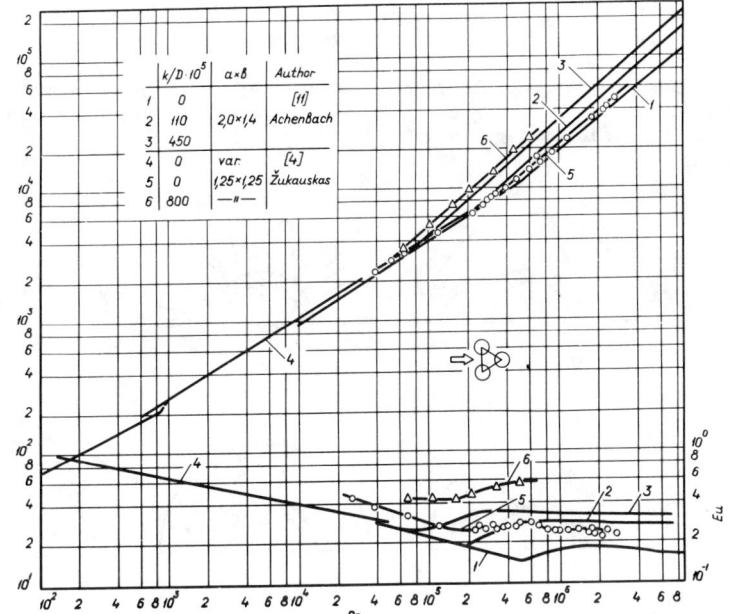

Fig.27. A Comparison of the Average Heat Transfer and the Pressure Drop of Staggered Banks of Smooth and Rough-Surface Tubes

## 4. FINNED TUBES

Although finned tube exchangers have been in existence just for 60 years, there have been numerous developments in the technology. Figure 28 shows two basic finned tube geometries; the plate fin and the helical fin. These are the most common configurations.

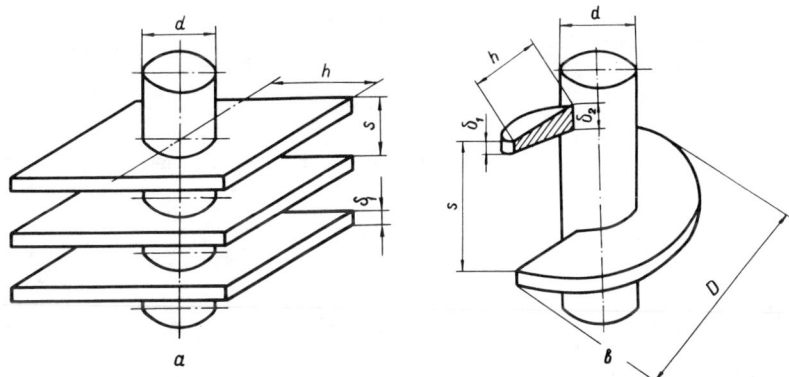

Fig.28. Basic Finned Tube Geometries: a - Plate Fin; b - Helical Fin

The technical literature contains performance and experimental data of plate fin, wavy plate fin, fluted fin heat exchangers.

Helical finned tubes technology encompasses a wider range of geometric variables and materials of construction than that for plate fins. A variety of fin geometries is possible: Fig. 29.

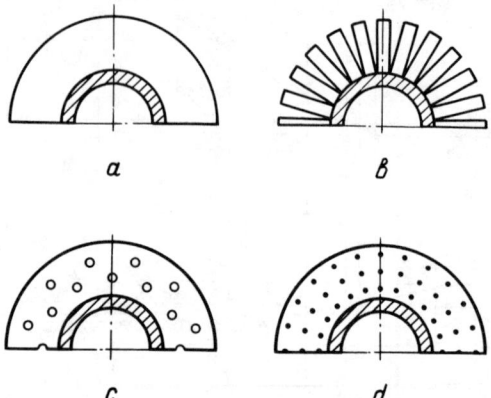

Fig.29. Helical Fin Geometries: a - Plain, b - Segmented, c - Slotted, d - Punched

Finned surfaces are introduced to increase heat transfer facilities and power efficiencies of recuperative heat exchangers, predominantly for the shell-side gaseous fluids. Fins are usually located on the outer surfaces of tubes. The heat being supplied from the tube, fin temperatures decrease towards their tips. Maximum fin-foot heat flux corresponds to the limiting condition, when its tip temperature approaches that of the flow. Any further increase of the fin height gives no effect on the heat flux density at the base.

The relation between the average heat transfer coefficient of a finned tube and the reduced heat transfer coefficient is expressed by

$$\alpha_c = \bar{\alpha}_T ( \frac{A_F}{A_{tot}} E \psi \xi + \frac{A_T}{A_{tot}} ) \qquad (19)$$

where

$E = \frac{\tanh(mh)}{mh}$ - efficiency for annular fins of constant thickness

$m = \sqrt{\frac{2\alpha}{t\lambda_w}}$

$A_F$ and $A_T$ - the surfaces of fins and unfinned places of the tube.

The coefficient, taking into account the trapezoidic profile of the fin, is as follows

$\xi = 1 + (0.125 - 0.125 \sqrt{t_1/t_2}) mh$

# AIR-COOLED HEAT EXCHANGERS

For radial fins of rectangular profiles coefficient $\xi$ is equal to 1.

To determine the value of the correction factor for non-uniform heat transfer for transverse annular and helical fins, use Fig. 30 (see Kuznecov and Pshenisnov [12], and equation $\psi = 0.97 - 0.056\ mh$) in the ranges of variables: $0 < mh < 2.0$, $5 \times 10^3 < Re < 5 \times 10^5$.

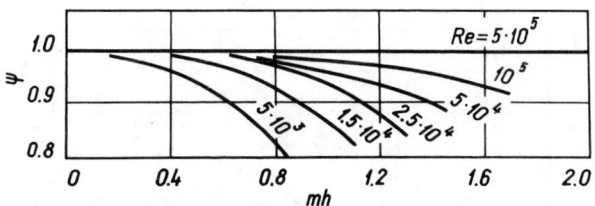

Fig.30. Relation of the Non-Uniformity Factor to mh and Re

## 4.1. Local Heat Transfer and Pressure Drop of a Finned Tube

Fluid dynamics for a finned tube, because of its complex geometry, involves variations of the heat transfer coefficient both on the circle and on the height of a separate fin. The determining factors are tube finning, tube location in a bank and flow regime.

Publication ([13] Fig. 31) presents the results of distribution of the local heat transfer coefficients on the height of a fin, and on the circle, for a single finned tube and a finned tube in a staggered bank. The maximum heat transfer coefficient is observed at $\varphi = 90°$ and $270°$ on the least inter-tube space.

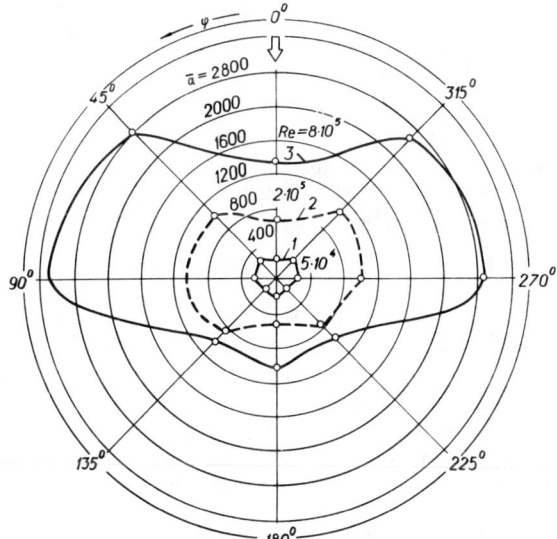

Fig.31. Distribution of the Local Heat Transfer Coefficient on a Finned Tube at Different Values of Re. Solid Line -a Single Finned Tube. Dotted Line - a Finned Tube in a Staggered Bank. 1-Re=4.8x10⁴, 2-Re=1.8x10⁵, 3-Re=7.6x10⁵

A comparison of the local heat transfer coefficient on a single tube and on a tube in a bank shows a more uniform distribution in the latter case. Spiral fins give moderate effects on the heat transfer around a tube.

The same data interpreted as the relative heat transfer coefficient values $\bar{\alpha}_h/\bar{\alpha}_F$ ($\bar{\alpha}_F$ - average heat transfer coefficient of fin height and circle), are shown on Fig. 32.

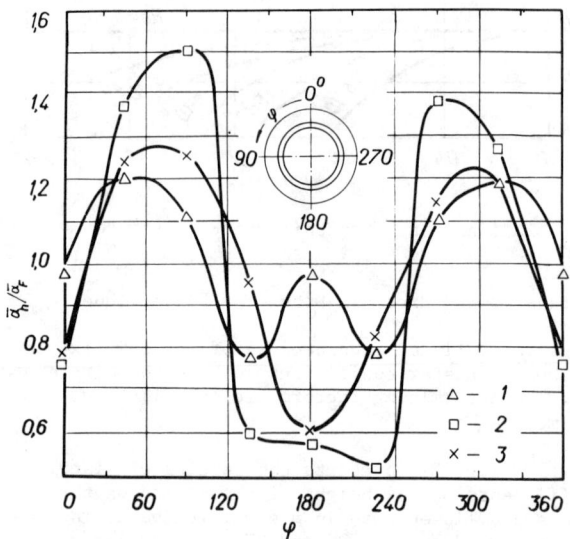

Fig. 32. Variation of the Relative Heat Transfer Coefficient on a Finned Tube at Different Re. 1 - Re=4.8x10$^4$ - Single Tube; 2 - Re=7.6x10$^5$ - Single Tube; 3 - Re=1.8x10$^5$ - in a Bank

The relative heat transfer coefficient of the tube increases downstream from $\varphi$ =0 to $\varphi$ =70-90° because of the increase of the velocity on the surface. The subsequent separation of the boundary layer causes a sharp decrease of the relative heat transfer coefficient.

On the rear part of a separate finned tube, an increase of the relative heat transfer coefficient is observed at low Re. The increase is hardly noticeable at high Re or in the inner rows of a bank. A more uniform distribution of the heat transfer coefficient on a tube is observed at lower Re, than at higher ones.

Fig. 33 [14] presents the curves of the heat transfer coefficient in a separate row of a five-row in-line bank for indentical Re. In the first row, the heat transfer coefficient distribution is similar for in-line and staggered banks. In the second row the heat transfer coefficient increases from $\varphi$ =0 to $\varphi$ =65-70° for an in-line bank, but decreases from $\varphi$ =0° to $\varphi$ =180° in a staggered one. In the first row the heat transfer is determined by the main flow parameters, but on the inner rows, the turbulence effects of the preceding rows prevail.

The study of fluid dynamics on finned surfaces which define the heat

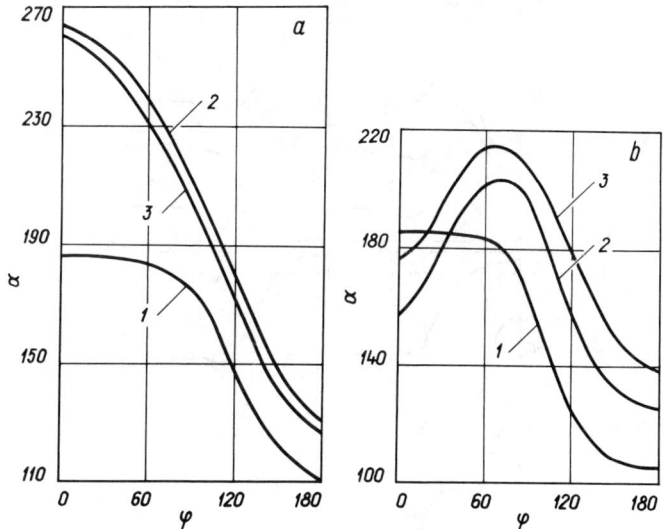

Fig.33. Distribution of the Local Heat Transfer Coefficient on a Finned Tube in Different Rows of a Staggered Bank (a) and an In-Line Bank (b). 1, 2, 3 I, II-IV, V Rows of Tubes Respectively

transfer deserves a special attention. The structure of the flow on a tube may yield explanation for the distribution of the local heat transfer coefficient and for the heat transfer mechanism as a whole.

One of the methods which help to study the fluid dynamics is the pressure distribution on a finned tube [15] Fig. 34. The pressure distribution on the fin top and on the base tube in the first row of a staggered bank and in the in-line bank are similar. But in the second row as well as in the inner ones no increase in pressure of the flow direction from $\varphi = 0$ is observed. It is necessary to note that hydrodynamic picture of the flow on the tubes as described [15] fully corresponds the description of the heat process [14].

4.2. Heat Transfer and Pressure Drop in Banks of Finned Tubes in Crossflow

Heat transfer and pressure drop in banks of tubes are functions of multiple factors, including those of fin shape, bank geometry, number of rows, physical properties and velocity of the fluid.

Design calculations are encumbered by the numerous suggestions rate calculation models for different bank arrangements.

We present here the experimental results on the effects of the above factors on the heat transfer of finned tubes in staggered banks for Re from 20 to $1.4 \times 10^6$ (Fig. 35) and in-line banks for Re up to $10^5$.

Our analysis of all available data resulted in following heat transfer relations for staggered banks; we suggest, on the basis of data from [16, 17].

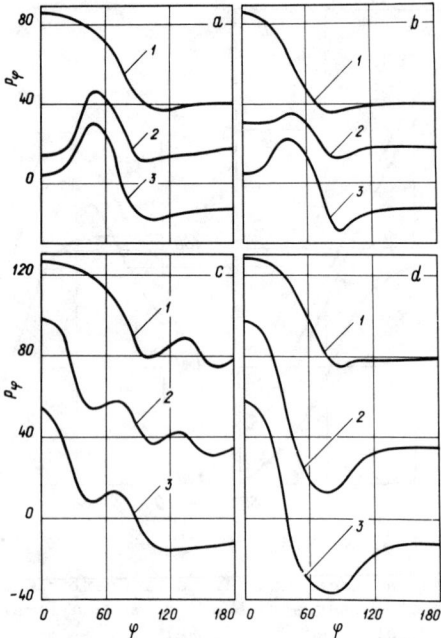

Fig.34. Pressure Distribution on Finned Tubes in Banks (a,b - In-Line Bank; c, d - Staggered Bank) a,c - Fin Tip; b,d - Fin Base

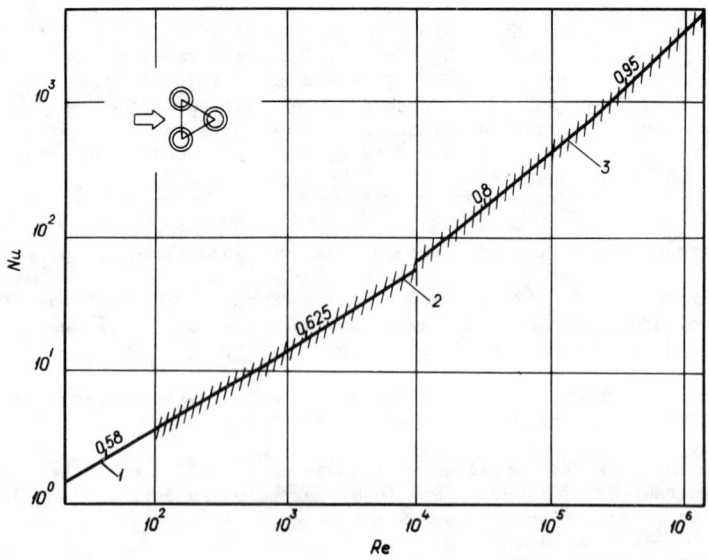

Fig.35. Heat Transfer of Finned Tubes in Staggered Banks ($20 < Re < 1.4 \times 10^6$)

$$Nu = 0.245 \, Re^{0.58} \tag{20}$$

for Re from 20 to $5 \times 10^2$. Similarly, on the basis of [16, 19].

$$Nu = 0.4 \, Re^{0.625} \cdot \varepsilon^{-0.375} \tag{21}$$

for Re from $5 \times 10^2$ to $10^4$ and $\varepsilon$ from 5 to 12.

According to [13]

$$Nu = 0.043(a/b)^{0.2}(s/d)^{0.18}(h/d)^{-0.14} Re^{0.8} \tag{22}$$

for Re from $10^4$ to $2 \times 10^5$ and

$$Nu = 0.0069(a/b)^{0.2}(s/d)^{0.18}(h/d)^{-0.14} Re^{0.95} \tag{23}$$

for Re over $2 \times 10^5$. These relations apply to staggered banks with a from 2.17 to 4.13, b from 1.27 to 2.14, s/d from 0.125 to 0.28 and h/d from 0.125 to 0.59.

For in-line banks from data presented in [16]:

$$Nu = 0.266 \, Re^{0.625} \, \varepsilon^{-0.375} \tag{24}$$

for Re from $5 \times 10^3$ to $10^5$ and $\varepsilon$ from 5 to 12.

Relations (20 through 24) refer to $\bar{\alpha}_T$, which is determined from the average temperature of the total finned surface.

A similar analysis of the known data on the pressure drop (Fig.36) suggest for staggered banks from [17, 18]

$$Eu = 67.6 \, \varepsilon^{0.5} \, Re^{-0.69} \, z \, c_z \, a^{-0.55} \, b^{-0.5} \tag{25}$$

for Re from 20 to $10^3$ and $\varepsilon = 9.12$, $a = 2$, $b = 1.73$.

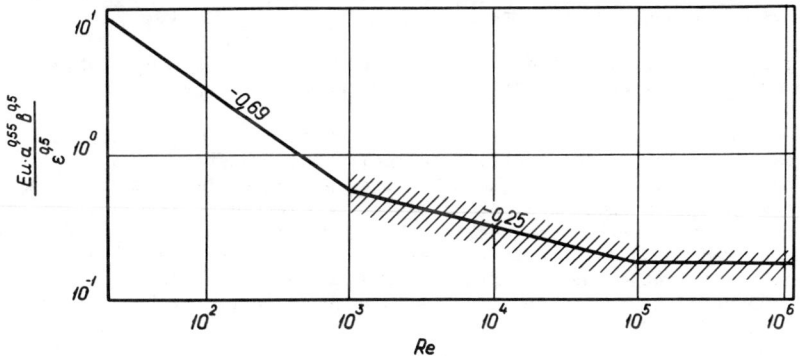

Fig.36. Pressure Drop on Finned Tubes in Staggered Banks ($20 < Re < 1.2 \times 10^6$)

$$Eu = 3.2\ \varepsilon^{0.5} Re^{-0.25} z\ c_z^{-0.55} a\ b^{-0.5} \tag{26}$$

for Re from $10^3$ to $10^5$ and

$$Eu = 0.18\ \varepsilon^{0.5} z\ c_z^{-0.55} a\ b^{-0.5} \tag{27}$$

for Re from $10^5$ to $1.4 \times 10^6$. Relations (26) and (27) apply for staggered banks when $\varepsilon$ is from 1.9 to 16.3, a from 1.6 to 4.13 and b from 1.2 to 2.35.

For in-line banks, on the basis of data in [18] we suggest

$$Eu = 0.068\ \varepsilon^{0.5} z\ c_z\ \eta^{-0.4} \tag{28}$$

for Re from $10^3$ to $10^5$, $\varepsilon$ from 1.9 to 16.3, a from 2.83 to 3.13 and b from 1.2 to 2.35, where $\eta = (a-1)/(b-1)$.

Judging by (20 through 23) and by Fig. 35 to determine the heat transfer of a specific finned tube heat exchanger, the knowledge of the intended range of Re for its operation is obligatory.

In the laminar-turbulent flow regime, the value of m, the power index of Re, is 0.58 for Re from 20 to $5 \times 10^2$. In the subcritical flow regime, the value of m is 0.625 for Re from $5 \times 10^2$ to $10^4$. The range of Re from $10^4$ to $2 \times 10^5$ is critical at which a predominantly turbulent flow regime is established, so that the value of m should be increased to 0.8.

We note here, that on plain tubes the laminar-turbulent transition is observed at higher values of Re.

At Re about $2 \times 10^5$, the supercritical, completely turbulent flow regime, is established. In the heat transfer relations for banks of finned tubes m = 0.95 is applied for Re over $2 \times 10^5$.

Our comparison of the results by different authors suggests considerable advantages of staggered banks, as compared to in-line ones.

Analytical descriptions of the pressure drop also depend on the flow regime. At Re about $10^5$, with the onset of the supercritical flow regime, the pressure drop acquires on autonomous character.

Comparisons of different banks to evaluate the effect of the relative pitches a and b revealed higher thermal efficiencies of compact staggered banks.

As to the fins they ought to be as thin, as possible, but neither their shape, nor thickness have a predominant effect on their heat transfer. Some condiseration on the geometry of finning-fin height h and fin spacing s, revealed a definite effect of the inter-fin space h/s. At specific values of h/s, stagnation regions may be formed in the inter-fin spaces, so that certain parts of the surface become excluded from the active heat transfer. To avoid this, $h/s \leq 1.9$ is recommended where the amount of heat is proportional to the finned surface. With h/s > 1.9, the amount of heat transfer reduces in proportion to $(h/s)^{-0.7}$. This applies to Re from $5 \times 10^3$ to $5 \times 10^4$ and t from $0.5 \times 10^{-3}$ to $0.8 \times 10^{-3}$. There exists a close relation between the ratio h/s and the finning factor $\varepsilon$. The heat transfer coefficient decreases both with the increase of the fin height & with the decrease of the fin spacing, that is with the growth of the finning factor $\varepsilon$.

An interesting relation referring the effect of the finning factor $\varepsilon$ was observed; the higher the value of $\varepsilon$, the more pronounced is the advantage of staggered banks, as compared to in-line ones. An increase of $\varepsilon$ is reflected by different changes of the heat transfer in staggered and in-line banks.

An explanation to this lies on the fact, that separated streams and wakes are formed downstream the separate fins, thus affecting the heat transfer. In in-line banks the wake of a preceding tube causes the exclution of the front part of a following tube from the active heat transfer and makes finned tubes less efficient in arrangements of the in-line type. The coefficients of convective heat transfer are the same for banks of different finning geometries (h/s and h/d), but with the same finning factors $\varepsilon$. Thus finning factor $\varepsilon$ is a value, which accounts with sufficient accuracy both for fin height, and for fin spacing, and its relation to the coefficient of convective heat transfer would be welcome.

The values of Nu and Eu from (20 through 28) correspond to average values for one row in a bank. The equations for the leading rows are

$$Nu_z = Nu \ c_z \qquad (29)$$

$$Eu_z = Eu \ c_z' \qquad (30)$$

where $c_z$, $c_z'$ - correction factors from Fig. 37.

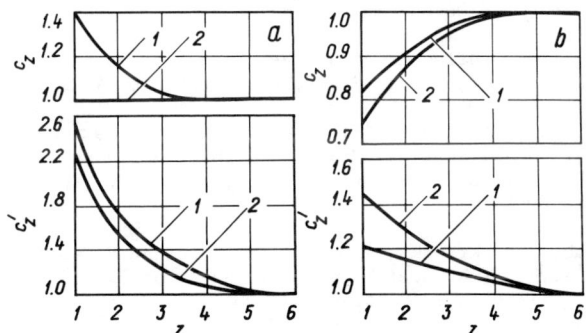

Fig.37. The Relative Correction Factor for the Heat Transfer and Pressure Drop of Small Banks, a) In-line Bank; b) Staggered Bank. (1-Re=1.2 x $10^3$, 2- Re = 5 x $10^4$)

REFERENCES

1. Kays, W.M., London, A.L. 1958. Compact Heat Exchangers. McGraw-Hill, New York.

2. Fraas, A.P., Özışık, M.N. 1965. Heat Exchanger Design. John Willey and sons, Inc., New York.

3. Burkov, V.V., Indeikin, A.I. 1978. Avtotraktornye Radiatory. Automobile-and Tractor Radiators. Mashynostroenije, Leningrad. (In Russian).

4. Žukauskas, A., Mararevićius, V., Šlančiauškas, A. 1968. Teplootdacha Puchkov Trub v Poperechnom Potoke Zhidkosti. Heat Transfer in Banks of Tubes in Crossflow of Fluid. Mintis, Vilnius.(In Russian).

5. Žukauskas, A. 1972. Vol.8. Advances in Heat Transfer. Academic Press. New York, London, pp. 93-160.

6. Žukauskas, A., Žiugžda, J. 1979. Teplootdacha Tsilindra v Poperechnom Potoke Zhidkosti. Heat Transfer of Cylinder in Crossflow of Fluid. Mokslas, Vilnius (In Russian).

7. Žukauskas, A., Ulinskas, R.V., Sipavićius, Č.-S.I. 1978. Average Heat Transfer and Pressure Drop in Cross Flow of viscous Fluids over Tube Bundles at Low Re, Heat Transfer Soviet Research, Vol.10, N 6, pp. 90-101.

8. Žukauskas, A., Ulinskas, R.V., Martsinauskas, K.F. 1977. Influence of the Geometry of the Tube Bundle on the Local Heat Tranfser Rate in the Critical Region of Stream-Line Flow International Chemical Engineering, Vol.17, N 4, pp. 744-751.

9. Poškas, P., Survila, V., Žukauskas, A. 1979. Srednyaia Teplootdacha Truby v Puchkakh, Poperechno Obtekaemykh Potokom Vozdukha pri Bolschykh Re. Mean Heat Tranfser of Tube in Banks in Cross-Flow of Air at High Re. -Lietuvos TSR Mokslu Akademijos Darbai, Ser. B, 2(111), pp. 69-79.

10. Kostić, Z. 1968. Heat Transfer from a Cylinder in Turbulent Wake of a Preceeding Cylinder, Int. Seminar "Heat and Mass Transfer in Flows with Separated Regions and Measurement Techniques", Herceg-Novi, Yugoslavia.

11. Achenbach, E. 1971. Influence of Surface Roughness on the Flow Through a Staggered Tube Bank, Warme-und Stoffübertragung, Vol.4, N 1, pp. 120-126.

12. Kuznetsov, N.V., Pshenisnov, I.F. 1974. O Vlyanii Neravnomernosti Teplootdachi po Poverkhnosti Kruglogo Rebra na yego Effektivnost. The Effect of Non-Uniform Heat Transfer of a Fin on its Efficiency. Teploenergetika, 8, pp. 42-45. (In Russian).

13. Stasiulevićius, J., Skrinska, A. 1974. Teplootdacha Poperechno Obtekaemykh Puchkov Rebristykh Trub. Heat Transfer in Banks of Finned Tubes in Crossflow, Mintis, Vilnius. (In Russian).

14. Kuntysh, V.B., Yokhvedov, F.M. 1977. Eksperimentalnoe Issledovanijie Mestnykh Koeffitsientov Teplootdachi Trub so Spiralnymi Rebrami v Poperechno Obtekaemykh Rebristykh Puchkakh. An Experimental Study Banks of Tubes with Helical Fins in Crossflow. Izvestia Vuzov SSSR. Energetika, 2, pp.105-110.(In Russian)

15. Yokhvedov, F.M., Kuntyah, V.F., Genin, S.M. 1972. Eksperimentalone Izuchenye Raspredelenya Davlenya po Rebru v Puchkakh iz Orebrennykh Trub. An Experimental Study of Pressure Distribution of Fins in Banks of Finned Tubes. Trudy Arkhangelskogo Lesotekn. Inst., Arkhangelsk, 31, pp.60-63 (In Russian).

16. Schmidt, Th. E. 1963. Der Warmeübergang an Rippenrohren und die Berechnung von Rohrbundel, Warmeaustauschern.-Kaltetechn., Bd.15, Heft 4, S.98-102; Heft 12, S. 370-378.

17. Koshmarov, Yu.A., Svirschevskii, S.B.,Inozemtseva, E.N. 1978. Teploobmen

i Soprotivlenye Orebrennykh Trub pri Nizkikh Chislakh Reynoldsa. Heat Transfer and Pressure Drop of Finned Tubes at Low Re. Tematicheskii Sbornik Moskovskogo Aviatsionnogo Instituta. Moscow, 463, pp.33-37. (In Russian).

18. Lokshin, V.A., Fomina, V.M. 1978. Obobshchenye Materyalov po Eksperimental-nomu Issledovanyu Rebristykh Puchkov. An Analysis of Experimental Data on Banks of Finned Tubes. Teploenergetika, N 6, pp.36-39. (In Russian).

19. Yudin, V.F., Tokhtarova, L.S., Lokshin, V.A., Tulin, S.N. 1968. Obobshchenye Opytnykh Dannykh o Konvektivnom Teploobmene pri Poperechnom Omyvanii Puckov Trub a Poperechnym Lentochnym i Shaibovym Orebrenyem. An Analysis of Experimental Data of Convective Heat Transfer of Banks of Tubes with Plate-and Disc Fins in Crossflow. Trudy Tsentralnogo Kotlo-turbinnogo Instituta, Moscow, 82, pp. 108-134. (In Russian).

# Total and Local Heat Transfer and Pressure Drop of Staggered and In-Line Tube Bundles

**E. ACHENBACH**
Institut für Reaktorbauelemente
Kernforschungsanlage Jülich GmbH
Jülich, Germany

ABSTRACT

Heat transfer and pressure drop of a smooth staggered and in-line tube bundle have been measured in the range of Reynolds number $4 \times 10^4 < \text{Re} < 7 \times 10^6$. By means of local heat transfer results the phenomena occurring in the boundary layer of the tubes are described. Beyond the critical Reynolds number the heat transfer of both types of heat exachngers increases due to the occurrence of a turbulent boundary layer around the tubes. The unexpected variation of the pressure drop coefficient near the critical Reynolds number is discussed.

NOMENCLATURE

| | | |
|---|---|---|
| a | - | ratio of transversal pitch to tube diameter |
| b | - | ratio of longitudinal pitch to tube diameter |
| d | m | tube diameter |
| k | m | surface roughness height |
| n | - | exponent of the Reynolds number |
| Nu | - | Nusselt number |
| p | $N/m^2$ | pressure |
| Re | - | Reynolds number |
| $s_l$ | m | longitudinal pitch |
| $s_t$ | m | transversal pitch |
| $u_c$ | m/s | velocity in the smallest cross section |
| z | - | number of rows |

GREEK SYMBOLS

| | | |
|---|---|---|
| $\alpha$ | $W/(m^2 K)$ | heat transfer coefficient |
| $\zeta$ | - | pressure drop coefficient |
| $\eta$ | kg/(ms) | dynamic viscosity |
| $\lambda$ | $W/(mK)$ | thermal conductivity |
| $\rho$ | $kg/m^3$ | density |
| $\psi$ | - | angular position measured from front stagnation point |
| $\psi_s$ | - | angular position of boundary layer separation |
| $\psi_t$ | - | angular position of boundary transition |

1. INTRODUCTION

Cross flow heat exchangers applied to nuclear gas-cooled reactors operate at high Reynolds numbers. Hammeke et al. [1] found that at a distinct Reynolds number heat transfer and pressure drop change their dependence upon the Reynolds number: Both quantities are observed to increase considerably, at least for staggered tube arrangements.

In own flow investigations it was demonstrated previously [2], [3], [4], that this phenomenon is associated with the occurrence of a critical Reynolds number at which the boundary layer of the tubes indicates transition from laminar to turbulent flow in the rearward part of the tubes. The transition point shifts upstream in direction of the front stagnation point with increasing Reynolds number. Thus the augmentation of the heat rate transferred can be explained.

In the papers mentioned above it was additionally pointed out that the critical Reynolds number is dependent on the surface roughness of the tubes. Increasing roughness causes a decrease of the critical Reynolds number. Thus an enhancement of heat transfer can be generated already at lower Reynolds number using rough surfaces of the heat exchanger tubes. Corresponding results are found, for instance, in a paper by Groehn and Scholz [5].

The aim of the present paper is to demonstrate that cross flow tube bundles behave like a bluff body. Particularly, analogous phenomena as observed for the single cylinder in cross flow occur. Thus in contradiction to the common opinion, the boundary layer of the tubes is laminar up to very high Reynolds numbers and undergoes transition to turbulent flow for a smooth tube bundle only for $Re > 4.5 \times 10^5$.

In this paper the heat transfer and pressure drop results of only one smooth staggered and one in-line-tube arrangement are discussed. The description is started with results obtained for a single row of tubes which evidently indicates the close relationship between tube bundle and single cylinder. This evidence should be emphasized so this report is predominantly not a matter of production of experimental data.

The definition of the dimensionless groups used for the presentation of the results are as follows: The Reynolds number is formed with the tube diameter d and the velocity $u_c$ in the narrowest cross section between neighbouring tubes:

$$Re = \frac{u_c d \rho}{\eta} \qquad (1)$$

Correspondingly the length scale of the Nusselt number is given by the tube diameter

$$Nu = \frac{\alpha \, d}{\lambda} \qquad (2)$$

The pressure drop coefficient $\zeta$ is referred to the velocity $u_c$ and the flow resistance per row.

$$\zeta = \frac{\Delta p}{z \frac{\rho}{2} u_c^2} \qquad (3)$$

The presentation of the local heat transfer results is made in terms of $Nu/\sqrt{Re}$, since the heat transfer in the laminar boundary layer increases with the square root of Re. Thus $\sqrt{Re}$ acts as a normalizing function.

## 2. EXPERIMENTAL ARRANGEMENT

A staggerd and an in-line heat exchanger model consisting of seven rows and three tubes per row were tested. The ratio of transversal pitch to tube diameter $a = s_t/d$ was $a = 2.04$, that of the longitudinal pitch $b = s_1/d = 1.43$. The tubes had a diameter of $d = 0.147$ m and were highly polished. Thus the ratio of roughness height to diameter was $k/d < 10^{-5}$.

The test section was rectangular ($0.9 \times 0.5$ m$^2$). It could be installed in a wind tunnel operating with air at ambient conditions as well as in a high pressure circuit, which was also run with air, but up to a pressure of 40 bars. Thus Reynolds numbers up to $Re = 7 \times 10^6$ could be verified. Details about the high pressure wind tunnel are found in [6].

The heat transfer rate was determined by electrically heating a single tube of the bundle. This tube could be installed at arbitrary positions of the bank to evaluate the effect of tube position on heat transfer. Concerning the measurement of local heat transfer a small plug was provided which was mounted flush with the surface and separately heated. It was controlled at the same temperature as its surroundings. As the cylinder could be rotated around its axis the local probe could be placed at any angular position requested. The method has carefully been checked during tests on a single circular cylinder in cross flow. The details of the device are described and the difficulties of this measurement technique pointed out in a previous paper [7].

Since the test cylinder a sketch of which is shown in Figure 1 was manufactured from copper the boundary condition of constant temperature of the surface could nearly be verified. At highest Reynolds numbers a maximum variation of $\pm 5\%$ of the mean temperature difference between wall and gas was expected.

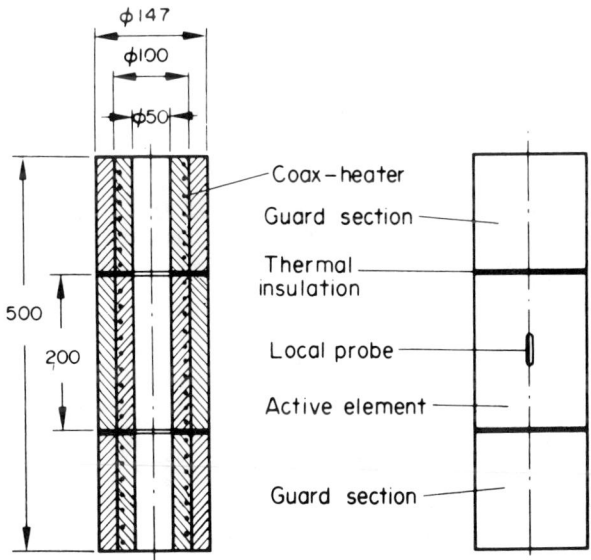

Fig.1. Test Cylinder

## 3. RESULTS

### 3.1 SINGLE ROW

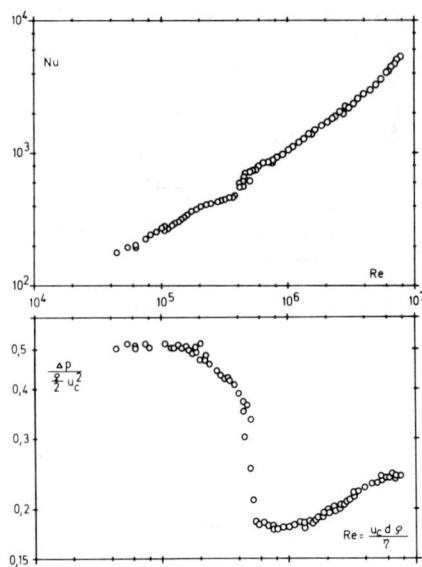

Fig.2. Total Heat Transfer and Pressure Drop of a Single Row of Smooth Tubes, a = 2.04

Figure 2 represents the experimental results of total heat transfer and pressure drop of a single row of smooth tubes for a = 2.04 as a function of the Reynolds number. The pressure drop coefficient $\zeta$ exhibits the typical evidence of a bluff body subjected to a flow near and beyond critical flow conditions. Similar to the single circular cylinder in cross flow four flow ranges can be distinguished [3]: At subcritical flow $\zeta$ is independent of Re. The boundary layer laminarly separates from the wall upstream of the narrowest cross section at about $\psi = 80°$. In the range $2 \times 10^5 < Re < 4.5 \times 10^5$ $\zeta$ decreases drastically down to the minimum value at Re $4.5 \times 10^5$ which is denoted the critical Reynolds number $Re_{crit}$. The steep drop of $\zeta$ is caused by the transition of the boundary layer from laminar to turbulent flow associated with the occurrence of separation bubbles on both sides of the tube and turbulent reattachment of the flow. The rearward turbulent boundary layer which is characterized by strong interchange of energy normal to the main flow direction is able to follow the contour of the wall over a larger distance against a pressure rise than the laminar boundary layer. Thus the local static pressure in the rear of the tube increases causing a decrease of the flow resistance. With further increase of Re the existence of the intermediate laminar separation and turbulent reattachment ceases and immediate transition laminar – turbulent occurs. While the transition point shifts upstream with increasing Re the separation point also moves upstream causing a decrease of the static pressure in the rear and hence an increase of the pressure drop. At transcritical flow conditions the pressure drag coefficients seems to level out to a constant value. A more detailed description of the boundary layer phenomena has been given in [9] for the single cylinder in cross flow. That representation can analogously be applied to the flow past a single row of tubes.

The total heat transfer is also affected by the boundary layer phenomena described above (Figure 2). At the beginning of the critical flow range the heat transfer diminishes as the length of the laminar boundary layer increases, which represents a remarkable heat resistance. The occurrence of the turbulent reattachement of the boundary layer instantaneously leads to an augmentation of the heat transfer. For $Re > 4 \times 10^6$ the curve $Nu = f(Re)$ again exhibits a steeper slope because the boundary layer transition point begins to shift upstream so that the contribution of the turbulent boundary layer to the total length of the boundary layer increases.

The details of the boundary layer phenomena can be observed in Figure 3 which illustrates the local heat transfer of a single row of smooth tubes at variable Reynolds number. At $Re = 1.8 \times 10^5$ the flow is still subcritical and laminar separation occurs at about $\psi = 85°$. At $Re = 4.4 \times 10^5$ the separation bubble is immediately being formed. It is fully developed for $Re = 1.2 \times 10^6$ where near $\psi = 105°$, i.e. the point of intermediate laminar separation, the heat transfer coefficient becomes very small due to the strong increase of the laminar boundary layer. The downstream turbulent boundary layer causes a steep increase of the heat transfer which shows a second minimum near the separation point of the turbulent boundary layer separation point at $\psi = 140°$. The curve valid for $Re = 3 \times 10^6$ exhibits that direct transition laminar - turbulent occurs, but on the rear part of the tube. Finally, the transition takes place at $\psi = 40°$ for $Re = 6.9 \times 10^6$ indicated by the steep increase of Nu.

The experimental results for the heat transfer at the stagnation point collapse at a value of about $Nu/\sqrt{Re} = 0.7$. This value is consistent with that of the single cylinder ($Nu/\sqrt{Re} \simeq 1$) since the Reynolds number of the tube row is by a factor of 2 larger compared with the single cylinder.

The evaluation of local heat transfer results yields the position of boundary layer separation which is plotted dependent on Re in Figure 4. The

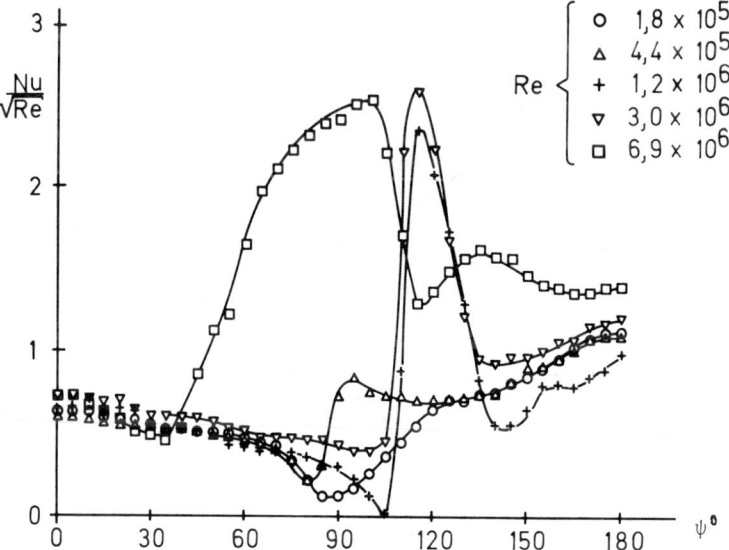

Fig.3. Local heat transfer of a single row of smooth tubes, a = 2.04

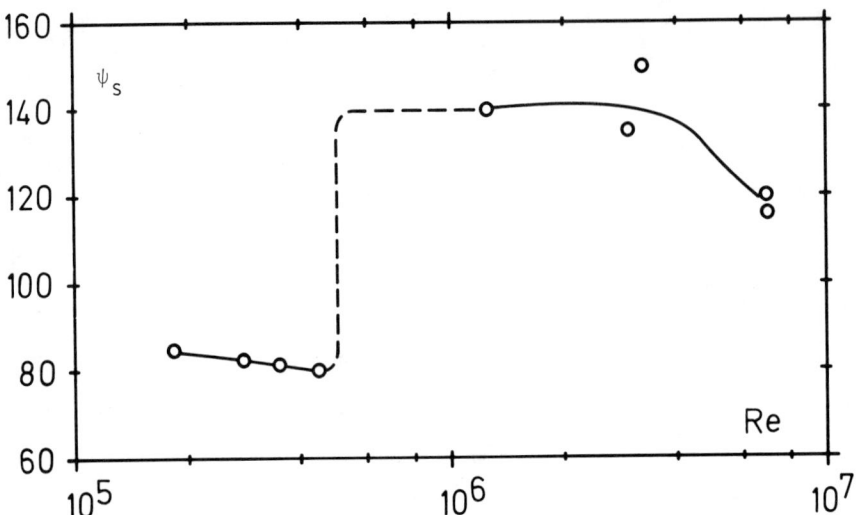

Fig.4. Position of boundary layer separation of a single row of tubes

critical Reynolds number is recognized by the step in the curve which indicates the sudden shift of the separation into the rear of the tubes.

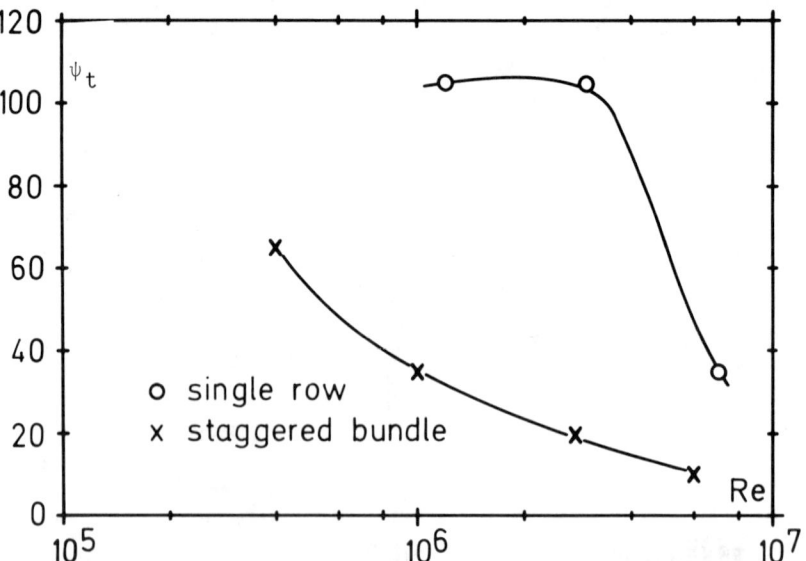

Fig.5. Position of boundary layer transition laminar-turbulent of a single row of tubes

Figure 5 also results from local heat transfer measurements. The curve representing the transition laminar - turbulent of the boundary layer of a single row of tubes demonstrates the upstream shifting of $\psi_t$ at the beginning of the transcritical flow range. The additional curve of Fig. 5 is referred to below.

## 3.2 STAGGERED TUBE BUNDLE

The heat transfer and the pressure drop of a staggered tube arrangement essentially exhibit the same behaviour as seen for the single row. Due to the geometrical conditions of the tube array and to the increased turbulence level of the flow through the bundle the pressure drop coefficient does not fall down abruptly (Figure 6). However, as an indication of the beginning of transition from laminar to turbulent boundary layer $\zeta$ shows a minimum at $Re_{crit}$ = $4.5 \times 10^5$. At this value the curve Nu = f (Re) takes a steeper slope as with growing Reynolds number the heat transfer is increased due to the turbulent exchange of heat through the boundary layer normal to the wall.

With a view to Figure 7, which represents the local heat transfer of the central tube in the fifth row of a staggered bundle at variable Reynolds number it becomes evident that between Re = $4 \times 10^5$ and Re = $3 \times 10^6$ the transition point laminar - turbulent shifts upstream causing the total heat transfer to increase nearly proportional to Re. This range is, however, transitional, and it is succeeded by the transcritical regime where Nu $\sim$ Re$^{0.8}$. In Figure 5 the dependency of the location $\psi_t$ of boundary layer transition on the Reynolds number is illustrated. It is obvious that compared with the single row of tubes premature transition occurs due to the high tubulence intensity within the bundle.

Coming back to Figure 7 it is seen that the boundary layer is laminar even at the highest Reynolds number near the vicinity of the stagnation point. The peak values of the turbulent boundary layer increase with a o.8-power of the Reynolds number.

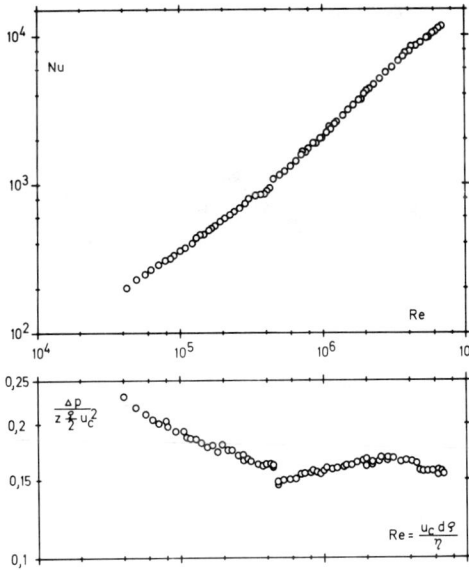

Fig.6. Total heat transfer and pressure drop of a smooth staggered tube bundle

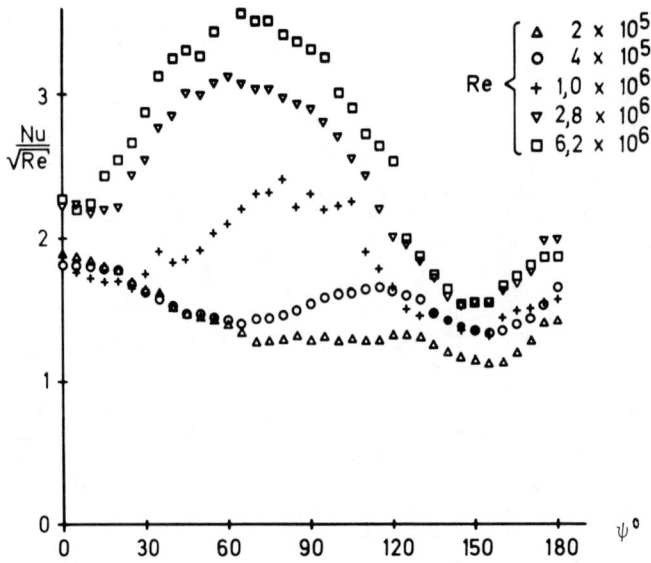

Fig.7. Local heat transfer of a smooth staggered tube bundle at variable Reynolds number

## 3.3 IN-LINE TUBE BUNDLE

A tube of an in-line bundle is situated in the wake of the preceding tube and at the same time it is affected by the succeeding tube with respect to the mechanism of the boundary layer separation. Therefore the geometrical

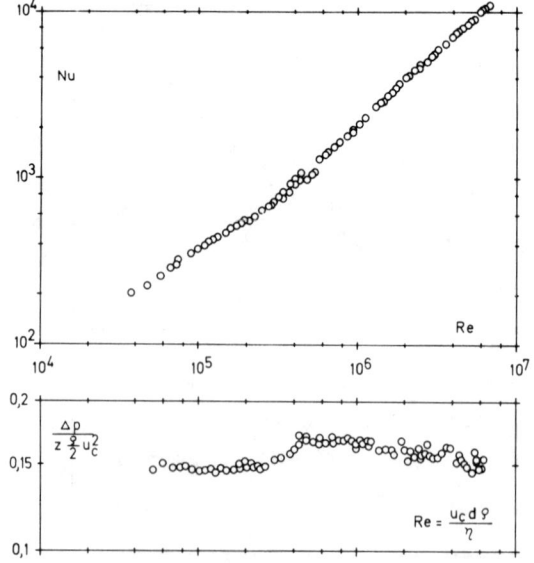

Fig.8. Total heat transfer and pressure drop of a smooth in-line tube bundle

conditions, above all the longitudinal pitch, have an important effect on the flow. In spite of this the typical phenomena of the flow past bluff bodies occur in such an array as for instance the existence of the critical Reynolds number (see Figure 8). Its value is the same as for the staggered bundle of the same transversal pitch. Considering the total heat transfer, beyound $Re_{crit}$ the Nusselt number is increasing with a larger power of Re than for $Re < 4 \times 10^5$. Similar to the staggered bundle the curve levels out to a power of o.8 of Re for Re $3 \times 10^6$.

The pressure drop coefficient exhibits a maximum value at $Re = Re_{crit}$. On a first view this is surprising, particularly with respect to the evidence observed for the staggered tube bundle. However, in a previous paper [4] in which experimental data on local static pressure and skin friction are published it is shown that the downstream shifting of the boundary layer separation at critical flow conditions causes a decrease of the angle of impact at the front of the succeeding tube which leads to an increase of static pressure in the front part of the tube. Thus the total flow resistance is increased. With further growing of Re the separation point moves upstream again and initiates a gradual decrease of $\zeta$.

The local heat transfer is plotted at various Reynolds number in Fig. 9 for the in-line tube bundle. An established boundary layer is found only in the angular position $30° < \psi < 120°$. It is laminar for $Re < Re_{crit}$. The sudden increase of Nu indicates the transition to the turbulent boundary layer. The peak values depend on Re with a o.8-power.

The comparison of the heat transfer data obtained for the staggered and in-line bundles, respectively, yield the surprising result that within a few per cent the heat transfer of both bundles obeys the same relationship with respect to the Reynolds number. Figure 1o represents the own results compared with those of other authors [1], [1o], [11]. It is obvious that the data of

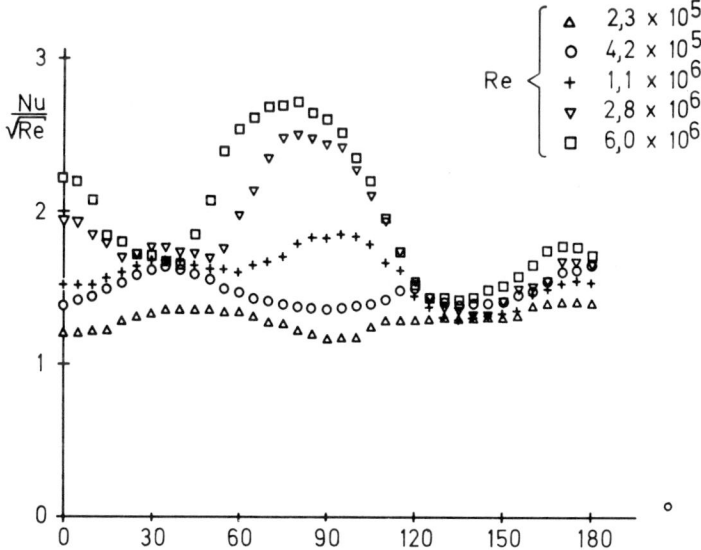

Fig.9. Local heat transfer of an in-line tube bundle at variable Reynolds number

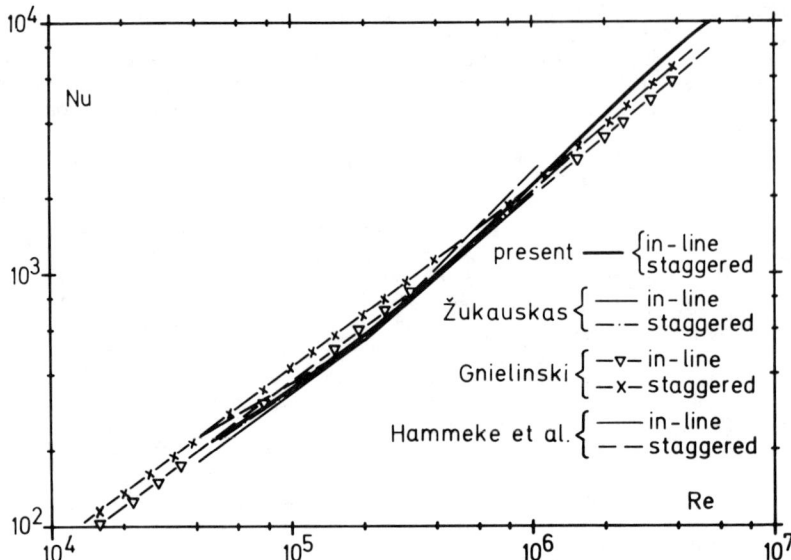

Figure 1o: Total heat transfer of tube bundles, comparison with other authors

all experimenters as well as those of the semi-empirical solution by Gnielinski [11] collapse within 15 per cent, independent whether the arrangement is staggered or in-line.

3.4 ENTRANCE EFFECT

As the total heat transfer was measured by heating a single tube in different positions of the bundle the entrance effect on heat transfer can be elaborated. Taking the heat transfer of the central tube in the fifth row, $Nu_5$, of a bundle consisting of seven rows as reference the comparison can be made for each row. In Figure 11 the experimental data are plotted for the in-line and staggered arrangement at various Reynolds numbers. The general trend is an increase of $Nu_z$ from the first row, the contribution of which lies between 40 and 60%, up to the third row, where the values of an infinitely extended tube bundle are approximately reached. The low values of the first row are due to the low velocity of the incident flow associated with a low tubulence level. In contradiction to the staggered bundle the heat transfer of the second row of an in-line array comes close to that of the infinite bundle. This is why the second tube row is already in the wake of the first row whereas the incident flow to the second row of a staggered tube bank has still a low turbulence level.

4. CONCLUSION

The pressure drop and the heat transfer of staggered and in-line tube bundles show a close similiarity to the corresponding quantities of a single circular cylinder in cross flow. Up to the critical Reynolds number which was determined to be $Re_{crit} = 4.5 \times 10^5$ for a smooth bundle of the transversal pitch a = 2.04, the boundary layer is completely laminar. Beyond $Re_{crit}$ it becomes turbulent downstream of the main span. With further increase of the

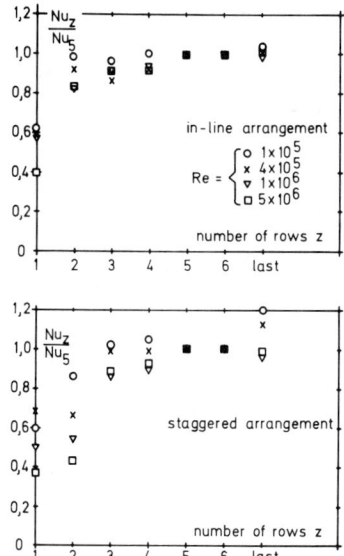

Figure 11: Entrance effect on heat transfer of tube bundles

Reynolds number the location of transition laminar-turbulent shifts upstream. Thus the total heat transfer is enhanced. The pressure drop, too, changes its dependence upon the Reynolds number beyond $Re_{crit}$ where the staggered and in-line bundle exhibit a different behaviour.

The own experimental results exceed the data already available by half an order of magnitude concerning the maximum Reynolds number. In the lower range of Reynolds number the agreement between the particular heat transfer results of different authors is good. The present data indicate that in the transition range between the critical and transcritical flow regime the heat transfer is nearly proportional to the Reynolds number. The slope of Re levels out to $n = 0.8$ in the transcritical regime at $Re > 3 \times 10^6$.

REFERENCES

1. Hammeke, K.; Heinecke, E.; Scholz, F. Wärmeübergangs- und Druckverlustmessungen an querangeströmten Glattrohrbündeln, insbesondere bei hohen Reynolds-Zahlen, Int. J. Heat Mass Transfer 1o (1967), pp. 427-446.

2. Achenbach, E. Investigations on the flow through a staggered tube bundle at Reynolds numbers up to $Re = 10^7$. Wärme- und Stoffübertragung 2 (1969), pp. 47-52.

3. Achenbach, E. Influence of surface roughness on the flow through a staggered tube bank. Wärme- und Stoffübertragung 4 (1971), pp. 12o-126.

4. Achenbach, E. On the cross flow through in-line tube banks with regard to the effect of surface roughness. Wärme- und Stoffübertragung 4 (1971) pp. 152-155.

5. Groehn, H.G.; Scholz, F. Wärme- und strömungstechnische Untersuchungen an fluchtenden Rohrbündel-Wärmeaustauschern aus pyramidenförmig aufgerauhten Rohren. JüL-1437-(1977).

6. Grosse, H.; Scholz, F. Der Hochdruck-Gaskanal. Kerntechnik 4 (1965) pp. 150-158.

7. Achenbach, E. Total and local heat transfer from a smooth circular cylinder in cross-flow at high Reynolds number. Int. J. Heat Mass Transfer 18 (1975) pp. 1387-1396.

8. Achenbach, E. Influence of surface roughness on the cross-flow around a circular cylinder. J. Fluid Mech. 46 (1971) pp. 321-335.

9. Achenbach, E. Distribution of local pressure and skin friction around a circular cylinder in cross-flow up to $Re = 5 \times 10^6$. J. Fluid Mech. 34 (1968) pp. 625-639.

10. Žukauskas, A. Heat transfer from tubes in cross-flow Advances in Heat Transfer 8 (1972) pp. 93-160.

11. Gnielinski V. Wärmeübergang bei Querströmung durch einzelne Rohrreihen und Rohrbündel. VDI-Wärmeatlas, Abschnitt Ge, Düsseldorf (1977) VDI-Verlag.

# Thermal Hydraulic Investigation of Yawed Tube Bundle Heat Exchangers

**H.-G. GROEHN**
Kernforschungsanlage Jülich GmbH
Institut für Reaktorbauelemente

ABSTRACT

In connexion with developing of heat exchangers for High-Temperature-Gas-Cooled-Reactor (HTGR) Plants a program was started to study the hydraulic resistance and heat transfer coefficient of 28 tube bundles with yawed tubes. The pitches are: $S_T$ = 1.25 - 1.5; $S_L$ = 1.09 - 1.5, the arrangement of the tubes are in-line, staggered, and crossed, the angle of attack is varied between $15° \leq \varphi \leq 90°$. After a description of the test arrangement and an estimate on the influence of the yaw angle results obtained from 12 tube bundles at subcritical Reynolds numbers are given.

NOMENCLATURE

| | | |
|---|---|---|
| $c_p$ | Ws/(kg K) | specific heat at constant pressure |
| $D$ | m | tube diameter |
| $p$ | N/m$^2$ | static pressure |
| $\Delta p$ | N/m$^2$ | difference in static pressure over the tube bank |
| $s_T$ | | transversal pitch; see Fig. 4 |
| $s_L$ | | longitudinal pitch, see Fig. 4 |
| $T$ | K | temperature |
| $U$ | m/s | = $V \cdot \sin \varphi$, velocity component normal to the tube axis |
| $V$ | m/s | = $V_\infty \cdot (s_T - 1)/s_T$, velocitiy in the narrowest cross section between the tubes |
| $V_\infty$ | m/s | velocity in the empty channel |
| $W$ | m/s | = $V \cdot \cos \varphi$, velocity parallel to the tube axis |
| $Z$ | | tube row number |

GREEK SYMBOLS

| | | |
|---|---|---|
| $\alpha$ | W/(m$^2 \cdot$ K) | heat transfer coefficient |
| $\nu$ | m$^2$/s | kinematic fluid viscosity |
| $\eta$ | kg/(m · s) | fluid viscosity |
| $\theta$ | degree | angular position at circumference |
| $\lambda$ | W/(m · K) | heat conductivity |
| $\rho$ | kg/m$^3$ | fluid density |
| $\varphi$ | degree | yaw angle, see Fig. 4 |

SUBSCRIPTS

$\varphi_0$      yaw angle
$90°$      cross flow
$\infty$      infinite deep bundle; empty channel

CHARACTERISTIC NUMBERS

$\zeta_U$      $= \Delta p/(\rho/2 \cdot U^2 \cdot Z)$, hydraulic resistance coefficient
$\zeta_V$      $= \Delta p/(\rho/2 \cdot V^2 \cdot Z)$, hydraulic resistance coefficient
Nu      $= \alpha \cdot D/\lambda$ , Nusselt number
Pr      $= \eta \cdot c_p/\lambda$ , Prandtl number
Re      $= U \cdot D \cdot \rho/\eta$ , Reynolds number
$Re_V$      $= V \cdot D \cdot \rho/\eta$ , Reynolds number

INTRODUCTION

The problem of oblique flow arises in two types of HTGR heat exachangers; firstly in the entrance and exit regions of bundles with mainly axial flow direction. In this kind of tube arrangements the angle of attack varies from cross-flow ($90°$) to axial flow ($0°$), Fig. 1, right. Secondly in helical type heat exchangers with a large lead of the tubes, Fig. 1, left. In this case the yaw angle is constant along the whole length of the tube bundle. As a speciality of this heat exchanger type a design can be considered the neighboured tube cylinders of which are coiled in a contrary direction.

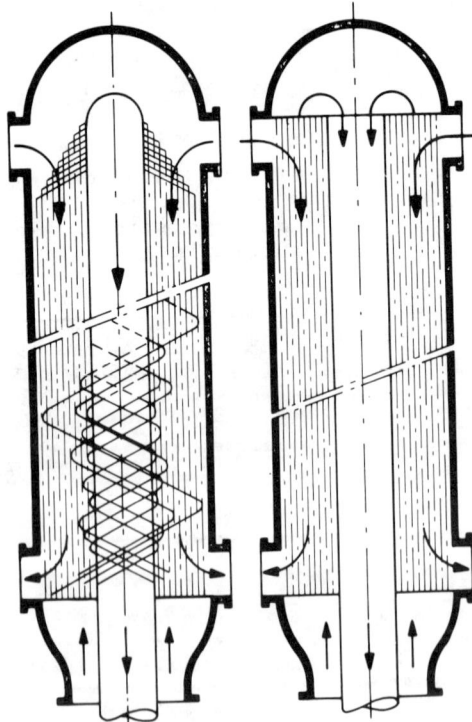

Fig.1. Types of heat exchangers with oblique flow paritions

Besides specific thermal hydraulic problems which must be solved for each particular heat exchanger there are some general questions of oblique flow which are only insufficient or partly inconsistent answered in the literature. For instance there is no reference to the influence of the tube pitches on heat transfer and pressure drop of inclined heat exchangers. Furthermore there is almost no information about the range of Reynolds number for which the data are valid. On the other hand there is no doubt that the yaw angle influences the critical Reynolds number. To clarifiy these and some other questions an experimental investigation has been started. A large number of tube bundles with different tube arrangements and tube pitches will be tested at subcritical as well as supercritical flow conditions.

FUNDAMENTALS OF OBLIQUE FLOW

The flow pattern around a yawed cylinder is three dimensional. However, a theoretical solution is essentially simplified by the fact that the flow is independent of the axial extension of the cylinder. Fig. 2 illustrates the flow conditions around a yawed cylinder. $V_\infty$ is the velocity in the flow direction. It can be splitted into the components $U_\infty$ normal and $W_\infty$ parallel to the tube axis. As yaw angle $\psi$ the angle between the main flow direction and the tube axis is defined. The coordinates x and z are in the tube wall normal and parallel to the tube axis. y is the coordinate perpendicular to the tube wall. $\theta$ denotes the angular position at the circumference counted from the point of impact.

The invicied velocity outside the boundary layer $V(x)$ can be regarded as a superimposition of the constant velocity component $W_\infty$ and of the velocity component $U(x)$ normal to the tube axis. If u, v, w mean the boundary velocity components in x-, y-, and z-direction the momentum equations can be written applying the boundary layer simplifications as

$$u \, \partial u/\partial x + v \, \partial u/\partial y = U dU/dx + \nu \, \partial^2 u/\partial y^2 \tag{1}$$

$$u \, \partial w/\partial x + v \, \partial w/\partial y = \nu \, \partial^2 w/\partial y^2 \tag{2}$$

Additionally the equation of continuity yields

$$\partial u/\partial x + \partial v/\partial y = 0 \tag{3}$$

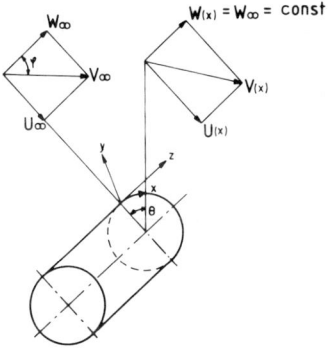

Fig.2. Flow conditions around a yawed tube

As remarked above the velocity components are independent of the z-direction. Consequently the velocity components u and v normal to the tube axis can be calculated independent of the axial velocity component w. This fact is called the principle of independence.

Equations (1) and (3) describe the two-dimensional flow around a circular cylinder. Their solution is given, for instance, by Blasius [1]. His method has been extended by Sears [2] and Görtler [3] to solve equation (2) that means to find out the velocity component w.

As the temperature distribution is also independent of the z-direction the energy equation for an inclined cylinder is equal to that of a circular cylinder in cross flow.

$$u\partial T/\partial x + v\partial T/\partial y = \lambda/(\rho c_p)\partial^2 T/\partial y^2 \qquad (4)$$

From the principle of independence follows, for instance, that the point of boundary layer separation of a yawed cylinder is unaffected by the axial velocity component. In the same way the pressure distribution around the cylinder should be not influenced by the yaw angle. Measurements from Bursnal [4], for instance, and also our own results, Fig. 3, confirm the validity of the principle of independence for a single cylinder at subcritical Reynolds numbers. The normalized pressure distribution is nearly independent of the yaw angle. Accordingly the hydraulic resistance coefficient $c_d$ differs unsignificantly, too.

The question arises if the principle of independence is appropriate to describe the thermalhydraulic behaviours of inclined tube bundles, too. Experiments should clarify down to which yaw angles the flow through an inclined tube bundle can be related to a bundle in pure cross flow.

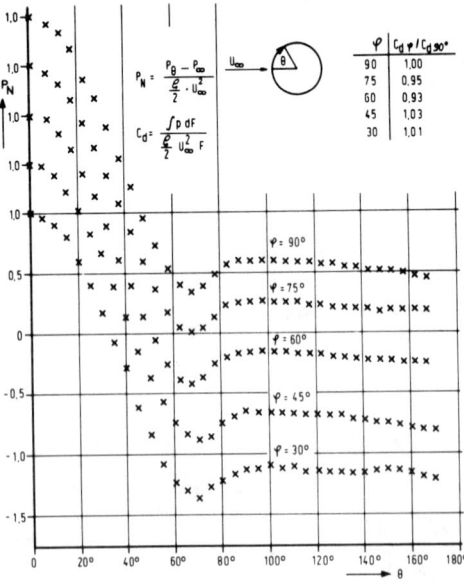

Fig. 3. Pressure distribution around a circular cylinder at different yaw angles

## APPARATUS AND TESTS

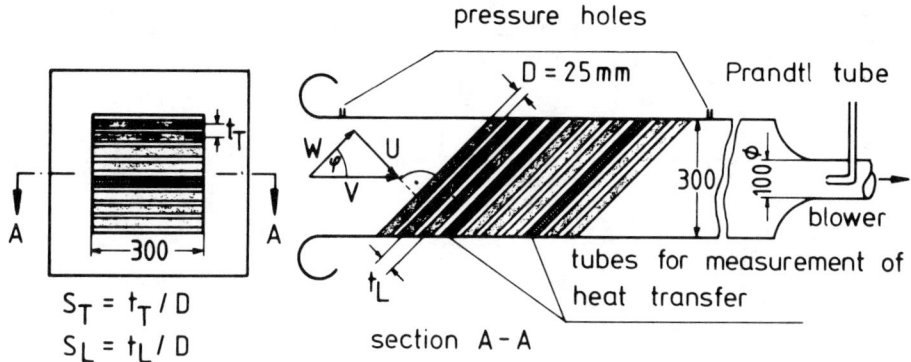

Fig.4. Experimental set-up

To cover the Reynolds number range of interest measurements are necessary at atmospheric conditions as well as with pressurized air up to 40 bars. 28 tube bundles with different tube pitches and different tube arrangements will be tested at first in an open atmospheric wind tunnel. From the results of these measurements it will be decided which of the tube bundles have to be tested at higher pressure, too. The dimensions of the test section result from the data of the available blowers and from the necessity to arrange tube bundles with a representative number of tubes across and along the flow direction. The test set-up is sketched in Fig. 4. The dimensions chosen allow to arrange 8 tubes per row normal to the flow direction at a transversal tube pitch of $S_T$ = 1,5 and 9,5 tubes at $S_T$ = 1,25. In flow direction 1o rows are used. The mass flow was determined means of a Prandtl tube. It was calibrated at different assemblies.

For heat transfer measurements electrical heated tubes are applied. A coax heater is inserted in to a copper rod. It is pressed into one of the steel tubes of the bundle. Thus the test tube has the same surface conditions as the remainder dummy tubes. The temperature of the test tube is measured at the inner wall of the steel tube because the tube diameter is relatively small and the surface of the tube should not be hurt. The temperature decrease through the tube wall is calculated. Three thermocouples are distributed over the active length of o,1 m of the heater and four over the circumference. The heaters are three divided. The outer parts act as guard heater while the inner piece represents the active element. The particular sections are thermally insulated from each other and separately heated.

## RESULTS

### ROW TO ROW MEASUREMENTS

The flow through oblique and cross flow tube bundles exhibits to a great extend similarities which can be revealed by carrying out row to row experiments of heat transfer and pressure drop. Initially it was not clear whether, for instance, the mass flow distribution is uniform along the tubes of a bundle consisting only of a few rows. Therefore the most of the tube arrays

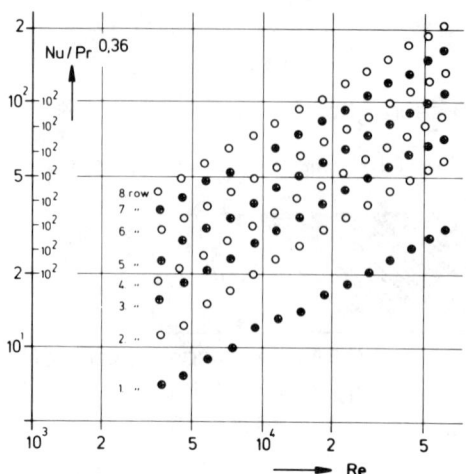

Figure 5 to Figure 7:
Entrance effect on heat transfer of yawed tube banks

**Fig.5.** $\varphi = 45°$; in-line; $S_T = 1,25$; $S_L = 1,25$

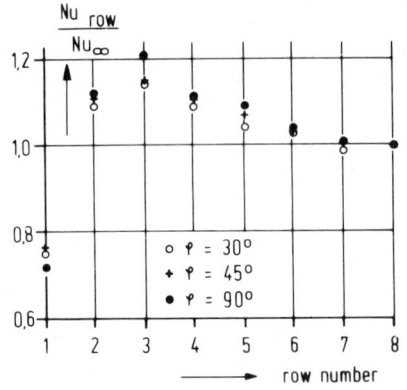

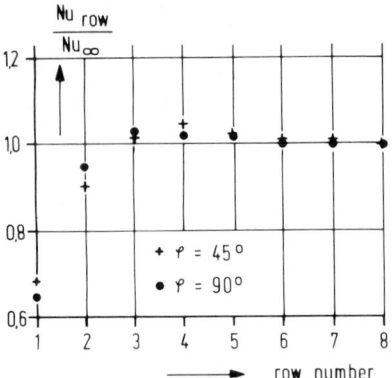

**Fig.6.** $Re = 2 \cdot 10^4$; in-line; $S_T = 1,25$; $S_L = 1,25$

**Fig.7.** $Re = 2 \cdot 10^4$; staggered; $S_T = 1,5$; $S_L = 1,3$

were investigated with variabel row numbers. The heat transfer coefficient was determined from the first up to the eighth row. In each case at least two rows were mounted downstream of the test tube. As an example Fig. 5 represents the dimensionless heat transfer coefficient as a function of the Reynolds number for an in-line tube bank at a yaw angle of $\varphi = 45°$. The term $Pr^{0,36}$ was taken from Zukauskas [6] while the Prandtl number was constant for our own experiments. To provide a clear representation of the results the ordinate scale in Fig. 5 is displaced by 25% for each row. As can be seen the slope of imagined curves through the data points is the same for all rows in the investigated Reynolds number range except those for the first and

second row which differ slightly. This corresponds to results from tube
bundles in cross flow. In Fig. 6 and Fig. 7 the heat transfer coefficient
of each row at constant Reynolds number is related to that of the eighth
row which stands for the heat transfer coefficient of an infinitely deep
tube bundle. Fig. 6 shows experimental data for an in-line tube bank at
different yaw angles. The increase of heat transfer from the first to the
third row, well-known from corss flow banks is characteristic also for
smaller yaw angles down to $\varphi = 30°$. However, the maximum at the third row
decrease with smaller yaw angles. Opposing to Fig. 6 the increase of heat
transfer for the staggered tube banks occurs over four rows and is weaker.
The influence of the yaw angle is insignificant. Differences between Fig. 6
and Fig. 7 probably are not only caused by the different tube arrangement
but to a high degree by the different tube pitches.

A similar representation as given in Fig. 5 for the heat transfer has
been made for the hydraulic resistance coefficient of a yawed tube bundle,
Fig. 8. Analoguesly to the result of Fig. 5 the pressure drop coefficient
exhibits the same dependence on the Reynolds number for all row numbers. In
order to compare the change of the hydraulic resistance over the depth of a
tube bundle for different angles of inclination the representation of Fig. 9
is suitable. The product $\zeta \cdot z$ means the actual dimensionless pressure drop
of the total tube bundle. For more than four rows the pressure loss increases
linearly.

From this evidence it can be concluded that the entrance and exit losses
are negligable for $z > 4$ at all yaw angles. The slope of the curves can be
related to the pressure drop coefficient by $\quad \zeta \approx \tan \alpha$ . This equation
is fairly satisfied if applied to a bundle consisting of ten rows. Summarizing one can say that the cross flow character of inclined tube bundles is
kept down to yaw angles of $\varphi = 30°$.

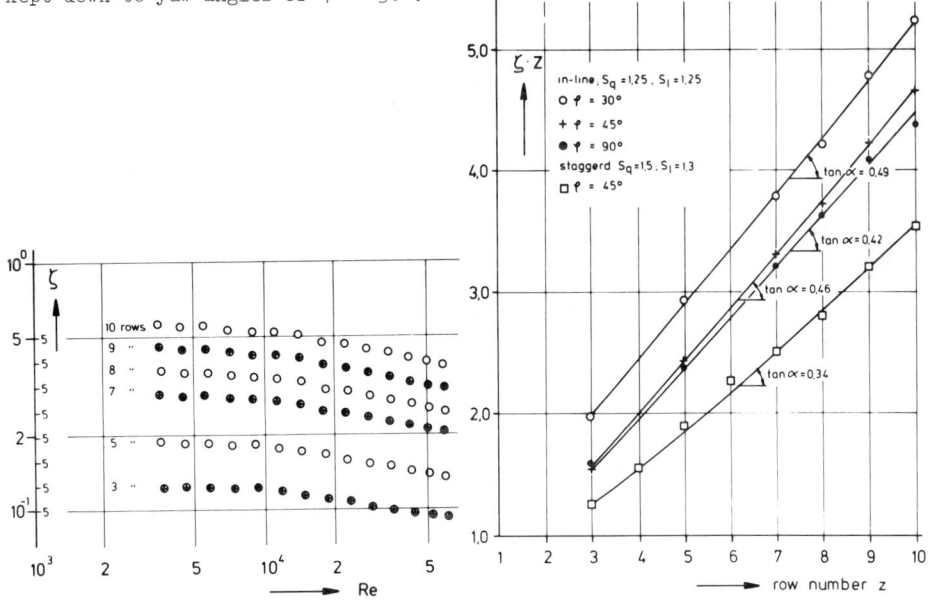

Fig.8. $\varphi = 45°$; in-line;
$S_T = 1,25$; $S_L = 1,25$

Fig.9. Re $= 2 \cdot 10^4$

Entrance effect on hydraulic resistance of yawed tube banks

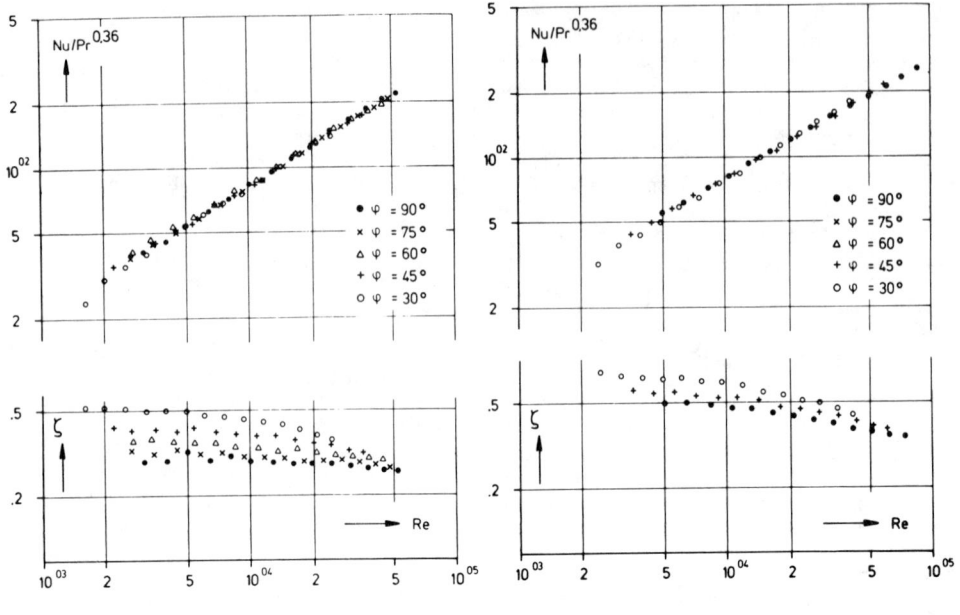

Fig.10. in-line; $S_T = 1{,}5$; $S_L = 1{,}5$

Fig.11. in-line; $S_T = 1{,}25$; $S_L = 1{,}25$

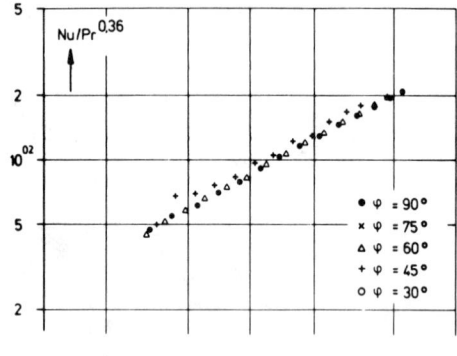
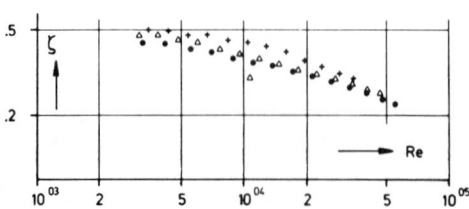

Fig.12. staggered; $S_T = 1{,}5$; $S_L = 1{,}3$

Figure 1o to Figure 12:

Influence of the yaw angle on heat transfer and hydraulic resistance of tube banks

# YAWED TUBE BUNDLE HEAT EXCHANGERS

## INFLUENCE OF INCLINATION ON HEAT TRANSFER AND PRESSURE DROP

From the principle of independence it is suggested that heat transfer and pressure drop of yawed tube bundles can be predicted using results of cross flow tube bundles. For this purpose the velocity component normal to the tube axis must be taken as reference velocity of the characteristic numbers. In this representation which has been used in the Fig. 1o through Fig. 14 the experimental data should coincide for all angles of inclination. As can be seen from Fig. 1o the heat transfer data are in very good agreement to the theory. However, the hydraulic resistance deviates systematically for different yaw angles. Generally the hydraulic resistance coefficient increases with smaller yaw angles in the subcritical Reynolds number range. This is in opposite to the results from the single cylinder shown in Fig. 3. Besides, the dependence on the Reynolds number varies with the yaw angle. The reason for that probably is the approach to the critical Reynolds number which decreases for smaller angles of inclination. Similar results were obtained for in-line tube bundles with the pitches $s_T = 1,25$ and $s_L = 1,25$, Fig. 11. In difference to the greater tube pitches the influence of the yaw angle on the hydraulic resistance is smaller by about one half. The smallest investigated yaw angle of a staggered tube bundle is $\varphi = 45°$. In contrary to the in-line banks the measured heat transfer data for $\varphi = 45°$ are about 7% greater than for cross flow. The dependence of the yaw angle on the $\xi$ -value is similar as for in-line tube banks.

## PARALLEL AND CROSSED YAWED TUBE BUNDLES

As mentioned above tube bundles with crossed tube layers are mainly used as helical type heat exchangers where the neighboured tube cylinders are coiled contrarywise. Information on thermalhydraulic data is not yet available for this type. In the course of this test series two tube bundles with straight tubes at yaw angles of $\varphi = 60°$ and $\varphi = 75°$ were examined. The most important change in the flow when crossing the tubes results from the fact that the narrowest transverse cut of the parallel tube arrangement exists only at the crossing points of the tube layers. The same is true for the maximum velocity between the tubes which was chosen for the characteristic numbers. Theoretical the maximum velocity must be halved but this doesn't seem to be realistic. In the following diagrams the maximum velocity as defined for parallelly arranged tubes will be maintained for comparison in the dimensionless terms. Fig. 13 shows heat transfer and pressure drop of the two types of tube bundles with an angle of inclination of $\varphi = 60°$.

As expected the heat transfer and pressure drop are reduced for the crossed tube bundle due to the actual decreased velocity. The decrease in heat transfer is about 14% over the whole Reynolds number range examined, whereas the decrease in pressure drop runs up to amounts to 5o% at $Re = 3 \cdot 10^3$ and to about 7o% at $Re = 5 \cdot 10^4$. That means that a rather small loss of heat transfer contrast with a considerable amount of pressure drop. For a yaw angle of $\varphi = 75°$ the influence resulting from the fact that the tubes are crossed is rather small. In Fig. 14 these results are given together with those of $\varphi = 60°$ and those of a tube bundle in cross flow. An estimated calculation for heat transfer and hydraulic resistance coefficient of crossed tube bundles in relation to yawed banks with parallel flow can be obtained if the characteristic velocity is reduced by the factor $(\sin \varphi)^2$. In this case the data points for heat transfer of parallel and crossed tube bundles coincide rather well. For the flow resistance this estimate is applicable only in the lower Reynolds number range. At higher Reynolds number the slope of the curves differs for parallel and crossed tubes and a deviation up to 2o% occurs.

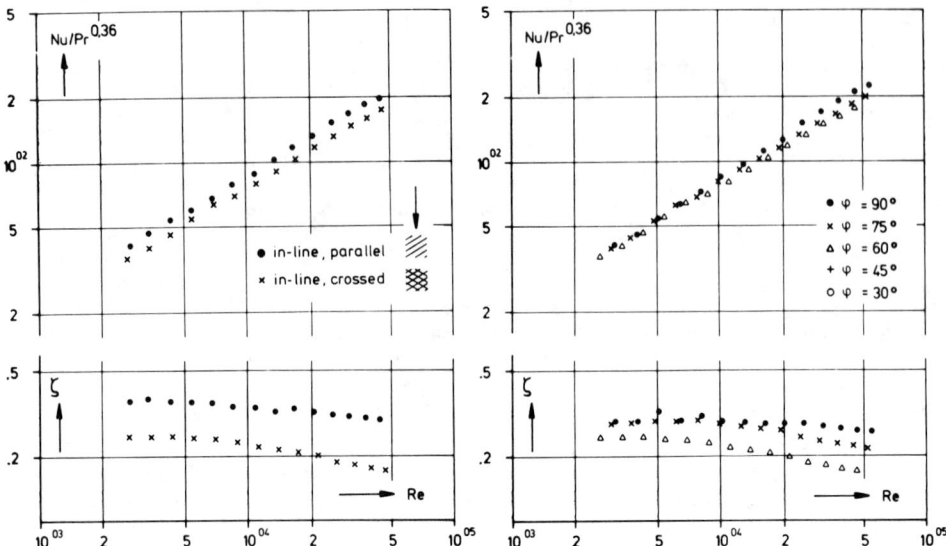

Heat transfer and hydraulic resistance of tube banks with crossed tubes;
$S_T = 1,5$; $S_L = 1,5$

Fig.13. comparison between in-line and crossed arrangement

Fig.14. Influence of the yaw angle

COMPARISON WITH DATA IN LITERATURE

In the literature the influence of the yaw angle on heat transfer and pressure drop is accounted for by partition factors. The heat transfer coefficient $\alpha$ and the pressure drop $\Delta p$ are related to the data of tube banks in cross flow. This representation is equivalent to that one where the Reynolds number and the hydraulic resistance coefficient are related to the velocity in the main flow direction. Fig. 15 shows the experimental data taken from Fig. 1o in such a representation. An estimate for the partition factors can be derived from the principle of independence. Among other things it follows from this principle that the hydraulic resistance coefficient should be independent of the yaw angle.

$$\zeta_\varphi = \zeta \qquad (5)$$

To describe the dependence of $\zeta$ on the Reynolds number a power relation of the form

$$\zeta = A \cdot Re^b \qquad (6)$$

is used. Where A and b are constants.

Introducing the definitions of the $\zeta$ -value and the Reynolds number as well as the geometrical relationship between the total velocity V and the component U normal to the tube axis the partition factor is

# YAWED TUBE BUNDLE HEAT EXCHANGERS

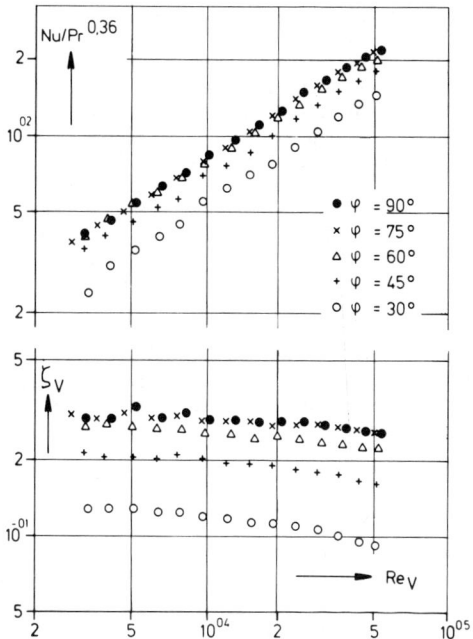

Fig.15. Influence of the yaw angle on heat transfer and hydraulic resistance of an in-line tube bank; $S_T = 1,5$; $S_L = 1,5$

Characteristic numbers refered to the velocity V in main flow direction

$$\Delta P_\varphi / \Delta P_{90} = (\sin \varphi)^{2+b} \tag{7}$$

In a similar way one obtains

$$\alpha_\varphi / \alpha_{90} = (\sin \varphi)^d \tag{8}$$

with the relationship

$$Nu = C \cdot Re^d \tag{9}$$

where C and d are constants.

Equations (7) and (8) are drawn in Fig. 16 together with predicted results for heat transfer and pressure drop taken from Michejew [7], VDI-

Wärmeatlas [8], and Idel'chick [9]. The constants of equation (7) and (8) are established from our own results obtained for cross flow tube bundles. The predictions in [7], [8], [9] are not specified as to the Reynolds number range, the tube pitches, or tube arrangements. Only Idel'chick differentiates between in-line and staggered tube arrangements. The explanation given for Fig. 1o through Fig. 12 has already shown that the principle of independence is not suitable for describing the pressure drop of yawed tube bundles, as one can be see when comparing equation (7) and the measured data. The best agreement reached with Michejew's [7] curve. However, the dependence on the Reynolds number is different for inclined and cross flow tube banks. If one considered the partition factor for $\varphi = 30°$ at a Reynolds number of $Re_v = 3 \cdot 10^3$ which are $\Delta p_{30}/ \Delta p_{90} = 0.47$ one would obtain a deviation of 24% in relation to [7].

The heat transfer data are in accordance to equation (8). Deviation to the predictions of the authors named above is unessential. As an additional piece of information the heat transfer coefficient of a tube bundle in axial flow is given in Fig. 16. The value is calculated for the same tube pitches used here with Presser's [1o] prediction.

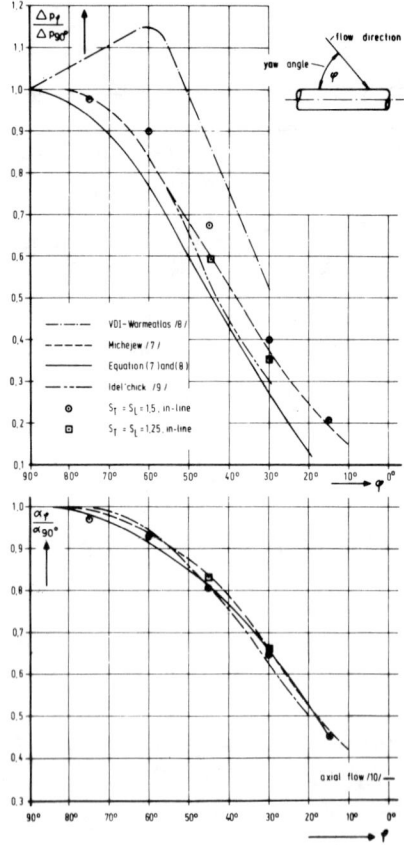

Fig.16. Comparison of experimental results with several predictions; $Re = 2 \cdot 10^4$

## CONCLUDING REMARKS

Tests at yawed tube bundles in the subcritical Reynolds number range have shown that the principle of independence can be applied to the prediction of heat transfer. This means that heat transfer equations concerning tube bundles in cross flow can be transferred to oblique flow if the velocity component normal to the tube axis is taken to form the Reynolds number. This is proved for various in-line tube arrangements and has to be confirmed for staggered tube bundles in further tests.

At the moment the hydraulic resistance is described most favourable by Michejew's [7] prediction. However, the dependence on the Reynolds number varies with the yaw angle and the tube pitches. Down to $\varphi = 30°$ deviations up to 24% occur.

First data for the calculation of crossed yawed tube bundles are given. In comparison to yawed tube bundles with parallel tubes a considerable decrease of the pressure loss is observed, whereas the decrease of the heat transfer coefficient is relatively small.

The investigations will be extended to yaw angles of $\varphi = 15°$ and supercritical Reynolds numbers.

## REFERENCES

1. Schlichting, H., Grenzschichttheorie, Verlag G. Braun, Karlsruhe 1965.

2. Sears, W.R., The boundary layer of yawed cylinders, Journal of the Aeronautical Sciences, Vol. 15, No.1, Jan. 1948, pp. 49-52.

3. Görtler, H., Zur laminaren Grenzschicht am schiebenden Zylinder, Teil I, Arch. Math. Bd. III, 1952, S. 216-231.

4. Bursnal, W.J. and Loftin, L.K., Jr., Experimental investigation of the pressure distribution about a yawed circular cylinder in the critical Reynolds number range NACA TN-2463, September 1951.

5. Chin, W.S. and Lienhard, J.H., On real fluid over yawed circular cylinders, ASME Paper 67 - WA/FE-M. 1967.

6. Žukauskas, A., Makarewitschius, V. and Slanciauskas, A., Heat transfer in banks of tubes on cross flow of fluid (Russian) Verlag Mintis, Vilnius, 1968.

7. Michejew, M.A., Grundlagen der Wärmeübertragung, VEB Verlag Technik, Berlin, 1961.

8. VDI-Warmeatlas, 3. Auflage, VDI-Verlag, Düsseldorf 1977.

9. Idel'chik, I.E., Handbook of hydraulic resistance, AEC-TR-6630, 1966.

10. Presser, K., Wärmeübergang und Druckverlust an Reaktorbrennelementen in Form längsdurchströmter Rohrbündel, Jül-486-RB, 1967.

# Compact Heat Exchangers

**RAMESH K. SHAH**
Harrison Radiator Division
General Motors Corporation
Lockport, New York 14094 USA

ABSTRACT

The unique features, subtle characteristics, and problems associated with compact heat exchangers are discussed in this paper. Starting with the definitions of important nondimensional groups, experimental methods are described to obtain single-phase heat transfer and flow friction characteristics of compact heat exchanger surfaces. Theoretical solutions and experimental correlations for surface characteristics of plate-fin, tube-fin and regenerative exchanger surfaces are then summarized. Header design being equally important for compact heat exchangers is briefly discussed along with the quantitative analysis of degradation of performance due to malflow distribution. Finally, the important characteristics and problems associated with highly compact surfaces are discussed.

1. INTRODUCTION

Starting with the definition of a compact heat exchanger and compact heat transfer surface, commonly used construction types, surface geometries, and flow arrangements are described later in this section for a compact heat exchanger. The $\varepsilon$-$N_{tu}$ method of the exchanger analysis [1] is generally employed for the thermal design of a compact heat exchanger. Since this method as well as the pressure drop analysis is well documented in the literature [1], they will not be repeated here.

1.1 Definitions

Loosely defined, a compact heat exchanger is one which incorporates a heat transfer surface having a high "area density". That is, a high ratio of heat transfer surface to volume. The heat exchanger is not necessarily of small bulk and mass. However, if it did not incorporate a surface of high area density, it would be much more bulky and massive.

Somewhat arbitrarily, we will specify that a compact surface has an area density $\beta$ greater than 700 $m^2/m^3$ (213 $ft^2/ft^3$). A spectrum of surface area densities of heat exchanger surfaces is shown in Fig. 1. The description and discussion of this figure is provided in [2]. The motivation for using compact surfaces is to gain specified heat exchanger performance, $q/\Delta t_m$, within acceptably low mass and box volume constraints. As

$$\frac{q}{\Delta t_m} = U \beta V \qquad (1)$$

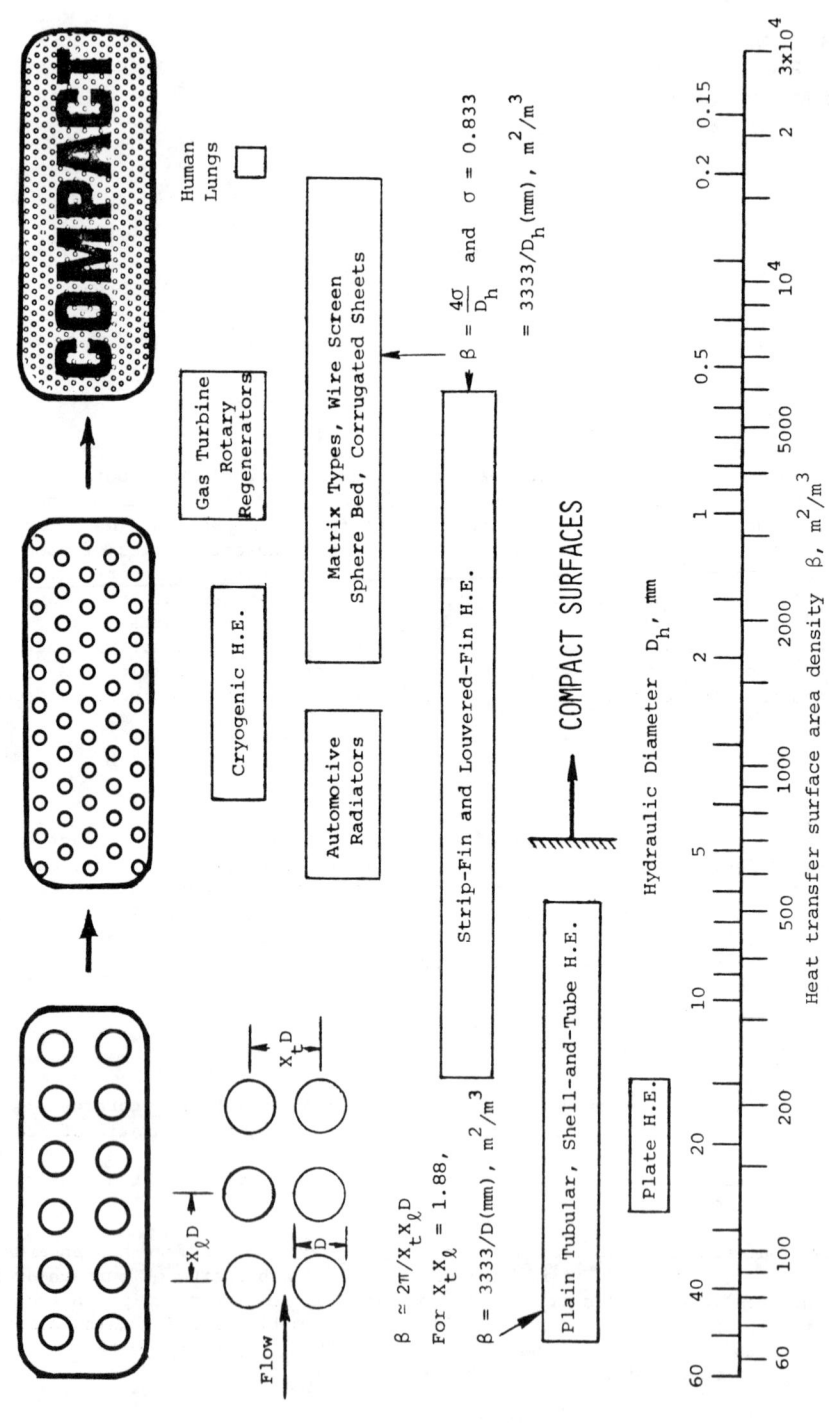

Fig. 1 Heat transfer surface area density spectrum of exchanger surfaces.

clearly, a high β minimizes the exchanger volume. Moreover, compact surfaces generally result in a higher overall conductance U, again a contribution to a smaller volume. As compact surfaces can achieve structural stability and strength with a thinner section, the gain in lower exchanger mass is even more pronounced than the gain in lower volume. Various techniques employed to make heat transfer surfaces compact are: fins between plates, finned circular tubes, or densely packed continuous or interrupted cylindrical flow passages of various shapes.

A heat exchanger of any structural construction is considered as compact if it employs a compact surface on either one or more sides of a two-fluid or a multifluid heat exchanger.† The convective heat transfer coefficient for gaseous fluids is generally one or two orders of magnitude lower than water, oil and other liquids. Thus to reduce the size and weight of a gas-to-liquid heat exchanger, the heat transfer surface on the gas side needs to be much more compact than can be practically realized with circular tubes. Hence, for a somewhat "balanced" design, a compact surface is employed on the gas side. Thus, major applications of compact heat exchangers are gas-to-gas, gas-to-liquid and gas-to-condensing or evaporating fluid heat exchangers.

The uniquenesses of compact heat exchangers are: (1) many surfaces available having different order of magnitudes of surface area density; (2) flexibility in distributing the area on the hot and cold sides as desired by design considerations; and (3) generally substantial cost, weight or volume savings.

The subtle characteristics of compact heat exchangers are: (1) usually at least one of the fluids is a gas; (2) fluids must be clean and relatively non-corrosive; (3) the fluid pumping power (i.e., pressure drop) is often of equal importance to the heat transfer rate; (4) operating pressures and temperatures are somewhat limited compared to shell-and-tube exchangers due to the brazing, mechanical expansion, etc., construction features; (5) with the use of increasingly more compact surfaces, the resultant shape of the exchanger is of a thin disk having large frontal area and short flow lengths; the header design of a compact heat exchanger is thus equally important for a uniform flow distribution; (6) the market potential must be large enough to warrant the sizeable manufacturing, research and development costs.

1.2 Construction Types and Surface Geometries

As mentioned earlier, a compact heat exchanger on the gas side requires a significantly greater amount of surface area for a specified heat transfer rate than for a liquid as a working fluid. The increase in surface area is achieved by employing surfaces that have high heat transfer surface area density β. The basic construction types employed in the design of a compact heat exchanger are: extended surface heat exchangers employing fins on one or more sides, regenerators employing small hydraulic diameter surfaces, and tubular exchangers employing small diameter tubes.

Some of the basic criteria for selecting a particular construction type are the cost, operating pressure and temperature, fouling, fluid contamination, and ruggedness. From the viewpoint of the cost per unit heat transfer surface area, the tubular exchanger in general is the most expensive followed by extended surface exchanger, and the regenerative exchanger in general is the least expensive. Extended surface and regenerative exchangers are generally designed for low pressure applications, with operating pressures limited to about 1,000

---

†Note the distinction between the terminologies "compact heat exchanger" and "compact heat transfer surfaces".

kPa gage (150 psig).[†] In this low pressure application, the metal extended surface exchangers are considered for maximum operating temperatures up to about 540-650 °C (1,000-1,200 °F); for higher temperatures up to 2,000 °C (3,600 °F), ceramic regenerative exchangers are considered. Fouling is generally not so severe a problem with gases as with liquids. Extended surface and small hydraulic diameter passages are hence used on the gas side of a compact heat exchanger. This exchanger is generally not designed for the application involving heavy fouling. Regenerators have self-cleaning characteristics because the hot and cold gases flow in the opposite directions periodically through the same passage. And as a result, compact regenerators have minimal fouling problems and usually have very small hydraulic diameter passages. Carryover and bypass leakages from the hot fluid to the cold fluid (or vice versa) occur in the regenerator. Where this leakage and subsequent fluid contamination is not permissible, the regenerator is not used.

Two most common types of extended surface exchangers are the plate-fin and tube-fin types. In a plate-fin exchanger, fins or spacers are sandwiched between parallel plates (referred to as parting plates or parting sheets) or formed tubes [2]. Fins are attached to the plates by brazing, soldering, gluing, welding, mechanical fit or extrusion. The plate fins are categorized as (1) plain (uncut surfaces) and straight fins such as plain triangular and rectangular fins; (2) plain but wavy fins; and (3) interrupted fins such as strip, louver, perforated and pin fins [2]. The heat transfer coefficient and friction factor in the developing flow region are considerably higher than those in the fully developed region. This is because the developing boundary layers are thinner and offer lower thermal and hydrodynamic resistances compared to those for the thick boundary layers associated with the fully developed flows. Wavy and interrupted fins have boundary layers developing after each interruption. With a proper design, the resultant heat transfer coefficients and heat transfer rates are significantly higher at the same pressure drops for wavy and interrupted fins compared to those for the plain fins. Thus these fins employ the materials of construction more efficiently than plain fins and are therefore used when allowed by the design constraints.

In a tube-fin exchanger, tubes of round, rectangular or elliptical shapes are generally used. When fins are used, they are employed either outside, inside or inside and outside of the tubes depending upon the application. They are attached to the tubes by a tight mechanical fit, tension winding, soldering, brazing, welding, gluing or extrusion. Fins outside the tubes may be categorized as (1) normal fins on individual tubes; (2) longitudinal fins on individual fins; and (3) continuous (plain, wavy or interrupted) fins on an array of tubes. Fins inside the tubes are categorized as integral or attached fins [2]. Tube-fin exchangers can withstand high pressures on the circular tube side. The tube-fin exchanger usually has a lower compactness compared to the plate-fin unit.

In a regenerator, any of the plate-fin surface geometries can be used in the matrix. However, the interrupted surfaces are not used in a rotary regenerator application because of a transverse flow leakage. In a fixed-matrix regenerator, all of the surface geometries used in a rotary regenerator, can be employed. In addition, also used are the quartz pebbles, steel or copper or lead shots, copper wool, randomly packed woven screens, and crossed rods.

---

[†]Some cryogenic plate-fin heat exchangers have been designed for the operating pressures up to about 8,300 kPa gage (1,200 psig).

# COMPACT HEAT EXCHANGERS

## 1.3 Flow Arrangements

For an extended surface compact heat exchanger, crossflow is the most common flow arrangement. This is because it greatly simplifies the header design at the entrance and exit of each fluid. If the desired heat exchanger effectiveness is high (say greater than 80%), the size of a crossflow unit may become excessive. In such a case, an overall cross-counterflow multipass unit or a counterflow unit may be preferred. However, there are manufacturing difficulties associated with a true counterflow arrangement as it is necessary to separate the fluids at each end and the headering problem is more complex. Some header configurations are presented in [2]. Multipassing retains the header and ducting advantages of the simple crossflow heat exchanger, while it is possible to approach the thermal performance for counterflow. For high temperature applications, different materials may be used in different passes to reduce material cost and increase the life of the heat exchanger. The structural temperature differences are reduced significantly by multipassing relative to a single-pass crossflow design. The counterflow unit has the least structural temperature differences. The parallel flow arrangement having the lowest exchanger effectiveness for a given $N_{tu}$ and highest structural temperature differences is seldom used as a compact heat exchanger. Single-pass and multipass crossflow arrangements and counterflow arrangements are used in plate-fin exchangers, while generally only the single-pass and multipass crossflow arrangements are employed in a tube-fin exchanger. The therodynamically superior counterflow arrangement is usually employed in a regenerator.

## 1.4 Other Aspects

In the following sections, after introducing basic dimensionless groups and describing experimental procedures, presented are the single-phase heat transfer and flow friction characteristics of compact heat exchanger surfaces. As mentioned earlier, the header design and flow distribution problems are very important for compact heat exchangers and are discussed briefly. Peculiarities and subtle characteristics of highly compact surfaces, referred to as laminar flow surfaces, are then summarized.

## 2. SURFACE BASIC HEAT TRANSFER AND FLOW FRICTION CHARACTERISTICS

The nondimensional heat transfer and fluid flow friction (pressure drop) characteristics of a heat transfer surface are simply referred to as the surface basic characteristics or surface basic data. These characteristics are primarily obtained by experiments and for some simple geometries by analytical means. A wealth of information is available in the literature for hundreds of surface geometries. An excellent source of design information on the compact exchanger surfaces is a monograph by Kays and London [1]. Theoretical solutions for laminar flow through constant cross section ducts are summarized by Shah and London [3]. In addition, numerous technical papers are published in the literature every year. Some of these will be summarized later. When the basic data are not available in the literature and are difficult to predict analytically, they are obtained experimentally.

In this section, first the nondimensional groups used in presenting the basic data are summarized. Next, experimental techniques to determine the basic data for a variety of surfaces are described. Finally, the theoretical solutions and available correlations for compact heat exchanger surfaces are presented.

## 2.1 Nondimensional Groups

Heat transfer characteristics of compact heat exchanger surfaces are generally presented in terms of Nusselt number, Stanton number or Colburn factor versus Reynolds number. Flow friction characteristics are generally presented in terms of friction factor versus Reynolds number. These and other related nondimensional groups are defined next.

<u>Nusselt number Nu.</u> It is defined as the ratio of the convective conductance $h$ to the pure molecular thermal conductance $k/D_h$

$$Nu = \frac{h}{k/D_h} = \frac{hD_h}{k} = \frac{q''D_h}{k(t_w - t_m)} \tag{2}$$

The Nusselt number is strongly dependent upon the thermal boundary conditions and flow passage geometry in laminar flow, and weakly dependent upon these parameters in turbulent flow. The Nusselt number is constant for thermally and hydrodynamically fully developed laminar flow. It is dependent upon $x^* = x/(D_h Pe)$ for developing laminar temperature profiles, and is dependent upon $x^*$ and Pr for simultaneously developing laminar velocity and temperature profiles. The Nusselt number is dependent upon Re and Pr for fully developed turbulent flows. The Nusselt number is related to Stanton, Prandtl and Reynolds numbers as

$$Nu = St \, Pr \, Re = St \, Pe \tag{3}$$

<u>Stanton number St.</u> It is defined as the ratio of wall heat flux per unit temperature difference between the wall and the fluid to the heat capacity rate of fluid per unit of flow area.

$$St = h/Gc_p \tag{4}$$

The behavior of St with Re parallels that of the Fanning friction factor $f$ versus Re. The Stanton number is also directly related to the number of heat transfer units on one side of the exchanger.

$$St = N_{tu,1}(D_h/4L) \tag{5}$$

where $N_{tu,1} = (hA/C)_1$ and subscript 1 refers to one side. The Stanton number is dependent upon the fluid Prandtl number, Reynolds number and the flow passage geometry.

<u>Colburn factor j.</u> It is defined as

$$j = (h/Gc_p)Pr^{2/3} = St \, Pr^{2/3} = (Nu Pr^{-1/3})/Re \tag{6}$$

Since the constant property Stanton number is dependent upon the fluid Prandtl number, Colburn [4] proposed this modulus to take into account the moderate variations in the Prandtl number. The constant property j versus Re characteristic for a given surface is nearly independent of the flowing fluid for $0.5 \leq Pr \leq 10.0$ from laminar to turbulent flow conditions.

<u>Reynolds number Re.</u> It is defined as

$$Re = \frac{\rho u_m D_h}{\mu} = \frac{GD_h}{\mu} \tag{7}$$

For internal flow, Re is proportional to the ratio of flow momentum rate ("inertia force") to viscous force for a specified duct geometry. Thus the

Reynolds number is a flow modulus.

<u>Prandtl number Pr.</u> It is defined as the ratio of momentum diffusivity to thermal diffusivity of the fluid.

$$Pr = \frac{\nu}{\alpha} = \frac{\mu c_p}{k} \qquad (8)$$

The Prandtl number is solely a fluid property modulus. $Pr \leq 0.03$ for liquid metals, $0.2 \leq Pr \leq 1$ for gases, $1 \leq Pr \leq 10$ for water, $5 \leq Pr \leq 50$ for light organic liquids, and $Pr > 30$ for oils and other viscous liquids.

<u>Friction factor f.</u> It is defined as the ratio of equivalent shear stress (wall shear force per unit of heat transfer/friction area) in the flow direction to the flow kinetic energy per unit volume, $\rho u_m^2/2g_c$.

$$f = \frac{\tau_w}{\rho u_m^2/2g_c} \simeq \frac{\Delta p}{\rho u_m^2/2g_c} \frac{D_h}{4L} \qquad (9)$$

The friction factor is dependent upon Re, $L/D_h$ and the passage geometry of the exchanger in laminar flow. In turbulent flow, it is a function of Re and surface roughness and is a weak function of the passage geometry and $L/D_h$.

As an example of the foregoing correlating groups, the basic heat transfer and flow friction characteristics for fully developed air flow in a circular tube are presented in Fig. 2. Notice the similarity in j and f versus Re characteristics. They are almost parallel. Also note three flow regimes, laminar, transition, and turbulent, in this figure. The dip in j and f vs. Re characteristics in the transition region is usually a behavior of long continuous flow passages. Generally, the interrupted surfaces do not have such a sharp dip in the transition region.

It should be emphasized at this point that since the j and f vs. Re characteristics are nondimensional, they are valid for any surface geometrically similar to the original surface.[†] However, as soon as one or more geometrical dimensions are changed (such as a change in the fin pitch, fin height, or fin thickness), the surface is no more geometrically similar to the original surface, and the j and f data of the original surface may not be applicable to this surface. This is because the hydraulic diameter is not a universal geometric parameter for correlation, the other geometric parameters may have a significant influence on the surface performance.

2.2 Experimental Methods

Primarily, three different test techniques are used to determine the surface heat transfer characteristics. These techniques are based on the steady state, transient and periodic nature of heat transfer modes through the test sections. The selection of a particular method for testing will be evident from the following description and discussion.

<u>Steady state test technique.</u> This is one of the most common test techniques used to establish the j vs. Re characteristic of a surface primarily used in

---

[†] A geometrically similar surface may be visualized as the surface photographically enlarged or reduced in size.

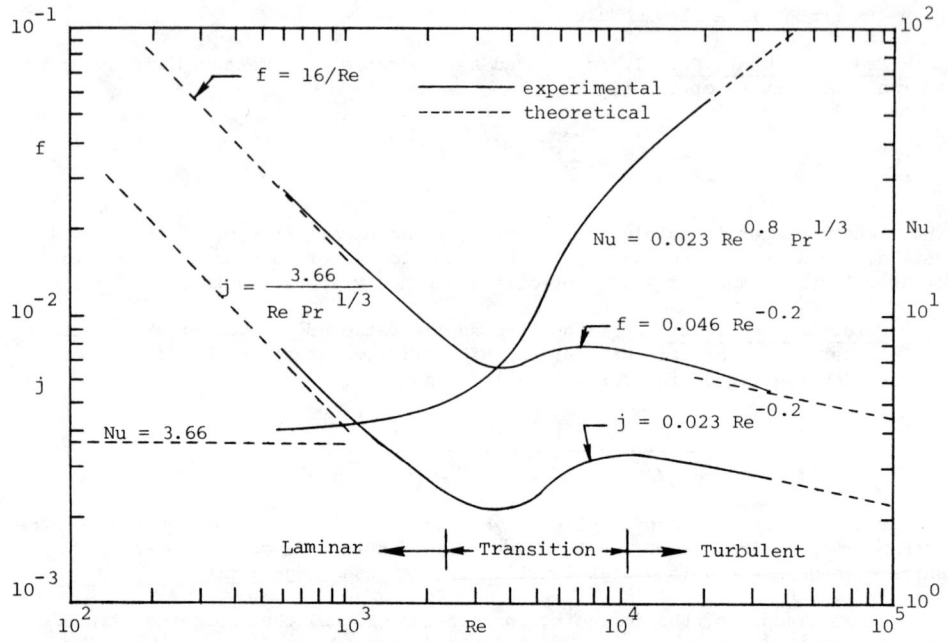

Fig. 2  Basic heat transfer and flow friction characteristics for airflow through a long circular tube.

a recuperator. Generally, a crossflow heat exchanger is employed as a test section. On one side, a surface for which the j vs. Re characteristic is known, is employed; a fluid with high heat capacity rate flows on this side. On the other side of the exchanger, a surface for which the j vs. Re characteristic is to be determined, is employed; the fluid which flows over this "unknown" surface, is the one which is used in a particular application of the unknown side surface. Generally, air is used on the unknown side, and steam, hot water, chilled water or oils are used on the known side. A typical test setup used by Kays and London [1] is shown in Fig. 3 to provide some ideas on the airside (unknown side) components of the test rig. For further details, refer to [5].

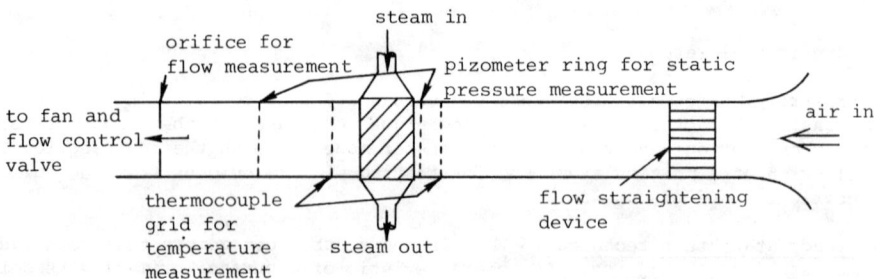

Fig. 3  Schematic of steam-to-air steady state heat transfer test rig.

In the experiments, the fluid flow rates on both sides of the exchanger are set constant at predetermined values. Once the steady state conditions are achieved, fluid temperatures upstream and downstream of the test section on both sides are measured, as well as all pertinent measurements for the determination of the fluid flow rates. The upstream pressure and pressure drop across the core on the unknown side are also recorded to determine the "hot" friction factors.[†] The tests are repeated with different flow rates on the unknown side to cover the desired range of the Reynolds number.

In order to determine the $j$ factor on the unknown side, the exchanger effectiveness is determined from the temperature measurements, and the heat capacity rate ratio is determined from the flow measurements and specific heats. $N_{tu}$ is subsequently computed from the appropriate $\varepsilon$-$N_{tu}$ relationship for the test core flow arrangement. In a test section, generally the fouling resistances are neglected so that $(\eta_o hA)$ on the unknown side is determined from the following equation where UA is found from $N_{tu}$.

$$\frac{1}{UA} = \frac{1}{(\eta_o hA)_{\text{unknown side}}} + R_w + \frac{1}{(\eta_o hA)_{\text{known side}}} \qquad (10)$$

Knowing the surface area, and the geometry of the extended surface, if any, h and $\eta_o$ are computed iteratively. Once h is known, the computation of the $j$ factor is straightforward from Eq. (6). The Reynolds number on the unknown side for the test point is determined from its definition of Eq. (7) for the known flow rate.

The test core is designed with two basic considerations in mind to reduce the experimental uncertainty in the $j$ factors: (1) the thermal resistances on each side as well as the wall; and (2) the range of $N_{tu}$.

The thermal resistances in a heat exchanger are related by Eq. (10). To reduce the uncertainty in the determination of the thermal resistance of the unknown side (with known overall thermal resistance, 1/UA), the thermal resistances of the exchanger wall and the known side should be kept minimum. The wall thermal resistance is usually negligible and may be further minimized through the use of a thin material with high thermal conductivity. The resistance of the surface on the known side is minimized by achieving a high conductance h, by employing a highly compact interrupted surface, and by the use of liquid at high flow rates or condensation of vapor. Generally, hot (or cold) water at high flow rates or condensing steam is used as a working fluid. Because of the foregoing reasons, the thermal boundary condition achieved during the testing is generally a constant wall temperature condition.

The $N_{tu}$ range for testing is generally restricted to 3 or less as explained now. The above designed test section has an approximately zero heat capacity rate ratio ($C^* \simeq 0$). To arrive at the desired range of $N_{tu}$ for the test section, consider $C^* = C_{min}/C_{max} = 0$ case. $N_{tu}$ is then found from the $\varepsilon$-$N_{tu}$ relationship for $C^* = 0$ [1].

$$N_{tu} = \ln\left(\frac{1}{1-\varepsilon}\right) = \ln\left(\frac{t_{h,i} - t_{c,i}}{t_{h,i} - t_{c,o}}\right) \simeq \ln\left(\frac{t_{h,o} - t_{c,i}}{t_{h,o} - t_{c,o}}\right) \qquad (11)$$

Here, it is assumed that the heat is transferred from the known to the unknown

---

[†] The friction factor determined from the $\Delta p$ measurement taken during the heat transfer testing is referred to as the "hot" friction factor.

side. The temperature change of the fluid on the known side is very small, so that $t_{h,i} \simeq t_{h,o}$. Thus, one can realize that the important temperature difference for the determination of $N_{tu}$ is $(t_{h,o} - t_{c,o})$ since it decreases with increasing $N_{tu}$. If the core is too effective (high $N_{tu}$), then the accurate measurement of this small temperature difference will be difficult. So generally, the testing is restricted to $\varepsilon$ of about 95% or less or $N_{tu}$ of 3 or less. The temperature measurements at low air or gas flows become less accurate because of significant conduction and radiation losses associated with thermocouples or resistance thermometers. Care must be exercised for the accurate temperature measurements at low gas flows. Otherwise, a "drop-off" in j vs. Re will be experienced due to an error in $(t_{h,o} - t_{c,o})$ measurement, even with the maximum value of $N_{tu}$ as low as 1.5 to 2.

The experimental uncertainty in the j factor for the steady-state method is ±5% when the temperatures are measured accurately to within ±0.1°C. The uncertainty in the Reynolds number is ±2% when the flow is measured accurately within ±0.7%.

<u>Transient test technique.</u> This is one of the most common test techniques used to establish the j vs. Re characteristic of a matrix type or a high $N_{tu}$ surface. The test section is a single-fluid exchanger (matrix) built up from the heat transfer surface for which the j vs. Re characteristic is to be determined. Generally, air is used as a working fluid. In the test facilities, an electric heater (resistance heating screen) is employed upstream of the test section to obtain a step change in air temperature at the inlet of the test section.

In the experiments, the air flow rate is set constant at a predetermined value. The air is heated with the resistance heating screen to about 10°C (20°F) above the ambient temperature which in turn heats the matrix. The heating is continued until the core reaches a uniform temperature exhibited by a negligible difference between the air temperature at inlet and exit of the matrix. Once the stable condition is reached, the power to the heating screen is turned off. The temperature-time history of the air leaving the matrix is continuously recorded during the matrix cooling period.

This temperature-time history of air depends upon the heat transfer rate from the matrix and is thus a function of the matrix $N_{tu}$. There are several theoretical methods to determine $N_{tu}$, such as: (1) maximum slope data reduction method; (2) zero intercept method; (3) direct curve matching method; and (4) first moment of area method. For further details on these methods, refer to [6-8]. The maximum slope data reduction method is most commonly used for high $N_{tu}$ surfaces.

During the experiments, sometimes both the cooling and heating temperature-time history of the exit air is recorded and the average of the pertinent information is then used in the data reduction. The experiment is repeated at different flow rates to cover the desired range of Reynolds number. This method is also referred to as the "single-blow" transient test technique. The experience shows that the thermal boundary condition achieved during the single-blow testing is in between the constant wall temperature and constant wall heat flux conditions. When the maximum slope data reduction method is used, the experimental uncertainty in the j factors is ±13% for $N_{tu} \geq 3.5$.

<u>Periodic test technique.</u> This method is similar to, but a variant of, the single-blow technique. In the single-blow method, a step change in the fluid temperature is achieved at the inlet of the matrix. In the periodic method, the temperature of the inlet air is continuously periodically changed by a periodic

(sinusoidal) power input to the electric heater. The phase shift and/or the amplitude change between inlet and outlet air temperatures is used to determine h and hence j factors. This method is applicable to a much wider range of $N_{tu}$ than the previous two methods. The inlet air temperature does not need to be sinusoidal, but then a Fourier analysis is performed to the inlet and exit temperature-time history. This method is described in detail by Stang and Bush [9].

Flow friction characteristics. The experimental determination of flow friction characteristics of heat exchanger surfaces is relatively straightforward. Regardless of the core construction and the method of heat transfer testing, the determination of f is made under steady fluid flow rates with or without heat transfer. At a given fluid flow rate on the unknown side, the following measurements are made: core pressure drop, core inlet pressure and temperature, core outlet temperature for "hot" friction data, fluid flow rate, and the core geometrical properties. The Fanning friction factor f is then determined from the following equation.

$$f = \frac{r_h}{L} \frac{1}{(1/\rho)_m} \left[ \frac{2g_c \Delta p}{G^2} - \frac{1}{\rho_i}(1-\sigma^2+K_c) - 2\left(\frac{1}{\rho_o} - \frac{1}{\rho_i}\right) + \frac{1}{\rho_o}(1-\sigma^2-K_e) \right] \qquad (12)$$

This equation is an inverted form of the core pressure drop equation of Kays and London [1]. Here $K_c$ is the sudden contraction loss coefficient at the exchanger core inlet and $K_e$ is the sudden expansion loss coefficient at the exchanger core exit. They are available from Kays and London [1]. For the isothermal pressure drop data, $\rho_i = \rho_o = 1/(1/\rho)_m$. The friction factor thus determined includes the effects of skin friction, form drag, and local flow contraction and expansion losses, if any, within the core. Tests are repeated with different flow rates on the unknown side to cover the desired range of the Reynolds number. The experimental uncertainty in the f factors is ±5%.

Generally, the Fanning friction factor f is determined from isothermal pressure drop data (no heat transfer across the core). The hot friction factor vs. Re curve should be close to the isothermal f vs. Re curve, particularly when the variations in the fluid properties are small, i.e., the average fluid temperature for the hot f data is not significantly different from the wall temperature. Otherwise, the hot f data must be corrected to account for the temperature-dependent fluid properties [10] during the data reduction.

2.3 Theoretical Solutions and Correlations for Simple Geometries[†]

Heat exchangers employ surfaces having either continuous flow passages or flow passages with frequent boundary layer interruptions. The velocity and temperature profiles across the flow section are generally fully developed in the continuous flow passages; while they are developing at each boundary layer interruption in an interrupted surface. The heat transfer and flow friction characteristics are generally substantially different for fully developed flows and developing flows. The analytical results are discussed separately next for developed and developing flows for simple flow passage geometries. For complex surface geometries, the surface basic characteristics are primarily obtained by experiments. They are discussed in Section 2.4.

---

[†]All the results presented in this section are for constant fluid properties. The influence of temperature-dependent properties is taken into account by the property ratio index method presented in [10].

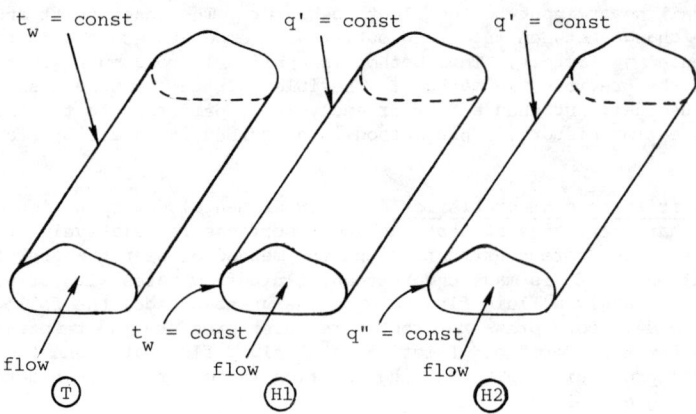

Fig. 4  Thermal boundary conditions for the pipe flow: ⓣ constant wall temperature; Ⓗ1 constant axial wall heat flux with constant peripheral wall temperature; Ⓗ2 constant wall heat flux axially and peripherally.

Fully developed laminar flow. The constant property fully developed laminar flow Nusselt numbers are constant, independent of Re and Pr, but dependent upon the flow passage geometry and thermal boundary conditions. The constant property product of Fanning friction factor and Reynolds number is also constant, independent of Re and Pr, but dependent upon the flow passage geometry, for fully developed laminar flow. Before summarizing the analytical results, three important thermal boundary conditions ⓣ, Ⓗ1, and Ⓗ2 are defined first. Shah and London [3] describe these and other thermal boundary conditions in detail.

The ⓣ boundary condition refers to the constant wall temperature both axially and peripherally throughout the duct (or passage) length as shown in Fig. 4. This boundary condition is approximated in condensers, evaporators, and liquid-to-gas heat exchangers with high liquid flows. The Nusselt numbers for this boundary condition have a subscript T. The $Nu_T$ is always less than $Nu_{H1}$ [3].

The wall heat transfer rate is constant in the axial direction while the wall temperature at any cross section is constant in the peripheral direction for the Ⓗ1 boundary condition. The wall heat transfer rate is constant in the axial direction as well as in the peripheral direction for the Ⓗ2 boundary condition. These boundary conditions are shown in Fig. 4. The Ⓗ1 and Ⓗ2 boundary conditions are realized in gas turbine regenerators, counterflow exchanger with $C^* \simeq 1$, nuclear heating and electric resistance heating. In these applications, the Ⓗ1 boundary condition is realized for highly conductive materials for which the temperature gradients in the peripheral direction are minimum; the Ⓗ2 boundary condition is realized for very low conductive materials for which temperature gradients exist in the peripheral direction. For intermediate values of thermal conductivity, the boundary condition will be in between Ⓗ1 and Ⓗ2. The Nusselt numbers for these boundary conditions have subscripts H1 and H2 respectively. $Nu_{H2}$ is lower than $Nu_{H1}$ for noncircular flow passages [3].

The Nusselt numbers and fRe for some of the practically important flow passage geometries are presented in Table 1. Shah and London [3] summarize the laminar flow solutions for 40 duct geometries. These results provide a valuable baseline for simple passage geometries. Due to the entrance length effect, actual Nusselt numbers should be higher than that for the fully developed flow. However, actual matrix passages are never ideal and uniform. Passage-to-assage nonuniformity to be discussed later reduces heat transfer more than the gain by the thermal entrance effect. The friction factors are generally higher than those for fully developed flow because of a significant effect of hydrodynamic entrance length. Passage-to-passage nonuniformity reduces the friction factor (and $\Delta p$) only slightly. Thus, generally experimental j (or Nu) is lower and

Table 1. Solutions for heat transfer and friction for fully developed laminar flow through specified ducts [3]

| GEOMETRY ($L/D_h > 100$) | $Nu_{H1}$ | $Nu_{H2}$ | $Nu_T$ | fRe | $\dfrac{j_{H1}}{f}$ † | $\dfrac{Nu_{H1}}{Nu_T}$ |
|---|---|---|---|---|---|---|
| $2b$, $2a$, $\dfrac{2b}{2a} = \dfrac{\sqrt{3}}{2}$ | 3.014 | 1.474 | 2.39 | 12.630 | 0.269 | 1.26 |
| $2b$, $2a$, 60°, $\dfrac{2b}{2a} = \dfrac{\sqrt{3}}{2}$ | 3.111 | 1.892 | 2.47 | 13.333 | 0.263 | 1.26 |
| $2b$, $2a$, $\dfrac{2b}{2a} = 1$ | 3.608 | 3.091 | 2.976 | 14.227 | 0.286 | 1.21 |
| hexagon | 4.002 | 3.862 | 3.34 | 15.054 | 0.299 | 1.20 |
| $2b$, $2a$, $\dfrac{2b}{2a} = \dfrac{1}{2}$ | 4.123 | 3.017 | 3.391 | 15.548 | 0.299 | 1.22 |
| circle | 4.364 | 4.364 | 3.657 | 16.000 | 0.307 | 1.19 |
| $2b$, $2a$, $\dfrac{2b}{2a} = \dfrac{1}{4}$ | 5.331 | 2.930 | 4.439 | 18.233 | 0.329 | 1.20 |
| $2b$, $2a$, $\dfrac{2b}{2a} = \dfrac{1}{8}$ | 6.490 | 2.904 | 5.597 | 20.585 | 0.355 | 1.16 |
| $\dfrac{2b}{2a} = 0$ | 8.235 | 8.235 | 7.541 | 24.000 | 0.386 | 1.09 |
| $\dfrac{b}{a} = 0$, insulated | 5.385 | - | 4.861 | 24.000 | 0.253 | 1.11 |

†$j_{H1}/f = Nu_{H1} Pr^{-1/3}/fRe$ with $Pr = 0.7$.

f is higher than the theoretical solutions for fully developed flow. Also, the thermal boundary condition for heat transfer may not exactly correspond to any of the previously described boundary conditions. Hence, accurate j and f vs. Re characteristics are generally experimentally determined even for simple geometries used in highly compact heat exchangers.

As an illustration, these characteristics are presented below for two simple geometries. London et al. [11] present the following correlations for gas flow through triangular passages ($40 < Re < 800$)

$$f = 14.0/Re \qquad j = 3.0/Re \qquad (13)$$

London and Shah [12] present the following correlation for gas flow through hexagonal passages ($80 < Re < 800$)

$$f = 17.0/Re \qquad j = 4.0/Re \qquad (14)$$

**Fully developed turbulent flow.** Even though the turbulent flow is seldom realized in compact heat exchangers, the correlations are included here for completeness. The Fanning friction factor for fully developed turbulent flow through a pipe is dependent upon the pipe wall surface roughness and Reynolds number. For smooth circular pipes, it is expressed by Kármán-Nikuradse equation.

$$\frac{1}{\sqrt{f/2}} = 2.46 \ln(Re\sqrt{f/2}) + 0.30 \qquad (15)$$

Since this equation is implicit in f, it is approximated as follows.

$$f = \begin{cases} 0.079 \, Re^{-0.25} & \text{for} \quad 5{,}000 < Re < 30{,}000 \\ 0.046 \, Re^{-0.2} & \text{for} \quad 30{,}000 < Re < 10^6 \end{cases} \qquad (16)$$

Petukhov and Kirillov [13] proposed the following equation for f that fits well with the test data over the range $10^4 < Re < 5 \times 10^6$.

$$f = (1.58 \ln Re - 3.28)^{-2} \qquad (17)$$

For noncircular tubes, Eqs. (14)-(16) also yield accurate f factors if the hydraulic diameter $D_h$ is used as a characteristic dimension in Re. There is only a weak influence of passage shape in turbulent flow. It is now discussed for some noncircular passages.

For smooth rectangular tubes, the friction factor is also dependent upon the aspect ratio, although not as severely as in laminar flow. It monotonically increases with decreasing aspect ratio from 1 to 0. Jones [14] correlated all available experimental data, and arrived at an equivalent diameter $D_e$ for the rectangular passage that is related to the hydraulic diameter and the aspect ratio $\alpha^*$.

$$D_e = \phi \, D_h \qquad (18)$$

where

$$\phi = \frac{2}{3} + \frac{11}{24} \alpha^*(2 - \alpha^*) \qquad (19)$$

# COMPACT HEAT EXCHANGERS

Subsequently, the Reynolds number is calculated using $D_e$ instead of $D_h$. The friction factors are then computed from Eqs. (15)-(17). Note that this friction factor is used in conjunction with the hydraulic diameter $D_h$ to calculate $\Delta p$ from Eq. (9) or (12).

For isosceles triangular ducts, even when the flow is fully developed turbulent near the base, it is laminar near the apex. This results into an overall lower friction factor. Carlson and Irvine, as summarized by Kays and Perkins [15], provide the following correlation.

$$f_D = 4f = C \, Re^{-0.25} \qquad (20)$$

where C varies from 0.24 for the apex angle 0 to 0.31 for the apex angle of 60°.

For concentric annuli, the dependence of the radius ratio on the friction factor is small. Experimental f data that are reported in the literature for concentric annuli are about 10% higher than those for a circular tube. Kays and Perkins [15] recommend the following correlation for $6{,}000 < Re < 300{,}000$.

$$f = 0.085 \, Re^{-0.25} \qquad (21)$$

The hydraulic diameter $D_h = d_o - d_i$ is used in the definition of Re.

In fully developed turbulent flow, the <u>constant-property</u> Nusselt number is independent of thermal boundary conditions for $Pr > 1$, but it is dependent upon both Re and Pr. For $Pr < 1$, the turbulent flow Nusselt number is also dependent upon the thermal boundary condition. The ratio of $Nu_H/Nu_T$ is provided by Kays and Perkins [15] as a function of Re and Pr.

Webb [16] critically compared analytical solutions for $Pr > 0.7$ with experimental data corrected for constant properties condition.[†] He concluded that the following equation proposed by Petukhov and Kirillov [13] fits the experimental data within ±10% for $10^4 < Re < 5 \times 10^6$ and $0.5 < Pr < 2{,}000$.

$$Nu = \frac{(f/2) Re Pr}{1.07 + 12.7(Pr^{2/3} - 1)\sqrt{f/2}} \qquad (22)$$

where the Fanning friction factor f is obtained from Eq. (15) or (17). Based on a comparison with the mass transfer data, this equation provides Nu for high Pr ($10^3 < Pr < 10^5$) accurately within about ±15%. Webb also concluded that the foregoing correlation fits the experimental data very well compared to the Dittus-Boelter or Colburn equation.

Dittus-Boelter Equation: $\quad Nu = 0.023 \, Re^{0.8} Pr^{0.4} \qquad (23)$

Colburn Equation: $\quad Nu = 0.023 \, Re^{0.8} Pr^{1/3} \qquad (24)$

The Nusselt numbers predicted by these equations are too low compared to Eq. (22) for $Pr < 600$.

Sleicher and Rouse [17] propose an empirical formula that is much simpler than Eq. (22) and having less than 5% difference in Nu as

---

[†] The Nusselt numbers are obtained for several decreasing values of heat flux at a constant Reynolds number. Extrapolation of Nu vs. q" to zero heat flux yields the Nusselt number for the constant properties condition.

$$Nu = 5 + 0.015 \, Re^a Pr^b \tag{25}$$

where
$$a = 0.88 - 0.24/(4 + Pr) \tag{26}$$

$$b = 0.333 + 0.5 \exp(-0.6 \, Pr) \tag{27}$$

The range of applicability of Eq. (25) is $10^4 < Re < 10^6$ and $0.1 < Pr < 10^4$. For gases ($0.6 < Pr < 0.9$), Eq. (25) simplifies within 4% to

$$Nu = 5 + 0.012 \, Re^{0.83} (Pr + 0.29) \tag{28}$$

For liquid metals ($Pr < 0.1$), Sleicher and Rouse [14] recommend the following correlations for the Ⓗ and Ⓣ boundary conditions.

$$Nu_H = 6.3 + 0.0167 \, Re_f^{0.85} \, Pr_w^{0.93} \quad \text{for } Pr < 0.1 \tag{29}$$

$$Nu_T = 4.8 + 0.0156 \, Re_f^{0.85} \, Pr_w^{0.93} \quad \text{for } Pr < 0.1 \tag{30}$$

Here $Re_f$ denotes the Reynolds number with the fluid properties evaluated at the film temperature $(t_m + t_w)/2$ and $Pr_w$ denotes the Prandtl number with the fluid properties evaluated at the wall temperature. Note that thus the influence of temperature-dependent fluid properties is included in Eqs. (29) and (30).

Webb [18] has provided an excellent summary on correlations for turbulent flow through rough pipes.

Developing laminar flow. Shah and London [3] have summarized the developing flow solutions for circular tube, parallel plates, rectangular ducts, isosceles triangular ducts, and concentric and eccentric annular ducts. Shah [19] has correlated the $f_{app} Re$ factors by the following equation.

$$f_{app} Re = 3.44(x^+)^{-0.5} + \frac{K(\infty)/(4x^+) + fRe - 3.44(x^+)^{-0.5}}{1 + C(x^+)^{-2}} \tag{31}$$

Here $f_{app}$ is the apparent Fanning friction factor that takes into account both the skin friction and the change in momentum rate in the hydrodynamic entrance region. It is based on the static pressure drop from $x = 0$ to $L$. It is defined by

$$f_{app} \frac{L}{r_h} = \frac{\Delta p}{\rho u_m^2 / 2g_c} \tag{32}$$

$K(\infty)$, $f_{app} Re$ and $C$ of Eq. (31) for rectangular, triangular and concentric annular ducts are presented in Table 2.

Although Eq. (31) may provide some guidelines for friction factors for the interrupted surfaces, it includes only the effect of skin friction. The form drag associated with the blunt (smooth and burred) edges of the surface may contribute significantly to the pressure drop. Hence, analytical values of apparent friction factors are generally not used in designing exchangers. But, as a rule of thumb, $f = 4j$ or such relationship is used to predict $f$ factors for interrupted surfaces for which $j$ factor is already known either from the theory or from experiments.

Thermal entry length solutions with developed velocity profiles are summarized by Shah and London [3] for a large number of practically important flow passage geometries. They proposed the following correlations for thermal

Table 2. $K(\infty)$, $f_{app}Re$ and C for use in Eq. (27), from Shah [19]

|  | $K(\infty)$ | fRe | C |
|---|---|---|---|
| $\alpha^*$ | Rectangular ducts | | |
| 1.00 | 1.43 | 14.227 | 0.00029 |
| 0.50 | 1.28 | 15.548 | 0.00021 |
| 0.20 | 0.931 | 19.071 | 0.000076 |
| 0.00 | 0.674 | 24.000 | 0.000029 |
| $2\phi$ | Equilateral triangular duct | | |
| 60° | 1.69 | 13.333 | 0.00053 |
| $r^*$ | Concentric annular ducts | | |
| 0 | 1.25 | 16.000 | 0.000212 |
| 0.05 | 0.830 | 21.567 | 0.000050 |
| 0.10 | 0.784 | 22.343 | 0.000043 |
| 0.50 | 0.688 | 23.813 | 0.000032 |
| 0.75 | 0.678 | 22.967 | 0.000030 |
| 1.00 | 0.674 | 24.000 | 0.000029 |

entrance solutions for circular and noncircular ducts having laminar developed velocity profiles and developing temperature profiles.

$$Nu_{x,T} = 0.427 \, (fRe)^{1/3} \, (x^*)^{-1/3} \tag{33}$$

$$Nu_{m,T} = 0.641 \, (fRe)^{1/3} \, (x^*)^{-1/3} \tag{34}$$

$$Nu_{x,H1} = 0.517 \, (fRe)^{1/3} \, (x^*)^{-1/3} \tag{35}$$

$$Nu_{m,H1} = 0.775 \, (fRe)^{1/3} \, (x^*)^{-1/3} \tag{36}$$

where f is Fanning friction factor for fully developed flow, Re is the Reynolds number and $x^* = x/(D_h RePr)$. For interrupted surfaces, $x = \ell$. The above equations are recommended for $x^* < 0.001$.

Since developing laminar flow exists up to $Re \simeq 10^5$ for short ducts having $x^+ = x/D_h Re < 0.001$ [3], and Re seldom exceeds $10^4$ for compact heat exchanger surfaces, the interrupted surfaces have only developing flows in the operating Re range. That is why no dip is found in the j vs. Re characteristic of an interrupted surface such as that in Fig. 5b. The functional relationship of $Nu \propto (x^*)^{-1/3}$ of Eqs. (33)-(36) is then quite useful for predicting the j data for a new surface if they are known for another surface of the same family and $\ell/D_h$ are known for both surfaces.

For simultaneously developing laminar velocity and temperature profiles, refer to [3] for available analytical solutions.

<u>Developing turbulent flow.</u> In the entrance region, the turbulent flow friction factors are higher than those in fully developed flow. However, the entrance length is very short, generally less than 10 tube diameters, and other sources of pressure drop at the tube entrance are far more important in most

applications. Hence, the influence of developing turbulent flow region is generally neglected in the pressure drop evaluation.

The thermal entry length solutions for the developed or developing velocity profile have been summarized by Kays and Perkins [15], and will not be repeated here.

## 2.4 Analytical Solutions and Correlations for Extended Surfaces[†]

One of the most comprehensive sources of the basic data for extended surfaces is by Kays and London [1] published in 1964. They have presented $j$ and $f$ data for 56 plate-fin surfaces and 21 tube-fin surfaces in a unified format. A partial list of additional basic data since 1964 are as follows: (1) Plain fins [20], Wavy fins [21-24][‡], Strip (offset) fins [25-30], Louver fins [27,31,32], Perforated fins [6,8,33-37], and Pin fins [38]. An extensive bibliography and basic data for a variety of tube-fin geometries are summarized by Rosenman et al. [39] and Webb [40]. These basic data will not be presented here due to the space limitations.

However, as an illustration, the $j$ and $f$ data are presented for two compact plate-fin surfaces in Fig. 5. Although one surface having plain triangular passages is slightly more compact than the other surface having strip fins, they are of the same order of magnitude for the following comparisons: (1) While a dip exists in the $j$ vs. Re characteristic for the plain triangular passage, no such dip exists for the strip fin surface. (2) Both $j_2/j_1$ and $f_2/f_1 \simeq 4$ at $Re = 1,000$ where subscripts 1 and 2 are for the plain and strip fins respectively. While both surfaces are compact, the plain triangular fin is hardly a "high performance" surface. The strip fin surface is a "high performance" surface. (3) The $j/f$ ratio for the plain and strip fins is 0.25 and 0.26, respectively. (4) The Reynolds number range for these surfaces is typical for many compact exchanger applications. Compact heat exchanger surfaces seldom operate beyond $Re \simeq 10,000$. For highly compact surfaces, the operating Reynolds number can be as low as 50-100.

Before summarizing the published correlations for interrupted surfaces, let us briefly discuss some of the important characteristics of the plate-fin surfaces. A variety of methods have been employed to increase the heat transfer coefficient for the exchanger surfaces. An undesirable consequence of this augmentation is an increase in the friction factor. Since friction power expenditure is equally very important for compact heat exchanger applications, only a limited number of augmentation techniques is employed in practice. Various turbulators that increase the turbulence level are generally not employed because they are very inefficient from the viewpoint of a relative increase of the heat transfer coefficient over the friction power expenditure. Similarly, active mechanical devices for augmentation are not used for compact heat exchangers due to the required mechanical and frictional powers. The passive technique of interrupting the surface in the flow direction is the most common method of augmentation for compact heat exchanger surfaces. The plate-fin surfaces are well suited for the surface interruptions. If the fin area of such a surface (for example,

---

[†]All the results presented in this section are for constant fluid properties. The influence of temperature-dependent properties is taken into account by the property ratio index method presented in [10].

[‡]In Ref. [21,22], the data are presented for corrugated plates of plate heat exchangers. The qualitative information from these references may be applicable to the wavy plate-fin surfaces.

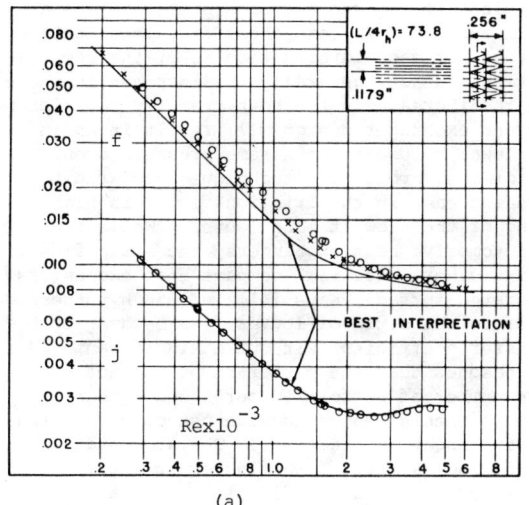

**Surface Geometry**

Fin pitch = 6.68 per cm, 16.96 per in.
Plate spacing, b = 6.50 mm, 0.256 in.
Fin length flow direction = 127 mm, 5.00 in.
Flow passage hydraulic diameter, $4r_h$ = 0.001723 m, 0.005652 ft
Fin metal thickness = 0.152 mm, 0.006 in., aluminum
Total heat transfer area/volume between plates, $\beta$ = 1994 $m^2/m^3$, 607.81 $ft^2/ft^3$
Fin area/total area = 0.861

(a)

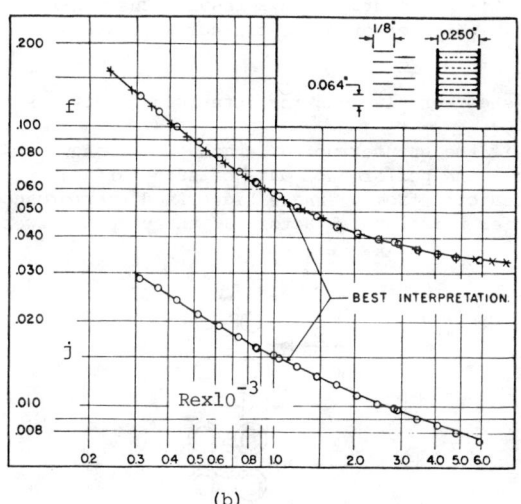

**Surface Geometry**

Fin pitch = 6.15 per cm, 15.61 per in.
Plate spacing, b = 6.35 mm, 0.250 in.
Fin offset length (flow direction), $\ell$ = 3.175 mm, 1/8 in.
Flow passage hydraulic diameter, $4r_h$ = 0.002383 m, 0.007817 ft
Fin metal thickness = 0.102 mm, 0.004 in.
Total heat transfer area/volume between plates, $\beta$ = 1548 $m^2/m^3$, 471.9 $ft^2/ft^3$
Fin area/total area = 0.923

(b)

Fig. 5 Surface basic characteristics for: (a) a plain plate-fin surface 16.96T [1], and (b) an offset plate-fin surface 104 [25].

wavy, louver, strip, perforated, and pin fins) is substantially large compared to the prime surface area, a marked increase in the overall performance can be achieved. Also many tube-fin geometries have an inherent feature of surface interruption. If an exchanger is designed properly, even a large friction factor will result in an equal pressure drop by an appropriate reduction in the flow length and an increase in the core frontal area.

The most common interrupted surfaces are the strip fins and louver fins.

The increase in heat transfer coefficient at a given Reynolds number may be as much as 2 to 4 times over a plain fin surface. In both these fins, the contribution of the form drag due to blunt edges of each interruption is of the same order of magnitude as the skin friction. Hence, generally the increase in the friction factor is even higher than the increase in the heat transfer coefficient. However, as mentioned earlier, a proper exchanger design will result in an increase in heat transfer (or overall lower volume) at the same pressure drop. The louver fin is easily adaptable for mass production techniques compared to the strip fins. However, a relative increase in the friction factor is higher for the louver fin due to the bending of the flow at each louver. While the louver fins are most common in the automotive heat exchangers, the strip fins are most common in aircraft, cryogenics, and other industrial compact heat exchangers. Based on the basic data of one perforated surface, Kays [41] in 1958 hypothesized that the perforated fins are superior to the strip or louver fins because of a substantial increase in the heat transfer coefficient without a significant increase in the friction factor. He reasoned that the perforations do not add the form drag. A testing of a large number of perforated cores since then revealed the results contrary to this hypothesis [35]. No wide use of perforated fins has been found in compact heat exchanger applications. The augmentation characteristic of wavy fins is not so good as that of the louver or strip fin and hence wavy fins are used in a limited number of applications. In a pin-fin plate-fin exchanger, a new boundary layer develops on each individual pin, however, the pins introduce a substantial form drag. As a result, the pin-fins are also used in those applications in which other plate-fin surfaces are not suitable.

Now we will summarize the published correlations for interrupted surfaces.

LaHaye et al. [42] showed that if the uninterrupted length $\ell$, as shown in Fig. 6, is a characteristic dimension, the performance data of such surface geometries could be approximately predicted from a single "idealized" performance plot of Fig. 7. Here the heat transfer performance factor $J$ and pumping power

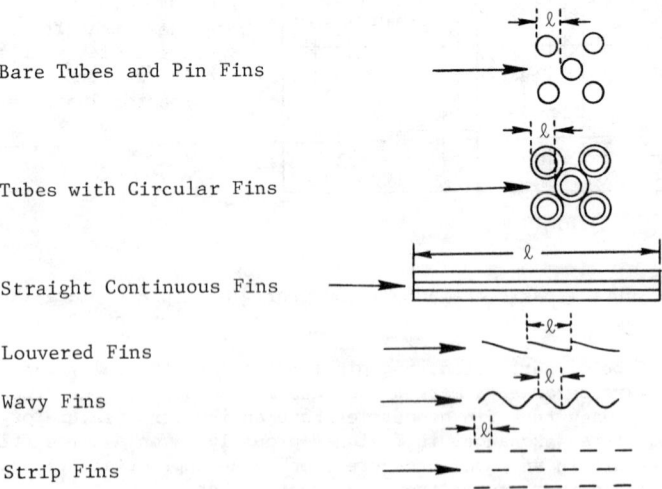

Fig. 6 Characteristic flow length between major boundary layer disturbances for various heat exchanger surfaces.

# COMPACT HEAT EXCHANGERS

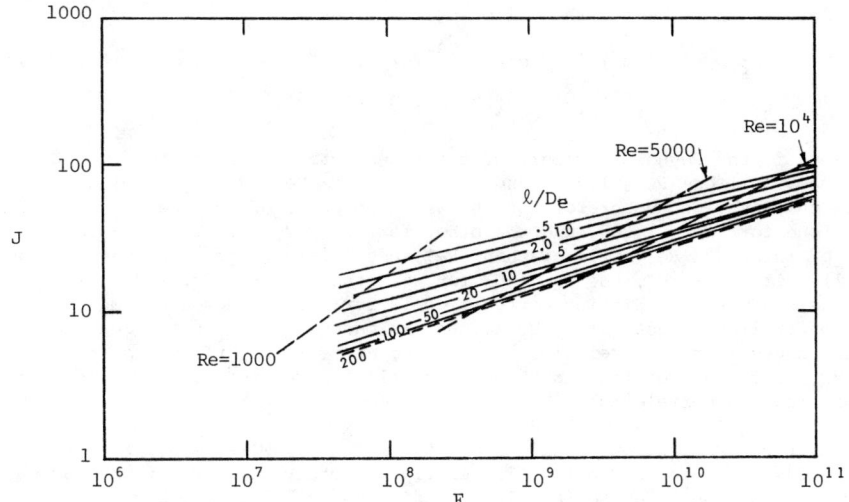

Fig. 7 Idealized dimensionless performance plot of J versus F.

factor F are defined as

$$J = jRe = h \frac{Pr^{2/3}}{c_p \mu} D_h = h\beta \frac{Pr^{2/3}}{c_p \mu} \frac{D_h^2}{4\sigma} \tag{37}$$

$$F = fRe^3 = E \frac{2g_c \rho^2}{\mu^3} D_h^3 = E\beta \frac{2g_c \rho^2}{\mu^3} \frac{D_h^4}{4\sigma} \tag{38}$$

Figure 7 is prepared for the turbulent flow performance. From this figure, J versus F at a given Reynolds number, and subsequently, j and f versus Re characteristics for a specified $\ell/D_h$ surface could be determined.

Wieting [43] correlated available data for 22 strip (offset) fin surfaces (Fig. 6) and proposed the following set of equations.

For Re ≤ 1,000,

$$f = 7.661 \, (\ell/D_h)^{-0.384} \, \alpha^{*-0.092} \, Re^{-0.712} \tag{39}$$

$$j = 0.483 \, (\ell/D_h)^{-0.162} \, \alpha^{*-0.184} \, Re^{-0.536} \tag{40}$$

For Re ≥ 2,000,

$$f = 1.136 \, (\ell/D_h)^{-0.781} \, (\delta/D_h)^{0.534} \, Re^{-0.198} \tag{41}$$

$$j = 0.242 \, (\ell/D_h)^{-0.322} \, (\delta/D_h)^{0.089} \, Re^{-0.368} \tag{42}$$

Here $\ell$ is the strip length or uninterrupted fin flow length, $\delta$ is the fin thickness, $D_h$ is the hydraulic diameter of the passages, and $\alpha^*$ is the ratio of width to height of the passage. 85% of all data correlated are within ±15% for f and ±10% for j and a few data have a discrepancy as high as 40%.

To obtain the f and j factors for a transitional Reynolds number, Wieting suggested the following procedure. Determine the reference Reynolds number for f and j from the following equations.

$$Re_f^* = 41 \ (\ell/D_h)^{0.772} \ \alpha^{*-0.179} \ (\delta/D_h)^{-1.04} \tag{43}$$

$$Re_j^* = 61.9 \ (\ell/D_h)^{0.952} \ \alpha^{*-1.1} \ (\delta/D_h)^{-0.53} \tag{44}$$

Here $Re_f^*$ is the Reynolds number for the intersection point of the two f vs. Re curves, one for $Re \leq 1,000$, and the other for $Re \geq 2,000$. Similarly, $Re_j^*$ is the Reynolds number for the intersection point of the two j vs. Re curves, one for $Re \leq 1,000$, and the other for $Re \geq 2,000$. If the Reynolds number of interest Re is lower than $Re_f^*$, use Eq. (39) for f, otherwise use Eq. (41). If $Re < Re_j^*$, use Eq. (40) for j, otherwise use Eq. (42). It should be emphasized that the Wieting correlation is strictly based on a limited amount of available test data. No account is made of burrs on the leading and trailing edges in the correlation. Care must be exercized in extrapolating data for the fin geometries that have geometrical parameters outside the range of those for the correlation.

A careful examination of all good data that are published has revealed the ratio $j/f \leq 0.25$ for strip fin, louver fin and other similar interrupted surfaces. This can be approximately justified as follows. The flows are developing along each interruption in such a surface. Based on the Reynolds analogy, in absence of the form drag, j/f should be 0.5. Since the contribution of the form drag is of the same order of magnitude as the skin friction for such an interrupted surface, j/f will be about 0.25. Published data for strip and louver fins are questionable if $j/f > 0.3$ and such is the case for the results of Mochizuki and Yagi [28]. All the measurements and possible sources of leaks and losses must be checked thoroughly to verify those basic data having $j/f > 0.3$ for strip and louver fins.

## 3. HEADER DESIGN AND FLOW DISTRIBUTION

### 3.1 Header Design

One disadvantage in the use of highly compact surface geometries is that the resulting core shapes are characterized by large flow frontal areas and short flow lengths for the gas flow path. The common automobile "radiator" is a close-at-hand example -- a frontal area of approximately 0.3 $m^2$ and an air flow length of 3 cm for a 160 kW automobile engine! In this application, this large frontal area can be accommodated by a front mounting on the vehicle. However, heat exchangers in other engines must be located in ducting and the flow header[†] configuration has a definitive influence on the engine system envelope geometry. This situation is illustrated in Fig. 8 where we can see that the addition of a regenerator to the gas turbine changes a cigar shape "jet-engine" configuration into a box configuration.

Important objectives for the header design are: (1) flow should be distributed as uniformly as possible at the core face; (2) pressure drop in the headers should be as minimum as possible by minimizing flow separation and flow impact effects; and (3) flow nonuniformity in the headers should be minimized to avoid high velocity region that can produce localized erosion particularly

---

[†]An inlet header is a transition "duct" joining the inlet face of the exchanger core or matrix to the inlet pipe. The outlet header joins the outlet face of the exchanger core to the outlet pipe. The header is variously referred to as a tank, manifold, box, or distributor.

# COMPACT HEAT EXCHANGERS

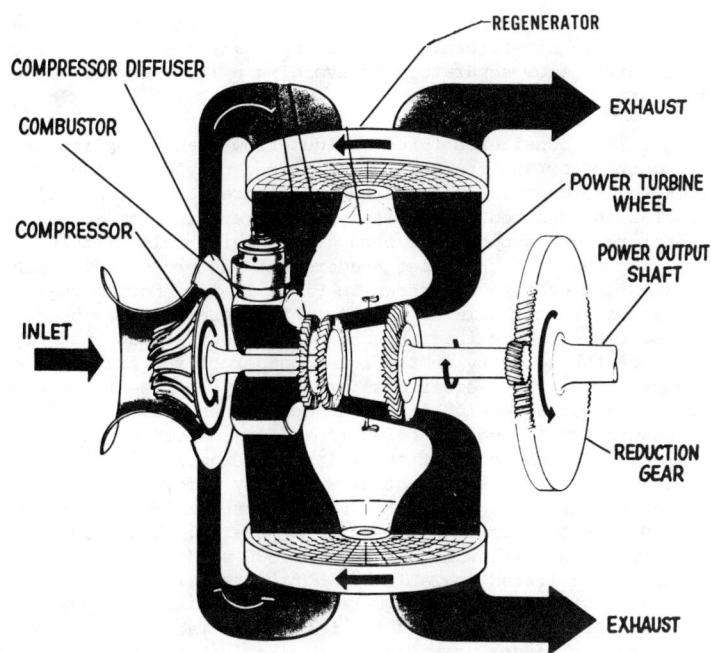

Fig. 8  A vehicular gas turbine engine with regenerators.

for liquid flows.

Headers may be categorized as normal-flow, turning-flow, and oblique-flow headers. In normal-flow headers, the inflow is perpendicular to the heat transfer core face as shown in Fig. 9a. The inlet normal-flow header acts as a diffuser. An extensive analysis and correlations for various flow regimes in a diffuser have been reported by Kline [44] and Miller [45]. In turning-flow headers, the flow is turned 90°, 180° or any other angle as necessary. To minimize the losses and provide the uniform flow distribution, the ideal turning-flow headers are shaped in a streamline fashion or guide vanes are used. In oblique-flow headers, the inflow is either parallel or at an angle to the

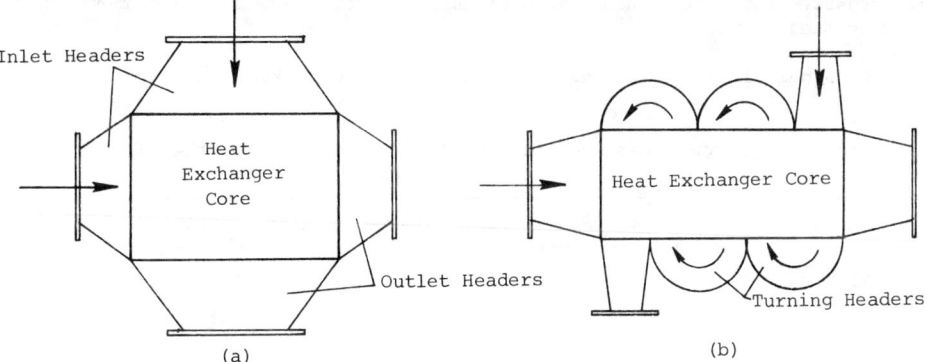

Fig. 9  (a) Normal inlet and outlet headers, (b) turning headers.

heat transfer core face as shown in Figs. 10-12. Since the oblique-flow headers reduce the header volume significantly (as is important for compact heat exchangers) and minimizes flow separation by avoiding a large expansion, we will summarize the theoretical results here.

London et al. [46] considered three oblique-flow header configurations: parallel flow, counterflow and free-discharge. They provided a theory to specify the inlet header shapes and sizes (for a box configuration of the outlet header) that yields the desired uniformity of the pressure drop across the core. Their experimental results support the predictions quite well (within ±5% flow nonuniformity). The shape of the inlet headers and the pressure distributions in the inlet and outlet headers are shown in Figs. 10-12. Notice that the flow continuously accelerates in the inlet header of the parallel flow headers, flow continuously decelerates in the inlet header of the counterflow headers, and flow does not change the velocity in the inlet header of the free-discharge header. London et al. provided equations for the shape of the inlet headers and pressure losses associated with inlet and outlet headers. In order to assure the matching pressure profile needed for uniform flow distribution and for a minimum header loss, they showed that the inlet header dimension $z_i$ can be either larger or smaller than the outlet dimension $y_o$ for a parallel flow header; and $z_i$ must be $0.636\ y_o\ (\rho_o/\rho_i)^{1/2}$ for a counterflow header. They provided numerical results for pressure losses in a parallel flow header ($z_i = y_o$, $\rho u_i^2/2g_c = \rho u_o^2/2g_c$), a counterflow header [$z_i = 0.636\ y_o\ (\rho_o/\rho_i)^{1/2}$], and a free-discharge header as listed in Table 3. Notice that the outlet header loss is largely associated with the nonuniform velocity distribution shown as $u_o(y)$ in Figs. 10 and 11. From the review of the results of Table 3, if an option is available, a counterflow header arrangement is clearly the preferred geometry.

While the foregoing results are based on a uniform inlet velocity $u_i$, as shown in Figs. 10,11, experimental results reported in [46] demonstrate that header performance is relatively insensitive to the inlet velocity maldistribution.

Kutchey and Julien [47] considered radial and circumferential flow nonuniformity in the inlet headers for a vehicular gas turbine regenerator. Based on the experiments and analysis, they showed that the mismatched airside and gas-side radial flow variations reduce the exchanger effectiveness significantly. Circumferential flow variations result in second order effects for the high effectiveness regenerator.

In practice, a variety of other oblique-flow headers are used in heat exchangers. The design theory is available only for those simple geometries as discussed above.

As a guideline, fluid does not turn in the direction one would believe.

Table 3. Pressure loss coefficient $K = \Delta p/(\rho u_i^2/2g_c)$ for the inlet, outlet and inlet plus outlet headers [46]

| Header Type | Inlet Header | Outlet Header | Inlet + Outlet Headers |
|---|---|---|---|
| Parallel flow | 1.822 | 0.645 | 2.467 |
| Counterflow | 0.333 | 0.262 | 0.595 |
| Free-discharge | 1.000 | 0 | 1.000 |

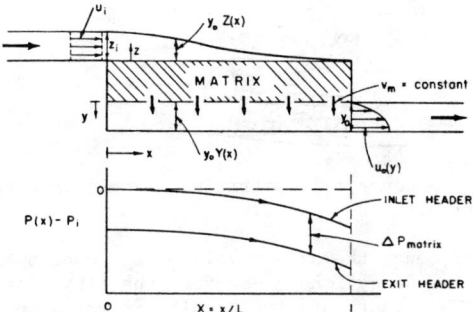

Fig. 10  Parallel flow headers and core, and pressure distribution in inlet and exit headers.

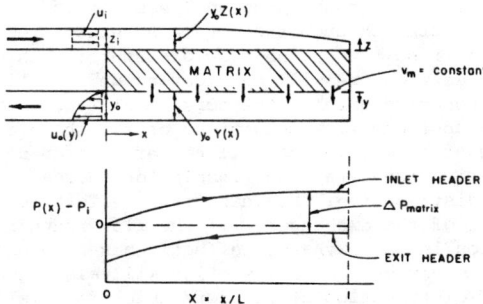

Fig. 11  Counterflow headers and core, and pressure distribution in inlet and exit headers.

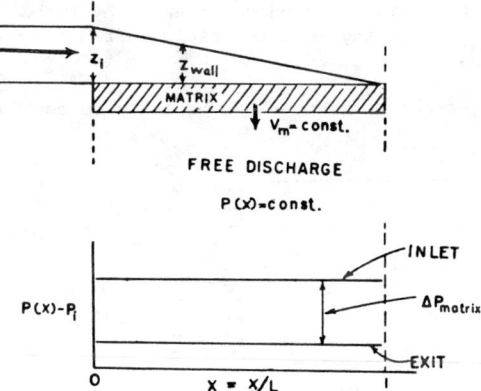

Fig. 12  Free discharge header and core, and pressure distribution at core inlet and outlet.

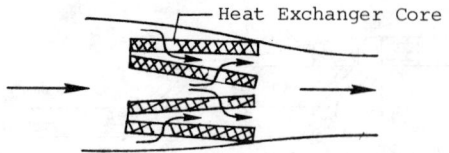

Fig. 13  A folded core concept illustrating an oblique-flow header configuration.

A folded core concept to reduce the header volume is shown in Fig. 13. The arrows show the desired fluid flow directions. However, the fluid does not turn and flow perpendicular to core. This configuration results in very nonuniform flow distribution through the core and a serious reduction in heat exchanger performance.

3.2  Gross Malflow Distribution

In all heat transfer and pressure drop analyses, it is presumed that the fluid is uniformly distributed through the core. A serious reduction in heat exchanger performance may result when the flow distribution through the core is nonuniform. In this section, we will consider nonuniform flow distribution within the core due to nonuniformity at the core inlet face. This nonuniformity is the result of either upstream flow conditions or the shape of the inlet header. Reduction in the exchanger effectiveness will be derived for several flow arrangements, followed by a derivation of a relationship for increase in pressure drop due to nonuniform flow distribution. Nonuniform flow distribution will be considered only on one side of the exchanger. If the flow nonuniformity occurs on both sides, the reduction in the exchanger effectiveness can only be determined numerically. The results derived in this section will also be applicable to the case where the flow distribution at the core inlet face is ideally uniform, but there is a gross blockage in the heat exchanger core. Gross blockage may be due to brazing, soldering, or due to other manufacturing considerations.

<u>Counterflow heat exchanger.</u>  Consider a counterflow heat exchanger in which flow distribution is uniform on the cold side, and nonuniform on the hot side.[†] Also consider nonuniform flow distribution represented by two different uniform velocities, $u_{max}$ and $u_{min}$, flowing into the minimum free flow area $A_{o,1}$ and $A_{o,2}$, respectively, as shown in Fig. 14.

The performance of the heat exchanger can be analyzed considering two heat

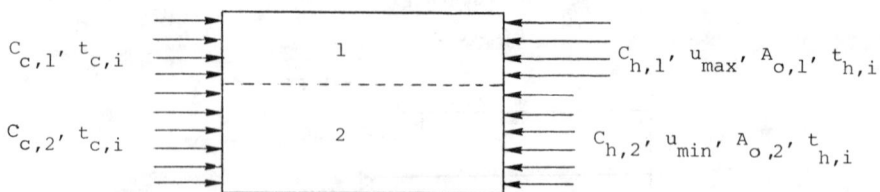

Fig. 14  Idealized flow nonuniformity on one side of a counterflow heat exchanger.

---

[†]The results derived will be generalized to the case when flow distribution is uniform on the hot side, and nonuniform on the cold side.

exchangers (#1 and 2) in parallel, each having uniform velocity at the entrance. Let us first write down the exchanger effectiveness for the case of uniform flow distribution on both sides. For this case,

$$C_h = C_{h,1} + C_{h,2} \qquad C_c = C_{c,1} + C_{c,2} \tag{45}$$

and the mean velocity on the hot side, for constant $\rho$, is

$$u_m = (u_{max} A_{o,1} + u_{min} A_{o,2})/A_o \tag{46}$$

The temperature effectiveness on the hot side when the flow is uniform on both sides is [1]

$$\varepsilon_{h,o} = \frac{t_{h,i} - t_{h,o}}{t_{h,i} - t_{c,i}} = \frac{1 - \exp[-(N_{tu})_h (1 - C_h/C_c)]}{1 - (C_h/C_c) \exp[-(N_{tu})_h (1 - C_h/C_c)]} \tag{47}$$

Here the range of $C_h/C_c$ is $0 \le C_h/C_c \le \infty$, and $(N_{tu})_h = UA/C_h$.

For the case of nonuniform flow distribution on the hot side, as shown in Fig. 14, the temperature effectiveness of Exchanger 1 will be

$$\varepsilon_{h,1} = \frac{(t_{h,i} - t_{h,o})_1}{t_{h,i} - t_{c,i}} = \frac{1 - \exp[-(N_{tu,1})_h (1 - C_{h,1}/C_{c,1})]}{1 - (C_{h,1}/C_{c,1}) \exp[-(N_{tu,1})_h (1 - C_{h,1}/C_{c,1})]} \tag{48}$$

Similarly $\varepsilon_{h,2}$ can be obtained by replacing the subscript 1 of Eq. (48) with 2. Here

$$\frac{C_{h,1}}{C_{c,1}} = \frac{C_{h,1}}{C_h} \frac{C_h}{C_c} \frac{C_c}{C_{c,1}} = \frac{u_{max}}{u_m} \frac{C_h}{C_c} \tag{49}$$

$$\frac{C_{h,2}}{C_{c,2}} = \frac{u_{min}}{u_m} \frac{C_h}{C_c} \tag{50}$$

$$N_{tu,1} = (UA)_1/C_{h,1} \tag{51}$$

where
$$\frac{1}{(UA)_1} = \frac{1}{(\eta_o h_1 A_1)_h} + \frac{a}{A_{w,1} k_w} + \frac{1}{(\eta_o h A_1)_c} \tag{52}$$

Similarly $(N_{tu,2})_h$ and $1/(UA)_2$ are defined by replacing the subscript 1 with 2. On the hot side, the velocities in Exchangers 1 and 2 are $u_{max}$ and $u_{min}$, and are different from the mean velocity $u_m$ if the flow were completely uniform. This change in the velocity will affect the heat transfer coefficient. Therefore a proper heat transfer coefficient $h_1$ must be used in Eq. (52) for $(UA)_1$. Similarly, a proper $h_2$ must be used for the $(UA)_2$ expression. If the scale or fouling is considered, proper resistances must be added in Eq. (52).

The exchanger temperature effectiveness for the case of nonuniform flow distribution on the hot side is then obtained from the following definition.

$$\varepsilon_h = \frac{C_{h,1} \varepsilon_{h,1} + C_{h,2} \varepsilon_{h,2}}{C_h} \tag{53}$$

Note that this definition is based on $q = q_1 + q_2$. The reduction in the temperature effectiveness on the hot side is then

$$\Delta\varepsilon = \frac{\varepsilon_{h,o} - \varepsilon_h}{\varepsilon_{h,o}} \tag{54}$$

where $\varepsilon_{h,o}$ is given by Eq. (47). It can be shown rigorously that $\Delta\varepsilon$ represents a reduction in the <u>exchanger</u> effectiveness regardless of whether $C_h$ is $C_{min}$ or $C_{max}$.

The foregoing derivation for the reduction in the exchanger effectiveness was made for flow nonuniformity on the hot side. For the case of flow nonuniformity on the cold side, Eqs. (47)-(54) are applicable after replacing the subscript h with c and c with h. Also the flow nonuniformity was considered to be a two-step function velocity distribution. This concept can be extended to an n-step function velocity distribution.

<u>Crossflow heat exchanger, nonuniform flow on unmixed side, uniform flow on mixed side.</u> Consider a crossflow heat exchanger with hot fluid unmixed and cold fluid mixed. Also consider nonuniform flow distribution represented by two different uniform velocities on the hot side, and uniform flow distribution on the cold side as shown in Fig. 15. The maximum and minimum velocities are $u_{max}$ and $u_{min}$, and the respective free flow areas are $A_{o,1}$ and $A_{o,2}$.

In this case, also the performance of the exchanger can be analyzed by considering two heat exchangers (#1 and 2) in parallel, each having uniform velocity at the entrance. Now let us first express the temperature effectiveness on the hot side (that is unmixed) when the flow is uniform on <u>both</u> sides.

For $C_h$ unmixed, $C_c$ mixed [1]

$$\varepsilon_{h,o} = \frac{t_{h,i} - t_{h,o}}{t_{h,i} - t_{c,i}} = \frac{C_c}{C_h}\left[1-\exp\left\{-\frac{C_h}{C_c}\left[1-\exp\left(-\frac{UA}{C_h}\right)\right]\right\}\right] \tag{55}$$

Here the range of $C_h/C_c$ is $0 \leq C_h/C_c \leq \infty$. And $C_h$ and $C_c$ represent the heat capacity rates based on the total flow rate on the respective sides. For the case of nonuniform flow distribution on the hot side, the temperature effectiveness of Exchangers 1 and 2 are

$$\varepsilon_{h,1} = \frac{(t_{h,i} - t_{h,o})_1}{t_{h,i} - t_{c,i}} \qquad \varepsilon_{h,2} = \frac{(t_{h,i} - t_{h,o})_2}{t_{h,i} - t_{c,I}} \tag{56}$$

Here $t_{c,I}$ is the outlet temperature of the cold fluid from Exchanger 1 as shown in Fig. 15. Now applying Eq. (55) for Exchanger 1, we have

$$\varepsilon_{h,1} = \frac{C_c}{C_{h,1}}\left[1-\exp\left\{-\frac{C_{h,1}}{C_c}\left[1-\exp\left(-\frac{(UA)_1}{C_{h,1}}\right)\right]\right\}\right] \tag{57}$$

Similarly $\varepsilon_{h,2}$ can be obtained by replacing the subscript 1 with 2 in this equation. Here $(UA)_1$ is given by Eq. (52) in which proper heat transfer coefficient $(h_1)_h$ should be used as discussed just after Eq. (52). Similarly, a proper $(h_2)_h$ should be used in the $(UA)_2$ expression.

The temperature $t_{c,I}$ in Eq. (56) is determined by an energy balance in

# COMPACT HEAT EXCHANGERS

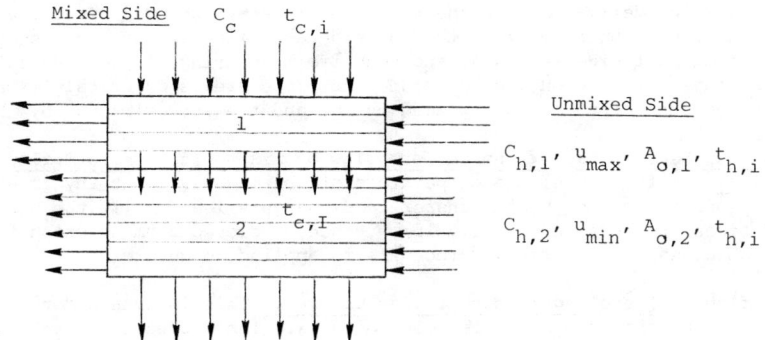

Fig. 15 Idealized flow nonuniformity on the unmixed side of an unmixed-mixed crossflow heat exchanger.

Exchanger 1. The heat transfer rate $q_1$ in this exchanger is

$$q_1 = \varepsilon_{h,1} C_{h,1} (t_{h,i} - t_{c,i}) = C_c (t_{c,I} - t_{c,i}) \qquad (58)$$

so that

$$t_{c,I} = t_{c,i} + \frac{\varepsilon_{h,1} C_{h,1}}{C_c} (t_{h,i} - t_{c,i}) \qquad (59)$$

Now the heat transfer rate in Exchanger 2 is given by

$$q_2 = \varepsilon_{h,2} C_{h,2} (t_{h,i} - t_{c,I}) \qquad (60)$$

Substituting the value of $t_{c,I}$ from Eq. (59) into Eq. (60) and simplifying,

$$q_2 = \varepsilon_{h,2} C_{h,2} \left(1 - \frac{\varepsilon_{h,1} C_{h,1}}{C_c}\right) (t_{h,i} - t_{c,i}) \qquad (61)$$

The overall temperature effectiveness on the hot fluid side is defined by

$$q = q_1 + q_2 = \varepsilon_h C_h (t_{h,i} - t_{c,i}) \qquad (62)$$

Using Eqs. (58)-(62), the temperature effectiveness on the hot fluid side with the nonuniformity model of Fig. 15 is

$$\varepsilon_h = \frac{1}{C_h} \left[ \varepsilon_{h,1} C_{h,1} + \varepsilon_{h,2} C_{h,2} \left(1 - \frac{\varepsilon_{h,1} C_{h,1}}{C_c}\right) \right] \qquad (63)$$

The reduction in the temperature effectiveness on the hot side is then

$$\Delta\varepsilon = \frac{\varepsilon_{h,o} - \varepsilon_h}{\varepsilon_{h,o}} \qquad (64)$$

where $\varepsilon_{h,o}$ is given by Eq. (55). As mentioned earlier, $\Delta\varepsilon$ of this equation represents the reduction in the exchanger effectiveness regardless of $C_h$ being $C_{min}$ or $C_{max}$.

The foregoing derivation for the reduction in the exchanger effectiveness was made for flow nonuniformity on the hot side as unmixed. For the case of cold side as unmixed, hot side as mixed, and nonuniform entering flow on the cold side and uniform entering flow on the hot side, the procedure is identical, except everywhere the subscript h is replaced by c, and the subscript c by h.

Crossflow heat exchanger, nonuniform flow on mixed side, uniform flow on unmixed side. In this case, the temperature of the mixed side at any location will be dependent upon the heat transfer in Exchangers 1 and 2 (if the unmixed and mixed sides are reversed in Fig. 15). Hence, a closed form solution for $\Delta\varepsilon$ cannot be obtained. An iterative procedure is applied to determine $\Delta\varepsilon$.

Crossflow heat exchanger, both fluids unmixed. This is even a more complicated arrangement from the analysis point of view. The decrease in exchanger effectiveness can be determined by a numerical analysis. Fortunately, a method has been described and some results of practical interest have been obtained by Chiou [48]. He considered two-dimensional nonuniformity either on one side or on both sides of a crossflow exchanger.

Increase in pressure drop due to gross malflow distribution. Consider a two-step function velocity distribution at the core inlet on the hot fluid side as shown in Fig. 14. Exchangers 1 and 2 in this figure are considered in parallel. Evaluate the pressure drop $\Delta p_1$ and $\Delta p_2$ in Exchangers 1 and 2, respectively, by the method of Kays and London [1]. Also evaluate $\Delta p$ considering flow as uniform at the core inlet in Fig. 14. Because of the nonuniform flow distribution, the static pressures at the core inlet and outlet faces will also be nonuniform, i.e., they will not be constant. Therefore, as a conservative approach, the higher of the two $\Delta p$'s (i.e., $\Delta p_1$) will be the pressure drop on the hot side of the exchanger with imposed flow nonuniformity of Fig. 14. The increase in pressure drop due to flow nonuniformity is then

$$(\Delta p)_{increase} = \Delta p_1 - \Delta p \tag{65}$$

It should be emphasized that the entrance and exit losses, in addition to the core friction contribution, will be higher (in the evaluation of $\Delta p_1$) compared to those for uniform flow.

If the flow nonuniformity occurs on both sides of an exchanger, the same procedure is applied to both sides. Since the pressure drops on both sides of a two-fluid exchanger are relatively independent of each other, the flow arrangement of the two-fluid exchanger does not come into the picture. Hence, the foregoing procedure is applicable to any flow arrangement, however, an experimental verification of the proposed procedure is needed.

## 4. LAMINAR FLOW SURFACES

A heat transfer surface is designated as a laminar flow surface[†] when predominantly thermally developed or developing laminar boundary layers[†] develop on the surface. Somewhat arbitrarily, we will define a laminar flow surface as the heat transfer surface having area density $\beta$ greater than 3,300 $m^2/m^3$ (1,000 $ft^2/ft^3$).

In a gas-to-gas heat exchanger, the friction power expenditure is generally

---

[†]Because of our major concern for heat transfer in an exchanger, we have considered thermal boundary layers. In laminar flow surfaces, the velocity boundary layers will also be laminar developing or developed types.

COMPACT HEAT EXCHANGERS                                                                141

significant and a controlling factor. As a result, for a reasonable pressure
drop, the flow velocity must be kept low. This may lower the effective heat transfer coefficient h. In addition, the heat transfer coefficient h on the gas
side has generally 1/10 or a lower value compared to that for a liquid. Hence,
when a large surface area is required, and if the heat exchanger volume is not
to be excessive, compact surfaces must be employed.

As more compact surfaces are developed, the hydraulic diameter $D_h$ becomes
smaller, and with low velocities, the design Reynolds number $GD_h/\mu$ is also
smaller. Hence, highly compact surfaces are laminar flow surfaces. For an automobile radiator, the airside Re is in the range 1,000-3,000. In contrast,
150 is the design Re for the highly compact ceramic matrix ($\beta \simeq 6,000$ m$^2$/m$^3$,
1,830 ft$^2$/ft$^3$) used in the Ford 707 gas turbine engine regenerator. At these
low Reynolds numbers, the flow mixing associated with turbulent eddies is minimal,
and a laminar type boundary layer forms over the heat transfer surface. Heat
transfer away from the wall and <u>through</u> the fluid is then largely dominated by
the heat conduction mechanism.

4.1 Characteristics

Laminar flow surfaces are classified into two categories: those surfaces
having predominantly thermally developed (low Re) laminar flow, and those surfaces
having predominantly thermally developing laminar flow. Surfaces having continuous cylindrical passages fall into the first category. Compact strip, louver,
perforated and wavy fins fall into the second category.

Thermally developed laminar flow problems can be treated mathematically as
a heat conduction problem in the fluid. Theoretical solutions are available for
a large number of duct cross-section geometries [3]. Some of these are summarized
in Table 1. The following interesting characteristics are observed from these
solutions.

1. There is a strong influence of thermal boundary conditions (T), (H1), and
   (H2) on the convective behavior. Depending on the flow cross-section geometry,
   variations of j of the order of 50% or more can be imposed by the wall heat
   flux or temperature boundary condition.

2. As Nu = $hD_h/k$, a constant Nu implies a convective coefficient h that is
   independent of the flow velocity.

3. An increase in h is best achieved by reducing $D_h$, or by a change in the
   type of geometry, e.g., a change from a triangular to an 8:1 rectangular
   geometry.

4. The friction factor varies inversely with velocity so that the flow pressure
   drop tends to vary with the first power of the velocity.

The foregoing low Re laminar flow characteristics are rather different
from what we have learned to expect from the high Re turbulent flow situation
characteristic of "not so compact" tubular heat exchangers. Also note that the
transition from fully developed laminar to turbulent flow occurs at Re $\simeq$ 2,300.
For continuous passage geometries, usually there is a dip in j and f vs.
Re curves in the range Re ~ 2,000-5,000 as shown in Fig. 2.

Thermally developing laminar flow problems have also been analyzed for a
number of duct cross-section geometries [3]. Some correlations are presented
in Eqs. (33)-(36). The following characteristics are observed from these
solutions.

1. Influence of thermal boundary conditions on the convective behavior appears to be of the same order as that for the fully developed based on the available solutions.

2. Since $Nu$ is proportional to $(x^*)^{-1/3} = (x/D_h \text{RePr})^{-1/3}$, $Nu$ is proportional to $\text{Re}^{1/3}$ and hence, h varies as $u_m^{1/3}$.

3. Influence of the duct shape on thermally developing $Nu$ is not so great as that for the fully developed $Nu$.

With interrupted surfaces, there is considerable form drag associated with the blunt (burred or deburred) edges. This affects both the friction factor and heat transfer coefficient. Hence, firm design data (j and f vs. Re) are largely derived from physical model tests rather than from mathematical modeling. Transition from thermally developing laminar to thermally developed turbulent flow for interrupted exchanger surfaces occurs at Re ~ 4,000-10,000. In contrast to continuous passage geometries, there is no dip in j and f vs. Re curves in the transition region as shown in Fig. 5b.

## 4.2 Design Problems

The nature of a highly compact surface introduces some additional problems which may not be so severe in less compact surfaces. Some of these are flow nonuniformity, influence of brazing, and influence of fouling. They result into a degradation of the exchanger heat transfer performance.

*Flow nonuniformity.* Ideally, the flow should be uniform across the entire inlet face of the heat exchanger for maximum performance. However, the flow may not be uniform either at the core face or within the matrix. Two types of flow nonuniformities (malflow distribution) associated with highly compact surfaces are: gross (or bulk) flow nonuniformity and passage-to-passage (or small scale) nonuniformity.

As mentioned earlier, the shape of a compact heat exchanger is generally awkward having a large frontal area and a short flow length, a thin disk-type shape. The large frontal area results in the necessity for turning the flow streams in a system of headers. The turning and a marked area change in the headers may possibly lead to a very uneven flow distribution through the heat exchanger core, even though all flow passages may be uniformly constructed. Thus, this gross malflow distribution is caused by the variation in the core upstream (with or without the downstream) static pressure distribution and hence different driving potential $\Delta p$ for flow passages. The gross malflow distribution is thus independent of the heat exchanger surface geometry configuration and uniformity. The inevitable result is a decrease in the performance of the heat exchanger as discussed in Section 3.2. The header design problem is thus equally very important for the highly compact heat exchanger and has been discussed in Section 3.1.

In a highly compact surface, a loss of performance occurs when flow passages do not have identical cross-section areas. The matrix surfaces for rotary regenerators are particularly susceptible to manufacturing tolerances because it is difficult to control precisely the passage size when small dimensions are involved. The highly compact plate-fin type surfaces have tight fin spacing. A slight misalignment or buckling of fins can easily occur in the manufacturing process. This could substantially change the aspect ratio or the shape of flow passages. For the same driving potential $\Delta p$, since the flow resistance may be different across differently shaped passages, it results into a passage-to-

# COMPACT HEAT EXCHANGERS

passage (or small scale) malflow distribution. This malflow distribution is thus solely dependent upon the heat exchanger core or matrix geometrical non-uniformity. This malflow distribution does not produce a substantial loss in performance in either turbulent flow or for highly interrupted laminar flow surfaces, but can have a marked influence in fully developed laminar flow. For this reason, the passage-to-passage nonuniformity is of special interest for highly compact surfaces. This nonuniformity reduces the j factors substantially with a slight reduction in f factors. The theoretical analysis for this non-uniformity for low Re laminar flow surfaces has been presented in detail by London [49] and Shah and London [50]. The analysis was carried out for the passage-to-passage nonuniformity associated with the rectangular and triangular passages. This analysis will not be repeated here, but some illustrative results are presented in Figs. 16-18 for two and three passage nonuniformity for rectangular passages that are highly recommended for very compact heat exchanger applications.

In Fig. 16, a reduction in $N_{tu}$ for rectangular passages is shown when 50% of the passages are large ($c_2 > c_r$) and 50% of the passages are small ($c_1 < c_r$) compared to the reference or nominal passages. The results are presented for the passages having a nominal aspect ratio $\alpha_r^*$ of 1, 0.5, 0.25 and 0.125. Both the constant heat flux boundary condition (H1) and the constant wall temperature boundary condition (T) are considered. The channel deviation parameter $\delta_c$ is defined as

$$\delta_c = 1 - c_1/c_r \tag{66}$$

and the $N^*_{tu,cost}$ is defined as

$$N^*_{tu,cost} = \left[1 - \frac{N_{tu,e}}{N_{tu,r}}\right] \times 100 \tag{67}$$

Here $N_{tu,e}$ is the effective $N_{tu}$ when the 2-passage model passage-to-passage nonuniformity is present, and $N_{tu,r}$ is the reference or nominal $N_{tu}$. For $N_{tu,r} = 5.0$, it can be seen from Fig. 16 that a 10% channel deviation (which is common for a highly compact surface) results in 10% and 21% reduction in $N_{tu,H1}$ and $N_{tu,T}$ respectively for $\alpha_r^* = 0.125$. In contrast, the gain in the pressure drop due to the passage-to-passage nonuniformity is only 2.5% for $\delta_c = 0.10$ and $\alpha_r^* = 0.125$, as found from Fig. 17. Here $\Delta p^*_{gain}$ is defined as

$$\Delta p^*_{gain} = \left[1 - \frac{\Delta p_{actual}}{\Delta p_{nominal}}\right] \times 100 \tag{68}$$

Figure 18 shows the influence of passage-to-passage nonuniformity for a three-passage model with the results from the two-passage model superimposed as solid lines for $\alpha_r^* = 1$. Here three-passage model consists of a small, a nominal and a large rectangular passage. Two variables of nonuniformity are the passage size (selected as $c_r - c_1 = c_2 - c_r$) and the number of passages [selected as $\chi_1 = \chi_2$ and $\chi_3 = \chi_r = 1 - (\chi_1^2 + \chi_2^2)$]. And $\chi_i$ is the fractional distribution of the i-th shaped passage. The data point in Fig. 18 are for each $(1 - c_1/c_r)$ and for $\chi_1 = \chi_2 = 0.10, 0.20, 0.30, 0.333, 0.40$ and $0.50$. The channel deviation parameter $\delta_c$ for three-passage and higher number passage model is defined as

$$\delta_c = \sum_{i=1}^{n} [\chi_i (1 - c_i/c_r)^2]^{1/2} \tag{69}$$

The following observations may be made from Fig. 18: (1) The channel

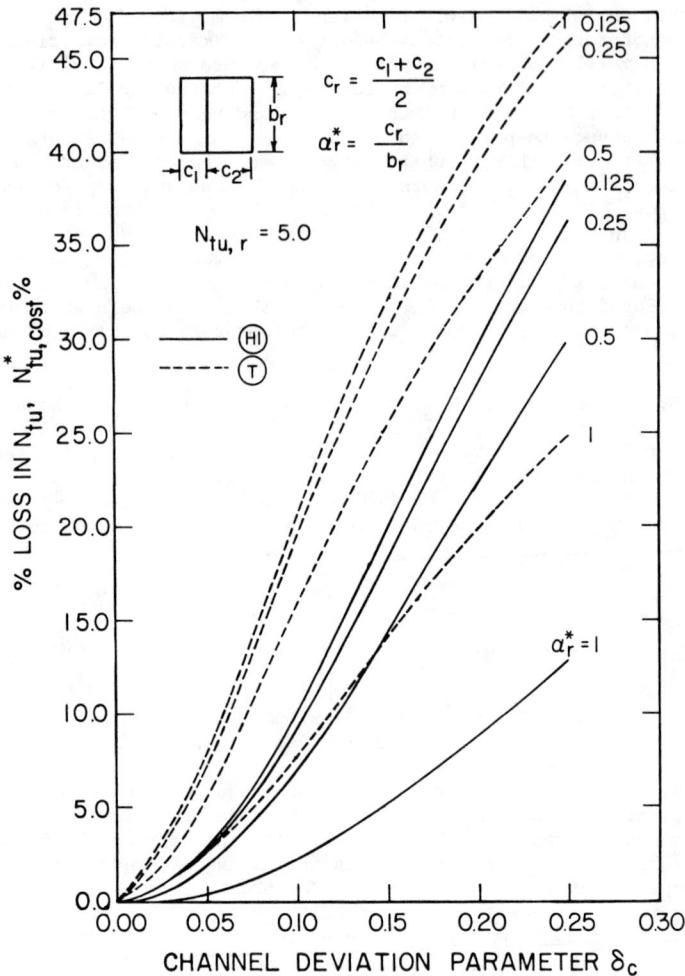

Fig. 16 Two-passage nonuniformity in rectangular passages -- % loss in $N_{tu}$ as a function of $\delta_c$ and $\alpha_r^*$.

deviation parameter $\delta_c$ correlates the deterioration in performance very well for two and three-passage model symmetrical passage-to-passage nonuniformity. The results have been extended to the Gaussian passage-to-passage nonuniformity and the foregoing conclusion is equally valid [50]. Hence, for any symmetrical passage-to-passage nonuniformity for rectangular passages, once $\delta_c$ is determined, Fig. 16 can be used to compute the reduction in $N_{tu}$, and Fig. 17 can be used to compute the reduction (gain) in $\Delta p$. (2) The loss in $N_{tu}$ is significant for the Ⓣ boundary condition compared to the Ⓗ1 boundary condition. (3) The loss in $N_{tu}$ increases with higher nominal $N_{tu}$. (4) The loss in $N_{tu}$ is much more significant compared to the gain (or reduction) in $\Delta p$ at a given channel

# COMPACT HEAT EXCHANGERS

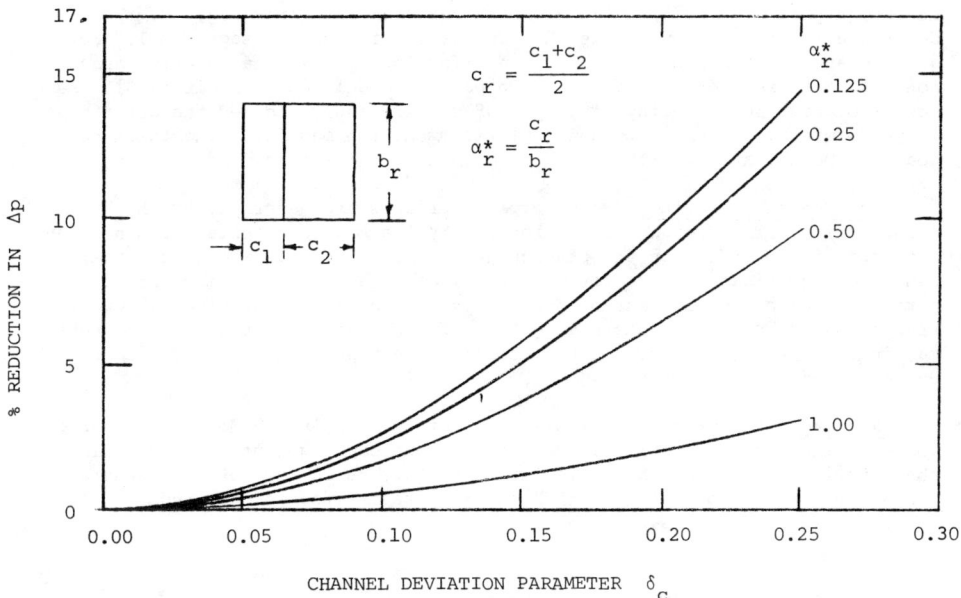

Fig. 17  Two-passage nonuniformity in rectangular passages -- % reduction in $\Delta p$ as a function of $\delta_c$ and $\alpha_r^*$.

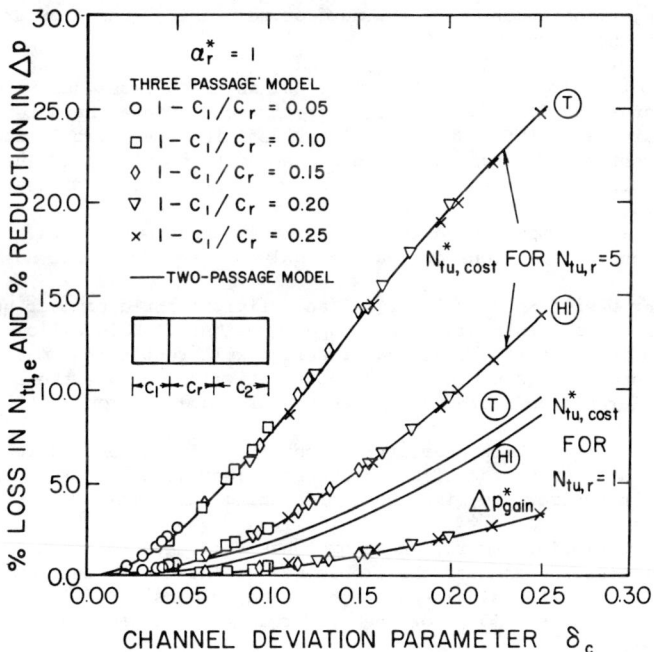

Fig. 18  Three passage nonuniformity in rectangular passages.

deviation parameter. (5) Although not clearly obvious from Fig. 18, the deterioration in performance is the highest for the two-passage model. For a given channel deviation $(1 - c_1/c_r)$, $\delta_c$ for the Gaussian passage-to-passage nonuniformity is 0.48 that for the two-passage model nonuniformity [50]. As seen from Fig. 18, reducing $\delta_c$ by 0.48 significantly reduces the deterioration in performance. Shah and London [50] recommend the manufacturing tolerance goals to be set at $\delta_c \leq 5\%$.

**Influence of brazing.** Very compact surfaces are generally of the brazed construction. The "roughness" introduced by the brazed material does not have any appreciable effect on the passage geometry, if the passages are large; but can have significant effect for very small passage sizes. It may introduce partial or complete blockage of some of the passages. Gross blockage due to brazing may reduce j factors slightly. In contrast, the brazing roughness may increase the f factors substantially. This subject is considered by Shah and London [26].

**Fouling.** The problem of fouling is more severe in the heat exchanger of very small passage geometries. Unless the heat exchanger has self-cleaning characteristics due to a periodic flow reversal, as in the rotary regenerators, the use of highly compact heat exchangers is restricted to only those applications involving "clean" fluids.

## 5. SUMMARY

Starting with a definition, unique features and subtle characteristics of compact heat exchangers are described. A brief discussion is then presented on the exchanger construction types, surface geometries, and flow arrangements used for compact heat exchangers.

Each of a large number of compact exchanger surfaces has unique dimensionless heat transfer and flow friction characteristics. These characteristics are presented in many different ways. After defining the commonly used dimensionless groups, experimental techniques are presented to obtain these surface basic characteristics. Theoretical solutions and correlations for heat transfer and friction factors are then presented for simple geometries. These surface geometries are commonly used in highly compact heat exchangers. Analytical solutions and correlations are subsequently presented for extended surfaces.

Header design is very important and critical for a compact heat exchanger due to its awkward thin disk-type shape. Design of oblique-flow header is summarized in the text. Poor header design will result in malflow distribution. An analysis is presented to quantitatively evaluate the degradation in heat transfer and an increase in pressure drop due to gross malflow distribution.

Finally, the subtle characteristics of highly compact surfaces, referred to as laminar flow surfaces, are discussed along with the problems of passage-to-passage nonuniformity, influence of brazing, and fouling.

As a concluding remark, an attempt has been made to describe unique characteristics and some of the problems associated with the compact heat exchangers. Only the single-phase phenomena is considered. The reader is referred to Robertson [51] for the performance of compact heat exchangers with two-phase fluids.

# COMPACT HEAT EXCHANGERS 147

NOMENCLATURE

| | |
|---|---|
| $A$ | total heat transfer surface area (both primary and secondary, if any) on one side of a direct transfer type exchanger, total heat transfer surface area of a regenerator, $m^2$, $ft^2$ |
| $A_o$ | minimum free flow (or open) area on one side of the exchanger, $m^2$, $ft^2$ |
| $A_w$ | total wall area for heat conduction from hot fluid to cold fluid, $m^2$, $ft^2$ |
| $a$ | wall thickness, m, ft |
| $C$ | flow stream heat capacity rate, $Wc_p$, W/°C, Btu/hr °F |
| $C_{min}$ | minimum of $C_c$ or $C_h$, W/°C, Btu/hr °F |
| $c_p$ | specific heat of fluid at constant pressure, J/kg °C, Btu/lbm °F |
| $D_h$ | hydraulic diameter of flow passages, $4A_o L/A$ or $4\sigma/\beta$, m, ft |
| $E$ | fluid pumping power per unit surface area, $W\Delta p/\rho A$, $W/m^2$, $hp/ft^2$ |
| $f$ | Fanning friction factor, defined by Eq. (9), dimensionless |
| $f_D$ | Darcy friction factor, $4f$, dimensionless |
| $G$ | mass velocity, $W/A_o$, $kg/m^2 s$, $lbm/hr\ ft^2$ |
| $g_c$ | proportionality constant in Newton's second law of motion, $g_c = 1$ and dimensionless in SI units, $g_c = 32.174\ lbm\ ft/lbf\ sec^2$ |
| Ⓗ | thermal boundary condition referring to constant axial as well as peripheral wall heat flux, also constant peripheral wall temperature; boundary condition valid only for the circular tube, parallel plates, and concentric annular ducts when symmetrically heated |
| Ⓗ1 | thermal boundary condition referring to constant axial wall heat flux with constant peripheral wall temperature |
| Ⓗ2 | thermal boundary condition referring to constant axial wall heat flux with constant peripheral wall heat flux |
| $h$ | heat transfer coefficient, $W/m^2$ °C, Btu/hr $ft^2$ °F |
| $j$ | Colburn factor, $StPr^{2/3}$, dimensionless |
| $K_c$ | flow contraction loss coefficient at exchanger entrance, dimensionless |
| $K_e$ | expansion loss coefficient for flow at exchanger exit, dimensionless |
| $K(\infty)$ | incremental pressure drop number for fully developed flow, see [3] for the definition, dimensionless |
| $k$ | fluid thermal conductivity, W/m °C, Btu/hr ft °F |
| $L$ | fluid flow (core) length on one side of the exchanger, m, ft |
| $\ell$ | effective flow length between major boundary layer disturbances, distance between interruptions, m, ft |
| $Nu$ | Nusselt number, defined by Eq. (2), dimensionless |
| $N_{tu}$ | number of heat transfer units, $UA/C_{min}$, dimensionless |
| $(N_{tu})_h$ | number of heat transfer units based on $C_h$, $UA/C_h$, dimensionless |
| $Pr$ | Prandtl number, $\mu c_p/k$, dimensionless |
| $\Delta p$ | fluid static pressure drop on one side of a heat exchanger core, Pa, $lbf/ft^2$ |
| $q$ | heat transfer rate in the exchanger, W, Btu/hr |

| | |
|---|---|
| $q''$ | heat transfer rate per unit surface area, $q/A$, W/m$^2$, Btu/hr ft$^2$ |
| $R_w$ | wall thermal resistance, $a/k_w A_w$ for a plate, °C/W, hr °F/Btu |
| Re | Reynolds number based on the hydraulic diameter, $GD_h/\mu$, dimensionless |
| $r_h$ | hydraulic radius, $A_o L/A$ or $D_h/4$, m, ft |
| St | Stanton number, $h/Gc_p$, dimensionless |
| ⓣ | thermal boundary condition referring to constant wall temperature, both axially and peripherally |
| $t$ | fluid static temperature to a specified arbitrary datum, °C, °F |
| $t_m$ | fluid bulk mean temperature, °C, °F |
| $\Delta t_m$ | true mean temperature difference, defined by Eq. (1), °C, °F |
| U | overall heat transfer coefficient, W/m$^2$ °C, Btu/hr ft$^2$ °F |
| $u_m$ | fluid mean axial velocity, m/s, ft/sec |
| V | heat exchanger total volume, m$^3$, ft$^3$ |
| W | fluid mass flow rate, $\rho u_m A_o$, kg/s, lbm/hr |
| x | Cartesian coordinate along the flow direction, m, ft |
| $x^+$ | axial distance, $x/D_h$ Re, dimensionless |
| $x^*$ | axial distance, $x/D_h$ RePr, dimensionless |
| $\alpha$ | fluid thermal diffusivity, $k/\rho c_p$, m$^2$/s, ft$^2$/sec |
| $\alpha^*$ | ratio of small to large side of a rectangular duct, dimensionless |
| $\beta$ | heat transfer surface area density: a ratio of total transfer area on one side of a plate-fin heat exchanger to the volume between the plates on that side; a ratio of total area on one side of the tube-fin heat exchanger to total volume of the exchanger; a ratio of total (hot and cold side) surface area of a bare bank of tubes or of a regenerator to total volume of the exchanger, m$^2$/m$^3$, ft$^2$/ft$^3$ |
| $\delta$ | fin thickness, m, ft |
| $\delta_c$ | channel deviation parameter, defined by Eq. (66) or (69), dimensionless |
| $\varepsilon$ | heat exchanger effectiveness, refer to [1], dimensionless |
| $\varepsilon_h$ | temperature effectiveness of the hot fluid, defined similar to $\varepsilon_{h,o}$, dimensionless |
| $\varepsilon_{h,o}$ | temperature effectiveness of the hot fluid when flow is uniform on both sides of a two-fluid heat exchanger, defined by Eq. (47), dimensionless |
| $\eta_o$ | temperature effectiveness of total heat transfer area on one side of the extended surface heat exchanger, see [1] for the definition, dimensionless |
| $\mu$ | fluid dynamic viscosity coefficient, Pa·s, lbm/hr ft |
| $\rho$ | fluid density, kg/m$^3$, lbm/ft$^3$ |
| $\sigma$ | ratio of free flow area to frontal area, $A_o/A_{fr}$, also the volumetric porosity for regenerators, dimensionless |
| $\chi_i$ | fractional distribution of i-th shaped passage, dimensionless |

Subscripts

| | |
|---|---|
| c | cold fluid side |

| | |
|---|---|
| h | hot fluid side |
| H, H1, H2 | constant axial wall heat flux boundary conditions |
| i | inlet to the exchanger |
| o | outlet to the exchanger |
| r | reference or nominal passage |
| T | constant wall temperature boundary condition |
| w | wall |

## REFERENCES

1. W.M. Kays and A.L. London, Compact Heat Exchangers, Second Edition, McGraw-Hill Book Co., New York (1964).
2. R.K. Shah, Classification of heat exchangers, in Heat Exchangers - Thermohydraulic Fundamentals and Design, edited by S. Kakac, A.E. Bergles, and F. Mayinger, Hemisphere Publishing Corp., New York (1981).
3. R.K. Shah and A.L. London, Laminar Flow Forced Convection in Ducts, Supplement 1 to Advances in Heat Transfer, Academic Press, New York (1978).
4. A.P. Colburn, A method of correlating forced convection heat transfer data and a comparison with fluid friction, Trans. AIChE, Vol. 29, 174-210 (1933); reprinted in International Journal of Heat and Mass Transfer, Vol. 7, 1359-1384 (1964).
5. W.M. Kays and A.L. London, Heat transfer and flow friction characteristics of some compact heat exchanger surfaces - Part I: Test system and procedure, Trans. ASME, Vol. 72, 1075-1085 (1950); also Description of test equipment and method of analysis for basic heat transfer and flow friction tests of high rating heat exchanger surfaces, TR No. 2, Dept. Mech. Eng., Stanford Univ., Stanford, CA (1948).
6. P.F. Pucci, C.P. Howard and C.H. Piersall, Jr., The single blow transient testing technique for compact heat exchanger surfaces, Trans. ASME, Journal of Engineering for Power, Vol. 89, Series A, 29-40 (1967).
7. A.J. Wheeler, Single-blow transient testing of matrix-type heat exchanger surfaces at low values of $N_{tu}$, TR No. 68, Dept. Mech. Eng., Stanford Univ., Stanford, CA (1968).
8. J.R. Mondt and D.C. Siegla, Performance of perforated heat exchanger surfaces, Trans. ASME, Journal of Engineering for Power, Vol. 96, Series A, 81-86 (1974).
9. J.H. Stang and J.E. Bush, The periodic method of testing compact heat exchanger surfaces, Trans. ASME, Journal of Engineering for Power, Vol. 96, Series A, 87-94 (1974).
10. R.K. Shah, Compact heat exchanger design procedures, in Heat Exchangers - Thermohydraulic Fundamentals and Design, edited by S. Kakac, A.E. Bergles, and F. Mayinger, Hemisphere Publishing Corp., New York (1981).
11. A.L. London, M.B.O. Young, and J.H. Stang, Glass ceramic surfaces, straight triangular passages -- heat transfer and flow friction characteristics, Trans. ASME, Journal of Engineering for Power, Vol. 92, Series A, 381-389 (1970).
12. A.L. London and R.K. Shah, Glass-ceramic hexagonal and circular passage surfaces -- heat transfer and flow friction design characteristics, SAE Trans., Vol. 82, Section 1, 425-434 (1973).
13. B.S. Petukhov and V.V. Kirillov, The problem of heat exchange in the turbulent flow of liquids in tubes, Teploenergetika No. 4, 63-68 (1958).
14. O.C. Jones, Jr., An improvement in the calculation of turbulent friction in rectangular ducts, Trans. ASME, Journal of Fluids Engineering, Vol. 98, Series I, 173-181 (1976).
15. W.M. Kays and H.C. Perkins, Forced convection, internal flow in ducts, in Handbook of Heat Transfer, edited by W.M. Rohsenow and J.P. Hartnett, Chapter 7, McGraw-Hill Book Co., New York (1973).

16. R.L. Webb, A critical evaluation of analytical solutions and Reynolds analogy equations for turbulent heat and mass transfer in smooth tubes, Wärme-und Stoffübertragung, Vol. 4, 197-204 (1971).
17. C.A. Sleicher and M.W. Rouse, A convenient correlation for heat transfer to constant and variable property fluids in turbulent pipe flow, International Journal of Heat and Mass Transfer, Vol. 18, 677-683 (1975).
18. R.L. Webb, Toward a common understanding of the performance and selection of roughness for forced convection, in Studies in Heat Transfer: A Festschrift for E.R.G. Eckert, edited by J.P. Hartnett, T.F. Irvine, Jr., E. Pfender, and E.M. Sparrow, Hemisphere Publishing Corp., New York, pp. 257-272 (1979).
19. R.K. Shah, A correlation for laminar hydrodynamic entry length solutions for circular and noncircular ducts, Trans. ASME, Journal of Fluids Engineering, Vol. 100, Series I, 177-179 (1978).
20. Z.V. Tishchenko, V.N. Bondarenko, and L.I. Golechek, Heat transfer and pressure drop in gas-carrying ducts formed by smooth-finned plate-type heat-exchange surfaces, Heat Transfer - Soviet Research, Vol. 11, No. 5, 117-124 (1979).
21. K. Okada, M. Ono, T. Tomimura, T. Okuma, H. Konno and S. Ohtani, Design and heat transfer characteristics of new plate type heat exchanger, Heat Transfer-Japanese Research, Vol. 1, No. 1, pp. 90-95 (1972).
22. G. Rosenblad and A. Kullendorf, Estimating heat transfer rates from mass transfer studies on plate heat exchanger surfaces, Wärme-und Stoffübertragung, Vol. 8, 187-191 (1975).
23. L. Goldstein, Jr., and E.M. Sparrow, Mass transfer experiments on secondary-flow vortices in a corrugated wall channel, International Journal of Heat and Mass Transfer, Vol. 19, 1337-1339 (1976).
24. L.J. Goldstein and E.M. Sparrow, Heat/mass transfer characteristics for flow in a corrugated wall channel, Trans. ASME, Journal of Heat Transfer, Vol. 99, Series C, 187-195 (1977).
25. A.L. London and R.K. Shah, Offset rectangular plate-fin surfaces -- heat transfer and flow friction characteristics, Trans. ASME, Journal of Engineering for Power, Vol. 90, Series A, 218-228 (1968).
26. R.K. Shah and A.L. London, Influence of brazing on very compact heat exchanger surfaces, ASME Paper No. 71-HT-29 (1971).
27. E.V. Dubrovskii and A.I. Fedotva, Investigation of heat-exchanger surfaces with plate fins, Heat Transfer-Soviet Research, Vol. 4, No. 6, 75-79 (1972).
28. S. Mochizuki and Y. Yagi, Heat transfer and friction characteristics of strip fins, Heat Transfer-Japanese Research, Vol. 6, No. 3, 36-59 (1977).
29. E.M. Sparrow, R.R. Baliga, and S.V. Patankar, Heat transfer and fluid flow analysis of interrupted wall channels, with application to heat exchangers, Trans. ASME, Journal of Heat Transfer, Vol. 99, Series C, 4-11 (1977); discussion by R.K. Shah, Trans. ASME, Journal of Heat Transfer, Vol. 101, Series C, 188-189 (1979).
30. E.M. Sparrow and C.H. Liu, Heat-transfer, pressure-drop and performance relationships for in-line, staggered and continuous plate heat exchangers, International Journal of Heat and Mass Transfer, Vol. 22, 1613-1625 (1979).
31. M.C. Smith, Performance analysis and model experiments for louvered fin evaporator core development, Society of Automotive Engineers, SAE Paper No 720078 (1972).
32. L.T. Wong and M.C. Smith, Airflow phenomena in the louvered-fin heat exchanger, Society of Automotive Engineers, SAE Paper No. 730237 (1973).
33. H.L. Miller and C.A. Leeman, Heat transfer and pressure drop characteristics of several compact plate surfaces, Heat Transfer: Fundamentals and Industrial Application, AIChE Symp. Ser. 131, Vol. 39, 63-71 (1973).
34. R.K. Shah, Perforated heat exchanger surfaces. Part 1 - Flow phenomena, noise and vibration characteristics, ASME Paper No. 75-WA/HT-8 (1975).
35. R.K. Shah, Perforated heat exchanger surfaces. Part 2 - Heat transfer and flow friction characteristics, ASME Paper No. 75-WA/HT-9 (1975).

36. C.Y. Liang and W.J. Yang, Heat transfer and friction loss performance of perforated heat exchanger surfaces, Trans. ASME, Journal of Heat Transfer, Vol. 97, Series C, 9-15 (1975).
37. C.P. Lee and W.J. Yang, Augmentation of convective heat transfer from high-porosity perforated surfaces, Heat Transfer 1978, Vol. 2, 589-594, Hemisphere Publishing Corp., New York (1978).
38. G. Theoclitus, Heat transfer and flow friction characteristics of nine pin-fin surfaces, Trans. ASME, Journal of Heat Transfer, Vol. 88, Series C, 383-390 (1966).
39. T. Rosenman, S.K. Momoh, and J.M. Pundyk, Heat transfer and pressure drop characteristics of dry tower extended surfaces. Part I: Heat transfer and pressure drop data; Part II: Data analysis and correlation, Report No. PFR7-100 and PFR7-102, Battelle Memorial Institute, Richland, Washington (1976).
40. R.L. Webb, Air-side heat transfer in finned-tube heat exchangers, Heat Transfer Engineering, Vol. 1, No. 3, 33-49 (1980).
41. W.M. Kays, The heat transfer and flow friction characteristics of six high performance heat transfer surfaces, Trans. ASME, Journal of Engineering for Power, Vol. 82, Series A, 27-34 (1960); also as The heat transfer and flow friction characteristics of a wavy fin, a strip fin and a perforated fin heat transfer, TR No. 39, Dept. Mech. Eng., Stanford Univ., Stanford, CA (1958).
42. P.G. LaHaye, F.J. Neugebauer, and R.K. Sakhuja, A generalized prediction of heat transfer surfaces, Trans. ASME, Journal of Heat Transfer, Vol. 96, Series C, 511-517 (1974).
43. A.R. Wieting, Empirical correlations for heat transfer and flow friction characteristics of rectangular offset-fin plate-fin heat exchangers, Trans. ASME, Journal of Heat Transfer, Vol. 97, Series C, 488-490 (1975).
44. S.J. Kline, Diffusers -- flow phenomena and design, a Technical Report, Mechanical Engineering Department, Stanford University, Stanford, CA (July 1978).
45. D.A. Miller, Internal Flow, British Hydraulic Research Association (1978); available through Air Science Co., PO Box 143, Corning, NY 14830.
46. A.L. London, G. Klopfer and S. Wolf, Oblique-flow headers for heat exchangers, Trans. ASME, Journal of Engineering for Power, Vol. 90, Series A, 271-286 (1968).
47. J.A. Kutchey and H.L. Julien, The measured influence of flow distribution on regenerator performance, SAE Trans. Section 1, 743-752 (1974).
48. J.P. Chiou, The advancement of compact heat exchanger theory considering the effects of longitudinal heat conduction and flow nonuniformity, in Symposium on Compact Heat Exchangers -- History, Technological Advancement and Mechanical Design Problems, edited by R.K. Shah, C.F. McDonald, and C.P. Howard, Book No. G00183, ASME, New York (1980).
49. A.L. London, Laminar flow gas turbine regenerators - the influence of manufacturing tolerances, Trans. ASME, Journal of Engineering for Power, Vol. 92, Series A, pp. 46-56 (1970).
50. R.K. Shah and A.L. London, Effects of nonuniform passages on compact heat exchanger performance, Trans. ASME, Journal of Engineering for Power, Vol. 102, Series A, pp. 653-659 (1980).
51. J.M. Robertson, Review of boiling, condensing and other aspects of two-phase flow in plate-fin heat exchangers, in Symposium on Compact Heat Exchangers-- History, Technological Advancement and Mechanical Design Problems, edited by R.K. Shah, C.F. McDonald, and C.P. Howard, Book No. G00183, ASME, New York (1980).

# Heat Transfer in Concurrent Flow Double Pipe Heat Exchangers

**M. D. MIKHAILOV and B. K. SHISHEDJIEV**
Applied Mathematics Centre
P.O. Box 384, Sofia 1000, Bulgaria

ABSTRACT

An algorithm is described for the calculation of heat transfer coefficients in concurrent flow double pipe heat exchangers if only the velocity distribution is known. The conventional piston and parabolic flow cases as well as the pseudoplastic and Binghamplastic fluid flows are calculated. It is established, that: (1) the normalized Nusselt number for pseudoplastic and Bingham flows is practically identical with the one for Newtonian fluids, and (2) when heat capacity flow rate ratio increases, the fully developed effectiveness coefficient may either increase or decrease, passing through a maximum.

NOMENCLATURE

$a_i$ = width or radius of channel i
$b$ = wall thickness
$c_i$ = specific heat of fluid i
$h_i$ = heat transfer coefficient for channel i
$k_i$ = thermal conductivity of fluid i
$k_w$ = thermal conductivity of circular tube wall
$r$ = radial distance from center of circular tube
$T_i$ = local temperature of fluid i
$T_{io}$ = inlet temperature of fluid i
$u_i$ = local velocity of fluid i
$\bar{u}_i$ = average velocity of fluid i
$x$ = distance normal to heat transfer surface of annular space measured from insulated wall

$\alpha_i$      thermal diffusivity of fluid i

$\rho_i$      density of fluid i

$U_i$      $u_i/\bar{u}_i$      velocity of fluid i

H      $(1+n)(1+b/a_1)^n(c_2\,\rho_2\bar{u}_2 a_2/c_1\,\rho_1\bar{u}_1 a_1)$      heat capacity flow rate ratio

n      0 or 1 for parallel plate channel or circular tube respectively

K      $(k_1/k_2)(a_2/a_1)(1+b/a_1)^{-n}$      relative thermal resistance of fluid

$Nu_i$      $2a_i h_i/k_i$      Nusselt number for channel i

X      $x/a_2$, $R = r/a_1$, $Z = z\,\alpha_1/\bar{u}_1 a_1^2$, $\theta_i = (T_i - T_{io})/(T_{2o} - T_{1o})$

$K_w$      $(k_1/k_w)\int_j^{1+b/a_1} R^{-n}\,dR$ = relative thermal resistance of wall

## INTRODUCTION

In [1,2] Stein presented an introductory analytical investigation of heat transfer coefficients in concurrent flow double pipe heat exchangers. His analysis is based on a two-region Sturm-Liouville system consisting of two equations coupled at a common boundary. The solutions of this system form an infinite sequence of eigenfunctions with corresponding eigenvalues.

If the velocity distributions are assumed to be uniform, eigenfunctions are the familiar tabulated functions and eigenvalues are given by the positive nonzero roots of an eigenvalue transcendental equation. These plug flow models of the heat exchanging fluids were utilized in [1].

A perturbetion method is used for the laminar flow case in [2]. The approximations for eigenfunctions are presented as a linear sum of plug flow eigenfunctions. The coefficients multiplying each term in the sum are determined so that the Sturm-Liouville equations are approximately satisfied. In [2] it was discussed, that the convergence was not sufficiently rapid or uniform. In most cases the approximations exhibited oscillatory behaviour with the increase of the terms of the sum, making it difficult to judge the accuracy of the result. For this reason the range of parameters in [2] was limited.

In [3], one of the authors presents a general solution of the diffusion equations coupled through general boundary conditions, including Stein's solutions, in the references mentioned [1,2], as a very special case. For numerical evaluation of the one-dimensional case of this general solution the authors developed an algorithm.

The present contribution uses this algorithm for computing the fully developed Nusselt numbers and effectiveness coefficients for double pipe hear exchangers. Stein's and the author's results are compared and discussed. Results are also given for non-Newtonian fluids flowing in the central tube.

## FORMULATION OF THE PROBLEM

The usual double pipe heat exchanger consists of two channels (parallel plane channels or concentric circular pipes) separated by a common wall with fluid flowing through the two channels, as illustrated in Fig. 1.

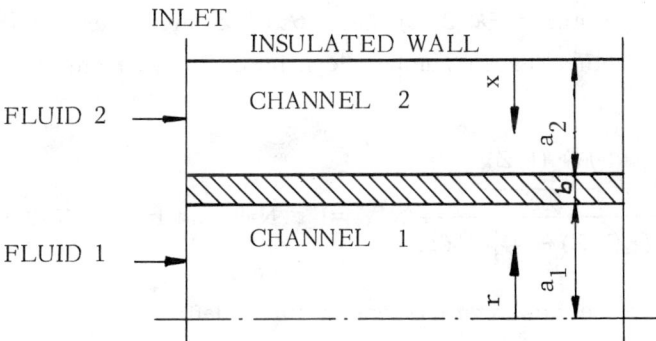

Fig. 1. Double pipe heat exchanger geometry

Stein [1,2] accepts the following idealizations: (1) At the inlet to the duct the temperature distribution within the fluids is uniform, (2) Physical properties are independent from temperature, (3) Viscous dissipation is negligible, (4) Longitudinal heat conduction in the heat exchanger wall is negligible, (5) Longitudinal heat conduction in the fluids is negligible, (6) The velocity distribution within the fluids corresponds to a fully developed laminar flow, and (7) The annular space is approximated by the space between infinitely wide parallel planes.

With the above simplifications, the problem is defined as follows:

$$R^n U_1(R) \frac{\partial \theta_1(R,Z)}{\partial Z} = \frac{\partial}{\partial R} \left( R^n \frac{\partial \theta_1(R,Z)}{\partial R} \right) \qquad (1\ a)$$

$$\frac{KH}{n+1} U_2(X) \frac{\partial \theta_2(X,Z)}{\partial Z} = \frac{\partial^2 \theta_2(X,Z)}{\partial X^2} \tag{1 b}$$

$$\theta_1(R, 0) = 0 \quad , \quad \theta_2(X, 0) = 1 \tag{1c, d}$$

$$\frac{\partial \theta_1(0, Z)}{\partial R} = 0 \quad , \quad \frac{\partial \theta_2(0, Z)}{\partial X} = 0 \tag{1e, f}$$

$$K \frac{\partial \theta_1(1, Z)}{\partial R} + \frac{\partial \theta_2(1, Z)}{\partial X} = 0 \tag{1 g}$$

$$K_w \frac{\partial \theta_1(1, Z)}{\partial R} + \theta_1(1, Z) - \theta_2(1, Z) = 0 \tag{1 h}$$

Once the temperatures $\theta_1(R, Z)$ and $\theta_2(X, Z)$ are determined from the problem (1), the Nusselt numbers can be evaluated from the equations

$$Nu_i(Z) = \frac{2 \frac{\partial \theta_i(1, Z)}{\partial N}}{\theta_i(1, Z) - \theta_{i, av}(Z)} \quad , \quad i=1, \; N=R \text{ and } i=2, \; N=X \tag{2 a, b}$$

where $\theta_{i, av}(Z)$ are the flow average temperatures defined as

$$\theta_{1, av}(Z) = (n + 1) \int_0^1 R^n U_1(R) \, \theta_1(R, Z) \, dR \tag{3 a}$$

$$\theta_{2, av}(Z) = \int_0^1 U_2(X) \, \theta_2(X, Z) \, dX \tag{3 b}$$

The overall heat transfer coefficient is also defined in the usual manner as [2]

$$Nu_1^o(Z) = \frac{2 \frac{\partial \theta_1(1, Z)}{\partial R}}{\theta_{2, av} - \theta_{1, av}} \tag{4 a}$$

It is a relatively simple mater to show that [2]

$$1/Nu_1^o = 1/Nu_1 + K_w/2 + K/Nu_2 \tag{4 b}$$

Stein defined in [2] the heat exchanger effectiveness as the ratio of the actual rate of heat transfer between fluids for a particular heat exchanger to the rate of heat transfer for a similar exchanger with an infinite heat transfer area

$$E(Z) = \theta_{1,av}(Z)\,(1+H)/H \tag{5}$$

The dimensionless temperature distributions and, consequently, Nusselt numbers and heat exchanger effectiveness depend on the velocity distributions within the fluids and the three dimensionless groups H, K and $K_w$.

## SOLUTION OF THE PROBLEM

The formulated problem (1) is a very special case of the general problem treated in [3,4]. To solve eqs. (1) it is necessary, at first, to obtain the solution of the two-region boundary-value problem stated by [4]:

$$y'_{1,i}(R) = -\mu_i^2\, R^n\, U_1(R)\, y_{2,i}(R) \tag{6 a}$$

$$y'_{2,i}(R) = R^{-n}\, y_{1,i}(R) \tag{6 b}$$

$$y'_{3,i}(R) = -\mu_i\, R^n\, U_1(R)\left\{\mu_i\, y_{4,i}(R) + 2\, y_{2,i}(R)\right\} \tag{6 c}$$

$$y'_{4,i}(R) = R^{-n}\, y_{3,i}(R) \tag{6 d}$$

and

$$y'_{5,i}(X) = -\mu_i^2\, \frac{KH}{n+1}\, U_2(X)\, y_{6,i}(X) \tag{6 e}$$

$$y'_{6,i}(X) = y_{5,i}(X) \tag{6 f}$$

$$y'_{7,i}(X) = -\mu_i\, \frac{KH}{n+1}\, U_2(X)\left\{\mu_i\, y_{8,i}(X) + 2\, y_{6,i}(X)\right\} \tag{6 g}$$

$$y'_{8,i}(X) = y_{7,i}(X) \tag{6 h}$$

with the boundary conditions

$$y_{1,i}(0) = y_{3,i}(0) = y_{4,i}(0) = 0 \quad , \quad y_{2,i}(0) = 1 \tag{7 a}$$

$$y_{5,i}(0) = y_{7,i}(0) = y_{8,i}(0) = 0 \quad , \quad y_{6,i}(0) = 1 \tag{7 b}$$

and

$$y_{2,i}(1)\, y_{5,i}(1) + K\, y_{1,i}(1)\, y_{6,i}(1) + K_w\, y_{1,i}(1)\, y_{5,i}(1) = 0 \tag{8}$$

Solution for dimensionless temperature distribution can be written as

$$\theta_1(R,Z) = H/(1+H) + \sum_{i=1}^{\infty} A_i\, y_{5,i}(1)\, y_{2,i}(R)\, \exp(-\mu_i^2\, Z) \qquad (9\text{ a})$$

$$\theta_2(X,Z) = H/(1+H) - K \sum_{i=1}^{\infty} A_i\, y_{1,i}(1)\, y_{6,i}(X)\, \exp(-\mu_i^2\, Z) \qquad (9\text{ b})$$

where

$$A_i = \frac{2}{\mu_i} \cdot \frac{y_{1,i}(1)\, y_{5,i}(1)}{y_{5,i}^2(1)\begin{vmatrix} y_{4,i}(1) & y_{3,i}(1) \\ y_{2,i}(1) & y_{1,i}(1) \end{vmatrix} + K\, y_{1,i}^2(1) \begin{vmatrix} y_{8,i}(1) & y_{7,i}(1) \\ y_{6,i}(1) & y_{5,i}(1) \end{vmatrix}} \qquad (9\text{ c})$$

Application of eqs. (9) to the defining relation for the dimensionless average temperature gives the expressions

$$\theta_{1,av}(Z) = H/(1+H) - \sum_{i=1}^{\infty} B_i\, \exp(-\mu_i^2\, Z) \qquad (10\text{ a})$$

$$\theta_{2,av}(Z) = H/(1+H) - (n+1)/H \sum_{i=1}^{\infty} B_i\, \exp(-\mu_i^2\, Z) \qquad (10\text{ b})$$

where

$$B_i = (1/\mu_i^2)\, A_i\, y_{1,i}(1)\, y_{5,i}(1) \qquad (10\text{ c})$$

For sufficiently large values of $Z$ all but the $i=1$ terms of eqs. (9) and (10) can be neglected. Application of these large $Z$ solutions gives the following fully developed heat transfer coefficients:

$$Nu_1(\infty) = \frac{2\mu_1^2\, y_{1,1}(1)}{2\, y_{1,1}(1) + \mu_1^2\, y_{2,1}(1)} \qquad (11\text{ a})$$

$$Nu_2(\infty) = \frac{2\mu_1^2\, y_{5,i}(1)}{y_{5,1}(1) + \mu_1^2\, y_{6,1}(1)\, K\, H/(n+1)} \qquad (11\text{ b})$$

$$Nu_1^o(\infty) = \mu_1^2\, H/(1+H) \qquad (11\text{ c})$$

Introducing eq. (9a) into eq. (5), after multiplying and dividing by quantities related to the first eigenvalue, and using eq. (11 c) we get

$$E(Z) = 1 - \emptyset(Z) \exp/-(\frac{1+H}{H} Nu_1^o(\infty))Z/ \qquad (12\,a)$$

where

$$\emptyset(Z) = 2\frac{1+H}{H} \sum_{i=1}^{\infty} B_i \exp/-(\mu_i^2 - \mu_1^2)Z/ \qquad (12\,b)$$

For sufficiently large Z, eq. (12 b) gives the following asymptotic values

$$\emptyset(\infty) = 2\frac{1+H}{H} B_1 \qquad (12\,c)$$

called in [2] a fully developed effectiveness coefficient.

For the plug flow case, $U_1(R) = 1$ and $U_2(X) = 1$. For the laminar flow case $U_1(R) = (1-R^2)(n+3)/2$ and $U_2(X) = (1-X)X\,6$. For this two special cases the solutions presented here give Stein's results [1,2].

Except these two cases, we considered also certain classes of non-Newtonian fluids flowing in a central circular tube (i. e. n=1):

(1) pseudoplastic fluids [5] with

$$U_1(R) = (1-R^{m+1})(m+3)/(m+1) \qquad (13)$$

and (2) Bingham-plastic fluids [5] with

$$U_1(R) = 2(1 - ((R-R_o)/(1-R_o))^2)(1-R_o)^2(1-R_o^4 - \tfrac{4}{3}R_o(1-R_o^3))^{-1} \qquad (14\,a)$$

for $R_o \leq R \leq 1$ and

$$U_1(R) = (1-R_o)^2(1-R_o^4 - \tfrac{4}{3}R_o(1-R_o^3))^{-1} \qquad (14\,b)$$

for $0 \leq R \leq R_o$.

Note that equations (13) and (14) for m=1 and $R_o=0$ lead to a laminar flow case.

## ALGORITHM FOR SOLVING THE PROBLEM

The working scheme for the digital computation is as follows [4]:

(1). Calculate the first eigenvalue $\mu_1^*$ by numerical integration (Runge-Kutta method) of eqs. (6 a, b, c, d) at the boundary conditions (7a), so that with the Newton's iterative method equation $y_{11}(1) = 0$ may be identically satisfied. The calculation starts with the initial approximation

$$\mu_1^* = \frac{4\pi}{3\int_0^1 \sqrt{U_1(R)}\, dR}$$

(2). Similarly calculate the eigenvalue $\mu_1^{**}$, solving numerically eqs. (6 e, f, g, h) at the boundary conditions (7 b) so that equation $y_{5,1}(1)=0$ will be identically satisfied. The calculation begins with

$$\mu_1^{**} = \frac{13\pi}{12\frac{KH}{n+1}\int_0^1 \sqrt{U_2(X)}\, dX}$$

(3). Determine the interval where the system (6) eigenvalue $\mu_1$ should be found. It is closed between zero and the lesser of the above obtained values for $\mu_1^*$ and $\mu_1^{**}$.

(4). The eigenvalue and eigenfunctions are determined by a direct numerical solution of eqs. (6) at the boundary conditions (7) so that eq. (8) is identically satisfied.

(5). Using formulas (11) and (12), the fully developed Nusselt numbers and the effectiveness coefficient are calculated.

(6). Calculate the normalized Nusselt [2] number by dividing the tube side Nusselt number $Nu_1(\infty)$ by the uniform flux value of

$$Nu_{uf} = (2\int_0^1 R^{-1}(\int_0^R R\, U_1(R)\, dR)^2\, dR)^{-1} \tag{15}$$

DISCUSSION OF THE NUMERICAL RESULTS

Using the above algorithm the computer program COUPFLOW [4] was used and calculations were made of $\mu_1$, $Nu_1(\infty)$, $Nu_1(\infty)/Nu_{uf}$, $Nu_2(\infty)$, $Nu_1^o(\infty)$ and $\varphi(\infty)$ for plug flow, Newton's laminar flow (m=1), pseudoplastic fluid flow (m=2 and 3) and Bingham-plastic fluid flow ($R_o$ from 0. to 0.5). The range of the parameters explored include the following values: H from 0.1 to 2; K from 0.1 to 2, and $K_w$ from 0 to 1.

The numerical results obtained by us with a higher accuracy compared with Stein's results, showed that the values in [2] have the declared degree of accuracy. Therefore we are showing in nomograms only a part of our results leading to important conclusions.

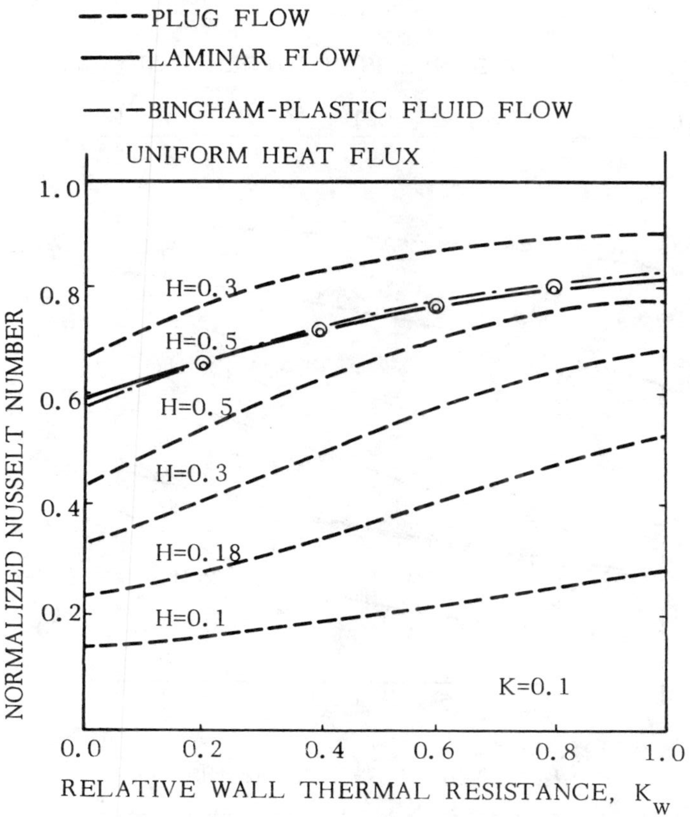

- - - PLUG FLOW
——— LAMINAR FLOW
—··— BINGHAM-PLASTIC FLUID FLOW

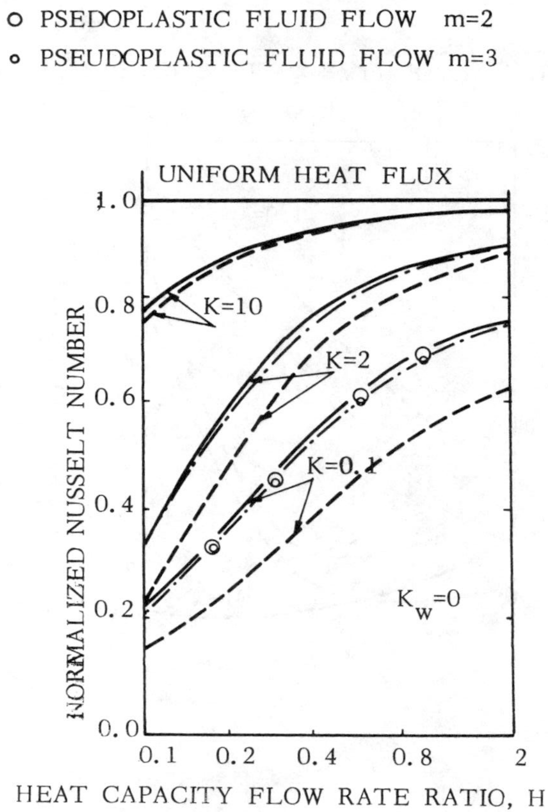

○ PSEDOPLASTIC FLUID FLOW  m=2
○ PSEUDOPLASTIC FLUID FLOW m=3

Fig. 2.

Fig. 3

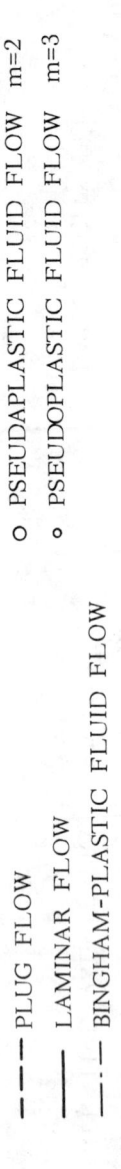

—— PLUG FLOW  
——— LAMINAR FLOW  
—··— BINGHAM-PLASTIC FLUID FLOW  

○ PSEUDAPLASTIC FLUID FLOW  m=2  
∘ PSEUDOPLASTIC FLUID FLOW  m=3  

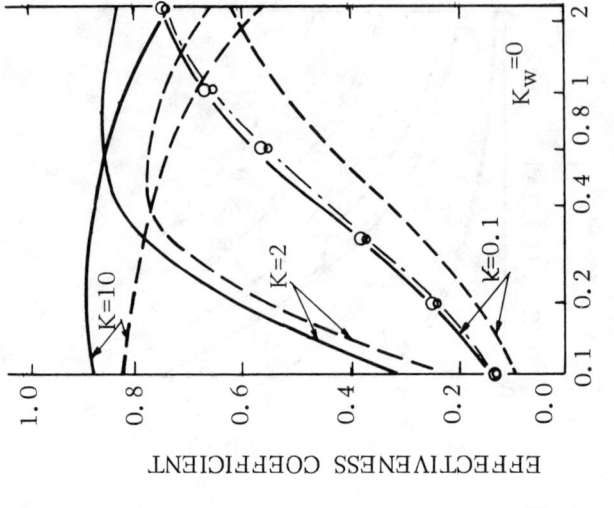

Fig. 5  
EFFECTIVENESS COEFFICIENT vs HEAT CAPACITY FLOW RATE RATIO, H

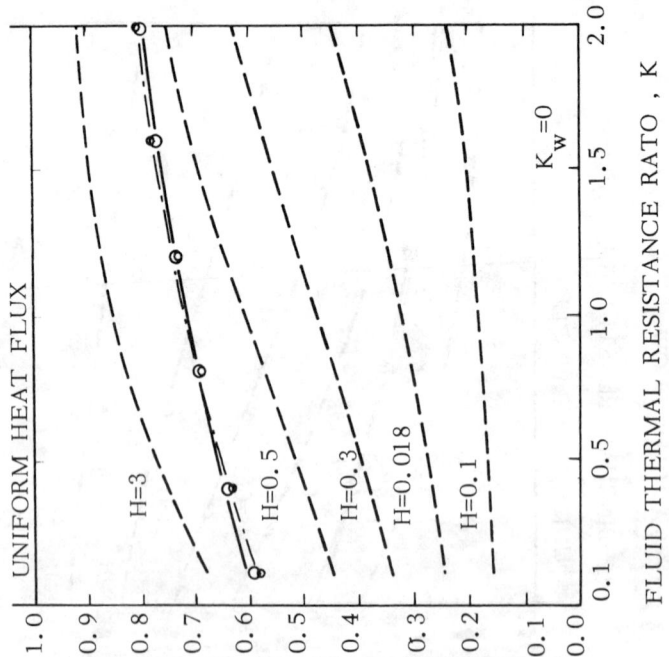

Fig. 4.  
NORMALIZED NUSSELT NUMBER vs FLUID THERMAL RESISTANCE RATIO, K

Figures 2, 3 and 4 show graphs of the normalized Nusselt number $Nu_1(\infty)/Nu_{uf}$ as a function of the parameters H, K and $K_w$ for parabolic, pseudoplastic, and Bingham-plastic fluid flow in central circular tube by Newtonian laminar flow through the annular space. Also shown on the graphs are the results for the plug flow case.

It is of great interest to note that the ratio $Nu_1(\infty)/Nu_{uf}$ for pseudoplastic and Bingham-plastic flows is very near to the ratio $Nu_1(\infty)/Nu_{uf}$ for a Newtonian fluid. This is valid only for the normalized Nusselt number and is explained with the change of $Nu_1(\infty)$ being approximately proportional to that of $Nu_{uf}$ as defined by eq. (15). The latter is easily calculated, so that it is sufficient to have the data for the parabolic flow case in order to evaluate $Nu_1(\infty)$ for a non-Newtonian fluid.

Fig. 5 shows graphs of the fully developed effectiveness coefficient $\emptyset(\infty)$ as a function of the parameters H and K for $K_w=0$. The graphs show explicity, that when H increases, $\emptyset(\infty)$ may either increase or decrease, passing through a maximum, that could not be observed by Stein, as his method does not provide the possibility to calculate $\emptyset(\infty)$ for a larger interval of parameters H and K.

REFERENCES

1. R. P. Stein 1965. Heat transfer coefficient in liquid metal concurrent flow double pipe heat exchangers. Chem. Engng. Progr. Symp. Ser. 59: 64-75.
2. Stein, R. P. The Graetz problem in concurrent flow double pipe heat exchangers, Chem. Engng. Progr. Symp. Ser. 59 :78-87 (1965).
3. Mikhailov, M. D. 1973. General solutions of the diffusion equations coupled at boundary conditions. Int. J. Heat Mass Transfer 16: 2155-2164.
4. Mikhailov, M. D. and Shishedjiev, B. K. 1976. Coupled at boundary mass or heat transfer in entrance concurrent flow. Int. J. Heat Mass Transfer 19: 553-557.
5. Wilkinson, L. W. 1964. Non-Newtonian Liquids (Russian translation)

# Plate Heat Exchangers

**KENNETH J. BELL**
School of Chemical Engineering
Oklahoma State University
Stillwater, Oklahoma USA

ABSTRACT

The three types of plate heat exchangers are briefly surveyed and their major construction features described. The plate and frame type is discussed in more detail and several applications are presented.

1. INTRODUCTION

Under the general term, "plate heat exchangers", at least three major heat exchanger configurations are commonly understood: the plate and frame or gasketed plate exchanger (which is the most important one and generally the one referred to when one speaks only of plate heat exchangers), the spiral plate heat exchanger, and the lamella heat exchanger. The common factor among these three configurations is that the heat transfer surface is composed of parallel metal plates kept separated by surface embossing or spacer tabs and the heat exchanging fluids flow in the thin channels thus formed.

This section starts with a brief consideration of the three basic types and identifies their major features and chief areas of application. There follows a more detailed examination of the particular features of the plate and frame heat exchanger.

2. TYPES AND APPLICATIONS OF PLATE HEAT EXCHANGERS

2.1 Plate and Frame Heat Exchangers

A typical plate and frame heat exchanger is shown in an exploded view in Figure 1. The heat transfer surface is composed of a series of plates with ports for fluid entry and exit in the four corners. Fluid flow into a given channel is controlled by the presence or absence of gaskets around the port for each plate. The plates are embossed in a herringbone or rectangular pat-

---

Many of the figures used in this lecture are taken from the technical literature of Alfa Laval, Lund, Sweden, and are used with their permission. The author greatly appreciates this assistance.

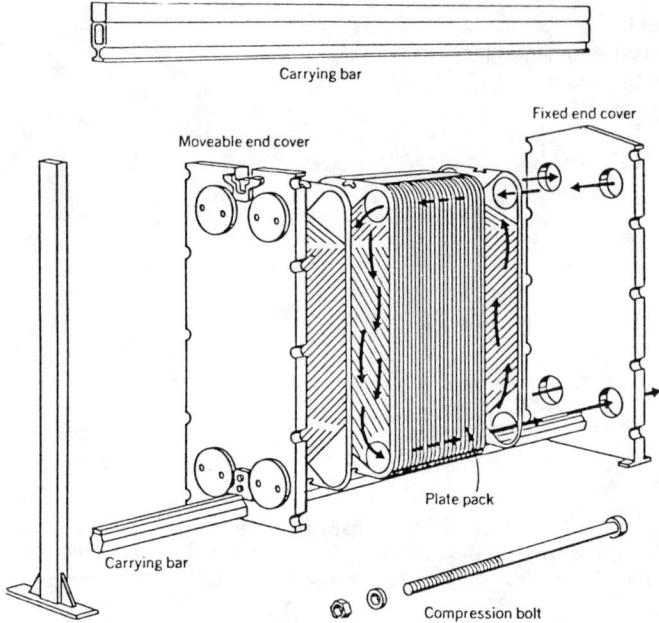

Figure 1. Typical Plate and Frame Heat Exchanger (Exploded View). The plate pack is clamped together in the frame between the frame and pressure plate by means of lateral bolts, hanging in the upper carrying bar and guided by the lower. (Courtesy Alfa-Laval AB, Sweden.)

tern in such a way as to provide a multiplicity of support points against collapse of the channel by high pressures on the opposite side of the bounding plates. The plates can be stacked up to several hundred in a frame and held together by the bolts which hold the stack in compression. Leakage from the channels between the plates to the surrounding atmosphere is prevented by the gasketing around the exterior of the plate.

2.2  Sprial Plate Heat Exchanger

Typical spiral plate heat exchanger configurations are shown in Figure 2. In this case, the plates are kept separated by embossed or welded-on tabs on one of the plates. The plates are arranged in a spiral form to give a generally circular outline of the external configuration. Ports admit fluid to one or the other of the two spiral channels. The cover plates prevent fluid from leaking out of either channel into the external atmosphere, or short-circuiting of the fluid from one spiral to the next. It is possible to leave off one of the plate covers to admit, for example, vapors into the exposed heat transfer surface; heat exchangers with such an arrangement are used for internal column reflux, thereby avoiding the necessity to take the vapors outside the column for condensation.

# PLATE HEAT EXCHANGERS

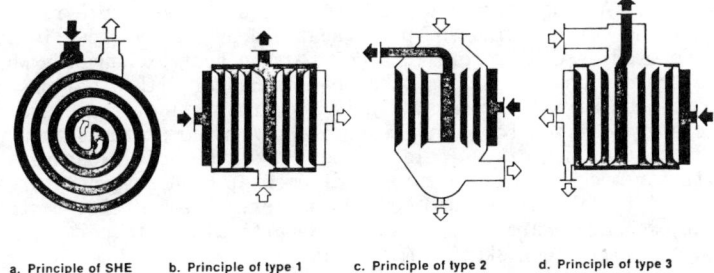

a. Principle of SHE    b. Principle of type 1    c. Principle of type 2    d. Principle of type 3

Figure 2. Typical Spiral Plate Heat Exchangers (Courtesy Alfa-Laval AB, Sweden.)

## 2.3 The Lamella Heat Exchanger

The Ramen lamella heat exchanger is diagrammed in Figure 3. It consists essentially of a series of metal plates roll-formed in such a way that there are periodic ribs (which can be welded together if desired) that mutually support the surface against collapse from external pressure. One fluid flows through the channels formed in these plates and the other fluid flows on the shell side of the exchanger, on the outside of the plates. A packing gland is used at the bottom of the plate array to allow sealing of the plate-side fluid and yet permit the removal of the entire plate bundle from the top of the heat exchanger when the split flanges are removed at the bottom.

## 2.4 Areas of Application for Plate Heat Exchangers

The plate and frame heat exchanger and the spiral heat exchanger are especially limited by gasket material characteristics to relatively low temperatures and pressures. The lamella heat exchanger is able to work over a substantially larger range, but eventually the packing gland limits the maximum

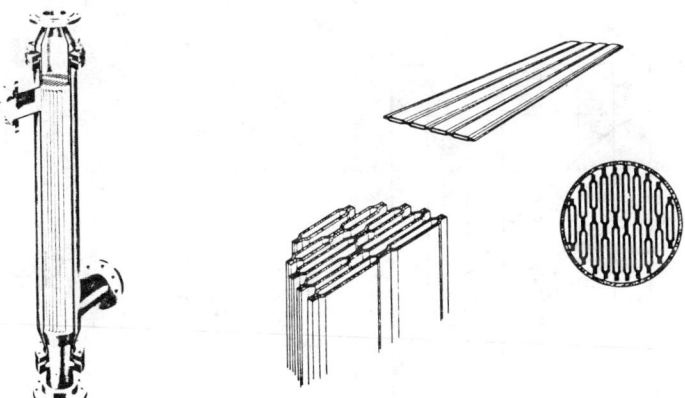

Figure 3. Ramen Lamella Heat Exchanger Details (Courtesy Alfa-Laval AB, Sweden.)

pressure possible. A rough comparison of the pressure temperature ranges of application for the various kinds of plate heat exchangers is shown in Figure 4. The line denoted as CTHE refers to a "close tube heat exchanger", which is essentially a conventional shell and tube heat exchanger with a bellows arrangement to absorb thermal stresses. It is seen that all of the plate type heat exchangers are limited to lower pressures and/or temperatures than typical shell and tube heat exchangers. In fact, shell and tube heat exchangers are commonly built for much higher pressures and somewhat higher temperatures than shown in Figure 4 for the close tube heat exchanger. The limiting factors for all plate heat exchangers are the pressure and particularly temperature limitations upon available gasketing materials.

However, within the pressure-temperature limits alluded to above, the plate heat exchanger types represent generally extremely efficient heat transfer surface (on a volumetric basis). Heat transfer coefficients tend to be quite high, due to the repeated expansion and contraction of the flow through the embossed plates of the plate and frame type, or due to the eddying secondary flow of the spiral heat exchanger. Also, pressure drops are correspondingly rather high, compared to shell and tube type exchangers.

The heat transfer surface for plate and frame type exchangers can be readily formed out of a wide variety of alloy materials. Since the plates themselves are thin, the plate exchangers give relatively very high heat transfer coefficients for the mass of alloy material required. Therefore, if it is necessary to use alloy metals for heat transfer for corrosion protection, they can compete very well economically with more conventional heat exchanger designs, especially shell and tube exchangers. Plate heat exchangers are seldom, if ever, made out of low carbon steel.

Some idea of the comparative heat transfer and pressure drop characteristics of plate heat exchangers, compared to flow inside tubes, can be gained by examining Figures 5 and 6. Figure 5 shows the j factors as a function as Reynolds number for several different plate heat exchanger configurations as compared to the corresponding values for flow inside tubes. It is seen that at a given Reynolds number, the value for the plate exchanger can be up to 10 times that for the flow inside tubes. Figure 6 shows the friction factor as a function of Reynolds number for similar cases. Once again, it is seen that

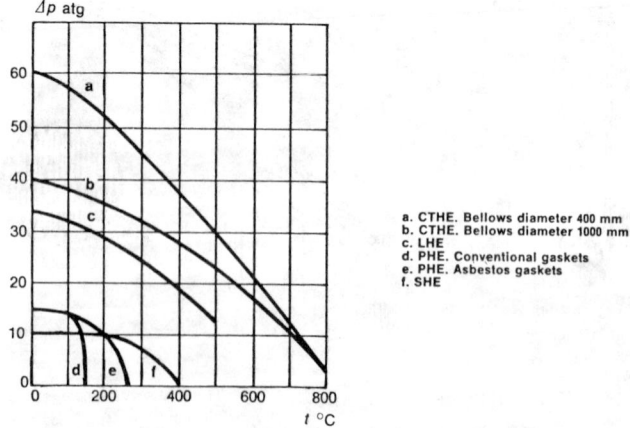

Figure 4. Typical Pressure-Temperature Ranges for Heat Exchanger Application (Courtesy Alfa-Laval AB, Sweden.)

# PLATE HEAT EXCHANGERS

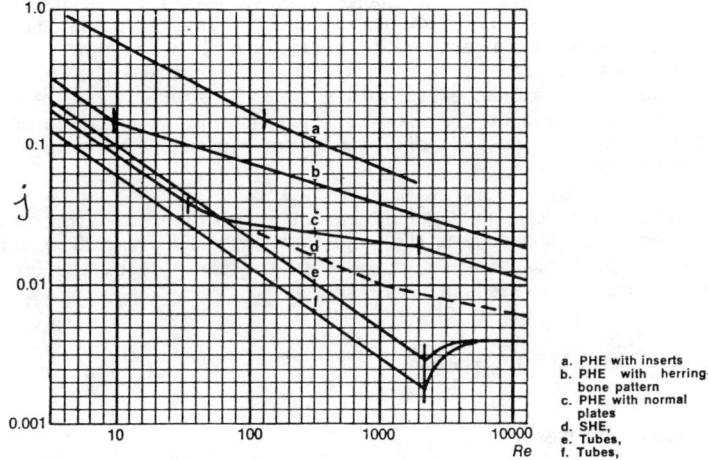

Figure 5. j Factors for Several Heat Transfer Surfaces (Courtesy Alfa-Laval AB, Sweden.)

a. PHE with inserts
b. PHE with herringbone pattern
c. PHE with normal plates
d. SHE,
e. Tubes,
f. Tubes,

Defining equations:

$$Re = \frac{d_h \rho V}{\mu}$$

$$d_h = \frac{4 \text{ (cross-sectional area for flow)}}{\text{wetted perimeter}}$$

$$j = \left(\frac{h}{C_p \rho V}\right) \left(\frac{C_p \mu}{k}\right)^{2/3}$$

$$f = \frac{2 \Delta p d_h}{4 \rho V^2 L}$$

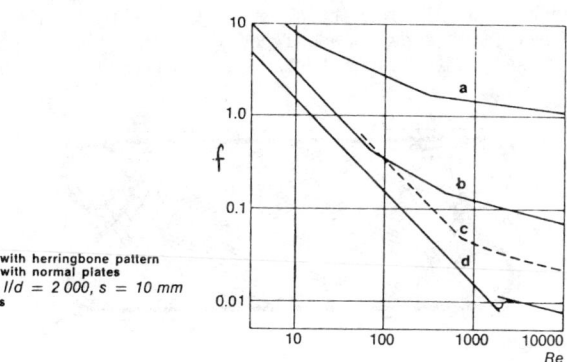

a. PHE with herringbone pattern
b. PHE with normal plates
c. SHE; $l/d = 2\,000, s = 10$ mm
d. Tubes

Figure 6. f Factors for Several Heat Transfer Surfaces (Courtesy Alfa-Laval AB, Sweden.)

the pressure drop for plate heat exchangers is substantially higher (by a factor up to 20) than that for flow inside tubes. This supports the statement above that plate heat exchangers are essentially high heat transfer, high pressure drop devices. It is, of course, possible to lower the velocity through the plate heat exchanger to obtain lower pressure drops, resulting in a large decrease in the heat transfer coefficient that is obtained and a corresponding loss of volumetric efficiency and economic advantage of the plate heat exchanger.

Most plate heat exchangers permit easier inspection, cleaning and repair procedures than shell and tube units, and the plate and frame type allows easy expansion of the heat transfer surface (up to the capacity of the frame). These features are summarized in Table I.

## 3. PLATE AND FRAME HEAT EXCHANGERS

### 3.1 Mechanical Features

The plates may be constructed of any material that can be pressed to the desired shape. Common materials include types 304, 316, 317 and 18/26/6 stainless steels, titanium, titanium with 0.2% palladium, Incoloy 825, Inconel 625, Hastelloy C-276 and B, Carpenter 20, monel 400, cupronickels, and aluminum brass. Plates have also been pressed on a nonstandard basis from silver, copper and tantalum. A typical plate design is shown in Figure 7.

Gasket materials commonly specified include nitryl and butyl rubbers, Hypalon, Viton, Teflon, and Fluon. Compressed asbestos gaskets are used for higher temperatures and pressures. A gasket cement is used with the organic gaskets which allows the gasket to be removed during stripping of the plate exchanger and reused several times if care is taken in the operation.

The frames are made out of carbon steel and the bolts are made out of high tensile steel. Plate heat exchangers can be manufactured to the usual pressure vessel codes, e.g., ASME. Connections are made out of the same material as the plates or out of rubber-lined carbon steel.

The maximum allowable pressure depends upon the plate corrugation pattern, material thickness, material strength, the gasket system and the strength of the frame and bolts. All industrial types will withstand a design pressure of 0.7 MPa (100 psig), most can be designed to 1 MPa (150 psig), some can go to 1.6 MPA (225 pisg), a few can withstand 2.0 MPa (300 psig) and one or two are

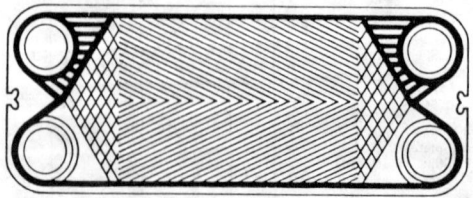

Figure 7. Typical Herringbone Plate Design  (Courtesy Alfa-Laval AB, Sweden.)

TABLE I

INSPECTION, CLEANING, REPAIR AND EXTENSION
OF VARIOUS HEAT EXCHANGER TYPES

| Requirement | | Heat Exchanger Type | | | |
|---|---|---|---|---|---|
| | | Plate Heat Exchanger | Spiral Plate Heat Exchanger | Lamella Heat Exchanger | Close Tube Heat Exchanger | Shell and Tube |
| Inspection for Fouling: | one side | a | a | a | a | a |
| | both sides | a | a or d | c | c | b or d |
| Inspection for Leakage: | one side | b | a | a | a | a |
| | both sides | b | a | a | a | a |
| Inspection for Corrosion: | one side | a | a or c | b | a or b | a or b |
| | both sides | a | b or c | d | c | a or d |
| Chemical Cleaning: | one side | a | a | a | a | a |
| | both sides | a | a | a | a | b or c |
| Manual Cleaning: | one side | a | a or c | b | a | a |
| | both sides | a | a or c | d | d | a or d |
| Extension | | a | d | d | d | d |
| Repair | | a | c | c | b | b |

a, very good; b, acceptable for most types; c, poor; d, impossible

acceptable up to 2.5 MPa (360 psig). All of these will withstand a test pressure of 1.5 times the design pressure. If ASME VIII requirements are to be met, the maximum design pressure is probably 1.6 MPa (225 psig).

As of July 1980, the largest plate manufactured measured 3.63 square meters (39 square feet) of heat transfer surface. Up to 599 plates can be put into a single frame, giving the maximum area per frame of 2,174 square meters or 23,400 square feet. Plates will probably not go much beyond about 4 square meters per plate because larger plates are difficult to handle. If larger heat transfer areas are required, multiple frames in series or in parallel are specified. The smallest plate available is 0.032 square meters or 0.34 square feet.

The maximum connection size presently available is 400 millimeter diameter or 16 inches. At 6 meters per second (20 feet per second), the maximum liquid flow rate is thus about 0.75 cubic meters per second (or about 12,000 U.S. gallon gpm). For vapor or gas flow, the maximum velocity is dependent upon the density. Approximately $u_{max} = 60\rho^{-1/3}$ where u is in meters per second and $\rho$ is in kilograms per cubic meter.

### 3.2 Flow Arrangements

Some typical flow arrangements are shown in Figure 8. Plate exchangers are very flexible in the way in which channels may be arranged. A blank port on a plate cause the fluid to flow through that channel rather than pass

A. A series-parallel arrangement

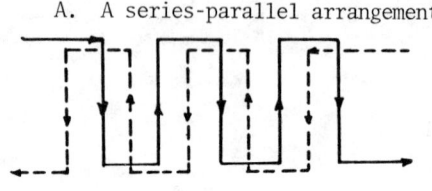

B. All channels in parallel

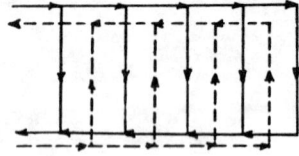

C. All channels in series

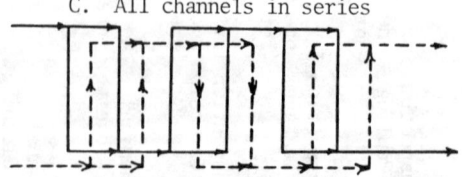

Figure 8. Typical Flow Arrangements

through that port and enter into the next channel. Among the common flow arrangements, all of the channels may be in series; this arrangement is used when two fluids of relatively low flow rate need to exchange heat over a wide temperature range. The arrangement with all channels in parallel, on the other hand, is usually designed for streams with large flow rates but relatively short ranges of temperature change required. Series-parallel arrangements deal with intermediate cases of flow and temperature range. In cases in which the flow pattern is not purely countercurrent, corrections need to be made to the logarithmic mean temperature difference for the departure from pure countercurrent flow. Different plate configurations can be used in the same frame in order to more nearly match the thermal-hydraulic characteristics of the two streams and to make maximum use of the allowable pressure drops.

3.3 Applications

Typical applications are mainly liquid-to-liquid turbulent flow situations. One example of very great importance, especially in at-sea or coastal locations, is shown in Figure 9. In this case, a freshwater system is used within the plant to remove heat from process streams and equipment such as compressor intercoolers; this stream then gives up its heat in a plate heat exchanger system to cooling water from the surrounding ocean. Corrosion and fouling problems are transferred from the process plant equipment to the plate heat exchanger system. Equipment in the plant can be fabricated out of low carbon steel, while the plate exchangers can be made out of titanium to withstand seawater corrosion. Fouling problems can be ultimately solved by the ability to tear down the plate heat exchanger system for cleaning if necessary. Package cooling systems including pumps, pipework, valves, chlorination and back-flushing systems or filters are available commercially.

Another application is shown in Figure 10 where plate heat exchangers are used for heat recovery in a MEA (monoethanolamine) gas absorption system. In this case, the plate heat exchanger is exchanging heat between the hot lean effluent MEA solution from the stripper going into the absorber and the cold rich MEA solution going from the absorber into the stripper. As much as 90-95 percent heat recovery is possible, considerably reducing the areas of the reboiler and condenser as well as plant service stream requirements.

Plate heat exchangers are used as conventional process heaters and coolers, as well as condensers for relatively high density vapors, such as ammonia or propylene.

Two phase flows are possible in plate exchangers, usually under conditions of a low vapor fraction. Crude streams with small amounts of natural gas, re-boiled vapor-liquid streams for columns, and streams which generate gas during heating are candidates for these kinds of applications.

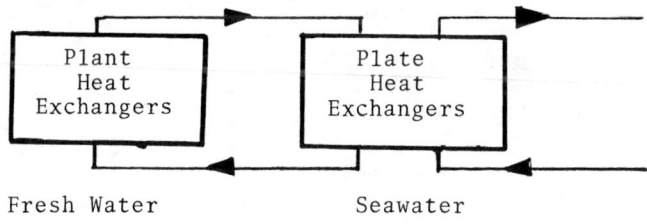

Figure 9. Central Cooling System Using Sea Water as Heat Sink

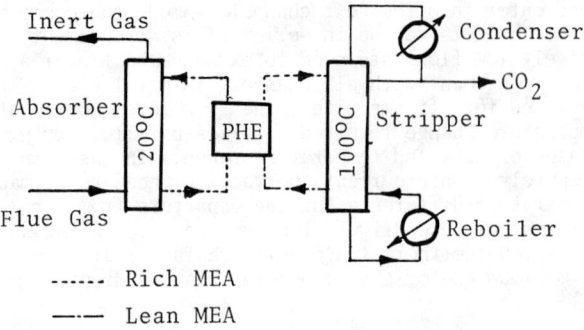

Figure 10. Heat Recovery System for an Acid Gas Absorber

Plate heat exchangers are also used extensively in the food and dairy industries because of the relative ease for disassembling the heat exchanger for cleaning and sterilization to meet health and sanitation requirements.

### 3.4 Heat Transfer and Pressure Drop Characteristics

There are very few published data on specific channel characteristics. All of the major manufacturers have run extensive laboratory tests on their own proprietary plates and have these data available for design. However, these data are considered very proprietary by the companies and are usually available only to major customers for checking designs. Computer design programs are used by the major manufacturers to make optimum use of available pressure drop. Some data are available in reports from the National Engineering Laboratory, East Kilbride, Scotland [2] and implicit data are included in a study of fouling and plate heat exchangers by HTRI [3]. A typical heat transfer result is published by NEL for Alfa-Laval's plate P 22:

$$Nu \cdot Pr^{-1/3} = 0.520 \cdot Re^{0.61} \cdot L(\mu/\mu_w)^{0.14}$$

at $Re > 40$

where $Re = m \cdot d_\mu/(\mu \cdot b \cdot s)$

$m$ = kg/s per channel

$d_h$ = 0.00588 m

$b$ = 0.215 m

$s$ = 0.00294 m

$$f = 10.5 \cdot Re^{-0.33}$$

$$\Delta p = f \cdot 2 \left(\frac{m}{b \cdot s}\right)^2 \cdot \frac{L}{d_h} \cdot \rho^{-1}$$

$L$ = 0.605 m

(Probably not valid at $Re > 500$, where $f = 2.5 \, Re^{-0.1}$ would be more expected)

However, comparison of this correlation with proprietary data from Alfa Laval shows that the NEL correlation for heat transfer seems to be about 10 percent too low.

In general terms, for water at 40°C and at an allowable pressure drop of 100 kilopascals (kpa) per plate, the typical plate heat exchangers will produce a film heat transfer coefficient of about 21,000-25,000 watts w/m$^2$°C. These figures bear out the earlier statement that plate heat exchangers should be regarded predominantly as high pressure drop/high heat transfer devices.

REFERENCES

1. Thermal Handbook, Alfa-Laval AB, Lund, Sweden.

2. National Engineering Laboratory, Reports No. 283, 284, 285, and 286, East Kilbride, Scotland (1967).

3. Cooper, A., Suitor, J. W., and Usher, J. D., "Cooling Water Fouling in Plate Heat Exchangers", Heat Transfer Engineering, 1, No. 3, 50-55 (1980).

# Heat Transfer between Packed, Agitated and Fluidized Beds and Submerged Surfaces

E. U. SCHLUNDER
Institut für Thermische Verfahrenstechnik
University of Karlsruhe
Kaiserstr. 12
D-7500 Karlsruhe 1 / FRG

ABSTRACT

Heat transfer between packed, agitated and fluidized beds and submerged surfaces is treated by a common theoretical concept. The thermal properties appearing in this concept may be predicted a priori. The remaining problem is how to describe the particle motion. It is shown that the penetration model, which is well known from gas liquid systems, also applies to gas-solid systems. The results of the theoretical approach are compared with numerous experimental data obtained from literature.

1. INTRODUCTION

Heat transfer between packed, agitated and fluidized beds and submerged surfaces may be the rate controlling step in numerous processes, such as coal gasification, fluidized bed combustion, contact drying etc. In all these cases heat is to be transferred from submerged surfaces to particle systems either to heat the particles or to cover latent heat consumption due to physical (drying) or chemical (combustion) reactions.

The variety of according equipment design is fairly large. However, certain general statements can be made likewise applying to different types of equipment. These statements concern particular heat transfer fundamentals as well as interdependencies between heat transfer and particle motion. The particular heat transfer fundamentals are fairly clear today. Therefore, the remaining problem in many cases is a reliable description of the particle motion. At present, however, such a description implies the introduction of more or less arbitrary hypotheses and models. Those models are specific for each type of equipment. At this point the general treatment branches into specified treatments fitting to the respective special type of equipment. Also at this point a proper combination of the theoretical approach with empirical knowledge seems to be reasonable in order to achieve useful results applicable to equipment design.

The purpose of this paper is to summarize the particular heat transfer fundamentals, and to point out, which consequences they have as to equipment design and performance. Based upon this

it shall be shown what kind of models are to be introduced in
order to combine the fundamentals with empirical knowledge on
equipment performance. Eventually, fundamentals and models shall
be compared with experimental data obtainable from literature.

## 2. INTERDEPENDENCE OF HEAT TRANSFER AND PARTICLE MOTION

Heat transfer to a bed of particles occures in a sequence
of two steps. In the first step heat is transferred from the
submerged surface onto the surface of the adjacent layer of
particles. The respective transfer coefficient is the "wall-to-
particle" heat transfer coefficient $\alpha_{wp}$. In the second step the
heat is transferred into the bed by heat conduction. Internal
heat conduction, however, does not play any role, as long as the
bed is isothermal.

The bed is always isothermal, if

a) the residence time of the bed at the submerged surface
is sufficiently short,

b) there is only isothermal latent heat consumption in
the bed close to the submerged surface,

c) the bed is perfectly mixed by any kind of mixing device.

In case a) mixing of the bed is definitely insignificant,
in case b) it is practically insignificant. In case c) it is
unimportant whether there is sensible or latent heat consumption
in the bed.

Since in practice the description of the particle motion is
often very difficult, it is of great interest to know under which
circumstances case a) and b) appear in practical applications.
Therefore, it is necessary to quantify, what a "sufficiently
short time" means in the case a). As to case b) it must be
explained, what the distance is which defines the expression
"close to the submerged surface". Eventually, the condition of a
"perfectly mixed bed" is to be defined and the consequences are
to be shown.

All the three cases a), b) and c) have in common, that the
over all heat transfer resistance is only related to the wall-to-
particle heat transfer coefficient $\alpha_{wp}$. Therefore, the prediction
of this coefficient will be treated in the following chapter.

Most practical cases are in between the three limiting
cases a), b) and c); i.e. there is partly sensible and likewise
partly latent heat consumption in the bed during longer residence
times while the mixing of the bed is not at all perfect. In these
cases also heat conduction within the bed becomes important. This
mechanism, however, is strongly affected by the particle motion.
Therefore, the subsequent chapters deal with a suitable description
of these phenomena.

## 3. HEAT TRANSFER BETWEEN A BED OF PARTICLES AND SUBMERGED SURFACES

The fundamental phenomenon of heat transfer between a bed of particles and a submerged surface is the heat conduction through the gaseous gap which separates the particle from the surface. This phenomenon has been analysed by Schlünder /1/ in a semi-rigorous way. According to /1/ the local heat transfer coefficient between a smooth submerged surface and the smooth surface of an adjacent spherical particle is given by

$$\alpha_{WP,local} = \frac{\lambda_g}{s+\sigma} + \alpha_{RD} \qquad (1)$$

where

$$\sigma = \sigma_o \, 2 \, \frac{2-\gamma}{\gamma} \qquad (2)$$

$\lambda_g$ is the heat conductivity of the continuum gas, $\sigma_o$ is the mean free path of the gas molecules, $\gamma$ is the accomodation coefficient and s is the local width of the gaseous gap. The mean free path can be predicted by the formula

$$\sigma = \frac{1}{\sqrt{2} \, n \, \pi \, d^2_{mol} \, Lo} \qquad (3)$$

$$n \cong \frac{P}{\tilde{R}T} \qquad (4)$$

is the molar density of the gas, $d_{mol}$ the molecule diameter and $Lo = 6,02 \cdot 10^{26}$ 1/kmol the Loschmidt number. P is the gas pressure, $\tilde{R}$ the molar gas constant and T the absolute temperature. The accomodation coefficient $\gamma$ may be estimated as given by Martin /2/

$$\lg \left[ \frac{1}{\gamma} - 1 \right] = A' - \left[ \frac{1000 \, K}{T} + B' \right] \frac{1}{C'} \qquad (5)$$

with $A' = 0.6$; $B' = 1.0$ and $C' = 2.8$ for air. For other gases A' and B' are the same, however, C' is 50 for He, 6 for Ne, 3 for Ar, 2.5 for Kr and 2.25 for Xe. Fig. 1 is a graphical representation of eq. 5.

The term $\alpha_{RD}$ represents the heat transfer by radiation and may be predicted by

$$\alpha_{RD} = 0,04 \, C_{12} \left[ \frac{T_m}{100 \, K} \right]^3 \qquad (6)$$

where $C_{12} \cong 5 \, W/m^2 K$. T is the mean temperature between the submerged surface and the particles.

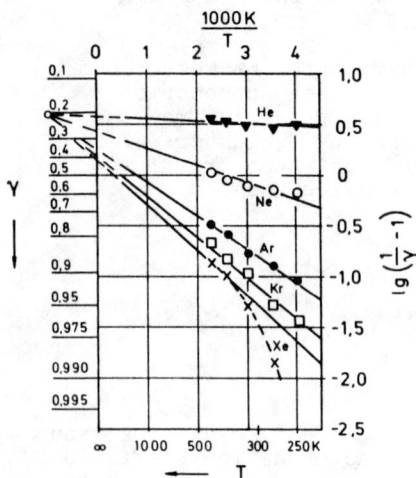

Fig. 1
Accomodation-coefficient for nobel gases contacting Pt-surface acc.Eq.5. Experimental data from Reiter, Camposilvan, Nehren; Wärme- und Stoffübertragung 5 (1972).

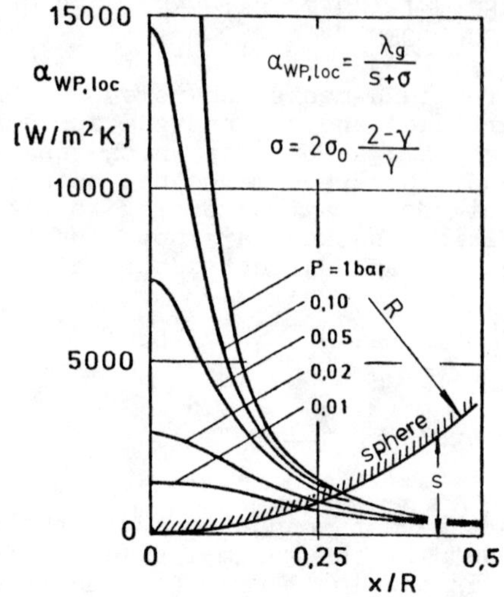

Fig. 2
Local heat transfer coefficient $\alpha_{WP,loc}$ between a smooth wall and a contacting sphere with a smooth surface at various gas pressures P. Sphere-diameter 2R = 1 mm, air at 25°C.
$\sigma_0$ = 0,07 µm at P = 1 bar
$\lambda_g$ = 0,025 W/mK at 25°C
$\gamma$ = 0,90 ; $\alpha_{RD}$ = 5 W/m²K

Fig. 2 shows the local heat transfer coefficient, $\alpha_{WP,loc}$ versus the relative distance from the contact point x/R for a spherical particle at various gas pressures P. The maximum value appears at the contact point and follows from Eq. 1

$$\max \alpha_{WP,loc} = \frac{\lambda_g}{\sigma} + \alpha_{RD} \tag{7}$$

With increasing distance x/R the local heat transfer coefficient drops drastically. By integration one obtains the average coefficient

$$\alpha_{WP} = \frac{1}{\pi R^2} \int_0^R \alpha_{WP,loc} \, 2\pi x \, dx \tag{8}$$

Evaluation of Eq. 8 yields

$$\alpha_{WP} = \frac{2\lambda_g}{R} \left\{ \left(1 + \frac{\sigma}{R}\right) \ln \left(1 + \frac{R}{\sigma}\right) - 1 \right\} + \alpha_{RD} \tag{9}$$

This heat transfer coefficient is related to the projection area of the sphere. With respect to the total area of the submerged surface one obtains

$$\alpha_{WS} = \varphi\, \alpha_{WP} + (1-\varphi)\, \alpha_{RD} \tag{10}$$

where $\varphi$ is in the order of 0.75 to 0.85.

Eq. 9 is based on the assumption that both the particles as well as the submerged bodies have a smooth surface and that there is a pointwise contact. This might not always be true in reality. In case that the particles are partly flattened one would expect higher heat transfer coefficients $\alpha_{WP}$ than predicted by Eq. 9. On the other hand, if the particles and/or the submerged bodies have a rough surface, the heat transfer coefficient $\alpha_{WP}$ should be lower than predicted by Eq. 9. Both deviations have been observed experimentally. However, it may be stated, that these deviations remain within relatively narrow limits and that Eq. 9 applies in many practical cases without any further correction.

## 4. HEAT CONDUCTION WITHIN THE BED

Heat conduction within the bed is insignificant, if the bed is isothermal. This is true, if

a) the residence time is very short

b) there are strong heat sources or sinks

c) the bed is mixed very well

Therefore, heat conduction becomes important, if

α) the residence times are sufficiently long

ß) there is only sensible heat consumption in the bed

γ) the particle motion is zero

Presupposed that the latter conditions hold, the packed bed may be considered as a quasi-homogeneous medium, to which Fourier's theory of heat conduction applies. Then the heat flux $\dot{q}$ is given by Fourier's law

$$\dot{q} = -\lambda_{so}\, \frac{\partial \vartheta}{\partial z} \tag{11}$$

where $\lambda_{so}$ is the apparent heat conductivity of the quasi-homogeneous packed bed. Today, $\lambda_{so}$ may be predicted fairly accurate as shown in the literature /3/ to /12/. For packed beds filled with air under normal pressure $\lambda_{so}$ is roughly 10 to 20 times higher than the heat conductivity of the air.

For a semi-infinite packed bed contacting a plane surface the Fourier theory yields for the **instantaneous** "internal heat transfer coefficient" $\alpha_{so}$ to the expression

$$\alpha_{so} = \frac{1}{\sqrt{\pi}} \sqrt{\lambda_{so}\, \rho_{so}\, c_{so}} \; \frac{1}{\sqrt{t}} \qquad (12)$$

where $\rho_{so}$ and $c_{so}$ are the apparent density and heat capacity, resp., of the packed bed, while t is the residence time. Eq.12 holds for constant surface temperature and sufficiently long, but not too long residence times, so, that the temperature profiles within the packed bed are still non-developed. Integration with respect to time gives for the time averaged coefficient

$$\bar{\alpha}_{so} = \frac{2}{\sqrt{\pi}} \sqrt{\lambda_{so}\, \rho_{so}\, c_{so}} \; \frac{1}{\sqrt{t}} \qquad (13)$$

For fully developed temperature profiles at long residence times one obtains

$$\bar{\alpha}_{so} = \frac{\pi^2}{2} \frac{\lambda_{so}}{H} \qquad (14)$$

where H is the height of the packed bed when resting one a horizontal plane surface.

5. SYNTHESIS OF WALL HEAT TRANSFER RESISTANCE $1/\alpha_{ws}$ AND INTERNAL HEAT TRANSFER RESISTANCE $1/\alpha_{so}$ FOR PACKED AND PERFECTLY MIXED BEDS

When a <u>packed bed</u> contacts a plane surface of constant temperature, the heat transfer is subsequently controlled first by Eq. 9, 10; then by Eq. 13 and eventually by Eq. 14. Therefore, the heat transfer coefficient between the plane surface and the packed bed drops from $\alpha_{ws}$ at the beginning to $\bar{\alpha}_{so} = (\pi^2/2)(\lambda_{so}/H)$ as the residence time t increases. This is depicted in Fig. 3 by the lower curve, which thus represents the limiting case of zero particle motion.

The other limiting case is the <u>perfectly mixed bed</u>, which is always isothermal, so that the heat transfer is permanently controlled by Eq. 9, 10, independent of the residence time. This case is depicted in Fig. 3 by the upper curve, which simultaneously represents the maximum heat transfer coefficient.

The heat transfer coefficient for non perfectly mixed beds is expected to be found between the upper and the lower limit as indicated in Fig. 3. It is of practical interest that below a so called critical residence time $t_{crit}$ the upper and the lower limit coincide, so that the degree of mixing, i.e. the velocity of the particle motion has no influence on the heat transfer. The critical residence time follows from Eq. 10 and 13

$$t_{crit} = \frac{4}{\pi} \frac{\lambda_{so}\, \rho_{so}\, c_{so}}{\alpha_{WS}^2} \qquad (15)$$

At normal pressure and with particle diameters of some millimeters this critical time is in the order of seconds. Under vacuum, however, it may be in the order of hours.

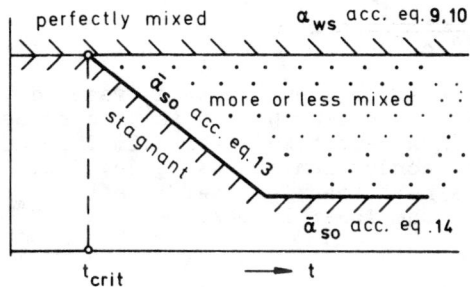

Fig. 3  Heat transfer coefficient α between a plane surface and an adjacent packed bed or perfectly mixed bed of spherical particles

Eq. 15 says quantitatively, what the above mentioned condition a) "very short residence time" means.

The above mentioned condition b) is fulfilled, if there is a heat sink at the particle surface, which isothermally absorbes all the heat transferred from the submerged surface to the adjacent first layer of particles. Then the whole bed is isothermal, whether it is mixed or not. However, once the capacity of the heat sink is exhausted, there must be a minimum particle motion replacing exhausted particles from the layer adjacent to the submerged surface by fresh particles from the bulk. One example, which will be treated below, is the process of contact drying.

6.  SYNTHESIS OF WALL HEAT TRANSFER RESISTANCE $1/\alpha_{WS}$ AND INTERNAL HEAT TRANSFER RESISTANCE $1/\alpha_{SO}$ FOR NON PERFECTLY MIXED BEDS - THE PENETRATION MODEL

To predict the heat transfer to a non isothermal and non perfectly mixed bed raises the problem how to predict the particle

motion. Since no rigorous theory of the particle motion is
available, one has to rely on more or less realistic models.
One model, which has been successfully applied in gas-liquid
systems is the so called "Penetration model". It shall be adopted
also in this case.

It is assumed that the condition of non perfect mixing can be
described by a periodic sequence of two steps. In the first step
the particle motion is zero and the bed is equal to a packed bed.
This step is lasting for a certain time $t_R$ at which the bed gains
heat through transient conduction. The instantaneous heat transfer
resistance during this period may be written approximately as the
sum of the wall heat transfer resistance and the internal heat
transfer resistance

$$\frac{1}{\alpha_{mom}} = \frac{1}{\alpha_{WS}} + \frac{\sqrt{\pi\, t}}{\sqrt{\rho_{so}\, c_{so}\, \lambda_{so}}} \qquad^{+)} \qquad (16)$$

The second step occurs after the time t has reached the fictive
contact time $t_R$. Then the packed bed is assumed to become
perfectly mixed within a zero time intervall. Thereafter the heat
transfer by transient conduction starts again. Fig. 4 illustrates
how the instantaneous heat transfer coefficient $\alpha_{mom}$ oscillates
between the upper and the lower limits.

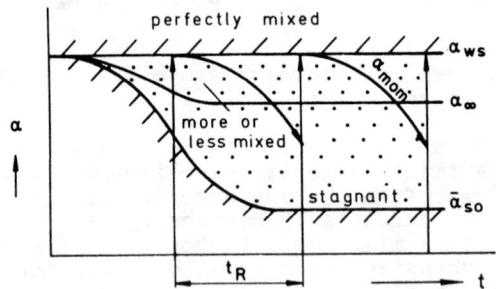

Fig. 4  The instantaneous and average heat transfer
coefficients $\alpha_{mom}$ and $\alpha_\infty$, resp., according
to the penetration model

---

$^{+)}$ Footnote:
A rigorous calculation including the influence of the heat
capacity of the heated plate has been given by Muchowski /26/.
However, for most technical applications the approximate equation
16 should be sufficient.

Integrating Eq. 16 with respect to the time gives the steady state heat transfer coefficient $\alpha_\infty$ for a non perfectly mixed bed:

$$\frac{\alpha_\infty}{\alpha_{WS}} = \frac{2}{\sqrt{\pi\tau}} \left\{ 1 + \frac{1}{\sqrt{\pi\tau}} \ln \frac{1}{1+\sqrt{\pi\tau}} \right\} \quad (17)$$

where $\tau$ is the dimensionless fictive contact time according to

$$\tau = \frac{\alpha_{WS}^2}{\rho_{so} c_{so} \lambda_{so}} t_R \quad (18)$$

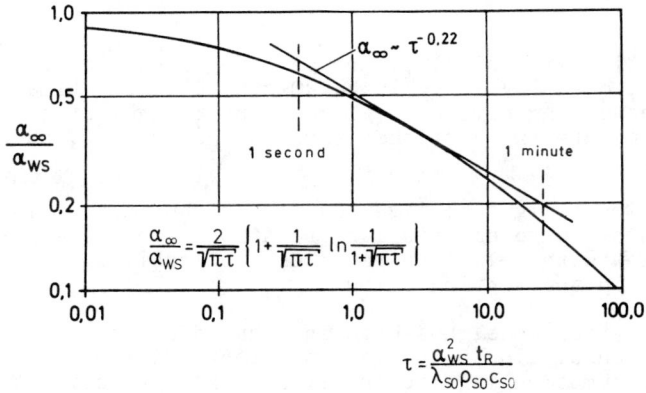

Fig. 5   Relative heat transfer coefficient $\alpha_\infty/\alpha_{WS}$ as a function of the fictive contact time $\tau$ according to the penetration model, see Eq. 17

Fig. 5 represents $\alpha_\infty/\alpha_{WS} = f(\tau)$ according to Eq. 17. One recognizes that $\alpha_\infty$ approaches to the upper limit as $\tau$ goes to zero, while the lower limit is approached, if $\tau$ is sufficiently long.

Eq. 17 does not really solve the problem, since a new variable, the fictive contact time $t_R$, has been introduced. However, a certain progress can be stated, if the fictive contact time turns out to be a <u>pure mechanical</u> property of the non perfectly mixed bed. If this holds, we have been able to synthesize the heat transfer phenomenon from thermal and mechanical properties of the system.

The advantage of this concept is that at least dimensionless groups can be established which are useful for scaling up and down even if the exact correlations between these groups are not known.

In this context Eq. 17 might be generalized in the following way

$$N_{heat} = f(N_{therm} \cdot N_{mech}) \tag{19}$$

where

$$N_{heat} = \frac{\alpha_\infty}{\alpha_{WS}} \tag{20}$$

$$N_{therm} = \frac{\alpha_{WS}^2 \, t_{mech}}{\rho_{so} \, c_{so} \, \lambda_{so}} \tag{21}$$

$$N_{mech} = \frac{t_R}{t_{mech}} \tag{22}$$

are the respective dimensionless groups being connected by the function f. The quantity $t_{mech}$ in Eq.20 and 21 is the time constant of the mixing device, e.g. given by the number of rotations per minute (rpm) of the stirrer, the scraper, the rotary drum etc.

The function f might be given by Eq. 17, but could also be given by any other pure empirical correlation, such as e.g. a simple power law with a limited range of applicability. This will be referred to in the section below.

The penetration model has been introduced in connection with sensible heat consumption only. In many practical applications latent heat consumption is also involved if not predominant. The penetration model may be extended by replacing the heat capacity $c_{so}$ by the total enthalpy change of the particulate material:

$$\tau = \frac{\alpha_{WS}^2 \, t_R}{\rho_{so} \, \lambda_{so} (dh/d\vartheta)_{so}} \tag{23}$$

where

$$\left(\frac{dh}{d\vartheta}\right)_{so} = c_{so} + \left(\frac{\partial h}{\partial X}\right)_{so} \frac{dX}{d\vartheta} \tag{24}$$

with dX being the change of mass to which the latent heat pertains (e.g. moisture content in case of drying). If the latent heat consumption is predominating $d\vartheta$ is zero and $\tau$ approaches zero, regardless how long $t_R$ is. This means that in case of predominant latent heat absorption the heat transfer coefficient $\alpha_\infty$ always equals the maximum heat transfer coefficient $\alpha_{WS}$ regardless how perfect the bed is mechanically mixed.

## 7. APPLICATION OF THE PENETRATION MODEL TO ROTATING DRUM APPARATUSSES

Indirectly heated drums are used as dryers, kilns, etc. Usually near the feed of the drum the material is warmed up. In the middle of the apparatus latent heat absorption may prevail, while near the end the solid might become superheated. This is typical, e.g., for a rotary dryer performance.

If we identify the time constant of the mixing device $t_{mech}$ with the reciprocal of the drum revolutions per second $\omega$:

$$t_{mech} = \frac{1}{\omega} \qquad (25)$$

we obtain

$$\tau = \frac{\alpha_{WS}^2}{\rho_{so} \lambda_{so} (dh/d\vartheta)_{so} \omega} \cdot N_{mech} \qquad (26)$$

Let $N_{mech}$ be any given number, $\tau$ still will vary from the feed towards the end of the dryer because of the variation of $(dh/d\vartheta)_{so}$ with the moisture content X. Near the feed within the warming up zone $\tau$ will be fairly long. In the middle, where the drying zone lies, $\tau$ will become very short. Towards the end, where the solid material becomes superheated, $\tau$ will increase again and superceed the initial value because $\lambda_{so}$ is much lower for the dry material as for the wet one. Fig. 6 shows the corresponding variation of $\alpha_\infty/\alpha_{WS}$ qualitatively.

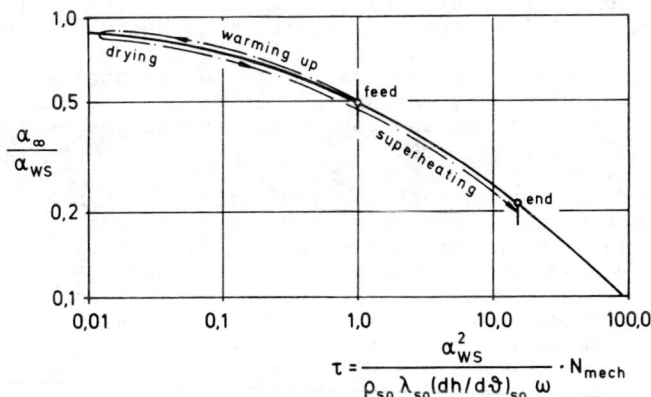

Fig. 6 Heat transfer in an indirectly heated rotary dryer

Only few experimental heat transfer data for rotary dryers have been reported in the literature so far, /13/, /14/. These data indicate, that the heat transfer coefficient is proportional to $\omega^m$, where m lies around 0.2 to 0.3. Comparison with Eq. 17 yields that $\tau$ should have been in the order of 0.01 to 1.0. These values seem to be not too unrealistic as shown in section 9.

## 8. APPLICATION OF THE PENETRATION MODEL TO FLUIDIZED BEDS WITH IMMERSED BODIES

Heat transfer from immersed bodies to fluidized beds occurs along three parallel paths namely Particle Convection ($\alpha_{PC}$), Gas Convection ($\alpha_{GC}$) and Radiation ($\alpha_{RD}$):

$$\alpha = \alpha_{PC} + \alpha_{GC} + \alpha_{RD} \qquad (27)$$

Gas convection is only important as the particles are larger than say 1 mm. Radiation contributes considerably only at high temperatures. For smaller particles under normal conditions the particle convection mechanism predominates.

Until recently only empirical correlations did exist for the prediction of $\alpha_{PC}$. One of the most widely used is the correlation of Zabrodsky /14/:

$$\alpha_{PC} = 35.8 \, \rho_P^{0.2} \, \lambda_g^{0.6} \, d_P^{-0.36}$$

where $\rho_P$ is the particle density, $\lambda_g$ is the heat conductivity of the gas and $d_P$ is the particle diameter, all in SI units. Meanwhile Martin /2/ has presented a new physically based theory, which not only comprises Zabrodsky's equation but also applies to a much wider range of involved parameters. This theorey starts out from the previous work of Heyde and Klocke /15/ and Schlünder /1/ and predicts many experimental data obtained under various conditions with only one adjustable parameter. Since this theory is also based on the penetration model the essential features shall be repeated here.

At first it is assumed that the fluidized bed is perfectly mixed which means that $\tau$ approaches zero and therefore $\alpha_{PC}$ equals $\alpha_{WS}$. The next question is, how the heat transfer coefficient between the submerged surface and the fluidized bed $\alpha_{WS}$ is interconnected with the heat transfer coefficient between the surface and the single particle $\alpha_{WP}$. Since in a fluidized bed the particles do not permanently touch the submerged surface, the relationship between $\alpha_{WS}$ and $\alpha_{WP}$ will not only be affected by the void fraction of the fluidized bed but also by the contact time $t_C$ of the particles with the submerged surface. Again we see, that the missing information $t_C$ is a pure mechanical one, which must be found by some data fitting procedure similar as to the fictive contact time $t_R$ for dense beds.

Since the particle contact time $t_C$ is closely related to the average random particle velocity $\bar{w}_P$, Martin /2/ adopted as a starting point the basic energy equation as set forth in the kinetic theory of gases:

$$\dot{q}_{PC} = \frac{1}{6} \rho_P (1-\psi) c_p \bar{w}_p \gamma_P (\vartheta_W - \vartheta_B) \tag{28}$$

where $\dot{q}_{PC}$ is the heat flux due to particle convection, $\rho_P (1-\psi) c_P$ is the heat capacity of the particle collective per unit volume, $\bar{w}_P$ is the average random particle velocity, $\gamma_P$ is the "particle accommodation coefficient" and $(\vartheta_W - \vartheta_B)$ is the temperature difference between the submerged surface and the bulk of the fluidized bed. $\gamma_P$ is defined as

$$\gamma_P = \frac{\vartheta_P - \vartheta_B}{\vartheta_W - \vartheta_B} \tag{29}$$

where $\vartheta_P$ is the particle temperature after departure from the surface. If the contact time is sufficiently long $\vartheta_P$ will equal $\vartheta_W$. While the sufficiently short contact times $\vartheta_P$ will remain at $\vartheta_B$.

$\gamma_P$ can be calculated by the following equation

$$\gamma_P = 1-\exp\left\{-\frac{\alpha'_{WP} A_P}{\rho_P c_P V_P} t_C\right\} \tag{30}$$

$$\alpha'_{WP} = \frac{2\lambda_g}{R}\left\{(1+\frac{\sigma}{R}) \ln(1+\frac{R}{\sigma}) - 1\right\} \tag{31}$$

$$A_P = \pi R^2 \tag{32}$$

$$V_P = \frac{4\pi}{3} R^3 \tag{33}$$

Eq. 30 implies that heat conduction within the particles may be neglected compared to the heat transfer through the gaseous gap by $\alpha'_{WP}$ which can be seen when forming a Biot number

$$Bi = \frac{\alpha'_{WP} R}{\lambda_P} = 2 \frac{\lambda_g}{\lambda_P}\left\{(1+\frac{\sigma}{R}) \ln(1+\frac{R}{\sigma}) - 1\right\} \tag{34}$$

In general Bi is much less than unity. With the definition

$$\alpha_{PC} = \frac{\dot{q}_{PC}}{\vartheta_W - \vartheta_B} \tag{35}$$

we obtain

$$\alpha_{PC} = \frac{1}{6} \rho_P (1-\psi) c_P \bar{w}_P \left\{ 1 - \exp\left( -\frac{\alpha'_{WP}}{\frac{1}{6} \rho_P c_P \bar{w}_P} \cdot \frac{\bar{w}_P t_C}{4 d_P} \right) \right\} \quad (36)$$

This equation contains besides the still unknown contact time $t_C$ the also unknown random particle velocity $\bar{w}_P$. Within the frame of the kinetic theory of gases this velocity is given by the thermodynamic temperature. As to the fluidized bed Martin /2/ postulated that the kinetic energy of the randomly moving particles is generated by conversion of potential energy being released during the free fall of the particles between two collisions:

$$\frac{1}{2} \bar{w}_P^2 = g \, l_{free} \quad (37)$$

g is the acceleration due to gravity and $l_{free}$ is the mean free path of the particles in the fluidized bed. $l_{free}$ may be calculated exactly analogous to the kinetic theory of gases from the number density and the particle diameter. As a result one obtains

$$\bar{w}_P = \sqrt{g \, d_P \, F(\psi, \psi_L)} \quad (38)$$

wherein

$$F(\psi, \psi_L) = \frac{\psi - \psi_L}{5 (1-\psi_L)(1-\psi)} \quad (39)$$

$\psi_L$ is the void fraction at the minimum fluidization point.

The remaining question is how to determine the contact time $t_C$. It sould be brought to mind that $t_C$ as it appears in Eq. 30 is a fictive quantity, since heat conduction through the gaseous gap between the submerged surface and the particle does not only occur as long as the particles sit on the surface, but also during the periods of approach to and departure from the surface. Therefore Martin /2/ postulated that the "effective" contact time required to displace the particle by one particle diameter:

$$c'_C \, t_C = \frac{d_P}{\bar{w}_P} \quad (40)$$

Then Eq. 36 may be rewritten as

$$\boxed{\alpha_{PC} = \alpha'_{WP} \frac{(1-\psi)}{4 \, C'_C} \frac{1-\exp(-N)}{N}} \quad (41)$$

wherein

$$N \equiv \frac{3\, \alpha'_{WP}/C'_C}{2\, \rho_P\, c_P\, \sqrt{g d_P F(\psi, \psi_L)}} \qquad (42)$$

with $\alpha'_{WP}$ according to Eq. 31 and $F(\psi, \psi_L)$ according to Eq. 39.

The Eq. 41 has two limiting cases, given by $N \to 0$ and $N \to \infty$ resp. We obtain

$$\lim_{N \to 0} \alpha_{PC} = \alpha'_{WP}\, \frac{1-\psi}{4C'_C} \qquad (43)$$

and

$$\lim_{N \to \infty} \alpha_{PC} = \frac{1}{6}\, \rho_P\, (1-\psi)\, c_P\, \bar{w}_P \qquad (44)$$

In the first case ($N \to 0$), that occurs with larger particles, the controlling mechanism is the heat conduction through the gaseous gap between surface and particle during the collision with the wall.

In the latter case ($N \to \infty$), that occurs with smaller particles, the heat transfer is limited by the heat capacity and the velocity of the particles.

Furthermore Eq. 41 yields an optimum void fraction $\psi_{OPT}$ at which the heat transfer coefficient $\alpha_{PC}$ reaches a maximum value. This optimum void fraction can be determined by putting $\partial \alpha_{PC}/\partial \psi = 0$. One obtains

$$\psi_{OPT} = \psi_L + \frac{1}{2}(1-\psi_L)\left(1 - \frac{N}{\exp N - 1}\right) \qquad (45)$$

One realizes that for $N \to 0$ (larger particles) $\psi_{OPT}$ equals $\psi_L$, while for $N \to \infty$ (smaller particles) $\psi_{OPT}$ equals $1/2\,(1+\psi_L)$, which is half way in between minimum and maximum fluidization point. Eventually Eq. 41 says that there is an absolute maximum of $\alpha_{PC}$ with respect to a particle diameter variation, which follows from the condition $(\partial \alpha_{PC}/\partial d_P)_{\psi_{OPT}} = 0$. Fig. 7 shows that the optimum particle diameter is 40 μm, at which the heat transfer coefficient $\alpha_{PC}$ reaches an absolut maximum value of $\approx 800$ W/m$^2$K in case that the fluidization gas is air at 25°C. If the fluidization gas would be hydrogen, the optimum particle diameter would be 110 μm and the maximum heat transfer coefficient would be $\approx 1400$ W/m$^2$K, as shown in Fig. 8.

These maximum heat transfer coefficients at optimum void fractions $\psi_{OPT}$ are due to the fact that $\alpha_{PC}$ increases with decreasing particle diameter as long as Eq. 43 is rate controlling, however, decreases when Eq. 44 becomes rate controlling. It is obvious that the latter mechanism only applies in the case of sensible heat absorption through the particles. In case of isothermal latent heat absorption with the heat sink being located at the surface of the particles this limiting mechanism does not exist, since $c_P = (\partial h/\partial \vartheta)_p \to \infty$. In this case $\alpha_{PC}$ increases monotonously with decreasing particle diameter up to the absolute upper limit, which follows from Eq. 43 and 31 putting $d_P \to 0$ as

$$\alpha_{PC} < \frac{\lambda_g}{\sigma} \frac{1-\psi_L}{4\, C_C'} \tag{46}$$

Martin has determined the "collision factor" $C_C'$ by comparison with experimental data taken from Wunder and Mersmann /16/, Baskakov /17/ and Wicke and Fetting /22/. He found

$$C_C' = \frac{3}{4} \tag{47}$$

This means that the fictive contact time $t_C$ of the particle is of the same order of magnitude as the travelling time of the particle along a path of one particle diameter, see Eq. 40.

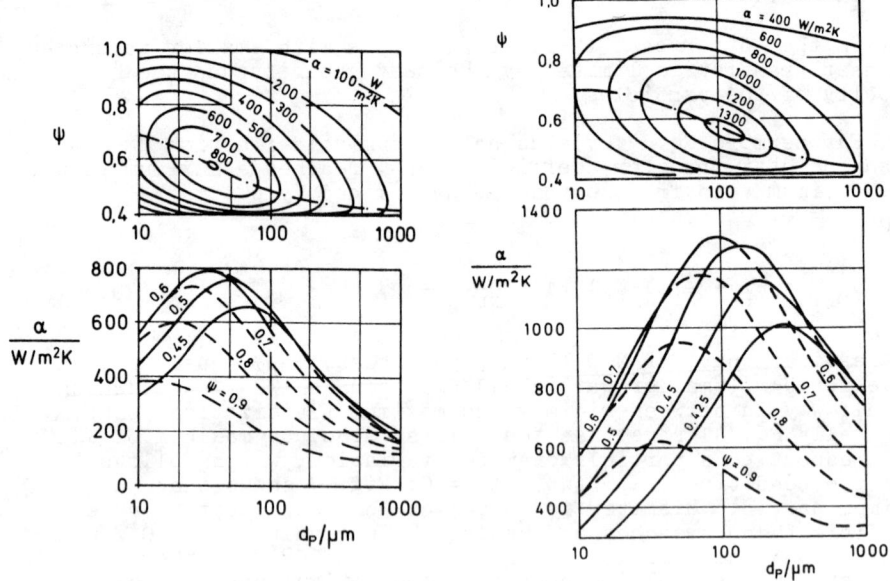

Fig. 7
Wall heat transfer coefficient for an air fluidized bed of ceramic particles at various void fractions $\psi$ (bed expansions) depending on the particle diameter
$\rho_P = 2500$ kg/m³
$\rho_P c_P = 2$ MJ/m³K
$\psi_L = 0.40$

Fig. 8
Wall heat transfer coefficient for a hydrogen fluidized bed. Other data as in Fig. 7.

For comparison with experimental data, which partly might include heat transfer through radiation (high temperatures) and gas convection (large particles) additional equations for $\alpha_{RD}$ and $\alpha_{GC}$ (see Eq. 27) are needed. For the contribution through radiation Dammköhler's Formula

$$\alpha_{RD} = 0.04 \; C_{12} \left[\frac{T_m}{100 \; K}\right]^3 \tag{48}$$

should be sufficient with $C_{12} \cong 5 \; W/m^2 K$.

The contribution through gas convection has been studied by Baskakov /17/. He found

$$\alpha_{GC} = \frac{\lambda_g}{d_P} \; 0,009 \; Pr_g^{1/3} \; \sqrt{Ar} \tag{49}$$

with

$$Pr_g = \nu_g / \kappa_g \tag{50}$$

and

$$Ar = \frac{g \; d_P^3}{\nu_g^2} \cdot \frac{\rho_P - \rho_g}{\rho_g} \tag{51}$$

## 9. COMPARISON BETWEEN THEORY AND EXPERIMENT

Wunschmann and Schlünder /18/ have measured instantaneous heat transfer coefficients under transient conditions while heating packed and mechanically agitated beds resting on a horizontal heating plate at normal pressure and at vacuum. With these experiments only sensible heat absorption was involved.

Experiments with latent heat absorption through the particles have been carried out by Günes and Schlünder /19/ by contact drying of wetted porous particles under vacuum. Numerous experimental data with sensible heat absorption in fluidized beds have been evaluated by Martin /2/. These experimental data shall now be compared with the theoretical predictions as given above.

### 9.1 Heat Transfer Between a Horizontal Plate and Packed as well as Mechanically Agitated Beds with Sensible Heat Absorption

Fig. 9 shows experimental data according to Wunschmann and Schlünder /18/ for a packed bed of 3.1 mm glas beads resting on an electrically heated horizontal plate. The time averaged heat transfer coefficient is plotted against the residence time t for various gas pressures. On the left hand side $\alpha = \alpha_{WS}$ according to Eq. 9, 10, is depicted, while on the right hand side $\alpha = \alpha_{SO}$ according to Eq. 13 is shown. The experimental data seem to confirm the theoretical predictions. This figure also confirm that the critical residence time $t_{crit}$ according to Eq. 15 is in the order of hours at vacuum. At very low pressure (P<0,001 bar) the only heat transfer mechanism is radiation.

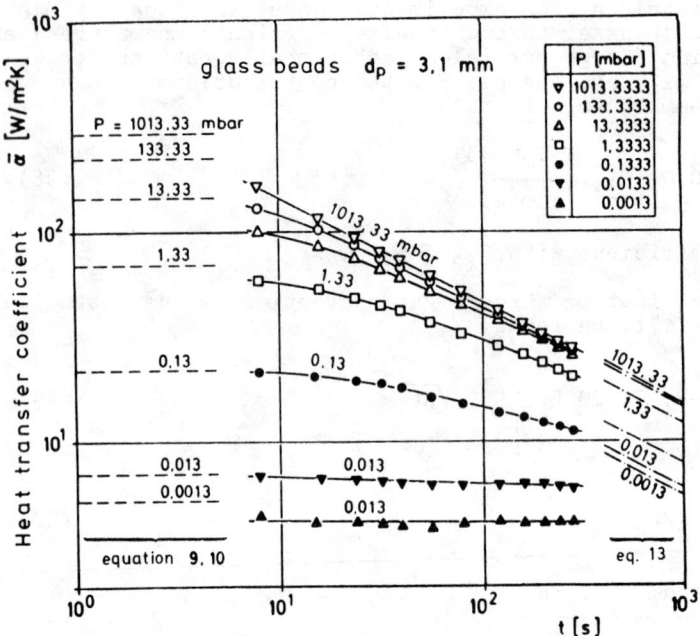

Fig. 9. Time averaged heat transfer coefficient $\bar{\alpha}$ between an electrically heated horizontal plane and a packed bed of glass beads as a function of the residence time t at various gas pressures P. Bed height 50 mm. Particle diameter $d_p$ = 3,1 mm, temperature $20°C$.

Fig. 10 shows analogous results for a mechanically mixed bed at two different gas pressures. The mixing device was an ordinary stirrer with adjustable speed ω. One recognizes steady state terminal heat transfer coefficients $\alpha_\infty$ within the lower and the upper limit depending on the stirrer speed ω.

Plotting these terminal values $\alpha_\infty$ against the dimensionless group $N_{therm}$ as defined by Eq. 21 and 25 one obtains a descending sequence of data points with increasing $N_{therm}$ as shown in Fig. 11. The higher $\alpha_\infty$-values appear at low pressure. These sequence of data can be fitted by Eq. 17 when putting

$\tau = 15\ N_{therm}$

Thus we obtain $N_{mech}$ = 15, which means that 15 stirrer revolutions were necessary as an equivalent for a perfect (thermal) mixing of the bed in the context with the penetration model.

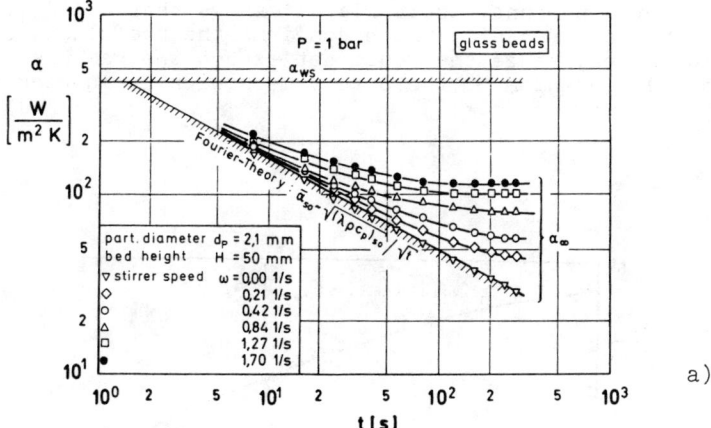

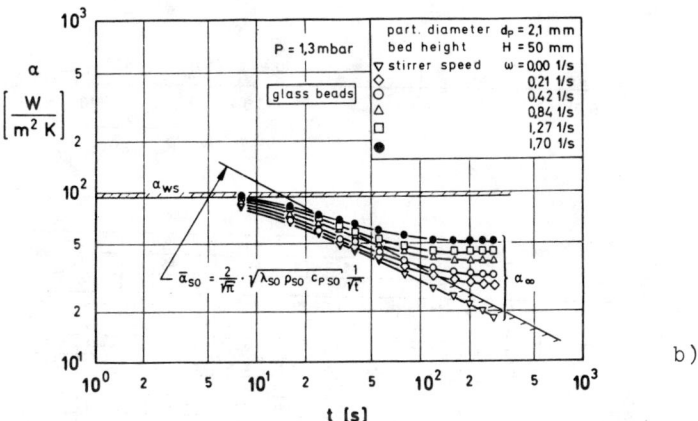

Fig. 10. Heat transfer coefficients as in Fig.9, however, for a mechanically agitated bed of glass spheres. Particle diameter $d_P = 2,1$ mm

    a) Gas pressure 1 bar
    b) Gas pressure 1,3 mbar

As a very important result Fig. 11 shows that $N_{mech}$ does not depend on the gas pressure, which confirms the usefulness of the penetration model, since one would not expect the gas pressure to affect the mechanism of the purely mechanically induced mixing process.

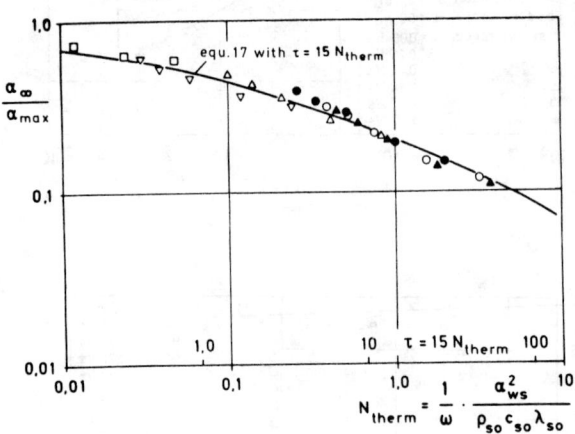

| p [mbar] | 1000 | | 520 | | 130 | | 13 | | 1,3 | | 0,13 | |
|---|---|---|---|---|---|---|---|---|---|---|---|---|
| $\alpha_{ws}$ [W/m²K] | 400 | | 370 | | 300 | | 185 | | 85 | | 22 | |
| $\lambda_{so}$ [W/mK] | 0,177 | | 0,177 | | 0,176 | | 0,165 | | 0,111 | | 0,041 | |
| $\omega$ [s⁻¹] | $N_{therm}$ | $\alpha_\infty/\alpha_{ws}$ | $N_{therm}$ | $\alpha_\infty/\alpha_{ws}$ | $N_{therm}$ | $\alpha_\infty/\alpha_{ws}$ | $N_{therm}$ | $\alpha_\infty/\alpha_{ws}$ | $N_{therm}$ | $\alpha_\infty/\alpha_{ws}$ | $N_{therm}$ | $\alpha_\infty/\alpha_{ws}$ |
| 1,70 | 0,45 | 0,29 | 0,39 | 0,31 | 0,26 | 0,38 | 0,10 | 0,48 | 0,030 | 0,60 | 0,006 | 0,72 |
| 1,27 | 0,60 | 0,25 | 0,52 | 0,27 | 0,34 | 0,33 | 0,14 | 0,42 | 0,039 | 0,52 | 0,008 | 0,72 |
| 0,85 | 0,91 | 0,20 | 0,78 | 0,22 | 0,51 | 0,28 | 0,21 | 0,35 | 0,059 | 0,45 | 0,012 | 0,72 |
| 0,43 | 1,80 | 0,14 | 1,54 | 0,15 | 1,02 | 0,19 | 0,41 | 0,26 | 0,118 | 0,36 | 0,024 | 0,62 |
| 0,21 | 3,66 | 0,11 | 3,13 | 0,12 | 2,07 | 0,15 | 0,84 | 0,21 | 0,237 | 0,31 | 0,048 | 0,59 |
| | ▲ | | ○ | | ● | | △ | | ▽ | | □ | |

| $\tau$ | 0,10 | 0,20 | 0,50 | 1,00 | 2,00 | 5,00 | 10,0 | 20,0 | 50,0 |
|---|---|---|---|---|---|---|---|---|---|
| $\alpha_\infty/\alpha_{ws}$ | 0,735 | 0,665 | 0,56 | 0,48 | 0,39 | 0,30 | 0,24 | 0,18 | 0,13 |

Fig. 11. Relative heat transfer coefficient $\alpha_\infty/\alpha_{ws}$ between a horizontal plate and a mechanically agitated bed of glass spheres at various gas pressures and various revolutions per second of the stirrer, acc. /15/

$d_p$ = 2,1 mm, $\rho_{so}$ = 1800 kg/m³, $c_{so}$ = 652 J/kg,
$\alpha_{WS} = \alpha_{WP}$

## 9.2 Heat Transfer Between a Horizontal Plate on Mechanically Agitated Beds with Latent Heat Absorption

Contact drying of moist porous particles under vacuum have been investigated by Günes and Schlünder /19/. As long as the entire bed is isothermal which is true during the constant rate period there is only latent heat absorption. Then the energy balance yields

$$\dot{m}_{H_2O} \, \Delta h_v = \dot{q} \tag{52}$$

where $\dot{m}_{H_2O}$ is the drying rate (kg/m²s) and $\Delta h_v$ is the latent heat of evaporation. As long as evaporation takes place at the particle surface the heat flux is entirely controlled by $\alpha_{WS}$ according to Eq. 9, 10:

$$\dot{q} = \alpha_{WS} \, (\vartheta_H - \vartheta_S) \tag{53}$$

$\vartheta_H$ is the temperature of the heated plate and $\vartheta_S$ of the bed, resp. Combining Eq. 53 and 52 gives the maximum drying rate

$$\max \dot{m}_{H_2O} = \alpha_{WS} \, \frac{\vartheta_H - \vartheta_S}{\Delta h_v} \tag{54}$$

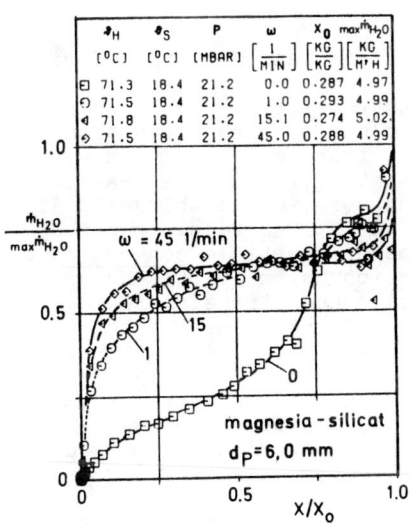

Fig. 12. Normalized drying rates $\dot{m}_{H_2O}/\max \dot{m}_{H_2O}$ versus normalized moisture content $X/X_0$ for contact drying of a mechanically agitated bed of magnesia-silicat particles under vacuum at various revolutions per second of the stirrer. $d_P$ = 6 mm

Fig. 12 shows experimental data according to Günes and Schlünder /19/. The relative drying rate $\dot{m}_{H_2O}/max\dot{m}_{H_2O}$ is plotted against the relative particle moisture content $X/X_o$ with $X_o$ being the initial moisture content. The pressure $P^o$ and therefore also $\vartheta_S$ as well as $\vartheta_H$ have been kept constant, while the stirrer speed $\omega$ has been varied. One recognizes that the maximum drying rate was reached only at the beginning of the drying process, followed by an almost constant rate period at about 60 % of the maximum value. This reduction may be explained be the dry out of rough surface layer of the particles, see /19/. The interesting fact is, that within the constant rate period the stirrer speed $\omega$ has practically no effect on the drying rate since the bed is isothermal anyway. However, the stirrer speed zero is to be excluded, since a certain minimum speed is necessary to remove dry particles from the surface of the plate.

Towards the end of the drying process the stirrer speed becomes more effective as to the drying rate. This is because at low moisture contents there is not only latent but also sensible heat absorption due to the rising particle temperature. In this case developing temperature profiles in the bed will be flattened by the stirrer thus giving rise to steeper temperature gradients at the plate surface and therefore also to higher heat transfer rates. This effect can be well described by the penetration model, if the dimensionless fictive contact time $\tau$ is formed according to Eq. 23 thus taking into account simultaneous latent and sensible heat absorption. This has been shown in detail by Günes and Schlünder /20/. So, one may conclude that the penetration model is a useful tool to describe the heat transfer from submerged surfaces to mechanically agitated beds with sensible, latent as well as combined sensible and latent heat absorption.

## 9.3 Heat Transfer Between Fluidized Beds and Submerged Surfaces

Fig. 13 shows the maximum heat transfer coefficients $\alpha_{max}$ between fluidized beds and submerged surfaces as a function of the particle diameter $d_P$ at optimum void fraction $\psi_{OPT}$ with air as fluidizing agent. The experimental data for sensible heat absorption are taken from various authors, while the solid line is predicted according to the Eqs. 27 to 51. One recognizes that the controlling heat transfer mechanisms are <u>particle convection</u> as the particles are small, and <u>gas convection</u> as the particles are larger. Within the particle convective heat transfer regime one has to distinguish between sensible and latent heat absorption. In the latter case there is not such a limitation as by the heat capacity of the particles (lower solid line in Fig. 13), so that the heat transfer coefficient follows Eq. 43 with $\psi = \psi_L$, see Eq. 45, (upper dotted line in Fig. 13). In practical cases with simultaneous sensible and latent heat absorption, $\alpha_{max}$ should fall into the shaded zone fenced by the two limiting curves.

Fig. 14 shows $\alpha_{max}$ versus $d_P$ with hydrogen as fluidizing agent. One recognizes that the shaded zone is shifted towards larger particle diameters. This might be important, e.g. for processes like coal gasification.

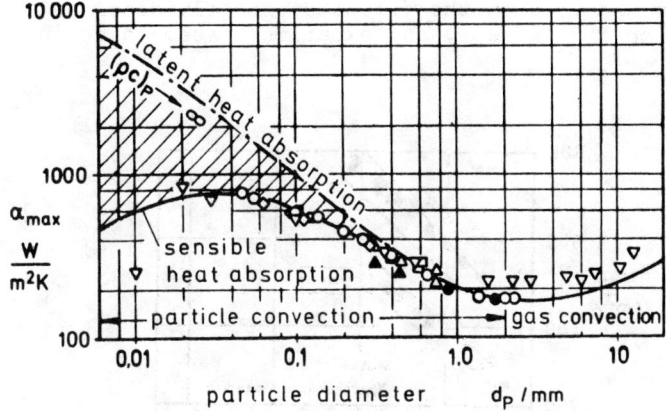

| Wunder a. Mersmann | Baskakov | Wicke a. Fetting |
|---|---|---|
| ○ glass beads<br>● aluminum<br>◇ sharp particles<br>□ wide particle diameter distrib | ▽ korund | △ sand<br>▲ aluminum |

Fig. 13. Maximum heat transfer coefficient $\alpha_{max}$ between fluidized beds and submerged surface at optimum void fraction $\psi_{OPT}$. Data according to various authors. Full and dotted lines calculated according to equations 27 to 51 for air, 1 bar, 25°C

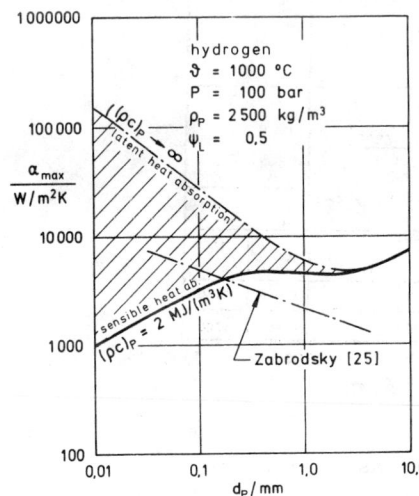

Fig. 14.
Maximum heat transfer coefficients as in Fig.13, however for hydrogen as fluidizing agent at 100 bar and 1000°C

Fig. 15 and 16 show the effect of the fluidized bed temperature on $\alpha_{max}$. Theory and experiment seem to be in good agreement.

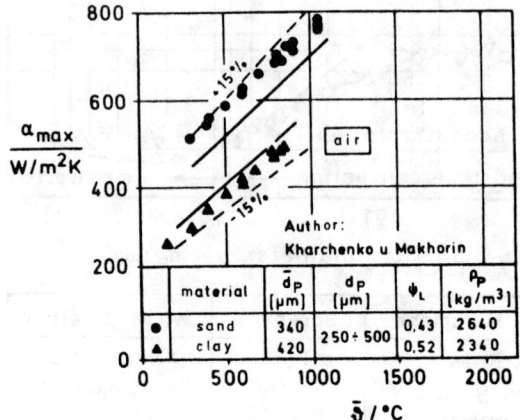

Fig. 15. Effect of temperature on the maximum heat transfer coefficient. Full lines theoretical. Data taken from /23/.

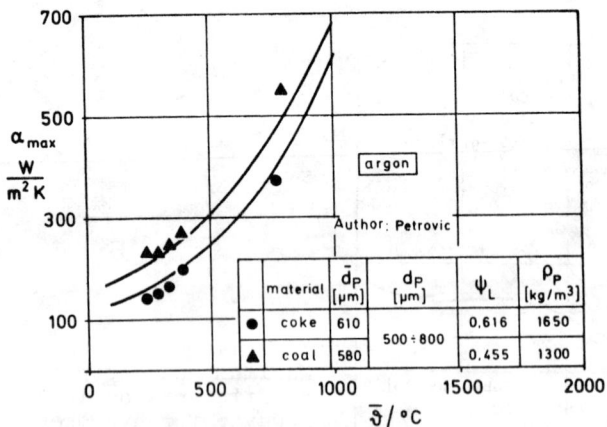

Fig. 16. Effect of temperature as in Fig. 15. Data taken from /21/.

Fig. 17 shows the effect of the gas pressure on $\alpha_{max}$. The theoretical predictions (with $C_K' = 3/4$) fall 30 % below the experimental data of Botterill /23/, however, the pressure effect is correctly predicted.

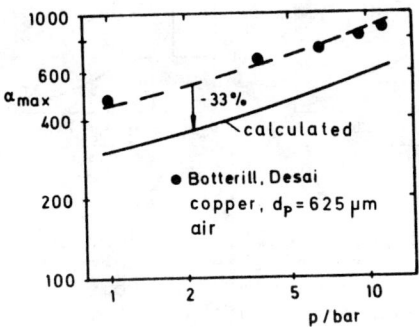

Fig. 17. Effect of pressure on the maximum heat transfer coefficient. Full lines theoretical. Data taken from /24/.

Since the void fraction $\psi$ of the fluidized bed as a function of the gas velocity $u_g$ is well known (see appendix), the theory of Martin /2/ allows to predict not only the maximum heat transfer coefficients $\alpha_{max}$ at the optimum void fraction $\psi_{OPT}$ corresponding to an optimum gas velocity $u_{g,OPT}$, but also the whole function $\alpha = \alpha(u_g)$ for any given particle diameter and given fluid properties. This is shown in Fig. 18 and 19 where $\alpha(u_g)$ is depicted for air operated fluidized beds and various gas pressures as well as various particle diameters as parameter. Experimental data of Wunder and Mersmann /16/ seem to be in good agreement with the theoretical predictions [+]. The data of Botterill and Desai /23/ deviate by 30 %, however, the effect of gas velocity and pressure is precisely predicted.

---

[+] Footnote. The slight deviations are due to the fact that Baskakov correlated the contribution by gas convection only at the optimum gas velocity. A (unpublished) generalisation of Baskakov's correlation by Martin results in almost perfect agreement between theory and experiment.

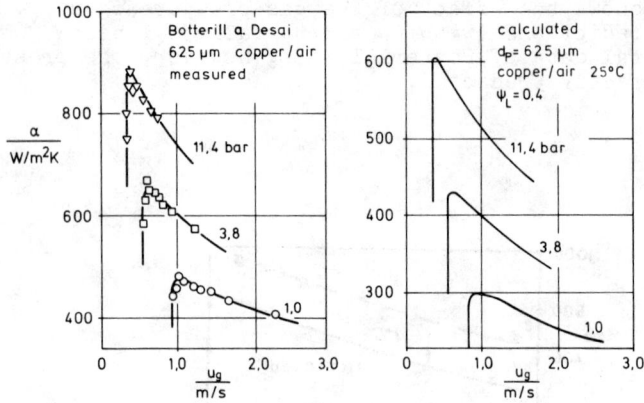

Fig. 18. Effect of gas velocity and pressure on the heat transfer coefficient between fluidized beds and immersed surfaces. Comparison between experimental and predicted data.

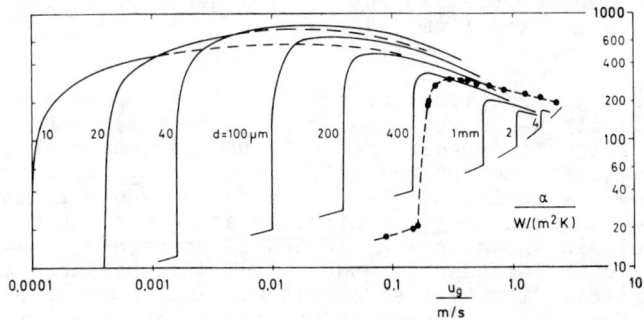

Fig. 19. Effect of gas velocity and particle diameter on the heat transfer coefficient between fluidized beds and immersed surfaces. Full lines theoretical. Data taken from /16/.

## 10. CONCLUSIONS

The heat transfer between packed, mechanically agitated and gas fluidized beds and submerged surfaces can be described by one unique concept. This concept contains three essential parameters:

1. The heat transfer coefficient between a plane surface and a smooth spherical particle $\alpha_{WP}$.
2. The heat conductivity of the quasi-homogeneous packed bed $\lambda_{so}$.
3. The fictive contact time of the packed bed $t_R$ or the single particle, $t_C$ [+], respectively.

The quantities $\alpha_{WP}$ and $\lambda_{so}$ are nowadays predictable with satisfactory accuracy. As long as the fictive contact time is identical with the residence time of the bed at the submerged surface (packed and "en bloc" moving meds), the heat transfer is predictable through $\alpha_{WP}$ and $\lambda_{so}$ without any further information.

If, however, the fictive contact time differs from the residence time (mechanically agitated and gas fluidized beds) additional hypotheses are necessary. It has been shown that the so called "Penetration Model", which much earlier has successfully been applied to gas-liquid systems, is also successfully applicable to gas-solid systems as described here.

In case of mechanically agitated beds (dense beds) <u>one additional parameter</u> which is the relationship between the fictive contact time and the time constant of the mixing device, is needed. The same is true for the heat transfer between fluidized beds (dilute beds) and submerged surfaces. Only <u>one additional parameter</u>, the so called collision factor, which correlates the fictive contact time of a single particle with the submerged surface and the particle velocity, is needed.

After all, it seems to be possible to synthesize the phenomenon of heat transfer on to beds of particles by two <u>physically independent</u> phenomena such as heat conduction and particle motion. Thus a status may have been reached, as it is well known from the treatment of transport phenomena in single phase gas or liquid flow. In particular, the establishment of typical dimensionless groups seems to alleviate the solution of scale up problems in many practical applications. Moreover, it could be shown that in the presence of strong heat sources or sinks (combustion, gasification, drying etc.) no additional information about the particle motion is necessary at all, because - besides of intraparticle transport phenomena - the a priori predictable parameter $\alpha_{WP}$ is the solely rate controlling one. Further research in this direction seems to be desirable.

---

[+] Footnote: Not to be mistaken with the "residence time" t of the packed, agitated or fluidized bed upon the submerged surface!

## 11. APPENDIX

Relationship between the void fraction of the fluidized bed $\psi$ and the gas velocity $u_g$.

Minimum fluidization point

$$Re_L = \frac{A}{2B}(1-\psi_L)\left\{\sqrt{1+\frac{\psi_L^3}{(1-\psi_L)^2}\frac{Ar}{A^2/4B}}-1\right\} \qquad (55)$$

Maximum fluidization point

$$Re_A = 18\left\{\sqrt{1+\frac{1}{9}\cdot\sqrt{Ar}}-1\right\}^2 \qquad (56)$$

(with homogeneous fluidization)

$$Re_A = \sqrt{\frac{4}{3}Ar} \qquad (57)$$

(with heterogeneous fluidization)

Fluidized bed expansion

$$\frac{u_g}{u_{g,A}} = \frac{Re}{Re_A} = \psi^n \qquad (58)$$

$$n = \frac{\ln Re_L/Re_A}{\ln \psi_L}$$

With the help of these equations and Eq. 41 $\alpha_{PC}$ can also be calculated in terms of the gas velocity $u_g$ as shown in Fig. 18. Constants in Eq. 55: A = 150; B = 1.75.

## Symbols

| | | |
|---|---|---|
| $A$ | $m^2$ | surface area |
| $c_p$ | J/kg K | heat capacity |
| $C_{12}$ | $W/m^2 K^4$ | radiation constant |
| $d$ | m | diameter |
| $g$ | $m/s^2$ | gravity constant |
| $h$ | kJ/kg | enthalpy |
| $\Delta h_v$ | kJ/kg | heat of vaporization |
| $H$ | m | bed height |
| $l_{free}$ | m | mean free path of the particles |
| $n$ | $kmol/m^3$ | molar gas density |
| $\dot{m}_{H_2O}$ | $kg/m^2 s$ | drying rate |
| $P$ | Pa | pressure |
| $\dot{q}$ | $W/m^2$ | heat flux |
| $R$ | m | particle radius |
| $\tilde{R}$ | J/kmol K | gas constant |
| $s$ | m | gap width |
| $t_R, t_C$ | s | contact time |
| $t$ | s | residence time |
| $T$ | K | absolute temperature |
| $u_g$ | m/s | gas velocity in the empty cross section |
| $V$ | $m^3$ | volume |
| $\bar{w}_P$ | m/s | particle velocity |
| $x$ | m | distance from the contact point |
| $X$ | - | mass fraction, moisture content |
| $\alpha$ | $W/m^2 K$ | heat transfer coefficient |
| $\gamma$ | - | accomodation coefficient |
| $\vartheta$ | °C | temperature |
| $\lambda$ | W/m K | heat conductivity |
| $\sigma_o$ | m | mean free path of the gas molecules |
| $\nu$ | $m^2/s$ | gas viscosity |
| $\rho$ | $kg/m^3$ | density |
| $\tau$ | - | dimensionless contact time |
| $\psi$ | - | void fraction |
| $\varphi$ | - | relativ surface area covered with particles |
| $\kappa$ | $m^2/s$ | thermal diffusivity |

Subscripts

| | |
|---|---|
| A | at maximum fluidization point |
| L | at minimum fluidization point |

| | |
|---|---|
| g | gas |
| GC | gas convective |
| loc | local |
| mom | instantaneous |
| PC | particle convective |
| RD | radiative |
| so | packed bed (stagnant) |
| ∞ | terminal |
| B | bulk |
| WP | wall-to-particle |
| WS | wall-to-bed |
| W | wall |
| H | heating surface |
| P | particle |
| m | molecular |

Dimensionless Groups
―――――――――――――――

$$Ar = \frac{gd_P^3}{\nu_g^2} \cdot \frac{\rho_P - \rho_g}{\rho_g} \quad \text{Archimedes number}$$

$$Pr = \nu_g/\kappa_g \quad \text{Prandtl-number}$$

$$Re = \frac{ud_P}{\nu_g} \quad \text{Reynolds-number}$$

REFERENCES
――――――――――

1. Schlünder, E.U. 1971. Wärmeübergang an bewegte Kugelschüttungen bei kurzfristigem Kontakt. Chem.Ing.Techn., Vol. 43, pp. 651-654.

2. Martin, H. 1980. Wärme- und Stoffübertragung in der Wirbelschicht. Chem.Ing.Techn., Vol. 52, pp. 199-209.

3. Zehner, P. and Schlünder, E.U. 1970. Wärmeleitfähigkeit von Schüttungen bei mäßigen Temperaturen. Chem.Ing.Techn., Vol.42, pp. 933-941.

4. Zehner, P. and Schlünder, E.U. 1972. Einfluß der Wärmestrahlung und des Druckes auf den Wärmetransport in nichtdurchströmten Schüttungen. Chem.Ing.Techn., Vol. 44, pp. 1303-1308.

5. Zehner, P. and Schlünder, E.U. 1973. Die effektive Wärmeleitfähigkeit durchströmter Kugelschüttungen bei mäßigen und hohen Temperaturen. Chem.Ing.Techn., Vol. 45, pp. 277-284.

6. Zehner, P. 1973. Experimentelle und theoretische Bestimmung der effektiven Wärmeleitfähigkeit durchströmter Kugelschüttungen bei mäßigen und hohen Temperaturen. VDI-Forsch.heft, Düsseldorf.

7. Bauer, R. and Schlünder, E.U. 1976. Effektive radiale Wärmeleitfähigkeit gasdurchströmter Schüttungen aus Partikeln unterschiedlicher Form. Chem.Ing.Techn., Vol. 48, pp. 227-228.

8. Bauer, R., Muchowski, E. and Schlünder E.U. 1976. Wärmetransport in ruhenden und bewegten Kornschichten. Fortschritte der Verfahrenstechnik 1975, Bd. 14, Abschnitt A 3.3. VDI-Verlag, pp. 54 - 59.

9. Bauer, R. and Schlünder E.U. 1977. Effektive radiale Wärmeleitfähigkeit gasdurchströmter Schüttungen. verfahrenstechnik, Vol. 11, pp. 605-614.

10. Bauer, R. and Schlünder, E.U. 1977. Effektive radiale Wärmeleitfähigkeit gasdurchströmter Schüttungen aus Partikeln unterschiedlicher Form und Größenverteilung. VDI-Forschungsheft, 582.

11. Bauer, R. and Schlünder, E.U. 1978. Effective Radial Thermal Conductivity of Packings in Gas Flow. Part I. Convective Transport Coefficient. Part II. Thermal Conductivity of the Packing Fraction without Gas Flow. Int.Chem.Engng., Vol. 18, pp. 181-188 and 189-204.

12. Schlünder, E.U. 1978. Transport Phenomena in Packed Bed Reactors. ACS Symposium Series, No. 72, Chemical Reaction Engineering Reviews, Houston.

13. Baunack, F. 1956. Freiberger Forschungsheft A 47, Berlin.

14. Keßler, H.G. 1969. Die Konstakttrocknung rieselfähiger Güter bei Normaldruck und bei Vakuum. Chem.Ing.Techn., Vol. 41, pp. 463-472.

15. Heyde, M. and Klocke, H.J. 1979. Wärmeübergang zwischen Wirbelschichten und Einbauten - ein Problem des Wärmeüberganges bei kurzfristigem Kontakt. vt-verfahrenstechnik, Vol. 13, pp.886-892.

16. Wunder, R. and Mersmann, A. 1978. Wärmeübergang zwischen Gaswirbelschichten und senkrechten Austauschflächen. Vortrag auf dem Jahrestreffen der Verfahrensingenieure in Aachen.

17. Baskakov, A.P. et al. 1973. Heat Transfer to Objects immersed in fluidized beds. Powder Technology, Vol. 8, pp. 273-282.

18. Wunschmann, J. and Schlünder, E.U. 1975. Wärmeübergang von beheizten Flächen an Kugelschüttungen. vt-verfahrenstechnik, Vol. 9.

19. Günes, S., Schlünder, E.U. and Gnielinski, V. 1980. Kontakttrocknung von grobkörnigem Granulat im Vakuum. vt-verfahrenstechnik, Vol. 14, pp. 31-39.

20. Günes, S. and Schlünder E.U. 1980. Über den Einfluß der mechanischen Durchmischung auf die Trocknungsgeschwindigkeiten bei der Kontakttrocknung von grobkörnigem Granulat. vt-verfahrenstechnik, Vol. 14.

21. Petrovich, V. 1978. Messung und Berechnung des Wärmeüberganges von einem Heizrohr an eine Kohlewirbelschicht in Wasserdampf und Inertgas. Dissertation TH Aachen.

22. Wicke, E. and Fetting, F. 1954. Wärmeübertragung in Gaswirbelschichten. Chem.Ing.Techn., Vol. 26, pp. 301-456.

23. Karchenko, N.V. and Makhorin, K.E. 1964. The rate of heat transfer in fluidized bed and an immersed body at high temperatures. Int.Chem.Engng., Vol. 4, pp. 650-654.

24. Botterill, J.S.M. and Desai, M. 1972. Limiting factors in gas fluidized bed heat transfer. Powder Technology, Vol. 6, pp. 231-238.

25. Zabrodsky, S.S. 1966. Hydrodynamics and Heat Transfer in Fluidized Beds. The M.I.T. Press.

26. Muchowski, E. 1979. Transient Heat Transfer Between a Perfect Conductor with Heat Generated in it and a Semi-infinite Solid Including a Contact Resistance. Wärme- und Stoffübertragung, Vol. 12, pp. 161-164.

# Fluidized Bed Heat Exchangers: Theory and Practice

BIROL İ. KILKIŞ
Mechanical Engineering Department
Middle East Technical University
Ankara, Turkey

ABSTRACT

The performance of fluidized bed heat exchangers depend upon the quality of fluidization which is closely related with the bubbling phenomena. The associated hydrodynamical instabilities taking place especially around the heat exchanging bundles draw much attention of the designers. Furthermore, local fluidization phenomena as observed and numerically analysed by Kılkış [1] helps to predict the performance of a fully fluidized bed. This paper presents experimental and numerical investigations of the Author about bubble behaviour and local fluidization characteristics. After a short review of the relevant theory, bubble behaviour at different conditions are discussed. The mechanism of local fluidization is explained and sample computer results are presented in relation to typical applications.

NOMENCLATURE

| | |
|---|---|
| $Al_f$ | Dimensionless local fluidization thickness ($u/D_o$) |
| $A_o$ | Dimensionless diameter of local instability bubble ($D_b/D_o$) |
| C | Pressure drop constant |
| $C_p$ | Heat capacity |
| D,d | Diameter |
| f | Frequency |
| $g; g_c$ | Acceleration of gravity; conversion factor |
| h | Heat transfer coefficient |
| k | Thermal conductivity |
| $L_s$ | Depth of obstacle center from the top surface of the fluidized bed |
| $L_f$ | Bed height |
| $L_p$ | Pipe length |
| Nu | Nusselt number |
| P | Pressure |
| Pr | Prandtl number |
| Re | Reynolds number |
| T | Temperature |
| t | Time |
| $t_p$ | Residence time of particles in the first row against wall |
| $t_o$ | Periodicity of bubble transit |
| $t_b$ | Bubble transit time |
| $U_o$ | Superficial velocity |
| u | Maximum thickness of local fluidized region |
| V | Velocity |

Subscripts

| | |
|---|---|
| B | Bed |
| b | Bubble |
| f | Fluid |
| g | Gas |
| i;f | Initial, final |
| L | Local |
| m | Emulsion phase |
| mf | Minimum fluidization |
| n | Normal, normalized |
| o | Obstacle, particle contacting |
| p | Particle, packed bed |
| s | Solid |
| t | Terminal |
| v | Vertical, vessel |
| w | Wall |

Greek Symbols

| | |
|---|---|
| $\alpha$ | Total heat transfer coefficient |
| $\beta;\gamma;\theta$ | Geometrical parameters of locally fluidized region |
| $\mu$ | Viscosity |
| $\nu$ | Kinematic viscosity |
| $\delta$ | Average space between the wall and first row of facing particles |
| $\varepsilon$ | Void fraction of the bed |
| $\phi_s$ | Sphericity of solid particles |
| $\rho$ | Density |
| $\gamma_p$ | Specific weight of the particle. |

## 1. INTRODUCTION

Fluidized beds are used extensively for heat exchanging purposes in both physical applications and chemical processes mainly because of their unique ability to rapidly transport heat and maintain a nearly uniform temperature. Fluidization is the condition that results when a fluid is blown through a grid at a rate sufficient to overcome the static weight of the solids atop the grid to a point where the mass of these solid particles, referred to as the bed behaves like a liquid. One of the most important features of fluidized beds is the high heat transfer rate occuring between the interior of the bed and the walls of contacting surfaces. Apart historical applications such as gasification of powdered coal (After Winkler, 1922), rising fuel prices, depletion of primary energy sources, and the demand for better operations and processes have drawn attentions to fluidized bed combustors and inherently fluidized bed heat exchangers. Recent advances in material technology have also helped in commercializing such systems. As an example; adoption of gas turbine nickel-base alloys to fluidized bed steam turbine propulsion systems resulted in compatible designs. In this system fluidization technique is both employed at the combustor section itself and other components such as coolers and high temperature chemical reactors. Regardless of whatever the process or operation is, the problem of heat exchange is the major challenging subject which is attacked by many investigators both theoretically and experimentally. The advantages of this technique have usually been listed as:

i- Extremely large area of contact between solids and the fluid.
ii- The comparative ease with which solids can be handled.
iii- Nearly uniform temperature distribution.
iv- High rates of heat transfer taking place between the solid particles and an

immersed surface.

There are of course certain limitations inherent within the fluidization technique:
i- The process demands extra expenditure of power for fluidization.
ii- Some solids can not be fluidized and usually there exists a limit to the size of fluidizable solid particles.
iii-The dynamics of the bed is not sufficiently well understood so as to satisfactorily scale-up the laboratory scale data to large scale units. The bed performance depends upon various parameters which in turn these may vary so much that a theory could not been unified yet. Available equations can not be readily adopted to specific design problems.
i- In order to increase particle mixing and agitation, the operation regime is generally selected so as to achieve a "boiling bed" where it means in a gas fluidized bed that excess gas passes through the bed vessel in the form of practically particle free voids (bubbles).Bubbles behave in very varied ways and general bubbling behaviour is much affected by the presence of immersed heat exchanging probes (tube bundles). Although faster particle agiation generally means a further increase in the heat exchange rate, bubble transit across the tube surfaces reduce the time average of this heat exchange. Pressure and temperature fluctuations and variations thus occuring over the tube surfaces create non-uniform thermal and mechanical stresses and vibrations whose mechanisms and effects has not been yet sufficiently exploited.

In this paper, it has been shown both experimentally and theoretically that the gas flow rate which is in fact the main process controlling variable, plays also an important role on the bubble behaviour.The general hydrodynamical behaviour of a bubbling bed is analysed in certain extent so as to provide useful information for predicting the bed performance.In this paper a numerical solution is presented which helps to position and align heat exchanging pipes correctly for optimum performance.These recent achievements lead one to state that fluidized beds offer excellent media for heat exchanging purposes provided that care is taken during the design phase.

## 1.1. Principles of Fluidization

Figure.1 shows different regimes of a typical bed.Starting from "No Flow" conditions,particles are in physical contact with each other and are densely packed.The same is also true for very low flow rates.The flow rate is generally identified by Superficial Velocity($U_o$) defined at the inlet as follows:[1]

$U_o$ : Volumetric flow rate / Inlet area     (1)

As the superficial velocity is increased,particles start to unlock.If superficial velocity reaches a specific value ( $U_{mf}$ ),weight of the particles are practically counterbalanced by the drag force exerted through the upward moving gas.This condition is known as the "Minimum Fluidization Condition ( MFC ),and the bulk of the bed starts to exhibit liquid-like behaviour.There are basic differences in the dynamic behaviour of gas and liquid fluidized beds beyond MFC.In gas fluidized beds,a further increase in $U_o$ imposes large instabilities in the form of so called "bubbles" as shown in the third drawing in Figure.1 (Heterogenous Fluidization).In liquid fluidized beds no bubbling is observed.

---

1).Symbols are explained in the following section.

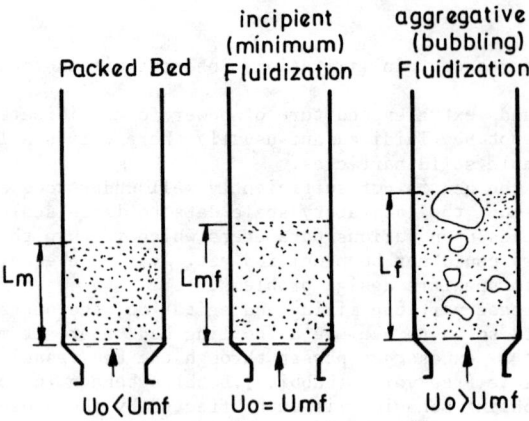

Fig.1. Different bed regimes

Minimum fluidization velocity $U_{mf}$ can be predicted as follows:

$$U_{mf} \cong \frac{(\phi_s d_p)^2}{150} \cdot \frac{\rho_s - \rho_g}{\mu} \cdot g \cdot \left(\frac{\epsilon_{mf}^3}{1-\epsilon_{mf}}\right) \qquad Re_p < 20 \qquad (2\text{-a})$$

$$U_{mf}^2 \cong \frac{\phi_s d_p}{1.75} \cdot \frac{\rho_s - \rho_g}{\rho_g} \cdot g \cdot \epsilon_{mf}^3 \qquad Re_p > 1000 \qquad (2\text{-b})$$

These expressions are derived by extrapolating Ergun's data [2] which is as follows:

$$\frac{\Delta P}{L} g_c = 150 \frac{(1-\epsilon_m)^2}{\epsilon_m^3} \cdot \frac{\mu U_o}{(\phi_s d_p)^2} + 1.75 \frac{1-\epsilon_m}{\epsilon_m^3} \cdot \frac{\rho_g U_o^2}{\phi_s d_p} \qquad (3)$$

and using the following condition:

$$\frac{\Delta P}{L_{mf}} = (1-\epsilon_{mf})(\rho_s - \rho_g)\frac{g}{g_c} \qquad (4)$$

For constant bed properties, the unit pressure drop is constant over the bed height.

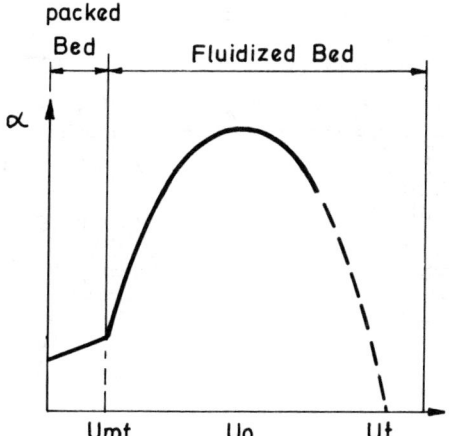

Fig.2. Variation of Heat Transfer Coefficient with $U_o$.

Bubbles behave in very varied ways in gas fluidized beds. Although generally they are observed to grow as gas flows into them from the surrounding continuous phase and as a lower bubble is caught and coalesces with one at above, one may be observed to diminish. They may divide spontaneously or on contact with an obstruction. Bubbles also tend to break when encountering horizontal tubes. However this tendency depends upon the relative size of tubes with respect to bubbles and their configuration as will be discussed later. Another factor which strongly effects the quality of bubbling fluidization is the distributor type which primarily effects the bubble incipience and formation [3].

Figure 2 shows the typical dependence of bed-wall heat transfer for a homogenously fluidized bed. After reaching $U_{mf}$ the initial rapid rise in heat transfer coefficient is explained by the increased solid circulation whereas the decrease beyond the maximum is attributed to the lowered solid concentration which finally reaches down to zero at terminal speed ($U_t$). A similar dependence may be expected for heterogenous beds with certain differences which strictly depend upon the bubble behaviour.

1.2. Heat Exchanging Applications

Heat exchanging applications of fluidized beds can be grouped in two categories.One of them is the physical operations such as direct heat exchanging and the other is chemical processes such as solid heating/cooling.Figure.3 shows a fluidized bed cooler for hot alumina particles.Water passing through the heat exchanging tubes is the primary coolant and cool fluidization air accomplishes a partial extra cooling.Cool particles are taken out from the unit at the cold water inlet side.Figure.4 illustrates a fluidized batch used for reactors which need constant and high temperature operations.Typical applications employ hot combustion gases for fluidization.Air leaving the bed from the cyclone enters a secondary heat exchanger in order to recover the useful heat.

The use of solid fuels to generate power and process steam is becoming the principal strategy to reduce oil consumption and imports.Fluidized bed combustion which inherently covers also a fluidized bed heat exchanging medium is hoped to play a major role by providing for the increased combustion of solid fuels in an environmentally acceptable manner at comparable costs.It has been predicted

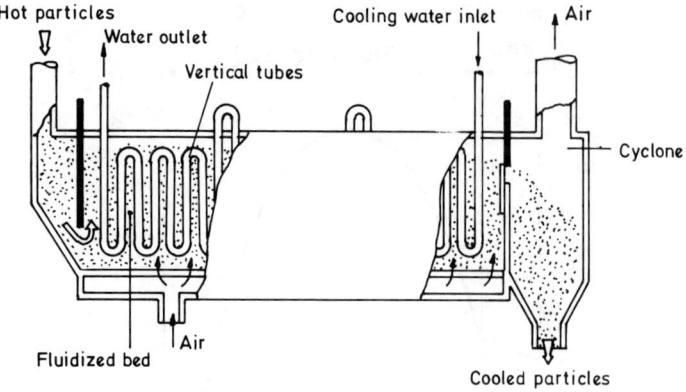

Fig.3. Fluidized Bed Cooler[3]

that this method will rapidly become the preferred method for firing solid fuels.To control the bed temperature at desired levels,the heat-transfer surface is submerged in the bed.Units with 45000 Kg/H steam generators delivering 45 Atm. of saturated steam have already been marketed commercially.Recently it has been reported that a marine steam turbine plant designed for the lower grade solid fuels is on the way of commercialization.It is characterised by high efficiency,higher steam temperatures and new principles.Lightweight machinery and shorter engine rooms are reported to be additional features[4].There are other heat exchanger applications for continuous chemical operations which are generally termed as either countercurrent or crosscurrent multistage heat-exchangers depending upon the principle of operation.However,true countercurrent operation is unobtainable because of the high degree of mixing that usually occurs within a fluidized bed.

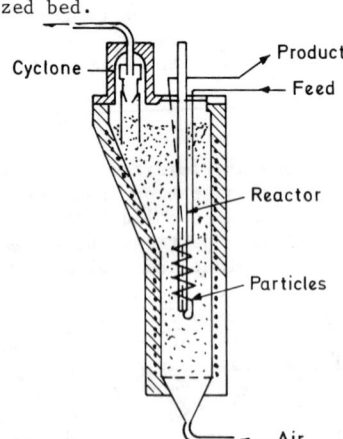

Fig.4. A Fluidized Chemical Reactor[5]

## 1.3. Hydrodynamics of Fluidized Beds

Extensive researches have been made for the assessment of the hydrodynamical behaviour of fluidized beds and their effect on the heat exchange. Considering the case of a bubbling bed at the presence of immersed heat exchanging pipes two complications are encountered. The first one is the local fluidization and the other one is the complete disturbance of flow field in a bubbling bed:

1.3.1. Local fluidization. The Author [1] had performed extensive experiments in a two-dimensional bed in order to assess local fluidization phenomena which takes place in the close vicinity of heat exchanging pipes as will be explained in more detail at later sections of this paper. Local fluidization is principally the incipience of fluidization in a small area around a portion of pipe surfaces although the bulk of the bed is not fluidized yet. This region is so peculiar that there occurs minute bubbling in those regions as shown in Figure 5. This figure shows a direct photograph of the close vicinity of a cylindrical obstacle taken by a high speed camera.

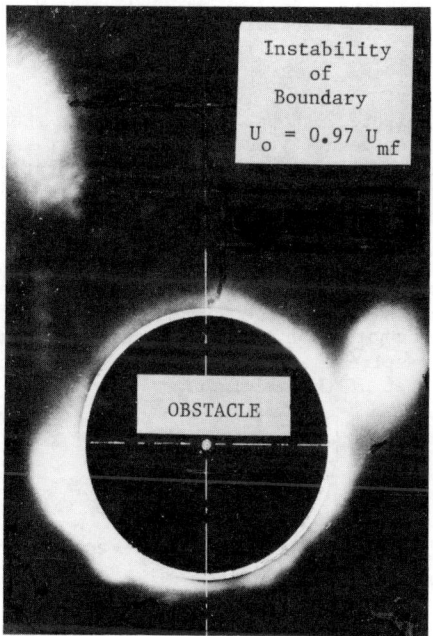

Fig.5. Local Fluidization Zone Around a Cylindrical Obstacle [1]

($U_o = 0.97\ U_{mf}$)

At superficial velocities close to $U_{mf}$, even small bubbles departing from the obstacle surface can be observed as can be seen at the upper left hand side of Figure 5. Regarding to rapid increase of heat transfer coefficient as shown in Figure 2, it may be concluded that an early increase in local heat transfer coefficient at the bottom of pipes may be expected due to this local fluidization phenomena. Local fluidization can be directly attributed to the flow disturbance around the pipes. Figure 6 shows the cross-sectional view of a simple pipe. Assuming a potential flow in the bed, local increase in the vertical gas velocity due to flow deflection can be easily expressed as follows:

$$\frac{a^2}{r^2} \cos 2\theta - 1 = \frac{U_{mf}}{U_v} \qquad (U_o < U_{mf}) \qquad (5)$$

The dotted lines show the locus of points where $U_v$ is equal to $U_{mf}$. The region bounded by this line is the local fluidization region as depicted by the simple potential flow equation.

Local fluidization is observed to start as early as $U_o$ is only 60% of $U_{mf}$. When $U_o$ reaches $U_{mf}$ the bulk of the bed is fluidized also. However there always exists a dead (locked) zone above heat exchanging pipes as shown in Figure 10. This zone fluctuates as bubbles pass but never diminishes. When the bed starts to bubble the flow field is furthermore disturbed and large instabilities occur:

1.3.2. Bubble behaviour. Figure 7 shows a typical bubble photographed as rising through the bed vessel after being injected at MFC. Incipience of bubbles over the distributor plate in a real system is completely a statistical problem and only certain generalizations about the bubble behaviour can be made. Harrison and Leung [6] found that the frequency of bubble formation at an orifice depended primarily on the gas flow rate to the orifice and was largely independent of the overall bed height, fluidizing conditions and the nature of the particles. The bubble volume flow rate is roughly equal to the excess gas flow rate. Bubbles generally grow as they rise. There are instants where growing bubbles split. Figure 8 outlines a photograph taken just at the incipience of such a separation. When a bubble splits there occurs a decrease in the size of the two formed as compared with that of the original.

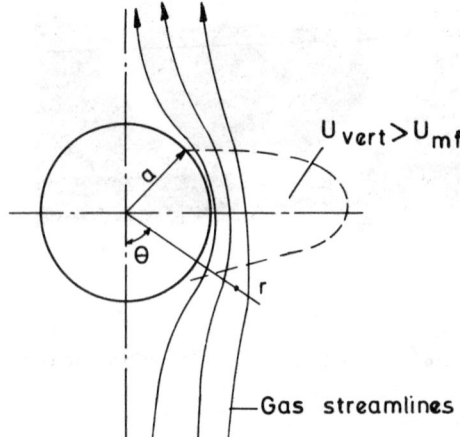

Fig.6. Flow Field Around a Cylinder

# FLUIDIZED BED HEAT EXCHANGERS: THEORY AND PRACTICE

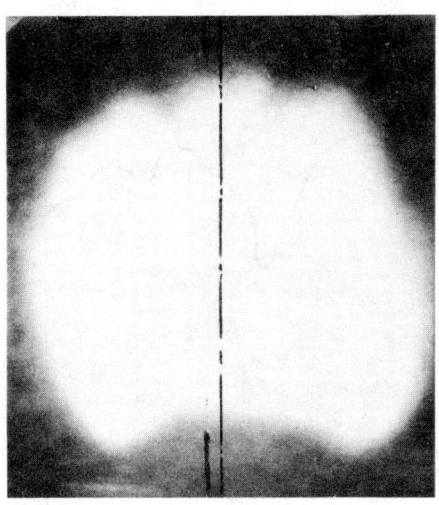

Fig.7. A Typical Single Rising Bubble [1]

Presence of "obstacles" within the bed completely change the bubble behaviour. First of all such obstacles act as collectors and bubbles tend to coalesce with those pipes. Bubbles coalescing with these pipes show varying attitudes as will be discussed at later sections; depending upon their relative sizes with respect to the pipe size. During residence of bubbles along the pipes, the heat exchange is violated and suddenly drops near to zero, because bubbles are practically particle free. Therefore the analysis of bubble coalescence, bubble residence and bubble frequency becomes very important factors in predicting the time average heat exchange rate in a bubbling bed. Finally the entire picture becomes too complicated to hope for an analytical solution, therefore approximate numerical methods still backed up by experimental data is in order.

In order to obtain a simple approach, some assumptions has to be made. One of them may be the assumption that heat transfer suddenly drops down to zero during bubble transit at the region where direct contact with the bubble occurs. When one remembers that more than one bubble may be in contact with a pipe along its length, it becomes evident that the problem can not be so easily reduced to a two dimensional case.

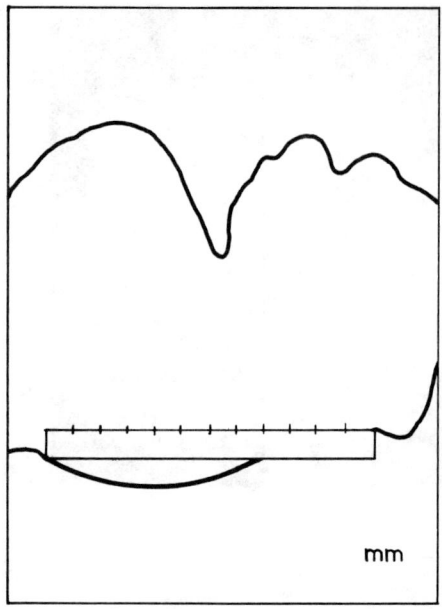

Fig.8. A Single Bubble at the Incipience of Seperation

## 2. THEORY

Bed-wall coefficients in gas fluidized beds have been found to be 20 to 40 times that for gases alone, and since the bed represents a complex interaction of gas and solid, many factors will enter into any generalized correlation. Usually the radiation effects are excluded. For a packed bed the mode of heat transfer is both conduction and convection ($U_o>0$), for a fluidized bed the mode of heat transfer is practically convection. The variables to be considered in the convection process are:
1. Properties of gas: $\rho_g, \mu, C_{pg}, k_g$
2. Properties of solid: $d_p, f_s, \phi_s, C_{ps}$
3. Conditions at minimum fluidization: $U_{mf}, \varepsilon_{mf}$
4. Flow conditions: $U_o, \varepsilon_f$
5. Geometrical parameters: Bed dimensions (vessel diameter $d_v$ for cylindrical bed), pipe dimensions ($D_o$), pipe layout, $L_s, L_f, L_p$ (pipe length)
6. For a bubbling bed: f(frequency of bubbles), $D_b$, $t_o$ (residence time) and $V_b$ (velocity of bubble).

The above variables in relation with bubbles can be reduced to a single variable namely the well-known $U_o$ with certain approximations as will be discussed in Section 4.

In practice, despite encouraging laboratory results, attainable heat transfer coefficients when radiant heat transfer is insignificant have seemed to be limited to about 400 W/m²°C and may drop down to 60 W/m²°C. In other instances marked temperature gradients have occured within fluidized bed reactors when uniform temperatures were required. The reason for these divergences from expected behaviour lie in the complexing of bubbling fluidized beds. Although many empirical correlations relating bed to surface heat transfer coefficients have been proposed, they are of doubtful validty because they can not make adequate reference for the dynamic behaviour of a gas-fluidized system. For this reason the rest of this paper is devoted to a deeper assessment for the dynamical behaviour of bubbles in relation to the heat exchange rate.

## 2.1. Heat Transfer in a Homogenously Fluidized Bed

Before starting to analyse fluidization region, a short review of packed bed heat transfer will be made:

The heat transfer coefficient $h_p$ between the surface of a sphere of diameter $d_p$ and a fluid through which it is moving with relative velocity $U_o$ is given by Ranz and Marshall [7] as:

$$Nu_p = \frac{h_p \cdot d_p}{k_g} = 2 + 0.6 \, Pr^{1/3} \cdot Re_p^{1/2} \tag{6}$$

In a system of packed beds, the average $Nu_p$ depends upon the type of packing of particles [8]. However Equation (6) can be used with a slight modification to express heat transfer in beds of coarse solids:

$$Nu_p = 2 + 1.8 \, Pr^{1/3} \cdot Re_p^{1/2} \qquad Re_p > 100 \tag{7}$$

Joly [8] made extensive simulative experiments and have concluded that $Nu_p$ can be expressed in the following form:

$$Nu_p = C \cdot Pr^b \cdot Re_p^a \tag{8}$$

He gives the corresponding approximate constants as follows:

$C = 1,5$

$b = 0$

$a = 0,38$

Experimental findings of many investigators for plug flow conditions at the onset of fluidization show a systematic trend at lower Reynolds numbers. All the available data reported on this basis in correlated by the following equation [3]:

$$Nu_p = 0.3 \, Re^{1/3} = C \cdot Re^a \tag{9}$$

From the definition of $Re_p$:

$$\alpha = \frac{\dot{Q}}{(T_p - T_g)} = h_p \times \Pi \times d_p^2 \times (T_p - T_g)/(T_p - T_g)$$

$$= \Pi \times d_p \times k_g \times C \times \left[\frac{U_o \cdot d_p}{\nu}\right]^a \tag{10}$$

For a given system and known operating conditions, Equation (10) reduces to:

$$\alpha = A' \times [U_o]^a \tag{11}$$

This simple relationship can be observed in Figure 2.

An expression for the local heat transfer coefficient ($\alpha_L$) is given by Zabrodzky [9] as follows:

$$\alpha_L = \frac{\dot{Q}_L}{(T_W-T_B)} \frac{C_p \cdot \gamma_p \cdot d_p}{3.9\, t_p} \cdot (1-\varepsilon)^{2/3} \cdot [1-e^{-\frac{1.2 \times k_g}{d_p \cdot \delta_W \cdot \gamma_p \cdot c_p} t_p}] \cdot \left(\frac{T_W-T_{pi}}{T_W-T_B}\right) \qquad (12)$$

Figure 9 shows the temperature history of the particles where:

$t_p$ = Average residence time of particles in the first row facing a unit surface heated up element of the wall (exchange surface)

$T_{pi}$ = Average initial temperature of particles in the first row

$T_{pf}$ = Average final temperature of particles leaving the first row.

In this Figure, two typical cases are shown. Case A shows the time history of a particle whose residence time is small. Such a particle has an initial temperature $T_{pi}$ which is higher than $T_B$ due to the fact that particle A is already close to the surface element. The heat flux depends upon the temperature difference ($T_W-T_p$) at any given instant. Due to this high temperature gradient, the temperature time curve is steep. Another particle coming from the bulk of the bed (particle B) has slightly lower $T_{pi}$. ($T_{pi}=T_B$). For such a particle with longer residence time, the temperature-time curve will be more steep and eventually $T_{pf}$ will be higher than that of particle A. However in terms of the local heat transfer coefficient ($\alpha_L$), the heat transfer rate will increase as residence time decreases. Similarly $\alpha_L$ will increase with increasing ($T_W-T_{pi}$). These conclusions exemplifies the relationship between $\alpha_L$ and $t_p$, $T_W$ and $T_{pi}$ as depicted in Equation (12). This equation is valid for a smooth fluidization and a plane wall. For such a case the local heat transfer coefficient becomes identical with its global value for a contacting medium. However the same equation can be taken as a basis for the development of the theory for a bubbling bed as will be given in Section 3.2.

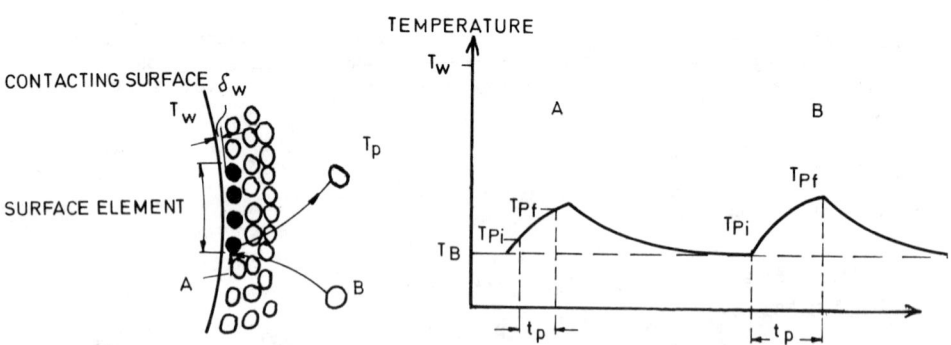

Fig.9. Time History of Particles

A number of experimental studies exist for immersed vertical tubes in homogenously fluidized beds. Following equation is proposed both for heating and cooling purposes [3].

$$\frac{h_w d_p}{k_g} = 0.01844 \, C_R(1-\varepsilon) \cdot \left(\frac{C_{pg}\rho_g}{k_g}\right)^{0.43} \cdot \left(\frac{d_p \rho_g U_o}{\mu}\right)^{0.23} \cdot \left(\frac{C_{ps}}{C_{pg}}\right)^{0.8} \cdot \left(\frac{\rho_s}{\rho_g}\right)^{0.66} \quad (13)$$

$$10^{-2} < \frac{d_p \rho_g U_o}{\mu} < 10^2$$

$C_R$ represents the correction factor for non-axial tube location. $C_R$ varies between 1 and 2 depending upon dimensionless position from the center of vessel [10]. Average deviation is ± 20 % for the 423 data points used in the correlation. ($C_{pg} \rho_g/k_g$) has units of $sec/cm^2$. Vreedenberg [10] investigated the effect of tube diameter, particle size, shape and density, and gas velocity on the heat transfer coefficient on immersed horizontal tubes, including data on large-scale beds:

$$\frac{h_w d_o}{k_g} = 0.66 \left(\frac{C_{pg} \cdot \mu}{k_g}\right)^{0.3} \cdot \left[\left(\frac{d_o \cdot f_g \cdot U_o}{\mu}\right)\left(\frac{\rho_s}{\rho_g}\right)\left(\frac{1-\varepsilon}{\varepsilon}\right)\right]^{0.44} \quad (14)$$

2.2. Local Fluidization

When a solid obstacle (as tube) is immersed in a fluidized bed, the flow-fields on both gas and particles are modified. For the case of a packed bed ($U_o \leq U_{mf}$) an important result can be noted. Around such an obstacle, the vertical component of the gas increases and can eventually exceeds the minimum velocity even    the bulk of the bed has not been fluidized yet. This local fluidization zone primarily depends upon the obstacle geometry as shown in Figures 10, and 11. A careful analysis of these zones for a bed with multiple bundles yield useful information for the performance characteristics in the fluidized regime.

3.3. Complications in a Bubbling Gas Fluidized Bed

An expression similar to Equation (12) can be derived for the local heat transfer coefficient at the presence of bubbles [11] with following assumptions:
 i- The heat transferred during the bubble transit is negligible.
 ii- The "local" heat transfer is considered. So a particle enters or leaves the "first row" facing the surface element by sliding along the wall (particle A on Figure 9). As a consequance $T_{pi}$ can be slightly greater than the average bulk temperature of the bed.
 iii- Void fraction ($\varepsilon$) for heterogenous fluidization is essentially constant and equal to the void fraction of the emulsion phase ($\varepsilon \simeq \varepsilon_{mf}$). The expansion of the bed as a function of superficial velocity is taken into account by the variation of $t_b/t_o$. Here $t_b$ is the bubble transit time across the obstacle and $t_o$ is the periodicity of bubble passage:

$$\alpha_L = \frac{C_p \cdot \gamma_p \cdot d_p}{3.9 \, t_p} \cdot (1-\varepsilon)^{2/3} \cdot \left[1 - e^{-\frac{1.2 \cdot k_g}{d_p \cdot \delta_w \cdot \gamma_p \cdot c_p} t_p}\right] \cdot \left(\frac{T_w - T_{pi}}{T_w - T_B}\right) \cdot \left(\frac{t_o - t_b}{t_o}\right) \quad (15)$$

Here the last term $\left(\frac{t_o - t_b}{t_o}\right)$ exemplifies a linear time proportioning for the local heat transfer rate.

For vigorous agitation the exponent (e) is small ($t_p$ is small), and the expression $\alpha_L$ can be written as:

$$\alpha_L \simeq (1-\varepsilon)^{2/3} \frac{k_g}{\delta_w} \left(\frac{T_w-T_{pi}}{T_w-T_B}\right) \cdot \left(\frac{t_o-t_b}{t_o}\right) \qquad (16)$$

A further simplification can be made if the particles close to the wall are frequently replaced and have the opportunity to give off their excess heat before returning to the surface. In this case the factor $(T_w-T_{pi}/T_w-T_B)$ approaches its upper limit 1 if $T_{pi} \simeq T_B$ [11].

## 3. ANALYSIS

Although equation (16) seems to be simple, the term $(t_o-t_b/t_o)$ is discouragingly difficult to predict for a rigorously bubbling bed with immersed tube bundles. Experimental and theoretical investigations have been performed in order to assess this difficult problem. Experiments yielded the general functional trends and indicated certain boundary conditions which were found to be useful for the theoretical work. All experiments were performed in a two-dimensional bed with inside dimensions of 30 x 555 x 1500 mm. Glass beads of 0.5 mm diameter are fluidized by air sucked through the bed at atmospheric pressure. A number of photodiodes are imbedded in the bed wall and obstacle surface in order to record and determine bubble speed and population (see Figure 11).

### 3.1. Experimental Foundings

Typical results for a single tube with diameters varied between 5 and 10 cm were tested for assessing local fluidization characteristics. A short cylindrical obstacle ($D_O$=5 and 10 cm) is immersed. Figure 10 shows the development of local fluidization with respect to increasing $U_O$. A small region starts to develop at $U_O$=0.7 $U_{mf}$. First, a distinct zone is formed on each side. Small bubbles are formed and move along the obstacle surface. They disappear at the upper boundary of fluidized region. As $U_O$ is increased, the two zones expand more and merge at the upstream (lower) part of the obstacle. At the same time fluidization occurs into two channels up to the top of the bed. Bubbles detach from the obstacle and move upwards. There remains a dead region on the upper surface of the obstacle. The size of this "wake" is only weakly dependent upon $U_O$. Angle $\gamma$ depends upon the obstacle (tube) diameter ($D_O$), and the depth of the tube from the top surface of the bed ($L_s$). As $L_s$ increases, $\gamma$ angle also increases. The same is also true for $D_O$.

Although the present results are not sufficiently complete to describe all details of the local fluidization, a number of valuable conclusions can be drawn for the conditions which are of practical interest($U_O \simeq U_{mf}$) and for the particle behaviour close to the obstacle surface. As one can see from Figure 10, fluidization is confined to a region extending from the lower part of the obstacle up to 40-45 degrees above the horizontal plane of symmetry. As a result, a relatively high heat transfer rate is expected over 3/4 of the heat exchanger surface. This fraction depends on the cross-sectional geometry. However, this fraction is almost independent of $U_O$ for $U_O/U_{mf}$ ratios greater than 0.9. The frequency of bubble passage is in the order of 10 per second as can be seen in Figure.11. A distinction must be made between different points on the obstacle. At point 1, the pressure fluctuations are rather erratic, while at points 2 and 3, a definite periodicity can be observed. Small blocks in Figure.11 shows pressure recordings at points 1,2,and 3 respectively. The pressure variation at point 3 shows an alter-

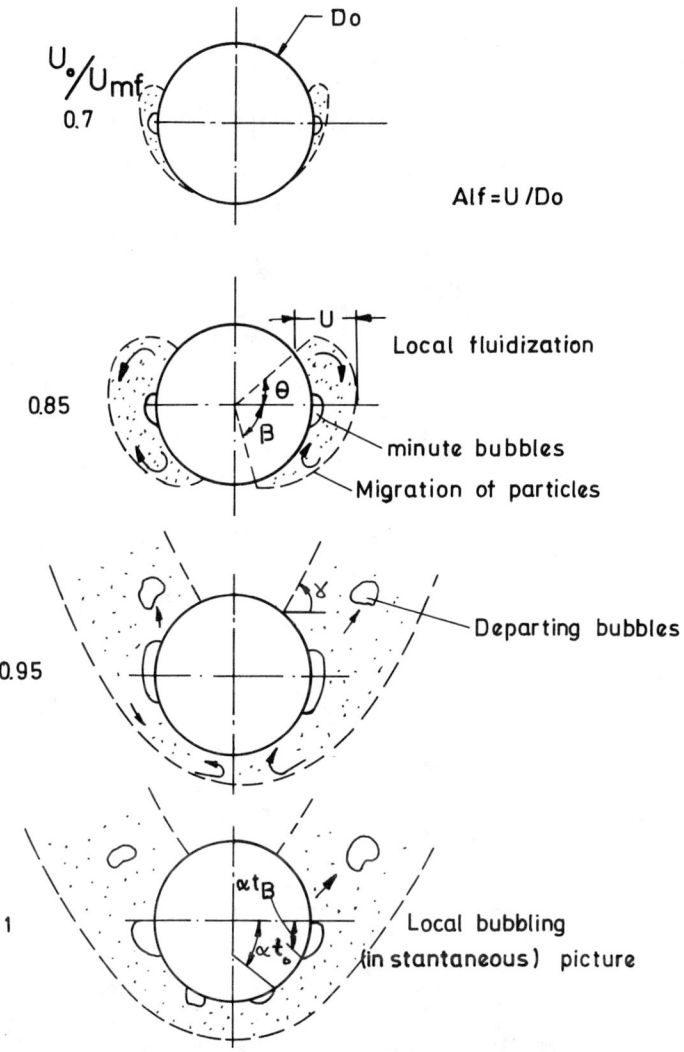

Fig.10. Development of Local Fluidization Around a Cylindrical Obstacle.

nation of smaller and larger peaks.This indicates that a second bubble is formed before the first one leaves the obstacle.Bubbles grow during their motion at a rate faster than could be explained by the expansion of the gas.Thus,it can be concluded that gas from the emulsion phase flows into the bubble.An estimate can be made of the factors $(t_o-t_b/t_o)$ and $t_p$, from Figure.11 in relation to Equation No.15. $(t_o-t_b/t_o)$ varies in the range 1/2(point2)to 1/3(point3) due to the growth of the bubble.A similar variation in heat transfer can be expected.The upper limit for $t_p$ is given by $t_o-t_b$ and varies in the range 1/2x1/10 to 1/3x1/10 seconds.The true value of $t_p$ is smaller because of the agitation in the wake of each

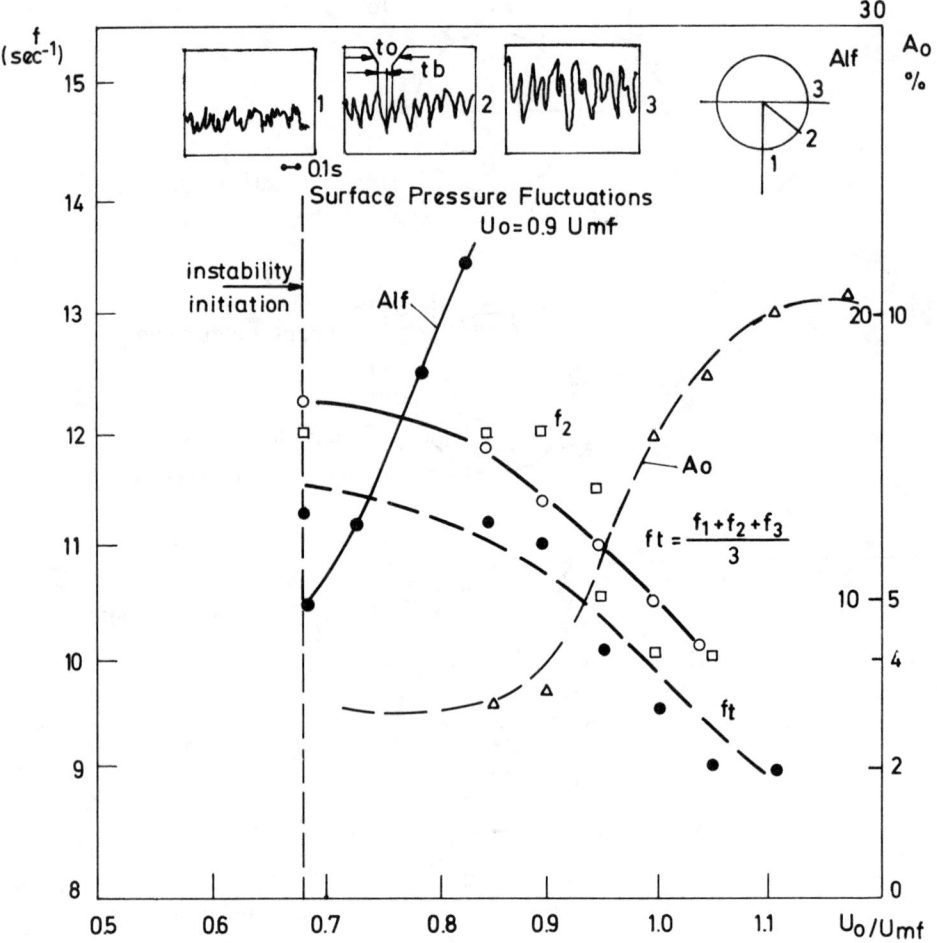

Fig.11. Amplitude, Frequency and Pressure Fluctuations Around a Cylindrical Obstacle

bubble. The wake occupies a large fraction of the emulsion phase in between two successive bubbles. This figure shows that the average frequency ( $f_t$ ) decreases as $U_o$ increases. Conversely, $A_o$ and $A_{1f}$ increase with $U_o$.

These cylindrical obstacles were also used in the fluidization regime. First, the bed was just fluidized for a value of $U_o \simeq 1.1\ U_{mf}$, and a single bubble was injected from an orifice placed at the middle of the distributor plate. For the present case, only a bubble moving along the centerline of the bed is considered. The diameter of the bubble is the only parameter which was changed. Secondly, vigorously bubbling bed was analysed to derive some conclusions concerning with coalescence and mixing.Pressure variations at a point on the obstacle surface were recorded by an appropriate transducer.Figure.12 shows a typical recording. As the bubble rises from the bottom and approaches to the obstacle,pressure field near the obstacle is distorted and pressure on any point on the obstacle start to increase.When the bubble touches the obstacle,nearly a constant rate pressure decrease is observed until the bubble leaves the obstacle completely. This constant rate is directly related to the bubble pressure which is constant in the bubble at a given time but decreases with the advent of rising in the bed. The bubble pressure is equal to the static bed pressure at the bed height corresponding to the current bubble position defined by its center of gravity.Consequently as the bubble rises,bubble pressure decreases as with the bed.This constant decreasing period directly gives the bubble residence time ( $t_b$) on the obstacle.The slope indicates the bubble velocity which is nearly constant here. As the bubble leaves the obstacle,nominal pressure starts to assume again the static bed pressure at obstacle level.Small ripples on the curve shows that local fluidization is always present except when the bubbles are actually in contact with the obstacle.The residence time is observed to be strongly dependent upon the bubble diameter.Amount of delay which is a ratio between the actual residence time and the time that the bubble would cover the same distance at the absence of the obstacle is:

DELAY : $t_{o_m} / t_{o_h}$ (17)

Here $t_{o_m}$ is the measured (actual) residence time and $t_{o_h}$ is the transit time at the absence of an obstacle.The transit time can be expressed in terms of bubble length end the bubble velocity.The effective bubble length along its vertical axis is approximately 0.97 $D_b$ ,and $V_b$ : $0.57( g.D_b)^{1/2}$.

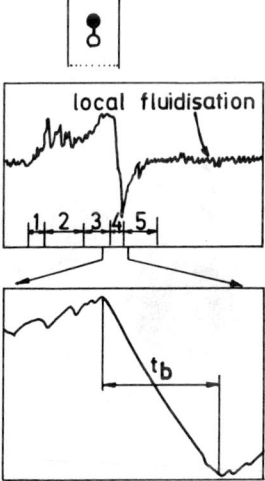

Fig.12. Surface Pressure Variation at the Vicinity of an Obstacle.

Then:

$$t_{o_h} : 0.97 \cdot (D_b)^{1/2} / 0.57 \cdot (g)^{1/2} \qquad (18)$$

It is found that as bubble size increases, bubbles "feel" the presence of an obstacle more. Small bubbles, simply split into two and these are rapidly swept upwards along the obstacle, whereas large bubbles tend to remain practically as a single unit and are deflected more from their original path and reside more on the obstacle. Figure 13 shows a photograph of a large bubble at the vicinity of a cylindrical obstacle. Figure 14 indicates a linear relationship between the delay and relative bubble size ($D_n$).

The measured residence time of bubble remains practically the same over the entire obstacle surface provided that $D_n$ is greater or equal to unity. In other words, the bubble grasps the obstacle instead of just slipping and bouncing away from the obstacle. This situation is clearly shown in Figure 15. The only exception for this situation is that very large bubbles tend to move along one side of the obstacle. In the view of experimental results, the following generalizations can be made:

i. During the bubble transit, heat transfer is suppressed. This is an unfavorable effect. It becomes more important with an increase in $U_o$, because bubble diameter and frequency also increases with $U_o$. Therefore a reduction in the peak $\alpha_L$ value as given in Figure 2 should be expected. The situation becomes worse due to the fact that delay period increases as bubble diameter increases. One can therefore conclude that $U_o$ should not be much greater than $U_{mf}$.

ii. The stationary particles in the wake above the obstacle are pushed aside by the bubble and replaced by fresh emulsion. So this part of the surface also becomes partially active in the heat transfer process. This favourable effect is more pronounced for larger bubbles which have a tendency to move along one side of the obstacle.

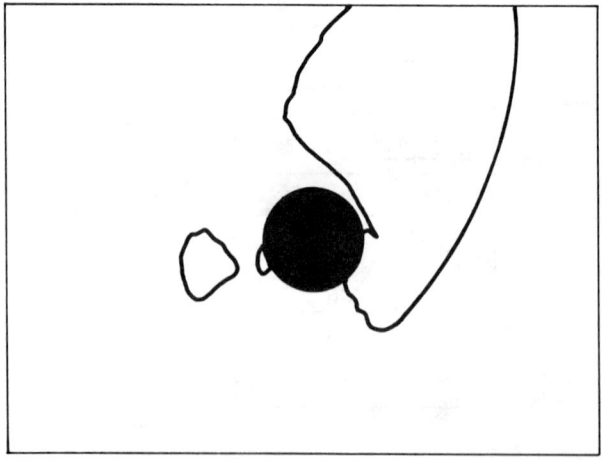

Fig.13. Behaviour of a Large Bubble Near an Obstacle

# FLUIDIZED BED HEAT EXCHANGERS: THEORY AND PRACTICE

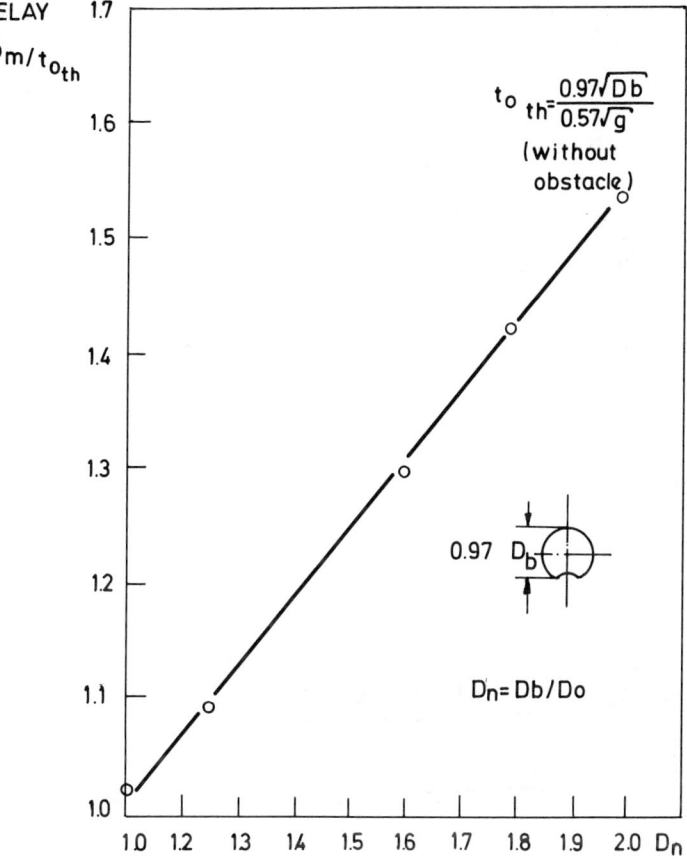

Fig.14. Bubble Delay at the Vicinity of an Obstacle

The above two items show that it is necessary to optimize the $U_o$ value. However the problem is not so simple because the actual problem consists of multiple obstacles and bubbles. However Equation (16) can be further simplified for a simple rising bubble also considering delay.

$$\alpha_L \simeq (1-\varepsilon)^{2/3} \frac{k_g}{\delta_w} \left( \frac{t_o - t_b}{t_o} \right) \qquad (19)$$

$\delta_w$ is a function of obstacle geometry and particle size. As an example; $\delta_w$ is approximately $d_p/6$ for moderately sized cylindrical obstacles. The term $t_o-t_b/t_o$ can be related to $U_o$ as follows:

$$\frac{t_o - t_b}{t_o} = g(t_{o_m}) = g'(D_n) = g'(D_b/D_o) \qquad (20)$$

$D_b$ is obviously a function of $U_o$. For normal operating conditions and medium sized particles [1,3] the following relationship can be used:

$$D_b \simeq 0.15\, U_{mf}(U_o/U_{mf}-1) \qquad [cm] \tag{21}$$

Therefore

$$t_{o_m} = f'(U_o) \tag{22}$$

Then

$$\alpha_L \simeq (1-\varepsilon)^{2/3} \frac{k_g}{6 \cdot d_p} f(t_{o_m}) = f''(U_o) \tag{23}$$

for given system parameters.

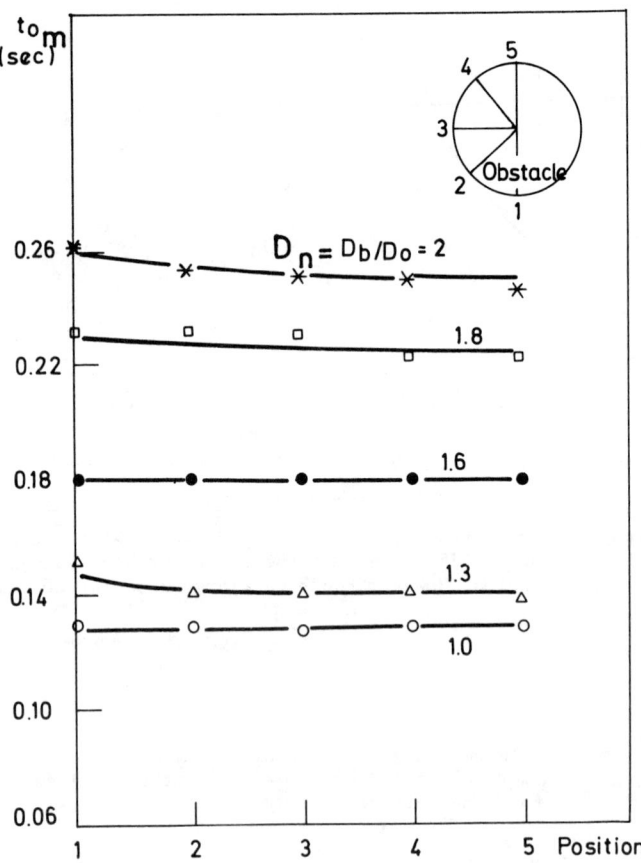

Fig.15. Variation of Residence Time with $D_n$ and Position

# FLUIDIZED BED HEAT EXCHANGERS: THEORY AND PRACTICE

## 3.2. Numerical Solution

When a solid obstacle is immersed in a fluidized bed, the flow fields of both particles and gas are modified. The Davidson model for the flowfield around a single rising bubble has been, in spite of its simple assumptions, a powerful tool for the prediction of bed behaviour. For the case of an immersed obstacle an analoque model can worked out:

Postulates: i- The gas flow in the emulsion phase satisfies Darcy's law
ii- Particles move around the obstacle as would an incompressible inviscid fluid.

Boundary conditions:
i- Far from the obstacle, the undisturbed pressure gradient exists.
ii- Neither gas nor solids can cross the obstacle surface.

Figure 16 shows the assumed boundary conditions.

Within the bulk of the bed:

$$\nabla^2 P = 0 \qquad (24)$$

From Darcy model;

$$U_g = U_p - C \text{ grad } P \qquad (25)$$

with $U_p \simeq 0$ (Davidson model):

$$U_g|_y = -C \frac{\partial P}{\partial y} \qquad (26)$$

$$U_g|_x = -C \frac{\partial P}{\partial y} \qquad (27)$$

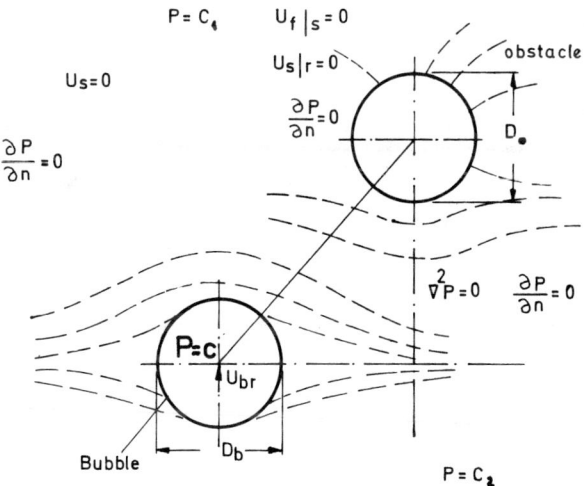

Fig.16. Boundary Conditions [12]

Where C can be solved at the bed inlet conditions at incipience of fluidization:

$$C = -\frac{\varepsilon^3 \cdot (\phi_s \cdot d_p)^2}{150\mu(1-\varepsilon)^2} \cdot g_c$$

A two dimensional finite element program was developed [12] for axi-symmetric problems to solve the pressure field and derive necessary relationships for a bubbling regime.

The same program was also used to test different tube bundles at the local fluidization regime. Figure 17 shows two rectangular tubes. As seen from this figure, local fluidization starts at $U_o/U_{mf}=0.75$ on the side walls of the tubes. Locally fluidized region expands as the $U_o/U_{mf}$ ratio increases. However there always exist a dead zone (wake) at the top surface and between the tubes.

Different tube geometries were tested using the computer program and certain rules were derived:

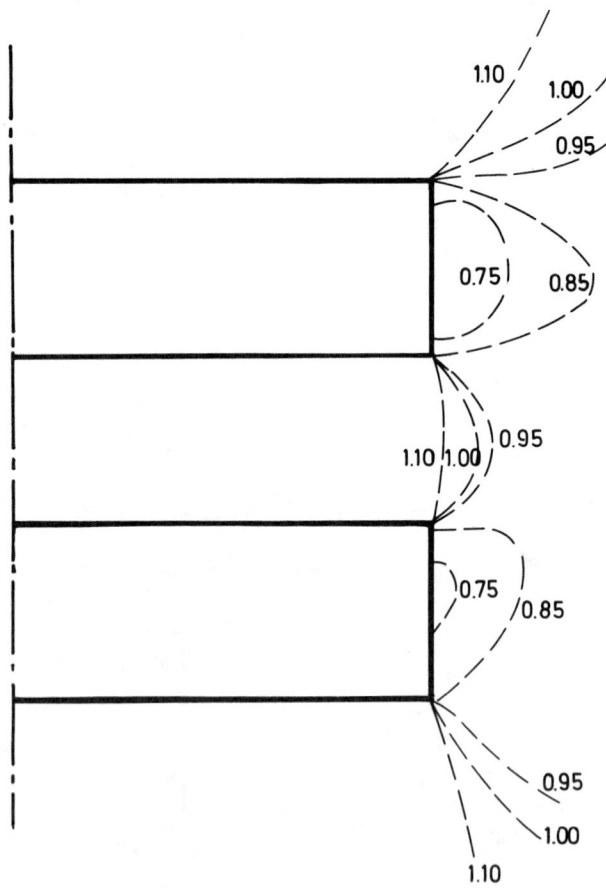

Fig.17. Incipience and Expansion of Local Fluidization Region Around Two Parallel Rectangular Tubes

1. Tubes should be placed as high as possible within the bed. Yet care should be paid, because bubble diameter increases with the bed height and large bubbles reduce the heat transfer rate.
2. At a fluid bed height, simple tube sections can be ranked for their efficiencies as follows; square, ellipse, finned tube, circle.
3. Vertical tubes rather than horizontal tubes should be preferred.
4. In order to eliminate the dead zones in between the tubes, they should be appropriately positioned.

## 4. CONCLUSIONS

In this paper it is attempted to asses the possibilities of gaining more insight to the hydrodynamics of fluidized bed heat exchangers. Experimental and theoretical studies were carried out to correlate the heat transfer coefficient with the bed hydrodynamics. Generalizations derived from this study well agree with already existing ones showing that this numerical algorithm can be used as an efficient tool in designing fluidized bed heat exchangers. By an iterative approach, the designer can select the design parameters to suit his requirements best. The nature of the results and the type of variables in the governing heat transfer equations imply that these data may be easily adopted for predicting heat transfer coefficients.

## REFERENCES

1. Kılkış, B. 1972. An investigation into the hydrodynamics of fluidized bed. Von Karman Institute Project Rp. 1972-318.

2. Ergun, S. 1952. Chem. Eng. Progr.(48) p.89.

3. Kunii, D. and Levenspiel, O. 1969. Fluidization Engineering. John Wiley and Sons, Inc.

4. Bredin, H.W. 1979. Fluidized bed combustion. Mechanical Engineering, Jan. pp. 30-32.

5. Adams, F.C., Gernand, M.O., and Kimberlin, Jr..1954. Ind. Eng. Chem.,(46), 2485.

6. Harrison, D. and Leung, L.S. 1961. Trans. Inst. Chem. Eng.(39). p. 409.

7. Ranz, W.E. and Marshall, Jr..1951. Chem.Eng.Progr.(48). p. 141.

8. Joly, C. and Ginoux, J.J., 1972. Influence of a local obstruction on heat transfer in packed beds. Von Karman Institute. TN.80.

9. Zabrodsky, S.S. 1966. Hydrodynamics and heat transfer in fluidized beds. M.I.T. Press.

10. Vreedenberg, H.A. 1952. J.Appl.Chem.(London).2. Suppl.1. 526.

11. Kılkış, B.1973. Hydrodynamics of a two dimensional fluidized bed in the vicinity of a cylinder with horizontal axis. Congrés. Int. La Fluidisation et ses Applications.

12. Kılkış, B. 1979. A two dimensional finite element analysis of hydrodynamical behaviour of fluidized beds.Num.Met.in thermal problems.pp. 840-849.

# THERMAL-HYDRAULIC FUNDAMENTALS: TWO-PHASE

# Boiling and Evaporation—I

**J. G. COLLIER**
United Kingdom Atomic Energy Authority
Harwell, UK

ABSTRACT

The basic processes involved in boiling and evaporation are described including the thermodynamics of vapour formation, evaporation at planar interfaces, homogeneous and heterogeneous nucleation, sizing of active nucleation sites, bubble growth and bubble detachment and frequency. The various heat transfer regimes in natural convection pool boiling are then identified and practical correlations for each regime given. This discussion leads logically on to the case where the flow past the heater tube or cylinder is by forced rather than by natural convection. Finally, the limited information about the processes which occur during boiling inside a tube bundle are reviewed and recommendations made.

1. INTRODUCTION

Boiling occurs in a wide range of heat exchange equipment. This includes water-tube boilers, pipe stills, refrigeration equipment, water-cooled nuclear reactors, evaporators and many other major items of chemical and power plant equipment. Boiling can be defined as the addition of heat to a liquid in such a way that generation of vapour occurs. The subject of boiling and evaporation in heat exchangers will be considered in two separate parts. In this first Chapter we will consider the basic processes of boiling and evaporation in terms of the thermodynamics of vapour formation, evaporation at planar interfaces, homogeneous and heterogeneous nucleation, sizing of active nucleation sites, bubble growth and bubble detachment and frequency. We will then go on to discuss the various heat transfer regimes in natural convection pool boiling and to present practical correlations to estimate heat transfer rates for each regime. This discussion will lead logically to the case where the flow past the heater or cylinder is by forced rather than by natural convection. Finally, the limited information about the processes which occur during boiling inside a tube bundle are reviewed.

2. VAPOUR FORMATION

Vapour may be generated from a liquid in one of three ways, each corresponding to departure from a thermodynamic equilibrium liquid-vapour state:

(a) formation of vapour at a planar interface corresponds to departure from a *stable equilibrium* state and occurs when the liquid temperature is increased above or the local vapour pressure reduced below the corresponding saturation condition. The term *evaporation* is usually reserved for this process.

(b) formation of vapour nuclei within a liquid held at a temperature above or at a static pressure below the saturation condition corresponds to departure from a *metastable equilibrium* state. The term *homogeneous nucleation* is used to describe this situation.

(c) formation of vapour at a pre-existing vapour nucleus within a liquid corresponds to departure from an *unstable equilibrium* state. The term *heterogeneous nucleation* is used to describe this process.

### Evaporation

Phase change at a planar interface can be viewed from the standpoint of kinetic theory as the difference between two molecular fluxes – a rate of arrival of molecules from the vapour space towards the interface and a rate of departure of molecules from the surface of the liquid into the vapour space. This approach leads to the following equation for the interfacial heat transfer coefficient:

$$\alpha_i = \left(\frac{\tilde{M}}{2\pi \tilde{R}}\right)^{\frac{1}{2}} \frac{\Delta h_v^2}{T^{3/2}(v_g - v_1)} \qquad (1)$$

Whether liquids evaporate, at the maximum rate calculated by equation (1) has been hotly debated for many years. For water, for example, it is commonly reported that the evaporation rate and therefore the interfacial heat transfer coefficient is 0.03 – 0.05 of the maximum value given by equation (1). However, more recent experiments have indicated that for both non-polar and polar liquids (Lednovick and Fenn[1]), even water (Bonacci et al[2], Maa[3]) the maximum rate is achieved and that experimental uncertainty about the liquid interface temperature is responsible for the reported very low values. The rate of evaporation falls rapidly as the time of exposure of the evaporating interface is increased. Temperature gradients are set up in the liquid phase adjacent to the interface in order to supply the latent heat of the net vapour flux. A simple transient conduction analysis (Saha[4]) yields the following equation for the "effective" interface heat transfer coefficient after an exposure time (t)

$$\alpha_i = \frac{\sqrt{3}\,\lambda_1}{\sqrt{\pi \kappa t}} \qquad (2)$$

### Homogeneous Nucleation

Initial considerations relating to the formation of vapour from a metastable liquid or an unstable equilibrium state invariably start from the equation defining the mechanical equilibrium of a spherical vapour nucleus (radius r*) in a liquid at a constant temperature ($T_g$) and pressure ($p_1$). Thus

$$p_g - p_l = \frac{2\sigma}{r^*} \tag{3}$$

where $p_g$ is the vapour pressure inside the nucleus and $p_l$ is the imposed liquid pressure corresponding to a saturation temperature $T_{SAT}$. To calculate the liquid superheat $(T_g - T_{SAT})$ corresponding to the pressure difference $(p_g - p_l)$ use can be made of the Clausius-Clapeyron equation and the perfect gas law. The resulting equation is

$$(T_g - T_{SAT}) = \frac{\tilde{R}\, T_{SAT}\, T_g}{\Delta h_v \, \tilde{M}} \ln\left[1 + \frac{2\sigma}{p_l r^*}\right] \tag{4}$$

If $(2\sigma/p_l r^*) \ll 1$ then equation (4) simplifies to

$$(T_g - T_{SAT}) = \frac{\tilde{R}\, T_{SAT}^2}{\Delta h_v \, \tilde{M}} \frac{2\sigma}{p_l r^*} \tag{5}$$

As the superheat is increased so the size of the equilibrium nucleus reduces and approaches molecular dimensions. Thermal fluctuations occur in the metastable liquid and there is a small but finite probability of a cluster of molecules with vapour-like energies coming together to form a vapour embryo of the size of the equilibrium nucleus. This probability is given in terms of the number of nuclei of radius $(r^*)$ formed per unit volume in unit time

$$\frac{dn}{dt} = N \left[\frac{2\sigma}{\pi m}\right]^{\frac{1}{2}} \exp\left[\frac{-16\pi\sigma^3}{3kT_g (p_g - p_l)^2}\right] \tag{6}$$

The rate of homogeneous nucleation (dn/dt) is an extremely sensitive function of the superheated liquid temperature $(T_g)$. At low superheats the rate is insignificantly small but it increases very rapidly as the superheat is increased. Significant nucleation occurs for values of (dn/dt) between $10^9$ and $10^{13}$ m$^{-3}$s$^{-1}$.

For water at atmospheric pressure the liquid temperature corresponding to (dn/dt) of $10^{13}$m$^{-3}$s$^{-1}$ is 320.7°C, i.e. a superheat of 220.7°C. This value is much higher than any experimental measurement for water even under very carefully controlled conditions. For water, at least, homogeneous nucleation can be discounted as a mechanism for vapour formation. However, homogeneous nucleation can and does occur in organic liquids. The review by Blander and Katz[5] is recommended for further details.

Heterogeneous Nucleation

Non-condensible gas bubbles or foreign bodies held in suspension in the liquid, together with gas- or vapour-filled cracks or cavities in container surfaces (known as *nucleation sites*) normally provide ample nuclei to act as centres for vapour formation. The presence of dissolved gas reduces the superheat required to maintain a bubble of radius r* in unstable equilibrium.

On any container surface there will be pits or cavities having relatively small included angles. Consideration of the successive positions of the interface within such a narrow cavity shows that the minimum radius of curvature of the interface, which in turn dictates the superheat for a vapour bubble to form at this cavity is always that determined by the mouth of the cavity provided $\beta < 90°$ i.e. the liquid wets the surface. This was verified experimentally by Griffith and Wallis[18].

### Sizing of active nucleation sites

Only a very small fraction of the crevices and cavities in a surface are able to act as effective nucleation sites. A cavity that is completely filled with liquid cannot act as a nucleation site (Figure 1b). If, however, the walls of the cavity are poorly wetted or irregular in shape, the curvature of the interface may reverse inside the cavity so that surface tension forces are now resisting further penetration of the liquid even when the vapour pressure in the cavity is negligible. Stabilisation of the interface within the crevice may occur as a result of an internal enlargement (Figure 1c), as a result of a non-wetting inclusion in, say, a metal surface (Figure 1d) or as a result of a non-wetting surface or deposit (Figure 1e). In this latter case the entire surface may be covered by such a film or deposit. When the surface is flooded liquid is forced into the cavity by the imposed liquid pressure. If heat is applied to the surface subsequently, "wetting" may occur as a result of dissolution of grease films in a solvent liquid or in the case of liquid metals, a chemical reaction between a non-wetting surface oxide and the liquid metal.

Lorentz, Mikic and Roksenow[7] have developed a model which takes into account both the wettability of the surface, in terms of the contact angle $\beta$, and the geometrical shape of the cavity, in terms of the included angle $\psi$. The model assumes a conical cavity being inundated by an advancing liquid front (Figure 2a). Once the vapour is trapped by the liquid, the interface readjusts to form a vapour embryo of radius r (Figure 2b). Conservation of vapour volume requires that r is a function of $\beta$ and $\psi$ (Figure 2c). This model is useful in that if the size of active sites on a surface is known for one liquid, then the equivalent value of r for the other liquids with different contact angles can be derived.

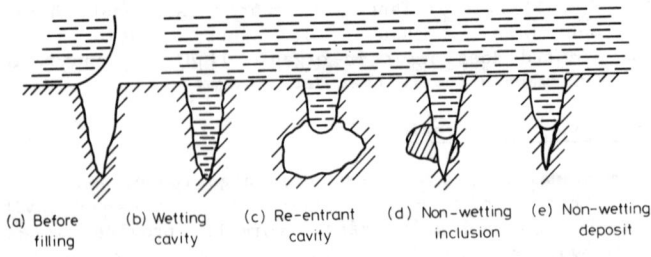

(a) Before filling  (b) Wetting cavity  (c) Re-entrant cavity  (d) Non-wetting inclusion  (e) Non-wetting deposit

Figure 1   Formation of an active site

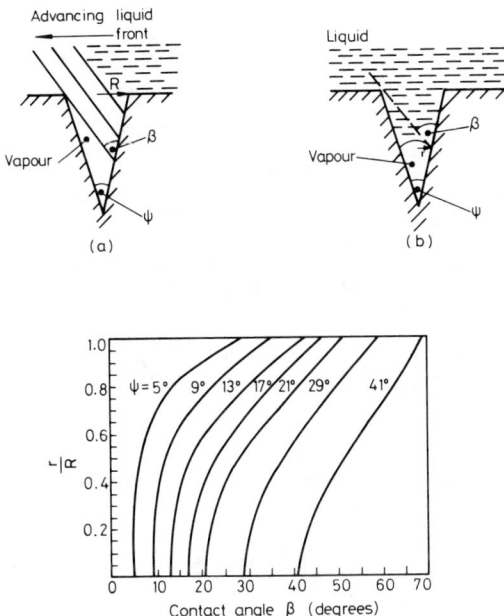

Figure 2    Vapour Trapping Model of Lorentz, Mikic and Rohsenow for sizing active cavities

As a rough guide the following figures may be taken for the maximum size of active nucleation sites on smooth metallic surfaces; for water ~5 µm, for organics and refrigerants ~0.5 µm, for cryogenic fluids on aluminium or copper ~0.1-0.3 µm. These figures may be considerably increased on rough surfaces, specially prepared porous surfaces or porous deposits caused by fouling.

Bubble growth

Once the vapour nucleus in the superheated liquid has attained a size greater than that for unstable equilibrium it will grow spontaneously. The growth rate is limited in the early stages by the inertia of the surrounding liquid and in the later stages by the rate at which the latent heat of vaporisation can diffuse to the vapour-liquid interface. Mikic et al[8] have examined analytically vapour bubble growth close to surface and within a liquid. For a bubble growing in an initially uniformly superheated liquid from an initial radius greater than the critical radius (r*) given by equation (4) they give the following expression

$$R^+ = \frac{2}{3}\left[(t^+ + 1)^{3/2} - t^{+3/2} - 1\right] \qquad (7)$$

where

$$R^+ = RA/B^2 \; ; \; t^+ = tA^2/B^2$$

$$A = \left[b \frac{\Delta T \Delta h_v \rho_g}{T_{SAT} \rho_l}\right]^{\frac{1}{2}} \; ; \; B = \left[\frac{12}{\pi} Ja^2 \kappa_l\right]^{\frac{1}{2}}$$

For bubble growth in an infinite medium b = 2/3 and $\Delta T = (T_g - T_{SAT})$. In the inertia-controlled region $t^+ \ll 1$ Equation (7) becomes

$$R^+ = t^+ \qquad (8)$$

which is the well known Rayleigh[9] solution. For the diffusion-controlled region $t^+ \gg 1$ Equation (7) becomes

$$R^+ = (t^+)^{\frac{1}{2}} \qquad (9)$$

which is the asymptotic solution derived by Plesset and Zwick[10] and others. Mikic et al[8] extended their analysis to the case of vapour bubbles growing in a non-uniform temperature field adjacent to a heated surface.

Bubble detachment and frequency

The size and shape of vapour bubbles departing from a heated surface are a strong function of the way in which they are formed. The prime forces acting on a vapour bubble during the later phases of its growth are buoyancy and hydrodyamic drag forces attempting to detach it from the surface and surface tension and liquid inertia forces acting to prevent detachment. The growth velocity of a bubble and hence the inertial force is a strong function of the liquid superheat which, in turn, is inversely proportional to the size of the active cavity. A small cavity thus forms a bubble with a faster growth rate than from a large cavity. When surface tension forces dominate the departing bubbles tend to be spherical. When inertial forces dominate the bubble tends to be hemispherical and when both forces are significant the bubble has an oblate shape (Johnson et al[11]).

Individual nucleation sites emit bubbles with a constant frequency, the value of which varies from site to site. Jakob[12] observed that the product of bubble frequency (f) and departure diameter ($D_b$) was a constant. More recently Ivey[13] has shown that this is an approximation and that the form of the product depends on whether the bubble growth is inertia- or diffusion-controlled. In the former case the relationship becomes $D_b f^2$ = const while for the region where bubble growth is limited by the diffusion of heat $D_b f^{\frac{1}{2}}$ = const. For water at atmospheric pressure, bubble diameters at departure are in the range 1-2.5mm and bubble frequencies in the range 20-40 $s^{-1}$.

# 3. POOL BOILING

Pool boiling is defined as boiling from a heated surface submerged in a large volume of stagnant liquid. This liquid may be at its boiling point in which case the term *saturated pool boiling* is employed or below its boiling point when the term *subcooled pool boiling* is used. The results of investigations into heat transfer rates in pool boiling are usually plotted on a graph of surface heat flux ($\dot{q}$) against heater wall surface temperature ($T_W$) - the boiling curve. An alternative presentation might use the wall superheat ($T_W-T_{SAT}$) rather than the wall temperature itself.

The component parts of the boiling curve are:

(a) *the natural convection region* where temperature gradients are set up in the pool and heat is removed by natural convection to the free surface and thence by evaporation to the vapour space.

(b) *the onset of nucleate boiling* (ONB) where the wall superheat becomes sufficient to cause vapour nucleation at the heating surface. This may occur close to the point where the natural convection and nucleate boiling curves meet as is usually the case for water at atmospheric pressure and above. Alternatively, it may occur at much larger superheats than those required to support fully developed nucleate boiling, resulting in a sharp drop in surface temperature for the case of a constant surface heat flux when vapour is first generated. This latter behaviour is associated with fluids at very low reduced pressures, e.g. water at below atmospheric pressure and liquid metals in particular.

(c) *the nucleate boiling region* where vapour nucleation occurs at the heating surface. Starting with a few individual sites at low heat fluxes the vapour structure changes as the heat flux is increased, as a result of bubble coalescence and finally, at high heat fluxes, vapour patches and columns are formed close to the surface.

(d) *the critical heat flux* (CHF) marks the upper limit of nucleate boiling where the interaction of the liquid and vapour streams causes a restriction of the liquid supply to the heating surface.

(e) *the transition boiling region* is characterised by the existence of an unstable vapour blanket over the heating surface that releases large patches of vapour at more or less regular intervals. Intermittent wetting of the surface is believed to occur. This region can only be studied under conditions approximating to a constant surface temperature.

(f) *the film boiling region* where a stable vapour film covers the entire heating surface and heating is released from the film periodically in the form of regularly-spaced bubbles. Heat transfer is accomplished principally by conduction and convection through the vapour film with radiation becoming significant as the surface temperature is increased.

<u>The onset of nucleate boiling</u>

As the surface heat flux is increased, so the surface temperature exceeds the saturation temperature. Establishing the onset of nucleate boiling requires a criterion for bubble nucleation within the non-uniform liquid temperature field adjacent to the heating surface.

Hsu[14] postulated that the criterion for nucleation from a surface cavity with a hemispherical bubble cap is that the temperature of the liquid surrounding the top of the bubble should exceed that necessary for the nucleus to remain in equilibrium (Equation (4)). The equation for the critical size of nucleation site derived by Han and Griffith[15] based on this assumption was

$$r_c = \frac{\delta(T_W - T_{SAT})}{3(T_W - T_{1\infty})} \left[ 1 \pm \sqrt{1 - \frac{12(T_W - T_{1\infty})T_{SAT}\sigma}{(T_W - T_{SAT})^2 \delta \rho_g \Delta h_v}} \right] \qquad (10)$$

where $\delta$ is the thickness of the thermal boundary layer. The size of cavity satisfying Equation (10) is approximately 50-250 μm for water at atmospheric pressure. The superheat corresponding to this range of cavity size is less than 1°C for water. In practice, some 10-15°C of superheat is normally required to initiate boiling of water from a flat metallic surface at atmospheric pressure.

The discrepancy lies in the fact that cavities of radius 50-250 μm are not normally active nucleation sites (that is, they do not contain a vapour embryo). If an active site of size $r_c$ does not exist on the heating surface then the wall temperature must be increased until the liquid temperature at the top of the bubble reaches the equilibrium temperature for the smaller active cavities that do exist (see *sizing of active nucleation sites*).

Nucleate boiling

In the nucleate boiling region the surface temperature increases only slowly for relatively large changes in the surface heat flux. The surface temperature variation can be expressed in the form of a simple power law relationship

$$T_W = T_{SAT} + \psi \dot{q}^m \qquad (11)$$

where $\psi$ and m are constants depending upon the physical properties of the liquid and vapour and upon the nucleation properties of the surface; m will normally be in the range 0.25-0.5 with a most probable value around 0.3-0.33.

The important effect that surface condition has on nucleate pool boiling must be stressed immediately. As an example, Figure 3 shows experimental data taken by Berenson[16] for n-pentane boiling on copper. Different surface finishes cause the wall superheat $\Delta T_{SAT}$ (=$T_W - T_{SAT}$) for a given heat flux to vary by nearly a factor of five.

It is usual to express heat transfer rates in terms of a heat transfer coefficient ($\alpha$). In convective processes this represents the ratio between the heat flux ($\dot{q}$) and the temperature difference between surface and fluid ($\Delta T$). Rearranging Equation (11) to give an explicit expression for $\alpha$, it is noted that in pool boiling the heat transfer coefficient is itself a function of the heat flux ($\dot{q}$). Thus

$$\alpha = \frac{1}{\psi} \dot{q}^{(1-m)} = A\dot{q}^n \qquad (12)$$

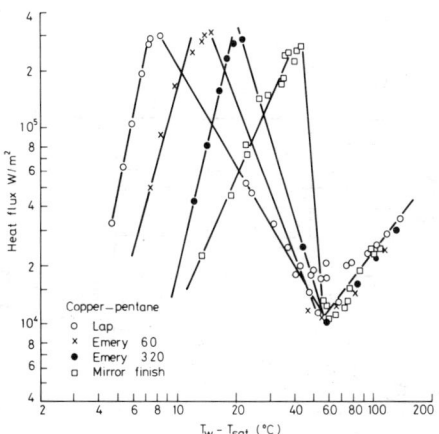

Figure 3  Effect of surface finish on pool boiling curve (Berenson[16]):
● emery 320; x emery 60
o lap;  □ mirror finish

where n ≃ 0.7 and A is a function of liquid and vapour physical properties and of surface condition.

Many of the available correlations for nucleate pool boiling heat transfer start from a simplified model of the boiling process and are expressed in terms of a number of dimensionless groups. Appropriate characteristic dimensions and velocities, such as the bubble departure diameter or the bubble growth velocity are used in the formulation of these dimensionless groupings. One such well known correlation is that proposed by Rohsenow[17].

Another well tried correlation based on similar precepts is that of Forster and Zuber[18].

Such dimensionless correlations have a number of disadvantages; they require accurate physical properties, they are complicated to evaluate and the inherent uncertainty induced by the condition of the surface is considerable. As an alternative, simple dimensional equations can be prescribed for individual fluids using data from experimental studies. One basis for such equations is the work of Borishanski [19] which makes use of the law of corresponding states. The heat transfer coefficient is evaluated from

$$\alpha = A^* \dot{q}^{0.7} F(p) \tag{13}$$

where $F(p)$ is a function of reduced pressure, $p_r$ as shown in Figure 4 and $A^*$ is a constant evaluated at the reference reduced pressure $p_r^* = 0.0294$.

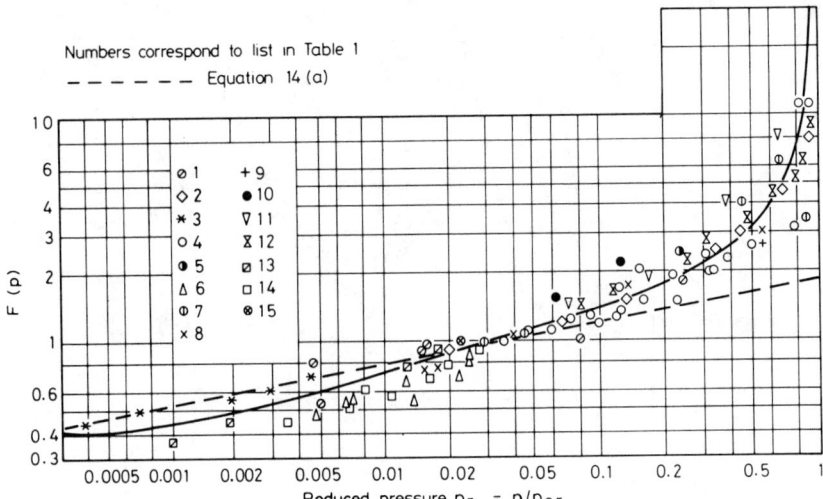

Figure 4  Variation of F(p) with reduced pressure (Borishanski[19])

A variation of this method has been given by Mostinski[20] who has proposed the following expression for A* and F(p):

$$A^* = 0.1011\, p_{cr}^{0.69}$$

$$F[p]_1 = 1.8\, p_r^{0.17} + 4\, p_r^{1.2} + 10\, p_r^{10}$$

(14)

Values of A* evaluated from Equation (14) at $p_r^* = 0.0294$ are given in Table 1 for comparison with values derived from experimental measurements by Borishanski and others. Palen et al[21] report that the Mostinski correlation was superior to any of the physical property based correlations that they had tested. However, for application to pool boiling type reboilers a safe design curve is recommended which is obtained by dropping the last two terms in the polynomial of Equation (14):

$$F[p]_2 = 1.8\, p_r^{0.17}$$

(14a)

This recommendation is due to the great scatter and generally increasing uncertainty in the boiling behaviour as the critical point is approached.

German workers at the University of Karlsruhe[22] have found that for the refrigerants (R11, R12, R113, R115 etc.) the function F[p] is somewhat different than given by Equation (14) and proposed by Borishanski. They propose a relationship for F[p] such that:

TABLE 1  Simple dimensional equations for nucleate pool boiling heat transfer (after Borishanski[19])

$$\alpha = A\dot{q}^{0.7} \text{ ; } \alpha \text{ in W/m}^2{}^\circ\text{K, } \dot{q} \text{ in W/m}^2$$

| Liquid | Pressure range (bar) | Reference pressure p* ($p_r$ = 0.0294) (bar) | A* from exp study | A* from Equation (14) | Critical pressure $p_{cr}$ (bar) | Reference | No. in Fig.4 |
|---|---|---|---|---|---|---|---|
| Water | 1 - 70 | 6.50 | 4.09 | 4.21 | 221.2 | (23) | 1 |
| Water | 1 - 196 | 6.50 | 4.02 | 4.21 | 221.2 | (24)(25) | 2 |
| Water | 0.09 - 1 | 6.50 | 5.80 | 4.21 | 221.2 | (26) | 3 |
| Water | 1 - 72.5 | 6.50 | 4.47 | 4.21 | 221.2 | (24) | 4 |
| Water | 1 - 170 | 6.50 | 4.44 | 4.21 | 221.2 | (26) | 5 |
| Water | 1 - 5.25 | 6.50 | 5.74 | 4.21 | 221.2 | (27) | 6 |
| Pentane | 1 - 28.6 | 0.98 | 1.09 | 1.14 | 33.4 | (23) | 7 |
| Heptane (80%) | 0.45 - 14.8 | 0.78 | 1.18 | 0.967 | 26.4 | (23) | 8 |
| n-heptane | 0.45 - 21.7 | 0.78 | 1.63 | 0.967 | 26.4 | (23) | 9 |
| Benzene | 1 - 44.4 | 1.44 | 1.06 | 1.48 | 49.0 | (23) | 11 |
| Benzene | 0.9 - 20.7 | 1.44 | 1.32 | 1.48 | 49.0 | (28) | - |
| diphenyl | 0.9 - 8 | 0.91 | 1.12 | 1.08 | 31.0 | (28) | - |
| Methanol | 0.08 - 1.39 | 2.33 | (0.69) | 2.07 | 79.5 | (29) | 13 |
| Ethanol | 1 - 7.9 | 1.87 | 1.83 | 1.78 | 63.8 | (23) | 10 |
| Ethanol | 1 - 59 | 1.87 | 2.59 | 1.78 | 63.8 | (24)(25) | 12 |
| Butanol | 0.17 - 1.38 | 1.31 | (0.44) | 1.39 | 44.7 | (29) | 14 |
| R11 ($CFCl_3$) | 1 - 3 | 1.28 | 1.95[1.73] | 1.37 | 43.7 | (31) | - |
| R12 ($CF_2Cl_2$) | 1 - 4.87 | 1.21 | 2.43 | 1.31 | 41.1 | (30) | 15 |
| R12 ($CF_2Cl_2$) | 6 - 40.5 | 1.21 | 3.49[2.57] | 1.31 | 41.1 | (33) | - |
| R13 ($CF_3Cl$) | 2.8 - 10.5 | 1.13 | 1.79 | 1.26 | 38.6 | (22) | - |
| R13B1 ($CF_3Br$) | 17 - 39 | 1.17 | 4.43[2.48] | 1.29 | 39.9 | (33) | - |
| R22 ($CHF_2Cl$) | 0.4 - 2.15 | 1.45 | [2.39] | 1.49 | 49.36 | (22) | - |
| R113 ($C_2Cl_3F_3$) | 1 - 3 | 1.00 | 1.24 | 1.15 | 34.1 | (34) | - |
| R115 ($C_2F_5Cl$) | 8 - 31 | 0.92 | 3.79[2.38] | 1.08 | 31.2 | (33) | - |
| RC318 ($C_4F_8$) | 3.6 - 27 | 0.82 | 3.13[2.50] | 1.00 | 27.8 | (33) | - |
| Methylene chloride | 1 - 4.5 | 1.79 | (1.91) | 1.72 | 60.8 | (31) | - |
| Ammonia | 1 - 8 | 3.32 | 3.90 | 2.64 | 113 | (31) | - |
| Methane | 1 - 42 | 1.37 | 2.68 | 1.43 | 46.5 | (32) | - |

Values shown in round brackets ( ) are uncertain

Values shown in square brackets [ ] relate to the use of Equation (15) for F[p]

TABLE 2   Values of F[p] as a function of reduced pressure

| $p_r$ | $F[p]_1$ from Equation (14) | $F[p]_2$ from Equation (14a) | $F[p]_3$ from Equation (15) |
|---|---|---|---|
| 0.0001 | 0.3760 | 0.3760 | 0.7010 |
| 0.0002 | 0.4231 | 0.4231 | 0.7020 |
| 0.0005 | 0.4948 | 0.4944 | 0.7050 |
| 0.002  | 0.5572 | 0.5562 | 0.7100 |
| 0.002  | 0.6281 | 0.6258 | 0.7200 |
| 0.005  | 0.7382 | 0.7313 | 0.7500 |
| 0.01   | 0.8386 | 0.8227 | 0.8000 |
| 0.2    | 0.9622 | 0.9256 | 0.9008 |
| 0.0294 | 1.0464 | 0.9883 | 0.9958 |
| 0.05   | 1.1915 | 1.0817 | 1.2053 |
| 0.1 | 1.4693 | 1.2169 | 1.722 |
| 0.2 | 1.9489 | 1.3691 | 2.800 |
| 0.5 | 3.3508 | 1.5999 | 6.700 |
| 0.8 | 5.8670 | 1.7330 | 15.100 |
| 1.0 | 15.8000 | 1.8000 | |

$$F[p]_3 = 0.7 + 2 p_r \left[ 4 + \frac{1}{(1-p_r)} \right] \qquad (15)$$

This function is compared with that proposed by Mostinski in Table 2. It is recommended that for refrigerants $F[p]$, together with the values of $A^*$ in square brackets in Table 1, be used in Equation 13.

A large number of independent variables influence heat transfer rates in nucleate boiling. The following is a partial check list of those variables which have the most important influence.

  (i)    the system pressure
  (ii)   the surface condition
  (iii)  the presence of non-condensible gases
  (iv)   hystersis in the boiling curve
  (v)    the size and orientation of the surface
  (vi)   the subcooling of the liquid
  (vii)  the wettability of the surface
  (viii) the gravitational acceleration

Critical heat flux (CHF)

The pool boiling curve can only be obtained in its entirety under circumstances where the temperature of the heating surface is controlled to a specified value. In many practical cases however, the surface heat flux is the independently controlled variable. In this case the boiling curve in the natural convection and nucleate boiling regions remains basically unaltered. However, if an attempt is made to increase the value of the surface heat flux indefinitely, at a particular value, the surface temperature will jump from that corresponding to nucleate boiling to an operating point in the film boiling region. In many practical cases this large temperature jump is sufficient to cause failure of the heating surface. Hence, the colloquial name *burnout* which is often used to refer to this phenomenon.

It is now generally accepted that the critical heat flux in pool boiling occurs as a result of a hydrodynamic flow pattern transition close to the heating surface. The mechanism is one in which insufficient liquid is able to reach the heating surface due to the rate at which vapour is leaving the surface. Zuber[35] idealised the flow pattern as a square array of vapour jets leaving the heated surface with a spacing corresponding to the most rapidly growing Taylor instability wavelength $\lambda_d$. Interaction of adjacent vapour jets due to Helmholtz instability limits the vapour volumetric flux from the surface to

$$(j_g)_{MAX} = \frac{\pi}{24} \left[ \frac{g_n(\rho_1 - \rho_g)}{\rho_g^2} \right]^{\frac{1}{4}} \qquad (16)$$

For the case where the liquid is at the saturation temperature and all the vapour is produced at the heating surface (for liquid metals and also for certain two-component systems this latter assumption may not be valid) then

$$j_g = \left(\frac{\dot{q}}{\rho_g \Delta h_v}\right) \qquad (17)$$

and thus

$$\dot{q}_{cr} = K \Delta h_v \rho_g^{\frac{1}{2}} \left[g_n(\rho_l - \rho_g)\right]^{\frac{1}{4}} \qquad (18)$$

where according to Zuber, $K = \pi/24$. This theory thus confirmed the earlier correlation proposed by Kutateladze[36] in which $K$ was $0.16 \pm 0.03$. More recently, Lienhard and Dhir[37] have re-examined the Zuber theory in the light of up-to-date experimental evidence and have concluded that the constant $K$ given by Zuber should be increased by a factor 1.14, i.e., $K$ should be 0.149. The value given by equation (18) may be affected by a variety of independent variables including

(a) the geometry and orientation of the heating surface
(b) the viscosity of the liquid
(c) the subcooling of the liquid
(d) the condition of the surface.

Transition boiling

In this little-studied region of the boiling curve, liquid periodically contacts the heating surface with the result that the formation of large amounts of vapour forces the liquid away from the surface and an unstable vapour film or blanket is formed. This, in turn, collapses allowing the liquid to contact the surface once more. The region is normally only obtainable by controlling the surface temperature to a predetermined value. For water at atmospheric pressure this temperature is about 140°-250°C. Because of the periodic nature of the process the surface heat flux and temperature undergo large variations with both time and position on the heater. However, the average heat flux decreases as the temperature increases because the proportion of the time and heater surface which the liquid spends in contact with the surface continually decreases. A discussion of some of the factors which are important in transition boiling was given by Bankoff and Mehra[38]. No adequate theory or model of this region exists at this time. However, it is usually reasonable to interpolate linearly on log-log paper between the critical and minimum heat flux points on a curve of heat flux versus temperature difference.

Minimum heat flux

The minimum heat flux $\dot{q}_{min}$ is reached when the heat flux matches the minimum rate of vapour formation which will sustain a stable vapour film over the heating surface. Vapour is released from this film in the form of bubbles produced regularly in both time and space. If the heat flux is less than $\dot{q}_{min}$, the interface will collapse onto the heating surface cooling it and re-establishing nucleate boiling. The vapour release rate can be considered as the product of the volume of a released bubble, the number of bubbles per cycle per unit area and a minimum wave frequency. This leads to the expression for $\dot{q}_{min}$ on a flat plate first derived by Zuber[35] and subsequently modified by Berenson[39].

$$\dot{q}_{min,F} = \text{const } \rho_g \Delta h_v \left[ \frac{\sigma g_n (\rho_1 - \rho_g)}{(\rho_1 + \rho_g)^2} \right]^{\frac{1}{4}} \qquad (19)$$

Because of uncertainty in the minimum wave frequency the constant in Equation (19) must really be determined experimentally. Zuber proposed various values for the constant which are mostly too high and Berenson's later experiments resulted in 0.09 and this remains the current best estimate for a flat plate surface. The value of $\dot{q}_{min}$ for a vertically-orientated heater is lower than for a horizontally mounted heater in a similar ratio to that found for the critical heat flux, i.e. ~0.75.

Film boiling

At high temperature differences a continuous vapour film blankets the heater surface. The major resistance to heat transfer is confined to this vapour film and because there is lack of liquid-solid contact, this region is the most tractable to analyse. The relationships for the heat transfer coefficient in laminar or turbulent film boiling in various geometrical situations can be established by direct analogy with the identical relationships derived from film wise condensation.

Berenson[40] analysed film boiling on a *large horizontal plate* and produced the following equation which was compared with data for pentane carbon tetrachloride, benzene and ethyl alcohol

$$\alpha = 0.425 \left[ \frac{\lambda_g^3 \rho_g (\rho_1 - \rho_g) g_n \Delta h_v'}{\mu_g \Delta T \left( \frac{\lambda_c}{2\pi} \right)} \right]^{\frac{1}{4}} \qquad (20)$$

where $\lambda_c$ is the "critical" (or shortest unstable) wavelength for Taylor instability given by

$$\lambda_c = 2\pi \left[ \frac{\sigma}{g_n (\rho_1 - \rho_g)} \right]^{\frac{1}{2}} \qquad (21)$$

and $\Delta h_v'$ is an effective latent heat of vaporisation allowing for the effect of superheat

$$\Delta h_v' = \Delta h_v \left[ 1 + \xi \left( \frac{c_{pg} \Delta T}{\Delta h_v} \right) \right] \qquad (22)$$

For use in Equation (22) $\xi$ should be taken as 0.4.

For laminar film boiling from a *horizontal tube*, Bromley[41] gives

$$\alpha = 0.62 \left[ \frac{\lambda_g^3 \rho_g (\rho_1 - \rho_g) g_n \Delta h_v'}{\mu_g \Delta T \, D} \right]^{\frac{1}{4}} \qquad (23)$$

where $\xi$ in Equation (23) should be taken as 0.68.

For very large tubes, horizontal flat surfaces and very thin wires Equation (23) is inaccurate. For large diameter tubes and flat horizontal surfaces, the characteristic length is more correctly $\lambda_c$ than the tube diameter D. Thus, Breen and Westwater[42] modified Equation (23)

$$\alpha = \left[0.59 + 0.069 \frac{\lambda_c}{D}\right] \left[\frac{\lambda_g^3 \rho_g (\rho_l - \rho_g) g_n \Delta h_v'}{\mu_g \Delta T \lambda_c}\right]^{\frac{1}{4}} \qquad (24)$$

where $\Delta h_v'$ is now given by

$$\Delta h_v' = \Delta h_v \left[1 + 0.34 \left(\frac{c_{pg} \Delta T}{\Delta h_v}\right)\right]^2 \qquad (25)$$

The effect of liquid sub-cooling on film boiling coefficients can be considerable. As an example, for organic liquids such as n-hexane, benzene, methanol, carbon tetrachloride and ethyl alcohol, coefficients will be increased fourfold by subcooling the liquid 40°C.

Surface temperatures are usually high in film boiling and heat may be transferred by radiation. Bromley[41] proposes the following approximation for combining the effects of convection and radiation.

$$\alpha = \alpha_c + 0.75 \alpha_r \qquad (26)$$

where $\alpha_c$ is the convective coefficient (for example, Equation (20) or (24) and $\alpha_r$ is the radiation coefficient. For simplicity the radiation coefficient $\alpha_r$ can be calculated by assuming radiation between infinite parallel planar surfaces with the liquid acting as a perfect black body

$$\alpha_r = C_s \varepsilon \left[\frac{T_W^4 - T^4}{(T_W - T)}\right] \qquad (27)$$

where $C_s$ is the Stefan-Boltzman constant and $\varepsilon$ is the emissitivity of the heating surface.

Comparisons of the equations given in this section for film boiling with experimental data sets covering a wide range of fluids, viz nitrogen, oxygen, water, methane, ethane, ethylene, ethylene mixtures, propane, n-butane, methanol, ethyl alcohol, benzene, carbon tetrachloride, suggest an average standard deviation of between 30-40%. For fuller details the reader is referred to the reviews by Clements and Colver[43].

## 4. BOILING OUTSIDE TUBES AND TUBE BUNDLES

### Boiling outside single tubes in cross flow

When a saturated liquid flows upwards across a uniformly heated cylinder a vapour cavity forms in the cylinder's wake. Initially this cavity is not continuous along the length of the cylinder but as the heat flux is increased so the cavity transforms into a very uniform vapour sheet. Under these circumstances the only liquid reaching the top half of the cylinder is that supplied between the vapour bubbles and the heater surface as the bubbles enter the cavity wake near the horizontal diameter.

Figure 5 shows experimental data by Beecher[44] for saturated water flowing at 1 m/s normal to an electrically heated stainless steel cylinder 1.22mm diameter. The data are compared with a curve for water boiling from a similar cylinder under natural convection pool boiling conditions. In the region of high heat flux the two curves are nearly coincident whilst at low temperature differences, the individual curves are in agreement with the respective single-phase forced convection and natural convection predictions. The forced convection boiling curve may therefore be established by an interpolation procedure.

Figure 6 shows the critical heat flux for water for boiling from heated cylinders under cross flow conditions. The critical heat flux is plotted against cylinder diameter with velocity as parameter. Lienhard and Eickhorn[50] have correlated the available experimental critical heat flux data for cross flow of liquids over single cylinders and tubes in relation to the various hydrodynamic flow patterns. The only exception is the data of McKee and Bell[45] for relatively large diameter cylinders. However, comparison is made difficult by the short length of the test specimen (L/D ~2.5) and by the design of the mountings used in this particular study.

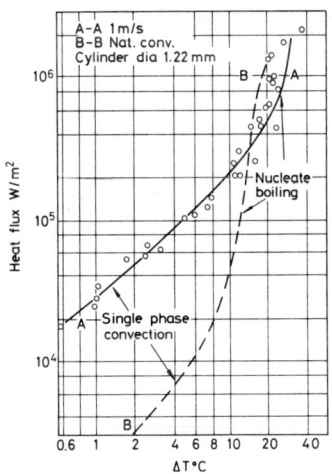

Figure 5    Effect of cross flow on nucleate boiling curve for water (1 bar) (Beecher[44])

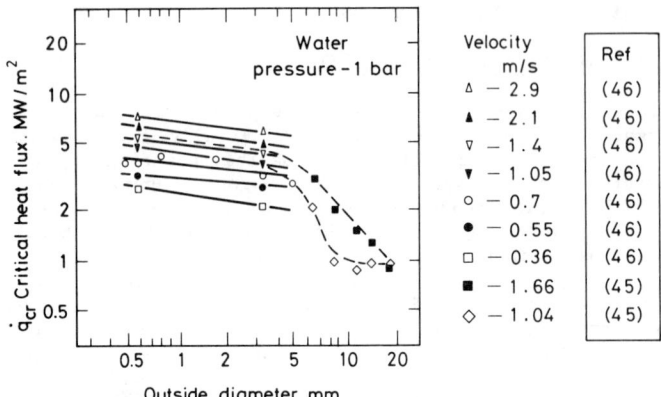

Figure 6    Critical heat flux in cross flow - effect of diameter and velocity (McKee and Bell[45])

Film boiling during upwards cross flow over a horizontal cylinder has been studied by Bromley et al[47]. Equation (23) is satisfactory for low velocities but for higher velocities

$$\frac{u_\infty}{\sqrt{g_n D}} > 2.0 \tag{28}$$

the heat transfer coefficient is given by

$$\alpha_c = 2.7 \left( \frac{u_\infty \lambda_g \rho_g \Delta h_v'}{D \, \Delta T_{SAT}} \right)^{\frac{1}{2}} \tag{29}$$

For subcooled liquids ($\Delta T_{SUB}$ up to 45°C) and high velocities (up to 2.5m/s) the values of heat transfer coefficient in film boiling numerically approach the values expected for nucleate boiling for organic liquids (2-3000 W/m² K).

<u>Boiling on the outside of tubes within a horizontal tube bundle</u>

Although it is known that the rate of heat transfer from a heated tube within an evaporator tube bundle is different from that for an isolated single tube, there have been few published studies quantifying the differences. Horizontal tube bundles are used in refinery kettle reboilers and in water chillers and this lack of published information is surprising. There are a number of essential differences between the curve for boiling outside a horizontal tube and for boiling on a tube within a horizontal tube bundle. Agitation, particularly due to the vapour within the bundle increases the heat transfer coefficient in the nucleate boiling region particularly at low heat fluxes. At higher values of wall superheat $\Delta T_{SAT}$ the presence of large amounts of vapour within the bundle causes vapour blanketting of heat transfer surfaces to occur at heat fluxes below the critical heat flux for an isolated single tube. This reduction is greater the larger the physical size of the tube bundle and the closer the packing density of the tubes. The variation of heat flux with wall superheat in the region of the critical heat flux also tends to be much flatter than for a single tube. Finally the heat transfer coefficient in the film boiling region may be enhanced by the increased levels of turbulence in the tube bundle although to a somewhat lesser extent than in the nucleate boiling region.

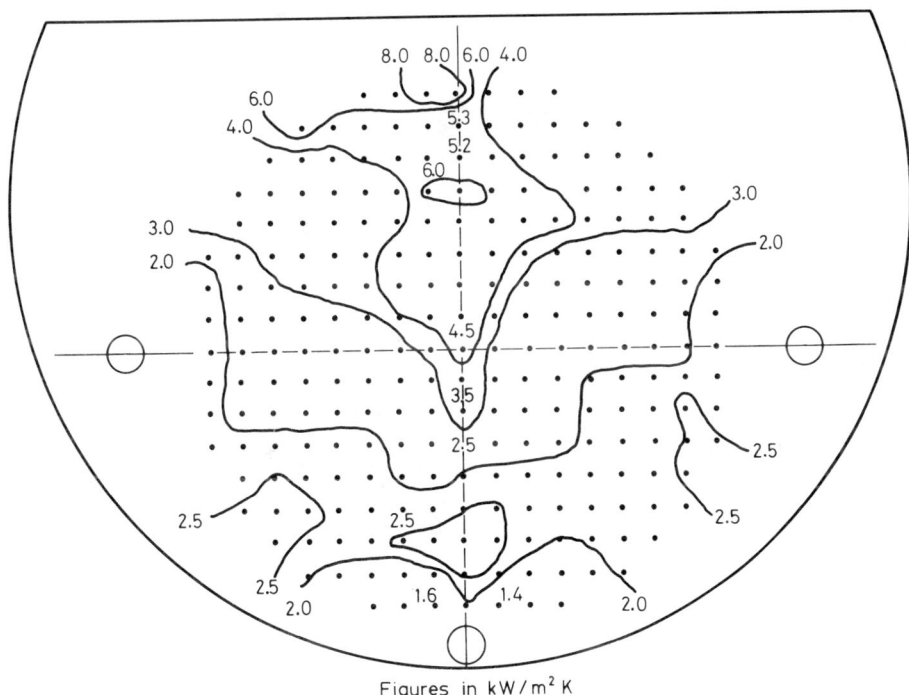

Figures in kW/m² K

Figure 7    Contours of heat transfer coefficient in evaporation in a kettle reboiler (Leong and Cornwall[48])

Experiments on a thin slice through a kettle reboiler tube bundle using a refrigerant have recently been carried out by Leong and Cornwall[48]. These experiments show (Figure 7) that significant variations of heat transfer coefficient occur with the value increasing with bundle height. The pattern is complex and methods for predicting this variation are not yet available. A number of mechanisms are possible whereby a limiting heat flux is reached within a tube bundle. They may be briefly described as follows (Figure 8).

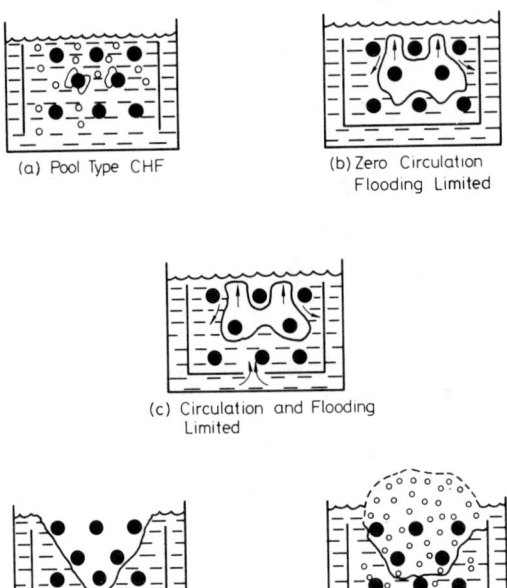

Figure 8   Sketches of possible dryout mechanisms inside tube bundles

(a) *pool boiling type critical heat flux* - this mechanism will occur in small bundles with widely spaced tubes. In this case the flow passages within the bundle are essentially filled with liquid and the limiting process is the same as that for an isolated single cylinder.

(b) *zero circulation - flooding limited* - this mechanism occurs when there is no net circulation through the tube bundle. Liquid entering the bundle can only do so from above and is hindered in doing so by the vapour being released from the bundle. A flooding condition is reached where the vapour release is such as to prevent sufficient down-flow of liquid to reach all the heating surfaces within the bundle. This mechanism is more likely to occur in large bundles with closely spaced tubes. Downflow can occur in some lanes and with upflow in others, giving an apparent internal circulation.

(c) *circulation and flooding limited* - this mechanism is similar to (b) except that some small circulation into the bundle from the sides and base occurs. This low in-flow of liquid is evaporated within the bundle and the resulting vapour passes upwards to join that generated from liquid passing into the bundle from above. Again, the limiting condition is reached when this total vapour flow is such as to prevent a sufficient liquid in-flow from above. As with (b), this mechanism occurs in large bundles with closely spaced tubes.

(d) *circulation limited* - as the circulation through the bundle increases, a condition is reached where the vapour flow produced from this circulation alone is sufficient to prevent any liquid entering the bundle from above. The limiting condition is then complete evaporation of the liquid feed, starving the upper tubes in the bundle of liquid. Such a condition might occur in narrow but tall tube bundles.

In practice it is difficult to distinguish between mechanisms (b), (c) and (d) in actual tube bundle geometries.

(e) *entrainment limited* - as the circulation through the bundle is further increased, so a substantial amount of liquid may be entrained as droplets within the vapour flow and thus dryout will occur on the upper tubes on the bundle at vapour qualities considerably less than 100%.

Palen and Small[49] have given a correlation for the critical heat flux within a tube bundle under conditions where the liquid circulation is limited. It was arrived at by modifying the isolated single tube value given by Zuber, $\dot{q}_{cr,z}$ (equation (18)) by means of a dimensionless tube density factor and a dimensionless physical property factor.

## 5. CONCLUDING REMARKS

Although the study of the basic physics of nucleation and bubble growth are helpful in gaining a good understanding of the physical processes occurring in boiling, the results of such studies are not directly useable by the heat exchanger designer.

Likewise, the all embracing dimensionless heat transfer correlation is also to be viewed with suspicion, not because of the factors it includes but rather for the unspecified variables it leaves out. For example, the most intractable problem in the design of heat exchangers involving boiling and evaporation is that of the surface condition. Indeed, because under boiling conditions, fouling of heat transfer surfaces is particularly prevalent the surface condition may vary significantly with operating time leading to corresponding changes in performance of the heat exchanger.

REFERENCES

(1) Lednovick, S.L. and Fenn, J.B. "Absolute Evaporation Rates for Some Polar and Non-Polar Liquids", A.I.Ch.E.J. Vol.23, No.3, pp.454-459, 1977.

(2) Bonacci, J.C., Myers, A.L., Nongbri, G. and Eagleton, L.C. "The evaporation and condensation coefficient of water, ice and carbon tetrachloride", Chem.Eng.Sci. Vol.31, pp.609-617, 1976.

(3) Maa, J.R. "Rates of evaporation and condensation between pure liquids and their own vapours", Ind.Eng.Chem Vol.9, No.2, pp.283-287, 1970.

(4) Saha, P. "Development of constitutive relations - Effect of pressure change", Reactor Safety Research Programs QPR (April 1st - June 30th 1977) BNL-NUREG-50683, 1977.

(5) Blander, M. and Katz, J.L. "Bubble nucleation in liquids", A.I.Ch.E.J. Vol.21, No.5, pp.833-848, 1975.

(6) Griffith, P. and Wallis, J.D. "The role of surface conditions in nucleate boiling", Chem.Engng.Prog.Symp.Series No.30, 49, 1960.

(7) Lorentz, J.J., Mikic, B.B. and Rohsenow, W.M. "The effect of surface conditions on boiling characteristics", Heat Transfer 1974, Vol.4, Proceedings of the 5th Int. Heat Transfer Conference, Tokyo, Japan.

(8) Mikic, B.B., Rohsenow, W.M. and Griffith, P. "On bubble growth rates", Int.J.Heat Mass Transfer Vol.13, pp.657-665, 1970.

(9) Rayleigh, L. "Pressure due to collapse of bubbles", Phil.Mag. Vol.34, pp.94-98, 1917.

(10) Plesset, M.A. and Zwick, S.A. "The growth of vapor bubbles in superheated liquids", J.Appl.Phys. Vol.25, No.4, pp.493-500, 1954.

(11) Johnson, M.A., de la Pera, J. and Mesler, R.B. "Bubble shapes in nucleate boiling", Chem.Eng.Prog.Symp. Series Vol.62, p.1, 1966.

(12) Jakob, M. "Heat Transfer Vol.1", published by John Wiley, New York, 1958.

(13) Ivey, H.J. "Relationship between bubble frequency, departure diameter and rise velocity in nucleate boiling", Int.J.Heat Mass Transfer, Vol.10, No.8, pp.1023, 1967.

(14) Hsu, Y.Y. "On the size range of active nucleation cavities on a heating surface", J. Heat Transfer Vol.1, 84C(3), pp.207-216 (1962).

(15) Han, C.Y. and Griffith, P. "The mechanism of heat transfer in nucleate pool boiling, Pt.I Bubble initiation growth and departure", Int.J.Heat Mass Transfer, Vol.8(6), pp.887-904 (1965).

(16) Berenson, P.J. "Experiments on pool boiling heat transfer", Int.J. Heat Mass Transfer, Vol.5, pp.958 (1962).

(17) Rohsenow, W.M. "A method of correlating heat transfer data for surface boiling of liquids", Trans. ASME Vol.74, pp.969 (1952).

(18) Forster, H.K. and Zuber, N. "Dynamics of vapour bubble growth and boiling heat transfer", A.I.Ch.E.J., Vol.1, No.4, pp.531-535 (1955).

(19) Borishanski, V.M. "Correlation of the effect of pressure on the critical heat flux and heat transfer rates using the theory of thermodynamic similarity", Problems of Heat Transfer and Hydraulics at Two-Phase Media, pp.16-37, Published by Pergamon Press (1969).

(20) Mostinskii, I.K. Teploenergetika 4, 66 (1963); English Abstract, Brit. Chem. Engng., 8(8), p.580 (1963).

(21) Palen, J.W., Yarden, A. and Taborek, J. "Characteristics of boiling outside large-scale horizontal multi-tube bundles". "Heat Transfer - Tulsa", A.I.Ch.E. Symp. Series, Vol.68, No.118, pp.50-62 (1972).

(22) Bier, K., Engelhorn, H. and Gorenflo, D. Wärmeübergang an tiefsiedende Halogenkältemittel Ki Klima and Kälte Ingenieur 11 pp.399-406 (1976).

(23) Cichelli, M. and Bonilla, C. Trans. A.I. Chem. Eng. 41, p.755 (1945).

(24) Borishanski, V.M.  (a) Energomaschinostroyeniye No.7 (1958)
                       (b) Zh. tekh. Fiz. No.2 (1956)
                       (c) Kotloturbostroyeniye No.4 (1952)

(25) Borishanski, V.M., Bobrovich, G.I. and Minchenko, F.P. "Heat transfer from a tube to water and to ethanol in nucleate pool boiling", pp.85-106 in Problems of Heat Transfer and Hydraulics of Two-Phase Media, editor S.S. Kutateladze (Pergamon Press 1969).

(26) McAdams, W. Heat Transmission, McGraw Hill, 1954.

(27) Kutateladze, S.S. Trud. ZKTI Book 2, Mashig 1947.

(28) Huber, D.A. and Hoehne, J.C. "Pool boiling of benzene, diphenyl and benzene-diphenyl mixtures under pressure". J. of Heat Transfer, pp.215-220, August 1963.

(29) Cryder, D.S. and Finalborgo. Trans. Am.I.Chem.Engrs. 33, p.346 (1937).

(30) Myers, J. and Katz, D. "Boiling coefficients outside horizontal tubes", Chem. Eng. Prog. Symp. Series No.5 "Heat Transfer Atlantic City", p.49 (1953).

(31) Danilova, G.N. "Heat transfer to boiling refrigerants", pp.107-130 in Problems of Heat Transfer and Hydraulics of Two-Phase Media, editor S.S. Kutateladze (Pergamon Press 1969).

(32) Sciance, C.T., Colver, C.P. and Sliepcevich, C.M. Advances in Cryogenic Engng. 12, pp.395-409 (1967).

(33) Bier, K., Gorenflo, D. and Wickenhäuser, G. "Pool boiling heat transfer at saturation pressures up to critical", Chapter 7, Heat Transfer in Boiling, edited by E. Hahne and U. Grigull, Academic Press/Hemisphere Publishing Corp. (1977).

(34) Gorenflo, D. "Influence of pressure on heat transfer from horizontal tubes to boiling refrigerants". Proc. of the XII Int. Congress of Refrigeration, Vol.II S-587.599 Madrid (1969).

(35) Zuber, N. On the stability of boiling heat transfer. Trans. ASME Vol.80, p.711 1958; see also Zuber, N., Tribus, M. and Westwater, J.W. "The hydrodynamic crisis in pool boiling of saturated and subcooled liquids", Int. Dev. Heat Trans. Vol.2, p.230 (1961).

(36) Kutateladze, S.S. "A hydrodynamic theory of changes in the boiling process under free convection conditions", Izv. Akad. Nauk SSSR, Otd. Tekh. Nauk No.4, pp.529-536 (1951).

(37) Lienhard, J.H. and Dhir, V.K. "Extended hydrodynamic theory of the peak and minimum pool boiling heat fluxes", Nat. Aero Space Admin. Report NASA-CR-2270, 194 pages, (July 1973).

(38) Bankoff, S.G. and Mehra, V.S. "A quenching theory for transition boiling", Ind. Eng. Chem. Fundamentals, Vol.1, pp.39-40 (1962).

(39) Berenson, P.J. "Transition boiling heat transfer from a horizontal surface", Mass Inst. Tech. Heat Transfer Laboratory Technical Report No.17, (1960).

(40) Berenson, P.J. "Film boiling heat transfer from a horizontal surface", J. Heat Transfer, Vol.83, pp.351-358 (August 1961).

(41) Bromley, L.A. "Heat Transfer in Stable Film Boiling", Chem. Engng. Prog. Vol.46, No.5, pp.221-227 (May 1950).

(42) Breen, B.P. and Westwater, J.W. "Effect of diameter of horizontal tubes on film boiling heat transfer", A.I.Ch.E. - ASME 5th Nat. Heat Transfer Conf., Houston, A.I.Ch.E. Preprint No.19 (August 1962).

(43) Clements, L.D. and Colver, C.P. "Natural Convection Film Boiling Heat Transfer", Ind. Engng. Chem., Vol.62 (9), pp.26-46 (September 1970).

(44) Beecher, N. M.S. Thesis in Chemical Engineering, Mass. Inst. of Technology, 1948.

(45) McKee, H.R. and Bell, K.J. "Forced convection boiling from a cylinder normal to the flow", Chem. Engng. Symp. Series - "Heat Transfer - Philadelphia", Vol.65, No.92, pp.222-230 (1969). See also Ph.D. Thesis by McKee, "Forced convection boiling from a cylinder normal to the flow", Oklahoma State University, April 1967.

(46)  Vliet, G.C. and Leppert, G. "Critical heat flux for nearly saturated water flowing normal to a cylinder", Paper presented at the Winter Annual Meeting, ASME, New York, Nov. 25-30 1962, Paper No.62-WA-173.

(47)  Bromley, L.A., Le Roy, N.R. and Robbers, J.A. "Heat transfer in forced convection film boiling", Ind. Eng. Chem. Vol.45, No.22, pp.2639-2646, 1953.

(48)  Leong, L.S. and Cornwell, K. "Heat transfer coefficients in a reboiler tube bundle". The Chemical Engineer, No.343, pp.219-221, April 1979.

(49)  Palen, J.W. and Small, W.M. "A new way to design kettle and internal reboilers", Hydrocarbon Processing Vol.43, No.11, pp.199-208, November 1964.

(50)  Lienhard, J.H. and Eickhorn, R. "Peak boiling heat flux on cylinders in a cross flow", Int. J. Heat Mass Transfer, Vol.19, pp.1135-1142, 1976.

# Boiling and Evaporation—II

**J. G. COLLIER**
United Kingdom Atomic Energy Authority
Harwell, UK

ABSTRACT

This lecture is devoted to a description of forced convection boiling of single component liquids in vertical tubes. The various heat transfer regimes which occur are identified and described with the aid of a three-dimensional representation - the "boiling surface". Appropriate correlations for each heat transfer regime are given so that this "surface" may be constructed for any particular fluid or geometry.

1. INTRODUCTION

In the previous lecture we identified some of the fundamental thermodynamic aspects of liquid-vapour phase change and went on to consider what happens when a liquid is boiled in a container in which the only means of circulation is that induced by the vapour bubbles rising from the heated surface. In many industrial boilers and evaporators the liquid being evaporated is at a significant pressure and in this case evaporation is conveniently accomplished within a heated tube. This chapter is therefore devoted to a discussion of heat transfer during forced convection boiling inside tubes.

Research work from many fields of engineering has contributed to our knowledge of forced convection boiling. Notably, work has been published in the technical journals covering mechanical, chemical, regrigeration, petroleum, boiler plant and nuclear engineering amongst others. Studies in some of these fields date back to the nineteenth century. However, it is only in the last thirty years that important contributions have been made towards understanding the processes involved in progressive vaporisation along a tube. The lack of progress in the earlier work arises, in part, from the large number of variables which are present but, in the main, it is the consequence of the fact that early workers in the field reported only effects on the performance of the evaporator resulting from independent variables altered by the operator. It is clear from a brief inspection of the complex flow patterns which exist when evaporation takes place along a tube that it is essential to consider the local or point conditions in the evaporator. No such study was published prior to 1952.

Another major difficulty comes from the fact that the addition of heat to a two-phase flow causes variations in the amount and distribution of each phase and the flow pattern or topology of the flow. These changes, in turn, induce variations in the local heat transfer processes. Because of the continuous change of all the thermal and hydraulic properties of the flow, the situation at any axial point in the channel can never be fully developed either thermally or hydrodynamically. Since the local situation is not an equilibrium situation, it is necessary to understand the variation of the flow and thermal properties upstream of the point being considered. This information is required to define the magnitude and direction of the departure from equilibrium. Add to these difficulties the fact that certain situations involve time-varying properties and others involve departures from thermodynamic equilibrium and one begins to realise the magnitude of the problem being tackled.

Although a complete understanding of evaporation within a tube is not yet available, Hewitt and his co-workers[28,29,30] have been able to develop a quantitative closed form theory for the prediction of non-equilibrium evaporating flows. However, for the purpose of heat exchanger design something simpler is usually required.

In this chapter therefore, the various regions of single-phase and two-phase heat transfer encountered in boilers and evaporators are identified with the aid of a three dimensional diagram - the "heat transfer (or boiling) surface". The boundaries between the various regions are discussed and correlations given to allow the appropriate heat transfer rates to be established.

2. HEAT TRANSFER REGIONS IN A VERTICAL TUBE[1]

The heat transfer regions encountered when a liquid is evaporated in vertical heated tube are shown in Figure 1. Boiling cannot occur until the tube wall temperature exceeds the saturation temperature. The amount by which the wall temperature exceeds the saturation temperature is known as the "degree of superheat" $\Delta T_{SAT}$ and the difference between the saturation and the local bulk fluid temperature is known as the "degree of subcooling" $\Delta T_{SUB}$. In the two-phase region where net vapour generation occurs, the variable characterising the heat transfer mechanism is the thermodynamic vapour "quality" (x). At any distance, z along the tube this is given by:

$$x(z) = \frac{i(z) - i_f}{i_{fg}} \qquad (1)$$

It is useful at this stage to describe, at least qualitatively, the progressive variation of the local surface temperature (or local heat transfer coefficient) along the length of the tube as evaporation proceeds. The local heat transfer coefficient can be established by dividing the surface heat flux (constant over the tube length) by the difference between the wall temperature and the bulk fluid temperature. Typical variations of these two temperatures with length along the tube are shown in Figure 1. The variation of heat transfer coefficient with length along the tube for the conditions represented in Figure 1 is given in Figure 2 (curve (i), solid line). In the *single phase convective heat transfer region* (region A) the wall temperature is displaced above the bulk fluid temperature by a relatively constant amount, (the heat transfer coefficient is approximately constant) and is modified only slightly by the influence of temperature on the liquid physical properties. In the *sub-cooled nucleate boiling region* (region B) the temperature difference between the wall and the bulk fluid decreases linearly with length up to the point where x = 0. The heat transfer coefficient, therefore, increases linearly with length in this region. In the *saturated nucleate boiling region*

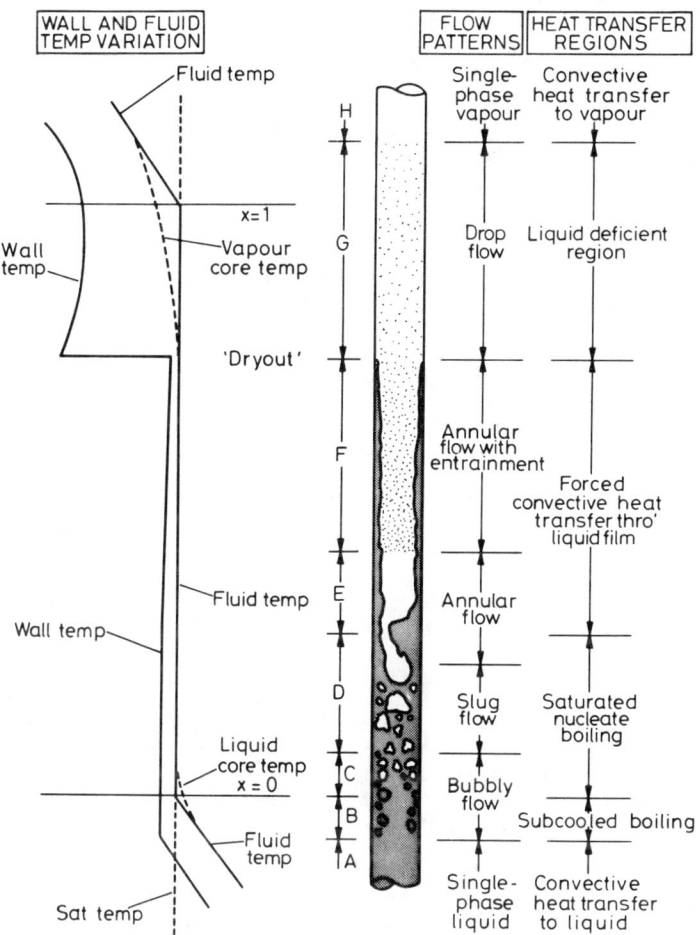

Fig.1  Regions of Heat Transfer in Convective Boiling

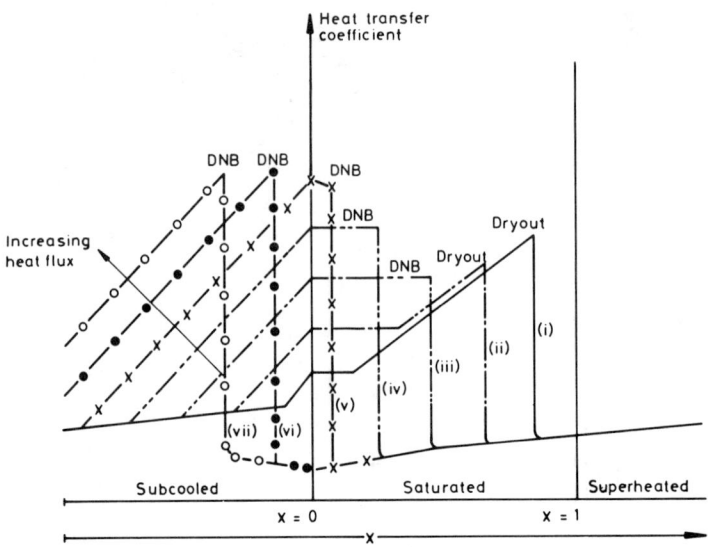

Fig.2 Variation of Heat Transfer Coefficient with Quality with increasing Heat Flux as Parameter.

(regions C and D) the temperature difference and, therefore, the heat transfer coefficient remains constant. Because of the reducing thickness of the water film in the *two-phase forced convective region* (regions E and F) the difference in temperature between the surface and the saturation temperature reduces and the heat transfer coefficient increases with increasing length or steam quality. At the dryout point the heat transfer coefficient is suddenly reduced from a very high value in the forced convective region to a value near to that expected for heat transfer by forced convection to dry saturated steam. As the quality increases through the *liquid deficient region* (region G) so the steam velocity increases and the difference in temperature between the surface and the saturation value decreases with the corresponding rise in heat transfer coefficient. Finally, in the single-phase steam region (x >1) the wall temperature is once again displaced by a constant amount above the bulk fluid temperature and the heat transfer coefficient levels out to that corresponding to convective heat transfer to a single phase steam flow.

The above comments have been restricted to the case where a relatively low heat flux is supplied to the walls of the tube. The effect of progressively increasing the surface heat flux whilst keeping the inlet flow-rate constant, will now be considered with reference to Figures 2 and 3. Figure 2 shows the heat transfer coefficient plotted against steam quality with increasing heat flux as parameter (curves (i)-(vii)). Figure 3 shows the various regions of two-phase heat transfer in forced convective boiling on a three-dimensional diagram with heat flux, steam quality and temperature as co-ordinates - "the heat transfer surface". Curve (i) relates to the conditions shown in Figure 1.

for a low heat flux being supplied to the walls of a tube. The temperature pattern shown in Figure 1 will be recognised as the projection in plan view (temperature-quality co-ordinates) of curve (i). Curve (ii) shows the influence of increasing the heat flux. Sub-cooled boiling is initiated sooner, the heat transfer coefficient in the nucleate boiling region is higher but is unaffected in the two-phase forced convective region. Dryout occurs at a lower steam quality. Curve (iii) shows the influence of a further increase in heat flux. Again, sub-cooled boiling is initiated earlier and the heat transfer is again higher in the nucleate boiling region. As the steam quality increases, before the two-phase forced convective region is initiated, and while bubble nucleation is still occurring, an abrupt deterioration in the cooling process takes place. This transition is essentially similar to the critical heat flux phenomenon in saturated pool boiling and will be termed *departure from nucleate boiling* (DNB).

The mechanism of heat transfer under conditions where the critical heat flux (DNB or dryout) has been exceeded is dependent on whether the initial condition was the process of "boiling" (i.e. bubble nucleation in the sub-cooled or low steam quality regions) or the process of "evaporation" (i.e. evaporation at the water film-steam core interface in the higher steam quality areas). In

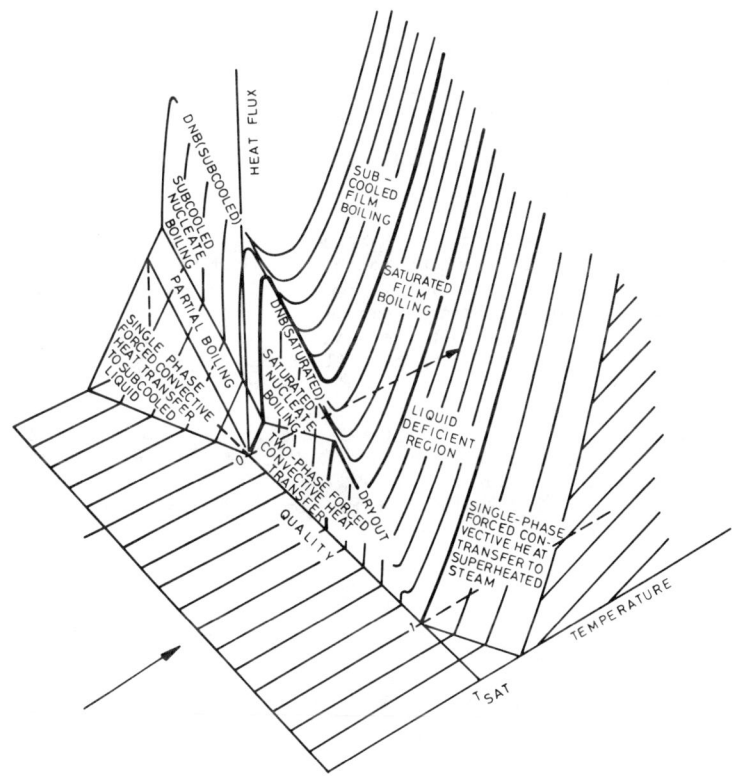

Fig.3 The Heat Transfer (or Boiling) Surface

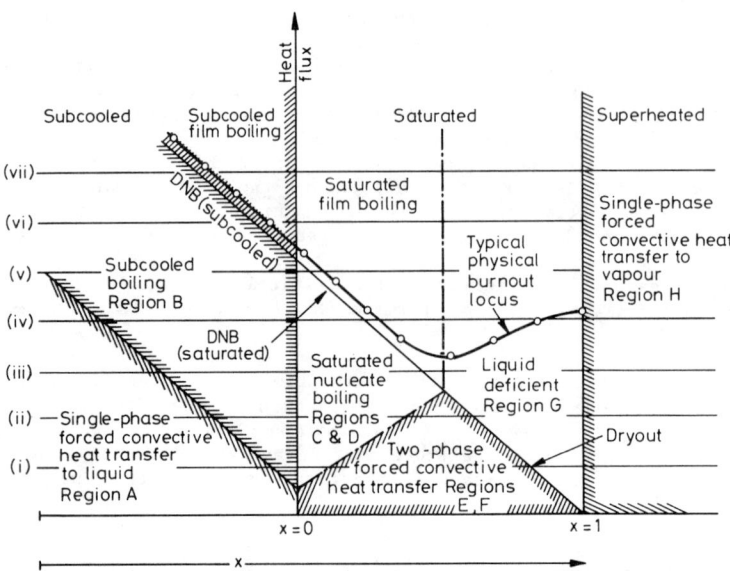

Fig.4 Regions of Two-Phase Forced Convective Heat Transfer as a Function of Quality with increasing Heat Flux as ordinate

the latter case, the liquid deficient region is initiated; in the former case the resulting mechanism is one of *film boiling* (Figure 3).

Returning to Figures 2 and 3, it can be seen that further increases in heat flux (curves (vi) and (vii)) cause the condition of "departure from nucleate boiling" (DNB) to occur in the sub-cooled region with the whole of the saturated or "quality" region being occupied by, firstly, "film boiling" and, in the latter stages, the "liquid deficient region" - both relatively inefficient modes of heat transfer. In Figure 3 the film boiling surface has been arbitrarily divided into two regions: *sub cooled film boiling* and *saturated film boiling*. Film boiling in forced convective flow is essentially similar to that observed in pool boiling. An insulating steam film covers the heating surfae through which the heat must pass. The heat transfer coefficient is orders of magnitude lower than in the corresponding region before the critical heat flux was exceeded, due mainly to the lower thermal conductivity of the steam adjacent to the surface.

A *transition boiling* region occupies the reverse slope (partly obscured) in Figure 3. In the transition boiling region the vapour film next to the heating surface becomes unstable and there is intermittent contact between the liquid phase and the surface.

Figure 4 shows the regions of two-phase forced convective heat transfer as a function of "quality" with increasing heat flux as ordinates (an elevation view of Figure 2).

In the following sections criteria will be given whereby the boundaries delineated in Figures 3 and 4 - the "heat transfer surface" - can be established. In addition, methods for calculating heat transfer rates in each heat transfer region will be discussed briefly.

## 3. SINGLE-PHASE LIQUID HEAT TRANSFER

The tube surface temperature in region A, convective heat transfer to single-phase liquid, is given by

$$T_W = T_f(z) + \Delta T_f \tag{2}$$

and

$$\Delta T_f = \phi/h_{fo} \tag{3}$$

where $\Delta T_f$ is the temperature difference between the tube inside surface and the mean bulk liquid temperature at a length z from the tube inlet, $h_{fo}$ is the heat transfer coefficient to single-phase liquid under forced convection. The liquid in the channel may be in laminar or turbulent flow. For laminar flow a variety of theoretical relationships are available, depending on the boundary conditions, i.e. constant surface heat flux or surface temperature, developing velocity profile or fully-developed flow. The following empirical equation based on experimental data takes into account the effect of varying physical properties across the flow stream and the influence of natural convection.

$$\left[\frac{h_{fo} D}{k_L}\right] = 0.17 \left[\frac{GD}{\mu_L}\right]^{0.33} \left[\frac{c_p \mu}{k}\right]_L^{0.43} \left[\frac{Pr_L}{Pr_W}\right]^{0.25} \left[\frac{D^3 \rho_L^2 g \beta \Delta T}{\mu_L^2}\right]^{0.1} \tag{4}$$

This relationship is valid for heating in vertical upflow or cooling in vertical downflow for $z/D > 50$ and $GD/\mu_L < 2000$.

For turbulent flow the well-known Dittus-Boelter equation has been found satisfactory

$$\left[\frac{h_{fo} D}{k_L}\right] = 0.023 \left[\frac{GD}{\mu_L}\right]^{0.8} \left[\frac{c_p \mu}{k}\right]_L^{1/3} \tag{5}$$

This relationship is valid for heating in vertical up flow and $z/D > 50$ and $\{GD/\mu_L\} > 10,000$.

## 4. THE ONSET OF SUBCOOLED NUCLEATE BOILING

Consider the variations of temperature of the tube inner surface at a point, z, from the inlet as the heat flux is steadily increased at a given inlet subcooling and mass velocity. Figure 5 shows the relationship in a qualitative form. Three regions are shown as the single-phase (sub-saturation) region AB, the "partial boiling" region BCDE and the fully-developed subcooled boiling region EF. Figure 5 will be easily recognised as a part-section in the subcooled region through the surface illustrated in Figure 3.

As the heat flux is increased at constant subcooling, the relationship between the surface temperature, $T_W$, and the heat flux will follow the line ABD' until the first bubbles nucleate. A higher degree of superheat is necessary to initiate the first bubble nucleation sites at a given heat flux then indicated by the curve ABCDE. When nucleation first occurs the surface temperature drops from D' to D and, for further increases in heat flux, follows the line DEF. The criterion for the onset of boiling can crudely be established as the intersection of line ABD' and the fully-developed boiling curve C'' EF.

A more refined treatment of the onset of nucleation can be derived by considering the temperature profile in the region adjacent to the heated wall. This treatment was originally proposed by Hsu for pool boiling and has been used to predict both the onset of subcooled boiling and the suppression of saturated nucleate boiling. Consider Figure 6; if the liquid is at a uniform temperature, $T_G$, then bubble nuclei of radius $r^*$ will grow if this temperature exceeds that given by

$$(T_G - T_{SAT}) = \frac{R \, T_{SAT} \, T_G}{i_{fg} \, M} \ln \left[ 1 + \frac{2\sigma}{p \, r^*} \right] \qquad (6)$$

In a heated system, there is a temperature gradient away from the wall which is approximated in the linear form

$$T_L(y) = T_W - \left( \frac{\phi y}{k} \right) \qquad (7)$$

The postulate of Hsu is that the bubble nuclei on cavities in the heated wall will only grow if the lowest temperature on the bubble surface (i.e. that furthest away from the wall) is greater than $T_G$. Allowing for the distortion of the liquid temperature profile by the bubble by plotting $T_G$ against $nr^*$ and $T_L$ against y on the same ordinates (Figure 6) it will be seen that only that range of bubbles for which $T_L$ is greater than $T_G$ (i.e. radii, $r_{MIN}$ to $r_{MAX}$) can grow. If, over the whole field, $T_L$ is less than $T_G$, then no nuclei will grow. When equation (7) is just tangent to equation (9) then nuclei of a critical radius, $r_{CRIT}$ will grow. Thus, for nucleation of the critical nucleus (n = 1)

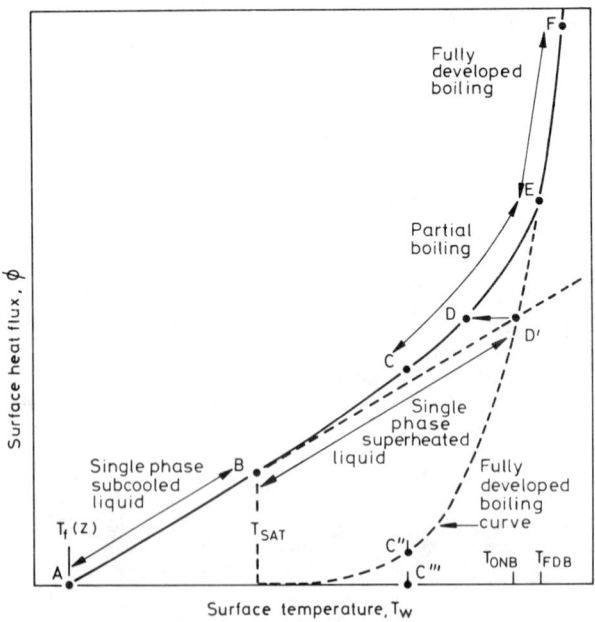

Fig.5 The Forced Convection Boiling Curve

$$T_L(y) = T_G \text{ and } \frac{dT_L(y)}{dy} = \frac{dT_G}{dr} \qquad (8)$$

Analytical solution of the equations was carried out by Davis and Anderson[2] who obtained the following expressions

$$\phi_{ONB} = \frac{k_L}{4B}(\Delta T_{SAT})^2_{ONB} \qquad (9)$$

$$\text{and } r_{CRIT} = \sqrt{\frac{Bk_L}{\phi}} \qquad (10)$$

$$\text{where } B = \left[\frac{2\sigma T_{SAT}(v_G - v_L)}{i_{fg}}\right] \qquad (11)$$

The above treatments can only predict the onset of nucleate boiling accurately if there is a sufficiently wide range of "active" cavity sizes available. At low fluxes and low pressures the values of $r_{CRIT}$ predicted from equation (10) may be so large that no "active" sites of this size are present on the heating surface. In this case an estimate of the largest "active" cavity size available on the heating surface must be made. Reasonable agreement with experimental data for water was found when a maximum "active" cavity size of 1 μm radius was used.

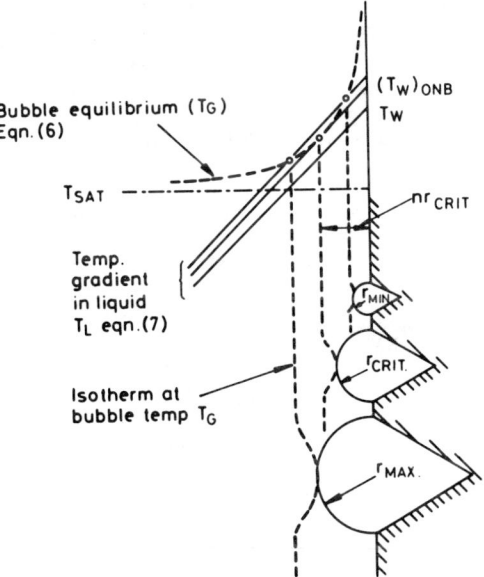

Fig.6 Onset of Nucleation in Subcooled Boiling

For the subcooled boiling region, equation (9) must be solved simultaneously with the heat transfer equation

$$\phi = h_{fo} (\Delta T_{SAT} + \Delta T_{SUB}(z)) \tag{12}$$

to give the heat flux $\phi_{ONB}$ and $(\Delta T_{SAT})_{ONB}$ required for the onset of boiling. The boundary between region A and region B shown in Figure 4 can be derived from such a simultaneous solution. The treatment may also be extended to cover the influence of dissolved gases upon the onset of nucleation.

## 5. SUBCOOLED NUCLEATE BOILING

Once boiling has been initiated, only comparatively few nucleation sites are operating initially so that a proportion of the heat will be transferred by normal single-phase processes between patches of bubbles. As the surface temperature increases, so the number of bubble sites also increases and the area for single-phase heat transfer decreases. Finally, the whole surface is covered by bubble sites, boiling is "fully-developed" and the single-phase component reduces to zero. In the "fully-developed" boiling region, velocity and subcooling, both of which have a strong influence on single-phase heat transfer, have little or no effect on surface temperature as observed experimentally.

Jens and Lottes[3] summarised experiments on subcooled boiling of water flowing upwards in vertical electrically-heated stainless steel or nickel tubes, having inside diameters ranging from 3.63 to 5.74 mm. System pressures ranged from 7 bars to 172 bars, water temperatures from 115°C to 340°C, mass velocities from 11 to 1.05 x $10^4$ kg/m²s, and heat fluxes up to 12.5 MW/m². These data were correlated by a dimensional equation <u>valid for water only</u>.

$$\Delta T_{SAT} = 25\phi^{0.25} e^{-p/62} \tag{13}$$

where p is the absolute pressure in bar, $\Delta T_{SAT}$ is in °C and $\phi$ is in MW/m².

More recently, Thom[4] reported that the values of $T_{SAT}$ estimated from equation (13) were consistently low over the range of his experiments. A modified equation of the same form and valid only for water was suggested

$$\Delta T_{SAT} = 22.65\phi^{0.5} e^{-p/87} \tag{14}$$

in SI units. This equation is recommended for calculational purposes. A number of studies of the basic mechanisms of subcooled boiling have been carried out but no generally accepted theory has so far emerged.

## 6. SATURATED NUCLEATE BOILING REGION

Here, the mechanism of heat transfer is essentially identical to that in the subcooled region. A thin layer of liquid near to the heated surface is superheated to a sufficient degree to allow nucleation. Just as with all other characterising variables, such as coolant temperature, void fraction, pressure gradient, etc., the heat transfer coefficient and heater surface temperature variation is smooth and continuous through the thermodynamic boundary (x = 0) marking the onset of saturated boiling. The methods and equations used to correlate experimental data in the subcooled region remain valid for this region with the provision that $T_f(z) = T_{SAT}$. Just as the heat transfer mechanism in the subcooled region is independent of the subcooling and, to a large degree, the mass velocity, so it may be inferred that the heat transfer process in this region is independent of the steam quality $x(z)$, and the mass velocity. This, indeed, is found to be experimentally correct for the case of "fully-developed nucleate boiling". Thus, because the bulk temperature is constant in this region the heat transfer coefficient is also constant since $\Delta T_{SAT}$ is fixed.

## 7. SUPPRESSION OF SATURATED NUCLEATE BOILING

To maintain nucleate boiling on the surface, it is necessary that the wall temperatures exceeds the critical value for a specified heat flux. If the wall superheat is less than that given by equation (10) for the imposed surface heat flux, then nucleation does not take place; the value of $\Delta T_{SAT}$ for comparison with this equation is calculated from the ratio $(\phi/h_{TP})$, where $h_{TP}$ is the two-phase heat transfer coefficient in the absence of nucleation.

$$\phi_{ONB} = \frac{4B}{k_L} h_{TP}^2 \qquad (15)$$

Equation (15) represents an equation for the boundary between the saturated nucleate boiling and the two-phase forced convective regions (Figure 4). This relationship is only valid on the basis of a complete range of "active" cavities on the heating surface. As with subcooled boiling, there is a similar "partial boiling" region between fully-developed saturated nucleate boiling and the two-phase forced convective region. In this transition region, both the forced convective and nucleate boiling mechanisms are significant.

Direct observations of nucleation in annular flow have been reported by Hewitt et al[5]. The results qualitatively confirmed the expected trends. However, the values of $\phi_{ONB}$ calculated from equation (15) are very much lower than those measured. This discrepancy is consistent with the findings of Davis and Anderson mentioned above. Equation (15) predicts that nucleation will persist at wall superheat down to 3°C but it is very unlikely that the large sizes of cavity required to be "active" down to this superheat were present on the stainless steel heating surface.

## 8. THE TWO-PHASE FORCED CONVECTION REGION

The two-phase forced convective region is most likely to be associated with the annular flow pattern. Heat is transferred by conduction or convection through the water film and steam is generated continuously at the interface. Extremely high heat transfer coefficients are possible in this region; values can be so high as to make accurate assessment difficult. Typical figures for water of up to 200 kW/m$^2$ °C have been reported.

Many workers have correlated their experimental results of heat transfer rates in the two-phase forced convective region in the form

$$\frac{h_{TP}}{h_f} = fn\left(\frac{1}{X_{tt}}\right) \qquad (16)$$

where $h_f$ is the value of the single-phase liquid heat transfer coefficient based on the liquid component flow, and $X_{tt}$ is the Matinelli parameter.

A number of relationships of the form of equation (16) have been proposed and in some cases these have been extended to cover the saturated nucleate boiling region also. However, these correlations do have a high mean error ($\pm$ 30%) and a more satisfactory correlation has been proposed by Chen[6]. The correlation covers both the "saturated nucleate boiling region" and the "two-phase forced convective region". Both mechanisms occur to some degree over the entire range of the correlation and that the contributions made by the two mechanisms are additive.

$$h_{TP} = h_{NcB} + h_c \tag{17}$$

where

$$h_{NcB} = 0.00122 \left[ \frac{k_L^{0.79} c_{pL}^{0.45} \rho_L^{0.49}}{\sigma^{0.5} \mu_L^{0.29} i_{fg}^{0.24} \rho_G^{0.24}} \right] \Delta T_{SAT}^{0.24} \Delta p^{0.75} (S) \tag{18}$$

$$\text{and } h_c = 0.023 \left[ \frac{G(1-x)D}{\mu_L} \right]^{0.8} \left[ \frac{\mu c_p}{k} \right]_L^{0.4} \left( \frac{k_L}{D} \right) (F) \tag{19}$$

The parameter S, the nucleate boiling suppression factor, and the parameter F, are graphical functions which may be approximated by the following curve-fits for S[13]

$$S = \begin{bmatrix} [1 + 0.12 \ (Re'_{TP})^{1.14}]^{-1} & , & Re'_{TP} < 32.5 \\ [1 + 0.42 \ (Re'_{TP})^{0.76}]^{-1} & , & Re'_{TP} > 32.5 \end{bmatrix} \tag{20}$$

and for F

$$F = \begin{bmatrix} 1.0 & , & \frac{1}{X_{tt}} \leq 0.10 \\ 2.35 \ (\frac{1}{X_{tt}} + 0.213)^{0.736} & , & \frac{1}{X_{tt}} > 0.10 \end{bmatrix} \tag{21}$$

where

$$\frac{1}{X_{tt}} = \left( \frac{x}{1-x} \right)^{0.9} \left( \frac{\rho_L}{\rho_G} \right)^{0.5} \left( \frac{\mu_G}{\mu_L} \right)^{0.1} \tag{22}$$

and

BOILING AND EVAPORATION-II

$$\text{Re}'_{TP} = \left(\frac{G(1-x)D}{\mu_L}\right) F^{1.25} \times 10^{-4} \quad (23)$$

## 9. THE CRITICAL HEAT FLUX CONDITION

The heat flux in the nucleate boiling regions cannot be increased indefinitely. At some critical value sufficient liquid is unable to reach the heating surface to cool it due to the rate at which vapour is leaving th surface and an abrupt deteriortion in the cooling process takes place — "departure from nucleate boiling" (DNB). Likewise, at some critical value of the "quality", complete evaporation of the liquid film may occur — "dryout". The term "critical heat flux" will be used to cover both these separate and distinct processes. A distinction should also be made between the precise point at which a more or less rapid surface temperature rise from a temperature close to the saturation value is initiated and the point at which failure of the heating surface due to rupture or melting occurs — "physical burnout". The term "burnout heat flux", which also suggests this latter condition, has been abused by many, who use it to note the heat flux at which the rapid deterioration of the cooling processes occurs. This ambiguity can be illustrated by reference to Figures 3 and 4. Low heat transfer coefficients coupled with the relatively high heat flux values required to initiate film boiling in the subcooled or low quality regions results in extremely high temperature differences at the critical heat flux condition. Failure of the heating surface usually occurs and thus the heat flux to initiate DNB is often identical with that to cause "physical burnout". However, this is not the case in the liquid deficient region where higher heat transfer coefficients and lower critical heat fluxes cause only modest temperature excursions at "dryout". In this region the physical burnout locus denotes a particular isotherm representing the failure criterion for the chosen heating surface.

Experiments to determine the critical heat flux for water in vertical uniformly-heated and non-uniformly heated round tubes have been carried out in many countries over the past 30 years or so. Thompson and Macbeth[7] published a compilation of 4389 separate experimental results in 1964 and more recently the USSR Academy of Sciences[31] has provided a series of standard tables of critical heat flux as a function of the local bulk mean water condition for various pressures and mass velocities for a fixed tube diameter of 8mm.

The critical heat flux is a complex function of a large number of independent variables, thus

$$\phi_{CRIT} = \text{fn } [G, i(z) \, p, \, D, \, z] \quad (24)$$

or

$$\phi_{CRIT} = \text{fn } [G, (\Delta i_{SUB})_i, \, p, \, D, \, z] \quad (25)$$

When the influence of one particular variable on the critical heat flux is considered, it is essential to specify which of the other independent (or dependent) variables is held constant. Thus,

(i) for fixed p, D, and z the critical heat flux increases linearly with $(\Delta i_{SUB})_i$ the inlet subcooling for fixed G and increases with G, the mass velocity at fixed $(\Delta i_{SUB})_i$.

(ii) for fixed p, D and z the critical heat flux decreases linearly with increasing exit mass quality (x(z)) for fixed G. In the subcooled region $\phi_{CRIT}$ increases with G at constant exit conditions whilst in the saturated region the reverse is the case.

(iii) for fixed p, D and G, the critical heat flux decreases with increasing tube length z at fixed inlet subcooling $(\Delta i_{SUB})_i$.

(iv) for fixed p, D and G the effect of tube length (z) on $\phi_{CRIT}$ for fixed exit quality (x(z)) is small.

(v) for fixed p, z, and G an increase in tube diameter (D) increases $\phi_{CRIT}$ for a fixed value of $(\Delta i_{SUB})_i$.

(vi) for fixed p, z and G, the critical heat flux decreases as the tube diameter is increased for fixed values of exit quality.

(vii) for fixed D, z and G the critical heat flux falls rapidly as the pressure is increased for a fixed exit quality (x(z)). If the inlet subcooling is held constant $\phi_{CRIT}$ passes through a maximum at low pressure and then falls as the pressure is increased (Figure 7).

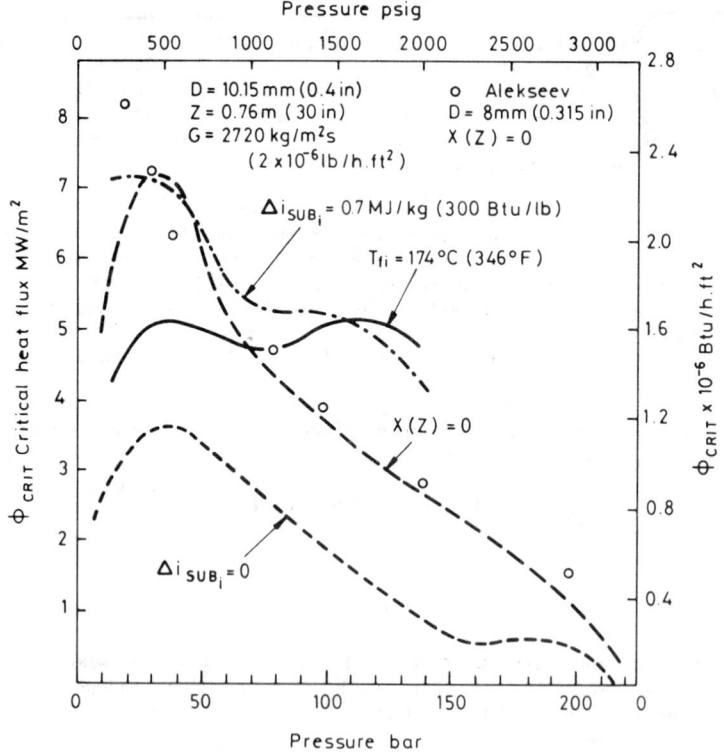

Figure 7    The influence of system pressure on the critical heat flux

TABLE 1

SELECTION OF CRITICAL HEAT FLUX CORRELATIONS

| AUTHOR | GEOMETRY | CORRELATION |
|---|---|---|
| Macbeth[7]<br>Bowring[8] | Round Tubes<br>Rod Bundles | $\phi_{CRIT} = \dfrac{A + CDG(\Delta i_{SUB})_i/4}{1 + Cz}$<br>where $A = fn(G,D,p)$ and $C = fn(G,D,p)$ |
| CISE[9] | Round Tubes | $x_{CRIT} = a \left[ \dfrac{z_{SAT}}{z_{SAT} + b} \right]$    where $a = fn(G,p)$<br>$b = fn(G,D,p)$ |
| Barnett[10] | Annuli,<br>Rod Bundles | $\phi_{CRIT} = \dfrac{A + B(\Delta i_{SUB})_i}{C + z}$    where $A = fn(G,D_h, D_e, p)$<br>$B = fn(G,D_h)$<br>$C = fn(G,D_e)$ |
| Hewitt[11] | Round Tubes,<br>Annuli,<br>Rod Bundles | $x_{CRIT} \left[ k_1(G) k_1(p) \right] = fn \left[ k_2(G) k_2(D) z_{SAT} \right]$<br>where $k_1(G)$, $k_1(p)$, $k_2(G)$ and $k_2(D)$ are multiplying factors (graphical functions of $G$, $D$, and $p$) |
| Biasi et al[12] | Round Tubes | $\phi_{CRIT} = \dfrac{1.883 \times 10^3}{D^n G^{1/6}} \left[ \dfrac{f(p)}{G^{1/6}} - x(z) \right]$<br>for low quality<br>$\phi_{CRIT} = \dfrac{3.78 \times 10^3 h(p)}{D^n G^{0.6}} [1 - x(z)]$    cgs units<br>for high quality<br>$n = 0.4$ for $D \geqslant 1$ cm    $f(D)$ and $h(p)$ are<br>$n = 0.6$ for $D < 1$ cm    empirical functions of pressure |
| Becker[14] | Round Tubes<br>Annuli<br>Rod Clusters | $\phi_{CRIT} = G^{-0.5} f_1[x,p] f_2(D)$<br>where $f_1[x,p]$ and $f_2(D)$ are functions presented graphically. Additional corrections are added to account for heated perimeter effect. |

Experimental CHF data are usually summarised in the form of correlations in which the critical heat flux is expressed as a function of the various independent and sometimes dependent variables. Table 1 summarises a selection of the better known correlations, particularly those in use in the UK. A complete compilation would run to well over 100 such correlations. Each correlation is usually optimised within a well-defined range of variables and extrapolation outside this range is not recommended.

An alternative to the use of correlations is the quantitative closed form theory for dryout in annular flow developed by Hewitt and his co-workers[28,29,30]. The construction of this general model was made possible by the development of correlations for the rate of droplet deposition from the vapour core onto the annular liquid film and for the rate of droplet entrainment. Dryout is predicted to occur when the flowrate in the liquid film on the heated surface becomes zero. The importance of this phenomenological model is its ability to describe all the important dependent variables viz channel pressure drop, void fraction, heat transfer coefficient and critical heat flux <u>without</u> the use of empirical correlations.

Experimental data for the critical heat flux for fluids other than water is relatively sparse and what there is tends to be very limited in the ranges of the independent variables covered. Attempts have been made to derive a set of scaling laws which would relate the extensive information available for the critical heat flux for water to that for non aqueous fluids. One of the more successful approaches is that of Ahmad[32]. Three dimensionless groups are commonly matched in the scaling of $\phi_{CRIT}$; $[(\Delta i_{SUB})_i/i_{fg}]$, $(\rho_f/\rho_g)$, $(z/D)$. Ahmad proposed a further modelling parameter $\psi_{cr}$ as follows:

$$\psi_{cr} = \left[\frac{GD}{\mu_f}\right]\left[\frac{\mu_f}{\sigma D \rho_f}\right]^{2/3}\left[\frac{\mu_f}{\mu_g}\right]^{1/8} \qquad (26)$$

or alternatively

$$\psi_{cr} = \left[\frac{GD}{\mu_f}\right]\left[\frac{\gamma^{0.5}\mu_f}{D\rho_f^{0.5}}\right]^{2/3}\left[\frac{\mu_f}{\mu_g}\right]^{1/8} \qquad (27)$$

where

$$\gamma = \frac{d(\rho_f/\rho_g)}{dp}$$

Equation (26) is the more basic version but equation (27) gives nearly identical results over a wide range of pressure and temperature. The usefulness of this scaling law may be judged from Figure 8 where data for water, Freon 12 and carbon dioxide are compared[33].

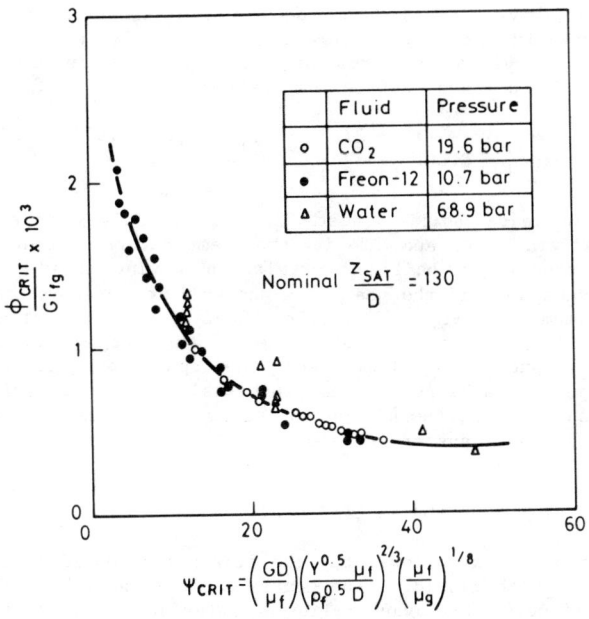

Figure 8   Application of Ahmad scaling method for water, Freon 12 and Carbon dioxide   (Hauptmann et al[33])

## 10. HEAT TRANSFER BEYOND THE CRITICAL HEAT FLUX

There is some uncertainty about the heat transfer rates beyond the critical heat flux condition at low steam qualities and under subcooled conditions. This is because with water, at least, it is impossible to study such situations experimentally except under transient conditions. With water in the pressure range up to 100 bar, the critical heat flux values are in excess of 3 MW/m$^2$ for subcooled boiling, while typical values for the film boiling heat transfer coefficients in this region are 150-500 W/m$^2$ °C. Any attempt to pass through the critical heat flux under steady state conditions would produce heater surface temperatures in excess of 2000°C and would cause failure of most common metal surfaces - physical burnout.

## 11. TRANSITION BOILING

The use of experimental techniques[15,16,17] where the surface temperature rather than the surface heat flux is the controlling variable, has established quite definitely the existence of a "transition boiling" region in forced convection conditions as well as for pool boiling. A comprehensive review of the published information on transition boiling under forced convection conditions has been prepared by Groeneveld and Fung[18]. The Groeneveld and Fung review lists those studies carried out prior to 1976. Since that date, in response to the urgent demands of the nuclear industry, further studies have been initiated.

Various attempts have been made to produce correlations for the transition boiling region. Probably the most useful currently available is that by Tong and Young[19] which is given in terms of the heat flux in the transition boiling region ($\phi_{TB}$) for a given surface temperature ($T_W$)

$$\phi_{TB} = \phi_{NB} \exp\left[-0.001 \frac{x^{2/3}}{(dx/dz)} \left(\frac{\Delta T}{55.5}\right)^{(1 + 0.0029 \Delta T)}\right] \qquad (28)$$

where $\phi_{NB}$ is the nucleate boiling heat flux (presumably equated with the critical heat flux) in W/m$^2$ and $\Delta T$ is the temperature difference ($T_W - T_{SAT}$) in °C. A transition boiling correlation having a wider range of application may be developed if the heat flux and wall temperature difference at the points of maximum and minimum on the "heat transfer surface" can be predicted with confidence. the present state of the art allows an accurate prediction of $\phi_{CRIT}$ and also $\Delta T_{CRIT}$ but the conditions at the minimum point are still subject to a large degree of uncertainty. The experimental data suggest that the conditions at the minimum point are a complex function of mass quality, flowrate and heat transfer surface properties[20].

## 12. SUBCOOLED AND SATURATED FILM BOILING

At low qualities and mass flow rates the flow regime would appear to be an "inverted annular" one with liquid in the centre and a thin vapour film adjacent to the heating surface. The vapour-liquid interface is not smooth, but irregular. These irregularities occur at random locations, but appear to retain their identity to some degree as they pass up the tube with velocities of the same order as that of the liquid core. The vapour in the film adjacent to the heating surface would appear to travel at a higher velocity.

Various analytical models have been used to predict film boiling heat transfer coefficients. A selection of these correlations is given in Table 2. The evidence at the present time is that laminar film boiling with a smooth interface only occurs over relatively short distances (7-10 cms) downstream of the dryout or "rewet" front. At longer distances the coefficient becomes independent of distance and takes on a value considerably (approximately a factor of 2) higher than the laminar solution would indicate. Figure 9 shows some experimental results for film boiling inside a vertical heated tube for water at low pressure. These data were taken using a transient technique[21] and show a considerable effect of flow direction (upflow or downflow) for saturated film boiling. This effect is absent for subcooled film boiling.

TABLE 2

FILM BOILING HEAT TRANSFER CORRELATIONS

| Geometry | Correlation |
|---|---|
| Flat plate, vertical, laminar flow | $\left[\dfrac{\bar{h}(z)z}{k_G}\right] = C \left[\dfrac{z^3 g \rho_G (\rho_L - \rho_G) i'_{fg}}{k_G \mu_G \Delta T}\right]^{1/4}$<br>$C = 0.943$ for zero interfacial shear stress ($\tau_i = 0$)<br>$C = 0.667$ for zero interfacial velocity ($u_i = 0$) |
| Flat plate, vertical, turbulent flow | $\left[\dfrac{h(z)z}{k_G}\right] = 0.056\, Re_G^{0.2} \left[\left(\dfrac{c_p \mu}{k}\right)_G \dfrac{z^3 g \rho_G (\rho_L - \rho_G)}{\mu_G^2}\right]^{1/3}$ |
| Flat plate, horizontal (Berenson) | $h = 0.425 \left[\dfrac{k_G^3 g \rho_G (\rho_L - \rho_G) i'_{fg}}{\mu_G \Delta T \left(\dfrac{\lambda c}{2\pi}\right)}\right]^{1/4}$<br>where $\left(\dfrac{\lambda c}{2\pi}\right)$ is the characteristic bubble spacing = $\left[\dfrac{\sigma}{g(\rho_L - \rho_G)}\right]^{1/2}$ |
| Cylinder, external surface, vertical (Bailey) | $h = 0.40 \left[\dfrac{k_G^3 g \rho_G (\rho_L - \rho_G) i'_{fg}}{\mu_G \Delta T\, r}\right]^{1/4}$<br>$r$ = cylinder (fuel rod) radius |
| Cylinder, external surface, horizontal, stagnant (Bromley) | $h = 0.62 \left[\dfrac{k_G^3 g \rho_G (\rho_L - \rho_G) i'_{fg}}{\mu_G \Delta T\, D}\right]^{1/4}$<br>Bromley EQN (1) |
| Cylinder, external surface, horizontal, flowing (Bromley) | $h = 2.7 \left[\dfrac{u\, k_G\, \rho_G\, i'_{fg}}{D\, \Delta T}\right]^{1/2}$<br>Bromley EQN (2) |

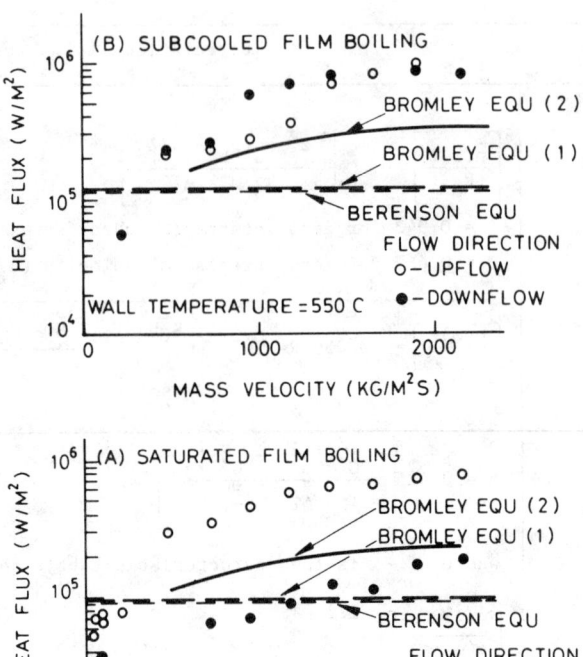

Fig.9 Comparison of experimental film boiling data for water[21] with correlations

## 15. LIQUID DEFICIENT REGION

Heat transfer rates in the liquid deficient region are bounded by two limiting situations, viz.

(1) <u>complete departure from equilibrium</u>. The rate of heat transfer from the steam phase to the entrained water droplets is so slow that their presence is simply ignored and the steam temperature $T_G(z)$ downstream of the

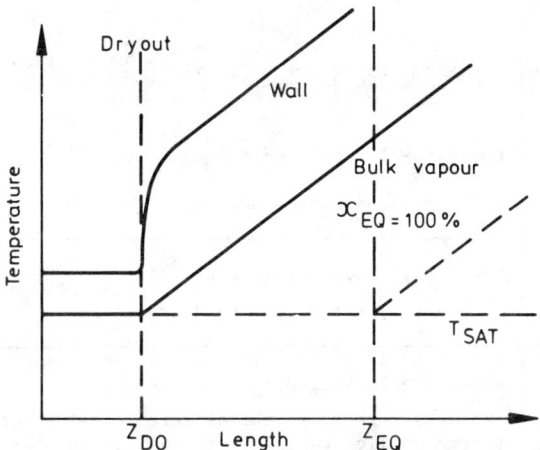

(a) COMPLETE LACK OF THERMODYNAMIC EQUILIBRIUM.

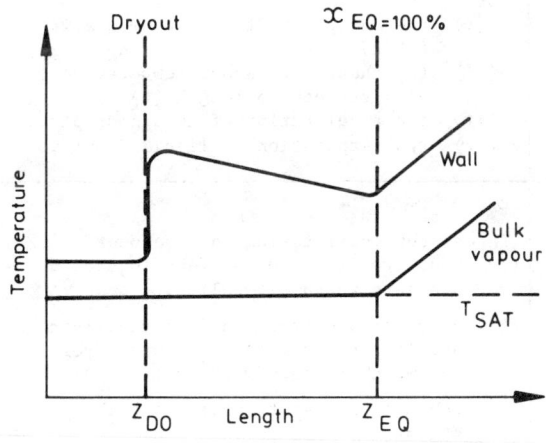

(b) COMPLETE THERMODYNAMIC EQUILIBRIUM

Fig.10 Limiting Conditions for Post-dryout Heat Transfer

TABLE 3

LIQUID DEFICIENT REGION HEAT TRANSFER CORRELATIONS

| Author | Correlation | (%) |
|---|---|---|
| Groeneveld[22] Slaughterback[23] (empirical) | $Nu_G = a \left[ Re_G \left\{ x + \frac{\rho_G}{\rho_L}(1-x) \right\} \right]^b Pr_{G,w}^c Y^d$ <br> where $Y = 1 - 0.1 \left( \frac{\rho_L}{\rho_G} - 1 \right)^{0.4} (1-x)^{0.4}$ <br> and a,b,c and d are indices | 11.5% |
| Mattson et al[27] (empirical) | $h = C \exp(-0.5 \sqrt{\Delta T}) + a\, Re_G^b\, Pr_{G,w}^c\, D_e^f\, k_G^h\, x^j$ <br> where C, a,b,c,f,h and j are indices | 17.28% |
| Groeneveld and Delorme[15] (taking into account non-equilibrium) | $x(z) - x^*(z) = \exp(-\tan \psi)$ <br> where $\psi = fn(Re_{TP}, P, x_e)$ <br> and $x(z) - x^*(z)$ is the difference between the thermodynamic "equilibrium" quality and the "true" steam quality | 6.7% |
| Bennett et al[36] | Simultaneous solution of differential equations for <br> (i) the change of "true" steam quality ($x^*(z)$) with respect to length (z) <br> (ii) the change of vapour temperature ($T_G$) with respect to length (z) <br> (iii) the acceleration of liquid droplets <br> (iv) the evaporation of liquid droplets | – |
| Iloeje et al[26] | $\phi = \phi_{dc} + \phi_{dw} + \phi_c$ <br> Three step model taking into account <br> (a) heat transfer from surface to liquid droplets which hit wall ($\phi_{dc}$) <br> (b) heat transfer from surface to liquid droplets which enter boundary layer but which do not "wet" wall ($\phi_{dw}$) <br> (c) convective heat transfer to bulk steam ($\phi_c$) | – |

dryout point is calculated on the basis that all the heat added to the fluid goes into superheating the steam. The wall temperature $T_W(z)$ is calculated using a conventional single-phase heat transfer correlation (Figure 10).

(2) **complete thermodynamic equilibrium.** The rate of heat transfer from the steam phase to the entrained water droplets is so fast that the steam temperature $T_G(z)$ remains at the saturation temperature until the energy balance indicates all the droplets have evaporated. The wall temperature $T_W(z)$ is again calculated using a conventional single-phase heat transfer correlation, this time with allowance made for the increasing steam velocity resulting from droplet evaporation (Figure 10).

It is known that liquid deficient heat transfer behaviour tends towards situation (1) at low pressures and low velocities, whilst at high pressure (approaching the critical condition) and high flow rates ( $> 3000$ kg/m$^2$s) situation (2) pertains.

A selection of correlations and models for the liquid deficient region is given in Table 3. The best of the empirical correlations are those of Groeneveld[22] and Slaughterback[23]. Figure 11 shows heat transfer coefficients predicted by various empirical correlations as a function of pressure. The Slaughterback correlation is to be preferred at low pressures because it does not predict artificially high heat transfer coefficients under these conditions.

The correlation of Groeneveld and Delorme[24] is representative of the type of correlation which recognises departures from thermodynamic equilibrium and produces considerable improvement in the prediction of wall temperature compared with the empirical correlation. However, such correlations assume that the extent of the departure from thermodynamic equilibrium is characterised by the "local" conditions, i.e. is unaffected by what has happened preceding the point being considered. This is obviously not the case.

A comprehensive model of heat transfer in the post-dryout region must take into account the various paths by which heat is transferred from the surface to the bulk vapour phase. Six separate mechanisms can be identified.

(i) heat transfer from the surface to water droplets which impact the wall ("wet" collisions)

(ii) heat transfer from the surface to liquid droplets which enter the thermal boundary layer but which do not "wet" the surface ("dry" collisions)

(iii) convective heat transfer from the surface to the bulk steam

(iv) convective heat transfer from the bulk steam to suspended liquid droplets in the steam core

(v) radiation heat transfer from the surface to the liquid droplets

(vi) radiation heat transfer from the surface to the bulk steam.

One of the first phenomenological models proposed was that of Bennett et al[25] which is a one-dimensional model starting from known equilibrium conditions at the dryout point. This model, however, did not consider mechanisms (i) and (ii). However, good agreement with experimental wall temperature profiles was seen. More recently, Iloeje et al[15] have proposed a three step model which takes account of mechanisms (i), (ii) and (iii).

## 14. SINGLE-PHASE VAPOUR HEAT TRANSFER

Based on experimental data in tubes and annuli, the following correlation of Bishop et al[26] is recommended for steam at moderately high pressures and Reynolds numbers:

$$\left(\frac{hD}{k_G}\right)_f = 0.0073 \left(\frac{DG}{\mu}\right)_f^{0.886} \left(\frac{c_p \mu}{k}\right)_f^{0.61} \left[1 + \frac{2.76}{z/D}\right] \quad (29)$$

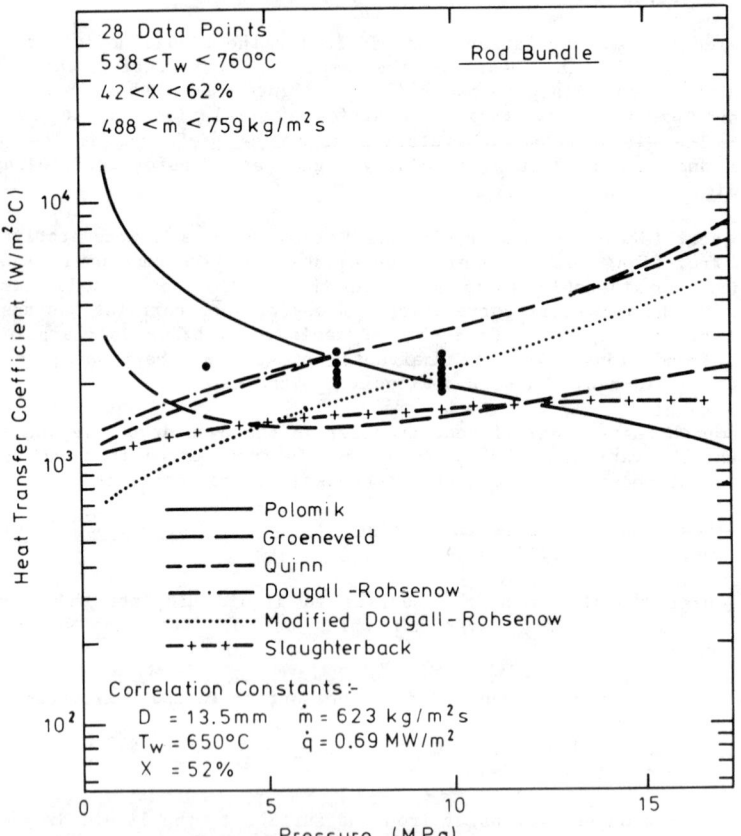

Fig.11 Heat Transfer Coefficients in Post-dryout Region as a Function of pressure (Slaughterback[23])

where subscript f refers to film conditions evaluated at $T = (T_W+T_b)/2$.

## 15. NON-EQUILIBRIUM EFFECTS

Many of the difficulties which arise from the use of correlations involving just local parameters are concerned with non-equilibrium effects. It is necessary to consider both departures from hydrodynamic equilibrium (i.e. not "fully developed" flow) and from thermodynamic equilibrium (i.e. vapour and liquid phases at different temperatures). By way of an example, consider an isothermal air-water flow in which all the water is injected at the channel entrance close to the wall and the air flows down the centre of the channel. After a finite distance the flow will have rearranged itself such that the disposition of liquid and gas does not change further with distance. Under boiling conditions, however, such "fully developed" flows can never be achieved because of the progressively increasing vapour flow with distance along the heated channel. However, analytical methods have now been proposed in order to calculate such hydrodynamically developing two-phase flows[28,29,30]. Turning to departures from thermodynamic equilibrium, these can become apparent - as compared with the "one-dimensional equilibrium model" used in most sub-channel codes - as a result of non-uniformities of temperature across the section, transient boundary layers around bubbles or droplets and kinetic theory limits on evaporation and condensation rates. Thus, vapour bubbles can exist in sub-cooled liquids and liquid droplets in superheated vapours. Departures from equilibrium where the two-phases exist at different mean temperatures can also occur in post-dryout heat transfer, in flashing and critical flows.

## 16. CONCLUDING REMARKS

In this chapter we have tried to identify the various heat transfer regimes which occur during single phase and two-phase flows in boilers and evaporators. The concept of a complete "heat transfer surface" has been developed. The premise upon which this "heat transfer surface" is based is that the heat transfer rates can be described by the instantaneous local hydrodynamic conditions. Thus, it is assumed that the heat flux calculation can be undertaken on the basis of the local conditions at the point being considered. It has been shown that this is not strictly true because of (a) flow history effects, i.e. developing flow states upstream of the point being considered and (b) lack of thermodynamic equilibrium, i.e. unequal temperatures for the liquid and vapour phases. Models are being developed to incorporate some of these effects into the thermohydraulic analysis of heat exchangers.

## REFERENCES

(1) Collier, J.G. "Convective Boiling and Condensation" (2 ed) published by McGraw Hill Book Co. (UK) Ltd., (1980).

(2) Davis, E.J. and Anderson, G.H. "The Incipience of Nucleate Boiling in Forced Convection Flow". AIChE Journal $\underline{12}(4)$, 774-780 (July 1966).

(3) Jens, W.H. and Lottes, P.A. "Analysis of heat transfer burnout, pressure drop and density data for high pressure water". ANL-4627 (May 1951).

(4)  Thom, J.R.S., Walker, W.M., Fallon, T.A. and Reising, G.F.S. "Boiling in subcooled water during flow up heated tubes or annuli". Paper 6 presented at the Symposium on Boiling Heat Transfer in Steam Generating Units and Heat Exchangers held in Manchester, 15-16 September 1965 by Inst. of Mech. Eng. (London).

(5)  Hewitt, G.F., Kearsey, H.A., Lacey, P.M.C. and Pulling, D.J. "Burnout and nucleation in climbing film flow". AERE-R4374 (1963).

(6)  Chen, J.C. "A correlation for boiling heat transfer to saturated fluids in convective flow". Paper presented to 6th National Heat Transfer Conference, Boston, 11-14 August 1963, ASME preprint 63-HT-34.

(7)  Thompson, B. and Macbeth, R.V. "Boiling water heat transfer – burnout in uniformly heated round tubes: a compilation of world data with accurate correlations". AEEW-R356 (1964).

(8)  Bowring, R.W. "A simple but accurate round tube uniform heat flux dryout correlation over the range 0.7 – 17 $MN/m^2$ (100-2500 psia)" AEEW-R789 (1972).

(9)  Bertoletti, S. et al. "Heat Transfer Crisis with Steam-Water Mixtures". Energia Nucleare $\underline{12}$(3), March 1965. See also Gaspari, G.P. et al. "Some considerations on CHF in rod clusters in annular dispersed vertical upward two-phase flow". 4th Int. Conf. Heat Transfer, Paris (1970).

(10) Barnett, P.G. "A correlation of burnout data for uniformly heated annuli and its use for predicting burnout in uniformly heated rod bundles". AEEW-R463 (1966).

(11) Hewitt, G.F. and Kearsey, H.A. "Correlation of critical heat flux for vertical flow of water in uniformly heated tubes, annuli and rod bundles". AERE-R5590 (1970).

(12) Biasi, L. et al. "Studies on Burnout" Part 3, Energia Nucleare $\underline{14}$(9) 530-536 (1967).

(13) Butterworth, D. Private Communication (1978).

(14) Becker, K.M. "A burnout correlation for flow of boiling water in vertical rod bundles". AE276 AB Atomenergi (1967).

(15) Iloeje, O.C., Plummer, D.N., Rohsenow, W.M. and Griffith, P. "A study of wall rewet and heat transfer in dispersed vertical flow". MIT Dept. of Mech. Engng. Report 72718-92 (September 1974).

(16) Plummer, D.N., Iloeje, O.C., Rohsenow, W.M., Griffith, P. and Ganic, E. "Post critical heat transfer to flowing liquid in a vertical tube". MIT Dept. of Mech. Engng. Report 72718-91 (September 1974).

(17) Groeneveld, D.C. "Effect of a heat flux spike on the downstream dryout behaviour". J. of Heat Transfer, pp.121-125 (May 1974).

(18) Groeneveld, D.C. and Fung, K.K. "Forced Convection Transition Boiling – a review of literature and comparison of prediction methods". AECL-5543 (1976).

(19) Tong, L.S. and Young, J.D. "A phenomenological transition and film boiling heat transfer correlation". Paper B 39 Vol.3, Proc. of 5th Int. Heat Transfer Conference, Tokyo, September 1974.

(20) Iloeje, O.C., Plummer, D.N., Rohensow, W.M. and Griffith, P. "An investigation of the collapse and surface rewet in film boiling in forced vertical flow". J. of Heat Transfer, pp.166-172, (May 1975).

(21) Ralph, J.C., Sanderson, S. and Ward, J.A. "Post dryout heat transfer under low flow and low quality conditions". Symp. on Thermal and Hydraulic Aspects of Nuclear Safety, 1977 ASME Winter Meeting, Atlanta, Georgia, Nov.27 - Dec.2 (1977).

(22) Groeneveld, D.C. "Post-dryout heat transfer at reactor operating conditions". AECL-4513, Nat. Topical Meeting on Water Reactor Safety, ANS Salt Lake City, Utah, March 26-28 (1973).

(23) Slaughterback, D.C. et al. "Statistical regression analysis of experimental data for flow film boiling heat transfer". ASME-AIChE Heat Transfer Conference, Atlanta, August 1973.

(24) Groeneveld, D.C. and Delorme, G.G.J. "Prediction of thermal non-equilibrium in the post-dryout region". Nuc. Engng. and Design $\underline{36}$, 17-36, (1976).

(25) Bennett, A.W., Hewitt, G.F., Kearsey, H.A. and Keeys, R.K.F. "Heat transfer to steam-water mixtures flowing in uniformly heated tubes in which the critical heat flux has been exceeded". Thermodynamics and Fluid Mechanics Convention, Bristol, 278-29 March 1968, I. Mech. E. See also AERE-R5373 (1967).

(26) Bishop, A.A. et al. "Forced convection heat transfer to superheated steam at high pressure and high Prandtl numbers". ASME Paper 65-WA/HT-35 (1965).

(27) Mattson, R.J. et al. "Regression analysis of post CHF - flow boiling data". Paper B3.8 Vol.4, Proc. of 5th Int. Heat Transfer Conference, Tokyo, September (1974).

(28) Whalley, P.B., Hutchinson, P. and Hewitt, G.F. "The calculation of critical heat flux in forced convection boiling". Paper B6.11, 5th Int. Heat Transfer Conference, Tokyo, Japan (1974).

(29) Whalley, P.B., Hutchinson, P. and Hewitt, G.F. "Prediction of annular flow parameters for transient conditions and for complex geometries". European Two-Phase Flow Group Meeting, Haifa, Israel, June 1975.

(30) Whalley, P.B. "The calculation of dryout in a rod bundle". AERE-R8319. European Two-Phase Flow Group Meeting, Erlangen, Germany, June 1976 (Paper A9).

(31) "Tabular data for calculating burnout when boiling water in uniformly heated round tubes", Working Party of the Heat and Mass Transfer Section of the Scientific Council of the USSR Academy of Sciences, Teploenergtikia, Vol.23 (9) pp.9=-92 (1976) - or in Thermal Engineering pp.77-79 September (1977).

(32) Ahmad, S.Y. "Fluid-to-fluid modelling of the critical heat flux: A compensated distortion model", <u>Int. J. Heat Mass Transfer</u>, Vol.16 pp.641-661 (1973).

(33) Hauptmann, E.G., Lee, V. and McAdam, D. "Two-phase fluid modelling of the critical heat flux", Proc. of Int'l Meeting, Reactor Heat Transfer, Karlsruhe, pp.557-576, 9-11 October 1973.

# Condensers: Basic Heat Transfer and Fluid Flow

**D. BUTTERWORTH**
Heat Transfer and Fluid Flow Service (HTFS)
AERE Harwell, Oxon, UK

ABSTRACT

The various modes of condensation (dropwise, filmwise, etc.) are described and the resistances to heat transfer illustrated. Methods of calculating the heat transfer coefficient for condensing inside and outside tubes at various orientations and with low and high vapour velocities are presented. The prediction of heat and mass transfer in multicomponent mixtures is discussed. Pressure drop calculation methods for both the shell side and tube side are given.

1.  INTRODUCTION

With very low rates of condensation, or when the heat-transfer surface is non-wetting, dropwise condensation may occur as illustrated in Fig.1a. The predominant mode of condensation in practical exchangers is, however, filmwise which is illustrated in Fig.1b. Dropwise condensation is therefore not discussed in these lectures. When condensing mixtures which form immiscible liquid phases, a variety of flow patterns may occur and one of these is illustrated in Fig.1c. Homogeneous condensation may also occur in condensers: that is, condensation of droplets in the gas/vapour phase instead of on the wall (see Fig.1d). This can occur in partial condensers (ie. those with incondensable gas) when the gas is cooled faster than the dew point is able to fall by vapour condensation. Some condensers are designed without a wall between the vapour and the coolant. These so-called direct-contact condensers are described further in the lecture on thermohydraulic design.

In a condenser, there are a number of temperature drops between the bulk gas/vapour and the wall and we usually associate these with heat-transfer coefficients. Figure 2a illustrates these temperature differences for a pure vapour of temperature $T_G$ which we will take as being at the saturation temperature $T_S$. A small temperature difference occurs at the liquid-vapour interface due to molecular-kinetic effects but the main temperature difference arises due to the thermal resistance of the condensate layer. When an incondensable gas is present (Fig.2b) a further temperature difference arises in the gas phase due to mass transfer limiting effects. Similar problems arise when condensing vapour mixtures. The calculation of heat-transfer coefficients associated with these resistances are discussed below.

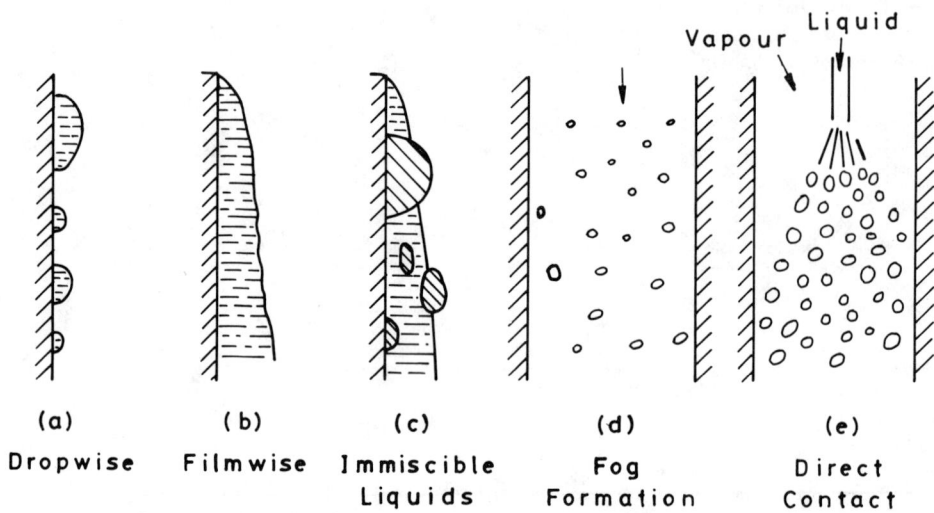

Fig.1. Modes of condensation

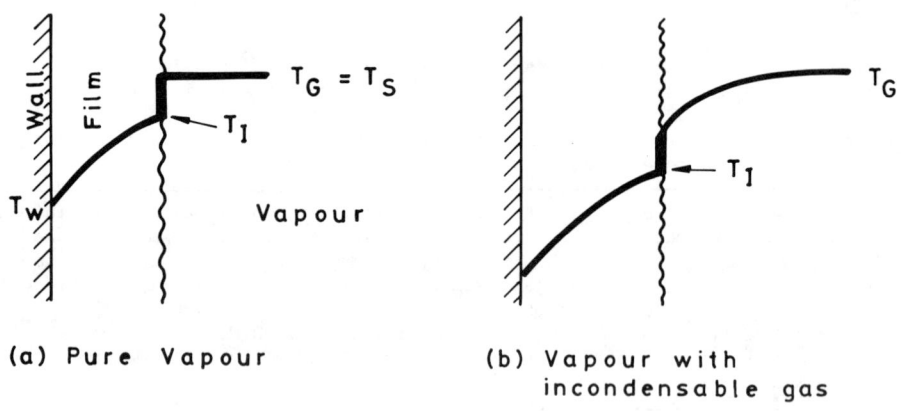

Fig.2. Resistances to condensation

# CONDENSERS: BASIC HEAT TRANSFER AND FLUID FLOW

2. INTERFACIAL (MOLECULAR-KINETIC) RESISTANCE

This resistance is usually negligible but may become important at very low pressures (say less than 0.01 bar). It may be calculated from [1]

$$\alpha_I = \frac{2\sigma_c}{2-\sigma_c} \left(\frac{\tilde{M}}{2\pi RT}\right)^{\frac{1}{2}} \frac{\lambda^2 p \tilde{M}}{RT^2} .\qquad(1)$$

There is some uncertainty about the value to use for $\sigma_c$ which is the fraction of molecules bombarding the liquid which sticks (ie. do not bounce off again). Recent work has suggested that $\sigma_c$ is slightly below unity and a value of 0.8 is recommended for design.

3. CONDENSATE FILM RESISTANCES

3.1 Outside Horizontal Tubes

The coefficient for an isolated horizontal tube may be given satisfactorily by Nusselt's [2] equation provided that the vapour velocity and temperature drop across the film are both low:

$$\alpha_c = 0.725 \left\{\frac{k_L^3 \rho_L (\rho_L - \rho_G) g \lambda}{\mu_L D_o (T_S - T_W)}\right\}^{\frac{1}{4}} .\qquad(2)$$

Corrections to this equation for large temperature drops across the film are given by Chen [3] but these are not usually very important and are therefore neglected here. A convenient alternative form of equation (2) is

$$\frac{\alpha_c}{k_L} \left\{\frac{\mu_L^2}{\rho_L(\rho_L - \rho_G)g}\right\}^{1/3} = 1.51 \left(\frac{4\Gamma'}{\mu_L}\right)^{-1/3} .\qquad(3)$$

The mean coefficient $\bar{\alpha}_N$ for N tubes arranged in a vertical column is usually below the coefficient predicted by equation (2) because of condensate from higher tubes falling onto lower ones. This effect is often accounted for by the following form of equation

$$\bar{\alpha}_N / \alpha_1 = N^{-s} \qquad(4)$$

where $\alpha_1$ is the coefficient for the top tube as given by equation (2). Theory based on a continuous laminar layer flowing from tube to tube (as in Fig.3a) gives s = 1/4. The actual condensate flow is, however, illustrated in Fig.3b which gives higher values of $\bar{\alpha}_N$ than with s taken as 1/4. Kern [4] from "experience" suggested s = 1/6 and this value has since been shown by Grant [5] to agree with the experimental data of Grant and Osment [6].

Equation (4) may be recast in a form which gives the heat-transfer coefficient for the N'th tube from the top, $\alpha_N$, as

$$\frac{\alpha_N}{\alpha_1} = N^{1-s} - (1-N)^{1-s} .\qquad(5)$$

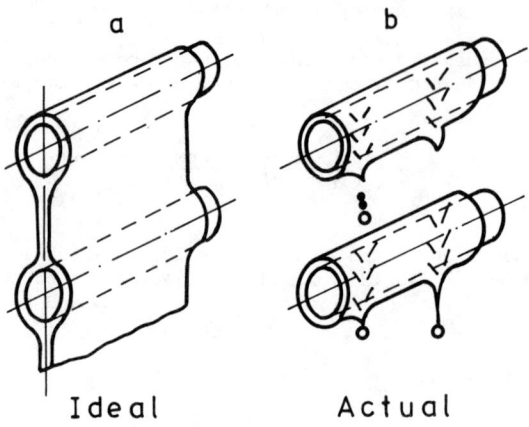

Fig.3. Condensation outside horizontal tubes

An alternative method of calculating the coefficient for the N'th tube is to use equation (3) but with $\alpha_c$ replaced by $\alpha_N$ and with $\Gamma'$ taken as the condensate flow rate per unit length from this N'th tube. This method was demonstrated experimentally by Short and Brown [7] and Butterworth [8] has shown the method to be compatible with the Kern [4] result.

The equations given so far are for laminar condensate films, which are those with film Reynolds numbers less than about 1600. The group $4\Gamma'/\mu_L$ is twice the film Reynolds number for one side of the tube, which means that equation (3) only applies for $4\Gamma'/\mu_L < 3200$. There are no data for turbulent film condensation on horizontal tubes but, by analogy with vertical plates, we can approximate the heat transfer coefficient for the N'th tube as [8]

$$\frac{\alpha_N}{k_L}\left\{\frac{\mu_L^2}{\rho_L(\rho_L-\rho_G)g}\right\}^{1/3} = 0.023\left(\frac{2\Gamma'}{\mu_L}\right)^{1/4} Pr_L^{1/2} \qquad (6)$$

where $\Gamma'$ is here the condensate flow per unit length from the N'th tube. This equation applies for $2\Gamma'/\mu_L > 1600$.

There have been a number of recent papers [9, 10, 11, 12, 13] on the effect of crossflow on the condensate heat transfer coefficient for a horizontal tube. The paper by Fujii et al [9] presents a semi-empirical relationship which appears both soundly based as well as predicting a wide range of data obtained by different experimentors. This relationship may be written as

$$\frac{\alpha_c}{\alpha_o} = 1.4 \left\{ \frac{u_G^2 (T_S-T_W) k_L}{g D_o \lambda \mu_L} \right\}^{0.05} \qquad (7)$$

where $\alpha_o$ is the heat transfer coefficient for no vapour shear (ie. that calculated from equation (2)) and $u_G$ is the vapour velocity. This equation is only used if $\alpha_c/\alpha_o > 1$, otherwise equation (2) is used. The data used in producing equation (7) are limited to cases where $\alpha_c/\alpha_o < 1.7$. The equation should therefore be used with care for higher values. It may be used for horizontal and downward vapour flows but not for upward flows.

There is considerable uncertainty about applying equation (7) to rod bundles (as distinct from isolated tubes). The first problem is in choosing $u_G$. Is it the superficial velocity or the maximum velocity as the vapour squeezes between the tubes, or indeed some other velocity? The safest velocity to use is the superficial velocity which is that obtained if the tubes were not present. The second problem is how to allow for the effect of condensate dripping from one tube to the next. The deterioration due to this effect is not so severe when vapour shear is present and it is probably therefore reasonable to ignore it.

It is possible for the coefficient to be decreased by vapour velocity [14] and the reason for this can only be guessed at. It is tentatively suggested that this arises from local pressure reductions as the vapour flows over the tube. This gives a corresponding reduction in local saturation temperature and consequent loss in temperature difference which may be avoided by ensuring that $T u_G^2/(T_S-T_W)\lambda$ is small (say less than 0.05) where $T$ is the absolute temperature.

### 3.2 Outside Vertical Tubes

Nusselt's [2] theory for laminar, wave-free condensate gives the mean coefficient over the tube length, L, as

$$\bar{\alpha}_c = 0.943 \left\{ \frac{k_L^3 \rho_L (\rho_L-\rho_G) g \lambda}{\mu_L (T_S-T_W) L} \right\}^{\frac{1}{4}} \qquad (8)$$

An alternative form of this equation is

$$\frac{\bar{\alpha}_c}{k_L} \left\{ \frac{\mu_L^2}{\rho_L (\rho_L-\rho_G) g} \right\}^{\frac{1}{3}} = 1.47 \, Re_{LB}^{-\frac{1}{3}} \qquad (9)$$

where $Re_{LB} = 4\Gamma_B/\mu_L$ which is the film Reynolds number at the bottom of the plate.

In terms of the local coefficient, equation (9) may be written as

$$\frac{\alpha_c}{k_L} \left\{ \frac{\mu_L^2}{\rho_L (\rho_L-\rho_G) g} \right\}^{\frac{1}{3}} = 1.1 \, Re_L^{-\frac{1}{3}} \qquad (10)$$

where $Re_L$ is the local film Reynolds number $(4\Gamma/\mu_L)$. For values of $Re > 30$ waves on the surface of the film may increase the coefficient and a correction for this effect has been proposed by Kutateladze [15] as

$$\text{Correction} = 0.8(Re_L/4)^{0.11}. \tag{11}$$

Combining this with equation (10) gives

$$\frac{\alpha_c}{k_L}\left\{\frac{\mu_L^2}{\rho_L(\rho_L-\rho_G)g}\right\}^{1/3} = 0.756\, Re_L^{-0.22}. \tag{12}$$

For values of $Re_L > 1600$, turbulence in the film becomes important and the following equation obtained by Labuntsov [16] should be used

$$\frac{\alpha_c}{k_L}\left\{\frac{\mu_L^2}{\rho_L(\rho_L-\rho_G)g}\right\}^{1/3} = 0.023\, Re_L^{1/4}\, Pr^{1/2}. \tag{13}$$

It is sometimes more convenient to have the average coefficient over the height of the tube rather than the local value. This average may be obtained from the following equation provided that $T_W$ is constant:

$$\frac{Re_{LB}}{\overline{\alpha}_c} = \int_0^{Re_{LB}} \frac{dRe_L}{\alpha_c}. \tag{14}$$

This may be used with equations (10), (12) and (13) to give

$$\frac{\overline{\alpha}_c}{k_L}\left\{\frac{\mu_L^2}{\rho_L(\rho_L-\rho_G)g}\right\}^{1/3} = \frac{Re_{LB}}{1.08\, Re_{LB}^{1.22} - 5.2} \tag{15}$$

for $30 < Re_{LB} < 1600$, and

$$\frac{\overline{\alpha}_c}{k_L}\left\{\frac{\mu_L^2}{\rho_L(\rho_L-\rho_G)g}\right\}^{1/3} = \frac{Re_{LB}}{8750 + 58Pr_L^{-1/2}(Re_{LB}^{3/4} - 253)} \tag{16}$$

for $Re_{LB} > 1600$. Of course, equation (9) applies for $Re_{LB} < 30$.

Downward and horizontal vapour velocities may increase the coefficient above that predicted from the above equations (see section 3.4).

3.3  Inside Horizontal Tubes

Two distinct models for intube condensation are usually found in the literature. These are the annular and stratifying models illustrated in Figure 4. Annular flow occurs when shear forces predominate and stratifying flow when gravity forces predominate.

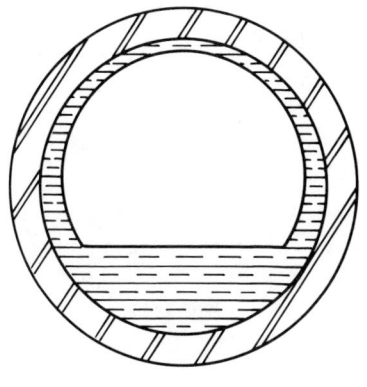
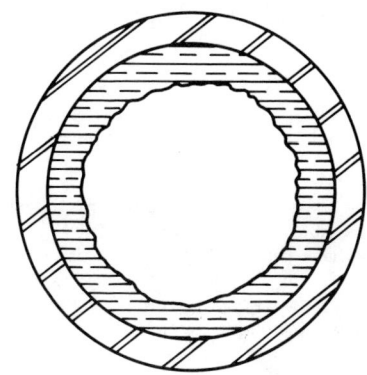

Stratifying         Annular

Fig.4. Models for condensation in a horizontal tube

The stratifying models are based on modifications to Nusselt's theory for condensation outside tubes:

$$\alpha_c = 0.725 \, \Omega \left\{ \frac{k_L^3 \rho_L (\rho_L - \rho_G) g \lambda}{\mu_L D_i (T_S - T_W)} \right\}^{\frac{1}{4}} \quad (17)$$

where the correction factor $\Omega$ allows for the reduction in coefficient caused by liquid accumulation in the bottom of the tube. Typically it is about 0.8. Jaster and Kosky [17] give the following simple yet reasonable equation for $\Omega$:

$$\Omega = \varepsilon_G^{\frac{3}{4}} \quad (18)$$

where the void fraction, $\varepsilon_G$, is calculated from the Zivi [18] equation

$$\varepsilon_G = 1 \Big/ \left\{ 1 + \frac{1-x}{x} \left(\frac{\rho_G}{\rho_L}\right)^{2/3} \right\} \quad (19)$$

There is a wide variety of methods of calculating $\alpha_c$ in annular flow. Two of the more simple methods are those of Boyko and Kruzhilin [19] and Palen et al [20], which are respectively

$$\alpha_c = \alpha_{LO}\left\{1 + x\left(\frac{\rho_L}{\rho_G} - 1\right)\right\}^{\frac{1}{2}} \quad (20)$$

$$\alpha_c = \alpha_L \phi_L \quad (21)$$

where $\alpha_{LO}$ is the coefficient if the total flow were flowing as liquid and $\alpha_L$ is the coefficient if the liquid phase were flowing alone. $\phi_L$ is the two phase multiplier given in section 5.1.

The more complicated models for annular flow use the Martinelli analogy for heat transfer to express $\alpha_c$ as

$$\alpha_c = \frac{c_L(\rho_L \tau_W)^{\frac{1}{2}}}{T^+} \quad (22)$$

where $T^+$ is a function of the condensate Reynolds number, $Re_L$, and condensate Prandtl number, $Pr_L$. There are a number of expressions for $T^+$ ranging from the simple relationship proposed by Carpenter and Colburn [21], which is

$$T^+ = 23.3\, Pr_L^{\frac{1}{2}}, \quad (23)$$

to the more complete method of Kosky and Staub [22], which involves the following equations:

$$T^+ = Pr_L \delta^+, \text{ for } \delta^+ \leq 5 \quad (24a)$$

$$T^+ = 5\left[Pr_L + \ln\left\{1 + Pr_L\left(\frac{\delta^+}{5} - 1\right)\right\}\right], \text{ for } 5 < \delta^+ \leq 30 \quad (24b)$$

$$T^+ = 5\left[Pr_L + \ln(1 + 5Pr_L) + \frac{1}{2}\ln\frac{\delta^+}{30}\right], \text{ for } \delta^+ > 30 \quad (24c)$$

where $\delta^+$ is a dimensionless film thickness given by

$$\delta^+ = (Re_L/2)^{\frac{1}{2}}, \text{ for } Re_L < 1250 \quad (25a)$$

$$\delta^+ = 0.0504\, Re_L^{\frac{7}{8}}, \text{ for } Re_L > 1250 \quad (25b)$$

The wall shear stress, $\tau_W$, in equation (22) is most easily determined from the definition of the two-phase frictional pressure gradient, $dp_F/dz$, defined on the basis of the momentum equation of two-phase flow:

$$\tau_W = \frac{D_i}{4}\left(-dp_F/dz\right) \quad (26)$$

Methods of calculating $dp_F/dz$ are outlined in section 5.1 below.

At first sight, these various methods look very different but Butterworth [23] has shown that equation (20) may be transformed to

$$T^+ = 8.5 \, Re_L^{0.1} \, Pr_L^{0.57} \tag{27}$$

It is also a simple matter to show that equation (21) transforms to

$$T^+ = 6.6 \, Re_L^{0.1} \, Pr_L^{0.6} \tag{28}$$

Some of these methods of determining $T^+$ are compared in Fig.5.

The question now arises as to which model one uses, annular or stratified, for a given problem. A simple rule of thumb is to calculate the coefficient, $\alpha_c$, by both methods and then use the higher of the two values thus determined. Alternative criteria have been proposed by Jaster and Kosky [17] and Palen et al [20]. The method of Palen et al is the simplest and is therefore given here. They propose calculating a dimensionless gas velocity as follows:

$$V_G^* = \left[ \frac{\rho_G V_G^2}{(\rho_L - \rho_G) g D_i} \right]^{\frac{1}{2}} \tag{29}$$

They suggest that the stratifying model applies for $V_G^* < 0.5$ and the annular flow model applies for $V_G^* > 1.5$. In between, the following interpolation equation would seem reasonable:

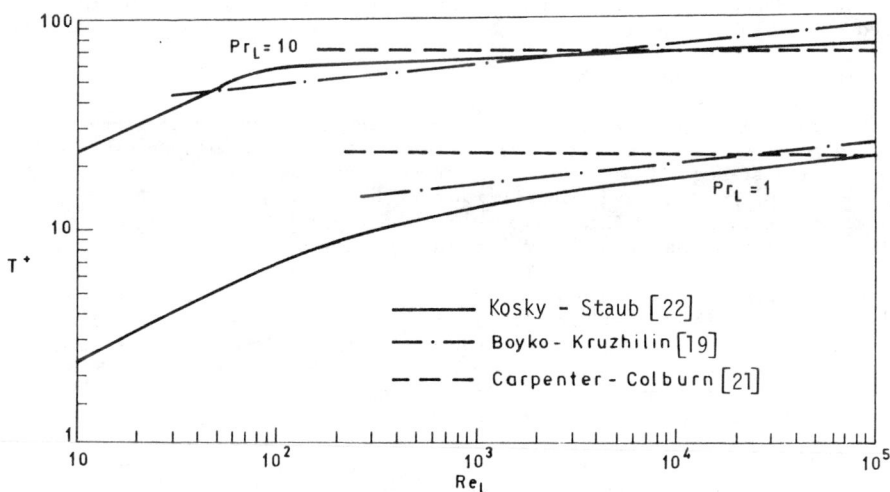

Fig.5. Comparison of some methods for calculating $T^+$

$$\alpha_c = \alpha_{an} + (V_G^* - 1.5)(\alpha_{an} - \alpha_{str}) \qquad (30)$$

where $\alpha_{an}$ and $\alpha_{str}$ are the coefficients calculated from the annular and stratifying models respectively.

## 3.4 Inside Vertical Tubes

Let us consider first vapour downflow. If the vapour velocities are low, the methods described in section 3.2 for condensation outside vertical tubes apply. If, on the other hand, the vapour velocities are very high, the annular flow model described above for horizontal tubes will apply. Both methods should be tried and the higher of the two coefficients used in preliminary design. Slightly greater accuracy is achieved by taking the root-mean-square of the coefficients obtained by the two methods.

For vapour upflow, however, we must keep the vapour velocities low enough to avoid flooding. This is the condition when the vapour upflow prevents the smooth downward drainage of condensation. Flooding may be avoided by using the Wallis [25] criterion

$$(V_G^*)^{\frac{1}{2}} + (V_L^*)^{\frac{1}{2}} < C \qquad (31)$$

where

$$V_L^* = \left[ \frac{\rho_L V_L^2}{(\rho_L - \rho_G) g D_i} \right]^{\frac{1}{2}}. \qquad (32)$$

Although C is not a true constant, it seldom falls below about 0.6, so that this value will usually be safe. After having ensured that flooding is avoided, the vapour shear on the surface of the film is too small to effect its flow. Hence, the coefficient, $\alpha_c$, is predicted satisfactorily by the methods given in section 3.1.

## 3.5 With Immiscible Liquid Phases

The case discussed here is that of condensing a binary vapour mixture whose composition is the eutectic composition. Other compositions can give gas-phase resistance effects similar to those discussed below. Coefficients for immiscible-liquid condensate films are difficult to predict accurately because of the variety of flow patterns which are possibly dependent on the fluid, the surface, the geometry, and the condensation rate.

For organic water mixtures, two main flow patterns have been observed [25]. The first is channelling flow in which the organic and the water form fairly distinct streams on the condenser surface (although each stream can contain droplets of the other phase). The second is where large standing droplets of water form on the surface surrounded by a flowing film of organic. The droplets roll off the surface fairly infrequently. The standing drop mechanism appears most likely on surfaces which are not wet easily by water and it gives rise to lower coefficients. Although not certain at the moment, channelling flow is possibly most likely to occur in operating heat exchangers.

The most suitable methods currently available for predicting the coefficient for the condensate resistance with immiscible phases, ignore the flow patterns just discussed and correlate all the data by the same general form of equation. The method of Bernhard et al [26] gives the mean coefficient as

$$\bar{\alpha}_c = x_{v1}\bar{\alpha}_{c1} + x_{v2}\bar{\alpha}_{c2} \tag{33}$$

where $\bar{\alpha}_{ci}$ is the coefficient for the pure i component condensing on the same surface as predicted by equation (2) or (8), dependant on the geometry. $x_{vi}$ is the volumetric fraction in the condensate. Alternatively, the method of Ackers and Turner [27] may be used to give the coefficient as

$$\frac{\bar{\alpha}_c}{k_{av}} \left\{ \frac{\mu_{L1}}{\rho_{av}(\rho_{av}-\rho_G)g} \right\}^{1/3} = 1.51 \left( \frac{4\Gamma'}{\mu_{L1}} \right)^{-1/3} \tag{34}$$

for a horizontal tube, or

$$\frac{\bar{\alpha}_c}{k_{av}} \left\{ \frac{\mu_{L1}}{\rho_{av}(\rho_{av}-\rho_G)g} \right\}^{1/3} = 1.47 \left( \frac{4\Gamma_B}{\mu_{L1}} \right)^{-1/3} \tag{35}$$

for a vertical surface. The condensate thermal conductivity, $k_{av}$, is volume fraction averaged and the density, $\rho_{av}$, is mass fraction averaged. The liquid viscosity, $\mu_{L1}$, is that of the film forming component which would invariably be the organic in an organic-water mixture.

4. GAS PHASE RESISTANCES

4.1 Simple Models

Simple, general models are often used for multicomponent/partial condensers because the more precise models are extremely complicated. One of the more soundly based of these models is that of Silver [28] and Bell and Ghaly [29]. The assumption is made in this method that the condensation path follows the equilibrium condensation curve which is expressed in the form of the condensing temperature, T, plotted against either the cumulative heat release rate, $\dot{Q}$, or the mixture specific enthalpy, h. The quantities $\dot{Q}$ and h are related by

$$\dot{Q} = (h_{in} - h)\dot{M} \tag{36}$$

where $h_{in}$ is the inlet enthalpy and $\dot{M}$ is the total mass flow rate of the condensing stream.

Bell and Ghaly give the effective condensing side coefficient as

$$\frac{1}{\alpha_{eff}} = \frac{1}{\alpha_c} + \frac{Z}{\alpha_G} \tag{37}$$

where

$$Z = x\, c_G \frac{dT}{dh} \tag{38}$$

and $\alpha_G$ is the heat transfer coefficient between the gas phase and the condensate surface. The coefficient $\alpha_G$ is usually calculated by assuming that the gas phase is flowing alone since the effect of the liquid is uncertain. Price and Bell [30] have suggested a method of allowing for the effect of the liquid film in $\alpha_G$ but this has not been confirmed by data. McNaught [31] has devised a correction for the condensing mass flux.

There have been a few published comparisons of the Silver, Bell and Ghaly method with more precise methods and these have shown that it can under-predict the heat transfer area by 30% under some circumstances, yet predict double the area in others.

### 4.2 Colburn-Hougen Method for Single Vapour with Incondensable Gas

The Colburn-Hougen [32] method consists of solving the following equations for the interface temperature $T_I$:

$$\alpha^*(T_I - T_o) = \alpha_G \phi (T_G - T_I) + \lambda \beta \rho_V \ln \frac{p - p_{AI}}{p - p_{Ab}} \qquad (39)$$

where $T_o$ is the coolant temperature, $\rho_V$ is the density of the condensing vapour at the system pressure (p) and $p_{AI}$ and $p_{Ab}$ the partial pressure of the condensing component at the interface and in the bulk gas respectively. The interface partial pressure is the vapour pressure corresponding to $T_I$. The coefficient $\alpha^*$ is that between the coolant and the condensate surface:

$$\frac{1}{\alpha^*} = \frac{1}{\alpha_c} + r + \frac{1}{\alpha_o} . \qquad (40)$$

The factor $\phi$ was introduced by Ackerman [33] to account for the effect of mass transfer on $h_G$, and is given by

$$\phi = \eta/(1 - e^{-\eta}) \qquad (41)$$

where

$$\eta = \dot{m}_c c_V / \alpha_G . \qquad (42)$$

The condensation mass flux $\dot{m}_c$ is given by

$$\dot{m}_c = \beta \rho_V \ln \frac{p - p_{AI}}{p - p_{Ab}} . \qquad (43)$$

The mass transfer coefficient, $\beta$, may be obtained from the Chilton-Colburn [34] analogy between heat and mass transfer

$$\beta = \left( \frac{Pr_G}{Sc_G} \right)^{2/3} \frac{\alpha_G}{\rho_G c_G} . \qquad (44)$$

For known $T_G$, $T_o$ and $p_{Ab}$, equation (39) may be solved by trial and error to give the temperature $T_I$. An effective local overall coefficient U may then be calculated from

CONDENSERS: BASIC HEAT TRANSFER AND FLUID FLOW

$$U = \frac{\alpha^*(T_I - T_o)}{T_G - T_o} \cdot \qquad (45)$$

Frequently, to simplify the calculations, $T_G$ and $p_{Ab}$ are taken as the equilibrium values from the condensation curve. They should, however, be calculated from heat and mass balances in the gas phase which give respectively

$$\frac{dT_G}{dA} = - \frac{\alpha_G(T_G - T_I)}{\dot{M}_G c_G} \left( \frac{\eta}{e^\eta - 1} \right) \qquad (46)$$

$$\frac{d\dot{M}_{GA}}{dA} = - \dot{m}_c \cdot \qquad (47)$$

The partial pressure of the vapour in the bulk is given from $\dot{M}_{GA}$ by

$$p_{Ab} = \frac{p \, \dot{M}_{GA}/\tilde{M}_A}{\dot{M}_{GA}/\tilde{M}_A + \dot{M}_{GB}/\tilde{M}_B} \cdot \qquad (48)$$

Equations (46) and (47) have to be integrated numerically while, at the same time, determining $T_I$ iteratively from equation (39). This is complicated!

4.3 Fogging

Dependent on circumstances, the above calculations may predict gas-phase temperatures $T_G$ which fall below the dew point temperature. This is more likely for higher molecular weight vapours, with increasing temperature difference and with reduced initial superheat [35]. If the vapour is sub-cooled sufficiently, fog may form which is undesirable since it can give loss of product through the vent line. Fog formation occurs when a critical supersaturation ratio, S, is exceeded; (S is the partial pressure of condensable divided by the vapour pressure at gas temperature $T_G$). The critical supersaturation ratio may be estimated from [35]

$$S = \exp\left[ 1.7 \times 10^7 \frac{\tilde{M}_A}{\rho_L} \left( \frac{\sigma}{T_G} \right)^{3/2} \right] \qquad (49)$$

when $T_G$ is the gas-phase temperature in Kelvin.

Methods of avoiding fog formation are discused by Steinmeyer [34]. If it cannot be avoided some form of mesh demister should be inserted in the vent line. Special demisters are available for the fine droplets occurring in fog (0.1 - 40 µm).

4.4 Colburn-Drew Method for Binary Vapour Mixtures [36]

For condensing binary vapour mixtures (with no incondensables) equation

(39) above becomes more complicated:

$$\alpha^*(T_I - T_o) = \alpha_G \phi (T_G - T_I) + \dot{n}_T \{z\tilde{\lambda}_A + (1-z)\tilde{\lambda}_B\} \tag{50}$$

where $\dot{n}_T$ is the total molar condensation flux given by

$$\dot{n}_T = \beta \tilde{\rho}_G \ln \frac{z - \tilde{y}_{AI}}{z - \tilde{y}_{Ab}} \tag{51}$$

where $z = \dot{n}_A / \dot{n}_T$. The value of $\eta$ needed in equation (41) in order to calculate $\phi$ is now

$$\eta = \dot{n}_T \{z\tilde{c}_A + (1-z)\tilde{c}_B\}/\alpha_G \tag{52}$$

If the compositions in the bulk gas and in the liquid at the interface are known, equation (50) may be solved iteratively for $T_I$ and $z$ by using vapour-liquid equilibria data for $\tilde{y}_{AI}$. Assuming ideality, the vapour-liquid equilibria relationships are

$$\tilde{y}_{AI} = \frac{p_{AS}}{p} \tilde{x}_{AI} \tag{53a}$$

$$\tilde{y}_{BI} = 1 - \tilde{y}_{AI} = \frac{p_{BS}}{p} \tilde{x}_{BI} \tag{53b}$$

where $\tilde{x}_{AI}$ and $\tilde{x}_{BI}$ are the liquid mole fractions at the interface and $p_{AS}$ and $p_{BS}$ are the saturation pressures evaluated at temperature $T_I$. Two extreme methods are usually used to calculate the liquid mole fractions at the interface. The first is to assume excellent mixing in the liquid which gives

$$\tilde{x}_{AI} = \tilde{x}_{Ab}, \quad \tilde{x}_{BI} = \tilde{x}_{Bb} \tag{54}$$

and the second is to assume no mixing in the liquid which gives

$$\tilde{x}_{AI} = z, \quad \tilde{x}_{BI} = 1-z . \tag{55}$$

As with the Colburn-Hougen method described in section 4.2 above, differential heat and mass balances are used to progress the above calculations through the condenser. The gas phase heat balance equation is the same as before (ie. equation (46)) while the gas phase mass balances are

$$\frac{d\dot{M}_{GA}}{dA} = -z\dot{n}_T \tilde{M}_A , \quad \frac{d\dot{M}_{GB}}{dA} = -(1-z)\dot{n}_T \tilde{M}_B \tag{56}$$

Having determined $\dot{M}_{GA}$ and $\dot{M}_{GB}$, the liquid component flow rates are readily calculated from overall mass balances.

# CONDENSERS: BASIC HEAT TRANSFER AND FLUID FLOW

## 4.5 Multicomponent Models

Section 4.2 and 4.4 above deal with binary systems. The condensation rates in these models are derived from equations of the form

$$\dot{n}_i = D \, \tilde{\rho}_G \frac{d\tilde{y}_i}{ds} + \dot{n}_T \, \tilde{y}_i \qquad (57)$$

where s is the distance from the interface. Equations (43) and (51) above are obtained by integrating this equation across a fictitious laminar film and manipulating the result. An alternative simplified integrated form which is valid for low mass transfer rates is

$$\dot{n}_i = \tilde{\rho}_G \, \beta (\tilde{y}_{ib} - \tilde{y}_{iI}) + \dot{n}_T \, \tilde{y}_{ib} . \qquad (58)$$

For binary systems, equation (58) may easily be used since there is only one value of $\beta$ for each component. For multicomponent mixtures with m components, equation (58) becomes

$$\dot{n}_i = \tilde{\rho}_G \sum_{j=1}^{m} K_{ij} (\tilde{y}_{jb} - \tilde{y}_{jI}) + \dot{n}_T \, \tilde{y}_{ib} \qquad (59)$$

which involves a matrix of coefficients. Recent works have been directed towards methods of evaluating the $K_{ij}$ [37].

## 4.6 Gas Phase Effects with Immiscible Liquids

Figure 6 shows the phase diagram for a totally immiscible binary mixture. When condensing vapour with a eutectic composition, the only resistance to heat transfer is in the liquid phase and the coefficient for this is calculated by the method given in section 3.5 above. When condensing a non-eutectic mixture, there are two possibilities. The first is when the interface temperature is above the eutectic temperature. In that case, only one vapour condenses and the other acts as a non-condensable gas. The calculation method for this is therefore the Colburn-Hougen method which is that described in section 4.2 above. The second case is when the Colburn-Hougen method predicts that $T_I$ is less than $T_E$. Since this is not possible, a more complicated calculation is necessary which gives binary condensation. Deakin [38] has developed such a calculation method which is a modified Colburn-Drew method. In his method, equation (50) above is solved iteratively for z with $T_I$ set to $T_E$. This is a simpler calculation to that for miscible liquids since $T_I$ is known.

## 5. PRESSURE DROP

### 5.1 Tube Side

Tube side pressure drop calculations for two phase flows are discussed in detail in standard text books [39, 40]. The subject is only therefore covered in outline here. The local pressure gradient for flow in a horizontal tube is given by

$$-\frac{dp}{dz} = -\frac{dp_F}{dz} + \dot{m}^2 \frac{d}{dz} \left\{ \frac{x^2}{\varepsilon_G \, \rho_G} + \frac{(1-x)^2}{(1-\varepsilon_G) \rho_L} \right\} . \qquad (60)$$

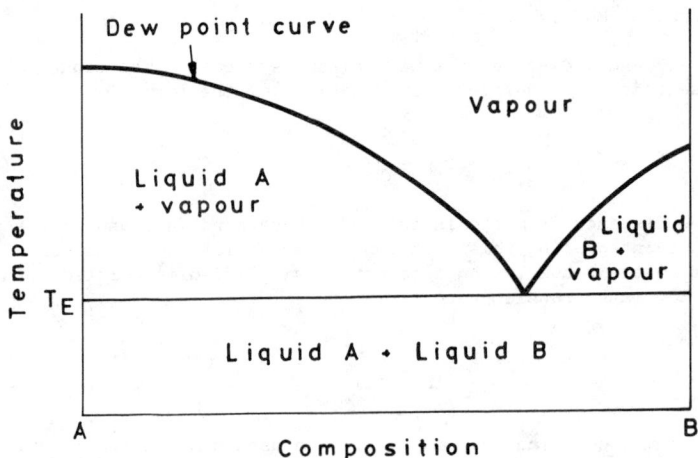

Fig.6. Temperature composition diagram for a totally immiscible binary system

This may be integrated to give the pressure drop over the tube length as

$$\Delta p = \Delta p_F - \dot{m}^2 \left\{ \frac{x_1^2}{\varepsilon_{G1}\rho_{G1}} + \frac{(1-x_1)^2}{(1-\varepsilon_{G1})\rho_{L1}} - \frac{x_2^2}{\varepsilon_{G2}\rho_{G2}} - \frac{(1-x_2)^2}{(1-\varepsilon_{G2})\rho_{L2}} \right\} \quad (61)$$

where the subscripts 1 and 2 refer to the start and end of the length. The two terms on the right hand side of equation (61) have opposite signs. The first gives a fall in pressure due to wall friction, while the second gives a rise in pressure due to pressure recovery from the decelerating flow. There are many correlations for the void fraction $\varepsilon_G$. That suggested most frequently in the literature for condensation in tubes is by Zivi [18] (see equation (19) above), but this is probably because of its ease of use rather than its accuracy.

The frictional pressure drop term $\Delta p_F$ must be calculated in a stepwise manner. The channel should be divided into a number of short lengths, $\Delta z$, over which the conditions change only moderately. The pressure drop over one of these lengths is thus calculated from

$$\Delta p_{Fj} = \left( -\frac{dp_F}{dz} \right) \Delta z \quad (62)$$

where the pressure gradient $dp_F/dz$ is evaluated using conditions in the middle of the length. This is given by

$$-\frac{dp_F}{dz} = -\left( \frac{dp_F}{dz} \right)_L \phi_L^2 \quad (63)$$

where $(dp_F/dz)_L$ is the pressure gradient if the liquid were flowing alone in the tube and $\phi_L$ is the correction factor given by

$$\phi_L^2 = 1 + \frac{C_o}{X} + \frac{1}{X^2} \qquad (64)$$

The parameter X is given by

$$X^2 = (dp_F/dz)_L / (dp_F/dz)_G \qquad (65)$$

where $(dp_F/dz)_G$ is the pressure gradient if the gas-vapour were flowing alone in the tube. Fig.7 gives $C_o$ as determined by Chisholm and Sutherland [41]. For high rate of condensation, the frictional pressure gradient may be higher than is predicted from equation (63) because of the increased interfacial shear generated by the condensing vapour.

5.2 Shell Side

The pressure drop in the window regions and the crossflow regions should be calculated separately using mean conditions in each section. The pressure drop $\Delta p$ is given by [42]

$$\frac{\Delta p}{\Delta p_{LO}} = 1 + (\gamma^2 - 1) \left\{ Bx^{\frac{2-n}{2}} (1-x)^{\frac{2-n}{2}} + x^{2-n} \right\} \qquad (66)$$

where n is the index in the Blasius relationship for single-phase flow ($f \propto 1/Re^n$) in that geometry. For the window zone, n is taken as zero.

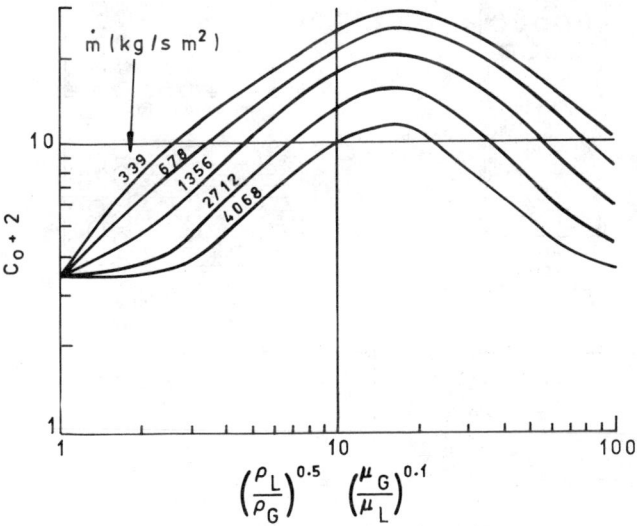

Fig.7. Curves for $C_o$ obtained by Chisholm and Sutherland [41]

The parameter γ is given by

$$\gamma^2 = \Delta p_{GO}/\Delta p_{LO} = \frac{\rho_L}{\rho_G}\left(\frac{\mu_G}{\mu_L}\right)^n \qquad (67)$$

where $\Delta p_{GO}$ is pressure drop if the total flow were flowing with gas-vapour-phase properties and $\Delta p_{LO}$ is the pressure drop if the total flow were flowing with the liquid-phase properties. The parameter B depends on the conditions as shown in Table 1.

Table 1: Values of B for Use in Equation (66)

|  |  | Horizontal (side to side) | Vertical (up and down) |
|---|---|---|---|
| Cross flow: | Spray and bubble | 0.75 | 1.0 |
|  | Stratified and stratified spray | 0.25 | - |
| Window |  | $2/(\gamma+1)$ | $(\rho_H/\rho_L)^{0.25}$ |

The flow patterns indicated in this table are illustrated in Figure 8, and a map indicating when they occur is given in Figures 9 and 10. Up-and-down flow (with horizontally cut baffles) is unusual in condensers. Intermittent flow should be avoided since it gives severe vibration.

As with the tube side case, some pressure recovery due to flow deceleration is possible but is difficult to estimate.

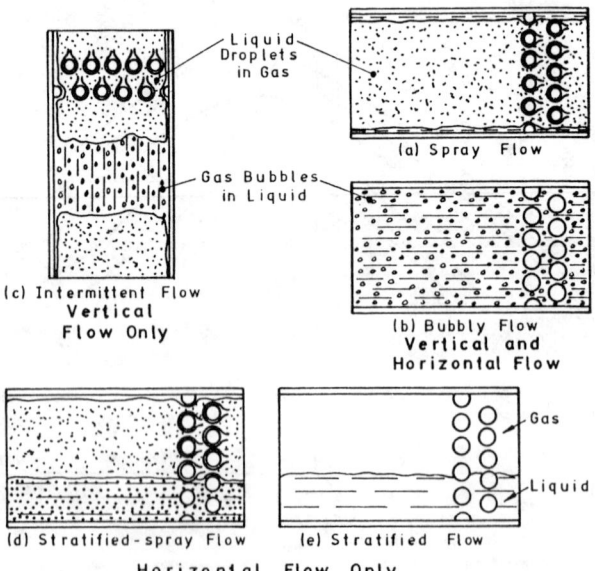

Fig.8. Flow patterns defined by Grant and Chisholm [42]

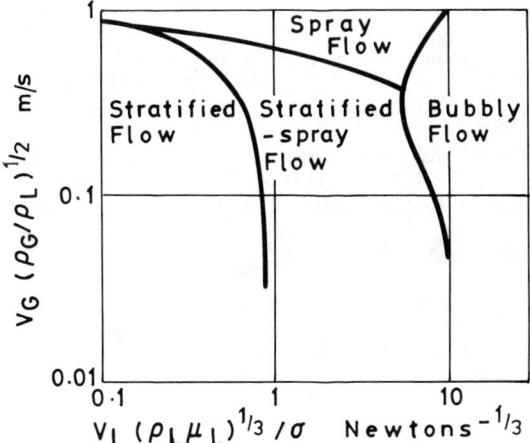

Fig.9. Flow pattern map for horizontal cross flow

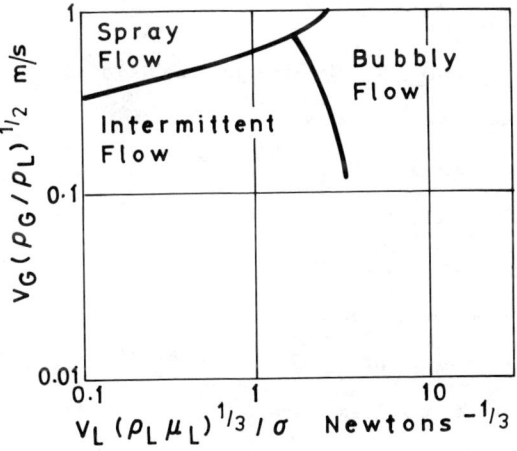

Fig.10. Flow pattern map for vertical cross flow

## NOMENCLATURE

| | |
|---|---|
| B | Parameter in equation (66), given by table 1. |
| $C_o$ | Parameter in equation (64), given by figure 7. |
| c | Specific heat capacity (J/kg K) |
| $\tilde{c}$ | Molar specific heat capacity (J/kmol K) |
| $D_o$ | Tube outside diameter (m) |
| $D_i$ | Tube inside diameter (m) |
| $D$ | Diffusion coefficient (m$^2$/s) |
| f | Friction factor |
| g | Gravitational acceleration (9.81 m/s$^2$) |
| h | Mixture specific enthalpy (J/kg) |
| k | Thermal conductivity (W/m K) |
| $K_{ij}$ | Multicomponent mass transfer coefficients (m/s) |
| L | Tube length (m) |
| $\dot{m}$ | Total mass flow per unit area in tube (kg/m$^2$s) |
| $\dot{m}_c$ | Condensation mass flux (kg/m$^2$s) |
| $\dot{M}$ | Mass flow rate (kg/s) |
| $\tilde{M}$ | Molecular weight (kg/kmol) |
| $\dot{n}$ | Condensing molar flux (kmol/s m$^2$) |
| p | Pressure or partial pressure (Pa) |
| Pr | Prandtl number (c μ/k) |
| $\dot{Q}$ | Cumulative heat release rate (W) |
| r | Thermal resistance of tube wall and fouling (m$^2$K/W) |
| R | Gas constant (8314 J/kmol K) |
| $Re_L$ | Liquid Reynolds number ($4\Gamma/\mu_L$ or, for flow in tubes, $(1-x)\dot{m}D_i/\mu_L$) |
| $Re_{LB}$ | Film Reynolds number at bottom of surface |
| s | Index in equations (4) and (5) |
| s | Distance from interface (m) |
| S | Supersaturation ratio |

# CONDENSERS: BASIC HEAT TRANSFER AND FLUID FLOW

| | |
|---|---|
| $Sc_G$ | Gas-phase Schmidt number ($k_G/\rho_G c_G D$) |
| T | Temperature (K) |
| u | Velocity (m/s) |
| V | Superficial velocity (m/s) |
| V* | Dimensionless velocity defined by equations (29) and (32) |
| x | Vapour-gas phase mass flow fraction |
| $\tilde{x}_A, \tilde{x}_B$ | Mole fractions in liquid phase |
| X | Parameter given by equation (65) |
| $\tilde{y}_A, \tilde{y}_B, \tilde{y}_j$ | Mole fraction in gas/vapour phase |
| z | $\dot{n}_A/\dot{n}_T$ in section 4. Distance along tube in section 5 (m) |

**Greek**

| | |
|---|---|
| $\alpha$ | Heat transfer coefficient (W/m$^2$K) |
| $\alpha^*$ | Heat transfer coefficient given by equation (40) (W/m$^2$K) |
| $\beta$ | Mass transfer coefficient (m/s) |
| $\gamma$ | Parameter given by equation (67) |
| $\Gamma$ | Condensate flow rate per unit perimeter of tube (kg/m s) |
| $\Gamma_B$ | Condensate flow rate per unit perimeter of tube at the bottom of the tube (kg/m s) |
| $\Gamma'$ | Condensate flow rate per unit length of tube (kg/m s) |
| $\delta^+$ | Dimensionless film thickness predicted by equation (25) |
| $\Delta$ | Indicates change in some parameter |
| $\eta$ | Parameter given by equation (42) |
| $\lambda$ | Latent heat (J/kg) |
| $\mu$ | Viscosity (Ns/m$^2$) |
| $\rho$ | Density (kg/m$^3$) |
| $\tilde{\rho}$ | Molar density (kmol/m$^3$) |
| $\sigma$ | Surface tension (N/m) |
| $\sigma_c$ | Accommodation coefficient |
| $\tau$ | Shear stress (N/m$^2$) |
| $\phi$ | Correction factor (various, as defined) |

Ω         Correction factor in equation (18)

## Subscripts

| | |
|---|---|
| A | Component in binary mixture |
| an | Annular flow |
| b | Phase bulk value |
| c | Condensation or condensing |
| B | Component in binary mixture |
| B | Bottom of vertical tube |
| eff | Effective value |
| F | Frictional value |
| G | Gas-vapour phase (alone) |
| GO | Indicates that the total flow is flowing with gas-vapour-phase properties |
| H | Homogeneous flow |
| I | Interface |
| in | Inlet |
| L | Liquid phase (alone) |
| LO | Indicates total flow having liquid properties |
| N | For N tubes arranged above one another |
| o | Coolant |
| S | Saturated |
| str | Stratifying flow |
| T | Total condensing |
| V | Vapour |
| W | Value at wall |

## Superscripts

| | |
|---|---|
| — | Average over height of surface or over N tubes |
| ~ | Molar value |

REFERENCES

1. Berman, L.D. 1967. On the effect of molecular-kinetic resistance upon heat transfer with condensation. Int. J Heat Mass Transfer, Vol. 10, p 1463.

2. Nusselt, W. 1916. Surface condensation of water vapour, Z. Ver dt. Ing. Vol.60, No. 27, pp 541-546, and Vol. 60, No.28, pp 569-575.

3. Chen, M.M. 1961. An analytical study of laminar film condensation: Part 1 - Flat plate; Part 2 - single and multiple horizontal tubes, Trans. ASME, J. Heat Transfer, Vol. 83c, pp 48-60.

4. Kern, D.Q. 1958. Mathematical development of tube loading in horizontal condensers, AICh.E. J. Vol. 4, No. 2, pp 157-160.

5. Grant, I.D.R. 1972. Film condensation of pure vapours. Notes for condensation and condensers course. Birniehill Institute, National Engineering Laboratory, East Kilbride, Glasgow.

6. Grant, I.D.R. and Osment, B.D.J. 1968. The effect of condensate drainage on condenser performance. National Engineering Laboratory Report No. NEL 350, East Kilbride, Glasgow.

7. Short, B.E. and Brown, H.E. 1951. Condensation of vapours in vertical banks of horizontal tubes. Inst. Mech. Engnrs. Proc. General Discussion on Heat Transfer, pp 27-31.

8. Butterworth, D. 1980. Contribution on inundation without vapour shear, Workshop on Modern Developments in Marine Condensers, Naval Postgraduate School, Monterey, California, 26-28 March.

9. Fujii, T., Honda H. and Oda, K. 1979. Condensation of steam on a horizontal tube - the influence of oncoming velocity and thermal conductivity at the tube wall. 10th National Heat Transfer Conference, San Diego, California, ASME/AIChE, 6-8 August.

10. Fujii, T, Uehara, H. and Kurata, C. 1972. Laminar filmwise condensation of flowing vapour on a horizontal cylinder, Int. J. Heat Mass Transfer, Vol. 15, pp 235-246.

11. Honda, H. and Fujii, T. 1974. Effect of direction of oncoming vapour on laminar and filmwise condensation on a horizontal cylinder. Heat Transfer 1974, Scripta Book Co, Vol. 3, pp 299-303.

12. Kutateladze, S.S., Gogonin, N.I, Dorokhov, A.R. and Susonov, V.I. 1979. Film condensation of flowing vapour on a bundle of plain horizontal tubes, Thermal Engng, Vol. 26, No. 5, pp 270-273.

13. Berman, L.D. 1979. Influence of vapour velocity on heat transfer with filmwise condensation on a horizontal tube. Thermal Engineering, Vol. 26, No. 5, pp 274-278.

14. Hawes, R.I., 1976. Effect of vapour crossflow velocity on condensation; NEL Report No. 619, pp 55-58, National Engineering Laboratory, East Kilbride, Glasgow.

15. Kutateladze, S.S. 1963. Fundamentals of Heat Transfer, Academic Press Inc.

16. Lubuntsov, D.A. 1957. Heat transfer in film condensation of pure steam on vertical surfaces and horizontal tubes. Teploenergetika, Vol. 4, No. 7, pp 72-80.

17. Jaster, H. and Kosky, P.G. 1976. Condensation heat transfer in a mixed flow regime. Int. J. Heat Mass Transfer, Vol. 19, pp 95-99.

18. Zivi, S.M. 1964. Estimation of steady state steam void fraction by means of the principle of minimum entropy production. J. Heat Transfer, Trans. ASME, Series C, Vol. 86, pp 247-252.

19. Boyko, L.D. and Kruzhilin, G.N. 1967. Heat transfer and hydraulic resistance during condensation of steam in a horizontal tube and a bundle of tubes. Int. J. Heat Mass Transfer, Vol.10, pp 361-373.

20. Palen, J.W, Breber, G. and Taborek, J. 1980. Prediction of flow regimes in horizontal tube side condensation. Heat Transfer Engineering Vol. 1, No. 2, pp 47-57.

21. Carpenter, E.F and Colburn, A.P. 1951. The effect of vapour velocity on condensation inside tubes. Inst. Mech. Engrs. Proc. General Discussion on Heat Transfer, pp 20-26.

22. Kosky, P.G. and Staub, F.W. 1971. Local condensation heat transfer coefficients in the annular flow regime. A.I.Ch.E. J. Vol. 17, No.5, pp 1037-1043.

23. Butterworth, D. 1977. Developments in the design of shell and tube condensers, ASME Preprint 77-WA/H7-24, ASME Winter Annual Meeting, Atlanta, Georgia, Nov. 27-Dec. 2.

24. Wallis, G.B. 1961. Flooding velocities for air and water in vertical tubes, UKAEA Report No. AEEW-R123.

25. Deakin, A.W. 1976. The condensation of vapours of binary immiscible liquids. PhD Thesis, University of Birmingham.

26. Bernhardt, S.H, Sheridan, J.J and Westwater, J.W. 1972. Condensation of immiscible mixtures, AIChE Symposium Series, Vol. 68, No. 118, pp 21-37.

27. Ackers, W.W and Turner, M.M. Condensation of vapours of immiscible liquids. AIChE J, Vol. 2, No. 5, pp 587-589.

28. Silver, L. 1947. Gas cooling with aqueous condensation, Trans Inst. Chem. Engrs. Vol. 25, pp 30-42.

29. Bell, K.J and Ghaly, M.A. 1972. An approximate generalized design method for multicomponent/partial condensers. AIChE Symposium Series, Vol. 69, No. 131, pp 72-79.

30. Price, B.C and Bell, K.J. 1974. Design of binary vapour condensers using the Colburn-Drew equations. AIChE Symposium Series, Vol. 7, No. 138, pp 267-272.

31. McNaught, J.M. 1979. Mass transfer correction terms in design methods for multicomponent/partial condensers. 18th National Heat Transfer Conference, San Diego, California, ASME/AIChE 6-8 August.

32. Colburn, A.P and Hougen, O.A 1934. Design of cooler condensers for mixtures of vapours with non-condensing gases. Ind. Engng. Chem. Vol. 26, pp 1178-1182.

33. Ackermann, G. 1937. Simultaneous heat and molecular mass transfer by overall temperature and partial pressure differences. Forschungshaft, No. 382, pp 1-16.

34. Chilton, T.H and Colburn, A.P. 1934. Mass transfer (absorption) coefficients. Ind. Eng. Chem. Vol. 26, pp 1183-1187.

35. Steinmeyer, D.E. 1972. Fog formation in partial condensers, Chemical Engineering Progress, Vol. 68, No.7, pp 64-68.

36. Colburn, A.P and Drew, T.B. 1937. The condensation of mixed vapours, Trans. Amer. Inst. Chem. Engrs, Vol.33, pp 197-215.

37. Krishna, R, Panchal, C.B, Webb, D.R and Coward, I. 1976. An Ackermann-Colburn-Drew type analysis for condensation of multicomponent mixtures. Letters on Heat and Mass Transfer, Vol. 3, No. 2, pp 163-172.

38. Deakin, A.W. 1975. Design of condensers for binary mixtures of vapours of immiscible liquids. UKAEA Report No. AERE-R8118.

39. Collier, J.G. 1972. Convective boiling and condensation, McGraw Hill.

40. Butterworth, D and Hewitt, G.F. 1977. Two phase flow and heat transfer, Oxford University Press, Oxford.

41. Chisholm, D and Sutherland, L.A. 1969-70. Prediction of pressure gradients in pipeline systems during two-phase flow. Proc. Instn. Mech. Engrs. Vol. 184, No. 3C, pp 24-32.

42. Grant, I.D.R and Chisholm, D. 1977. Two phase flow on the shell side of a segmentally baffled shell and tube heat exchanger. ASME paper 77-WA/HT-22, ASME Winter Annual Meeting, Atlanta, Georgia.

# Heat Transfer wtih Natural Convection Boiling in Multicomponent Mixtures

K. STEPHAN
Institut für Technische Thermodynamik
und Thermische Verfahrenstechnik
Universität Stuttgart
FRG

ABSTRACT

Heat transfer in natural convection boiling of multicomponent mixtures is usually lower than that of the pure components. Most of the hypotheses to explain this effect originate from the different equilibrium compositions of the vapor and the liquid phases. Based on this effect a model for an approximative prediction of heat transfer coefficients could be derived, representing experimental data with a satisfactory accuracy in such cases where the thermal properties may be approached by a linear function of the mole fraction. In order to establish correlations with wide application also to fluids where the thermal properties are an arbitrary function of mole fraction heat transfer coefficients can be predicted from a correlation based on a critical review of the nearly 5000 existing experimental data points for natural convection boiling heat transfer of pure substances. An additional term in this correlation describes the influence of mass transport on heat transfer in multicomponent systems.

1. INTRODUCTION

From existing experiments it is well known that heat transfer in pure liquids may be considerably reduced when small amounts of impurities are added. In boiling water for instance 50 ppb of micro-particles may cause a remarkable decrease in heat transfer coefficients. Pure liquids therefore are interesting for the study of heat transfer phenomena because of the lower number of parameters exerting an influence on the heat transfer mechanism. All real liquids however contain small amounts of additional other components and may not be regarded as single component systems.

The purpose of this paper is to study the behaviour of soluble binary liquid mixtures with respect to boiling heat transfer as compared to pure liquids and to explain how heat transfer coefficients may be evaluated. Though our knowledge in this area is still very limited, sufficient experimental data are available at present for this purpose.

The earliest experiments performed by Jakob and Linke [1] de-

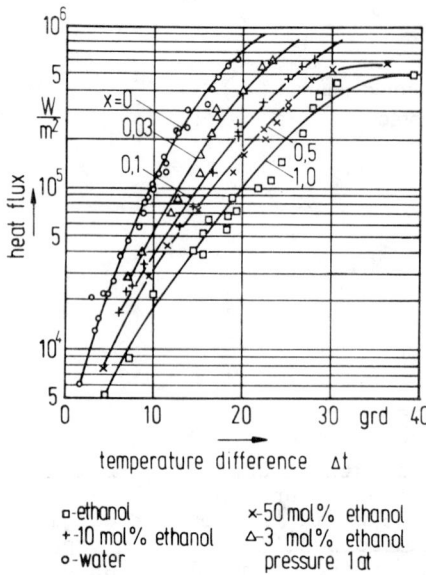

Fig. 1. Heat transfer in boiling water-ethanol

monstrated an increase of heat transfer coefficients if a well-wetting liquid is added to water. Extensive experiments were done by Bonilla and Perry [2]. Fig. 1 shows an example from their experiments on boiling ethanole-water-mixtures. The heat flux density does not change linearly with the mole fraction but lies between those of the pure components. This behaviour applies to many mixtures. For mixtures with no surface-active components heat transfer coefficients were always found to lie below the values, which might be obtained assuming that heat transfer coefficient varies linearly with the mole fraction.

Van Wijk, Vos and van Stralen [3] provide the following explanation for this effect: Individual components of a mixture pass from the liquid to the vapour phase in different proportions, and the faster evaporation of the more volatile component results in an enrichment of the bubble forming boundary layer with the less volatile component, so that the local boiling temperature increases. Thus, the effective excess temperature of the heating surface is lower than the measured value which is based on the temperature of the liquid at some distance from the heating surface. Other workers [4-7] attribute the decrease in heat transfer to the additional resistance to mass diffusion, which exists in mixtures and slows down bubble growth.

According to Grigoryev [8], the reduction in heat transfer is due to increased critical bubble radii of the mixtures, resulting in a smaller bubble population at the same degree of wall superheating. A detailed and critical examination of Grigoryev's hypothesis is given by Shock [9]. On the basis of thermodynamic considerations, Stephan and Körner [1o] found that the reversible isothermal work required to generate a single bubble in a binary mixture is greater than that needed for a pure component of the same physical properties. They assume that for a given temperature difference between the wall and the boiling liquid there are fewer bubbles generated, and consequently heat transfer decreases.

The above mentioned behaviour changes drastically if the liquid contains a surface-active component. In practically all such cases heat transfer coefficients increase as compared to the pure liquid. This may be explained by the decrease of energy which is absorbed for bubble formation, which is caused by an increase of bubble sites.

The critical heat flux of binary mixtures also does not depend linearily on the mole-fraction of the components. According to several authors [3,5,11,12,13] it appears to be higher than that of pure liquids. Probably this effect may also be explained by the lower mole-fraction of the volatile component near the wall, which prevents vapor formation and a rapid formation of a vapor film near the wall. As experiments by Hovestreijdt [28] indicate the critical heat flux is higher in mixtures with higher surface tension of the volatile component than in mixtures where the volatile component has a lower surface tension.

All the hypotheses mentioned definitely originate from the fact, that equilibrium compositions of vapour and liquid phases of boiling mixtures are usually different. Vapourliquid equilibria of mixtures consisting of more than two components can often be estimated from binary data, and it might, therefore, be assumed that also heat transfer coefficients of multi-component mixtures could be estimated with sufficient accuracy, if heat transfer data for pure components and for the relevant binary subsystems were available. Investigation of the validity of this assumption it also the subject of this paper.

## 2. THERMODYNAMIC THEORY OF BUBBLE FORMATION IN MIXTURES

Though a study of the thermodynamic processes of bubble formation in liquids will not lead to general relations for heat transfer, we can draw from it some information on the different behaviour of bubbles forming in pure and in binary mixture and we can learn about the additional parameters which might be important in mixtures.

We consider a single bubble (phase") in a binary liquid mixture (phase '). If the sign $\sigma$ stands for the surface tension, the sign $o$ for bubble surface, we have for the free energy F of the system

$$F(T,V,N_1,N_2) = F'(T,V',N_1',N_2') + F''(T,V'',N_1'',N_2'') + \sigma o \quad (1)$$

The temperature is denoted by $T$; $V = V'' + V'$ stands for the volume and $N_1 = N_1' + N_1''$, $N_2 = N_2' + N_2''$ for the mole number of component 1 and 2. In thermodynamic equilibrium we have $dF = 0$. From there we find after some mathematical and thermodynamical operations [for further details see: K. Stephan, M. Körner, Chem.

Ing. Techn. 41, 1969, p. 409/417] the work which is necessary to form a viable nucleus in a binary mixture

$$\Delta G^+ = \frac{16\pi}{3} \sigma^3 \frac{\overline{V}''^2}{\Delta T^2 \{\frac{\Delta \overline{H}}{T} + (x''-x') \frac{\partial^2 \overline{G}'}{\partial x^2} \frac{\Delta x'}{\Delta T}\}^2} \qquad (2)$$

where $\Delta \overline{H}$ is the molar enthalpy of transformation which is the difference between the molar vapor enthalpy $x'' H_1'' + (1-x'')H_2''$ and the molar enthalpy $x'' H_1' + (1-x'')H_2'$ of a liquid of the same composition. It is clearly different from the molar enthalpy of evaporation. For multicomponent systems the corresponding equation reads [14]:

$$\Delta G^+ = \frac{16 \pi \sigma^3 \overline{V}''^2}{3\Delta T^2 \{\frac{\Delta \overline{H}}{T} + \frac{(\Delta x_n'/\Delta T)}{x_n'' - x_n'} \sum_{i=1}^{n-1} \sum_{j=1}^{n-1} (\frac{\partial^2 \overline{G}}{\partial x_i \partial x_j})(x_i''-x_i')(x_j''-x_j')\}^2} \qquad (3)$$

For one-component systems we get from equ.(2) or (3)

$$G^+ = \frac{16\pi}{3} \sigma^3 \frac{\overline{V}''^2}{\Delta T [\Delta \overline{H}/T]^2} \qquad (4)$$

where $\Delta \overline{H}$ stands now for the molar enthalpy of evaporation and $\sigma$ for the surface tension.

## 2.1 Conclusions from the thermodynamic considerations

Comparing equations (2), (3) and (4) we can get the work necessary for the formation of a viable nucleus in a binary mixture. Starting from equ. (4) we only require to replace there the surface tension $\sigma$, the molar volume $\overline{V}''$ and the transformation enthalpy by the corresponding quantity of the mixture and by adding to the denominator the term $(x''-x')(\partial^2 \overline{G}/\partial x^2)'(\Delta x'/\Delta T)$, which is responsible for the transfer of matter, and thus obtain equ.(2).

As we know heat transfer in evaporation is governed by the number of nuclei and therefore by the work needed for producing a viable bubble. From this fact and from the above statement we can deduce, that it does not suffice to replace the thermodynamic properties in the equation for pure component systems by the properties of a mixture in order to get the heat transfer coefficient of a mixture. Moreover one has to introduce an additional term, containing the molar difference $x''-x'$, thermodynamic properties $\partial^2 \overline{G}/\partial x^2$ and the quantities $\Delta x'$, $\Delta T$ responsible for the process of mass and heat transfer. In aceotropic mixtures ($x''=x'$) the additional term vanishes. Furthermore equs. (2) and (3) clearly indicates that in non-superheated systems ($\Delta T \to 0$) and in aceotropic mixtures the term $\Delta G^+$ is infinite. In this case bubble formation is impossible. In non-aceotropic mixtures, however, bubble formation is possible even if the liquid is not superheated. The concentration difference $\Delta x'$, which stands for the difference between the concentration of a liquid with bubbles within it and the concentration of the same liquid without bubbles, acts then as a driving force for bubble formation.

Some statements are possible regarding the sign of the term $(x''-x')(\partial^2 \bar{G}/\partial x^2)'(\Delta x'/\Delta T)$ in equ. (2) and the corresponding term in equ. (3). As we know from thermodynamics $(\partial^2 G/\partial x^2) > 0$ is one of the stability conditions for binary systems. Furthermore after the first rule of Konowalow [15] the boiling point in isobaric mixtures increases by addition of that component which has a lower concentration in the vapor than in the liquid phase.

Therefore, if we have $x''-x' > 0$, we get $(\Delta x'/\Delta T) < 0$ and in the reverse case if $x''-x' < 0$ we get $(\Delta x'/\Delta T) > 0$. The sign of the second term in the denominator of equs. (2) and (3) therefore is always negative. Hence we may conclude that heat transfer coefficients in multicomponent mixtures are smaller than that in pure liquids or in aceotropic mixtures. In the neighbourhood of aceotropic points a decrease of heat transfer coefficients must be expected also when departing from the poor liquid. According to this explanation the heat transfer coefficients of a mixture plotted over the concentration may show therefore maxima or minima.

The phenomena described above, as we can see from equs. (2) and (3) are also influenced by the concentration dependent surface tension and by the other thermodynamic properties in equs. (2) and (3). At present no general relations on the surface tension as function of concentration exist. We know only that its value is between those of the pure components. The mixing rules, which say that surface tension of a mixture is a linear function of the mole fraction, seem to fail in many cases [15]. Surface-active components accumulate along the surface of the liquid mixture and heat transfer increases. In cases where the volatile component is extremely surface-active as compared to the other component, we observe formation of foam provided that heat flux densities are large. This effect may be explained if we consider that the liquid near the bottom of a bubble contains less of the volatile component than at the top of the bubble. Therefore a concentration field builds up around the bubble sur-

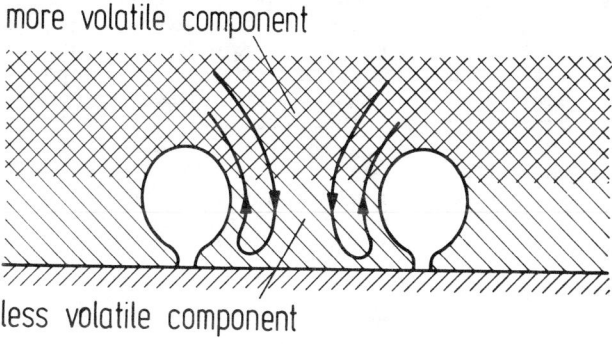

Fig. 2. Flow pattern due to Marangoni-effect

face. Near the bottom of the bubble we find higher surface tensions than near its top. The gradient in surface tension provokes shear stresses, which are known as Marangoni-effect, and which try to compensate each other. There develops a secondary flow pattern as shown in fig. 2, which may lift the bubble and cause foaming. This explanation is in agrement with several observations [16] on boiling mixtures of oil and refrigerants and with experiments [17] on bubble formation in distilling columns.

The different mixtures studied in the present paper, did not contain extremely surface-active components. Therefore we may presume that for all mixtures regarded here, surface tension gradients had only a negligible effect on heat transfer. This may also be justified in so far as heat resistances in boiling of non-surface-active liquids depend weaker than linearily on the surface tension. But it should be emphasized that the following results are only applicable to liquids without surface-active-components.

## 3. COMPARISON BETWEEN THERMODYNAMIC THEORY AND EXPERIMENTS

According to the thermodynamic theory a decrease of heat transfer in mixtures is to be expected depending on the difference between equilibrium concentrations $x_i''-x_i'=y_i-x_i$. Heat transfer coefficients will decrease the greater this difference.

Evidently the sign of the difference is unimportant. This theory is based on the assumption of thermodynamic equilibrium which, however, in boiling liquids does not exist. It is therefore doubtful, if the theoretical results agree with experimental observations. However, analyzing the results of measurements on boiling mixtures shows, that the theoretical results are confirmed. The diffusional effects, therefore, may be expected to cause not too far a departure from equilibrium.

In the following the good qualitative agreement between thermodynamic theory and experiments shall be discussed. The experimental results are taken from a paper by Körner [18] and by Happel and Stephan [19].

The boiling and the dew lines of n - heptane - methylcyclohexane practically coincide in the boiling diagram, as shown in fig. 3. According to equs. (2) and (3) the term responsible for the transfer of matter in the denominator is unimportant. As fig. 4 demonstrates heat transfer coefficients depend only slightly on the concentration. The error is very small, if one assumes a linear dependence of heat transfer coefficients on the coefficients of the pure components. Studying a mixture with larger differences between liquid and vapor composition, as for example a mixture consisting of acetone and n-butanole shown in fig. 5, we observe a pronounced decrease of heat transfer coefficients in the range of large concentration differences. This effect is clearly shown in fig. 6. It is very conspicuous in an aceotropic mixture. As an example let us regard a methanole-benzene-mixture fig. 7 and fig. 8. We observe a remarkable reduction of heat transfer coefficients in the region of large concentration differences y-x. Approaching the aceotropic point

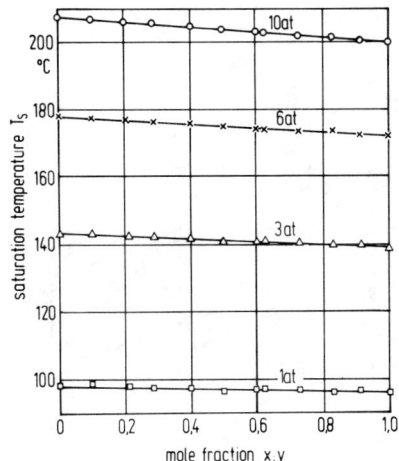

Fig. 3. Saturation temperature of n-heptane-methylcyclohexane

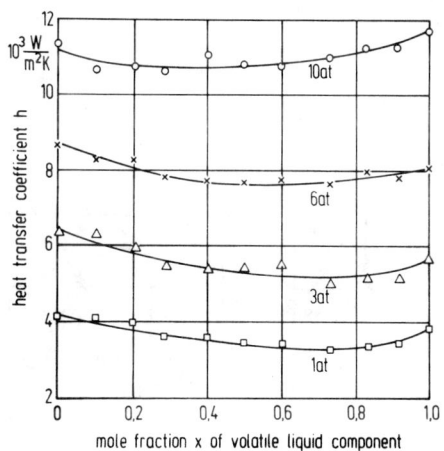

Fig. 4. Heat transfer coefficient for boiling-n-heptane-methylcyclohecane, $q = 10^5$ W/m$^2$

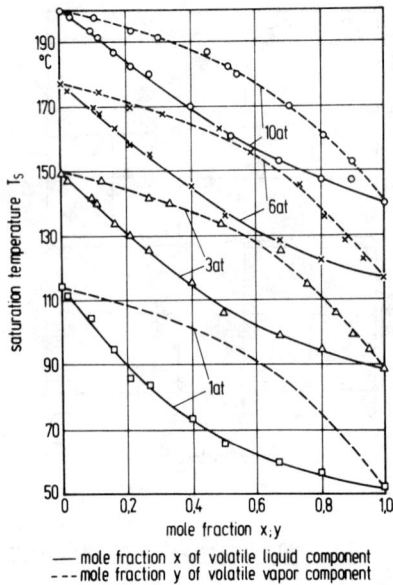

Fig. 5. Saturation temperature of acetone-n-butanol

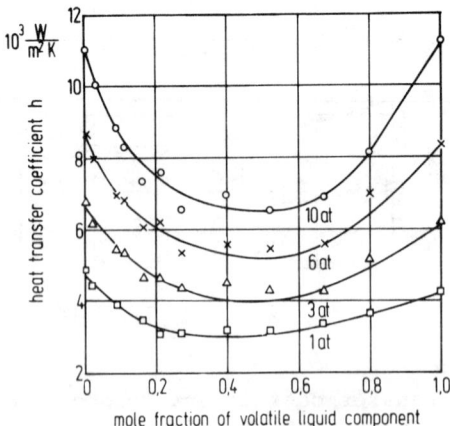

Fig. 6. Heat transfer coefficient for boiling acetone-n-butanol, $\dot{q} = 10^5$ W/m$^2$

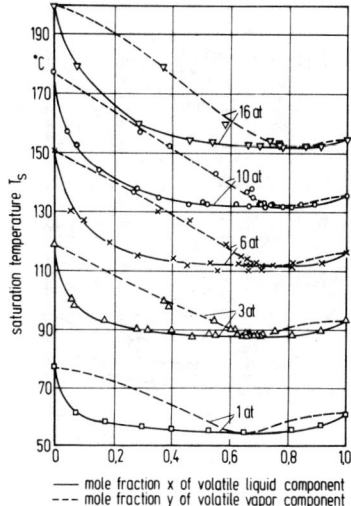

Fig. 7. Saturation temperature of methanol benzene

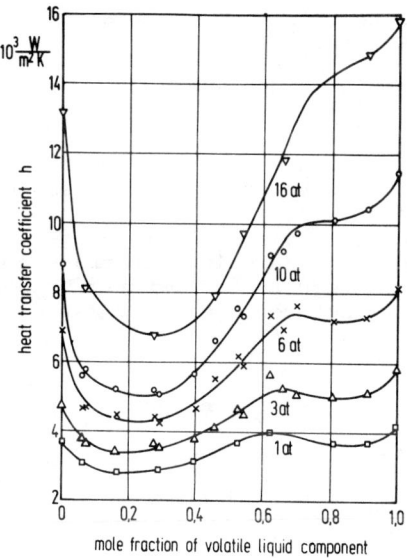

Fig. 8. Heat transfer coefficient of methanol-benzene, $\dot{q} = 10^5$ W/m$^2$

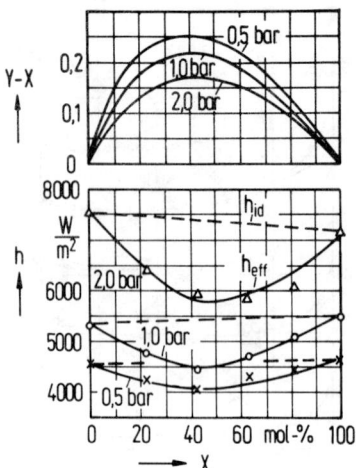

Fig. 9. Vapour-liquid-concentration differenze and heat transfer coefficient of benzene-toluene

heat transfer coefficients increase. Hence the result of the thermodynamic theory is confirmed. It predicted a decrease of heat transfer coefficients near aceotropic points, depending only on the absolute value of the concentration difference. As fig. 6 and 8 indicate, the influence of concentration on the reduction of heat transfer coefficient increase with pressure. This may be explained by the number of bubbles per square inch; it increases with pressure.

The free liquid area between the bubbles at the wall, therefore, becomes smaller with increasing pressure and the mass flow of liquid from the bulk to the wall will be restrained. The subsequent delivery of the volatile component is damped, which results in a further decrease in heat transfer.
Finally fig. 9 gives results on mixtures of benzene and toluone boiling at pressures between o,5 and 2 bar with a heat flux of $10^5 W/m^2$.

## 4. CONCLUSIONS AND PREDICTION OF HEAT TRANSFER COEFFICIENTS

The results from equilibrium thermodynamics do not lead to a theory of heat transfer, which is a nonequilibrium process. However, they indicate a reduction of heat transfer coefficients as compared to those of the reference liquid, defined as a liquid in which the heat transfer coefficients at a given heat flux varies linearly with the mole fraction. The reduction of heat transfer coefficients obviously depends on the difference in mole fractions. The diffusion resistance, which by some authors is regarded as the decisive factor for the reduction in heat transfer, also increases with the difference in mole fractions. Thus whatever the prevailing mechanism of heat transfer be, either diffusion resistance or decrease in bubble population, we may conclude, that heat transfer coefficients in multicomponent mixtures are smaller than those of their reference mixture and that the reduction depends mainly on the difference in mole fractions. The question however arises, how to determine, for a given heat flux, the heat transfer coefficients or the temperature difference $\Delta T$. Since the reference mixture is a hypothetical one, the driving temperature difference $\Delta T$ cannot be found by experiments. If there existed a general theory for pool boiling heat transfer in pure liquids, the temperature difference could be calculated from the heat transfer equations. As is well known, such a theory does not exist and most of the empirical correlations are only applicable to those liquids for which they were established. Equations of this kind, as useful as they are for heat transfer calculations, may not be employed for the determination of the temperature difference of the reference mixture. However, in some other fields of heat transfer, it proved to be quite successfull to determine the heat transfer coefficients of a mixture from those of the pure components according to their mole or volume fractions. As an example we may mention heat transfer in total condensation of miscible binary mixtures [2o] or an equation proposed for condensation of a vapour with components immiscible in the liquid phase [21]. A similar procedure introduced by Stephan and Körner [22] for pool boiling of binary mixtures started from a reference mixture whose temperature difference for a given heat flux is defined as

$$\Delta T_{ref} = \Delta T_{id} = \Sigma x_i' \Delta T_i.$$ (5)

Introducing of a reference state of this kind avoids the evaluation of the decisive properties of the mixture. They evidently yield fairly accurate results for mixtures with thermal properties depending linearly on the mole fraction or for mixtures with very similar thermal properties. They are to be expected less accurate for mixtures where the decisive properties are strong function of the mole fraction. However, all existing experiments on pool boiling of binary mixtures demonstrated that heat transfer coefficients for a given heat flux may be correlated in a first approximation by using a reference temperature according to equ. (5). The deviation from the reference state due to the different mole fractions in liquid and vapour phase may be included in an additional term containing the mole fractions. Equations of this type were proposed by Afgan [6], Körner and Stephan [23] and by Calus and Leonidopoulos [23]. A very simple but accurate procedure is that by Körner and Stephan [22]. They represented the temperature difference $\Delta T$ from their experiments for a given heat flux by

$$\Delta T / \Delta T_{id} = 1 + |\Delta T^E| \Delta T_{id}|$$ (6)

where $T_{id}$ is defined by equ.(5) and the ratio of "excess-temperature" $\Delta T^E$ over $\Delta T_{id}$ proved to be a function of the difference in mole fraction. The value $\Delta T^E/\Delta T_{id}$ proved to be always positive, due to the reduction of heat transfer. Experiments with 12 different mixtures of one component i with another component j of hydrocarbons or of hydrocarbons with water could be represented by a linear equation

$$\Delta T^E/\Delta T_{id} = K_{ij} (x_i'' - x_i') \, , \qquad (7)$$

where $K_{ij}$ is a positive factor independent from the mole fraction. An extension of these results to multicomponent mixtures suggests to use the temperature difference of the reference mixture according to equ.(5), and furthermore to introduce for the excess-temperature following equ.:

$$\Delta T^E/\Delta T_{id} = \left| \sum_{i=1}^{n-1} K_{in} (x_i'' - x_i') \right| \qquad (8)$$

$\Delta T^E/\Delta T_{id}$ is always positive and varies linearly with the (n-1) independent variable concentration differences. The $K_{in}$ are individual constants which have to be determined by experiment. Subscript 1 indicates the component with the lowest, and n that with the highest boiling point at the operating pressure p.

For a ternary mixture, equ.(8) may be simplified to

$$\Delta T^E/\Delta T_{id} = |K_{13}(x_1''-x_1') + K_{23}(x_2''-x_2')| \; . \qquad (9)$$

A ternary mixture includes the binary systems as limiting cases. If $x_1' = x_1'' = 0$, the ternary system reduces to a binary one, consisting of components 2 and 3; if $x_2' = x_2'' = 0$, the resultant binary system consists of components 1 and 3. Consequently, the coefficients of Eq.(9), which are determined from ternary mixture experiments should agree with those obtained from measuring heat transfer to the respective binary systems. Thus heat transfer to boiling ternary mixtures could be predicted from the binary data.

Coefficients $K_{in}$ were estimated by the method of least squares, based on experimental heat transfer coefficients of 32 binary and 50 ternary compositions in the heat flux range of 50 kW/m² ≤ q̇ ≤ 200 kW/m². The results are presented in Table 1.

Table 1. Proportionality coefficients in Relation between Excess Temperature and Concentration Differences Obtained from Experiments on Binary and Ternary Mixtures.

| Mixtures | $K_{in}$ | Binary | Ternary |
|---|---|---|---|
| Acetone(1)-methanol(2)-water(3) | $K_{12}$ | 1.19 | - |
|  | $K_{13}$ | 0.81 | 0.58 |
|  | $K_{23}$ | 0.56 | 0.54 |
| Methanol(1)-ethanol(2)-water(3) | $K_{12}$ | 1.39 | - |
|  | $K_{13}$ | 0.56 | 0.31 |
|  | $K_{23}$ | 0.71 | 0.23 |

With the aid of Eqs(5) to (9) and coefficients $K_{in}$ listed in Table 1, experimental heat transfer data were predicted with a mean error of $\pm 4.6$ % for binary and $\pm 7.8$ % for the ternary mixtures. However, it is clear from Table 1 that the coefficients $K_{in}$ obtained from experiments on ternary mixtures and on the relevant binaries do not show a satisfactory agreement. The ternary coefficients $K_{in}$ are in all cases lower than the binary ones. On the average, the predicted heat transfer to the ternary mixtures is too low if the binary coefficients $K_{in}$ are substituted in Eq.(9). This is illustrated in Fig. 10 which presents the plots of experimental heat transfer coefficients of the ternary mixtures against the values, estimated by using binary data. However, the deviations rarely exceed 20 %. Therefore, when experimental results for ternary systems are not available, the calculation of heat transfer coefficients for ternary mixtures from binary data, by the above procedure, may be recommended as a first approximation.

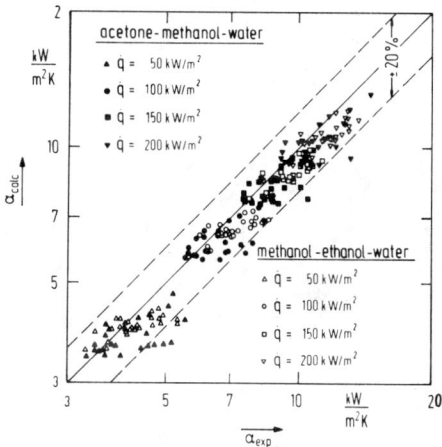

Fig.10. Comparison between heat transfer coefficients ($\alpha_{calc}$) in the ternary mixtures obtained from binary data and experimental values ($\alpha_{exp}$).

## 5. A SIMPLE MODEL FOR THE DEVIATION BETWEEN HEAT TRANSFER COEFFICIENTS FROM EXPERIMENTS AND FROM THE PROPOSED CORRELATIONS

There are two explanations as to why the reduction in heat transfer coefficients of ternary mixtures is less than previously assumed. The first follows from the hypothesis of van Wijk, Vos and van Stralen [3], which postulates that the reasons for reduced heat transfer to binary mixtures are the different composition and a higher boiling point of the boundary layer liquid, compared to that of the bulk liquid. At small differences in composition, the resulting change in the boiling point of a mixture of n components can be approximately represented by

$$\Delta T_s = \sum_{i=1}^{n-1} \Delta T_{si} \qquad (10)$$

where

$$\Delta T_{si} \sim (x_i'' - x_i')(\partial T_s/\partial x_i')_{\tilde{x}_j', p} \qquad (11)$$

Hence, the local change in boiling point of a binary mixture depends not only on the concentration difference $(x_i'' - x_i')$ between the vapour and liquid phases, but also on the slope $(\partial T_s/\partial x_i')_p$ of the boiling point curve (Fig.11). The partial derivatives $(\partial T_s/\partial x_i')_{x_j', p}$ of ternary mixtures, are often substantially smaller,

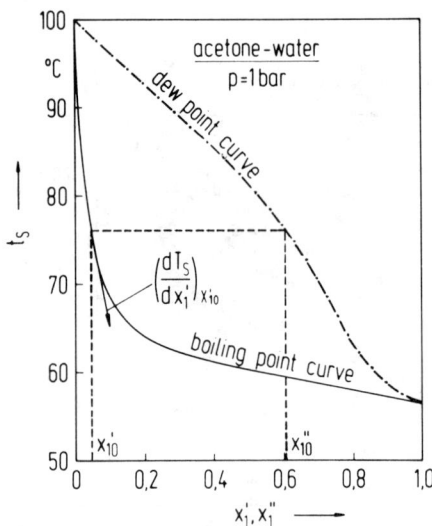

Fig.11. Isobaric vapour-liquid equilibrium diagram of the system acetone-water.

than those of corresponding binary compositions. This was also confirmed by van Stralen and Sluyter [23], who investigated the critical heat flux of ternary mixtures consisting of organic compounds and water, and may be illustrated by plotting boiling curves $t_s(x_1')$ for different $x_2'$, and $t_s(x_2')$ for different $x_1'$ as shown in Fig. 12. It may be observed, for example that the curve $t_s(x_2')$ would become considerably flatter, if small quantities of acetone were added to the binary mixture methanol-water. Hence, the difference between the boiling points of the boundary layer and the bulk liquid is less than might be expected, if the concentration differences $(x_i'' - x_i')$ alone were considered.

The second explanation is based on the deviation of the

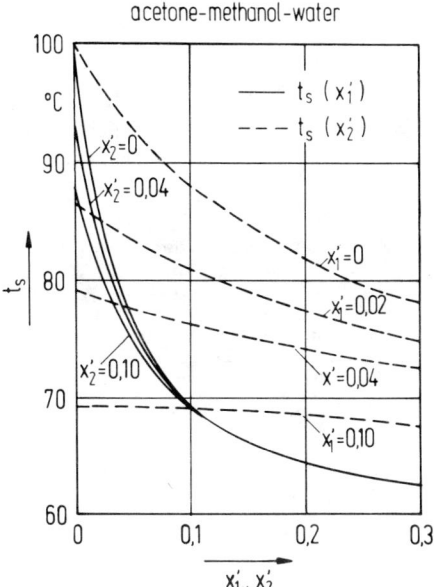

Fig.12. Boiling point curves of the mixture acetone-methanol-water, p = 1 bar.

vapour composition at the heating surface, from the value it would have if at equilibrium with the bulk liquid. The result of this deviation is an additional diffusional resistance.

In a binary mixture the difference between the two vapour compositions is approximately expressed by the linear relation

$$\Delta x_1'' \sim (x_1'' - x_1')(\partial x_1''/\partial x_1')_p \quad . \tag{12}$$

For a mixture of n components, there are (n-1) independent deviations of vapour mole fractions

$$\Delta x_i'' \sim (x_i'' - x_i')(\partial x_i''/\partial x_i')_{x_j',p} \quad . \tag{13}$$

For binary mixtures, the partial derivatives $(\partial x_1''/\partial x_1')_p$ represent the slopes of the $x_1''(x_1')$-equilibrium curves, shown in Fig. 13 for the systems acetone-methanol and acetone-water. As confirmed by calculation, these curves flatten out considerably on addition of small quantities of another organic component to the binary aqueous solutions.

In Fig. 14 the term $(x_1''-x_1')(\partial x_1''/\partial x_1')_p$ is plotted as a function of mole fraction for the binary systems acetone-methanol and acetone-water. The experimental values of $\Delta T^E/\Delta T_{id}$, which indicate the reduction in the heat transfer coefficients according to Eq.( 8 ), are also shown. For the binary systems consisting of only organic components, the corresponding curves are similar. The decrease in the heat transfer coefficient may be explained by deviation from the equilibrium concentration. However, the behaviour of aqueous solutions is different, as shown by the acetone-

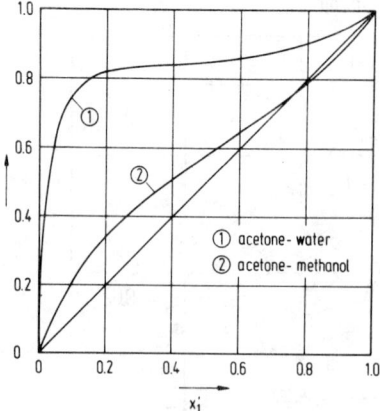

Fig.13. Equilibrium compositions of mixtures acetone-methanol and acetone-water, p = 1 bar.

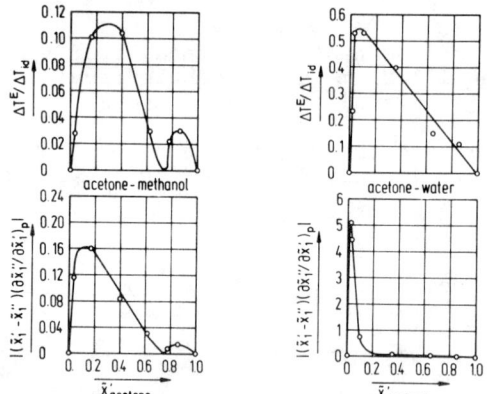

Fig.14. Comparison between curves $\Delta T^E/\Delta T_{id}$ and $(x_1''-x_1')\cdot(\delta x_1''/\delta x_1')$ over mole fraction

water mixtures. Only in the range of very small mole fractions of the organic components, is such an explanation applicable. Even van Wijk's et al. [3] hypothesis of higher local boiling temperature does not predict a distinct decrease in the heat transfer outside this narrow range. The explanations, as to why heat transfer to ternary mixtures is less reduced than that to binary ones, cannot be valid unless the relatively poor heat transfer to aqueous solutions can be partly attributed to other phenomena, such as e.g. variation of physical properties with concentration.

## 6. GENERALIZED CORRELATIONS

So far little has been known about the influence of physical properties on boiling heat transfer. The numerous correlations which do exist for calculating nucleate boiling heat transfer to pure liquids, often produce contradictory results. All of the existing equations are based on models which prove nonsuitable for some substances, apparently none take into account all of the processes important to boiling heat transfer, and present knowledge is not sufficient for the building up of a valid general model. Consequently, it did not seem appropriate to start from a given model, but correlate instead the existing experimental data by means of more mathematical methods.

This has been done recently by Stephan and Abdelsalam [24], who applied the methods of regression analysis to the existing experimental data. They collected all the existing experimental data and critically reviewed them. This was done under the following criteria:

Only data concerning pool boiling on horizontal surfaces in the range of fully established nucleate boiling under the influence of the gravity field were considered and, in order to permit conclusions on the influence of wall material on heat transfer, the data were further limited to those, for which the heating surface material was indicated. These conditions were fulfilled by about 5000 data from 72 papers. Only a few of the above mentioned 5000 data from the literature gave information on the roughness of the heating wall. In cases where these specifications were missing, a mean surface roughness of 1 μm, as is often met in technical applications, was assumed. Very often the experimenters reported their results only with fitted curves accompanied by some of their raw data. In order to have a common basis for the analysis therefore, all the experimental data were fitted by curves $\alpha(\dot{q})$ and each of these curves represented by a certain number (usually four) of characteristic points. Thus the total of 5000 original measuring points were replaced by about 1553 characteristic points. A great number of experimental data are available for water, hydrocarbons, cryogenic liquids and for refrigerants. For each of these substances, or respectively, groups of substances there exists approximately the same number of about 400 characteristic points. It seemed reasonable therefore to study these groups separately, all the more so, as the accuracy of measurement is different for the groups. Heattransfer data on pool boiling of water are, for instance, more reliable than those on pool boiling of cryogenic liquids. Also, the transport properties of one group of substances may differ considerably from those of another group, whereas the deviations between substances within one group are usually smaller, so some of the dimensionless parameters are different for the various groups of substances. Some of them, important for one group of substances, are expected to be unimportant for another group, an effect which indeed was confirmed by the regression analysis. By considering the groups of substances separately first, equations can be developed with a minimum number of dimensionless variables representing the experimental data within the scope of their accuracy. Eventually an overall-correlation valid for all substances of the four groups was established. These equations permit prediction of nucleate boiling heat transfer to most of the industrially important liquids with sufficient accuracy over a wide pressure range, between $10^{-4} \le p/p_c \le 0.97$. It was finally rewritten in the form $\alpha = c \dot{q}^n$, where the value c was plotted as a function of pressure for the different substances. Diagrams c(p) are given in [24].

Experiments with mixtures were only done at moderate pressures and mainly with hydrocarbons and water. The above mentioned overall correlation was therefore slightly changed in order to adapt it to these substances in the moderate pressure range; thus improving the accuracy, however, in a limited pressure range and for a smaller number of substances. This equation for pure substances had to be extended by addition of an appropriate term for boiling of mixtures. As already explained this term should reflect the local rise in boiling temperature or departure from the equilibrium vapour concentration. In fact, experimental results for pure components, as well as for all binary and ternary mixtures are well approximated by the equation

$$\frac{\alpha d_o}{\lambda'} = c \left(\frac{\dot{q} d_o}{\lambda' T_s}\right)^{0.674} \left(\frac{\rho''}{\rho'}\right)^{0.156} \left(\frac{rd_o^2}{a'^2}\right)^{0.371} \left(\frac{a'^2 \rho'}{\sigma d_o}\right)^{0.350} \times$$

$$\left(\frac{\eta' c_p'}{\lambda'}\right)^{-0.162} \left[1 + \left| \sum_{i=1}^{n-1} (x_i'' - x_i')(\partial x_i''/\partial x_i')_{x_j',p} \right| \right]^{-0.0733} \quad (14)$$

where the departure diameter is $d_o = 0.0146 \; \beta \sqrt{2\sigma/[g(\rho' - \rho'')]}$ and $c = 0.0871$. The mean deviation of heat transfer coefficients calculated with the aid of Eq.(14) from the experimental values is $\pm 8.7$ %. Contact angles ß were assumed to be 45° for pure water and 35° for the organic components and for all mixtures. The factor c depends on the properties of the heating surface which are not accounted for in Eq.(14).

Eq.(14) differs from Stephan's and Abdelsalam's correlation mainly in the exponents on the Prandtl number and on the term ($\rho''/\rho'$) which were fitted to heat transfer coefficients of pure substances acetone, methanol, ethanol, and water, determined experimentally during the present investigation. This adaptation produced an improved reproduction of experimental results in the range of moderate pressures. Eq.(14) furthermore includes in addition the last term, which is a measure of deviation from equilibrium concentration. This additional term is equal to unity for pure components and azeotropic mixtures.

Fig. 15 presents the heat transfer coefficients calculated from Eq.(14) for pure components as well as for all binary and ternary mixtures, plotted against the experimental results. 92 % of the points are accurate to within 15 %.

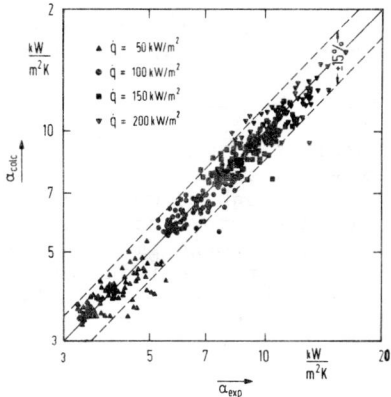

Fig. 15. Heat transfer coefficients of pure components and of binary and ternary mixtures, calculated from Eq.(14) plotted against experimental values.


## 7. PRACTICAL EVALUATION OF HEAT TRANSFER IN NUCLEATE BOILING OF MULTICOMPONENT MIXTURES

Based on the preceding results, heat transfer in nucleate boiling of multicomponent mixtures may be calculated as follows: To begin with, it ought to be ascertained in what way the presumably most important physical properties, i.e., thermal conductivity, surface tension, and viscosity vary with liquid composition. If the variations are approximately linear, heat transfer should be calculated employing the correlation of Stephan and Körner, which has now been extended to mixtures of more than two components. The required heat transfer coefficients of pure components are either known or may be calculated from an appropriate heat transfer correlation. For estimation of heat transfer to mixtures of more than two components, the coefficients $K_{in}$ of the constituent binary systems may be used in Eq.(8). If these are not available, the mean value 1.53, proposed by Stephan and Körner [22] can be inserted for all $K_{in}$. From the hypotheses of local increase in boiling temperature and of difference in concentrations between the generated vapour and its equilibrium value relative to the bulk of the liquid, it may be concluded that calculation of heat transfer to multicomponent mixtures from binary data will be accurate enough provided the boiling point and dew point curves of the binary mixtures do not show excessive curvature and the boiling points of pure components do not differ too widely. From experiments on aqueous solutions, it may be concluded that the above procedure can produce satisfactory results even for mixtures, with physical properties which do not vary linearly with composition. This is due to the fact that the non-linear variations in physical properties are at least partially taken into account by the empirical constants of Eq.(8). Hence, Eqs(5) to (8) can be recommended for a rough estimate of heat transfer to multicomponent mixtures, when sufficiently accurate values of physical properties are not available and provided that the applicability of Eqs(5) to (8) to the constituent binary systems has been confirmed experimentally. The resulting errors lead to design of equipment with heat transfer surfaces larger than required. Systematic failure of this correlation may, however, occur for aqueous solutions when predicting multicomponent system heat transfer from binary mixture data and at azeotropic points of binary aqueous solutions. Although concentration differences between the vapour an liquid phases disappear, the actual heat transfer coefficients are found to be smaller than the ideal values. This may be attributed to the reduced thermal conductivity and increased viscosity. For this reason, Eq.(14) is considered to be more reliable for aqueous solutions boiling at pressures near to atmospheric.

NOMENCLATURE

| | |
|---|---|
| $a = \lambda/(\rho c_p)$ | thermal diffusivity |
| c | empirical constant |
| $c_p$ | specific heat capacity at constant pressure |
| $d_o$ | bubble diameter at detachment |
| g | gravitational acceleration |
| K | empirical constant |
| M | molar mass |
| p | pressure |
| q | heat flux |
| r | latent heat of evaporation |
| T | temperature |
| $\Delta T = T_W - T_S$ | difference between wall and liquid temperatures |
| x | mole fraction |
| $Nu = \alpha d_o/\lambda'$ | Nusselt number |
| $\alpha = \dot{q}/\Delta T$ | heat transfer coefficient |
| ß | contact angle |
| η | viscosity |
| λ | thermal conductivity |
| ρ | density |
| σ | surface tension |

Subscripts

| | |
|---|---|
| c | critical |
| id | ideal |
| s | saturated |

Superscripts

| | |
|---|---|
| E | excess quantity |
| ' | liquid phase |
| " | vapour phase |

# REFERENCES

1. Jakob, M. and Linke, W. Phys.Z. 36, 26780 (1935).

2. Bonilla, C.F. and Perry, C.W. Trans.Amer.Inst.chem.Engr. 37, 685/705 (1941).

3. van Wijk, W.R., Vos, A.S., van Stralen, S.J.D., Chem.Eng. Sci.5 (1956) pp.68-80.

4  Sternling, C.V., Tichacek, C.J., Chem.Eng Sci.16 (1961) pp. 297-337.

5. Huber, D.A., Hoehne, J.C., Trans.ASME Ser. C 85 (1963) Nr. 8, pp. 215-220.

6. Afgan, N.H., Int. Heat Transf. Conf., Chicago 1966, vol.III, pp. 175-185.

7. Valent, V., Afgan, N.H. Waerme-Stoffuebertrag 6(1973) pp. 235-240.

8. Grigoryev, L.N., Teplo i Massoperenos 2 (1962) pp.120-127.

9. Shock, R.A.W., Int.J.Heat Mass Transfer 20(1977) pp.701-709.

1o. Stephan, K., Körner, M., Chem.-Ing.-Tech. 41 (1969) Nr. 7 pp. 409-417.

11. Dunskus, T. and Westwater, W.J., Chem. Engng. Progr., Sympos. Ser. 57, 173/81 (1960).

12. van Wijk, W.R. and van Stralen, S.J.D., Chem.Ing.Techn.37, 509/17 (1965).

13. Vos, A.S. and van Stralen, S.J.D., Chem.Engng.Sci.5, 50/56 (1956).

14. Stephan, K. and Preußer, P., Reprints 6th Internat. Heat Transf.Conf., Toronto (1978), Hemisphere Publ.Comp., Vol.1, 187/192.

15. Plank, R., Handbuch der Kältetechnik, Vol.4, Die Kältemittel, Berlin, Göttingen, Heidelberg 1956, S. 78.

16. Stephan, K., Kältetechn. 16, 162/66 (1964).

17. Zuiderweg, F.J. and Harmens, A., Chem.Engng.Sci.9, 89/103, (1958).

18. Körner, M., Diss. Aachen 1967.

19. Happel, O. and Stephan, K., Reprints 5th Internat. Heat Transf. Conf., Tokyo (1974), Jap. Soc. Mech. Engs. and Soc. Chem. Engrs, Japan, Vol. IV, 340/344.

20. Pressburg, B.J. and Todd, J.B., Amer.Inst.Chem.Engrs. J.3, (1957), 348.

21. Bernhardt, S.H., Sheridan, J.J. and Westwater, J.W., Amer. Inst.Chem.Engrs., Symp. 118, (1971), 21.

22. Stephan, K. and Körner, M., Chem.-Ing.-Techn. 41, (1969),409.

23. van Stralen, S.J.D. and Sluyter, W.M., Int.J.Heat Mass Transfer 12 (1969) Nr. 11, pp. 1353-1384.

24. Stephan, K. and Abdelsalam, M., Int.J. Heat Mass Transf., 23 (1980), 73/87.

# Heat Transfer with Condensation in Multicomponent Mixtures

K. STEPHAN
Institut für Technische Thermodynamik
und Thermische Verfahrenstechnik
Universität Stuttgart
FRG

ABSTRACT

Condensation of multicomponent mixtures can either be homogeneous or consist of several immiscible phases. Condensation with a homogeneous liquid phase is most often met in practical applications. In this case the condenser design usually is based on the assumption that total condensation occurs, which is equivalent to the assumption that the temperature of the vapor-liquid interface is constant and given by the boiling temperature. The temperature of the vapor-liquid interface, however, may change considerably according to the rate of condensation, an effect which is mostly neglected. Also the usually adopted film theory by Ackermann can only be applied if the mass fluxes normal to the cooling wall are induced by diffusion. Since this condition mostly is not fulfilled when condensing vapor mixtures, an approximative procedure is proposed taking into account the influence of suction on condensation heat transfer.

In cases where the liquid consists of several phases very different phenomena are to be observed. Heat transfer coefficients may then also be calculated approximatively. When the vapor condenses at an immiscible liquid interface an additional heat resistance, however, must be taken into account.

1. INTRODUCTION

Heat transfer in condensation of mixtures depends on how the liquid phases are formed near a cooled wall. They can either be homogeneous or consist of several immiscible phases.

In practical applications condensation with a homogeneous liquid phase is most often met. The classical liquid film theory by Nusselt applies, as is well known to homogeneous condensation, however, only to the condensation of pure vapors.

However, in chemical engineering applications it is almost always necessary to deal with multicomponent mixtures, the components of which distribute themselves in different ways among the phases. In addition, in some cases thermal resistance of considerable magnitude can appear at the interfaces between the phases; in the classical theories such resistances are assumed to be not significant. The calculation of heat transfer by means of these theories therefore often leads to inaccurate values, and hence to the faulty dimensioning of equipment.

2. PENOMENA IN MULTICOMPONENT CONDENSATION WITH IMMISCIBLE LIQUID PHASES

In technical equipment it is seldom that pure gases are liquified; instead, vapor mixtures are usually encountered, or the vapors contain admixtures of so-called inert gases, which by definition are not condensed under the prevailing conditions. While heat transfer during the condensation of two-component mixtures has been investigated many times in the past, heat transfer during the condensation of mixtures of three or more components has been studied very little. However, the basic processes are similar to those in the condensation of binary mixtures, so that the special features of this case should first be considered.

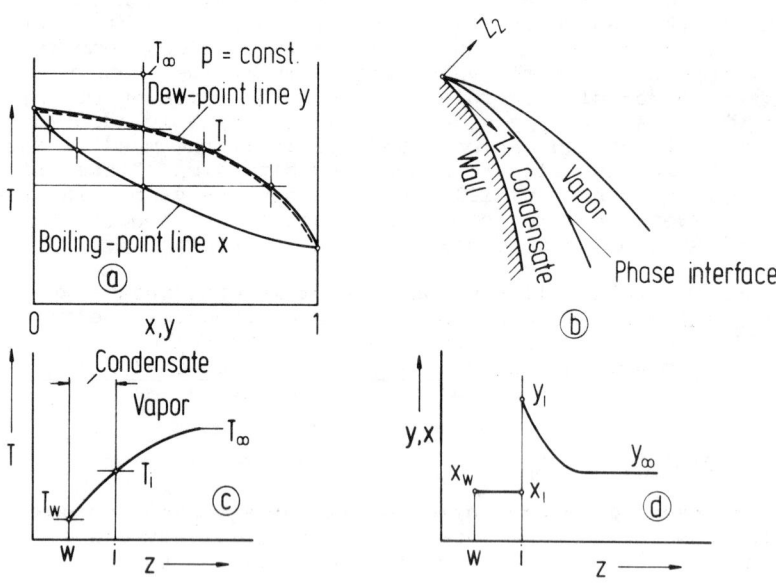

Fig. 1  Temperature concentration diagram for a binary mixture, and changes of concentrations in vapor and condensate.
Subscripts: w cooling wall; i phase interface; ∞ core flow.

If a two-component mixture, whose boiling point and dew point curves are shown in Figure 1a is condensed on a wall of temperature $T_w$, condensate is formed, as shown in Figure 1b. Figure 1c shows the temperature distribution in the condensate and vapor. The component with the higher boiling point transfers preferentially into the condensate at the interface. In consequence, the vapor mixture at the phase interface contains, in general, more of the lower-boiling volatile component than at a greater distance from the interface, as shown by the concentration distribution $y_i - y_\infty$ in Figure 1d. An increase in concentration of the lower-boiling component therefore results toward the interface, and as a result of this, there is a back-diffusion of this component into the vapor phase, by means of which in the steady state the molecules of the lower-boiling more volatile component, which do not condense at the phase interface, are transported back into the vapor phase. Because of the small flow velocities in the condensate compared with the vapor, energy, momentum and mass transfer by convection can be ignored in the condensate as long as the Prandtl or Schmidt numbers are not very small. In this case, the assumptions of Nusselt are valid, according to which the flow in the condensate is determined only by viscous and field forces (gravity force), and the temperature profile is governed mainly by thermal conduction [1,2]. In these cases the concentration in the liquid is constant, since the wall is impermeable to material, and convective mass transfer in the liquid film is negligible. If the condensate and vapor are in counter-current flow, as in falling film columns or reflux condensers, the condensate formed by condensation from the moving vapor is richer in the more volatile components than in the case of co-current flow. If equilibrium is assumed at the phase interface, then, because of the larger mole fraction $x_i$ in counter-current flow, the mole fraction $y_i$ at the phase interface is also larger than in co-current flow (see Figure 1d). Hence, the concentration difference driving force $y_i - y_\infty$ is larger, and it is possible for more of the lower-boiling, more volatile molecules to be transferred in the vapor phase than in co-current flow (see Figure 2).

As sketched in Figure 2, the composition of the counterflowing liquid is influenced by the vapor phase: mass exchange occurs between the liquid and the vapor by means of rectification, so that the condensate leaving the system (downwards in Figure 2) contains more of the lower-boiling, more volatile component than in co-current flow. If it is desired to obtain a condensate which is greatly enriched in the less volatile components, a counter-current flow system is therefore preferable; this effect was first pointed out by Claude [3] more than 50 years ago, and it has been recently shown for the first time [4] that it markedly influences the mass transfer resistance, particularly in the vapor phase. In general it is true that, as a result of such additional mass transfer resistances, the mass flux to the phase interface at a given value of the temperature driving force $T_\infty - T_w$ is smaller than in the condensation of a pure vapor, and hence, the mass flux density which is transferred is also smaller.

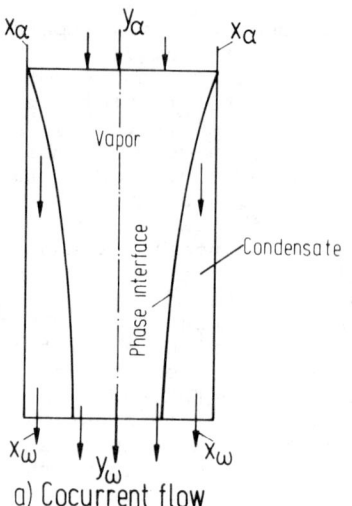 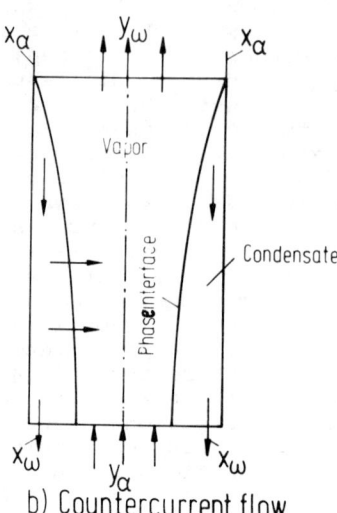

Fig. 2 Concentrations at the phase interface in cocurrent and countercurrent flow. In case (b) the downward-flowing falling film is richer in the more volatile components than in case (a). In case (a) the fraction of the more volatile components decreases downwards and molecules of the more volatile component pass into the vapor.

These facts are shown by Figure 3a which is taken from a paper by Lucas [5], in which the reduced heat flux density $\dot{q}/\dot{q}_o$ is plotted against the temperature $T_\infty$, which in this case should be equal to the saturation temperature. The curves which are plotted are valid for condensation of a methanol/water vapor with negligible effect of the gravity forces (i.e., on a horizontal plate, or when the vapor velocity is large, so that the Froude number $gx/u_\infty^2$ is very small).

As is known, the heat flux density $\dot{q}$ which is actually transferred is smaller than the heat flux density $\dot{q}_o$, which would be transferred if there were no transport resistance in the vapor, and hence the temperature $T_\infty$ prevailed at the phase interface. The dashed curves according to Taitel and Tamir [6] which are shown for comparison are valid for condensation at large Froude numbers and, therefore, for example, at a vertical wall with free flow of the vapor. Figure 3 shows clearly the effect of the temperature driving force $T_\infty - T_W$. The heat flux density $\dot{q}$ transferred increases with the temperature driving force. This can be shown clearly by means of a boiling diagram according to

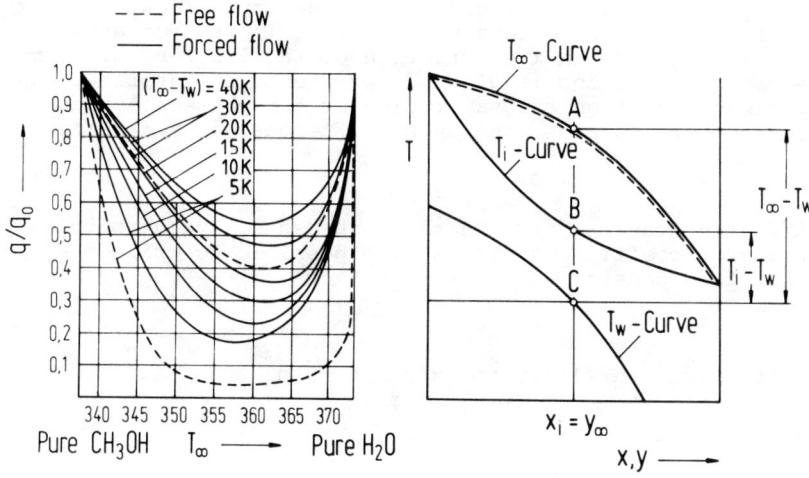

Fig. 3  Reduction of heat flux in the condensation of methanol-water.

Tamir et al. [7] if sufficiently large temperature differences $T_\infty - T_W$ are assumed, so that the vapor is completely condensed from point A onwards in Figure 3b, leaving only liquid, which is characterized by point B in Figure 3b. The wall temperature is characterized by point C. $\overline{BC}/\overline{AC} = (T_i - T_W)/(T_\infty - T_W)$ is a monotonically increasing function of $\dot{q}/\dot{q}_0$. If $T_\infty - T_W$ is now held constant and the temperature $T_\infty$ is increased, in Figure 3a one will therefore follow the path of a curve with $T_\infty - T_W =$ const in the direction of increasing temperature $T_\infty$, then, in Figure 3b the distance $\overline{BC}/\overline{AC}$ can again be measured off if a line parallel to the dew-point line is drawn through the point C. The distance $AC = T_\infty - T_W$ thereby remains unchanged, while BC first becomes smaller and then larger again. $\dot{q}/\dot{q}_0$ varies also in the same ratio. On the other hand, if $T_\infty$ is held constant and $T_\infty - T_W$ is reduced, so that the wall temperature is increased, then, in Figure 3b the $T_W$-curve must be shifted upwards. A smaller value of $\dot{q}/\dot{q}_0$ is therefore obtained.

The preceding explanation is only valid when it is assumed that the temperature at the phase interface is given by the point B in Figure 3b. However, actually this temperature is only attained if the condensation rate is infinitely large (local total condensation). If this is not the case, a higher temperature $T_i$ results at the phase interface, as can be shown starting from the material balance for the lower-boiling, more volatile component of a binary mixture. Hence,

$$\dot{m}w_i' = \dot{m}w_i'' + j \qquad (1)$$

where $\dot{m} = \dot{m}_V$ is the mass flow density of the vapor flowing toward the phase interface and condensing there. It is removed as condensate $\dot{m}_L = \dot{m}_V$, $w_i'$ is the mass fraction of the lower-boiling, more volatile component on the liquid side of the phase interface i, while $w_i''$ denotes that on the vapor side of the phase interface. In the vapor there is additionally a diffusional flux density

$$j = D(\partial w/\partial z_2)_i \tag{2}$$

while diffusion plays no part in the condensate. By introducing a mass transfer coefficient, it is also possible to write the diffusional flux density as

$$j = -\beta(w_\infty'' - w_i'') . \tag{3}$$

As a result, Equation (1) is transformed into

$$\dot{m} = -\rho\beta \frac{w_i'' - w_\infty''}{w_i'' - w_i'} . \tag{4}$$

The mass flux density is negative since it is directed in the opposite direction to the $z_2$-axis as shown in Figure 1. Only two limiting cases can occur in Equation (4):

a) In the case of local total condensation all the vapor flows in the boundary layer toward the phase interface. In this case $\dot{m} \to \infty$ by definition. The material balance for a vapor stream of the more volatile component directed perpendicularly to the phase interface yields $\dot{m}(w_\infty'' - w_i'') = -\rho\beta(w_\infty'' - w_i'')$, i.e., we have $\dot{m} = -\rho\beta$ and $w_i' = w_\infty''$. In the corresponding case for heat transfer, $\dot{m}c_p = -\alpha$. (For a control volume extending up to the interface and including the boundary layer, we have for the case in which the more volatile component flows only perpendicularly to the interface: $\dot{m}_\infty w_\infty'' - \dot{m}w_i'' = -\rho\beta(w_\infty'' - w_i'')$, with $\dot{m}_\infty = \dot{m}$, from which the equation above follows.)

b) If the locally resulting mass flux density of the condensate is vanishingly small, $\dot{m} \to 0$, it follows from Equation (4) that $w_i'' = w_\infty''$.

At rates of condensation $0 < \dot{m} < \infty$, such as actually occur in practice, the concentrations and temperatures according to Figure 1 fall between the two extreme cases for $\dot{m} \to \infty$ and $\dot{m} \to 0$. For calculating the temperature of the phase interface, the energy balance at the phase interface is required as a further balance equation:

$$\dot{q}_L = \dot{q}_D - \dot{m}\Delta h_v \tag{5}$$

where $\dot{q}_L$ is the heat flux density conveyed by conduction in the condensate layer and $\dot{q}_D$ is that conveyed by conduction in the vapor, and $\Delta h_v$ is the enthalpy of vaporization of the mixture. The minus sign in Equation (5) takes into account that the mass flux density is negative according to the sign convention being used. Then

$$\dot{q}_L = \alpha_L(T_i - T_w) = \frac{\alpha_L}{(\alpha_L/k')+1}(T_i - T_k) \tag{6}$$

where $\alpha_L$ is the heat transfer coefficient in the condensate and $k'$ is the partial heat transfer coefficient between the wall and the cooling medium, the temperature of which is $T_K$. In the case of constant wall temperature $T_W$, $k' \to \infty$ and $T_K = T_W$. The following relationship is valid in addition:

$$\dot{q}_D = \alpha_D(T_\infty - T_i) . \qquad (7)$$

Substitution of Equations (4),(6) and (7) into Equation (5) gives for the energy balance

$$\frac{\alpha_L}{\alpha_L/k'}(T_i - T_K) = \alpha_D(T_\infty - T_i) + \rho\beta \frac{w_i'' - w_\infty''}{w_i'' - w_i'} \Delta h_v . \qquad (8)$$

Since the concentrations $w_i'$, $w_i''$ depend only on $T_i$ for binary mixtures with the assumption of equilibrium at the phase interface, $T_i$ can be determined from this relationship, assuming that the interdependence between $\alpha_L$, $\alpha_D$, $\beta$ and $T_i$ is known. Since at least the natural flow occurs in the vapor, the term $\alpha_D(T_\infty - T_i)$ in Equation (8) can usually not be neglected. In vapor-gas mixtures also, ignoring the heat flux $\alpha_D(T_\infty - T_i)$ leads to low heat flux densities at the phase interface, as shown by Akers et al. [8] and by Sparrow and Eckert [9], as a result of natural convection. The mass transfer coefficients $\beta$ can be calculated by use of the analogy between heat and mass transfer, from which it is found that

$$\rho\beta = (\alpha_D/c_{po})(Le)^{2/3} . \qquad (9)$$

The mass transfer coefficients calculated in this way still need to be multiplied by a correction factor taking into account the mass flux directed toward the phase interface by convection [10-13]. The actual heat and mass transfer coefficients must also be larger than those obtained from the known correlations as a result of the wavy nature of the film surface. The roughness of the liquid film can be taken into account approximately by multiplying the heat and mass transfer coefficients for a smooth pipe by the friction factor for the roughness of the phase interface, which has been determined from measurements by several authors [13-18]. In the case of large temperature differences $T_\infty - T_K$, and hence also of large rates of condensation, the heat transfer coefficient $\alpha_D$ must be replaced approximately by a larger value $\alpha_D'$ which takes into account the transport of energy as a result of mass transfer, according to the relationship of Ackermann [13] and Greiner [12]:

$$\alpha_D' = \alpha_D \frac{a_o}{1 - e^{-a_o}} \text{ with } a_o = jc_{pD}/\alpha_D .$$

In general, the temperature $T_i$ can only be calculated iteratively from Equation (8) for a given position or a section of the condenser. The calculation must then be repeated for other sections.

## 3. SOME REMARKS ON HEAT EXCHANGER DESIGN

Processes applied up to now for the design of condensers for mixtures introduce to some extent further drastic simplifications, in order to shorten the calculations. The film theory developed by Ackermann [13] in 1937 offers an approximate method of analysis. The influence of the mass flux normal to the boundary layer is taken into account by a correction factor with which the heat and mass transfer coefficient for pure conduction and diffusion are multiplied. Because the correction factor is calculated on the basis of the assumption that "a one-directional mass flux by diffusion" takes place [13], the film theory fails when a strong convective flow (Reynolds flow) to the wall occurs. Nevertheless because of its computational simplicity the film theory approach or modifications of it is often used [10,14,15]. By far most often used is the procedure by Kern [20], which is also recommended in slightly modified form in the VDI-Wärmeatlas [16]. It is assumed that at the beginning of condensation $\dot{m} = 0$, by means of which the temperature at the phase interface at the beginning of condensation is established. It is further assumed that there is complete mixing of the liquid and gas phases, and constant velocities of the two phases parallel to the phase interface.

Another procedure developed by Roetzel [15,17] also assumes that the liquid and vapor are completely mixed and exist in equilibrium. However, the boiling-point and dew-point lines are replaced in the individual sections of the condenser by straight lines parallel to each other, which simplifies the calculations. Other procedures start from the assumption of local total condensation [18-22]; they are therefore valid for quite large temperature differences. The interfacial temperature can then be taken with sufficient accuracy as the boiling temperature corresponding to the concentration of the vapor.

For calculating the area required, use is made, according to a proposal of Pressburg and Todd [18] of the mean heat transfer coefficient calculated from the mole fractions and the heat transfer coefficients of the pure components:

$$\alpha_L = \alpha = x_1 \alpha_1 + (1-x_1)\alpha_2 . \qquad (10)$$

Of course, it is still quite unclear when local total condensation can be assumed in full-scale equipment. According to a study by Lucas [5] on the laminar film condensation of binary mixtures on flat surfaces, the error in the heat flux density which is made by assuming local total condensation depends on the properties of the mixture, and hence not only on the temperature difference $T_\infty - T_W$, but also on the absolute values of $T_\infty$ or $T_W$. For example, if an error of 5 % in heat flux density is permitted in the condensation of methanol/water vapors, then, at high absolute values of the temperature $T_\infty$ ($T_\infty > T_{S\,methanol}$), a temperature difference $T_\infty - T_W$ of up to 30 K is required, while at small absolute values of the temperature $T_\infty$ ($T_\infty \approx T_{S\,methanol}$) a temperature difference of only 7 K is sufficient in order that the assumption of local total condensation is satisfied. Numerous approximate theories have been developed for the calculation of partial condensation in falling films. A theory given by Hausen [23] is well supported by experiments, and has recently been extended by Begemann [4]. For further details reference should be made to the literature.

4. AN APPROACH FOR CONDENSATION ON A FLAT PLATE

An approach, which yields more accurate results for condensation on a flat plate is based on the following model. We consider laminar film condensation of a binary vapor on a flat plate which has an angle of inclination $\varphi$ to the direction of gravity (Figure 4).

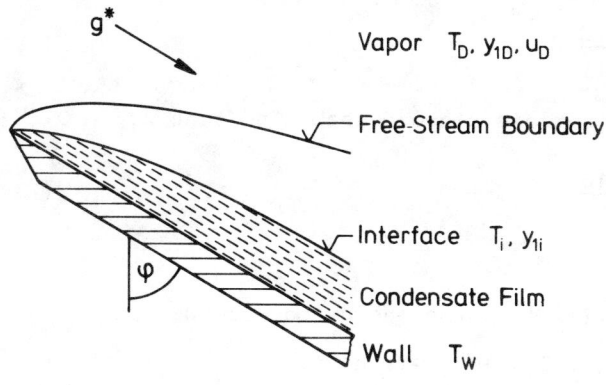

Fig. 4 Physical Model.

The vapor is assumed to be saturated with a free-stream equilibrium composition as characterized by the mass species-concentration of the low boiling component $y_{1D}$. Its free-stream temperature is $T_D$ and its free-stream velocity is $u_D$. Changes of these free-stream values caused by pressure drop or decrease of the vapor mass flux along the plate are not considered. The condensate forms a homogeneous film of liquid at the surface of the plate which is at temperature $T_w$. At the interface between vapor and condensate the two phases have equilibrium compositions $y_{1i}$ and $x_{1i}$, as fixed by the interface temperature $T_i$.

The common characteristic of all condensation processes is the rapid decrease in specific volume which accompanies the phase change. This creates a "suction effect", which results in the transport of vapor molecules to the interface. The strength of the effect depends on the temperature difference $\Delta T$ between the vapor free stream and the surface of the plate. This "suction force" is inhibited by a mass transfer resistance which occurs whenever the vapor consists of components with different boiling points. Mainly because of these effects, the interface temperature $T_i$ is lower than the vapor temperature $T_D$. The temperature difference $T_i-T_w$ must be known to calculate the heat transfer coefficients. The wall temperature $T_w$ often is known or can be calculated from the heat transfer rate to the cooling liquid, so this becomes a matter of determining $T_i$. The approximative calculation of temperature $T_i$ starts with mass and energy balances at the interface. All fluxes from the vapor to the interface are taken to be positive. Thus the local mass flux due to condensation is given by

$$\dot{m} = \rho \beta_k \cdot Y \quad . \tag{11}$$

In Eq.(11), $\rho\beta_k$ is the conductivity of the interface, where the index k indicates that the mass transfer coefficient has been corrected for the suction effect. Factor Y is a concentration-difference ratio and may be considered as the driving force of the process. For the general case in which both components are condensable, it is defined by

$$Y = \frac{y_{1i} - y_{1D}}{y_{1i} - x_{1i}} \quad . \tag{12}$$

If one of the components is noncondensable so that $x_{1i} = 0$, this reduces to

$$Y = \frac{y_{1i} - y_{1D}}{y_{1i}} \tag{13}$$

for that component.

Introducing Eq.(1) into the energy balance we get

$$\alpha_L (T_i - T_w) = \alpha_k (T_D - T_i) + \rho\beta_k \cdot Y \cdot \Delta h_v \quad . \tag{14}$$

The vapor heat transfer coefficient $\alpha_k$ must also be corrected for the energy transport by the condensing mass flux. The heat transfer coefficient in the condensate film $\alpha_L$ is defined in the usual manner

$$\alpha_L = \frac{\lambda_L}{\delta} \quad . \tag{15}$$

With this the energy balance becomes

$$\frac{\lambda_L}{\delta} (T_i - T_w) = \alpha_k (T_D - T_i) + \rho\beta_k \cdot Y \cdot \Delta h_v \quad . \tag{16}$$

The interface temperature $T_i$ is obtained by an iterative solution of Eq.(16). An iterative procedure is necessary because the film thickness $\delta$, the heat transfer coefficient between vapor and condensate $\alpha_k$, and the corresponding mass transfer coefficient $\beta_k$, are all complicated functions of $T_i$.

The work of Lucas [5] offers some valuable guidance in the approach to the solution of Eq.(16). It has shown that condensation heat transfer is influenced mainly by the kind of the vapor flow along the interface. It can be either free convection or forced convection, or combined free and forced convection. This is characterized by the Froude number

$$\xi = \frac{g^* \cdot x}{u_D^2} \quad , \tag{17}$$

which ranges between two extremes:
- the limiting case of forced convection, when $\xi \to 0$ and the inertia forces predominate and gravity forces may be neglected;

- the limiting case of free convection, when $\xi \to \infty$ and gravity forces are predominant.

Both cases $\xi \to \infty$ and $\xi \to 0$ were studied in detail by Stephan and Lasecke [24].

The Froude number has a finite value when the condensation takes place under the simultaneous influence of both body and inertia forces. The calculation of this case is reduced to a reasonable amount of work when further simplifications are made. Because of the great density difference between liquid and gas, the thickness and the motion of the condensate film are mainly determined by gravity. The interface shear stress has only little effect. Thus the Nusselt model is a good approximation for determining the film thickness $\delta$ in Eq.(16) when the acceleration of gravity g is replaced by its component in the direction of flow

$$g^* = g \cdot \cos \varphi \; . \tag{18}$$

The film thickness according to Nusselt is then given by

$$\delta = \left( \frac{4 \cdot \lambda_L \cdot \nu_L \cdot (T_i - T_w) \cdot x}{g^* \cdot \rho_L \cdot \Delta h_v} \right)^{0.25} . \tag{19}$$

Introduction into the dimensionless condensate film thickness

$$\frac{1}{\eta_\delta} = \frac{1}{\delta} \sqrt{\frac{\nu_L \cdot x}{u_D}} \tag{20}$$

yields the dimensionless form

$$\frac{1}{\eta_\delta} = \left( \frac{\xi \cdot \rho_L \cdot \Delta h_v \cdot \nu_L}{4 \cdot \lambda_L \cdot (T_i - T_w)} \right)^{0.25} \tag{21}$$

where the Froude number $\xi$ is given by Eq.(17).

To arrive at convenient expressions for the heat and mass transfer coefficients $\alpha_k$ and $\beta_k$ in Eq.(16) it proved effective to consider the transport processes in the vapor boundary layer as consisting in each case of two parts: that due to conduction as given by $\alpha_o$, or diffusion as given by $\beta_o$, and that due to the corresponding suction effect. Then

$$\alpha_k = \alpha_o + k \cdot \dot{m}_{therm} \cdot c_p \tag{22}$$

$$\rho \beta_k = \rho \beta_o + k \cdot \dot{m}_{therm} \tag{23}$$

with

$$\dot{m}_{therm} = \rho \cdot v_i \; . \tag{24}$$

Factor k takes into account that the suction-effect flow is inhibited by the presence of the vapor boundary layer; when such a layer exists ($\alpha_0, \beta_0 \neq 0$), k<1, and in the absence of such layer k = 1. The value of k for forced convection along a horizontal plate with suction can be determined from the results of Hartnett and Eckert [25]. These are reproduced well by Eqs. (22) and (23), if one takes k = 0.9 for the mixtures of water-methanol and water-air that are treated as examples.

$\alpha_0$ and $\beta_0$ are determined from the well known Pohlhausen solution for laminar flow

$$Nu_o = 0.332 \cdot Re_x^{0.33} \tag{25}$$

$$Sh_o = 0.332 \cdot Re_x^{0.5} \cdot Sc^{0.33} . \tag{26}$$

Written in dimensionless form Eqs.(22) to (26) become

$$\frac{Nu_k}{Re_x^{0.5}} = 0.332 \cdot Pr^{0.33} + k \cdot f_i \cdot Pr \tag{27}$$

$$\frac{Sh_k}{Re_x^{0.5}} = 0.332 \cdot Sc^{0.33} + k \cdot f_i \cdot Sc \tag{28}$$

with the suction parameter

$$f_i = \frac{v_i}{u_D} Re_x^{0.5} . \tag{29}$$

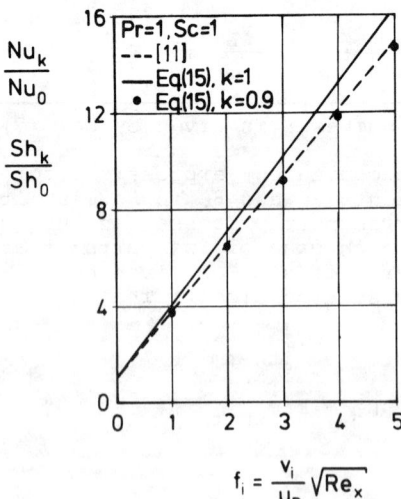

Fig. 5  Improvement of forced convection heat and mass transfer by wall suction.

It can be seen from Figure 5 that Eqs.(27) and (28) provide a good approximation for the exact solution, justifying the simple but physically plausible approach of Eqs.(22) and (23). In order to use Eqs.(27) to (29) and determine the interface temperature $T_i$ in eq.(16) some further information on the suction velocity $v_i$ or the suction parameter is required. Both depend on the driving temperature difference $T_i-T_w$. In order to obtain a relation between $v_i$ and the driving temperature difference Sparrow et al. [26] started from the differential eqs. of liquid and vapor phase and introduced a few neglections: Heat conduction from the vapor to liquid film was assumed to be negligible in comparison to the enthalpy of condensation, furthermore for the vapor boundary layer the interface velocity was assumed to be negligiliby small. Under these reasonable assumptions the suction parameter $f_i$ turned out to be given by

$$f_i = \frac{R \cdot E}{\eta_\delta} \qquad (30)$$

with

$$R \cdot E = \sqrt{\frac{\rho_L \mu_L}{\rho \mu}} \cdot \left| \frac{\lambda_L (T_i - T_w)}{\mu_L \cdot \Delta h_v} \right| . \qquad (31)$$

Inserting now the film thickness $\delta$ from Eq.(19) and the heat and mass transfer coefficients Eqs.(27) and (28) under consideration of Eqs.(29) to (31) into the energy balance Eq.(16) the unknown interface temperature $T_i$ in combined body force and forced-convection condensation may be determined.

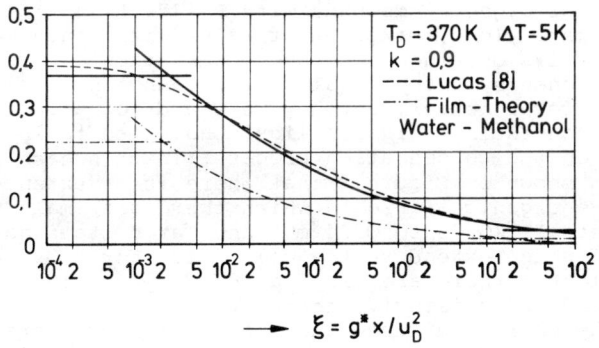

Fig. 6   Reduced temperature difference in combined free- and forced-convection condensation.

Results are shown in Figures 6 and 7 in which the ratio of the temperature difference $(T_i-T_w)$ is given as a function of the Froude number $\xi$. Between the horizontal asymptotes for both forced- and free-convection flow, the curves represent the general case considered in this section. The agreement with the complete solution by Lucas [5] is seen to be acceptable.

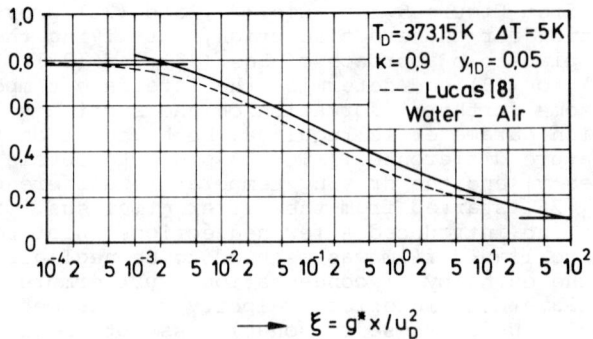

Fig. 7  Reduced temperature difference in combined free- and forced-convection condensation.

The improvement over the film-theory results, also shown in Figure 6, comes from taking into account the suction effect.

The simplicity of the mathematical requirements for using the approximate procedure is a further advantage.

## 5. IMMISCIBLE LIQUID PHASES

In industrial practice it is often necessary to condense mixtures which are not miscible in the liquid phase. Examples are mixtures of organic cmpounds with R 112, R 113, perchloroethylene or paraxylene, mixtures of water with benzene, phenol, etc. The results obtained for the condensation of vapors with miscible components in the liquid phase cannot be applied to this case. However, experiments by Bernhardt et al. [27] using water and a nonmiscible organic liquid have led to interesting information on the condensation process. In this case it is observed, as shown in Figure 8 that there is an extended liquid film I of the organic liquid in which there are small moving water droplets II (0.03 to 0.04 mm) and large stationary water droplets III which extend to the wall (diameter 0.05 to 4 mm), and on the latter there are tiny moving droplets IV of the organic liquid. Surprisingly, therefore, the organic liquid does not cover the entire surface. The small water droplets move with velocities between 2,5 and 127 mm/sec, while the tiny droplets of organic liquid of about 0.02 mm diameter dance back and forth with velocities of 25 to 50 mm/sec.

Heat transfer during the condensation of immiscible liquids has been calculated from the experiments. However, most of the equations which have been produced are suitable only for correlating the measured data on which they are based, and do not agree with other measured data. This is not surprising, since obviously a very complicated phenomenon is being dealt with. However, all the measured values known up to know can be

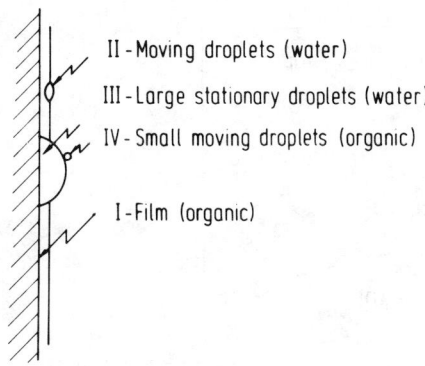

Fig. 8 Condensation of immiscible liquids.

described adequately according to a proposal by Bernhardt et al. [27] if it is assumed that each liquid phase occupies a fraction of the surface which is proportional to its volumetric fraction. Although this assumption is not in fact the case, as Bernhardt et al. themselves pointed out, it nevertheless leads to useful results. We can therefore recommend calculating the heat transfer coefficients by the equation

$$\alpha = \alpha_1 v_1 + \alpha_2 v_2 \qquad (32)$$

where $v_1$ and $v_2$ are the volume fractions of the pure components in the liquid phase. The heat transfer coefficients $\alpha_1$ and $\alpha_2$ should be calculated from Nusselt's formula at the pressure p and the temperature T of the mixture. The difference between the saturation and wall temperatures is substituted in this equation as the temperature difference, where the saturation temperature is taken to be equal to the eutetic temperature of the mixture at the prevailing pressure. This follows directly from Gibbs' phase rule, since a two-component mixture consisting of two components and three phases (two liquid and one gaseous phase) has only one definite temperature (the so-called eutetic temperature) at which it can exist at equilibrium at a given pressure. As shown in Figure 9, Equation (32) reproduces the measured values of Bernhardt et al. very well.

Completely different results are obtained if, in contrast to the experiments noted above, the liquid phase is not obtained first by condensation, but the vapor is condensed on a liquid film which is already present, where again the condensate formed is not miscible with the liquid film. A more careful investigation shows that in these phenomena the thermal resistance $1/\alpha_i$ between the condensate surface and the vapor may have a decisive effect. Thus, Tamir et al. [28,29] have shown that the thermal resistance at the phase interface between a water film and n-pentane, methyl-chloride, R 113 or 1-1-dichlorethane condensing on it can be of

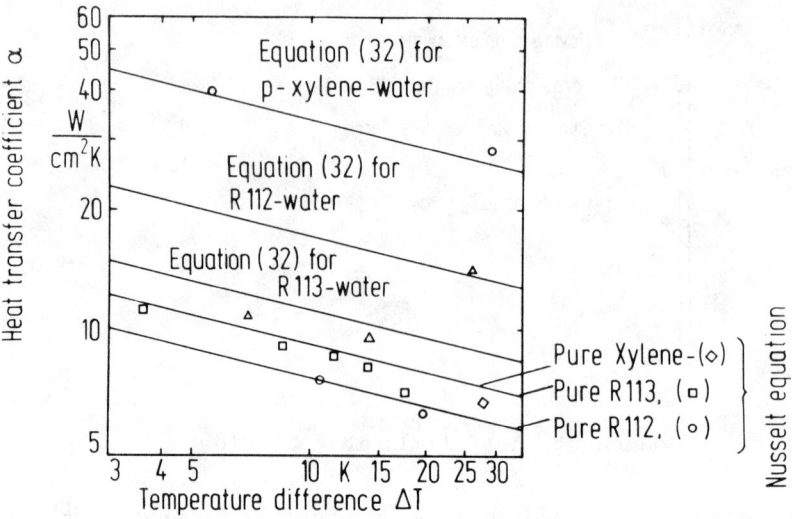

Fig. 9  Correlation of measured values with Equation (32). After Bernhardt et al.[27].

the same size as the thermal resistance of the water film itself. As an empirical correlation for the mean heat transfer coefficient at the phase interface, Tamir and Rachmilev recommend the following relationship

$$\bar{\alpha}_i = 0.225 \cdot 10^{-10} \exp\left(\frac{6375}{T_\infty}\right) \qquad (33)$$

where $T_\infty$ is the saturation temperature of the vapor, and $\bar{\alpha}_i$ is given in cal/(cm$^2$)(sec)(K). The total thermal resistance is then calculated from

$$\frac{1}{\bar{\alpha}} = \frac{1}{\bar{\alpha}_i} + \frac{1}{\bar{\alpha}_{L\infty}} \qquad (34)$$

where $\bar{\alpha}_{L\infty}$ is the mean heat transfer coefficient of the condensate film, and is calculated by means of the Nusselt water-film theory. The additional resistance at the interface can be explained by the fact that condensation is only possible through nucleation. The rate of nucleus formation $\dot{n}$ is given by $\dot{n} = \nu \exp(-\Delta G/kT_\infty)$, where $\nu$ is a frequency factor, $\Delta G$ is the free enthalpy, and k is the Boltzmann constant. The derivation of Equation (33) can be obtained approximately as indicated by Sykes and Marchello [30]. However, no proof is given in [30].

# NOMENCLATURE

| | |
|---|---|
| $c_p$ | specific heat capacity |
| $D_{12}$ | binary diffusion coefficient |
| $f_i$ | suction parameter at the interface |
| $g$ | gravitational acceleration |
| $g^*$ | component of gravitational acceleration in flow direction |
| $\Delta h_v$ | latent heat of vaporization |
| $k$ | correction factor |
| $M$ | mass per mol |
| $\dot{q}$ | heat flux |
| $\dot{q}_o$ | heat flux with vanishing mass transfer resistance |
| $T$ | temperature |
| $\Delta T = T_D - T_w$ | temperature difference |
| $u$ | velocity in x-direction |
| $v_i$ | velocity in y-direction at the interface |
| $x$ | coordinate along the plate |
| $x_1$ | mass species concentration of low boiling component in the liquid |
| $y_1$ | mass species concentration of low boilimg component in the vapor |
| $Y$ | reduced concentration difference |

Greek symbols

| | |
|---|---|
| $\alpha$ | heat transfer coefficient |
| $\beta$ | mass transfer coefficient |
| $\delta$ | thickness of condensate film |
| $\delta'$ | modified thickness of condensate film |
| $\eta_\delta$ | reduced thickness of condensate film |
| $\lambda$ | thermal conductivity |
| $\mu$ | dynamic viscosity |
| $\nu$ | kinematic viscosity |
| $\xi$ | dimensionless coordinate, see Eq.(7) |
| $\rho$ | density |
| $\varphi$ | inclination angle |

Dimensionless numbers

| | |
|---|---|
| $Nu = \alpha \cdot x / \lambda$ | Nusselt number |
| $Pr = \nu \cdot \rho \cdot c_p / \lambda$ | Prandtl number |
| $Re_x = u_D \cdot x / \nu$ | Reynolds number of vapor bulk flow |
| $Sc = \nu / D_{12}$ | Schmidt number |
| $Sh = \beta \cdot x / D_{12}$ | Sherwood number |

Subscripts

| | |
|---|---|
| D | vapor bulk |
| i | interface |
| k | corrected |
| L | liquid |
| therm | thermal suction |
| w | wall |
| o | vanishing suction |
| 1 | component 1 |
| 2 | component 2 |

The vapor properties are not subscripted.

## REFERENCES

1. Denny, V.E. and Mills, A.F. 1969. Int.J.Heat Mass Transf.12. pp. 965-979.

2. Denny, V.E. and Jusionis, V.J. 1972. Int.J.Heat Mass Transf. 15. pp. 2143-2153.

3. Claude, G. 1926. "Air Liquide Oxygène, Azote, Gas Rares" (Liquid Air Oxygen, Nitrogen, Rare Gases), 2nd ed. Dunod, Paris.

4. Begemann, E. 1972. VDI-Forsch.-Heft 553. VDI-Verlag Düsseldorf.

5. Lucas, K. 1974. Habil.-Schrift, Ruhr-Universität Bochum.

6. Taitel, Y. and Tamir, A. 1974. Int.J. Multiphase Flow 1. pp. 697-714.

7. Tamir, A., Taitel, Y. and Schlünder, U.E. 1974. Int.J.Heat Mass Transf. 17. pp. 1253-1260.

8. Akers, W.W., Davis, S.H., and Crawford, J.E. 1960. J.E.Chem. Eng.Progr., Symp.Ser. 56, No.30. pp. 139-144.

9. Sparrow, E.M. and Eckert, E.R.G. 1961. AIChE Journal 7, No.3. pp. 473-477.

10. Colburn, A.P. and Drew, T.B. 1937. Trans.Am.Inst.Chem.Eng.33. pp. 197-212.

11. Eckert, E. and Lieblein, V. 1949. Forsch.Ing.-Wes. 16. pp. 33-42.

12. Greiner, M. 1977. Dissertation, TU München (Munich Tech.Univ.)

13. Ackermann, G. 1937. VDI-Forsch.-Heft 382. Berlin. pp. 1-16.

14. Röhm, H.-J. 1978. Wärme- und Stoffübertragung 11. pp. 63-72.

15. Roetzel, W. 1975. Wärme- und Stoffübertragung 8. pp. 211-218.

16. "VDI-Wärmeatlas". 1974. 2nd ed., Section 7b, VDI-Verlag Düsseldorf.

17. Roetzel, W. 1974. Wärme- und Stoffübertragung 7. pp. 60-64.

18. Pressburg, B.S. and Todd, J.B. 1975. AIChe Journal 3. pp. 348-352.

19. Wallace, J.L. and Davison, A.W. 1938. Ind.Eng.Chem.30. pp. 948-953.

20. Haselden, G.G. and Prosad, S. 1949. Trans.Inst.Chem.Eng. 27. p. 195.

21. Mirkovich, V.V. and Missen, R.W. 1963. Can.J.Chem.Eng. 41. p. 73.

22. Haselden, G.G. and Platt, W.A. 1960. Brit.Chem.Eng. 5. p. 37.

23. Hausen, H. 1957. "Handbuch der Kältetechnik"(Refrigeration Handbook), Vol.8. pp 154-160. Springer-Verlag Berlin-Göttingen-Heidelberg; also Hausen, H. and Schlatterer, R. 1949. Chem.Ing.Techn. 21. pp. 23-24. 453-460.

24. Stephan, K. and Laesecke, A. 1980. Wärme- und Stoffübertragung 13. pp. 115-123.

25. Hartnett, J.P. and Eckert, E.R.G. 1957. Trans.ASME 79. pp. 247-254.

26. Sparrow, E.M., Minkowycz, W.J., and Saddy, M. 1967. Int.J. Heat Mass Transfer 10. pp. 1829-1845.

27. Bernhardt, S.H., Sheridan, J.J., and Westwater, J.W. 1971. AIChE. Symp.Ser. 118. pp. 21-37.

28. Tamir, A., Taitel, Y., and Schlünder, E.U. 1974. Int.J.Heat Mass Transfer 17. pp. 1253-1260.

29. Tamir, A. and Rachmilev, I. 1974. Int.J.Heat Mass Transfer 17. pp. 1241-1251.

30. Sykes, J.A. and Marchello, J.M. 1970. Ind.Eng.Chem., Proc. Des.Dev. 9. pp. 63-71.

# Film Evaporation and Condensation in Desalination

**SAMUEL SIDEMAN**
Technion-Israel Institute of Technology
Haifa, Israel

ABSTRACT

The advantages of film evaporator/condensers are detailed and evaporation and condensation of laminar and turbulent films, as manifested in various water desalination schemes, are reviewed with special emphasis on the more recent and promising horizontal-evaporator-condenser units. The importance of the interacting effects of the phase change phenomena occurring across the metal wall is noted and the importance of the thermal conductivity of the wall for operation with intermittent nucleate boiling and enhanced heat transfer areas are discussed. Recent work on the tube shapes yielding highest heat transfer rates is presented.

NOMENCLATURE

$C$    specific heat of liquid [J/Kg°K]

$g$    gravitation constant [m/s$^2$]

$h$    heat transfer coefficient [w/m$^2$°K]

$h^*$    dimensionless heat transfer coefficient [$=h(\bar{\nu}^2/k^3 g)^{1/3}$]

$k$    conductivity of liquid [w/m°K]

$L$    length of plane

$q$    heat flux [J/s]

$Re$    film Reynolds Number ($4\Gamma/\mu$)

$T$    local temp. [°K]

$T_s$    saturation temp. of steam at the condensate free surface

$U_o$    overall heat transfer coefficient [w/m$^2$°K]

$u^+$    dimensionless velocity

$y$    normal coordinate [m]

$y^+$    dimensionless film thickness

$y_w$    wall thickness [m]

$z$    axial coordinate of the tube [m]

Greek letters

$\alpha$    Eddy thermal diffusivity

$\Gamma$    liquid mass flow rates per unit width [Kg/ms]

$\delta^+$    dimensionless film thickness

$\delta$    film thickness

$\bar{\delta}$    time averaged film thickness

$\lambda$    heat of evaporation and condensation [J/Kg]

$\tau$    shear stress [N/m$^2$]

$\rho$    density [Kg/m$^3$]

$\nu$    kinematic viscosity, ($\equiv \mu/\rho$) [m/s$^2$]

Subscripts

o    outside

f    fluid

i    inside

w    wall

s    saturated, steam

v    evaporation side, vapor

## 1. INTRODUCTION

This section deals with film evaporation and film condensation associated with water desalination plants. Of particular interest are those schemes whereby the evaporation which takes place on one side of the tube is thermally sustained by the condensation of a saturated vapor on the other surface of the tube. Here, however, we shall mainly treat each phenomenon separately and elaborate on the simultaneous phenomena in the next chapter. In a limited available space, one can only hope to attune the learned reader to some points of interest and introduce the novice to the basic characteristics associated with this mode of operation.

Horizontal-tube evaporator-condensers, where condensation takes place inside the tube bundle while the cooling evaporating film cascades outside the tubes, dripping downwards from tube to tube, were recently shown to have economical advantage over the classical vertical arrangements. This relatively new mode of evaporation is of immediate relevance and will therefore be elaborated upon in more detail.

Also, of particular economic importance is the vapor recompression scheme which reduces the operating costs as compared with the more conventional single or multiple effect evaporators. Both the vertical and horizontal tube evaporator/condenser arrangements compete for priorities within the recompressing scheme. The schematics of the vertical tube falling film evaporator/condenser are shown in Figs.1 and 2. A schematic of the single stage horizontal evaporator/condenser vapor compression unit is shown in Fig.3. It is to be noted that though the multiple effect system, say the MSF, is the most frequently used plant today, it does not approach the equivalent of the mechanical recompression unit which approximates some 30 effects.

Some indication as to the relative advantages of the horizontal tube evaporator (THA) and the MSF desalination scheme is given in Table I. Clearly, the combination of the horizontal tube evaporator with the vapor recompression mode of operation is the most promising scheme for water desalination.

Some operating details may be instructive. As compared with the violently turbulent flashing MSF, the horizontal tube multiply effect evaporator (THA) is a quiet process. This minimizes corrosion (protective films are not eroded) and brief acid treatments do not affect the tubes. For example, the velocities of the flashing brine in the MSF are up to 10 m/sec whereas the boiling brine flows at about 0.7 m/sec over the tubes. The evaporation of the films outside the horizontal tubes lead to heat transfer coefficients twice or thrice larger than those realized in the MSF, while scale formation on these tubes is practically non-existent. These advantages lead to steam economy (or performance ratios) of 6.5 to 7.5 Kg/1000 K joule (15-17 lb/1000 BTU) steam as compared with a maximum of about 5.2 (12.5 lb/1000 BTU) in the MSF. (Actual performance ratios are quite lower). Furthermore, because of the nature of their operation, the horizontal units can be easily shut off, or started and brought to full capacity in minutes. Capital construction costs for a 4000 $m^3$/day unit in Aruba were 15% less than for an equivalent MSF plant, and units of up to $13.5 \times 10^3$ $m^3$/day are being offered by Sasakura, Japan. A 4600 $m^3$/day (aluminum) horizontal tube plant is being operated by IDE (Israel Desalination Engineering, Tel Baruch, Israel) for some years now and a 46000 $m^3$/day plant is presently being built in Ashdod, Israel. Other interested manufacturers are Hamon Sobelco, Aqua Chem, Aiton Ltd.

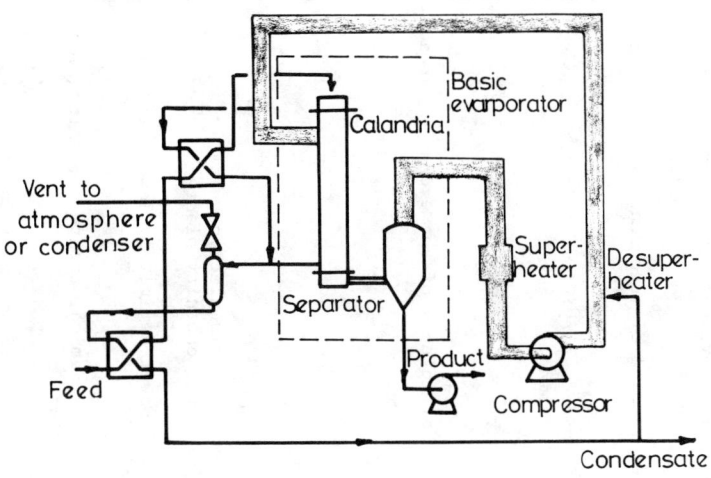

Fig.1. Schematic of a vertical falling film evaporator/condenser recompression unit.

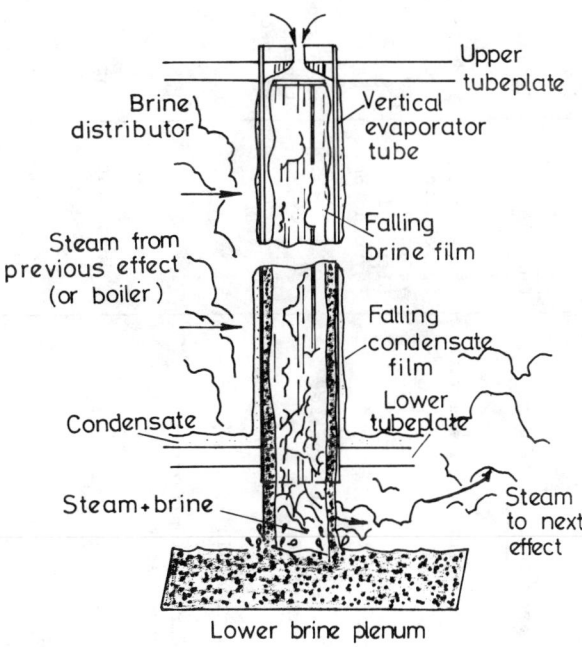

Fig.2. Evaporating/condensing films on a vertical tube

Table I: Comparison between the horizontal tube evaporator (THA) and MSF distillation schemes (1)

| Object | MSF | THA | Remarks |
|---|---|---|---|
| 1. Bundle of tubes | Horizontal | Horizontal | THA allows for a vertical conception as well as a horizontal one |
| 2. Vapor condensation in the tubes | Outside | Inside | THA: better removal of incondensable material |
| 3. Flow of brine in sea water | Thick layer forced convection. Elevated Reynolds No. | In thin films around the tubes. Low Reynolds No. | MSF: antifoam necessary; low efficiency due to incomplete flashing. |
| 4. Relative heat transfer coefficients | 1 | x1.5 smooth tubes x3 special tubes | THA: maximum efficiency of the exchange surface |
| 5. Irreversibilities | High | Low | |
| 6. Circulation flow rate/ produced water flow rate | 7 to 12 | 1.5 | |
| 7. Recycling pump | 1 important | Not | MSF: planning of recirculation pumps is delicate for large capacity plants |
| 8. Pumps interaction | | None in the vertical | THA: Adsorbed power definitely lower as compared to MSF |
| 9. Sea water/product water flow rate ratio | 2 | 1.5 | THA: degassing, chemical treatment, filtration, reduced chlorination |
| 10. Brine to sea water concentration ratio | 1.7 to 2 | 2 to 3 according to cycle | THA: flow rate of exit brine reduced; brine is used more economically |

contd.....

(Table 1: contd.)

| | Object | MSF | THA | Remarks |
|---|---|---|---|---|
| 11. | Influence of deposition tubes on: production; performance ratio | Lowers<br>Lowers | Lowers<br>Little influence | THA: the calories consumption is proportional to the production rate |
| 12. | Primary vapor pressure versus that of sea water to heat | Lower in the final heater of the brine | Higher in the evaporator | THA: it is impossible to pollute the distillate (heating circuit and fresh water production) |
| 13. | Adaptation to variations in loading | Difficult | Automatic | THA: easy coupling to electric power stations |
| 14. | Performance ratio (coefficient) | 10 to 12 | 14 to 18 depending on cycle | THA: important advantage on the site when the density of installation is high |
| 15. | Crowding on the ground | Large | Small in the vertical structure | |
| 16. | The salinity of the distillate | 10-100 ppm | 1-10 ppm | THA: very low salinity |
| 17. | Adaptation to vapor compression cycles | Interesting | Very interesting | THA: allows high energetic yields |

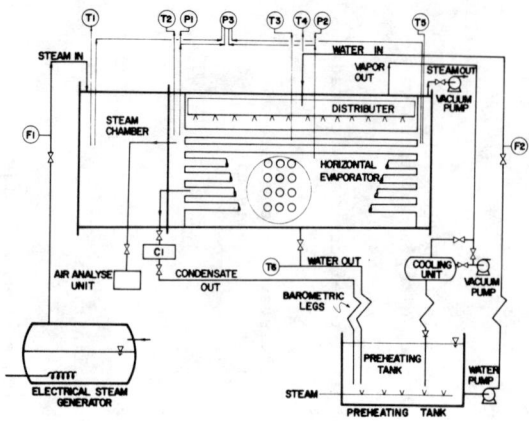

Fig.3. Schematic of horizontal tube film evaporator/
condenser recompression unit.

## 2. SOME BASIC FILM CHARACTERISTICS

### 2.1 The Film Thickness

A simple momentum balance on a film falling down a vertical plane which accounts for the momentum flux across the film and the gravity force acting upon it yields $\delta$, the average film thickness:

$$\delta = \left(\frac{\mu \Gamma}{\rho^2 g}\right)^{1/3} \tag{1}$$

where $\Gamma$ is the mass flow rate per unit width of wall and $\mu$ is the viscosity, $\rho$ is the liquid density and $g$ is the gravitational acceleration.

Film thickness, in dimensionless units $y^+ = y(\tau_w/\rho)^{1/2}/\nu$, where $\tau_w = \rho g \delta$ and $\nu$ is the kinematic viscosity, is given by [2]:

$$\delta^+ = (0.75 \text{ Re})^{1/2} \quad \text{laminar film} \tag{2a}$$

$$\delta^+ = 0.102 \text{ Re}^{0.8} \quad \text{turbulent film} \tag{2b}$$

where $Re = 4\Gamma/\mu$. A graphical presentation of the dimensional and dimensionless film thickness $\delta^+$ is given in Fig.4 [3]. Time average local thickness $\bar{\delta}$ of a water film (at 24°C) falling along a vertical plane were measured for $145 < Re < 4030$ by laser scattering from suspended latex particles. The results are given as [3a]

$$\bar{\delta} = a \text{ Re}^n \; ; \quad a = \begin{Bmatrix} 0.0362 \text{ mm} \\ 0.0247 \text{ "} \\ 0.0068 \text{ "} \end{Bmatrix} \; n = \begin{Bmatrix} 0.325 \\ 0.374 \\ 0.554 \end{Bmatrix} \quad \begin{array}{l} Re < 600 \\ \text{transitional} \\ Re > 2000 \end{array} \tag{3}$$

A review of some other studies on the hydrodynamics of the falling film is given by Seban [3b]. It is noted that a substantial film length - of order of 1 m - is required before equilibrium values of the maximum and minimum film thickness are realized, though the mean film thickness reaches its equilibrium

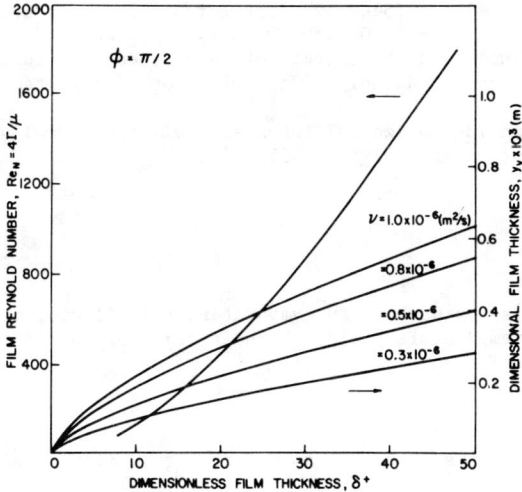

Fig.4. Dimensional and dimensionless film thickness.

value very quickly. The velocity of the waves which produce the variation in film thickness is of order of 1.6 times the mean velocity of the turbulent film of constant thickness.

2.2 Condensing Films

Nusselt's analysis of film condensation on a vertical plate [29] is still the basis for most later developments. Nusselt assumed the temperature distribution in the liquid film to be linear, that the vapor is pure saturated at $T_s$ and that no vapor shear acts at the liquid vapor interface. The resulting equation for the average heat transfer coefficient over a plate of length L is given by:

$$h_{vert} = 0.943 \left( \frac{g\rho(\rho-\rho_v)k^3 \lambda'}{L\mu(T_s-T_w)} \right)^{0.25} \tag{4a}$$

or

$$h^*_{vert} = 1.47 \, Re^{-1/3} \quad ; \quad h^* \equiv h \left( \frac{\nu^2}{k^3 g} \right)^{1/3} \tag{4b}$$

where $\rho_v$ is the density of the vapor and $\lambda'$ is the latent heat of condensation corrected for sensible heat $[=\lambda+(3C/8)(T_s-T_w)]$. q is the heat flux, $T_w$ is the wall temperature; $T_s$ is the steam temperature. A similar analysis yields, for condensation outside a horizontal tube of diameter D.

$$h_{oriz} = 0.728 \left( \frac{g\rho(\rho-\rho_v)k^3 \lambda'}{D\mu(T_s-T_w)} \right)^{1/4} \tag{5a}$$

or

$$h^*_{oriz} = 1.514 \, Re^{-1/3} \tag{5b}$$

Comparison of Equations (4) and (5) shows that for otherwise identical conditions $h_{vert}=h_{oriz}$ when $L=2.78$ D. Obviously $h_{oriz}>h_{vert}$ when $L/D>2.78$. The detailed derivations, including the effects of vapor shear and turbulence, is given in detail by Rohsenow and Choi [4] and will not be repeated here.

For condensation <u>inside</u> horizontal tubes at low ($Re_v<35,000$) vapor velocities, the constant in Eq.(5a) is 0.555 [5] or

$$h^*_{oriz}=1.154\ Re^{-1/3} \tag{6}$$

## 2.3 Evaporating Films

Utilizing Nusselt's approach to an evaporating film flowing down a vertical plane of constant temperature yields the heat transfer coefficient analogous to Eq.(4) [6] i.e.:

$$h^*=1.10\ Re^{-1/3} \tag{7}$$

where, for simplicity, one neglects the amount of liquid evaporated along the path.

Films flowing down a vertical plane are characterized by wave formation at $Re>20$. These are gravitational waves which are accompanied by capillary waves. The enhancement in heat transfer due to the waves at low laminar Reynolds numbers is accounted empirically by [7]:

$$h_{wave}/h=Re^{0.04} \tag{8}$$

Note that the distortion of the film waves is promoted by increasing the heat flux q and, at the extreme, leads to the film breakdown.

Another empirical correction accounting for the increase of the film heat transfer coefficient due to rippling was suggested by Chun and Seban [8]

$$h_{ripples}=0.8\left(\frac{\Gamma}{\mu}\right)^{0.11} \tag{9}$$

Reasonable agreement with evaporating film data was obtained for $Re<350$ [9].

Empirical heat transfer coefficients for evaporation at low heat fluxes ($q<0.7\times10^5$ w/m$^2$) are given by [2]:

$$h^*=1.76\ Re^{1/3} \quad \text{laminar, subcooled liquid film} \tag{10}$$

$$h^*=0.9\ Re^{-0.22} \quad \text{laminar, evaporation} \tag{11}$$

$$h^*=0.006\ Re^{0.4} \quad \text{turbulent (Re>3200)} \tag{12}$$

Eqs.(11),(12) are similar, but some 10% higher, than those suggested by Chun & Seban [8] for evaporation of water downflowing outside a stainless steel tube. Different models of turbulence yield somewhat different predictions. A short discussion of some of these models is given in [3b].

At higher heat fluxes, bubble formation is quite evident, and the effect of the film flow rate on the heat transfer coefficients disappears. Under these conditions of fully developed nucleate boiling [2]:

$$h=1.24\ q^{0.741} \tag{13}$$

which is essentially similar to the values realized in fully developed flow-boiling in a channel.

It is interesting to note that Eqs. (11) and (12) fall within the compilation of vapor <u>condensation</u> data on vertical tubes [7], indicating that $h^*=f(Re)$.

Data up to $Re\simeq 400$ agrees well with Nusselt's equation (modified by Eq.(8)). However, in the wide <u>transition range</u> of 400 Re 4000 the heat transfer rate is practically constant, i.e. $h^*\simeq 0.22$, with the evaporation data of [2] falling some 10% lower. Total condensate flow rate is to be used for the vertical tube as well as for the horizontal tube pack. Of related interest we note the experimental evaporation studies carried out by Markatos [8a] in a (2.44 m long) horizontal wind tunnel. The net increase of the Nusselt numbers due to the waves is about 28%, as compared to data for smooth flow.

## 3. VERTICAL TUBE EVAPORATOR/CONDENSER

The most common water distillation schemes are based on evaporation inside and condensation outside the tube [10-19]. A major feature of the film evaporators is the attainment of the economically desired low temperature difference between the evaporating brine and the heating steam. This entails that the evaporator to be used has to be capable of stable operation. Falling film evaporators, rather than the rising film type, have the advantage that the film is established at the top of the tube by means of a distributor and is not dependent on the evolution of vapor to form and maintain it. The latter type is obviously advantageous with high heat flux evaporators. Though the surface area in this mode of operation is relatively small due to high transfer rates, the area costs are nevertheless appreciably high due to the small temperature driving forces involved.

A scheme for internal condensation with external evaporation has also been suggested [20,21]. A falling film plate evaporator has also been successfully used for smaller duties and special applications, mainly under vacuum, for heat sensitive products. Low vapor velocities and pressure drops have been measured with the APV - Rosco unit in which the heating unit is made of welded plates inside a cylindrical vessel, with condensation taking place inside the plates and a liquor film evaporating on the outside.

Closely related is the study dealing with condensation inside vertical tubes [22], and the numerous studies of condensation on, and evaporation from, inclined [23] and vertical surfaces [24-26].

As a practical comment we note that the vertical tube multiple-effect distillation units presently operating in the Virgin Islands (at about 8500 m$^3$/day) do not show any advantage over the present MSF operation. However, newer designs utilizing fluted tubes are promising, and, based upon their experience with a 500 m$^3$/day unit in Innoshima, Hitachi Zosen is presently advertising the vertical tube multi-effect unit, claiming 80-90% of the MSF operation cost as well as great savings in the capital costs.

## 4. HORIZONTAL TUBES

### 4.1 Condensation Outside Horizontal Tubes

Condensation outside horizontal tubes is a basic element in the multistage flash (MSF) evaporation process [27,28], and as such received great attention. At low temperature driving forces and low vapor velocities, laminar film condensation occurs and the Nusselt assumptions [29] hold up to $Re\approx 400$. Modifications of Nusselt's model include the effects of the sensible heat, high pressures, large temperature driving forces [30] and convection in the flow direction [31]. Other improved analyses were presented by Chen [32] and Sparrow et al [33,34,35]. A compilation of experimental studies and correlations is presented by McAdams [36] and Rohsenow [6]. Some additional related references are given in [23]. Rohsenow et al [37], Danny and Mills [38] and Lee [39] considered the effect of vapor-velocity and turbulence in the condensate film. Forced convection was found to improve the condensation rates and the effect of noncondensables under such conditions is less than in gravity flow [34]. Similar conclusions were noted by Grant [40], who studied the effect of different tube arrangements for venting noncondensing gases on condenser performance, and Tamir and Taitel [41] who studied ways to improve (direct) condensation rate by forced convection and interfacial suction of noncondensable gases. Closely related are the studies of condensation on horizontal [42,43] and inclined tubes [44,45] and flat surfaces [46-48].

The heat transfer coefficient decreases as the number of tubes in a vertical row in the tube bundle increases [36]. However, experimental data with a pack of horizontal tubes shows that h* is practically constant (0.2-0.3) over the transition range of Reynolds Numbers, 400<Re<4000, independent of wavelength of the falling condensate film and the spacing S between the tubes, provided S is larger than the falling drop size. The advantages and limitations of condensation on horizontal tubes in a tube bundle as compared to vertical tubes are discussed by Chaddock [49]. An analysis of condensation in a vertical bank of horizontal tubes was presented by Kern [50] for different tube layouts bounded by a circle.

## 4.2 Condensation Inside Horizontal Tubes

As seen in Fig.3, the horizontal tube evaporator/condenser [51-58] is characterized by condensation inside the tube and evaporation outside. Condensation inside horizontal tubes, which is commonly encountered in refrigeration systems, results in a two-phase flow system [59,60]. Of immediate practical interest are the stratified flow regimes (at $Re_v < 3.5 \times 10^4$) and the annular condensate flow regime at higher vapor flow rates [5]. In the stratified flow regime, the condensation mechanism is predominantly affected by the viscous and body forces [61,62], while the process of condensation in the annular flow regime is controlled by the dynamics of the vapor and the condensate and their interaction [63-65]. As shown experimentally [66-69], the condensation heat transfer coefficient increases with steam flow rate and tube length to diameter ratio. A presentation of the condensation rates at various steam flow rates is given by Rohsenow [6]. The effect of the accumulated condensate at the bottom of a slightly inclined tube on the heat transfer rate is presented by Chato [5] and Chaddock [49], the latter including comparisons between condensation outside a vertical tube and condensation outside and inside horizontal tubes.

## 4.3 Evaporation Outside Horizontal Tubes

Evaporation, as distinct from boiling, denotes surface evaporation due to relatively low temperature driving forces and/or low heat fluxes. Thin film evaporators have long been desired for their better performance and designs ranged from free falling films to mechanically wiped films [70] where transfer coefficients as high as 45 KW/m$^2$°C were reported [70]. As the heat fluxes and/or temperature driving forces increase, nucleate boiling may occur, increasing the evaporation side coefficient [71-73]. The proactical distinction between these two mechanisms of phase change is rather interesting. The heat flux in the evaporation process is usually uniform, hence the thermal properties of the wall should not affect the transfer rate. However, when bubble formation occurs, local temperature gradients are distributed along the wall and the average heat flow rate will depend on the thermal conductivity of the wall.

Evaporation from films flowing over electrically heated horizontal tubes, at relatively high heat fluxes, were studied by Fletcher et al [74,75]. A fairly similar experimental technique was utilized by Edwards [76] in his study of film evaporation on a grooved horizontal tube. Closely related are his studies of [26] on evaporation from vertically falling saline water films in laminar transitional flow. Also related is the study of evaporation from a film flowing on a near horizontal plane [77,78,79] and the studies of heat transfer, without evaporation, to a film flowing over a horizontal tube [80,81].

## 5. SIMULTANEOUS EVAPORATION AND CONDENSATION

As indicated in the introduction, we limit our discussion to the horizontal-tube evaporator/condenser mode of operation which presently seems to have the economic and operational advantages.

Saturated steam enters the tube at one end and condenses on the inside surface of the tube. The condensate film flows tangentially, driven by gravity, along the periphery of the tube, and forms a layer at the bottom of the tube so that the flow of steam and liquid along the tube is stratified. The external surface of the tube is continually wetted by saturated (sea) water dripping from above, flowing by gravity along the external perimeter, draining at the bottom of the tube and dripping on top of the tube below.

Depending on the external liquid feeding rate and the imposed thermal driving force, the evaporating and condensate films can each be either laminar or turbulent. Accordingly, distinction should be made whether molecular transport (of momentum and energy), macro-eddies transport, or both, play a role in the combined evaporation-condensation process.

Difficulties may arise in calculation of mixed regimes in a tube where, over different sections both turbulent and laminar steam and condensate flow regimes occur simultaneously. For instance, when steam condenses in the horizontal tube one has to account for gravity forces as well as the dynamic effect of the steam. Fortunately, in the operation range considered here of $Re_v < 35,000$, the shear stress on the condensing film is negligible [52]; steam and accumulated condensate at the bottom move in stratified flow.

### 5.1 Overall Coefficients Based on Average Individual Values

Wilkes [51] analyzed the evaporating film outside a horizontal tube and the condensate film inside the tube as two independent phenomena, based on the measured average value of the tube wall temperature. Thus, the overall heat transfer coefficient $U_o$ is given by

$$\frac{1}{U_o} = \frac{1}{h_o} + \frac{y_w}{k_w}\frac{A_o}{A_w} + \frac{1}{h_i}\frac{A_o}{A_i} \tag{14}$$

where $A_o, A_i$ denote the outer and inner surface areas of the tube, respectively, $k_w$ is the thermal conductivity of the tube wall and $y_w$ is the wall thickness.

Newson [58] has analyzed published heat transfer studies which reported experimental measurements of $U_o$. Utilizing Harwell's program to calculate $h_i$, he solved for $h_o$.

Some experimental observations are of interest:

a) Once the tubes were wetted the overall heat transfer coefficients was found to be independent of the (sea) water flow rate;
b) The values of $U_o$ were independent of the heat load between 6300 to 38000 w/m$^2$. Larger heat fluxes showed a 5-10% reduction in $U_o$ ;
c) $h_i$ increases with vapor velocity inside the tube due to the vapor shear on the condensing film, but the overall effect on $U_o$ is about 5%-10% (for 1.5 to 11 m/sec);
d) The value of $U_o$ is practically independent of tube diameter (for 16-38-67 mm $\phi$ tubes);
e) Both $h_i$ and $h_o$ (hence $U_o$) increase with increasing temperatures, the latter mainly due to changes in the physical properties of the brine;
f) Brine salinity has not effect on $U_o$;
g) The evaporating side heat transfer coefficient $h_o$ is the controlling one, being some 70% of the magnitude of $h_i$.

## 5.2 Overall Coefficients Based on Individual Local Values

The above analyses relate to a constant boundary condition at the wall, and do not consider the interacting evaporation-condensation process along the periphery of the conduit. Sideman et al analyzed this simultaneous process, accounting for the peripheral temperature variation of the wall of a circular tube [82,86] an ellipse [3,83] and conduits with various cross sections [9,84]. The analysis also accounts for the presence of noncondensables, which results in a temperature change along the tube, in the axial direction, as the inert concentration builds up. However, for sufficiently high steam flow rates and/or initially low concentration of the noncondensables and low temperature driving forces, a homogeneous distribution of the inerts can be reasonably assumed [82]. Details of the theoretical analysis are summarized elsewhere [85].

The theoretical analysis of the simultaneous process with smooth tubes can be simplified by stipulating that the thickness of the conduit's wall may be neglected. This introduces an error of only about 5% [83]. This assumption, however, may lead to significant errors when applied to enhanced heat transfer areas such as grooved or fluted tubes.

## 5.3 Enhanced Heat Transfer: Profiled Tubes

Augmentation of the heat transfer by utilizing profiled tubes is of interest in all forms of saline water distillation processes [87]. A listing of over 450 investigations and surveys dealing with augmentation of convective heat and mass transfer is given by Bergles and Webb [88], following an earlier extensive discussion of the subject [89]. A review dealing with surface condensers was presented by Williams [90]. The following comments are therefore, just aimed at highlighting this rapidly developing mode of operation, with particular emphasis on enhanced transfer rates to or from condensing and evaporating films.

Until recently, attention has been directed towards vertical fluted tubes suitable for multi-effect evaporators. Evaporation [91-93] and condensation [94-96] on enhanced vertical planes and tubes indicated improvements of up to 300% in the overall transfer rate.

Nabavian and Bromely [98] used a Gregorig type surface [94] on copper
tubes, and measured <u>condensation</u> transfer coefficients of up to 80,000 W/m$^2$°K.
Withers [99] reported an increase of up to twice the smooth tube value.
Karkhu and Borovkov [100] obtained 50 to 100% enhancement with brass tubes,
while Mills et al [101] reported enhancements of up to 550% with a brass tube
and even higher values with a copper tube. Theoretical analyses of condensation
on horizontal grooved tubes have also been presented [76,100].

An internally grooved tube yielded condensation coefficients of 17,600
to 35,800 W/m$^2$°K, [52], which yielded $U_o$ some 40 to 60% longer than with
a smooth tube.

Attempts have also been made to devise horizontal tube shapes which would
enhance evaporation rates. Sideman [3,83,97] suggested that the application of
elliptically shaped tubes would enhance $U_o$ by some 25%. An analysis of the
experimental data obtained with various horizontal evaporator-condenser units
was recently presented by Newsom [58,102], indicating that knurling can increase
the evaporating side coefficient by 40 to 70% over smooth tubes. Edwards [76]
and Prince [103] studied evaporation from films flowing over grooved horizontal
tubes, and reported significant, 300-400%, enhancements. Based on the convective
mechanism indicated by the study of Davies [104], turbulent promoting,
longitudinally transverse, ribbed horizontal tubes, with a distance to height
ratio of about 7, were suggested [58] for enhanceing the evaporation side
transfer coefficients. However, experiments in these laboratories showed a
50% reduction of $U_o$ when axially square-grooved tubes were employed in the
horizontal tube evaporator/condenser. A 100% enhancement was obtained in these
laboratories utilizing aluminum tubes with square circumferential 1x1x1 mm
grooves [105-107].

A theoretical analysis [108] of the effect of the shape of the external
grooves of various configurations Fig.5 on the film evaporation rate and the
overall transfer coefficient is of particular importance. In agreement with
the earlier studies of square grooves [105-107], the study shows that the amount
of heat transferred through the film in the grooved surface depends on the water
flow rate in the groove which, in turn, depends on the size and shape of the groove.

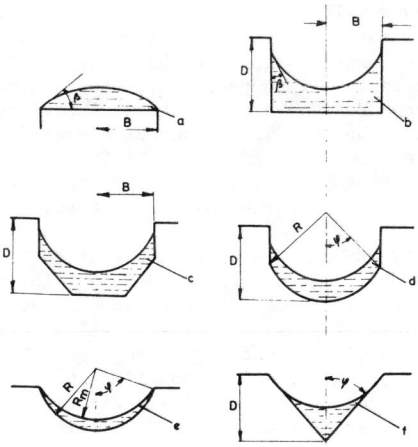

Fig.5. Shapes of grooves studied.

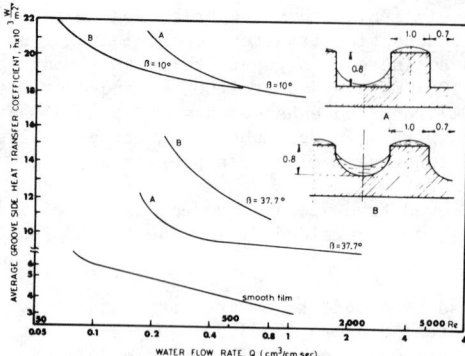

Fig.6. Average groove side heat transfer coefficients for square-bottom and circular-bottom squared grooves.

In general, the analysis yields that the square-edged grooves, with either straight or cut-edged bottom, are most advantageous for Re>500. The circular-bottom square-groove seems advantageous for 250 Re 500, yielding high transfer rates with relatively little sensitivity to the flow rate. The triangular groove is useful only at the low, Re 500, flow range while the circular groove is not advantageous at all. Comparison of the results is shown in Fig.6. Note the large effect of the solid-liquid contact angle which indicates that the heat transfer rates increase as the contact angle decreases. This is consistent with the well-known advantage of adding surfactants to sea water.

Unlike the case of the smooth tube, the tube material in profiled tubes greatly affects the transfer rates. Profiled copper tubes yield higher condensation coefficients than similar brass tubes, and much higher coefficients than realized with similar cupro-nickel tubes [101]. Copper gives 4 times higher transfer coefficient than titanium [76]. Furthermore, each tube material shows a different dependence of the transfer coefficient on the temperature driving force [101]. These characteristics are due to the different two dimensional temperature fields developed in the profiled tube walls of different conductivity. It is noted that the metal thermal conductivity (hence the wall resistance) is of two orders of magnitude larger than that of water. This can justify the commonly used constant wall temperature boundary condition when relatively thick films are considered. However, the wall resistance may not be neglected when, as is usually the case with grooved or fluted tubes, parts of the film are extremely thin, i.e. with a very small film resistance. Under these conditions the film resistance is, locally, of the order of the resistance of the wall and variations of the latter will affect the overall performance.

Profiled surfaces have also proved advantageous in nucleate and flow boiling systems [109-113]. Withers and Habdas [110] reported significant improvements in the heat transfer coefficients and burnout heat flux for tube boiling with internal helical ridging. Golovinskii [111] reviewed the mechanisms of boiling heat transfer from finned or corrugated tubes. Obviously, significant improvement of the overall transfer coefficients of the evaporation condensation process can be realized if the relatively efficient condensing surface is complimented with enhanced evaporation or boiling.

6. CONCLUSION

The apparent advantages of the film evaporator/condenser unit, and particularly the horizontal-tube type unit, are discussed. The numerous studies dealing independently with either film condensation or film evaporation, inside or outside the tube, are recalled. The overall transfer coefficients, $U_o$, in the horizontal tube evaporator/condenser are then reviewed, with reference to the experimental and theoretical studies. Distinction is made between $U_o$ calculated from the individual film coefficients and $U_o$ calculated for the simultaneous phase change occurring on the two sides of the metal wall. The latter method of calculation is the required one, particularly when profiled tubes are utilized to enhance the heat transfer rates.

## 7. ACKNOWLEDGEMENT

We acknowledge with thanks the financial support of the Israel National Council for R&D.

## 8. REFERENCES

1. Deronzier,J.C.1979. Desalination 31:115-124.

2. Fujita,T. and Ueda,T. 1978. Intl.J.Heat Mass Transfer 21:97-118.

3. Semiat,R.,Sideman,S.and Moalem-Maron,D.1978. Proc.6th Int.Heat Transfer Conf.,Toronto,HX-30:361-366.

3a. Salazar,R.P.and Marschall,E. 1978.Int.J.Multiphase Flow 4:405-412.

3b. Seban,R.A. 1978. Proc.6th Int.Heat Transfer Conf.Toronto,Canada 6:417-428.

4. Rohsenow,W.M. and Choi,H.Y. 1961. Heat and Mass and Momentum Transfer; Chapter 10, Prentice Hall,N.Y.

5. Chato,J.C. 1962. J.Am.Soc.Refrig.Air Cond.Eng. p.52.

6. Rohsenow,W.M. and Hartnett,J.P. 1973. Handbook of Heat Transfer; Section 12 pp 12-19, McGraw-Hill, N.Y.

7. Kutateladze,S.S. and Gogonin,I.I. 1979. Int.J.Heat Mass Transfer 22:1593-1599.

8. Chun,K.R. and Seban,R.A.1971. J.Heat Transfer 93:391.

8a. Markatos,N.C.G. 1977. Chem.Ing.Tech. 49:989.

9. Sideman,S.,Moalem-Maron,D. and Semiat,R. 1977. Desalination 21:221-233.

10. Hoffman,H.W.,Alexander,L.G. and Bundy,R.D 1966. USAEC Rept.ORNL-CF-66-7-4, Oak Ridge National Lab.

11. Alexander,L.G.,Hoffman,H.W. and Holz,P.P. 1967. USAEC Rept.ORNL-CF-67-7-22, Oak Ridge National Lab.

12. First VTE pilot plant report. O.S.W. Rept. No.367, July (1967).

13. Evaluation of the VTE and multistage flash desalination processes. O.S.W. Rept. No. 580, Aug.(1970).

14. Performance characteristics of advanced evaporator tubes for LTV evaporators. O.S.W. Rept. No. 644, Jan (1971).

15. VTE pilot plant annual report for the period ending July 1,1968, O.S.W. Rept. No.646, Dec.(1970).

16. Development report No.9, V.T.E. process development. O.S.W. Rept.No.739, Octo.(1971).

17. Van Der Mast, V.C. and Bromely, L.A. 1976. AIChE J. 22(3):533.

18. Jones,H.H.M.1960. The industrial chemist, Dec.p.599.

19. Fong,C., King,C.J. and Sephton,H.H. 1975. AIChE-ASME Heat Transfer Conf. Aug. San Francisco.

20. Ishigai,S., Nakanisi,S.,Takehara,M. and Oyabu,Z. 1974. Bulletin of the ASME, 17(103):106.

21. Domanskii,I.V. and Sokolov,V.N. 1967. J.Appl.Chem.of the USSR.40(1):56.

22. Ueda,T.,Kubo,T. and Inoue,M. 1974. 5th Int.Heat Transfer Conf. Tokyo.,Sept.

23. Suryanarayana,N.V. and Malchom,G.L. 1975. J.of Heat Transfer,Trans ASME p.79.

24. Jones,W.P. and Renz,U. 1974. Int.Heat Mass Transfer 17:1019.

25. Kroll,J.E. and McCutcham,J.W. 1968. J.Heat Trans.p.201.

26. Unterberg,W. and Edwards,D.K. 1965. AIChE J. 11:1073.

27. Weinberg,J.,Gazit,E. and Koren,A. 1976. Int.Symp. on fresh water from the sea, May, Alghero, Sardinia, Italy.

28. Takada,M. 1976. 5th Int.Symp.on fresh water from the sea, May, Alghero, Sardinia, Italy.

29. Nusselt,W.Z., 1916. V.O.I. 60:541-569.

30. Bromely,L.A. 1952. Ind.Eng.Chem. 44(12):2966.

31. Rohsenow,W.M. 1954. ASME Paper No.54-A-144.

32. Chen,M.M. 1961. J.of Heat Transfer, 83C:48-55.

33. Sparrow,E.M. and Gregg,J.L. 1959. J.Heat Transfer 81C:pp 13,291.

34. Sparrow,E.M.,Mincowycz,W.J. and Saddy,M. 1967. Int.J.Heat Mass Transfer 12:1157-1169.

35. Koh,J.C.Y.,Sparrow,E.M. and Hartnett,J.P. 1969. Int.J.Heat Mass Transfer 12:69.

36. McAdams,W.H. 1954. Heat transmission. McGraw-Hill, N.Y. 3rd ed.

37. Rohsenow,W.M.,Weber,J.H. and Ling,A.T. 1956. Trans. ASME 78:1637.

38. Denny,V.E. and Mills,A.F. 1969. J. Heat Transfer 41:495.

39. Lee,J. 1964. AIChE J. 10:540.

40. Grant,I.D.R. 1969. Br.Chem.Eng. 14:1709-1712.

41. Tamir,A. and Taitel,Y. 1971. Israel J. of Tech. 9:69-81.

42. Glicksman,L.R., Mikic,B.B. and Snow,D.F. 1973. AIChE J. 19(3):636.

43. Berman,L.D. 1973. Teploenergetika, 20(8):76-77, 103.

44. Sheynkman,A.G. and Linetsky,V.N. 1969. Heat Trans.Sov.Res. 1(3):90.

45. Hassan,K.E. and Jakob,M. 1958. Trans.ASME 80:887.

46. Moalem-Maron,D.,Sideman,S. and Semiat,R. 1977. Desalination, 21:51-58.

47. Nimmo,B.G. and Leppert,G. 1970. 4th Int.Heat Trans.Conf. Paris,p.22.

48. Gerstmann,J. and Griffith,P. 1967. Int.J.Heat Mass Transfer, 10(5):567.

49. Chaddock,J.B. 1957. Refrig.Engng. 65:36-41, 90-94.

50. Kern,D.Q. 1958. AIChE J. 4(2):157.

51. Aqua Chem.Inc.,Research and development on the horizontal spray film evaporator; O.S.W. Rept. No. 209, Nov. (1966).

52. Universal Desalting Corp., Pilot plant tests and design study of a 2.5 MGD horizontal-tube multiple-effect plant, O.S.W. Rept. No. 492, Oct. (1969); No. 592 May (1970).

53. Third report on horizontal tube multiple-effect process pilot-plant tests, O.S.W. Rept. No. 740, Oct.(1971).

54. Operation of the multi-effect multi-stage flash distillation plant (Clair Engle), third report (annual), San Diego, California, O.S.W. Rept. No. 668, April (1971).

55. Design studies on multistage flash distillation vessels, 1971. O.S.W. Rept. No. 687, July.

56. Cannizzaro,C.J. et al. 1974. Fourth report on horizontal-tube multi-effect (HTME) process pilot-plant test program, O.S.W. R&D Rept. No. 967.

57. Cannizzaro,C.J. et al. 1974. Fifth report on horizontal tube multiple-effect (HTME) program pilot plant test program, O.S.W. Rept. No. 968.

58. Newson,I.H. 1978. Proc.6th Int.symp.fresh water from the sea,2:113-124.

59. Baker,O. 1954. Oil Gas J. 53:185.

60. Bell,K.J.,Taborek,J. and Fenoglio,F. 1970. Heat transfer, Minneapolis, 1969, CEP Symp. Ser.No. 102, 66:150.

61. Rufer,C.E. and Kezios,S.O. 1965. J.Heat Transfer, 88C:265.

62. Sarma,A.S.P., Sarma,P.K. and Apparao,K. 1972. Can.J.Chem.Engng. 50:541.

63. Akers,W.W,, Deans,H.A. and Corsser,O.K. 1959. Chem.Engng.Prog.Symp.Ser. 55(99):171.

64. Akers,W.W. and Rossom,H.F. 1960. Chem.Engng.Prog.Symp.Ser. 56(30):145.

65. Murthy,V. and Sarma,P.K. 1972. Can.J.Chem.Engng. 50:546.

66. Rosson,H.F, and Myers,J.A. 1965. Chem.Eng.Prog.Symp.Ser. 61(59):1980.

67. Myers,J.A. and Rosson,H.F. 1961. Chem.Engng.Prog.Symp.Ser. 57(32):150.

68. Anaiev,E.P., Boyko,L.D. and Kruzhelin,G.N. 1961. Int.Heat Trans.Conf. Part II, p.290.

69. Razavi,M.D. and Clutterbuck, E.K. 1974. Chem. and Ind. 2:205.

70. Lustenader,E.L., Richter,R. and Neugebauer,F.J. 1959. J.Heat Transfer, Nov.297-307.

71. Agrawal,S.C. and Brahmin,B. 1972. Indian J. of Technol. 10:299.

72. Alam,S.S. and Varshney,B.S. 1972. Indian J. of Technol. 10:172.

73. Slesarenko,V.N. and Yakubovshiy,Y.V. 1974. Heat Transfer Sov.Res.6(4):158.

74. Fletcher,L.S.,Sernas,V. and Galowin,L.S. 1974. Ind.Eng.Chem.Process Res.Dev. 13(2):265.

75. Fletcher,L.S., Sernas,V. and Parker,W.H. 1975. Ind.Eng.Chem.Process Res.Dev. 14(4):411.

76. Edwards,D.K.,Gier,K.D., Aygasmamy,P.S. and Cotton,L.I. 1973. ASME-AIChE Heat Transfer Conf. Atlanta, Ga., Aug.ASME paper 73-HT-25.

77. Lis,J. and Strickland,J.A. 1970. 4th Int.Heat Transfer Conf. B.4.6,Paris.

78. Riedle,K. and Purcupile,J.C. 1973. ASHRAE Meeting, Chicago.

79. Gollan,A. and Sideman,S. 1968. Int.J.Heat Mass Transfer, 11:1761.

80. Zfati,A. 1971. M.Sc. thesis (in Hebrew), Dept. of Mech.Eng. Technion-Israel Institute of Technology, Haifa,Israel.

81. Adams,F.W., Broughton,G. and Conn, A.L. 1936. Ind. Eng.Chem. 23:537.

82. Moalem-Maron,D. and Sideman,S. 1976. Int.J.Heat Mass Trans. 19:259.

83. Moalem-Maron,D. and Sideman,S. 1975. ASME J. Heat Trans. 97:352.

84. Sideman,S.,Moalem-Maron,D. and Semiat,R. 1975. Desalination 17:167.

85. Sideman,S. and Moalem-Maron,D. 1979. Proc.Int.Seminar ICHMT Dubrovnik; F.Durst, N.Afgan Eds., Hemisphere Publ.Comp. N.Y.

86. Semiat,R.,Moalem-Maron,D. and Sideman,S. 1978. Physical chemistry and hydrodynamics, Levich Birthday Conf.,Oxford,B.D.Spalding, Ed.Advance Publ. U.K.

87. Symposium on enhanced tubes for desalination plants, Dept. of the Interior, U.S.A. (1970).

88. Bergles,A.E. and Webb,R.L. 1970. American Soc. of Mechanical Engineers, N.Y. p.1-15.

89. Bergles,A.E. 1969. Prog. in Heat and Mass Transfer, 1:331.

90. Williams,A.G. Nandapurkar, S.S. and Holland,F.A. 1968. Trans. I.Ch.E. The Chemical Engineer, 46:367.

91. Simpson,H.C., Beggs,G.C. Lewis,J.S. and Linstrum,A. 1976. 5th Int.Symp. on fresh water from the sea, 2:289.

92. Johnson,B.M.,Jansen,G. and Owzanski,P.C. 1971. ASME-AIChE Heat Trans. Conf. Tulsa, Okla.

93. Kays,D.D. and Chia, W.S. 1971. ASME-AIChE Heat Transfer Conf. Tulsa, Okla.

94. Gregorig,R. 1954. Zeitschrift fur angewandte Mathematik und Physik,5:36.

95. Thomas,G.D. 1967. I&EC Fundamentals, 6:97.

96. Thomas,G.D. 1968. AIChE J., 14:644.

97. Sideman,S.,Semiat,R. and Moalem,D. 1976. 5th Int.Symp. on fresh water from the sea, 2:315-324.

98. Nabavian,K. and Bromely, L.A. 1963. Chem.Eng.Sci. 18:651.

99. Withers,J.C. 1970. Symp. on enhanced tubes for desalination plants, U.S. Dept. of the Interior, p.119.

100. Karkhu,V.A. and Borovkov,V.P. 1971. Heat Transfer Sov. REs. 3:183.

101. Mills,D.F., Hubbard,G.L., James,R.K. and Tam,C. 1975. Desalination 16:121.

102. Newson,I.H. and Hodgson,T.D. 1973. AERE Rept. No. R7318 - UKAEA.

103. Prince,W.J. 1971. Enhanced tubes for horizontal evaporator desalination processes, M.S. Thesis, UCLA School of Engineering.

104. Davies,J.T. and Shawk,P.M. 1976. 59th Annual AIChE Meeting, Chicago.

105. Semiat,R.,Moalem-Maron,D. and Sideman,S. 1978. Transfer characteristics of convex and concave rivulet flow on inclined surfaces with straight-edged grooves, Technion R*D Rept. No.51, Proj. 071-082, March.

106. Moalem-Maron,D., Semiat,R. and Sideman,S. 1978. Enhanced heat transfer in horizontal evaporator-condensers with straight-edged grooved tubes, Technion R&D Rept. No. 52, Proj.071-082, March.

107. Sideman,S.,Semiat,R. and Moalem-Maron,D. 1978. Enhanced transfer in horizontal evaporator-condenser conduits with circumferential grooves, Proc. 6th Int.Symp. Fresh water from the sea, Las Palmas, 1:183-192.

108. Sideman,S. and Levin, A. 1979. Desalination 31:7-18.

109. Czikk, A.N. and O'Neill,P.S. 1969. Symp. on enhanced tubes for desalination plants, Washington, D.C.

110. Withers, J.C. and Habdas, E.P. 1973. 47th National Meeting AIChE, New Orleans, LA.

111. Golovinskii, G.P. 1975. Int.Chem.Eng., 15(2):258.

112. Danilora, G.N. and Dyundin, V.A. 1972. Heat Transfer - Sov.Res. 4;48.

113. Carnavos, T.C. 1964. O.S.W. Symp. Enhanced tubes for distillation plants, Washington D.C.

# Experimental and Theoretical Investigation of Deposition Motion of Liquid Droplets in Two-Phase Flow through a Vertical Tube

E. N. GANIĆ AND K. MASTANAIAH
Department of Energy Engineering
University of Illinois at Chicago Circle
Box 4348
Chicago, Illinois 60680 USA

ABSTRACT

Measurements of deposition rates are reported for air-water droplet system in a 12.7mm I.D acrylic tube at Re = 52 500 and 94 600. A theory has also been formulated for the deposition motion of large sized particles from a turbulent gas stream. The theory shows that the primary resistance to transport of large particles exists in the turbulent core. Satisfactory agreement is obtained between the proposed theory and the present measurements for particles within the Stokes regime. Deposition data of Agarwal [6] for uniform sized particles, and of Cousins & Hewitt [12] involving drop size distribution are also described by the present deposition model.

1. INTRODUCTION

Accurate description of the mechanism of purely turbulence-controlled deposition is essential for the analysis of deposition motion of particles under more complex conditions where other mechanisms such as gravitational settling and electrostatic effects are also present simultaneously. A critical examination of the deposition measurements for a wide range of particle sizes from submicron to several hundreds of microns has been reported by McCoy & Hanratty [1]. As shown in figure 1 they have presented the dimensionless deposition velocity $k_d^+$ ($=k_d/u^*$) for vertical systems vs. the dimensionless particle relaxation time $\tau^+$ ($=d_p^2 \rho\rho_p u^{*2}/18\mu^2$) based on volume median diameter. Here $k_d = N_o/\bar{c}$, $N_o$ is the mass rate of deposition of particles per unit area, $\bar{c}$ is the bulk concentration of particles across the tube, $u^*$ is the friction velocity, $d_p$ and $\rho_p$ are the particle diameter and density, and $\rho$ and $\mu$ are the density and dynamic viscosity of the fluid respectively. The following observations have been made in view of the data depicted in Fig. 1.

For particles in the submicron range, $\tau^+ < 0.15$, particles follow the streamlines of fluid motion, and Brownian diffusion is the mechanism reponsible for deposition, suggesting that $k_d^+$ is independent of $\tau^+$ and is a function of Schmidt number $Sc(=\nu/\bar{D})$ only, where $\nu$ is the kinematic viscosity of the fluid and $\bar{D}$ is the Brownian diffusivity.

When $\tau^+ > 0.15$, $k_d^+$ is however found to be independent of Brownian motion, and Sc is no longer an important dimensionless group. According to the theory proposed by Friedlander & Johnstone [2], particles in this range diffuse towards the wall due to radial velocity fluctuations of the turbulent eddies up to one stopping distance from the wall, and then deposit on the wall by a free-flight

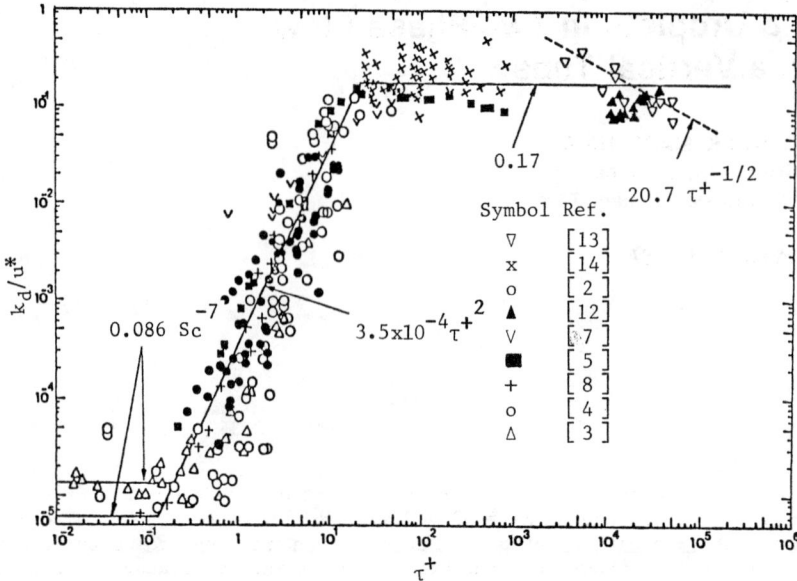

Fig. 1. Summary of literature deposition data.

(inertial) mechanism through the viscous boundary layer, owing to the initial momentum imparted to them by the fluid eddies. A stopping distance, S, is defined as the distance a particle would travel through a stagnant fluid with a given initial velocity under the conditions of Stokesian drag, and is given by $S = (d_p^2 \rho_p v_{po}/18\mu)$, where $v_{po}$ is the initial particle velocity at the start of free flight. The data indicate that $k_d^+$ is approximately proportional to $\tau^{+2}$ up to $\tau^+$ about 20 to 30. The experimental deposition data in the above range of $\tau^+$ are reported in [2-8]. In order to better explain the observed deposition data, the theory of Friedlander & Johnstone [2] is later modified by a number of investigators [9, 10, 4, 11].

However, as the particle size increases beyond $\tau^+ > 20$ to 30, none of the above models is able to predict the experimental data. There is observed a marked change in the behavior of $k_d^+$, suggesting a deviation in the mechanism of particle deposition. Indeed, this range of $\tau^+$ arises in most practical applications involving two-phase dispersed or drop flow. For vertical flow systems, the deposition data in this range are reported in [12,13,14,5,6]. Unfortunately, the accumulated data show a large amount of scatter, thereby masking the separate effects of Re and $\rho/\rho_p$, where Re is the Reynolds number ($=\bar{U} D/\nu$), D being the tube diameter and $\bar{U}$, the superficial gas velocity. Different investigators using different techniques have obtained different particle deposition rates under similar conditions.

On the other hand, very few theoretical analyses exist in the literature for predicting the deposition rates in this region of particle size. Hutchinson et al. [15] consider that particle motion associated with the eddy interaction in the turbulent core, and in the wall region of highly damped turbulence is random. The deposition rate is found to depend on Re, $\rho_p/\rho$, $d_p/D$ and L/D, where L is the length. From the best fit of the theory with the available data, they find that the wall region is described by $o < y^+ < 1.25$, where $y^+ = y u^*/\nu$ and y is the radial distance from the wall.

Some criticism of the above theory has been reported by McCoy [16]. It is mentioned that if the final free flight begins at $y^+ = 1.25$, the initial particle velocity should be the one that corresponds to local fluctuating velocity of the fluid, instead of the velocity that exists in the core region. Another inconsistency in the theory is that $y^+ = 1.25$ cannot be considered universal in the sense that the larger particles would be able to traverse a much larger region of damped turbulence than the smaller particles, since the inertia of the particle increases as $d_p^3$ while the fluid Stokesian drag increases as $d_p$. Also Gardner [17] observes that the theory of [15] predicts maximum possible $k_d^+$ as 0.81, which is about six times higher than that measured by Liu & Agarwal [5].

Forney & Spielman [14] have presented deposition data and developed an expression for the calculation of deposition velocity for particles having $S^+>30$. Their theory employs the stopping distance concept and considers that the location where free flight starts can vary from $y^+ = 30$ to $y^+ = r_o^+$, depending on the particle size. The boundary condition employed is $c = o$ at $y^+ = S^+$ for $30 < S^+ < r_o^+$.

The main difficulty with the theory of Forney & Spielman [14] is that the assumption of a zero particle concentration at locations corresponding to the start of free flight in the turbulent core is incorrect, and is in direct contrast with the reported measurements of Hagiwara et al. [18]. The error in such an assumption becomes increasingly serious when the start of free flight becomes closer to the tube center. Another limitation of their theory is that their analysis fails to describe the deposition process when the stopping distance exceeds the tube radius, in which case turbulent diffusion with free flight as the final transport totally breaks down.

The prediction procedure of Namie & Ueda [19] is based on the assumption that turbulent diffusion is important up to $y^+ \simeq 100$, since the motion for $y^+ < 100$ consists of eddies of small sizes and high frequency, and that the particle responds to large scale eddies having low frequency. They assume that all particles that diffuse up to $y^+ \simeq 100$ will reach the wall and deposit, and consider gradient type boundary conditions for the particle concentration. The theory is not compared with the data of others.

Reeks & Skyrme [20] have published a stochastic model to express the dependence of deposition rate on particle inertia. As in other models, the deposition process is assumed to be based on turbulent diffusion followed by inertial projection through the boundary layer. Good agreement is obtained with the data of Liu & Agarwal [5] when the stopping distance is taken as $y^+ = 5$ instead of 1.25 as employed in [15].

Cleaver & Yates [21] have proposed a deposition model based on the recent observations of the structure of the wall region of a turbulent boundary layer. It is assumed that particles move to a certain height above the surface by turbulent diffusion and are convected to the wall in the downsweep of fluid towards the wall, while some of them are carried back into the turbulent core by the turbulent burst (upsweep). The analysis for $\tau^+ \sim 0(1)$ display an interesting dependence of deposition flux on $\rho_p/\rho_g$ as well as on $\tau^+$, thereby partly explaining some of the scatter in the reported data. However, their theory predicts an assymptote as $\tau^+ \to 1000$, a result that is in contrast with data for large particles [6].

From the foregoing considerations, it is evident that there exists no satisfactory theory that can explain the mechism of droplet transport when the particle relaxation time is in excess of about 30. The purpose of the present work is to experimentally and theoretically study the mechanism of turbulence-controlled deposition in the higher range of particle size.

## 2. ANALYSIS

### 2.1 Proposed Physical Model

The present model is concerned with particles which have stopping distance $S^+$ in excess of the combined thickness of the viscous sublayer and the buffer layer ($y^+ = 30$). The intensity of radial turbulence from the center of the tube ($y^+ = r_o^+$) to the periphery of the turbulent core ($y^+ = 30$) is high and practically uniform as indicated by Laufer's data [22]. Therefore, the particle is subjected to radial velocity fluctuation regardless of $S^+$ and thereby diffuses to the periphery of the turbulent core. The radial velocity fluctuations outside the turbulent core ($0 < y^+ < 30$) are, however, small and non-uniform compared to those in the core region [22], and therefore particle transport towards the wall by turbulent diffusion is questionable. However, all particles with $S^+ > 30$ have already enough momentum, imparted to them by the turbulent eddies at the perifery of the turbulent core, to penetrate the boundary layer by inertial flight. The above arguments form the physical basis for our consideration that $y^+ = 30$ separates the eddy diffusivity transport from passive transport, independent of the particle size, for $S^+ > 30$. The idea is in fact consistent with the conjecture of Friedlander & Johnstone [2] that "for $S^+ > 30$, the particles need to diffuse only to the edge of the sublayer" although they have not considered any experimental data to examine the validity of their supposition. It is to be pointed out that for particles with $S^+ > 30$, Friedlander and Johnstone [2] have assumed equality of particle and fluid diffusivities and employed Reynolds analogy to obtain the deposition velocity that is independent of particle size. But this assumption is not realistic for large particles since their response to fluid fluctuations becomes imperfect and uncertain [23].

The present work therefore considers the effect of particle inertia on the deposition rates, and this marks an important difference between the present model and that proposed by Friedlander and Johnstone [2].

### 2.2. Deposition Velocity

Based on the proposed physical model described above, the analysis for drop deposition is performed for particle $\tau^+ > 40$. The following assumptions are made.

1. The flow is fully developed.
2. The concentration profile is developed so that entrance effects are not present.
3. Drop concentration is small enough to consider that the fluid turbulence characteristics are unaltered.
4. The size of the particles is small enough that their motion relative to the fluid obeys Stokes law of resistance.
5. There is no wall rebound or reentrainment of drops, once the drops deposit on the wall.
6. The local mass flux of droplets in the turbulent core varies linearly from zero at the tube center to a value of $N_o$ at $y^+ = 30$ corresponding to the edge of the buffer layer, and remains constant, equal to $N_o$ in the region $0 < y^+ < 30$. This assumption, however, requires plug flow idealization. Because the turbulent flow velocity profile is relatively flat, the above approximation introduces little error, as mentioned by Kays [24].

The rate law for the diffusion of droplets due to concentration gradients is given by

$$N = \varepsilon_p \frac{dc}{dy} \tag{1}$$

where N is the local rate of mass flux of droplets in the radial direction, c is the concentration of the droplets in the units of mass/volume and $\varepsilon_p$ is the particle diffusivity equal to $\delta\varepsilon_f$, $\varepsilon_f$ being the fluid eddy diffusivity.

As is well known, the eddy diffusivity of fluid is not constant across the turbulent core [24]. The eddy viscosity expression proposed by Reichardt [24] is considered here, and is given by

$$\varepsilon_f = \frac{Kr_o u^*}{6}\left[1 - \left(\frac{r}{r_o}\right)^2\right]\left[1 + 2\left(\frac{r}{r_o}\right)^2\right] \tag{2}$$

Here K is the mixing length constant equal to 0.4. Integration of (2) between $y^+ = 30$ and a point $y^+$ in the core leads to the concentration distribution as

$$c - c_b = \frac{N_o}{u^* \delta K}\left[\ln\left(\frac{1 + 2x}{1 - x}\right)\right] \quad x = (1 - y^+/r_o^+)^2 \;\; (1 - 30/r_o^+)^2 \tag{3}$$

At $y^+ = r_o^+$, we have

$$c_c - c_b = \frac{N_o}{u^* \delta K} \ln\left[\frac{1 + 2(1 - 30/r_o^+)^2}{1 - (1 - 30/r_o^+)^2}\right] \tag{4}$$

where $c_c$ and $c_b$ are the concentrations at the tube center and the edge of the buffer layer respectively. The mass transfer coefficient or the so-called "deposition velocity" is defined as

$$k_d = N_o / \bar{c} \tag{5a}$$

with the bulk concentration $\bar{c}$ given by

$$\bar{c} = \int_0^{r_o} cur\, dr \Big/ \int_0^{r_o} ur\, dr, \tag{5b}$$

where u(r) is the radial velocity distribution in the tube.

In evaluating (5b), it is assumed that $c = c_b$ for $0 < y^+ < 30$. This has become necessary in view of the uncertainties regarding the phase distribution in the immediate vicinity of the wall. However $\bar{c}$ is seen to be relatively insensitive to the type of concentration profile assumed in the sublayer, as would be expected.

In order to complete the solution of the problem, an additional relation for $c_b$ is necessary. This is obtained here by invoking an auxiliary boundary relation [10,11]:

$$N_o = v_{pb} c_b \tag{5c}$$

where $v_{pb}$ is the velocity of the particle at the edge of the buffer layer, wherefrom free flight is considered to start. From the relations to be derived later in the text, it can be shown that

$$v_{pb} = v_{fb} \sqrt{\delta} \tag{5d}$$

where $v_{fb}$ is the rms radial velocity of the fluid at $y^+ = 30$, which can be taken equal to $0.75\ u^*$. Therefore

$$v_{pb} = 0.75\ u^* \sqrt{\delta} \tag{5e}$$

Equating (5a) and (5c), we obtain a relation for $c_b$ as

$$c_b / \bar{c} = k_d / (0.75\ u^* \sqrt{\delta}) \tag{5f}$$

The value of $\bar{c}/c_c$ is determined from (3), (5b) and (5f). In (5b), the following distribution for $u^+$ ($= u/u^*$) is considered.

$$u^+ = y^+, \qquad\qquad 0 < y^+ < 5 \tag{6a}$$

$$u^+ = 5 \ln y^+ - 3.05, \qquad 5 < y^+ < 30 \tag{6b}$$

$$u^+ = 5.5 + 2.5 \ln \left[ \frac{y^+ + 1.5(1 + r/r_o)}{1 + 2(r/r_o)^2} \right], \qquad 30 < y^+ < r_o^+ \tag{6c}$$

Equation (6c) is due to Reichardt [24], and is in excellent agreement with the experimental data up to the tube center.

From Equations (4), (5a) and (5f), it is shown that the dimensionless deposition velocity is given by

$$\frac{k_d}{u^*} = \frac{\delta(c_c / \bar{c})}{2.5 \ln\left[\frac{1 + 2(1 - 30/r_o^+)^2}{1 - (1 - 30/r_o^+)^2}\right] + (\sqrt{\delta}/0.75)} \tag{7}$$

## 2.3 Particle to Fluid Diffusivity Ratio $\delta$

The diffusivity of a particle in turbulent flow has been analysed by Tchen [25]. The details and limitations of his theory, and further considerations towards improvement have been given in Soo [26] and Hinze [22].

For a solid particle or water droplets in a gas stream ($\rho_p \gg \rho$), in the absence of external forces like gravity and neglecting the forces due to Basset term, additive mass and pressure gradient, the equation of motion subjected to radial velocity fluctuations of the fluid under Stokesian drag is written as

$$\frac{dv'_p}{dt} = a(v'_f - v'_p) \tag{8}$$

where

$$a = 36\mu / \left[(2\rho_p + \rho) d_p^2\right], \tag{9}$$

$v'_f$ and $v'_p$ are the fluid and the particle velocities respectively, and t is the diffusion time or the time of travelling of the particle.

For shear flow through tubes of finite dimensions, the radial diffusion time is of the same scale as the characteristic Lagrangian integral time scale $T_L$ which is of the order of $r_0/u^*$, where $r_0$ is the tube radius. In such situations where $t \sim T_L$, the dispersion of the particle can be rather well approximated as a small time diffusion process (Monin & Yaglom [27]), due to the fact that the diffusivity is only very weakly dependent on the specific form of the correlation function. The assumption of the dispersion via a small time diffusion is also consistent with Tchen's assumption of no over-shooting (Hinze 1975, p. 462 & 470) which implies that the displacement of the particles relative to the fluid element is smaller than the Kolmogoroff's microscale. Tchen's theory then leads to

$$\varepsilon_p/\varepsilon_f = \int_0^\infty E_{fL}(n) \, dn / \int_0^\infty E_{pL}(n) \, dn = \overline{v_p^2} / \overline{v_f^2} \tag{10}$$

where $E_{fL}(n)$ and $E_{pL}(n)$ are the Lagrangian energy-spectrum functions for the particle and the fluid respectively. Assuming an exponential relation for the correlation coefficient, it can be shown that [28]

$$\delta = \varepsilon_p/\varepsilon_f = \frac{1}{1 + \frac{\rho_p}{18} \frac{d_p^2}{\mu} \frac{1}{T_{fL}}} \tag{11}$$

where $T_{fL}$ is the Lagrangian integral time scale of turbulence. It may be remarked that (11) is similar to the one derived by Friedlander [29].

Unfortunately there does not exist sufficient experimental information about $T_{fL}$ for turbulent pipe flows because of the difficulty of measurements. $T_{fL}$ is a measure of the longest connection between two fluid particles. It is generally related to the Eulerian integral scale $T_E$ by (see Monin & Yaglom [27], p. 577):

Where

$$T_{fL} = \beta \, T_E, \tag{12}$$

$$T_E = \ell_E / v, \tag{13}$$

v and $\ell_E$ being the radial velocity scale and Eulerian length scale of turbulence respectively. For the turbulent core flow in a tube, v can be considered uniform and taken equal to 0.8 $u^*$. Equation (12) is based on the consideration that the Lagrangian and the Eulerian autocorrelations have similar shape but differ only in time scale [27].

The measurements of $\ell_E$ obtained by Martin & Johanson [30] will be adopted here, and are represented by

$$\ell_E/r_o = 5.028 \times 10^{-4} \, Re^{.509} \tag{14}$$

There exists considerable degree of uncertainty in the present state of knowledge concerning the important quantity $\beta$ which relates the Lagrangian and the Eulerian integral time scales. As pointed out in Monin & Yaglom [27], the reported values of $\beta$ are extremely scattered about a mean value of $\beta = 4$ [31]. More recently, Snyder & Lumley [32] have interpreted a value of $\beta = 3$ based on their measurements in a grid generated turbulence. It should be noted that $\beta$ is not a universal constant; but in view of the absence of reliable and consistent value of $\beta$ at the present time, a value of $\beta = 3$ is employed in the present work as a first approximation. Therefore we have

$$T_{fL} = \alpha \beta r_o / u^* \tag{15a}$$

where

$$\alpha = 6.285 \times 10^{-4} \, Re^{.509} \tag{15b}$$

$$\beta = 3 \tag{15c}$$

From (10,11) and (15), the diffusivity ratio $\delta$ is obtained as

$$\delta = \frac{1}{1 + \frac{1}{9} \frac{\rho_p}{\rho} \left(\frac{d_p}{D}\right)^2 Re\sqrt{f/2} \left(\frac{1}{\alpha\beta}\right)} \tag{16}$$

where f is the friction factor. Expressed in dimensionless relaxation time, (16) becomes

$$\delta = \frac{1}{1 + \frac{2}{Re\sqrt{f/2}} \tau^+/(\alpha\beta)} = \frac{1}{1 + (\tau^+/r_o^+)/(\alpha\beta)} \tag{17}$$

The friction factor is calculated from the well-known correlations for smooth tubes as

$$\begin{aligned} f &= 0.0791 \, Re^{-.25}, & 3 \times 10^3 < Re < 10^5 \\ &= 0.046 \, Re^{-0.2}, & 10^5 < Re < 10^6 \end{aligned} \tag{18}$$

Thus (17) shows that $\delta$ depends on $\tau^+$ and Re only. The trend of $\delta$ given by (17) is similar to that obtained recently by Soo [33] who has deduced three different expressions for the diffusivity ratio depending on the particle size. However, the present result (17) offers the advantage of being continuous for all $\tau^+$ within the Stokes regime.

It may be remarked that $\delta$ as given by (18) is spatially uniform across the core. This appears to be reasonable since the intensity of radial turbulence,

# LIQUID DROPLETS IN TWO-PHASE FLOW THROUGH A VERTICAL TUBE

which is responsible for the particle radial migration, is nearly homogeneous in the core region [22], and since the Eulerian length scale as mentioned earlier is also nearly independent of radial position.

From (7) and (17) the final expression for the deposition velocity is obtained in dimensionless form as

$$\frac{k_d}{u^*} = \frac{c_c / \bar{c}}{\left\{ 2.5 \ln \left[ \frac{1+2(1-30/r_o^+)^2}{1-(1-30/r_o^+)^2} \right] + \frac{1}{0.75} \sqrt{\frac{1}{1+(\tau^+/r_o^+)/(\alpha\beta)}} \right\}} \cdot \frac{1}{\left[ 1+(\tau^+/r_o^+)/(\alpha\beta) \right]} \quad (19)$$

where $\alpha = 6.285 \times 10^{-4} \, Re^{.509}$

and $\beta = 3$,
as given by (15b) & (15c) respectively. Note that $r_o^+ = Re\sqrt{f/2}/2$.

Equation (19) above suggests that the dimensionless deposition velocity $k_d/u^*$ is a function of Re and $\tau^+$ only.

On the other hand, according to Friedlander & Johnstone [2], for $s^+ > 30$ the deposition velocity is given by

$$k_d / \bar{U} = f/2 \quad (20a)$$

which can be written as

$$k_d / u^* = \sqrt{f/2} = 2 \, r_o^+ / Re \quad (20b)$$

Equation (20b) implies that $k_d/u^*$ is independent of particle size but depends on Re only. The improvement of the present result (19) over (20b) will be illustrated later in the text.

In many practical situations, the suspended droplets may not be of uniform size but have a size distribution. In such cases an appropriate mean diameter has to be specified for evaluating $\tau^+$. The arithmetic mean drop size is taken as the effective droplet diameter in the present calculations. Although most of the mass is carried by larger drops, an arithmetic mean diameter is chosen here for the following reason: Eqn. (19) suggests that the deposition velocity varies as $1/d_p^2$ for larger drops, and since the mass of the drops varies as $d_p^3$, the length mean diameter appears to be a reasonable characteristic drop size in the calculation of deposition flux $N_o = k_d \bar{c}$. In other words, the probability of larger drops being deposited is relatively small compared to that of the small drops, because larger drops are less influenced by the eddy motion. This is consistent with the fact that the radial transport of small particles is associated with the momentum exchange with the turbulent eddies, in which case, as mentioned by Soo [26], a length mean diameter $d_{10}$ would be appropriate.

It should be recognized that the present result cannot be applied to arbitrarily large $\tau^+$ in view of the restriction on the Stokes resistance law, which is valid for $Re_p < 1.0$. In fact there exists a value of $\tau^+ = \tau_m^+$ above which the condition that $Re_p < 1.0$ will not be satisfied.
By definition,

$$Re_p = d_p |v_p - v_f| / \nu \quad (21a)$$

In (21a), $v_f$ is taken as 0.8 $u^*$ and $v_p$ is given by (10) and (11) as

$$v_p = v_f \sqrt{\delta} \qquad (21b)$$

where $\delta$ is given by (16). After some algebraic manipulation, it can easily be shown that the maximum value of $\tau^+$ is given by the transcendental relation

$$Re_p = 0.75 \sqrt{18\tau_m^+ (\rho/\rho_p)} \left[1 - \sqrt{\delta}\right] \qquad (21c)$$

In (21c), $Re_p$ is taken equal to 1.0. Thus it is seen that

$$\tau_m^+ = f(Re, \rho_p/\rho) \qquad (21d)$$

## 3. COMPARISON WITH EXPERIMENTAL DATA

The accuracy of the proposed theory is now tested by comparing the calculated results with the existing experimental data as well as the present measurements.

### 3.1 Data for Concentration Distributions (Hagiwara et al. [18])

To the authors' knowledge, the only data available for the distribution of droplet concentration in a vertical tube are those recently reported by Hagiwara et al. [18]. In their experiments using downward flow of air-water droplets in a 26.4 mm I.D. vertical acrylic pipe, the droplets are generated by seven equally spaced atomizing nozzles similar to those employed by Nukiyama and Tanasawa [34]. No droplet entrainment is observed in their experiments.

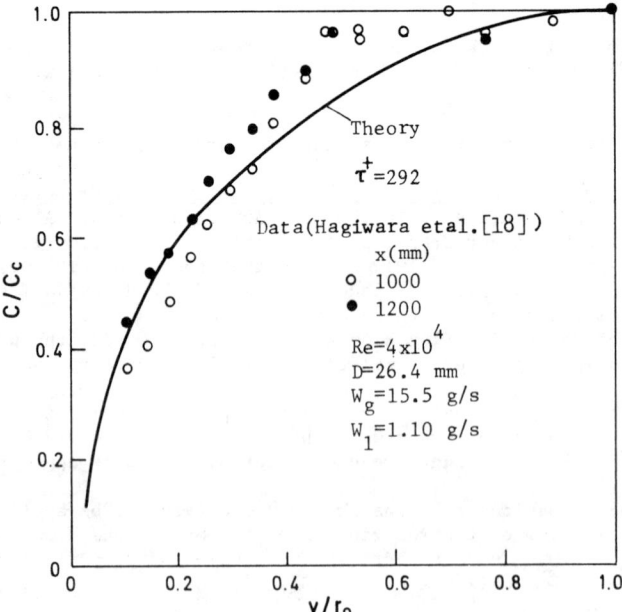

Fig. 2. Comparison of predicted concentration profile with the data of Hagiwara et al. [18].

The data of Hagiwara et al. [18] for concentration distribution at Re = 40 000 is presented in dimensionless form in figure 2. The data shown pertain to two axial locations in the test section, i.e. at locations x = 1000 mm and 1200 mm downstream of the nozzle inlet, where entrance effects may not be present. The data are taken down to $y/r_o$ = 0.11 from the tube center. The measured arithmetic mean diameter $d_{10}$ is about 32μm. The corresponding value of $\tau^+$ based on $d_{10}$ is 292. The predicted values of $c/c_c$ from the present analysis is shown in figure 2 for $\tau^+$ = 292.

The measured concentration profile at x = 1200 mm lie somewhat below the data corresponding to x = 1000 mm. It is seen from figure 2 that the predicted concentration distribution is in good agreement with the data, except for some region near the tube center where the measured concentration is relatively flat. It is unfortunate that no data appear to exit for concentrations near the perifery of the turbulent core, with which to compare the present predictions. The buffer layer for the conditions of the data presented in figure 2 corresponds to $y/r_o$ = 0.0284. Calculations have shown that the value of $c_b$ is of the order of $k_d/u^*$, and is almost always less than 10 percent of the center-line concentration. The data also clearly suggest that the primary resistance to particle transport exists in the turbulent core region, thus confirming the soundness of the physical model employed here.

3.2 Deposition Data of Agarwal [6]

Agarwal [6] has obtained deposition data for uniform-size uranine-tagged olive aerosol in vertical down-flow of air in 0.327cm I.D. glass tube with L/D = 91.7 at Re = 6000, and in 1.38cm I.D. copper tube with L/D = 73.9 at Re = 50 000. The test section tubes are smooth. The drops are generated by means of a vibrating orifice monodisperse aerosol generator. The maximum size of the droplet used is 21 μm, and the droplet to fluid density ratio is about 713.

The calculated deposition velocity $k_d^+$ vs. $\tau^+$ using the present analysis is compared in figure 3 with the above data. The data for Re = 6000 covers a maximum $\tau^+$ of 291 corresponding to d = 21 μm, while a maximum of $\tau^+$ = 449 corresponding to d = 17.9 μm is obtained for the conditions at Re = 50 000. It is to be noted that the data for $\tau^+$ > 40 only are presented in the figure, since that represents the scope of our present work. The calculated results from the theory of Friedlander & Johnstone [2] are also displayed.

It is seen that the theory of Friedlander & Johnstone [2] is unable to predict the trends of the deposition velocity. Even at $\tau^+$ = 40, their theory considerably under-predicts the deposition velocity. On the other hand the present theory is able to represent the trends of the data, although the theory slightly under-predicts the measurements. The minor deviation between the theory and the data can be partly attributed to the uncertainities associated with the value of $\ell_E$ and $\beta$ in (14) and (15c) respectively. Nevertheless the agreement should be considered satisfactory at the present time in view of the apparent complexity of the phenomena involved.

The calculated value of particle Reynolds number $Re_p$ at $\tau^+$ = 291 for Re = 6000 is 1.8, while that at $\tau^+$ = 449 for Re = 50 000 is 1.1. This implies that the present theory appears to apply for $Re_p$ as large as about 2.0. That is, the theory is seen to be satisfactory for particle size having $\tau^+$ somewhat greater than $\tau_m^+$ given by (21c). This is to be expected since Stokes drag coefficient is not very much in error for $Re_p$ even up to about 2.0, while a value $Re_p$ = 1.0 is used in (21c).

3.3 Deposition Data of Cousins & Hewitt [12]

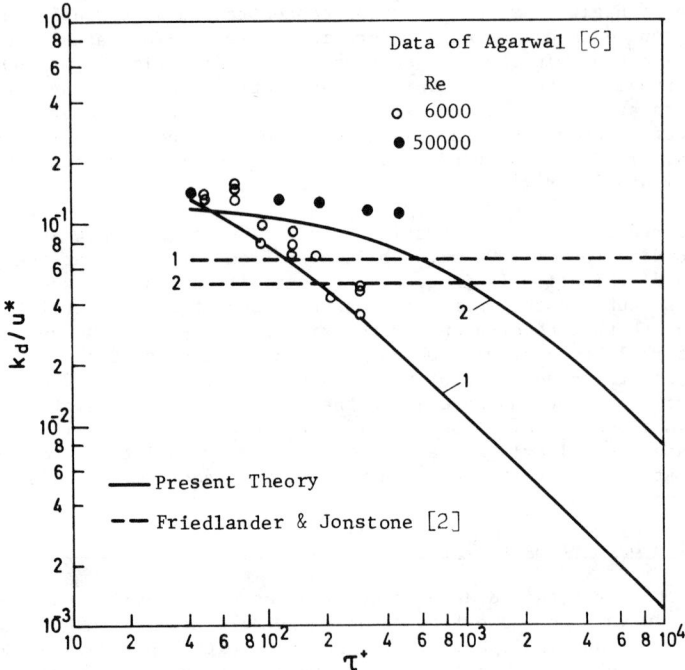

Fig. 3. Comparison of predicted deposition velocity with the experimental data of Agarwal [6].

Cousins & Hewitt [12] have measured deposition velocities in vertical upflow of air and water in 0.009525 m and 0.0318 m I.D. tubes. Water is introduced as an annulus through a porous sinter and entrainment of water droplets is generated due to vapor shear. The liquid film deposited in the test section is again removed, and the fraction of liquid deposited, F, is calculated directly. A wide range of deposition lengths and liquid loading are considered. Drop size is measured using photographic technique. The measured mean Sauter diameter, $d_{32}$, ranges from 40-70 μm in the smaller tube, and 70-110 μm in the larger diameter tube.

The $d_{10}$ values are computed from the average of measured values of $d_{32}$ and the relationship $d_{32}/d_{10} = 4.667$, as developed by Tatterson, Dallman & Hanratty [35]. This relationship is based on drop size distributions measured by many investigators.

Table 1 depicts the comparison of calculated deposition velocities with the measured values for the two tube diameters. The data shown in Table 1 for the 0.009525 m dia. tube are averaged for the different test section lengths excluding the shortest section for which entrance effects are observed, and also for the different liquid loadings. The measured deposition velocity for the 0.0318m pipe, however, corresponds to a deposition length of 1.94 m. No reentrainment of droplets is observed for the data selected here for comparison. It is seen from Table 1 that the present theory is in good agreement with the data. The maximum error for the lower Re is about 17%. At higher Re, the error is about 11%. On the other hand, the theory of Friedlander & Johnstone [2] is in error of about 40% compared to the data. It is interesting to note from Table 1 that the particle $Re_p$ has not exceeded about 0.85, thus indicating the validity of the present theory.

Table 1. Comparison of predicted deposition velocities with the experimental data of Cousins & Hewitt [12]

| D,m | Re | $\tau^+$ | $k_d$, m/s | | | % Error | | $Re_p$ |
|---|---|---|---|---|---|---|---|---|
| | | | Expt. | Present Theory | F&J* | Present Theory | F&J | |
| .009525 | 37,380 | 164 | .165 | .178 | .098 | 7.9 | 40.6 | .54 |
| | 56,051 | 334 | .186 | .218 | .133 | 17.0 | 28.6 | .85 |
| .0318 | $1.4 \times 10^5$ | 354 | .137 | .129 | .070 | 5.8 | 48.8 | .36 |
| | $1.96 \times 10^5$ | 642 | .179 | .159 | .091 | 10.9 | 42.6 | .54 |

*F & J - Theory of Friedlander & Johnstone [2]

3.4 Present Deposition Measurements

Experimental apparatus and procedure. In order to provide a further check on the proposed deposition model, the theory is also compared with the present deposition data for air-water system in a vertical tube at near atmospheric pressure. For details of the experiment, reference [28] may be consulted. In the experiments, air from a compressor passes first through a refrigerated air dryer to remove moisture and later through an oil filter to eliminate oil content. Air flow is metered by means of calibrated variable area rotameters with a rating of 39.2 scfm. Distilled water from a water tank is circulated by a variable speed Master Flex pump, and is metered through a calibrated flow meter. A flow integrator is used at the discharge end of the water pump to eliminate any flow pulsations.

The liquid droplets are generated by an atomizer with a sharp-edged air orifice and a cylindrical liquid nozzle. Secondary air is used for the atomizer while the primary air stream is used to vary the bulk flow rate in the test section. The atomizer and the deposition test section are illustrated in figure 4. The major dimensions of the atomizer are similar to those employed by Nukiyama & Tanasawa [34] in some of their pioneering experiments. The air orifice has a diameter of 1.5 mm. The Sauter mean diameter of the droplet size distribution is calculated from the Nukiyama & Tanasawa correlation [34]. The characteristic arithmetic mean diameter $d_{10}$ are obtained from $d_{32}$, using the relations developed by Tatterson et al. [35]. It is possible to vary drop size and droplet concentration by adjusting the secondary and the primary air flow rates.

As shown in figure 4, an entrance section having 12.7mm I.D. and L/D of 60 is employed to insure that the flow is fully developed at the inlet of the test

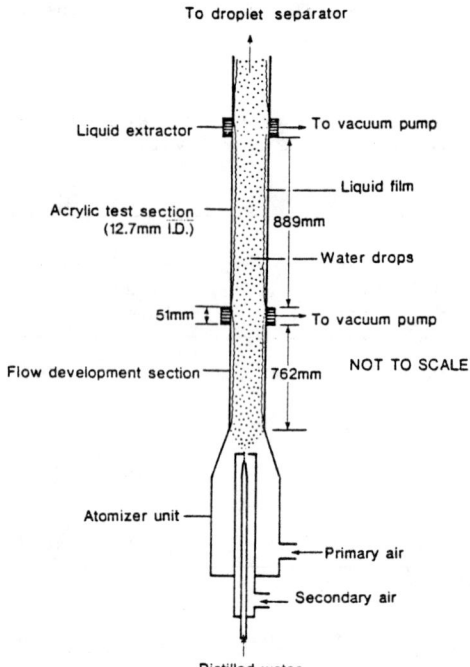

Fig. 4. Experimental system for deposition studies.

section. A converging section having length of 50.8 mm is used to connect the atomizer unit to the test section. The test section is a smooth acrylic tube 12.7 mm I.D. and 889 mm long. A pressure measurement is made at the inlet of the test section, and the pressure drop across the test section is measured using a water manometer. Other necessary pressure and temperature readings are observed in the experimental loop. Extraction sections, 50.8 mm long, are employed to remove the liquid film formed on the wall at the two ends of the test section. A major part of the liquid flowing through the nozzle collects into the wall layer as film in the flow development section, and is removed at the inlet extraction unit. By adjusting the pump speed, it is possible to completely remove the liquid film formed at the inlet and outlet of the test section, which is verifed by visual examination.

The droplets from the two-phase mixture at the exit of the test section are separated by a cyclone separator followed by a filter having a retention efficiency of 99%. The technique of measuring the deposition rate is essentially the same as applied by Cousins & Hewitt [12]. During the experiment, the liquid layer formed at the inlet of the test section is completely removed. The drops then migrate towards the wall and form a thin liquid film that grows continuously along the test section. The liquid film formed at the test section exit is also removed completely. The flow rate of liquid collected at the test section exit gives the amount of liquid deposited over the entire test section length. The flow rate of liquid present in the air-water droplet mixture flowing past the top extractor unit is measured by collecting the liquid drains form the cyclone separator and the filter. This serves as a check on the overall mass balance. All the data are taken in a steady state condition.

The fractional deposition F is given by the ratio of the flow rate of the liquid collected at the test section outlet extractor to the flow rate of droplets entering the test section. The droplet flow rate at the test section inlet is given by the difference between the total liquid flow rate $W_\ell$ entering the liquid nozzle and the flow rate of liquid collected at the inlet extractor unit. With a knowledge of F, the deposition velocity $k_d$ is determined from a mass balance given as $F = 1-\exp[-4(k_d/\bar{U})(L/D)]$ (22)

During the experiments, visual observation has indicated that the test section wall is covered with a thin liquid film usually smooth but sometimes with small ripples at the larger water flow rates. However no significant wave on the water film surface has been noted so that the droplet entrainment is believed not to have occured in the experiments. The entrainment E is defined as the amount of liquid entrained expressed as a fraction of the liquid film flow rate. The correlation of Truong Quang Minh & Huyghe [36] suggests entrainment of less than 2 percent for Re = 94 600, and less than 0.5 percent for Re = 52 500. In view of these factors the entrainment effect is not taken into account in deducing the deposition velocities.

Through the extractor unit, a small amount of air is also blow off together with the liquid film. This is evident from the visual observation of air bubbles present in between the liquid masses flowing in the acrylic tubing connecting the extractor to the vacuum pump. The flow rate of air removed through the extractor is however measured to be less than about one percent of the total air flow rate through the test section, and is therefore considered to have no significant effect on the measured deposition data.

An error analysis has indicated that the measured deposition velocities are within about 12 percent accuracy. The reproducibility of the data has been assured by repeating some of the runs. The data in general are consistent without significant scatter and are therefore considered reliable.

<u>Comparison of theory with present data.</u> The measured dimensionless deposition velocities for Re = 52 500 and 94 600 are depicted in figures 5 & 6 respectively. For the data at Re = 52 500, the test section inlet pressure is 1.15 bar, the drop diameters $d_{10}$ are varied from 12 - 46 µm and the droplet concentration is in the range of $(8.5-62) \times 10^{-3}$ kg/m³. The corresponding dimensionless drop relaxation time $\tau^+$ ranges from 264 - 3520. The conditions for Re = 94 600 are as follows: the test section inlet pressure is 1.37 bar, the $d_{10}$ values range from 8-45 µm, and the droplet concentration is varied from $(22-75) \times 10^{-3}$ kg/m³, while $\tau^+$ has a range of 295-8670. The room temperature is about 290 K. It is interesting to note that the present data pertain to relatively large $\tau^+$ compared to those of Agarwal [6] and of Cousins & Hewitt [12] which are limited to $\tau^+$ of about 40 - 640, while the present measurements yield data for $\tau^+$ in the range of 260 - 8700. Therefore the present data are believed to be a valuable supplement to the existing body of deposition data.

It is seen from figures 5 & 6 that $k_d/u^*$ decreases with $\tau^+$ up to some value of $\tau^+$ beyond which it remains nearly independent of $\tau^+$. This appears to be an important new information since this trend has not yet been reported earlier. The results also reveal that $k_d/u^*$ is not quite sensitive to Re. The dependence of $k_d/u^*$ on concentration is not found to be significant. This is a characteristic trend at low droplet concentrations ($\bar{c}$ typically less than about 1.0 kg/m³) as observed earlier by Cousins & Hewitt [12] for a vertical

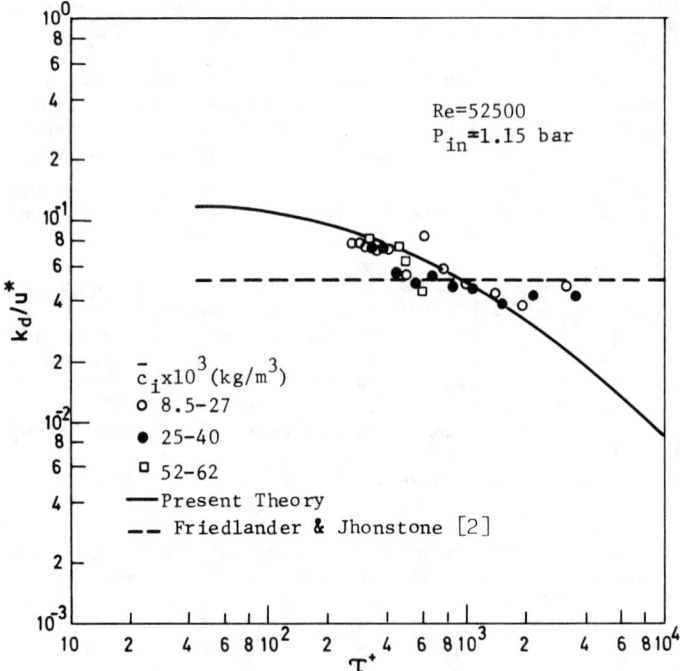

Fig. 5. Comparison of predicted deposition velocity with the present experimental data at Re = 52,500.

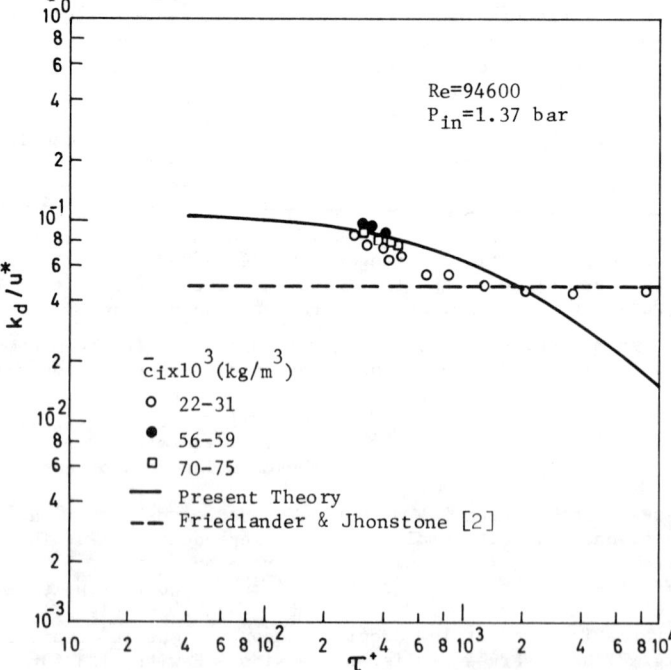

Fig. 6. Comparison of predicted deposition velocity with the present experimental data at Re = 94,600.

system. At high concentration, $k_d/u^*$ is generally found to decrease (Hagiwara et al. [18]) due to changes in the turbulence characteristics of the gas phase.

The predictions of $k_d/u^*$ from the present theory and the theory of Friedlander & Johnstone [2] are also compared with the data in figures 5 & 6. It is observed from figure 5 that the present theory is in good agreement with the data for Re = 52 500 up to $\tau^+$ of about 1500 above which the theory under-predicts the data, the deviation increasing with increasing $\tau^+$. Near the point of deviation, the value of $Re_p$ is about 3.1. Similar trend is noticed for Re = 94 600 also. Figure 6 suggests that the present analysis agrees with the data for Re = 94 600 up to $\tau^+$ of 2000 although the theory slightly overestimates the data for $\tau^+$ of 600 - 1500. Above $\tau^+ = 2000$, the theory deviates with the data. The value of $Re_p$ at the point of deviation is 3.0. The deviation of the theory with the data at relatively large $\tau^+$ is therefore attributable to inadequacy of the assumption of Stokesian drag for large size particles.

On the other hand figure 5 shows that for Re = 52 500, the theory of Friedlander & Johnstone [2] is not able to represent the trend of the data up to $\tau^+$ of about 400, but there is seen to be a surprisingly good agreement for $\tau^+$ of 400 - 3000. Similar trend is observed for Re = 94 600 also. However, since their theory is not able to describe the deposition velocity for $\tau^+ < 400$, the seemingly good comparison obtained for $\tau^+ > 400$, as in the case of Re = 52 500, does not seem to have significance.

In view of all the above comparisons, it is evident that the present theory offers a considerable improvement over the theory of Friedlander & Johnstone [2].

## 4. CONCLUSION

The deposition motion of droplets from a turbulent gas stream in a vertical tube has been studied theoretically and experimentally. It has been shown that the dispersion in the bulk plays a principal role in governing the resistance characterizing the rate of deposition. The proposed theory satisfactorily represents the existing data and the present measurements of deposition rates in the Stokes regime.

ACKNOWLEGMENTS

This work is supported by the National Science Foundation under grant ENG 78 - 06211.

REFERENCES

1. McCoy, D.D. and Hanratty, T.J. 1977. Rate of deposition of droplets in annular two-phase flow. Int. J. Multiphase flow, Vol. 3, pp. 319-331.

2. Friedlander, S.K. and Johnstone, H.F. 1957. Deposition of suspended particles from turbulent gas streams. Ind. Engng. Chem.,Vol. 49, pp. 1151-1156.

3. Wells, A.C. and Chamberlain, A.C. 1967 Transport of small particles to vertical surfaces. Brit. J. Appl. Phys., Vol. 18, 1793-1799.

4. Sehmel. G.A. 1970. Particle deposition from turbulent air flow. J. Geophys. Res., Vol. 75, pp. 1776-1781.

5. Liu, B.Y.H. and Agarwal, J.K. 1974. Experimental observation of aerosol de-

position in turbulent flow. J. Aerosol Sci., Vol. 5, pp. 145-155.

6. Agarwal, J.K. 1975. Aerosol sampling and transport. Ph.D. thesis, Univ. Minnesota, Particle Tech. Lab Publ. No. 265.

7. Ilori, T.A. 1971. Turbulent deposition of coarse aerosols from turbulent flow. J. Aerosol Sci., Vol. 5, pp. 257-271.

8. Schwendimen, L.C. and Postma, A.K. 1961. Turbulent deposition in sampling lines. USAEC R. HW-65309.

9. Davies, C.N. 1966. Deposition of aerosols from turbulent flow through pipes. Proc. Roy. Soc. A., Vol. 289, pp. 235-246.

10. Beal, S.K. 1970. Deposition of particles in turbulent flow on channel or pipe walls. Nuclear Sci. and Eng., Vol. 40, pp. 1-11.

11. Liu, B.Y.H. and Ilori, T.A. 1974. Aerosol deposition in turbulent pipe flow. Environmental Sci. and Tech., Vol. 8, pp. 351-356.

12. Cousins, L.B. and Hewitt, G.F. 1968. Liquid phase mass transfer in annular two-phase flow: Droplet deposition and liquid entrainment. AERE-R 5657.

13. Farmer, R.A., Griffith, P. and Rohsenow, W.M. 1970. Liquid droplet deposition in two-phase flow. Trans. A.S.M.E., J. Heat Transfer, Vol. 92, pp. 587-594.

14. Forney, L.J. and Spielman, L.A. 1974. Deposition of coarse aerosols from turbulent flow. J. Aerosol. Sci., Vol. 5, pp. 257-271.

15. Hutchinson, P., Hewitt, G.F. and Dukler, A.E. 1971. Deposition of liquid or solid dispersions from turbulent gas streams; a stochastic model. Chem. Engng. Sci., Vol. 26, pp. 419-439.

16. McCoy, D.D. 1975. Particle deposition in turbulent flow. M.S. thesis, Dept. Chem. Engng., Univ. of Illinois at Urbana-Champaign.

17. Gardner, G.S. 1975. Deposition of particles from a gas flowing parallel to a surface. Int. J. Multiphase flow, Vol. 2, pp. 213-218.

18. Hagiwara, Y. and Sato, T. 1979. An experimental investigation on liquid droplets diffusion in annular-mist flow. Second Multi-phase flow and Heat Transfer Symposium-Workshop, Miami Beach, Florida.

19. Namie, S. and Ueda, T. 1973. Droplet transfer in two-phase annular mist flow (Part 2, Prediction of droplet transfer rate). Bull. J.S.M.E., Vol. 16, pp. 752-764.

20. Reeks, M.W. and Skyrme, G. 1976. The dependence of particle deposition velocity on particle inertia in turbulent pipe flow. J. Aerosol. Sci., Vol. 4, pp. 485-495.

21. Cleaver, J.W. and Yates B. 1975. A sublayer model for the deposition of particles from a turbulent flow. Chem. Engng. Sci., Vol. 30, pp. 983-992.

22. Hinze, J.O. 1975. Turbulence, 2nd edn. McGraw-Hill.

23. Owen, P.R. 1969. Pneumatic transport. J. Fluid Mech., Vol. 39, pp. 407-432.

24. Kays, W.M. 1966. Convective Heat and Mass Transfer, McGraw-Hill.

25. Tchen, C.M. 1947. Mean value and correlation problems connected with the motion of small particles suspended in a turbulent fluid. Martinus Nijhoff, The Hague.

26. Soo, S.L. 1967. Fluid Dynamics of Multiphase Systems, London: Blaisdell Press.

27. Monin, A.S. and Yaglom, A.M. 1971. Satistical Fluid Mechanics: Mechanics of Turbulence (Edited by J.L. Lumley), MIT Press.

28. Ganic, E.N. and Mastanaiah, K. 1979. Studies of liquid drop deposition, drop carry-over velocity and heat transfer in two-phase drop flow. Report No. TR-E-79-1, Prepared for the National Science Foundation, Dept. of Energy Engg. Univ. of Illinois at Chicago Circle.

29. Friedlander, S.K. 1957. Behavior of suspended particles in a turbulent fluid. A.I.Ch.E.J., Vol. 3, pp. 381-385..

30. Martin, G.Q. and Johanson, L.N. 1965. Turbulence characteristics of liquids in pipe flow. A.I.Ch.E.J., Vol. 11, pp. 29-33.

31. Hay, J.S. and Pasquill, F. 1957. Diffusion from a fixed source at a height of a few hundred feet in the atmosphere. J. Fluid Mech., Vol. 2, pp. 299-310.

32. Snyder, W.H. and Lumley, J.L. 1971. Some measurements of autocorrelation functions in a turbulent flow. J. Fluid Mech., Vol. 48, pp. 41-71.

33. Soo, S.L. 1978. Diffusivity of spherical particles in dilute suspensions. A.I.Ch.E. J., Vol. 74, pp. 184-185.

34. Nukiyama, S. and Tanasawa, Y. 1938 & 1939. An experiment on the atomization of liquid by means of an air stream. Reports 1 and 4; Trans. J.S.M.E., Vol. 4, p. 86, and Vol. 5, p. 68.

35. Tatterson, D.F., Dallman, J.C. and Hanratty, T.J. 1977. Drop sizes in annular gas-liquid flows. A.I.Ch.E J., Vol. 23, pp. 68-76.

36. Troung Quang Minh and Huyghe, J. 1965. Some hydrodynamical aspects of annular dispersed flow: entrainment and film thickness. Symposium on Two-phase Flow, Paper C2, Exeter, England.

# RADIATIVE HEAT TRANSFER IN HEAT EXCHANGERS

# Radiative Heat Transfer in Heat Exchangers

**M. NECATI ÖZIŞIK**
Department of Mechanical and Aerospace Engineering
North Carolina State University
Raleigh, North Carolina 27650 USA

ABSTRACT

The analysis of radiation heat transfer in heat exchanger applications is presented with particular emphasis to: Radiating extended surfaces; forced convection to transparent fluid with radiation boundary conditions; forced convection to absorbing and emitting fluid; and forced convection to absorbing, emitting and scattering fluid.

NOMENCLATURE

| | |
|---|---|
| $A$ | area |
| $C_p$ | specific heat |
| $F$ | view factor |
| $dF$ | elemental view factor |
| $h$ | heat transfer coefficient |
| $h^* = \dfrac{h}{q_w}\left(\dfrac{q_w}{\bar{\sigma}}\right)^{1/4}$ | |
| $I(\tau,\mu)$ | radiation intensity |
| $L$ | characteristic dimension |
| $N_c = \dfrac{kt}{L^2 \bar{\sigma} T_b^3}$ | , conduction-to-radiation parameter |
| $N = \dfrac{k\kappa}{4n^2 \bar{\sigma} T^3}$ | , conduction-to-radiation parameter |
| $n$ | refractive index |
| $q$ | heat flux |
| $q^r$ | radiation heat flux |
| $q^c$ | conduction heat flux |
| $Q^r = \dfrac{q^r}{4n^2 \bar{\sigma} T^4}$ | |

| | |
|---|---|
| $R$ | radiosity |
| $S = \dfrac{4h}{\rho U_m C_p}$ | |
| $T$ | temperature |
| $T_b$ | bulk temperature |
| $T_w$ | wall temperature |
| $t$ | thickness |
| $\beta = \dfrac{R}{\sigma T_b^4}$, | dimensionless radiosity |
| $\beta = \dfrac{R}{q_w}$, | dimensionless radiosity |
| $\beta$ | extinction coefficient |
| $\kappa$ | absorption coefficient |
| $\sigma$ | scattering coefficient |
| $\bar{\sigma}$ | Stefan-Boltzmann constant |
| $\xi$ | dimensionless axial variable |
| $\eta = \dfrac{Q}{Q_{ideal}}$, | fin efficiency |
| $\mu$ | direction cosine |
| $\nu$ | frequency |
| $\theta = \dfrac{T}{T_b}$, | dimensionless temperature |
| $\theta = \left(\dfrac{\bar{\sigma}}{q_w}\right)^{1/4} T$, | dimensionless temperature |
| $\varepsilon$ | emissivity |
| $\rho$ | reflectivity |
| $\tau$ | optical variable |
| $\tau_o$ | optical thickness |

CONTENTS

1. INTRODUCTION
2. RADIATIVE HEAT EXCHANGE BETWEEN SURFACES
3. HEAT TRANSFER FROM RADIATING FINS
    3.1 Radiating longitudinal fins
    3.2 Radiating plate fin with interaction between the fin base and fin surface

4. FORCED CONVECTION TO A TRANSPARENT GAS WITH RADIATION BOUNDARY CONDITIONS
5. FORCED CONVECTION TO AN ABSORBING AND EMITTING GAS WITH RADIATION
   5.1 Radiation heat flux in an absorbing, emitting, plane parallel medium
   5.2 Absorbing, Emitting, couette flow
6. RADIATION HEAT FLUX IN AN ABSORBING, EMITTING, SCATTERING PLANE-PARALLEL MEDIUM
   6.1 Absorbing, emitting, scattering, thermally developing slug flow between parallel-plates
REFERENCES

1. INTRODUCTION

Radiation is an important mechanism of heat transfer at high temperatures, when the fourth-power law of heat dissipation becomes significant in comparison to convection and conduction. For example, in heat rejection system for space vehicles, the waste heat from the power plant, electronic equipment, and various other components is carried by a coolant fluid to the space radiator which employ some form of extended surface. The waste heat reaching the radiator is conducted to the fin surface and then from the fin surface is dissipated to the atmosphere-free space by thermal radiation. Since the radiators are probably among the heaviest components in the cooling system of a space vehicle, and the reduction of weight and size is of prime importance in the design of space vehicles, it is desirable to know the heat transfer characteristics of the radiator as accurately as possible in order to minimize the weight and size. In such applications, the basic mechanism of heat transfer from the radiator into the space is conduction through the fin combined with radiation in a non-participating medium.

In engineering applications involving heat transfer from a flowing fluid, the role of radiation becomes important in comparison to convection when the temperatures of the fluid and of the conduit become sufficiently high. For a transparent gas (i.e., a gas that does not absorb, emit or scatter radiation) flowing over a surface subjected to a prescribed heat flux, the radiation between the wall surface and an external environment alters both the wall surface temperature and the gas temperature rise.

For a gas that absorbs and emits radiation, the presence of radiation alters the temperature distribution within the fluid, hence a simultaneous treatment of convection and radiation is necessary in such situations. That is the energy equation for heat transfer in the fluid contains the divergence of the radiative heat flux, which must be obtained from the solution of the equation of radiative transfer. As a result the convection and radiation problems are coupled.

In situations when the gas contains suspended particles which scatters radiation in addition to absorption and emission of radiation by the medium itself, the analysis of the combined convection and radiation becomes more involved because of the difficulties associated with the solution of the equation of radiative transfer containing the scattering term.

In this paper, the methods of analysis of heat transfer when radiation effects are important are presented with particular emphasis to heat exchanger applications. In section 2, basic relations are developed for radiative heat exchange between surfaces needed in the analysis of radiative exchange from extended surfaces as well as for combined convection and radiation for a transparent gas. In section 3, the analysis of radiative heat exchange for extended surfaces, including the interaction between the adjacent fins, is presented. In section 4, combined convection and radiation for the flow of transparent gas through conduits subjected to a prescribed heat flux at the walls is treated. In section 5, heat transfer by combined convection and radiation to an absorbing and emitting gas is given. Finally, in section 6, the effects of scattering of radiation by the suspended particles in the gas on heat transfer are discussed and the methods of analysis of heat transfer for such situations are illustrated.

## 2. RADIATIVE HEAT EXCHANGE BETWEEN SURFACES

We first examine radiative heat exchange among the surfaces of an <u>enclosure</u> that contains a <u>nonparticipating</u> medium. The term <u>enclosure</u> is used to designate a region completely surrrounded by a set of surfaces that are characterized by their radiative properties and temperatures (or heat fluxes), so that a full account can be made of the incoming and outgoing radiation at any of these surfaces. The term <u>nonparticipating</u> medium refers to an environment that does not absorb, emit or scatter radiation.

A convenient way of analyzing the radiation problem among the surfaces is to separate the entire surface of the enclosure into a finite number of zones. Several investigators utilized this approach and introduced very restrictive assumptions in order to reduce the analysis of radiative heat transfer problem for an enclosure to the solution of a <u>set of algebraic equations</u>. Such methods of analysis are well documented in the literature [1-5]. A scrutiny of all these methods reveals that, for a given physical system each will provide the same answer, despite the apparent difference between the formulation of the problem. We refer to these methods as <u>Simplified Zone Analysis</u> [6]. The utility of the Simplified Zone Analysis is severely restricted because of the assumption that the radiosity is uniform over the surface of each zone.

Therefore, we present here the approach called <u>Generalized Zone Analysis</u> in which the radiative heat transfer problem for an enclosure is transformed to the solution of a set of <u>integral equations</u>. In this method, the enclosure surface is divided into a finite number of zones, $i = 1, 2, \ldots, N$, and the following conditions are assumed satisfied at the surface of each zone:

1. The radiative properties are uniform and independent of direction.
2. Either the temperature or the heat flux distribution is specified over the surface of each zone.
3. The surfaces are diffuse emitters and diffuse reflectors.

We also assume that the entire energy spectrum is divided into a finite number of frequency bands, $\Delta \nu_k$, $k = 1, 2, \ldots, K$, such that the radiative properties of the surfaces are uniform over each band, $\Delta \nu_i$. Then the integral equations governing the radiative heat exchange between the zones, for each frequency band $\Delta \nu_k$, are given by [6]

$$R_{i,k}(\underline{r}_i) = \varepsilon_{i,k} \pi I_{b,k}[T_i(\underline{r}_i)] + \rho_{i,k} \sum_{j=1}^{N} \int_{A_j} R_{j,k}(\underline{r}_j) dF_{dA_i - dA_j} \tag{1}$$

$$q_i(\underline{r}_i) = \sum_{k=1}^{K} q_{i,k}(\underline{r}_i) \tag{2a}$$

where

$$q_{i,k}(\underline{r}_i) = R_{i,k}(\underline{r}_i) - \sum_{j=1}^{N} \int_{A_j} R_{j,k}(\underline{r}_j) dF_{dA_i - dA_j} \qquad (2b)$$

or

$$q_{i,k}(\underline{r}_i) = \frac{1}{\rho_{i,k}} \left\{ \varepsilon_{i,k} \pi I_{b,k}[T_i] - (1 - \rho_{i,k}) R_{i,k}(\underline{r}_i) \right\} \text{ for } \rho_{i,k} \neq 0 \qquad (2c)$$

Here, $R_{i,k}(\underline{r}_i)$ is the <u>radiosity</u> at the location $\underline{r}_i$ on the zone $A_i$ for the frequency band $\Delta\nu_k$, $I_{b,k}[T_i(\underline{r}_i)]$ the Planck function integrated over the band $\Delta\nu_k$, $dF_{dA_i-dA_j}$ is the differential view factor between the elementary surfaces $dA_i$ and $dA_j$ over the zones $A_i$ and $A_j$, $q_{i,k}(\underline{r}_i)$ is the partial radiative heat flux over the band $\Delta\nu_i$ and $q_i(\underline{r}_i)$ is the total radiative heat flux over the entire frequency band at the location $\underline{r}_i$ on the zone $A_i$. The positive value of $q_i$ means that the radiative energy is <u>leaving the surface</u>. The quantities $\varepsilon_{i,k}$ and $\rho_{i,k}$ are the average emissivity and reflectivity of the zone $A_i$ over the frequency band $\Delta\nu_i$.

Equations (1) and (2) provide the complete mathematical formulation of the radiation problem considered above for the band approximation.

In the case of a <u>gray</u> enclosure the radiative properties of the surfaces are assumed to be independent of frequency. Then we omit the subscript k, and equations (1) and (2) for a <u>gray enclosure</u> reduce to

$$R_i(\underline{r}_i) = \varepsilon_i \bar{\sigma} T_i^4(\underline{r}_i) + \rho_i \sum_{j=1}^{N} \int_{A_j} R_j(\underline{r}_j) dF_{dA_i - dA_j} \qquad (3)$$

$$q_i(\underline{r}_i) = R_i(\underline{r}_i) - \sum_{j=1}^{N} \int_{A_j} R_j(\underline{r}_j) dF_{dA_i - dA_j} \qquad (4a)$$

or

$$q_i(\underline{r}_i) = \frac{1}{\rho_i} [\varepsilon_i \bar{\sigma} T_i^4(\underline{r}_i) - (1 - \rho_i) R_i(\underline{r}_i)] \text{ for } \rho_i \neq 0 \qquad (4b)$$

$i = 1, 2, \ldots, N$, and $\bar{\sigma} \equiv$ Stefan-Boltzmann constant.

Equations (3) and (4) are the complete mathematical formulation of radiative heat exchange for an N zone <u>gray enclosure</u>.

To illustrate the physical significance of these equations in the solution of radiation problems, we consider an N zone gray enclosure and assume that the temperatures $T_i(\underline{r}_i)$, $i = 1, 2, \ldots, N$, are prescribed over the surface of each zone. We wish to determine the distribution of radiative heat fluxes, $q_i(\underline{r}_i)$, over the zones. Then, equations (3) provide N coupled integral equations for the N unknown radiosity functions $R_i(\underline{r}_i)$, $i = 1, 2, \ldots, N$. Once the radiosity functions are known from the solution of such a system, the distribution of radiative heat flux $q_i(\underline{r}_i)$ for any of these zones is determined from equation (4a) or (4b).

Equations (3) and (4) are also applicable to a gray enclosure in which <u>temperatures are prescribed for some of the zones and heat fluxes for the others</u>. Then the problem becomes one of determination of temperatures $T_i(\underline{r}_i)$ for the zones for which the heat fluxes are prescribed, and the determination of heat fluxes $q_i(\underline{r}_i)$ for the zones for which temperatures are prescribed. The reader should consult reference [6] for further discussion and the details of this matter.

It is to be noted that the equations for the <u>simplified zone</u> analysis are immediately obtainable from the foregoing equations by introducing an additional assumption that the radiosity is uniform over the surface of each zone. The temperatures and heat fluxes are also regarded uniform over the surface of each zone. For example, for a gray enclosure, if we assume that the radiosities $R_j$ are uniform over each zone, then $R_j$ is taken out of the integral and the differential view factor $dF_{dA_i-dA_j}$ is integrated over the surface. Then equations (3) and (4) respectively reduce to

$$R_i = \varepsilon_i \bar{\sigma} T_i^4 + \rho_i \sum_{j=1}^{N} R_j F_{i-j} \tag{5}$$

$$q_i = R_i - \sum_{j=1}^{N} R_j F_{i-j} \tag{6a}$$

or

$$q_i = \frac{1}{\rho_i} [\varepsilon_i \bar{\sigma} T_i^4 - (1 - \rho_i) R_i] \quad \text{for} \quad \rho_i \neq 0 \tag{6b}$$

$$i = 1, 2, \ldots, N.$$

Here, $F_{i-j}$ is the diffuse view factor between the zones $A_i$ and $A_j$.

Clearly, the analysis of radiative heat exchange is significantly simplified under the additional assumption of uniform radiosity over the surface of each zone. Suppose the temperatures are prescribed for each zone. Then equations (5) are N algebraic equations for the N radiosities $R_j$. Once the radiosities are determined from the solution of such a system, the radiative heat flux at any one of the zones is given by equation (6a) or (6b).

In all the expressions given above it is considered that the diffuse view factor $F_{i-j}$ is known or an analytical expression is available for the elemental view factor $dF_{dA_i-dA_j}$.

## 3. HEAT TRANSFER FROM RADIATING FINS

The heat exchangers for space vehicles for the rejection of waste heat into atmosphere-free space usually employ some form of extended surfaces as illustrated in Fig. 1. The analysis of heat transfer from a radiating fin, by neglecting the effects of mutual radiation between a fin and other surfaces of the radiator (i.e., the adjacent fins, the fin base) has been investigated for different fin geometries and the resulting fin efficiencies are well documented [7]. The analysis become more involved when the effects of mutual radiation between the adjacent surfaces are to be included [3,6,8-11]. To illustrate the analysis of heat transfer for such situations we consider the following two cases.

### 3.1 Radiating Longitudinal Plate Fins

Figure 2 illustrates the geometry and the coordinates for a longitudinal plate fin to be investigated here. This arrangement represents a repeating

# RADIATIVE HEAT TRANSFER IN HEAT EXCHANGERS

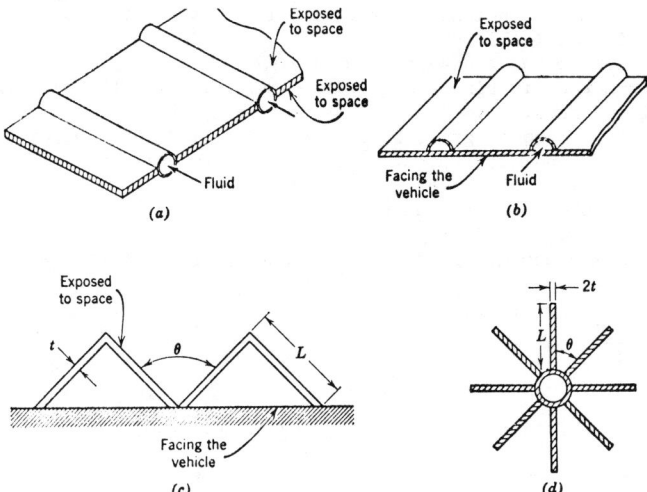

Fig. 1. Radiator Configurations for Space Vehicles

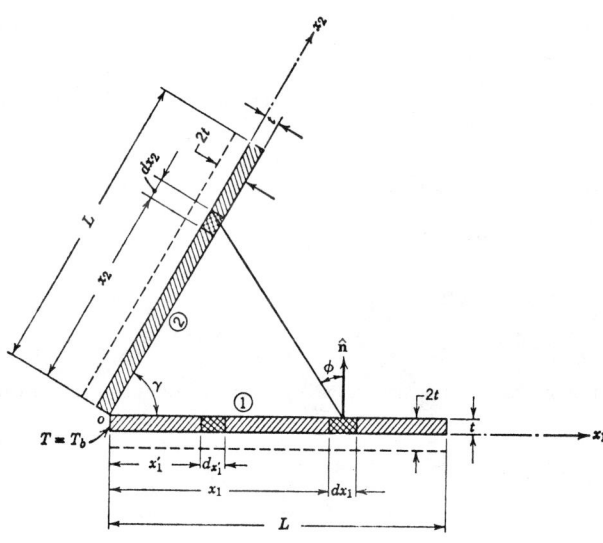

Fig. 2. Longitudinal Plate Fin

pattern in the fin configuration Fig. 1d for a plate fin of thickness 2t. The following assumptions are made.

1. The fin thickness is small so that the temperature gradients across the fin thickness can be neglected.
2. Temperature is a function of the axial coordinates $x_1$ and $x_2$ only.

3. The radiative energy incident on the fin surface from the external environment is negligible.
4. The temperature at the fin base is uniform, i.e., $T_b$.
5. Heat loss from the fin tip is negligible compared with that leaving from the lateral surfaces.
6. The surfaces are opaque, gray, diffuse emitters and have uniform emissivity $\varepsilon$. The thermal conductivity, k, of the plates is uniform.
7. Kirchhoff's law is applicable.
8. The surfaces are diffuse reflectors.

The steady-state energy equation for the plate 1 is written as

$$\frac{d^2 T_1(x_1)}{dx_1^2} = \frac{1}{k \cdot t} q_1(x_1) \tag{7}$$

and the radiative heat flux $q_1(x_1)$ is determined from (4a) as

$$q_1(x_1) = R_1(x_1) - \int_{x_2=0}^{L} R_2(x_2) \, dF_{dx_1-dx_2} \tag{8}$$

From equations (7) and (8), we obtain the following integrodifferential equation for the temperature distribution $T_1(x_1)$ in plate 1:

$$\frac{d^2 T_1(x_1)}{dx_1^2} = \frac{1}{kt} \left[ R_1(x_1) - \int_{x_2=0}^{L} R_2(x_2) \, dF_{dx_1-dx_2} \right] \quad \text{in} \quad 0 \leq x_1 \leq L \tag{9a}$$

subject to the boundary conditions

$$T_1(x_1) = T_b \quad \text{at} \quad x_1 = 0 \tag{9b}$$

$$\frac{dT_1(x_1)}{dx_1} = 0 \quad \text{at} \quad x_1 = L \tag{9c}$$

and the equation for the radiosity $R_1(x_1)$ is obtained from equation (3) as

$$R_1(x_1) = \varepsilon \bar{\sigma} T_1^4(x_1) + (1 - \varepsilon) \int_{x_2=0}^{L} R_2(x_2) \, dF_{dx_1-dx_2} \tag{9d}$$

where $\rho$ is replaced by $(1 - \varepsilon)$.

A set of equations similar to equations (9) can be written for the temperature distribution $T_2(x_2)$ in plate 2. However, the problem possesses symmetry, that is, $R_1(x_1) = R_2(x_2)$ and $T_1(x_1) = T_2(x_2)$ for $x_1 = x_2$. Then subscripts 1 and 2 can be removed from the radiosity and temperature functions; then equations (9) represent the complete mathematical formulation of the problem and the equations in the dimensionless form are given by

$$\frac{d^2\theta(\xi_1)}{d\xi_1^2} = \frac{1}{N_c}\left[\beta(\xi_1) - \int_{\xi_2=0}^{1} \beta(\xi_1)\, dF_{d\xi_1-d\xi_2}\right] \quad \text{in} \quad 0 \le \xi_1 \le 1 \tag{10a}$$

$$\theta(\xi_1) = 1 \quad \text{at} \quad \xi_1 = 0 \tag{10b}$$

$$\frac{d\theta(\xi_1)}{d\xi_1} = 0 \quad \text{at} \quad \xi_1 = 1 \tag{10c}$$

and

$$\beta(\xi_1) = \varepsilon\, \theta^4(\xi_1) + (1-\varepsilon)\int_{\xi_2=0}^{1} \beta(\xi_2)\, dF_{d\xi_1-d\xi_2} \tag{11}$$

where the dimensionless quantities are defined as

$$\theta \equiv \frac{T}{T_b}, \quad \beta \equiv \frac{R}{\sigma T_b^4}, \quad N_c \equiv \frac{kt}{L^2 \sigma T_b^3}, \quad \xi_1 \equiv \frac{x_1}{L} \quad \text{and} \quad \xi_2 \equiv \frac{x_2}{L} \tag{12}$$

Here, the quantity $N_c$ is the conduction-to-radiation parameter which signifies the relative importance of conduction compared with radiation. Clearly, the problem simplifies to that of pure conduction for $N_c \to \infty$ and to that of pure radiation for $N_c \to 0$. Analytical expression can readily be obtained for the elemental diffuse view factor $dF_{d\xi_1-d\xi_2}$ in terms of the dimensionless coordinates $\xi_1$ and $\xi_2$, and the angle of inclination $\gamma$ between the fin plates.

The distribution of radiative heat flux $q(x_1)$ along the fin surface given by equation (8) is now written in the dimensionless form as

$$\frac{q(\xi_1)}{\sigma T_b^4} = \beta(\xi_1) - \int_{\xi_2=0}^{1} \beta(\xi_2)\, dF_{d\xi_1-d\xi_2} \tag{13}$$

The net rate of heat dissipation by radiation $Q$, from the surface of one of the fin plates per unit width normal to the plane of the Fig. 2, is given by

$$Q = \int_{x_1=0}^{L} q(x_1)\, dx_1$$

or

$$\frac{Q}{\sigma T_b^4} = L\int_{\xi_1=0}^{1}\left[\beta(\xi_1) - \int_{\xi_2=0}^{1} \beta(\xi_2)\, dF_{d\xi_1-d\xi_2}\right] d\xi_1 \tag{14}$$

In order to define the radiative effectiveness $\eta$ for the fin, we now consider an ideal situation in which the fin surfaces are black ($\varepsilon = 1$) and maintained at a uniform temperature $T_b$ (i.e., the base temperature) everywhere. The rate of ideal heat dissipation by radiation from the surface of the fin plate per unit width normal to the plane of Fig. 2 is given by

$$Q_{ideal} = \bar{\sigma} T_b^4 \left[ L \sin \frac{\gamma}{2} \right] \tag{15}$$

Then, the radiative effectiveness $\eta$ for the fin is defined as

$$\eta \equiv \frac{Q}{Q_{ideal}} = \frac{1}{\sin\left(\frac{\gamma}{2}\right)} \int_{\xi_1=0}^{1} \left[ \beta(\xi_1) - \int_{\xi_2=0}^{1} \beta(\xi_2) \, dF_{d\xi_1 - d\xi_2} \right] d\xi_1 \tag{16}$$

The mathematical formulation of heat transfer by radiation from a longitudinal plate fin for a radiator of the type shown in Fig. 1d is now complete. Once the dimensionless radiosity function $\beta(\xi_1)$ is determined from the solution of the two coupled integrodifferential equations (10) and (11), the radiative effectiveness $\eta$ is computed from equation (16). Knowing the radiative effectiveness $\eta$, the heat dissipation rate by radiation Q, from the surface of one of the fin plate per unit width perpendicular to the plane of the figure is calculated from

$$Q = \eta \, Q_{ideal} = \eta \, \bar{\sigma} \, T_b^4 \, L \sin \frac{\gamma}{2} \tag{17}$$

Figure 3, obtained from reference [8] shows the radiative effectiveness

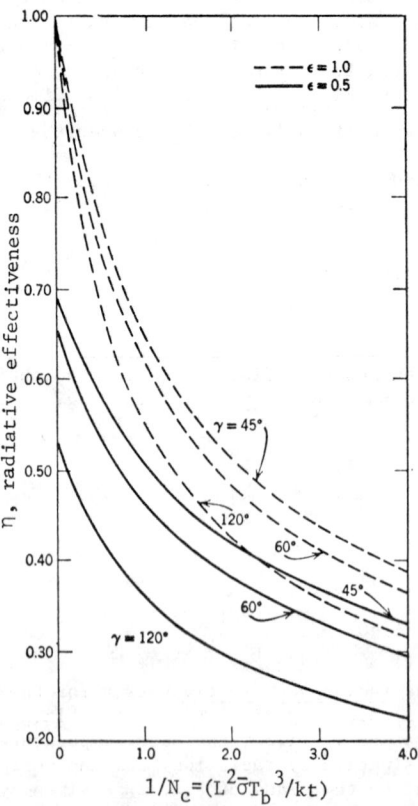

Fig. 3. Radiative Effectiveness for Longitudinal Plate Fins

for longitudinal plate fins plotted against $1/N_c$ for $\varepsilon = 1.0$ and $0.5$, for different values of the opening angle $\gamma$. For $\varepsilon = 1$, the curves converge to the greatest possible heat loss as $N_c \to \infty$ (i.e., as thermal conductivity becomes very high). As expected, the fin effectiveness decreases as the emissivity $\varepsilon$ decreases. For any given $N_c$ the fin effectiveness is always greater for smaller opening angles. Fin effectiveness decreases as $N_c$ decreases (i.e., radiation effects increase).

The presence of fins on a radiator surface increases not only the heat transfer area but also the radiator weight. Therefore, it is important to know the conditions that will yield maximum heat dissipation for a fixed profile, i.e., $A = Lt = $ constant, and for given values of $\varepsilon$ and $N_c$.

When the values of $\varepsilon$, $\gamma$ and $A$ are fixed, equation (17) can be written as

$$Q = \bar{\sigma} T_b^4 \frac{A}{t} \eta \sin \frac{\gamma}{2} \qquad (18)$$

where

$$N_c = \frac{kt^3}{A^2 \bar{\sigma} T_b^3} \qquad (19)$$

and the fin effectiveness $\eta$ becomes a function of $N_c$, that is, $\eta \equiv \eta(N_c)$; then, in equation (18), the fin thickness $t$ is the only independent variable. To maximize $Q$ with respect to $t$, we differentiate equation (18) with respect to $t$, equate the resulting expression to zero, and eliminate the term $\partial N_c/\partial t$ by means of the differentiation of equation (19). Then, the value of the conduction-to-radiation parameter, $(N_c)_{opt}$, that will give the maximum heat transfer rate, $Q_{max}$, for fixed values of $\varepsilon$, $\gamma$ and $A$ is determined as

$$\left(N_c\right)_{opt} = \frac{1}{3} \frac{\eta}{\left(\frac{\partial \eta}{\partial \eta_c}\right)} \qquad (20)$$

## 3.2 Radiating Plate Fins with Interaction Between the Fin Base and Fin Surface

We now consider a rectangular plate fin geometry as shown in Fig. 4, in which the fin base has appreciable exposed surface. In the formulation of this radiative heat transfer problem, the interaction between the fin and the fin base should be taken into account in addition to that between the adjacent fin surface. In order to illustrate the mathematical formulation of the problem for such a situation, we make the same assumption as those given in the previous example regarding the fin surfaces. In addition, we assume that the fin base is opaque, gray, diffuse emitter, diffuse reflector and have the same emissivity as the fin surfaces. Because of symmetry, we consider only half the thickness of the fin plate.

The one-dimensional, steady-state energy equation and the boundary conditions for plate 1 are given by

$$\frac{d^2 T_1(x_1)}{dx_1^2} = \frac{1}{kt} q_1(x_1) \qquad \text{in} \quad 0 \leq x_1 \leq L \qquad (21a)$$

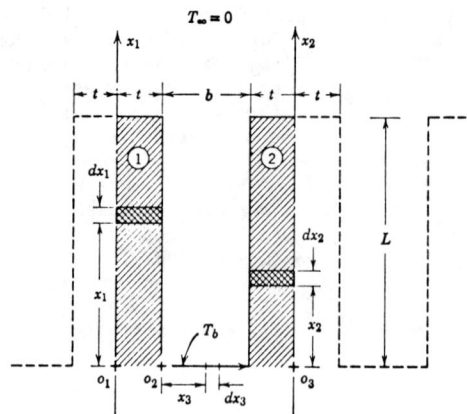

Fig. 4. Rectangular Plate Fin with Exposed Base Surface

$$T_1(x_1) = T_b \quad \text{at} \quad x_1 = 0 \quad (21b)$$

$$\frac{dT_1(x_1)}{dx_1} = 0 \quad \text{at} \quad x_1 = L \quad (21c)$$

and the radiative heat flux $q_1(x_1)$ is determined in accordance with equation (4a) as

$$q_1(x_1) = R_1(x_1) - \int_{x_2=0}^{L} R_2(x_2)\, dF_{dx_1-dx_2} - \int_{x_3=0}^{b} R_3(x_3)\, dF_{dx_1-dx_3} \quad (22)$$

Here, the second and third terms on the right-hand side represents the radiation received from the fin plate 2 and the fin base respectively. The equations for the radiosities are written in accordance with equation (3) as

$$R_1(x_1) = \varepsilon \bar{\sigma} T_1^4(x_1) + (1-\varepsilon)\int_{x_2=0}^{L} R_2(x_2) dF_{dx_1-dx_2} + (1-\varepsilon)\int_{x_3=0}^{b} R_3(x_3) dF_{dx_1-dx_3}$$
(23a)

$$R_3(x_3) = \varepsilon \bar{\sigma} T_b^4 + (1-\varepsilon)\int_{x_1=0}^{L} R_1(x_1) dF_{dx_3-dx_1} + (1-\varepsilon)\int_{x_2=0}^{L} R_2(x_2) dF_{dx_3-dx_2}$$
(23b)

Similar set of equations can be written for the fin plate 2; but they are not needed because of symmetry considerations, that is, $T_1(x_1) = T_2(x_2)$ and $R_1(x_1) = R_2(x_2)$ for $x_1 = x_2$. Then the subscripts 1 and 2 can be removed from $T_1$, $R_1$ and $R_2$, and the complete mathematical formulation of the problem can be written in the dimensionless form as

$$\frac{d^2\theta_1(\xi_1)}{d\xi_1^2} = \frac{1}{N_c}\left[\beta(\xi_1) - \int_{\xi_2=0}^{1}\beta(\xi_1)dF_{d\xi_1-d\xi_2} - \int_{\xi_3=0}^{1}\beta_3(\xi_3)dF_{d\xi_1-d\xi_3}\right], \quad (24a)$$

$$\text{in} \quad 0 \leq \xi_1 \leq 1 \quad (24b)$$

$$\theta(\xi_1) = 0 \quad \text{at} \quad \xi_1 = 0 \quad (24c)$$

$$\frac{d\theta(\xi_1)}{d\xi_1} = 0 \quad \text{at} \quad \xi_1 = 1 \quad (24d)$$

$$\beta(\xi_1) = \varepsilon\theta^4(\xi_1) + (1-\varepsilon)\int_{\xi_2=0}^{1}\beta(\xi_2)dF_{d\xi_1-d\xi_2} + (1-\varepsilon)\int_{\xi_3=0}^{1}\beta_3(\xi_3)dF_{d\xi_1-d\xi_3} \quad (25)$$

$$\beta_3(\xi_3) = \varepsilon + (1-\varepsilon)\int_{\xi_1=0}^{1}\beta(\xi_1)dF_{d\xi_3-d\xi_1} + (1-\varepsilon)\int_{\xi_2=0}^{1}\beta(\xi_2)dF_{d\xi_3-d\xi_2} \quad (26)$$

where various dimensionless quantities are defined as

$$\beta = \frac{R_1}{\sigma T_b^4} = \frac{R_2}{\sigma T_b^4}, \quad \beta_3 = \frac{R_3}{\sigma T_b^4}, \quad N_c = \frac{kt}{L^2\sigma T_b^3} \quad (27)$$

$$\theta = \frac{T_1}{T_b}, \quad \xi_1 = \frac{x_1}{L}, \quad \xi_2 = \frac{x_2}{L}, \quad \xi_3 = \frac{x_3}{b}$$

Equations (24), (25) and (26) are three simultaneous equations for the three unknowns $\theta(\xi_1)$, $\beta(\xi_1)$ and $\beta_3(\xi_3)$.

Once the radiosities are determined from the solution of the above system, the radiative heat flux at the fin surface is determined from the dimensionless equation (22).

## 4. FORCED CONVECTION TO A TRANSPARENT GAS WITH RADIATION BOUNDARY CONDITIONS

When a transparent gas flows over a surface, the coupling between convection and radiation occurs only through the fourth-power thermal radiation boundary condition at the surface. However, one should distinguish between the situations in which the boundary surface temperature is prescribed and the surface heat flux is prescribed. In the former case the radiation and convection are treated separately, since the surface temperature is fixed and radiation with the external environment does not alter the surface temperature. In the latter case, however, the radiation alters the surface temperature, hence the radiation and convection are coupled through the boundary condition.

The problem of combined convection and radiation in forced flow of a transparent gas has been treated in several references [3,5,12-19]. To illustrate the basic approach in the analysis of such problems we now examine forced convection through a circular tube with radiation boundary conditions.

### 4.1 Forced Convection Through a Circular Tube

We consider the steady, fully developed flow of a transparent gas inside a circular tube subjected to uniformly applied wall heat flux, $q_w$. Figure 5 illustrates the geometry and the coordinates. The heat supply to the wall is dissipated by convection to the fluid and radiation through the two open ends of the tube to outer environments at temperatures $T_1$ and $T_2$. The gas enters the tube at a uniform temperature $T_{g1}$ and heated by heat supplied from the tube wall.

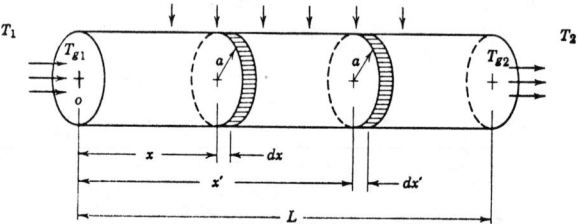

Fig. 5. Forced Convection Through a Circular Tube with Radiation Boundary Condition

In this problem we are interested in the determination of the tube surface temperature and the gas temperature rise as a function of the axial distance along the tube.

We assume that the tube surface is opaque, gray, diffuse emitter, diffuse reflector, has uniform emissivity $\varepsilon$ and that the Kirchhoff law is valid. To simplify the analysis we consider only the radially averaged gas temperature, $T_g(x)$, and consequently assume that <u>a priori</u> knowledge of the heat transfer coefficient, h, for the flow of gas through the tube is available.

The energy balance equation for the radially averaged gas temperature is written as

$$\frac{dT_g(x)}{dx} = \frac{2h}{\rho U_m C_p a} [T_w(x) - T_g(x)] \tag{28a}$$

with the boundary condition

$$T_g(x) = T_{g1} \quad \text{at} \quad x = 0 \tag{28b}$$

where $U_m$ is the mean velocity, $\rho$ is the density, $C_p$ is the specific heat of the gas, $T_w(x)$ is the wall temperature, h is the heat transfer coefficient and "a" is the tube radius.

The heat supplied to the wall, $q_w$, is dissipated by convection and radiation, so that

$$q_w = q^c(x) + q^r(x) \tag{29}$$

where $q^c(x)$ and $q^r(x)$ are the convective and radiative heat fluxes at the location x at the wall surface. Here the convective heat flux is given by

$$q^c(x) = h[T_w(x) - T_g(x)] \tag{30}$$

and the radiative heat flux is determined by taking into consideration the radiative heat exchange from the band of radius "a" width dx at the location x to the complete interior surface of the tube and to the openings at x = 0 and x = L. The appropriate expression for this radiative heat flux is determined according to the expression given by equation (4a) as

$$q^r(x) = R(x) - \left[ \bar{\sigma} T_1^4 \cdot F(x) + \bar{\sigma} T_2^4 \cdot F(L-x) + \int_{x'=0}^{L} R(x') dF_{dx-dx', |x-x'|} \right] \tag{31}$$

where

$R(x)$ = radiosity at the cylindrical surface at x,
$F(x)$ = diffuse view factor from the band (a,dx) at x to the opening at x = 0 at a distance x apart,
$F(L-x)$ = diffuse view factor from the band (a,dx) at x to the opening at x = L at a distance (L-x) apart,
$dF_{dx-dx', |x-x'|}$ = elemental diffuse view factor from the band (a,dx) at x to the band (a,dx) at x' at a distance $|x-x'|$ apart.

In equation (31), the first, second and third terms inside the bracket on the right-hand side are respectively the radiation from the openings at x = 0 and x = L and from the entire inner surface of the tube to the band (a,dx) at x.

Finally, the expression for the radiosity $R(x)$ is determined according to equation (4b) as

$$q^r(x) = \frac{1}{\rho} [\varepsilon \bar{\sigma} T_w^4(x) - (1-\rho)R(x)]$$

or by setting $\varepsilon = 1 - \rho$ and rearranging we find

$$R(x) = \bar{\sigma} T_w^4(x) - \frac{1-\varepsilon}{\varepsilon} q^r(x) \tag{32}$$

Equations (28) through (32) give the complete mathematical formulation of this convection and radiation problem.

These equations can be written in the dimensionless form by defining the following dimensionless quantities.

$$\theta = \left( \frac{\bar{\sigma}}{q_w} \right)^{1/4} T \text{ , dimensionless temperature}$$

$$\beta = \frac{R}{q_w} \text{ , dimensionless radiosity}$$

$$S = \frac{4h}{\rho U_m C_p}, \text{ Stanton number}$$

$$h^* = \frac{h}{q_w}\left(\frac{q_w}{\bar{\sigma}}\right)^{1/4}, \text{ dimensionless heat transfer coefficient}$$

$$\xi = \frac{x}{2a}, \quad \xi' = \frac{x'}{2a} \quad \text{and} \quad \xi_L = \frac{L}{2a}$$

Then equations (28) become

$$\frac{d\theta_g(\xi)}{d\xi} = S[\theta_w(\xi) - \theta_g(\xi)] \tag{33a}$$

$$\theta_g(\xi) = \theta_{gl} \quad \text{at} \quad \xi = 0 \tag{33b}$$

Equations (29), (30) and (31) are combined as

$$1 = h^*[\theta_w(\xi) - \theta_g(\xi)] + \beta(\xi) - \left[\theta_1^4 F(\xi) + \theta_2^4 F(\xi_L - \xi) + \int_{\xi'=0}^{\xi} \beta(\xi') dF_{d\xi-d\xi',|\xi-\xi'|} + \int_{\xi'=\xi}^{\xi_L} \beta(\xi') dF_{d\xi-d\xi',|\xi'-\xi|}\right] \tag{34}$$

where $\theta_1$ and $\theta_2$ are the environment temperatures at $\xi = 0$ and $\xi = \xi_L$.

Equation (32) is combined with equations (29) and (30) to give

$$\beta(\xi) = \theta_w^4(\xi) - \frac{1-\varepsilon}{\varepsilon}\left\{1 - h^*[\theta_w(\xi) - \theta_g(\xi)]\right\} \tag{35}$$

Equations (33), (34) and (35) are three simultaneous equations for the three unknown functions $\theta_w(\xi)$, $\theta_g(\xi)$ and $\beta(\xi)$.

The diffuse view factors appearing in the foregoing equations can be determined by utilizing the rules for the view factor determination [6]. We find

$$F(z) = \frac{\frac{1}{2} + z^2}{\sqrt{1+z^2}} - z \tag{36a}$$

where

$$z = \xi \quad \text{or} \quad (\xi_L - \xi).$$

and

$$dF_{d\xi-d\xi',|z|} = \left[1 - |z|\frac{z^2 + \frac{3}{2}}{\left(z^2+1\right)^{3/2}}\right] d\xi' \tag{36b}$$

with $z \equiv \xi - \xi'$.

Although analytical solution cannot be found for the problem defined by equations (33) through (36), direct numerical integration is possible. However, the utility of direct numerical integration is restricted to short tube length ($\xi_L$ = 5 to 10), because too many subdivisions are required to approximate the integral for long tubes.

We present in Fig. 6 a plot of the tube surface temperatures as a function of the axial distance along the tube as obtained from the calculations of reference [15]. The results are given for the case h* = 0.8, S = 0.01, $\theta_{g1} = \theta_1 = \theta_2 = 1.5$ and $\theta_{g2} = \theta_2$. Included on this figure is the temperature distribution for the case of pure convection (i.e., without radiation effects), shown by the dotted lines. The effects of wall emissivity and of tube length on temperature distribution are illustrated. For short tubes, the radiation heat loss from the tube surface to the external environment becomes important, hence the tube temperature distribution is flattened. As the emissivity of the tube is reduced from 1 to 0.01, the radiation effects becomes less important because heat cannot be dissipated by radiation effectively.

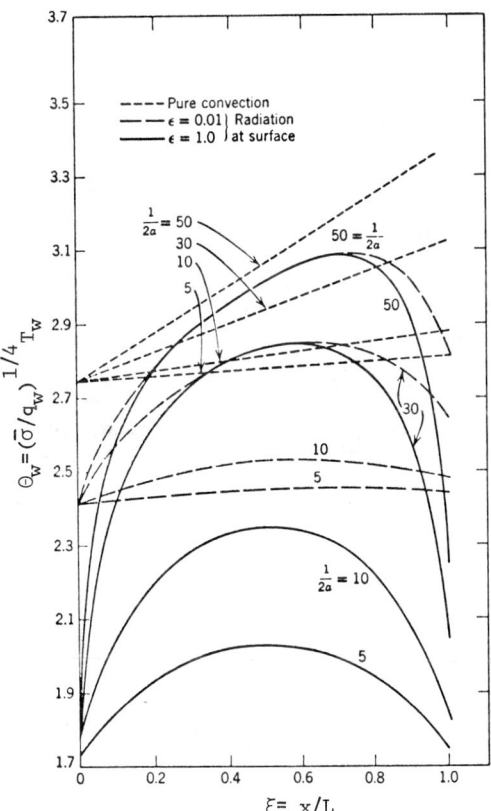

Fig. 6. Effects of Radiation on Tube Surface Temperature

## 5. FORCED CONVECTION TO AN ABSORBING AND EMITTING GAS WITH RADIATION

When a gas absorbs and emits radiation, at sufficiently high temperatures the radiation effects become important and the presence of radiation within the gas alters the temperature distribution in the fluid, which in turn affects the heat transfer at the wall. In such situations convection and radiation are coupled because the temperature distribution within the fluid is affected by both convection and radiation; as a result a simultaneous treatment of heat transfer by both mechanism is necessary. The problem of combined convection and radiation has been studied by several investigators [20-29]. When the medium absorbs and emits radiation, the energy equation for the fluid is modified as a result of the inclusion into the energy equation of the divergence of the radiation heat flux term. The continuity and momentum equations, however, remain unaltered by radiation for most engineering applications.

For a participating, incompressible fluid, for example, the energy equation, including the radiation effects, can be written in the form

$$\rho C_p \frac{DT}{Dt} = \nabla \cdot k \nabla T - \nabla \cdot q^r + \mu \Phi \qquad (37)$$

where

$\frac{D}{Dt}$ = the substantial derivative

$q^r$ = radiative heat flux vector

$\Phi$ = viscous energy dissipation function

$k$ = thermal conductivity

Clearly, the radiation effects enters the energy equation as a source term, $\nabla \cdot q^r$, where the radiative heat flux should be obtained from the solution of the equation of radiative transfer and depends on the fourth power of the temperature. Because of the difficulties associated with the solution of the equation of radiative transfer for multidimensional cases, the studies of combined convection and radiation are generally restricted to situations in which the radiation part of the problem is treated as one-dimensional in the direction where the temperature gradients are greatest. Before proceeding to the analysis of combined convection and radiation, we present a brief discussion of the determination of radiative heat flux term $q^r$ for the one-dimensional absorbing, emitting plane-parallel medium.

### 5.1 Radiation Heat Flux in an Absorbing, Emitting, Plane-Parallel Medium

We consider an absorbing, emitting, plane-parallel, gray medium of optical thickness $\tau_o$ confined between two diffusely reflecting, diffusely emitting, opaque, gray boundaries at $\tau = 0$ and $\tau = \tau_o$. Let $T(\tau)$ be the temperature distribution in the medium, $T_1$ and $T_2$ be the temperatures, $\rho_1$ and $\rho_2$ be the diffuse reflectivities, and $\varepsilon_1$ and $\varepsilon_2$ be the emissivities of the boundaries at $\tau = 0$ and $\tau = \tau_o$ respectively. Figure 7 shows the geometry and the coordinate. The radiation problem for this particular case, comprising the one-dimensional equation of radiative transfer and the two boundary conditions are given by

$$\mu \frac{\partial I(\tau,\mu)}{\partial \tau} + I(\tau,\mu) = \frac{n^2 \bar{\sigma} T^4(\tau)}{\pi} \quad \text{in} \quad 0 \leq \tau \leq \tau_o, \; -1 \leq \mu \leq 1 \qquad (38a)$$

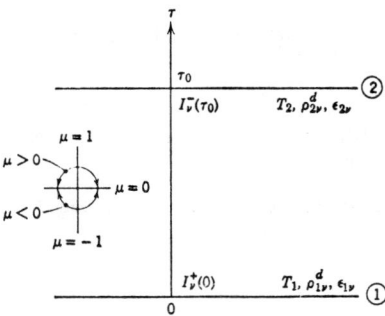

Fig. 7. Geometry and Coordinate for Radiation Problem

$$I(0) = \epsilon_1 \frac{n^2 \bar{\sigma} T_1^4}{\pi} + 2\rho_1 \int_0^1 I(0,-\mu')\mu' d\mu', \quad \text{at} \quad \tau = 0, \ \mu > 0 \tag{38b}$$

$$I(\tau_o) = \epsilon_2 \frac{n^2 \bar{\sigma} T_2^4}{\pi} + 2\rho_2 \int_0^1 I(\tau_o,\mu')\mu' d\mu', \quad \text{at} \quad \tau = \tau_o, \ \mu < 0 \tag{38c}$$

where $I(\tau,\mu)$ is the radiation intensity, n is the refractive index for the medium, $\mu$ is the cosine of the angle between the direction of the radiation beam and the positive $\tau$ axis, and the optical variable $\tau$ is related to the physical space variable y by $\tau = \kappa y$, where $\kappa$ is the absorption coefficient for the medium. Once the radiation intensity $I(\tau,\mu)$ is determined from the solution of the above radiation problem, the radiative heat flux $q^r(\tau)$, anywhere in the medium, is determined from its definition

$$q^r(\tau) = 2\pi \int_{-1}^1 I(\tau,\mu)\mu d\mu \tag{39}$$

It is to be noted that in the problems of interaction of radiation by convection, the radiation problem is always taken as steady-state because the transients associated with the propagation of radiation through the medium dies out fast compared to that associated with convection. Therefore, the time dependence enters the radiation problem merely as a parameter through the temperature distribution for the medium. The radiation problem defined by equations (38) can be solved for the radiation intensity $I(\tau,\mu)$ and the net radiation heat flux, $q^r(\tau)$, in the $\tau$ direction is determined according to its definition given by equation (39). The resulting expressions for the net radiative heat flux $q^r(\tau)$ and the derivative of the radiative heat flux $dq^r(\tau)/d\tau$ are given by [6, chapter 8]

$$q^r(\tau) = 2\pi [I^+(0) E_3(\tau) - I^-(\tau_o) E_3(\tau_o - \tau)] +$$

$$+ 2\pi \left\{ \int_0^\tau I_b[T(\tau')]E_2(\tau - \tau')d\tau' - \int_\tau^{\tau_o} I_b[T(\tau')]E_2(\tau' - \tau)d\tau' \right\} \tag{40}$$

$$\frac{dq^r(\tau)}{d\tau} = 4\pi I_b[T(\tau)] - 2\pi[I^+(0)E_2(\tau) + I^-(\tau_o)E_2(\tau_o - \tau)] -$$

$$-2\pi \left\{ \int_o^\tau I_b[T(\tau')]E_1(\tau-\tau')d\tau' + \int_\tau^{\tau_o} I_b[T(\tau')]E_1(\tau'-\tau)d\tau' \right\} \quad (41)$$

and the boundary surface intensity functions $I^+(0)$ and $I^-(\tau_o)$ are

$$I^+(0) = \frac{a_1 + b_1 a_2}{1 - b_1 b_2} \quad (42a)$$

$$I^-(\tau_o) = \frac{a_2 + b_2 a_1}{1 - b_1 b_2} \quad (42b)$$

where

$$b_1 = 2\rho_1 E_3(\tau_o) \quad (43a)$$

$$b_2 = 2\rho_2 E_3(\tau_o) \quad (43b)$$

$$a_1 = \varepsilon_1 I_b(T_1) + 2\rho_1 \int_o^{\tau_o} I_b[T(\tau')]E_2(\tau')d\tau' \quad (43c)$$

$$a_2 = \varepsilon_2 I_b(T_2) + 2\rho_2 \int_o^{\tau_o} I_b[T(\tau')]E_2(\tau_o - \tau')d\tau' \quad (43d)$$

here the exponential integral function, $E_n(z)$, is defined as

$$E_n(z) = \int_o^1 e^{-\frac{z}{\mu}} \mu^{n-2} d\mu \quad (44)$$

Clearly several special cases are obtainable from the expressions given above for $q^r(\tau)$ and $dq^r(\tau)/d\tau$. For example, for black boundaries at both walls, we set $\rho_1 = \rho_2 = 0$ and $\varepsilon_1 = \varepsilon_2 = 1$.

The coupling between the radiation problem and the energy equation of fluid dynamics now becomes apparent. To determine $dq^r(\tau)/d\tau$ or $q^r(\tau)$ one needs the distribution of temperature in the medium; whereas to determine the temperature distribution from the solution of the energy equation one needs the divergence of the radiative heat flux vector.

## 5.2 Absorbing, Emitting Couette Flow

To illustrate the application of the foregoing expressions for the radiative heat flux in the analysis of combined convection and radiation, we consider the steady-state heat transfer to an absorbing, emitting, gray couette flow between two parallel plates. Figure 8 shows the geometry and coordinates.

# RADIATIVE HEAT TRANSFER IN HEAT EXCHANGERS

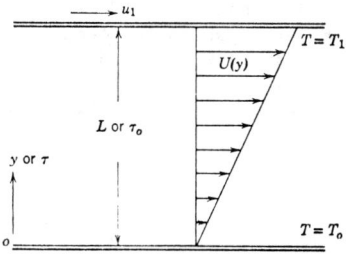

Fig. 8. Absorbing, Emitting Couette Flow

The upper and lower plates are black surfaces at temperatures $T_1$ and $T_2$ respectively. The steady-state energy equation, neglecting the viscous energy dissipation term, is given in the dimensionless form as

$$\frac{d}{d\tau}\left[N\frac{d\theta(\tau)}{d\tau} - Q^r(\tau)\right] = 0 \quad \text{in} \quad 0 \leq \tau \leq \tau_0 \quad (45a)$$

$$\theta(\tau) = \theta_1 \quad \text{at} \quad \tau = 0 \quad (45b)$$

$$\theta(\tau) = 1 \quad \text{at} \quad \tau = \tau_0 \quad (45c)$$

where

$N = \dfrac{k\kappa}{4n^2\bar{\sigma}T_2^3}$ = conduction-to-radiation parameter

$Q^r(\tau) = \dfrac{q^r(\tau)}{4n^2\bar{\sigma}T_2^4}$ = dimensionless radiative heat flux

$\theta(\tau) = \dfrac{T(\tau)}{T_2}$ = dimensionless temperature

$\tau = \kappa y$ = optical variable

$\tau_0 = \kappa L$ = optical distance between the plates

$\kappa$ = absorption coefficient.

The radiative heat flux $q^r(\tau)$ is obtained from equation (42) and the dimensionless radiative flux $Q^r(\tau)$ appearing in equation (43a) is given by

$$Q^r(\tau) = \frac{1}{2}\left[\theta_1^4 E_3(\tau) - E_3(\tau_0 - \tau)\right] +$$

$$+ \frac{1}{2}\left[\int_0^\tau \theta^4(\tau')E_2(\tau-\tau')d\tau' - \int_{\tau'=\tau}^{\tau_0} \theta^4(\tau')E_2(\tau'-\tau)d\tau'\right] \quad (46)$$

A simultaneous solution of equations (43) and (44) yields the temperature distribution in the medium. Figure 9 shows the temperature distribution in the gas for $\theta_1 = 0.5$, optical thickness $\tau_o = 1$ and for different values of the conduction-to-radiation parameter, N. As the value of N is reduced, the radiation effects are more pronounced; the case N = 0 corresponds to heat transfer by pure radiation. Conversely, as the value of N is increased, the conduction effects become dominant; the curves for $N \geq 10$ are indistinguishable from that of pure conduction.

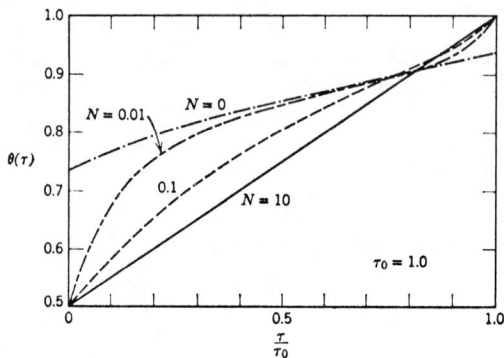

Fig. 9. Effects of Radiation on Temperature Distribution in Couette Flow

## 5. RADIATION HEAT FLUX IN AN ABSORBING, EMITTING, SCATTERING PLANE-PARALLEL MEDIUM

When the medium scatters radiation, in addition absorption and emission, the solution of the radiation problem becomes more involved because of the presence of the scattering term in the equation of radiative transfer. To illustrate this matter we consider an absorbing, emitting, isotropically scattering, plane-parallel, gray medium of optical thickness $\tau_o$ confined between two diffusely reflecting, diffusely emitting, opaque, gray boundaries at $\tau = 0$ and $\tau = \tau_o$. Let $T(\tau)$ be the temperature distribution in the medium, $T_1$ and $T_2$ be the temperatures, $\rho_1$ and $\rho_2$ be the diffuse reflectivities, and $\varepsilon_1$ and $\varepsilon_2$ be the emissivities of the boundary surfaces at $\tau = 0$ and $\tau = \tau_o$ respectively. The mathematical formulation of this radiation problem is given by

$$\mu \frac{\partial I(\tau,\mu)}{\partial \tau} + I(\tau,\mu) = (1-\omega) \frac{n^2 \bar{\sigma} T^4(\tau)}{\pi} + \frac{\omega}{2} \int_{-1}^{1} I(\tau,\mu')d\mu', \quad \text{in} \quad 0 \leq \tau \leq \tau_o,$$

$$-1 \leq \mu \leq 1 \qquad (47a)$$

$$I(0) = \varepsilon_1 \frac{n^2 \bar{\sigma} T_1^4}{\pi} + 2\rho_1 \int_0^1 I(0,-\mu')\mu'd\mu', \quad \text{at} \quad \tau = 0, \quad \mu > 0 \qquad (47b)$$

$$I(\tau_o) = \varepsilon_2 \frac{n^2 \bar{\sigma} T_2^4}{\pi} + 2\rho_2 \int_0^1 I(\tau_o,\mu')\mu'd\mu', \quad \text{at} \quad \tau = \tau_o, \quad \mu < 0 \qquad (47c)$$

where, $\omega = \frac{\sigma}{\beta}$, is the single scattering albedo. Clearly, for a nonscattering medium we have $\omega = 0$ and the problem given by equations (47) reduces to that given by equation (37) for an absorbing and emitting medium. Once the problem given by equations (47) is solved for the radiation intensity, $I(\tau,\mu)$, for a specified temperature distribution, $T(\tau)$, in the medium, the net radiative heat flux, $q^r(\tau)$, anywhere in the medium is determined according to its definition given by equation (39).

Various approximate and exact methods of analysis have been reported in the literature for the solution of the radiation problem given by equations (47). For example, the lowest order analysis of the spherical harmonics method, called the "P-1 approximation" transforms the integrodifferential equation (47a) into a simple second degree ordinary differential equation; but the results obtained with the P-1 approximation are not very accurate.

The radiation problem (47) can be transformed into a set of coupled ordinary differential equations by approximating the integral term by Gaussian quadratures and the resulting system can be solved numerically. The use of this approach in the interaction problems require a large amount of computer time for the solution of the problem; hence is not so convenient.

The normal-mode expansion technique [6,30,31] has been applied to solve the radiation problem exactly in the problems of interaction of radiation with convection [29].

A purely numerical scheme employing cubic splines as approximating functions and Gaussian quadrature points as the collocation points has been used for the solution of a combined convection and radiation problem [32].

To solve the combined convection and radiation problem by an iterative scheme between the energy and radiation problems, it is desirable to have very efficient and sufficiently accurate method of analysis for the radiation part of the problem. Recently two such methods have appeared in the literature: The so-called "F-N" method [33-35] and the "Source Function Expansion" method [36-37]. The former derives its basis from the normal mode expansion technique and the latter makes use of a Legendre polynomial expansion in the space variable. Although these two methods have not yet used in the analysis of the interaction problem, they require little computing time for the solution of the radiation part of the problem and give sufficiently accurate results for the net radiative heat flux. They appear to be promising for use in such applications.

## 6.1 Absorbing, Emitting, Scattering, Thermally Developing Slug Flow Between Parallel-Plates

To illustrate the effects of radiation on convection, we consider combined convection and radiation for an absorbing, emitting, isotropically scattering, gray, thermally developing slug flow between infinite parallel plates separated by a distance 2L. The fluid enter the channel with a uniform temperature $T_o$ at the origin of the axial coordinates $x = 0$. In the region $x > 0$, the walls are kept at a uniform temperature $T_w$. The energy equation for a radiating, incompressible, constant property fluid in steady slug flow, neglecting viscous dissipation and axial conduction, is given in the dimensionless form as

$$\frac{\partial^2 \theta(\tau,\xi)}{\partial \tau^2} - \frac{1}{N}\frac{\partial Q^r(\tau,\xi)}{\partial \tau} = \frac{\partial \theta(\tau,\xi)}{\partial \xi}, \quad \text{in } 0 \leq \tau \leq \tau_o, \quad \xi \geq 0 \qquad (48a)$$

with the boundary conditions

$$\theta(\tau,\xi) = 1 \quad \text{at} \quad \tau = 0, \quad \xi > 0 \tag{48b}$$

$$\frac{\partial \theta(\tau,\xi)}{\partial \tau} = 0 \quad \text{at} \quad \tau = \tau_o, \quad \xi > 0 \tag{48c}$$

$$\theta(\tau,\xi) = \theta_o \quad \text{at} \quad \xi = 0, \quad 0 \leq \tau \leq \tau_o \tag{48d}$$

where the radiation flux $Q^r(\tau,\xi)$ is related to the dimensionless radiation intensity $\psi(\tau,\mu,\xi)$ by

$$Q^r(\tau,\xi) = 2\pi \int_{-1}^{1} \psi(\tau,\mu',\xi)\mu'\,d\mu' \tag{49}$$

and $\psi(\tau,\mu,\xi)$ is the solution of the following radiation problem

$$\mu \frac{\partial \psi(\tau,\mu,\xi)}{\partial \tau} + \psi(\tau,\mu,\xi) = (1-\omega)\frac{1}{4\pi}\theta^4(\tau,\xi) + \frac{\omega}{2}\int_{-1}^{1}\psi(\tau,\mu',\xi)d\mu',$$

$$\text{in } 0 \leq \tau \leq \tau_o, \quad -1 \leq \mu \leq 1 \tag{50a}$$

subject to the boundary conditions

$$\psi(\tau,\mu,\xi) = \varepsilon + \rho^s \psi(\tau,-\mu,\xi) + 2\rho^d \int_{o}^{1}\psi(\tau,-\mu',\xi)\mu'\,d\mu', \tag{50b}$$

$$\text{at} \quad \tau = 0, \quad \mu > 0 \tag{50b}$$

$$\psi(\tau,-\mu,\xi) = \psi(\tau,\mu,\xi) \quad \text{at} \quad \tau = \tau_o, \quad \mu < 0 \tag{50c}$$

Here the radiation boundary condition (50b) allows the reflectivity of the plate to be partly specular, $\rho^s$, and partly diffuse, $\rho^d$, such that $\rho^s + \rho^d + \varepsilon = 1$. The boundary condition (50c) merely states the symmetry about the central plane in the channel.

An iterative technique was applied to solve the combined convection and radiation problem defined by equations (48) - (50). The normal mode expansion technique was used to solve the radiation part of the problem exactly. Once the temperature distribution in the flow field is determined, the local Nusselt number $Nu(\xi)$ is computed from

$$Nu(\xi) = \frac{4\tau_o}{1 - \theta_m(\xi)}\left[-\frac{\partial \theta(\tau,\xi)}{\partial \tau} + \frac{1}{N}Q^r(\tau,\xi)\right]_{\tau=0} \tag{51}$$

where $\theta_m(\xi)$ is the mean temperature at the axial location $\xi$.

We present in Table 1 the local Nusselt number, $Nu(\xi)$, at different axial locations for the values of conduction and radiation parameter, $N = 0.5$, the dimensionless inlet and outlet temperatures, $\theta_o = 0$ and $\theta_\omega = 1$.

Increased scattering of radiation, characterized by larger $\omega$, reduces the effects of radiation; that is the radiation effects are greatest for nonscattering fluid (i.e., $\omega = 0$). The local Nusselt number with diffusely reflecting surfaces is slightly higher than with specularly reflecting surfaces. The local Nusselt number increases with increasing optical thickness of the fluid. The cases with optical thickness $\tau_o \leq 0.1$ are almost identical to those with no radiation. The case $\omega = 1$ corresponds to no interaction between convection and radiation; hence it represents the nonradiating case for slug flow.

Table 1. Effects of $\omega$ and $\tau_o$ and the wall reflectivities on the local Nusselt number Nu at different axial locations for $N = 0.5$, $\theta_o = 0$ and $\theta_w = 1$, for slug flow between parallel plates.

| $\tau_o$ | $\omega$ | $\varepsilon$ | $\rho^s$ | $\rho^d$ | \multicolumn{5}{c}{$(\frac{x}{D_e})/(\text{Re Pr})$} |
|---|---|---|---|---|---|---|---|---|---|
| | | | | | $3.125 \times 10^{-4}$ | $6.25 \times 10^{-4}$ | $3.125 \times 10^{-3}$ | $6.25 \times 10^{-3}$ | $3.125 \times 10^{-2}$ |
| | | | | | \multicolumn{5}{c}{The Local Nusselt Numbers} |
| 1 | 0 | 1 | 0 | 0 | 36.5569 | 27.3605 | 15.5963 | 13.4641 | 15.2006 |
| | | 0.5 | 0.5 | 0 | 35.7727 | 26.5570 | 14.7730 | 12.5808 | 13.4970 |
| | | 0.5 | 0 | 0.5 | 35.7766 | 26.5610 | 14.7808 | 12.5911 | 13.5316 |
| | | 0.1 | 0.9 | 0 | 35.1089 | 25.8768 | 14.0811 | 11.8379 | 12.0934 |
| | | 0.1 | 0 | 0.9 | 35.1176 | 25.8859 | 14.0985 | 11.8609 | 12.1631 |
| | 0.5 | 1 | 0 | 0 | 36.0605 | 26.8500 | 15.1069 | 12.9606 | 14.0129 |
| | | 0.5 | 0.5 | 0 | 35.4748 | 26.2498 | 14.1439 | 12.2761 | 12.6938 |
| | | 0.5 | 0 | 0.5 | 35.4888 | 26.2641 | 14.4852 | 12.2837 | 12.7071 |
| | | 0.1 | 0.9 | 0 | 34.8698 | 25.6298 | 13.8254 | 11.5758 | 11.3789 |
| | | 0.1 | 0 | 0.9 | 34.8980 | 25.6586 | 13.8484 | 11.5915 | 11.4041 |
| | 1 | 1 | 0 | 0 | 34.6827 | 25.4379 | 13.4983 | 10.0947 | 9.8701 |
| 0.5 | 0 | 1 | 0 | 0 | 35.4795 | 26.2552 | 14.4065 | 12.1151 | 12.0325 |
| 0.1 | 0 | 1 | 0 | 0 | 34.7432 | 25.5001 | 13.5676 | 11.1719 | 10.0228 |

REFERENCES

1. Eckert, E. R. and Drake, R. M. 1959. Heat and Mass Transfer. McGraw-Hill Book Co., New York.

2. Hottel, H. C. 1954. "Radiant heat transmission" in Heat Transmission, by W. H. McAdams, McGraw-Hill Book Co., New York.

3. Sparrow, E. M. and Cess, R. D. 1965. Radiant Heat Transfer. Brooks/Cole Publishing Co., Belmont, Calif.

4. Sarofim, A. F. and Hottel, H. C. 1967. Radiative Transfer. McGraw-Hill Book Co., New York.

5. Özişik, M. N. 1977. Basic Heat Transfer. McGraw-Hill Book Co., New York.

6. Özişik, M. N. 1973. Radiative Transfer. John Wiley & Sons, New York.

7. Kern, D. Q. and Kraus, A. D. 1972. Extended Surface Heat Transfer. McGraw-Hill Book Co., New York.

8. Sparrow, E. M., Eckert, E. R. G. and Irvine, T. F. 1961. Effectiveness of radiating fins with mutual irradiation. J. Aerospace Sci., Vol. 28, pp. 763-772.

9. Sparrow, E. M. and Eckert, E. R. G. 1962. Radiant interaction between fin base and base surfaces. J. Heat Transfer, Vol. 84C, pp. 12-18.

10. Donovan, R. C. and Rohrer, W. M. 1969. Radiative and conductive fins on a plane wall, including mutual irradiation. ASME paper No. 68-WA/HT-22.

11. Tien, C. L. 1967. Approximate solutions of radiative exchange between conducting plates with specular reflection. J. Heat Transfer, Vol. 89C, pp. 119-120.

12. Cess, R. D. 1962. The effects of radiation upon forced-convection heat transfer. Appl. Sci. Res., Vol. A10, pp. 430-438.

13. Sparrow, E. M. and Lin, S. H. 1965. Boundary layers with prescribed heat flux - Application to simultaneous convection and radiation. Int. J. Heat Mass Transfer, Vol. 8, 437-448.

14. Perlmutter, M. and Siegel, R. 1962. Heat transfer by combined forced convection and thermal radiation in a heated tube. J. Heat Transfer, Vol. 84C, pp. 301-311.

15. Siegel, R. and Perlmutter, M. 1962. Convective and radiant heat transfer for flow of transparent gas in a tube with gray wall. Int. J. Heat Mass Transfer, Vol. 5, pp. 639-660.

16. Chen, C. J. 1966. Laminar heat transfer in a tube with nonlinear radiant heat-flux boundary condition. Int. J. Heat Mass Transfer, Vol. 9, pp. 433-440.

17. Liu, S. T. and Thorsen, R. S. 1970. Combined forced convection and radiation heat transfer in asymmetrically heated parallel plates. Proc. Heat Transfer and Fluid Mechanics Institute, pp. 32-44, Stanford University Press, Palo Alto, Calif.

18. Keshock, E. G. and Siegel, R. 1964. Combined radiation and convection in an asymmetrically heated parallel plate flow. J. Heat Transfer, Vol. 86C, pp. 342-350.

19. Campo, A. and Auguste, J. C. 1976. Laminar heat transfer in ducts with viscous dissipation and convective-radiative exchange at the walls. ASME paper 76-WA/HT-59.

20. Goulard, R. and Goulard, M. 1959. Energy transfer in the Couette flow of a radiant and chemically reacting gas. Proc. of Heat Transfer and Fluid Mechanics Institute, pp. 126-129, Stanford University Press, Palo Alto, Calif.

21. Viskanta, R. and Grosh, R. J. 1962. Boundary layers in thermal radiation absorbing and emitting media. Int. J. Heat Mass Transfer, Vol. 5, pp. 795-806.

22. Viskanta, R. 1964. Heat transfer in a radiating slug flow in a parallel-plate channel. Appl. Sci. Res., Vol. 13, pp. 291-311.

23. Einstein, T. H. 1963. Radiant heat transfer to absorbing gas enclosed between parallel flat plates with flow and conduction. NASA Tech. Rept. TR-R-154.

24. Chen, J. C. 1964. Simultaneous radiative and convective heat transfer in an absorbing, emitting and scattering medium in slug flow between parallel plates. A.I.Ch.E.J., Vol. 2, pp. 253-259.

25. DeSoto, S. 1968. Coupled radiation, conduction and convection in entrance region flow. Int. J. Heat Mass Transfer, Vol. 11, pp. 39-53.

26. Pearce, B. E. and Emery, A. F. 1969. Heat transfer by thermal radiation and laminar forced convection to an absorbing fluid in the entry region of a pipe. ASME paper No. 69-WA/HT-16.

27. Landram, C. S., Grief, R. and Habib, I. S. 1969. Heat transfer in turbulent pipe flow with optically thin radiation. J. Heat Transfer, Vol. 91C, pp. 330-336.

28. Thorsen, R. S. 1971. Combined conduction, convection, and radiation effects in optically thin tube flow. ASME paper No. 71-HT-17.

29. Lii, C. C. and Özişik, M. N. 1973. Heat transfer in an absorbing, emitting and scattering slug flow between parallel plates. J. Heat Transfer, Vol. 95-C, 538-540.

30. Özişik, M. N. and Siewert, C. E. 1969. On the normal-mode expansion technique for radiative transfer in scattering, absorbing and emitting slab with specularly reflecting boundaries. Intern. J. Heat Mass Transfer, Vol. 12, pp. 611-620.

31. Beach, H. L., Özişik, M. N. and Siewert, C. E. 1971. Radiative transfer in linearly anisotropic-scattering, conservative and non-conservative slabs with reflecting boundaries. Intern. J. Heat Mass Transfer, Vol. 14, pp. 1551-1565.

32. Chawla, T. C. and Chan, S. H. 1979. Combined radiation convection in thermally developing flows with scattering. 18th National Heat Transfer Conference, San Diego, California, August 1979.

33. Siewert, C. E. and Benoist, P. 1979. The $F_N$ method in neutron transport theory. Part I. Theory and applications. Nucl. Sci. Eng., Vol. 69, pp. 156-160.

34. Grandjean, P. and Siewert, C. E. 1979. The $F_N$ method in neutron transport theory. Part II. Applications and numerical results. Nucl. Sci. Eng., Vol. 69, pp. 161-168.

35. Siewert, C. E., Maiorino, J. R. and Özişik, M. N. 1980. The use of $F_N$ method for radiative transfer problems with reflective boundary conditions. JQSRT (in press).

36. Özişik, M. N. and Sutton, W. H. 1980. A source function expansion in radiative transfer. 19th National Heat Transfer Conference, Orlando, Florida, July 27-30, 1980.

37. Özişik, M N. and Shouman, S. M. 1980. A source function expansion method for radiative transfer in a two-layer slab. (To be published).

# HEAT EXCHANGER DESIGN: RATING, SIZING, AND OPTIMIZATION

# Heat Exchangers—Basic Methods

WARREN M. ROHSENOW
Department of Mechanical Engineering
Massachusetts Institute of Technology
Cambridge, Massachusetts 02139 USA

ABSTRACT

In sizing heat exchangers or predicting their performance the local thermal resistances must first be calculated. A brief study of the order of magnitudes of these resistances shows which of them are negligible in various situations. Then the temperature distribution in the heat exchanger is predicted. In special cases a logarithmic mean is the appropriate mean temperature and an NTU method may be used. Reviewed here also are single phase heat transfer coefficients and friction factors. One example of elementary sizing methods is presented for a shell-and-tube counter flow exchangers.

I.2.1 INTRODUCTION

A heat exchanger is a piece of equipment in which heat is transferred from a hot fluid to a colder fluid. In its simplest form, the two fluids mix and leave at an intermediate temperature determined from conservation of energy. This device is not truly a heat exchanger but rather a mixer. In most applications the fluids do not mix but transfer heat through a separating wall which takes on a wide variety of geometries. Perhaps the most common geometries are shell-and-tubes and stacks of parallel plates with hot and cold fluids in alternate passages between plates. Fins may be present on the surfaces.

Heat exchanger design presents many stress and corrosion problems as well as problems of fluid mechanics and heat transfer. In most cases design aims at minimum cost, balancing the cost of pumping the fluids and initial cost of the exchanger against the savings resulting from the heat transfer. In some cases — such as in missile, aircraft or shipboard applications — design may be governed by the necessity of minimizing either volume or weight.

I.2.2 OVER-ALL HEAT TRANSFER COEFICIENT AND FOULING FACTORS

Heat exchanger walls are usually single materials, although sometimes as a protection against corrosion the wall may be bimetallic (steel with aluminium cladding) or coated with a plastic. Most heat exchanger surfaces tend to acquire an additional heat transfer resistance which increases with time. This may be either a very thin surface oxidation layer or, at the other extreme, may be a thick crust deposit, such as that which results from salt water coolant in steam condensers. We define a scale coefficient of heat tranfer $h_s$ in terms of thermal resistance of this scale as follows:

$$\frac{\Delta T}{q} = R_s = \frac{1}{Ah_s} \qquad (I.2.1)$$

where the area $A$ is the original heat transfer area of the surface before scaling began.

The local over-all coefficient of heat transfer for a single wall material is then expressed as follows:

$$\frac{T_h - T_c}{dq/dA_r} = \frac{1}{U_r} = \frac{1}{h_c}\frac{dA_r}{dA_c} + \frac{1}{h_{s_c}}\frac{dA_r}{dA_c} + R_w\frac{dA_r}{dA_w} + \frac{1}{h_{sh}}\frac{dA_r}{dA_h} + \frac{1}{h_h}\frac{dA_r}{dA_h} \qquad (I.2.2)$$

where reference area $A_r$ is usually selected either as $A_c$ or $A_h$. For the flat plate with $A_c = A_h = A_r$, $R_w = x_w/k_w$, and for a cylinder,

$$R_w \frac{dA_r}{dA_w} = \frac{\ln(r_o/r_i)}{2\pi k_w}\frac{dA_r}{dL}$$

The temperature drops and corresponding resistances across a wall are shown in Fig. I.2.1. Table I.2.1 gives magnitudes of $1/h_s$, called fouling factor, recommended for inclusion in $U_r$ for calculating the required heat transfer surface of an exchanger. Looking at the order of magnitude of resistances, it is seen that the scale resistance is usually significant only when there are liquids on both sides of the exchanger wall.

A separating wall may be finned differently on each side, Fig. I.2.2. On either side, the heat transfer takes place from the fins (subscript f in the equations that follow) as well as from the unfinned portion of the wall (subscript u); introducing the fin efficiency, $\eta$, the total $q$ is given by

$$q = (\eta A_f h_f + A_u h_u) \Delta T \qquad (I.2.3)$$

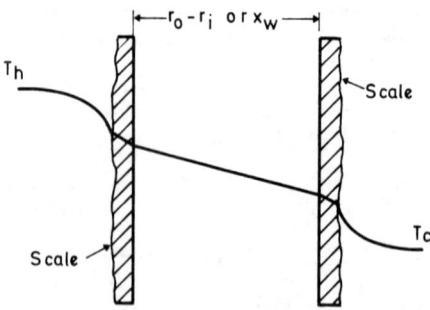

Fig. I.2.1 Temperature Distribution Across Exchanger Wall

# HEAT EXCHANGERS—BASIC METHODS

where $\Delta T$ is either $(T_h - T_1)$ or $(T_2 - T_c)$. Taking $h_u = h_f = h$, an overall surface efficiency $\varepsilon$ may be defined as follows:

$$\varepsilon \equiv \frac{q}{Ah\,\Delta T} = \eta\,\frac{A_f}{A} + \frac{A_u}{A} \tag{I.2.4}$$

TABLE I.2.1 Heat Exchanger Fouling Factors*

$(r_s = 1/h_s \; \text{oC m}^2/\text{w})$

| | up to 115 C | | 115 – 205 C | |
|---|---|---|---|---|
| Temperature of heating medium ........ | | | | |
| Temperature of water ............... | 50 C or less | | above 50 C | |
| Water velocity, m/sec ............... | 1 and less | over 1 | and less | over 1 |
| Distilled water ..................... | 0.0001 | 0.0001 | 0.0001 | 0.0001 |
| Sea water........................... | 0.0001 | 0.0001 | 0.0002 | 0.0002 |
| City or well water .................. | 0.0002 | 0.0002 | 0.0004 | 0.0004 |
| Treated boiler feed water ........... | 0.0002 | 0.0001 | 0.0002 | 0.0002 |
| Mississippi River water ............. | 0.0006 | 0.0004 | 0.0008 | 0.0006 |

| | |
|---|---|
| Liquid gasoline, organic vapors ....................................... | 0.0001 |
| Refrigerating liquids, cooling brine, oil-bearing steam ............. | 0.0002 |
| Refrigerating vapors, distillate bottoms above 20° API, air .......... | 0.0004 |
| Fuel oil, salty crude oil, residual bottoms less than 20° API ........ | 0.0010 |
| Diesel exhaust gas, coke-oven gas, cracking unit residuum ............ | 0.0020 |

where $A \equiv A_u + A_f$. An over-all heat transfer coefficient for the entire wall is then given by

$$\frac{T_h - T_c}{dq/dA_r} \equiv \frac{1}{U_r} = \frac{1}{\varepsilon_c h_{eff_c}}\frac{dA_r}{dA_c} + R_w \frac{dA_r}{dA_w} + \frac{1}{\varepsilon_h h_{eff_h}}\frac{dA_r}{dA_h} \tag{I.2.5}$$

where

$$\frac{1}{h_{eff}} \equiv \frac{1}{h_{fluid}} + \frac{1}{h_{scale}}$$

Here the reference area $A_r$ is the total surface area $A = A_u + A_f$ for either the hot or cold sides, and $A_w$ is the total primary wall area. For straight fins of uniform thickness

---

* From Standards of Tubular Exchanger Manufacturers' Assoc., 3rd Ed., N.Y., 1952

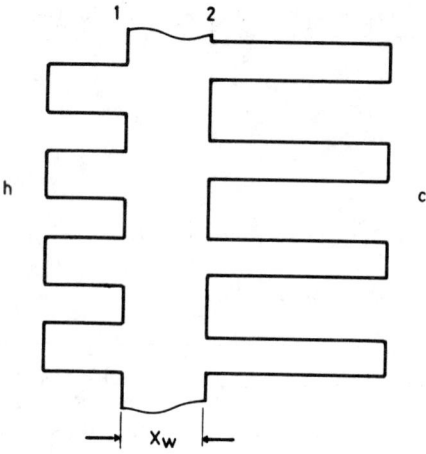

Fig. I.2.2 Finned Wall

$$\eta = \frac{\tanh BL}{BL}$$

where $B \equiv \sqrt{\dfrac{2h}{\delta k}}$ (I.2.6)

where $\delta$ is the fin thickness and L is the fin length.

## I. 2.3 ORDER OF MAGNITUDE OF THERMAL RESISTANCES

For a flat wall of thickness $x_w$ and $h_1$ and $h_2$ on either side with scale $h_s$, only on one side Equation (I.2.2) becomes

$$\frac{1}{U} = \frac{1}{h_1} + \frac{1}{h_s} + \frac{x_w}{k_w} + \frac{1}{h_2} \qquad (I.2.7)$$

The orders of magnitude and ranges of various magnitudes of h are shown in Table I.2.2.

TABLE I.2.2  Order of Magnitude of  $h$, w/m$^2$C

| | |
|---|---|
| Gases (Natural convection) | 5 - 25 |
| Flowing Gases | 10 - 250 |
| Flowing Liquids (non-metal) | 100 - 10,000 |
| Flowing Liquid Metals | 5,000 - 250,000 |
| Boiling Liquids | 1,000 - 250,000 |
| Condensing Vapors | 1,000 - 25,000 |

For a wall, the equivalent $h_{eq} = k/x$. For example, a steel wall 0.0025 m thick with k = 50/wm C, $h_{eq}$ = 50/0.0025 = 20000. For a 0.25 m thick asbestos wall with k = 0.25 w/m C, $h_{eq}$ = 0.25/0.25 = 1.0.

# HEAT EXCHANGERS—BASIC METHODS

For various combinations of resistances in Equation (I.2.7), some may be negligible.

Example I.2.1: What resistances are negligible for a 0.0025 m thick steel plate (k = 50) with flowing liquid ($h_1$ = 5000) and flowing gas ($h_2$ = 50) with $h_s$ = 5000?

Substitute in Equation (I.2.7):

$$\frac{1}{U} = \frac{1}{5000} + \frac{1}{5000} + \frac{0.0025}{50} + \frac{1}{50} \approx \frac{1}{50} = \frac{1}{h_2}$$

In this case only the gas side resistance is significant.

Example I.2.2: In Example I.2.1 replace the flowing gas by condensing steam ($h_2$ = 5000).

$$\frac{1}{U} = \frac{1}{5000} + \frac{1}{5000} + \frac{1}{20,000} + \frac{1}{5000}$$

In this case none of the resistances is negligible.

Example I.2.3: In Example I.2.1 replace the flowing liquid by another flowing gas ($h_1$ = 25).

$$\frac{1}{U} = \frac{1}{25} + \frac{1}{5000} + \frac{1}{20,000} + \frac{1}{50} \approx \frac{1}{25} + \frac{1}{50} = \frac{1}{h_1} + \frac{1}{h_2}$$

Here the wall and scale resistances are negligible.

## I.2.4 TEMPERATURE DISTRIBUTION IN HEAT EXCHANGERS

The preceeding discussions of heat transfer from one fluid to another separated by a wall have been limited to the heat transfer processes occurring at a particular place in a heat exchange system where the fluid temperatures have particular values.

In a heat exchanger the fluid temperatures change as the fluids flow along the heat exchanger length. If the fluid condenses or evaporates along the length, the temperature remains constant. Figure I.2.4 shows typical temperature distributions which may be obtained in heat exchangers. For the element of length, dx, with associated heat transfer area, dA, Figure I.2.3,

$$dq = U(T_h - T_c) \, dA \tag{I.2.8}$$

Also from the steady flow energy equation for each fluid,

$$dq = w_c di_c = w_h di_h \tag{I.2.9}$$

where w is the flow rate, i is the enthalpy; the kinetic potential energy changes of the fluids are neglected.

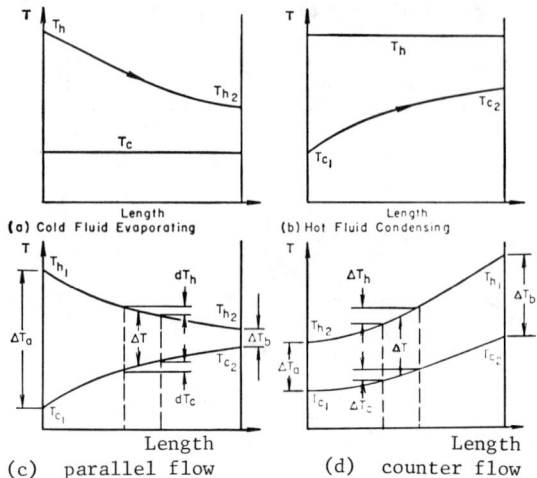

Fig. I.2.3 Axial Temperature Distribution in Heat Exchangers

If the cold fluid evaporates and the hot fluid condenses and $T_h$ and $T_c$ are each constant along the exchanger length, Equation (I.2.8) when integrated with U constant becomes

$$q = AU(T_h - T_c) \qquad (I.2.10)$$

For the cases of parallel flow and counterflow of fluids without phase change, Figures I.2.3c and I.2.3d, the enthalpy change may be written as $di = c\,dt$; then

$$dq = w_c c_c\, dT_c = \pm w_h c_h\, dT_h \qquad (I.2.11)$$

where the (+) sign refers to counterflow since $dT_h/dx$ is positive and the (-) sign to parallel flow since $dT_h/dx$ is negative. Then from Equation (I.2.11)

$$d(T_h - T_c) = dT_h - dT_c = dq\left(\pm \frac{1}{w_h c_h} - \frac{1}{w_c c_c}\right) \qquad (I.2.12)$$

Substituting for dq from Equation (I.2.8)

$$\frac{d(T_h - T_c)}{(T_h - T_c)} = U\left(\pm \frac{1}{w_h c_h} - \frac{1}{w_c c_c}\right) dA \qquad (I.2.13)$$

# HEAT EXCHANGERS—BASIC METHODS

which, when integrated with constant values of $U$, $w_h c_h$ and $w_c c_c$ between limits $\Delta T_a$ and $\Delta T_b$ results in

$$\ln \frac{\Delta T_b}{\Delta T_a} = UA \left( \pm \frac{1}{w_h c_h} - \frac{1}{w_c c_c} \right) \tag{I.2.14}$$

Similarly, integration of Equation (I.2.11) results in

$$q = w_c c_c (T_{c2} - T_{c1}) = w_h c_h (T_{h1} - T_{h2}) \tag{I.2.15}$$

Solving this for $w_c c_c$ and $w_h c_h$ and substituting in Equation (I.2.14)

$$q = AU \frac{\Delta T_a - \Delta T_b}{\ln(\Delta T_a / \Delta T_b)} \tag{I.2.16}$$

for either parallel or counterflow. The quantity

$$\frac{\Delta T_a - \Delta T_b}{\ln(\Delta T_a / \Delta T_b)}$$

is the logarithmic mean value, $\Delta T_{lm}$, of $\Delta T$ between $\Delta T_a$ and $\Delta T_b$, so $q = AU \, \Delta T_{lm}$. It should be noted that $\Delta T$, for example is the difference in temperature of the fluids at a particular place in the heat exchanger; $\Delta T_a = (T_{h1} - T_{c1})$ for parallel flow, but $\Delta T_a = (T_{h2} - T_{c1})$ for counterflow.

In the case of counterflow with $w_c c_c = w_h c_h$, the quantity $\Delta T_{lm}$ is indeterminate since $(T_{h1} - T_{h2}) = (T_{c2} - T_{c1})$ and $\Delta T_a = \Delta T_b$. Equation (I.2.11) shows that for any increment of area $dA$, $dT_c = dT_h$ for this case; therefore, $\Delta T$ is uniform along the heat exchanger and Equation (I.2.10) represents the over-all heat transfer performance with $(T_h - T_c) = \Delta T_a = \Delta T_b$. This may also be shown from Equation (I.2.16) by applying the calculus of limits.

The integration of Equation (I.2.8) for these other flow arrangements results in a form of an integrated mean temperature difference $\Delta T_m$ such that

$$q = AU \, \Delta T_m \tag{I.2.17}$$

where $\Delta T_m$ is a complex function of $T_{h1}$, $T_{h2}$, $T_{c1}$, and $T_{c2}$. Generally this function $\Delta T_m$ can be determined in terms of the following quantities:

$$\Delta T_{lmc} = \frac{(T_{h2} - T_{c1}) - (T_{h1} - T_{c2})}{\ln \frac{(T_{h2} - T_{c1})}{(T_{h1} - T_{c2})}}$$

$$\frac{C_c}{C_h} = \frac{w_c c_c}{w_h c_h} = \frac{T_{h1} - T_{h2}}{T_{c2} - T_{c1}} \tag{I.2.18}$$

$$\varepsilon_c = \frac{T_{c2} - T_{c1}}{T_{h1} - T_{c1}}$$

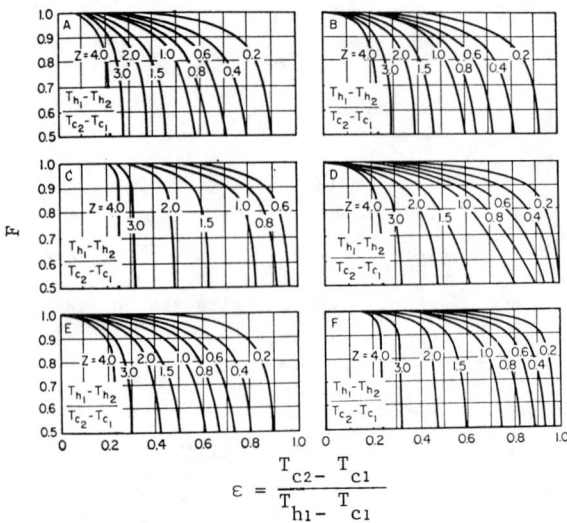

$$\varepsilon = \frac{T_{c_2} - T_{c_1}}{T_{h_1} - T_{c_1}}$$

Fig. I.2.4. Mean temperature difference in heat exchangers with various flow arrangements [Bowman, Mueller and Nagle (1)]. (A) One shell pass and 2,4,6, etc., tube passes. (B) One shell pass and 3,6,9, etc., tube passes. (C) Four shell passes and 8,16,24, etc., tube passes. (D) Crossflow, both fluids unmixed, one tube pass. (E) Crossflow, shell fluid mixed, one tube pass. (F) Crossflow, shell fluid mixed, two tube passes, shell fluid flows over first and second passes in series.

These quantities may be interpreted as follows: $\Delta T_{1cm}$ is the logarithmic mean temperature difference for a counterflow arrangement with the same fluid inlet and outlet temperatures: $C_h/C_c$ is the ratio of wc products of the two fluids, and $\varepsilon_c$ is called the effectiveness of the heat exchanger on the cold fluid side because it is a measure of the ratio of the heat actually transferred to the cold fluid to the heat which would be transferred if the same fluid were to be raised to the temperature of the hot inlet fluid.

Charts (Fig. I.2.4) for convenience of solution have been prepared for various flow arrangements with $F \equiv \Delta T_m/\Delta T_{1mc}$ as a function of $\varepsilon_c$ and $C_c/C_h$. Then

$$q = AUF \; \Delta T_{1mc} \tag{I.2.19}$$

In a multipass or a crossflow arrangement, the fluid temperature may not be uniform at a particular distance into the exchanger unless the fluid is well mixed along the path length. For example, in crossflow (Fig. I.2.5) the hot and cold fluids may enter at uniform temperatures, but if there are channels in the flow path to prevent mixing, the exit temperature distributions will be as shown. If such channels are not present, the fluids may be well mixed along the path length and the exit temperatures are more nearly uniform. A similar stratification of temperatures occurs in the shell- and-tube multipass exchanger. A series of baffles may be required if mixing

# HEAT EXCHANGERS—BASIC METHODS

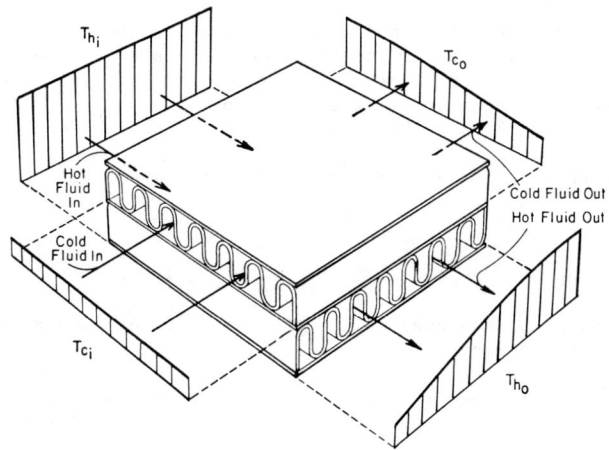

Fig. I.2.5 Temperature Distribution in Crossflow Heat Exchanger

of the shell fluid is to be obtained. Charts are presented for both mixed and unmixed fluid in Fig. I.2.4.

The preceeding analysis assumed U to be uniform throughout the heat exchanger. If U is not uniform the heat exchanger calculations may be made by subdividing the heat exchanger into sections over which U is nearly uniform and applying the previously developed relations to each subdivision.

## I.2.5 NTU DESIGN METHOD

The heat exchanger heat transfer equations such as Equations (I.2.14) and (I.2.15) may be written in dimenesionless form (2) resulting in the following dimensionless groups:

(1) Capacity rate ratio,

$$C_{min}/C_{max} \qquad \text{(I.2.20a)}$$

where $C_{min}$ and $C_{max}$ are respectively the smaller and larger of the two magnitudes $C_h$ and $C_c$. Recall that C is the product of the fluid flow rate and its specific heat.

(2) Exchanger heat transfer effectiveness,

$$\varepsilon \equiv \frac{C_h(T_{h1} - T_{h2})}{C_{min}(T_{h1} - T_{c1})} = \frac{C_c(T_{c2} - T_{c1})}{C_{min}(T_{h1} - T_{c1})} \qquad \text{(I.2.20b)}$$

which is the ratio of the actual heat transfer rate in the exchanger to the thermodynamically limited maximum prossible heat transfer rate which could be realized only in a counterflow heat exchanger of infinite heat transfer area. The first definition in Equation (I.2.20b) is for $C_h = C_{min}$ and the second for $C_c = C_{min}$.

(3) Heat transfer area number,

$$N_A = \frac{AU}{C_{min}} \qquad (I.2.20c)$$

This dimensionless group has been called the number of exchanger heat transfer units, NTU, by A.L. London.

In the equations of the preceeding section assume that $C_c > C_h$; so $C_h = C_{min}$ and $C_c = C_{max}$. Equation (I.2.14) may the be written

$$\frac{\Delta T_a}{\Delta T_b} = \exp[-N_A(\pm 1 - C_{min}/C_{max})] \qquad (I.2.21)$$

where the (+) sign is for counterflow and (−) for parallel flow. With Equations (I.2.20) and (I.2.15) and Figure I.2.3, the following identities are obtained:

Counterflow:

$$\frac{\Delta T_a}{\Delta T_b} = \frac{T_{h_2} - T_{c_1}}{T_{h_1} - T_{c_2}} = \frac{1 - \varepsilon}{1 - (C_{min}/C_{max})\varepsilon} \qquad (I.2.22a)$$

Parallel flow:

$$\frac{\Delta T_a}{\Delta T_b} = \frac{T_{h_1} - T_{c_1}}{T_{h_2} - T_{c_2}} = \frac{1}{1 - \varepsilon(1 + C_{min}/C_{max})} \qquad (I.2.22b)$$

Combining Equations (I.2.21) and (I.2.22)

Counterflow:

$$\varepsilon = \frac{1 - \exp[-N_A(1 - C_{min}/C_{max})]}{1 - (C_{min}/C_{max})\exp[-N_A(1 - C_{min}/C_{max})]} \qquad (I.2.23a)$$

Parallel flow:

$$\varepsilon = \frac{1 - \exp[-N_A(1 + C_{min}/C_{max})]}{1 + C_{min}/C_{max}} \qquad (I.2.23b)$$

These same results are obtained if it is assumed that $C_c < C_h$; then $C_c = C_{min}$ and $C_h = C_{max}$.

Two limiting cases are of interest — $C_{min}/C_{max}$ equal to zero and unity. For $C_{min}/C_{max} = 1.0$, Equation (I.2.23a) is indeterminate, but the following result may be obtained directly from (Equations (I.2.10) (I.2.15) and (I.2.20) or by applying the calculus to Equation (I.2.23a).

# HEAT EXCHANGERS—BASIC METHODS

For $(C_{min}/C_{max}) = 1.0$,

Counterflow:

$$\varepsilon = \frac{N_A}{1 + N_A} \tag{I.2.24a}$$

Parallel flow:

$$\varepsilon = \frac{1}{2}(1 - e^{-2N_A}) \tag{I.2.24b}$$

For $(C_{min}/C_{max}) = 0$, Figs. 1.2.3(a) and I.2.3(b), for parallel or counterflow, Equations (I.2.23) become

$$\varepsilon = 1 - e^{-N_A} \tag{I.2.24c}$$

These relations are shown graphically in Figs. I.2.6(a) and I.2.6(b). Similar relations and curves have been obtained for crossflow and for cross-counterflow exchangers and are shown in Figs. I.2.6(c),(d),(e). In a multipass cross-counterflow arrangement if each fluid is "mixed" between passes and if $N_A$ is equally distributed among the passes, the over-all effectiveness was derived (2) to be

$$\varepsilon = \frac{\left[\frac{1 - \varepsilon_p(C_{min}/C_{max})}{1 - \varepsilon_p}\right]^n - 1}{\left[\frac{1 - \varepsilon_p(C_{min}/C_{max})}{1 - \varepsilon_p}\right]^n - \frac{C_{min}}{C_{max}}} \tag{I.2.25a}$$

and if $C_{min}/C_{max} = 1.0$,

$$\varepsilon = \frac{n\varepsilon_p}{1 + (n - 1)\varepsilon_p} \tag{I.2.25b}$$

where $n$ is the number of identical passes and $\varepsilon_p$ is the effectiveness of each pass, a function of $N_A/n$.

The curves of Fig. I.2.6 show the asymptotic character of the $\varepsilon$ vs. $-N_A$ relation. The curves become quite flat beyond $N_A \cong 3$; hence, it is usually found that the most economical exchanger of a particular flow arrangement and $C_{min}/C_{max}$ will have a magnitude of $N_A$ in the range of approximately 1 to 3. Figure I.2.6 shows that a close approach to counterflow can be obtained by multipassing and that using more than three of four passes does not increase significantly the effectiveness. Figure I.2.6(f) compares the performance of various flow arrangements for $C_{min}/C_{max} = 1.0$.

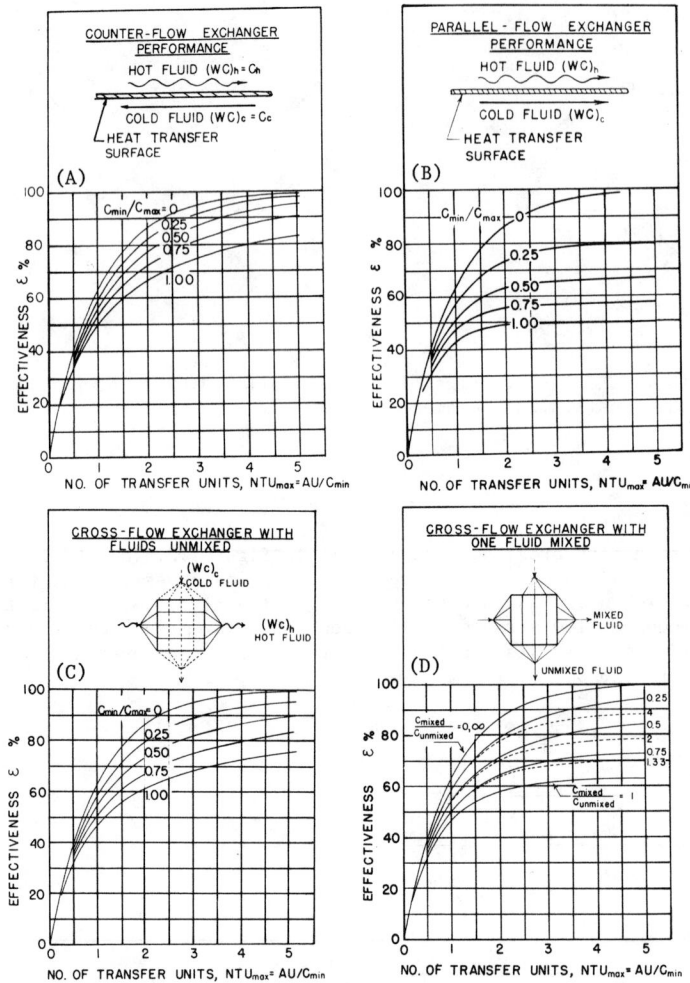

Fig. I.2.6  Effectiveness vs. NTU for Various Types of Heat Exchanger Flow Arrangements [Kays, London and Johnson (2)].

## I.2.6  FLOW IN CLOSED CONDUITS — FRICTION FACTOR

In fully developed flow in a pipe — either laminar or turbulent, we assume $\Delta p$ is proportional to the length $L$ and that the following functional relationship is valid:

$$\frac{\Delta p}{L} = \phi(V, D, \rho, \mu, e) \tag{I.2.26}$$

The quantity $e$ is a statistical measure of surface roughness of the pipe and has the dimensions of length. With force $F$, mass $M$, length $L$, and time $\theta$ as fundamental dimensions, and $V, D, \rho$ as the set of maximum number of

# HEAT EXCHANGERS—BASIC METHODS 441

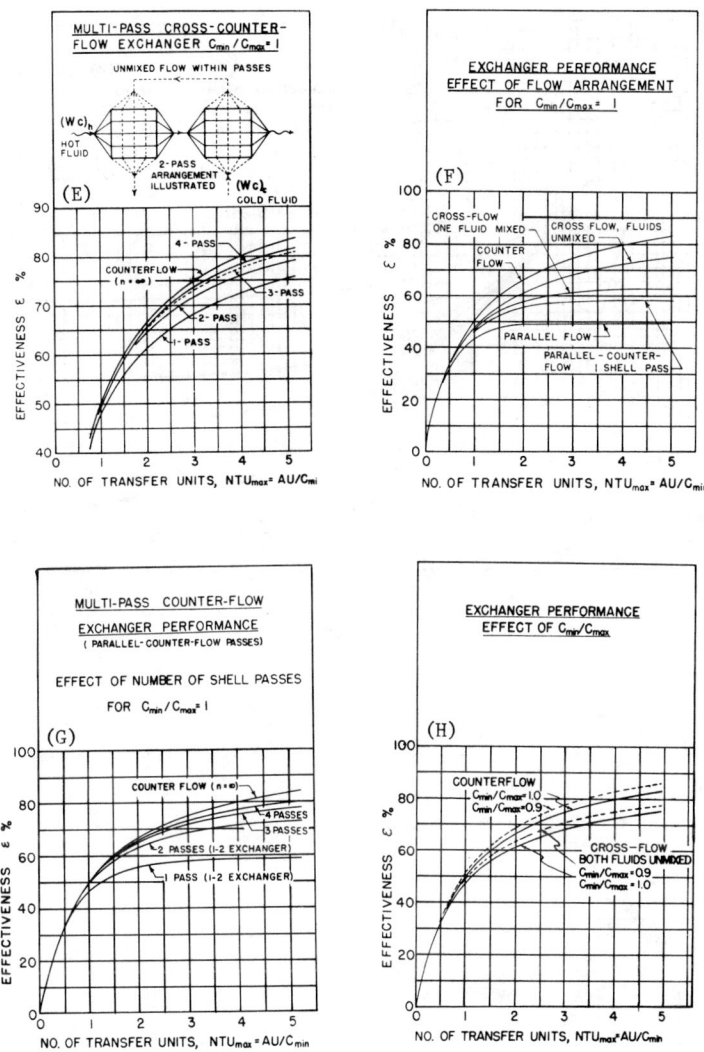

Fig. I.2.6 Continued

quantities which in themselves cannot form a dimensionless group, the pi theorem leads to

$$\frac{\Delta p}{4(L/D)(\rho V^2/2)} = \psi \left( \frac{V D \rho}{\mu}, \frac{e}{D} \right) \qquad (I.2.27)$$

where the dimensionless numerical constants 4 and 2 are added here for convenience.

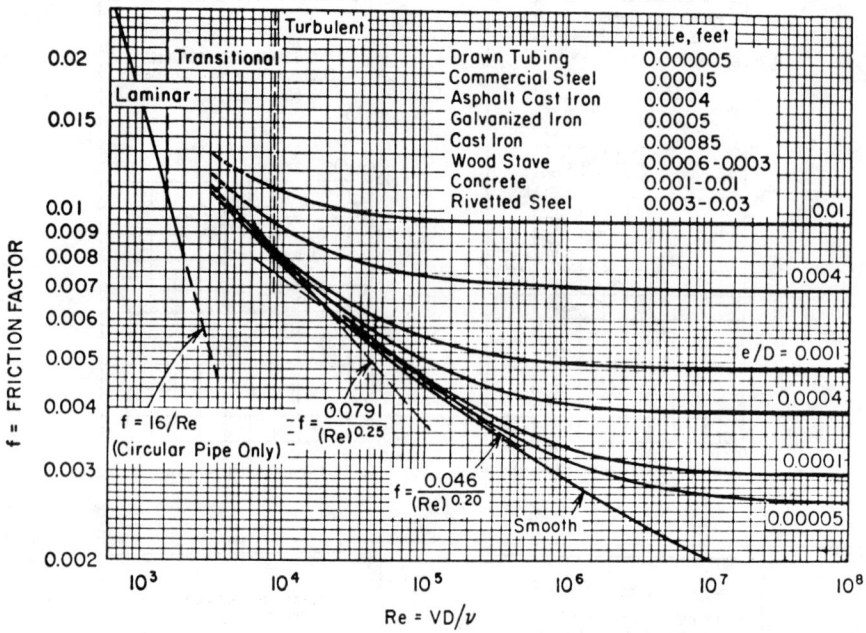

Fig. I.2.7 Friction Factor for Flow in Circular Pipes [Moody (5)]

The above dimensionless group involving $\Delta p$ has been defined as a friction factor, $f^*$.

$$f \equiv \frac{\Delta p}{4(L/D)(\rho V^2/2)} \tag{I.2.28}$$

The, Equation (I.2.27( becomes

$$f = \psi(\text{Re}, e/D) \tag{I.2.29}$$

Figure I.2.7. shows this relationship as deduced by Moody (5) from experimental data for fully developed flow. In the laminar region, existing empirical data on pressure drop within round pipes can all be correlated by a simple relationship between $f$ and $\text{Re}$, independent of the surface roughness.

Laminar:

---

* Other definitions of $f$ appear in the literature. Some multiply the right side of Eq. (I.2.28) by 2 or 4, changing the numerical value by a factor.

$$f = \frac{16}{Re} \tag{1.2.30}$$

The transition from laminar to turbulent flow is somewhere in the neighborhood of 2300 - 4000 for Re. Values of $f$ for commercial pipes are difficult to determine precisely; even for s0-called smooth pipes $f$ may vary as much as ± 5 cen cent. In addition, the change of roughness with age is difficult to predict.

The $f$-vs.-$Re$ relation for smooth pipes in turbulent flow has a slight curvature on a log-log plot, Fig. I.2.7, and is given as

Turbulent:

$$\frac{1}{\sqrt{f}} = 4.0 \log_{10}(Re \sqrt{f}) - 0.40 \tag{I.2.31}$$

Two linear approximations, shown dotted in Fig. I.2.7 are:

Turbulent:

$$f \cong \frac{0.046}{Re^{0.2}} \tag{I.2.32}$$

$$f \cong \frac{0.0791}{Re^{0.25}} \tag{I.2.33}$$

For calculation purposes $f$ should be read from the graph, but these approximate equations are often useful in showing functional relationships of various quantities.

For fully developed flow in a tube, a simple force balance yields

$$\Delta p \frac{\pi}{4} D^2 = \tau_0 (\pi DL)$$

which may be combined with Equation (I.2.28) to get an equivalent form for the friction factor:

$$f = \frac{\tau_0}{\rho V^2/2} \tag{I.2.34}$$

## I.2.7 NONCIRCULAR CROSS SECTIONS

A duct of noncircular cross section is not geometrically similar to a circular pipe; hence, dimensional analysis does not relate the performance of these two geometries. However, in turbulent flow, $f$ for noncircular cross sections (annular spaces, rectangular and triangular ducts, etc.) may be evaluated from the data for circular pipes if $D$ is replaced by an "equivalent diameter," $D_e$, defined by

$$D_e = \frac{4A}{P} = \frac{4 \text{ (flow area)}}{\text{wetted perimeter}} \tag{I.2.35}$$

using the equivalent diameter in turbulent flow gives results for $f$ within about ± 8% of measured results, Brundrett [3].

The equivalent diameter of an annulus of inner and outer diameter $D_i$, $D_o$ is

$$D_e = \frac{4(\pi/4)(D_o^2 - D_i^2)}{\pi(D_o + D_i)} = (D_o - D_i)$$

For a circular pipe, Equation (I.2.35) reduces to $D_e = D$.

The transition Reynolds number $VD_e\rho/\mu$ is also found to be approximately 2300, as for circular ducts.

For laminar flow, however, the results for noncircular cross sections are not universally correlated. In a thin annulus in which spacing $Z$ is very much less than the mean diameter of the annulus, the flow has a parabolic distribution perpendicular to the wall and has this same distribution at every circumferential position. If we treat this as flow between parallel flat plates

$$\frac{\Delta p}{\Delta x} = \frac{12\mu V}{Z^2} \tag{I.2.36}$$

Here $D_e = 2Z$ and Equation (I.2.36) can be written in the form

$$f = \frac{24}{Re} \tag{I.2.37}$$

with $D_e$ replacing $D$ in the definitions of $f$ and $Re$. This equation is obviously different from Equation (I.2.30) which applied to laminar flow in circular pipes.

Flow in a rectangular duct (dimensions $Z_1 \times Z_2$) in which $Z_2 \ll Z_1$ is similar to this annular flow. For rectangular ducts of other aspect ratios $(Z_1/Z_2)$,

$$f = \frac{16}{\phi\, Re} \tag{I.2.38}$$

where

$$D_e = \frac{2(Z_1 Z_2)}{(Z_1 + Z_2)} \tag{I.2.39}$$

and $\phi$ is given by Fig. I.2.8.

For laminar flow in ducts of triangular and trapezoidal cross section, Nikuradse [4] showed that $f$ is approximated by $16/Re$ with $D_e$ given by Equation (I.2.35), and transition occurs at approximately $Re = 2300$.

## I.2.8 ONE-DIMENSIONAL FLOW — CORRECTION FOR NONUNIFORM VELOCITY

Flow in closed conduits, such as a heat-exchanger tube can often be analyzed as a steady, one-dimensional flow. Consider a control volume shown in Fig. I.2.9.

Continuity of matter requires simply

# HEAT EXCHANGERS—BASIC METHODS

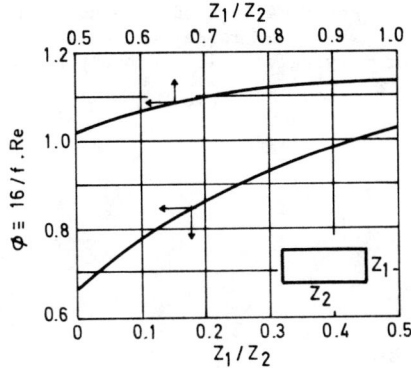

Fig. I.2.8 Values of $\phi$ for rectangular ducts

$$d(\rho AV) = 0 \quad \text{or} \quad \rho AV = \text{const.} \tag{I.2.40}$$

where $\rho$ is the density, $A$ is the control surface area normal to flow, $V$ is the average velocity.

A momentum balance in the flow or streamline direction involves (a) net average momentum flux associated with mass flow, (b) pressure and viscous forces acting on the control surface, and (c) gravity body force acting on the control volume. At steady state,

$$\rho AV \frac{dV}{dl} \Delta l + A \frac{dp}{dl} \Delta \ell + \tau_o (\text{per.}) \Delta l + g\rho A \Delta l \sin \theta = 0 \tag{I.2.41}$$

or

$$-dp = \rho V \, dV + \frac{4\tau_o \, dl}{D_e} + \rho g \, dZ \tag{I.2.42}$$

Equation (I.2.42) is exactly true only if the velocity is uniform over the flow cross section. If the velocity varies in some arbitrary manner as shown in

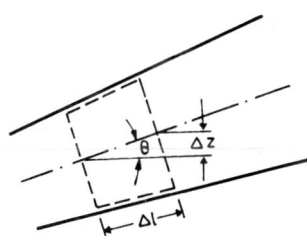

Fig. I.2.9 One-dimensional Flow

Fig. I.2.10, the average momentum flux entering section 1 is not equal to $\rho AV^2$, but may be expressed as $\rho AV^2/\alpha$ where $\alpha$ is defined by Equation (I.2.43) to account for the effect of varying velocity across the section. Referring to Fig. I.2.10, we can write

$$\frac{1}{\alpha}(\rho AV^2) = \int_A v(\rho v \, dA)$$

or, with $\rho$ uniform,

$$\frac{1}{\alpha} = \frac{1}{A}\int_A \left(\frac{v}{V}\right)^2 dA \qquad (I.2.43)$$

For uniform velocity, e.g., highly turbulent sluglike flow, $\alpha = 1$; for parabolic velocity distribution in a circular pipe (Hagen-Poiseuille flow), $\alpha = 0.75$.

For flow of fluid in a tube of uniform cross section, Equation (I.2.42), modified by factor $\alpha$ if necessary, can be readily integrated to find the pressure drop between any two sections 1 and 2 along the tube. We can replace $\tau_0$ in Equation (I.2.42) by the friction factor $f$ using Equation (I.2.34), noting further that $\rho$ in Equation (I.2.34) may be evaluated at

$$\frac{1}{\rho_m} = \frac{1}{2}\left(\frac{1}{\rho_1} + \frac{1}{\rho_2}\right)$$

Letting $G = \rho V$, Equation (I.2.42) when integrated for $\alpha \neq 1$ becomes

$$P_1 - P_2 = \frac{G^2}{\alpha}\left(\frac{1}{\rho_2} - \frac{1}{\rho_1}\right) + 4f \frac{L}{D_e} \frac{G^2}{2\rho_m} + g\rho_m(Z_2 - Z_1) \qquad (I.2.44)$$

For liquids $\rho_1 \cong \rho_2 \cong \rho_m$ making the first term on the right-hand side of Equation (I.2.44) essentially zero.

For a fluid with nearly uniform velocity ($\alpha = 1$) at any cross section, Equation (I.2.42) becomes

$$d(H) = d\left(\frac{p}{\rho} + \frac{v^2}{2}\right) + gZ = dF \qquad (I.2.45)$$

where $dF$ is the frictional effect. The quantity $[(p/\rho + (V^2/2) + (gZ)$ is known as the Bernoulli head, $H$, and Equation (I.2.45) for idealized frictionless flow ($dF = 0$) is known as the Bernoulli equation.

Similar to $\alpha$, a factor $\beta$ which we shall find useful in the next section may be defined to account for the fact that if velocity is nonuniform at a section, the average kinetic energy of fluid at that section is not simply $\rho AV \cdot V^2/2$ but may be expressed as $(1/\beta)(\rho AV \cdot V^2/2)$. Referring again to Fig. I.2.10,

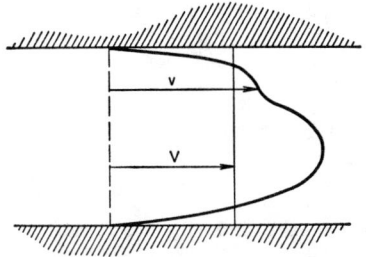

Fig. I.2.10 Velocity Distribution

$$\frac{1}{\beta} \rho A V \frac{V^2}{2} = \int_A (\rho v \, dA) \frac{v^2}{2}$$

uniform $\rho$ :

$$\frac{1}{\beta} = \frac{1}{A} \int_A \left(\frac{v}{V}\right)^3 dA \qquad (I.2.46)$$

For uniform velocity, $\beta = 1$ , and for a parabolic distribution, $\beta = 0.5$.

## I.2.9 SUDDEN ENLARGEMENT AND CONTRACTION

The pressure drop associated with sudden enlargements and sudden contractions, Fig. I.2.11 are reported in terms of a decrease in Bernoulli head H , Equation (I.2.45), and a loss coefficient K referred to the kinetic energy of the flow in the smaller cross section. Since $\Delta Z = 0$ for each of the cases,

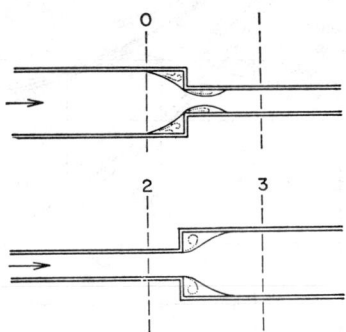

Fig. I.2.11 Sudden Contraction and Enlargement

$$\frac{p_o - p_1}{\rho_{o1}} + \frac{v_o^2 - v_1^2}{2} = K_c \frac{v_1^2}{2} \qquad (I.2.47)$$

$$\frac{p_2 - p_3}{\rho_{23}} + \frac{v_2^2 - v_3^2}{2} = K_e \frac{v_2^2}{2} \qquad (I.2.48)$$

From continuity since $\rho AV = \text{const.}$, and $\rho_1 \cong \rho_2$ and $\rho_2 \cong \rho_3$,

$$A_o V_o = A_1 V_1 \quad \text{and} \quad A_2 V_2 = A_3 V_3 \qquad (I.2.49)$$

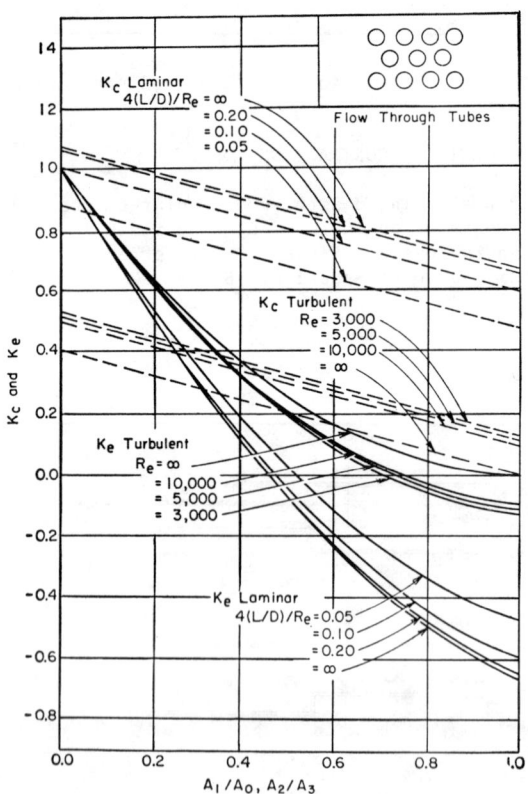

Fig. I.2.12 Values of $K_c$ and $K_e$ for a Tube Bundle [Kays (50)].

HEAT EXCHANGERS—BASIC METHODS

Values of $K_c$ and $K_e$ have been determined by Kays [5] for a number of different geometries and checked by his test data. Fig. I.2.12 presents $K_c$ and $K_e$ for a bundle of tubes with headers at either end where the velocity distribution is essentially uniform.

I.2.10 HEAT TRANSFER

In fully developed laminar flow hydraulic diamter, $D_e$, does not correlate the heat transfer results. The Nu for any geometry reaches a uniform value in fully developed flow. The results are usually presented as a Nu based on $D_e$, in Table I.2.3.

TABLE I.2.3 NUSSELT NUMBER FOR FULLY DEVELOPED LAMINAR FLOW

| Geometry | Velocity distribution | Condition at wall | $\frac{hD_e}{k}$ |
|---|---|---|---|
| Circular tube | Parabolic | $(q/A)_o$ Uniform | 4.36 |
| Circular tube | Parabolic | $T_o$ Uniform | 3.66 |
| Circular tube | Slug | $(q/A)_o$ Uniform | 8.00 |
| Circular tube | Slug | $T_o$ Uniform | 5.75 |
| Parallel plates | Parabolic | $(q/A)_o$ Uniform | 8.23 |
| Parallel Plates | Parabolic | $T_o$ Uniform | 7.60 |
| Triangular duct | Parabolic (ref.2) | $(q/A)_o$ Uniform | 3.00 |
| Triangular duct | Parabolic (ref.2) | $T_o$ Uniform | 2.35 |

In the entrance region the magnitudes of Nu are higher as the boundary layers develop. The average Nu approaches these fully developed values at approximately,

$$\frac{L}{D} \cong \frac{Re\, Pr}{20} \tag{I.2.50}$$

In fully developed turbulent flow the following three relations are used by various designers:

Mc Adams:
$$\left(\frac{hD}{k_b}\right) = 0.023 \left(\frac{GD}{\mu_b}\right)^{0.8} \left(\frac{\mu c_p}{k}\right)_b^{0.4} \tag{I.2.51}$$

Colburn:
$$\frac{h}{c_{pb} G} \left(\frac{\mu_f c_{pb}}{k_b}\right)^{2/3} = \frac{0.023}{(GD/\mu_f)^{0.2}} \tag{I.2.52}$$

Sieder and Tate:
$$\frac{h}{c_{pb} G} \left(\frac{\mu c_p}{k}\right)_b^{2/3} = \frac{0.027}{(GD/\mu_b)^{0.2}} \cdot \left(\frac{\mu_b}{\mu_w}\right)^{0.14} \tag{I.2.53}$$

For non-circular ducts use the hydraulic diameter $D_e$ in the equations. The results will be within ± 8% of the data.

In the entrance region for gases

$$\frac{Nu_{avg}}{Nu_\infty} \simeq 1 + \frac{1.4}{L/D} \qquad (I.2.54)$$

## I.2.11 SIZING COUNTERFLOW SHELL-AND-TUBE HEAT EXCHANGERS FOR A GIVEN JOB

Bare tube shell-and-tube heat exchangers find many applications in gas-to-gas or liquid-to-liquid service.

The "job" of the heat exchanger here will be defined as transfering a specified amount of heat between two fluids at given flow rates and inlet temperatures with specified amounts of pumping power on each side. This then sets magnitudes for the following quantities:

$w_1$, $T_{in,1}$, $\delta T_1$, $\Delta P_1$

$w_2$, $T_{in,2}$, $\delta T_2$, $\Delta P_2$

where subscript 1 refers to the hotter fluid inside the tubes and subscript 2 to the colder fluid in the shell, Fig. I.2.13.

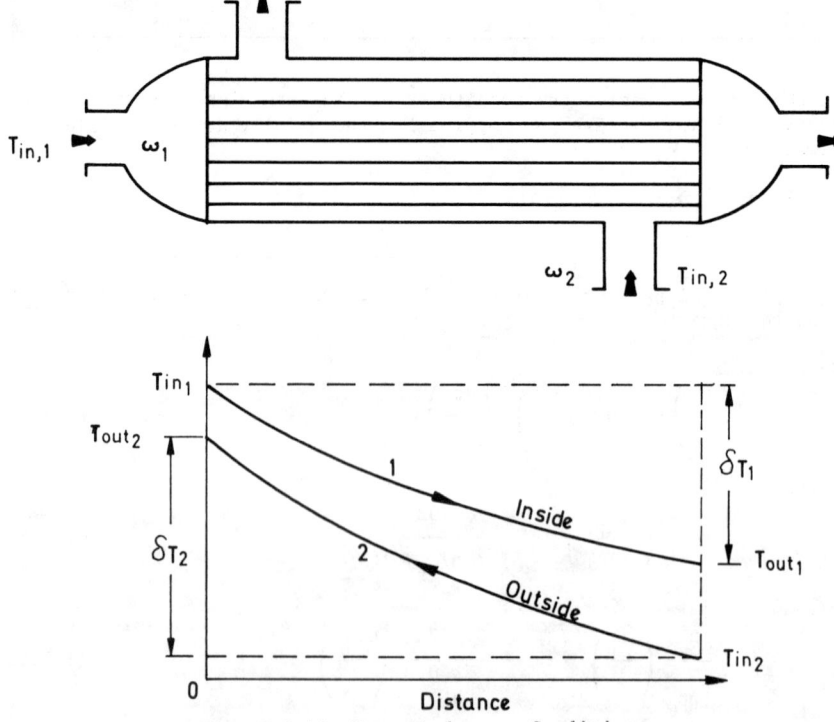

Fig. I.2.13 Heat Exchanger Conditions

## HEAT EXCHANGERS—BASIC METHODS

It will be shown that by defining the "job" the magnitudes of $G_1$ and $G_2$ are then established independent of tube size selected and the exchanger length, $L$, and number of tubes, $n$, are determined by the tube size selected. The procedure described here is a modification of one given by Fraas and Ozisik [10].

The following basic equations are applicable in turbulent flow approximating the friction factor as $f = 0.046/Re_D^{0.2}$ which is defined by $\Delta P = 4f(L/D)G^2/2g_o\rho$ for both sides. Here $D$ is $D_1$ for the inside and $D_2$ (equivalent or hydraulic diameter) for the outside.

$$\Delta P_1 = \left(\frac{0.092\,\mu_1^{0.2}}{g_o\rho_1}\right) \frac{G_1^{1.8} L}{D_1^{1.2}} = K_1 \frac{G_1^{1.8} L}{D_1^{1.2}} \qquad (I.2.55)$$

$$\Delta P_2 = \left(\frac{0.092\,\mu_2^{0.2}}{g_o\rho_2}\right) \frac{G_2^{1.8} L}{D_2^{1.2}} = K_2 \frac{G_2^{1.8} L}{D_2^{1.2}} \qquad (I.2.56)$$

$$h_1 = \left(0.023\,k_1 \frac{Pr_1^{0.4}}{\mu_1^{0.8}}\right) \frac{G_1^{0.8}}{D_1^{0.2}} = K_3 \frac{G_1^{0.8}}{D_1^{0.2}} \qquad (I.2.57)$$

$$h_2 = \left(0.023\,k_2 \frac{Pr_2^{0.4}}{\mu_2^{0.8}}\right) \frac{G_2^{0.8}}{D_2^{0.2}} = K_4 \frac{G_2^{0.8}}{D_2^{0.2}} \qquad (I.2.58)$$

$$\frac{1}{U_1} = \frac{1}{h_1} + \frac{D_1}{D_o h_2} = \frac{1}{h_1}\left(1 + \frac{D_1 h_1}{D_o h_2}\right) \qquad (I.2.59)$$

$$A_{x_1} = \frac{\pi}{4} D_1^2 n; \quad A_{x_2} = \frac{\pi}{4} D_2 D_o n \qquad (I.2.60)$$

$$A_1 = \pi D_1 n L \qquad (I.2.61)$$

$$q = A_1 U_1 \Delta T_{lm} = w_1 c_1 \delta T_1 = w_2 c_2 \delta T_2 \qquad (I.2.62)$$

Here $D_o$ is the outside diameter of the tubes. In Equations (I.2.55) through (I.2.58) the $K_1$, $K_2$, $K_3$ and $K_4$ are defined as the quantities in the parentheses and represent known quantities.

The following are some equations derived from the above equations. From Equations (I.2.55) and (I.2.56)

$$\frac{\Delta P_1}{\Delta P_2} = \left(\frac{K_1}{K_2} \frac{G_1}{G_2}\right)^{1.8} \left(\frac{D_2}{D_1}\right)^{1.2} \equiv K_5 \qquad (I.2.63)$$

From Equation (I.2.60)

$$\frac{G_1}{G_2} \equiv \frac{w_1}{A_{x_1}} \frac{A_{x_2}}{w_2} = \frac{w_1}{w_2} \frac{D_2 D_o}{D_1^2} \quad (I.2.64)$$

From Equations (I.2.63) and (I.2.64)

$$\frac{G_1}{G_2} = \left(\frac{w_1}{w_2} \frac{D_o}{D_1}\right)^{0.4} \left(\frac{K_2 K_5}{K_1}\right)^{1/3} \equiv K_6 \quad (I.2.65)$$

From Equations (I.2.57), (I.2.58), (I.2.64) and (I.2.65),

$$\frac{h_1}{h_2} = \frac{K_3}{K_4}\left(\frac{G_1}{G_2}\right)^{0.8}\left(\frac{D_2}{D_1}\right)^{0.2} = \frac{K_3}{K_4}\left(\frac{K_2 K_5}{K_1}\right)^{1/3}\left(\frac{D_o}{D_1}\right)^{0.2}\left(\frac{w_1}{w_2}\right)^{0.2} \equiv K_7 \quad (I.2.66)$$

From Equations (I.2.57), (I.2.59) and (I.2.66),

$$\frac{1}{U_1} = \frac{D_1^{0.2}}{K_3 G_1^{0.8}}\left(1 + \frac{D_1}{D_o}K_7\right) = K_8 \frac{D_1^{0.2}}{G_1^{0.8}} \quad (I.2.67)$$

where

$$K_8 \equiv \frac{1}{K_3}\left(1 + \frac{D_1}{D_o}K_7\right) \quad (I.2.68)$$

From Equations (I.2.61) and (I.2.62),

$$q = n\pi D_1 U_1 L \Delta T_{1m} = \frac{\pi}{4} D_1^2 n G_1 c_1 \delta T_1$$

or

$$L = \left(\frac{c_1 \delta T_1}{4 \Delta T_{1m}}\right) \frac{G_1 D_1}{U_1} \equiv K_9 \frac{G_1 D_1}{U_1} \quad (I.2.69)$$

Combining Equations (I.2.67) and (I.2.69)

$$L = K_9 K_8 D_1^{1.2} G_1^{0.2} \quad (I.2.70)$$

Note L and D appear in the same way in both Equations (I.2.55) and (I.2.70). Combining these equations eliminates L and D to obtain

$$G_1 = \left(\frac{\Delta P_1}{K_1 K_8 K_9}\right)^{1/2} = K_{10} \quad (I.2.71)$$

From $w_1 = G_1 A_{x_1}$ and Equation (I.2.60)

HEAT EXCHANGERS—BASIC METHODS

$$n = \frac{w_1}{G_1} \frac{4}{\pi D_1^2} = \frac{K_{11}}{D_1^2} \tag{I.2.72}$$

Then from Equations (I.2.61), (I.2.70) and (I.2.72),

$$A = \pi D_1 nL = \pi K_8 K_9 K_{10}^{0.2} K_{11} D_1^{0.2} = K_{12} D_1^{0.2} \tag{I.2.73}$$

With heat exchanger job defined by fixing the flow rates, temperatures, pressure drops and the properties of the two fluids, the magnitudes of all of the quantities $K_1 \ldots K_{12}$ are determined provided a magnitude for $D_o/D_1$ is selected. For thin walled tubing $D_o/D_1$ is approximately unity, and it seldom is greater than 1.2. The final design size is not very sensitive to this magnitude of $D_o/D_1$.

To proceed with the design define the flow rates, temperature, pressure drops, fluid properties and pick an expected value of $D_o/D_1$. Then any of the twelve K's may be evaluated. Not all of them need be calculated.

First the magnitude of the ratio $G_1/G_2$ is determined form $K_6$, Equation (I.2.65). From Equation (I.2.71) or $K_{10}$, $G_1$ is caculated, hence, $G_2$ is determined. At his point the tube diameter $D_1$ is selected and L, n and A calculated from Equations (I.2.70), (I.2.72) and (I.2.61), thus fixing the size of the heat exchanger.

Defining a heat exchanger job by fixing the flow rates, temperatures, pumping powers and fluids of each stream immediately establishes the required magnitude of $G_1$ and $G_2$, Equations (I.2.65) and (I.2.71). Then upon deciding on a tube diameter $D_1$, the magnitudes of L, n and A are established, Equations (I.2.70), (I.2.72) and (I.2.61).

NOMENCLATURE

| | |
|---|---|
| A | area |
| $A_x$, S | flow cross-sections area |
| C | capacity rate wc |
| $C_p$, c | specific heat at constant pressure |
| D, d | diameter |
| $D_e$ | equivalent diameter, defined by Eq. (I.2.35) |
| e | roughness of inner surface of pipe |
| f | friction factor |
| G | mass flow rate, (G = w/S = $\rho$V) |
| g | gravitational acceleration |
| h | heat transfer coefficient |
| k | thermal conductivity |
| L, $\ell$ | length |
| n | number of tubes |
| P, p | pressure |
| P | wetted perimeter |
| q | rate of heat transfer |
| q/A, q" | heat flux |
| T | temperature |
| U | over-all heat transfer coefficient |
| V | velocity |
| w | flow rate |

$\varepsilon$        exchanger heat transfer effectiveness
$\varepsilon$        surface efficiency
$\eta$        fin efficiency
$\mu$        absolute viscosity
$\nu$        kinetic viscosity $\mu/\rho$
$\rho$        mass density

# REFERENCES

1. Bowman, R.A., Mueller A.C., and Nagle, W.M. 1940. Trans. ASME, 62, 283-94.

2. Kays, W.M., London A.L., and Johnson, D.W. 1951. Gas Turbine Plant Heat Exchangers, ASME, New York.

3. Brundrett, E. 1978. "Modified Hydraulic Diameter for Turbulent Flow", Proc. NATO Adv. Study Inst., Istanbul, Turkey, Hemisphere Publ. Co., Washington, D.C.

4. Nikuradse, J. 1930. Ingenieur-Archiv, 1, 306.

5. Kays, W.M. 1950. Trans. ASME, 72:8, 1067.

6. Kays, W.M., and London, A.L. 1952. Trans., ASME. 74:7, 1179 (Oct.)

7. McAdams, W.H. 1954. Heat Transmissionm 3rd Ed., McGraw-Hill, New York.

8. Colburn, A.P. 1933. Trans. Am. Inst., Chem. Engrs., 29, 174.

9. Siedler, E.N., and Tate, G.E. 1936. Ind. Eng. Chem., 28, 1429-36.

10. Fraas, A.P, and Ozisik, M.N. 1963. Heat Exchanger Design", Wiley, 161.

# Heat Exchanger Design Methodology—An Overview

**RAMESH K. SHAH**
Harrison Radiator Division
General Motors Corporation
Lockport, New York 14094 USA

ABSTRACT

An overview is presented here on various quantitative and qualitative steps involved in arriving at the optimum heat exchanger design. These steps include thermal and hydraulic design, mechanical design, evaluation procedure and costing.

HEAT EXCHANGER DESIGN METHODOLOGY

Heat exchanger design methodology is illustrated in Fig. 1 which is modified from that presented by Kays and London [1] and Shah [2].[†] This design procedure may be characterized by a case study method. It is a complex procedure because of the many qualitative judgments (in addition to the quantitative calculations) that must be introduced. Let us overview the overall methodology now.

The specified problem for heat exchanger design may contain information as little as just the flow rate and temperatures of the fluid stream of concern; or it may contain as much as all details on operating pressures, temperatures, flow rates, surfaces on each side, required pressure drops and heat transfer, size, weight, and other design constraints including cost, materials to be used, and alternative heat exchanger types and arrangements. The designer requires less judgment with more information on problem specifications. However, if there are too many constraints specified, there may not be a feasible design, and some compromises may be needed for a feasible solution. The designer provides missing information based on his experience, judgment and discussions with the customer.

Based on the problem specifications and experience, the exchanger construction type and flow arrangement are first selected. Next selected are the core or surface geometry and material. The core geometry is selected for a shell-and-tube exchanger while the surface geometry is chosen for plate, extended surface and regenerative exchangers. There are several quantitative and qualitative criteria for surface selection. Some of the quantitative criteria for compact heat exchanger surfaces are discussed [2]. The qualitative criteria for surface selection are the operating temperature and pressure levels, the

---

[†]Even though Fig. 1 was originally prepared for single-phase heat exchangers, it is also applicable in general for two-phase heat exchangers.

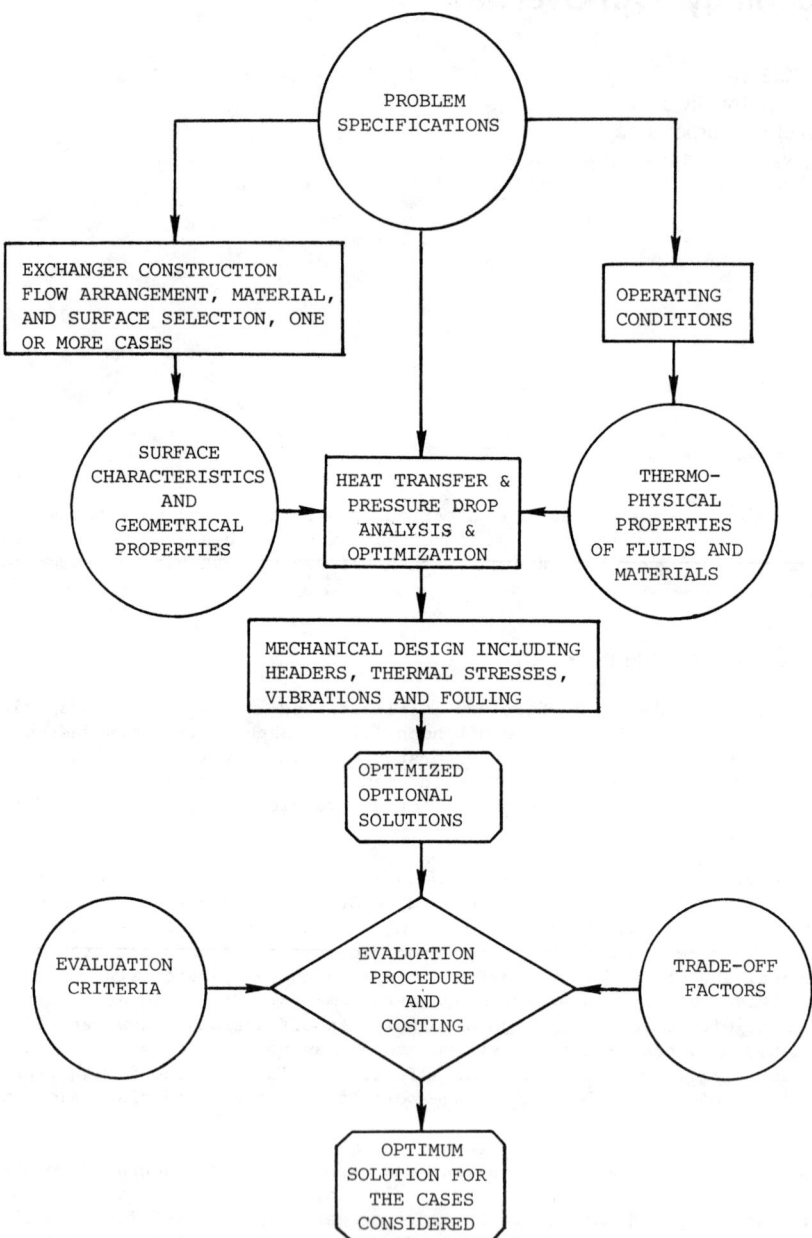

Fig. 1 Methodology of heat exchanger design.

designer's experience and judgment, cost, availability of surfaces, manufacturability, maintenance requirements, reliability, safety, and so on. For shell-and-tube exchangers, the criteria for selecting core geometry or configuration are the desired heat transfer performance within specified pressure drop, operating pressures and temperatures, thermal stresses, allowable leakage rate, corrosion characteristics of fluids, fouling, cleanability, maintenance, minimal operational problems (vibrations, freeze-up, instability, etc.), and cost. Some of these are discussed by Mueller [3].

Heat exchanger analysis procedures, generally available as a computer program, involve quantitative heat transfer and pressure drop evaluation or exchanger sizing. These are discussed by [1,4,5]. Inputs to the theoretical procedures are the surface heat transfer and flow friction characteristics (such as $j$ and $f$ versus $Re$ curves) and thermophysical properties of fluids, in addition to the problem specifications. Mathematical optimization techniques may be employed at this stage to optimize heat exchanger design. The output from the heat exchanger theoretical analysis may be a large number of optimized optional solutions depending upon the design variables considered for each surface. The surface basic heat transfer and flow friction characteristics and heat exchanger optimization techniques are discussed in [6] and [7], respectively.

Heat exchanger thermal design problems may be categorized primarily as rating and sizing problems. Inputs to the rating problem are: the exchanger construction, flow arrangement and overall dimensions, complete details on the materials and surface geometries including their heat transfer and pressure drop characteristics, fluid flow rates, inlet temperatures and fouling factors. The fluid outlet temperatures, total heat transfer capability, and pressure drops on each side of the exchanger are then determined in the rating problem. The rating problem is also sometimes referred to as the performance problem. In contrast, in the sizing problem, the core lengths, surface areas and core dimensions are to be determined. Inputs to the sizing problem are surface geometries (including their heat transfer and pressure drop characteristics), fluid flow rates, inlet and outlet fluid temperatures, fouling factors, and pressure drops on each side. The sizing problem is also sometimes referred to as the design problem.[†] Further details on the rating and sizing problems are discussed later in this lecture series.

Next, the mechanical design is pursued to ensure the mechanical integrity of the exchanger under steady state and transient operating conditions. Pressure stress calculations are performed to determine the fin, plate, tube or shell thicknesses. A proper selection of the material and the method of bonding fins to plates or tubes is made depending upon the operating temperatures, pressures and fluids. A proper selection of the particular sides of a two-fluid (or n-fluid) exchanger for working fluids is made depending upon the operating temperatures, pressures, available pressure drops, and cleaning requirements. A proper design of headers, tanks, manifolds, nozzles or pipes is made to ensure uniform flow distribution through the exchanger passages. Thermal stress calculations are performed to ensure the durability and desired life of the exchanger for expected startup and shutdown periods and for part-load operating conditions. A check is made to eliminate or minimize any flow-induced vibrations. Flow velocities are checked to eliminate or minimize erosion, corrosion

---

[†]The design problem in the literature is variously referred to as rating or sizing or both problems. To avoid confusion of the terminology "design problem," we will distinctly refer to the heat exchanger thermal design problems as rating and sizing problems.

and fouling. At this stage, considerations are also given to other operating problems, if existing, such as dynamic instability, freezing, etc. In the mechanical design, considerations are also given to maintenance requirements (such as cleaning, repair and serviceability), shipping limitations for overall size, etc. Thus a mechanical design of the exchanger is equally important and probably more difficult than the thermal design. Many of the mechanical design criteria are considered simultaneously with the thermal design.

Several optional solutions may be available when the thermal and mechanical designs are completed. The designer then considers the evaluation procedure (evaluation criteria and trade-off factors) and cost estimating to arrive at an optimum solution.

The evaluation criteria are in a large measure qualitative. Some of the factors that may influence the design to be chosen are: the manufacturing limitations of die, tools, machines and furnace, manufacturability from individual parts, ease of handling by altering the shape or by adding tabs and slots, mounting of the exchanger in the system, shipping limitations, shop work load, delivery dates, company policy, an estimate of the strength of the competition, etc.

The trade-off factors may be developed to quantitatively weigh the relative costs of pressure drop, heat transfer performance, weight, envelope size, leakage, initial cost versus life of the exchanger for fouling, corrosion and fatigue failures, cost of tubes versus fin material or shell, and cost of one-of-a-kind versus a large volume of the exchanger.

An overall optimum design in most cases is the one which meets the performance requirements at a minimum cost. The overall cost associated with a heat exchanger may be categorized as the capital and operating costs. The capital cost includes the costs associated with design, materials, manufacturing (machinery, labor and overhead), testing, shipping, installation and depreciation. The operating cost consists of the costs associated with fluid pumping powers, maintenance, repair, and cleaning. Some of the cost estimates are difficult to obtain and are of qualitative nature.

The final output is an <u>optimum design</u>, or possibly, several such designs to submit to the customer. In many cases, the final output is used to formulate a new problem input statement for a parametric study. This is because the heat exchanger may be only one component of a thermodynamic system and trade-offs between fluid pumping power and heat transfer power may be necessary. The parametric study leads to an optimum overall system rather than just an optimum heat exchanger based on a somewhat arbitrary initial specification of requirements. With this new problem, a redesign of an optimum heat exchanger is carried out. If the heat exchanger incorporates new design features or if it is a critical part of the system, a model heat exchanger is built and tested in the laboratory to confirm its heat transfer and pressure drop performance, fatigue characteristics due to vibrations, pressure and temperature cycling, corrosion characteristics, and burst pressure limit.

The problem of heat exchanger design is very intricate. Only a part of the total design consists of quantitative analytical evaluation. Because of a large number of qualitative judgments, trade-offs and compromises, the heat exchanger design is more of an art at this stage. In general, no two engineers will come up with the same heat exchanger design for a given application. Most probably a "better" design will be arrived at by an experienced engineer since he has a "feel" for the qualitative considerations.

## REFERENCES

1. W. M. Kays and A. L. London, Compact Heat Exchangers, Second Edition, McGraw-Hill Book Co., New York, p. 11 (1964).

2. R. K. Shah, Compact heat exchanger surface selection methods, Heat Transfer 1978, Vol. 4, 193-199, Hemisphere Publishing Corp., New York (1978).

3. A. C. Mueller, Heat exchangers, in Handbook of Heat Transfer, Edited by W. M. Rohsenow and J. P. Hartnett, McGraw-Hill Book Co., New York, Chapter 18 (1973).

4. W. M. Rohsenow, Heat exchangers-basic methods, in Heat Exchangers-Thermohydraulic Fundamentals and Design, Edited by S. Kakac, A. E. Bergles, and F. Mayinger, Hemisphere Publishing Corp., New York (1981).

5. R. K. Shah, Compact heat exchanger design procedures, in Heat Exchangers-Thermohydraulic Fundamentals and Design, Edited by S. Kakac, A. E. Bergles, and F. Mayinger, Hemisphere Publishing Corp., New York (1981).

6. R. K. Shah, Compact heat exchangers in Heat Exchangers-Thermohydraulic Fundamentals and Design, Edited by S. Kakac, A. E. Bergles, and F. Mayinger, Hemisphere Publishing Corp., New York (1981).

7. R. K. Shah, Heat exchanger optimization, Heat Transfer 1978, Vol. 4, 185-191, Hemisphere Publishing Corp., New York (1978).

# Finite Element Analysis of Heat Exchangers

**M. D. MIKHAILOV**
Applied Mathematics Center
P.O. Box 384
Sofia 1000, Bulgaria

**M. N. ÖZIŞIK**
North Carolina State University
P.O. Box 5246
Raleigh, North Carolina 27650 USA

ABSTRACT

A finite element method of analysis is adapted for heat exchanger calculations, to compute the total effectiveness of a complex assembly of heat exchangers and the heat transfer through an array of extended surfaces. The application is illustrated with examples.

NOMENCLATURE

| | |
|---|---|
| A | heat transfer surface of the exchanger ($m^2$) |
| $\Delta A$ | heat transfer surface of a cell ($m^2$) |
| [A], [B] and {F} | matrices defined by eqs. (3) |
| e | element number |
| E | number of elements |
| h | heat-transfer coefficient (W/$m^2$ °C) |
| k | thermal conductivity (W/m°C) |
| $k_{nn}^{(e)}$, $k_{nm}^{(e)}$, $k_{mn}^{(e)}$ and $k_{mm}^{(e)}$ | thermal influence coefficients |
| $[K]^{(e)}$ | expanded matrix of thermal influence coefficients |
| [K] | global matrix |
| L | second order linear operator |
| M | thermal transfer matrix defined by eq. (10a) |
| n | node number |
| N | number of nodes |
| $Q_n^{(e)}$ | heat flow rate entering the element "e" at the node "n" (W) |
| $\{Q\}^{(e)}$ | extended column vector of nodal heat transfer rates for element "e" |

{Q}     column vector of external nodal heat flow rates
ΔQ      rate of heat exchange in a cell (W)
r       fluid heat capacity rate ratio
$T_n$   temperature at the node "n"
{T}     column vector of nodal temperatures
U       overal heat transfer coefficient
$W_1$   thermal capacity rate for fluid with lower thermal capacity rate (W/°C)
$W_2$   thermal capacity rate for fluid with higher thermal capacity rate (W/°C)
E       effectiveness of the exchanger

## INTRODUCTION

The finite element method (FEM) is one of the newest and most popular numerical techniques applied by engineers and scientists. This method originated some 20 years ago in the aircraft industry as an effective matrix method for structural analysis, but now it is recognized as a powerful and versatile tool for obtaining computer solutions of a wide variety of engineering problems [1, 2].

In this contribution we will consider several heat exchangers' problems for which we can identify network systems consisting of a collection of elements interconnected to each other at nodal points. Direct physical consideration was used to establish matrix relationships between the heat fluxes applied to the individual element and to its nodal temperatures. Then the so-called assembly procedure was used to combine all these "local" matrices and to form the "global" matrix for the complete system. Thus the analysis of some heat exchanger problems is reduced to the solution of a system of algebraic equations.

The application of FEM is illustrated with the following examples: (a) temperature distribution in multipass shell-and-tube exchangers with baffles, (b) analysis of complex assemblies of heat exchangers, and (c) heat transfer through an array of extended surfaces.

## THE FINITE ELEMENT METHOD

Let us consider the heat flow network system composed of many individual elements (e= 1, 2, .. E), connected through the nodes (N= 1, 2, .. N). The FEM renders the solution of this complex problem by subdividing it into a series of simpler interrelated problems. We focus our attention on an individual element shown in Fig. 1. We suppose that the relationships between the temperatures $T_n$, $T_m$ and the heat flow rates $Q_n^{(e)}$ and $Q_m^{(e)}$ entering the element at the nodes "n" and "m" can be written as

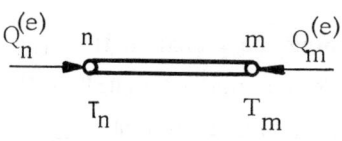

Fig. 1. An individual element.

$$\begin{Bmatrix} Q_n^{(e)} \\ Q_m^{(e)} \end{Bmatrix} = \begin{bmatrix} k_{nn}^{(e)} & k_{nm}^{(e)} \\ k_{mn}^{(e)} & k_{mm}^{(e)} \end{bmatrix} \begin{bmatrix} T_n \\ T_m \end{bmatrix} \quad (1\text{ a})$$

or in the expanded form, as

$$\begin{Bmatrix} 0 \\ \cdot \\ Q_n^{(e)} \\ \cdot \\ 0 \\ \cdot \\ Q_m^{(e)} \\ \cdot \\ 0 \end{Bmatrix} = \begin{bmatrix} 0 & . & 0 & . & 0 & . & 0 & . & 0 \\ . & & . & & . & & . & & . \\ 0 & . & k_{nn}^{(e)} & . & 0 & . & k_{nm}^{(e)} & . & 0 \\ . & & . & & . & & . & & . \\ 0 & . & 0 & . & 0 & . & 0 & . & 0 \\ . & & . & & . & & . & & . \\ 0 & . & k_{mn}^{(e)} & . & 0 & . & k_{mm}^{(e)} & . & 0 \\ . & & . & & . & & . & & . \\ 0 & . & 0 & . & 0 & . & 0 & . & 0 \end{bmatrix} \begin{bmatrix} T_1 \\ \cdot \\ T_n \\ \cdot \\ \cdot \\ \cdot \\ T_m \\ \cdot \\ T_N \end{bmatrix} \quad (1\text{ b})$$

or more concisely as

$$\{Q\}^{(e)} = [K]^{(e)} \{T\} \quad (1\text{ c})$$

We note that: (1) at the nodes where the elements are connected the temperature is the same for all elements forming the node, and (2) the algebraic sum of the heat flow rates at each node must be equal to the external heat flow at that node. Assuming also that we have found numerically, by some means, the coefficients $k_{nn}^{(e)}$, $k_{nm}^{(e)}$, $k_{mn}^{(e)}$ and $k_{mm}^{(e)}$, equations (1 c) are summed for all e= 1, 2,.. E to obtain

$$\{Q\} = [K]\{T\} \qquad (2\,a)$$

where $\{Q\}$ is a column vector of external nodal heat flow rates and $[K]$ is the global matrix defined as

$$[K] = \sum_{e=1}^{E} [K]^{(e)} \qquad (2\,b)$$

The computer assembly procedure forming the global matrix $[K]$ consists of the following steps: (1) Set up N x N null master matrix (all zero entries), (2) Starting with one element insert his coefficients $k_{nn}^{(e)}$, $k_{nm}^{(e)}$, $k_{mn}^{(e)}$ and $k_{mm}^{(e)}$ into the master matrix in the locations designated by their indices. Each time a term is placed in a location where another term has already been placed, it is added to whatever value is there, (3)Return to step 2 and repeat this procedure for one element after another until all elements have been treated. The result will be the global matrix $[K]$. This assembly procedure is an essential part of FEM.

At each node a boundary condition can be written as

$$A_n T_n + B_n Q_n = F_n \qquad (3\,a)$$

where $Q_n$ is the external nodal heat flow rate ; $A_n$, $B_n$ and $F_n$ are known constants. The cases $A_n=1$, $B_n=0$ and $A_n=0$, $B_n=1$ correspond to a prescribed temperature and heat flow rate, respectively. The nodal boundary conditions (3 a) can be writen in the matrix form as

$$[A]\{T\} + [B]\{Q\} = \{F\} \qquad (3\,b)$$

where

$$[A] = \begin{bmatrix} A_1 & & & \\ & A_2 & & \\ & & \ddots & \\ & & & A_N \end{bmatrix},\; [B] = \begin{bmatrix} B_1 & & & \\ & B_2 & & \\ & & \ddots & \\ & & & B_N \end{bmatrix} \text{ and } \{F\} = \begin{Bmatrix} F_1 \\ F_2 \\ \vdots \\ F_N \end{Bmatrix} \qquad (3\,c,d,e)$$

Elements of $[A]$ and $[B]$ not shown are zero.

At last eq. (2 a) is introduced into the node condition (3 b) to give the solution

$$T = ([A] + [B][K])^{-1}\{F\} \qquad (4)$$

FINITE ELEMENT ANALYSIS OF HEAT EXCHANGERS    465

The generality of the finite element method described offers a definite advantage: once a computer program has been developed for the solution of the heat flow network system it may be used for the solution of different classes of problems. Such program, developed in The Appl. Math. Cntr. (Sofia) was used to compute temperature distribution in multipass shell-and-tube exchangers with baffles, the total effectiveness of an complex assembly of heat exchangers and the heat transfer through an array of extended surfaces.

FINITE-ELEMENTS' MODELS FOR HEAT EXCHANGER CALCULATIONS

In [3] Saryal proposed lumped resistance, resistance-capacitance and hybrid models for heat exchanger calculations. Since the FEM described in the previous section solves easily lumped resistance model our first attempt was to apply the Saryal's model to recalculate data for the temperature distribution in the multipass shell-and-tube exchangers with baffles given in [4]. But the results obtained differ from these reported by Gaddis and Schlünder in [4].

In this section a new network model for heat exchanger calculation is introduced. The predictions of our model fully coincide with the results mentioned and given in [4].

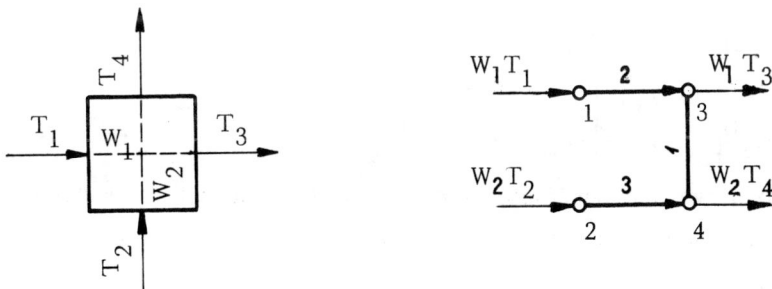

Fig. 2.    (a) a cell            (b) finite element model of the cell

The exchanger shown shematically in [4] is treated as a cascade of cells with mixing taking place in each fluid. The following assumptions are made: (1) Constant physical properties and (2) Constant overall heat transfer coefficient. Then from a steady-state heat balance in the cell shown at Fig. 2a, the following equations are formulated [4]:

$$\Delta Q = W_1 (T_3 - T_1) \tag{5 a}$$

$$\Delta Q = -W_2 (T_4 - T_2) \tag{5 b}$$

$$\Delta Q = U \Delta A (T_4 - T_3) \tag{5 c}$$

Substituting eq. (5 c) into eqs. (5 a, b) we obtain

$$W_1 (T_3 - T_1) = U \Delta A (T_4 - T_3) \tag{6 a}$$

$$W_2 (T_2 - T_4) = U \Delta A (T_4 - T_3) \tag{6 b}$$

Here we introduce the finite element model of the cell shown in Fig. 2b with the following element's matrices

$$\begin{bmatrix} Q_3^{(1)} \\ Q_4^{(1)} \end{bmatrix} = U \Delta A \begin{bmatrix} 1 & -1 \\ -1 & 1 \end{bmatrix} \begin{bmatrix} T_3 \\ T_4 \end{bmatrix} \tag{7 a}$$

$$\begin{bmatrix} Q_1^{(2)} \\ Q_3^{(2)} \end{bmatrix} = W_1 \begin{bmatrix} 1 & 0 \\ -1 & 0 \end{bmatrix} \begin{bmatrix} T_1 \\ T_3 \end{bmatrix} \tag{7 b}$$

and

$$\begin{bmatrix} Q_2^{(3)} \\ Q_4^{(3)} \end{bmatrix} = W_2 \begin{bmatrix} 1 & 0 \\ -1 & 0 \end{bmatrix} \begin{bmatrix} T_2 \\ T_4 \end{bmatrix} \tag{7 c}$$

To prove that our finite element model of the cell is equivalent to eqs. (6) we consider the heat balance for nodes 1, 2, 3 and 4:

$$Q_1^{(2)} = W_1 T_1 \tag{8 a}$$

$$Q_3^{(2)} + Q_3^{(1)} = - W_1 T_3 \tag{8 b}$$

$$Q_2^{(3)} = W_2 T_2 \tag{8 c}$$

$$Q_4^{(1)} + Q_4^{(3)} = -W_2 T_4 \tag{8 d}$$

Determining $Q_n^{(e)}$ from eqs. (7) and substituting these results in

# FINITE ELEMENT ANALYSIS OF HEAT EXCHANGERS

eqs. (8) we obtain the eqs. (6) i.e. our finite element model of the cell is correct.

The finite element model of the shell-and-tube exchanger shown shematically in Fig. 3a is given in Fig. 3b.

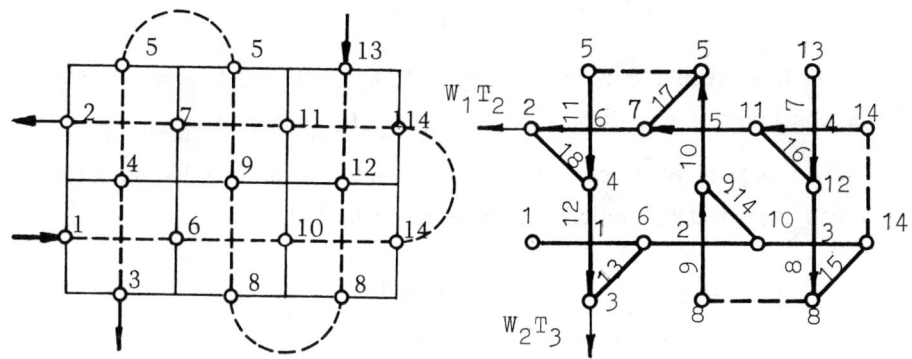

Fig. 3. (a) Subdividing the heat exchanger into cells  (b) Finite element model of the heat exchanger

The input data corresponding to this problem are listed in Table 1 and Table 2

Table 1. Topology of the Grid

| ELEMENT'S NUMBER | FIRST NODE | SECOND NODE | ELEMENT'S TYPE |
|---|---|---|---|
| 1 | 1 | 6 | 1 |
| 2 | 6 | 10 | 1 |
| 3 | 10 | 14 | 1 |
| 4 | 14 | 11 | 1 |
| 5 | 11 | 7 | 1 |
| 6 | 7 | 2 | 1 |
| 7 | 13 | 12 | 1 |
| 8 | 12 | 8 | 1 |
| 9 | 8 | 9 | 1 |
| 10 | 9 | 5 | 1 |
| 11 | 5 | 4 | 1 |
| 12 | 4 | 3 | 1 |
| 13 | 3 | 6 | 2 |
| 14 | 9 | 10 | 2 |
| 15 | 8 | 14 | 2 |
| 16 | 11 | 12 | 2 |
| 17 | 5 | 7 | 2 |
| 18 | 2 | 4 | 2 |

Table 2

| NODE'S NUMBER | A | B | F |
|---|---|---|---|
| 1 | 1 | 0 | 0 |
| 2 | 1 | 1 | 0 |
| 3 | 1 | 1 | 0 |
| 4 | 0 | 1 | 0 |
| 5 | 0 | 1 | 0 |
| 6 | 0 | 1 | 0 |
| 7 | 0 | 1 | 0 |
| 8 | 0 | 1 | 0 |
| 9 | 0 | 1 | 0 |
| 10 | 0 | 1 | 0 |
| 11 | 0 | 1 | 0 |
| 12 | 0 | 1 | 0 |
| 13 | 1 | 0 | 1 |
| 14 | 0 | 1 | 0 |

For the example considered ( $W_1 = W_2 = 1$ and $U \Delta A = 1/6$ ) there are two types of elements, the matrices of which have the standard form

$$c \begin{bmatrix} k_{nn} & k_{nm} \\ k_{mn} & k_{mm} \end{bmatrix}, \text{ namelly :}$$

(1) first type ( $c=1$, $k_{nn}=1$, $k_{nm}=0$, $k_{mn}=-1$ and $k_{mm}=0$) and
(2) second type( $c=1/6$, $k_{nn}=1$, $k_{nm}=-1$, $k_{mn}=-1$ and $k_{mm}=1$ .

The results obtained are given in Table 3. Our results coincide with these reported by Gaddis and Schlünder in [4].

Table 3. Solution

| Nodes | Temperature | Flux |
|---|---|---|
| 1 | 0.000 | 0.0000 |
| 2 | 0.4368 | -0.4368 |
| 3 | 0.5632 | -0.5632 |
| 4 | 0.6437 | -0.0000 |
| 5 | 0.6782 | 0.0000 |
| 6 | 0.0805 | 0.0000 |
| 7 | 0.4023 | 0.0000 |
| 8 | 0.8161 | -0.0000 |
| 9 | 0.7241 | 0.0000 |
| 10 | 0.1724 | 0.0000 |
| 11 | 0.3563 | 0.0000 |
| 12 | 0.9080 | 0.0000 |
| 13 | 1.0000 | 1.0000 |
| 14 | 0.2644 | 0.0000 |

## ANALYSIS OF COMPLEX ASSEMBLIES OF HEAT EXCHANGERS

In many practical cases combination of heat exchangers are used in series, parallel or series-parallel. The exchangers are often of non-identical type (or size) and, therefore, a multiplicity of cases are possible.

A general matrix method for obtaining the total effectiveness and intermediate temperatures of an assembly of heat exchangers in terms of individual effectiveness and fluid capacity rate ratio was reported by Domingos [5]. Closed form expression for total effectiveness was obtained for overal parallel or counter flow association. But for mixed flow association the final expressions can be very complex, particularly if the values of effectiveness are different for each exchanger.

In the present section the finite element approach is adopted for calculating the total effectiveness of complex assemblies of heat exchangers.

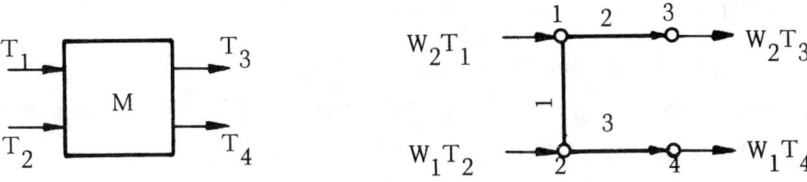

Fig. 4. (a) Parallel flow heat exchanger as a "black box"  (b) Finite element model of heat exchanger

Domingos [5] regards a parallel flow heat exchanger as a "black box" (shown in Fig. 4a), its behaviour is equivalent to a static thermal transfer matrix

$$M = \begin{bmatrix} 1-Er & Er \\ E & 1-E \end{bmatrix} \quad (10\text{ a})$$

which transforms two input temperatures (one of each stream) to two output temperature

$$\begin{Bmatrix} T_3 \\ T_4 \end{Bmatrix} = M \begin{Bmatrix} T_1 \\ T_2 \end{Bmatrix} \quad (10\text{ b})$$

where $r = W_1/W_2$ is the fluid heat capacity rate ratio and $E = (T_4-T_2)/(T_1-T_2)$ is the heat exchanger effectiveness.

Instead of M we introduce the finite element model shown in Fig. 4b. with the following element matrix

$$\begin{Bmatrix} Q_1^{(1)} \\ Q_2^{(1)} \end{Bmatrix} = W_1 E \begin{bmatrix} 1 & -1 \\ -1 & 1 \end{bmatrix} \begin{Bmatrix} T_1 \\ T_2 \end{Bmatrix} \quad (11\text{ a})$$

$$\begin{Bmatrix} Q_1^{(2)} \\ Q_3^{(2)} \end{Bmatrix} = W_2 \begin{bmatrix} 0 & 1 \\ 0 & -1 \end{bmatrix} \begin{Bmatrix} T_1 \\ T_3 \end{Bmatrix} \quad (11\text{ b})$$

and

$$\begin{Bmatrix} Q_2^{(3)} \\ Q_4^{(3)} \end{Bmatrix} = W_1 \begin{bmatrix} 0 & 1 \\ 0 & -1 \end{bmatrix} \begin{Bmatrix} T_2 \\ T_4 \end{Bmatrix} \qquad (11\ c)$$

To prove that our finite element model of a heat exchanger is equivalent to the matrix operator M introduced by Domingos, consider the heat ballance for nodes 1, 2, 3 and 4

$$Q_1^{(1)} + Q_1^{(2)} = W_2\ T_1 \qquad (12\ a)$$

$$Q_2^{(1)} + Q_2^{(3)} = W_1\ T_2 \qquad (12\ b)$$

$$Q_3^{(2)} = -W_2\ T_3 \qquad (12\ c)$$

$$Q_4^{(3)} = -W_1\ T_4 \qquad (12\ d)$$

Determining $Q_n^{(e)}$ from eqs. (11) and substituting these results in eqs. (12) we obtain eqs. (10) i.e. our finite element model of the heat exchanger is correct.

Consider now as an illustrative example the overal parallel flow association of five exchangers (shown in Fig. 5a), each characterized by its effectiveness $E_i < 1/(1+r)$ and the same heat capacity rate ratio r. This assembly of five exchanger is equivalent to only one with an effectiveness

$$E_t = (T_{12} - T_2)/(T_1 - T_2) \qquad (13\ a)$$

given by [5]

$$E_t = (1 - \prod_1^5 (1 - E_i(r+1)))/(r+1) \qquad (13\ b)$$

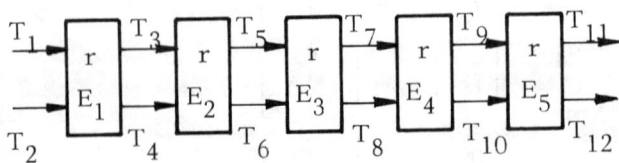

Fig. 5 (a) Overall parallel flow association

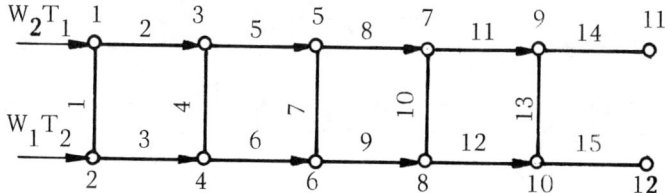

Fig. 5 (b) Finite element model of parallel flow association

The finite element model of this overall parallel flow association is shown in Fig. 5b. The input data corresponding to this problem are listed in Table 4 and Table 5. For the example considered ($W_1=1$, $W_2=r$) there are the following types of matrices:

type 1    : $c=1$, $k_{nn}=0$, $k_{nm}=1$, $k_{mn}=0$ and $k_{mm}=-1$

type 2    : $c=r$, $k_{nn}=0$, $k_{nm}=1$, $k_{mn}=0$ and $k_{mm}=-1$

type 3 to 7 : $c=E_i r$, $k_{nn}=1$, $k_{nm}=-1$, $k_{mn}=-1$ and $k_{mm}=1$ (i=1 to 5)

Our numerous numerical calculations indicate that: (1) the effectiveness does not depend on the temperature $T_{11}$ and $T_{12}$, and (2) the effectiveness calculated by the finite element model and the closed form solution, eq. (13 b), coincide. A part of our results are given in Table 10.

Table 4

| ELEMENT'S NUMBER | FIRST NODE | SECOND NODE | ELEMENT'S TYPE |
|---|---|---|---|
| 1 | 1 | 2 | 3 |
| 2 | 1 | 3 | 1 |
| 3 | 2 | 4 | 2 |
| 4 | 3 | 4 | 4 |
| 5 | 3 | 5 | 1 |
| 6 | 4 | 6 | 2 |
| 7 | 5 | 6 | 5 |
| 8 | 5 | 7 | 1 |
| 9 | 6 | 8 | 2 |
| 10 | 7 | 8 | 6 |
| 11 | 7 | 9 | 1 |
| 12 | 8 | 10 | 2 |
| 13 | 9 | 10 | 7 |
| 14 | 9 | 11 | 1 |
| 15 | 10 | 12 | 2 |

Table 5

| NODE'S NUMBER | A | B | F |
|---|---|---|---|
| 1 | 1.00 | -1.00 | 0.00 |
| 2 | 1.00 | -1.00 | 0.00 |
| 3 | 0.00 | 1.00 | 0.00 |
| 4 | 0.00 | 1.00 | 0.00 |
| 5 | 0.00 | 1.00 | 0.00 |
| 6 | 0.00 | 1.00 | 0.00 |
| 7 | 0.00 | 1.00 | 0.00 |
| 8 | 0.00 | 1.00 | 0.00 |
| 9 | 0.00 | 1.00 | 0.00 |
| 10 | 0.00 | 1.00 | 0.00 |
| 11 | 1.00 | 0.00 | 1.00 |
| 12 | 1.00 | 0.00 | 0.00 |

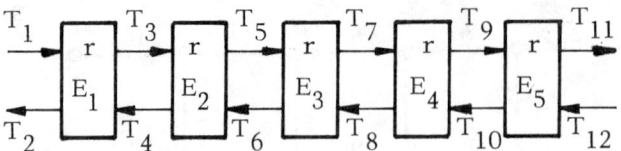

Fig. 6 (a) Overal counter-flow association.

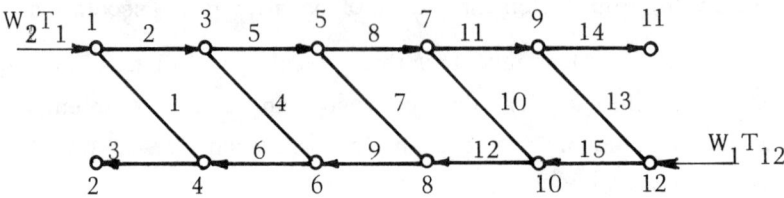

(b) Finite element model of counter-flow association.

Consider now as an illustrative example the overal counter-flow association of five exchangers (Shown in Fig. 6a ), each characterised by its effectiveness $E_i < 1$ and the same heat capacity rate ratio r. This assembly has an equivalent effectiveness

$$E_t = (T_2 - T_{12})/(T_1 - T_{12}) \qquad (14\text{ a})$$

given by [5]

$$E_t = \frac{1 - \prod_1^5 ((1-E_i r)/(1-E_i))}{r - \prod_1^5 ((1-E_i r)/(1-E_i))} \qquad (14\text{ b})$$

The finite element model of this overal counter-flow association is shown in Fig. 6b. The input data corresponding to this problem are listed in Table 6 and Table 7. The type of matrices is the same as these used in Table 4.

Our numerical calculations indicate again that: (1) the effectiveness does not depend on the temperature $T_2$ and $T_{11}$, and (2) the effectiveness calculated by the finite element model and the closed form solution, eq. (14 b), coincide.

## Table 6. Topology of the Grid

| ELEMENT'S NUMBER | FIRST NODE | SECOND NODE | ELEMENT'S TYPE |
|---|---|---|---|
| 1 | 1 | 4 | 3 |
| 2 | 1 | 3 | 1 |
| 3 | 4 | 2 | 2 |
| 4 | 3 | 6 | 4 |
| 5 | 3 | 5 | 1 |
| 6 | 6 | 4 | 2 |
| 7 | 5 | 8 | 5 |
| 8 | 5 | 7 | 1 |
| 9 | 8 | 6 | 2 |
| 10 | 7 | 10 | 6 |
| 11 | 7 | 9 | 1 |
| 12 | 10 | 8 | 2 |
| 13 | 9 | 12 | 7 |
| 14 | 9 | 11 | 1 |
| 15 | 12 | 10 | 2 |

## Table 7

| NODE'S NUMBER | A | B | F |
|---|---|---|---|
| 1 | 1.00 | -1.00 | 0.00 |
| 2 | 1.00 | 0.00 | 0.00 |
| 3 | 0.00 | 1.00 | 0.00 |
| 4 | 0.00 | 1.00 | 0.00 |
| 5 | 0.00 | 1.00 | 0.00 |
| 6 | 0.00 | 1.00 | 0.00 |
| 7 | 0.00 | 1.00 | 0.00 |
| 8 | 0.00 | 1.00 | 0.00 |
| 9 | 0.00 | 1.00 | 0.00 |
| 10 | 0.00 | 1.00 | 0.00 |
| 11 | 1.00 | 0.00 | 1.00 |
| 12 | 1.60 | -1.00 | 0.00 |

## Table 8

| ELEMENT'S NUMBER | FIRST NODE | SECOND NODE | ELEMENT'S TYPE |
|---|---|---|---|
| 1 | 1 | 12 | 3 |
| 2 | 1 | 3 | 1 |
| 3 | 12 | 2 | 2 |
| 4 | 3 | 6 | 4 |
| 5 | 3 | 5 | 1 |
| 6 | 6 | 4 | 2 |
| 7 | 5 | 8 | 5 |
| 8 | 5 | 7 | 1 |
| 9 | 8 | 6 | 2 |
| 10 | 7 | 4 | 6 |
| 11 | 7 | 9 | 1 |
| 12 | 4 | 10 | 2 |
| 13 | 9 | 10 | 7 |
| 14 | 9 | 11 | 1 |
| 15 | 10 | 12 | 2 |

## Table 9

| NODE'S NUMBER | A | B | F |
|---|---|---|---|
| 1 | 1.00 | -1.00 | 0.00 |
| 2 | 1.00 | 0.00 | 0.00 |
| 3 | 0.00 | 1.00 | 0.00 |
| 4 | 0.00 | 1.00 | 0.00 |
| 5 | 0.00 | 1.00 | 0.00 |
| 6 | 0.00 | 1.00 | 0.00 |
| 7 | 0.00 | 1.00 | 0.00 |
| 8 | 0.60 | -1.00 | 0.00 |
| 9 | 0.00 | 1.00 | 0.00 |
| 10 | 0.00 | 1.00 | 0.00 |
| 11 | 1.00 | 0.00 | 1.00 |
| 12 | 0.00 | 1.00 | 0.00 |

## Table 10

| INPUT DATA | | | | | | Parallel flow | | Counter flow | | Mixed flow |
|---|---|---|---|---|---|---|---|---|---|---|
| | | | | | | eq(13a) | eq(13b) | eq(14a) | eq(14b) | eq(15) |
| r | $E_1$ | $E_2$ | $E_3$ | $E_4$ | $E_5$ | $E_t$ | $E_t$ | $E_t$ | $E_t$ | $E_t$ |
| 0.6 | 0.5 | 0.2 | 0.3 | 0.4 | 0.6 | 0.624 | 0.624 | 0.869 | 0.869 | 0.768 |
| 0.8 | 0.1 | 0.2 | 0.3 | 0.4 | 0.5 | 0.552 | 0.552 | 0.745 | 0.745 | 0.554 |

A part of our results are given in Table 10.

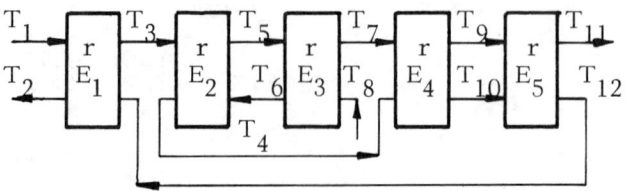

Fig. 7 (a) Mixed flow association.

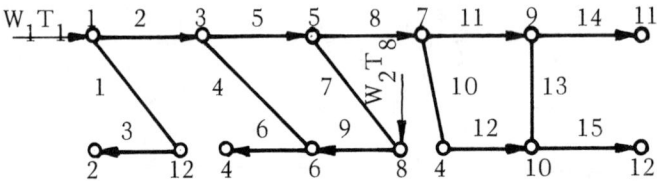

(b) Finite element model of mixed flow association.

Consider now as an illustrative example the mixed flow association (shown in Fig. 7a) where some exchangers are in overal counter flow and others in parallel flow. Domingos remarked that all expressions for obtaining the equivalent effectiveness

$$E_t = (T_2 - T_8)/(T_1 - T_8) \tag{15}$$

have been presented in [5], but the final formula can be very complex.

The finite element model of this mixed flow association is shown in Fig. 7b. The input data corresponding to this problem are listed in Table 8 and Table 9. The type of matrices is the same as of these used in Table 4. Our numerical calculation indicate again that the effectiveness does not depend on the temperature $T_2$ and $T_{11}$. A part of our results are given in Table 10.

## HEAT TRANSFER THROUGH AN ARRAY OF EXTENDED SURFACE

The heat transfer analysis of a single fin has been extensively studied and the solutions for various fin geometries are well documented [6]. Such results are applicable to study heat flow through a finned matrix whose repeating section is a single fin.

In certain types of compact heat exchangers, the repeating section of a finned matrix is not a single fin, but is an array of fins. The heat transfer analysis of an array of fins is a complicated matter; a limited number of works on this sybject are available only. The method of analysis in [6] involves a simultaneous solution of a set of ordinary differential equations associated with each of the fins in the array. If the differential equations can be integrated, the problem becomes one of solving a system of linear algebraic equations associated with the determination of the integration constants. The application of the method for the solution of practical problems is quite sophisticated. To alleviate this difficulty the procedure described in reference [7] treats each fin as a lumped parameter and cascades all the fins in the array via matrix operations.

In this section, each fin in the array is considered as a finite element whose characteristics are determined by the solution of a one-dimensional fin problem. Explicit expressions are developed for the determination of the thermal influence coefficients.

The one-dimensional steady state heat equation for a individual element can be written in the general form

$$L_e T_e(x) = 0 \qquad (16)$$

Eq. (15) holds from the base at $x=x_n$ to the tip $x=x_m$.

Let $u_e(x)$ and $v_e(x)$ be two fundamental solutions of the differential equation (16) for a given problem. The solution $T_e(x)$ is constructed by taking a linear combination of these two fundamental solutions as

$$T_e(x) = C_e u_e(x) + D_e v_e(x) \qquad (17\ a)$$

The evaluation of this equation at the nodes "n" and "m" (i. e. $x=x_n$ and $x_m$) yields

$$\begin{Bmatrix} T_n \\ T_m \end{Bmatrix} = \begin{bmatrix} u_e(x_n) & v_e(x_n) \\ u_e(x_m) & v_e(x_m) \end{bmatrix} \begin{Bmatrix} C_e \\ D_e \end{Bmatrix} \qquad (17\ b)$$

Consider the one-dimensional heat flow through an individual element shown in Fig. 8. Let $Q_n^{(e)}$ and $Q_m^{(e)}$ be the heat flow rates entering the element at the nodes "n" and "m" (i.e., $x=x_n$ and $x=x_m$), through the surfaces $A(x_n)$ and $A(x_m)$, respectively. Let the temperature distribution $T_e(x)$ in the element be governed by eq. (17 a). The heat flow rates $Q_n^{(e)}$ and $Q_m^{(e)}$ are determined according to Fourier's law and the result is written in matrix notation as

$$\begin{Bmatrix} Q_n^{(e)} \\ Q_m^{(e)} \end{Bmatrix} = \begin{bmatrix} -k_e(x_n)A(x_n)u_e'(x_n) & -k_e(x_n)A(x_n)v_e'(x_n) \\ +k_e(x_m)A(x_m)u_e'(x_m) & +k_e(x_m)A(x_m)v_e'(x_m) \end{bmatrix} \begin{Bmatrix} C_e \\ D_e \end{Bmatrix} \quad (17\text{ c})$$

where primes denote differentiation with respect to x.

The column vector of constants $C_e$ and $D_e$ is determined from eq. (17 b) and substituted into eq. (17 c). We obtain

$$\begin{Bmatrix} Q_n^{(e)} \\ Q_m^{(e)} \end{Bmatrix} = \frac{1}{u_e(x_n)v_e(x_m) - u_e(x_m)v_e(x_n)} \begin{bmatrix} k_{nn}^{(e)} & k_{nm}^{(e)} \\ k_{mn}^{(e)} & k_{mm}^{(e)} \end{bmatrix} \begin{Bmatrix} T_n \\ T_m \end{Bmatrix} \quad (18\text{ a})$$

where the thermal coefficients are defined as

$$k_{nn}^{(e)} = -k_e(x_n)A(x_n)\left[u_e'(x_n)v_e(x_m) - u_e(x_m)v_e'(x_n)\right] \quad (18\text{ b})$$

$$k_{nm}^{(e)} = -k_e(x_n)A(x_n)\left[u_e(x_n)v_e'(x_n) - u_e'(x_n)v_e(x_n)\right] \quad (18\text{ c})$$

$$k_{mn}^{(e)} = +k_e(x_m)A(x_m)\left[u_e'(x_m)v_e(x_m) - u_e(x_m)v_e'(x_m)\right] \quad (18\text{ d})$$

$$k_{mm}^{(e)} = +k_e(x_m)A(x_m)\left[u_e(x_n)v_e'(x_m) - u_e'(x_m)v_e(x_n)\right] \quad (18\text{ e})$$

We recognize that eqs. (18) have the standard form and therefore they completly define the heat conduction properties of an individual element.

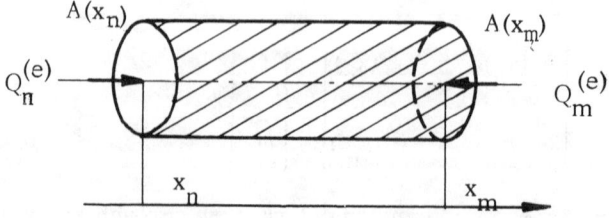

Fig. 8. A schematic representation of an element of the fin-array

# FINITE ELEMENT ANALYSIS OF HEAT EXCHANGERS

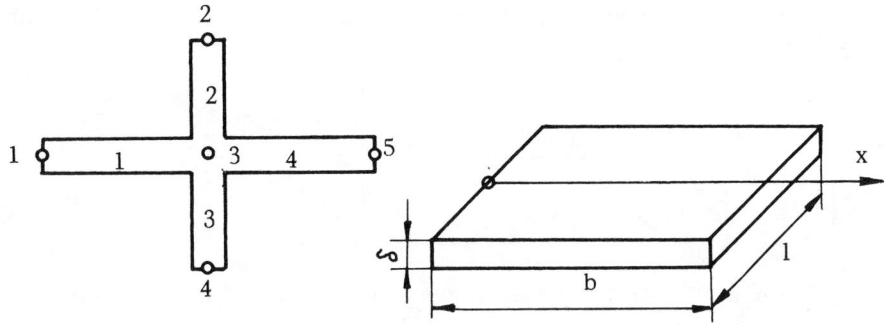

Fig. 9. (a) Nomenclature for the analysis of fin-array  (b) Nomenclature for fin geometry considered in the example

Consider now as an illustrative example the fin matrix configuration of Fig. 9a to represent the repeating section of a plate fin compact heat exchanger [7]. Let the element "2" and "3" represent the splitter plates; then there is no heat transfer at the nodes and we have $Q_2 = Q_4 = 0$. The nodes "1" and "5" are subjected to prescribed heat flow rates, hence $Q_1$ and $Q_5$ are known. The fin geometry is illustrated in Fig. 9b. The pertinent dimensions of the elements and the magnitudes of various other quantities are chosen as [7]:

Fins "1" and "4": $b_1 = b_4 = 6.34$ mm, $\delta_1 = \delta_4 = 0.152$ mm.
Fins "2" and "3": $b_2 = b_3 = 1.16$ mm, $\delta_2 = \delta_3 = 0.254$ mm.
Depth of the array : $l = 0.3048$ m.
$k = 173$ W/m°C, $h = 56.77$ W/m$^2$ °C.
Heat flow rates : $Q_1 = 2.93$ W, $Q_5 = 2.344$ W

The fin equation (15) for this particular case takes the form

$$\frac{d^2 \theta_e(x)}{dx^2} - m_e^2 \theta_e(x) = 0 \qquad (19)$$

where $m_e^2 = 2h/k\delta_e$ and $\theta_e(x)$ is the temperature in excess of the surrounding coolant temperature. The elementary solution $u_e(x)$ and $v_e(x)$ of eq. (16) are

$$u_e(x) = \cosh(m_e x) \qquad\qquad v_e(x) = \sinh(m_e x) \qquad (20\ a, b)$$

These solutions are now introduced into eqs. (18); the resulting fin matrix becomes

$$\begin{bmatrix} k_{nn}^{(e)} & k_{nm}^{(e)} \\ k_{mn}^{(e)} & k_{mm}^{(e)} \end{bmatrix} = \frac{k \cdot \delta_e \cdot l \cdot m_e}{\sinh(m_e b_e)} \begin{vmatrix} \cosh(m_e b_e) & -1 \\ -1 & \cosh(m_e b_e) \end{vmatrix}$$

(21)

The solution for the node temperature in excess of the surrounding coolant temperature was obtained by utilizing the numerical values given above. The results are given in Table 11.

SOLUTION

| NODES | TEMPERATURE | FLUX |
|---|---|---|
| 1 | 11.2010 | 2.9300 |
| 2 | 9.7652 | 0.0000 |
| 3 | 9.7825 | -0.0000 |
| 4 | 9.7652 | 0.0000 |
| 5 | 10.7570 | 2.3440 |

Our results correspond to these reported in reference [7] for the solution of the same problem by the "cascading process".

CONCLUSION

Finite element method permits easily to calculate: (a) temperature distribution in multipass shell-and-tube exchangers with baffles, (b) total effectiveness of any complex assembly of heat exchangers, and (c) heat transfer through an array of extended surfaces.

ACKNOWLEDGEMENT

Authors gratefully acknowledge the hospitality extended to them by the Institut für Verfahrens-und Kältetechnik, ETH- Zürich .

REFERENCES

1. Brebbia C. A., and Ferrante A. J. 1978. Computational methods for the solution of engineering problems, Pentech Press Limited.
2. Huebner K. H. 1975. The finite element method for engineers, John Wiley.
3. Saryal N. 1974. Electro-analog models for heat exchangers and a simplified method for heat exchanger calculations, Int. J. Heat Mass Transfer 17: 971-980.

4. Gaddis E. S., and Schlünder E. U. 1979. Temperature distribution and heat exchange in multipass shell-and-tube exchangers with baffles. Heat transfer engineering 1: 43-52.
5. Domingos J. D. 1969. Analysis of complex assemblies of heat exchangers. Int. J. Heat Mass Transfer 12 : 537-548.
6. Kern D. Q., and Kraus A. D. 1972. Extended surface heat transfer, McGraw-Hill, New York.
7. Kraus A. D., Snider A. D., and Doty L. F. 1978. An efficient algorithm for evaluating arrays of extended surface, J. Heat Transfer 100: 288-293.

# Exact Explicit Equations for Some Two- and Three-Pass Cross-Flow Heat Exchangers Effectiveness

**BRANISLAV S. BAČLIĆ and DUŠAN D. GVOZDENAC**
Institute of Fluid, Thermal and Chemical Engineering
Faculty of Technical Sciences, Mechanical Eng. Dept.
University of Novi Sad, 21000 Novi Sad, Yugoslavia

ABSTRACT

All of the previously unsolved cases of two- and three-pass cross-flow heat exchangers with passes coupled in identic order were solved analytically. In this paper, the final results of the analytical solutions for all possible combinations of these types of cross-flow heat exchangers, with one fluid unmixed throughout and the other mixed between passes and unmixed in each pass, are presented. These final results are given in the explicit forms of effectiveness - number of transfers units - capacity rate ratio relationships ($\varepsilon = \varepsilon$ (NTU,$\omega$)) as well as the mean mixed fluid temperatures between passes as the functions of the same arguments.

A useful term "ineffectiveness" for a single-pass crossflow heat exchanger is introduced for concise notation of the results.

The new relations are particularly helpful for computer programs.

1.  INTRODUCTION

The need for investigating the cross-flow heat transfer equipment with fluids unmixed on both sides arises as a result of their wide use in many branches of industry. A number of papers have dealt with the development of methods of designing such type of heat exchangers, but the exact analytical solutions of the mean-temperature difference and/or the thermal effectiveness as a function of number of transfer units and heat capacity ratio, are avalable only for a number of the possible cases.

According to Stevens et al.|6| the cases of multy-pass crossflow heat exchangers when both fluids are unmixed in each pass are "not susceptible to analytical solution" even if one fluid is unmixed between passes. That is not the case. The case is that the problem of determining the thermal effectiveness of the cross-flow heat exchanger arrangements with several passes, when both fluids are unmixed, is coupled with certain mathematical difficulties, but it does not mean that these difficulties can not be overcome.

The problem of establishing the $\varepsilon$-NTU-$\omega$ relationship is, in fact, that of evaluating the mean exit temperature for the unmixed flow. To reach this goal one needs the solution to the cross-flow heat transfer equation, viz. the temperature difference between the two streams as a function of position. The solutions of this kind, for the case of a single-pass exchanger, were obtained by

Nusselt [1,2], Mason [4] and Smith [3]. Recently a simplified and convinient formula for the single-pass cross-flow heat exchanger effectiveness was proposed by Baclic [5].

The solutions to all of the previously analytically unsolved cases of the multy-pass cross-flow heat exchangers are solved by the authors of this paper and will be published subsequently elsewhere. Due to the space limitations, in this paper, only the final results for the cases of two- and three-pass cross-flow heat exchangers with passes coupled in identic order when one fluid is unmixed throughout and the other is mixed between passes and unmixed in each pass, are presented. These final results are given in the form of explicit formulas for $\varepsilon = \varepsilon(NTU, \omega)$, as well as for mean mixed fluid temperatures between passes as the functions of the same arguments.

## 2. A USEFUL TERM: HEAT EXCHANGER INEFFECTIVENESS

Making the usual restrictive assumptions for any heat exchanger flow arrangement [7], and applying them to the two- and three-pass cross flow exchangers as it was done in [6], where the numerical solutions are obtained for many cases, one can be led to the mathematical formulation of the problem of determining the temperature field in the exchanger. By means of Laplace transform, similarly as it was done by Mason [4], the solution to the problem is obtainable for any multy-pass flow arrangement, wherefrom the formula for heat exchanger effectiveness is to be found by simple integration.

For the case of single-pass exchanger this procedure leads to the result [5]:

$$\varepsilon = 1 - e^{-(1+\omega)NTU} \left[ I_0(2NTU\sqrt{\omega}) + \sqrt{\omega} I_1(2NTU\sqrt{\omega}) - \frac{1-\omega}{\omega} \cdot \sum_{n=2}^{\infty} \omega^{n/2} I_n(2NTU\sqrt{\omega}) \right]. \tag{1}$$

The complementary value of the effectiveness (which is in fact the dimensionless mean exit temperature of the stream $W_1$):

$$\nu = 1 - \varepsilon = \frac{T_1'' - T_2'}{T_1' - T_2'} \tag{2}$$

may be termed the heat exchanger ineffectiveness. In the case of single-pass cross-flow heat exchanger with both fluids unmixed it follows that the interpretation of formula (1) may be written as:

$$\nu = \exp[-(1+\omega)NTU] \left[ I_0(2NTU\sqrt{\omega}) + \sqrt{\omega} I_1(2NTU\sqrt{\omega}) - \frac{1-\omega}{\omega} \sum_{n=2}^{\infty} \omega^{n/2} I_n(2NTU\sqrt{\omega}) \right]. \tag{3}$$

It may be shown that for the general case of multy-pass cross-flow heat exchangers, the function

$$\nu(a,b) = e^{-(a+b)} \left[ I_0(2\sqrt{ab}) + \sqrt{(b/a)} I_1(2\sqrt{ab}) - (\frac{a}{b} - 1) \sum_{n=2}^{\infty} (b/a)^{n/2} I_n(2\sqrt{ab}) \right] \tag{4}$$

often arises in the problems on coupling the passes in identic order when at least one fluid is mixed between passes. For that reason we introduce the notation simplifying the results to be given under next section. When we write $\nu$ without fractional subscripts, as well as without any arguments, namely as it is given by (2), it refers to the entire heat exchanger arrangement. However, when we denote $\nu_{\alpha/\beta}$ it should be regarded as the function defined by (4), where the integers $\alpha$ and $\beta$ are to be recognised from the arguments a and b. For the passes coupled in identic order, the latter are generally of the form

$$a = NTU/\beta \; ; \quad b = \alpha\omega \, NTU/\beta . \tag{5}$$

$\beta$ is usually the number of passes in the multy-pass cross-flow arrangement, while $\alpha = 1, 2, \ldots, \beta$.

Particularly, the ineffectiveness (3) of a single-pass cross-flow exchanger may be denoted as

$$\nu = \nu_{1/1} = \nu(NTU, \omega NTU) . \tag{6}$$

In the case of two-pass cross-flow exchangers we find useful two expressions:

$$\nu_{1/2} = \nu(NTU/2, \omega NTU/2) \tag{7}$$

$$\nu_{2/2} = \nu(NTU/2, \omega NTU) .$$

Further examples (for three-pass exchangers) are:

$$\nu_{1/3} = \nu(NTU/3, \omega NTU/3)$$

$$\nu_{2/3} = \nu(NTU/3, 2\omega NTU/3) \tag{8}$$

$$\nu_{3/3} = \nu(NTU/3, \omega NTU) .$$

Thus, in general case under $\nu_{\alpha/\beta}$ one should understand:

$$\nu_{\alpha/\beta} = \nu(NTU/\beta, \alpha\omega NTU/\beta) = \exp[-(1+\alpha\omega)NTU/\beta] \, [I_0(\frac{2NTU}{\beta}\sqrt{\alpha\omega}) +$$
$$+ \sqrt{\alpha\omega} \, I_1(\frac{2NTU}{\beta}\sqrt{\alpha\omega}) - (\frac{1}{\alpha\omega} - 1) \sum_{n=2}^{\infty} (\alpha\omega)^{n/2} I_n(\frac{2NTU}{\beta}\sqrt{\alpha\omega})]. \tag{9}$$

This notation is proposed as a simplification to the presentation of the results in the following section.

It should be noted that $\nu_{\alpha/\beta}$ may be seen as the ineffectiveness of a single-pass exchanger whose number of transfer units is equal to one $\beta$-th part of total NTU of multy-pass exchanger and whose capacity ratio is $\alpha$ times greater than the corresponding $\omega$ in de facto situation.

## 3. RESULTS

When one fluid is unmixed throughout and the other is mixed between passes but unmixed in each pass, the coupling of two passes in identic order permits only two possible cross-flow arrangements, viz. the conter-current and the co-current one. If three passes are to be coupled under these conditions, there are six possible cross-flow arrangements varying from pure co-current to pure counter-current flow scheme. The use of some of these arrangements may be dictated by the installation conditions.

Two possible two-pass and six possible three-pass combinations are presented in figures 1 through 8 by a scheme in the upper-left corners. The same figures contain a sketch in a coordinate sistem Oxy as well. These are given to provide the necessary notation as well as to recognize the difference in coupling in identic order. Identic order, for mathematical decription, means that the stream which is mixed between passes enters each pass at $x = 0$. The flud 2 is always assumed to be unmixed throughout, while the fluid 1 is one that is mixed between passes. To distinguish between possible combinations we denote the types of coupling by order in which the stream $W_1$ is entering some pass. An alphabetic order of passes is adopted. Thus, for the three-pass exchanger, a pure co-current scheme is denoted by type ABC, while the pure counter-current arrangement is of the type CBA. The other possible cases are then differentiated by simple permutations covering two feasible co-counter and two counter-co-current schemes.

In this manner we obtain the types AB, BA, ABC, and CBA identical to the cases presented by Stevens et al. |6| in their figures 16, 6, 21 and 11, respectively. Types ACB, BAC, BCA and CAB have not been considered in the literature previously. As a matter of fact, for any of eight types presented here the analytical solutions have not been reported.

Notation adopted in figures 1-8 for mean mixed (fluid 1) dimensionless temperatures between passes is as follows. One bar denotes the first interpass (always between passes A and B): $\bar{\Theta}_1 = (\bar{T}_1 - T_2')/(T_1' - T_2')$. Two bars denote the second interpass (always between passes B and C): $\bar{\bar{\Theta}}_1 = (\bar{\bar{T}}_1 - T_2')/(T_1' - T_2')$.

Each figure 1-8 contains the explicit formulas for the effectiveness and the mean temperatures in the interpasses of the corresponding type of exchanger. It may be seen that all of the formulas involve the ineffectiveness functions $\nu_{\alpha/\beta}$ described under previous section.

This permits the development of a simple general software for any information desired on the behaviour of any type of the exchanger presented. Formulas are convenient for programing even on a modest type of desk-top computers (few hundred programable steps are sufficient). A fast computation is provided due to the simplicity and good convergence of the formulas in any case. For example, all curves presented in the figures of this paper may be obtained on HP 9815 A calculator in approximately two hours.

On the basis of the results presented we made a check of the correction factors given by Stevens et al. in figs. 6, 11, 16, 21 of |6| and found tham correct within the tolerance acceptable in drawing the curves. For example the correction factors in figure 6 |6| for capacity ratios 0.4 and 0.7 and NTU > 4 are a little bit overpredicted.

In figure 9 the comparison between the effectiveness curves for well balanced flows ($\omega=1$) in all types of the two- and three-pass arrangements under consideration, is presented.

Studying the figures one can draw many conclusions on usefulness of some particular flow arrangements. The curves for the mixed temperatures in the interpasses may indicate the operation conditions under which some scheme may be used in practice. The duty of some passes, as well as the possible cases of heat flow inversion may be evaluated from the same curves. Thus, the region of senseless coupling of some passes and even of use of entire arrangement may be recognised. This is particularly valuable for schemes with identic effectivenesses (types BCA and CAB, and types ACB and BAC, respectively). For example, types BCA and CAB may be of some practical interes (see fig.9) due to relatively high values of $\epsilon$, however, the figures 6 and 7 indicate that preference is to be

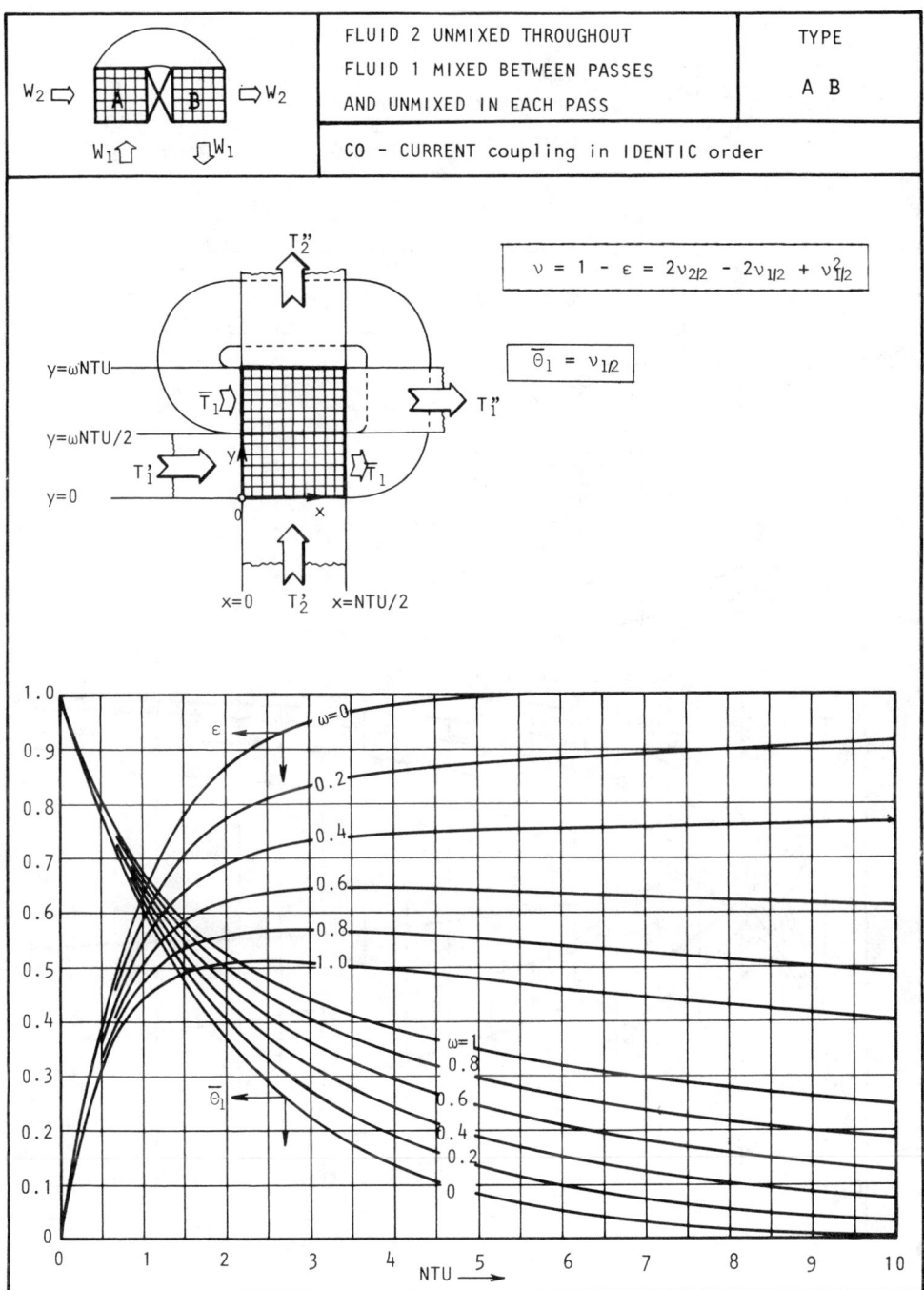

Fig. 1  Performances of AB type exchanger

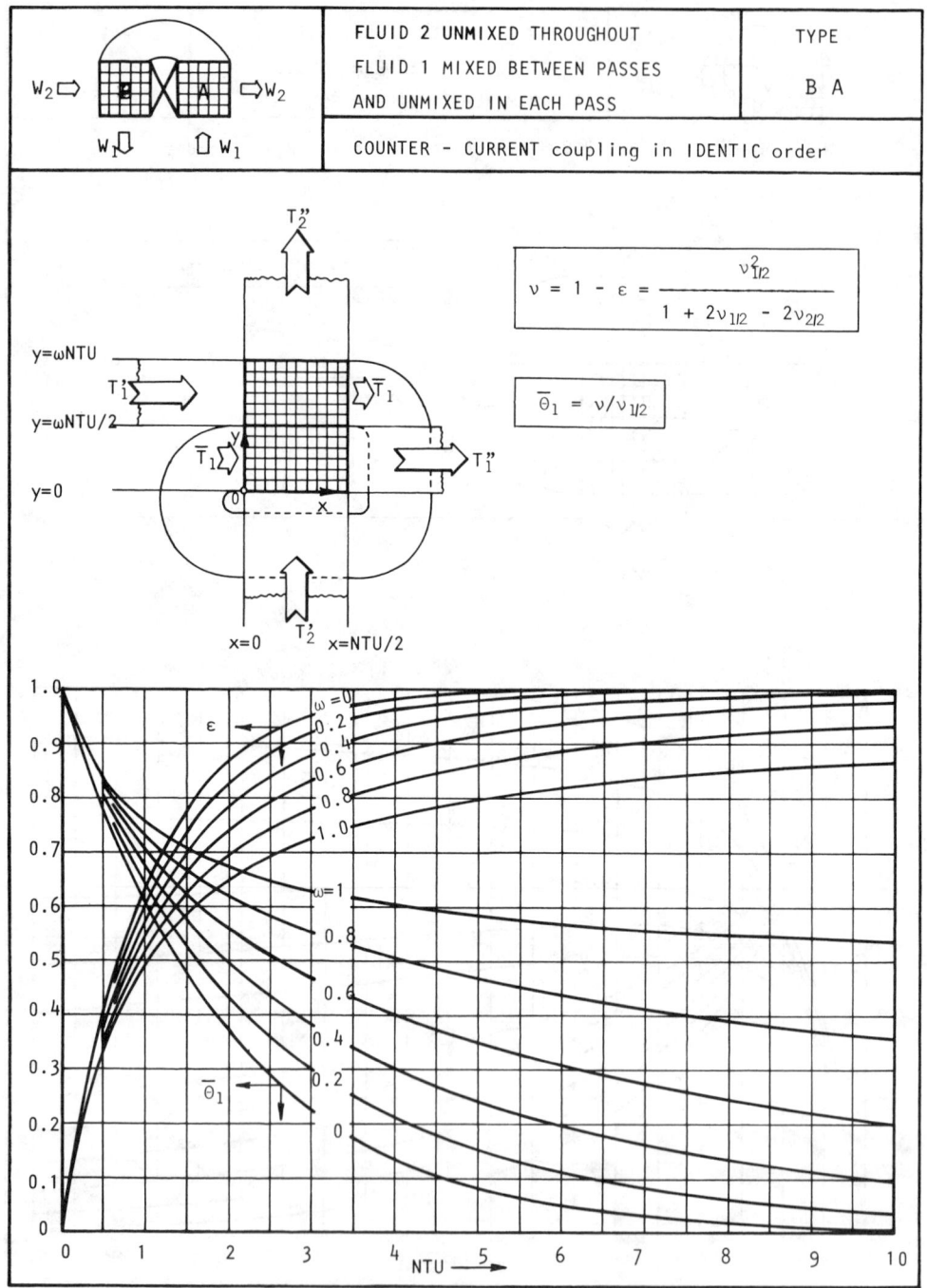

Fig. 2  Performances of BA type exchanger

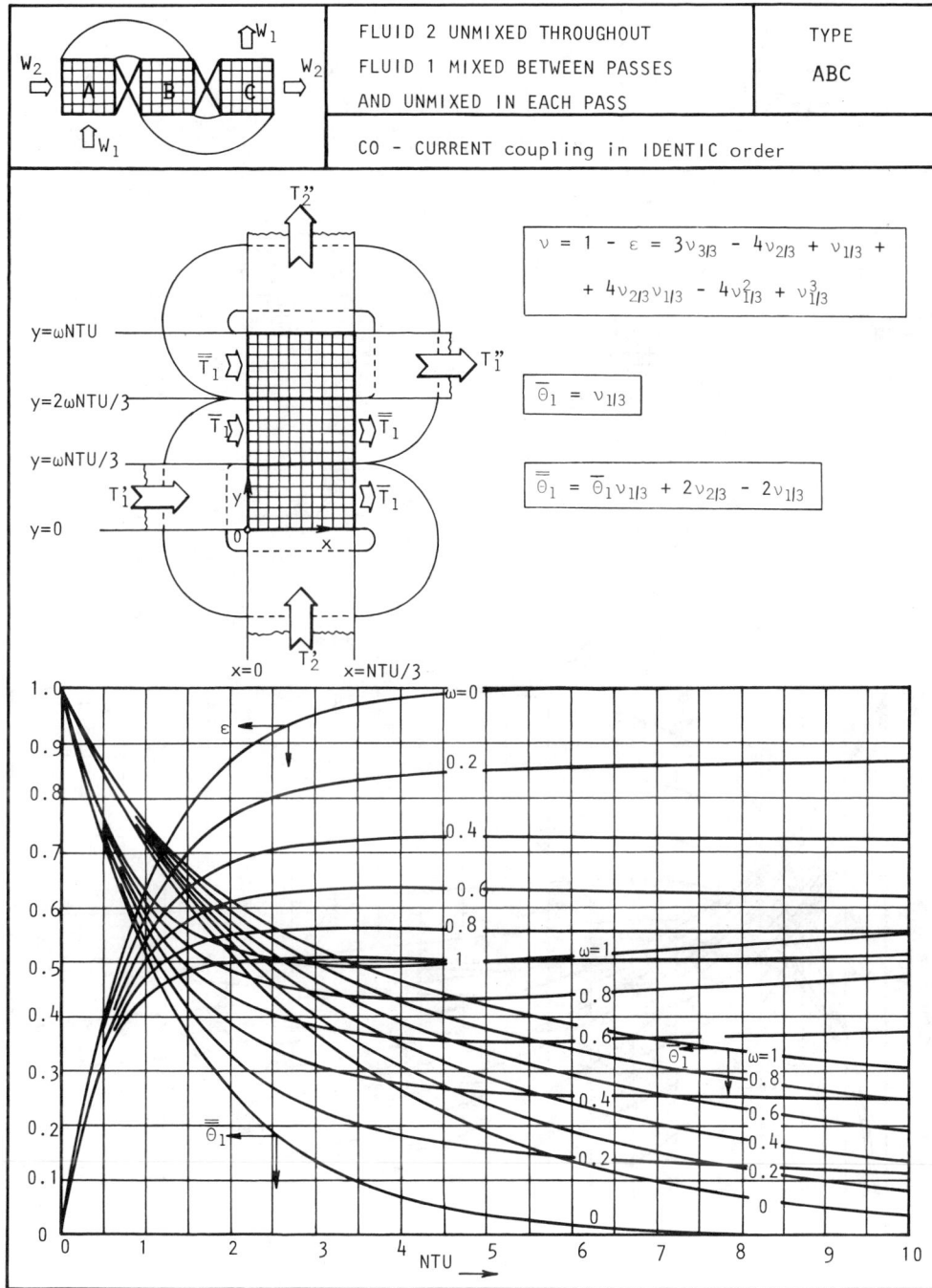

Fig. 3 Performances of ABC type exchanger

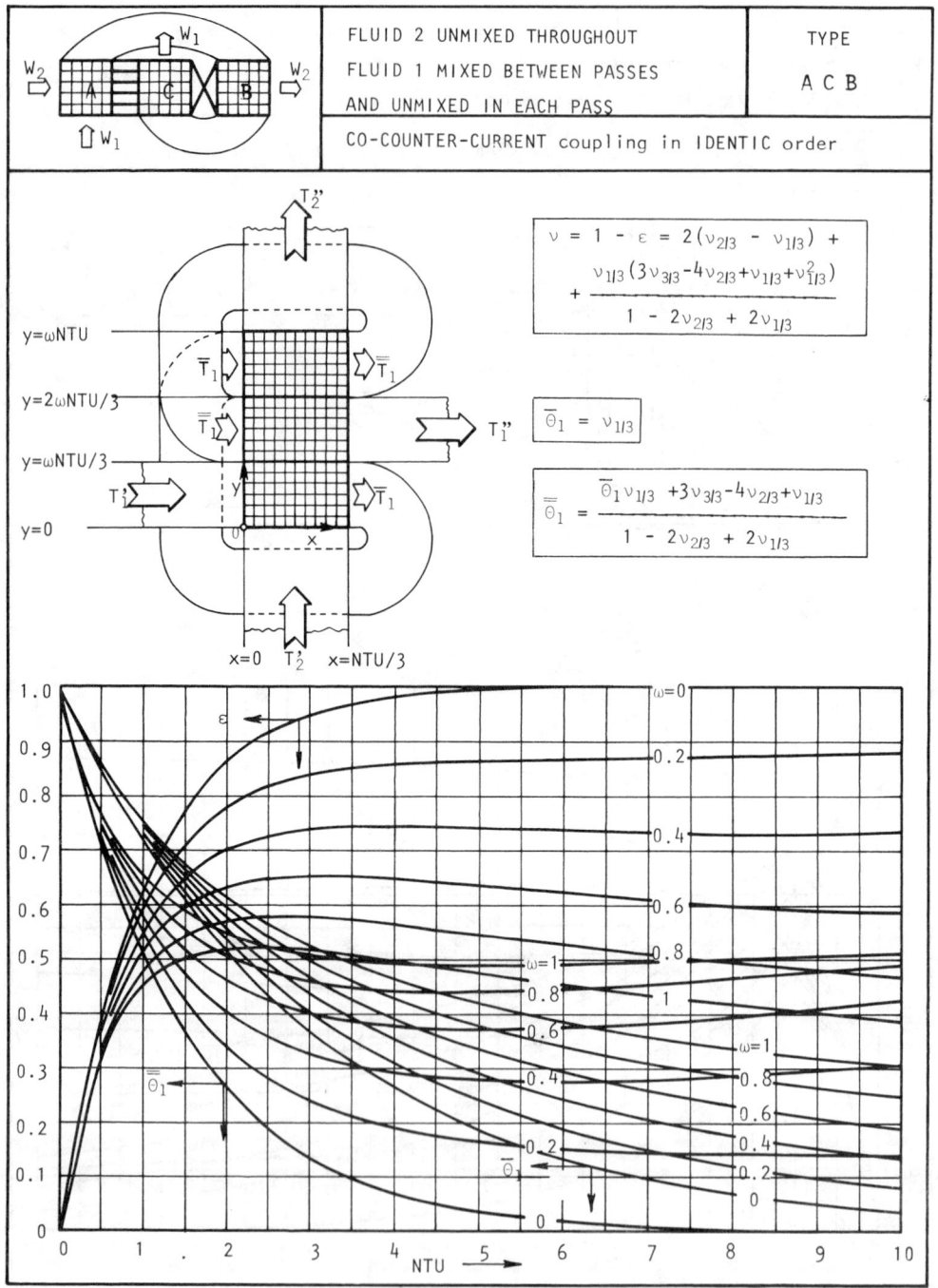

Fig. 4   Performances of ACB type exchanger

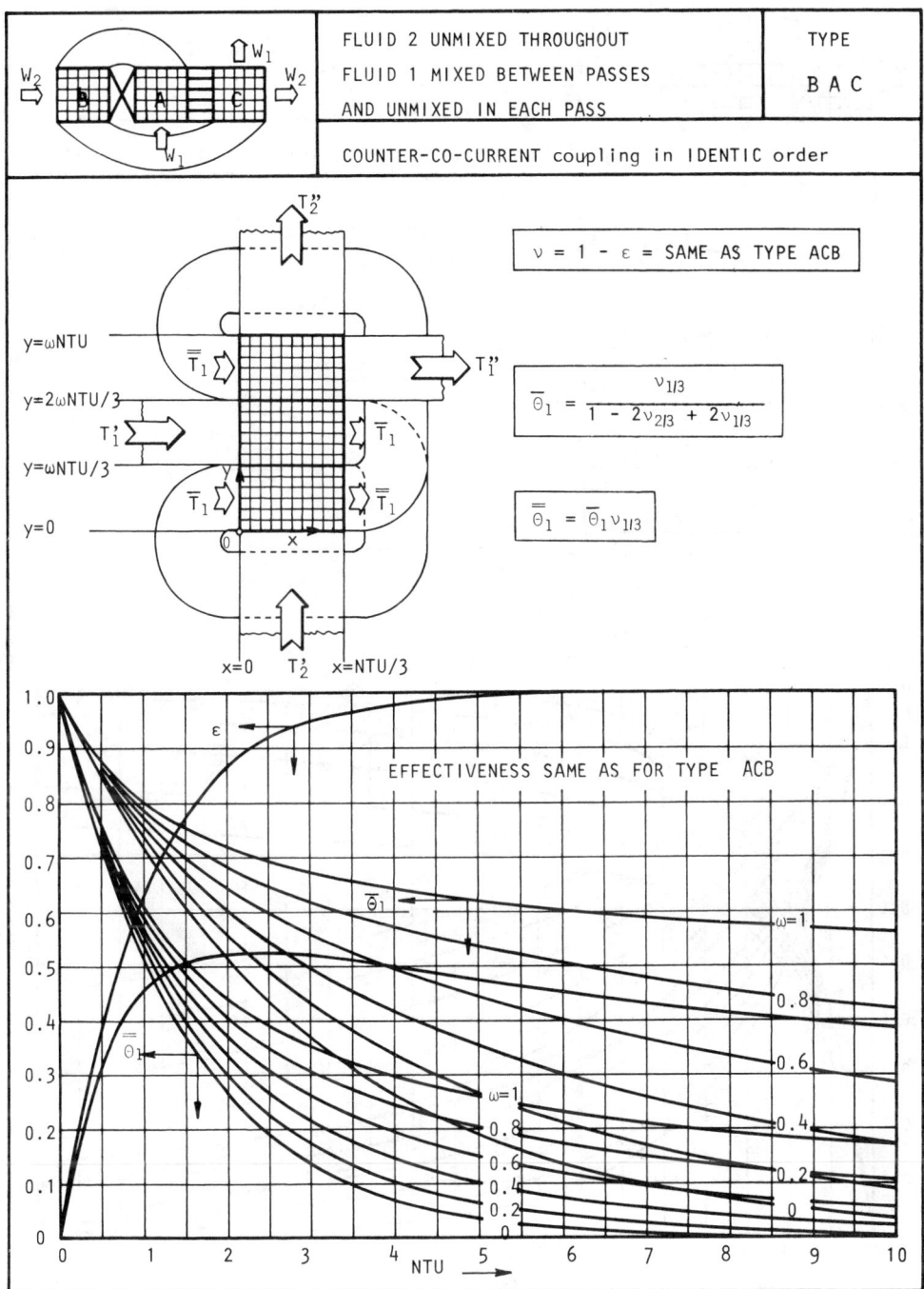

Fig. 5   Performances of BAC type exchanger

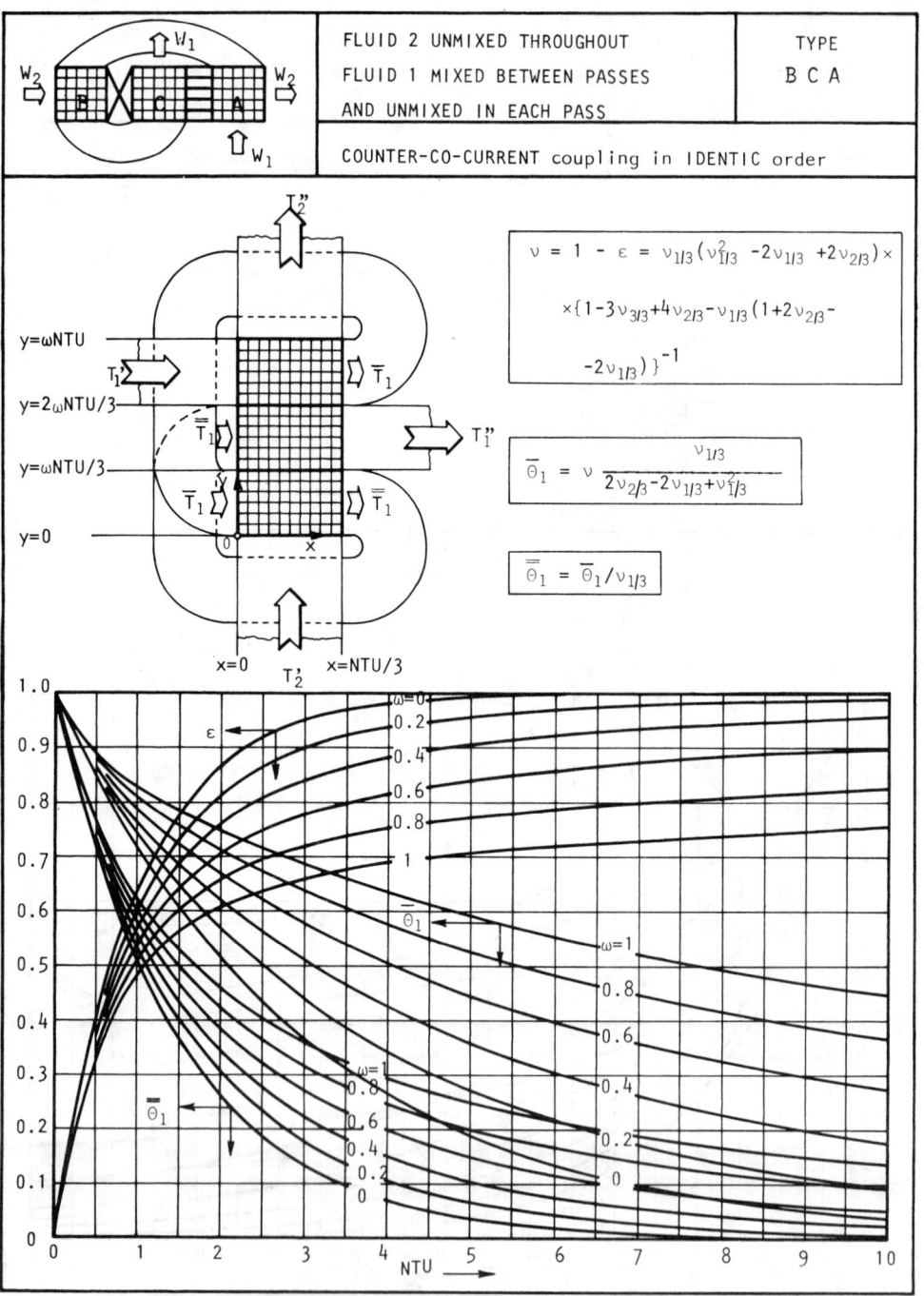

Fig. 6   Performances of BCA type exchanger

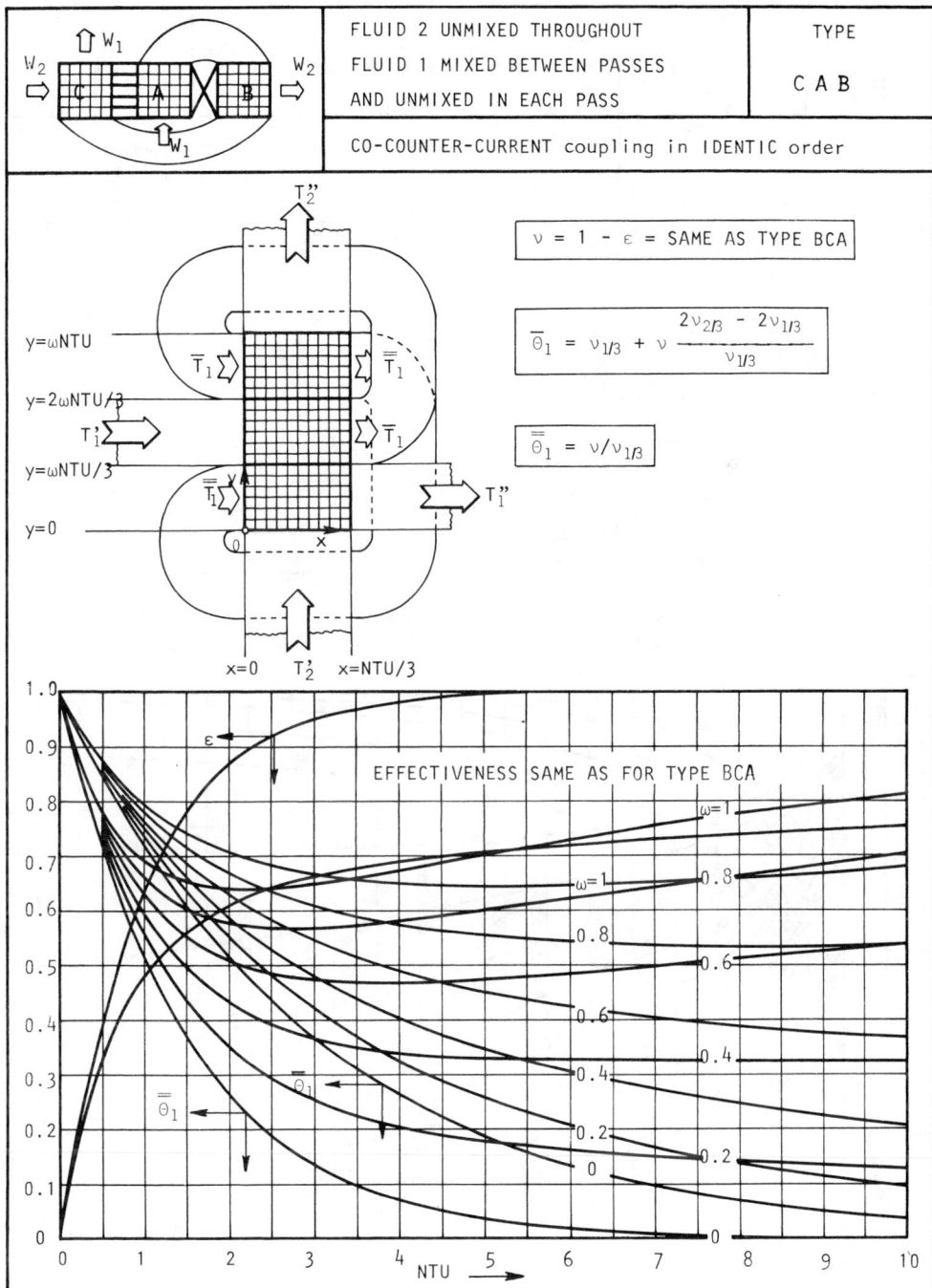

Fig. 7  Performances of CAB type exchanger

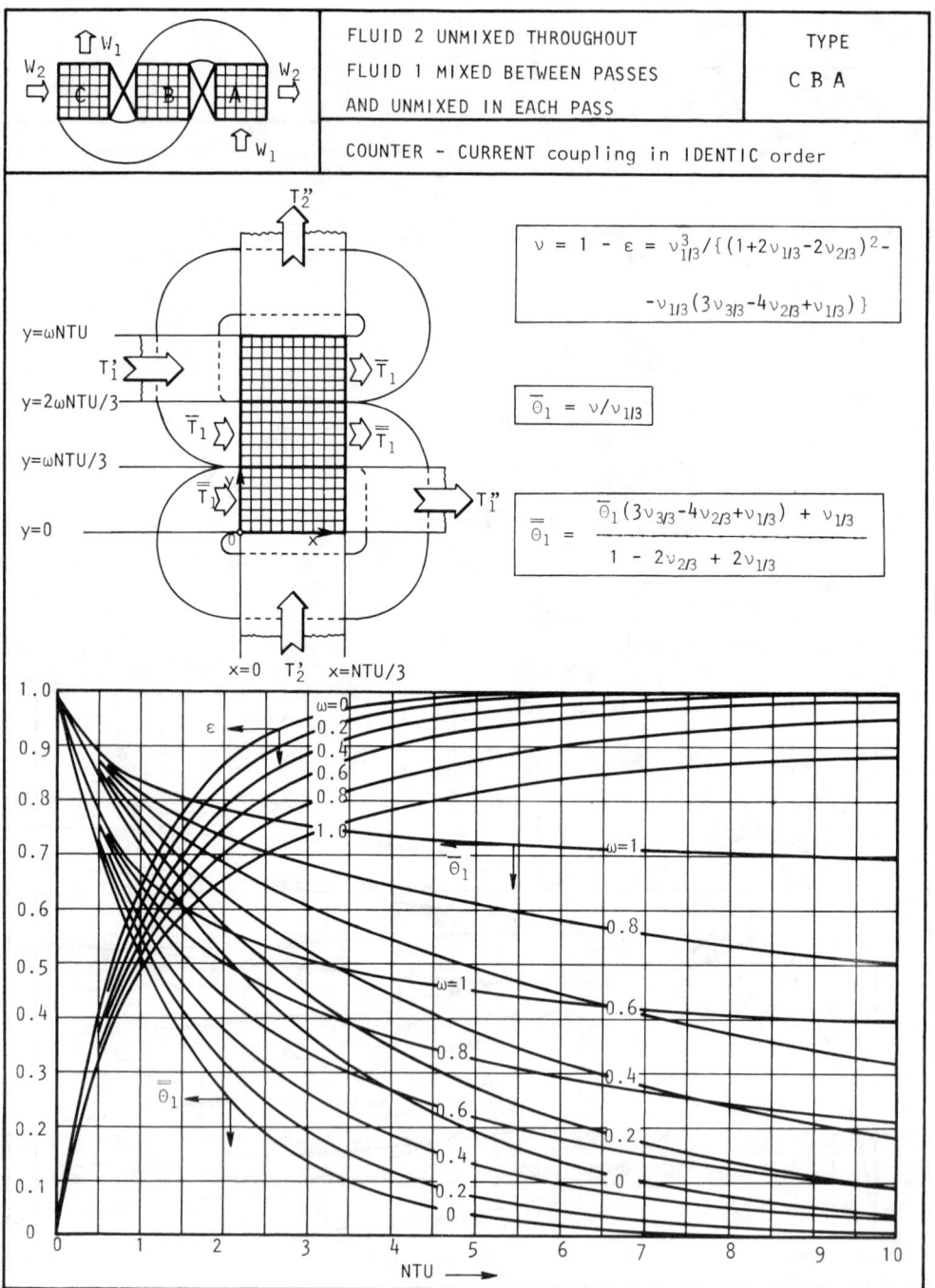

Fig. 8  Performances of CBA type exchanger

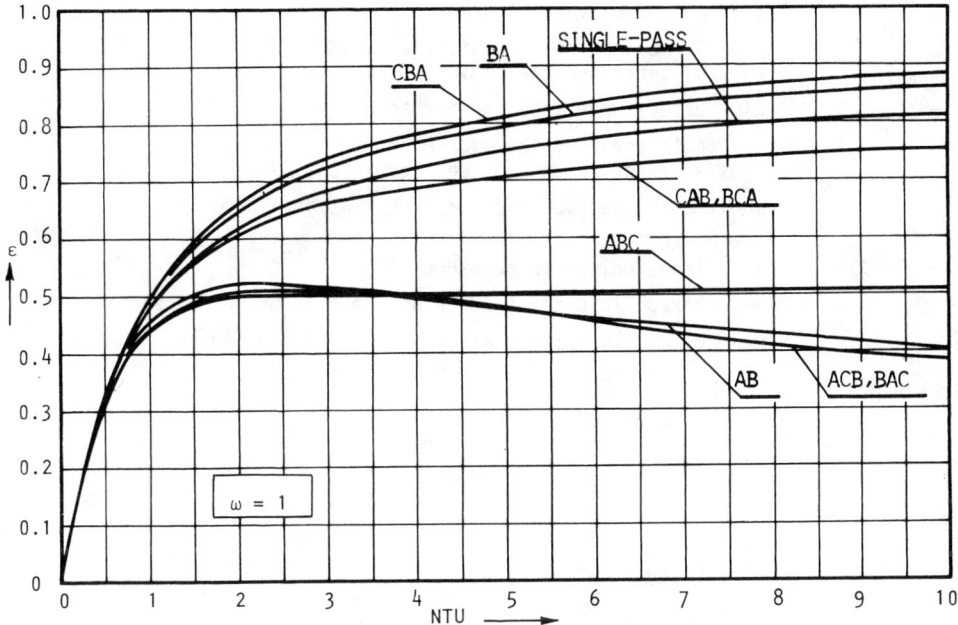

Fig. 9  Comparison of $\varepsilon = \varepsilon(NTU,1)$ for various types of cross-flow heat exchanger arrangements

given to the B$\underline{C}$A scheme up to higher values of NTU. (The intersections of the curves $\overline{\Theta}_1$ and $\overline{\overline{\Theta}}_1$ for corresponding values of $\omega$, form the upper limit of NTU up to which the use of third pass may have sense.)

Due to the simplicity of making an analysis of the figures in more details, we leave it to the reader, since the primary aim of this paper was to present the explicit formulas.

## 4. CONCLUSIONS

The exact explicit formulas for two- and three- in identic order coupled cross-flow heat exchanger passes effectiveness and mean temparatures in the interpasses may prove to be useful to the designers. Having the formulas an adequate parametric study may be performed not only from the quantitative standpoint but from the qualitative aspects as well.

Various optimization tasks on cross-flow heat exchangers may be fulfilled now by the use of the "ready" formulas particularly on computers.

## NOMENCLATURE

A  - heat transfer area, m$^2$
$c_p$ - specific heat at constant pressure, J/kgK
$\dot{M}$ - mass flow rate, kg/s
NTU = UA/W$_1$ - number of transfer units

$W = \dot{M}c_p$ — thermal capacity rate, W/K
$U$ — overall heat transfer coefficient, W/m²K
$\omega = W_1/W_2$ — thermal capacity rate ratio
$\varepsilon = (T_1'-T_1'')/(T_1'-T_2')$ — effectiveness of the exchanger
$\nu = 1 - \varepsilon$ — ineffectiveness of the exchanger
$\nu_{\alpha/\beta}$ — ineffectiveness function defined in eq. (9)
$I_n(\cdot)$ — modified Bessel function of n-th (integer) order
$T$ — temperature, K
$\Theta = (T-T_2')/(T_1'-T_2')$ — dimensionless temperature
$\overline{\Theta}$ — mean mixed fluid temperature at the exit of the first pass
$\overline{\overline{\Theta}}$ — mean mixed fluid temperature at the exit of the second pass

Subscripts:
1 — refers to fluid with $(\dot{M}c_p)_{min}$
2 — refers to fluid with $(\dot{M}c_p)_{max}$

Superscripts:
' — at exchanger inlet
" — at exchanger outlet

## REFERENCES

1. Nusselt, W., "Der Wärmeübergang im Kreuzstrom", Zeitschrift des Vereines deutscher Ingenieur, Vol. 55, 1911, pp. 2021-2024.
2. Nusselt, W., "Eine neue Formel für den Wärmedurchgang im Kreuzstrom", Technische Mechanik und Thermodynamik", Vol. 1, 1930, pp. 417-422.
3. Smith, D.M., "Mean Temperature Difference in Cross-Flow", Engineering, Vol, 138, November 2, 1934, pp. 479-481; Vol. 138, November 30, 1934, pp. 606-607.
4. Mason, J.L., "Heat Transfer in Cross-Flow", Proceedings of the Second U.S. National Congress of Applied Mechanics, ASME, New York, 1955, pp. 801-803.
5. Baclic, B.S., "A Simplified Formula for Cross-Flow Heat Exchanger Effectiveness", Trans. ASME, Journal of Heat Transfer, Vol. 100, 1978, pp. 746-747.
6. Stevens, R.A., Fernandez, J. and Woolf, J.R., "Mean-Temperature Difference in One, Two- and Three-Pass Crossflow Heat Exchangers", Trans. ASME, Vol. 79, 1957, pp. 287-297.
7. Gardner, K. and Taborek, J., "Mean Temperature Difference: A Reappraisal", AIChE, Vol. 23, No. 6, 1977, pp. 777-786.

# Compact Heat Exchanger Design Procedures

**RAMESH K. SHAH**
Harrison Radiator Division
General Motors Corporation
Lockport, New York 14094 USA

ABSTRACT

Two most common heat exchanger design problems are the rating and sizing problems. For an existing exchanger, the performance evaluation problem is referred to as the rating problem. To arrive at a design of a new exchanger to meet the specified performance is referred to as a sizing problem. A detailed procedure is outlined, with a specific example, to obtain solutions to these problems for a direct transfer type two-fluid compact heat exchanger. A general methodology of heat exchanger optimization is then presented.

1. INTRODUCTION

An overview on the heat exchanger design methodology has been presented in [1]. The details on the specific phases of the design have been covered throughout this lecture series. Considered here are two specific heat exchanger design problems, the rating and sizing problems for a two-fluid compact heat exchanger. A detailed procedure is outlined for the solution.

The rating problem, also referred to as the performance problem, is analyzed for predicting the performance of an already built exchanger. The objective here is either to verify vendor's specifications or to determine the performance at off-design conditions. In this problem, the following quantities are specified: the exchanger construction type, flow arrangement, overall core dimensions, complete details on the material and surface geometries on both sides including their heat transfer and pressure drop characteristics, fluid flow rates, inlet temperatures, and fouling factors. The designer's task is to predict the fluid outlet temperatures, heat transfer rate, and pressure drop on each side.

The sizing problem, also referred to as a design problem, involves a design of a new heat exchanger for specified performance within known constraints and minimum objective function. In this problem, the fluid inlet and outlet temperatures and flow rates are generally specified as well as the pressure drop on each side. The designer's task is to select construction type, flow arrangement, materials and surfaces on each side, and to determine the core dimensions to meet the specified heat transfer and pressure drop.

The more the information specified for either the rating or the sizing problem, generally the easier is the task for the designer for a solution. If the problem specification does not contain even minimum necessary information, the designer arrives at the necessary information based on the discussions with the customer, his experience and engineering judgements.

In this paper, a detailed solution procedure is outlined separately for the rating and sizing problems for a two-fluid direct transfer type compact heat exchanger.[†] This exchanger has gas as a working fluid at least on one side. The surface employed on the gas side of this exchanger has high surface area density usually attained by extended surfaces or by a prime surface having a small hydraulic diameter. Customarily, the $\varepsilon$-$N_{tu}$ method of heat exchanger analysis [3] is employed by the designers and manufacturers of compact heat exchangers. Hence, the solution procedure is outlined using the $\varepsilon$-$N_{tu}$ method.[††] Specific examples of rating and sizing problems are also included as an illustration of the design procedure. The solution to the sizing problem in general is not adequate for the design of a new exchanger since other constraints in addition to $\Delta p$ are imposed on the design; and the objective of the design is to minimize weight, volume, or other considerations in addition to meeting the required heat transfer. This is achieved by heat exchanger optimization. A general methodology of this optimization procedure is also presented.

## 2. THE RATING PROBLEM

The basic steps involved in the analysis of a rating problem are the determination of: surface geometrical properties, fluid physical properties, Reynolds numbers, surface basic characteristics, corrections to the basic characteristics due to temperature-dependent properties, heat transfer coefficients and fin temperature effectivenesses, wall resistance and overall conductance, $N_{tu}$, $C^*$, exchanger effectiveness, heat transfer rate, outlet temperatures, and pressure drop on each side. These steps are outlined in detail now.

### 2.1 Surface Geometrical Properties

The following geometrical properties on each side are needed for the heat transfer and pressure drop analysis: the primary and secondary surface area (if any) A, minimum free flow area $A_o$, frontal area $A_{fr}$, hydraulic diameter $D_h$, flow length L for the $\Delta p$ calculations, ratio of free flow area to frontal area $\sigma$ for exchanger entrance and exit losses, the fin length $\ell$ and other pertinent dimensions for the fin. Whatever information not specified directly is computed based on the specified surface geometries and core dimensions. The following are some of the basic relationships between the surface and core geometries on one side of a recuperator.

$$D_h = \frac{4A_o L}{A} \quad \text{or} \quad A_o = \frac{D_h A}{4L} \tag{1}$$

If $L_1$ and $L_2$ are the flow lengths and $N_p$ and $N_p+1$ are the number of flow passages on sides 1 and 2 respectively, then the volume between plates on each side is

---

[†] In a direct transfer type exchanger, two fluids are separated by a thin wall (parting sheets or tube walls) through which heat flows. This exchanger is also referred to as a recuperator. Because of time and space limitations, the design procedure for regenerators will not be outlined here. However, the pressure drop analysis for the regenerator is similar to the one outlined in this paper. For the regenerator heat transfer analysis, refer to [2].

[††] Since the dimensionless groups of the $\varepsilon$-$N_{tu}$ method and the log-mean temperature difference (LMTD) method are uniquely related [3], the solutions to the rating and sizing problems by the LMTD method will be identical to those by the $\varepsilon$-$N_{tu}$ method, although the detailed steps of calculations in the two methods will be different.

# COMPACT HEAT EXCHANGER DESIGN PROCEDURES

$$V_{p,1} = L_1 L_2 (b_1 N_p) \qquad V_{p,2} = L_1 L_2 b_2 (N_p+1) \qquad (2)$$

Heat transfer areas on each side are

$$A_1 = \beta_1 V_{p,1} \qquad A_2 = \beta_2 V_{p,2} \qquad (3)$$

Here $\beta_1$ and $\beta_2$ are the surface area density on each side per unit volume between plates. The ratio of minimum free flow area to frontal area on side 1 is

$$\sigma_1 = \frac{A_{o,1}}{A_{fr,1}} = \frac{A_{o,1} L_1}{A_{fr,1} L_1} = \frac{A_1 D_{h,1}/4}{V} = \frac{V_{p,1} \beta_1 D_{h,1}/4}{V}$$

$$= \frac{L_1 L_2 (b_1 N_p) \beta_1 D_{h,1}/4}{[b_1 N_p + b_2 (N_p+1) + 2a(N_p+1)] L_1 L_2} = \frac{b_1 N_p \beta_1 D_{h,1}/4}{b_1 N_p + b_2 (N_p+1) + 2a(N_p+1)}$$

$$\approx \frac{b_1 \beta_1 D_{h,1}/4}{b_1 + b_2 + 2a} \qquad (4)$$

Here $a$ is the thickness of parting plates. Similarly

$$\sigma_2 = \frac{b_2 (N_p+1) \beta_2 D_{h,2}/4}{b_1 N_p + b_2 (N_p+1) + 2a(N_p+1)} \approx \frac{b_2 \beta_2 D_{h,2}/4}{b_1 + b_2 + 2a} \qquad (5)$$

Here the last approximate equality is for the case when $N_p \gg 1$ or the number of passages on each side are the same. The heat transfer surface area on one side divided by the total volume $V$ of the exchanger, designated as $\alpha_1$, is

$$\alpha_1 = \frac{A_1}{V} = \frac{A_1}{L_1 A_{fr,1}} = \frac{(A_1/L_1)}{A_{fr,1}} = \frac{4 A_{o,1}/D_{h,1}}{A_{fr,1}} = \frac{4\sigma_1}{D_{h,1}} = \frac{b_1 \beta_1}{b_1 + b_2 + 2a} \qquad (6)$$

where the last equality is due to Eq. (4). Similarly,

$$\alpha_2 = \frac{A_2}{V} = \frac{b_2 \beta_2}{b_1 + b_2 + 2a} \qquad (7)$$

The two most commonly used plate-fin geometries are rectangular and triangular passages shown in Fig. 1. The actual rectangular fin geometry has rounded corners instead of the sharp corners shown in Fig. 1a. The fin surface area and primary surface area associated with this fin are also shown in Fig. 1a. From the review of Fig. 1, the fin length $\ell$ for heat conduction (up to the adiabatic plane) is

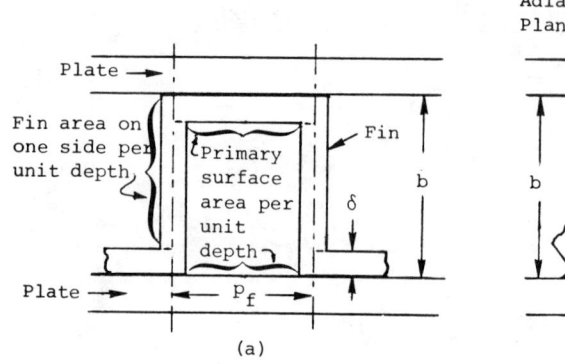 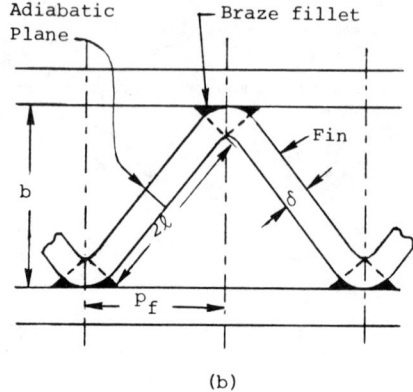

Fig. 1  (a) Idealized plain rectangular fin, and (b) plain triangular fin.

$$\ell = \frac{b-\delta}{2} \simeq \frac{b}{2} \tag{8}^\dagger$$

Note that no effect of the fin inclination is taken into account in Eq. (8) for the triangular fin. This represents a good approximation due to the presence of the braze fillet. However, a more accurate value of $\ell$ can be determined if the detailed measurements are available on the fin shape and braze fillet. For the fin geometries other than those of Fig. 1, an appropriate value of $\ell$ is determined for the fin temperature effectiveness calculations.

## 2.2 Fluid Properties and Bulk Mean Temperature

Exchanger transfer rate and fluid pumping power are strongly dependent upon the type of fluid flowing through the exchanger. It is essential that fluid properties be established accurately for the appropriate temperature and pressure conditions. Fluid properties needed are density, specific heat, viscosity, thermal conductivity and Prandtl number. These properties are available for a large number of fluids in the literature [4,5]. These properties are determined at an appropriate temperature on each side of the exchanger.

For the $\varepsilon$-$N_{tu}$ or LMTD method, we need to obtain a "single" temperature value to represent the flow of fluid through each side of the exchanger. For counterflow and parallel flow exchangers, the bulk fluid temperature varies in the flow direction as well as over the cross section of the individual flow passages. In a crossflow heat exchanger, it also varies in the flow direction of the other side fluid. In more complex arrangements, the fluid temperature variation is generally three-dimensional. The temperature changes in the flow direction affect the fluid bulk properties for the exchanger analyses. The temperature changes across the individual flow passages influence the velocity and temperature profiles, and thereby influence the friction factor and the

---

$^\dagger$An alternative approximation is $\ell = (b-2\delta)/2$. If $\ell \gg \delta$, then either of the approximations is good for engineering calculations. If $\ell$ is not very large compared to $\delta$, then $\eta_f \geq 0.95$, and either of the approximations will not significantly affect the value of $\eta_f$.

# COMPACT HEAT EXCHANGER DESIGN PROCEDURES

convective heat transfer coefficient. If the fluid properties vary substantially within an exchanger, an "average" temperature may not be adequate to accurately determine the heat transfer rate and pressure drop. In that case, the exchanger is divided into several segments, and the conventional $\varepsilon\text{-}N_{tu}$ or LMTD analysis is carried out for each segment with an average temperature representative of each segment.

The influence of temperature variations across the individual flow passages is accounted for by the property ratio method as discussed in Section 2.4. Here the temperature is characterized by both the wall and bulk mean temperatures and corrections to Nu and f are applied by appropriate property ratio fractions.

In this section, methods are presented to evaluate the flow length "average" (or bulk mean) temperature for each fluid side of a two-fluid heat exchanger. In the determination of the heat transfer rate through the exchanger, the true mean temperature is either used directly (as in the LMTD method, $q = UA\Delta t_m$) or indirectly (as in the $\varepsilon\text{-}N_{tu}$ method). Now $\Delta t_m = \Delta t_{\ell m}$ for counterflow and parallel flow exchangers or $\Delta t_m = F\Delta t_{\ell m} \simeq \Delta t_m$ for a well designed exchanger of any other flow arrangement. Here F is the log-mean temperature difference correction factor [3]. Therefore, we use $\Delta t_{\ell m}$ to evaluate the log-mean average temperature of the $C_{min}$ fluid, when the temperature rise or drop of the $C_{max}$ fluid is small. The average temperature of the $C_{max}$ fluid in this case is the arithmetic average temperature. When $\Delta t_h$ and $\Delta t_c$ are of the same order of magnitude, the mean temperature on each fluid side is considered simply as the arithmetic average temperature. An approximate procedure is outlined next for evaluation of this temperature for two-fluid heat exchangers with some specific flow arrangements.

<u>Heat exchangers with $C^* \simeq 0$.</u> Typical temperature distributions for both fluids are shown in Fig. 2. The flow length average temperatures for Fig. 2a situation are:

$$t_{c,m} = (t_{c,i} + t_{c,o})/2 \tag{9}$$

$$t_{h,m} = t_{c,m} + \Delta t_{\ell m} \tag{10}$$

where

$$\Delta t_{\ell m} = \frac{(t_{h,i} - t_{c,m}) - (t_{h,o} - t_{c,m})}{\ln[(t_{h,i} - t_{c,m})/(t_{h,o} - t_{c,m})]} \tag{11}$$

The flow length average temperature for Fig. 2b situation are:

$$t_{h,m} = (t_{h,i} + t_{h,o})/2 \tag{12}$$

$$t_{c,m} = t_{h,m} - \Delta t_{\ell m} \tag{13}$$

where

$$\Delta t_{\ell m} = \frac{(t_{h,m} - t_{c,i}) - (t_{h,m} - t_{c,o})}{\ln[(t_{h,m} - t_{c,i})/(t_{h,m} - t_{c,o})]} \tag{14}$$

<u>Counterflow and crossflow heat exchangers.</u> Typical temperature distributions for these exchangers are shown in [6]. The true mean temperature

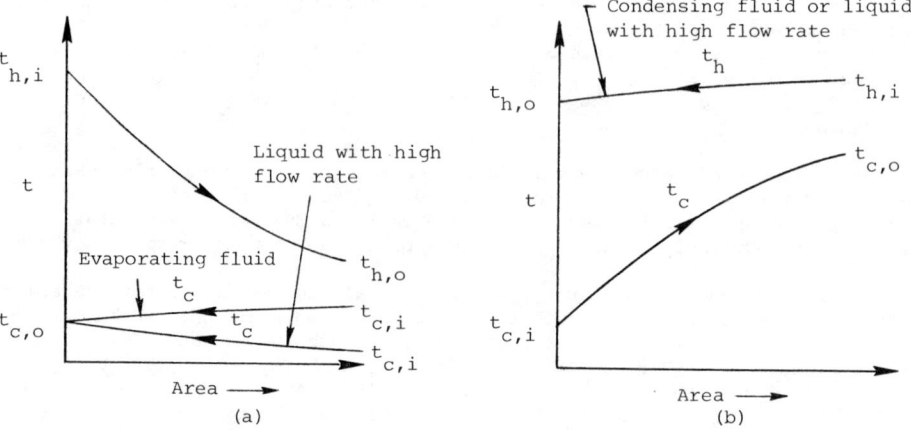

Fig. 2  Temperature distributions in a heat exchanger with (a) $C_h/C_c \simeq 0$, and (b) $C_c/C_h \simeq 0$.

difference for these exchangers is

$$\Delta t_m = \Delta t_{\ell m} \qquad \text{for counterflow} \tag{15}$$

$$\Delta t_m = F \Delta t_{\ell m} \qquad \text{for crossflow} \tag{16}$$

It appears that it may be more appropriate if $\Delta t_m$ is used to arrive at the average temperature of one of the two fluids of a crossflow exchanger. However, F generally varies between 0.75 and 0.98 and the resultant error in the average temperature, by considering F = 1, does not generally produce significant variations in the fluid properties. Hence, we evaluate the average temperature on each fluid side for crossflow by considering it as a counterflow heat exchanger.

Two different procedures are suggested for the determination of the average temperatures for a counterflow or crossflow exchanger. Somewhat arbitrarily, the mean temperatures are considered as the arithmetic averages for $C^* \geq 0.5$ and log-mean average for the $C^* < 0.5$ case. For the $C^* \geq 0.5$ case,

$$t_{h,m} = (t_{h,i} + t_{h,o})/2 \tag{17}$$

$$t_{c,m} = (t_{c,i} + t_{c,o})/2 \tag{18}$$

For the $C^* < 0.5$ case, the temperature change for the $C_{max}$ side is less than half compared to that for the $C_{min}$ side. Hence, the mean temperature on the $C_{max}$ side is taken as the arithmetic average.

$$t_m = (t_i + t_o)/2 \qquad \text{on the } C_{max} \text{ side} \tag{19}$$

# COMPACT HEAT EXCHANGER DESIGN PROCEDURES

For the $C_{min}$ side, the average temperature is considered as the log-mean average. The log-mean temperature difference is first evaluated from Eq. (11) or (14) using the mean temperature of Eq. (19) treated as the constant temperature. Subsequently, the log-mean average temperature on the $C_{min}$ side is determined from Eq. (10) or (13) as follows.

$$t_m \big|_{C_{min}} = t_m \big|_{C_{max}} \pm \Delta t_{\ell m} \qquad (20)$$

<u>Multipass heat exchangers.</u> Fluid temperatures are first determined after every pass so that inlet and outlet fluid temperatures are known for every pass. Then generally arithmetic averages of the terminal temperatures are taken to obtain the flow length average temperature for each fluid in each pass.

The effectiveness of each pass in terms of overall effectiveness and $C^*$ are as follows [3]. For overall counterflow arrangement,

$$\varepsilon_p = \frac{[(1-\varepsilon C^*)/(1-\varepsilon)]^{1/n} - 1}{[(1-\varepsilon C^*)/(1-\varepsilon)]^{1/n} - C^*} \qquad (21)$$

For overall parallel flow arrangement,

$$\varepsilon_p = \frac{1}{1+C^*} [1 - \{1 - (1+C^*)\varepsilon\}^{1/n}] \qquad (22)$$

where

$$\varepsilon_p = \frac{C_c(t_{c,o} - t_{c,i})_p}{C_{min}(t_{h,i} - t_{c,i})_p} = \frac{C_h(t_{h,i} - t_{h,o})_p}{C_{min}(t_{h,i} - t_{c,i})_p} \qquad (23)$$

This definition of $\varepsilon_p$ is then used to evaluate the mean outlet temperatures $t_{h,o}$ and $t_{c,o}$ for each pass of an overall parallel flow exchanger. For an overall counterflow arrangement, an iterative procedure will be required in order to determine the tempratures between passes. This is because inlet temperatures of both fluids are not known for any pass in the beginning.

For heat exchangers with other flow arrangements that are not considered here, use any one of the several methods outlined above. Engineering judgement is applied for the selection of a particular method for the determination of the flow length average temperature.

<u>Approximate outlet temperatures.</u> Since the outlet temperatures are not known for this problem, an approximate method is used first to best estimate them. For example, assume 60% effectiveness for shell-and-tube or plate heat exchangers, 75% effectiveness for most extended surface crossflow heat exchangers, and 85% effectiveness for regenerators and extended surface counterflow exchangers. If the designer knows more approximate value for his application, he uses that value of the effectiveness. Based on the assumed effectiveness, determine the outlet temperatures. Subsequently arrive at the mean temperature on both sides by the methods outlined above. Evaluate the physical properties of the fluids, $\mu$, $\rho$, $c_p$, $k$, $Pr$ on each side at these mean temperatures.

## 2.3 Reynolds Numbers and Surface Basic Characteristics

Once the surface geometrical and fluid physical properties are known, determine G and Re from their definitions.

$$G = W/A_o \qquad (24)$$

$$Re = GD_h/\mu \qquad (25)$$

The exchanger surface basic characteristics are generally presented in terms of j and f vs. Re or Nu vs. Re or x*, and $f_{app}$ Re vs. $x^+$ [7]. Thus depending upon the correlations, other independent parameters such as $L^* = L/D_h RePr$, $L^+ = L/D_h Re$, etc. may need to be evaluated before arriving at j, Nu or f. Some correlations may involve more complicated functions or geometrical parameters.

## 2.4 Corrections for Temperature-Dependent Properties

One of the basic idealizations made in the theoretical solutions for Nu and f is that the fluid properties remain constant throughout the flow field. Most of the experimental j and f data obtained also involve small temperature differences so that the fluid properties do not vary significantly. Since the transport properties of most fluids vary with temperature, they will vary over the flow cross section of a heat exchanger passage. The influence of heat transfer on the laminar velocity distribution is shown in Fig. 3 as an illustration [8]. Since the velocity profile is distorted due to heat transfer, the velocity gradients at the wall are different affecting both j and f factors. Thus the constant property solutions and results provide the best approximation for heat transfer problems involving small temperature differences. For those problems which involve large temperature differences, the constant property results would deviate substantially and need to be modified. In addition, j and f data obtained for a surface with air or water need to be modified, if they are to be used for the same surface with oil or other fluids. Influence of fluid property variations on j or Nu and f factors is presented in this section.

<u>Correction schemes.</u> For engineering applications, it is convenient to employ the constant property analytical solutions, or experimental data obtained with small temperature differences, and then to apply some kind of correction to account for property variations. Two schemes are commonly used for such correction: (1) the property ratio method, and (2) the reference temperature

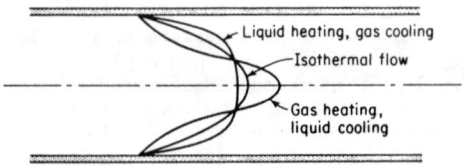

Fig. 3  Influence of heat transfer on laminar velocity distribution.

method. In the property ratio method, all properties are evaluated at the bulk mean temperature, and then all of the variable properties effects are lumped into a function. This function is a ratio of some pertinent property evaluated at the surface temperature to that property evaluated at the bulk temperature. In the reference temperature method, pertinent groups are evaluated such that the constant property results could be used directly to evaluate variable property performance. Typically, this may be the film temperature or the surface temperature. The _film_ temperature is an arithmetic average of the wall temperature and fluid bulk mean temperature.

The property ratio method is extensively used for the internal flow problem while the reference temperature method is the most common for the external flow problem. One of the reasons for the selection of the property ratio method for the internal flow problem is that the determination of $\rho u_m = G$ (used in calculating Re and St) is straightforward. It is computed from $G = W/A_o$ regardless of variations in $\rho$. In the reference temperature method, $\rho$ in $\rho u_m$ is evaluated at the reference temperature, while $\rho$ in determining $u_m$ ($u_m = W/A_o\rho$) is evaluated at the bulk mean temperature. Thus two densities are needed for the reference temperature method that leads to awkwardness and ambigious interpretations.

_Property ratio method correlations._ While variations in $\mu$, $k$ and $\rho$ are important for gases, the only property of major significance is $\mu$ for liquids. Thus gases and liquids incorporate different property groups for the correction.

For gases, the viscosity, thermal conductivity, and density are functions of the absolute temperature. This absolute temperature dependence is similar for different gases except for the temperature extremes. The temperature-dependent property effects for gases are adequately correlated by the following equations.

$$\frac{Nu}{Nu_{cp}} = \frac{j}{j_{cp}} = \left(\frac{T_w}{T_m}\right)^n \tag{26}$$

$$\frac{f}{f_{cp}} = \left(\frac{T_w}{T_m}\right)^m \tag{27}$$

where the suffix cp refers to the constant property variable and all temperatures are absolute. All of the properties in dimensionless groups of Eqs. (26) and (27) are evaluated at the bulk mean temperature. The values of the exponents n and m depends upon the flow regime (laminar or turbulent).

For liquids, viscosity is the only property of importance that varies greatly with temperature. Thus it is found that the temperature-dependent effects for liquids are adequately correlated by

$$\frac{Nu}{Nu_{cp}} = \frac{j}{j_{cp}} = \left(\frac{\mu_w}{\mu_m}\right)^n \tag{28}$$

$$\frac{f}{f_{cp}} = \left(\frac{\mu_w}{\mu_m}\right)^m \tag{29}$$

These exponents n and m for fully developed laminar and turbulent flows in a circular tube are summarized in Tables 1 and 2, respectively. These correlations are derived for the constant heat flux boundary condition (H). The variable property effects are generally not important for fully developed flow having constant wall temperature boundary condition, since $t_w$ approaches $t_m$ for fully developed flow. Hence, the correlations of Tables 1 and 2 are recommended for the (T) and (H) boundary conditions.

The j (or Nu) and f factors are also dependent upon the duct cross-sectional shape in laminar flow, and are practically independent of the duct shape in turbulent flow. The influence of variable fluid properties on Nu and f for fully developed laminar flow through rectangular ducts has been investigated by Nakamura et al. [12]. They concluded that the velocity profile is strongly affected by the $\mu_w/\mu_m$ ratio, and the temperature profile is weekly affected by the $\mu_w/\mu_m$ ratio. They found that the influence of the aspect

Table 1.  Property ratio method exponents of Eqs. (26), (27) or of Eqs. (28), (29) for laminar flow

| Fluid | Heating | Cooling |
|---|---|---|
| Gases | n = 0.0, m = 1.00 for $1 < T_w/T_m < 3$ [8] | n = 0.0, m = 0.81 for $0.5 < T_w/T_m < 1$ [8] |
| Liquids | n = -0.14, m = 0.58 for $\mu_w/\mu_m < 1$ [8] | n = -0.14, m = 0.54 for $\mu_w/\mu_m > 1$ [9] |

Table 2.  Property ratio method correlations or exponents of Eqs. (26), (27) or of Eqs. (28), (29) for turbulent flow

| Fluid | Heating | Cooling |
|---|---|---|
| Gases | Nu = $5+0.012$ Re$^{0.83}$(Pr+0.29)$(T_w/T_m)^n$<br>n = $-0.25 \log_{10}(T_w/T_m) + 0.3$<br>for $1 < T_w/T_m < 5$, $0.6 < Pr < 0.9$, $10^4 < Re < 10^6$ and $L/D_h > 40$ [10]<br>m = $-0.6 + 5.6(Re_w \rho_w/\rho_m)^{-0.38}$<br>for $1 < T_w/T_m < 3.7$ [11] | n = $-0.36$<br>for $0.37 < T_w/T_m < 1$ [11]<br><br><br>m = $-0.6 + 0.79(Re_w \rho_w/\rho_m)^{-0.11}$<br>for $0.37 < T_w/T_m < 1$ [11] |
| Liquids [11] | n = $-0.11$†<br>for $0.08 < \mu_w/\mu_m < 1$<br>$f/f_{cp} = (7-\mu_m/\mu_w)/6$‡<br>for $0.35 < \mu_w/\mu_m < 1$ | n = $-0.25$†<br>for $1 < \mu_w/\mu_m < 40$<br>m = $0.24$†<br>for $1 < \mu_w/\mu_m < 2$ |

†Valid for $2 \leq Pr \leq 140$, $10^4 \leq Re \leq 1.25 \times 10^5$.
‡Valid for $1.3 \leq Pr \leq 10$, $10^4 \leq Re \leq 2.3 \times 10^5$.

ratio on the correction factor $(\mu_w/\mu_m)^m$ for the friction factor is negligible for $\mu_w/\mu_m < 10$, and the values of $m$ for the circular tube are valid also for the rectangular ducts. The influence of the aspect ratio becomes important for $\mu_w/\mu_m > 10$. For the heat transfer problem, the Sieder-Tate correlation ($n = -0.14$) is valid only in the narrow range of $0.4 < \mu_w/\mu_m < 4$.

## 2.5 Heat Transfer Coefficient and Fin Effectiveness

Once the Colburn factor $j$ or the Nusselt number $Nu$ is known for each side and is corrected for the temperature-dependent fluid properties, determine the heat transfer coefficient from the definitions:

$$h = jGc_p Pr^{-2/3} \quad \text{or} \quad h = Nuk/D_h \tag{30}$$

If fins are used on any side, the temperature effectiveness of the fin, $\eta_f$, is evaluated using $h$ from Eq. (30), and the geometry and material of the fin. $\eta_f$ for the straight thin fin of uniform thickness $\delta$ of Fig. 1 is [3]

$$\eta_f = \frac{\tanh m\ell}{m\ell} \tag{31}$$

where 

$$m = \left[\frac{hP}{k_f A_k}\right]^{1/2} = \left[\frac{h(2L_f + 2\delta)}{k_f(L_f\delta)}\right]^{1/2} = \left[\frac{2h}{k_f\delta}\left(1 + \frac{\delta}{L_f}\right)\right]^{1/2} \simeq \left[\frac{2h}{k_f\delta}\right]^{1/2} \tag{32}$$

Here $P$ is the wetted perimeter of the fin, $A_k$ is the conduction cross-sectional area of the fin, and $L_f$ is the length of the fin in the fluid flow direction. Generally, $L_f \gg \delta$ and the expression on the extreme right-hand side is a good approximation for $m$. For the strip fin geometry $L_f$ is taken as the strip length $\ell_s$. For the temperature effectiveness of fins of other geometries, refer to Kern and Kraus [13]. The total surface temperature effectiveness $\eta_o$ for the extended surface is then calculated from

$$\eta_o = 1 - (1 - \eta_f)A_f/A \tag{33}$$

Here $A_f$ is the fin area.

The overall thermal conductance $UA$ is then computed from

$$\frac{1}{UA} = \frac{1}{(\eta_o hA)_h} + \frac{1}{(\eta_o h_s A)_h} + R_w + \frac{1}{(\eta_o h_s A)_c} + \frac{1}{(\eta_o hA)_c} \tag{34}$$

Here $h_s$ is the scale or fouling coefficient and $R_w$ is the wall thermal resistance.

$$R_w = \begin{cases} a/A_w k_w & \text{for plane parting sheets} \\ \ln(r_o/r_i)/(2\pi L k_w N_t) & \text{for circular tubes} \end{cases} \tag{35}$$

Here $A_w$ is the total wall area for heat conduction from the hot fluid to the cold fluid and $N_t$ is the total number of tubes in the exchanger.

## 2.6 C* and $N_{tu}$

Now first calculate the fluid heat capacity rate $C = Wc_p$ on each side, and then C* and $N_{tu}$ from their definitions

$$C^* = C_{min}/C_{max} \qquad (36)$$

$$N_{tu} = UA/C_{min} \qquad (37)$$

where $C_{min}$ is the minimum of $C_c$ and $C_h$.

## 2.7 Exchanger Effectiveness and Outlet Temperatures

Once $N_{tu}$ and C* are determined and the flow arrangement is known, the exchanger effectiveness is determined from the $\varepsilon$-$N_{tu}$ relationships of Table 3 or from the tabular data presented by Kays and London [3]. If the calculated effectiveness is greater than about 80%, the influence of longitudinal heat conduction must be evaluated. Longitudinal heat conduction reduces the exchanger effectiveness. A comprehensive review on the influence of longitudinal heat conduction on $\varepsilon$ has been provided by Chiou [15].

The outlet temperatures are subsequently calculated from the definition of $\varepsilon$.

$$t_{h,o} = t_{h,i} - \varepsilon(C_{min}/C_h)(t_{h,i} - t_{c,i}) \qquad (38)$$

$$t_{c,o} = t_{c,i} + \varepsilon(C_{min}/C_c)(t_{h,i} - t_{c,i}) \qquad (39)$$

If these outlet temperatures are significantly different from those assumed in the foregoing Section 2.2, the steps of Sections 2.3 through 2.7 are repeated. In most cases, one iteration with the outlet temperatures, Eqs. (38) and (39), from the first trial would be sufficient. The heat transfer rate is then computed from

$$q = \varepsilon C_{min}(t_{h,i} - t_{c,i}) \qquad (40)$$

## 2.8 Pressure Drops

Pressure drop on each side consists of core and manifold pressure drops. The core pressure drop is calculated from [3]

$$\frac{\Delta p}{p_i} = \frac{G^2}{2g_c} \frac{1}{p_i \rho_i} \left[ \underbrace{(1-\sigma^2 + K_c)}_{\text{Entrance effect}} + \underbrace{2\left(\frac{\rho_i}{\rho_o} - 1\right)}_{\text{Flow acceleration}} + \underbrace{f \frac{L}{r_h} \rho_i \left(\frac{1}{\rho}\right)_m}_{\text{Core friction}} - \underbrace{(1-\sigma^2 - K_e)\frac{\rho_i}{\rho_o}}_{\text{Exit effect}} \right] \qquad (41)$$

# COMPACT HEAT EXCHANGER DESIGN PROCEDURES

Table 3. $\varepsilon$-$N_{tu}$ relationships for common flow arrangements

| Flow arrangement | Equation |
|---|---|
| Counterflow [3] | $\varepsilon = \dfrac{1 - \exp[-N_{tu}(1 - C^*)]}{1 - C^* \exp[-N_{tu}(1 - C^*)]}$ |
| Parallel flow [3] | $\varepsilon = \dfrac{1 - \exp[-N_{tu}(1 + C^*)]}{1 + C^*}$ |
| Crossflow, both fluids unmixed [14] | $\varepsilon = 1 - \exp[-(1+C^*)N_{tu}][I_0(2N_{tu}\sqrt{C^*}) + \sqrt{C^*}\, I_1(2N_{tu}\sqrt{C^*}) - \dfrac{1 - C^*}{C^*} \sum_{n=2}^{\infty} C^{*n/2} I_n(2N_{tu}\sqrt{C^*})]$ |
| Crossflow, $C_{min}$ mixed, $C_{max}$ unmixed [3] | $\varepsilon = 1 - \exp[-\{1 - \exp(-N_{tu}C^*)\}/C^*]$ |
| Crossflow, $C_{max}$ mixed, $C_{min}$ unmixed [3] | $\varepsilon = \dfrac{1}{C^*}\left[1 - \exp\{-C^*[1 - \exp(-N_{tu})]\}\right]$ |
| Crossflow, both fluids mixed [3] | $\varepsilon = \left[\dfrac{1}{1 - \exp(-N_{tu})} + \dfrac{C^*}{1 - \exp(-N_{tu}C^*)} - \dfrac{1}{N_{tu}}\right]^{-1}$ |
| Parallel counterflow shell fluid mixed [3] | $\varepsilon = \dfrac{2}{(1 + C^*) + (1 + C^{*2})^{1/2}(1+e^{-\Gamma})/(1-e^{-\Gamma})}$ where $\Gamma = N_{tu}[1 + C^{*2}]^{1/2}$ |
| n-pass overall counterflow [3] | $\varepsilon = \dfrac{[(1 - \varepsilon_p C^*)/(1 - \varepsilon_p)]^n - 1}{[(1 - \varepsilon_p C^*)/(1 - \varepsilon_p)]^n - C^*}$ where $\varepsilon_p$ = effecitvness per pass = $\phi(N_{tu}/n, C^*)$ |
| n-pass overall parallel flow | $\varepsilon = \dfrac{1}{1 + C^*}[1 - \{1 - (1 + C^*)\varepsilon_p\}^n]$ where $\varepsilon_p$ = effectiveness per pass = $\phi(N_{tu}/n, C^*)$ |

Here $f$ is the Fanning friction factor, and is obtained from the known surface characteristics such as $f$ vs. Re or any similar correlation; $K_c$ and $K_e$ are abrupt contraction and expansion loss coefficients and are presented in

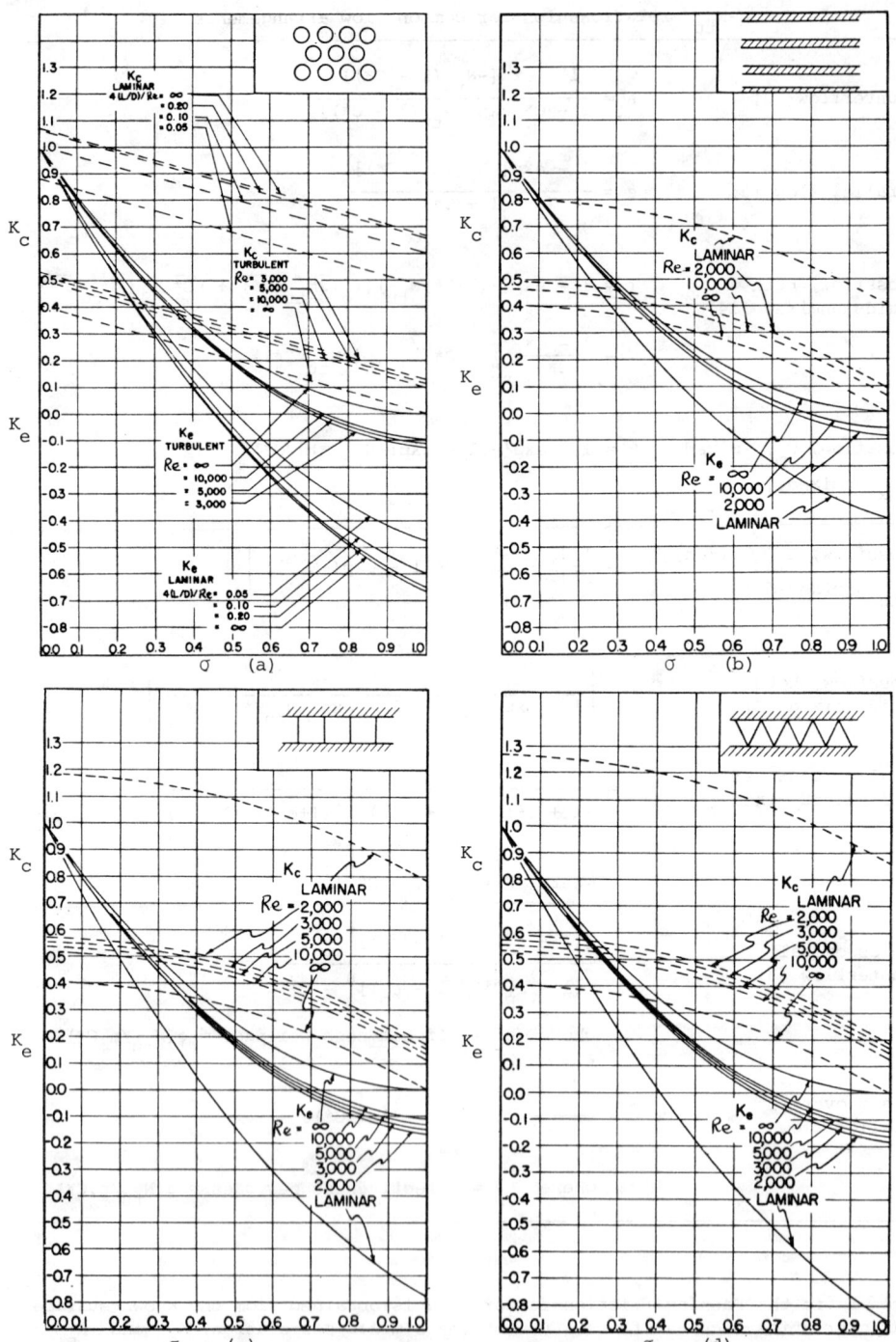

Fig. 4 Entrance and exit pressure loss coefficients: (a) circular tubes, (b) parallel plates, (c) square passages, and (d) triangular passages, from [3].

COMPACT HEAT EXCHANGER DESIGN PROCEDURES  509

Fig. 4 for four different entrance flow passage geometries; $(1/\rho)_m$ is the mean specific volume to be calculated from

$$\left(\frac{1}{\rho}\right)_m = \frac{1}{2}\left(\frac{1}{\rho_i} + \frac{1}{\rho_o}\right) \tag{42}$$

Generally, the core frictional pressure drop is the dominating term accounting for about 90% or more of $\Delta p$. The entrance and exit losses are important at low values of Re, $\sigma$ and L (short cores) and for gases. They are negligible for liquids. The magnitudes of $K_c$ and $K_e$ presented in Fig. 4 apply to long tubes for which the flow is fully developed at the exit. For partially developed flow, $K_c$ is lower and $K_e$ is higher than that for fully developed flow. For interrupted surfaces, the flow is never fully developed. For highly interrupted fin geometries, the flow is mixed very well and hence $K_c$ and $K_e$ for Re $\to \infty$ should represent a good approximation, because the entrance and exit losses for these surfaces are generally small compared to the core pressure drop.

The manifold pressure drops are strongly dependent upon the manifold geometry, flow velocity distribution, and working fluid. Some information on manifold $\Delta p$ is available in [8].

The solution procedure outlined in Sections 2.1-2.8 thus completes the details on solving the rating problem. As an illustration, an example of a rating problem is presented next.

2.9 An Example of a Rating Problem

A gas-to-air single-pass crossflow heat exchanger has overall dimensions of 0.300 m x 0.600 m x 0.898 m as shown in Fig. 5. Plain triangular fins and strip fins are employed on the gas and air sides, respectively. The geometrical properties and surface characteristics are provided in Fig. 6. Both fins and plates (parting sheets) are made from aluminum with k = 190 W/m°C. The plate thickness is 0.4 mm. The gas (process air) flows in at 1.2 m$^3$/s and 240°C. The makeup air on the other side flows in at 0.6 m$^3$/s and 4°C. Both fluids are at 110 kPa inlet pressure. Determine the heat transfer rate, outlet fluid temperatures and pressure drops on each side. Use the properties of air for the gas (process air).

Solution. We will follow the steps outlined in the text starting with the surface geometrical properties.

Surface geometrical properties. If the number of passages for the gas are $N_p$ and for the air are $N_p+1$,[†] then the noflow height is given by

$$L_3 = N_p b_g + (N_p+1)b_a + (2N_p+2)a$$

or

$$N_p = \frac{L_3 - b_a - 2a}{b_g + b_a + 2a} = \frac{898 - 6.36 - 2 \times 0.4}{6.35 + 6.35 + 2 \times 0.4} = 66$$

Here and in the following the subscripts g and a are used for the gas (process air) and the (makeup) air, respectively; b is the fin height (plate spacing) and a is the plate thickness. The frontal area on the gas and air

---
[†] This is a common practice to minimize the heat losses to ambient.

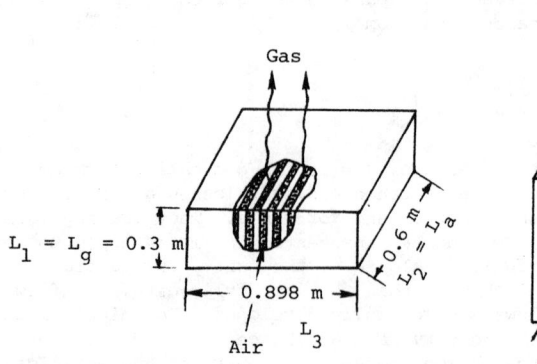
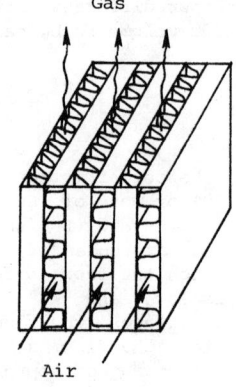

Fig. 5  A gas-to-air single-pass crossflow heat exchanger of the rating problem.

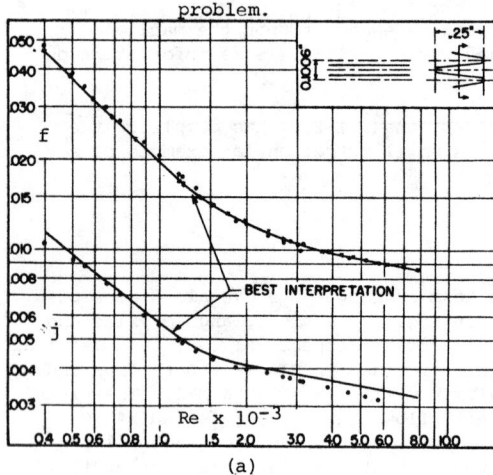

Gas Side Surface

Fin density = 0.782 per mm
Plate spacing, b = 6.35 mm
Hydraulic diameter, $D_h$ = 0.001875 m
Fin metal thickness, $\delta$ = 0.152 mm
Fin area/total area, $A_f/A$ = 0.849
Total heat transfer area/volume
between plates, $\beta$ = 1841 $m^2/m^3$

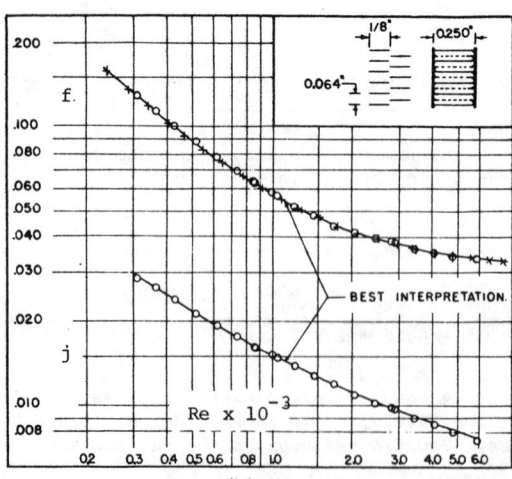

Air Side Surface

Fin density = 0.615 per mm
Plate spacing, b = 6.35 mm
Fin offset length, $\ell_s$ = 3.18 mm
Hydraulic diameter, $D_h$ = 0.002383 m
Fin metal thickness, $\delta$ = 0.152 mm
Fin area/total area, $A_f/A$ = 0.923
Total heat transfer area/volume
between plates, $\beta$ = 1548 $m^2/m^3$

Fig. 6  Surface basic characteristics for (a) plain plate-fin surface 19.86 [3], (b) rectangular strip offset fin surface 104 [16].

# COMPACT HEAT EXCHANGER DESIGN PROCEDURES

sides are

$$A_{fr,g} = L_2 L_3 = 0.6 \times 0.898 = 0.5388 \text{ m}^2$$

$$A_{fr,a} = L_1 L_3 = 0.3 \times 0.898 = 0.2694 \text{ m}^2$$

Heat exchanger volume between plates, on each side, is

$$V_{p,g} = L_1 L_2 (b_g N_p) = 0.300 \times 0.600 \times (6.35 \times 10^{-3}) 66 = 0.07544 \text{ m}^3$$

$$V_{p,a} = L_1 L_2 b_a (N_p + 1) = 0.300 \times 0.600 \times (6.35 \times 10^{-3}) \times 67 = 0.07658 \text{ m}^3$$

The heat transfer areas $A_g$ and $A_a$ are

$$A_g = \beta_g V_{p,g} = 1841 \times 0.07544 = 138.885 \text{ m}^2$$

$$A_a = \beta_a V_{p,a} = 1548 \times 0.07658 = 118.546 \text{ m}^2$$

The minimum free flow area is then calculated from the definition of the hydraulic diameter, $D_h = 4 A_o L / A$.

$$A_{o,g} = \frac{(D_h A)_g}{4 L_g} = \frac{0.001875 \times 138.885}{4 \times 0.300} = 0.2170 \text{ m}^2$$

$$A_{o,a} = \frac{(D_h A)_a}{4 L_a} = \frac{0.002383 \times 118.546}{4 \times 0.600} = 0.1177 \text{ m}^2$$

The ratio of minimum free flow area to frontal area is then

$$\sigma_g = A_{o,g} / A_{fr,g} = 0.2170 / 0.5388 = 0.403$$

$$\sigma_a = A_{o,a} / A_{fr,a} = 0.1177 / 0.2694 = 0.437$$

**Bulk mean temperatures and fluid properties.** In order to determine the bulk mean temperatures on each side, we need to calculate $C^*$. Since the flow rates are specified as volumetric at inlet temperatures, let us first calculate the gas and air density.

$$\rho_{g,i} = \frac{P_{i,g}}{\tilde{R} T_{g,i}} = \frac{(110 \times 10^3)}{287.041 \times (273.15 + 240.0)} = 0.7468 \text{ kg/m}^3$$

$$\rho_{a,i} = \frac{P_{i,a}}{\tilde{R} T_{a,i}} = \frac{(110 \times 10^3)}{287.041 \times (273.15 + 4.0)} = 1.3827 \text{ kg/m}^3$$

Hence the flow rates are

$$W_g = 1.2 \times 0.7468 = 0.8962 \text{ kg/s}$$

$$W_a = 0.6 \times 1.3827 = 0.8296 \text{ kg/s}$$

Since the air is at a lower temperature than the gas, its specific heat will also be lower. Hence, air will be the $C_{min}$ side. Now assume $\varepsilon = 0.75$ for the crossflow exchanger. Then using the definition of the exchanger effectiveness, Eqs. (39) and (38), we have

$$t_{a,o} = t_{a,i} + \varepsilon(t_{g,i} - t_{a,i}) = 4 + 0.75(240.0 - 4.0) = 181.0 \text{ °C}$$

$$t_{g,o} \simeq t_{g,i} - \varepsilon(W_a/W_g)(t_{g,i} - t_{a,i}) = 76.2 \text{ °C}$$

Note that we used $c_{p,a} \simeq c_{p,g}$ as a first approximation for determining $t_{g,o}$. Since $C^* \simeq W_a/W_g = 0.93$, we will use the arithmetic average temperature as the appropriate mean temperature on each side [Eqs. (17) and (18)].

$$t_{m,g} = (240.0 + 76.2)/2 = 158.1 \text{ °C}, \quad T_{m,g} = 431.25 \text{ °K}$$

$$t_{m,a} = (4.0 + 181.0)/2 = 92.5 \text{ °C}, \quad T_{m,a} = 365.65 \text{ °K}$$

In absence of information on the composition of the gas (process air), we will treat both the gas and makeup air as <u>dry</u> air. The properties of air are obtained from [4] as

|  | $\mu$, Pa·s | $c_p$, J/g °C | Pr | $Pr^{2/3}$ |
|---|---|---|---|---|
| Gas at 158.1 °C | $24.12 \times 10^{-6}$ | 1.022 | 0.687 | 0.779 |
| Air at 92.5 °C | $21.38 \times 10^{-6}$ | 1.013 | 0.694 | 0.784 |

<u>Mass velocities, Reynolds number, and j and f factors.</u>

$$G_g = (W/A_o)_g = 0.8962/0.2170 = 4.1300 \text{ kg/m}^2\text{s}$$

$$Re_g = (GD_h/\mu)_g = 4.1300 \times 0.001875/24.12 \times 10^{-6} = 321$$

$$G_a = (W/A_o)_a = 0.8296/0.1177 = 7.0484 \text{ kg/m}^2\text{s}$$

$$Re_a = (GD_h/\mu)_a = 7.0484 \times 0.002383/21.38 \times 10^{-6} = 786$$

From Fig. 6, we get the following information:

# COMPACT HEAT EXCHANGER DESIGN PROCEDURES

|     | Re  | j     | f     |
|-----|-----|-------|-------|
| Gas | 321 | 0.013 | 0.055 |
| Air | 786 | 0.017 | 0.065 |

Since Reynolds numbers indicate the flow as laminar on both gas and air sides, the correction to the j factor is unity because of $n = 0$ from Table 1. However, the correction to the f factor will not be unity since $m \neq 0$ from Table 1. We will determine this correction after calculating $T_w$.

Heat transfer coefficients and fin effectivenesses.

$$h_g = (jGc_p/Pr^{2/3})_g = 0.013 \times 4.1300 \times (1.022 \times 10^3)/0.779 = 70.44 \text{ W/m}^2 \text{ °C}$$

$$h_a = (jGc_p/Pr^{2/3})_a = 0.017 \times 7.0484 \times (1.013 \times 10^3)/0.784 = 154.82 \text{ W/m}^2 \text{ °C}$$

Now let us calculate the fin temperature effectivenesses for air and gas sides. Since the gas side has plain triangular fin,

$$m_g = (2h/k_f\delta)_g^{1/2} = [2 \times 70.44/(190 \times 0.152 \times 10^{-3}]^{1/2} = 69.84 \text{ m}^{-1}$$

Since the air side has strip fins, we will use Eq. (32) to take into account the strip edge exposed area,

$$m_a = \left[\frac{2h}{k_f\delta}\left(1 + \frac{\delta}{\ell_s}\right)\right]^{1/2} = \left[\frac{2 \times 154.82}{190 \times 0.152 \times 10^{-3}}\left(1 + \frac{0.152}{3.18}\right)\right]^{1/2} = 105.99 \text{ m}^{-1}$$

$$\ell_a = \ell_g \simeq b/2 - \delta = (6.35/2 - 0.152) \times 10^{-3} = 0.003023 \text{ m}$$

Thus,

$$\eta_{f,g} = \frac{\tanh(m\ell)_g}{(m\ell)_g} = \frac{\tanh(69.84 \times 0.003023)}{69.84 \times 0.003023} = 0.9854$$

$$\eta_{f,a} = \frac{\tanh(m\ell)_a}{(m\ell)_a} = \frac{\tanh(105.99 \times 0.003023)}{105.99 \times 0.003023} = 0.9671$$

The total surface temperature effectivenesses are

$$\eta_{o,g} = 1 - (1 - \eta_f)A_f/A = 1 - (1 - 0.9854) \times 0.849 = 0.9876$$

$$\eta_{o,a} = 1 - (1 - \eta_f)A_f/A = 1 - (1 - 0.9671) \times 0.923 = 0.9696$$

It should be pointed out that the fin conduction length $\ell$ for the end passages on the air side will be approximately b and not $(b/2 - \delta)$. This will result in a lower fin temperature effectiveness for the end passages. However, its influence will be smaller on the weighted average fin effectiveness considering all air passages. Hence, here we have neglected it. However, in a computer program, it can be easily incorporated.

<u>Wall resistance and overall conductance.</u> For the $R_w$ determination, the wall conduction area $A_w$ is

$$A_w = L_1 L_2 (2N_p+2) = 0.3 \times 0.6 \times 2 \times 67 = 24.12 \text{ m}^2$$

$$R_w = a/k_w A_w = 0.4 \times 10^{-3}/190 \times 24.12 = 8.728 \times 10^{-8} \text{ °C/W}$$

Since the influence of fouling is negligibly small for a gas-to-gas heat exchanger, we will neglect it. Then $1/UA$ from Eq. (34) is

$$\frac{1}{UA} = \frac{1}{0.9876 \times 70.44 \times 138.885} + 8.728 \times 10^{-8} + \frac{1}{0.9696 \times 154.82 \times 118.546}$$

$$= 1.0350 \times 10^{-4} + 8.728 \times 10^{-8} + 5.6195 \times 10^{-5} = 1.5978 \times 10^{-4}$$

$$UA = 6259 \text{ W/°C}$$

$N_{tu}$, <u>exchanger effectiveness and outlet temperatures.</u> In order to determine $N_{tu}$ and $\varepsilon$, first calculate $C_g$ and $C_a$.

$$C_g = (Wc_p)_g = 0.8962 \times (1.022 \times 10^3) = 915.9 \text{ W·s/°C}$$

$$C_a = (Wc_p)_a = 0.8296 \times (1.013 \times 10^3) = 840.4 \text{ W·s/°C}$$

$$C^* = \frac{C_{min}}{C_{max}} = \frac{C_a}{C_g} = \frac{840.4}{915.9} = 0.918$$

$$N_{tu} = UA/C_{min} = 6259/840.4 = 7.45$$

The effectiveness for the crossflow exchanger with both fluids unmixed, from the expression of Table 3, is

$$\varepsilon = 0.8235$$

This effectiveness is somewhat higher than normally used for a crossflow exchanger. Since this $\varepsilon > 80\%$, the decrease $\Delta\varepsilon$ if any in $\varepsilon$ due to longitudinal heat conduction should be evaluated. Longitudinal conduction parameters on the gas and air sides are

$$\lambda_h = \lambda_g = \left(\frac{k_w A_k}{LC}\right)_g = \frac{190 \times 0.03168}{0.3 \times 915.9} = 0.0219$$

# COMPACT HEAT EXCHANGER DESIGN PROCEDURES

$$\lambda_c = \lambda_a = \left(\frac{k_w A_k}{LC}\right)_a = \frac{190 \times 0.01608}{0.6 \times 840.4} = 0.00606$$

where the conduction cross-section area

$$A_{k,g} = 2N_p L_a a = 132 \times 0.6 \times 0.4 \times 10^{-3} = 0.03168 \text{ m}^2$$

$$A_{k,a} = (2N_p+2)L_g a = 134 \times 0.3 \times 0.4 \times 10^{-3} = 0.01608 \text{ m}^2$$

Other parameters needed for determining $\Delta\varepsilon$ are:

$$\frac{(\eta_o hA)_h}{(\eta_o hA)_c} = \frac{1.0350 \times 10^{-4}}{5.6195 \times 10^{-5}} = 1.84 \qquad \frac{\lambda_c}{\lambda_h} = \frac{0.00606}{0.0219} = 0.277$$

$$N_{tu} = 7.45 \qquad\qquad C_c/C_h = 0.918$$

From the interpolation of tabular results of Choiu [15], it is found that $\Delta\varepsilon/\varepsilon \approx 0.02$. Thus $\Delta\varepsilon = 0.0165$ and actual exchanger effectiveness is

$$\varepsilon_{actual} = 0.8235 - 0.0165 = 0.807$$

The heat transfer rate q is then

$$q = \varepsilon(t_{g,i} - t_{a,i})C_{min} = 0.807(240-4)840.4 = 160.1 \times 10^3 \text{ W}$$

The outlet temperatures are then

$$t_{g,o} = t_{g,i} - q/C_g = 240 - 160.1 \times 10^3/915.9 = 65.2 \text{ °C} = 338.35 \text{ °K}$$

$$t_{a,o} = t_{a,i} + q/C_a = 4 + 160.1 \times 10^3/840.4 = 194.5 \text{ °C} = 467.65 \text{ °K}$$

Since these outlet temperatures are different from those assumed for the initial determination of the fluid properties, a second iteration was carried out with fluid properties evaluated at the new average temperatures. It was found that the new outlet temperatures after the second iteration were the same as those calculated above.

<u>Pressure drops.</u> We will use Eq. (41) to compute the pressure drop on each side. The densities are evaluated using the perfect gas equation of state.

|     | $T_i$, °K | $T_o$, °K | $\rho_i$, kg/m$^3$ | $\rho_o$, kg/m$^3$ | $\rho_m$, kg/m$^3$ |
|-----|-----------|-----------|--------------------|--------------------|--------------------|
| Gas | 513.15    | 338.35    | 0.7468             | 1.1326             | 0.9001             |
| Air | 277.15    | 467.65    | 1.3827             | 0.8194             | 1.0290             |

Note that we have also considered the outlet pressures as 110 kPa since the pressure drop across the core is usually very small and hence is neglected in the first trial. The mean density in the last column is the harmonic mean value from Eq. (42).

Now let us determine $K_c$ and $K_e$. Since the gas side fin has Re = 321 and $L/D_h$ = 0.3/0.001875 = 160, the flow is fully developed laminar. For triangular passages, from Fig. 4d, for $\sigma_g$ = 0.40,

$$K_c = 1.2 \qquad K_e = 0.02$$

On the air side, strip fins are used. Because of the frequent boundary layer interruptions, the flow is well mixed and is treated as having Re = ∞. The aspect ratio of the rectangular passage, height/width = 6.35/(1/0.615 - 0.152) = 4.3. Since Re = ∞ curves for parallel plate and square passage geometries of Fig. 4 are identical, we could determine $K_c$ and $K_e$ from either geometry for $\sigma_a$ = 0.44 as

$$K_c = 0.33 \qquad K_e = 0.31$$

Before we compute the pressure drop, we need to correct the values of the isothermal friction factors by the method of Section 2.4 to account for the temperature-dependent properties. A review of Eq. (27) indicates that we need to calculate the fluid bulk mean temperatures and the wall temperature. The mean temperatures on the gas and air sides, based on the latest outlet temperatures, are

$$t_{g,m} = (240.0+65.2)/2 = 152.60\ °C, \qquad T_{g,m} = 425.75\ °K$$

$$t_{a,m} = (4.0+194.5)/2 = 99.25\ °C, \qquad T_{a,m} = 372.40\ °K$$

The thermal resistances on the gas and air sides are

$$R_g = \frac{1}{(\eta_o hA)_g} = 1.0350 \times 10^{-4}\ °C/W \qquad R_a = \frac{1}{(\eta_o hA)_a} = 5.6195 \times 10^{-5}\ °C/W$$

or
$$R_g/R_a = 1.84$$

If we neglect the wall resistance,

$$q = \frac{t_{g,m} - t_w}{R_g} = \frac{t_w - t_{a,m}}{R_a}$$

so that

$$t_w = \frac{t_{g,m} + (R_g/R_a)t_{a,m}}{1 + (R_g/R_a)} = \frac{152.60 + 1.84 \times 99.25}{1 + 1.84} = 118.04\ °C = 391.19\ °K$$

# COMPACT HEAT EXCHANGER DESIGN PROCEDURES

Since the gas is being cooled, using Eq. (27) and the exponent from Table 1,

$$f = f_{cp}(T_w/T_m)^m = 0.055(391.19/425.75)^{0.81} = 0.0514$$

Since the air is being heated, using Eq. (27) and the exponent from Table 1,

$$f = f_{cp}(T_w/T_m)^m = 0.065(391.19/372.40)^{1.00} = 0.0683$$

Now let us calculate the pressure drops using Eq. (41).

$$\left(\frac{\Delta p}{p_i}\right)_g = \frac{(4.1300)^2}{2\times110\times10^3\times0.7468}\left[(1-0.403^2+1.2)+2\left(\frac{0.7468}{1.1326}-1\right)\right.$$

$$\left.+\frac{0.0514\times0.3\times0.7468}{(0.001875/4)\times0.9001} - (1-0.403^2+0.02)\frac{0.7468}{1.1326}\right]$$

$$= 0.1038\times10^{-3}(2.0376-0.6813+27.2933-0.5655)$$

$$= 0.00292$$

Hence $(\Delta p)_g = 110\times0.00292 = 0.321$ kPa

$$\left(\frac{\Delta p}{p_i}\right)_a = \frac{(7.0484)^2}{2\times110\times10^3\times1.3827}\left[(1-0.437^2+0.33) + 2\left(\frac{1.3827}{0.8194}-1\right)\right.$$

$$\left.+\frac{0.0683\times0.6\times1.3827}{(0.002383/4)\times1.0290} - (1-0.437^2+0.31)\frac{1.3827}{0.8194}\right]$$

$$= 0.1633\times10^{-3}(1.1390+1.3749+92.4316-1.8883)$$

$$= 0.01520$$

Hence $(\Delta p)_a = 110\times0.01520 = 1.672$

Note that the pressure drop on the gas and air sides is 0.3% and 1.5% respectively. Our assumption of $p_o \simeq p_i$ to calculate $\rho_o$ is good. However, if the pressure drop is too high on either side, one more iteration, considering $p_o = p_i - \Delta p$, is recommended.

3. THE SIZING PROBLEM

The sizing or design problem is more difficult. Many early decisions to decide the construction type and basic geometries on each side are based on the experience, rules-of-thumb, engineering judgements, operating conditions, maintenance, manufacturing capability, and the expected life of the exchanger. We will first present some guidelines on the selection of construction types and flow arrangements.† We will then consider that the construction type and basic surface geometries on each side have been selected for the specified sizing problem. The problem then reduces to the determination of the core dimensions for the specified heat transfer and pressure drop performance. One could, of course, reduce this problem to the rating problem by tentatively specifying the dimensions, then predict the performance for comparison with that specified. This type of solution search is uncomfortable; not as bad as the search for a "needle in a haystack," but even with a computer program as an aid for the successive approximations involved, it has its aggravations. However, by reforming the surface characteristics input to include j/f versus Reynolds number, in addition to the separate j and f versus Reynolds number characteristics, a very substantial reduction of effort is involved. This coupling of heat transfer and flow friction will be made in the derivation of the core mass velocity equation that has been proposed by Kays and London [3]. Once the core mass velocity is determined, the solution to the sizing problem is carried out in a manner similar to the rating problem discussed in the preceding section.

3.1 Selection of Construction Type

A compact heat exchanger on the gas side requires a significantly greater amount of surface area for a specified heat transfer rate than for a liquid as a working fluid. This is because the heat transfer coefficient for the gas is 1/10 to 1/100 of that of a liquid. The increase in surface area is achieved by employing surfaces which have high heat transfer surface area density $\beta$. For example, fins are employed in an extended surface heat exchanger, or a small hydraulic diameter surface is employed in a regenerator or small diameter tubes are used in a tubular heat exchanger. The fourth basic construction type of Fig. 7, the plate exchanger, is generally not used as a compact heat exchanger

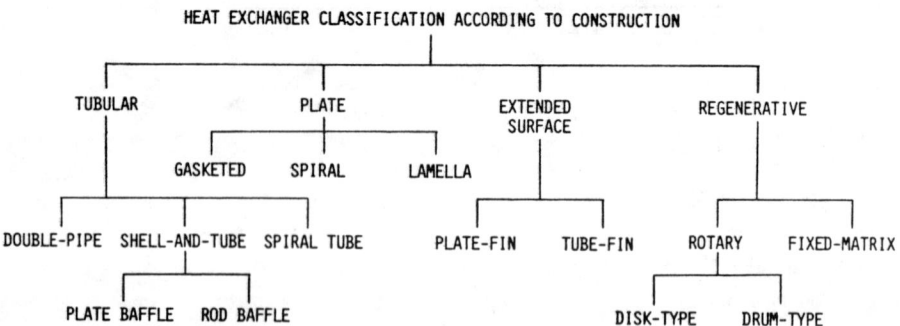

Fig. 7 Heat exchanger classification according to the construction type [6].

---

†The selection of material, surface geometries, fabrication methods, etc., is beyond the scope of this paper.

application because it may produce excessive pressure drop. The fluid pumping power is generally significant and a controlling factor in designing the gas side of a compact heat exchanger.

Some of the basic criteria for selecting a particular construction type are the cost, operating pressure and temperature, fouling, fluid contamination, and ruggedness. From the viewpoint of the cost per unit heat transfer surface area, the tubular exchanger in general is the most expensive followed by extended surface exchanger, and the regenerative exchanger in general is the least expensive. Extended surface and regenerative exchangers are generally designed for low pressure applications, with operating pressures limited to about 1,000 kPa gage (150 psig).[†] In this low pressure application, the metal extended surface exchangers are considered for maximum operating temperatures up to about 540-650 °C (1000-1200 °F); for higher temperatures up to 2000 °C (3600 °F), ceramic regenerative exchangers are considered. Fouling is generally not so severe a problem with gases as with liquids. Extended surface and small hydraulic diameter passages are hence used on the gas side of a compact heat exchanger. This exchanger is generally not designed for the application involving heavy fouling. Regenerators have self-cleaning characteristics because the hot and cold gases flow in the opposite directions periodically through the same passage. And as a result, compact regenerators do not have fouling problem and usually have very small hydraulic diameter passages. Carryover and bypass leakages from the hot fluid to the cold fluid (or vice versa) occur in the regenerator. Where this leakage and subsequent fluid contamination is not permissible, the regenerator is not used.

3.2  Selection of Flow Arrangement

Crossflow is the most common flow arrangement used for a compact heat exchanger. This is because it greatly simplifies the header design at the entrance and exit of each fluid. If the desired heat exchanger effectiveness is high (say greater than 80%), the size of a crossflow unit may become excessive. In such a case, an overall cross-counterflow multipass unit or a counterflow unit may be preferred. However, there are manufacturing difficulties associated with a true counterflow arrangement as it is necessary to separate the fluids at each end and the headering problem is more complex. Some header configurations are presented in [6]. Multipassing retains the header and ducting advantages of the simple crossflow heat exchanger, while it is possible to approach the thermal performance for counterflow. For high temperature applications, different materials may be used in different passes to reduce material cost and increase the life of the heat exchanger. The structural temperature differences are reduced significantly by multipassing relative to a single-pass crossflow design. The counterflow unit has the least structural temperature differences. The parallel flow arrangement having the highest structural temperature differences is seldom used in a compact heat exchanger because of its lowest effectiveness.

3.3  Core Mass Velocity Equation

In this subsection, an equation is derived to determine the core mass velocity $G$ on each side of a two-fluid exchanger such that the resultant core design will meet the specified heat transfer and pressure drop on each side.

The core pressure drop relationship is presented in Eq. (41) in which the

---

[†]Some cryogenic plate-fin heat exchangers have been designed for the operating pressures up to about 8300 kPa gage (1200 psig).

terms inside the bracket [ ] are identified as shown. The coefficient of this bracketed term on the right-hand side is the ratio of core inlet velocity head to pressure head.

$$\frac{G^2}{2g_c} \frac{1}{p_i \rho_i} = \frac{(u_i^2/2g_c)}{(p_i/\rho_i)} = \frac{\text{Velocity head}}{\text{Pressure head}} \qquad (43)$$

The dominant term in Eq. (41) is the core friction. The entrance and exit effects are generally relatively small and are of the opposite sign so their elimination is usually warranted in a "first approximation" type of calculation. The flow acceleration term is also relatively small in most heat exchangers, being generally less than 10% of the core friction term. The flow acceleration term will be positive for fluid heating ($\rho_i/\rho_o > 1$) and negative for fluid cooling ($\rho_i/\rho_o < 1$). It will also be neglected for the present purposes. With these approximations and $L/r_h = A/A_o$, Eq. (41) may be rearranged to the following form.

$$\frac{G^2}{2g_c p_i \rho_i} \approx \left(\frac{\Delta p}{p_i}\right)\left(\frac{A_o}{A}\right)\left(\frac{1}{\rho_i (1/\rho)_m}\right)\frac{1}{f} \qquad (44)$$

Now let us derive a simplified equation for heat transfer. From the required exchanger effectiveness $\varepsilon$ and known heat capacity rates on each side (known C*), the overall $N_{tu}$ is determined for the chosen flow arrangement. This $N_{tu}$ is related to $N_{tu,h}$ and $N_{tu,c}$ and the wall resistance by

$$\frac{1}{N_{tu}} = \frac{1}{N_{tu,h}(C_h/C_{min})} + C_{min} R_w + \frac{1}{N_{tu,c}(C_c/C_{min})} \qquad (45)$$

If the fouling is important, the corresponding resistances are added in Eq. (45). Generally, in a compact heat exchanger, the wall resistance term is small enough to neglect in a first approximation calculation. Hence, $N_{tu}$ on one side of interest (either hot or cold), designated as $N_{tu,1}$, may be estimated from the known overall $N_{tu}$ (without the subscript 1 or 2), if the designer is in a position to estimate the magnitudes of the component resistances on the hot and cold sides (as well as fouling, if any) relative to the overall resistance 1/UA. Lacking this information, if both fluids are gases, one could start with the estimate that the design is "balanced" by a selection of the hot- and cold-side surfaces so that $R_h \approx R_c = R_o/2$. Then

$$N_{tu,1} \approx 2 N_{tu} \qquad (46)$$

Alternatively, if we have liquid on one side and gas on the other, one might estimate

$$N_{tu,\text{gas side}} = 1.1 \, N_{tu} \qquad (47)$$

This argument is presented to support the claim that a reasonably knowledgeable designer is indeed in a position to estimate $N_{tu,1}$ from $N_{tu}$ required to achieve the specified performance. This $N_{tu,1}$ is related to Colburn factor j on side 1 by

$$N_{tu,1} = \left[\frac{\eta_o hA}{Wc_p}\right]_1 = \left[\eta_o \frac{h}{Gc_p}\frac{A}{A_o}\right]_1 = \left[\eta_o \, j \, Pr^{-2/3}\left(\frac{A}{A_o}\right)\right]_1 \qquad (48)$$

Eliminating $(A/A_o)$ from Eqs. (44) and (48), we get for one side

$$\frac{G^2}{2g_c P_i \rho_i} \simeq \left\{\frac{\Delta p/p_i}{N_{tu}}\right\}_{one\ side} \left\{\frac{Pr^{-2/3}}{\rho_i (1/\rho)_m}\right\} \eta_o \left(\frac{j}{f}\right) \qquad (49)$$

The core mass velocity G from this equation reduces to

$$G = \left[\left\{\frac{2g_c P_i \eta_o}{(1/\rho)_m Pr^{2/3}}\right\}\left\{\frac{\Delta p/p_i}{N_{tu}}\right\}_{one\ side} \frac{j}{f}\right]^{1/2} \qquad (50)$$

The feature that makes this equation so useful is that the ratio $j/f$ is a relatively flat function of the Reynolds number. Thus one can readily estimate an accurate magnitude of $j/f$ based on a "ball park" estimate of Re. A variety of compact surfaces are presented in Fig. 8 to show a relatively small variation of $j/f$ for a twofold range of the Reynolds number. Also notice that there is only a sixfold range of the $j/f$ for markedly different surface geometries. And the scale of the surface $(D_h)$ has only a minor influence on the required flow area as $D_h$ is involved only in the Reynolds number.

If there are no fins, $\eta_o = 1$ in Eq. (50). For a "good design," the fin geometry is chosen such that $\eta_o$ is in the range 70-90%. Therefore $\eta_o = 0.80$ is suggested as a first approximation to determine G from Eq. (50).

3.4 Solution to the Sizing Problem for a Counterflow or $C^* \simeq 0$ Exchanger

Now we will outline a detailed procedure for arriving at core dimensions for the counterflow exchanger or $C^* = 0$ exchanger for specified heat transfer and pressure drop. In a single-pass counterflow heat exchanger of any construction, if the core dimensions on one side are fixed, the core dimensions for the other side (except for the passage height) are also fixed. Therefore the design problem for this case is solved for the side that has more stringent $\Delta p/p$ specification. For the $C^* \simeq 0$ case, i.e., gas-to-liquid or phase changing fluid exchanger, the thermal resistance is primarily on the gas side and the pressure

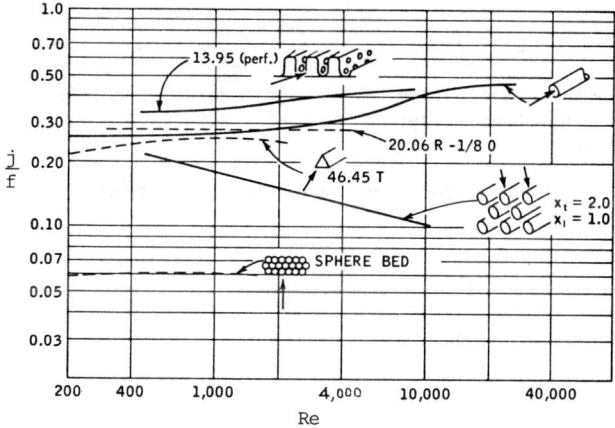

Fig. 8 Ratio of $j/f$ for some heat exchanger surfaces [3].

drop is more critical on the gas side. As a result, the core dimensions arrived are based on the gas side $\Delta p$ and $N_{tu,1}$. The dimensions on the other side are then chosen such that it meets the specified $\Delta p$. Thus either for counterflow or for $C^* = 0$ exchanger, the core dimensions are arrived for the side having the most stringent $\Delta p$. The following is a step-by-step procedure for the solution.

1. Determine the fluid outlet temperatures from the specified exchanger effectiveness. Obtain the fluid mean temperature on each side by the method of Section 2.2 and evaluate the fluid physical properties $\rho_i$, $\rho_o$, $(1/\rho)_m$, $c_p$, $\mu$, $k$, and $Pr$.

2. Calculate $C^*$ and $\varepsilon$, and determine $N_{tu}$ for the exchanger for the selected flow arrangement. The influence of longitudinal heat conduction, if any, is ignored in the first approximate calculation.

3. Determine $N_{tu}$ on one side by the approximations discussed with Eqs. (46) and (47).

4. For the selected surface on the critical side, plot $j/f$ vs. $Re$ curve and obtain a "ball park" value of $j/f$. If fins are employed on this side, also assume $\eta_o = 0.80$.

5. Evaluate $G$ from Eq. (50) using the information from Steps 1-4 and input value of $\Delta p/p$.

6. Calculate Reynolds number $Re$, and determine $j$ and $f$ for this $Re$.

7. Compute $h$, $\eta_f$ and $\eta_o$. In order to determine $U_1$, either lump the fouling resistances with the convective film resistances on the respective sides or ignore them. Similarly, either divide and lump the wall thermal resistance with the respective hot and cold side resistances or ignore it. $U_1$ from Eq. (34), after replacing the subscript $h$ and $c$ with $1$ and $2$ is

$$\frac{1}{U_1} = \frac{1}{(\eta_o h)_1} + \frac{(A_1/A_2)}{(\eta_o h)_2} \qquad (51)$$

where
$$\frac{A_1}{A_2} = \frac{\alpha_1}{\alpha_2} \qquad (52)$$

Here $\alpha_1$ and $\alpha_2$ (the ratios of heat transfer area on the respective side divided by the total exchanger volume) are known from the input geometry specifications.

8. Now calculate the core dimensions. Determine $A_1$ from $N_{tu}$ for this $U_1$ and known $C_{min}$.

$$A_1 = N_{tu} C_{min}/U_1 \qquad (53)$$

and $A_o$ from known $W$ and $G$

$$A_{o,1} = (W/G)_1 \qquad (54)$$

so that
$$A_{fr,1} = A_{o,1}/\sigma_1 \qquad (55)$$

where $\sigma_1$ is known as computed from Eq. (4). Finally evaluate the core length $L_1$ from known $A_{o,1}$, $A_1$ and $D_{h,1}$ as

$$L_1 = (D_h A/4A_o)_1 \qquad (56)$$

Once the frontal area is calculated, the core dimensions are chosen in reasonable size such that it will meet the pressure drop on the other side too.

9. Now correct the f factor for the variable property effects as mentioned in Section 2.4 and calculate $\Delta p/p$ from Eq. (41) or an appropriate equation.

10. If the calculated value of $(\Delta p/p)$ is within the input specification, the approximate solution to the sizing problem is finished. Finer refinements in the core dimensions may be carried out at this time.

    If the calculated value of $\Delta p/p$ is larger than the input value, compute an improved value of G from Eq. (41) using the specified $\Delta p/p$ and the values of f and L from the preceding step. With this new value of G, repeat Steps 6-10 until both heat transfer and pressure drops are met as specified.

    If the influence of longitudinal heat conduction is important, the longitudinal conduction parameter and other appropriate dimensionless groups are calculated based on the core geometry from the preceding iteration and input operating conditions. Subsequently, a new value of $N_{tu}$, accounting for longitudinal conduction, is calculated. This new value of $N_{tu}$ is then used in Step 8.

## 3.5 Solution to the Sizing Problem for a Crossflow Exchanger

For a crossflow exchanger, determining the core dimensions on one side ($A_{fr}$ and L) does not fix the dimensions on the other side. In such a case, the design problem is solved simultaneously on both sides. The solution procedure follows closely with that of Section 3.4 and is outlined next through detailed steps.

1. Determine G on each side by following Steps 1-5 of Section 3.4.

2. Follow Steps 6-7 of Section 3.4 and compute for each side the following: Re, j, f, h, $\eta_f$, and $\eta_o$. Subsequently, compute $U_1$ using Eq. (51).

3. The core dimensions are then calculated as follows: $A_1$, $A_{o,1}$, $A_{fr,1}$, and $L_1$ using Eqs. (53)-(56). Note that here $A_{fr,1} = L_2 L_3$. Similarly calculate $A_2$, $A_{o,2}$, $A_{fr,2}$ and $L_2$ using Eqs. (53)-(56) by replacing the subscript 1 to 2. Note that here $A_{fr,2} = L_1 L_3$. Thus the noflow (or stack) height $L_3$ can be determined from either $A_{fr,1}$ or $A_{fr,2}$ and known $L_2$ or $L_1$. It can be rigorously shown that $L_3$ calculated either from $A_{fr,1}$ or from $A_{fr,2}$ will be identical.

4. Now correct the f factor for the variable property effects as mentioned in Section 2.4 and calculate $\Delta p/p$ from Eq. (41) on each side.

5. If the calculated values of $(\Delta p/p)$ is within input specification on each side, the approximate solution to the sizing problem is finished. Finer refinements in the core dimensions may be carried out at this time.

    If the calculated value of $\Delta p/p$ is larger on one or both sides, compute an improved value of $G$ on each side using the specified $\Delta p/p$ and the values of $f$ and $L$ from the preceding step. With these new values of $G$, repeat Steps 2-5 until both heat transfer and pressure drops are met as specified.

If the influence of longitudinal heat conduction is important, the longitudinal conduction parameters and other appropriate dimensionless groups are calculated based on the core geometry from the preceding iteration and input operating conditions. Subsequently, a new value of $N_{tu}$, accounting for longitudinal conduction, is calculated. This new value of $N_{tu}$ is then used in Eq. (53).

Since we have not imposed any restrictions on the core dimensions, a unique set of core dimensions of a crossflow exchanger will be obtained by the aforementioned procedure that will meet heat transfer and pressure drops on each side.

3.6 An Example of a Sizing Problem

We will consider the same heat exchanger as that of the rating problem in Section 2.9. Design a gas-to-air single-pass crossflow heat exchanger operating at $\varepsilon = 0.823$ having gas and air inlet temperatures as 240 °C and 4 °C respectively, and gas and air flow rates as 0.8962 and 0.8296 kg/s, respectively. The gas and air side pressure drops are limited to 0.32 and 1.67 kPa, respectively. The inlet pressures on both sides are 110 kPa. The surfaces on the gas and air sides are selected as those of Fig. 6. Both fins and plates (parting sheets) are made from aluminum having $k = 190$ W/m °C. The plate thickness is 0.4 mm. Determine the core dimensions of this exchanger. Neglect the effect of longitudinal heat conduction and treat the gas as air for fluid property evaluation.

<u>Solution.</u> We will follow the steps outlined in Section 3.5 for a solution.

<u>Outlet temperatures.</u> In order to determine outlet temperatures from the known $\varepsilon$, we first need to know the $C_{min}$ side. Since $W_a < W_g$ and the specific heat of air (being the cold fluid) will be lower than that of the gas, the $C_{min}$ side will be the air side. Using Eqs. (39) and (38), we get

$$t_{a,o} = t_{a,i} + \varepsilon(t_{g,i} - t_{a,i}) = 4 + 0.823(240 - 4) = 198.2 \text{ °C}$$

$$t_{g,o} \simeq t_{g,i} - \varepsilon(W_a/W_g)(t_{g,i} - t_{a,i}) = 240 - 0.823(0.8296/0.8962) \times 236 = 60.2 \text{°C}$$

Since we do not know the specific heats, we have considered $c_{p,g} \simeq c_{p,a}$. This $t_{g,o}$ will be refined after we determine the fluid properties.

<u>Fluid properties.</u> Since $C^* \simeq W_a/W_g = 0.93$, the fluid properties will be evaluated at the arithmetic mean temperature.

$$t_{g,m} = (240 + 60.2)/2 = 150.1 \text{ °C} \qquad T_{g,m} = 423.25 \text{°K}$$

$$t_{a,m} = (4 + 198.2)/2 = 101.1 \text{ °C} \qquad T_{a,m} = 374.25 \text{°K}$$

$c_p$ of gas and air at these temperatures is 1.021 and 1.014 J/g °C, respectively. Hence, the correct $t_{g,o}$ will be

$$t_{g,o} = 240 - 0.823(1.014 \times 0.8296/1.021 \times 0.8962) \times 236 = 61.4 \text{ °C}$$

COMPACT HEAT EXCHANGER DESIGN PROCEDURES            525

Thus        $t_{g,m} = (240 + 61.4)/2 = 150.7\,°C$                $T_{g,m} = 423.85\,°K$

The specific heat of air at 423.85 °K is again 1.021 J/g °C and hence there is no further need of iterations. The air properties at $t_{g,m} = 150.7\,°C$ and $t_{a,m} = 101.1\,°C$ from [4] are as follows.

|  | $\mu$, Pa·s | $c_p$, J/g °C | Pr | $Pr^{2/3}$ |
|---|---|---|---|---|
| Gas at 150.7 °C | $23.82 \times 10^{-6}$ | 1.021 | 0.688 | 0.779 |
| Air at 101.1 °C | $21.75 \times 10^{-6}$ | 1.014 | 0.692 | 0.782 |

The inlet and outlet gas densities are evaluated at 110 and 109.7 Pa respectively. The inlet and outlet air densities are evaluated at 110 and 108.3 Pa respectively. The mean densities are evaluated using Eq. (42).

|  | $T_i$, °K | $T_o$, °K | $\rho_i$, kg/m$^3$ | $\rho_o$, kg/m$^3$ | $\rho_m$, kg/m$^3$ |
|---|---|---|---|---|---|
| Gas | 513.15 | 334.55 | 0.7468 | 1.1424 | 0.9032 |
| Air | 277.15 | 471.35 | 1.3827 | 0.8005 | 1.0140 |

**$C^*$ and $N_{tu}$.** From the foregoing values of $c_p$ and given flow rates

$$C_g = 0.8962 \times 1.021 \times 10^3 = 915.0\ W/°C$$

$$C_a = 0.8296 \times 1.014 \times 10^3 = 841.2\ W/°C$$

and          $C^* = C_{min}/C_{max} = 841.2/915.0 = 0.919$

In absence of longitudinal heat conduction, $N_{tu}$ for a crossflow exchanger with both fluids unmixed for $\varepsilon = 0.823$ and $C^* = 0.919$, from the expression of Table 3, is

$$N_{tu} = 7.45$$

Now we need to estimate $N_{tu,g}$ and $N_{tu,a}$ from the overall $N_{tu}$. The better the initial estimate, the closer will be the value of G as a first estimate. For a gas-to-gas heat exchanger, a good estimate would be equal resistances on each side considering a balanced design. This would correspond to Eq. (46).

$$N_{tu,a} = 2\ N_{tu} = 14.90$$

Then neglecting the wall thermal resistance, we get from Eq. (45),

$$N_{tu,g} = 2\ C^*\ N_{tu} = 2 \times 0.919 \times 7.45 = 13.69$$

**Core Mass Velocities.** In order to determine G from Eq. (50), we need to estimate the values of j/f and $\eta_o$. Since j and f versus Re characteristics are specified for the surfaces on the gas and air sides, j/f versus Re curves are constructed as shown in Fig. 9. However, we don't know Re yet. Hence, a "ball park" value of j/f is taken for each surface from this figure as

$$(j/f)_g \approx 0.3 \qquad\qquad (j/f)_a \approx 0.25$$

As a first approximation, we will assume $\eta_o$ on both the gas and the air side as 0.80. Now substituting all the values on the right-hand side of Eq. (50),

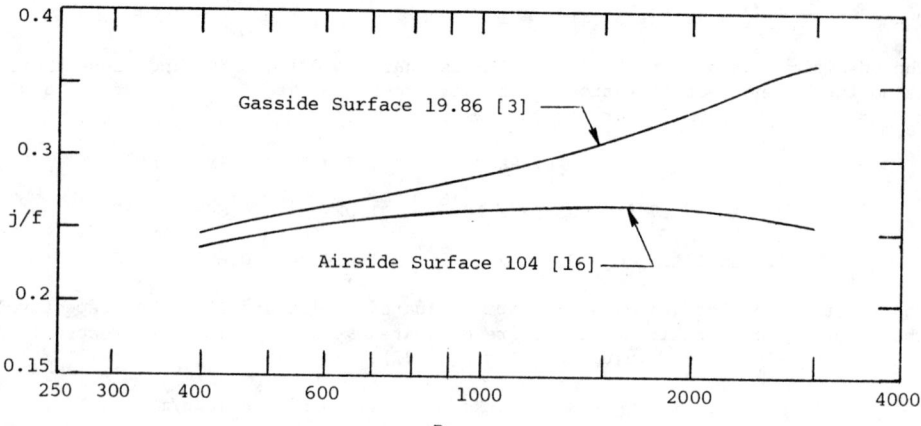

Fig. 9 j/f versus Re characteristics of surfaces of Fig. 6.

we get

$$G_g = \left[\frac{2 \times 0.8 \times (0.32 \times 10^3) \times 0.9032 \times 0.3}{0.779 \times 13.69}\right]^{1/2} = 3.61 \text{ kg/m}^2\text{s}$$

$$G_a = \left[\frac{2 \times 0.8 \times (1.67 \times 10^3) \times 1.014 \times 0.25}{0.782 \times 14.90}\right]^{1/2} = 7.62 \text{ kg/m}^2\text{s}$$

**Reynolds numbers, and j and f factors.** Compute the Reynolds number on each side from its definition as

$$Re_g = (GD_h/\mu)_g = 3.61 \times 0.001875/23.82 \times 10^{-6} = 284$$

$$Re_a = (GD_h/\mu)_a = 7.62 \times 0.002383/21.75 \times 10^{-6} = 835$$

From Fig. 6, determine j and f factors for these Reynolds number.

|     | Re  | j     | f     |
| --- | --- | ----- | ----- |
| Gas | 284 | 0.015 | 0.062 |
| Air | 835 | 0.016 | 0.063 |

Since Reynolds numbers indicate the flow as laminar on both gas and air sides, the correction to the j factor is unity because n = 0 from Table 1.

**Heat transfer coefficients, fin effectivenesses and overall U.**

$$h_g = (jGc_p/Pr^{2/3})_g = 0.015 \times 3.61 \times (1.021 \times 10^3)/0.779 = 70.97 \text{ W/m}^2 \text{ °C}$$

$$h_a = (jGc_p/Pr^{2/3})_a = 0.016 \times 7.62 \times (1.014 \times 10^3)/0.782 = 158.09 \text{ W/m}^2 \text{ °C}$$

Let us calculate $m_g$ and $m_a$ in order to calculate the fin effectiveness on each side. Since the gas side has plain triangular fins,

$$m_g = (2h/k_f\delta)_g^{1/2} = (2 \times 70.97/190 \times 0.152 \times 10^{-3})^{1/2} = 70.11 \text{ m}^{-1}$$

# COMPACT HEAT EXCHANGER DESIGN PROCEDURES

Since the air side has strip fins, we will use Eq. (32) to take into account the strip edge exposed area.

$$m_a = \left[\frac{2h}{k_f \delta}\left(1 + \frac{\delta}{\ell_s}\right)\right]^{1/2} = \left[\frac{2 \times 158.09}{190 \times 0.152 \times 10^{-3}}\left(1 + \frac{0.152}{3.18}\right)\right]^{1/2} = 107.10 \text{ m}^{-1}$$

$$\ell_a = \ell_g \simeq b/2 - \delta = (6.35/2 - 0.152) \times 10^{-3} = 0.003023 \text{ m}$$

Thus 
$$\eta_{f,g} = \frac{\tanh(m\ell)_g}{(m\ell)_g} = \frac{\tanh(70.11 \times 0.003023)}{70.11 \times 0.003023} = 0.9853$$

$$\eta_{f,a} = \frac{\tanh(m\ell)_a}{(m\ell)_a} = \frac{\tanh(107.10 \times 0.003023)}{107.10 \times 0.003023} = 0.9665$$

The total surface temperature effectivenesses are

$$\eta_{o,g} = 1 - (1-\eta_f)A_f/A = 1 - (1-0.9853) \times 0.849 = 0.9875$$

$$\eta_{o,a} = 1 - (1-\eta_f)A_f/A = 1 - (1-0.9665) \times 0.923 = 0.9691$$

In order to calculate $U_a$ from Eq. (51), we need to first calculate $\alpha_a$ and $\alpha_g$ using Eqs. (6) and (7).

$$\alpha_a = \frac{(b\beta)_a}{b_a + b_g + 2a} = \frac{6.35 \times 1548}{6.35 + 6.35 + 2 \times 0.4} = 728.1 \text{ m}^2/\text{m}^3$$

$$\alpha_g = \frac{(b\beta)_g}{b_g + b_a + 2a} = \frac{6.35 \times 1841}{6.35 + 6.35 + 2 \times 0.4} = 866.0 \text{ m}^2/\text{m}^3$$

Hence 
$$A_a/A_g = \alpha_a/\alpha_g = 728.1/866.0 = 0.841$$

Thus $U_a$ from Eq. (51) is

$$\frac{1}{U_a} = \frac{1}{0.9691 \times 158.09} + \frac{0.841}{0.9875 \times 70.97} = 0.01853$$

or $U_a = 53.97 \text{ W/m}^2 \text{ °C}$

**Surface area, free flow area, and core dimensions.** Since $N_{tu} = 7.45$, and $C_{min} = C_a = 841.2 \text{ W/°C}$,

$$A_a = N_{tu}C_a/U_a = 7.45 \times 841.2/53.97 = 116.1 \text{ m}^2$$

The minimum free flow area on the air side from the specified W and computed G is

$$A_{o,a} = (W/G)_a = 0.8296/7.62 = 0.1089 \text{ m}^2$$

The air flow length is then computed from the definition of the hydraulic diameter.

$$L_a = \left(\frac{D_h A}{4A_o}\right)_a = \frac{0.002383 \times 116.1}{4 \times 0.1089} = 0.635 \text{ m}$$

Since $A_a/A_g = 0.841$ and $A_a = 116.1 \text{ m}^2$, we get

$$A_g = 116.1/0.841 = 138.0 \text{ m}^2$$

Also

$$A_{o,g} = (W/G)_g = 0.8962/3.61 = 0.2483 \text{ m}^2$$

and

$$L_g = \left(\frac{D_h A}{4A_o}\right)_g = \frac{0.001875 \times 138.0}{4 \times 0.2483} = 0.261 \text{ m}$$

In order to calculate core frontal area on each side, we first need to determine $\sigma = \alpha D_h/4$ as

$$\sigma_a = \alpha_a D_{h,a}/4 = 728.1 \times 0.002383/4 = 0.434$$

$$\sigma_g = \alpha_g D_{h,g}/4 = 866.0 \times 0.001875/4 = 0.406$$

Hence

$$A_{fr,a} = A_{o,a}/\sigma_a = 0.1089/0.434 = 0.2509 \text{ m}^2$$

$$A_{fr,g} = A_{o,g}/\sigma_g = 0.2483/0.406 = 0.6116 \text{ m}^2$$

Since $A_{fr,a} = L_g L_3$ or $A_{fr,g} = L_a L_3$, we get

$$L_3 = A_{fr,a}/L_g = 0.2509/0.261 = 0.961 \text{ m}$$

or

$$L_3 = A_{fr,g}/L_a = 0.6116/0.635 = 0.963 \text{ m}$$

The difference in two values of $L_3$ is strictly due to the round off error.

**Pressure drops.** We will now use Eq. (41) to determine pressure drop on each side. The entrance and exit loss coefficients will be the same as those determined during the rating problem.

Gas side: $K_c = 1.2$   $K_e = 0.02$

Air side: $K_c = 0.33$  $K_e = 0.31$

In order to correct f factors for the temperature-dependent property effects, let us first calculate $t_w$. The thermal resistances on the hot and cold sides are

$$R_g = \frac{1}{(\eta_o hA)_g} = \frac{1}{0.9875 \times 70.97 \times 138.0} = 1.034 \times 10^{-4} \text{ °C/W}$$

$$R_a = \frac{1}{(\eta_o hA)_a} = \frac{1}{0.9691 \times 158.09 \times 116.1} = 5.622 \times 10^{-5} \text{ °C/W}$$

Therefore $R_g/R_a = 1.84$

Now $$t_w = \frac{t_{g,m} + (R_g/R_a) t_{a,m}}{1 + (R_g/R_a)} = \frac{150.7 + 1.84 \times 101.1}{1 + 1.84} = 118.56 \text{ °C} = 391.71 \text{ °K}$$

# COMPACT HEAT EXCHANGER DESIGN PROCEDURES

Since the gas is being cooled, using Eq. (27) and the exponent from Table 1,

$$f = f_{cp}(T_w/T_m)^m = 0.063(391.71/423.85)^{0.81} = 0.0591$$

Since the air is being heated, using Eq. (27) and the exponent from Table 1,

$$f = f_{cp}(T_w/T_m)^m = 0.063(391.71/374.25)^{1.00} = 0.0659$$

The pressure drops, using Eq. (41), are

$$\left(\frac{\Delta p}{p}\right)_g = \frac{(3.61)^2}{2 \times 110 \times 10^3 \times 0.7468}\left[(1-0.406^2+1.2) + 2\left(\frac{0.7468}{1.1424} - 1\right)\right.$$

$$\left. + \frac{0.0591 \times 0.261 \times 0.7468}{(0.001875/4) \times 0.9032} - (1-0.406^2+0.02)\frac{0.7468}{1.1424}\right]$$

$$= 7.932 \times 10^{-5}(2.0352 - 0.6926 + 27.2087 - 0.5590)$$

$$= 7.932 \times 10^{-5} \times 27.9923 = 0.00222$$

Hence $(\Delta p)_g = 0.00222 \times 110 = 0.244$ kPa

$$\left(\frac{\Delta p}{p}\right)_a = \frac{(7.62)^2}{2 \times 110 \times 10^3 \times 1.3827}\left[(1-0.434^2+0.33) + 2\left(\frac{1.3827}{0.8005} - 1\right)\right.$$

$$\left. + \frac{0.0659 \times 0.635 \times 1.3827}{(0.002383/4) \times 1.014} - (1-0.434^2+0.31)\frac{1.3827}{0.8005}\right]$$

$$= 1.9088 \times 10^{-4}(1.1416 + 1.4546 + 95.7822 - 1.9374)$$

$$= 1.9088 \times 10^{-4} \times 96.441 = 0.0184$$

Hence $(\Delta p)_a = 0.0184 \times 110 = 2.024$ kPa

Since the air side $\Delta p$ is higher than specified, a new value of G on both gas and air sides is determined again from Eq. (41) considering G as unknown.

$$\frac{0.32}{110} = \frac{G_g^2}{2 \times 110 \times 10^3 \times 0.7468} \cdot 27.9923 \rightarrow G_g = 4.13 \text{ kg/m}^2\text{s}$$

$$\frac{1.67}{110} = \frac{G_a^2}{2 \times 110 \times 10^3 \times 1.3827} \cdot 96.441 \rightarrow G_a = 6.92 \text{ kg/m}^2\text{s}$$

We can see that this new value of $G_g$ is already converged to the true value. However, because of too high a contribution of the core friction term, $G_a$ has been over corrected from 7.62 to 6.92 kg/m²s with the correct value from the rating problem being 7.048 kg/m²s. Repeating the calculations with the foregoing new values of G yield the following results.

$$L_g = 0.304 \text{ m} \qquad L_a = 0.587 \text{ m} \qquad L_3 = 0.909 \text{ m}$$

$(\Delta p)_g = 0.325$ kPa $\qquad (\Delta p)_a = 1.644$ kPa

Thus with the second iteration, the solution is almost converged to the specified pressure drops and heat transfer. The core dimensions are slightly different from those of the rating problem because of minor differences in the numbers used for the sizing problem. For example, $A_a/A_g = 0.841$ for the sizing problem and $0.854$ (= 118.546/138.885) for the rating problem, and so on.

The foregoing method clearly indicates how fast a solution to the sizing problem will converge to the core dimensions that will meet the heat transfer and pressure drops on both sides.

## 4. HEAT EXCHANGER OPTIMIZATION

The calculation procedure for the sizing problem has been outlined in the preceding section in which case no constraints except for the specified pressure drops are imposed on the design. The objective of that problem was to optimize the core dimensions to meet the required heat transfer.

Heat exchangers are designed for many different applications, and hence may involve many different performance criteria. These criteria for heat exchanger design may be minimum initial cost, minimum initial and operating costs, minimum weight or material, minimum volume or heat transfer surface area, minimum labor and so on.† When a single performance measure has been defined quantitatively and is to be minimized or maximized, it is called an "objective function" in a design optimization. A particular design may also be subjected to certain requirements such as required heat transfer, allowable pressure drop, limitations on height, width and/or length of the exchanger, and so on. These requirements are called "constraints" in a design optimization. A number of different surfaces could be incorporated in a specific design problem and there are many geometrical parameters that could be varied for each surface geometry.‡ In addition, operating flows and temperatures could also be changed. Thus a large number of "design variables" are associated with a heat exchanger design. The question arises as to how one can effectively adjust these design variables within imposed constraints and come up with a design having optimum objective function. This is what we mean by the most "efficient" design.

A complete mathematical optimization of heat exchanger design is presently neither practical nor possible. Many engineering judgements based on experience are involved in different stages of the design. However, once the general configuration and surfaces are selected, an optimized heat exchanger design may be arrived at if the objective function and constraints can be expressed mathematically, and if all of the variables are automatically and systematically changed on some statistical or mathematical basis.

A large number of optimization (search) techniques are available in the literature. A difficulty with the present state of the optimization art is that while any single search technique may work well on many problems, no single technique is able to solve every problem. A package of optimization computer

---

†Bergles et al. [17] present a list of performance evaluation criteria for enhanced heat transfer surfaces.

‡For a shell-and-tube exchanger, the geometrical variables are those associated with tube, baffles, shell, and front end and rear end heads. For an extended surface exchanger, the geometrical variables associated with a fin are the fin pitch, fin height, fin thickness, type of fin and other variables associated with each fin type.

programs is a convenient approach to overcoming this difficulty. Properly prepared, the programs can be used interchangeably and a new search can be called in where a previous one has quit thereby continuing or assuring progress toward optimum. In addition, when a package is used, various parts of the search procedure common to several optimization routines need be programmed only a single time. Heat exchanger optimization through a package of optimization computer programs has been suggested by Shah et al. [18]. They have discussed some details on the various optimization techniques, and outlined a procedure to arrive at an optimum heat exchanger design.

This procedure is repeated here with a flowchart of Fig. 10 for the completeness of the present paper. The procedure is referred to as the case study method.[†] In this method, each possible surface geometry and construction type is considered to be an alternative design as indicated in Fig. 10. In order to make a legitimate comparison of these alternatives, each design must be optimized for the specified application. Thus there may be several independent optimized solutions satisfying the problem requirements. Engineering judgment, comparison of objective function values, and other evaluation criteria are then applied to select a final optimum solution for implementation.

Consider a liquid-to-gas heat exchanger to be required for a specific application having minimum total cost. From the initial screening of surfaces, suppose two plate-fin type constructions (the louver-fin and strip-fin surfaces) and one flat-tube and wavy-fin type construction appear to be promising for the gas side. Then, for this problem, there are three alternate designs that need to be optimized.

As shown in Fig. 10, first formulate the total number of constraints for the problem. This includes customer's specified explicit constraints (such as fixed frontal area, the ranges of heat exchanger dimensions) and implicit constraints (such as required minimum heat transfer, allowable maximum pressure drop). Once the basic surface geometry for an alternate design is selected, the designer imposes some additional constraints such as the minimum and maximum values for fin height, fin thickness, fin pitch, fin thermal conductivity, flow length, number of finned passages, gas flow rate, etc. The designer wants to vary the above design variables within the ranges specified such that the exchanger will meet the required heat transfer, maximum pressure drop and other constraints with minimum total cost.

To optimize the heat exchanger, the designer starts with one set of heat exchanger surface geometrical dimensions which may not even satisfy all or some of the imposed constraints. Subsequently, the various geometrical properties (such as heat transfer area, free flow area, hydraulic diameter) and thermal properties are evaluated based on the input operating conditions. The heat transfer rate and pressure drop are then evaluated by the procedure outlined for the rating problem. Next, the output from heat exchanger calculations is fed to the optimization computer program package where the constraints and the objective function are evaluated. New values for the design variables are subsequently generated and heat exchanger calculations are repeated. The iterations are continued until the objective function is optimized within the accuracy specified and all the constraints are satisfied. In some situations, it may not be possible to satisfy all the constraints. The use of the "exterior penalty" method of optimization will essentially result in a design that will have constraints as close as possible to the specification. Engineering judgment will be needed to find out whether or not the optimum design is satisfactory.

One of the most important but least known input for the heat transfer and

---

[†]This method has been found most useful for heat exchanger optimization [18].

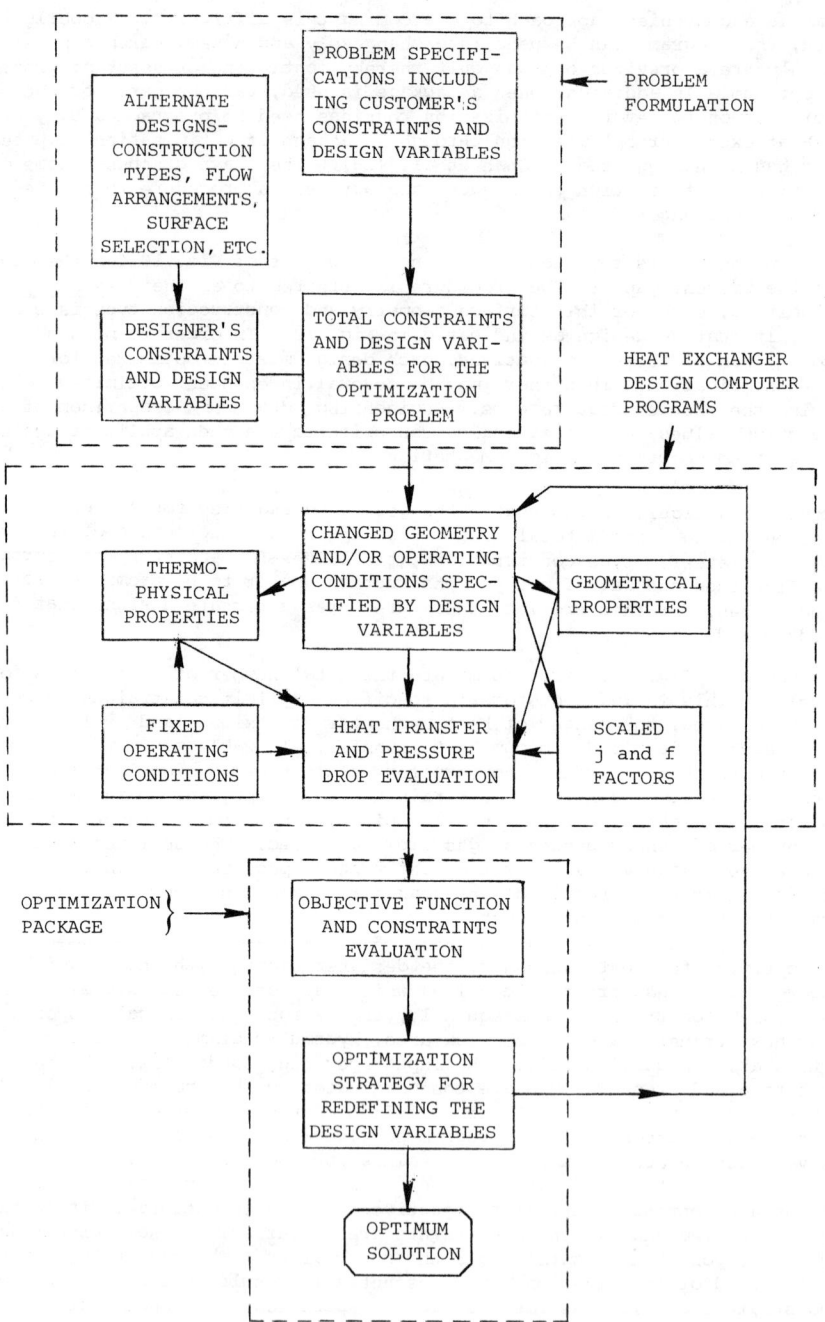

Fig. 10 Methodology for heat exchanger optimization [2].

pressure drop evaluation is the magnitude for "scaled" j and f factors. As soon as one of the surface geometrical dimensions is changed (such as fin pitch, height or thickness), the surface is no longer geometrically similar to the original surface for which experimental j and f data are available. In such cases, either theoretical or experimental correlations should be incorporated in the computer program to arrive "scaled" j and f factors for the new geometry. Some of these correlations as available in the literature are presented in [7]. The designer must use his experience and judgment regarding which correlations he should use to obtain the scaled j and f factors. In addition, a care must be exercised to avoid excessive extrapolations.

A review of Fig. 10 indicates two computer program packages needed for the optimization are the heat exchanger design programs and the optimization package. It should be emphasized that the heat exchanger programs must be working properly before any attempt is made to implement the optimization package. Otherwise, "garbage in gospel out" will result from the "black box." If the heat exchanger programs work properly, then any problem with the optimization package will not be serious, since the exchanger design will be an output from the heat exchanger programs, although that design may not be optimum.

## 5. SUMMARY

Two most common heat exchanger design problems, the rating and sizing problems, are first defined. A detailed procedure for the solutions of these problems is then outlined for a two-fluid compact heat exchanger (a recuperator) with a specific example. It is shown that a reasonably good solution to the rating or sizing problem can be obtained within two iterations. A general methodology is then discussed for design optimization that has been found very successful for compact heat exchanger applications, although the method is applicable to any type of heat exchanger.

## ACKNOWLEDGEMENT

The author is grateful to Mr. R. S. Johnson of Harrison Radiator Division for making constructive suggestions in the solution procedure for the sizing problem.

## NOMENCLATURE

| | |
|---|---|
| $A$ | total heat transfer surface area (both primary and secondary, if any) on one side of a two-fluid direct transfer type exchanger, $m^2$, $ft^2$ |
| $A_f$ | fin or extended surface area on one side of the exchanger, $m^2$, $ft^2$ |
| $A_{fr}$ | frontal or face area on one side of the exchanger, $m^2$, $ft^2$ |
| $A_k$ | total wall cross-sectional area for longitudinal conduction, $m^2$, $ft^2$ |
| $A_o$ | minimum free flow (or open) area on one side of the exchanger, $m^2$, $ft^2$ |
| $A_w$ | total wall area for heat conduction from hot fluid to cold fluid, $m^2$, $ft^2$ |
| $a$ | wall thickness, m, ft |
| $b$ | distance between two plates (fin height) in a plate-fin exchanger, m, ft |
| $C$ | flow stream heat capacity rate, $Wc_p$, W/°C, Btu/hr °F |
| $C^*$ | heat capacity rate ratio, $C_{min}/C_{max}$, dimensionless |

| | |
|---|---|
| $C_{max}$ | maximum of $C_c$ or $C_h$, W/°C, Btu/hr °F |
| $C_{min}$ | minimum of $C_c$ or $C_h$, W/°C, Btu/hr °F |
| $c_p$ | specific heat of fluid at constant pressure, J/kg °C, Btu/lbm °F |
| $D_h$ | hydraulic diameter of flow passages, $4r_h$ or $4A_oL/A$ or $4\sigma/\alpha$, m, ft |
| $F$ | log-mean temperature difference correction factor, dimensionless |
| $f$ | Fanning friction factor [3], dimensionless |
| $G$ | mass velocity, $W/A_o$, kg/m$^2$s, lbm/hr ft$^2$ |
| $g_c$ | proportionality constant in Newton's second law of motion, $g_c = 1$ and dimensionless in SI units, $g_c = 32.174$ lbm ft/lbf sec$^2$ |
| $h$ | heat transfer coefficient, W/m$^2$ °C, Btu/hr ft$^2$ °F |
| $I_n(\ )$ | modified Bessel function of the first kind and n-th order |
| $j$ | Colburn factor, StPr$^{2/3}$, dimensionless |
| $K_c$ | flow contraction loss coefficient at exchanger entrance, dimensionless |
| $K_e$ | expansion loss coefficient for flow at exchanger exit, dimensionless |
| $k$ | fluid thermal conductivity, W/m °C, Btu/hr ft °F |
| $k_f$ | thermal conductivity of the fin material, W/m °C, Btu/hr ft °F |
| $k_w$ | thermal conductivity of the wall material, W/m °C, Btu/hr ft °F |
| $L$ | fluid flow (core) length on one side of the exchanger, m, ft |
| $L_1$ | flow (core) length for Fluid 1 of a two-fluid heat exchanger, m, ft |
| $L_2$ | flow (core) length for Fluid 2 of a two-fluid heat exchanger, m, ft |
| $L_3$ | noflow height (stack height) of a two-fluid heat exchanger, m, ft |
| $\ell$ | fin length for heat conduction from primary surface to the midpoint between plates for symmetric heating, m, ft |
| $\ell_s$ | strip length of a strip (offset) fin, m, ft |
| $m$ | fin parameter, defined by Eq. (32), 1/m, 1/ft |
| $N_p$ | number of flow passages on one side in a plate-fin exchanger |
| Nu | Nusselt number, $hD_h/k$, dimensionless |
| $N_{tu}$ | number of heat transfer units, $UA/C_{min}$, dimensionless |
| $N_{tu,c}$ | number of heat transfer units on cold side, $(\eta_o hA)_c/C_c$, dimensionless |
| $N_{tu,h}$ | number of heat transfer units on hot side, $(\eta_o hA)_h/C_h$, dimensionless |
| $n$ | number of exchanger passes |
| Pr | Prandtl number, $\mu c_p/k$, dimensionless |
| $p$ | fluid static pressure, Pa, lbf/ft$^2$ |
| $\Delta p$ | fluid static pressure drop on one side of the exchanger core, Pa, lbf/ft$^2$ |
| $q$ | heat transfer rate in the exchanger, W, Btu/hr |
| $R$ | thermal resistance based on the surface area A, °C/W, hr °F/Btu |
| $\tilde{R}$ | gas constant for a particular gas, J/kg °K, lbf ft/lbm °R |
| $r_h$ | hydraulic radius, $A_oL/A$ or $D_h/4$, m, ft |
| St | Stanton number, $h/Gc_p$, dimensionless |
| $T$ | static temperature on the absolute scale, °K |

| | |
|---|---|
| $t$ | static temperature to a specified arbitrary datum, °C, °F |
| $U$ | overall heat transfer coefficient, see Eq. (34), W/m² °C, Btu/hr ft² °F |
| $u, u_m$ | fluid mean axial velocity, m/s, ft/sec |
| $V$ | heat exchanger total volume, m³, ft³ |
| $V_p$ | heat exchanger volume between plates on one side, m³, ft³ |
| $W$ | fluid mass flow rate, $\rho u_m A_o$, kg/s, lbm/hr |
| $\alpha$ | ratio of total heat transfer area on one side of the exchanger to the total volume of the exchanger, $A/V$, m²/m³, ft²/ft³ |
| $\beta$ | ratio of total heat transfer area on one side of a plate-fin exchanger to the volume between the plates on that side, $A/V_p$, m²/m³, ft²/ft³ |
| $\delta$ | fin thickness, m, ft |
| $\varepsilon$ | heat exchanger effectiveness, defined by Eq. (23) without the subscript p, dimensionless |
| $\varepsilon_p$ | heat exchanger effectiveness per pass, dimensionless |
| $\eta_f$ | temperature effectiveness of fins or an extended surface, dimensionless |
| $\eta_o$ | total surface temperature effectiveness, defined by Eq. (33), dimensionless |
| $\mu$ | fluid dynamic viscosity coefficient, Pa·s, lbm/hr ft |
| $\rho$ | fluid density, kg/m³, lbm/ft³ |
| $\sigma$ | ratio of free flow area to frontal area, $A_o/A_{fr}$, dimensionless |

Subscripts

| | |
|---|---|
| a | air side |
| c | cold fluid side |
| cp | constant properties |
| g | gas side |
| h | hot fluid side |
| i | inlet to the exchanger |
| m | mean or bulk mean |
| o | outlet to the exchanger |
| w | wall or properties at the wall temperature |
| 1 | Fluid 1 side |
| 2 | Fluid 2 side |

REFERENCES

1. R.K. Shah, Heat exchanger design methodology -- an overview, in Heat Exchangers - Thermohydraulic Fundamentals and Design, edited by S. Kakac, A.E. Bergles, and F. Mayinger, Hemisphere Publishing Corp., New York (1981).

2. R.K. Shah, Thermal design theory for regenerators, in Heat Exchangers - Thermohydraulic Fundamentals and Design, edited by S. Kakac, A.E. Bergles, and F. Mayinger, Hemisphere Publishing Corp., New York (1981).

3. W.M. Kays and A.L. London, Compact Heat Exchangers, Second Edition, McGraw-Hill Book Co., New York (1964).

4. K. Raznjevic, Handbook of Thermodynamic Tables and Charts, McGraw-Hill Book Co., New York (1976).

5. R.C. Reid and T.K. Sherwood, <u>The Properties of Gases and Liquids, Their Estimation and Correlation</u>, Second Edition, McGraw-Hill Book Co., New York (1966).

6. R.K. Shah, Classification of heat exchangers, in <u>Heat Exchangers - Thermohydraulic Fundamentals and Design</u>, edited by S. Kakac, A.E. Bergles, and F. Mayinger, Hemisphere Publishing Corp., New York (1981).

7. R.K. Shah, Compact heat exchangers, in <u>Heat Exchangers - Thermohydraulic Fundamentals and Design</u>, edited by S. Kakac, A.E. Bergles, and F. Mayinger, Hemisphere Publishing Corp., New York (1981).

8. W.K. Kays and H.C. Perkins, Forced convection, internal flow in ducts, in <u>Handbook of Heat Transfer</u> edited by W.M. Rohsenow and J.P. Hartnett, Chapter 7, McGraw-Hill Book Co., New York (1973).

9. R.L. Shannon and C.A. Depew, Forced laminar flow convection in a horizontal tube with variable viscosity and free-convection effects, <u>Trans. ASME, J. of Heat Transfer</u>, Vol. 91. Series C, 251-258 (1969).

10. C.A. Sleicher and M.W. Rouse, A convenient correlation for heat transfer to constant and variable property fluids in turbulent pipe flow, <u>International Journal of Heat and Mass Transfer</u>, Vol. 18, 677-683 (1975).

11. B.S. Petukhov, Heat transfer and friction in turbulent pipe flow with variable physical properties, in <u>Advances in Heat Transfer</u>, Vol. 6, 503-564 (1970).

12. H. Nakamura, A. Matsuura, J. Kiwaki, N. Matsuda, S. Hiraoka and I. Yamada, The effect of variable viscosity on laminar flow and heat transfer in rectangular ducts, <u>Journal of Chemical Engineering of Japan</u>, Vol. 12, No. 1, 14-18 (1979).

13. D.Q. Kern and A.D. Kraus, <u>Extended Surface Heat Transfer</u>, McGraw-Hill Book Co., New York (1972).

14. B.S. Baclic, A simplified formula for cross-flow heat exchanger effectiveness, <u>Trans. ASME, Journal of Heat Transfer</u>, Vol. 100, Series C, 746-747 (1978).

15. J.P. Chiou, The advancement of compact heat exchanger theory considering the effects of longitudinal heat conduction and flow nonuniformity, in Symposium on <u>Compact Heat Exchangers -- History, Technological Advancement and Mechanical Design Problems</u>, edited by R.K. Shah, C.F. McDonald, and C.P. Howard, Book No. G00183, ASME, New York (1980).

16. A.L. London and R.K. Shah, Offset rectangular plate-fin surfaces -- Heat transfer and flow friction characteristics, <u>Trans. ASME, Journal of Engineering for Power</u>, Vol. 90, Series A, 218-228 (1968).

17. A.E. Bergles, A.R. Blumenkrantz, and J. Taborek, Performance evaluation criteria for enhanced heat transfer surfaces, <u>Heat Transfer 1974</u>, Vol. II, 234-238 (1974).

18. R.K. Shah, K.A. Afimiwala, and R.W. Mayne, Heat exchanger optimization, <u>Heat Transfer 1978</u>, Vol. 4, 193-199, Hemisphere Publishing Corp., New York, (1978).

# Construction Features of Shell and Tube Heat Exchangers

**KENNETH J. BELL**
School of Chemical Engineering
Oklahoma State University
Stillwater, Oklahoma, USA

ABSTRACT

The major construction features of shell and tube exchangers are described. The flexibility of the basic design and the wide range of mechanical design options available to the designer are emphasized. Selection of features to meet a variety of service situations is illustrated.

1. INTRODUCTION

Shell and tube heat exchangers in their various construction modifications are, undoubtedly, the most widespread and commonly used basic heat exchanger configuration in the process industries. They are also used in conventional energy production as condensers, feed water heaters, and steam generators for pressurized water reactor plants. They are proposed for many alternative energy applications including ocean thermal and geothermal. And they are used in some refrigeration and air conditioning services.

The reasons for this near-universal acceptance are several. The shell and tube heat exchanger provides relatively large ratios of heat transfer area to volume and weight. It provides this surface in a form which is relatively easy to construct in a wide range of sizes and which is mechanically rugged enough to withstand normal shop fabrication stresses, shipping and field erection stresses, and normal operating conditions. The shell and tube exchanger can be reasonably easily cleaned, and those components most subject to failure - gaskets and tubes - can be easily replaced. The shell and tube exchanger offers great flexibility of mechanical features to meet almost any service requirement. Finally, good design methods are available, and the expertise and shop facilities for their successful design and construction are widespread.

This lecture will start by identifying the major components, then will discuss the various ways of dealing with thermal stresses between the tube

---

In this lecture, a number of figures of various heat exchanger configurations have been taken (with permission) from Manual No. 700-A, published by The Patterson-Kelly Company of East Stroudsburg, Pennsylvania. The author is most appreciative of Patterson-Kelly's consideration in this matter.

bundle and the shell, followed by a review of special service features, and will close with rules for allocating streams into the tubes vis-a-vis the shell.

2. BASIC COMPONENTS OF SHELL AND TUBE HEAT EXCHANGERS

While there is an enormous variety of specific design features that can be used in shell and tube exchangers, the number of basic components is relatively small. These components are shown in Fig. 1. In the following paragraphs, these basic components are listed together with a few words of description.

2.1 Tubes (Shown as "A", Figure 1)

The tubes are the basic component of the shell and tube exchanger, providing the heat transfer surface between one fluid flowing inside the tube and the other fluid flowing across the outside of the tubes. The tubes are generally drawn or extruded seamless metal, though welded tubes are becoming common. The metal is usually low-carbon steel, low alloy steel, stainless steel, copper, Admiralty, cupronickel, inconel, aluminum (in the form of various alloys), or titanium, though many other materials may be specified for special applications.

The tubes may be either bare or with low fins on the outside (see Figure 2). Low-fin tubes are used when the fluid on the outside of the tubes (the "shell-side" fluid) has a substantially lower heat transfer coefficient than the fluid on the inside of the tubes (the "tube-side" fluid). Low-fin surface provides 2-1/2 to 5 times as much heat transfer area on the outside as the corresponding bare tube, and this area ratio helps to offset the lower heat transfer coefficient. If low-fin tubes are used, it is necessary to consider the additional resistance to heat transfer caused by the longer conductive path; fin resistance or fin efficiency relationships have been developed for a number of important cases [1]. Fin efficiencies will typically be on the order of 0.9 for shell and tube applications.

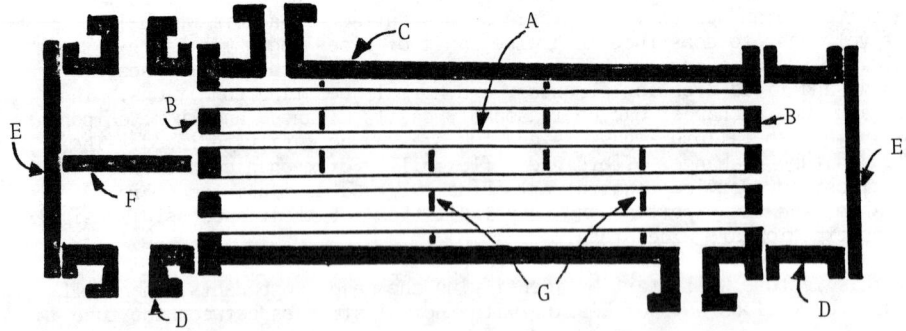

Figure 1. Diagram of a Typical Fixed Tube Sheet Shell and Tube Heat Exchanger. A: Tubes. B: Tube Sheets. C: Shell. D: Tube-side Channels and Nozzles. E: Channel Covers. F: Pass Divider. G: Baffles.

Figure 2. Cross-section of a Low-fin Tube Showing Full Diameter Plain End for Insertion in the Tube Sheet.

Low-fin tubes have a fin outside diameter slightly less than the unfinned ends so that they may be inserted into the bundle through the tube sheet holes. Unfinned lands along the tube may be provided to order, an important consideration when the tubes must pass through the baffle holes and be supported.

2.2 Tube Sheets (Component "B", Figure 1)

The tubes are held in place by being inserted into holes in the tube sheet and there either expanded into grooves cut into the holes (Figure 3) or welded to the tube sheet. Occasionally in low stress situations (e.g., power plant condensers), the tubes are simply expanded without grooving the tube sheet. Expansion into grooves provides an extremely strong joint, and welding is resorted to primarily to provide an additional barrier to leakage through the tube sheet. If welding is required, the tubes and tube sheets must usually be the same or very similar metals; for grooved joints, a wide variety of combinations is possible, dependent only upon reasonable limits on yield stress and corrosion compatibility.

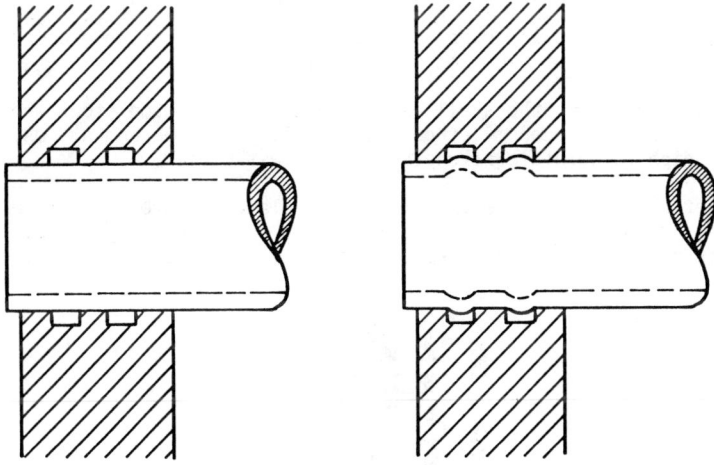

Figure 3. Diagram of Tube to Tube Sheet Expansion Joint Used in Process Heat Exchangers.

The tube sheet is usually a single round plate of metal that has been suitably drilled and grooved to take the tubes (in the desired pattern), the gaskets, the spacer rods (not shown), and the bolt circle where it is fastened to the shell. However, where mixing between the two fluids (by means of tube hole leaks) must be avoided, a double tube sheet such as is shown in Figure 4 may be provided (at considerable increase in cost).

The space between the tube sheets is open to the atmosphere so any leakage of either fluid should be quickly detected. Triple tube sheets (to allow each fluid to leak separately to the atmosphere without mixing) and even more exotic designs with inert gas shrouds and/or leakage recycling systems are used in cases of extreme hazard or high value of the fluid.

The tube sheet, in addition to its mechanical requirements, must withstand corrosive attack by both fluids in the heat exchanger and must be electrochemically compatible with the tube and all tube-side material. Tube sheets are sometimes made from low-carbon steel with a thin layer of corrosion-resisting alloy metallurgically bonded to one side; the bonding may be achieved by various detonation-bonding processes or by laying down a continuous weld deposit of the cladding metal.

2.3  The Shell and Shell-Side Nozzles (Item "C", Figure 1)

The shell is simply the container for the shell-side fluid, and the nozzles are the inlet and exit ports. The shell has a circular cross section and is commonly made by rolling a metal plate of the appropriate dimensions into a cylinder and welding the longitudinal joint ("rolled shells"). Small diameter shells (up to around 24 inches--0.6 m--in diameter) can be made by cutting pipe of the desired diameter to the correct length ("pipe shells"). The roundness of the shell is important in fixing the maximum diameter of the baffles (see below) that can be inserted and therefore the effect of shell-to-baffle leakage. Pipe shells are more nearly round than rolled shells unless particular care is taken in rolling. In order to minimize out-of-roundness, small shells are occasionally expanded over a mandrel; in extreme cases, the shell is cast and then bored out on a boring mill.

In large exchangers, the shell is made out of low carbon steel wherever possible for reasons of economy, though other alloys can be and are used when corrosion or high-temperature strength demands must be met.

The inlet nozzle often has an impingement plate (Figure 5A) set just inside the shell to prevent the incoming fluid jet from impacting directly at high velocity on the top row of tubes. Such impact can cause erosion, cavi-

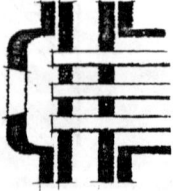

Figure 4:  Double Tube Sheet Design to Prevent Interleakage of Tube-side and Shell-side Fluids.

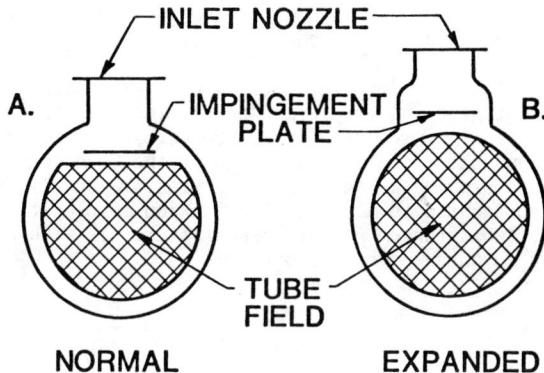

Figure 5: Inlet Shell-side Nozzles, Showing Impingement Plate Protection for the Tubes.

tation, and/or vibration. In order to put the impingement plate in and still leave enough flow area between the shell and plate for the flow to discharge without excessive pressure loss and excessive excape velocity, it may be necessary to omit some tubes from the full circle pattern. Alternatively, the nozzle may have an expanded section where it joins the shell (Figure 5B).

2.4 Tube-Side Channels and Nozzles (Item "D", Figure 1)

Tube-side channels and nozzles control the flow of the tube-side fluid into and out of the tubes of the exchanger. Since the tube-side fluid is often the more corrosive, these channels and nozzles will frequently be made out of alloy materials (compatible with the tubes and tube sheets, of course). They may be clad instead of solid alloy.

2.5 The Channel Covers (Item "D", Figure 1)

The channel covers are simply covers that bolt to the channel flanges and can be removed for tube inspection without disturbing the tube-side piping. In smaller heat exchangers, bonnets with flanged nozzles or threaded connections for the tube-side piping are often used instead of channels and channel covers.

2.6 Pass Divider (Item "F", Figure 1)

A pass divider is needed in one channel or bonnet for an exchanger having two tube-side passes, as the one illustrated in Figure 1, and they are needed in both channels or bonnets for an exchanger having more than two passes. If cast channels or bonnets are used, the dividers are integrally cast and then faced to give a smooth bearing surface on the gasket between the divider and the tube sheet. If the channels are rolled from plate or built up from pipe, the dividers are welded in place.

The pass dividers must fit tightly and cleanly into the grooves in the tube sheet and channel cover in order to minimize the possibility of leakage from one pass compartment to the next and a consequent serious deterioration of performance. The actual sealing is done by the gaskets which must be well-set and periodically checked and replaced.

The arrangement of the dividers in multiple-pass exchangers is somewhat arbitrary, the usual intent being to provide nearly the same number of tubes in each pass, to minimize the pressure difference across any one pass divider (to minimize leakage and therefore the violation of the Mean Temperature Difference derivation), to provide adequate bearing pressure on all parts of the sealing gaskets, and of course to minimize fabrication complexity and cost. For four tube passes, two arrangements are used (Figure 6), with neither one having an outstanding advantage over the other.

2.7  Baffles (Item "G", Figure 1)

Baffles serve two functions: Most importantly, they support the tubes in the proper position during assembly and operation and prevent vibration of the tubes caused by flow-induced eddies, and secondly, they guide the shell-side flow back and forth across the tube field, increasing the velocity and the heat transfer coefficient.

The most common baffle shape is the single segmental, shown in Figure 7. The segment sheared off must be less than half of the diameter in order to insure that adjacent baffles overlap at least one full tube row. For liquid flows on the shell side, a baffle cut of 20 to 25 percent of the diameter is common; for low pressure gas flows, 40 to 45 percent (i.e., close to maximum allowable cut) is more common, in order to minimize pressure drop. The baffle

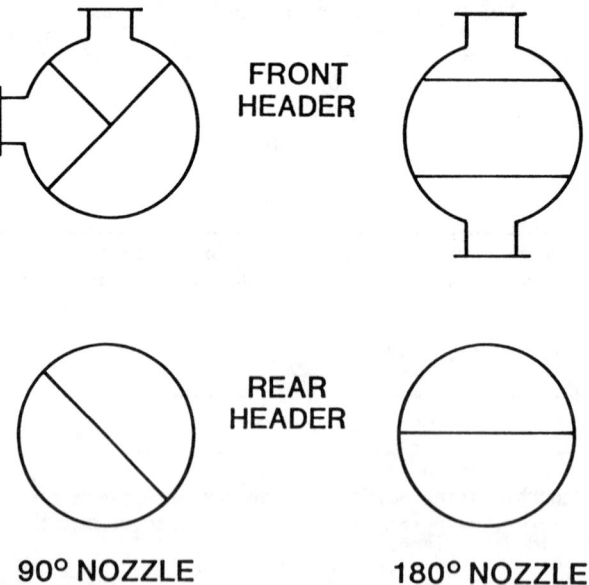

Figure 6: Two Possible Arrangements of Pass Dividers for a Four-Tube-Pass Exchanger.

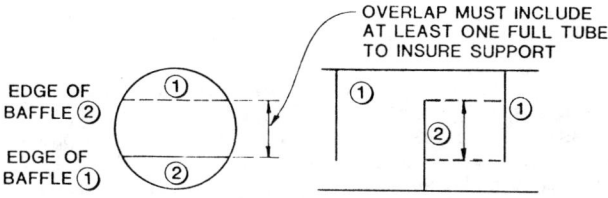

Figure 7. Diagram of Single Segmental Baffles.

spacing should be correspondingly chosen to make the free flow areas through the "window" (the area between baffle edge and shell) and across the tube bank roughly equal.

For many high velocity gas flows, the single segmental baffle configuration results in an undesirably high shell-side pressure drop. One way to retain the structural advantages of the segmental baffle and reduce the pressure drop (and, regrettably, to some extent, the heat transfer coefficient, too) is to use the double segmental baffle, shown in Figure 8. Exact comparisons must be made on a case-to-case basis, but the rough effect is to halve the local velocity and therefore to reduce the pressure drop by a factor of about 4 for a comparable size single segmental unit.

For sufficiently large units, it is possible to go to triple segmental arrangements and ultimately to strip and rod baffles, the important point being always to insure that every tube is positively constrained at periodic distances to prevent sagging and vibration.

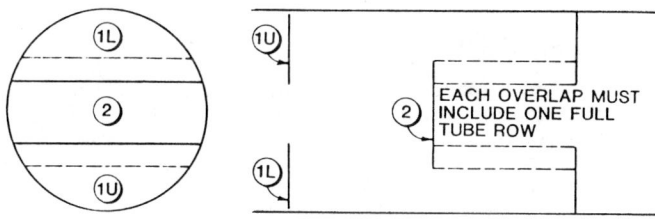

Figure 8. Diagram of Double Segmental Baffles.

Other baffle configurations such as disc-and-donut and orifice baffles have been used in the past but are seldom seen now.

A small clearance between tube outside diameter and baffle hole diameter is required to allow assembly of the tube bundle (and tube replacement, if desired). An excessive clearance provides too little tube support and possible vibration, as well as excessive leakage of fluid across the baffle. Too little clearance makes assembly and tube replacement difficult. Current TEMA Standards [2] call for 1/32 in. (0.8mm) diametral clearance in most cases, with 1/64 in. (0.4mm) diametral clearance under special considerations. In some air-conditioning applications, the tubes are expanded after assembly in order to eliminate the clearance altogether; this greatly complicates tube replacement, but might solve some potential tube vibration problems.

The outer diameter of the baffle must be less than the shell inside diameter to allow assembly, but the clearance should be as little as possible to minimize the shell-to-baffle leakage flow rate. (The shell-to-baffle leakage typically is the greatest penalty against the shell-side heat transfer coefficient.) TEMA [2] has set standards for this clearance, allowing an extra clearance for the out-of-roundness of rolled shells. The effect is significant, as will be seen clearly in later lectures on shell-side analysis.

## 3. TEMA NOTATION

As indicated above and developed at greater length below, a wide variety of shell and front and rear head configurations are available in shell and tube designs. To simplify specification, TEMA has developed a notation system for the major types, by which a basic configuration can be identified by three letters, the first identifying the front head, the second the shell, and the third the rear head. Figure 9 gives the TEMA notations. From this it may be seen that the exchanger shown in Figure 1 would be identified as AEL. Other examples will be given below.

## 4. AVOIDING THERMAL STRESS

### 4.1 The Thermal Stress Problem

Since, by its very purpose, the shell of the heat exchanger will be at a significantly different temperature than the tubes, differential expansion will result in stresses existing in both components and being transmitted through the tube sheets. The consequences of the thermal stress will vary with circumstances, but shells have been buckled or torn loose from supports and tubes have been buckled or pulled out of the tube sheet or simply pulled apart. The fixed tube sheet exchanger shown in Figure 1 is especially vulnerable to this kind of damage because there is no provision made for accommodating differential expansion.

There is a rough rule of thumb that says a simple fixed tube sheet configuration can only be used for cases where the inlet temperatures of the two streams do not differ by more than 50 or 60°C. Obviously, there must be many qualifications made to such a flat statement, recognizing the differences in materials and their properties, temperature level of operation, start-up and cycling operational procedures, etc., but clearly we need to consider means of solving this problem.

# CONSTRUCTION FEATURES OF SHELL AND TUBE HEAT EXCHANGERS

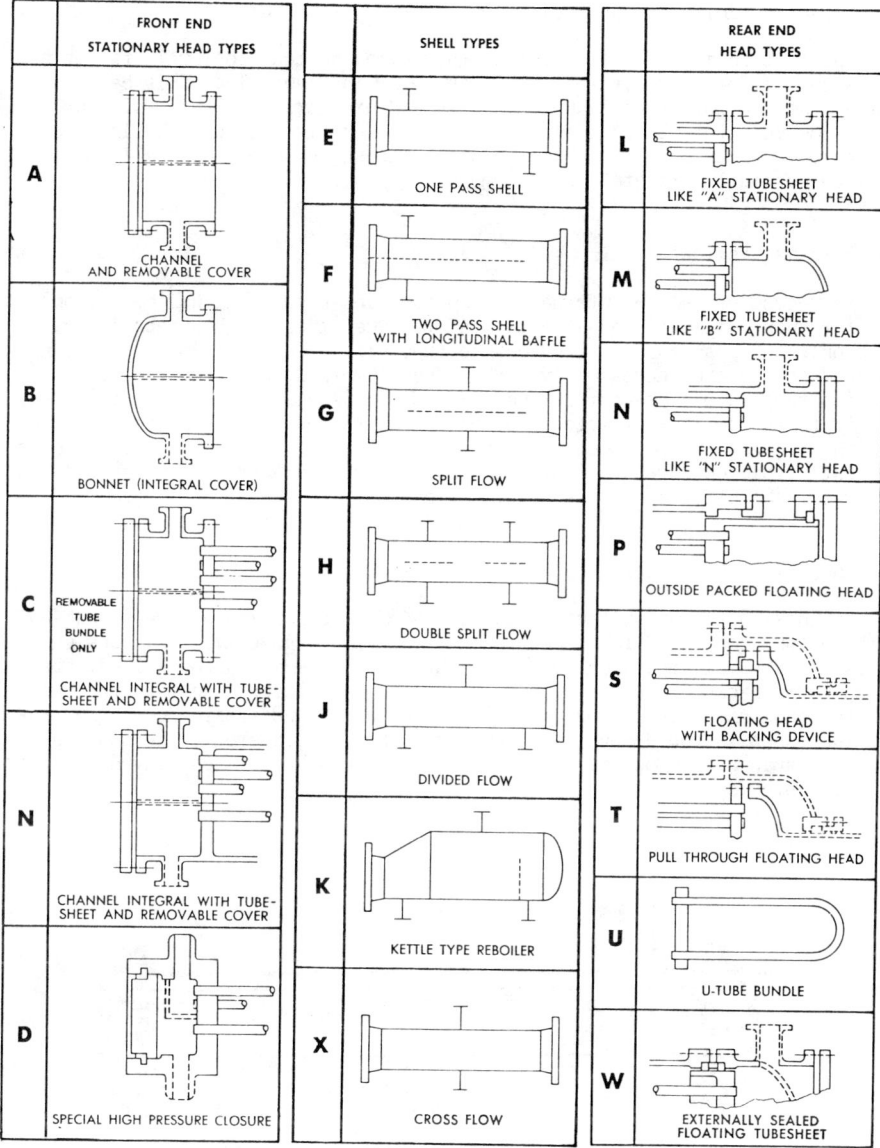

Figure 9. TEMA Notation for Shell and Tube Heat Exchangers. (Courtesy of Tubular Exchanger Manufacturers Association.)

## 4.2 Expansion Joint of the Shell

The most obvious solution to the thermal expansion problem is to put an expansion roll or joint in the shell, as shown in Figure 10. This becomes less attractive for large diameter shells and/or increasing shell-side pressure. However, very large diameter, near-atmospheric pressure shells have been designed with a partial ball-joint in the shell designed to allow the shell to partially "rotate" to accommodate stresses.

## 4.3 Internal Bellows

In recent years, an internal bellows design (Figure 11) has become popular for such applications as waste heat vertical thermosiphon reboilers, where only one pass is permitted on the tube side. These bellows have been designed to operate successfully with 7 MPa (1000 psi) water on the tube side and high temperature reactor effluent gas on the shell. There is, of course, a limit to the bellows diameter, decreasing as the pressures and temperatures go higher, but there has been much progress made in this area in recent years. Apparently, bellows are available to above 0.75m in diameter at pressures and temperatures typical of process applications.

## 4.4 The U-Tube Exchanger

One design variation that allows independent expansion of tubes and shell is the U-tube configuration shown schematically in Figure 12. This design solves the thermal expansion problem about as well as it can be solved, but it has some drawbacks. Individual tubes can not be replaced except in the outer row, the tube side can not be mechanically cleaned, and erosion can occur inside the tubes in the U-bend. In the larger bundle diameters, the long unsupported span of the outermost tubes in the U-bend has caused many destructive vibration problems.

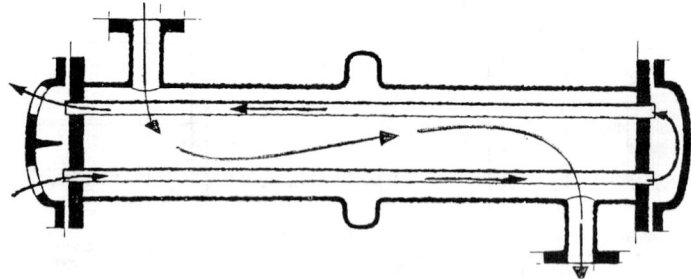

Figure 10. Fixed Tube Sheet Exchanger with Expansion Roll to Accommodate Thermal Stresses (Courtesy of Patterson-Kelley Co.)

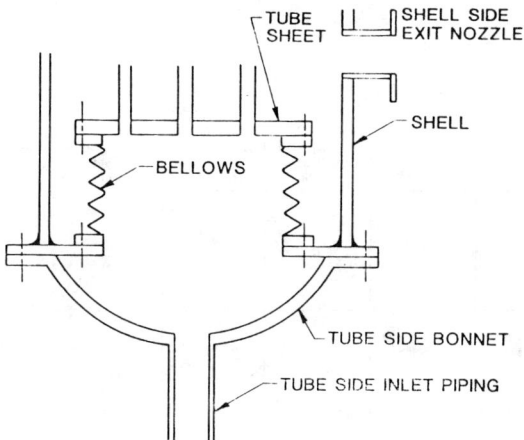

Figure 11. Use of an Internal Bellows to Accommodate Thermal Stresses

For easier comparison with other candidate bundle designs, the various features of this and the succeeding configurations are summarized in Table I. (The Table has also been taken from the Patterson-Kelley Manual 700-A.)

4.5 Floating Head Design

Several different designs of "floating head" shell and tube exchangers are in common use. The goal in each case, of course, is to solve the thermal stress problem and each design does accomplish that goal. Inevitably, however,

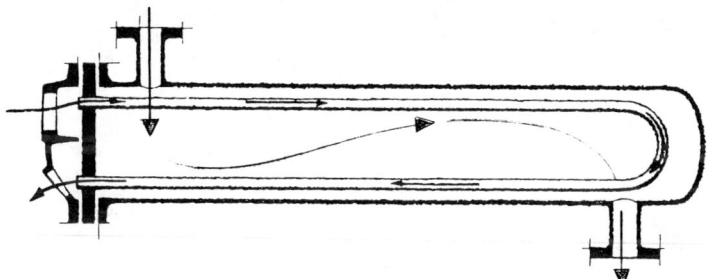

Figure 12. U-Tube Exchanger, Allowing Independent Expansion of Tubes

| Type of Design | "U"-Tube | Fixed Tubesheet | Floating Head Pull-Through Bundle | Floating Head Outside Packed Lantern-Ring | Floating Head Split Backing Ring | Floating Head Outside Packed Stuffing Box |
|---|---|---|---|---|---|---|
| Relative Cost Increases From (A) Least Expensive through (E) Most Expensive | A | B | C | C | D | E |
| Provision for Differential Expansion | individual tubes free to expand | expansion joint in shell | floating head | floating head | floating head | floating head |
| Removable Bundle | yes | no | yes | yes | yes | yes |
| Replacement Bundle Possible | yes | not practical | yes | yes | yes | yes |
| Individual Tubes Replaceable | only those in outside row | yes | yes | yes | yes | yes |
| Tube Interiors Cleanable | difficult to do mechanically can do chemically | yes, mechanically or chemically | yes, mechanically or chemically | yes, mechanically or chemically | yes, mechanically or chemically | yes, mechanically or chemically |
| Tube Exteriors With Triangular Pitch Cleanable | chemically only | chemically only | chemically only | chemically only | chemically only | chemically only |
| Tube Exteriors With Square Pitch Cleanable | yes, mechanically or chemically | chemically only | yes, mechanically or chemically | yes, mechanically or chemically | yes, mechanically or chemically | yes, mechanically or chemically |
| Double Tubesheet Feasible | yes | yes | no | no | no | yes |
| Number of Tube Passes | any practical even number possible | no practical limitations | no practical limitation (for single pass, floating head requires packed joint) | limited to single or 2 pass | no practical limitation (for single pass, floating head requires packed joint) | no practical limitation |
| Internal Gaskets Eliminated | yes | yes | no | yes | no | yes |

Table 1. Principal Features of Various Shell and Tube Bundle Types. (Courtesy Patterson-Kelley Co.)

CONSTRUCTION FEATURES OF SHELL AND TUBE HEAT EXCHANGERS

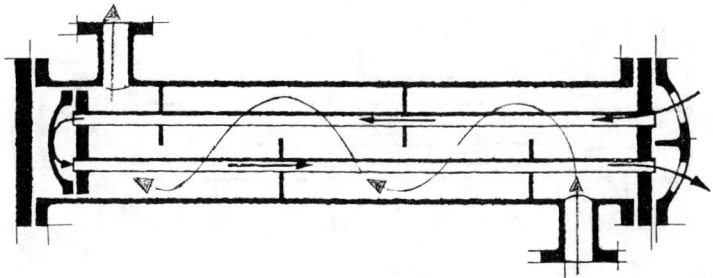

Figure 13. Pull-through Floating Head Heat Exchanger (Courtesty Patterson-Kelley Co.)

something must be given up, and each configuration has a somewhat different set of drawbacks to be considered when choosing one. These are summarized in Table I.

The simplest floating head design is the "pull-through bundle" type, shown in Figure 13. One of the tube sheets is made small enough that it and its gasketed bonnet may be pulled completely through the shell for shell-side inspection and cleaning. Unfortunately, many tubes must be omitted from the edge of the full bundle to allow for the bonnet flange and bolt circle. Also, the missing tubes around the periphery open up a low-resistance flow path for the shell-side fluid, the so-called "bundle by-pass stream". The reduced flow across the tube bundle has a significantly lower heat transfer coefficient than the ideal flow; additionally, the effective mean temperature difference is reduced. These problems can be reduced by the use of sealing strips put in longitudinally to partially block the bypass stream, as shown in Figure 14.

Another floating head design that partially offsets the above disadvantages is the "split ring floating head" type, Figure 15. Here the floating head bonnet is bolted to a split backing ring, rather than to the tube sheet. Therefore, no bolt circle has to be provided on the floating head tube sheet, though there still must be a gasket bearing surface. At some cost in added

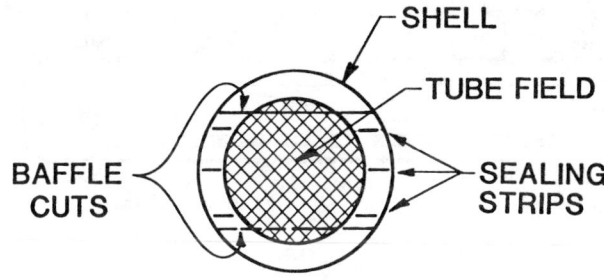

Figure 14. Diagram Showing Use of Sealing Strips Between Bundle and Shell.

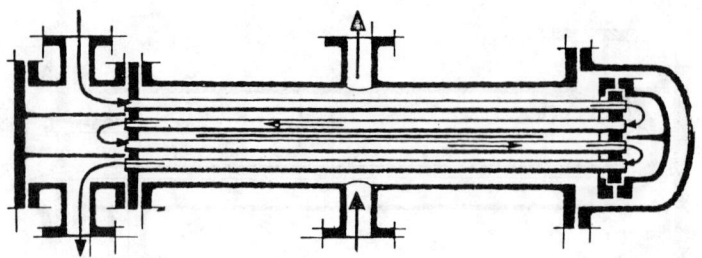

Figure 15. Split Ring Floating Head Heat Exchanger (Courtesy Patterson-Kelley Co.)

mechanical complexity, most of the tubes lost from the bundle in the pull-through design have been restored. The split ring floating head is widely used to meet the severe demands of the petroleum industry, though the pull-through floating head is still preferred when very high pressures or temperatures are involved or more positive gasketing between the two streams is needed.

Two other types, the "outside packed lantern ring", Figure 16, and the "outside packed stuffing box", Figure 17, are less positively sealed against leakage to the atmosphere than the foregoing types but have the advantage of allowing single tube-side pass construction. (The "pull through" and "split ring" types can be modified for single tube pass design, but they then give up the advantage of positive sealing so important in high pressure or hazardous fluid source.)

## 5. OTHER SHELL CONFIGURATIONS

Up to this point the discussion has emphasized the use of the TEMA E shell, with shell-side nozzles located at opposite ends of the shell. This is

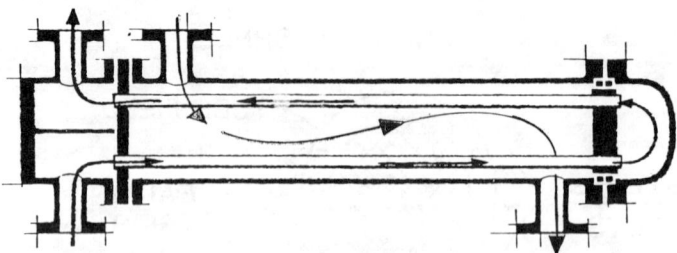

Figure 16. Outside Packed Lantern Ring Heat Exchanger (Courtesy Patterson-Kelley Co.)

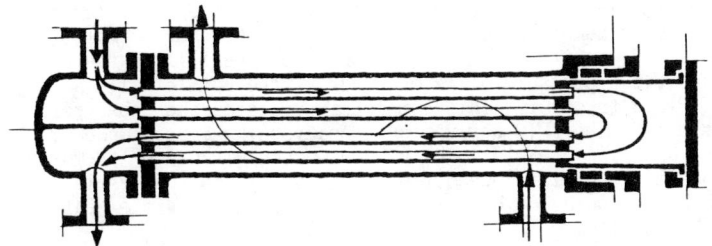

Figure 17. Outside Packed Stuffing Box Design for Thermal Stress Relief

the most common configuration, but a variety of other designs (F, G, H, J, K, and X) are used for specific purposes. The following paragraphs will discuss these types and their applications (and some cautionary considerations):

5.1 The F Shell

The F shell (Figure 18) has a longitudinal baffle in the shell, which provides in principle for two shell-side passes in series. The main reason for doing this is to provide (at least on paper) a countercurrent flow arrangement when one has two tube side passes, and therefore to allow the use of a configuration correction factor ($F_t$) of 1.0. If one traces through the flow paths, one sees that the two streams are always countercurrent to one another. The principle could be extended to multiple shell side passes to match multiple tube side passes but this is never (?) done in practice.

Even the provision of a single shell-side longitudinal baffle opens up a Pandora's box of fabrication, operation and maintenance problems. Without discussing all of the possibilities, we may observe that there will be - unless very special precautions are taken (i.e., an insulated baffle) - thermal leak-

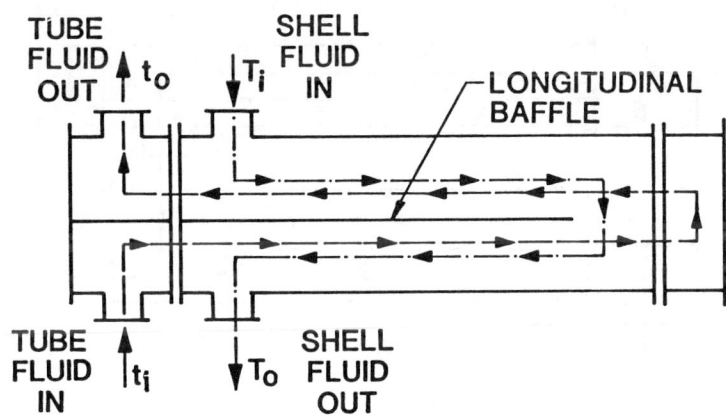

Figure 18. The F Shell, Providing Nominal Counterflow with Two Tubeside Passes

age from the hot shell-side pass through the baffle to the other (cold), which violates one of the assumptions of the LMTD derivation. Furthermore, unless the baffle is welded longitudinally to the shell or specially gasketed along the entire length, the pressure difference across the baffle will cause physical leakage of fluid to occur. This completely violates the "no-bypassing" assumption built into the LMTD derivation. Analyses have been made of the problem (Rozenman and Taborek [3]) which warn one when the penalty may become severe; but in general the use of multiple shell-side passes in heat exchangers is to be discouraged, especially when there is sensible heat transfer occurring on the shell side.

5.2 Multiple Shells in Series

If one needs to use multiple tube side passes (as one often does), and if the single shell pass configuration results in too low a value of $F_t$ (or in fact is thermodynamically inoperable), what can one do?

The usual solution is to use multiple shells in series, as diagrammed in Figure 19 for a very simple case.

More than two tube passes per shell may be used. The use of up to six shells in series is quite common, especially in heat recovery trains, but sooner or later pressure drop limits on one stream or the other limit the number of shells. Qualitatively, the overall flow arrangement of the two streams is countercurrent, even though the flow within each shell is still mixed. Since, however, the temperature change of each stream in one shell is only a fraction of the total change, the departure from true countercurrent flow is less. (A little reflection will show that as the number of shells in

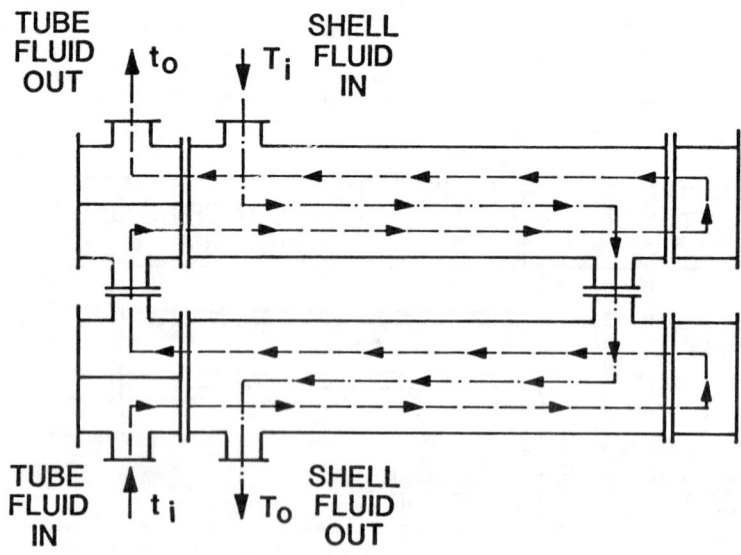

Figure 19. Two 1-2 Shells in Series to Improve the LMTD Configuration Correction Factor

series becomes infinite, the heat transfer process approaches true countercurrent flow and $F_t \to 1.0$.)

It is possible to analyze the thermal performance of a series of shells each having one shell pass and an even number of tube passes, by using heat balances and the $F_t$ equation applied to each shell. Such calculations quickly become very tedious, and it is much more convenient to use charts derived specifically for various numbers of shells in series. Such charts for 2 to 6 shells in series are given in Refs. [2] and [4] and other standard references.

5.3 The G and H Shells

The G and H shells, diagrammed in Figure 9, are structurally related to the F shell, being essentially a doubling and quadrupling of the basic F shell geometry. However, the flows are no longer purely countercurrent and $F_t$ no longer is even theoretically equal to unity in sensible heat transfer service. In fact, these shells are seldom used for sensible service, for the good reason that they have no obvious advantages and several drawbacks. They are used in phase-change service, particularly as horizontal thermosiphon reboilers and partial and total condensers. In these services, there is usually little penalty for thermal and physical leakage across the longitudinal baffle, whereas the baffle does help distribute the flow and flush out concentrations of non-condensable or non-volatile components that need to be continuously removed from the exchanger.

5.4 The J Shell

The J shell, or divided flow shell, is shown in Figure 9. It can be analyzed as two E shells placed end-to-end. The inlet flow can be either to the single nozzle at the center and exiting out the double nozzles at the end, or vice-versa. In fact, it is common practice to use the J shells in stacked pairs, usually with the single nozzles as the inlet and outlet for the array.

The J shell's main advantage is the sharply reduced shell-side pressure drop compared to an E shell of comparable dimensions--on the order of 1/4 to 1/8.

5.5 The K Shell

The K shell design (also termed a kettle reboiler or flooded chiller in the process and refrigeration industry, respectively) is employed when a portion of a stream needs to be vaporized, typically to a distillation column, Figure 9. The tube bundle has a diameter somewhat less than the smaller diameter shell, and the liquid being vaporized just covers the tube field. The large diameter shell provides a vapor-liquid separation space. The feed liquid enters the shell at the nozzle near the tube sheet, the nearly-dry vapor exits out the top nozzle, and the non-vaporized liquid overflows the end weir and exits through the right-hand nozzle. (It is essential that some liquid be removed continuously, or at least at frequent intervals from this nozzle to limit the buildup of non-volatile material.) Mesh demisters or centrifugal vane extractors are sometimes used at the vapor nozzle to minimize liquid carryover.

The tube bundle is commonly a U-tube configuration, but a floating tube sheet with tubeside header may also be used. In a few applications (e.g., the brine to isobutane vapor generators for the Raft River geothermal energy pilot plant), a double fixed tube sheet design is used with a single tube pass; this requires very careful startup and shutdown to minimize thermal stress.

K shells are an expensive construction, especially for high pressure vaporization, because of the large diameter shell required. However, it is the only reboiler design that will produce dry or nearly dry vapor without an external vapor-liquid separator.

## 5.6 The X Shell

The X, or crossflow, shell (Figure 9) is mostly used in condensing applications (including most central station steam power plant condensers), though it is occasionally employed in low pressure gas heating or cooling. It is a very low pressure drop arrangement (on the shell side), which is vital in vacuum condensing applications. This however does complicate getting good distribution of the vapor or vapor-gas mixture throughout the bundle, and it is important to provide a clear way free of tubes at the top of the bundle and running the length of the bundle (Figure 20). Alternatively, multiple inlet nozzles can be employed, or a "bathtub" nozzle--a large inverted trough-like manifold mounted on top of and running most of the length of the shell. Careful attention to removal of non-condensables is vital in X-shell condensers.

## 6. OTHER SHELL AND TUBE DESIGN FEATURES

### 6.1 The Rod Baffle Exchanger

The rod baffle shell and tube exchanger (Figure 21) can appear in any of the above shell types (though it is most commonly used in the E and X configurations), and is included here to illustrate another design option for the bundle [5].

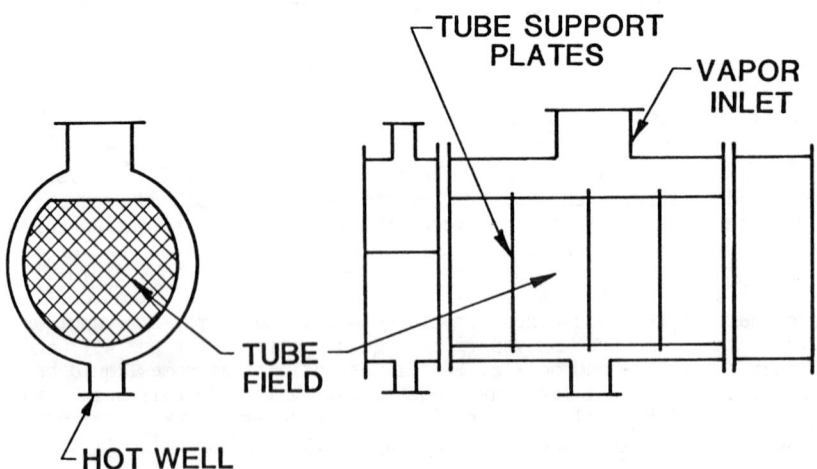

Figure 20. Schematic of an X Shell Heat Exchanger in Condensing Service

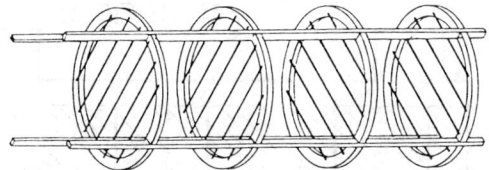

Figure 21. Baffle and Tie Rod Assembly for ROD Baffle Exchangers

As the name implies, the tube support is in the form of arrays of solid rods, passing transversely between the rows of tubes, which are arranged in a square layout. The rods are the same diameter as the tube-to-tube spacing so that each rod holds two rows of tubes apart so they can not vibrate or touch at that point. The rods are welded to an annular ring; each ring has a complete set of rods spaced two tube rows apart, with each successive ring having rods either filling the alternate spaces to the first ring, or rotated to the other axis. The pattern is repeated the length of the exchanger. Ring spacing is 6 to 8 in. (150-200mm). Thus each tube is supported on all four sides with each set of fan rings. A vibration analysis has not been done on this method of support, but the structure is very rigid and no rod baffle exchanger has even been reported as vibrating.

While originally developed to resist severe vibration problems, the rod baffle is also a very low pressure drop device. The low pressure drop feature has made it very interesting for condensers, especially if non-condensable gases are present and one is designing to very close temperature differences between coolant and condensing vapor. Rod baffle exchangers tend to be long and of small diameter and frequently can be made single pass on the coolant side and therefore countercurrent.

The rods, while there mainly to support the tubes, do improve the heat transfer somewhat over a pure longitudinal flow.

6.2 No-Tubes-in-the-Window Design

A schematic diagram of a shell and tube exchanger with no tubes in the window is shown in Figure 22. As the name implies, there are no tubes which pass through the window regions of the heat exchanger, the baffle cuts being somewhat smaller than they would be in a typical fully-tubed design. The advantage of the no-tubes-in-the-window design is that now every tube passes through and is supported by every baffle guiding the flow. Additionally, tube support plates may be inserted as required between the baffles; the purpose of the tube support plates is to support the tubes against vibration rather than to attempt to guide the flow. The no-tubes-in-the-window configuration is primarily of interest when shellside vibration problems are expected to be serious. While some fraction of the tubes is lost compared to the fully-tubed design, this loss can be minimized by having small baffle cuts and possibly by an increase in the shell side fluid velocity resulting in a significant increase in shellside heat transfer coefficient.

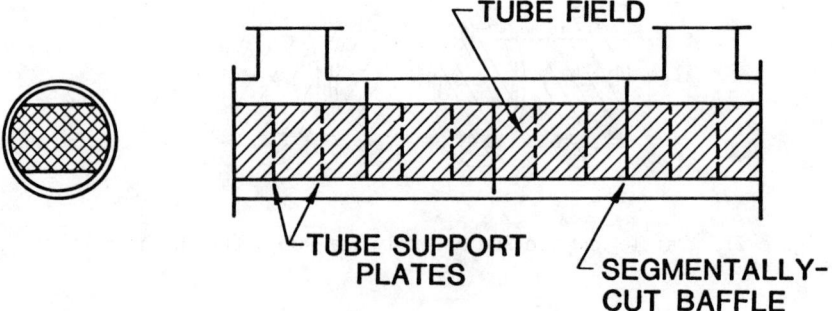

Figure 22. Diagram of a No-Tubes-in-the-Window Design

## 6.3 Annular Distributor

Especially when handling gases or vapors, pressure losses in the inlet and exit sections can become excessive and sometimes cause vibration problems. A very effective, if somewhat expensive, way to handle this problem is to use an annular distributor, or vapor belt, shown in Figure 23. In this design, the inlet (and sometimes the outlet) nozzle is welded to an enlarged annular shell section, or knuckle, with the actual exchanger shell forming the inner surface of the annulus. This shell is slotted around the periphery to allow the fluid to flow in through a much larger total cross section than would be possible with any normal-sized nozzle.

## 7. ALLOCATION OF STREAMS

An early decision that must be made in the design of a shell and tube exchanger is which fluid will go in the tubes and which in the shell. The following guidelines apply:

1. The high pressure fluid goes in the tubes, other things being equal. Because of their small diameter, normal thickness tubes are able to take quite high pressures and only the tube side channels and/or other

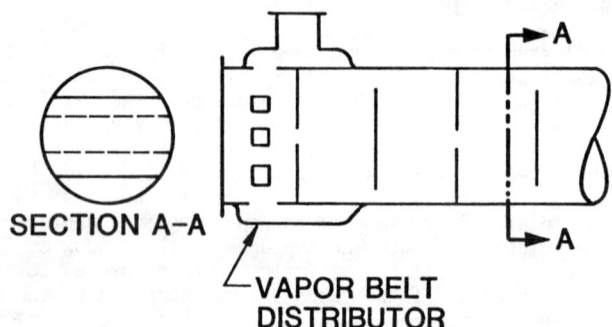

Figure 23. Annular or Vapor Belt Distributor for Minimizing Inlet Pressure Drop and Reducing Erosion and Vibration Problems

plumbing need normally be designed to withstand high pressure. If it is necessary to put the high pressure fluid in the shell, the exchanger will normally tend to a longer, smaller diameter configuration than otherwise.

2. The corrosive fluid goes in the tubes, other things being equal. Corrosion is resisted by using special--and therefore expensive--alloys, and it is much less expensive to avoid using alloy shells. The tube side of the tube sheet, the tube side channels, and the channel covers can usually be made of low carbon faced with the alloy either by detonation cladding or by depositing a continuous layer by welding. If the corrosion can not be effectively prevented but only slowed by choice of material, a design must be chosen in which corrodable components can be easily replaced (unless it is more economic to scrap the whole unit and start over.)

3. The more seriously fouling fluid goes in the tubes, other things being equal. The tube side is easier to clean, especially if mechanical cleaning (brushing, high velocity water jets, etc) is required.

4. The stream with the lower allowable pressure drop usually goes on the shell side, other things being equal. This is a rather curious anomaly, because the heat/momentum transfer analogies indicate that one can get more heat transfer per unit of pressure drop expended in a non-separating geometry (i.e., tubes) than in a separating one (i.e., crossflow on the shell side). However, the range of design variables on the shell side is so great compared to the tube side that it is usually possible to come up with a better mechanical design with the low pressure drop fluid on the shell side.

5. The stream with the lower heat transfer coefficient goes on the shell side, other things being equal. This allows the use of low finned tubing to partially offset the low coefficient with an increased area ration. However, tube side enhancement is now becoming available with acceptable engineering characteristics (availability, designability, reliability) and this rule is fading.

Problems arise when the above requirements are in conflict, e.g., one fluid is fouling and the other is at high pressure. Then the designer must estimate tradeoffs, and go with what seems the most economic choice, remembering the cardinal rule that the heat exchanger must perform its thermal duty and must do so with a very high degree of reliability and process availability--a cheap heat exchanger can become extremely expensive very quickly if it keeps shutting down a process.

REFERENCES

1. Kern, D. Q., and Kraus, A. D., Extended Surface Heat Transfer, McGraw-Hill, New York (1972).

2. Tubular Exchangers Manufacturers Association Standards, 6th Edition, New York (1978).

3. Rozenman, T., and Taborek, J., "The Effect of Leakage Through the Longitudinal Baffle on the Performance of Two-Pass Shell Exchangers", AIChE Symposium Series No. 118, "Heat Transfer--Tulsa", $\underline{68}$, 12-20 (1974).

4. Kern, D. Q., Process Heat Transfer, McGraw-Hill, New York (1950).

5. Small, W. M., and Young, R. K., "The RODbaffle Heat Exchanger", Heat Transfer Engineering 1, No. 2, 21-27 (1979).

# Preliminary Design of Shell and Tube Heat Exchangers

**KENNETH J. BELL**
School of Chemical Engineering
Oklahoma State University
Stillwater, Oklahoma, USA

ABSTRACT

The basic criteria for heat exchanger selection are presented and discussed. The structure of the design process is shown diagramatically and the differences between hand-based and computer-based design methods identified. A simple example of design logic for a computer is given. A preliminary estimation procedure for shell and tube exchangers is laid out.

1. CRITERIA FOR HEAT EXCHANGER SELECTION

The criteria that a process heat exchanger must satisfy are easily enough stated if we confine ourselves for the moment to rather broad statements:

First, of course, the heat exchanger must meet the process requirements. That is, it must effect the desired change in the thermal condition of the process stream within the allowable pressure drops, and it must continue to do this until the next scheduled shutdown of the plant for maintenance.

Second, the exchanger must withstand the service conditions of the plant environment. This includes the mechanical stresses of installation, startup, shutdown, normal operation, emergencies, and maintenance, and the thermal stresses induced by the temperature differences. It must also resist corrosion by the process and service streams (as well as by the environment); this is usually mainly a matter of choice of materials of construction, but mechanical design does have some effect. Desirably, the exchanger should also resist fouling, but there is not much the designer may do with confidence in this regard except keep the velocities as high as pressure drop and vibration limits permit.

Third, the exchanger must be maintainable, which usually implies choosing a configuration that permits cleaning - tubeside and/or shellside, as may be indicated - and replacement of tubes and any other components that may be especially vulnerable to corrosion, erosion, or vibration. This requirement may also place limitations on positioning the exchanger and in providing clear space around it.

Fourth, the exchanger should cost as little as is consistent with the above requirements; in the present listing, this refers to first cost or installed cost, since operating cost and the cost of lost production due to exchanger unavailability have already been considered by implication in the earlier and more important criteria.

Finally, there may be limitations on exchanger diameter, length, weight and/or tube specifications due to site requirements, lifting and servicing capabilities, or inventory considerations.

It is sometimes stated as a desirable feature that the exchanger design be selected with an eye to possible alternative uses in other applications. This has disturbing implications. Most heat exchangers are intended for projects having an expected life of five to twenty years - equal to or greater than the probable life of the exchanger. To suggest that a heat exchanger might become available sooner implies either that the exchanger or the process will prove unsatisfactory in its role. It is far better to labor under the positive compulsion that the only hope for success is by designing each item uniquely for the best performance in the task at hand.

## 2. LOGICAL STRUCTURE OF PROCESS HEAT EXCHANGER DESIGN

The basic logical structure of the process heat exchanger design procedure is shown in Figure 1. The basic structure is the same whether we use hand design methods or computer design; all that is different is the replacement of the very subtle and complicated human thought processes by a logical structure suited to a fast but inflexible computer. In fact, all of the most important decisions are made outside the dashed line bounding the computer's contribution to the solution.

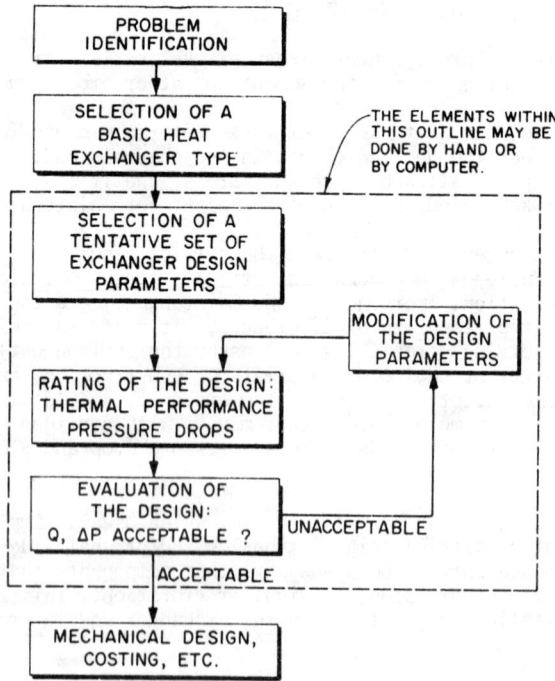

Figure 1. Basic Logical Structure for Process Heat Exchanger Design

# PRELIMINARY DESIGN OF SHELL AND TUBE HEAT EXCHANGERS

First, the problem must be identified as completely and unambigously as possible. Not only matters like flow rates and temperatures and compositions, but just as important: "What is the process engineer asking for?" and "Is that what he really needs?" and "Has he told me everything I need to know?". It is important to distinguish here between what must be done and how it is to be done. Not infrequently, the process engineer seeks to impose unwise design choices upon the exchanger designer; if the matter is of importance - if there is a chance that the choice will seriously adversely affect plant performance - the exchanger designer has no choice but to object strenuously.

The designer determines at the start of each problem the critical points of the problem - those which exercise a disproportionate effect upon the operational success, reliability, and/or cost of the exchanger. Each case is a little different in this respect - in one it will be the low coefficient associated with one stream, in another a close internal temperature approach between the streams, etc. These points are the ones upon which the designer will lavish his attention and skill.

Already by this point, the exchanger designer is making some very basic decisions. Right from the moment the problem hit his desk, the designer has some feel for whether this is a big exchanger (or exchanger system) or not, which directly translates into the amount of money involved. From this point, the designer will draw some early conclusions about the levels of engineering effort required and justified. The designer employed by an exchanger construction company may make a very different decision than one employed by a process company, to take the two extreme examples. The vendor can not afford to invest a great deal of engineering effort into an inexpensive heat exchanger (except in the unusual situation that his engineering work is separately contracted for), so the designer judges his level of effort by the selling price of the exchanger. The process company must make product, so the criterion for an engineering effort is the cost of the exchanger failing to work; a small but critical exchanger may warrant more effort than a much larger but less vital one. Recognition of this difference in constraints is in some measure responsible for the fact that most processors are now making detailed checks of the design of critical components in their new plants, even though they have engaged an engineering construction company to do the complete design.

It is at this point in the design process that the single most important design decision is made: the basic configuration of the heat exchanger, whether it is to be double-pipe, shell-and-tube, plate, etc. Possibly, an even more detailed decision may be made here, e.g., given that a shell and tube is chosen for a vaporizing service, the design may be further specified among the kettle, vertical or horizontal thermosiphon, or forced circulation types. Actually, the decision may have been made much earlier in the process engineer's mind, who recalls that a given configuration was used in the last instance of a similar application. And the process flow sheet from which the exchanger designer is working may already indicate by word or symbol a specific configuration. And in many applications (such as using atmospheric air to cool a process stream), there is really very little choice available as far as basic configurations are concerned. But, if the exchanger designer is to play his proper roles, he should have control over the choice of configuration within very broad guidelines and such physical constraints that exist in fact.

The next decision that needs to be made is what design method is to be used. Basically, these fall into two categories: hand design methods and computer design methods. Hand design methods as they exist in the most recent literature and as applied by a competent designer are still valid for at least half and perhaps as much as 90 per cent of all heat exchanger problems. On the other hand, anyone who tries to design a vertical thermosiphon reboiler by hand is asking for trouble. If one chooses to use a computer design method, one still has the task of selecting the level of the method. There are short-cut and detailed computer design methods available for most exchanger types.

Whether the chosen design method is by hand or by computer, the next step is to select a tentative set of exchanger geometrical parameters. Any experienced designer has a set of rules and procedures by which he can select a plausible set of candidate parameters for his first attempt; the better he is at estimating the starting design, the sooner he will come to the final design, and this is very important for hand calculation methods. Such a method is offered in Section 6 of this Chapter. On a computer, however, it is usually faster to let the computer select a starting point - usually a very conservative case - and use its enormous computational speed to move towards the desired design.

In either case, the initial design will be "rated"; that is, the thermal performance and the pressure drops for both streams will be calculated for this design. The rating program is described in more detail in the next section.

If the calculation shows that the required amount of heat can not be transferred or if one or both allowable pressure drops are exceeded, it is necessary to select a different, usually larger heat exchanger, and re-rate. Alternatively, if one or both pressure drops are much smaller than allowable, a better selection of parameters may result in a smaller and less costly heat exchanger, while utilizing more of the available pressure drop.

The selection of the modified design requires a very elegant piece of logic, but a very simple example will be described in Section 4.

## 3. THE RATING PROGRAM

The rating program shown schematically in Figure 2 is the core of the entire heat exchanger design program. In the rating program, the problem specifications and a preliminary estimate of the exchanger configurations are used as input data; the exchanger configuration given is tested for its ability to effect the required temperature change on the process streams within the pressure drop limitation specified.

The rating process carries out basically three kinds of calculations. First, it computes a number of internal geometry parameters - surface, flow areas, leakage areas, by-pass area, etc. - quantities that are required as further input into the heat transfer correlations and the pressure drop correlations. The other two basic calculations - heat transfer and pressure drop - are calculated for each stream in the configuration specified.

The results from the rating program are either the outlet temperatures of the streams if the length of the heat exchanger has been fixed or the length of the heat exchanger required to effect the necessary thermal change if the duty of the heat exchanger has been fixed. In either case, the rating program will also calculate the pressure drops for both streams in the exchanger.

The greatest amount of technical effort is required in generating the rating program because it is in this program that all of the correlations for heat transfer and pressure drop must be put in quantitative form by the heat exchanger designer. These correlations may come from theoretical analyses, or from experimental studies which have been correlated in suitable general terms,

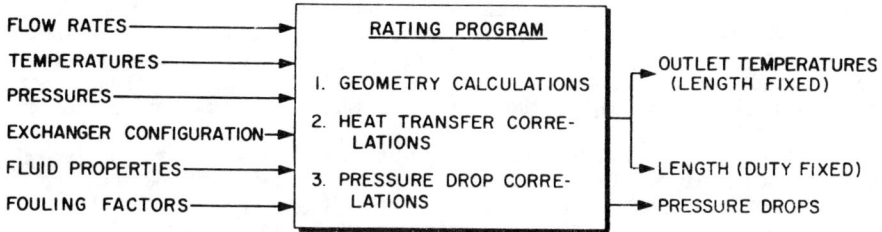

Figure 2. The Rating Program

or in some combination of these two methods. In addition, it is usually necessary to do a great deal of model building in order to correct the basic correlations to the real geometry factors and interaction effects that actually exist in such a complicated piece of equipment as a shell and tube heat exchanger.

The Delaware method for shell-side rating of shell and tube heat exchangers is given in the next Chapter. This method is suitable for both hand and computer use and has been found to be the best of the methods in the open literature. However, proprietary computer-based methods do predict exchanger performance more closely and are widely used.

## 4. THE DESIGN MODIFICATION PROGRAM

The design modification program takes the output from the rating program and modifies the configuration in such a way that the new configuration will do a "better" job of solving the heat transfer problem.

The configuration modification program is typically an extremely complex one logically because it must determine what limits the performance of the heat exchanger and what can be done to remove that limitation without adversely affecting either the cost of the exchanger or the operational characteristics of the exchanger which are satisfactory. If, for example, it finds that the heat exchanger is limited by the amount of heat that it can transfer, the program will try either to increase the heat transfer coefficient or to increase the area of the heat exchanger. To increase the tube side coefficient, one can increase the number of tube passes, thereby increasing the tube side velocity. If the shell-side heat transfer is limiting, one can decrease baffle spacing or decrease baffle cut. To increase area, one can increase the length of the exchanger, or increase the shell diameter, or go to multiple shells in series or in parallel.

If the exchanger is limited by a very low mean temperature difference or by a very low value of the correction factor on the mean temperature difference, the modification program can go to a countercurrent configuration or to multiple shells in series.

If the exchanger is limited by pressure drop on the tube side, the program can decrease the number of tube passes or it can increase the tube diameter; alternatively, it can decrease the tube length and increase the shell diameter and the number of tubes. If the exchanger is limited by pressure drop on the shell-side the program may increase the baffle cut, increase the baffle spacing, increase the tube pitch, or go to double or triple segmental baffles. There are other mechanical modifications which could be made, but these are the ones which are most commonly tested in the configuration modification programs.

A very simple but illustrative configuration modification program is diagrammed in Figure 3. In this particular program, it is assumed that the duty of the exchanger has been specified and the length required for the exchanger has been calculated by the rating program. The rating program has as its initial output the length of the largest diameter acceptable shell, the fewest tube passes consistent with the shell configuration, and the greatest allowable baffle spacing permitted by mechanical construction standards. The quantities with asterisks are the maximum allowable values of that parameter as specified in the input data.

The first question that arises is: Is the length calculated by the rating program less than the maximum allowable length of the exchanger that can be accommodated? If the answer to this question is "No", then it is necessary either to add another shell in parallel to accommodate the flow, or to adjust internal parameters in the heat exchanger in order to increase that heat transfer coefficient and also the pressure drop. The decision between these two choices is made on the basis of the calculated pressure drops as shown in Figure 3. The parameter adjusted depends upon the circumstances. For example, if the tube side coefficient and pressure drop are low compared to the limit allowed, one would increase the number of tube passes. In any case, after the adjustments have been made, it is necessary to go back to the rating program and re-rate the new design. The output of the rating program then comes back into the block on the left side of Figure 3 again.

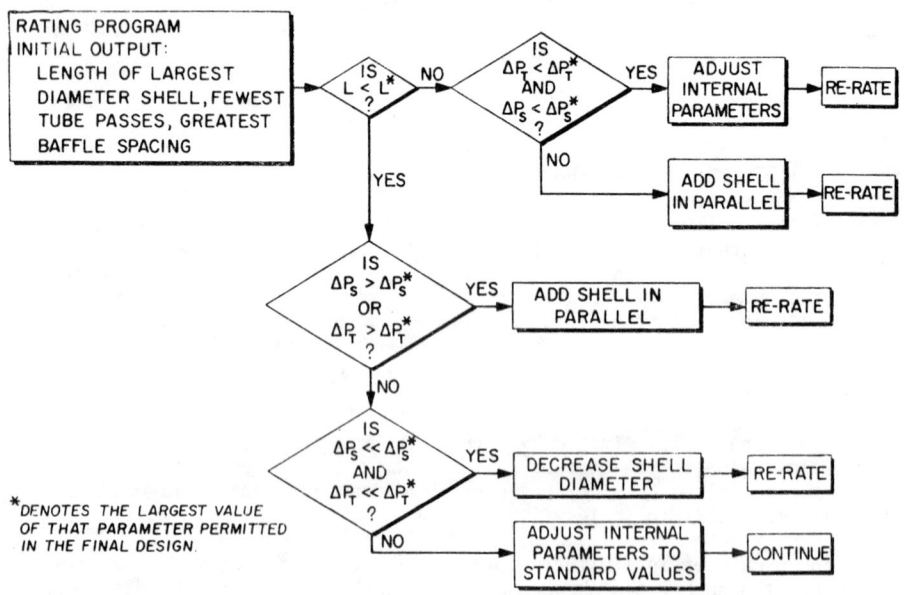

Figure 3. A Schematic Configuration Modification Program (Highly Simplified)

# PRELIMINARY DESIGN OF SHELL AND TUBE HEAT EXCHANGERS

If, on the other hand, the length of the exchanger is calculated to be less than the maximum allowable length (i.e., $L < L^*$), it is necessary to determine first whether the pressure drop limitations have been exceeded. If either pressure drop limitation is exceeded, it is necessary to add a shell in parallel and re-rate the new configuration. If both of the pressure drops are very much less than those allowed, it is probable that it will be possible to decrease the shell diameter one size and come up with a smaller and less expensive exchanger than the one previously rated. If the pressure drops calculated are only slightly less than the pressure drop allowed, then the exchanger designed is very close to the smallest acceptable exchanger (i.e., that will meet all of the operational requirements.) One then adjusts the design parameters to standard values for exchanger design and continues with the rest of the exchanger design program, including mechanical design, costing, optimization, statistical performance evaluation, etc.

## 5. VERIFICATION OF COMPUTER OUTPUT

Every key output item from a computer program should be personally inspected to verify its basic rationality. For heat exchanger designs, this inspection should be done by the heat exchanger designer who can very quickly determine whether the numbers presented are reasonable or if the design the computer has generated is a reasonable one.

It is important to do this every time. While any heat exchanger design program prepared by a reputable designer or programmer has been verified by running it against a number of test cases, the logical structure of a large program is so complex that there is no possibility of ever testing all of the possible logical circuits that might be encountered. For example, even a modest-sized design program might have 40 logical decision points in its structure, resulting in $2^{40} = 1.1 \times 10^{12}$ distinct ways through the logic. If we were to run a different logical path every second around the clock, it would take just under 35,000 years to check out all the possibilities. Therefore, in any given design problem, it is possible that a particular logical loop will be entered for the first time. If the logic of this loop is faulty, the results from it may be completely unrealistic or lead to further illogical steps.

Any questionable item in the output should be verified by analysis at whatever level of detail is required to establish its basic validity. Individual heat transfer coefficients that seem to be too high or too low can be verified by going back to the basic correlation and doing a very quick calculation of that correlation. Geometry factors can be checked very quickly against plausible values by simple geometrical relationships. Particular attention should be paid to those items which can substantially affect the performance and operating life of the plant. For example, if an overall heat transfer coefficient turns out to be especially sensitive to the choice of a fouling resistance, it is worth taking additional time to reflect upon experience as well as any applicable information from outside sources as to the probable value of that fouling resistance.

Any equipment designer, recognizing the inherent uncertainties in so much of what does, asks himself, "What is the penalty for being wrong?" And secondly, "What can be done about it?" There are many different ways of being wrong, but some of them result in massive penalties against the successful operation of the plant, while others result in only short-term or low-level deficiencies. The designer, faced with the fact that he can be and may be wrong on important matters, must ask himself what he can do to minimize the consequences. In some cases, this requires a second heat exchanger identical in configuration with the first and adequate to handle the entire load if the first exchanger must be taken off line for cleaning or maintenance (100 per cent standby). In other cases, it may simply consist of redesigning the exchanger so that the velocities in the exchanger are higher than normal practice if the problems are likely to be with fouling, or lower if erosion is involved; in this case it may be necessary to relax the usual pressure drop specifications.

## 6. PRELIMINARY DESIGN PROCEDURE

In the previous sections, the need for a preliminary estimate of the heat exchanger size was mentioned as a prerequisite for hand design methods of heat exchangers and useful for computer based methods. There are also a number of situations in which a quick approximate estimate is more useful than a delayed detailed design.

If, for example, a process designer is doing a very preliminary estimate of plant cost, layout and space requirements, he can probably estimate the approximate heat transfer area and shell length and diameter of a dozen exchangers in the length of time required to fill out the computer input data sheets for one, using a high level design program like HTRI's ST-4. Eventually, if the plant is to be built, he (or better yet, a heat exchanger specialist) will have to do the detailed calculations, but that task can profitably wait until the preliminary studies have ascertained whether the plant looks economically promising and given a pretty definite idea of how the components are to be strung together.

In this section, a preliminary design method is presented in some detail. This method covers most of the usual process applications for shell and tube exchangers. This method is essentially identical to that given in Ref. 4.

### 6.1 The Basic Design Equation

The basic design equation to be used in this section is

$$A_o = \frac{Q}{U_o (MTD)} = \frac{Q}{U_o F_T (LMTD)} \qquad (6.1\text{-}1)$$

where $A_o$ is the total heat transfer area required in the exchanger, ft$^2$, (calculated based on the outside diameter of the tube); Q is the heat duty of the exchanger, Btu/hr; $U_o$ is the overall heat transfer coefficient, Btu/hr ft$^2$°F (also based on the outside diameter of the tubes); LMTD is the logarithmic mean temperature difference calculated for countercurrent flow, °F; and $F_T$ is the configuration correction factor. Each of these terms will be discussed in greater detail below.

It should be noted that the validity of Eq. (6.1-1) is dependent upon a number of assumptions including constant specific heat of each stream (an isothermal stream such as condensing steam also satisfies this requirement) and constant overall heat transfer coefficient. These conditions are often not met, and in principle, a more elaborate formulation is required. However, that completely defeats the purpose of this procedure, and moreover, most of these departures from ideality introduce errors smaller than the probable error in the other approximations made in the method.

## 6.2 Estimation of Heat Load

The heat load can be quickly calculated for the sensible heat transfer case from

$$Q = WC_p (T_1 - T_2) = wc_p (t_2 - t_1) \quad (6.2\text{-}1)$$

where W and w are the mass flow rates (in $lb_m/hr$) of the hot and cold fluids, respectively, $C_p$ and $c_p$ the specific heats, (in $Btu/lb_m°F$) $T_1$ and $T_2$ the inlet and outlet temperatures of the hot stream (°F), and $t_1$ and $t_2$ the inlet and outlet temperatures of the cold stream. If one of the streams is an isothermal phase change stream (such as condensing steam).

$$Q = W\lambda \quad (6.2\text{-}2)$$

where W is the mass of the stream changing phase per unit time ($lb_m/hr$) and $\lambda$ is the latent heat of the phase change ($Btu/lb_m$).

More complex cases such as partial condensers require more elaborate analysis then space permits here, though the present method can be applied (with care!) to obtain at least rough estimates even in these cases.

## 6.3 Estimation of the Mean Temperature Difference

The first step in calculating the Mean Temperature Difference (MTD) is to find the Logarithmic Mean Temperature Difference (LMTD) for countercurrent flow. Using the temperatures defined in the previous section:

$$\text{LMTD} = \frac{(T_1 - t_2) - (T_2 - t_1)}{\ln \left( \frac{T_1 - t_2}{T_2 - t_1} \right)} \quad (6.3\text{-}1)$$

If $(T_1 - t_2) = (T_2 - t_1)$, then the equation reduces to

$$\text{LMTD} = (T_1 - t_2) = (T_2 - t_1) \quad (6.3\text{-}2)$$

For many purposes, the LMTD can be sufficiently approximated by the arithmetic mean temperature difference (AMTD):

$$\text{AMTD} = 1/2 \ [(T_1 - t_2) + (T_2 - t_1)] \quad (6.3\text{-}2)$$

The LMTD is always equal to or less than the AMTD. The difference between the LMTD and the AMTD increases with decreasing ratio of the smaller temperature difference to the larger. The ratio is given in Table 1.

The calculation of $F_T$ is somewhat more complicated, since it is necessary to use charts from, e.g., Refs (1) or (2), to obtain the exact value. However, there are certain practical considerations determining what that value may be. For the single tubeside pass, purely countercurrent heat exchanger, $F_T = 1.00$.

For preliminary design of a single shell with any even number of tube side passes, F may be estimated as 0.9, which is the average between the maximum possible value, 1.0, and the minimum recommended value, 0.8. This value may be

TABLE 1

Ratio of LMTD and AMTD as

Function of $\dfrac{\text{Smaller Terminal Temperature Difference}}{\text{Larger Terminal Temperature Difference}}$

| $(T_1 - t_2)/(T_2 - t_1)$* | LMTD/AMTD |
|---|---|
| 1.00 | 1.00 |
| 0.70 | 0.990 |
| 0.50 | 0.962 |
| 0.40 | 0.935 |
| 0.30 | 0.894 |
| 0.20 | 0.828 |
| 0.15 | 0.779 |
| 0.10 | 0.711 |
| 0.05 | 0.604 |
| 0.01 | 0.426 |

*If $(T_1 - t_2) < (T_2 - t_1)$. Otherwise, use the inverse ratio $(T_2 - t_1)/(T_1 - t_2)$ for this column.

shaded higher if the ratio of the (countercurrent) terminal temperature differences is near unity and lower if the outlet stream temperatures are similar. In the latter case - and more espcially if there is a temperature cross - the thermodynamic feasibility of the design should be checked before proceeding further. An absolute limit that may be quickly checked is

$$2T_2 \geq t_1 + t_2, \text{ hot fluid on shell} \qquad (6.3\text{-}3)$$

$$2t_2 \leq T_1 + T_2, \text{ cold fluid on shell} \qquad (6.3\text{-}4)$$

where the temperatures are as defined previously. If these limits are approached, it is necessary to use multiple shells in series.

6.4  Estimation of Number of Shells in Series

There is a rapid graphical technique for estimating a sufficient number of shells in series. The procedure is shown in Figure 4 and goes as follows:

a. The terminal temperatures of the two streams are plotted on the ordinates of ordinary arithmetic graph paper, the hot fluid inlet temperature and the cold fluid outlet temperature on the left hand ordinate and the hot fluid outlet and cold fluid inlet temperature on the right-hand ordinate. The distance between the ordinates is arbitrary, (corresponding to the total amount of heat exchanged between the two streams) and may be chosen to the convenience of the user.

# PRELIMINARY DESIGN OF SHELL AND TUBE HEAT EXCHANGERS

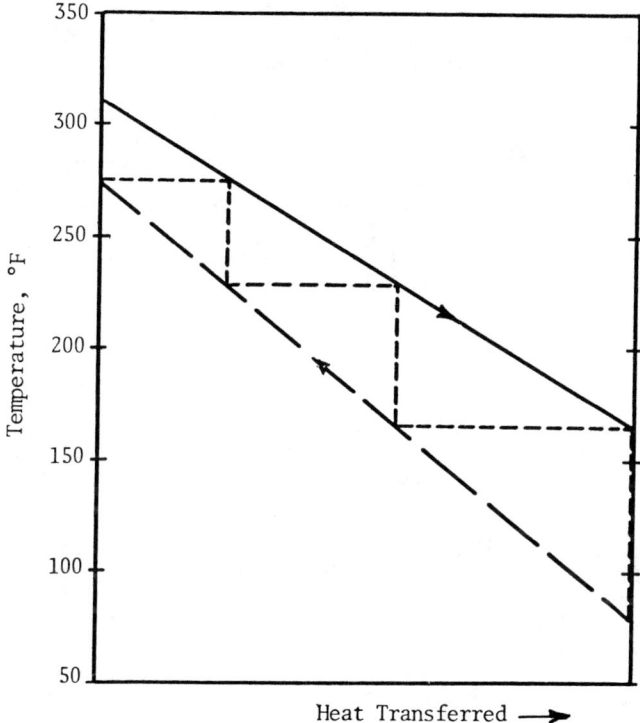

Figure 4. Estimation of Required Number of Shells in Series

b.  If the specific heat of each stream is constant, straight lines ("operating lines") are drawn from the inlet to the outlet temperature point for each stream.

    If the specific heat of one or both streams varies, it is necessary to calculate the temperature of that stream as a function of the amount of heat added or removed by the other, resulting in one or both operating lines being curved. In this case, the procedure given here for finding a sufficient number of shells in series will still be valid. However, an additional error will be introduced into the estimation procedure, the magnitude depending upon the curvature of the lines and the variation in $U_o$.

c.  Starting with the cold fluid outlet temperature (275°F in Figure 4), a horizontal line is laid off until it intercepts the hot fluid line. From that point a vertical line is dropped to the cold fluid line (reaching it in our example at a cold fluid temperature of 228°F). This operation defines a heat exchanger configuration in which the hot fluid temperature is never less than any temperature reached by the cold; that is, there is no temperature cross and we know where is no thermodynamic difficulty that can arise if this operation is carried out in one shell.

570  HEAT EXCHANGER DESIGN: RATING, SIZING, AND OPTIMIZATION

d. The process is repeated until a vertical line intercepts the cold fluid operating line at or below the cold fluid inlet temperature. (Alternatively, the process is continued until a horizontal line crosses the right-hand ordinate).

e. The number of horizontal lines (including the one that intersects the right hand ordinate) is equal to the number of shells in series that is clearly sufficient to perform the duty. In the case of the example problem, this number is three.

Following the above procedure will usually result in a number of shells having an overall $F_T$ between 0.8 and 0.9.

## 6.5 Estimation of $U_o$

The step with the greatest uncertainty in preliminary calculations is estimating the overall heat transfer coefficient. Tables of U's typical of various services are widespread; the drawback to their use is that, in trying to include the entire range ever encountered in practice, these tables give a spread of values so great as to be almost meaningless for more or less optimal design. A table of $U_o$ for various services is included here as Table 2. The values given are the plain tubes. They can usually be used with caution for low-finned tubes (e.g., Wolverine ST Trufin) if the controlling resistance is placed on the shell side; the value should be reduced by 10-30 per cent if the shell-side fluid is of medium or high viscosity, and by 50 per cent if the shell side fluid is of high viscosity and is being cooled.

A better procedure is to build up the value of $U_o$ from the individual h values, and wall, fin and fouling resistances using Eq. (6.5-1)

$$\frac{1}{U_o} = \frac{1}{h_o} + R_{fo} + R_{fin} + \frac{\Delta X}{k}\left(\frac{A_o}{A_m}\right) + \left(R_{fi} + \frac{1}{h_i}\right)\frac{A_o}{A_i} \qquad (6.5-1)$$

where $h_o$ is the outside (shellside) heat transfer coefficient; $R_{fo}$ is the shellside fouling resistance (in hr ft² °F/Btu); $R_{fin}$ is the fin resistance (used only with finned tubes and typically on the order of $1 - 5 \times 10^{-4}$ hr ft² °F/Btu for low finned tubes; Ref. (3) discusses the concept in greater detail); $\Delta X$ is the tube wall thickness (in ft); k is the thermal conductivity of the tube material; $A_m$ is the mean wall heat transfer area, usually taken to be the arithmetic mean, $1/2(A_o + A_i)$, in ft²; $R_{fi}$ is the inside fouling resistance and $h_i$ is the inside heat transfer coefficient, both based on the tube inside heat transfer area, $A_i$. A "contact resistance" term must also be added if mechanically-bonded finned tubes (as opposed to integrally-finned tubes) are used.

It will be generally found that one or at most two terms will dominate the value of $U_o$, and attention can be focused upon these controlling values. It will generally be found that the range of reasonable values is far smaller than the range of possible values. It will also prove useful to make some estimate of the basic uncertainties in each value, i.e., the uncertainty in the value that would still exist even if the best available correlation and physical properties were used. This will often indicate clearly the futility of worrying too much about the precise value of a coefficient for preliminary design purposes. Tabulation of probable values of the individual coefficients and their uncertainties is beyond the scope of this section.

# PRELIMINARY DESIGN OF SHELL AND TUBE HEAT EXCHANGERS

Table 2. Typical Overall Design Coefficients for Shell and Tube Heat Exchangers (Numbers in Parentheses Refer to Footnotes.)

| Fluid 1 | Fluid 2 | Total Fouling Resistance (1) hr ft$^2$°F/Btu | $U_o$ (1,2) Btu/hr ft$^2$°F |
|---|---|---|---|
| Water (3,4) | Water | 0.0015 | 250 - 300 |
| " | Gas, about 10psig | 0.001 | 15 - 20 |
| " | Gas, about 100psig | 0.001 | 30 - 40 |
| " | Gas, about 1000psig | 0.001 | 60 - 100 |
| " | Light organic liquids (5) | 0.0015 | 125 - 175 |
| " | Medium organic liquids (6) | 0.002 | 75 - 125 |
| " | Heavy organic liquids (7) | 0.0025 | 40 - 75 |
| " | Very heavy organic liquids (8) Heating Cooling | 0.004 | 10 - 40 5 - 15 |
| Steam | Gas, about 10psig | 0.0005 | 15 - 20 |
| " | Gas, about 100psig | 0.0005 | 35 - 45 |
| " | Gas, about 1000psig | 0.0005 | 70 - 110 |
| " | Light organic liquids (5) | 0.001 | 135 - 190 |
| " | Medium organic liquids (6) | 0.0015 | 80 - 135 |
| " | Heavy organic liquids (7) | 0.002 | 45 - 80 |
| " | Very heavy organic liquids (8) | 0.0035 | 15 - 45 |
| " (no non-condensables) | Water | 0.001 | 300 - 400 |
| Light organic liquids (5) | Light organic liquids (5) | 0.002 | 100 - 130 |
| " | Medium organic liquids (6) | 0.0025 | 70 - 100 |
| " | Heavy organic liquids (7) Heating Cooling | 0.003 | 40 - 75 25 - 50 |

Table 2. (cont.)

| | | | |
|---|---|---|---|
| Light organic liquids (5) | Very heavy organic liquids | 0.004 | |
| | Heating | | 20 - 50 |
| | Cooling | | 5 - 25 |
| Medium organic liquids (6) | Medium organic liquids (6) | 0.003 | 50 - 80 |
| " | Heavy organic liquids (7) | 0.0035 | |
| | Heating | | 30 - 50 |
| | Cooling | | 15 - 35 |
| " | Very heavy organic liquids (8) | 0.0045 | |
| | Heating | | 15 - 30 |
| | Cooling | | 5 - 25 |
| Heavy organic liquids (7) | Heavy organic liquids (7) | 0.005 | 10 - 30 |
| " | Very heavy organic liquids | 0.006 | 5 - 15 |
| Gas, about 10psig | Gas, about 10psig | 0 | 10 - 15 |
| " | Gas, about 100psig | 0 | 15 - 20 |
| " | Gas, about 1000psig | 0 | 15 - 25 |
| Gas, about 100psig | Gas, about 100psig | 0 | 20 - 30 |
| " | Gas, about 1000psig | 0 | 25 - 35 |
| Gas, about 1000psig | Gas, about 1000psig | 0 | 35 - 60 |
| Water | Condensing light organic vapors, pure component (5,9) | 0.001 | 150 - 200 |
| " | Condensing medium organic vapors, pure compenent (6,9) | 0.001 | 100 - 150 |
| " | Condensing heavy organic vapors, pure component (7,9) | 0.002 | 75 - 100 |

# PRELIMINARY DESIGN OF SHELL AND TUBE HEAT EXCHANGERS

General Notes on Table 2.

1. The total fouling resistance and the overall heat transfer coefficient are based on the total outside tube area.

2. Allowable pressure drops on each side are assumed to be about 10psi except for (a) low pressure gas and condensing vapor where the pressure drop is assumed to be about 5 per cent of the absolute pressure, and (b) heavy organics where the allowable pressure drop is assumed to be about 20 to 30 psi.

3. Aqueous solutions give approximately the same coefficients as water.

4. Liquid ammonia gives about the same results as water.

5. "Light organic liquids" include liquids with viscosities less than about 0.5 cp, such as hydrocarbons through $C_8$, gasoline, light alcohols and ketones, etc.

6. "Medium organic liquids" include liquids with viscosities between about 0.5 cp and 1.5 cp, such as kerosene, straw oil, hot gas oil, absorber oil, and light crudes.

7. "Heavy organic liquids" include liquids with viscosities greater than 1.5 cp, but not over 50 cp, such as cold gas oil, lube oils, fuel oils, and heavy and reduced crudes.

8. "Very heavy organic liquids" include tars, asphalts, polymer melts, greases, etc., having viscosities greater than about 50 cp. Estimation of coefficients for these materials is very uncertain.

9. These values may be used for vapor mixtures when the condensing range of the vapor is less than half of the temperature difference between the outlet coolant and the vapor. If the condensing range is greater than this, or if there are significant amounts of non-condensable gas present, the coefficient should be reduced towards the values shown for gas cooling; in these cases, the accuracy of the estimation is very uncertain.

6.6 Calculation of $A_o$

Once Q, MTD, and $U_o$ are known, the total outside heat transfer area (including fin area) $A_o$ is readily found from Eq. (6.1-1). The next question is, "What set of heat exchanger dimensions will accommodate that heat transfer area?" The following section presents a technique for answering that question.

7. ESTIMATION OF MAJOR EXCHANGER PARAMETERS

7.1 Heat Transfer Area for a Given Shell Diameter and Length

Figure 5 is the key to relating the heat transfer area in a shell and tube exchanger to the shell inside diameter and effective tube length. In this figure the ordinate is $A_o$, in $ft^2$, and the abscissa is the effective tube length (tube sheet to tube sheet for straight tube bundles, or length of a single straight section from tube sheet to tangent line for a U-tube bundle) in ft. The solid black lines in parameter are the commonly specified shell inside diameters in inches. Using standard tube count tables for 3/4 in. O.D. tubes on a 15/16 in. triangular pitch for a fixed tube sheet heat exchanger with one tube pass, the total outside tube heat transfer area than can be fitted into a shell has been calculated.

Therefore, once the required area $A_o$ is known, Figure 5 shows immediately the combinations of tube length and shell diameter that will provide that area in a single shell for an exchanger of the given tube size and layout. A technique of applying Figure 5 to other tube sizes, layouts, number of tube passes and shell-type will be given in the next section.

The dashed lines in parameter in Figure 5 (marked 3:1, 6:1, etc.) show the approximate locus of shells having a given effective tube length to shell diameter ratio. Shells shorter than three times the shell diameter (3:1) may suffer from poor fluid distribution and excessive entry and exit losses, and are likely to be more expensive than a longer, smaller diameter unit of the same area, especially if the shell side fluid is at high pressure. Shells longer that 15 times the shell diameter are likely to be difficult to handle mechanically, require a large clearway for bundle removal or retubing, and show the effect of diminishing returns on cost.

Many heat exchangers fall into the 6:1 to 8:1 range, with a pronounced trend towards the higher values as pressure drop prediction procedures have improved.

7.2 Extension of Figure 5

Figure 5 can be employed to estimate the required length and diameter of exchangers for other tube sizes and layouts, multiple tube side passes, and bundle constructions. Define an "effective" area, $A_o'$, by Eq. (7.2-1):

$$A_o' = A_o \, F_1 \, F_2 \, F_3 \qquad (7.2-1)$$

where $A_o'$ is the area on the ordinate of Figure 5:

$A_o$ is the required outside area of a heat exchanger as calculated from Eq. (6.1-1)

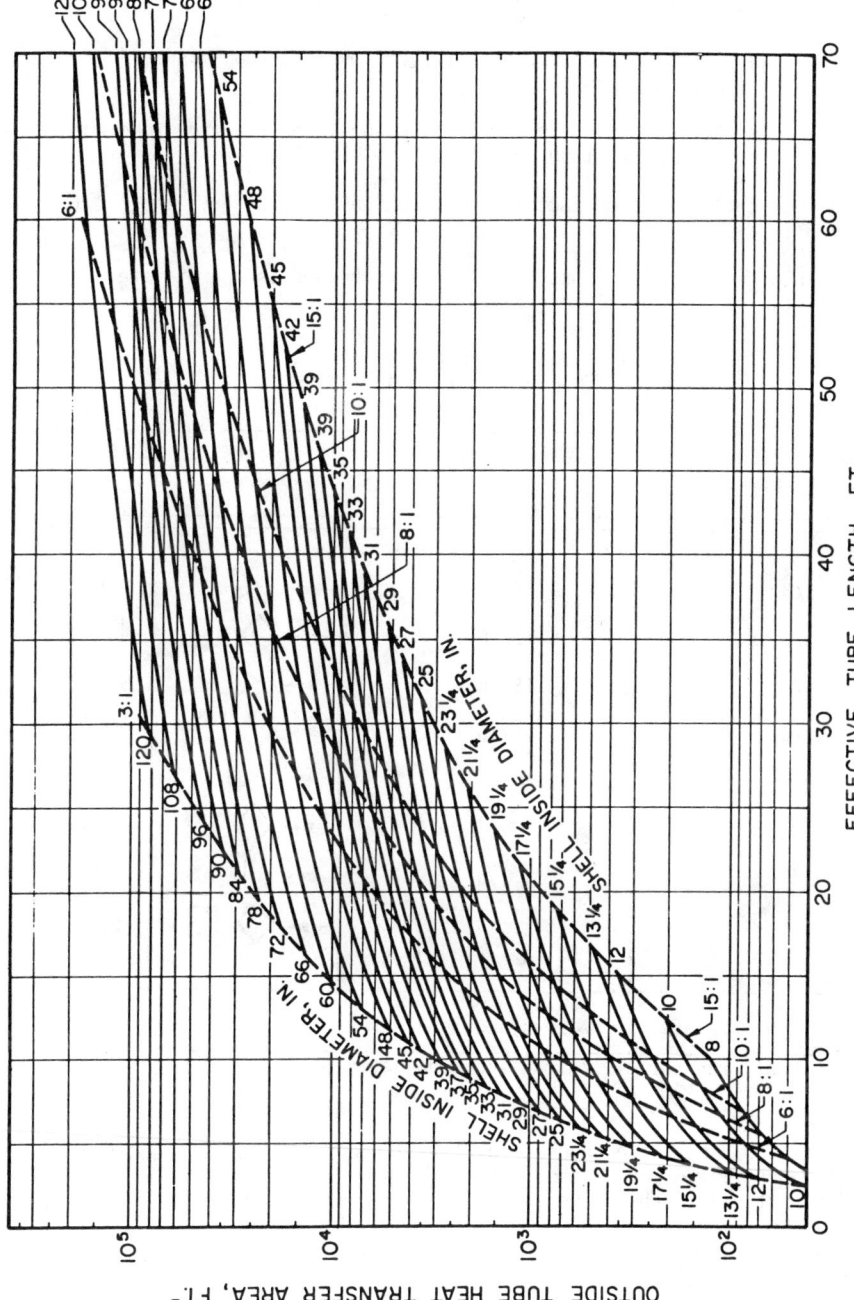

Figure 5. Heat Transfer Area as a Function of Shell Inside Diameter and Effective Tube Length for 3/4 in. Equilateral Triangular Tube Layout, Fixed Tube Sheet, One Tubeside Pass, Fully Tubed Shell.

$F_1$ is the correction factor for the unit cell tube array (= 1.00 for 3/4 in. tubes on a 15/16 in. triangular pitch).

$F_2$ is the correction factor for the number of tube passes (= 1.00 for one tube pass).

$F_3$ is the correction factor for the shell construction/tube bundle layout type (= 1.00 for fixed tube sheet).

Values of the correction factors are given in Table 3 ($F_1$), Table 4 ($F_2$) and Table 5 ($F_3$).

It should be emphasized that use of this technique is approximate, especially for the smaller diameter shells.

Table 3. Values of $F_1$ for Various Tube Diameters and Layouts

| Tube Outside Diameter, in. | Tube Pitch, in. | Layout | $F_1$ |
|---|---|---|---|
| 5/8 | 13/16 | △ | 0.90 |
| 5/8 | 13/16 | ◇ □ | 1.04 |
| 3/4 | 15/16 | △ | 1.00 |
| 3/4 | 15/16 | ◇ □ | 1.16 |
| 3/4 | 1 | △ | 1.14 |
| 3/4 | 1 | ◇ □ | 1.31 |
| 1 | 1 1/4 | △ | 1.34 |
| 1 | 1 1/4 | ◇ □ | 1.54 |

$$F_1 = \frac{\text{(Heat transfer area/cross-sectional area of unit cell)}_{\text{Reference}}}{\text{(Heat transfer area/cross-sectional area of unit cell)}_{\text{New case}}}$$

This table may also be used for low-finned tubing in the following way: The value estimated for $U_o$ should be based upon the total outside area (including fins) of the finned tube. This value will generally be somewhat less (10 to 30 per cent) than the plain tube values given in Table 2. Then the required value of $A_o$ is based upon the finned tube area, and the above values of $F_1$ are <u>divided</u> by the ratio of the finned tube area to the plain tube area (per unit length). Typically this value will be from 2.5 to 4.

Table 4. Values of $F_2$ for Various Numbers of Tube Side Passes*

| Inside Shell Diameter, in. | $F_2$ Number of Tube Side Passes | | | |
|---|---|---|---|---|
| | 2 | 4 | 6 | 8 |
| Up to 12 | 1.20 | 1.40 | 1.80 | - |
| 13-1/4 to 17-1/4 | 1.06 | 0.18 | 1.25 | 1.50 |
| 19-1/4 to 23-1/4 | 1.04 | 1.14 | 1.19 | 1.35 |
| 25 to 33 | 1.03 | 1.12 | 1.16 | 1.20 |
| 35 to 45 | 1.02 | 1.08 | 1.12 | 1.16 |
| 48 to 60 | 1.02 | 1.05 | 1.08 | 1.12 |
| Above 60 | 1.01 | 1.03 | 1.04 | 1.06 |

* Since U-tube bundles must always have at least two passes, use of this table is essential for U-tube bundle estimation. Most floating head bundles also require an even number of passes.

Table 5. $F_3$ for Various Tube Bundle Constructions

| Type of Tube Bundle Construction | $F_3$ Inside Shell Diameter, in. | | | | |
|---|---|---|---|---|---|
| | Up to 12 | 13¼-21¼ | 23¼-35 | 37-48 | Above 48 |
| Split Backing Ring (TEMA S) | 1.30 | 1.15 | 1.09 | 1.06 | 1.04 |
| Outside Packed Floating Head (TEMA P) | 1.30 | 1.15 | 1.09 | 1.06 | 1.04 |
| U-Tube* (TEMA U) | 1.12 | 1.08 | 1.03 | 1.01 | 1.01 |
| Pull-Through Floating Head (TEMA T) | --- | 1.40 | 1.25 | 1.18 | 1.15 |

* Since U-tube bundles must always have at least two tube side passes, it is essential to use Table 4 also for this configuration.

## 7.3 Example Problem

The use of the method can be best illustrated by an example:

High pressure recycle gas (1200 psig, with 40% hydrogen) is to be cooled from 550°F to 250°F, heating a mixed liquid aromatics stream from 80°F to 232°F. The gas flow rate is 380,000 lb/hr, with a specific heat of 1.8 Btu/lb°F and the aromatic flow rate is 2.49 x $10^6$ lb/hr with a specific heat of 0.54 Btu/lb°F.

A U-tube exchanger is to be used with the gas in the tubes. The tubes will be 1 in. O.D., 12 BWG on a 1-1/4 in. square pitch. Typical pressure drops of about 10 psi are allowed for each stream.

Estimate the effective tube length and diameter of an exchanger for this service.

Solution:

Check heat balance:

$Q_{gas}$ = 380,000 (1.8)(550-250)

= 2.05 x $10^8$ Btu/hr

$Q_{aro}$ = 2.49 x $10^6$ (0.54)(232 - 80)

= 2.04 x $10^8$ Btu/hr

Mean temperature difference:

$$\text{LMTD} = \frac{(550-232)-(250-80)}{\ln\left(\frac{550-232}{250-80}\right)}$$

= 236°F  (AMTD = 244°F)

From $F_T$ charts, $F_T$ = 0.85.

MTD = 0.85 (236) = 201°F

Estimation of $U_o$:

This particular case is not explicitly included in Table 2. However, inspection of the table shows that light to medium organic liquids give individual film transfer coefficients of about 250 Btu/hr ft²°F and gases at 1000 psig give individual coefficients of about 100 Btu/hr ft²°F. The presence of the hydrogen would be expected to increase this value substantially, to at least 150 Btu/hr ft²°F. Allowing a fouling resistance of 0.001 hr ft²°F/Btu for the aromatic gives, by Eq. (6.5-1):

$$\frac{1}{U_o} = \frac{1}{250} + 0.001 + \frac{0.109}{12(25)}\left(\frac{1.000}{0.891}\right) + \frac{1}{150}\left(\frac{1.000}{0.782}\right)$$

$U_o$ = 72 Btu/hr ft²°F

Calculation of Required Area:

$$A_o = \frac{2.05 \times 10^8}{72(201)} = 14{,}200 \text{ ft}^2$$

This is the actual area required in the exchanger. However, in order to use Figure 5, it is necessary to use Equation 7.2-1 and the associated tables to find $A_o'$:

From Table 3, for 1 in. tubes on a 1-1/4 in. square layout, $F_1 = 1.54$.

From Table 4, for two tube passes and a shell diameter about 60 in., $F_2 = 1.02$.

From Table 5, for U tube construction and a shell diameter of about 60 in., $F_3 = 1.01$.

Thus,

$$A_o' = 14{,}200 \ (1.54)(1.02)(1.01)$$

$$= 22{,}500 \text{ ft}^2$$

Entering Figure 5 at 22,500 ft$^2$ on the ordinate, we find the following shell diameter/tube length combinations will provide the required area:

| Shell, I.D., inches | Effective Tube Length, ft | Approximate Shell Length, ft |
|---|---|---|
| 72 | 22 | 27 |
| 66 | 27 | 32 |
| 60 | 33 | 37-1/2 |
| 54 | 40 | 44 |
| 48 | 50 | 54 |

Any of these would be considered a fairly standard job by a medium to large size heat exchanger manufacturer. Site considerations might limit the length, and tube side velocity, shell side pressure drop and vibration limits would certainly play an important role in further calculations. The author would probably select the 54 in. ID for further analysis.

# REFERENCES

1. Kern, D. Q., "Process Heat Transfer", McGraw-Hill Book Co., New York (1948).

2. Perry, R. H., and Chilton, C. H., Eds., "Chemical Engineers' Handbook", 5th Ed., McGraw-Hill Book Co., New York (1973).

3. Kern, D. Q., and Kraus, A. D., "Extended Surface Heat Transfer", McGraw-Hill Book Co., New York (1972).

4. Bell, K. J., "Estimate S & T Exchanger Design Fast", Oil and Gas Jour., Dec. 4, 1978, pp. 59-68.

# Delaware Method for Shell Side Design

**KENNETH J. BELL**
School of Chemical Engineering
Oklahoma State University
Stillwater, Oklahoma, USA

ABSTRACT

The logical structure of the Delaware Method for shell-side thermal-hydraulic design of shell and tube heat exchangers is described. Then the method itself is presented in an algorithmic form with the necessary tables and figures to carry out the calculations.

1. HISTORICAL DEVELOPMENT OF THE DELAWARE METHOD

From 1947 to 1963 the Department of Chemical Engineering at the University of Delaware carried out a comprehensive research program on shell-side fluid flow and heat transfer in shell and tube heat exchangers. Table 1 shows the high points during the period of the so-called Delaware project. In 1947, the project started under ASME sponsorship using funds from the Tubular Exchanger Manufacturers Association, The American Petroleum Institute, Andale Company, Davis Engineering Co., Downingtown Iron Works, E. I. duPont de Nemours and Company, Standard Oil Development Co. (Exxon), and York Corporation. The Principal Investigators were Professors Olaf Bergelin and Allan Colburn of the University of Delaware.

Table I

Milestones of the Delaware Project

| | |
|---|---|
| 1947: | Project started under ASME Sponsorship |
| 1947-1959: | Experimental program underway |
| 1950: | First report (Bulletin No. 2) (1) |
| 1958: | Second report (Bulletin No. 4) (2) |
| 1960: | Design method first published (3) |
| 1963: | Final report (Bulletin No. 5) (4) |

From 1947 to 1959, the experimental program was carried on, beginning with measurements of heat transfer and pressure drop during flow across ideal tube banks. These efforts were successively extended to introduce the various design features characteristic of shell and tube heat exchangers in commercial use. Sequentially, various baffle cut and spacing configurations were investigated inside a cylindrical shell with no baffle leakage and minimal bypass clearance. Baffle leakages between baffles and the shell and between the tubes and baffles were added in later studies. Finally, the bypass flow around the bundle between the outer tube limit and the shell inside diameter was studied, together with the effect of sealing devices. The first report was issued in 1950 (1) with a second report following in 1958 (2). In 1960, a preliminary design method for E shell exchangers was published (3). In 1963, the Final Report was published (4). Subsequent variations of the design method have appeared in the literature (e.g. (5)) and several new features are incorporated in the present version.

## 2. SIMPLIFIED MECHANISMS OF SHELL SIDE FLOW

In Figure 1, we show a diagram of the shell-side flow mechanisms in a highly idealized form. This diagram has been modified from Palen and Taborek (6) who in turn borrowed it and modified it from the original version shown by Tinker (7). We identify five different streams on the shell-side. Stream B is the main crossflow stream flowing through one window across the crossflow section and out through the opposite window. This is the stream that is desired on the shell-side of the exchanger.

However, because of the mechanical clearance required in a shell and tube exchanger, there are four other streams which compete with the B stream. First, there is the A stream leaking through the clearances between the tubes and the baffle, from one baffle compartment to the next. Then there is the C stream, the bundle bypass stream, flowing around the tube bundle between the outermost tubes in the bundle and the inside of the shell. The E stream is the shell-to-baffle leakage stream flowing through the clearance between the baffles and the inside diameter of the shell. The last of the identified major streams is the F stream, which flows through any channels within the tube bundle caused by the provision of pass dividers in the exchanger header (i.e., only in multiple tubepass configurations). (It should be noted that, for a two tubepass configuration as shown here, the pass divider ordinarily would be oriented perpendicular to the direction of the main crossflow stream, and would not provide an internal bypass stream; however, it is shown here because it can have a very serious effect in multiple tubepass configuration, where at least some of the pass lanes may be parallel to the direction of flow.)

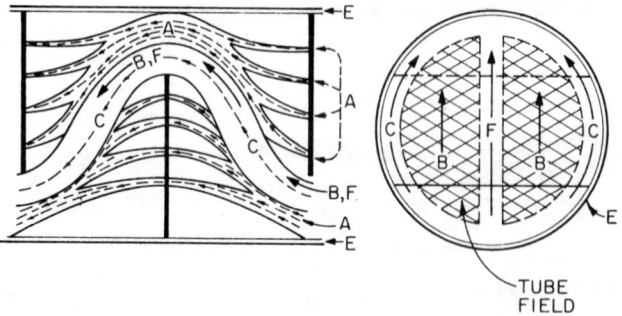

Figure 1. Idealized Diagram of Shell-side Flow Streams (Adapted from Tinker (7) and Palen and Taborek (8)).

These streams do not, of course, exist as precisely defined streams as shown in Figure 1. They form and mix and interact with one another, and a more complete mathematical analysis of the shell-side flow would take this into account. However, these analyses are also quite complicated (see Ref. (6)) and cannot be carried out exactly in any case, simply because of a lack of knowledge of the turbulent flow structures on the shell-side. Therefore, Figure 1 is an idealized representation but does allow us to talk in terms of the major effects modifying the idealized flow pattern.

In the Delaware method, the B stream is regarded as the essential stream in the exchanger with the other streams exerting various modifying effects upon the performance as predicted for the B stream alone. The various leakage and bypass streams affect the heat transfer rate in two separate ways: 1) They reduce the B stream and therefore the local heat transfer coefficient, and 2) they alter the shell-side temperature profile. The Delaware method in effect lumps these two effects together into a single correction.

Not all of the leakage and bypass streams have the same relative magnitude of effect and, of course, they respond differently to the various geometrical parameters of the shell-side. For example, the A stream (tube to baffle leakage) has only a relatively small effect upon the heat transfer coefficient and the pressure drop. The C stream has a relatively large effect, but there are mechanical ways of partially blocking this flow to minimize that effect. The E stream (shell to baffle leakage) has an extremely serious effect and unfortunately there is relatively little one can do to help. Finally, the pass divider bypass stream (F stream) has a moderate effect and responds to some of the same treatment that the bundle bypass stream does.

In the subsequent sections, we examine the basic structure of both the heat transfer and pressure drop correlations in the light of the Delaware program, and the formulation of the Delaware design method.

## 3. BASIC STRUCTURE OF THE DELAWARE METHOD

### 3.1 The Rating Problem

The Delaware method is a rating method, as indeed are most of the other so-called "design methods". "Rating" implies that a specific heat exchanger is fairly completely described geometrically (with the possible exception of the length) and the process specifications for the two streams are given. The major items of information that must be given are shown in Table II. With this information, the rating method will calculate the length of the heat exchanger required to perform the specified duty, or it will calculate the duty that can be performed in the heat exchanger if the length is specified. Additionally, the rating program will calculate the pressure drops for both the tube-side and shell-side.

### 3.2 Shell-side Heat Transfer Equation

The basic equation for calculating the effective average shell-side heat transfer coefficient is given in Equation (1).

$$h_{shell\text{-}side} = h_{ideal} \, J_c J_\ell J_b J_s J_r \qquad (3.2\text{-}1)$$

Table II

The Rating Problem

Given:  Process Specifications
  Flow rates
  Inlet temperatures
  Outlet temperatures (if length is to be found)
  Physical properties
  Fouling characteristics

  Geometry of Exchanger
  Shell inside diameter
  Outer tube limit
  Tube diameter and layout
  Baffle spacing and cut
  (Length)

Calculate: $\begin{cases} \text{Length (if not given)} \\ \text{Duty (if length is given)} \end{cases}$

  Pressure drops

---

$h_{ideal}$ is the heat transfer coefficient for pure crossflow in an ideal tube bank. This value is calculated assuming that the entire shell-side stream flows across the ideal tube bank formed by the tube array at the centerline of the exchanger. The Delaware ideal j-factor curves were obtained for a variety of geometries of industrial interest.

$J_c$ is the correction factor for baffle cut and spacing. This factor takes into account the heat transfer in the window and calculates the overall average for the entire heat exchanger. This correction factor is essentially a function of the fraction of the total tubes in the heat exchanger that are in crossflow (i.e., located between the baffle tips of adjacent baffles). This value is equal to 1.0 for a heat exchanger in which there are no tubes in the window, increases to a value as high as 1.15 for a design in which the windows are relatively small and the window velocity is very high, and decreases to a value of about 0.52 for very large baffle cuts. A typical value for a well-designed heat exchanger is near 1.0.

$J_\ell$ is the correction factor for baffle leakage effects, including both shell-to-baffle and tube-to-baffle leakage. This correction factor is a function of the ratio of total leakage area per baffle to the crossflow area between adjacent baffles and also of the ratio of the shell-to-baffle leakage area to the tube-to-baffle leakage area. $J_\ell$ weights the shell-to-baffle leakage more heavily than the tube-to-baffle leakage. It is also a function of the clearances between tube and baffle and shell and baffle so that credit is given for tighter constructional practices. The correlation for $J_\ell$ penalizes the design if the baffles are put too close together, leading to an excessive fraction of the flow being in the leakage streams compared to the crossflow. A typical value of $J_\ell$ is in the range of 0.7 to 0.8.

$J_b$ is the correction factor for the bundle bypass flow (C stream). $J_b$ accounts for differences in construction: For the relatively small difference between the outermost tubes and the shell for fixed tube sheet construction, $J_\ell \approx 0.9$, whereas for the much larger clearances required by pull-through floating head construction, $J_\ell \approx 0.7$. $J_b$ also considers the improvement made by sealing strips. These are typically longitudinal strips of metal between the outside of the bundle and the shell and fastened to the baffles; they force the bypass flow periodically back into the tube field, causing higher local velocities and heat transfer coefficients and mixing of the bypass stream with the main crossflow stream. Proper use of sealing strips in a split-ring or pull-through floating head can increase $J_\ell$ from 0.7 to 0.9.

$J_s$ is the correction factor for variable baffle spacing in the inlet and outlet sections. Nozzle dimensions frequently require that the nozzles be located far enough from the tube sheets that it may be necessary to increase the baffle spacing for the inlet and outlet sections. This correction factor allows for the change in the average shell-side coefficient caused by these locally lower velocities. $J_s$ will usually be between 0.85 and 1.0.

$J_r$ is the correction factor for adverse temperature gradient build-up. It is well-known that, in laminar flow, the heat transfer coefficient decreases with increasing distance from the start of heating because of the development of an adverse temperature gradient from the conduction process. This gradient resists further transfer of heat and therefore lowers the local and the average heat transfer coefficients with increasing distance. This correction has been worked out mathematically for flow in well-defined geometries such as inside round tubes, but it is also found experimentally to exist during flow across tube banks. For large heat exchangers in deep laminar flow, it can result in a decrease in the average heat transfer coefficient by a factor of two or more compared to what would have been predicted based on flow across a 10-row tube bank. This correction factor applies only if the shell-side Reynolds number is less than 100 and is fully effective only in deep laminar flow characterized by $Re_s < 20$.

The combined effect of all of these correction factors for a reasonably well designed shell and tube heat exchanger is typically on the order of 0.6; i.e., the effective mean shell-side heat transfer coefficient for the exchanger is on the order of 60 percent of that calculated if the entire flow took place across an ideal tube bank corresponding in geometry to one crossflow section. It is interesting to observe that this value has long been used as a rule of thumb.

3.3 Shell-side Pressure Drop

The shell-side pressure drop is built up in the Delaware method by summing the pressure drops for the inlet and exit sections. Before developing these individual terms, it is necessary to define a number of quantities:

$N_b$      Number of baffles in the exchanger

$N_c$      Number of rows of tubes in one crossflow section

$N_{cw}$      Number of rows of tubes in one window section, i.e., the number of tubes between the top of the baffle and the inside of the shell

$\Delta p_{b,i}$      Pressure drop in one crossflow section if there were no leakage or bypass

$\Delta p_{w,i}$   Pressure drop in one window section if there were no leakage or bypass

$R_\ell$   Correction factor for baffle leakage effects. This correction factor is different in magnitude from $J_\ell$ but depends upon the same ratios of total baffle leakage area to crossflow area. Typically, $R_\ell \approx 0.4$ to $0.5$ though lower values may be found in exchangers with closely spaced baffles.

$R_b$   Correction factor for bypass flow. Also different in magnitude from $J_b$, but similar in form. Typically, $R_b \approx 0.5$ to $0.8$, depending upon construction type and number of sealing steps. The lower value would be typical of a pull-through floating head with only one or two pairs of sealing strips, the higher value typical of a well tubed-out fixed tube sheet design.

$R_s$   Correction factor for the entrance and exit sections having a different baffle spacing than the internal sections.

From the very extensive tests at the University of Delaware in which the geometrical variables were varied systematically and independently over the practical range of construction, we know the following:

1. For both crossflow and window sections without leakage or bypass, we may calculate $\Delta p_{b,i}$ and $\Delta p_{w,i}$ fairly accurately from the friction factor curves for the given ideal tube bank lay-out.

2. For a "real" exchanger (that is, one that has both leakage and bypass), the pressure drop in the entrance and exit sections is affected by bypass but not by leakage. Additionally, there is an effect due to variable baffle spacing. Therefore, the combined pressure drop for the entrance and exit sections is:

$$2 \, \Delta p_{b,i} \, (1 + \frac{N_{cw}}{N_c}) \, R_b R_s \qquad (3.3\text{-}1)$$

The portion of the heat exchanger whose pressure drop is predicted by that term is shown by the shaded sections in Figure 2.

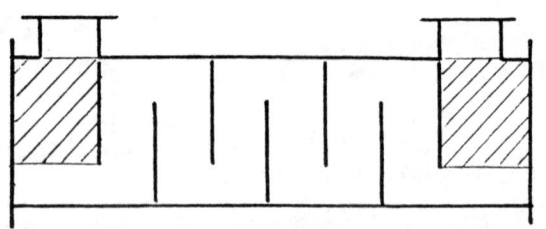

Figure 2.   Entrance and Exit Sections of an E Shell Exchanger

3. For a "real" exchanger, the pressure drop in the interior crossflow sections is affected by both bypass and leakage. Therefore, the combined pressure drop of all the interior crossflow sections is

$$(N_b - 1) \Delta p_{b,i} R_b R_\ell \qquad (3.3\text{-}2)$$

The portion of the heat exchanger whose pressure drop is given by the above term is shown in the cross-hatched portion of Figure 3.

4. For a "real" exchanger, the pressure drop in the windows is affected by leakage but not by bypass. Therefore, the combined pressure drop of all the window sections (indicated by the cross-hatched regions in Figure 4) is given by:

$$N_b \Delta p_{w,i} R_\ell \qquad (3.3\text{-}3)$$

5. Summing the above individual effects, we obtain the equation for the total nozzle-to-nozzle shell-side pressure drop:

$$\Delta p_s = [(N_b-1)\Delta p_{b,i} R_b + N_b \Delta p_{w,i}] R_\ell \qquad (3.3\text{-}4)$$

$$+ 2 \Delta p_{b,i} \left(1 + \frac{N_{cw}}{N_c}\right) R_b R_s$$

While each of the correction factors can vary over quite wide ranges, depending upon the configuration of the heat exchanger, the total shell-side pressure drop of a typical shell and tube exchanger is on the order of 20 to 30 per cent of the pressure drop that would be calculated for flow through the corresponding heat exchanger without baffle leakage and without tube bundle bypass effects. In fact, this is where most of the earlier correlations for heat exchanger design fell down; not knowing the very large effects on pressure drop due to leakage and bypass, it was not uncommon for some of them to overestimate the shell-side pressure drop by a factor of 5 or 10 or even more.

3.4 An Important Consequence of the Delaware and Subsequent Shell-side Design Methods

There is an interesting consequence of the ability of the Delaware method to predict shell-side pressure drops more accurately (and giving much lower values) than had been possible with earlier methods. Once it was established that the pressure drop in heat exchangers was better predicted by the Delaware

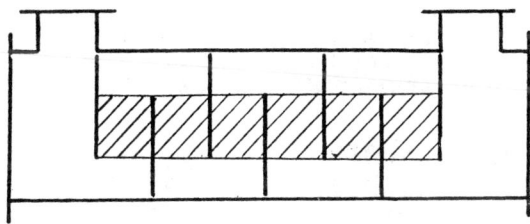

Figure 3. Internal Crossflow Sections of an E Shell

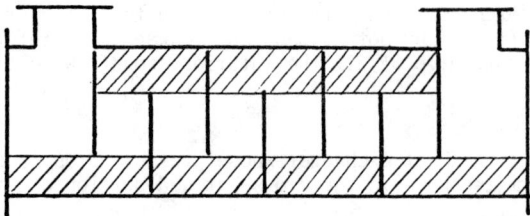

Figure 4. Window Sections in an E Shell Exchanger

method than by other methods available at that time, it was possible to design for higher shell-side velocities (i.e., using up more of the allowable pressure drop). This resulted in higher heat transfer coefficients and a smaller heat exchanger for a given service. An unforeseen consequence of this increase in velocities seems to have been the sudden appearance of significant numbers of tube vibration problems. This, in fact, has become a major design limitation in shell and tube exchangers and one which is not yet very well understood. It is outside the scope of this paper to deal with current efforts in this area, but it suffices to say that one must consider the possible downstream consequences of an improvement in design methods more carefully than in the past.

### 3.5 Accuracy of the Delaware Method

Finally, it should be remembered that this method, though apparently generally the best in the open literature, is not extremely accurate. An exhaustive study (6) tested various literature methods against 972 heat transfer data points and 1332 pressure drop data points covering a very wide range of fluids and geometrical parameters showed that this method predicted shell-side coefficients from about 50 per cent low to 100 per cent high, while the pressure drop range was from about 50 per cent low to 200 per cent high. The mean error for heat transfer was about 15 per cent low (conservative) for all Reynolds numbers, while the mean error for pressure drop was from about 5 per cent low (unsafe) at Reynolds numbers above 1000 to about 100 per cent high at Reynolds numbers below 10.

## 4. EXTENSION OF THE DELAWARE METHOD TO OTHER GEOMETRIES

The Delaware method, as originally developed and as it exists in the open literature, is more or less explicitly confined to the design of fully-tubed E shell configurations using plain tubes. However, there are many process reasons - better balance required between the shell-side and the tube-side heat transfer coefficients, vibration problems, more effective use of available shell-side pressure drop in low pressure drop cases, etc. - that lead to the importance of applying the method to variant configurations.

In the following sections, we will demonstrate how the existing method can be applied at least philosophically and sometimes with the aid of external information on shell-side performance, to the design of a few of these other designs. We shall by no means cover all of the possibilities, but these examples may serve to inspire others to develop their own procedures for using the basic structure of the Delaware method as applicable to configurations of interest to them. The applications that we shall specifically consider will be: 1) the use of low-finned tubes, in order to enhance the effective heat transfer rate on the shell-side; 2) the use of the no-tube-in-the-window configuration to minimize vibration problems; and 3) application to F shells, where it is possible by this means to improve the Mean Temperature Difference (MTD) substantially under the right circumstances.

4.1 Application to Low-Finned Tubes

Low-finned tubes are frequently used in shell and heat exchangers in order to increase the area in contact with the shell-side fluid. If that fluid has a relatively poor heat transfer coefficient such as a gas or a viscous liquid, it is often possible to reduce the overall size and cost of the exchanger significantly. It is possible to apply the Delaware method to the design of such exchangers in a fairly straightforward way making use of the results of Williams and Katz (8) and Briggs, Katz, and Young (9). While the data base in the open literature is limited, the results seem to be consistent with common sense.

Basically, the ideal tube bank heat transfer correlations have to be modified as shown qualitatively in Figure 5. For the heat transfer coefficient correlation, there is little or no effect on the $j_s$ factor at higher Reynolds numbers. At very low Reynolds numbers, apparently the fluid is retarded in passing through the spaces between the fins, leading to a lower value of the $j_s$ factor than would be predicted for a plain tube bundle. Nevertheless, because of the considerable increase in the area available, it is frequently economically advantageous to use low-finned tubes with viscous liquids. For gas-phase application, where ordinarily the Reynolds number is in turbulent flow regime, there is no penalty against the plain tube.

The friction factor data are less clear because of the uncertainties involved in reducing the data within the context of the Delaware method. Nevertheless, it seems that a conservative design basis can be obtained by using friction factor curves which are 1.5 times those for the corresponding plain tube bank.

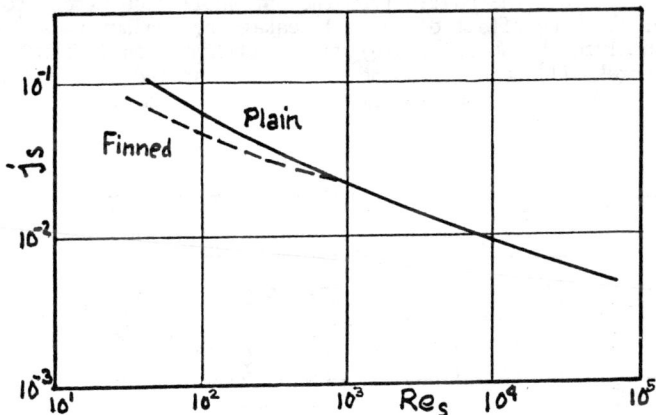

Figure 5. Correlation of $j_s$ vs $Re_s$ for Two Similar Shell and Tube Exchangers, One Finned Tube and One Plain Tube (9)

## 4.2 Application to No-Tubes-in-the-Window Configuration

A schematic diagram of a shell and tube exchanger with no-tubes-in-the-window is shown in Figure 6. As the name implies, there are no tubes which pass through the window regions of the heat exchanger, the baffle cuts being somewhat smaller than they would be in a typical fully-tubed design. The advantage of the no-tubes-in-the-window design is that now every tube passes through and is supported by every baffle guiding the flow. Additionally, tube support plates may be inserted as required between the baffles; the purpose of the tube support plates is to support the tubes against vibration rather than to attempt to guide the flow. The no-tubes-in-the-window configuration is primarily of interest when shell-side vibration problems are expected to be serious. While some fraction of the tubes is lost compared to the fully-tubed design, this loss can be minimized by having small baffle cuts and possibly by an increase in the shell side fluid velocity resulting in a significant increase in shell-side heat transfer coefficient.

The Delaware method can be applied with small modification to this configuration. One generally chooses a smaller baffle cut so that the free flow area through the window corresponds reasonably closely to the free crossflow area through the tube bank itself. The tube count for this configuration can be estimated from a fully-tubed exchanger tube count table multiplied by the fraction of the tubes that would be in crossflow at the selected baffle cut. The further calculations use $N_{cw} = 0$ and the baffle configuration correction factor $J_c = 1$. Otherwise the calculation proceeds essentially identically to a fully-tubed-out bundle.

## 4.3 Application to F Shells

The TEMA F shell configuration is shown in Figure 7. The F shell is commonly used with two tube passes on the tube-side as shown so that the flow arrangement is (at least on paper) entirely countercurrent, leading to a configuration correction factor of 1.0 on the logarithmic mean temperature difference (LMTD). It should be noted that it is easier to make this assumption in principle than it is to realize it in practice. The longitudinal baffle on the shell-side may, depending upon construction details, allow either thermal or physical leakage across the baffle from the hot shell-side stream to the cold (or vice versa). The effect of thermal leakage by conduction across the baffle has been analyzed by Whistler (10) and in nondimensional form by Rozenmann and Taborek, (11) and it can be reasonable handled in a quantitative fashion. For many operating conditions, the effect is a small one and quite acceptable to the designer.

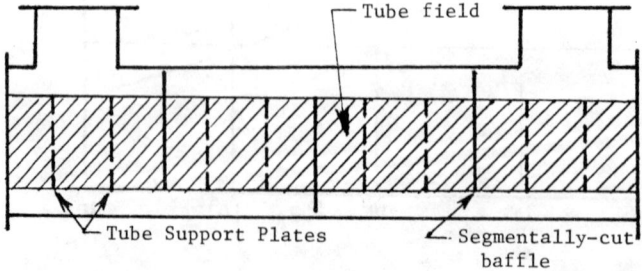

Figure 6. Diagram of a Shell and Tube Exchanger with No Tubes in the Window

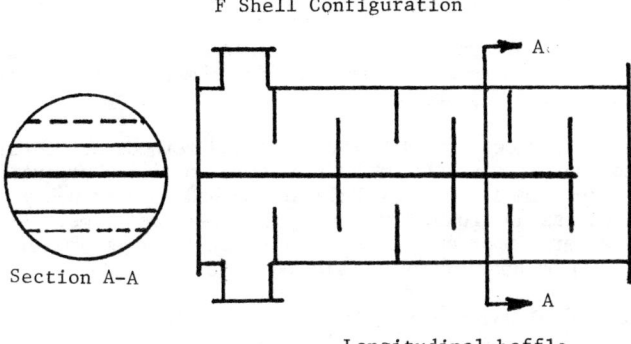

Figure 7. Diagram of a TEMA F Shell Configuration

Physical leakage of the fluid across the baffle is a different story. Even small clearances (which in any case are very difficult to predict accurately) can result in excessive leakage and a heat exchanger that is thermodynamically incapable of meeting terminal temperature specifications, no matter how much surface it has. Physical leakage can be avoided only if the baffle is welded into the shell, a construction which does not permit convenient withdrawal of the tube bundle for inspection or cleaning. Various sealing devices have been proposed for the longitudinal baffle but their long-term effectiveness is very dependent upon the care given to them during installation and maintenance procedures. Unless very particular attention is given to maintaining these seals in good shape, their use cannot be recommended if there is to be a reasonable guarantee of continued high thermal performance on the shell-side.

Assuming, however, that physical and thermal leakage across the longitudinal baffle is minimized, it is possible to adapt the Delaware method to the analysis of shell-side flow. Basically, all that is required is a minor modification of the geometrical parameters to recognize the fact that the shell-side is essentially split in two, reducing all of the areas for flow by half compared to the case for a single shell-side pass.

4.4 A Note on the Application of the Delaware Method

The Delaware method was originally intended for hand use in exchanger design. It is, however, easily adapted to computer use and can be programmed into a pocket-size computer/calculator. Palen and Taborek (6) showed that it was the most accurate of the methods in the open literature at that time by comparison against a very large data bank. The author has seen nothing since that time to indicate otherwise.

However, in terms of accuracy, the Delaware method can not compete against the large proprietary computer programs such as ST-4. In a world in which such programs are increasingly broadly available (if one has the price of admission to HTRI, HTFS, etc., which in terms of corporate dollars is almost negligible), is there a role for relatively simple methods?

The author thinks there is for the following reasons:

1. The proprietary programs are not available for general pedagogical use, so if one is trying to teach someone something about heat exchanger design, it is necessary to have an open literature method.

2. The input data sheets for the more advanced methods are getting so elaborate that it may take an hour or more to fill them out (more, the first few times out); the Delaware method in reasonably experienced hands has no trouble competing in this time frame. (In fact, there is a very important purpose served by the even faster, more simpleminded methods of the previous chapter.)

3. No one should be using one of the computer methods unless he/she has a pretty good idea ahead of time of what the answer should be, and has a feel for the magnitudes of the many variables. Some experience along these lines can be gained easily and independently from a hand method.

4. It is possible to make very rapid estimates of changes in process variables or exchanger geometry simply by looking at the Delaware correction factor charts. For example, one can estimate in a few minutes the probable change in pressure drop caused by fouling plugging up the baffle clearances.

## 5. SPECIFICATION/CALCULATION OF SHELL-SIDE GEOMETRY PARAMETERS

### 5.1 Input Data

The Delaware method assumes that the flow rate and the inlet and outlet temperatures (also pressures for a gas or vapor) of the shell-side fluid are specified and that the density, viscosity, thermal conductivity, and specific heat of the shell-side fluid are known or can be reasonably estimated as a function of temperature. The method also assumes that the following minimum set of shell-side geometry data are known or specified:

Tube outside diameter, $d_o$

Tube geometrical arrangement (unit cell)

Shell inside diameter, $D_i$

Shell outer tube limit (diameter), $D_{ot\ell}$

Effective tube length (between tube sheets), $\ell$

Baffle cut, $\ell_c$

Baffle spacing, $\ell_s$ (also the inlet and outlet baffle spacing, $\ell_{s,I}$ and $\ell_{s,o}$, if different from $\ell_s$)

Number of sealing strips/side, $N_{ss}$

From this geometrical information, all remaining geometrical parameters needed in the shell-side calculations can be calculated or estimated by methods given here, assuming that Standards of the Tubular Exchanger Manufacturers

Association (12) are met with respect to tube-to-baffle and shell-to-baffle clearances. However, if additional specific information is available (e.g., tube-baffle clearance), the exact values of certain parameters may be used in the calculation with some improvement in accuracy.

In order to complete the rating of a shell-and-tube exchanger, it is necessary to know also the tube material and wall thickness or inside diameter and to know or calculate the tube-side fluid flow rate and terminal temperatures (also pressures for gases and vapors) and heat transfer coefficient. In order to calculate the tube-side coefficient and pressure drop, the density, viscosity, thermal conductivity and specific heat of the tube-side fluid must generally be known or estimated as a function of temperature.

Not all of the fluid flow rates and temperatures can be independently specified, but are connected through the heat balance on the exchanger. Similarly, the overall rate equation, $Q = AU \, (MTD)$ must be satisfied, and it may well be that the U calculated by this design method does not equal that required by the heat balance and rate equation. If this happens when one is designing an exchanger to perform a given service, it is necessary to change one or more of the geometrical parameters (the tube length is a particularly popular choice because changing it does not require complete recalculation of the coefficient) until the calculated and required performances are in substantial agreement. If an existing exchanger is being rated, disagreement between calculated and required performance can only be resolved by changing flow rates and/or terminal specifications until agreement exists. The consequence often is that an existing unit cannot meet the specifications and a new unit must be designed and purchased.

## 5.2 Calculation of Shell-Side Geometrical Parameters

1. Total number of tubes in the exchanger, $N_t$: If not known by direct count, find in the tube count table, Table III, as a function of $D_{ot\ell}$, the tube pitch, p, and the layout. The shell diameter and outer tube limit are those for a conventional split-ring floating head design fully tubed out. For a given shell diameter, the value of $D_{ot\ell}$ will be greater than that shown for a fixed tube sheet design and smaller for a pull-through floating head. In any case, the tube count can be reasonably interpolated from the Table using the known or specified $D_{ot\ell}$, assuming that the tube count is proportional to $(D_{ot\ell})^2$. All tube count tables are only approximate since the actual number of tubes that can be fitted into a given tubesheet depends upon the pass position pattern, the thickness of the pass dividers and exactly where the drilling pattern is started relative to the dividers and the outer tube limit.

Additional tubes will be lost from the bundle for a U-tube design because the minimum bending radius prevents tubes from being inserted in some or all of the possible drilling positions near the centerline of the U-tube pattern. Tubes will also be lost if an impingement plate is inserted underneath the nozzle. For a no-tubes-in-the-window design, the actual number of tubes in the bundle is $F_c N_t$; see 5.2 (4) and Equation (5.2-2) below for the definition of $F_c$.

2. Tube pitch parallel to flow, $p_p$, and normal flow, $p_n$: These quantities are needed only for the purpose of estimating other parameters. If a detailed drawing of the exchanger is available, or if the exchanger itself can be conveniently examined, it is better to obtain these other parameters by direct count or calculation. The quantities are described by Figure 8 and read from Table IV for the most common tube layouts.

TABLE III
Tube Counts
(Adapted from Wolverine Tube Engineering Data Book and Perry's Handbook, 5th Ed.)

| Shell ID in. | Dia. of outer tube limit, in. | Tube OD in. | Tube pitch, in. layout | \multicolumn{5}{c}{Number of Tube Passes} | | | | |
|---|---|---|---|---|---|---|---|---|
| | | | | 1 | 2 | 4 | 6 | 8 |
| 8.071 | 6.821 | 3/4 | 15/16 △ | 38 | 32 | 26 | 24 | 18 |
| | | 3/4 | 1 □◇ | 32 | 26 | 20 | 20 | |
| | | 3/4 | 1 △ | 37 | 30 | 24 | 24 | |
| | | 1 | $1^1/_4$ □◇ | 21 | 16 | 16 | 14 | |
| | | 1 | $1^1/_4$ △ | 22 | 18 | 16 | 14 | |
| 10.02 | 8.77 | 3/4 | 15/16 △ | 62 | 56 | 47 | 42 | 36 |
| | | 3/4 | 1 □◇ | 52 | 52 | 40 | 36 | |
| | | 3/4 | 1 △ | 61 | 52 | 48 | 48 | |
| | | 1 | $1^1/_4$ □◇ | 32 | 32 | 26 | 24 | |
| | | 1 | $1^1/_4$ △ | 37 | 32 | 28 | 28 | |
| 12 | $10^3/_4$ | 3/4 | 15/16 △ | 109 | 98 | 86 | 82 | |
| | | 3/4 | 1 □◇ | 80 | 72 | 68 | 68 | 60 |
| | | 3/4 | 1 △ | 90 | 84 | 72 | 70 | 68 |
| | | 1 | $1^1/_4$ □◇ | 48 | 44 | 40 | 38 | 36 |
| | | 1 | $1^1/_4$ △ | 57 | 52 | 44 | 42 | 40 |
| $13^1/_4$ | 12 | 3/4 | 15/16 | 127 | 114 | 96 | 90 | 86 |
| | | 3/4 | 1 □◇ | 95 | 90 | 81 | 77 | 70 |
| | | 3/4 | 1 △ | 110 | 101 | 90 | 88 | 74 |
| | | 1 | $1^1/_4$ □◇ | 60 | 56 | 51 | 46 | 44 |
| | | 1 | $1^1/_4$ △ | 67 | 63 | 56 | 54 | 50 |
| $15^1/_4$ | 14 | 3/4 | 15/16 △ | 170 | 160 | 140 | 136 | 128 |
| | | 3/4 | 1 □◇ | 138 | 132 | 116 | 112 | 108 |
| | | 3/4 | 1 △ | 163 | 152 | 136 | 133 | 110 |
| | | 1 | $1^1/_4$ □◇ | 88 | 82 | 75 | 70 | 64 |
| | | 1 | $1^1/_4$ △ | 96 | 92 | 86 | 84 | 72 |
| $17^1/_4$ | 16 | 3/4 | 15/16 △ | 239 | 224 | 194 | 188 | 178 |
| | | 3/4 | 1 □◇ | 188 | 178 | 168 | 164 | 142 |
| | | 3/4 | 1 △ | 211 | 201 | 181 | 176 | 166 |
| | | 1 | $1^1/_4$ □◇ | 112 | 110 | 102 | 98 | 82 |
| | | 1 | $1^1/_4$ △ | 130 | 124 | 116 | 110 | 94 |
| $19^1/_4$ | 18 | 3/4 | 15/16 △ | 301 | 282 | 252 | 244 | 234 |
| | | 3/4 | 1 □◇ | 236 | 224 | 216 | 208 | 188 |
| | | 3/4 | 1 △ | 273 | 256 | 242 | 236 | 210 |
| | | 1 | $1^1/_4$ □◇ | 148 | 142 | 136 | 129 | 116 |
| | | 1 | $1^1/_4$ △ | 172 | 162 | 152 | 148 | 128 |

# DELAWARE METHOD FOR SHELL SIDE DESIGN

Table III (cont)

| Shell ID in. | Dia. of outer tube limit, in. | Tube OD in. | Tube pitch, in. layout | Number of Tube Passes ||||| 
|---|---|---|---|---|---|---|---|---|
| | | | | 1 | 2 | 4 | 6 | 8 |
| 21 | $19^1/_4$ | 3/4 | 15/16 △ | 361 | 342 | 314 | 306 | 290 |
| | | 3/4 | 1 □◇ | 276 | 264 | 246 | 240 | 234 |
| | | 3/4 | 1 △ | 318 | 308 | 279 | 269 | 260 |
| | | 1 | $1^1/_4$ □◇ | 170 | 168 | 157 | 150 | 148 |
| | | 1 | $1^1/_4$ △ | 199 | 188 | 170 | 164 | 160 |
| $23^1/_4$ | $21^1/_2$ | 3/4 | 15/16 △ | 442 | 420 | 386 | 378 | 364 |
| | | 3/4 | 1 □◇ | 341 | 321 | 308 | 296 | 292 |
| | | 3/4 | 1 △ | 381 | 369 | 349 | 326 | 328 |
| | | 1 | $1^1/_4$ □◇ | 210 | 199 | 197 | 186 | 184 |
| | | 1 | $1^1/_4$ △ | 247 | 230 | 216 | 208 | 202 |
| 25 | $23^3/_8$ | 3/4 | 15/16 △ | 531 | 506 | 468 | 446 | 434 |
| | | 3/4 | 1 □◇ | 397 | 391 | 370 | 360 | 343 |
| | | 3/4 | 1 △ | 470 | 452 | 422 | 394 | 382 |
| | | 1 | $1^1/_4$ □◇ | 250 | 248 | 224 | 216 | 210 |
| | | 1 | $1^1/_4$ △ | 294 | 282 | 256 | 252 | 242 |
| 27 | $25^3/_8$ | 3/4 | 15/16 △ | 637 | 602 | 550 | 536 | 524 |
| | | 3/4 | 1 □◇ | 465 | 452 | 427 | 418 | 408 |
| | | 3/4 | 1 △ | 559 | 534 | 488 | 474 | 464 |
| | | 1 | $1^1/_4$ □◇ | 286 | 275 | 267 | 257 | 250 |
| | | 1 | $1^1/_4$ △ | 349 | 334 | 302 | 296 | 286 |
| 29 | $27^3/_8$ | 3/4 | 15/16 △ | 721 | 692 | 640 | 620 | 594 |
| | | 3/4 | 1 □◇ | 554 | 542 | 525 | 509 | 500 |
| | | 3/4 | 1 △ | 630 | 604 | 556 | 538 | 508 |
| | | 1 | $1^1/_4$ □◇ | 348 | 340 | 322 | 314 | 313 |
| | | 1 | $1^1/_4$ △ | 397 | 376 | 354 | 334 | 316 |
| 31 | $29^3/_8$ | 3/4 | 15/16 △ | 847 | 822 | 766 | 722 | 720 |
| | | 3/4 | 1 □◇ | 633 | 616 | 590 | 586 | 570 |
| | | 3/4 | 1 △ | 745 | 728 | 678 | 666 | 640 |
| | | 1 | $1^1/_4$ □◇ | 402 | 390 | 366 | 360 | 348 |
| | | 1 | $1^1/_4$ △ | 472 | 454 | 430 | 420 | 400 |
| 33 | $31^3/_8$ | 3/4 | 15/16 △ | 974 | 938 | 872 | 852 | 826 |
| | | 3/4 | 1 □◇ | 742 | 713 | 687 | 683 | 672 |
| | | 3/4 | 1 △ | 856 | 830 | 774 | 760 | 732 |
| | | 1 | $1^1/_4$ □◇ | 460 | 453 | 430 | 420 | 414 |
| | | 1 | $1^1/_4$ △ | 538 | 522 | 486 | 470 | 454 |
| 35 | $33^3/_8$ | 3/4 | 15/16 △ | 1102 | 1068 | 1004 | 988 | 958 |
| | | 3/4 | 1 □◇ | 827 | 811 | 773 | 762 | 756 |
| | | 3/4 | 1 △ | 970 | 938 | 882 | 864 | 848 |
| | | 1 | $1^1/_4$ □◇ | 517 | 513 | 487 | 486 | 480 |
| | | 1 | $1^1/_4$ △ | 608 | 592 | 566 | 546 | 532 |

Table III (cont)

| Shell ID in. | Dia. of outer tube limit, in. | Tube OD in. | Tube pitch, in. layout | Number of Tube Passes ||||| 
|---|---|---|---|---|---|---|---|---|
| | | | | 1 | 2 | 4 | 6 | 8 |
| 37 | 35 1/4 | 3/4 | 15/16 △ | 1142 | 1200 | 1144 | 1104 | 1078 |
| | | 3/4 | 1 □◇ | 929 | 902 | 880 | 870 | 852 |
| | | 3/4 | 1 △ | 1090 | 1042 | 982 | 966 | 958 |
| | | 1 | 1 1/4 □◇ | 588 | 580 | 555 | 544 | 538 |
| | | 1 | 1 1/4 △ | 678 | 664 | 632 | 614 | 598 |
| 39 | 37 1/4 | 3/4 | 15/16 △ | 1377 | 1330 | 1258 | 1248 | 1212 |
| | | 3/4 | 1 □◇ | 1025 | 1012 | 984 | 964 | 952 |
| | | 3/4 | 1 △ | 1206 | 1176 | 1128 | 1100 | 1078 |
| | | 1 | 1 1/4 □◇ | 645 | 637 | 619 | 610 | 605 |
| | | 1 | 1 1/4 △ | 766 | 736 | 700 | 688 | 672 |
| 42 | 40 1/4 | 3/4 | 15/16 △ | 1611 | 1580 | 1498 | 1464 | 1456 |
| | | 3/4 | 1 □◇ | 1201 | 1171 | 1144 | 1109 | 1087 |
| | | 3/4 | 1 △ | 1409 | 1378 | 1314 | 1296 | 1280 |
| | | 1 | 1 1/4 □◇ | 745 | 728 | 708 | 686 | 680 |
| | | 1 | 1 1/4 △ | 890 | 878 | 834 | 808 | 800 |
| 44 | 42 1/4 | 3/4 | 15/16 △ | 1782 | 1738 | 1650 | 1624 | 1592 |
| | | 3/4 | 1 □◇ | 1349 | 1327 | 1286 | 1270 | 1252 |
| | | 3/4 | 1 △ | 1562 | 1535 | 1464 | 1422 | 1394 |
| | | 1 | 1 1/4 □◇ | 856 | 837 | 809 | 778 | 763 |
| | | 1 | 1 1/4 △ | 990 | 966 | 921 | 888 | 871 |
| 48 | 46 | 3/4 | 15/16 △ | 1965 | 1908 | 1834 | 1801 | 1766 |
| | | 3/4 | 1 □◇ | 1620 | 1598 | 1553 | 1535 | 1505 |
| | | 3/4 | 1 △ | 1827 | 1845 | 1766 | 1724 | 1690 |
| | | 1 | 1 1/4 □◇ | 1029 | 1010 | 975 | 959 | 940 |
| | | 1 | 1 1/4 △ | 1188 | 1163 | 1098 | 1076 | 1055 |
| 52 | 50 | 3/4 | 15/16 △ | 2347 | 2273 | 2178 | 2152 | 2110 |
| | | 3/4 | 1 □◇ | 1918 | 1890 | 1848 | 1826 | 1790 |
| | | 3/4 | 1 △ | 2212 | 2183 | 2092 | 2050 | 2010 |
| | | 1 | 1 1/4 □◇ | 1216 | 1196 | 1167 | 1132 | 1110 |
| | | 1 | 1 1/4 △ | 1405 | 1375 | 1323 | 1287 | 1262 |
| 56 | 54 | 3/4 | 15/16 △ | 2704 | 2660 | 2556 | 2526 | 2489 |
| | | 3/4 | 1 □◇ | 2241 | 2214 | 2167 | 2142 | 2110 |
| | | 3/4 | 1 △ | 2588 | 2545 | 2446 | 2409 | 2373 |
| | | 1 | 1 1/4 □◇ | 1420 | 1400 | 1371 | 1333 | 1307 |
| | | 1 | 1 1/4 △ | 1638 | 1605 | 1549 | 1501 | 1472 |
| 60 | 58 | 3/4 | 15/16 △ | 3399 | 3343 | 3232 | 3195 | 3162 |
| | | 3/4 | 1 □◇ | 2587 | 2556 | 2510 | 2485 | 2460 |
| | | 3/4 | 1 △ | 2987 | 2945 | 2827 | 2798 | 2770 |
| | | 1 | 1 1/4 □◇ | 1639 | 1615 | 1587 | 1553 | 1522 |
| | | 1 | 1 1/4 △ | 1889 | 1851 | 1797 | 1761 | 1726 |

# DELAWARE METHOD FOR SHELL SIDE DESIGN

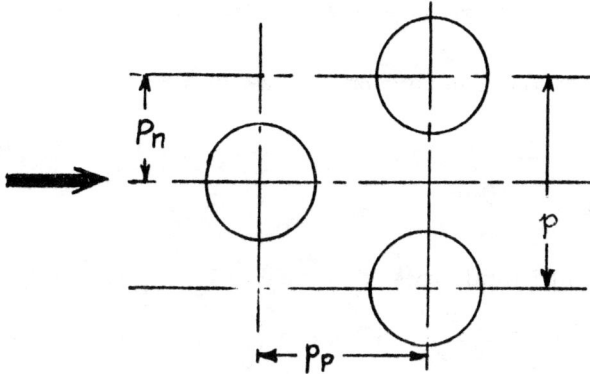

Figure 8. Tube Pitches Parallel and Normal to Flow (Equilateral Triangular Arrangement Shown)

TABLE IV

TUBE PITCHES PARALLEL AND NORMAL TO FLOW

| Tube, O. D. $d_o$, in. | Tube Pitch p, in. | Layout | $p_p$, in. | $p_n$, in. |
|---|---|---|---|---|
| 5/8 = 0.625 | 13/16 = 0.812 | → ◁ | 0.704 | 0.406 |
| 3/4 = 0.750 | 15/16 = 0.938 | → ◁ | 0.814 | 0.469 |
| 3/4 = 0.750 | 1 | → □ | 1.000 | 1.000 |
| 3/4 = 0.750 | 1 | → ◇ | 0.707 | 0.707 |
| 3/4 = 0.750 | 1 | → ◁ | 0.866 | 0.500 |
| 1 | 1-1/4 = 1.250 | → □ | 1.250 | 1.250 |
| 1 | 1-1/4 = 1.250 | → ◇ | 0.884 | 0.884 |
| 1 | 1-1/4 = 1.250 | → ◁ | 1.082 | 0.625 |

3. Number of tube rows crossed in one crossflow section (between baffle tips), $N_c$:

Count from exchanger drawing or estimate from

$$N_c = \frac{D_i [1 - 2(\frac{\ell_c}{D_i})]}{P_p} \tag{5.2-1}$$

4. Fraction of total tubes in crossflow, $F_c$. $F_c$ can be calculated from

$$F_c = \frac{1}{\pi} \left\{ \pi + 2(\frac{D_i - 2\ell_c}{D_{ot\ell}}) \sin [\cos^{-1}(\frac{D_i - 2\ell_c}{D_{ot\ell}})] - 2 \cos^{-1}(\frac{D_i - 2\ell_c}{D_{ot\ell}}) \right\} \tag{5.2-2}$$

where all the angles are read in radians. For convenience, $F_c$ has been plotted to an acceptable degree of precision in Figure 9 as a function of percent baffle cut, $[(\frac{\ell_c}{D_i})(100\%)]$, and shell diameter $D_i$. This figure is strictly applicable only to the $D_i - D_{ot\ell}$ combinations shown in Table III but may be used for other situations with minor error. For fixed tube sheet construction, $F_c$ is a little lower than that shown, especially for the smaller shell diameters; for pull-through floating head construction, $F_c$ is a little higher.

For no-tube-in-the-window designs, the actual number of tubes in the exchanger is $F_c N_t$ where $F_c$ is given by Equation (5.2-2) above; for all subsequent calculations for this design, $F_c$ is then taken as 1.00.

5. Number of effective crossflow rows in each window, $N_{cw}$. Estimate from

$$N_{cw} = \frac{0.8 \ell_c}{P_p} \tag{5.2-3}$$

This equation assumes that the shell-side fluid on the average crosses about half of the tube rows in the window (each such row twice) and the tube rows extend about 0.8 of the distance from the baffle tip to the shell inside diameter.

6. Number of baffles, $N_b$. Calculate from:

$$N_b = \frac{\ell - \ell_{s,I} - \ell_{s,o}}{\ell_s} + 1 \tag{5.2-4}$$

This equation considers that the entrance and/or exit baffle spacings may be different than the central baffle spacing. If the rating calculation is being carried out to determine the required length of the exchanger, this calculation is omitted until after the heat transfer calculations are completed (see Section 6); $N_b$ usually plays at most a minor role in calculating the shell-side heat transfer coefficient.

7. Crossflow area at or near centerline for one crossflow section, $S_m$. Estimate from:

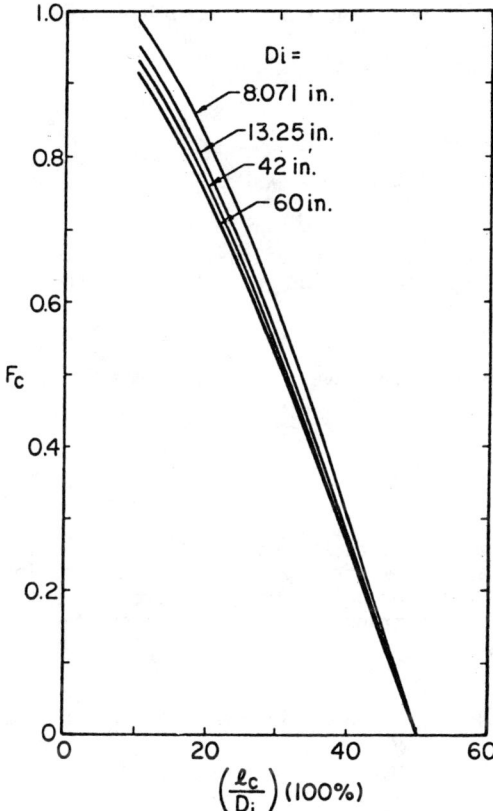

Figure 9. Estimation of Fraction of Tubes in Crossflow

$$S_m = \ell_s [D_i - D_{ot\ell} + (\frac{D_{ot\ell} - d_o}{P_n})(p - d_o)] \qquad (5.2\text{-}5a)$$

for rotated and inline square layouts;

$$S_m = \ell_s [D_i - D_{ot\ell} + (\frac{D_{ot\ell} - d_o}{p})(p - d_o)] \qquad (5.2\text{-}5b)$$

for triangular layouts.

These equations assume a nearly uniform tube field, except as required for tube pass partition lanes and the difference between the shell inside diameter and the diameter of the outer tube limit. (Those clearances are corrected for separately.) There is also no problem if the center line of the bundle normal to the crossflow is devoid of tubes, as required for U-tube or multiple tube pass construction; that is a minor disturbance of the uniformity of the tube field.

If low-finned tubes are used, the correct equations are

$$S_m = \ell_s [D_i - D_{ot\ell} + (\frac{D_{ot\ell} - d_r}{P_n})(p - d_r) - 2N_f h_f y_f] \quad (5.2\text{-}5c)$$

for rotated and inline square layouts and

$$S_m = \ell_s [D_i - D_{ot\ell} + (\frac{D_{ot\ell} - d_r}{p})(p - d_r) - 2N_f h_f y_f] \quad (5.2\text{-}5d)$$

for triangular layouts. In the above equations, $d_r$ is the root diameter of the finned tube, $N_f$ is the number of fins per inch, $h_f$ is the height of the fin and $y_f$ is the fin thickness.

8. Fraction of crossflow area available for bypass flow, $F_{Sbp}$. Estimate from

$$F_{Sbp} = \frac{[(D_i - D_{ot\ell} + 1/2 N_p W_p)]}{S_m} \quad (5.2\text{-}6)$$

where $N_p$ is the number of pass partition lanes through the tube field <u>parallel</u> to the direction of the crossflow stream. This term accounts for the <u>effect</u> of flow that can bypass wholly or partially the tube field, with a great reduction of contact with heat transfer surface and distortion of the temperature profile.

9. Tube-to-baffle leakage area for one baffle, $S_{tb}$. Estimate from:

$$S_{tb} = \pi d_o \delta_{tb} N_t (1/2)(1 + F_c) \quad (5.2\text{-}7)$$

where $S_{tb}$ is the diametral clearance between the tube and the baffle. TEMA Class R construction specifies a $\delta_{tb}$ of 1/32 in. where maximum unsupported tube length (normally $2\ell_s$) does not exceed 36 in. and a $\delta_{tb}$ of 1/64 in. otherwise. Values should be modified if extra tight or loose construction is specified, or if clogging by dirt is anticipated.

10. Baffle cut angle $\theta$. The baffle cut angle $\theta$ is the angle subtended by the intersection of the cut edge of the baffle with the inside surface of the shell. In terms of previously-defined quantities:

$$\theta = 2 \cos^{-1}(1 - \frac{2\ell_c}{D_i}) \quad (5.2\text{-}8)$$

This equation is shown graphically in Figure 10.

11. Shell-to-baffle leakage area for one baffle, $S_{sb}$. If diametral shell-baffle clearance, $\delta_{sb}$, is known, $S_{sb}$ can be calculated from

$$S_{sb} = \frac{\pi D \delta_{sb}}{2} [1 - \frac{\theta}{2\pi}] \quad (5.2\text{-}9)$$

# DELAWARE METHOD FOR SHELL SIDE DESIGN

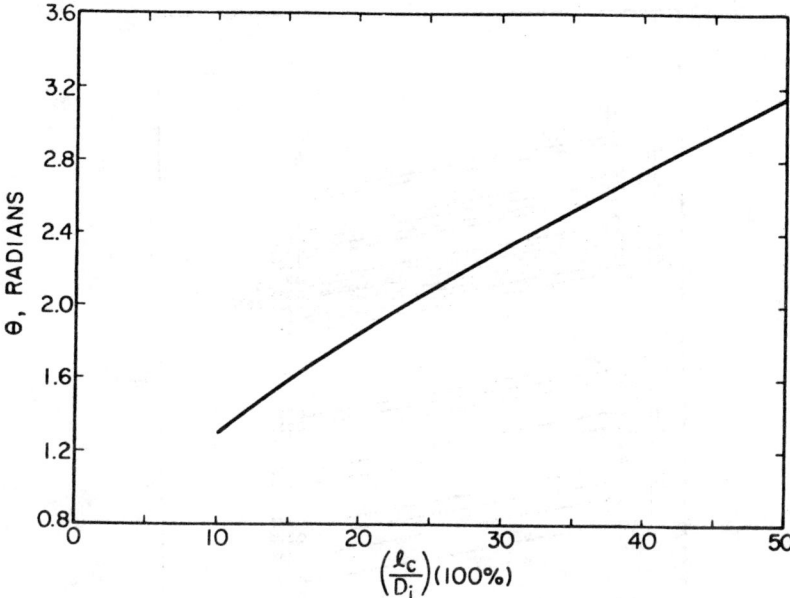

Figure 10. Baffle Cut Above

where the value of the term $\cos^{-1}(1 - 2\ell_c/D_i)$ is in radians and is between 0 and $\pi/2$. This area has been calculated and plotted in Figure 11 as a function of percent baffle cut, $[(\frac{\ell_c}{D_i})(100\%)]$, and inside shell diameter, $D_i$. Figure 11 is based on TEMA Class R standards:

| $D_i$, in. | Diametral shell-baffle clearance, in. | |
| --- | --- | --- |
| 8 - 13 | 0.100 | These values are |
| 14 - 17 | 0.125 | for pipe shells; |
| 18 - 23 | 0.150 | if rolled shells |
| 24 - 39 | 0.175 | are used, add |
| 40 - 54 | 0.225 | 0.125 in. |
| 35 - | 0.300 | |

Since pipe shells are generally limited to diameters below 24 in., the larger sizes are shown using the rolled shell specification.

Again, allowance should be made for especially tight or loose construction. In this connection, it may be noted that it is conservative to estimate a large clearance when calculating heat transfer, in the sense of calculating a lower heat transfer coefficient on the shell-side. However, when calculating pressure drop, one obtains a conservative estimate (higher $\Delta p$) if one assumes smaller clearances.

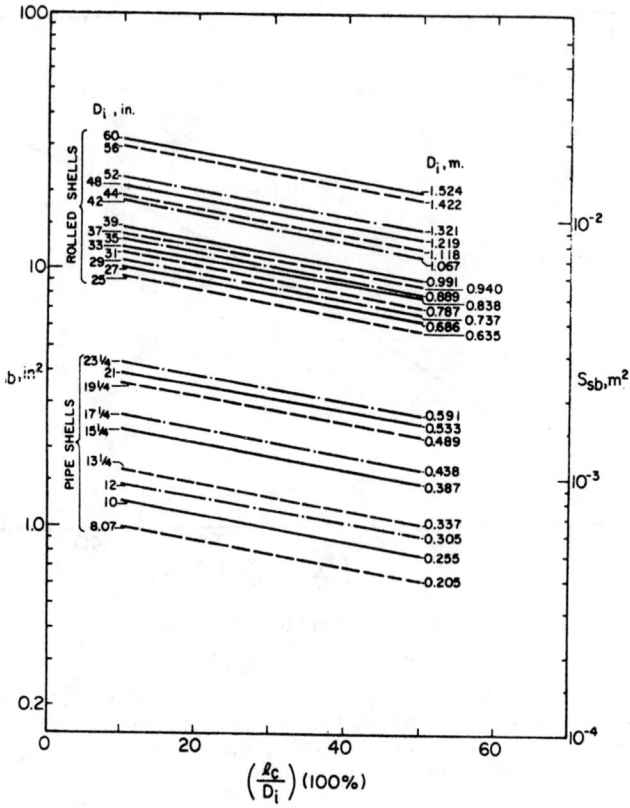

Figure 11. Estimation of Shell-to-Baffle Leakage Area

12. Area for flow through window, $S_w$. This area is obtained as the difference between the gross window area, $S_{wg}$ and the window area occupied by tubes, $S_{wt}$:

$$S_w = S_{wg} - S_{wt} \qquad (5.2\text{-}10a)$$

The value of $S_{wg}$ can be calculated from:

$$S_{wg} = \frac{D_i^2}{4} \left\{ \frac{\theta}{2} - [1 - 2(\frac{\ell_c}{D_i})]\sin(\frac{\theta}{2}) \right\} \qquad (5.2\text{-}10b)$$

For convenience, however, the values of $S_{wg}$ are plotted in Figure 12 as a function of $(\ell_c/D_i)$ and $D_i$.

The window area occupied by the tubes, $S_{wt}$, can be calculated from:

$$S_{wt} = \frac{N_t}{8}(1-F_c)\pi d_o^2 \qquad (5.2\text{-}10c)$$

# DELAWARE METHOD FOR SHELL SIDE DESIGN

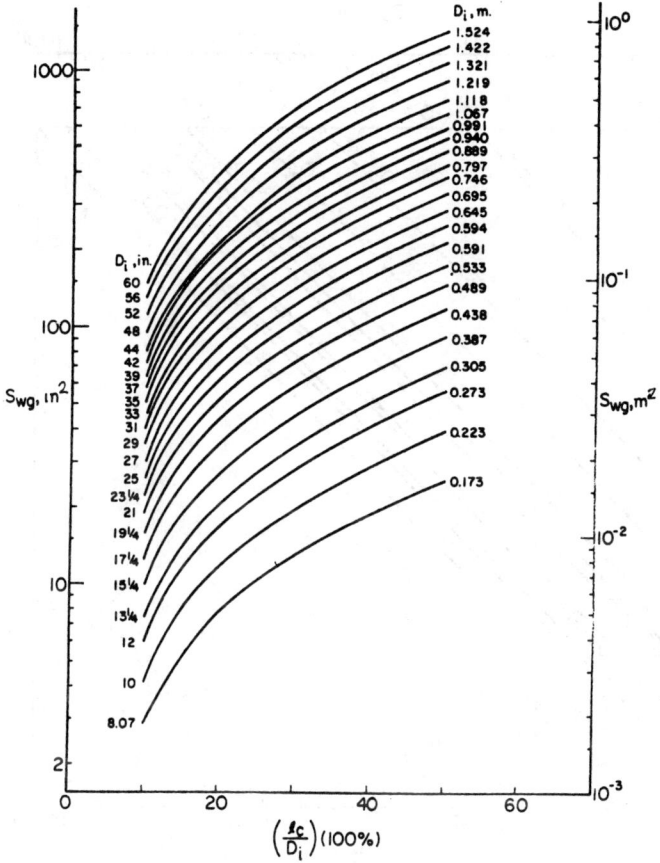

Figure 12. Estimation of Window Gross Flow Area

For convenience, the value of $S_{wt}$ can be found from Figure 13. To use this figure, enter the lower abscissa at the appropriate value of $N_t$ and proceed vertically to the solid line corresponding to $d_o$. Then proceed horizontally to the dashed line corresponding to the estimated value of $F_c$ and thence vertically to read the value of $S_{wt}$ on the upper abscissa.

13. Equivalent diameter of window, $D_w$. (Required only if laminar flow, defined as $Re_s \leq 100$, exists.) Calculate from

$$D_w = \frac{4 S_w}{\frac{\pi}{2} N_t (1-F_c) d_o + D_i \theta} \qquad (5.2\text{-}11)$$

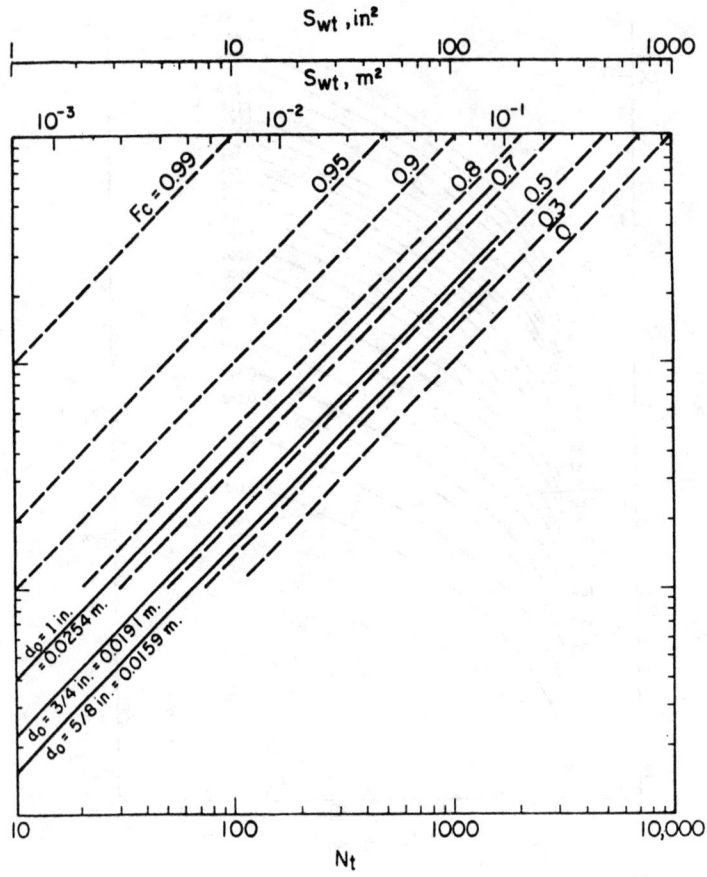

Figure 13. Estimation of Cross-Sectional Area of Tubes in Window

6. CALCULATION OF HEAT TRANSFER PERFORMANCE AND SHELL-SIDE PRESSURE DROP

6.1 Calculation of Shell-Side Heat Transfer Coefficient

1. Calculate shell-side Reynolds number, $Re_s$. The shell-side Reynolds number is defined as

$$Re_s = \frac{d_o W_s}{\mu_s S_m} \qquad (6.1\text{-}1a)$$

where $W_s$ = mass flow rate of shell-side fluid,

$\mu_s$ = bulk viscosity of shell-side fluid.

## DELAWARE METHOD FOR SHELL SIDE DESIGN

For finned tubes, the shell-side Reynolds number is defined as

$$Re_s = \frac{d_r W_s}{\mu_s S_m} \tag{6.1-1b}$$

where $d_r$ is the root diameter of the tube.

It is usually adequate to use the arithmetic mean bulk shell-side fluid temperature (i.e., halfway between the inlet and exit temperatures) to evaluate all bulk properties of the shell-side fluid. In the case of long temperature ranges or for a fluid whose viscosity is very sensitive to temperature change, special care must be taken (such as breaking the calculation into segments, each covering a more limited temperature range). Even then, the accuracy of the procedure is less than for more conventional cases.

2. Find $j_i$ from the ideal tube bank curve for a given tube layout at the calculated value of $Re_s$, using Figure 14. For finned tubes at low Reynolds numbers, refer to Figure 5 and the accompanying discussion for guidance in choosing $j_i$.

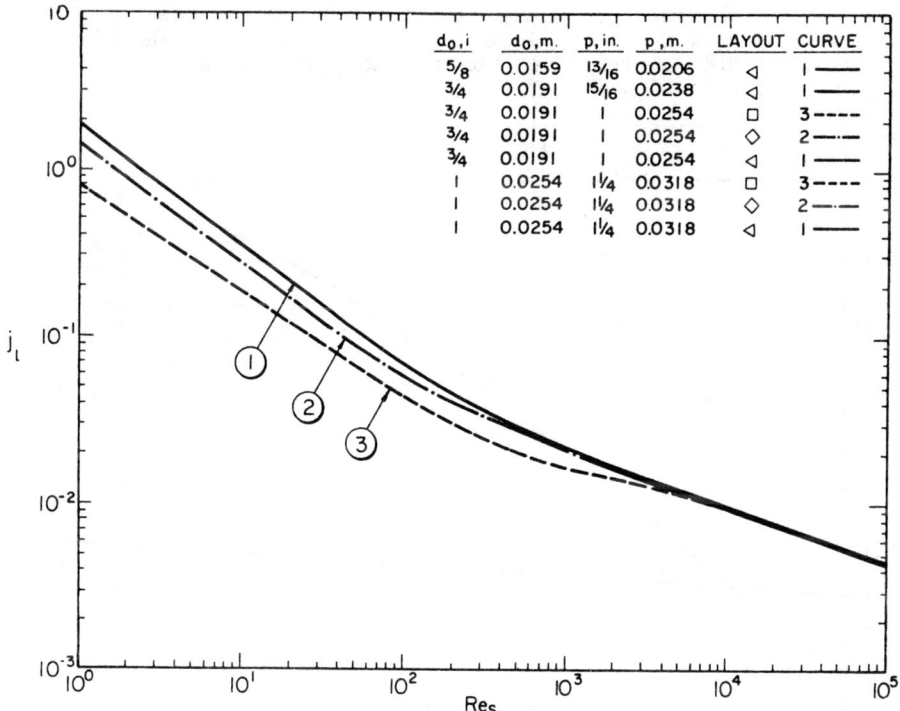

Figure 14. Correlation of j-Factor for Ideal Tube Bank

3. Calculate the shell-side heat transfer coefficient for an ideal tube bank, $h_{ideal}$

$$h_{ideal} = j_i c_s \left(\frac{W_s}{S_m}\right) \left(\frac{k_s}{c_s \mu_s}\right)^{2/3} \left(\frac{\mu_s}{\mu_{s,w}}\right)^{0.14} \qquad (6.1\text{-}2)$$

4. Find the correction factor for baffle configuration effects, $J_c$. $J_c$ is read from Figure 15 as a function of $F_c$. For no-tubes-in-the-window designs, $J_c = 1$.

5. Find the correction factor for baffle leakage effects, $J_\ell$. $J_\ell$ is found from Figure 16 as a function of the ratio of the total baffle leakage area, $(S_{sb} + S_{tb})$, to the crossflow area, $S_m$, and of the ratio of the shell-to-baffle leakage area $S_{sb}$ to the total baffle leakage area, $S_{tb} + S_{sb}$.

6. Find the correction factor for bundle bypassing effects, $J_b$. $J_b$ is found from Figure 17 as a function of $F_{Sbp}$ and of $N_{ss}/N_c$ (the ratio of the number of sealing strips per side to the number of rows crossed in one baffle crossflow section). The solid lines on Figure 17 are for $Re_s \geq 100$; the dashed lines for $Re_s \leq 100$. If there are pass divider lanes through the tube field parallel to the crossflow stream, it is assumed that equivalent steps will be taken to block that flow (the F stream) as for the bundle-shell bypass flow (C stream). This can be done by tie rods and spacers as well as by sealing strips.

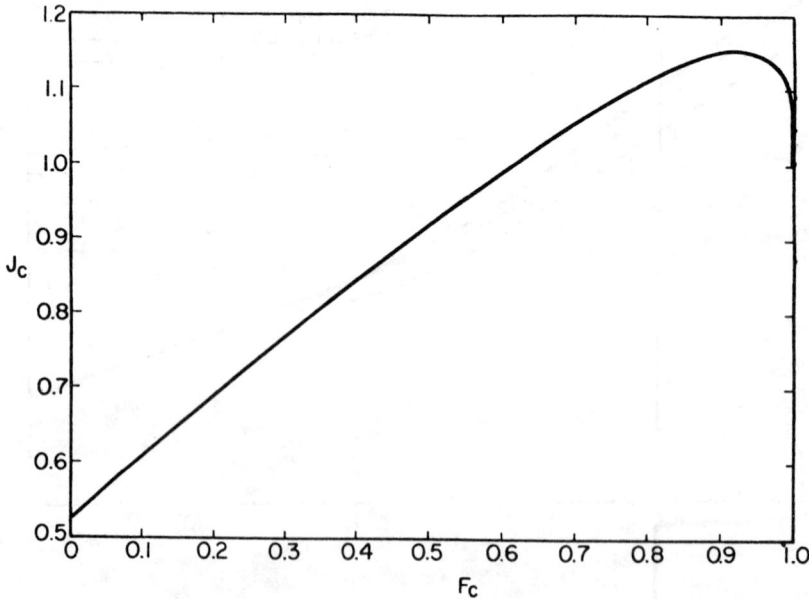

Figure 15. Correction Factor for Baffle Configuration Effects

# DELAWARE METHOD FOR SHELL SIDE DESIGN

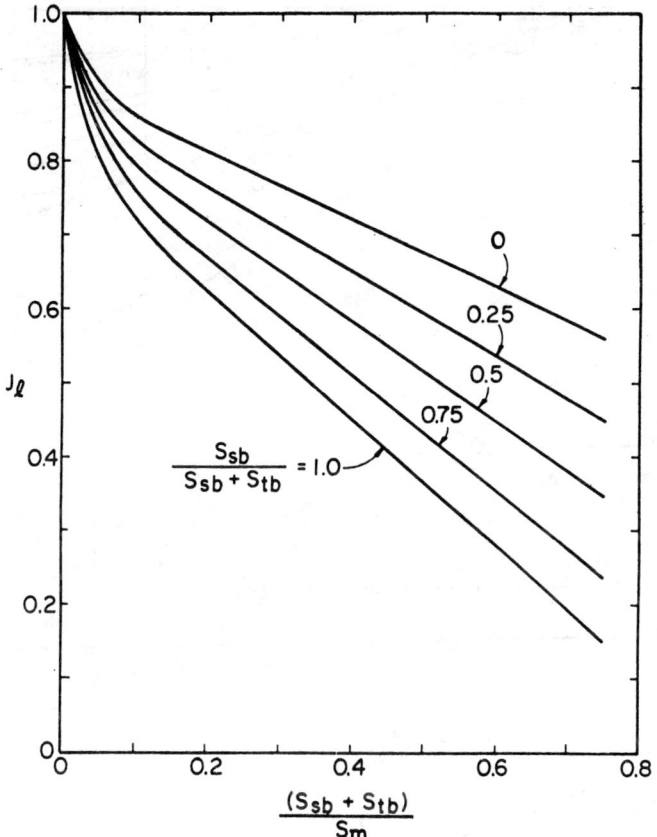

Figure 16. Correction Factor for Baffle Leakage Effects

7. Find the correction factor for adverse temperature gradient buildup at low Reynolds numbers, $J_r$. This factor is equal to 1.00 if $Re_s$ is equal to or greater than 100. For $Re_s$ equal to or less than 20, the correction factor is fully effective and a function only of the total number of tube rows crossed. For $Re_s$ between 20 and 100, a linear proportion rule is used.

Therefore:

A. If $Re_s < 100$, find $J_r^*$ from Figure 18, knowing $N_b$ and $(N_c+N_{cw})$.

B. If $Re_s \leq 20$, $J_r = J_r^*$.

C. If $20 < Re_s < 100$, find $J_r$ from Figure 19, knowing $J_r^*$ and $Re_s$.

8. Find the correction factor for unequal baffle spacing at inlet and/or outlet, $J_s$. The equation is:

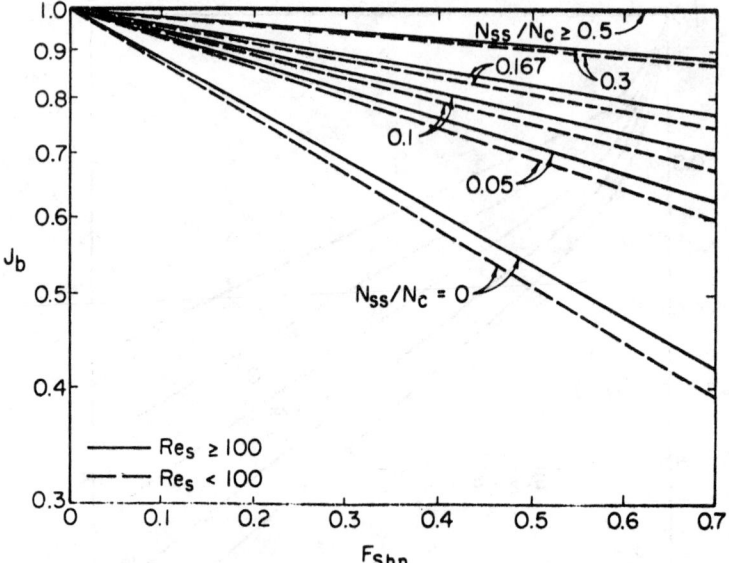

Figure 17. Correction Factor for Bypass Flow

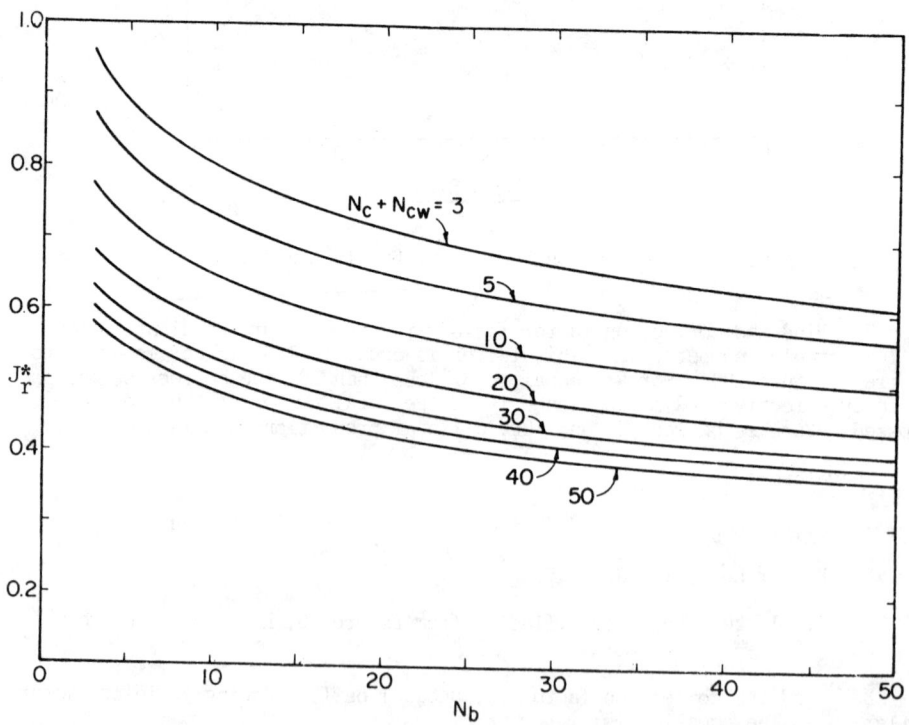

Figure 18. Basic Correction Factor for Adverse Temperature Gradient at Lo Reynolds Numbers

# DELAWARE METHOD FOR SHELL SIDE DESIGN

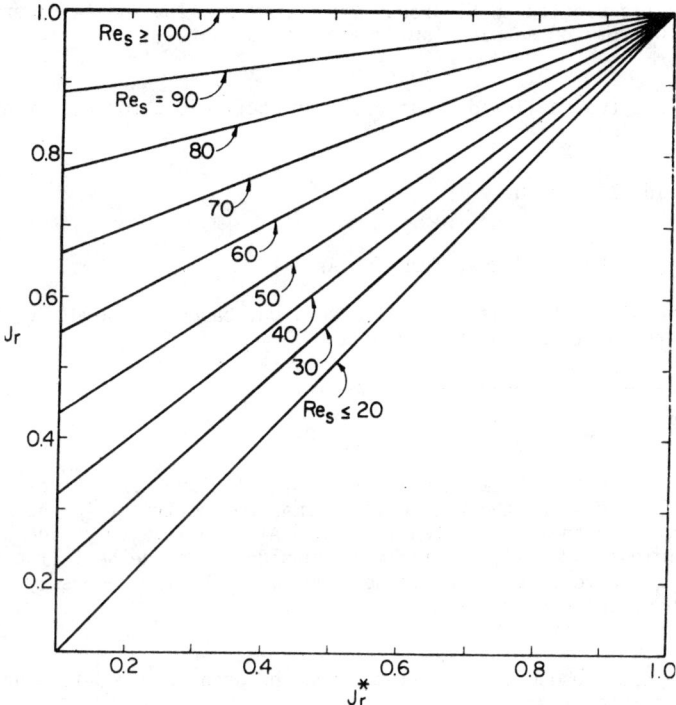

Figure 19. Correction Factor for Adverse Temperature Gradient at Intermediate Reynolds Numbers

$$J_s = \frac{(N_b-1)+(\ell^*_{s,I})^{1-n} + (\ell^*_{s,o})^{1-n}}{(N_b-1) + \ell^*_{s,I} + \ell^*_{s,o}} \qquad (6.1\text{-}3)$$

where

$N_b$ = number of baffles

$\ell^*_{s,I} = \ell_{s,I}/\ell_s$

$\ell^*_{s,o} = \ell_{s,o}/\ell_s$

$\ell_s$ = internal baffle spacing

$\ell_{s,I}$ = entrance baffle spacing

$\ell_{s,o}$ = exit baffle spacing

$n$ = 0.6 for turbulent flow ($Re_s > 100$)

= 1/3 for laminar flow ($Re_s < 100$)

Equation (6.1-3) is plotted in Figure 20 for turbulent flow and in Figure 21 for laminar flow, for the special (but common) case that $\ell_{s,I} = \ell_{s,o}$.

9. Calculate the shell-side heat transfer coefficient for the exchanger, $h_o$, from the equation:

$$h_o = h_{ideal} J_c J_\ell J_b J_r J_s \tag{6.1-4}$$

## 6.2 Calculation of Required Heat Transfer Area

1. The overall heat transfer coefficient, $U_o$, based on the shell-side heat transfer area $A_o$ is calculated from:

$$U_o = \frac{1}{\frac{1}{h_o} + R_{fo} + \frac{\Delta X_w}{k_w} \frac{A_o}{A_m} + R_{fi} \frac{A_o}{A_i} + \frac{A_o}{h_i A_i}} \tag{6.2-1}$$

where $R_{fo}$ and $R_{fi}$ are the shell-side and tube-side fouling resistances, $\Delta X_w$ the tube wall thickness, $k_w$ the tube wall thermal conductivity, $A_m$ the mean heat transfer area in the wall (often $1/2 [A_o + A_i]$), $h_i$ the tube-side heat transfer coefficient and $A_o/A_i$ the ratio of outside to inside heat transfer areas (= $r_o/r_i$ for bare tube). For finned tubes, it is also necessary to include the fin efficiency or a fin resistance term.

2. If the heat load Q is specified, then the area required to transfer the heat is calculated from

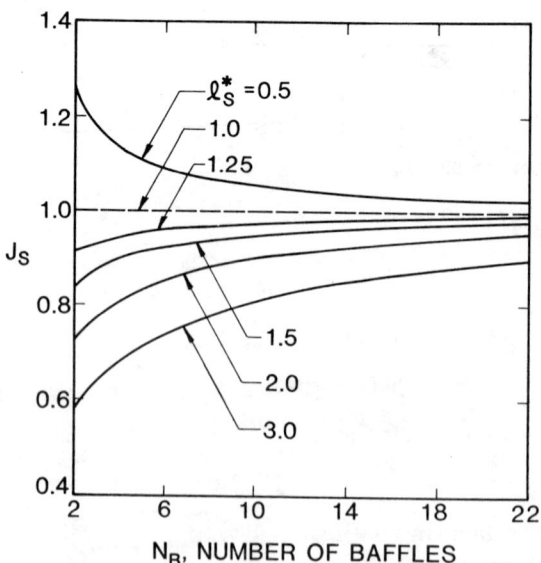

Figure 20. $J_s$ as a Function of $N_b$ for Turbulent Flow and Various Values of $\ell_s^* = \ell_{s,i}^* = \ell_{s,o}^*$

# DELAWARE METHOD FOR SHELL SIDE DESIGN

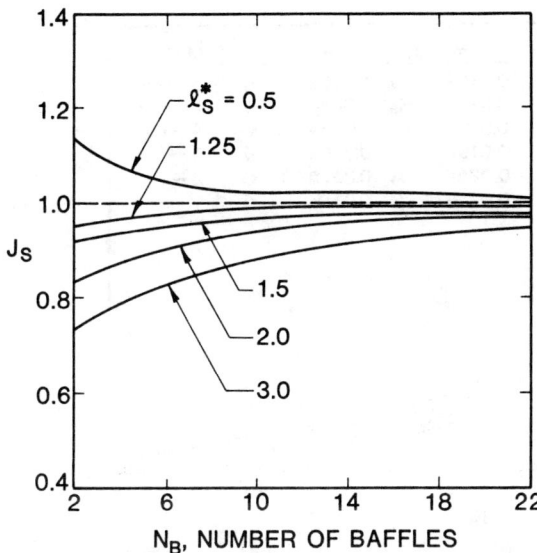

Figure 21. $J_S$ as a Function of $N_b$ for Laminar Flow and Various Values of $\ell_S^* = \ell_{S,I}^* = \ell_{S,o}^*$

$$A_o = \frac{Q}{U_o F_T (LMTD)} \qquad (6.2\text{-}2)$$

where LMTD is the logarithmic mean temperature difference for countercurrent flow and $F_T$ is the configuration correction factor (=1 for countercurrent flow, <1 for multiple tube-side pass arrangements). Consult References (5), (12), or (13) for these charts.

3. Then the required effective tube length of the exchanger, $\ell$, is calculated from

$$\ell = \frac{A_o}{\pi d_o N_t} \qquad (6.2\text{-}3)$$

and the number of baffles $N_b$ is found from Equation (5.2-4).

4. Alternatively, if the length of the exchanger and therefore the length of the tubes is specified, the amount of heat that can be transferred may be calculated by the NTU-effectiveness method. The details of this method are beyond the scope of this section, but References (13) and (14) provide charts and equations.

## 6.3 Calculation of Shell-Side Pressure Drop

1. Find $f_i$ from the ideal tube bank friction factor curve for the given tube layout at the calculated value of $Re_S$, using Figure 22 for triangular and

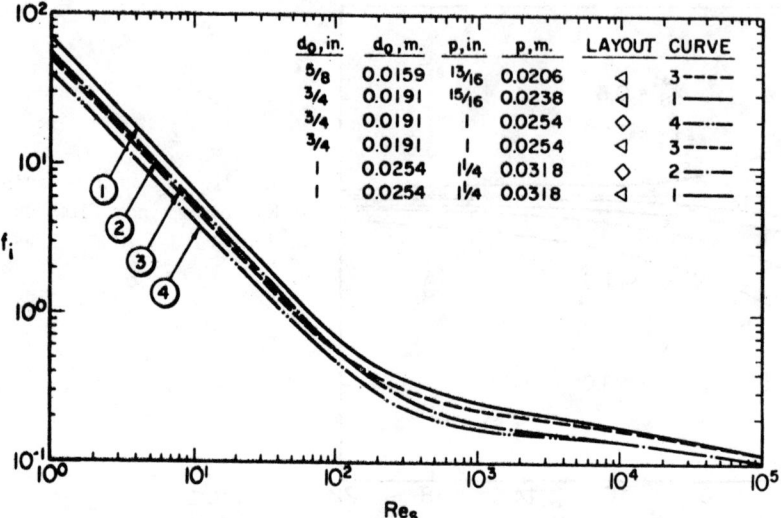

Figure 22. Correlation of Friction Factors for Ideal Tube Banks

rotated square arrays and Figure 23 for inline square arrays. For finned tubes, recall that $f_i$ should be increased by 50 percent.

2. Calculate the pressure drop for an ideal crossflow section, $\Delta p_{b,i}$:

$$\Delta p_{b,i} = \frac{4f_i W_s^2 N_c}{2\rho_s g_c S_m^2} \left(\frac{\mu_{s,w}}{\mu_s}\right)^{0.14} \qquad (6.2\text{-}4)$$

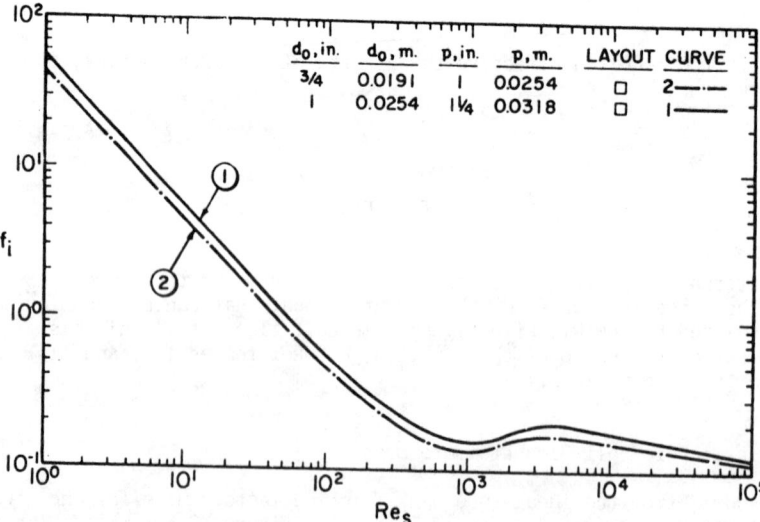

Figure 23. Correlation of Friction Factors for Ideal Tube Banks

# DELAWARE METHOD FOR SHELL SIDE DESIGN

3. Calculate the pressure drop for an ideal window section, $\Delta P_{w,i}$:

   A. If $Re_s \geq 100$:

   $$\Delta P_{w,i} = \frac{W_s^2 (2 + 0.6 N_{cw})}{2 g_c S_m S_w \rho_s} \quad (6.2\text{-}5a)$$

   B. If $Re_s < 100$:

   $$\Delta P_{w,i} = 26 \frac{\mu_s W_s}{g_c \, S_m S_w \rho_s} [\frac{N_{cw}}{p-d_o} + \frac{\ell_s}{D_w^2}]$$
   $$+ \frac{W_s^2}{g_c S_m S_w \rho_s} \quad (6.2\text{-}5b)$$

4. Calculate correction factor for effect of baffle leakage on pressure drop, $R_\ell$. Read from Figure 24 as a function of $(S_{sb}+S_{tb})/S_m$ with parameter of $S_{sb}/(S_{sb}+S_{tb})$. Curves shown are not to be extrapolated beyond the points shown.

5. Find the correction factor for bundle bypass, $R_b$. Read from Figure 25 as a function of $F_{Sbp}$ and $N_{ss}/N_c$. The solid lines are for $Re_s \geq 100$; the dashed lines are for $Re_s < 100$.

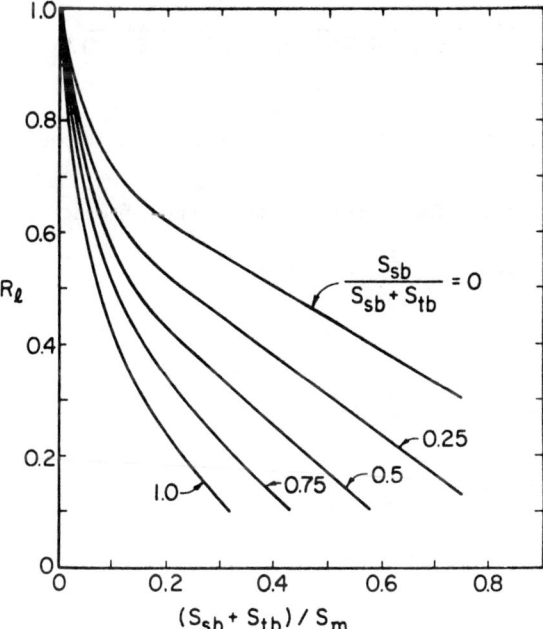

Figure 24. Correction Factor for Baffle Leakage Effect on Pressure Drop

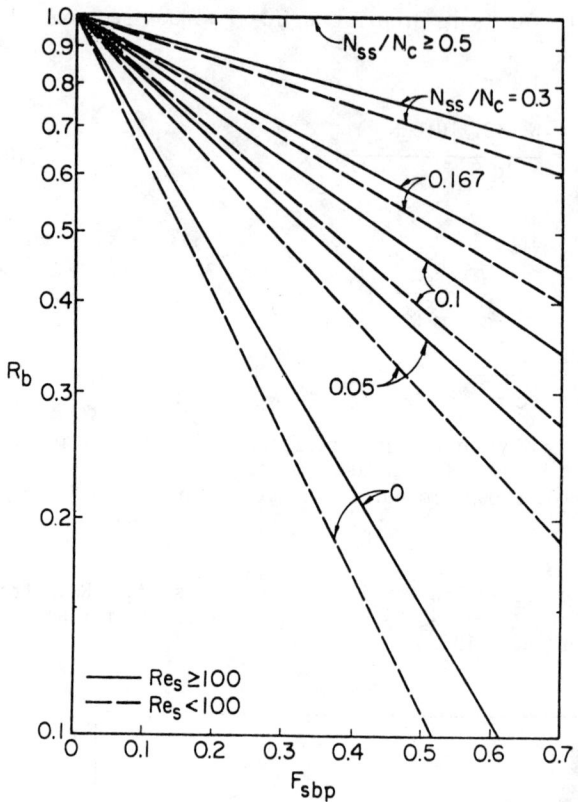

Figure 25. Correction Factor on Pressure Drop for Bypass Flow

6. Find the correction factor for unequal baffle spacing, $R_s$. The equation is:

$$R_s = \frac{1}{2} [(\ell^*_{s,I})^{-n'} + (\ell^*_{s,o})^{-n'}] \qquad (6.2\text{-}6)$$

where

$\ell^*_{s,I} = \ell_{s,I}/\ell_s$

$\ell^*_{s,o} = \ell_{s,o}/\ell_s$

$n' = 1.6$ for turbulent flow ($Re_s > 100$)

$\phantom{n'} = 1$ for laminar flow ($Re_s < 100$)

7. Calculate the pressure drop across the shell-side (excluding nozzles), $\Delta p_s$, from:

$$\Delta p_s = [(N_b - 1)(\Delta p_{b,i})R_b + N_b \Delta p_{w,i}] R_\ell + 2\Delta p_{b,i} R_b (1 + \frac{N_{cw}}{N_c}) R_s \qquad (6.2\text{-}7)$$

# DELAWARE METHOD FOR SHELL SIDE DESIGN

## NOMENCLATURE

$A_o, A_w, A_i$ = Heat transfer areas for shell-side, wall, and tube-side, respectively.

$c_s$ = Specific heat of shell-side fluid.

$D_i$ = Shell inside diameter.

$D_{ot\ell}$ = Diameter of the outer tube limit.

$D_w$ = Equivalent diameter of the window.

$d_o$ = Tube outside diameter.

$d_r$ = Root diameter of low-finned tube.

$F_c$ = Fraction of the total tubes that are in crossflow.

$F_{Sbp}$ = Fraction of total crossflow area that is available for bypass flow around tube bundle and through pass partition lanes.

$F_T$ = Configuration correction factor on LMTD.

$f_i$ = Friction factor for flow across an ideal tube bank.

$g_c$ = Gravitational conversion constant, $4.17 \times 10^8$ $lb_m ft/lb_f \, hr^2$.

$h_i$ = Tube-side heat transfer coefficiency.

$h_{ideal}$ = Shell-side heat transfer coefficient for ideal tube bank.

$h_o$ = Shell-side heat transfer coefficient for the exchanger.

$J_b$ = Correction factor on the shell-side heat transfer coefficient for bundle bypass effects.

$J_c$ = Correction factor on the shell-side heat transfer coefficient to account for baffle configuration effects.

$J_\ell$ = Correction factor on the shell-side heat transfer coefficient to account for baffle leakage effects.

$J_r$ = Correction factor on the shell-side heat transfer coefficient to account for buildup of adverse temperature gradient.

$J_r^*$ = Base correction factor on the shell-side heat transfer coefficient to account for buildup of adverse temperature gradient.

$J_s$ = Correction factor on the shell-side heat transfer coefficient to account for unequal baffle spacing.

$j_i$ = Colburn j-factor for an ideal tube bank.

$k_s$ = Thermal conductivity of shell-side fluid.

| | | |
|---|---|---|
| $k_w$ | = | Thermal conductivity of tube wall. |
| LMTD | = | Logarithmic mean temperature difference for countercurrent flow. |
| $\ell$ | = | Effective tube length (between tube sheets). |
| $\ell_c$ | = | Baffle cut distance from baffle tip to shell inside diameter. |
| $\ell_s$ | = | Baffle spacing, center-to-center of consecutive baffles. |
| $\ell_{s,I}, \ell_{s,o}$ | = | Baffle spacing at inlet and exit of the exchanger, respectively. $\ell^*_{s,I}$, and $\ell^*_{s,o}$ are the corresponding dimensionless values. |
| $N_b$ | = | Number of baffles in exchanger. |
| $N_c$ | = | Number of tube rows crossed during flow through one crossflow section. |
| $N_{cw}$ | = | Number of effective crossflow rows in each window section. |
| $N_{ss}$ | = | Number of sealing strips or equivalent obstructions to bypass flow encountered by the stream in one crossflow section. |
| $N_t$ | = | Total number of tubes in the exchanger. |
| $n, n'$ | = | Exponents for the relationship between $j_i$ and $Re_s$ and $f_i$ and $Re_s$, respectively. |
| $\Delta p_{b,i}$ | = | Pressure drop during flow across one ideal crossflow section. |
| $P_{w,i}$ | = | Pressure drop through one ideal window section. |
| $p$ | = | Tube pitch: distance between centers of nearest tubes in tube layout. |
| $p_n$ | = | Tube pitch normal to flow: distance between centers of adjacent tubes normal to the flow. |
| $p_p$ | = | Tube pitch parallel to flow: distance between centers of adjacent tube rows in the direction of flow. |
| $Q$ | = | Total heat duty of exchanger. |
| $R_b$ | = | Correction factor for effect of bundle bypass on pressure drop. |
| $R_\ell$ | = | Correction factor for effect of baffle leakage on pressure drop. |
| $R_s$ | = | Correction factor for effect of unequal baffle spacing on inlet and exit section pressure drops. |
| $R_{fi}, R_{fo}$ | = | Fouling resistance to heat transfer on tube-side and shell-side, respectively. |
| $Re_s$ | = | Reynolds number for shell-side. |
| $S_m$ | = | Crossflow area at or near centerline for one crossflow section. |

$S_{sb}$ = Shell-to-baffle leakage area for a single baffle.

$S_{tb}$ = Tube-to-baffle leakage area for one baffle.

$S_w$ = Area for flow through window.

$S_{wg}$ = Window gross area.

$S_{wt}$ = Window area occupied by tubes.

$U_o$ = Overall heat transfer coefficient based on shell-side heat transfer area.

$W_s$ = Mass flow rate of shell-side fluid.

$\delta_{sb}$ = Diametral clearance between shell and baffle.

$\delta_{tb}$ = Diametral clearance between shell and baffle.

$\mu_s$ = Viscosity of shell-side fluid at bulk stream temperature.

$\mu_{s,w}$ = Viscosity of shell-side fluid evaluated at surface temperature.

$\rho_s$ = Density of shell-side fluid.

$\theta$ = Baffle cut angle, radians.

REFERENCES

1. Bergelin, O. P., A. P. Colburn, and H. L. Hull, "Heat Transfer and Pressure Drop During Viscous Flow Across Unbaffled Tube Banks," Bulletin No. 2, University of Delaware Engineering Experiment Station, Newark, Delaware (1950).

2. Bergelin, O. P., M. D. Leighton, W. L. Lafferty, Jr., and R. L. Pigford, "Heat Transfer and Pressure Drop During Viscous and Turbulent Flow Across Baffled and Unbaffled Tube Banks," Bulletin No. 4, University of Delaware Engineering Station, Newark, Delaware (1958).

3. Bell, K. J., "Exchanger Design Based on the Delaware Research Program", Petro/Chem Engineer, 32 pp. C-26 - C-40c (October, 1960).

4. Bell, K. J., "Final Report of the Cooperative Research Program on Shell and Tube Heat Exchangers", Bulletin No. 5, University of Delaware Engineering Experiment Station, Newark, Delaware, (1963).

5. Perry, R. H., and Chilton, C. H., Eds., Chemical Engineers' Handbook, 5th Ed., Section 10, pp. 10-22 - 10-32, McGraw-Hill (1973).

6. Palen, J. W., and Taborek, J., "Solution of Shell-Side Flow Pressure Drop and Heat Transfer by Stream Analysis Method", CEP Symp. Ser. No. 92, 65, "Heat Transfer - Philadelphia" pp. 53-63 (1969).

7. Tinker, T., "Shell Side Characteristics of Shell and Tube Heat Exchangers", General Discussion on Heat Transfer, pp. 97-116, Institution of Mechanical Engineers, London, England (1951).

8. Williams, R. B., and Katz, D. L., "Performance of Finned Tubes in Shell and Tube Heat Exchangers", Trans. ASME 74, pp. 1307-1320 (1952).

9. Briggs, D. E., Katz, D. L., and Young, E. H., "How to Design Finned-Tube Heat Exchangers", Chem. Eng. Prog. 59, No. 11, pp. 49-59 (November, 1963).

10. Whistler, A. M., "Correction for Heat Conduction Through Longitudinal Baffle of Heat Exchanger", Trans. ASME 69 pp. 683-685 (1947).

11. Rozenmann, T., and Taborek, J., "The Effect of Leakage Through the Longitudinal Baffle on the Performance of Two-Pass Shell Exchangers", AIChE Symp. Ser. No. 118, 68, "Heat Transfer - Tulsa", pp. 12-20 (1972).

12. Tubular Exchanger Manufacturers Association, Standards, 6th Ed., New York (1978).

13. Rohsenow, W. M., and Hartnett, J. P., Eds., Handbook of Heat Transfer, Section 18, "Heat Exchangers", by A. C. Mueller. McGraw-Hill, New York (1973).

14. Kays, W. M., and London, A. L., Compact Heat Exchangers, McGraw-Hill, New York (1958).

# The Design of Boilers

J. G. COLLIER
United Kingdom Atomic Energy Authority
Harwell, UK

ABSTRACT

This lecture is concerned with the design of both fossil fired- and waste heat-boilers. It begins by reviewing the various types of equipment used in central station generating plant and in the process chemical industry. A special class of waste heat boiler is that of the steam generators used in nuclear power plant. The lecture then goes on to consider specific two-phase flow and boiling processes which are relevant to boiler design and discusses where special difficulties or problems may arise.

1. INTRODUCTION

Boilers are used to raise steam for a wide range of industrial uses. The steam may be used for space or process heating, for the generation of electricity, to drive compressors or pumps or in a variety of chemical processes. Boilers can be divided into two basic categories - those heated directly by the combustion of fossil fuels, e.g. coal, oil or gas and those heated indirectly by a hot gas or liquid which has received its heat from another energy source. The latter are usually referred to as *waste heat boilers*. Waste heat is often available from industrial chemical reactions which take place at high temperatures such as ammonia and methanol production. This heat is contained in either the process gases themselves or in the flue gases leaving the fired heater used to supply heat to the process. The economics of many chemical processes are dependent on the efficient and reliable generation of steam from this waste heat.

A special class of waste heat boilers are the *steam generators* used in central station nuclear power plants. These units are built in very large sizes up to 2000 MW(t) in a single heat exchanger. Heat energy is transferred from the fission reaction in the reactor core to the steam generators by means of either a high pressure gas, or by high pressure water or sometimes by a liquid metal such as sodium.

This lecture will review the various types of boiler used for the generation of electricity and in the process chemical industry. It will then go on to consider specific two-phase flow and boiling processes which are relevant to boiler design and will point out where special difficulties or problems may arise.

## 2. FOSSIL-FIRED EQUIPMENT

Boilers heated directly by combustion of fossil fuels can be broadly divided into two categories - *water-tube boilers* in which the steam/water mixture passes inside tubes with the flue gas combustion products on the outside or *fire-tube boilers* in which the flue gas passes inside the tubes ("smoke tubes") and boiling occurs on the outside surfaces of the tube. In order to remove the heat by boiling there must be a continuous circulation of water either inside the tubes, in the case of the water-tube boiler, or over the tubes in the case of the fire-tube boiler. This circulation may be induced by pumps, in which case the boiler is referred to as a *forced circulation boiler*, or alternatively by the driving head provided by the difference in density between water and steam, in which case the boiler is referred to as a *natural circulation boiler*.

### Central station fossil-fired boilers

A modern fossil station may be gas, oil or coal-fired. The steam generator consists of the following elements (Figure 1):

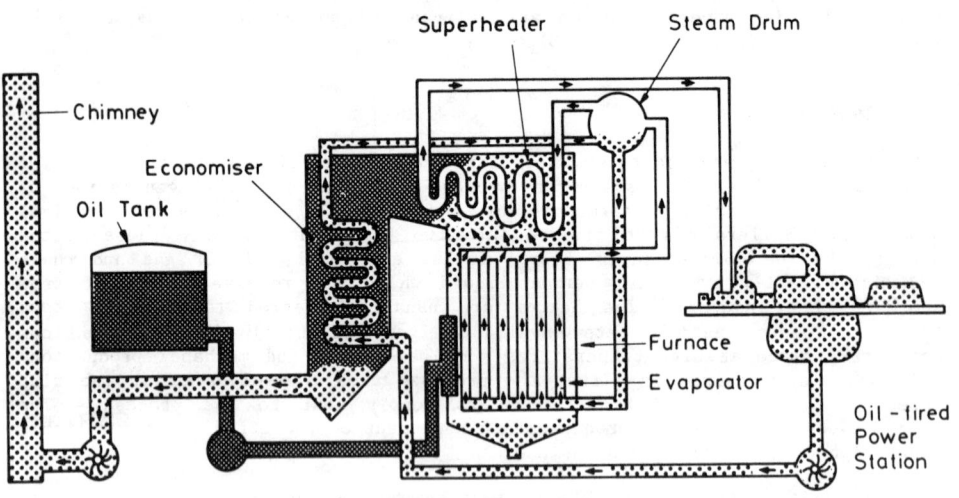

Figure 1   A modern fossil-fired boiler

# THE DESIGN OF BOILERS

(1) a water-cooled furnace

(2) a large superheater (with radiant superheater surface)

(3) a reheater (common but not essential)

(4) a convective heat transfer section (large or small) depending on design conditions

(5) an economiser or air heater or both.

The furnace, superheater, convection region and economiser are arranged to form an integral unit. The walls of the furnace are formed by a gas-tight welded membrane in which each boiler tube has an integrally formed fin which is welded to its neighbour using semi-automatic machines in the factory and erected in the form of large panels. The heat transfer situation is complicated by the fact that the heating is from one side only and because there are horizontal, inclined and curved tube lengths within the boiler configuration. Other problems encountered with this type of plant include instabilities (both parallel channel and full loop instabilities) and also separation and two-phase flow effects in the steam drums.

## Fluidised bed combustion

One of the more recent developments in fossil-fired steam plant is the concept of fluidised bed combustion (Thurlow (1978)). In this concept, fuel, usually coal, is fed into a turbulent bed of inert particles usually coal-ash. This bed of ash is kept turbulent by flowing up through it a stream of air evenly distributed over the whole base of the bed and is kept at a temperature below that at which the particles begin to soften or sinter, i.e. generally at a bed temperature of 800-850°C which is 250-300°C lower than the typical temperatures leaving conventional fossil fuel-fired furnaces. A further consideration is the high heat transfer coefficients obtained within the bed at ~300 $W/m^2K$ or 60 $Btu/hr^2ft$ °F. As a result, the heat can be extracted using less surface area than in a conventional furnace. Also, more of the heat transfer surface can be used since the entire circumference of the tube is available for heating. Generally, about 45-50% of the total boiler heat absorption occurs within the bed. One of the major problems is, however, that the boiler tubes must be horizontal.

It is possible to operate a fluidised combustion unit under pressure giving higher heat release rates and the possibility of increased thermal efficiency by combining with a gas turbine cycle.

## 3. NUCLEAR POWER PLANT

Nuclear power plants in common use around the world can be classified into four main types:

- the Light-Water Reactor (LWR)
- the Heavy-Water Reactor (HWR)
- the Gas-Cooled Reactor (GCR)
- the Liquid Metal-Cooled Reactor (LMFBR)

Two types of light water reactor are in operation viz

- the *Pressurised Water Reactor* (PWR) where the reactor core is cooled by ordinary (light) water at a pressure of 158 bar. Hot water from the reactor core is transferred via a series of two or more coolant loops to a number of steam generators.

- the *Boiling Water Reactor* (BWR) where boiling occurs directly within the core of the reactor and the water/steam mixture passes to separators where the steam is dried before going direct to the turbine. The separated water is returned to the core with the aid of a series of recirculating pumps.

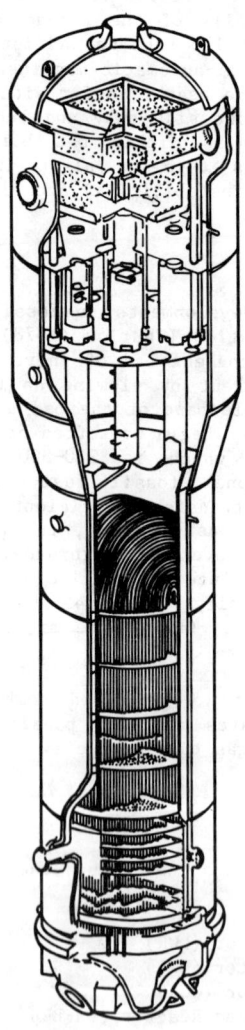

Figure 2    Inverted U tube type steam generator (Westinghouse)

## Steam generators for water-cooled reactors

Most steam generators used with PWRs and HWRs consist of a vertically-mounted shell containing a bundle of tubes in the form of an inverted U (Figure 2). The shell consists of two separate sections; an evaporator section containing the tube bundle and the larger diameter steam drum section where the steam is separated and dried.

The hot high pressure water from the reactor core flows into the channel head at the base of the unit, through the inside of the inverted U tube bundle containing some 45 miles of tubing and back to the channel head. A partition plate divides the channel head into inlet and outlet sections. The channel head is fabricated from ferritic steel and clad internally with stainless steel. The tubes are usually of Inconel or Incoloy and are mounted on a thick ferritic steel tube plate also clad on the primary side with Inconel. The tubes are rolled into the tube plate, welded to the primary side cladding and are supported at intervals by tube support plates. Feedwater enters the steam generator in the upper shell and mixes with water separated from the steam by the swirl vane separators. This water flows down the annulus between the steam generator shell and a baffle surrounding the tube bundle. When the water reaches the tube plate it flows radially across the upper surface of the tube plate into the tube nest. Boiling occurs on the outside surfaces of the tubes within the bundle and the steam-water mixture passes upwards into the swirl vane separators. Natural circulation is induced as a result of the density difference within the bundle and the annular downcomer. The steam from the separators passes through impingement type driers and exits from the top of the shell. Difficulties have been experienced with some designs of inverted U tube steam generator in maintaining the integrity of the boundary between the primary and secondary sides due to water-side corrosion on the secondary (shell) side.

Whilst the majority of PWR power plants are equipped with steam generators of the vertical shell, inverted U tube recirculating design, other steam generator designs are employed in some plants.

In the original Shippingport plant, (The Shippingport PWR, 1958) the steam generator equipment consisted of four units each comprising a heat exchanger, a steam drum and connecting piping. Two different types of heat exchanger, a horizontal U tube design supplied by Babcock & Wilcox and a horizontal straight tube design supplied by Foster Wheeler, were installed to evaluate the relative performance of the two designs (Figure 3). Each heat exchanger was of the shell and tube type. Primary coolant flowed through the tubes and steam was generated on the shell side. The steam-water mixture passed up the risers to the steam drum where standard separators and driers were used to separate the water from the steam. The water returned to the lower heat exchanger via the downcomers. The Babcock & Wilcox design was rated at 75 MW(t) and contained 921 x 19 mm stainless steel tubes 15 m long. The Foster Wheeler design was also rated at 75 MW(t) and contained 2096 x 12.7 stainless steel tubes 9.5 m long.

Horizontal natural circulation steam generators are also widely used in PWRs constructed in the USSR (Styrikovich, 1978). The units for the 440 MW(e) plant consist of a horizontal shell 11.5 m long and 3 m diameter. Vertical tubular headers located half way along the shell act as the inlet and outlet for the primary coolant. Horizontal bundles of U tubes mounted on these headers provide the heat transfer surface. These particular units shown in Figure 4 are rated at 250 MW(t) but units of 800 MW(t) have been manufactured for a 1000 MW(e) plant.

In the United States, one PWR supplier, Babcock & Wilcox, equips its reactors with a vertical shell, straight tube once-through steam generator (Figure 5). The primary coolant enters the header at the top of the unit and flows down through the tubes to exit at the base. On the secondary side, the feed water is boiled in the interspace between the tubes, totally evaporated and slightly superheated (by 30°C). The positioning of the feed nozzles and steam outlet on the shell and the use of some of the steam to preheat the feedwater in the annulus around the tube bundle overcomes the problem of differential thermal

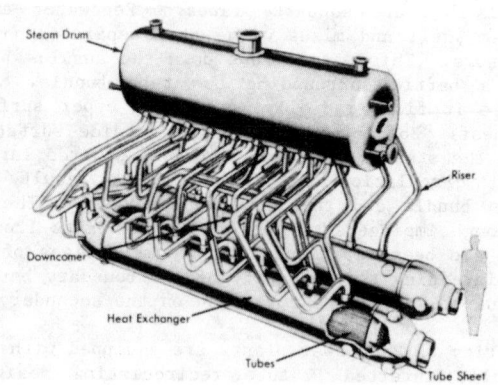

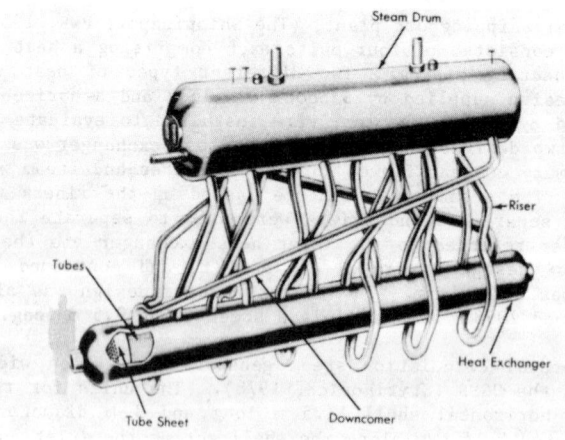

Figure 3  The steam generators used at Shippingport
(upper photo Babcock & Wilcox steam generator
lower photo Foster Wheeler steam generator)

# THE DESIGN OF BOILERS

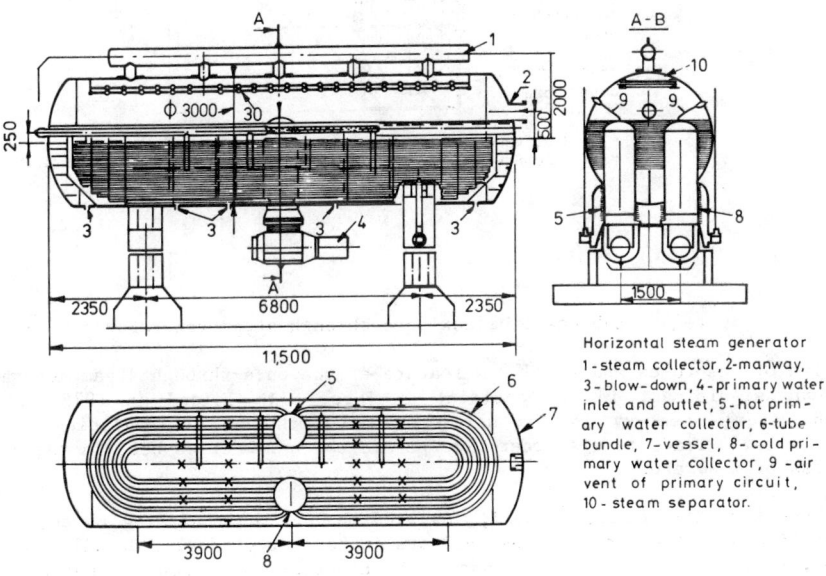

Figure 4  The steam generators used on the 440 MW(e) PWR constructed in the USSR (Stryrikovich)

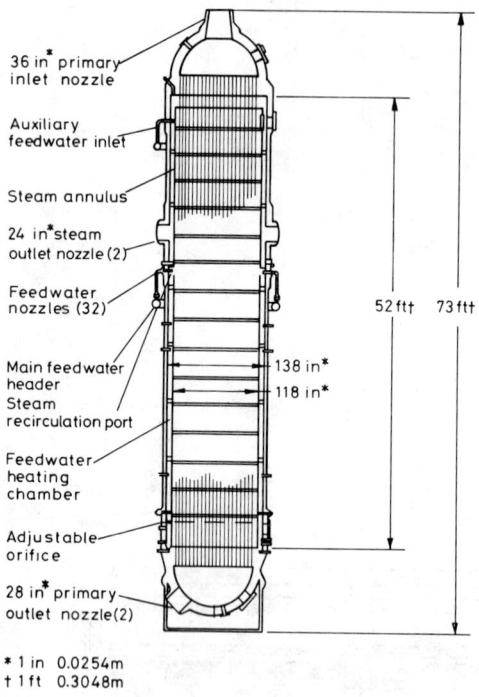

Figure 5   Babcock & Wilcox once-through steam generator

expansion of the tubes and shell. A feature of the once-through steam generator which was significant in the accident at Three Mile Island in 1979 is the reduced water inventory in the unit compared with the recirculating design which, in turn, leads to a shorter time before the unit "dries out" in the event of a loss of feedwater.

Even with the vertical shell, inverted U tube recirculating units, there are significant differences between vendors in respect of design details such as thermohydraulics parameters, methods of construction, tube supports and materials which profoundly influence their performance.

Some units are equipped with a feedwater preheating section or economiser located just above the tube plate on the cold leg side of the U tubes. The feedwater enters the preheater and is heated almost to saturation temperature by countercurrent heat transfer from the reactor coolant within the tubes.

# THE DESIGN OF BOILERS

In the design of unit offered by Foster Wheeler Limited (Davis and Hirst, 1979), the massive thick tube plate and channel head is dispensed with and is replaced by two cylindrical horizontal headers upon which the tube bundle is mounted directly. The primary reactor coolant passes through a vertical penetration in the steam generator shell to this horizontal header feeding the tube tank and exits by way of a similar header and shell penetration. The advantages claimed for this design include, avoidance of sludge deposition on the tube plate and the elimination of tube-to-tube plate crevices.

Tube support designs are particularly important because of the consequences of corrosion of the tube support material. Since the corrosion products of carbon steel occupy about twice the volume of the original metal it is possible, with some designs of tube support for the corrosion product to dent the tubes and to distort the support plate itself. Figure 6 shows a variety of tube support arrangements used in the steam generator units offered by the various PWR suppliers.

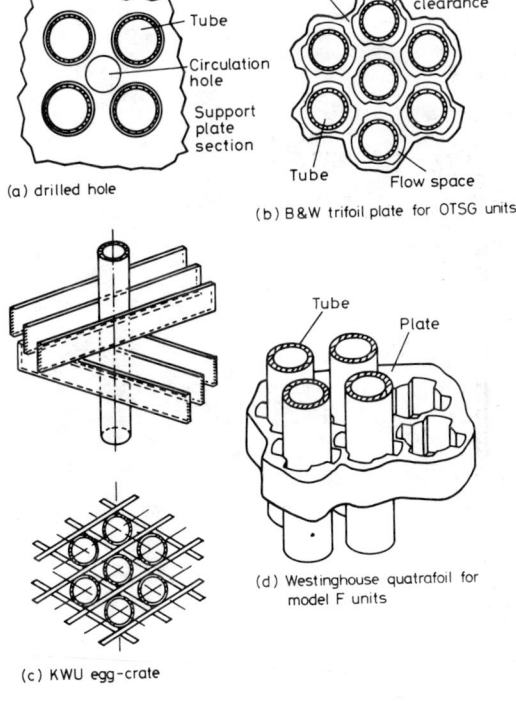

(a) drilled hole

(b) B&W trefoil plate for OTSG units

(c) KWU egg-crate

(d) Westinghouse quatrafoil for model F units

Figure 6   Tube support plate design for PWR steam generators (Garnsey)

Steam generators for gas-cooled and liquid metal-cooled reactors

The gas-cooled reactor and the liquid metal-cooled reactor use either a high pressure gas (carbon dioxide or helium) or sodium respectively to transfer heat from the reactor core to the steam generator. Because higher temperatures can be used with these types of reactor, higher steam pressures (up to 185 bar) and superheat/reheat cycles are used. Once-through steam generators are often employed in these designs. Details of some designs of gas-cooled reactor and liquid metal-cooled reactor steam generators are given in Table 1. Figure 7 shows an isometric view of the U tube design of evaporator used on the UK Prototype Fast Reactor (PFR).

4. WASTE HEAT BOILERS

A wide variety of designs of waste heat boiler have been constructed to recover heat from various chemical processes. Such boilers may have to accept process gases at temperatures up to 1000-1200°C and pressures up to 30 bar. As a result severe problems often arise due to local overheating sometimes associated with both fouling and corrosion. Typical of the designs at present in use are:

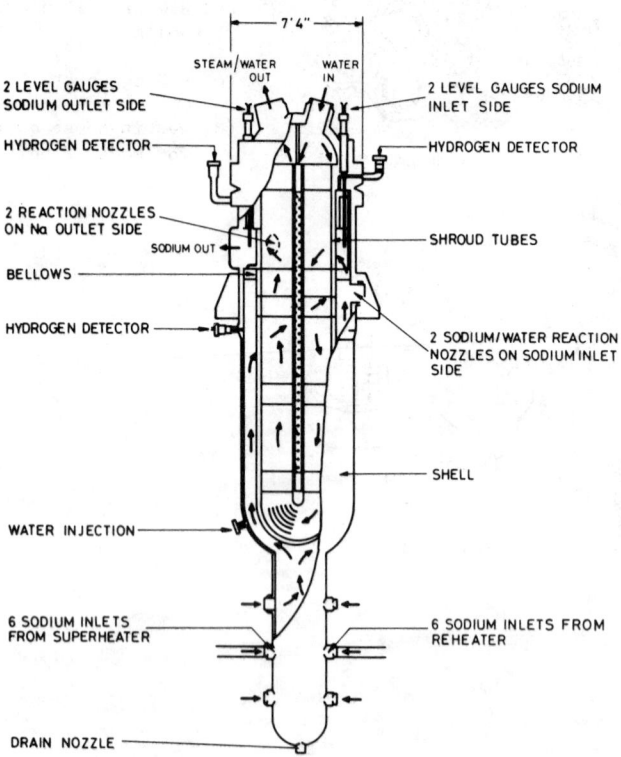

Figure 7   Cross-section view of evaporator on UK Prototype Fast Reactor (PFR)

TABLE 1. Main Design Parameters for Steam Generators of Gas-Cooled Reactors and Liquid Metal-Cooled Reactors

| | MAGNOX | | AGR | | HTR | PFR | LMFBR |
|---|---|---|---|---|---|---|---|
| | Trawsfynydd | Oldbury | Hinkley 'B' | Hartlepool | | | |
| Type of boiler | Drum assisted recirculation | Once-through | Once-through | Once-through | Once-through | Forced recirculation | Once-through or forced recirculation |
| Steam pressure bar | 65<br>20.6 | 96<br>52 | 165 | 165 | 165 | 165 | 165 |
| Tube geometry | Serpentine (finned) | Serpentine (finned) | Serpentine (finned) | Helical (finned) | Helical | U-tube | U-tube<br>Helical |
| Tube material | | | | | | | |
| Evaporator | Carbon steel | Carbon steel | Mild steel followed by 9Cr-1Mo | Mild Steel followed by 9Cr-1Mo | as for AGR | 2.25Cr-1Mo stabilized | – |
| Superheater primary | Carbon steel | Carbon steel | 9Cr-1Mo | 9Cr-1Mo | as for AGR | Austenitic stainless steel | Incoloy |
| Secondary | Carbon steel | Carbon steel | Austenitic stainless | Austenitic stainless | as for AGR | – | – |
| Tube dimensions | | | | | | | |
| Evaporator o.d. mm<br>thickness mm | 56<br>7 | 38<br>5 | 28.6<br>3.25 | 18(fin base)<br>3 | 20-30<br>3-5 | 25<br>2.3 | 15-25<br>2-4 |
| Superheater o.d. mm<br>thickness mm | –<br>– | 38<br>5 | 38<br>4.1 | 20<br>3 | 25-35<br>2.5 | 16<br>2 | 15-25<br>2-4 |
| Heat flux kW/m² | | | | | | | |
| Evaporator inlet<br>outlet<br>peak | 25<br>63<br>– | 54   19<br>190  110<br>–    – | 50<br>137<br>137 | 140<br>320<br>320 | 160<br>380<br>380 | 630<br>110<br>700 | R  OT<br>700 150<br>110 700 |
| Mass velocity kg/m²s | | | | | | | |
| Evaporator at full load | 575 | 410  540 | 1200 | 1650 | 2030 | 2580 | 3000 |

629

(a) *vertical calandria type units*

In this design of *fire-tube boiler*, a vertical calandria is formed by tubes located between two horizontal tube plates. The hot process gas passes upwards inside the tubes whilst water is fed to the shell side of the calandria at its base. The steam-water mixture generated within the shell is taken off at the top to a steam drum to separate the steam. The bottom tube sheet is exposed to the hot process gas and cooling by the water on the shell side may become inadequate if the circulation is impeded by the large volume of steam generated in this region. The use of ferrules at the entrance to the tubes can alleviate the problem but it is difficult to find materials which will withstand the arduous conditions.

(b) *bayonet-tube units*

A common design of waste heat boiler particularly on ammonia plants is the bayonet tube design (Figure 8). In this design the problems of the lower tube plate are eliminated by having a re-entrant design of water tube boiler. The water enters the unit from above, into an upper plenum

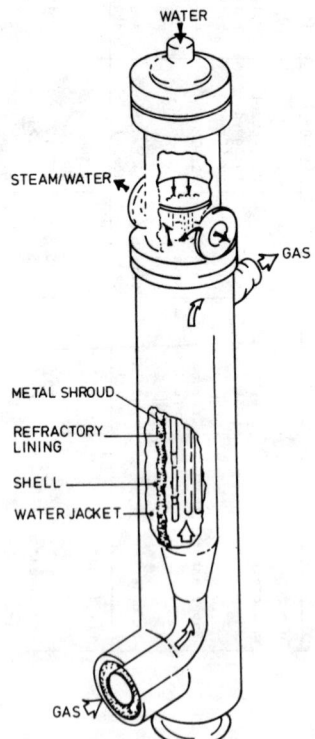

Figure 8   Bayonet tube waste heat boiler unit

and thence to a series of downcomer tubes. At the bottom of the downcomer tube the flow reverses and the steam-water mixture then flows upwards in the annulus between the inner downcomer tube and an outer tube which is in contact with the gas. The hot gas enters at the bottom of the vessel and flows across baffles leaving at the top of the unit. These units have in the past suffered from problems of overheating at the point where the flow reverses at the base of the bayonet tube, from erosion and the tube sheet design is rather complex.

(c) *horizontal U-tube units*

In this design a horizontal bundle of U-tubes is mounted in shell and tube type heat exchangers. Water is fed to the lower limb of each U-tube and the steam-water mixture generated within the unit passes out to a steam drum from the upper limb. Dryout heat fluxes are much lower in horizontal tubes compared with vertical tubes and the situation is made more complex by the presence of the U bend. Flow separation effects can occur downstream of the bend inducing premature dryout and overheating. The problem may be alleviated by the insertion of twisted tapes into the tubes but this increases the pressure drop. The problems of U-tube type waste heat boilers have been discussed by Robertson (1973).

(d) *horizontal cross flow units*

This design of *fire-tube* unit involves the hot process gases passing through a series of horizontal smoke tubes mounted between two vertically orientated tube plates (Figure 9). The steam-water mixture

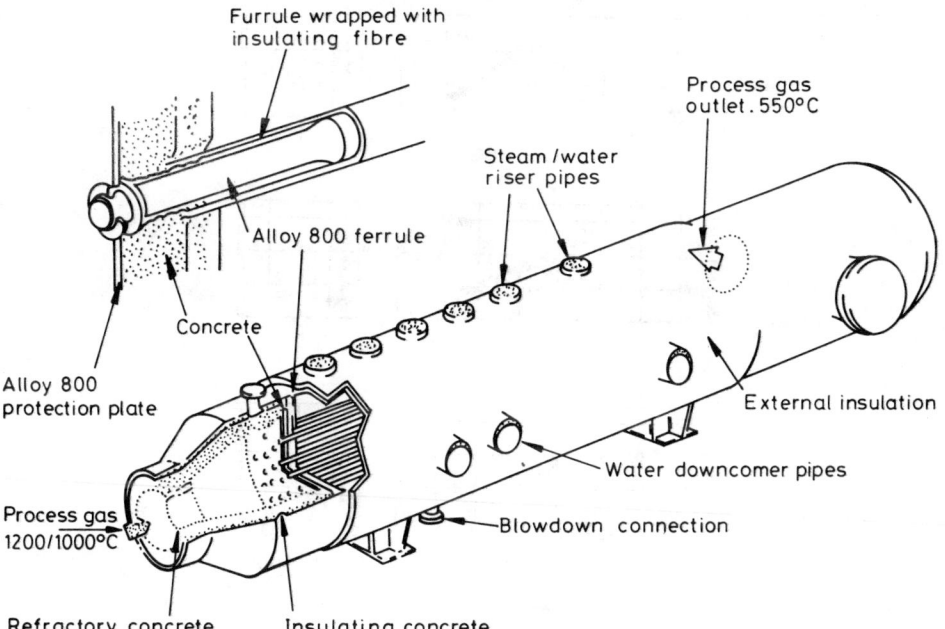

Figure 9   Horizontal cross flow fire-tube boiler

generated on the shell side is taken off through a series of riser pipes to a steam drum and the separated water is returned to the shell via downcomer pipes. The gas inlet is insulated using refractory and insulating concrete. Ferrules are inserted at the tube entrance but with the vertical tube plate less fouling occurs and the inside surface is swept by the recirculating steam-water mixture.

Further details of the various designs of waste heat boiler units has been given by Hinchley (1977, 1979).

## 5. HEAT TRANSFER AND FLUID FLOW PROCESSES OF RELEVANCE IN BOILER DESIGN AND OPERATION

### Dryout

A fossil-fired boiler tube is heated only over half the tube circumference. Conduction in the tube wall, however, ensures that some heat is transferred to the insulated half of the tube. It is important to establish to what extent this asymmetric heating around the circumference influences the dryout condition. Alekseev et al (1964) have examined the situation for a 10mm i.d. tube. Three ratios of the maximum heat flux ($\phi_{MAX}$) to the average heat flux ($\bar{\phi}$) were used, namely 1.12, 1.28 and 1.50. Allowance was made for the circumferential conduction in the tube wall when calculating the heat flux profile. The experimental results for each tube are shown in Figure 10 for

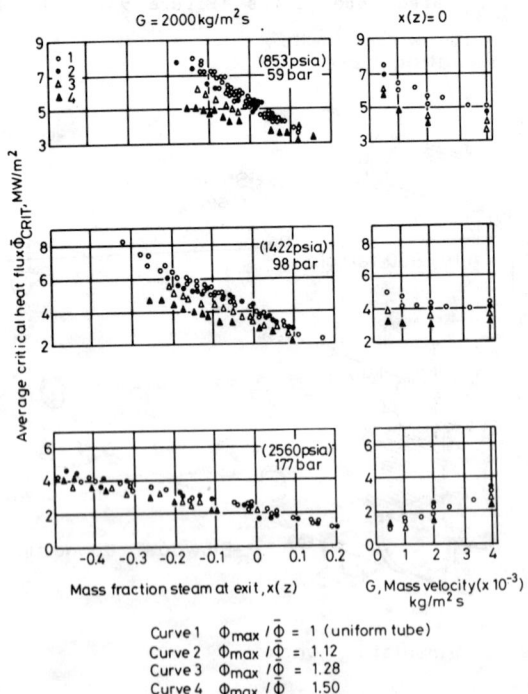

Curve 1   $\phi_{max}/\bar{\phi} = 1$ (uniform tube)
Curve 2   $\phi_{max}/\bar{\phi} = 1.12$
Curve 3   $\phi_{max}/\bar{\phi} = 1.28$
Curve 4   $\phi_{max}/\bar{\phi} = 1.50$

Figure 10    The influence of non-uniform heating around the tube circumference (Alekseev (1964))

# THE DESIGN OF BOILERS

three different pressures, together with the results for a uniform tube of the same dimensions. It is seen that as the circumferential heat flux profile becomes more asymmetric, so the average critical heat flux, $\phi_{CRIT}$ (evaluated from the total power to the tube) falls. At high subcoolings the value of $(\phi_{MAX})_{CRIT}$ for each of the asymmetric tubes approximately corresponds to the uniform tube value, $\phi_{CRIT}$. As the exit steam quality increases, so the <u>average</u> heat flux approaches the value for a uniform tube. These experiments confirm the "local" nature of the critical condition at high subcoolings and the tendency to a "fixed overall power" condition in the higher steam quality regions. It is recommended that the "fixed overall power" hypothesis be used to calculate the critical power which can be applied to a boiler tube with an asymmetric heating. The power calculated should be reduced by 10% to cover the worst condition for cases with $(\phi_{MAX}/\bar{\phi})$ between 1.0 and 1.5 and by 20% for cases with $(\phi_{MAX}/\bar{\phi})$ between 1.5 and 2.0.

In a once-through boiler the temperature rise at dryout is more important than the precise location of the dryout point. Figure 11 shows the wall temperature variation for a once-through boiler as determined from the measurements of Schmidt (1959). These diagrams show clearly the characteristic sharp increase in wall temperatures. Analysis of the extensive programme of work carried out by Herkenrath (1967), together with that of other workers, has shown that the temperature rise at dryout ($\Delta T_{MAX} = T_{W(MAX)} - T_{SAT}$) for a uniformly heated vertical tube can be expressed by the simple relationship:

$$\Delta T_{MAX} = C(\phi/G)^{2.5} \qquad (1)$$

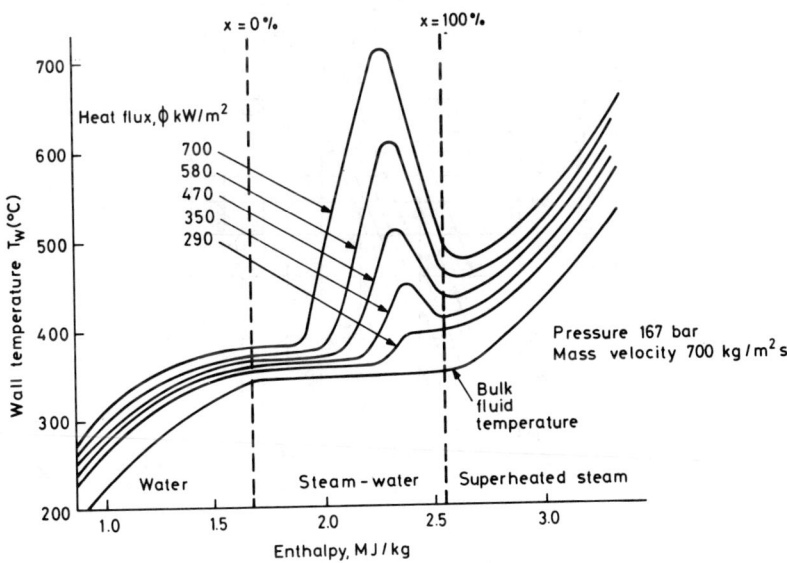

Figure 11   Tube wall temperatures in the liquid deficient region (Schmidt (1959))

The influence of tube diameter on $\Delta T_{MAX}$ is relatively small in the range 5-20 mm. Equation (1) is plotted in Figure 12 and this shows the strong dependence on system pressure. For example, with a constant mass velocity and heat flux the value of $\Delta T_{MAX}$ (and therefore C) decreases by a factor 4.5 as the pressure is increased in the range 140-190 bar. It might be expected that this decrease would extend all the way up to the critical pressure. This, in fact, is not the case, and above 210 bar the value of $\Delta T_{MAX}$ rises sharply. This singularity in the temperature rise at the critical heat flux has been investigated in detail by Bahr (1969). The explanation of the effect stems from the fact that as the pressure rises above 210 bar the mass quality corresponding to the critical heat flux falls with increasing speed from a positive value to a negative value; the mechanism of the critical heat flux changes from dryout to departure from nucleate boiling and the increase in $\Delta T_{MAX}$ results from the substitution of the relatively high heat transfer coefficients in the liquid deficient region with the much lower coefficients in the subcooled film boiling region.

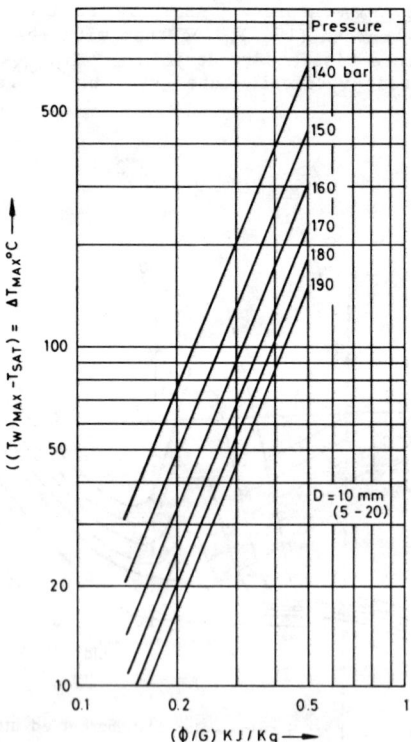

Figure 12  The influence of pressure on the temperature rise at dryout for a uniformly heated vertical tube cooled by high pressure water.

## Horizontal and inclined tubes

Evaporation and condensation within inclined and horizontal tubes is encountered frequently. Stratification of the flow occurs so that the majority of the liquid flows along the bottom of the tube. To a degree this is beneficial in the case of condensation but can cause problems during evaporation in that the liquid film at the top of the tube may dry out (Figure 13).

Dryout in a horizontal tube differs from that in a vertical tube in two ways:

(1) *Stratification* of the flow can occur at low velocities for both low quality and subcooled conditions. Such conditions can lead to overheating of steam boiler tubes at quite modest heat fluxes.

(2) *Dryout* of the tube at high vapour qualities occurs over a relatively long tube length, starting at the top of the tube where the film thickness and flowrate are lowest and ending up with the final evaporation of the rivulet running along the bottom of the tube. Under these conditions the vapour flow in the upper part of the tube may become superheated before dryout occurs at the base of the tube.

Styrikovich and Miropolskii (1950) reported the effects of stratification of a high pressure steam-water mixture in a horizontal pipe. These caused wide temperature differences between the top and bottom of the boiler tube. Experiments were carried out on a single 7.5 m, 56 mm i.d. tube at pressure between 10 bar and 220 bar with heat fluxes in the range 22-135 kW/m$^2$ and inlet velocities betwene 0.24 and 1 m/s. It was found that there was a critical two-phase velocity, j, below which stratification occurred and above which it did not, (Figure 14). Using an alcohol-water analogue for steam-water flow Gardner and Kubie (1976) established the following expression for the critial velocity, j:

$$j^2 = 6.6 \frac{[\sigma g \cos \alpha \, (\rho_f - \rho_g)]^{0.5}}{\rho_f \, f \, K^2} \quad (2)$$

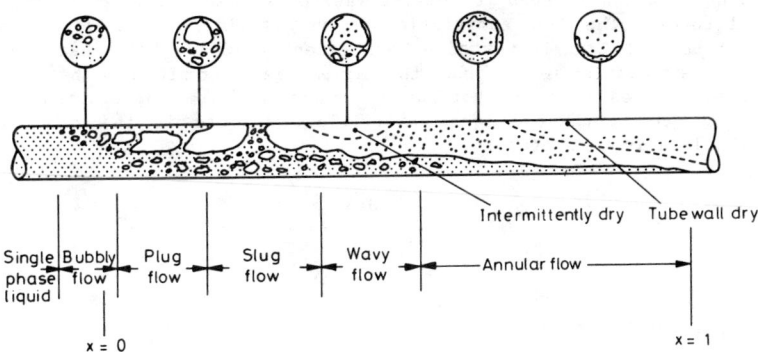

Figure 13   Flow patterns in a horizontal tube evaporator (Collier)

$\alpha$ is the angle over inclination of the tube to the horizontal, f is the friction factor and K is an empirical constant found to be 2.17. Substituting for f and K leads to the following equation for j:

$$j^{1.8} = 30.43 \frac{D^{0.2} [\sigma g (\rho_f - \rho_g)]^{0.5}}{\rho_f^{0.8} \mu_f^{0.2}} \quad (3)$$

Table 2 gives the values of j calculated from equation (3) for steam-water flows over a range of pressures and tube diameters. Excellent agreement is seen with the data in Figure 14 and equation (3) is recommended as giving the minimum single- or two-phase velocity below which stratification will occur in a horizontal tube.

TABLE 2
Values of j (m/s) to prevent stratification for steam/water flow in horizontal tubes

| D (mm) \ p (bar) | 33.5 | 64.2 | 112.9 | 146 | 165 | 187 |
|---|---|---|---|---|---|---|
| 20 | 2.71 | 2.47 | 2.11 | 1.82 | 1.63 | 1.37 |
| 40 | 2.92 | 2.67 | 2.28 | 1.97 | 1.77 | 1.48 |
| 60 | 3.06 | 2.80 | 2.38 | 2.06 | 1.85 | 1.55 |

Stratification may also occur in vertical bends and in helical coils. For the case of a vertical bend, Bailey (1977) showed that the gravitational force maintaining stratified flow is supplemented by centrifugal forces so that $g \cos \alpha$ in equation (2) is replaced by $(g + j^2/r)$. This neglects the secondary flow known to occur in both bends and coils.

Another problem relates to the estimation of heat transfer coefficients in horizontal tubes. The Chen correlation is only valid for situations where all surfaces of the tube remain wetted. More recently, Shah (1976) has proposed an alternative correlation which has the advantage that it can be applied to partially stratified flows in horizontal channels. The correlation is in the form of a graphical chart (Figure 15). The ordinate is the ratio ($h_{TP}/h_L$) and the abscissa is $C_o$ where:

$$C_o = \left(\frac{1-x}{x}\right)^{0.8} \left(\frac{\rho_g}{\rho_f}\right)^{0.5}$$

Two other parameters are used:

# THE DESIGN OF BOILERS

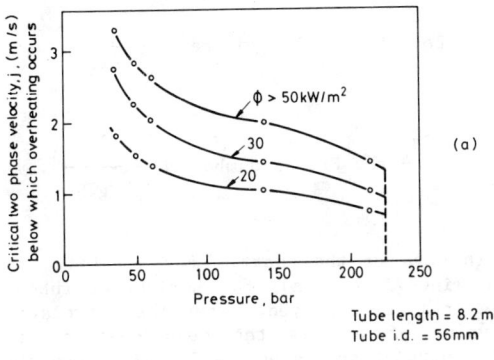

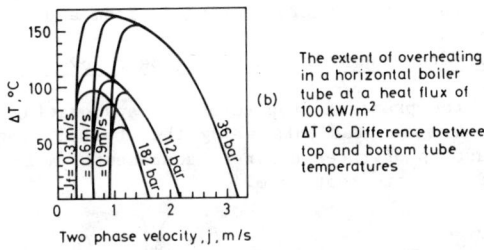

Figure 14  Stratification and overheating of horizontal boiler tubes (Styrikovich and Miropolskii (1950))

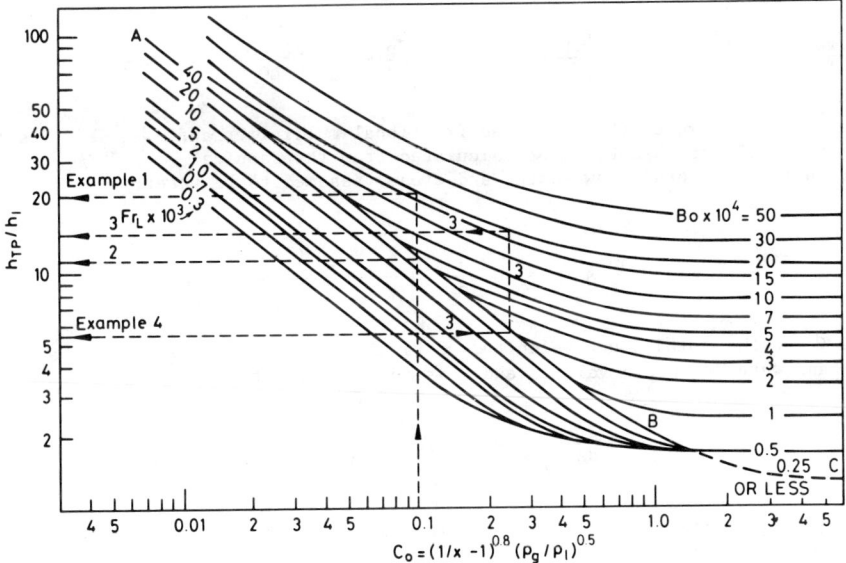

Figure 15  The correlation for two-phase heat transfer coefficient given by Shah (Shah 1976)

Bo - the boiling number $\left(\dfrac{\phi}{G\, i_{fg}}\right)$

and $Fr_f$ - the Froude number $\left(\dfrac{G^2}{\rho_f^2\, g\, D}\right)$

Figure 15 illustrates the use of the chart. For vertical tubes the value of $Fr_f$ is ignored and the line AB is used; for horizontal tubes different lines (according to the value $Fr_f$) are employed. When the correlation is used for a horizontal tube the value of $h_{TP}$ is the mean coefficient over the tube perimeter. The Shah correlation agrees well with experimental data for refrigerants but for water should only be used under conditions where the heat flux is a dependent variable, i.e. where heating is from a primary fluid.

## Two-phase pressure drop and void fraction ("Best Buys")

The evaluation of the pressure drop across an evaporating channel involves integrating the local pressure gradients along the channel length. This local pressure gradient is made up of three separate components; a frictional term, an accelerational term and a static heat term.

$$-\left(\dfrac{dp}{dz}\right)_{TP} = -\left(\dfrac{dp}{dz}F\right) - \left(\dfrac{dp}{dz}A\right) - \left(\dfrac{dp}{dz}z\right) \qquad (4)$$

Using the separated flow model these three components are respectively given by:

$$-\left(\dfrac{dp}{dz}F\right) = -\left(\dfrac{dp}{dz}F\right)_{LO} \phi_{LO}^2 \qquad (5)$$

where $\phi_{LO}^2$ is known as the two-phase frictional multiplier and $-\left(\dfrac{dp}{dz}F\right)_{LO}$ is the the frictional pressure gradient calculated from the Fanning equation for the total flow (water and steam) assumed to be water:

$$-\left(\dfrac{dp}{dz}A\right) = G^2 \dfrac{d}{dz}\left[\dfrac{x^2 v_G}{\alpha} + \dfrac{(1-x)^2 v_L}{(1-\alpha)}\right] \qquad (6)$$

where G is the total mass velocity, x is the quality and $v_G$ and $v_L$ are the steam and water specific volumes and $\alpha$ is the void fraction.

$$-\left(\dfrac{dp}{dz}z\right) = g \sin\theta \left[\dfrac{\alpha}{v_G} + \dfrac{(1-\alpha)}{v_L}\right] \qquad (7)$$

where g is the acceleration due to gravity, and $\theta$ is the angle of inclination of the tube to the horizontal.

## TABLE 3
### TWO-PHASE PRESSURE DROP CORRELATIONS FOR STEAM-WATER MIXTURES

| | CORRELATION |
|---|---|
| Homogeneous Collier (1972) | $\phi_{LO}^2 = \left[1 + x\left(\frac{\rho_L - \rho_G}{\rho_G}\right)\right]\left[1 + x\left(\frac{\mu_L - \mu_G}{\mu_G}\right)\right]^{-1/4}$ |
| Baroczy (1966) | $\phi_{LO}^2 = \Omega \phi_{LO}^2$ (G = 1356 kg/m²s) where $\phi_{LO}^2$ (G=1356)=$f([\mu_L/\mu_G)^{0.2}(\rho_G/\rho_L)],x)$<br>$\Omega = f([(\mu_L/\mu_G)^{0.2}(\rho_G/\rho_L)],x)$ |
| Chisholm (1973) | $\phi_{LO}^2 = 1 + (\Gamma^2 - 1)\left[Bx^{(2-n)/2}(1-x)^{(2-n)/2} + x^{2-n}\right]$ where $\Gamma = [(dp/dz)_{GO}/(dp/dz)_{LO}]^{\frac{1}{2}}$ and B = $f(G,\Gamma)$ |
| CISE Lombardi (1972) | $\left(\frac{dp}{dz}\right)_{TP} = \left[\frac{K G^n v^{0.86} \sigma^{0.4}}{D^{1.2}}\right]$ where $\begin{array}{l}v = [x/\rho_G + (1-x)/\rho_L]\\ K = f \text{ (geometry)}\\ n = f \text{ (geometry)}\end{array}$ |
| Martinelli-Nelson (1948) | $\phi_{LO}^2 = fn(p,x)$ |
| Smith-Macbeth Macbeth (1976) | $\phi_{LO}^2 = \left[\left\{\left(e(1-x) + \left(\frac{\rho_L}{\rho_G}\right) x\right)\left(e(1-x) + x\right)\right\}^{\frac{1}{2}} + (1-e)(1-x)\right]^2$ where e = 0.4 |

## TABLE 4
### TWO-PHASE PRESSURE DROP DATA FOR STEAM-WATER MIXTURES

| Data Bank | ESDU (1976) | | | Friedel (1977) | | | Idsinga (1975) | | | Harwell (1977) | | |
|---|---|---|---|---|---|---|---|---|---|---|---|---|
| Flow Direction | Upflow, downflow and horizontal | | | Upflow only | | | Upflow and horizontal | | | Upflow and horizontal | | |
| Correlation | n | e | σ | n | e | σ | n | e | σ | n | e | σ |
| Homogeneous Collier (1972) | 1709 | -13.0 | 34.2 | 2705 | -19.9 | 42.0 | 2238 | -26.0 | 22.8 | 4313 | -23.1 | 34.6 |
| Baroczy (1966) | 1447 | 4.2 | 30.5 | 2705 | -11.6 | 36.7 | 2238 | -8.8 | 29.7 | 4313 | -2.2 | 30.8 |
| Chisholm (1973) | 1536 | 19.0 | 36.0 | 2705 | -3.8 | 36.0 | 2238 | 0.5 | 40.5 | 4313 | 13.9 | 34.4 |
| CISE Lombardi (1972) | - | - | - | 2705 | 16.3 | 28.0 | 2225 | 22.6 | 28.9 | - | - | - |
| Martinelli-Nelson (1948) | 1422 | 16.3 | 36.6 | - | - | - | 2238 | 47.8 | 43.7 | - | - | - |
| Smith-Macbeth Macbeth (1976) | - | - | - | - | - | - | - | - | - | 4313 | -16.6 | 24.1 |

n = number of data points analysed

e = mean error (%) = $(\Delta p_{cal} - \Delta p_{exp}) \times 100 / \Delta p_{exp}$

σ = standard deviation of errors about the mean (%)

## TABLE 5
## VOID FRACTION CORRELATIONS FOR STEAM-WATER MIXTURES

| | CORRELATION |
|---|---|
| Lockhart & Martinelli (1949) | $\alpha = f(X)$ where $X = \left[\left(\frac{dp}{dz}\right)_L \Big/ \left(\frac{dp}{dz}\right)_G\right]^{\frac{1}{2}}$ |
| Hughmark (1962) | $\alpha = K\beta$ where $K = f[Re, Fr, (1-\beta)]$ |
| Smith (1969) | $S = e + (1-e)\sqrt{\dfrac{\rho_L/\rho_G + e(1/x - 1)}{1 + e(1/x - 1)}}$ where $e = 0.4$ |
| CISE Premoli (1970) | $S = f(G, D, \rho_L, \rho_G, \mu_L, \sigma, \beta) + 1$ |
| Chisholm (1973) | $S = \left[x\left(\dfrac{\rho_L}{\rho_G}\right) + (1-x)\right]^{\frac{1}{2}}$ |
| Thom (1964) | $S = f\left(\dfrac{\rho_L}{\rho_G}\right)$ |
| Bankoff-Jones (1961) | $S = \left[\dfrac{1 - \alpha}{A - \alpha + (1-A)\alpha^B}\right]$ where $A, B = f(p)$ |
| Bryce (1977) | $S = \left[\dfrac{1 - \alpha}{A - \alpha + (1-A)\alpha^B}\right]$ where $A = f(p, G, X, \rho_G, \rho_L)$, $B = f(p, \rho_G, \rho_L)$ |

TABLE 6

VOID FRACTION DATA FOR STEAM-WATER MIXTURES

| Data Bank | Analysis of mean density | | | | | | | Analysis of slip ratio | | | | | | |
|---|---|---|---|---|---|---|---|---|---|---|---|---|---|---|
| | Friedel (1977) | | | Bryce (1977) | | | | ESDU (1977) | | | | Bryce (1977) | | |
| Correlation | n | e | σ | | n | e | σ | n | e | σ | | n | e | σ |
| Lockhart & Martinelli (1949) | – | – | – | | – | – | – | 598 | -57.6 | 50.3 | | – | – | – |
| Hughmark (1962) | 484 | -10.8 | 33.0 | | – | – | – | 598 | – 9.1 | 29.2 | | – | – | – |
| Smith (1969) | 484 | 0.5 | 26.8 | | 639 | 8.6 | 31.5 | – | – | – | | 639 | 18.0 | 77.8 |
| CISE Premoli (1970) | 484 | 9.3 | 35.0 | | 639 | -1.4 | 22.7 | 598 | -23.7 | 27.2 | | 639 | - 1.2 | 68.6 |
| Chisholm (1973) | 484 | – 0.4 | 26.0 | | – | – | – | 598 | -14.5 | 30.8 | | – | – | – |
| Thom (1964) | 484 | 7.4 | 36.5 | | 639 | 43.3 | 61.7 | – | – | – | | 639 | 132.6 | 200.0 |
| Bankoff-Jones (1961) | – | – | – | | 639 | 9.32 | 31.6 | – | – | – | | 639 | 34.5 | 137.6 |
| Bryce (1977) | – | – | – | | 639 | 0.1 | 20.7 | – | – | – | | 639 | 6.1 | 86.9 |

n = number of data points analysed  
e = mean error (%) = (cal-exp)x100/exp  
σ = standard deviation of error about mean (%)

To evaluate the local pressure gradient, expressions are required for the functions $\phi_{LO}^2$ and $\alpha$. A very large number of correlations have been proposed for these functions and a summary of the better known correlations for $\phi_{LO}^2$ is given in Table 3. Recently, a number of workers have carried out systematic comparisons between these various correlations and data banks containing large numbers of experimental pressure drop measurements for steam-water mixtures. A summary of these various comparisons is given in Table 4. It can be seen that the various studies agree that the most accurate correlations for $\phi_{LO}^2$ are those of Baroczy, Chisholm and CISE but that in each case the standard deviation of errors about the mean is 30-35%.

To evaluate the changes in momentum (or kinetic energy) and also the mean density of a two-phase flow, it is necessary to be able to establish the local void fraction or fraction of the flow cross-section occupied by steam. Once again a large number of correlations have been proposed for the evaluation of void fraction ($\alpha$). Sometimes the correlation is expressed in terms of the slip ratio (S) which is related to the void fraction by the identity:

$$S = \left(\frac{x}{1-x}\right)\left(\frac{\rho_G}{\rho_L}\right)\left(\frac{1-\alpha}{\alpha}\right) \tag{8}$$

A summary of the better known correlations for S or $\alpha$ is given in Table 5. Similarly, systematic comparisons have been carried out between thee various void fraction correlations and data banks containing large numbers of experimental measurements of either void fraction or fluid density measurements for steam-water mixtures. A summary of these comparisons is given in Table 6. It can be seen that the various studies agreed that the most accurate void fraction correlations are those of Smith, CISE and Chisholm, with the latter having the added advantage of great simplicity. Again, the standard deviation of error on the mean density is about 20-30%.

6. CONCLUDING REMARKS

In this lecture it has only been possible to touch briefly upon the many problems of boiler design. The wide range of boiler types and configurations each bring with them their own specific design and operating problems. Fouling and corrosion are particularly important aspects of boiler design and operation and affect the integrity and reliability of all types of boilers. This topic is however considered elsewhere the Institute. Vibration is another cause of failure in boiler plant and that subject too is considered in other papers to the Institute.

# REFERENCES

Alekseev, G.V. et al (1964). "Burnout heat fluxes under forced flow", Paper presented at USSR at 3rd Int. Conf. on Peaceful uses of Atomic Energy, Geneva, A/CONF 28/P/327a.

Bagley, R. (1965). "The application of heat transfer due to the design of once-through boiler furnaces", Inst. of Mech. Engrs. Symp. on "Boiling Heat Transfer in Steam Generating Units and Heat Exchangers", Manchester.

Bahr, A. et al (1969). "Anomale Druckabhangegkeit der Warmeubertragung im Zwerphasengebeit bei Annaherung an der kritischen Druck". Brennstoff-Warme-Kraft, 21 (2), 631-633.

Bailey, N.A. (1977). "Dryout in the bend of a vertical U-tube evaporator", Personal Communication.

Baroczy, C.J. (1966). "A systematic correlation of two-phase pressure drop", Chem. Engng. Prog. Symp. Series 62 (64), pp.323.

Bryce, W.M. (1977). "A new flow dependent slip correlation which gives hyperbolic steam-water mixture flow equations", AEEW-R1099.

Chisholm, D. (1973). "Pressure gradients due to friction during flow of evaporating two-phase mixtures in smooth tubes and channels", Int. J. Heat Mass Transfer, 16, 347-358.

Chisholm, D. (1973). "Research note: void fraction during two-phase flow", J. Mech. Engng. Sci. 15 (3), 235-236.

Collier, J.G. (1972). "Convective boiling and condensation", Published by McGraw Hill Book Co. (UK) Ltd.

Davis, R.J. and Hirst, B. (1979). "Twin header bore welded steam generator for pressurised water reactors", Nucl. Energy Vol.18, April No.2 pp.133-140 (1979).

ESDU, (1976). "The frictional component of pressure gradient for two-phase gas or vapour/liquid flow through straight pipes", Engineering Sciences Data Unit (ESDU) London.

ESDU, (1977). "The gravitational component of pressure gradient for two-phase gas or vapour/liquid flow through straight pipes", Engineering Sciences Data Unit (ESDU) London.

Friedel, L. (1977). "Mean void fraction and friction pressure drop: Comparison of some correlations with experimental data", European Two-Phase Flow Group Meeting, Grenoble, Paper A7.

Gardner, G.C. and Kubie, J. (1976). "Flow of two liquids in sloping tubes: an analogue of high pressure steam and water", Int. J. Multiphase Flow, 2, 435-451.

Garnsey, R. (1979). "Corrosion of PWR steam generators", Nucl. Energy, Vol.18 April No.2, pp.117-132 (1979).

Herkenrath, H. et al (1967). "Heat transfer in water with forced circulation in 140-250 bar pressure range", EUR 3658d.

Hinchley, P. (1979). "The engineering of reliability into waste heat boiler systems", Proc. Inst. Mech. Engrs. Vol.193, No.8.

Hinchley, P. (1977). "Waste heat boilers: problems and solutions", Chem. Eng. Prog. 73, 90 and Chem. Ing. Tech. 49, 553.

Hughmark, G.A. (1962). "Hold-up in gas-liquid flow", Chemical Engineering Progress, 58 (4), 62-65.

Idsinga, W. (1975). "An assessment of two-phase pressure drop correlations for steam-water systems", M.Sc. Thesis, MIT.

Jones, A.B. (1961). "Hydrodynamic stability of a boiling channel", KAPL-2170.

Lockhart, R.W. and Martinelli, R.C. (1949). "Proposed correlation of data for isothermal two-phase, two component flow in pipes", Chemical Engineering Progress, 45 (1), 39-48.

Lombardi, C. and Peddrochi, E. (1972). "A pressure drop correlation in two-phase flow", Energia Nucleare, 19 (2).

Macbeth, R.V. (1976). Private Communication quoted in paper by Brittain, I. and Fayers, F.J. "A review of UK developments in thermal-hydraulic methods for loss of coolant accidents". Paper presented at CSNI Meeting on Transient Two-Phase Flow, Toronto.

Martinelli, R.C. and Nelson, D.B. (1948). "Prediction of pressure drop during forced circulation boiling of water", Trans. ASME 70, p.695.

Premoli, A., Di Francesco, D. and Prima, A. (1970). "An empirical correlation for evaluating two-phase mixture density under adiabatic conditions", European Two-Phase Flow Group Meeting, Paper B9, Milan.

Robertson, J.M. (1973). "Dryout in horizontal hair-pin waste heat boiler tubes", A.I.Ch.E. Symp. Series 69 (131), 55 (1973).

Schmidt, K.R. (1959). Warmetechnische Untersuchungen an hock belasteten kesselheizflacken. Mitteilungen der Vereinigung der grosskessel - bezitzer. 391-401.

Shah, M. (1976). "A new correlation for heat transfer during boiling flow through tubes", ASHRAE Trans. 82 (2), 66-86.

The Shippingport Pressurised Water Reactor (1958) - published by Addison Wesley Publishing Company (1958).

Smith, S.L. (1969). "Void fractions in two-phase flow. A correlation based on an equal velocity head model". Proc. Inst. Mech. Engrs. 184, Pt.1, (36), 647-664.

Styrikovich, M.A. and Miropolskii, Z.L. (1950). Dok. Akad. Nank. SSSR. 71 (2).

Styrikovich, M. (1978). "The role of two-phase flows in nuclear power plants", Int. Seminar "Momentum, Heat and Mass Transfer in Two-Phase Energy and Chemical Systems", Dubrovnik, Yugoslavia, 4-9 Sept. 1978, 20pp, ICHMT.

Thom, J.R.S. (1964). "Prediction of pressure drop during forced circulation boiling of water", Int. J. Heat Mass Transfer, 7, 709-724.

Thurlow, G.G. (1978). "The combustion of coal in fluidised beds", Inst. of Mech. Engrs. (London) Proceedings, 192 (15).

Ward, J.A. (1977). Private Communication.

# Condensers: Thermohydraulic Design

**D. BUTTERWORTH**
Heat Transfer and Fluid Flow Service (HTFS)
AERE Harwell, Oxon, UK

ABSTRACT

The various types of condenser equipment are described. One of the most important types is the shell and tube condenser. It is therefore shown how the methods in previous lectures on condensers are applied to the thermal and hydraulic design of these condensers. An important step in this is the calculation of local and mean temperature differences for multipass units and for fluid streams with non-linear temperature-enthalpy curves. This problem is therefore discussed in some detail. Spray condensers are also described and simple design methods for them outlined. Possible reasons why condensers fail to operate as expected are noted.

1. INTRODUCTION

Condensers are of two main types; those in which the coolant and condensing streams are separated by a solid surface, usually a tube wall, and those in which coolant and condensing vapour are brought into direct contact. Each of these types may be subdivided into further categories as illustrated in Figure 1. The direct contact type may consist of vapour which is bubbled into a pool of liquid or of liquid which is sprayed into vapour. Those in which the streams are separated may be subdivided into three main types; air-cooled, shell-and-tube, and plate. In the air-cooled type, condensation occurs inside tubes with cooling being provided by air which is blown or sucked across the tubes. Fins are usually provided on the air side to compensate for the low air-side coefficients by having a large surface area. The shell-and-tube type consists of a large cylindrical shell inside which there is a bundle of tubes. One fluid stream flows inside the tubes with the other on the outside or shell side. Condensation may occur outside or inside the tubes dependent on circumstances. Because of special circumstances in different applications, the design of these units may vary widely and it is therefore convenient to subdivide shell-and-tube condensers into process and turbine-exhaust types. A plate may be used instead of a tube wall to divide the coolant and condensing streams. In one design, the plates are corrugated to give rigidity and also to improve heat transfer. These are held together in a press or frame with gaskets between the plates to prevent fluid leakage. In another design, the plates are flat but corrugated metal sheets are sandwiched between them to act as fins. These units are made of aluminium and are used in cryogenic heat transfer applications.

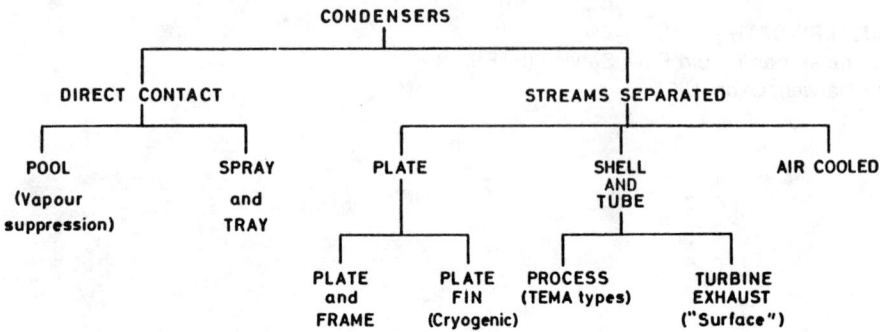

Fig.1. Subdivisions of Condenser types

Further details of some of the above types of unit are given below. The plate and frame type of unit is not described further here because it is not usually used as a condenser except when service steam is used to heat another stream. Information on this type of exchanger is, however, given by Alfa Laval [1], Butterworth [2] and elsewhere in these proceedings.

2. EQUIPMENT

2.1 Shell and tube condensers for chemical processing

As will be seen, there is a wide variety of units which fall into this category. The variety makes it possible to handle many and varied process streams. They are used for water, solvents, hydrocarbons, acids and many other fluids which may be condensing under vacuum or at high pressure. Single vapours, mixtures of vapours and vapours with non-condensable gas can all be handled with suitable choice of geometry. Standards for the design of these condensers are covered by the TEMA [3] standards, which also cover shell-and-tube exchangers for single-phase and boiling duties.

An example of a process condenser is shown schematically in Figure 2. In this instance, the hot (condensing) stream is on the shell side with the coolant on the tube side. The coolant here traverses the length of the exchanger once, thus giving a single tube side pass. Multipass arrangements are possible by, say, having U tubes or by partitioning the header boxes.

The baffles consist of circular plates with a segment cut away (as illustrated in Figure 3) and with slightly oversized holes drilled to take the tubes when the exchanger is constructed. These baffles perform two important duties: the first is to support the tubes and the second to direct the shell side stream back and forth across the tubes. For condensation on the shell side, it is usual to have the baffles cut vertically so that the condensate can fall easily to the bottom of the shell in all baffle compartments. The baffles are usually, however, cut horizontally if there is single-phase flow on the shell side. Notches are cut in the lowest position in the baffle to allow the unit to be drained. Other baffle types are possible and these are described elsewhere [3].

The choice of whether to put the condensing stream on the tube side or shell side is dictated by many factors. Dirty fluids are usually better on the tube side since it is easier to clean inside the tubes than outside

# CONDENSERS: THERMOHYDRAULIC DESIGN

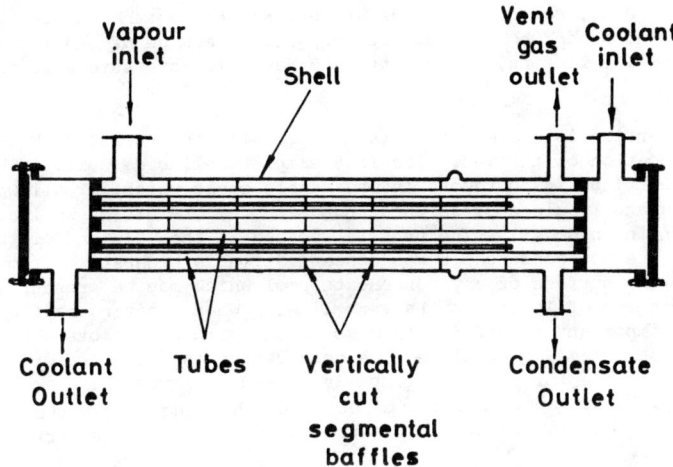

Fig.2. Example of shell side condenser: TEMA E type shell with single tube side pass

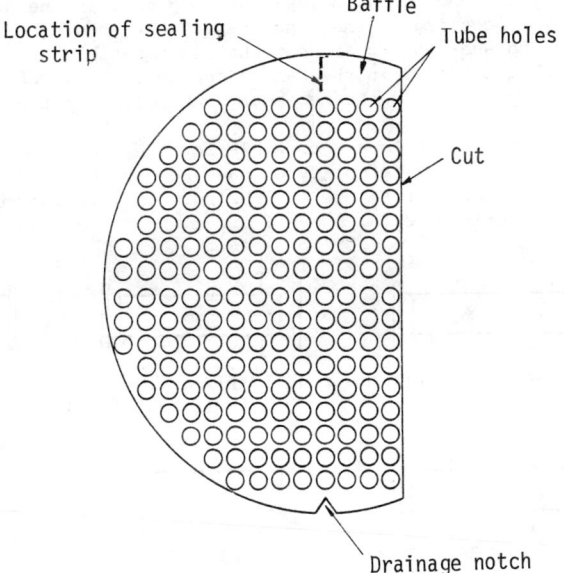

Fig.3 Single segmental baffle with vertical cut

them. Also, corrosive fluids are better in the tubes since then only the tubes and header boxes (but not the shell) need be of special materials. For similar reasons, it is better to have high-pressure streams on the tube side.

Venting is an important feature of all condensers and Figure 2 illustrates the location of the vent line in a simple shell side unit. The purpose of the vent line is to remove incondensable gases which, if allowed to accumulate, would depress the condensing temperature as well as hinder the condensing process. The vent is located in the cold end of the condenser where the non-condensable gas concentration is highest. It must be well above the surface of any condensate pool which may be present so as to avoid entrainment of liquid in the vent line. When locating baffles in a condenser it is important to arrange them so as to avoid dead spots where non-condensables may accumulate. A good design is one which directs the vapour flow over the tubes and continually towards the vent line. It is important to prevent inlet vapour flowing directly to the vent line without crossing any tubes. This may require the introduction of a sealing strip as indicated in figure 3.

It is important to ensure that a condenser may be drained properly. This means having a sufficiently large, constriction-free condensate outlet line which is located low down in the condenser and takes the condensate away downhill. If the condensate does not drain properly, a large fraction of the tube may become flooded thereby impairing the condenser performance. Shell side condensers are, however, often designed with some of the tubes deliberately flooded by having, say, a loop seal in the outlet line. This is to allow subcooling of the condensate. It should be noted, however, that condensing and sub-cooling are somewhat incompatible duties and they do not always go well together in the same exchanger. However, it is sometimes cheap and easy to obtain the necessary subcooling in the same exchanger.

Various arrangements are possible for the shell side of shell-and-tube condensers. These different types have been given letters of designation by TEMA [3] as shown in Figure 4. Using this lettering convention, the shell type described so far is an E-type.

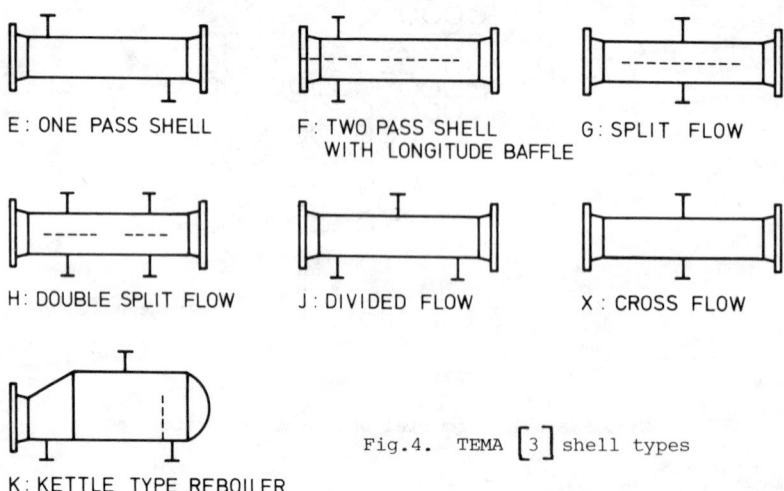

Fig.4. TEMA [3] shell types

# CONDENSERS: THERMOHYDRAULIC DESIGN

For shell side condensers, the E-type is the simplest and therefore the first choice. However, with low pressure or vacuum operation, the vapour velocities in the inlet region may be too high to allow this design. The high vapour velocities may lead to unacceptable erosion levels or tube vibration. Some avoidance of the high velocities is achieved by having a large inlet nozzle and by leaving out tubes in the vicinity of the inlet nozzle. If this does not lead to an acceptable reduction in vapour velocities, other shell types should be considered. The J shell could, for example, be used because it can have two inlet nozzles and one outlet. It is shown in Figure 4 as a divided flow shell but it can clearly be arranged in a combining-flow mode with two inlet nozzles at the top and one outlet at the bottom. Such shells are baffled in the same manner as has been already described for E shells. H shells also have the advantage of two inlet nozzles but are rather complicated by virtue of the longitudinal baffles shown as the broken line in Figure 4. The longitudinal baffles make it expensive to include a large number of cross baffles to support the tubes. The cross flow or X-type exchanger is a very useful unit for vacuum operation. Figures 5 and 6 show two possible arrangements for such units. It can be seen that these have many inlet nozzles and a distribution space above the tubes. A perforated plate is sometimes placed above the tubes to help distribute the inlet vapour. A particular advantage of the X shell is that the cross baffles no longer have the dual role of supporting the baffles and redirecting the vapour flow. In an E shell, unless of very special design, more baffles means higher vapour velocity. In an X shell, however, many baffles may be included without any significant effect on the vapour velocity. This is very useful when designing to avoid tube vibration since higher velocities tend to increase the chances of vibration whereas shorter tube spans reduce the chances. Crossflow condensers also give lower pressure drops for the same duty than is usual with other types.

Any of the TEMA shell types may be used for tube side condensers. The K type is a shell side boiler and the condensation on the tube side would be usually from service steam. It is normal to restrict the number of tube side passes to one in tube side condensers because of the difficulty of

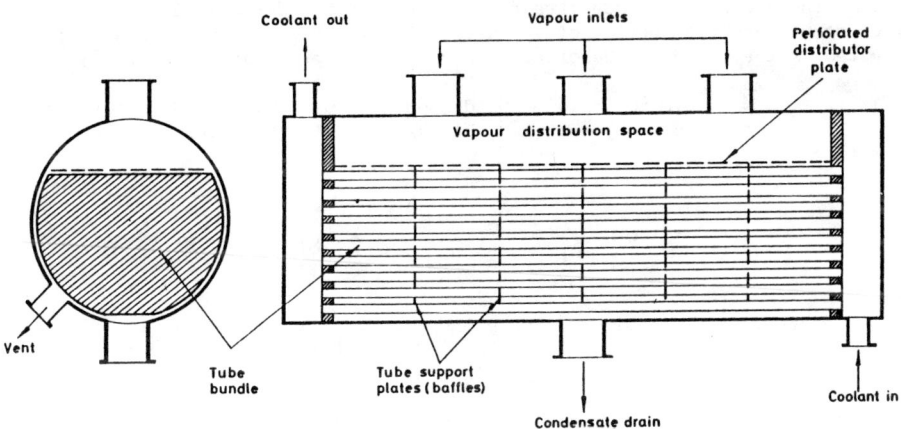

Fig.5. Main features of a crossflow condenser (TEMA X type)

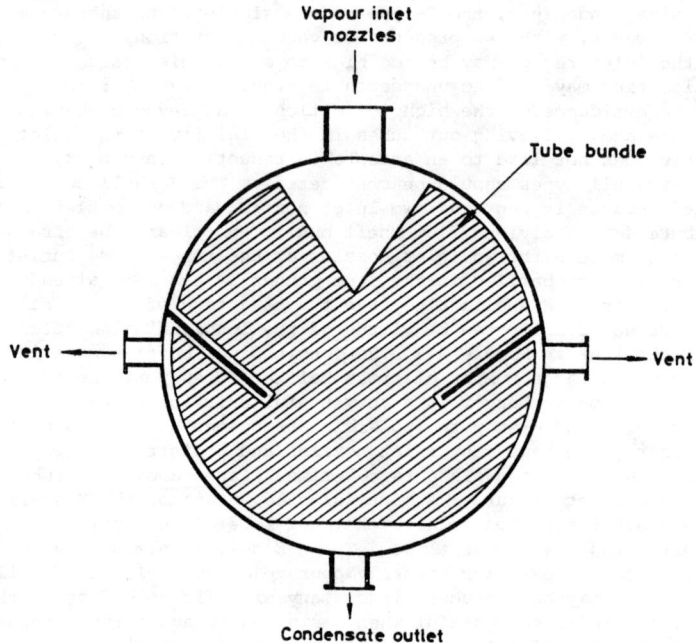

Fig.6. Alternative tube bundle layout
for a crossflow condenser

knowing the flow distribution in the second pass. However, the use of U tubes allows two tube passes without running into this problem.

It is usual to have tube side condensers horizontal but it is possible to have them vertical or inclined. When not horizontal, the possibility arises of having the vapour and condensate flowing in opposite directions, ie. down-flow of condensate with upflow of vapour. Clearly, the design of such units is limited by the flooding phenomenon discussed in a previous lecture.

It is also possible to design a condenser with upflow of both vapour and liquid. This can be done by ensuring that $V_G^*$ is well above unity at the top of the tube where

$$V_G^* = \left\{ \frac{\rho_G V_G^2}{(\rho_L - \rho_G) g D_i} \right\}^{\frac{1}{2}} \qquad (1)$$

It may, however, be difficult to ensure in practice that the liquid is dragged smoothly from the top of the tube for all likely condenser operating conditions. This mode is operation is not therefore recommended.

2.2 Shell and tube condensers for steam turbine exhausts

For historical reasons, these condensers are often referred to as surface condensers. In principle, there is no reason why they are any different from the shell side condensers just described. In practice, however, there are certain severe demands placed on them which have been overcome by special design features. These special demands arise from the large heat duties which they must perform and from the necessity to maintain a low condensing temperature in order to achieve the highest possible power-cycle efficiency.

The aim is to operate with a condensing temperature only a few degrees above the cooling-water temperature. Typically, for land based condensers, the cooling water will be about $20^{\circ}C$ with condensing taking place at around $30^{\circ}C$. The saturation pressure of water at this temperature is 0.04 bar absolute, which means that there is little pressure available for pressure drop through the unit. Equally clearly, there is little temperature difference to spare in order to overcome non-condensable gases. Hence, the design of surface condensers is governed by the need for good venting and low pressure drop.

Often, these condensers are very large. There may, for example, be two condensers serving a single 600 MW(e) turbine set. Hence, each condenser must handle around 250 kg/s of steam with approach velocities of up to 65 m/s. Surface areas are around 25000 $m^2$ which is achieved by having, say, 15000 tubes of 25 mm outer diameter with a length around 20 m. These very large condensers will often have box-shaped shells but the smaller ones, with surfaces of less than about 5000 $m^2$, can have cylindrical shells.

Surface condensers vary widely in their geometrical details and various types are described by the editors of Power [4], by Simpson [5] and by Sebald [6]. Standards for their design are given by the British Electrical and Allied Manufacturers' Association [7] and by the Heat Exchange Institute [8]. Despite the variations in construction, there are many features which are common to most designs and these are shown in Figure 7, which shows a relatively small surface condenser. This is not an actual unit but a concocted drawing to illustrate some of the main items.

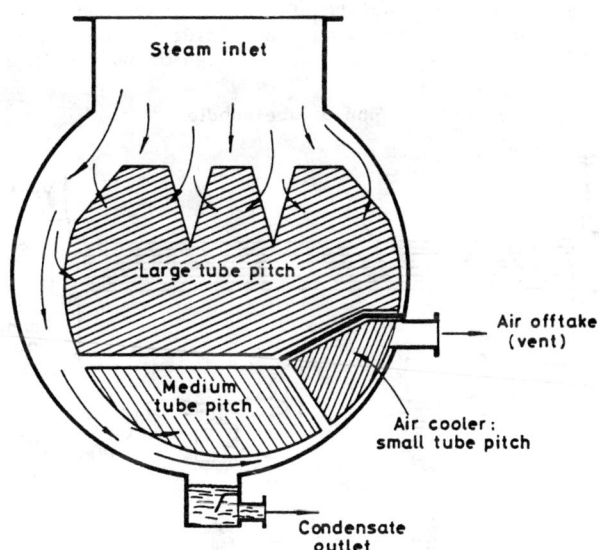

Fig.7. Small turbine exhaust condenser: areas occupied by tubes shown shaded

The vapour inlet velocity is very high because of the high thermal duty combined with the low pressure. The tubes near the inlet are therefore on a wider pitch than elsewhere and tubes are left out in places to provide paths or steam lanes which guide steam into the bundle. The combination of steam lanes and paths around the bundle means that there is a large bundle perimeter through which the steam may pass, thereby minimising the effects of the large inlet velocity. As the steam passes through the bundle towards the vent line, the steam flow rate reduces and the air concentration increases. Closer tubes and less superficial flow area are therefore used towards the exit in order to keep the steam velocities up. This improves the gas phase heat and mass transfer as well as reducing the danger of stagnant pockets of air forming. There is usually a separate compartment just by the vent line which has the smallest tube pitch and the coldest cooling water in the tubes. Since most of the steam has been extracted from the air by this stage, the compartment is called the air cooler. The purpose of the section is to extract the last possible moisture from the air which includes knocking out any entrained condensate. Some steam from the condenser inlet is allowed to pass right around the bundle through any condensate falling off the tubes. This helps to de-aerate the condensate. It is important, though, to prevent this vapour flowing straight to the vent without crossing any tubes.

2.3 Air-cooled heat exchangers

Many coolants are possible for process condensers: eg. air, cooling-tower water or a colder process stream which requires heating. In areas where there is a shortage of makeup water, air-cooled condensers may be favoured. They can also become economic if condensation is taking place at temperatures which are more than about 20°C above ambient. They suffer the disadvantage, however, of occupying a relatively large ground area and of generating noise from the fans.

Figure 8 illustrates a typical air-cooled heat exchanger which may be used as a condenser. It consists of a horizontal bundle of tubes with the air being blown across the tubes on the outside and condensation occurring inside the tubes. The unit shown is a forced draught unit since the air is

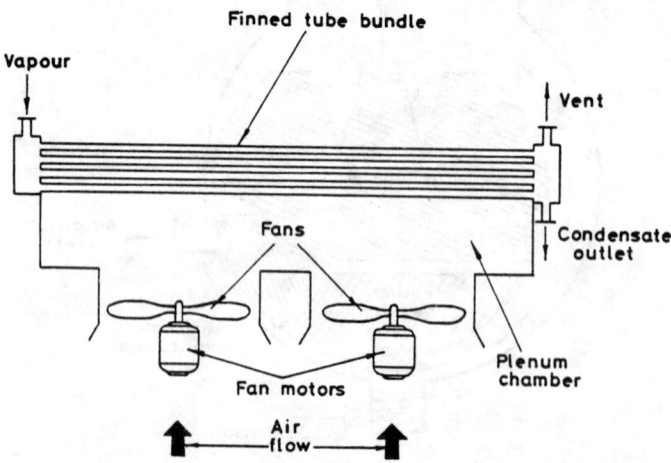

Fig.8. Forced draught air-cooled heat exchanger used as condenser

blown across. An alternative design is the induced draught unit which has
the fans on top which suck the air over the tubes. The tubes are finned
with transverse fins on the outside to overcome the effects of the low air-
side coefficients. There would normally be few tube rows and the process
stream may take one or more passes through the unit. With multipass
condensers, the problem arises with re-distributing the two-phase mixture
on entry to the next pass. This can be overcome in some cases by using
U tubes or by having separate passes just for subcooling or de-superheating
duties. In multipass condensers, it is important to have each successive
pass below the previous one to enable the condensate to continue downwards.
Further information in air-cooled heat exchangers is given by Ludwig [9],
by the American Petroleum Institute [10] and elsewhere in these proceedings.

2.4 Plate-fin heat exchangers

Figure 9 shows the general form of a plate-fin heat exchanger. The
fluid streams are separated by flat plates between which are sandwiched
corrugated fins. A more apt name for these exchangers is therefore a
finned-plate exchanger. They are often used in low temperature (cryogenic)
plant and where the temperature differences between the streams are small
(1 - 5°C). They are compact units having a heat transfer area per unit
volume of around 2000 $m^2/m^3$. The fins may be serrated or perforated to
improve heat transfer and help in the flow distribution across the plate.
Special manifold devices are provided at inlet and outlet to these
exchangers to provide good flow distribution across the plates and from
plate to plate. The plates are typically 0.5 - 1.0 mm thick and the fins
0.15 to 0.75 mm thick. The whole exchanger is made of aluminium alloy and
the various components are brazed together.

The flow channels in plate-fin exchangers are small and often contain
many interruptions to flow. This means that the flows have to be small

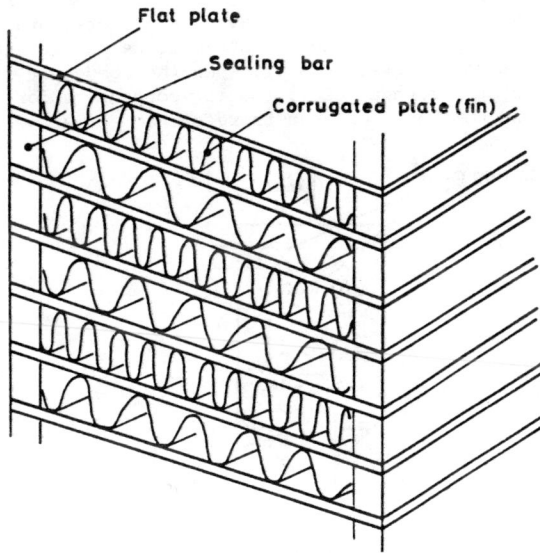

Fig.9. Basic construction of a plate-fin heat exchanger

(10 - 300 kg/m$^2$s) to avoid excessive pressure drops.  This can make the
channels prone to fouling which, when combined with the fact that they cannot
be mechanically cleaned, means that plate-fin exchangers are restricted to
clean fluids.  A useful feature of these exchangers is that they can exchange
heat between three or more fluid streams.  Further information on them is
given by Lenfestey [11].

2.5  Direct-contact condensers

Direct-contact condensers are cheap and simple devices but have limited
application because the processes streams and coolant are mixed.  The
removal of the intermediate wall means that they are not prone to fouling and
very high heat transfer rates per unit volume can be achieved.

Some direct-contact exchangers inject vapour into a pool of liquid.
This may be done to heat up a process fluid or to suppress vapour released
from a reaction vessel as a result of an accident or malfunction.  Two
difficulties arise with this method of condensation.  The first is that
the condensation front may move back into the vapour inlet line, causing the
liquid to be periodically ejected, often with some violence.  The second is
that a very large vapour bubble may form in the liquid pool and this may
collapse suddenly causing damage to the vessel.  These problems may be
avoided by having the vapour injected through a large number of small holes or
by using special ejectors which mix the incoming vapour with liquid in a
special mixing tube.

The most common type of direct-contact condenser is one in which subcooled liquid is sprayed into the vapour in a large vessel.  This
arrangement is illustrated in Figure 10.  Very often, these units are used
for condensing steam using water as coolant.  In these cases, the mixing
of water with condensate presents no major problem.  When condensing organics
with water, however, a separator is usually required.  Alternatively, the
condensate product may be cooled and some recycled as coolant spray.  At
first sight, there seems little benefit in having a direct-contact condenser
and a conventional single phase exchanger instead of using one shell-and-tube
condenser.  The advantage appears, however, when the condenser is operating
under vacuum.  As has been seen already, tubular condensers for vacuum
operation are large and complex.  It can therefore sometimes be economic to

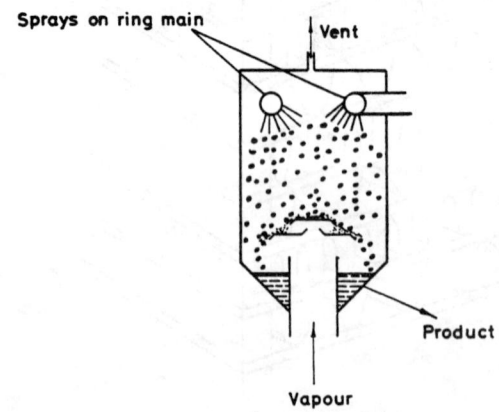

Fig.10.  Spray condenser

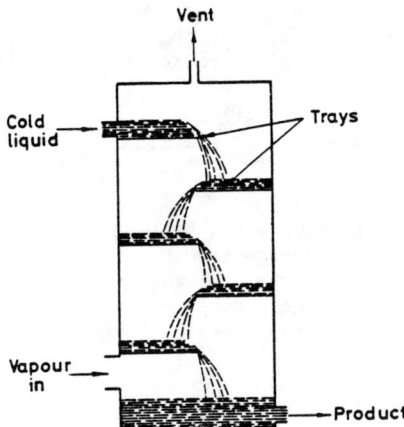

Fig.11. Tray condenser

replace such a condenser by a simple spray condenser and a compact single-phase cooler.

Spray condensers cannot be used with dirty coolants since the spray nozzle may block. In these circumstances, a tray condenser may be used, as illustrated in Figure 11.

## 3. THERMAL EVALUATION METHODS FOR SHELL AND TUBE CONDENSERS

### 3.1 Introduction and definition of terms

The term 'thermal evaluation method' is used to signify that the calculation process by which, for a known heat exchanger geometry, the thermal duty may be calculated or, alternatively, the required heat transfer area determined to suit the duty. In the latter case, the calculated heat transfer area may be incompatible with assumed geometry. These calculations fall short of a full design calculation which involved calculating stream pressure drops as well as repating the calculations for many different assumed geometries in order to find ones which satisfy all the imposed constraints. The final stage of design is to choose the best design on the basis of, say, capital cost.

The basis of thermal evaluation methods is an equation of the form

$$\frac{d\dot{Q}}{dA} = U\theta \qquad (2)$$

where $\dot{Q}$ is the heat transfer rate, A the heat transfer area, U the overall heat transfer coefficient and $\theta$ the temperature difference. It is important to appreciate that both U and $\theta$ can vary significantly throughout a condenser and hence equation (2) is based on the local values. The temperature difference $\theta$ may be defined in a number of ways provided that it is consistent with the definition of U. One definition often used for $\theta$ is that it is the difference between the *equilibrium* temperatures of the two streams if each were well mixed at the point in question. This would, for example, be the right definition of $\theta$ to use when determining U by the Silver [12], Bell and Ghaly [13] method described in previous lectures. It is the definition used in Sections 3.2 to 3.5 below.

The overall coefficient, $U$, may be regarded as being made up of a number of components as follows:

$$\frac{1}{U} = \frac{1}{\alpha_{hot}} + r_{hot} + \frac{\delta_w}{k_w} + \frac{1}{\alpha_{cold}} + r_{cold} \qquad (3)$$

where $\alpha_{hot}$ and $\alpha_{cold}$ are respectively the "film" coefficients for the hot and cold streams, $r_{hot}$ and $r_{cold}$ the respective fouling layer thermal resistances, $\delta_w$ the tube wall thickness and $k_w$ the tube wall thermal conductivity. For thick walled tubes, corrections are necessary to allow for the different surface areas inside and outside the tubes. A simplified form of equation (3) will be used in subsequent calculations:

$$\frac{1}{U} = \frac{1}{\alpha_{hot}} + r + \frac{1}{\alpha_{cold}} \qquad (4)$$

where $r$ is the combined thermal resistance of the tube wall and fouling. This resistance will be taken as a constant throughout the exchanger, whereas $\alpha_{hot}$ and $\alpha_{cold}$, and consequently $U$, may vary considerably. The determination of $\alpha_{hot}$ and $\alpha_{cold}$ is discussed elsewhere on this course, and it is therefore assumed in this lecture that we know how to calculate them locally. This lecture therefore concentrates on how to use this information in thermal evaluation.

Equation (2) can be rearranged and written in an integral form as follows:

$$\int_{\dot{Q}_T} \frac{d\dot{Q}}{\theta} = \int_{A_T} U dA \qquad (5)$$

where the subscript $T$ refers to the total value for the exchanger. Heat exchanger designs are usually summarised in terms of mean quantities which are related by an equation as follows:

$$\dot{Q}_T = U_m A \theta_m \qquad (6)$$

where $U_m$ is the mean overall coefficient and $\theta_m$ is the mean temperature difference. Comparing equations (5) and (6) suggests the following definitions for the mean quantities:

$$\frac{1}{\theta_m} = \frac{1}{\dot{Q}_T} \int_{\dot{Q}_T} \frac{d\dot{Q}}{\theta} \qquad (7)$$

$$U_m = \frac{1}{A_T} \int_{A_T} U dA . \qquad (8)$$

# CONDENSERS: THERMOHYDRAULIC DESIGN

In practice, it is unnecessary to evaluate both equations (7) and (8), since when either $\theta_m$ or $U_m$ has been determined the other may be calculated from equation (6).

Equation (2) may also be written as follows

$$A_T = \int_{\dot{Q}_T} \frac{d\dot{Q}}{U\theta} \tag{9}$$

Combining this with equation (4) gives

$$A_T = \int_{\dot{Q}_T} \frac{d\dot{Q}}{\alpha_{hot}\theta} + r \int_{\dot{Q}_T} \frac{d\dot{Q}}{\theta} + \int_{\dot{Q}_T} \frac{d\dot{Q}}{\alpha_{cold}\theta} \tag{10}$$

Dividing through by $A_T U_m$ and using equations (6) and (7) gives

$$\frac{1}{U_m} = \frac{1}{U_m A_T} \int_{\dot{Q}_T} \frac{d\dot{Q}}{\alpha_{hot}\theta} + r + \frac{1}{U_m A_T} \int_{\dot{Q}_T} \frac{d\dot{Q}}{\alpha_{cold}\theta} \tag{11}$$

which, on comparing with equation (4), suggests the following definitions for the mean "film" coefficients:

$$\frac{1}{\alpha_m} = \frac{1}{U_m A_T} \int_{\dot{Q}_T} \frac{d\dot{Q}}{\alpha\theta} \tag{12}$$

The derivation leading to this last equation was proposed by Smith [14].

Using equation (2), an alternative form of this definition is

$$\frac{1}{\alpha_m} = \frac{1}{U_m A_T} \int_{A_T} \left(\frac{U}{\alpha}\right) dA \tag{13}$$

Some special cases of equations (7) and (8) are useful. If $\theta$ varies linearly with $\dot{Q}$, equation (7) can be integrated to give

$$\theta_m = \theta_{ln} = \frac{\theta_a - \theta_b}{\ln \frac{\theta_a}{\theta_b}} \tag{14}$$

where $\theta_{ln}$ is the well-known logarithmic mean temperature difference and $\theta_a$ and $\theta_b$ are the end values of $\theta$. It is most unusual in a condenser for $\theta$ to vary linearly with $\dot{Q}$ over the whole exchanger but small portions of the exchanger can often be identified over which this assumption is well approximated. Examples of this will be seen below.

If U varies linearly with A, equation (8) may be integrated between $U_a$ and $U_b$ to give

$$U_m = \frac{1}{2}(U_a + U_b). \tag{15}$$

If both U and $\theta$ vary linearly with $\dot{Q}$, equation (8) may be integrated with the aid of (2) to give

$$U_m = \frac{U_a \theta_b - U_b \theta_a}{\theta_{ln} \ln \frac{U_a \theta_b}{U_b \theta_a}}. \tag{16}$$

This result was first obtained by Colburn [15]. If both $1/U$ and $\theta$ vary linearly with $\dot{Q}$, equation (8) may be integrated with the aid of (2) to give

$$\frac{1}{U_m} = \frac{1}{U_a} \frac{\theta_{ln} - \theta_b}{\theta_a - \theta_b} + \frac{1}{U_b} \frac{\theta_a - \theta_{ln}}{\theta_a - \theta_b}. \tag{17}$$

Again, these equations will not usually be valid over the whole of the condenser but may apply to small portions of it. It is not always clear which of the above equations is valid for a given set of circumstances. However, if $U_a$ and $U_b$ vary by only a small amount, equation (15) is preferred because of its simplicity. There is a long tradition in the use of equation (16) but with little justification. Equation (17) seems more in line with the variations observed in condensers and is hence recommended in those situations when equation (15) cannot be used due to the large difference between $U_a$ and $U_b$. Of course, any question about which equation is more accurate can always be avoided by dividing the exchanger into a large number of sections.

### 3.2 Co-current and counter-current condensers

The procedure given here applies to TEMA E type shells with a single tube side pass. It also applies to a J shell which can be divided down the middle and treated as two exchangers, one with co-current flow and the other with counter-current flow. Counter-current flow is more usual in E shells since it makes best use of the temperature difference between the streams. Indeed, some duties are not possible in co-current flow but can be handled without difficulty in counter-current flow. The following description is in terms of counter-current flow but the same approach can be used for co-current flow, and the differences in the results obtained are noted.

Figure 12 illustrates a counter-current flow exchanger. In this diagram and the subsequent discussion, the shell-side stream is denoted by a prime ($'$). Hence, the shell side stream enters with specific enthalpy $h'_{in}$ and leaves with specific enthalpy $h'_{out}$. The tube side specific enthalpy changes from $h_{in}$ to $h_{out}$. The shell side and tube side mass flows are respectively $\dot{M}'$ and $\dot{M}$. A heat balance over area A of the exchanger gives

$$h = h_{in} + \frac{\dot{M}'}{\dot{M}}(h' - h'_{out}) \tag{18}$$

# CONDENSERS: THERMOHYDRAULIC DESIGN

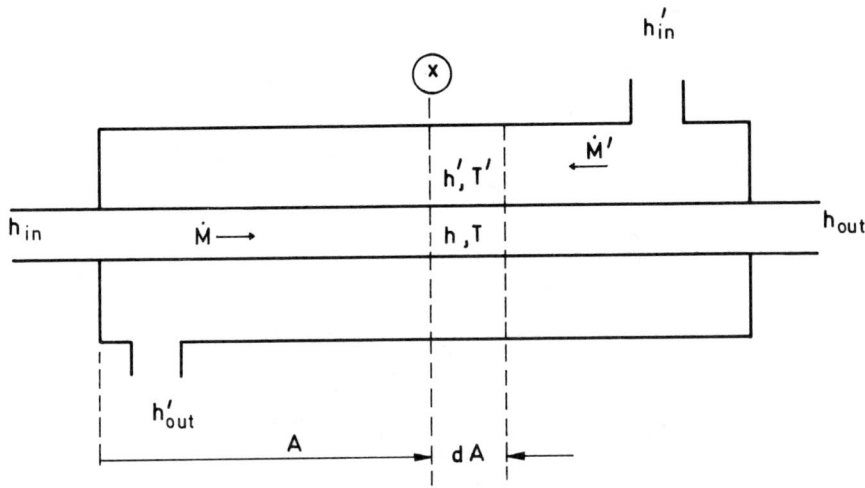

Fig.12. Counter flow heat exchanger

where $h'$ and $h$ are the shell side and tube side specific enthalpy, respectively, at position X on Figure 12. The corresponding equation for a co-current flow exchanger is obtained by replacing $h'_{out}$ by $h'_{in}$ and $\dot{M}'$ by $(-\dot{M}')$.

The first step in the thermal evaluation is to plot the equilibrium temperature, $T'$, against $h'$ for the shell side stream. The equilibrium temperature is used in accordance with our definition of the overall coefficient, U. Such a plot is shown in Figure 13. Using equation (18) and the temperature/specific enthalpy relationship for the tube side fluid, the corresponding tube side temperature, T, may be plotted on Figure 13 as shown. This diagram is extremely useful in condenser design and will be called here the "exchanger operating diagram". Figure 13 is typical of a condenser with a desuperheating zone and where condensation is occurring in the presence of non-condensable gas. The tube side curve shown would occur, say, if a pure liquid were being heated up and then boiled. The design is impossible if the two curves cross or touch anywhere.

The next step in the thermal evaluation is to divide this diagram into zones for which both the T curve and the $T'$ curve are linear. This is shown by the vertical broken lines in Figure 13. Now, over each zone, $\theta$ (ie $T' - T$) varies linearly with $h'$. This is the same as saying $\theta$ varies linearly with $\dot{Q}$, since

$$\dot{Q} = \dot{M}' (h'_{in} - h') . \qquad (19)$$

Thus, the logarithmic mean temperature difference, as defined by equation (14), applies for each zone and can be evaluated. The appropriate $\theta_a$ and $\theta_b$ for zone 2 are illustrated in Figure 13. Also, the overall coefficients at the zone boundaries may be calculated and a mean overall coefficient for each zone calculated using equations (15), (16) or (17), whichever is most

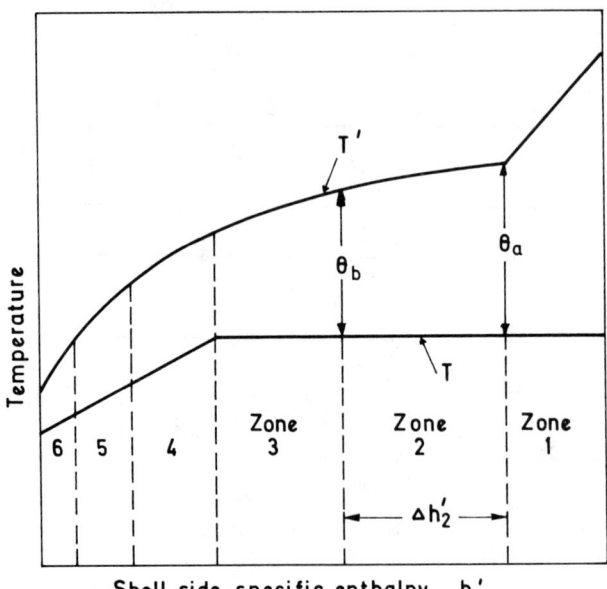

Fig.13. Example of exchanger operating diagram for a counter-flow heat exchanger

appropriate. Equation (6) may then be applied to each zone in the form

$$A_j = \frac{\dot{M} \Delta h'_j}{U_{m,j} \theta_{ln,j}} \qquad (20)$$

where the subscript j refers to the zone number and $\Delta h'_j$ is the specific enthalpy change of the shell side fluid in the j'th zone. Clearly, the total heat transfer area is given by

$$A_T = \sum_j A_j . \qquad (21)$$

Equations (7) and (8) may be expressed in summation form to give $U_m$ and $\theta_m$ for the whole exchanger:

$$\frac{1}{\theta_m} = \frac{1}{h'_{in} - h'_{out}} \sum_j \frac{\Delta h'_j}{\theta_{ln,j}} \qquad (22)$$

and

$$U_m = \frac{1}{A_T} \sum_j U_j A_j \qquad (23)$$

CONDENSERS: THERMOHYDRAULIC DESIGN

In the above calculation, the heat load on the exchanger is known and is given by

$$\dot{Q}_T = \dot{M}'(h'_{in} - h'_{out}) = \dot{M}(h_{out} - h_{in}). \qquad (24)$$

It is not, therefore, necessary to evaluate both equations (22) and (23) since $U_m$, $\theta_m$ and $\dot{Q}_T$ are related via equation (6). However, it is useful to evaluate equations (22) and (23) and substitute the results into equation (6) in order to crosscheck the arithmetic.

A useful feature of the above calculation procedure is that Figure 13 does not depend on the heat transfer coefficient and hence is independent of details in the geometry like the number of tubes, baffles etc. The same applies to the zonal and exchanger mean temperature differences. These quantities may need, therefore, only to be calculated when the number of passes is changed.

3.3 Shell side, E type condenser with two tube side passes

Figure 14 illustrates the operation of an E shell with two tube side passes. A heat balance over area A gives

$$\dot{M}'(h' - h'_{out}) = \dot{M}(h^I - h_{in} + h_{out} - h^{II}) \qquad (25)$$

where the superscripts I and II refer to the first and second tube side pass respectively. Combining this with equation (24) gives

$$h' = h'_{in} - \frac{\dot{M}}{\dot{M}'} (h^{II} - h^I) . \qquad (26)$$

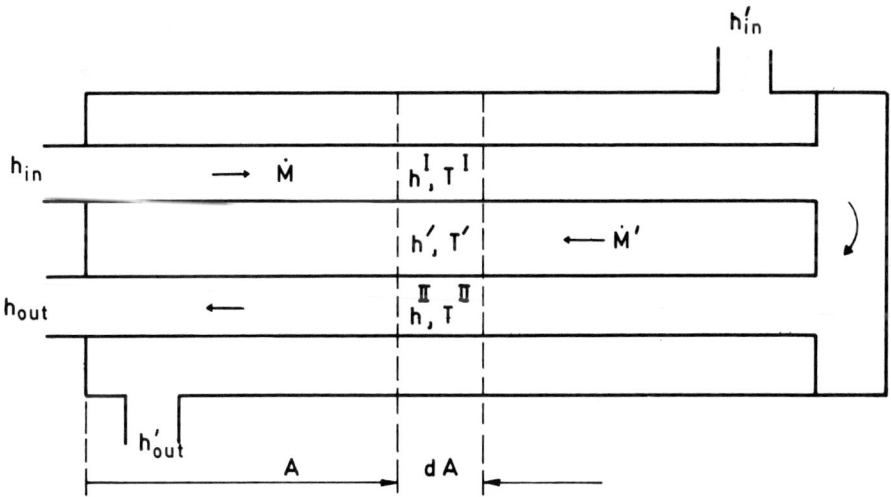

Fig.14. Exchanger with one shell pass and two tube passes

This result is for the first tube pass counter-current to the shell side flow. If, however, it is co-current, the result is

$$h^* = h^*_{out} + \frac{\dot{M}}{\dot{M}^*} (h^{II} - h^{I}) . \qquad (27)$$

Heat balances over area dA for passes I and II give respectively

$$\dot{M} dh^{I} = +U^{I}(T^* - T^{I}) \frac{dA}{2} \qquad (28)$$

and

$$\dot{M} dh^{II} = -U^{II}(T^* - T^{II}) \frac{dA}{2} . \qquad (29)$$

Dividing equation (29) by (28) gives

$$\frac{dh^{II}}{dh^{I}} = - \frac{U^{II}}{U^{I}} \frac{(T^*-T^{II})}{(T^*-T^{I})} . \qquad (30)$$

The same result is obtained if the shell side stream is co-current to the first tube side pass.

It is very convenient to simplify equation (30) by assuming $U^{II}/U^{I}$ is unity. This is often a reasonable approximation for shell side condensers provided that the tube side coefficient is constant or not controlling. With this assumption, equation (30) becomes

$$\frac{dh^{II}}{dh^{I}} = - \frac{T^*-T^{II}}{T^*-T^{I}} . \qquad (31)$$

The right hand side of equation (31) is a known function of $h^{II}$ and $h^{I}$, as becomes evident when one realises that $T^*$ is a known function of $h^*$, T a known function of h (whether superscripted I or II) and $h^*$ is related to $h^{II}$ and $h^{I}$ by equation (26) or (27). Hence, equation (31) can be integrated along the exchanger with the initial boundary conditions that $h^{I} = h_{in}$ when $h^{II} = h_{out}$.

For example, a simple numerical integration can be done by updating $h^{I}$ and $h^{II}$ as follows:

$$h^{I}_{new} = h^{I} + \delta h^{I} \qquad (32)$$

and

$$h^{II}_{new} = h^{II} - \frac{T^*-T^{II}}{T^*-T^{I}} \delta h^{I} \qquad (33)$$

# CONDENSERS: THERMOHYDRAULIC DESIGN

where $\delta h^I$ is a small change in $h^I$. There are, of course, more sophisticated integration methods for use with computers. This sort of integration may be used to construct the operating diagram shown in Figure 15. As with the counter-flow exchanger, this operating diagram is independent of detailed geometrical features and, therefore, applies to all two-pass E shells.

The heat leaving the shell side in area dA is

$$d\dot{Q} = U^I(T'-T^I)\frac{dA}{2} + U^{II}(T'-T^{II})\frac{dA}{2} \qquad (34)$$

which, since $U^I = U^{II} = U$ (say), gives

$$d\dot{Q} = (T'-\bar{T})\,U dA \qquad (35)$$

where

$$\bar{T} = \frac{T^I + T^{II}}{2} ; \qquad (36)$$

ie. $\bar{T}$ is the average temperature between the passes at a given point along the shell.

The curve for $\bar{T}$ can be plotted on the operating diagram as illustrated by the broken line in Figure 15. Equation (35) is now identical to equation (2), except that $\theta$ is replaced by $T'-\bar{T}$. The remainder of the thermal evaluation is now, therefore, the same as for the counter-flow exchanger except T is replaced by $\bar{T}$.

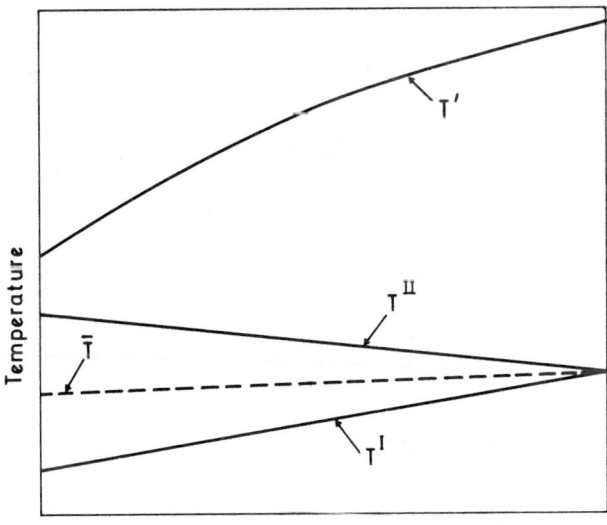

Fig.15. Example of operating diagram for shell side condenser (E type) with two tube passes

## 3.4 Shell side, E type condenser with four or more tube passes

An analysis of the type described above can be applied to exchangers with 4, 6, 8 or more even number of passes. The calculations become progressively more complicated, however, as the number of passes are increased. Furthermore, it is found in practice that the average tube side temperature $\bar{T}$ (now averaged over however many passes there are) does not change significantly as the number of passes are increased beyond 4, (see Butterworth, [16]).

Since the objective of the first part of the thermal evaluation is to construct an operating diagram containing the $\bar{T}$ curve, we could construct this for a four-pass exchanger and use this for any number of passes. There is, however, a convenient method of finding $\bar{T}$ for an infinite number of passes which we can use instead. This method is slightly less general than that given above since it only applies when there is a linear temperature-enthalpy curve for the tube side stream. Nevertheless, this covers the most important practical case of a single phase coolant.

The method is due to Emerson [17] who presents a more rigorous derivation than that given here although he did not spot some minor algebraic manipulations which could be used to simplify the calculation procedure. It is reasonable to postulate that the tube side stream sees a constant shell side temperature, $T'_{eff}$ say, as it traverses the length of the exchanger an infinite number of times. The logarithmic mean temperature difference then applies for heating the tube side stream since it has a linear temperature-enthalpy curve:

$$\theta_m = \frac{T_{out} - T_{in}}{\ln \dfrac{T'_{eff} - T_{in}}{T'_{eff} - T_{out}}} \quad (37)$$

We know also, for a large number of tube passes, that the mean tube side temperature, $\bar{T}$, is a constant, independent of $h'$. We can therefore write $\theta_m$ as

$$\theta_m = T'_{eff} - \bar{T} \quad (38)$$

Hence, combining equations (37) and (38) gives

$$\bar{T} = T'_{eff} - \frac{T_{out} - T_{in}}{\ln \dfrac{T'_{eff} - T_{in}}{T'_{eff} - T_{out}}} \quad (39)$$

But $\theta_m$ is also given by integrating equation (7).

The procedure for obtaining the operation diagram is therefore as follows:

(1) Plot the temperature versus specific enthalpy for the shell side stream.

# CONDENSERS: THERMOHYDRAULIC DESIGN

(2) Guess a value of $T'_{eff}$ (between $T'_{in}$ and $T'_{out}$).

(3) Calculate $\bar{T}$ from equation (39) and plot this as a horizontal line on the operating diagram.

(4) Determine $\theta_m$ by the methods already described for a counter-flow exchanger, ie. divide the diagram into linear zones, determine $\theta_{ln}$ for each zone and combine these using equation (22).

(5) Recalculate $T'_{eff}$ from equation (38) using the previously calculated $\theta_m$ and $\bar{T}$.

(6) Repeat the calculation from step 3 and continue the process until convergence is obtained. This usually takes two or three iterations.

The procedure described here will give reasonable results also for a multipass J shell.

## 3.5 Crossflow condensers

Let us first consider a single pass crossflow condenser as illustrated in Figure 16. A heat balance over area dA of this condenser gives

$$d\dot{Q} = U\theta dA \qquad (40)$$

For single phase coolant, and because the shell side temperature is constant in area dA, $\theta$ is given as

$$\theta = \frac{T_x - T_{in}}{\ln \frac{T' - T_{in}}{T' - T_x}} \qquad (41)$$

where $T_x$ is the temperature at exit to the tubes in question. Note that this is not the same on the coolant outlet temperature to the exchanger which is obtained after mixing the coolant from each tube. A heat balance over dN tubes sitting in area dA gives

$$\dot{m} c \frac{\pi D_i^2}{4} (T_x - T_{in}) dN = \pi D_o L u\theta dN \qquad (42)$$

where $\dot{m}$ is the tubeside mass flux, c the coolant specific heat, $D_i$ and $D_o$ the tube inside and outside diameters respectively, and L the tube length. This equation can be simplified to

$$T_x - T_{in} = \theta UB \qquad (43)$$

where

$$B = \frac{4 D_o L}{\dot{m} c D_i^2} \qquad (44)$$

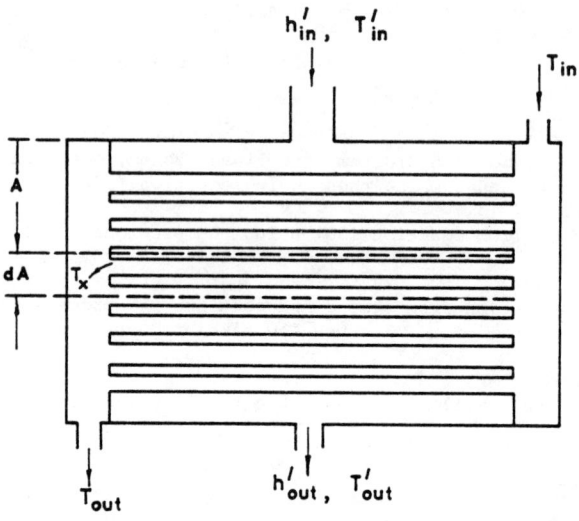

Fig.16. Crossflow condenser with single tube side pass

Combining equations (41) and (43) gives (after some manipulation)

$$\theta = (T' - T_{in}) \frac{1-e^{-UB}}{UB} \qquad (45)$$

Substituting this into equation (40) gives

$$\frac{d\dot{Q}}{dA} = \frac{1}{B}(T' - T_{in})(1 - e^{-UB}) \qquad (46)$$

The heat transfer area, or the heat load for a given area, can thus be determined by integrating equation (46). It must be borne in mind when doing this that both U and $T'$ vary with $\dot{Q}$.

A crossflow condenser with two tube passes may be treated as two crossflow units in series and hence the above method used again. This assumes, however, good mixing over the condenser length (which is not always the case) and that there is mixing of coolant between the passes. Mixing between the passes is, of course, not possible in a U-tube condenser. Multipass units are usually arranged with tube side passes layered and with the coolant flowing from lower to higher passes. Hence, a crossflow exchanger will approximate to a counter-flow unit when there are many passes.

3.6 Non equilibrium calculation methods

The methods presented so far are equilibrium ones in so far as the local stream temperatures are taken as their equilibrium values. The overall coefficient is then calculated in a way which is consistent with this.

# CONDENSERS: THERMOHYDRAULIC DESIGN

However, it becomes very difficult to estimate a suitable overall coefficient in some situations; particularly when condensing mixtures of fluids with large relative volatilities. Also, the equilibrium temperature method hides the actual liquid and gas-phase temperatures which may be crucial parts of the design. For example, the method will not tell one whether the gas phase is superheated or supersaturated. If the latter, there is danger of fog formation as has been discussed in a previous lecture [18]. The methods given so far, therefore, have severe limitations. Removing these limitations, unfortunately, makes the analysis much more complicated.

In order to illustrate some of the complexities of non equilibrium methods, and how to deal with these, we will take the case of condensing vapour in the presence of a non-condensable gas. We can recall some of the equations from a previous lecture [18].

$$\frac{dT_G}{dA} = -\frac{\alpha_G(T_G - T_I)}{\dot{M}_G c_G}\left(\frac{\eta}{e^\eta - 1}\right) \qquad (47)$$

$$\frac{d\dot{M}_V}{dA} = -\dot{m}_c . \qquad (48)$$

For a full explanation of these equations, the reader is referred to the previous lecture [18].

Let us assume for the moment that the coolant temperature $T_o$ is constant and that the condensing stream follows a single path (or identical parallel paths). The full thermal analysis therefore consists of integrating equations (47) and (48) along the condensing path. In doing this, it is necessary at each integration step to solve iteratively some non-linear equations in order to obtain values $T_I$, $\dot{m}_c$ and $\eta$ to use in the above equations. If now we let the coolant temperature vary, we have to integrate a further equation to obtain this temperature as the integration proceeds. This integration is straightforward if the coolant and condensing streams follow parallel paths which are either co- or counter-current. The appropriate equation is then

$$\frac{dT_o}{dA} = \pm \frac{\alpha^*(T_I - T_o)}{\dot{M}_o c_o} \qquad (49)$$

where $\dot{M}_o$ is the coolant mass flow rate and $c_o$ its specific heat capacity. The positive sign is for co-current flow and the negative sign for counter-current flow.

The integration of these equations, while not trivial, is certainly a feasible design approach given modern computers with standard library subroutines for integrating equations. A slight additional complication arises with counter-current flow because the coolant temperature may not be known at the start of the integration (because it is the coolant outlet temperature). However, in such cases, the outlet temperature may be guessed and the inlet temperature then determined by integration. If the calculated inlet temperature is different from its known value, a new guess has to be made for the inlet temperature.

So far we have only outlined the non-equilibrium calculation for a relatively trivial case. A slightly more complicated case is that of a TEMA E shell with shell side condensation and with two equal tube side passes. If we assume good radial mixing on the shell side, equations (47) and (48) now become

$$\frac{dT_G}{dA} = -\frac{1}{2}\left[\frac{\alpha_G(T_G - T_I)}{\dot{M}_G c_G}\left(\frac{\eta}{e^\eta - 1}\right)\right]^I$$

$$-\frac{1}{2}\left[\frac{\alpha_G(T_G - T_I)}{\dot{M}_G c_G}\left(\frac{\eta}{e^\eta - 1}\right)\right]^{II} \quad (50)$$

where the superscripts I and II refer to conditions pertaining to the tubes in pass I and II respectively. We also have to integrate separate equations for the pass I and II coolant temperatures:

$$\frac{dT_o^I}{dA} = \left[\frac{\alpha^*(T_I - T_o)}{\dot{M}_o c_o}\right]^I \quad (51)$$

$$\frac{dT_o^{II}}{dA} = -\left[\frac{\alpha^*(T_I - T_o)}{\dot{M}_o c_o}\right]^{II} \quad (52)$$

In setting the signs of the right hand sides of equations (50) - (52), it has been assumed that the integration is proceeding from the front-end header and that the shell side inlet nozzle is at the front end. In addition to these integrations, some iteration is normally required to match the calculated tube side temperatures, $T_o^I$ and $T_o^{II}$, at the rear-end header.

Clearly, the calculations just outlined for a two pass exchanger become even more involved when one goes to more passes or if some of the tubes are submerged in condensate. Nevertheless, such calculations can be handled economically and HTFS has a computer program, SCON, which uses similar methods for up to 16 tube passes.

3.7 Multidimensional shell side flows

The calculation method given in the previous sections, although quite complex, still contains a major simplifying assumption. This is that the flow paths are one-dimensional. In some situations, though, particularly that of steam power condensers, the multidimensional nature of the shell side flow is a key feature of the design. It is beyond the scope of this lecture to deal with this problem in depth. An outline only is therefore given, together with references to papers where further information of relevance to condenser design is given.

The first step in multidimensional analyses is to determine the flow pattern. This has been attempted by two methods. The first is a subchannel method [19] in which a flow network is set up with the nodes in that network being the regions between three adjacent tubes (in an equilateral triangle tube layout). Mass balance equations are set up at each node and momentum balance equations set up between each pair of adjacent nodes. This leads to a very large number of non-linear algebraic equations which must be solved for the flow pattern. The alternative method is the continuum method which essentially treats the rod bundle as an anisotropic porous medium with flow-dependent permeability. This leads to partial differential equations of mass and momentum continuity which are solved by finite difference or finite element methods. Butterworth [20, 21] has formulated this approach for single phase flow and discussed its application to steam condensers in reference [22]. Davidson [23, 24] has reviewed the application of this method to large power station condensers.

The second step in the multidimensional analysis is to determine the transport of non-condensable gas. This is difficult because the turbulent diffusion of the gas can be very important, especially in stagnant regions. The problem then centres on estimating the diffusion coefficient for turbulent, two-phase flow with tubes occupying some of the space. Of course, the art of good design is to avoid stagnant regions.

## 4. THERMAL EVALUATION METHOD FOR DIRECT-CONTACT CONDENSERS

### 4.1 Spray condensers

These are rather difficult to design with any degree of precision because of the uncertainties in droplet size, droplet trajectory, droplet coalescence and flow patterns in the gas phase. However, the simplicity of the basic geometry means that they can be oversized without much additional cost.

A preliminary step in spray condenser design is the selection of nozzle type and the determination of the nozzle behaviour for the liquid being used. Steinmeyer [25] describes various types of spray nozzles and discusses their method of operation, advantages and disadvantages. Spray condensers would usually operate with the simplest type of nozzle in which the source of atomisation is the pressure loss in the nozzle.

Nozzle types vary so widely from one manufacturer to another that it is necessary to use the manufacturers' data to determine such quantities as the pressure drop in the nozzle, the mean droplet size and the initial droplet velocity. Unfortunately the data given by manufacturers are usually limited to water at around 20°C spraying into ambient air. Often, however, there is enough information given to calculate the pressure drop and, from this, the inlet velocity, $u_1$, may be determined as

$$u_1 = K \left( \frac{2\Delta p}{\rho_L} \right)^{\frac{1}{2}} \tag{53}$$

where $\Delta p$ is the nozzle pressure drop and $\rho_L$ the liquid density. K is a coefficient which would be unity if there were no energy losses in the nozzle. A reasonable value of K for estimation purposes is 0.8. The mean droplet diameter for water can be determined from manufacturers' data and Steinmeyer [25] suggests an approximate equation for correcting this for other liquids:

$$\frac{d}{d_{wat}} = \left(\frac{\sigma}{\sigma_{wat}}\right)^{0.5} \left(\frac{\mu_L}{\mu_{wat}}\right)^{0.2} \left(\frac{\rho_{wat}}{\rho_L}\right)^{0.3} \quad (54)$$

when d is the droplet diameter, $d_{wat}$ the diameter for water sprayed with the same *volumetric* flow through the nozzle, $\mu_L$ is the liquid viscosity, $\mu_{wat}$ the viscosity of water at 20°C, $\rho_L$ the density of the liquid and $\rho_{wat}$ the density of water at 20°C. In reality, the dependence of droplet size on fluid properties is very complex and, hence, the above equation should be used with caution.

The problem of subcooled droplets injected into saturated vapour has been analysed by Brown [26]. He treated the droplets as solid spheres and solved the transient conduction equation in order to give the temperature rise in the droplet as a function of exposure time to the vapour. The droplet diameter is assumed to be independent of time which is reasonable since very little vapour can condense before the droplet is heated up to, or close to, saturation.

The droplet temperature rise is given by Brown as

$$\frac{T_{out} - T_{in}}{T_{sat} - T_{in}} = 1 - \frac{6}{\pi^2} \sum_{n=1}^{\infty} \frac{1}{n^2} \exp\left(-\frac{4n^2 \pi^2 \kappa_L t_c}{d^2}\right) \quad (55)$$

where $T_{out}$ is the mean outlet temperature, $T_{in}$ the inlet temperature, $T_{sat}$ the saturation temperature, $\kappa_L$ the liquid thermal diffusivity and $t_c$ the contact time. Figure 17 shows the results of this equation plotted for low pressure water. These results may be used to determine an effective mean coefficient $\bar{\alpha}$ for a given temperature rise. The results of such a calculation are given in Figure 18. This figure may be used in conjunction with the following equation in order to determine the desired contact time, $t_c$:

$$t_c = \frac{d \rho_L c_L}{6 \bar{\alpha}} \ln\left(1 - \frac{T_{out} - T_{in}}{T_{sat} - T_{in}}\right). \quad (56)$$

The coefficient $\bar{\alpha}$ can be very large and, indeed, comparable with the interfacial or molecular-kinetic coefficient. This is particularly so when, as is often the case, the condenser is operating at high vacuum. In such circumstances, equation (56) should be evaluated with $\bar{\alpha}$ replaced by $\alpha_{eff}$ which combines $\bar{\alpha}$ and the interfacial coefficient, $\alpha_I$, as follows:

$$\frac{1}{\alpha_{eff}} = \frac{1}{\bar{\alpha}} + \frac{1}{\alpha_I}. \quad (57)$$

Having determined the contact time $t_c$, it is necessary to estimate how far the droplets will travel in this time, thus enabling one to estimate the vessel size. A force balance on the droplets for vertical downward motion yields:

$$Mu\frac{du}{dz} = M\frac{du}{dt} = Mg - \xi\frac{\pi d^2}{4}\frac{\rho_G u^2}{2} \tag{58}$$

where u is the droplet velocity, M the droplet mass, z the distance, t the time, g the gravitational acceleration, $\xi$ the drag coefficient and $\rho_G$ the gas phase density. For horizontal flow, the same equation applies but g is zero. Also, g is often small compared with the other terms for high velocity droplets. Pita and John [27] have integrated this equation analytically

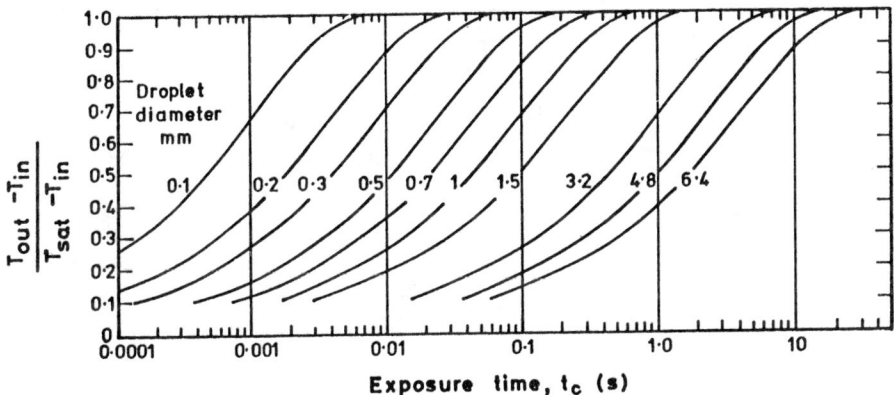

Fig.17. Temperature rise as a function of time for water droplets in saturated steam (Brown [26])

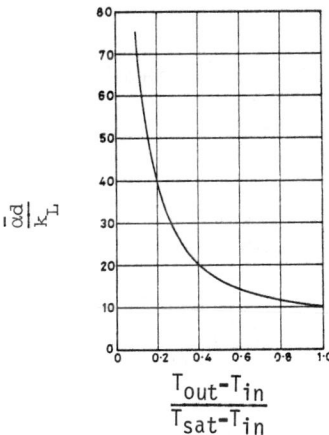

Fig.18. Effective mean heat transfer coefficient for droplet in pure vapour (Brown [26])

for the case when g is zero. In order to do this, they used the Ingabo [28] equation for $\xi$ which applies for $6 < Re < 400$:

$$\xi = 27 Re^{-0.84} \qquad (59)$$

where

$$Re = \frac{\rho_G u d}{\mu_G} \qquad (60)$$

and where $\mu_G$ is the gas phase viscosity. Peta and John obtained the total distance travelled, L, in time $t_c$, as

$$L = 0.06 \frac{d^{1.84}}{\Gamma} (u_1^{0.84} - u_2^{0.84}) \qquad (61)$$

where $u_1$ is the velocity at time $t_c$ given by

$$u_2 = \left( u_1^{-0.16} - 3.23 \frac{\Gamma t_c}{d^{1.84}} \right)^{-6.25} \qquad (62)$$

and where $\Gamma$ is a physical property grouping given by

$$\Gamma = \frac{\rho_G}{\rho_L} \nu_G^{0.84} \qquad (63)$$

As an alternative to the above calculation method, Fair [29] gives empirical calculation procedures based on the volumetric heat transfer coefficient.

The estimation of the effects of non-condensables is quite difficult and involves step by step calculations of the type described by Peta and John for droplet evaporation.

4.2 Tray Condensers

The uncertainties involved in the design of spray condensers become even more severe with tray condensers. It is possible to use the above heat transfer calculation method while using a large droplet diameter (say > 5 mm) but this can lead to error if the sheet does not break up into droplets before falling onto the next tray. The vertical distance travelled by the droplets can be determined by assuming that they fall freely under gravity. Alternatively, empirical calculation methods have been devised by Fair [29] which use volumetric coefficients.

5. REASONS FOR FAILURE OF CONDENSER OPERATION

Steinmeyer and Mueller [30] chaired a panel discussion session on why condensers do not operate as they are supposed to. Some of the main points arising from this discussion are noted here:

# CONDENSERS: THERMOHYDRAULIC DESIGN

(1) The tubes may be fouled more than expected - a problem not unique to condensers.

(2) The condensate may not be drained properly causing tubes to be flooded. This could mean that the condensate outlet is too small or too high.

(3) Venting of non-condensables may be inadequate. Remarks on the proper arrangement of vents are given in section 2 above.

(4) The condenser was designed on the basis of end temperatures without noticing that the design duty would involve a temperature cross in the middle of the range (see section 3 above).

(5) Flooding limits have been exceeded for condensers with backflow of liquid against upward vapour flow.

(6) Excessive fogging may be occurring. This can be a problem when condensing high molecular weight vapours in the presence of non-condensable gas.

## NOMENCLATURE

| | |
|---|---|
| $A$ | Heat transfer area ($m^2$) |
| $B$ | Parameter defined by equation (3) |
| $c$ | Specific heat capacity at constant pressure (J/kg K) |
| $d$ | Droplet diameter (m) |
| $D$ | Tube diameter (m) |
| $g$ | Gravitational acceleration ($m/s^2$) |
| $h$ | Specific enthalpy (J/kg) |
| $k$ | Thermal conductivity (W/m K) |
| $K$ | Parameter in equation (54) |
| $L$ | Length (m) |
| $\dot{m}$ | Mass flow per unit area ($kg/m^2 s$) |
| $M$ | Mass of a droplet (kg) |
| $\dot{M}$ | Mass flow rate (kg/s) |
| $N$ | Number of tubes |
| $p$ | Pressure (Pa) |
| $\dot{Q}$ | Heat flow through exchanger surface (W) |
| $r$ | Thermal resistance ($m^2 K/W$) |

| | |
|---|---|
| $t$ | Time (s) |
| $t_c$ | Contact time (s) |
| $T$ | Temperature (K) |
| $u$ | Velocity (m/s) |
| $U$ | Overall heat transfer coefficient (W/m²K) |
| $V$ | Superficial volumetric flowrate per unit area, ie. superficial velocity (m/s) |
| $z$ | Distance |

Greek

| | |
|---|---|
| $\alpha$ | Heat transfer coefficient (W/m²K) |
| $\alpha^*$ | Heat transfer coefficient from coolant to liquid surface (W/m²K) |
| $\Gamma$ | Physical property grouping defined by equation (3) (m$^{1.68}$/s$^{0.84}$) |
| $\Delta h$ | Enthalpy change (J/kg) |
| $\Delta p$ | Pressure drop (Pa) |
| $\eta$ | $\dot{m}_c \, c_v / \alpha_G$ |
| $\theta$ | Temperature difference between streams (K) |
| $\kappa$ | Thermal diffusivity (m²/s) |
| $\mu$ | Viscosity (Ns/m²) |
| $\nu$ | Kinematic viscosity (m²/s) |
| $\xi$ | Drag coefficient |
| $\rho$ | Density (kg/m³) |
| $\sigma$ | Surface tension (N/m) |

Subscripts

| | |
|---|---|
| a,b | Zone boundaries |
| cold | Cold stream |
| eff | Effective value |
| G | Gas phase (including vapour or gas vapour mixtures) |
| hot | Hot stream |
| i | Inside of tube |
| in | Inlet to exchanger |

| | |
|---|---|
| I | Interface value |
| ln | Logarithmic mean |
| L | Liquid phase (or condensate) |
| m | Mean value for exchanger |
| new | New value for next step |
| o | Outside of tube or, in section 3.6 only, the coolant |
| out | Outlet of heat exchanger |
| sat | Saturated value |
| T | Total for exchanger |
| v | Vapour |
| w | Wall |
| wat | For ambient water |
| x | Exit to tube |
| 1 | Initial value |
| 2 | Final value |

Superscripts

| | |
|---|---|
| ′ | Shell side |
| I | First pass |
| II | Second pass |
| — | Average between passes or over time of contact |

## REFERENCES

1. Alfa Laval, 1969. Thermal Handbook, Alfa Laval, Sweden.

2. Butterworth, D, 1977. Engineering Design Guide: Introduction to Heat Transfer, pp 19-20, Oxford University Press, Oxford.

3. TEMA 1978. Standards of Tubular Exchanger Manufacturers Association, Sixth Edition, New York.

4. Editors of Power, 1967. Power Generation Systems, pp 282-289, McGraw Hill, New York.

5. Simpson, H.C, 1969. Outline of Current Problems in Condenser Design, Proceedings of Symposium to Celebrate the Bicentenary of the James Watt Patent, University of Glasgow, September 1 & 2, pp 91-134.

6. Sebald, J.F, 1980. A History of Steam Surface Condensers for the Electricity Utility Industry. Heat Transfer Engineering, Vol 1 No 3, pp 80-87.

7. British Electrical and Allied Manufacturers' Association, 1967. Recommended Practice for the Design of Surface Type Steam Condensing Plant, London.

8. Heat Exchange Institute, 1970. Standards for Steam Surface Condensers, 6th edition, New York.

9. Ludwig, E.E, Applied Process Design for Chemical and Petrochemical Plants, Vol 3, pp 146-131, Gulf Publishing Co, Houston.

10. American Petroleum Institute, 1968. Air cooled heat exchangers for general refinery service, API Standard 661, Washington, USA.

11. Lenfestey, A.G, 1961. Low Temperature Heat Exchangers, Progress in Cryogenics, Vol 13, pp 25-47, Published Haywood & Co Ltd.

12. Silver, L. 1947. Gas Cooling with Aqueous Condensation, Trans. Inst. Chem. Engrs, Vol 25, pp 30-42.

13. Bell, K.J and Ghaly, M.A, 1973. An Approximate Generalized Design Method for Multicomponent/partial Condensers. Am.Inst. Chem. Engrs. Symp. Series, Vol 69, No 131, pp 72-79.

14. Smith, R.A, 1976. Private Communication, Heat Transfer Consultant, Middlesbrough.

15. Colburn, A.P, 1933. Mean Temperature Difference and Heat Transfer Coefficient in Liquid Heat Exchangers, Ind. Engng. Chem. Vol 25, pp 873-877.

16. Butterworth, D. 1973. A Calculation Method for Shell-and-Tube Heat Exchangers in which the Overall Coefficient Varies Along the Length. Conferences on Advances in Thermal and Mechanical Design of Shell-and-Tube Heat Exchangers, NEL Report No 590, 56-71, National Engineering Laboratory, East Kilbride, Scotland.

17. Emerson, W.H. 1973. Effective Tube-side Temperatures in Multi-pass Heat Exchangers with Non-uniform Heat-transfer Coefficients and Specific Heats. Conference on Advances in Thermal and Mechanical Design of Shell-and-tube Heat Exchangers. NEL Report No.590, 32-55, National Engineering Laboratory, East Kilbride, Scotland.

18. Butterworth, D. 1980. Condensers: Basic Heat Transfer and Fluid Flow. Advanced Study Institute on Heat Exchangers, August 4-15, Istanbul, Turkey.

19. Wilson, J.L. 1976. NEL Two Dimensional Condenser Computer Program. Meeting on Steam Turbine Condensers, NEL Report No. 619, pp 152-159, National Engineering Laboratory, East Kilbride, Scotland.

20. Butterworth, D. 1978. The Development of a Model for Three-Dimensional Flow in Tube Bundles, Int. J. Heat Mass Transfer, Vol.21, pp 253-256.

21. Butterworth, D. 1978. A Model for Heat Transfer during Three-dimensional Flow in Tube Bundles. Sixth Int. Heat Transfer Conference, Toronto, August, Vol.4, pp 219-223, Hemisphere Publishing Corporation, Washington DC.

22. Butterworth, D. 1980. Discussion of Paper by Davidson and Rowe on Simulation of Power Plant Condenser Performance by Computational Methods. Workshop on Modern Developments in Marine Condensers, Naval Postgraduate School, Monterey, California, 26-28 March.

23. Davidson, B.J. 1976. Computational Methods for Evaluating the Performance of Condensers. Meeting on Steam Turbine Condensers, NEL Report No.619, pp 152-159, National Engineering Laboratory, East Kilbride, Scotland.

24. Davidson, B.J. 1980. Simulation of Power Plant Condenser Performance by Computational Methods. Workshop on Modern Developments in Marine Condensers, Naval Postgraduate School, Monterey, California, 26-28 March.

25. Steinmeyer, D.E. 1973. Phase Dispersions: Liquid in Gas Dispersions, Chemical Engineers Handbook, 5th Edition, p 18/62, edited by Perry R.H and Chilton, C.H, McGraw Hill.

26. Brown, G. 1951. Heat Transmission by Condensation of Steam on a Spray of Water Drops. Inst. Mechanical Engrs. Proc. General Discussion on Heat Transfer, pp 49-52.

27. Pita, E.G. and John, J.E.A. 1970. The Effect of Forced Convection on Evaporative Cooling of Sprays in Air. Proc. Fourth Int. Heat Transfer Conf. Versailles, Vol 7, Paper CT3.12, Elsevier Publishing Co, Amsterdam.

28. Ingebo, R.D. 1951. Vaporization Rates and Heat Transfer Coefficients for Pure Liquid Drops, NACA, TN2368.

29. Fair, J.R. 1972. Design of Direct Contact Coolers/Condensers, Chemical Engineering, Vol 79, June 12, pp 91-100.

30. Steinmeyer, D.E. and Mueller, A.C. 1974. Why Condensers Don't Operate As They Are Supposed To. Chem. Engng. Progress, Vol 70, No 7, pp 78-82.

# Some Design Aspects of Thin Film Evaporators/Condensers in Water Desalinization

**SAMUEL SIDEMAN**
Technion, Israel Institute of Technology
Haifa, Israel

ABSTRACT

Thin film evaporators which are thermally sustained by simultaneous film condensation on the other side of the wall represent an efficient mode of operation aimed at economic production of vapor or concentrate. Other aspects involve thermal economics of the system, including heat recovery and unavoidable heat losses. These major aspects are highlighted via the review of presently operating desalination and wood pulping processes. Thus, while detailing the major concepts, important for the design of an efficient large-scale evaporator, an overview of the present state of the art is hopefully presented.

NOMENCLATURE

$a_s$ - water activity in solution
A - area of heat exchanger
Ar - Archimedes number, Eq.14
b - width of groove, m
$C_p$ - specific heat of liquid J/Kg°K
$d_p$ - diameter, particle, m
D - tube diameter, m
E - mechanical and electrical energy KWH/m$^3$
E - axis ratio elliptical tube
F - free energy
g - gravitation constant m/S$^2$
h - heat transfer coefficient w/m$^2$ °K
H - enthalpy KWH/m$^3$
k - conductivity of liquid w/m °K
m - moles of salt in solution
M - mass flow rate, Kg/hr
n - no. of effects
Nu - Nusselt number (hd/k), Eq.14
p - pressure
Pr - Prandtl number (Cp μ/k), Eq.14
$P_w$ - power KW/m$^3$
Q - heat input Kcal/Kg
R - performance ratio, Eq.8
$\hat{R}$ - gas constant

$R_{eq}$ - Equivalent circular tube radius, m
$Re_N$ - Reynolds number film $4\Gamma/\mu$
S - entropy
T - temperature °K
t - depth of groove, m
$\Delta T$ - temp. difference °K
u - velocity m/S
U - overall heat transfer coefficient w/m$^2$ °K
V - volume of water vapor, m$^3$/Kg
W - work KWH/m$^3$

α - thermal diffusivity
Γ - film flow rate Kg/mS
λ - latent heat, KWH/Kg
μ - viscosity N/Sm$^2$
ν - kinematic viscosity m$^2$/S
ρ - density Kg/m$^3$

## 1. INTRODUCTION

Thermally-based water desalination plants, and to a lesser degree pulping plants, represent some of the largest evaporators and condensers in practical use, and larger ones are under design. It is, therefore, important to review some of the basic characteristics of these exchangers and understand the major parameters involved in these large scale operations.

An overview of the various desalination processes is presented in Fig.1. Historically, the "heat consuming" processes were the preferred ones, but following the tremendous increase in energy costs in the second half of the last decade, the 'power consuming' processes are gaining economical importance. Here, however, we concentrate mainly on the thermal aspects associated with the desalination processes schematically represented in Fig.2.

Note the central position of the vapor-compression-evaporation mode of operation in Fig.1 which bridges and combines elements of the heat and power consuming processes. If one is entitled to some personal bias, the combination of this thermodynamic process together with the thermally efficient thin film evaporation (condensation) would presently get our vote as the most promising one for thermal water desalination. In general, it fulfills the following criteria for a good evaporator.

1. It allows for very large quantities of heat to be transferred from the heating side to the liquid to be evaporated.

2. It allows, for economical and ecological reasons, to separate the evaporated steam effectively from droplets that are carried over.

3. It is amenable to be joined together with similar apparatus in order to make the evaporator as economical as possible, both with regard to capital and operating costs.

4. It can easily cope with the physical (viscosity, boiling point rise, foaming) and chemical (scaling, crystallization, and corrosion) liquid properties.

5. If scaling occurs, it is possible to restore the scaled-up surfaces to the original condition effectively and simply.

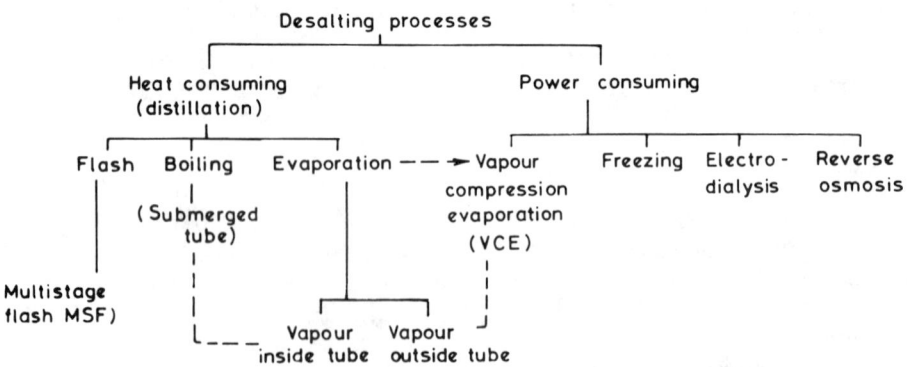

Fig.1. The major desalination processes.

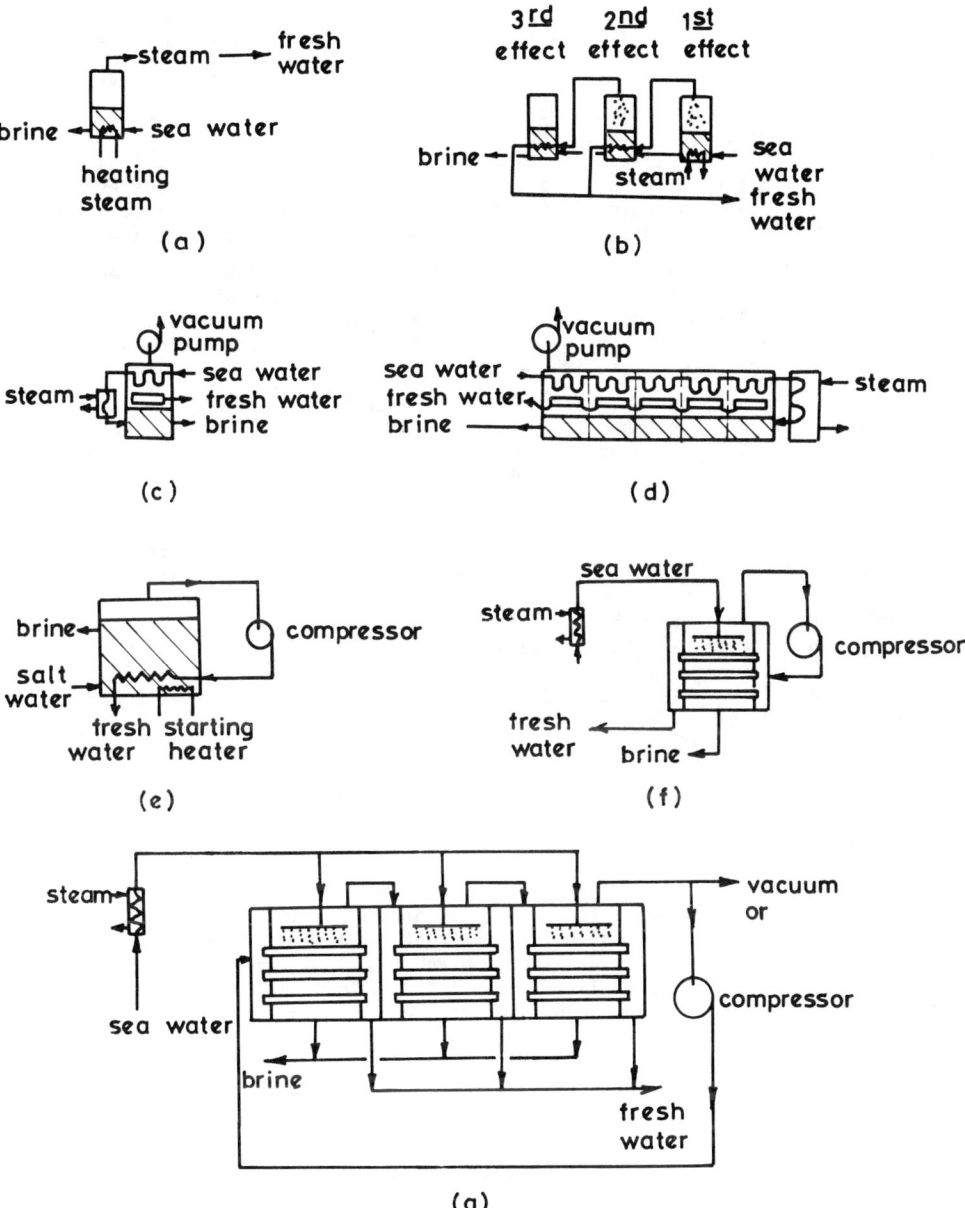

Fig.2. Schematic presentation of some thermal desalination processes.

(a) Single-effect evaporator; (b) Multi-effect pool boiling evaporator;
(c) Single-stage flash evaporator; (d) Multi-stage flash evaporator;
(e) Single-stage vapor-compression submerged evaporator; (f) Single-stage vapor-compression falling film horizontal tube evaporator;
(g) Multi-stage falling film horizontal tube evaporator.

Inspection of the various water 'distillation' processes presented in Fig.2 brings out a number of possible mechanism for vapor production. These include:

1. Submerged, pool, boiling. Vapor is formed by boiling on condensing steam-heated tubes submerged in the brine. The formed vapor is then channeled into a tube, compressed slightly and then condensed inside a tube submerged in the 'original' brine (single-effect vapor compression evaporator) or else channeled in tubes into a lower pressure compartment and condensing there while producing new vapor (multiple-effect evaporator).

2. Flash evaporation. Vapor is formed by allowing high pressure brine to 'expand' into a low pressure compartment. The vapor thus formed is usually allowed to condense on the outside surface of (cold brine carrying) tubes which pass above the flashed brine. The condensate, which drips onto trays, is collected and removed.

3. Thin film evaporation. The vaporization of brine films is thermally sustained by simultaneously condensing steam-vapor on the other side of the tube. Film evaporation can take place inside or outside vertical tubes or outside horizontal tubes.

It is noted that the vapor can be produced in one mode of heat transfer i.e., submerged boiling or film evaporation inside or outside the tube, and the condensation in another. Obviously, the combination of the most suitable thermal processes with the appropriate choice of materials, physical design and operating conditions, such as flow rates and temperatures, will lead to the most efficient and economical water desalination scheme.

No attempt is made here to evaluate the economics of the different processes. Rather, the basic ground rules and concepts associated with the various heat transfer with change of phase schemes will be elucidated, providing some background for understanding, design and economic evaluation. References [1-21] as well as the numerous papers published as proceedings of the various internationally held desalination symposia during the last two decades, should be helpful towards this end.

The relative advantages of the horizontal tube thin film evaporator as compared with the multi-stage flash (MSF) evaporator were listed elsewhere [22]. Clearly, the combination of the horizontal tube evaporator with vapor compression is a most promising water desalination scheme. Here, we shall review some of the basic thermal characteristics associated with the major desalination processes presently in use and thus demonstrate the important aspects associated with proper evaporator/condenser design. Particular notice will be given to the minimum energy required for desalination, the thermodynamics of vapor compression, the process heat flow of the MSF evaporation scheme, and, finally, to the more interesting film evaporator/condenser mode of operation. It is thus hoped that while the various design related 'trees' are detailed, one may also gain an overview of the forest.

2. THERMODYNAMIC CONSIDERATIONS

2.1. Minimum work of separation.

We shall relate here to the minimum mechanical work required to separate water vapor from a saline solution. The reversible, isothermal work for any process, regardless of mechanism, is given by

$$-W = \Delta H - T \Delta S = \Delta F \qquad (1)$$

where $-W$=work done upon the system to cause the process, $\Delta H$=enthalpy change, $\Delta S$=entropy change, $\Delta F$=free energy, change, and T=absolute temperature. In differential form

$$-\frac{dW}{dm} = \Delta F = \hat{R}T \ln a_s \qquad (2)$$

where dm=increment of water removed at any instant, $a_s$=activity of water in salt solution and R=universal gas constant.

Taking the standard state as that of pure water and assuming ideal gas relationships for the low pressure vapor, Eq.(2) becomes

$$-W = \int_{m_1}^{m_2} \hat{R}T \ln \frac{p_2}{p_1} \, dm \qquad (3)$$

where $m_1$ and $m_2$ are the initial and final number of moles of salt in solution, respectively, and $p_1$ and $p_2$ are the corresponding vapor pressures. Note that the minimum work is obtained at zero yield, i.e., when we actually have no product. At this very beginning of the separation process, where salt concentration in sea water is 35,000 ppm;

$$-W_{min} = 0.79 \text{ KWH/m}^3 \text{ at } 25°C$$

Another procedure to evaluate the minimum reversible work of separation is presented by Spiegler [8]. Assume that two air-evacuated reservoirs, one containing sea water and the other fresh water, are placed in a thermally insulated container. An equilibrium vapor pressure is established in each reservoir by evaporation of water. As is well-known, at a given temperature, the equilibrium vapor pressure of sea water is less than that of pure water. If we connect the two reservoirs, vapor from the pure water reservoir will naturally migrate to the sea water one. In order to reverse this phenomenon, it is necessary to install a pump or compressor between the vessels in order to increase the pressure of the 'sea water' vapor to a value just above that in the pure water vessel. The compression work done by this hypothetical pump in a flow system is given by

$$-W = \int_{p_0}^{p} V dp = V \Delta p \qquad (4)$$

where V=volume of water vapor per unit mass and $\Delta P$=pressure difference between reservoirs, $(p_0-p)$. Assuming sea water vapor pressure is 1.87% lower than that of pure water, and if V=43.4 lit/gr at 25°C and $p_0$=0.312 atm, then

$$-W = 0.70 \text{ KWH/m}^3$$

The difference in the results obtained by these two procedures is due to different values of vapor pressures utilized.

The actual power consumption of the compressor is obviously higher, because there is no ideal compressor which converts all of the driving energy into compression work. No process can be invented which will require less energy.

As distillation proceeds the salt concentration increases, the decrease of pressure of sea water with increasing salinity is given by [8]:

$$p = p_0(1 - 0.000537S) \qquad (5)$$

where S=gr solids per Kg sea water. Since $p_0$ is constant for a given temperature, $-W$ in Eq.(4) increases with increased salinity.

Equations (3) and (4) relate to zero product yield, which is unrealistic since it requires pumping an infinite amount of sea water as feed. As shown in

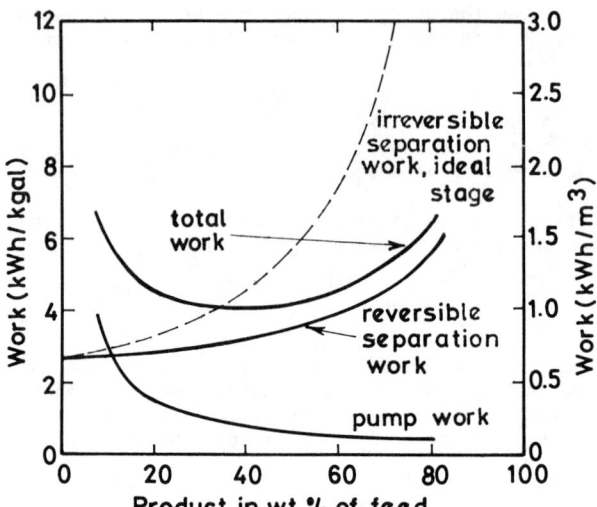

Fig.3. Comparison of minimum separation work, pump work and total energy requirement as a function of recovery of freshwater from sea water. 33 m head, 100% pump efficiency [20].

Fig.3, the required sea water pumping work decreases as recovery is increased, while the separation work increases. Depending on the assumed pumping head the total work will pass through a minimum at some particular recovery. [For order of magnitude feeling, the pumping work for 1 $m^3$ product, with a 50% recovery, overcoming a 16.4 m head is 0.12 KWH when a 70% pump efficiency is assumed. The separation work is then $-W=1.1$ KWH/$m^3$].

Equations (3) and (4) relate to reversible processes whereby work is done in a series of infinitesimal steps with complete equilibrium being maintained at all times. If one considers a continuous ideal single-stage process whereby the two outgoing streams (product water and concentrated brine) are at equilibrium, the minimum work is then simply given by [23]

$$-W = \hat{R}T \ln p/p_o \tag{6}$$

Eq.(6) is plotted for comparison as the dotted line in Fig.3. (For example, starting at 25°C with S=35 (=35,000 ppm), $p/p_o$=0.9606 for 50% recovery and $-W$=1.53 KWH/$m^3$. The effect of irreversibility is obvious.

2.2. Efficiency of real processes.

If we convert the minimum work of separation, of order of 1 KWH per $m^3$ product (or 3.786 KWH/1000 gal.) to its thermal equivalent, assuming a conversion (Carnot) efficiency of 40% we obtain $-W$=(860 Kcal/KWH)(100/40)=2150 Kcal/$m^3$ product.

The efficiency of real processes is determined by the following irreversible effects which affect the separation process:

1. Pressure drop due to fluid friction in lines and equipment.
2. Mechanical friction in pumps and/or compressors.
3. Heat leaks to, or from, the surroundings.
4. Finite temperature differences between fluids exchanging heat.

5. Heat conduction along the solids.
6. Miscellaneous.

These effects are unavoidable. To illustrate the significance of one of these parameters of irreversibility, consider the energy loss when the brine and the pure water product are discharged from the system at a temperature which is only 3°C above the incoming sea water. The wasted heat is 3000 Kcal/m$^3$ (water-brine). Now, if half of the incoming sea water is converted to fresh water (50% recovery), the heat wasted is 6000 Kcal/m$^3$ pure product. Thus, the energy loss due to the discharge temperature per unit m$^3$ of product is about 3 times the energy requirement for an ideally reversible desalination plant.

Other order of magnitude calculations demonstrate the actual efficiencies as compared to the theoretical minimum separation work. For instance, a metric ton of condensing steam will release 518,500 Kcal from which we can obtain (assuming a Carnot efficiency of 0.248) 518,500x0.248 = 128,500 (work) Kcal=149 KWH/m$^3$ condensate. This is about 150 times larger than the theoretical minimum work required for separating water from sea water, indicating that an ideal distillation plant should be able to produce about 150 m$^3$ of fresh water 1 m$^3$ (=1 ton) steam input. In practice, however, 10 m$^3$ product water per 1 m$^3$ of steam are considered reasonably good.

Consider a multiple effect evaporator in which 10 Kg of fresh water are produced per 1 Kg of steam input at 3.4 atm ($T_1$=399°K=126°C) and heat sink is at $T_2$=294°K=21°C). The heat requirement per m$^3$ product=(518,000 Kcal) (1/10 Kg steam/Kg product)= 51,800 Kcal/m$^3$ product. The equivalent work requirement $-W$=51,800x(105/399)=13,630 Kcal/m$^3$=15.85 KWH/m$^3$. Compared with $W_{min}$=1.1 KWH/m$^3$ (50% recovery), the energy efficiency of the process is 6.9%.

The energy efficiency of present plants operating with a 50% recovery is about 11 to 15%, depending on whether the differential (reversible) or single stage process is taken as a standard of comparison. (If one takes the $W_{min}$ for zero yield as a basis, the corresponding plant efficiency is about 8%). Clearly, the thermodynamic efficiency of saline water conversion is quite low, and efficiencies of about 20% are about the maximum expected in the foreseeable future.

2.3. Steam economy and performance ratio.

A simple calculation on a single effect evaporator shows that, approximately, the condensation of 1 Kg input steam will result in the production of 1 Kg vapor (at a lower pressure). It is possible to increase the amount vapor produced by a given amount of input steam by:

1. Placing several evaporators in series, with the pressures in the evaporating compartment being maintained at successively lower values to ensure heat flow. This is the basic concept behind the Multi-Effect Evaporator (MEE) and the Multi-Stage Flash Evaporator (MSF).

2. Utilizing Vapor Compression (VC) Evaporators, whereby the low pressure product-vapor is compressed (and hence heated) and reused as input steam.

The steam economy (SE), sometimes defined as "gained output ratio" (GOR), is defined as

$$SE \equiv \frac{\text{mass of distillate produced}}{\text{mass of high pressure input steam}} \quad (7)$$

Numerically, the SE will be slightly less than the number of effects. This is due to some heat losses (either to the heated sea water and/or to the surroundings) and to the fact that $\lambda_s < \lambda_{sol}$, i.e., the latent heat decreases as pressure increases, and less vapor is produced in each successive effect. A good approximation: SE=0.8 n where n is the number of effects. In comparing, performance of different plants, one must account for the equivalent heat input of mechanical and electrical inputs to the plants. The <u>performance ratio</u> R is defined as

$$R = \frac{\text{Kg distillate produced}}{1000 \text{ Kjoule heat input}} \qquad (8)$$

The utilization of R is preferred because it is independent of the conditions of the input stream, whereas the SE will vary with input steam pressure. (Note that because $\lambda \approx 1000$ BUT/lb the numerical value of $R \approx SE$ in British units system).

2.4. Heat-energy relationships.

Some problems are common to all distillation processes. Here, we shall only attempt to cover the heat-energy relationship and its associated heat rejection problem. Other common problems, which include pretreatment, entrainment, post treatment of distillate and proper disposal of waste brine and cooling water, are outside the scope of this presentation.

2.4.1. Heat rejection.

As shown in section 2.2, real systems operate at relatively low efficiency, and actual energy requirements exceed by far the minimum work required for isothermal separation of water from sea water. The excess energy used must be rejected from the plant if the temperatures within the plant are to remain constant.

For example, an efficient vapor compression plant will require about 12 KWH/m$^3$ of water produced, operating with a 50% recovery. From section 2, a minimum separation work of about 1.0 KWH/m$^3$ is required. The excess of 11 KWH/m$^3$ must be rejected from the plant. This translates to a temperature elevation of the outgoing brine and fresh water streams. To avoid thermal pollution by the reject brine, this temperature elevation above the incoming sea water must be controlled by increasing the flow rate of the cooling sea water.

2.4.2. Overall material and energy balances - cooling water requirement.

Consider an idealized distillation plant as a thermally insulated black box operating at steady state. Define:

$M_F$ - Kg/hr, feed, with enthalpy $H_F$ ($T_F$=25°C).
$M_{ci}$ - Kg/hr, cooling water, with enthalpy $H_{ci}$ ($T_{ci}$=25°C).
E  - Kcal/hr, mechanical and electrical energy ($-W_{min}$=3.0 KWH/m$^3$).
$Q_s$ - Kcal/hr, heat input ($\lambda$=664 Kcal/Kg).
$M_p$ - Kg/hr, product water, with enthalpy $H_p$ ($T_p$= 34°C).
$M_{co}$ - Kg/hr, cooling water, with enthalpy $H_{co}$ ($T_{co}$= 30°C).
$M_b$ - Kg/hr, brine, with enthalpy $H_b$, ($T_b$=47°C).

It is assumed that all the excess mechanical and electrical energy, regardless of source, dissipate as heat via the liquids. The material balance is given by

$$M_F + M_{ci} = M_b + M_{co} + M_p \quad ; \quad M_{co} = M_{ci} \qquad (9)$$

The energy balance is given by

$$Q_s + E = M_p H_p + M_b H_b + M_{co} H_{co} + M_p \cdot W_{min} - M_F H_F - M_{ci} H_{ci} \tag{10}$$

where $W_{min}$ =minimum energy of separation. Assuming for simplicity that enthalpy $H=C_p \Delta T = \Delta T$ (i.e., $C_p$=1 Kcal/Kg-°C), Eq.(14) reduces to:

$$\frac{Q_s + E}{M_p} = (T_p - T_F) + W_{min} + (T_{co} - T_{ci}) \frac{M_{ci}}{M_p} + (T_b - T_F) \frac{M_b}{M_p} \tag{11}$$

Assume that all heat input is due to high pressure steam coming in $M_s$ Kg/hr with a latent heat of 664 Kcal/kg; that $SE=M_p/M_s$=11.62, and that $M_b/M_p$=0.333, (i.e., the brine blow out rate is 0.333 lb per lb of distillate). Stipulating $(T_p-T_F)$=9°C, $(T_b-T_F)$=22°C and $(T_{co}-T_{ci})$=5°C. Eq.(15) yields $M_{ci}/M_p \approx 8.35$. Thus, we need circulate 8.35 m³ of cooling (sea) water for each m³ water product. It must be emphasized that the rejected heat, though quite large in quantity, is of low quality and, usually, cannot be utilized.

## 3. BASIC CHARACTERISTICS OF THE FLASHING PROCESS

### 3.1. The Flash Evaporator - Multi-Stage Flash (MSF) Evaporator.

The characteristic aspect of this process is that vapor is produced by throttling down the hot brine from a higher pressure to a lower one, step 3 to 4 on the enthalpy diagram, Fig.4. The thermodynamic process is simple: The energy stored in the water at its boiling point decreases as the water pressure is reduced. The excess energy produces vapor. The heat of condensation of the vapor released this way is used to preheat the sea water feed. This is the heat recovery part of the process. The higher the flashing temperature in the unit the higher the preheating temperature of the outgoing feed. Hence the desirability for many stages leading to the necessity of supplying the system only with the marginal heat required to bring the preheated feed to the pre-flash temperature. This is best demonstrated in Fig.5. Once the approach temperature ($=T_5-T_2$ in Fig.4) is specified, heat and material balances will fix the temperatures at all points.

A general characteristic of the MSF system is that the latent heat required for flashing is supplied by a sensible heat change of the liquid. Hence it is necessary to circulate in the system a large amount of liquid relative to the product. For the data given in Fig.4, the latent heat divided by the temperature gain in the external heater (116-74=42°C) yields 13.3 m³ circulating brine per m³ product. This high circulation distinguishes the MSF from the multi-effect-evaporation (MEE) process.

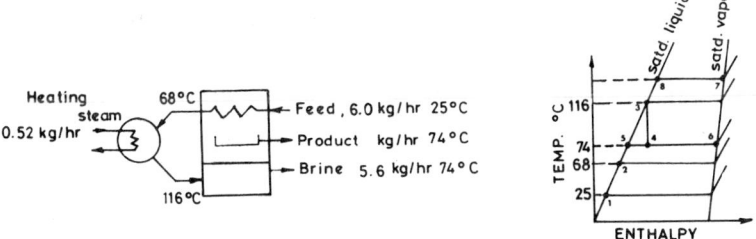

Fig.4. Thermodynamics of a single stage flash evaporator.

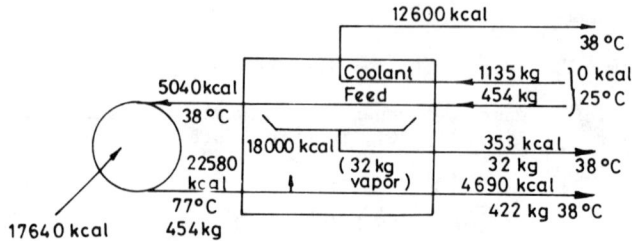

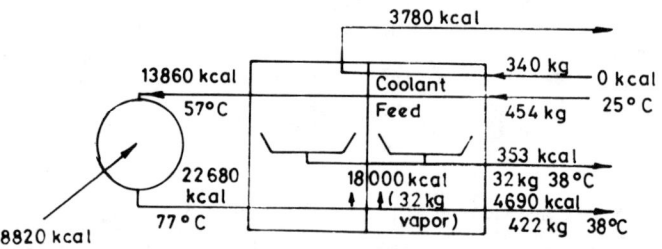

Fig.5. Comparison of the thermal requirements in one and two-stage flash evaporation units.

3.2. Advantages of MSF over the multi-effect evaporator (MEE).

The number of effects in the MEE is usually equivalent to the steam economy (SE) or performance ratio (R). The number of MSF stages, n, however, is determined mainly by the desired R and the allowable approach temperature. It is always larger than the R. Thus, an MSF evaporator designed to give a desired R will always require more heat transfer area, A, than the equivalent MEE. The area per unit product is approximated by

$$\frac{A}{M_p} \simeq \frac{\lambda n}{U \Delta T_F} \ln \frac{n}{n-R} \tag{12}$$

where $\Delta T_F$ is the flashing temperature range ($=T_3-T_5$). Assuming identical overall heat transfer coefficients, U, the area requirements of the MSF and MEE become equal when the MSF has an infinite number of stages, and Eq.12 reduces to

$$A_{min}/M_p = \lambda R/U \Delta T_F . \tag{13}$$

In spite of the increased pumping costs and area requirement of the MSF as compared to the MEE, it has superior characteristics to the submerged boiling MEE. These lead to cheaper product costs. Some of these advantages are:

a) Pool boiling evaporation requires a minimum $\Delta T$ of 8°C per effect. The larger the scale of evaporation, i.e. the recovery, the more serious becomes the effect of the boiling point elevation, thus placing limitations on the size of the units. The net result is that MEE with pool boiling is limited to relatively low performance ratios (R<6).

b) Flash evaporation can be realized with a very small, 1-2°C, temperature drop per stage. This permits to raise the temperature of the sea water feed (by the condensing vapors) and thus decrease the required energy input to the first stage. It allows design of units with high performance ratios thus giving cheaper energy costs.

c) The MSF evaporator is of a much simpler construction: a partitioned box-like enclosure with condenser tubes. Thus, although more heat transfer area is required, the product of area required and cost is much smaller than for the pool boiling MEE. (In any case, by taking a sufficiently large number of stages, the area requirement for MSF is only slightly larger than for MEE).

d) The MSF evaporator lends itself much more easily to scale-up than the MEE. All the energy in the MSF process is added to the liquid before the flashing begins while the energy in the MEE is added during the vapor production process.

3.3. The recycle and arrangement of MSF 'effects'.

In most MSF evaporators, a portion of the brine from the last stage is blended with the make-up brine and returned to the brine heater. This recirculation is in general no more than 8 parts to 1 part product. The advantage of recycle as compared to 'once through' is in reduced feed treatment cost. The disadvantage is in reduced maximum brine temperature due to increased salt concentration. In most cases the reduced chemical cost predominates.

The MSF processes, schematically shown in Fig.6, have several modifications in addition to the recirculation. The sea water feed first enters a group of flash stages that serves as a heat sink. This portion of the process is called the heat rejection section. The coolant sea water, upon leaving the heat rejection stage, is returned to the sea. The make-up water, on the other hand, is usually treated with one of the various sea water pretreating processes.

Equipment and operating cost studies indicate that maximum benefits are obtained with a recirculation to sea water make-up ratio of about 4/1. At higher ratios, pumping and equipment costs increase faster than the gain in thermal economy. To obtain near optimum conditions in all sections of the flash process more than one recycle stream can be utilized. To best accomplish this, the plant's flashing range can be divided into a series of "effects", each equipped with heat input, heat recovery, heat rejection section, and its own circulating pump. This permits the circulation rate to be independently set in each effect. The process composed of a series of such effects is called a Multiple Effect Multi-Stage Flash Evaporation Process (MEMSF).

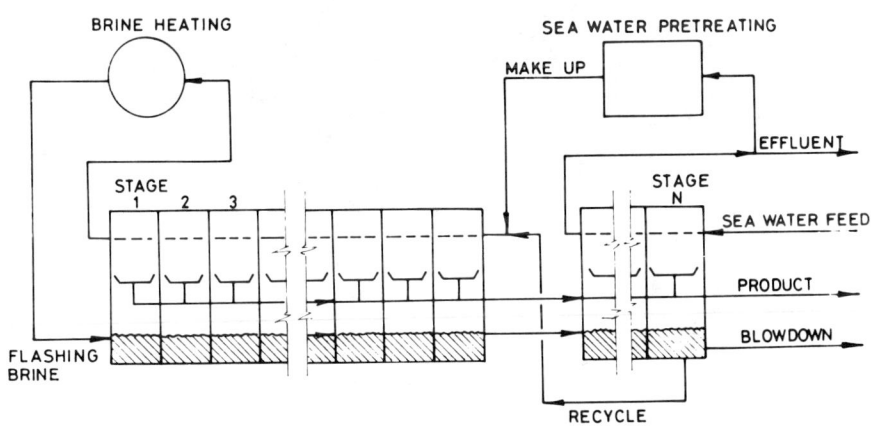

Fig.6. Sechmatic of a "single effect", multi-stage flash evaporator with recycle.

The basic characteristics of the MEMSF scheme are: Energy is supplied externally to the first effect. The heat from the first effect is rejected to the brine in the next effect. Thus, the heat rejection section of effect 1 is also the heat input section of effect 2, etc. The circulated brine flows through the tubes of the heat recovery section and then into the tubes of the combination heat rejection - heat input section. There the brine acquires sufficient energy to continue the flash process in the heat recovery section.

The increase in performance of these high capacities MEMSF plants may more than offset the additional capital and operating expenses due to the staging arrangement, controls, and pumps.

3.4. A novel vertical MSF scheme.

The Multi-Stage Flash/Fluidized Bed Evaporator (MSF/FBE) is a new novel type of a vertically assembled MSF unit [24, 25]. As shown schematically in Fig.7, the vertical column is made of the heat recovery section (the condensation section) and the flash column. In practice, the unit is of a circular structure with the condenser in the central shell surrounded by the annular flash section. The unit is assembled as a counter current heat exchanger: the raw sea water feed is heated while flowing upwards in the heat recovery section and fluidizing solid particles in the condenser tubes. After passing the heat input section (on top of the heat recovery section), the feed flows down by gravity and flashes via orifices in the interstage wall in all the vertically stacked stages. The gradual drop in the saturation pressure and temperature in each stage results in partial evaporation of the brine and the vapor, after being demisted, reaches the condenser tubes in the heat recovery section. The condensate cascades downwards via distillate siphons, and leaves at the bottom. Vacuum is maintained by means of steam ejectors.

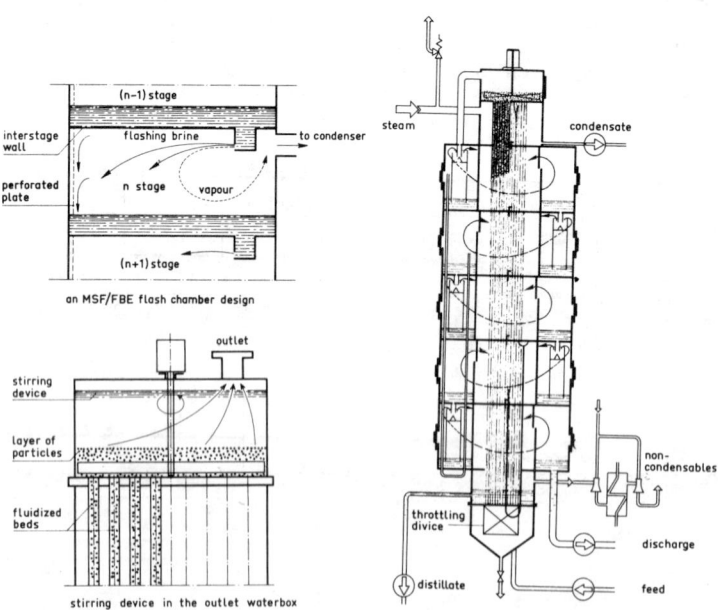

Fig.7. Schematics of the MSF/FBE and some details [24].

The unique feature of the MSF/FBE is the utilization of fluidized particles in the vertical condenser tubes in the heat recovery section, thus affecting the thermal boundary layer at the wall. This technique enhances the liquid side heat transfer coefficients in the tubes by some 6 to 7 folds over a comparable operation without fluidization, resulting with smaller column heights, and smaller (0.10 to 0.20 m/s) superficial velocities in the tubes. The wall to liquid heat transfer coefficients reported are best correlated by [26]

$$Nu_p = 0.067 \, Pr^{0.33} \, Re_p^{-0.237} \, Ar^{0.522} \quad ; \quad Re_p \cdot Ar^{-0.58} > 0.09 \tag{14}$$

where the dimensionless groups are defined by

$$Nu_p = \frac{h_\ell d_p}{k_\ell} \quad Pr = \frac{\mu_\ell C_{p\ell}}{k_\ell} \quad Re_p = \frac{\rho_\ell u_\ell d_p}{\mu_\ell} \quad Ar = \frac{g \rho_\ell^2 d_p^3}{\mu_\ell^2} \left(\frac{\rho_s - \rho_\ell}{\rho_\ell}\right)$$

and $d_p$ denotes particle diameter. Overall heat transfer coefficients of approx. 10,000 Kj/m²hr°C were reported [25].

The classical horizontal MSF is limited by the temperature difference over the stages to some 40 stages corresponding to a maximum gain ratio of approximately 12. This limitation does not exist for the suggested vertical MSF/FBE as stable brine transport is assured by a brine height of 10 to 15 cm, independent of the interstage temperature difference. This advantage allows design units with higher grain ratios. Furthermore, this allows stable operation at various brine loads. A 50 m³/day unit is presently operating in Holland.

## 4. VAPOR COMPRESSION DISTILLATION

### 4.1. Basic characteristics of process.

The vapor compression process is important because of its highly efficient usage of energy. A vapor compressor can be incorporated into any distillation system, the energy requirement being supplied in the form of mechanical energy instead of thermal energy. In a multiple effect plant, for example, the steam generated in the last effect can be compressed and used as the feed stream to the first effect. Multi-effect VC plant economy presently operating in Eilat is obtained through the use of a steam jet ejector which recompresses and recycles vapor from lower temperature (and pressure) effects to the first effect of the heat recovery section.

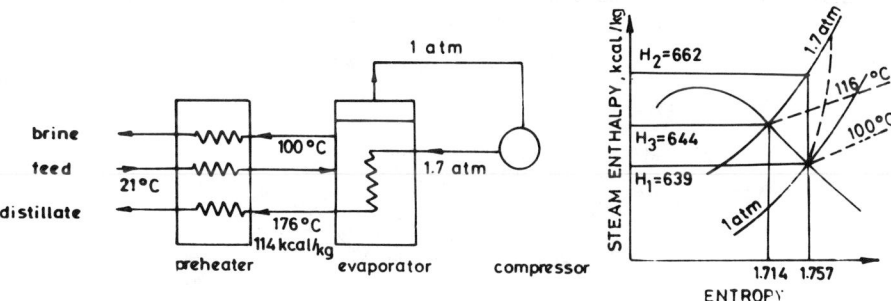

Fig.8. Thermodynamics of the vapor compression (VC) cycle.

As seen in the schematic diagram, Fig.8, the vapor compression process is based on recycling the latent heat from the evaporation side of the heat transfer surface to the condensing side of the same surface. The vapor produced is thus compressed (and hence heated) to the original pressure of the input steam, and thus its condensation will supply heat for the production of additional vapor from the same source of water. Fig.8 illustrates the use of a vapor compressor coupled to a single effect distillation plant. Feedwater is heated by counter current heat exchange with the product and blowdown streams. Usually, some compressed steam is also used in the feed preheater.

At steady state, heat rejection is equal to the heat input by the compressor, which is far less than the latent heat involved in the vapor production: the compressor supplies only the energy input necessary for the flashed water enthalpy drop plus losses. Thus, the energy input per unit mass of product is low, and very high performance ratios, up to 18.1, are realized.

In practice, the feed preheater can be incorporated into the evaporator shell, as is the case in the O.S.W. demonstration plant at Roswell, New Mexico, of unit capacity 0.83 m.g.d. This plant, the only large vapor compression unit, operates on a 15,800 ppm T.D.S. inland brackish water (cf. 35,000 ppm T.D.S. for sea water) and includes both ion-exchange and slurry recycle equipment for scale control.

As seen from Fig.8, we require about H=638-113 = 525 Kcal/Kg thermal energy to convert 1 Kg water to vapor, whereas only 638-632.5 = 5.5 Kcal/Kg (mechanical) energy is required to recreate the original state of the steam. The 3,750 m$^3$/day (0.83 m.g.d) VC plant at Roswell, New Mexico requires about 16.6 KWH/m$^3$ (63 KWH/Kgal). IDE's VC plants which range from 75 to 500 m$^3$/day in single units and up to 5000 m$^3$/day in a battery instillation, with a 40,000 m$^3$/day presently under construction, require similar power, i.e., 16-18 KWH/m$^3$ or about 4-4.5 Kgs fuel per m$^3$ product where power is derived from a diesel generator.

### 4.2. Some thermodynamic calculations.

The mechanical energy required for operating the compressor is given by

$$-W_s \left(\frac{Kcal}{hr}\right) = \frac{\Delta H_a \times M_D}{\eta} \quad ; \quad P_w = \frac{W_s}{860} \text{ KW} \quad (15)$$

where $\eta$=mechanical efficiency of the compressor, pumping $M_D$ Kg/hr vapor, and $\Delta H_a$, the true enthalpy difference due to non-ideality, is calculated from the adiabatic, theoretical, enthalpy change (line of constant entropy), and the isentropic efficiency, $\eta_{isent}$ by the relationship:

$$\Delta H_{act} = \frac{\Delta H_{theor.}}{\eta_{isent.}} \simeq \frac{\Delta H_{theor}}{0.7} = \frac{655-632.5}{0.7} \simeq 32.5 \frac{Kcal}{Kg} \quad (16)$$

The theoretical, adiabatic, work is given by

$$-W = \Delta H_{theor.} = \frac{k}{k-1} \hat{R} T_1 \left[\left(\frac{P_2}{P_1}\right)^{\frac{k-1}{k}} - 1\right] \quad (17)$$

$P_1$ and $T_1$ are the initial absolute pressure and temperature $p_2$ is the final pressure and k = 1.31. If the number of effects is n, the product $M_D$ will be n times the steam mass flow rate $M_F$. Hence the compressor energy requirement per lb of product will become -

$$-W = \Delta H_{theor.} = \frac{4.2 \hat{R} T_1}{n} \left[ (\frac{p_2}{p_1})^{0.237} - 1 \right] \text{Kcal/Kg product} \tag{18}$$

4.2.1. Comparison of single and multiple effect vapor compression units.

Consider a single effect VC plant operating between $T_2=93.3°C=366.3°K$ and $T_1=87.8°C=360.8°K$; $p_2=0.78$ atm (abs.); $p_1=0.64$ atm (abs.). By Eq.(18), the compressor energy requirement is = 8.4 Kcal/Kg product.

Assuming the temperature driving force per effect to be the same (=5.5°C) as in the single effect example above, the operating temperature range for a 10 effect plant will be 55°K. With $T_1=361°K$; $p_2=0.78$ atm (abs.) and $p_1=0.065$ atm (abs.) the compressor energy requirement will now be = 11.7 Kcal/kg product.

Thus, there is no energy advantage in going to multiple effect vapor compression units, providing the interstage temperature drop is held constant over the range of effects and the upper temperature $T_2$ is held constant. Mechanical and operating advantages, however, offset this disadvantage.

4.2.2. Heat transfer area.

The heat transfer area for evaporation in a single effect evaporator is defined by

$$A = \frac{Q}{U \Delta T_M} \tag{19}$$

where Q=evaporative heat load; U=heat transfer coefficient; $\Delta T_M$=mean temperature difference driving force (equal to the evaporation range less the temperature drop due to boiling point elevation and pressure loss). An economic balance between energy and capital costs must be carried out to determine the optimum design criteria.

The feed water to the VC unit must be heated to the evaporator operating conditions. The heat input is derived from three sources: (a) heat exchange with blowdown stream; (b) heat exchange with product stream; (c) heat exchange with steam. The design of the three feed heaters is not conducive to analytical treatment. However, it is clear that an optimum operating temperature $T_1$ exists such that the compressor operation costs are balanced against the feed heater costs. In achieving this balance and in considering the overall size of evaporators it may be economic to use a double effect evaporator despite the fact that energy requirements are higher for a VC multiple-effect plant.

Comparison of a VC unit with MEE or MSF plants is based solely on energy costs, and is given by Dodge [23]. For any $\Delta T$ less than about 7°C in the VC evaporator, the energy cost will be less than that of the MEE even if the latter had an infinite number of effects. It would require a 20-effect evaporator for equivalent energy costs to a VC unit with $\Delta T=8°C$.

5. THE FILM EVAPORATOR/CONDENSER

The elementary unit is the single effect evaporator, which, for well-known thermodynamic reasons is extended into the multiple effect evaporator. Typically, the heat transfer process in each effect involves the condensation of pure steam and the simultaneous evaporation of saltwater, separted by a solid wall. The overall temperature difference is of the order of 1-3°C. In view of the very small overall temperature difference, it is essential that both condensation and evaporation occur in thin films. These are realized in the vertical tube evaporator (VTE) or the horizontal tube evaporator (HTE). The rate of heat transfer clearly depends on the film thicknesses of the two films - evaporating liquor or brine and pure condensate. Here the main research effort is in the design of enhanced surfaces, in which swirling and/or surface tension reduces the film thickness over a part of the wall circumference to a fraction of the average thickness, leading to an enhancement of the overall heat transfer rate by up to 500%. The recent success of Rosenblad Corp.'s plate type falling film evaporator in concentrating relatively viscous liquors in the pulping industry [27] adds to the growing interest in film evaporators.

5.1. The Vertical Film Evaporator.

The film inside the tubes can be formed in two ways. One, by introducing the liquid at the bottom where boiling occurs. The rising vapor pulls the liquid upwards thus forming a film on the wall. This is the Rising Film Evaporator. Two, by distributing the liquid evenly at the top, allowing it to fall by gravity. This is the Falling Film Evaporator. Obviously, the latter is not limited only to tubes but, with proper liquid distribution at the top, is applicable to vertical walls of all shapes.

5.1.1. The Rising Film Evaporator.

This mode of operation is believed [27], to date back to Paul Kestner's invention in 1899: (Appareils & Evaporateurs Kestner still delivers evaporators of different types). The liquor is introduced at the bottom of the long vertical, 50 mm diameter tubes, and heat is supplied from the outside so that the liquor boils. The generated vapor bubbles rise vertically and carry the liquid upwards, forming a film along the outside tube-wall. Liquor and steam are separated after exiting through the top tube sheet. In a common variation, the liquor is introduced into half the tube bundle and thereafter is recirculated through the remaining tubes. In this manner the liquor flow per tube is doubled.

The main advantage of the rising film evaporator is that it is comparatively simple and therefore inexpensive to manufacture, especially in large units. Neither does it require any pumping power. It is, however, unsuitable for scaling liquor as, for example, sulfite liquors. Evaporators for highly scaling liquids usually use forced circulation, often with plate type heating surfaces.

The rising film evaporator is unsuitable for the pulping industry because of the low heat transfer coefficient at low specific heat loads (i.e. heat transferred per unit heating surface) or at relatively low temperatures and/or high liquor viscosities. An alternate evaporator type, useful for both sulfite and black liquor, is the Lockman Column, a multi-flash evaporator [27]. Here, the liquor is pumped through heat exchangers that are heated by the liquor's own flash steam. The liquor obtains its highest temperature in a heat exchanger supplied with live steam. On the liquor side the Lockman Column is comparable to a forced circulation evaporator.

5.1.2. The Falling Film Evaporator.

The major feature of the falling film evaporator is that the film is established at the top of the tube by means of a distributor, and is independent of the evolution of vapor. In the Vertical Tube Evaporator (VTE) depicted in Fig.9, the saline water, heated to near saturation, is introduced into the vertical tube while vapor from a preceding holter stage condenses outside the tubes. There is no static liquor pressure and no preheat zone. This may reduce the available temperature driving force if the liquid arrives below the saturation point. The falling film evaporator principle can be used both without and with circulation. The former gives the shortest retention time for the liquor,

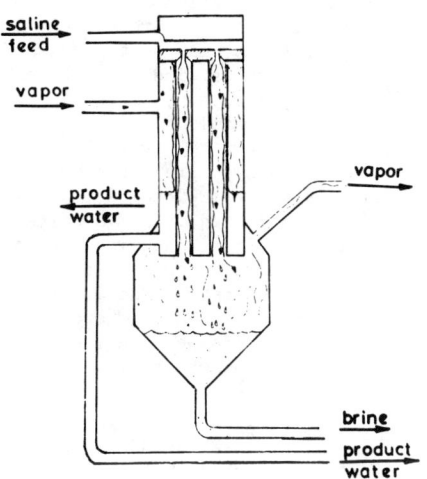

Fig.9. Schematic of the vertical tube falling film evaporator.

desirable during evaporation of heat sensitive liquids. In order to insure that the heating surface does not dry out, it is possible to recirculate the liquor, which also increases the HTCs. The drawback is of course, the need of pumping power. This power, however, is very small compared to that used in a forced circulation evaporator.

The falling film evaporators HTC is 50-100% higher than the forced circulation at high concentration and high temperatures [27]. At low temperature and medium concentration, the falling film and the forced circulation evaporator have the same HTCs, at a level which is about 50% higher than those obtained with the rising film evaporators.

The vertical tube falling film evaporators are useful for low viscosity liquids but unsuitable as sulfite and sulfate liquor concentrators. The latter are best treated with Rosenblad's (ROSCO) vertical wall, lamella type surfaces [27]. The heating surface consists of vertical "lamella mattresses" suspended within a pressure vessel, which in its entirety becomes both the heating element and the vapor body for the separation of liquor and vapor. The liquor therefore flows and boils on the outside of the lamellas, while the heat input takes place inside the elements. The liquor circulates with a pump to a distribution tray above the lamellas, and flows down over these through a large number of holes in the bottom of the distribution tray.

Similar design work was done in other countries and by other companies, mainly for the dairy industry. Especially important are the good characteristics at small temperature driving forces which explains why this type of evaporator can be used for fresh water evaporators, where the number of effects can exceed ten or more. For instance, Hitachi Zosen of Japan is advertising a multi-effect VTE, claiming great savings in both capital and operating costs as compared to the MSF plants of the same water capacity.

5.2. The Horizontal Tube Evaporator/Condenser.

One of the disadvantages of the VTE is that each effect requires its own pump to circulate the brine to the respective upper brine plenum. Also, the precise distribution of the incoming brine to form uniform films at the tube tops presents a problem. A possible arrangement which avoids these difficulties is the horizontal tube evaporator (HTE) Fig.10. The basic thermal principle is the same as in the VTE, but here the steam condenses inside the horizontal tubes, while the saline water to be evaporated cascades over the outside of these. The overall heat transfer rate is again dependent on the thickness of the two water films. Obviously, considerations must be given to condensate drainage, e.g., avoidance of tube blockage due to condensate backup [28] as well as to proper brine distribution on either smooth or grooved surfaces [29].

Horizontal tube falling film evaporators are presently utilized for water desalination in either single stage vapor compression (VC) units [30] or in multi-effect distillation (MED) units [31]. These processes are characterized

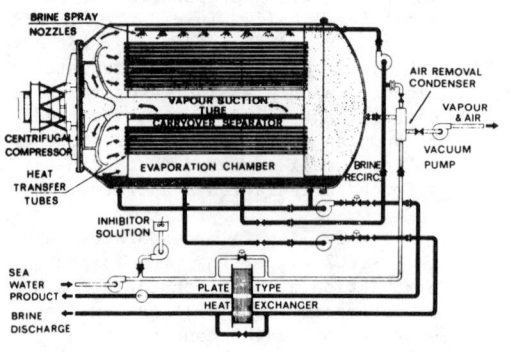

Fig.10. Schematic of the Aquaport vapor compression-horizontal tube evaporator.

by the utilization of low cost aluminum tubes and relatively low temperatures, low heat loads and low flow rates. The two schemes are complimentary, as the VC process is limited to the range of 20 to 1000 m$^3$/day (5x10$^3$ to 250x10$^3$ gpd) whereas the Med is suitable for 1000 to 50,000 m$^3$/day (0.25 to 13 mgd). The VC units utilize either electrical or mechanical energy. However, the MED plants are most attractive for dual purpose plants, where relatively inexpensive low pressure exhaust steam, below 75°C, is available.

As compared with the violently turbulent flashing MSF, the HTE is a quiet process. For example, the velocities of the flashing brine in the MSF are up to 10 m/sec whereas the boiling brine flows at about 0.7 m/sec over the tubes. The heat transfer coefficients are twice or thrice larger than those realized in the MSF. With low temperature driving forces, 2-4°C, and low pressure losses between adjacent effects, the total number of effects can be as high as 14 or 15. These advantages lead to performance ratios of 6.5 to 7.5 Kg/1000 Kjoule (15-17 lb/1000 BTU) steam as compared with a maximum of about 5.2 (12.5 lb/1000 BTU) in the MSF. Scaling problems are easily avoided at the low operating temperature range of the process through the use of polyphosphate or polyelectrolyte additives, thus avoiding the standard acid pretreatment step and minimizing corrosion of the vessel and the heat transfer surfaces. Because of the nature of their operation the horizontal units can be easily shut off, or started and brought to full capacity in minutes. Capital construction costs for a 4000 m$^3$/day unit in Aruba were 15% less than for an equivalent MSF plant, and units of up to 13.5x10$^3$/day are being offered by Sasakura, Japan. A 4600 m$^3$/day (aluminum) horizontal tube plant is being operated by IDE (Israel Desalination Engineering, Tel Baruch, Israel) for some years now and a 46000 m$^3$/day plant is presently being built in Ashdod, Israel. The design and operating features of a 10 mgd dual purpose plant is given in [30]. Other interested manufacturers are Hamon Sobelco, Aqua Chem., Aiton Ltd.

5.2.1. Some novel design ideas.

Recent research effort lies in the study of noncircular tubes [32-35], the underlying idea being that the vertical portion of the tube circumference is more efficient in providing a thin film than the inclined portions. This is clearly demonstrated by comparing the local overall heat transfer coefficients around circular and elliptical conduit. The theoretical results shown in Fig.11 were experimentally confirmed [29]. Note that the Elliptical configuration yields better results than any non-circular crossection conduit. However, increasing the axis ratio above the value of 4 does not yield any additional advantages, and may actually decrease the overall transfer coefficient by the formation of thicker condensation films along the wall. This effect can be reduced by inserting internal horizontal-partitions along the vertical part of the periphery, draining the condensate and thus forming 'new' clean metal surfaces for the vapor to condense on.

5.2.2. Enhanced heat transfer area.

Augmentation of the heat transfer by utilizing profiled tubes is of interest

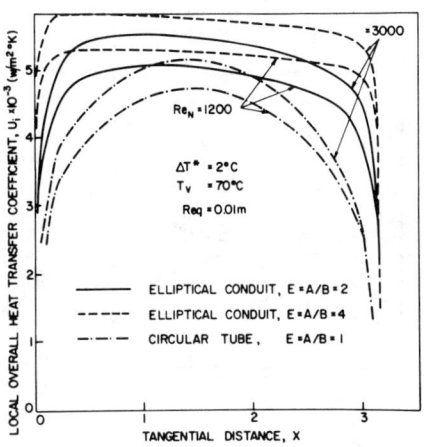 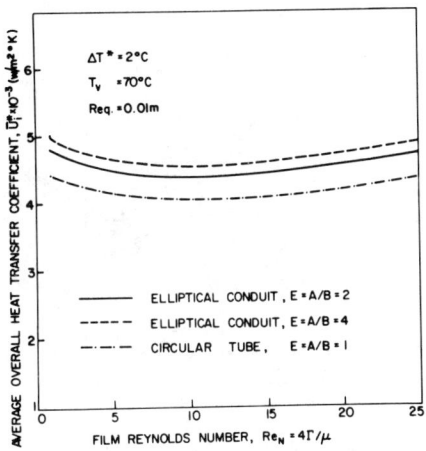

Fig.11. Comparison of local and average overall heat transfer coefficients for circular and elliptical horizontal tubes.

in all forms of heat transfer. References with particular emphasis on enhanced transfer rates to or from condensing and evaporating films is given elsewhere [22].

Until recently, attention has been directed towards vertical fluted tubes or surfaces for the falling film evaporators [36]. Evaporation and condensation on enhanced vertical planes and tubes indicated improvements of up to 300% in the overall transfer rate. More recently studies were directed to device tube shapes which would enhance evaporation rates from horizontal tubes [36-42].

Fig.12 represents the combined average HTCs calculated for the minimum and maximum (flooding) flow rates, as a function of the width of the groove's BOTTOM for various depths of the indentations. Nominal (tube-surface based) and actual wetted area HTCs are included. For a given width b, the combined average HTC increase with the decrease of TOP width. Not unexpected, the HTCs are smaller for deeper grooves. These results are in general accord with experimental data indicating the advantage of utilizing fine grooves as compared with coarse ones.

An elliptical tube with horizontal grooves (perpendicular to gravity flow) showed an appreciable decrease of the transfer rate as compared to smooth tubes. On the other hand, the experimental overall HTCs for the simultaneous evaporation-condensation process, with square-edged, circumferentially grooved, circular tubes (D=3.8 cm) show that narrower grooves (0.1/0.1 cm) enhance the transfer rate by some 80% whereas larger grooves (0.2/0.2 cm) improve the transfer rate only by 60%. As with smooth surfaces, the HTC is only slightly affected by the evaporation temperature change (from 68° to 82°C). Unlike smooth surfaces, the calculated effect of the feed flow rate is not monotonous, first increasing as the partially wetting rivulets grow with the flow rate to form a fully covered TOP and BOTTOM of the groove. With further increase of the feed flow rate, the film thickness of the concave rivulet at the BOTTOM increases and the HTC decreases. As seen in Fig.13, the agreement of the calculated curves with the experimental data is quite satisfactory, thus giving some credance to the suggested mode of calculation. The data for smooth circular and elliptical tubes are also included for comparison. These data demonstrate again the advantage of utilizing elliptical conduits. As already noted, finer grooves yield

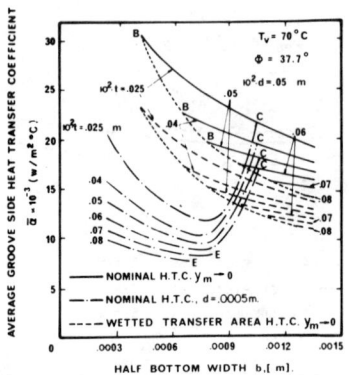 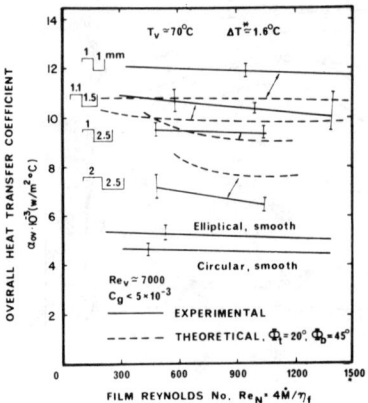

Fig.12. Average HTCs for minimum and flooding water flow rates on square-grooved horizontal tubes. Effects of groove's width and depth.

Fig.13. Comparison of theoretical and experimental HTC for square-edged grooved horizontal tube

higher transfer rates than large one. Also, widening the TOP negatively effects the transfer rate. This effect is larger than a comparable change in the width of the BOTTOM.

A theoretical analysis of the effect of the shape of the external grooves of various configurations on the film evaporation rate and the overall transfer coefficient [40] is of particular importance. In agreement with the earlier studies of square grooves, the study shows that the amount of heat transferred through the film in the grooved surface depends on the water flow rate in the groove which, in turn, depends on the size and shape of the groove.

In general, the analysis yields that the square-edged grooves, with either straight or cut-edged bottom, are most advantageous for Re>500. The circular bottom square-groove seems advantageous for 250 Re 500, yielding high transfer rates with relatively little sensitivity to the flow rate. The triangular groove is useful only at the low, Re 500, flow range while the circular groove is not advantageous at all. Note that the heat transfer rates increase as the contact angle decreases, consistent with the well-known advantage of adding surfactants to sea water. The effects of the thermal properties of the tube wall are discussed elsewhere [22]. The major interest shown here in the evaporation side is due to the relatively high resistance on this side as compared with the condensation side. However, significant improvements of the internal condensation side HTCs will certainly improve the overall performance. This point is discussed elsewhere in this meeting.

## 6. CONCLUSIONS

The various parameters involved in the thermal design of the large-scale evaporator/condenser units have been discussed. The unavoidable low efficiency of the thermal processes for water desalination as compared with the theoretical separation work of salt from water leave a lot to be desired, particularly, when energy cost increase beyond expectations. Known concepts and novel new ideas are considered, and the highly efficient and economically promising horizontal tube film-evaporator/condenser is discussed in detail. It is hoped with the details considered, one gains an overview required for a good design of efficient evaporators/condenser.

## 7. ACKNOWLEDGEMENT

Some of the work reported here was supported by the Israel National Council for R&D and the Vice President for Research Fund. Thanks are due to my colleague and friend D.M. for help and encouragement.

## 8. REFERENCES

1. OSW - Office of Saline Water Research and Development Progress Reports, USA - National Technical Information Service (NTIS), Order Dept. Springfield, VA.22151 USA.
2. Schamus,J.J. et al.1965. Bibliogrpahy of Saline Water Conversion Literature. Office of Saline Water (OSW) Report No.146.
3. Delyannis,A.,Delyannis,E. 1960. Handbook of Saline Water Conversion Bibliography, vol.1. Published by A.and E. Delyannis, P.O.Box 1199, Athens, Greece.
4. Indexed Bibliography of Nuclear Desalination Literature, Oak Ridge National Laboratory (ORNL), Oak Ridge, Tennessee, USA, National Desalination Information Center (NDIC), vol.1. 1968.
5. Desalination, International Journal on Science and Technology of Water Desalting, Published by Elsevier, P.O.Box 211, Amsterdam, Holland.
6. Desalination Abstracts, Issued by the Israel National Council for Research and Development, 1967. P.O.Box 20125, Tel-Aviv, Israel.
7. Popkin,R. 1968. Desalination-Water for the World's Future. Frederick A. Praeger, New York.
8. Spiegler,K.S. 1962. Salt Water Purification. John Wiley, New York.
9. Spiegler,K.S. 1966. Editor, Principles of Desalination. Academic Press, New York. 566 pp.
10. d'Orival,M. 1967. Water Desalting and Nuclear Energy. Verlag Karl Thiemig K.G. Muenchen. 197 pp.
11. Burley,M.J., Mawer,P.A. 1966. Desalination as a Supplement to Conventional Water Supply - Part 1. Water Research Association Technical Report No.50, Medemenham, Bucks., England. 270 pp.
12. Burley,M.J., Mawer,P.A. 1967. Desalination as a Supplement to Conventional Water Supply - Part 2. Water Research Association Technical Report No.60, 190 pp.
13. Levine,S.N. 1968. Selected Papers on Desalination and Ocean Technology. Dover Publication, New York, 437 pp.
14. Wilson,J.R. 1960. Demineralization by Electrodialysis. Butterworth, London,
15. Merten,U. 1966. Ed. Desalination by Reverse Osmosis. The M.I.T. Press, U.S. 290 pp.
16. Kellog, 1965. Saline Water Conversion Engineering Data Book. Report for the O.S.W.
17. Dukler,A.E. et al. 1971. Distillation Plant Data Book. Houston Research Institute, USA., for OSW.

18. Ralph M.Parsons Co. 1966. Guideline for Uniform Presentation of Desalting. Cost Estimates. OSW. Report No. 267, July 1967.

19. Post,R.G.,Seale,R.L. Eds.1966. Water Production Using Nuclear Energy. Univ. of Arizona Press, Tucson.

20. Howe,D. 1974. Fundamentals of Water Desalination. Marcel Dekker Inc. New York.

21. Porteous,A. 1975. Saline Water Distillation Processes. Longman.

22. Sideman, S. 1980. Film Evaporation and Condensation in Desalination in Heat Exchangers: Thermal-Hydraulic Fundamentals and Design. S.Kakac,Ed. Proc.NATO Adv. Study Inst. Bogazici Univ. Istanbul.

23. Dodge,B.F., Eshaya,A.M. 1961. Thermodynamics of Some Desalting Processes. In Advances in Chemistry Series.

24. Veenman,A.W. 1976. The MSF/FBE: An Improved Multistage Flash Distillation Process. Desalination 19: 1-14.

25. Veenman,A.W.1976. Construction and Initial Operating of the MSF/FBE. 5th Int.Symp.on Fresh Water from the Sea, 2:193.

26. Ruckenstein,E., Shorr,V. 1959. Despre transferul de caldura dintre un strat fluidizat cu lichid si peretele vasului care contine. Akad.Rep. Populare. Romini, 10:235.

27. von Matern,F., Jonsson,S.E. 1980. Fallfilmsindunstoning-nygammal taknik. Svensk Papperstidning 83:125-129.

28. Levin,A., Sideman,S. 1980. Comparison of Parallel and Counter-Current Flow of Steam and Condensate in Horizontal Tube Evaporators. Technion R&D Rept. No. 74, Proj. 071-225, Submitted to Israel Nat.Counc. for R&D. June 1980.

29. Sideman,S.,Moalem-Maron,D. 1979. Transport Characteristics of Thin Films: Evaporators, Condensation and Mass Transfer on Smooth and Grooved Horizontal Conduits. In. Two-phase Momentum Heat and Mass Transfer in Chemical Processes and Energy Engineering Systems. vol.2, pp.877-896, F.Durst, G.V. Tsiklauri., N.H.Afgan, Eds., Hemisphere Publ. Corp.N.Y.

30. Howarth,J.R.,Wood,F.C. 1973. Mechanical Vapor Compression Evaporators Incorporating the Horizontal Falling Film Principle. 4th Int.Symp. Fresh Water from the Sea. 1:327.

31. Barak,A.Z., Barnes,W.L. 1977. The Design, Construction, Testing and Operation of a Large-Scale Prototype Desaling Plant in Israel. Desalination 20:163.

32. Sideman,S.,Moalem-Maron,D., Semiat,R. 1975. Theoretical Analysis of Horizontal Condenser-Evaporator Conduits of Various Cross-Sections. Desalination J. 17:167-192.

33. Moalem-Maron,D.,Sideman,S. 1975. Theoretical Analysis of a Horizontal Condenser Evaporator Elliptical Tube. ASME J. Heat Transfer, 97:352.

34. Sideman,S.,Semiat,R.,Moalem-Maron,D. 1977. Performance Improvement of Horizontal Evaporator-Condenser Desalination Units. Desalination J. 21:221-233.

35. Semiat,R.,Sideman,S.,Moalem-Maron,D. 1978. Turbulent Film Evaporation on and Condensation in Horizontal Elliptical Conduits. Proc.6th Int.Heat Trans. Conf. Toronto, HX-30 361-366.

36. Sander-Beuermann,W.,Schroeder,J.J. 1978. Investigation of Isothermal Falling-Film-Flow on Vertically Profiled Surfaces. Proceedings 6th Intn. Symp. Fresh Water from the Sea. 1:173-182.

37. Semiat,R.,Moalem-Maron.D., Sideman,S. 1978. Transfer Characteristics of Convex and Concave Rivulet Flow on Inclined Surfaces with Straight-edged Grooves. Technion R&D Rept. No.51, Proj.071-082. Desalination, in press.

38. Moalem-Maron,D.,Semiat,R.,Sideman,S. 1978. Enhanced Heat Transfer in Horizontal Evaporator-Condensers with Straight-edged Grooved Tubes. Technion R&D Rept., No.52. Proj.071-082. Desalination, in press.

39. Sideman,S. 1978. Enhanced Transfer in Horizontal Evaporator-Condenser Conduits with Circumferential Grooves. Proc. 6th Int.Symp.Fresh Water from the Sea, Las Palmas. 1:183-192.

40. Sideman,S., Levin,A. 1979. Effect of the Configuration on Heat Transfer to Gravity Driven Films Evaporating on Grooved Tubes. Desalination 31:7-18.

41. Edwards,D.K., Gier,K.D.,Ayyaswamy,P.S.,Catton,I. 1973. Evaporation and Condensation in Circumferential Grooves, ASME paper No.73-HT-25, ASME-AIChE Heat Transfer Conf. Atlanta, Georgia.

42. Edwards,D.K., Balakrishnan,A.,Catton,I. 1974. Power-law Solutions for Evaporation from a Finned Surface, J.Heat Transfer. 96:423-426.

# On the Heat Transfer and Fluid Flow in Falling Film Shell-and-Tube Evaporators

**E. N. GANIĆ**
Department of Energy Engineering
University of Illinois at Chicago Circle
Chicago, Illinois 60680 USA

ABSTRACT

Recent developments in the area of heat transfer and fluid flow in falling film evaporators are summarized in this paper. Attention is focused primarily on a) heat transfer to falling film, b) falling film breakdown and c) vapor/liquid interaction and entrainment.

1. HEAT TRANSFER TO FALLING LIQUID FILMS

Falling liquid films have long been used in the heat transfer industry and in some chemical engineering operations. They are characterized by high heat transfer rates at low flow rates and temperature differences. The flow in a falling liquid film may be either (a) laminar with a constant thickness or (b) laminar with variable thickness due to waves or (c) turbulent with waves moving down the interface. Many experimental results and several analytical solutions for heat transfer to falling liquid films have been reported in the literature with relation to such cases as filmwise condensation and annular two-phase flow.

For a laminar fully developed liquid film, with constant heat flux along the wall, the analytical solutions are obtained [1,2] by introducing a film thickness derived by Nusselt. In this case when all of the heat transferred from the heating surface is absorbed in the liquid film,

$$h^* = 2.27 \, Re^{-1/3} \tag{1}$$

and in another case where there is heat transfer away from the film surface,

$$h^* = 1.76 \, Re^{-1/3}. \tag{2}$$

An extensive experimental study of heat transfer coefficient to subcooled liquid films was performed by Wilke [3]. The heat transfer coefficient to a liquid film flowing downwards on the outside of a 2400 mm long vertical tube was measured. A vertical tube was heated by hot water flowing through the inside of the tube. The experimental data were obtained over a wide range of liquid Prandtl numbers from $Pr = 5.4$ to $240$, by using water/glycol mixtures of various concentrations. In reference [4], the empirical equations derived by Wilke [3] were transformed into relations between $h^*$ and $Re$ using the film thickness derived by Nusselt for $Re < 1600$ and that given by Brauer [5] for $Re \geq 1600$. These relations for water are expressed as:

$$Re \leq 2460 \ Pr^{-0.646}, \quad h^* = 1.76 \ Re^{-1/3}; \tag{3}$$

$$2460 \ Pr^{-0.646} \leq Re < 1600, \quad h^* = 0.0323 \ Re^{1/5} \ Pr^{0.344}; \tag{4}$$

$$1600 \leq Re < 3200, \quad h^* = 0.00102 \ Re^{2/3} \ Pr^{0.344}; \tag{5}$$

$$3200 \leq Re, \quad h^* = 0.00871 \ Re^{2/5} \ Pr^{0.344}. \tag{6}$$

The experimental data obtained in reference [4] were compared with equations (3-6). In a low heat flux range, the experimental data showed a trend similar to that predicted by equations (3-4), although data were generally overpredicted as shown in Figure 4 (reference [4]). In a higher heat flux range, near the film breakdown (the film breakdown is discussed in the next section), the measured values of h are widely scattered and much lower than Wilke's predictions. The experimental data of heat transfer coefficient near film breakdwon obtained in references [4,6] are shown in Figure 1, and they are generally below the values given by equation (1). Similar data for a horizontal tube are shown in Figure 11 (reference [7]). Near the film breakdown condition, the distortion of the film (i.e. a film consists of crests and valleys) was noticed and the film thickness is far from uniform [7]. The distortion of the film reduces the heat transfer coefficient. Since the thickness of the film in the valleys is very small, the temperature of the liquid in these regions approaches the tube surface temperature. Also, the falling-film velocity in the valleys is lower than in the crests due to the gradients in surface tension and liquid density.

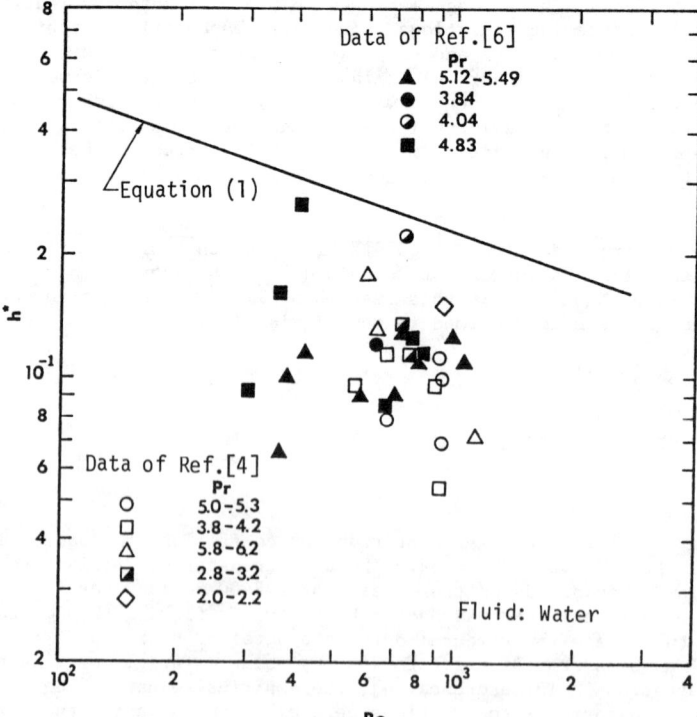

Fig. 1. Heat transfer coefficient vs. film Reynolds number for vertical tubes

Due to the effect of surface tension, the Weber number, We, affects the heat transfer coefficient as discussed in reference [6,4].

Heat transfer to saturated liquid films was studied in references [8-15]. Sernas [8] reported experimental data for thin film evaporation on horizontal tube and has correlated his data using common dimensionless groups:

$$\frac{\bar{h}}{k}\left(\frac{v^2}{g}\right)^{1/3} = C\left(\frac{4\Gamma}{\mu}\right)^{0.24}\left(\frac{v}{\alpha}\right)^{0.66} \qquad (7)$$

where C = 0.01925 for D = 25 mm and C = 0.01729 for D = 50 mm. This correlation is not tested outside the range of Sernas' data. Chun and Seban [10] developed a well-known correlation for heat transfer to evaporating liquid films on smooth vertical tubes, which reads:

$$\text{laminar:} \quad h_c = 0.821\left(\frac{v^2}{k^3 g}\right)^{-1/3}\left(\frac{4\Gamma}{\mu}\right)^{-0.22} \qquad (8a)$$

$$\text{turbulent:} \quad h_c = 3.8 \times 10^{-3}\left(\frac{v^2}{k^3 g}\right)^{-1/3}\left(\frac{4\Gamma}{\mu}\right)^{0.4}\left(\frac{v}{\alpha}\right)^{0.65} \qquad (8b)$$

Both equations give the local heat transfer coefficient as a function of film Reynolds number, $4\Gamma/\mu$. In the laminar flow range, equation (8a) includes the influence of waves and ripples which have the effect of increasing heat transfer by reducing the effective film thickness [16]. As mentioned in reference [16], equations (8a) and (8b) should apply equally well for either constant heat flux or constant wall temperature boundary conditions. In laminar flow $h_c$ is essentially equal to $k/\delta$, regardless of boundary conditions; and in turbulent flow the behavior should be essentially similar to that in pipes where it is known that Nusselt numbers for constant heat flux and constant wall temperature are approximately equal for Pr > 0.5. The interaction of correlation (8a) and (8b) yields the "transition" Reynolds number:

$$\left(\frac{4\Gamma}{\mu}\right)_{tr} = 5800\left(\frac{v}{\alpha}\right)^{-1.06}. \qquad (9)$$

As indicated in reference [16], this equation should not be regarded as an actual indication of the transition from laminar to turbulent regime, but only as the point of transition from one correlation to the other.

A thermal developing length $L_d$ is required for the falling film to be superheated from the saturation temperature to a fully developed linear profile. In the thermal developing region all the heat transferred from the wall goes into superheating the liquid film, and no evaporation occurs [16]. Assuming a fully developed laminar falling film velocity profile together with a linear temperature profile at the end of the developing region, the average heat transfer coefficient in the developing region was calculated [16] from energy balance as:

$$\bar{h}_d = \frac{3}{8} C_p \frac{\Gamma}{L_d} \qquad (10)$$

where

$$L_d = \frac{\Gamma^{4/3}}{4\pi\rho\alpha} \sqrt[3]{\frac{3\mu}{g\rho^2}} \tag{11}$$

The value of $L_d$, given by the above equation, was derived assuming a constant film thickness, given by the Nusselt expression and employing an approximate integral method [16]. The average heat transfer coefficient over the length L is then [16]:

$$\bar{h} = h_d \frac{L_d}{L} + h_c\left(1 - \frac{L_d}{L}\right) \tag{12}$$

where $h_d$, $L_d$ and $h_c$ are given by equations (10, 11, 8) respectively. The comparison of equation (12) with various experimental data is shown in Figure 2, together with other correlations. The variance between the Chun-Seban correlation and the correlation (12) reflects the influence of the thermal developing region. The large divergance is noted at high flow rates where the developing length is most important. The Dukler prediction is reasonably good (see Figure 2) at higher flow rates. The Chun-Seban relation was originally developed for falling films on vertical surfaces rather than on horizontal tubes.

If the liquid film superheat is sufficiently high, boiling may occur in the film. The greater heat fluxes are attainable with boiling in thin films than with boiling in pools [17]. The boiling heat transfer coefficient given by Rohsenow's correlation [18]:

$$h_b = \frac{\mu\, h_{fg}}{C_{sf}^3 \sqrt{\frac{g_o \sigma}{g\rho}}} \left(\frac{C_p}{h_{fg}\, Pr^s}\right)^3 \Delta T^2, \tag{13}$$

(where s = 1 for water, s = 1.7 for all other fluids) was suggested to be applicable for the film (in preference to nucleate boiling correlation for stagnant liquid layers [16]) when combined with equation (12) and assuming a weighting factor of unity, i.e.

$$\bar{h} = h_b + h_d \frac{L_d}{L} + h_c\left(1 - \frac{L_d}{L}\right). \tag{14}$$

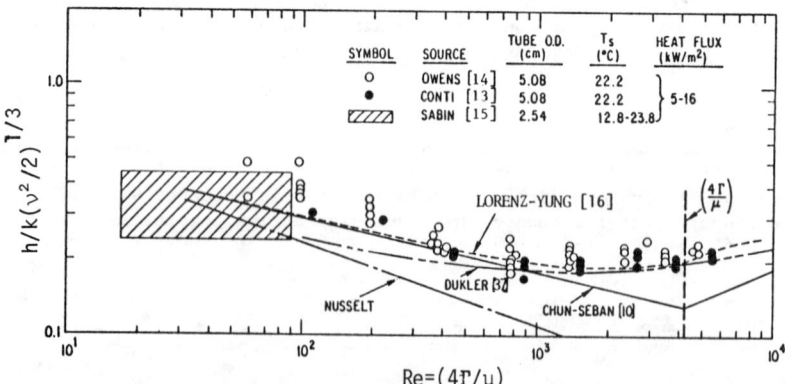

Fig. 2. Comparison of predictions with experimental data for ammonia evaporating on smooth horizontal tubes

The parameter $C_{sf}$ in equation (13) is a function of fluid-surface combination.

Table 1 summarizes the available experimental data in the literature concerning heat transfer to falling liquid films. The range of liquid Reynolds numbers and the type of heating are indicated in this table. A recent survey of a transport to falling films is given in references [19, 25, 26].

## 2. FALLING LIQUID FILM BREAKDOWN

When the flow rate of a thin film running over a heated surface is very low and/or the surface is very hot, the liquid film breaks down. That is, dry patches form on the surface. These dry patches can be rewetted by upstream disturbances at low heat fluxes, but at higher heat fluxes the dry areas eventually persist. This results in an abrupt decrease in the heat transfer coefficient, and it may also cause overheating of the surface.

One of the main difficulties encountered in designing falling film heat exchangers is maintaining the complete wettability of the tubes. The minimum wetting rate (i.e., the minimum liquid flow rate required to wet the tube surface at a given local heat flux ) should be known in order to determine the minimum liquid recycling ratio. This is especially important in the case of heat exchangers like those used for solar energy application. Such exchangers require very large heat transfer areas to compensate for the low $\Delta T$'s, and the pumping power needed to recycle the liquid affects the efficiency and feasibility of these systems.

The only experimental data for the liquid film breakdown on horizontal tubes are given in reference [7]. On the other hand, experimental investigations of film breakdown on vertical tubes have been widely conducted and the results reported in [4,6,11,27-30]. Since the wetting flow rates depend very much on the method of introducing the liquid film on the wall, the data for horizontal and vertical tubes will not be the same.

The experimental data for the film breakdown on horizontal tubes are shown in Figures 3 and 4. The film breakdown measurements were made after a steady state was achieved, and then the heat flux to the test section was increased in small steps (a steady state was achieved between steps) until dry patches were visually observed on the tube surface [7]. (This point is the lowest part of the bandad data, e.g., point A in Figure 3.) After a certain period of time, dry patches that did not rewet themselves were manually rewetted. The heat flux was again increased until the next dry patch (or patches) appeared. When manually rewetted dry spots continued to reform over a period of time, a permanent dry patch was defined (the upper limit of the banded data, i.e., point B in Figure 3). The same procedure was then repeated at a different flow rate. In some cases we were able to register only a permanent dry patch (e.g., point C in Figure 3). Also, a few data points for stable dry patches were obtained by decreasing the flow rate slightly from its initial value instead of increasing the heat flux (e.g., point D in Figure 3). The dry patches appear first on the sides of the tube. If q is increased, the dry patches will grow and eventually form a dry area around the tube - like a ring.

The effect of tube spacing (i.e., distance between horizontal tubes) on the breakdown heat flux is shown in Figure 4. It is reasonable to assume that the velocity of the liquid falling from the top tube to the next tube is approximately $(2gH)^{0.5}$. Therefore, increasing H increases the kinetic energy of the falling film and reduces the tendency of dry patches to form at a given surface

Table 1. Experimental Investigations of Heat Transfer to Falling Liquid Film

| Investigator | Liquid | Range of Re | Comments |
|---|---|---|---|
| 1. Bays & McAdams (1937) | oil | 2 - 2000 | Heating the wall by condensation $280 < Pr < 7000$ |
| 2. Wilke (1962) | mixture of water and glycol | 8 - 10,000 | Heating the wall by condensation $5.4 < Pr < 210$ |
| 3. Ponter (1968) | water with 0.15% Teepol L solution and 50% aqueous glycerol solution | 20 - 2800 | Heating the vertical tube internally by hot water flow $25°C < T_a < 70°C$ |
| 4. Chun & Seban (1971) | water | 320 - 21,000 | Electrically heated vertical tube $1.77 < Pr < 5.7$ |
| 5. Shah & Darby (1973) | water | 5825 - 8809 | Heating the wall by condensation, The effect of surface was considered. |
| 6. Gimbutis (1974) | water | 1800 - 45,000 | Electrically heated vertical tube $4.3 < Pr < 8.4$ |
| 7. Fletcher (1975) | sea water | 800 - 6,000 | Electrically heated horizontal tube $49°C < T_s < 126.7°C$ |
| 8. Faghre (1976) | water | 2000 - 21,000 | Electrically heated horizontal tube $1.77 < Pr < 6.335$ |
| 9. Sabin (1978) | ammonia | 20 - 90 | Electrically heated horizontal tube $12.8°C < T_s < 23.8°C$ |
| 10. Conti (1978) | ammonia | 100 - 8,000 | Electrically heated horizontal tube $T_s = 22.2°C$ |
| 11. Owens (1978) | ammonia | 120 - 10,000 | The same as Conti (1978) |
| 12. Fujita & Ueda (1978) | water | 500 - 5000 | Electrically heated vertical tube $2.2 < Pr < 6.2$ |
| 13. Sernas (1979) | water | 1151 - 6044 | Electrically heated horizontal tube $1.453 < Pr < 3.717$ |
| 14. Ganic & Roppo (1980) | water | 50 - 200 | Electrically heated horizontal tube $26.7°C < T_i < 50°C$ |
| 15. Ganic & Roppo (1980) | water | 300 - 1000 | Electrically heated vertical tube $3.84 < Pr < 5.49$ |

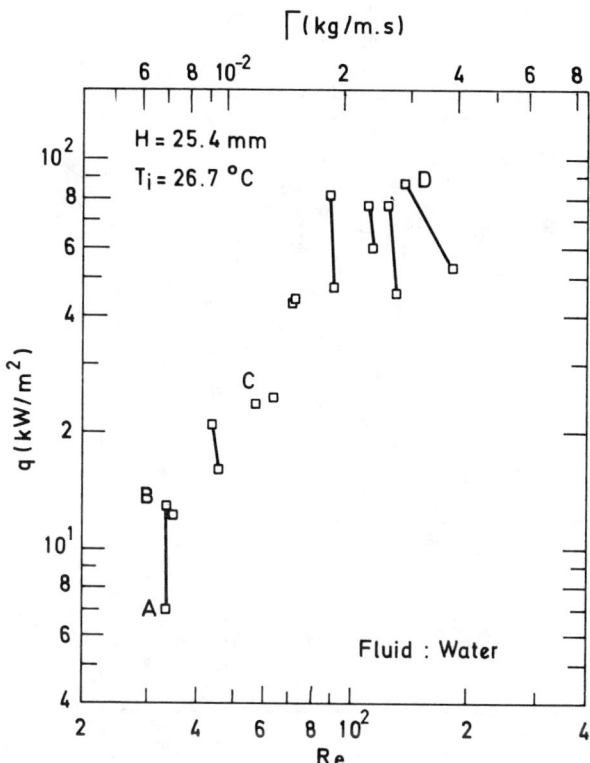

Fig. 3. Film breakdown heat flux vs. liquid film flow rate for horizontal tube [7].

heat flux. This is supported by the experimental data for the 12.7 mm (0.5 inch) and 25.4 mm (1.0 inch) tube spacing.

However, the breakdown data for 50.8 mm (2.0 inch) tube spacing appear to contradict this thinking, and this is discussed further in reference [7]. In any case, for most heat-exchanger applications the data on the two smaller spacings (25.4 and 12.7 mm) are more relevant.

The effect of film inlet temperature on the film breakdown is associated with the effect of surface tension. At higher inlet temperatures the surface tension is smaller and the liquid film on the horizontal tube has a more uniform thickness. It is a well-known phenomenon that on a wavy surface, which consists of valley and crest, the region of greater film thickness (i.e., the crest) has a lower surface temperature and subsequently a higher surface tension than the thinner region. Therefore, the surface tension force draws liquid from the valley into the crest. Eventually, a liquid deficiency occurs in the thinner regions, causing film breakdown, i.e., forming a dry patch on the surface. At higher flow rates this distortion of the liquid film, due to the temperature difference set up in the film, is significantly reduced by the inertial force of the film flow.

The experimental data on the breakdown heat flux for vertical tubes are summarized in Figure 5. As in the case of horizontal tubes, the film breakdown heat flux increases with flow rate. At higher flow rates the liquid film

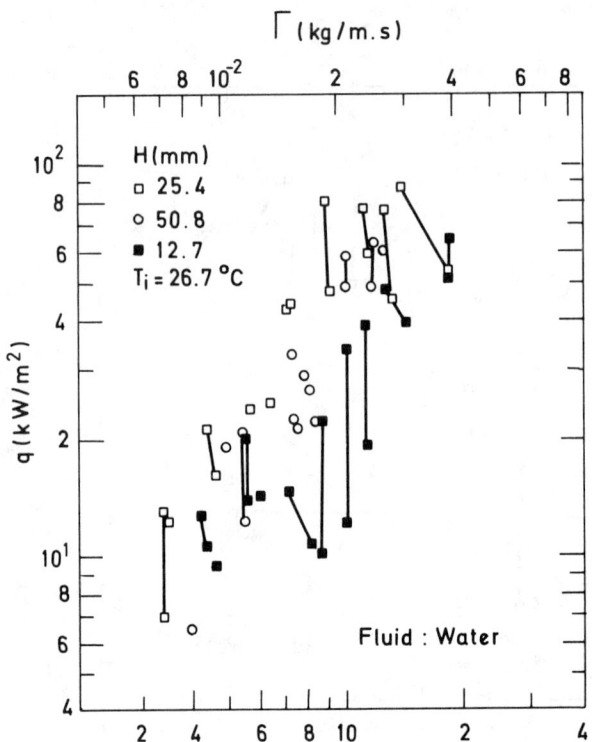

Fig. 4. Film breakdown heat flux vs. liquid film flow rate (effect of tube spacing [7]) for horizontal tube.

carries more energy (kinetic plus surface), and higher heat fluxes are required to prevent rewetting of dry patches. This is in qualitative agreement with the results of [31]. Most of the data on film breakdown and heat transfer coefficient are presented in terms of the film Reynolds number ($4\Gamma/\mu$). Some of the data (Figure 5) are presented in terms of the liquid film flow rate $\Gamma$ only, since quantity ($4\Gamma/\mu$) may obliterate the effect of inlet temperature or surface tension because $\mu$ strongly depends on T. This indicates that $4\Gamma/\mu$ is not always a good correlating parameter, at least by itself.

The stability of dry patches may be effected by the heating procedure of heat transfer surface. All data discussed here are related to the prescribed heat flux condition since the test section was heated electrically.

Falling film breakdown on an adiabatic vertical surface and subsequent rewetting of the dry patch were theoretically studied initially by Hartley and Murgatroyd [32], who investigated the equilibrium of forces acting at the upsteam stagnation point of the dry patch. The force balance included the inertial force due to bringing the upstream liquid to rest at the stagnation point, and the surface tension force due to a nonzero contact angle between the liquid and solid surface. By solving the force balance equation, the undisturbed film thickness (film thickness upstream of the dry patch) was obtained with the contact angle as a parameter. This work has been improved by many researchers who included the gravitational body force, the thermocapillary force (a force resulting from the variation of surface tension with temperature), the wall shear

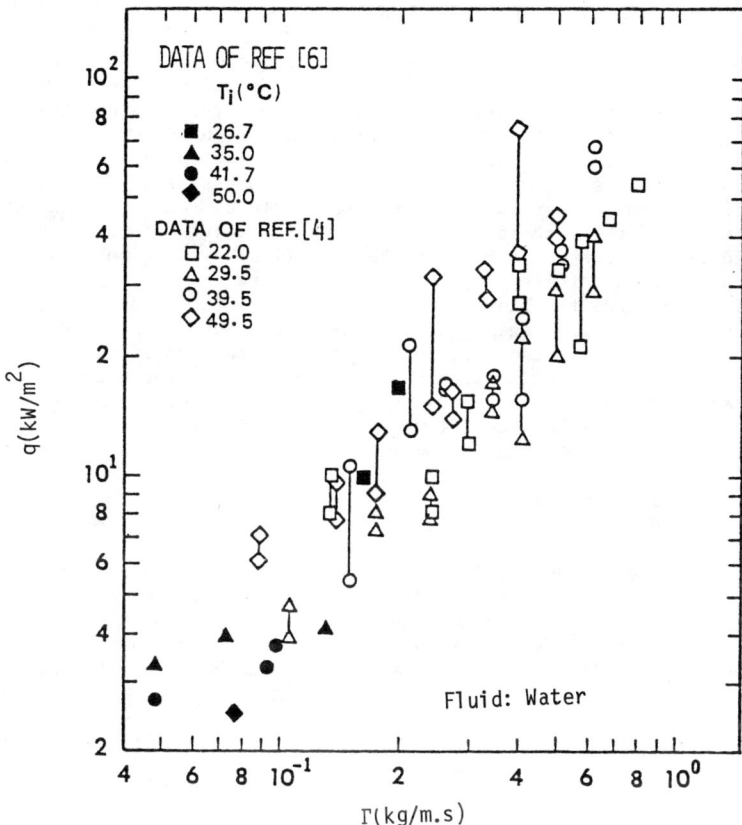

Fig. 5. Film breakdown heat flux vs. film flow rate for vertical tube [6].

force, the vapor thrust if there is evaporation at the film interface, and the interfacial shear force [28,30,33].

A second model, also reported in [32], was based on the assumption that a stable film configuration corresponds to a minimum power transmission by the film in the form of kinetic and surface energy. A substantial improvement of this criterion was made in [31].

A third approach employed in the film breakdown studies is classified as the small perturbation theory model and is summarized in [30].

Overall, the volume of film breakdown data and comparison to theories is small and inconclusive. A reasonable agreement of data with a model was reported in reference [30], providing a large contact angle was assumed. However, on the same subject an opposite conclusion to the above was given in reference [11] concerning the value of the contact angle. It appears that at the present time there is a need for systematic experimental data for a contact angle under dry patch conditions.

## 3. VAPOR/LIQUID INTERACTION AND ENTRAINMENT

The purpose of this section is to review the basic mechanisms of vapor/liquid interaction and entrainment on the shell side of falling film evaporators. The results presented here are basically drawn from our recent study [34]. Compared to the vertical configuration, the horizontal design is considerably more vulnerable to vapor/liquid interaction. In a typical horizontal tube falling film evaporator, fluid is fed on the top of vertical banks of horizontal tubes. In the absence of vapor crossflow the unevaporated fluid from any given tube will fall directly on the next lower tube. The qualitative structure of the falling liquid is dependent primarily on the feed flowrate. At relatively low feed flowrates the liquid falls in the form of droplets (Figure 6a), and at relatively high flowrates the liquid falls in the form of columns (Figure 6b). The transition from one flow regime to the other occurs over a relatively large range of $\Gamma$. From the high-speed pictures of the flow regimes it was concluded that the droplet regime can exist within what appears to be the column regime to the human eye [7]. At higher tube spacing values, the columns change to droplets [7]. At still higher feed flowrates the liquid falls as unstable sheets and columns. As mentioned in [34], these high flowrates are generally outside the range of interest for most falling film evaporators because the required pumping power becomes excessive, and the thermal performance remains practically unchanged.

As mentioned above, when the liquid flow rate is small the liquid falls from one tube to the next in droplet form (Figure 6a). The droplets are generated at discrete points along the underside of the horizontal tube. This problem on horizontal tubes were related to the classic hydrodynamics problem known as Taylor instability. Based on that, the column and droplet spacing is given as

$$\lambda = 2\pi \sqrt{n\sigma/\rho g} , \qquad (15)$$

and experimental data suggested n = 2.

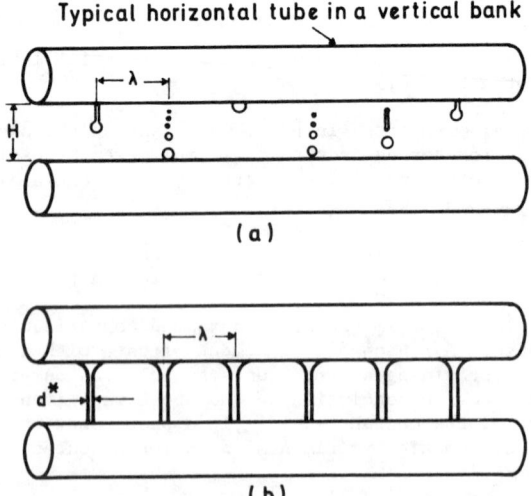

Fig. 6. Liquid falling in (a) droplet mode; and (b) stable column mode [34].

The liquid droplets falling from the underside of the tube were photographed in reference [34] by high speed camera at 200 frames/s. As a relatively large drop detached from the film, a long narrow tail was observed, which, by the well-known Rayleigh instability, eventually breaks up into four or five smaller droplets. It was estimated that the volume of small droplets (secondary droplets) compared to the volume of the large droplet (primary droplet) is about 19 percent. From the experimental data of water and ethyl alcohol the following correlations were deduced:

$$0.24 < d_s/d_p < 0.46, \qquad (16)$$

$$d_p = c_1 \sqrt{\sigma/\rho g} \qquad (17)$$

where $c_1 = 3$. Note that the general form of equation (17) is also applicable to to the bubble formation at an orifice [35].

As a result of vapor generation, a vapor crossflow exists on the shell side. The droplets and columns falling from a given tube may be partially or totally deflected away from the next tube by vapor crossflow, thereby causing liquid redistribution and incomplete wetting of lower tubes in the bank. Using a single droplet trajectory model it was shown in reference [34] that the minimum vapor crossflow velocity $U_g$ required to deflect the droplet of size d (so that the droplet will not hit the lower tube) is given as

$$U_g = \left(\frac{3}{2} \frac{\rho}{\rho_g} d\, g\right)^{1/2} \left[\frac{P}{D}\left(\frac{P}{D} - 1\right)\right]^{-1/4}. \qquad (18)$$

The minimum vapor crossflow velocity beyond which the liquid column will no longer impinge upon the next lower tube is given as [34]

$$U_g = \left[\left(\frac{\tan \alpha}{\cos \alpha}\right) \frac{(4\Gamma \lambda g)}{(2gH)^{0.5}\, C_d\, d^*\, \rho_g}\right]^{0.5} \qquad (19)$$

where:

$\lambda$ is given by equation (15); $C_d \approx 1$;

$$\alpha = \tan^{-1}\left\{0.5\left[\frac{P}{D}\left(\frac{P}{D} - 1\right)\right]^{-0.5}\right\}; \qquad (20)$$

$$d^* = \left(\frac{8\lambda \Gamma}{\pi \rho}\right)^{0.5} (2gH)^{-0.25} \qquad (21)$$

Equation (19) is derived making the force balance on the liquid column (i.e. a balance between the drag force due to the vapor crossflow and the weight of the liquid column). As pointed out in reference [34], the deflection of liquid away from the next lower tube does not necessarily imply a loss in heat transfer performance because the liquid may eventually experience a good "hit" on a tube in the adjacent column. However, it is expected that good wetting of the tubes, which is essential to good heat transfer performance, can be better achieved by a well-controlled liquid flow from one tube to the next rather than relying on random hits on tubes in adjacent columns.

In addition to these deflection mechanisms, a number of entrainment mechanisms can also occur. If nucleate boiling is present in the film, a mist of small droplets is generated as bubble burst through the film, and small droplets are readily entrained by the flowing vapor. Newitt et al. [36] have found that bubble size is an important parameter affecting entrainment. Droplet generation resulting from boiling in the film appears to be significant only at heat fluxes appreciably higher than those of interest for falling film evaporators.

An entrainment mechanism common also to both horizontal and vertical falling film evaporators is that of shearing or stripping of the film from the tube surface. It was shown in reference [34] that this type of entrainment mechanism can be important for vertical design and for horizontal design. However, the vapor velocity ($U_g$) given by equations (18) and (19) (which is an upper design limit for the crossflow velocity) is below the critical vapor velocity required for stripping.

## 4. CONCLUDING REMARKS

From this survey on the heat transfer and fluid flow in falling film evaporators the following major conclusions can be drawn:

1. The heat transfer coefficient at conditions prior to film breakdown should be studied further. The predicted values for a heat transfer coefficient are generally much higher than the experimentally determined values in this region.

2. The Lorenz-Yung correlation should be used if the thermal entrance region in a liquid film is important. Also, this correlation is suitable for predicting a heat transfer coefficient when a boiling within the film is present.

3. It appears that the Wilke's relations should be limited to the liquid films with a uniform thickness around the periphery and with small temperature differences in it (i.e. a small heat flux). This conclusion is consistent with results of reference [4].

4. There are basically two flow regimes associated with liquid film falling from one horizontal tube to the next: a) the droplet regime which is related to lower flow rates and b) the column regime related to higher flow rates. It is logical to expect that at extremely high film flow rates the columns will disappear, and the liquid will fall from one tube to the next as a sheet.

5. As the film flow rate increases, the heat flux needed to cause a film breakdown (i.e., a dry patch) becomes higher.

6. Equations (17), (18) and (19) are useful for evaluating vapor/liquid interaction and entrainment in falling film evaporators.

The results presented in this paper should prove to be useful tools for the thermal design and performance evaluation of falling film shell-and-tube evaporators.

## ACKNOWLEDGMENT

Figures 2 and 6 are provided by Drs. J. Lorenz and D. Yung of Argonne

National Laboratory. Miss Martha Gaarsoe has prepared the typescript of this lecture in record time.

NOMENCLATURE

- $C_p$ — specific heat, J/kg °C
- $C_d$ — vapor drag coefficient
- $d$ — droplet diameter, m
- $d^*$ — effective column diameter, equation (21), m
- $D$ — tube diameter, m
- $g$ — gravitational acceleration, m/s$^2$
- $h$ — local heat transfer coefficient, W/m$^2$°C
- $\bar{h}$ — average heat transfer coefficient over the tube circumference, W/m$^2$°C
- $h^*$ — dimensionless heat transfer coefficient, $\bar{h}(\nu^2/gk^3)^{1/3}$, W/m$^2$°C
- $h_{fg}$ — enthalpy of evaporation, J/kg
- $H$ — tube spacing, m
- $k$ — thermal conductivity, W/m°C
- $L$ — length, m
- $P$ — tube pitch, m
- $Pr$ — Prandtl number, $\mu C_p/k$
- $q$ — film breakdown heat flux, W/m$^2$
- $Re$ — film Reynolds number, $4\Gamma/\mu$
- $T$ — temperature, °C
- $\Delta T$ — temperature difference, °C
- $T_a$ — average film temperature, °C
- $T_i$ — film inlet temperature, °C
- $T_s$ — film saturation temperature, °C
- $\alpha$ — thermal diffusivity, m$^2$/s
- $\mu$ — dynamic viscosity, Ns/m$^2$
- $\nu$ — kinematic viscosity, m$^2$/s
- $\rho$ — density, kg/m$^3$
- $\sigma$ — surface tension, N/m
- $\Gamma$ — liquid film flow rate, per unit width, kg/m.s
- $\delta$ — film thickness, m

Subscripts

- b — boiling
- g — gas
- i — inlet
- p — primary
- s — secondary

REFERENCES

1. Ueda, T. and Tanaka, T. 1974. Studies of liquid film flow in two-phase annular and annular-mist flow regimes (Part 1, Downflow in a vertical tube), Bull. J.S.M.E., Vol. 17, No. 107, pp. 603-613.

2. Ueda, T. and Nose, S. 1974. Studies of liquid film flow in two-phase annular and annular-mist flow regimes (Part 2, Upflow in a vertical tube), Bull. J.S.M.E., Vol. 17, No. 107, pp. 614-624.

3. Wilke, W. 1962. Wärmeübergang und Rieselfilme, ForshHft. Ver. Dt. Ing. 490, B28.

4. Fujita, T. and Ueda, T. 1978. Heat transfer to falling liquid films and film breakdown - I, Int. J. Heat Mass Transfer, Vol. 21, pp. 97-108.

5. Brauer, H. 1956. Strömung und Wärmeübergang bei Rieselfilmen, ForschHft. Ver. Dt. Ing. 457, B22.

6. Ganic, E.N. and Roppo, N.M. 1980. A note on heat transfer to falling liquid films on vertical tubes, Letters in Heat Mass Transfer, Vol. 7, pp. 145-154.

7. Ganic, E.N. and Roppo, N.M. 1980. An experimental study of falling liquid film breakdown on a horizontal cylinder during heat transfer, J. Heat Transfer, Vol. 102, May (1980).

8. Sernas, V. 1979. Heat transfer correlation for subcooled water films on horizontal tubes, J. Heat Transfer, Vol. 101, pp. 176-178.

9. Liu, P. 1975. Ph.D. Thesis, Mech. Eng. Dept., Univ. of Wisconsin, Madison.

10. Chun, K.R. and Seban, R.A. 1971. J. Heat Transfer, Vol. 93, pp. 391-396.

11. Fujita, T. and Ueda, T. 1978. Heat transfer to falling liquid films and film breakdown - II, Int. J. Heat Mass Transfer, Vol. 21, pp. 109-118.

12. Fletcher, L.S., Sernas, V., and Parken, W.H. 1975. Ind. Eng. Chem., Process Des. Dev., Vol. 14, No. 4, pp. 411-416.

13. Conti, R.J. 1978. Experimental investigation of horizontal-tube ammonia film evaporations with small temperature differentials, Proc. Fifth Ocean Thermal Energy Conversion Conference, Miami Beach, Florida.

14. Owens, W.L. 1978. Correlation of thin film evaporation heat transfer coefficients for horizontal tubes, Proc. Fifth Ocean Thermal Energy Conversion Conference, Miami Beach, Florida.

15. Sabin, C.M. and Poppendiek, H.F. 1978. Film evaporation of ammonia over horizontal round tubes, Proc. Fifth Ocean Thermal Energy Conversion Conference, Miami Beach, Florida.

16. Lorenz, J.J. and Yung, D. 1979. A note on combined boiling and evaporation of liquid films on horizontal tubes, J. Heat Transfer, Vol. 101, pp. 178-180.

17. Nishikawa, K., Kusada, H., Yamasaki, K. and Tanaka, K. 1967. Bull. J.S.M.E. Vol. 10, No. 38, pp. 328-338.

18. Rohsenow, W.M. 1972. J. Heat Transfer, Vol. 94, pp. 255-256.

19. Seban, R.A. 1978. Transport to falling films, HEAT TRANSFER 1978, Vol. 6, pp. 417-428.

20. Bays, G.S. and McAdams, W.H., 1937. Ind. Eng. Chem., Vol. 29, p. 1240.

21. Gimbutis, G. 1974. Heat transfer of a turbulent vertically falling film, HEAT TRANSFER 1974, Vol. 2, pp. 85-89.

22. Faghri, A. 1976. Ph.D. Thesis, Univ. of Calif. at Berkeley, Berkeley, CA.

23. Shah, B.H. and Darby, R. 1973. The effect of surfactant on evaporative heat transfer in vertical film flow, Int. J. Heat Mass Transfer, Vol. 16, p. 1889.

24. Ponter, A.B. and Davies, G.A. 1968. Heat transfer to falling films, Chem. Engng. Sci. Vol. 23, pp. 664-665.

25. Faghri, A. and Payvar, P. 1979. Transport to thin falling films, Reg. J. Energy Heat Mass Transfer, Vol. 1, No. 2, pp. 153-173.

26. Rothfus, P.P. and Lavi, G.H. 1978. Vertical Falling Film Heat Transfer: a literature survey, Proc. Fifth Ocean Thermal Energy Conversion Conference, Miami Beach, Florida.

27. Hallett, V.A. 1966. Surface phenomena causing breakdown of falling liquid films during heat transfer, Int. J. Heat Mass Transfer, Vol. 9, pp. 283-294.

28. Simon, F.F. and Hsu, Y.Y. 1970. Thermocapillary induced breakdown of falling liquid films, NASA TN D - 5624.

29. Munakata, T., Watanabe, K. and Miyashita, K. 1975. Minimum wetting rate on wetted-wall column, Journal of Chem. Eng. of Japan, Vol. 8, No. 6, pp. 440-444.

30. Bankoff, S.G. and Chung, J. 1978. Dryout of a thin heated liquid film, Proc. of Int. Heat Mass Transfer Center Seminar - Dubrovnik, Hemisphere Publishing Co., 1978.

31. Mikielewicz, J. and Moszynski, J.R. 1976. Minimum thickness of a liquid film flowing vertically down a solid surface, Int. J. Heat Mass Transfer, Vol. 19, pp. 771-776.

32. Hartley, D.E. and Murgatroyd, W. 1964. Criteria for the break-up of thin liquid layers flowing isothermally over solid surfaces, Int. J. Heat Mass Transfer, Vol. 7, pp. 1003-1015.

33. Zuber, N. and Staub, F.W. 1966. Stability of dry patches forming in liquid films flowing over heated surface, Int. J. Heat Mass Transfer, Vol. 9, pp. 897-905.

34. Yung, D., Lorenz, J.J. and Ganic, E.N. 1980. Vapor/liquid interaction and entrainment in falling film evaporators, J. Heat Transfer, Vol. 102, No. 1, pp. 20-25.

35. Wallis, G.B. 1969. One-dimensional two-phase flow, McGraw-Hill, New York.

36. Newitt, D.M., Dombrowski, N. and Knelman, F.H. 1954. Liquid entrainment: 1. the mechanism of drop formation from gas or vapor bubbles, Trans. Inst. Chem. Eng. Vol. 32, pp. 244-261.

37. Dukler, A.E. 1960. Chem. Eng. Prog. Symposium Series, Vol. 56, No. 30, pp. 1-10.

# Thermal Design Theory for Regenerators

**RAMESH K. SHAH**
Harrison Radiator Division
General Motors Corporation
Lockport, New York 14094 USA

ABSTRACT

Starting with an introduction on the similarity and differences between rotary and fixed-matrix regenerators, an in-depth analysis is presented for thermal design that includes the following: exchanger variables, derivations and physical significance of dimensionless groups associated with the $\varepsilon$-$N_{tu,o}$ and $\Lambda$-$\Pi$ methods, results and correlations for counterflow regenerators using $\varepsilon$-$N_{tu,o}$ and $\Lambda$-$\Pi$ approaches, influence of rotation, influence of longitudinal wall heat conduction, influence of wall thermal resistance, and influence of bypass and carryover leakages on the regenerator effectiveness.

1. INTRODUCTION

In an earlier lecture, indirect contact type exchangers have been classified as direct transfer type, storage type and fluidized bed exchangers[1]. In a direct transfer type heat exchanger, two fluids are separated by a thin wall through which heat flows. Although a simultaneous flow of both fluids is required in the exchanger, there is no mixing of two fluids. There are generally no moving parts in the exchanger. This type of exchanger is referred to as a recuperator. In contrast, for a storage type exchanger, the same flow passages are occupied by one of the two fluids. The heat transfer surface is of cellular structure usually referred to as a matrix. During the hot gas flow through a passage, thermal energy is stored in the matrix wall. During the cold gas flow through the same passage later, the matrix wall delivers thermal energy to the cold fluid. Thus heat is not transferred through the wall as in a recuperator, but is alternately stored and rejected by the matrix wall. This storage type exchanger is generally referred to as a regenerator. It is exclusively used for gas-to-gas heat transfer applications.

In order to have a continuous operation in a regenerator, either gas flows must be diverted to and from the fixed matrices as in a fixed-matrix regenerator, or the matrix must be moved periodically in and out of the fixed streams of gases as in a rotary regenerator. Thus for a continuous operation, a fixed-matrix regenerator has at least two matrices operated in parallel, but usually three or four [1]. Proper opening and closing of valves at appropriate time intervals is needed to ensure the design performance. The outlet fluid temperatures vary with time in a fixed-matrix regenerator. A disk-type rotary regenerator is shown in Fig. 1a. The outlet fluid temperatures in this regenerator vary across the flow area and are independent of time. In spite of these subtle differences, if the elements of a regenerator (either rotary or fixed-matrix) are considered fixed in the space by choosing a proper reference coordinate system, the heat transfer analysis is identical for both regenerators for arriving at the regenerator effectiveness.

In this paper, we will describe the methods for heat transfer analysis of regenerators that have attained "steady-state" or regular <u>periodic-flow</u> conditions. The transient response will be considered in a paper presented later in this lecture series. We will start with the usual idealizations made in the design theory followed by the derivation of governing differential equations. These equations will be made dimensionless to arrive at the dimensionless groups for the $\varepsilon$-$N_{tu,o}$ and $\Lambda$-$\Pi$ methods. The solutions to the differential equations will then be presented in terms of the regenerator effectiveness as a function of the pertinent dimensionless groups for the counterflow and parallel flow regenerators. Finally the influence of longitudinal wall heat conduction, wall thermal resistance and carryover leakage on the regenerator heat transfer performance will be presented.

## 2. IDEALIZATIONS FOR THE REGENERATOR PERIODIC-FLOW THEORY

The following idealizations are built into the derivation of the governing differential equations presented in the next section.

1. Heat transfer between the exchanger and surroundings is negligible. There are no thermal energy sources within the exchanger. No phase change occurs in the exchanger.

2. The mass flow rates of both fluids, although may be different, do not vary with time during the respective flow periods. The fluids are uniformly distributed.

3. The velocity and temperature of each fluid at the inlet are uniform over the flow cross section and are constant with time.

4. The fluid velocity on each side is considered constant with position, temperature and time throughout the matrix.

5. The heat transfer coefficients ($h_h$ and $h_c$) between the fluids and the matrix wall are <u>constant</u> (with position, temperature, and time) throughout the exchanger.

6. The surface area of the matrix as well as the rotor mass is uniformly distributed.

7. The temperatures of both fluids and the matrix are dependent upon the axial coordinate $x$ and the time $\tau$.

8. The thermal properties of both fluids and the matrix wall material are constant, independent of time and position.

9. Heat conduction in the fluids in longitudinal and transverse directions is neglected.

10. The thermal conductivity of the matrix wall is zero in the flow direction (no longitudinal heat conduction) and infinite in the transverse (wall thickness) direction. Thus the temperature across the wall thickness is uniform at a cross section and the wall thermal resistance is treated as zero.

11. No mixing of the fluids occurs during the switch from hot to cold flows. Also the fluid carryover is negligible relative to the flow rates $W_h$ and $W_c$.

The first through sixth idealizations parallel those usually made in the design theory for direct transfer type exchangers (recuperators). In the third idealization, the fluid velocities and temperatures are considered uniform over the entering cross sections. A deterioration in heat transfer occurs for non-uniform entering velocity and temperature profiles. Generally the temperature is uniform; however, the velocity profile may be nonuniform due to the header design. The influence of the nonuniform velocity profile at the entrance is considered in [22].

The matrix and fluid temperatures depend upon $x$ and $\tau$ coordinates in a fixed-matrix regenerator. In a rotary regenerator, the fluid temperatures are functions of the axial coordinate $x$ and the angular coordinate $\theta$ for a stationary observer looking at the regenerator. For an observer riding on the matrix, the fluid temperatures in a rotary regenerator are also functions of $x$ and $\tau$. Thus we will consider $t_h$ and $t_c$ as functions of $x$ and $\tau$ for both types of regenerators.

Saunders and Smoleniec [2] investigated the eighth idealization. They found the error in the effectiveness less than 1% due to variations in fluid and matrix specific heats. However, if a significant influence of variations in specific heats of the gases and the matrix is anticipated, a numerical solution to the problem is suggested.

Heat conduction in the fluids is generally negligible for the Péclet number[†] $Pe = RePr > 50$ [3]. It is important primarily for the liquid metals having $Pe < 50$. Since the regenerators are exclusively used for gas-to-gas heat exchanger applications having $Pe > 50$, the idealization of negligible heat conduction in the fluids is excellent.

Longitudinal heat conduction in the wall may not be negligible, particularly for metal matrices having continuous flow passages. This effect will be considered in Section 5. The transverse conduction for thick ceramic walls may not be infinity. This effect will be considered in Section 6. The influence of the fluid carryover losses and the fluid mixing due to bypass flows is discussed in Section 7.

3. GOVERNING DIFFERENTIAL EQUATIONS

On the basis of the foregoing idealizations, let us derive the governing differential equations and boundary conditions. Consider the counterflow regenerator of Fig. 1.[‡] A disk-type rotary regenerator is shown in Fig. 1a. For clarity, only one regenerator elemental flow passage and the associated flow matrix are shown in Fig. 1b during the hot gas flow period and in Fig. 1c during the cold gas flow period. In fact in the derivation of the differential

---

[†] The Péclet number is proportional to the ratio of thermal energy convected to the fluid to the thermal energy axially conducted within the fluid. The inverse of Péclet number represents the relative importance of fluid axial heat conduction.

[‡] Although the differential equations are derived for a counterflow regenerator, the results are presented for both counterflow and parallel flow regenerators in Section 4.

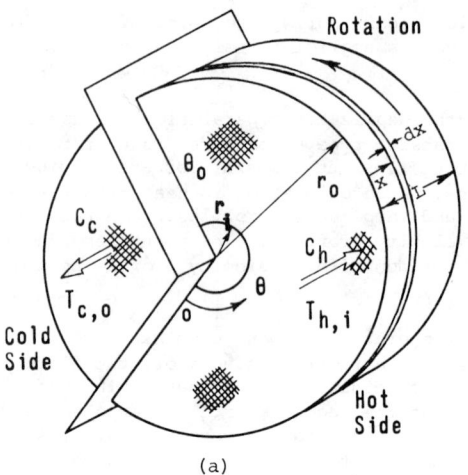

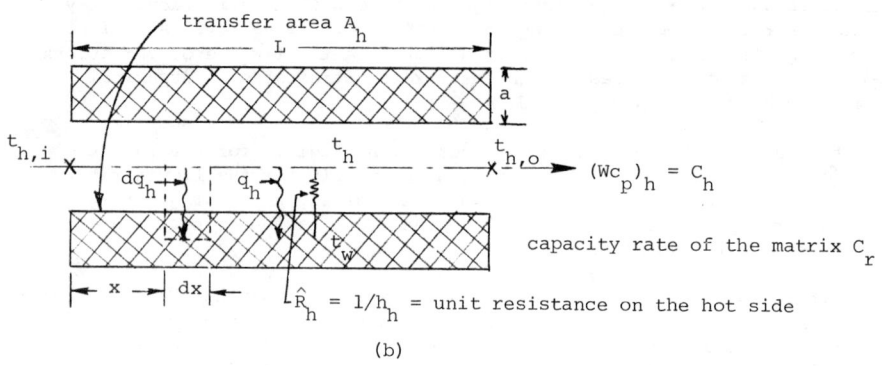

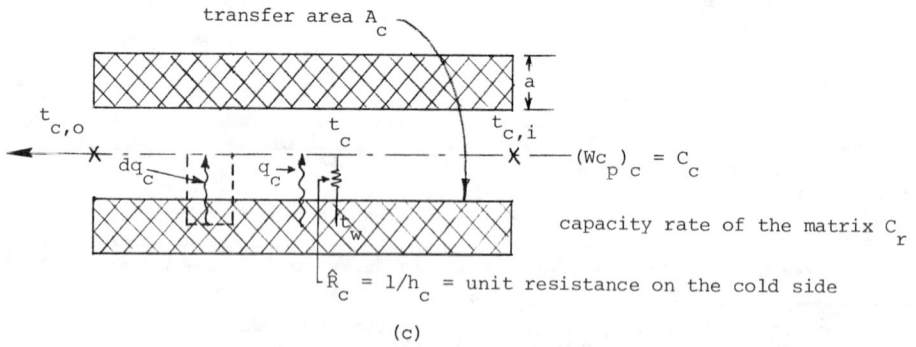

Fig. 1  (a) A rotary regenerator showing sections $x$ and $dx$, (b) regenerator elemental flow passage and associated matrix during the hot gas flow period, (c) the same as (b) during the cold gas flow period.

# THERMAL DESIGN THEORY FOR REGENERATORS

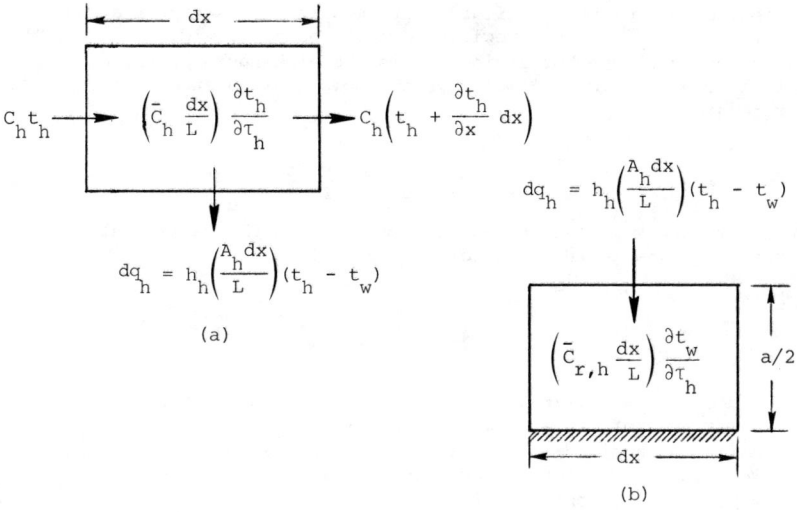

Fig. 2  Energy rate terms associated with the elemental passage dx (a) of fluid, and (b) of matrix at a given instant of time during hot gas flow period.

equations, considered are all quantities (surface area, flow area, flow rate, etc.) associated with a complete cross section of the regenerator at x and x+dx. The reference coordinate system considered is (x, τ) so that Figs. 1b and 1c are valid for a rotary regenerator having an observer riding on the matrix. Figures 1b and 1c are also valid for a fixed-matrix regenerator with the observer standing on the stationary matrix. To show clearly that the theoretical analysis is identical for rotary and fixed-matrix regenerators, we will consider variables and parameters associated with a complete regenerator in this section. That means we will consider the heat transfer surface area, flow rates, etc. associated with all matrices of a fixed-matrix regenerator.[†]

The differential fluid and matrix elements of the hot gas flow period are shown in Fig. 2 with the associated energy transfer terms at a given instant of time.

Let us first define the heat capacitance terms $\bar{C}_h$, $\bar{C}_c$, $\bar{C}_{r,h}$ and $\bar{C}_r$ for the fluids and the matrix, and the heat transfer areas $A_h$ and $A_c$ before setting up the energy balance and rate equations. $\bar{C}_h$ is the hot fluid heat capacitance within the exchanger and $C_h$ is the hot fluid heat capacity rate. Their definitions and relationships are as follows.

$$\bar{C}_h = M_h c_{p,h} \qquad C_h = W_h c_{p,h} = \frac{u_{m,h} \bar{C}_h}{L} = \frac{\bar{C}_h}{\tau_{d,h}} \qquad (1)$$

---

[†] It should be emphasized at this point that although all the matrices of a fixed-matrix regenerator are considered in the derivation of differential equations in this section, we will consider only one matrix for the analysis by the Λ-Π method.

Here $M_h$ is the mass of the hot fluid contained in the exchanger at any instant of time, $c_{p,h}$ is the specific heat of the hot fluid, $u_{m,h}$ is the mean axial velocity of the hot fluid, $L$ is the exchanger length, and $\tau_{d,h}$ is the hot fluid dwell time. Similarly, the cold fluid heat capacitance and capacity rate are

$$\bar{C}_c = M_c c_{p,c} \qquad C_c = W_c c_{p,c} = \frac{u_{m,c} \bar{C}_c}{L} = \frac{\bar{C}_c}{\tau_{d,c}} \tag{2}$$

The regenerator matrix wall heat capacitance $\bar{C}_r$ and the matrix wall heat capacity rate $C_r$ are defined and related as follows.

$$\bar{C}_r = M_w c_w \qquad C_r = \begin{cases} M_w c_w N = \bar{C}_r N \\ M_w c_w / P_t = \bar{C}_r / P_t \end{cases} \tag{3}$$

Here $M_w$ is the mass of all matrices, $c_w$ is the specific heat of the matrix material, $N$ is the rotational speed for a rotary regenerator and $P_t$ is the total period for a fixed-matrix regenerator. This $P_t$ is the interval of time between the start of two successive heating periods and is the sum of hot gas flow period $P_h$, cold gas flow period $P_c$, and reversal period $P_r$.

$$P_t = P_h + P_c + P_r \tag{4}$$

Since $P_r$ is generally small compared to $P_h$ or $P_c$, it is usually neglected.

The matrix heat capacitance $\bar{C}_{r,h}$ and $\bar{C}_{r,c}$ are related to the total matrix heat capacitance as

$$\bar{C}_{r,h} = \bar{C}_r \frac{P_h}{P_t} \text{ or } \bar{C}_r \frac{\theta_h}{\theta_t} \qquad \bar{C}_{r,c} = \bar{C}_r \frac{P_c}{P_t} \text{ or } \bar{C}_r \frac{\theta_c}{\theta_t} \tag{5}$$

Here $\theta_h$ and $\theta_c$ are the disk sector angles for a rotary regenerator through which hot and cold gases flow, and

$$\theta_t = \theta_h + \theta_c + \theta_s = 2\pi \tag{6}$$

with $\theta_s$ as the sector angle covered by the seals. The matrix wall heat capacity rates during the hot and cold periods, using Eq. (5), are then

$$C_{r,h} = \frac{\bar{C}_{r,h}}{P_h} = \frac{\bar{C}_r}{P_t} = \bar{C}_r N \qquad C_{r,c} = \frac{\bar{C}_{r,c}}{P_c} = \frac{\bar{C}_r}{P_t} = \bar{C}_r N \tag{7}$$

Thus
$$C_{r,h} = C_{r,c} = C_r \tag{8}$$

The heat transfer areas $A_h$ and $A_c$ are related to the total heat transfer area $A$ of all matrices of a fixed-matrix regenerator as

$$A_h = \frac{AP_h}{P_t} = \frac{\beta V P_h}{P_t} \qquad A_c = \frac{AP_c}{P_t} = \frac{\beta V P_c}{P_t} \tag{9}$$

and for a rotary regenerator,

$$A_h = \frac{A\theta_h}{\theta_t} = \frac{\beta V \theta_h}{\theta_t} \qquad A_c = \frac{A\theta_c}{\theta_t} = \frac{\beta V \theta_c}{\theta_t} \tag{10}$$

# THERMAL DESIGN THEORY FOR REGENERATORS

Here $\beta$ is the heat transfer surface area density and $V$ is the total volume of all matrices.

At this time, it may again be pointed out that we have selected the reference coordinate system as $(x, \tau)$ for both rotary and fixed-matrix regenerators. Hence, even for a rotary regenerator, we will use Eq. (9) and the pertinent expression in Eqs. (3), (5) and (7) in terms of $P_t$, $P_h$ and $P_c$.

Now let us review Fig. 2a. During its flow through the elemental passage, the hot gas transfers heat to the wall by convection resulting in reduction in its outlet enthalpy and internal stored thermal energy. Applying the energy balance, the first law of thermodynamics, to this elemental passage, we get

$$h_h \left(\frac{A_h dx}{L}\right)(t_h - t_w) + C_h\left(t_h + \frac{\partial t_h}{\partial x} dx\right) - C_h t_h + \left(\bar{C}_h \frac{dx}{L}\right)\frac{\partial t_h}{\partial \tau_h} = 0 \quad (11)$$

Substituting the value of $\bar{C}_h$ from Eq. (1) into this equation and simplifying, we get

$$\frac{\partial t_h}{\partial \tau_h} + \frac{L}{\tau_{d,h}} \frac{\partial t_h}{\partial x} = \frac{(hA)_h}{C_h \tau_{d,h}} (t_w - t_h) \quad (12)$$

Since the hot gas flows on both sides of the wall, the plane at the half wall thickness is adiabatic as shown in Fig. 2b. In the absence of longitudinal and transverse wall heat conduction, the heat transferred from the hot gas to the matrix wall is stored in the wall. An energy balance on the matrix wall elemental passage is

$$\left(\bar{C}_{r,h} \frac{dx}{L}\right)\frac{\partial t_w}{\partial \tau_h} = h_h\left(\frac{A_h dx}{L}\right)(t_h - t_w) \quad (13)$$

Combining with Eq. (7) and simplifying

$$\frac{\partial t_w}{\partial \tau_h} = \frac{(hA)_h}{C_{r,h} P_h} (t_h - t_w) \quad (14)$$

For the cold gas flow period, a pair of similar equations results:

$$\frac{\partial t_c}{\partial \tau_c} + \frac{L}{\tau_{d,c}} \frac{\partial t_c}{\partial x} = \frac{(hA)_c}{C_c \tau_{d,c}} (t_c - t_w) \quad (15)$$

$$\frac{\partial t_w}{\partial \tau_c} = \frac{(hA)_c}{C_{r,c} P_c} (t_w - t_c) \quad (16)$$

The boundary conditions are as follows. The inlet temperature of the hot gas is constant during the hot gas flow period, and the inlet temperature of the cold gas is constant during the cold gas flow period.

$$t_h(0, \tau_h) = t_{h,i} = \text{constant} \qquad \text{for } 0 \leq \tau_h \leq P_h \quad (17)$$

$$t_c(L, \tau_c) = t_{c,i} = \text{constant} \qquad \text{for } 0 \leq \tau_c \leq P_c \quad (18)$$

The periodic equilibrium conditions for the wall are

$$t_w(x, \tau_h = P_h) = t_w(x, \tau_c = 0) \qquad \text{for } 0 \leq x \leq L \quad (19)$$

$$t_w(x, \tau_h = 0) = t_w(x, \tau_c = P_c) \quad \text{for } 0 \leq x \leq L \quad (20)$$

Since the regenerator is in periodic equilibrium, Eqs. (17)-(20) are valid for $\tau = \tau + nP_t$, where n is an integer, $n \geq 0$.

The boundary conditions of Eqs. (17) and (18) are the simplest for the analysis. In application, the fluids inlet to the regenerator may have non-uniform temperature profiles. Then the solution can only be obtained by further numerical analysis.

Based on the foregoing differential equations and boundary conditions, the dependent fluid as well as matrix temperatures are functions of the following variables and parameters.

$$\underbrace{t_h, t_c, t_w}_{\text{Dependent variables}} = \phi \{ \underbrace{x, \tau_h, \tau_c}_{\text{Independent variables}}, \underbrace{t_{h,i}, t_{c,i}, C_h, C_c, \tau_{d,h}, \tau_{d,c}}_{\text{Operating conditions}},$$

$$\underbrace{C_{r,h}, (hA)_h, (hA)_c, L, P_h, P_c}_{\text{Parameters under designer's control}} \} \quad (21)$$

$C_{r,c}$ is not included in the foregoing list since $C_{r,c} = C_{r,h}$.

## 4. DIMENSIONLESS GROUPS AND SPECIFIC SOLUTIONS

Fifteen independent variables and parameters exist for the dependent regenerator fluid and wall temperatures. These are reduced by formulating appropriate dimensionless groups. The specific form of these groups is to some extent optional. Two such options have been used for the regenerator analysis. The effectiveness-number of transfer units ($\varepsilon$-$N_{tu,o}$) method is generally used for the rotary regenerators. The reduced length - reduced period ($\Lambda$-$\Pi$) method is generally used for the fixed-matrix regenerators. It will be shown that both methods are equivalent. Solutions to Eqs. (12), (14)-(20) for counterflow and parallel flow regenerators[†] are also presented in this section.

### 4.1 The $\varepsilon$-$N_{tu,o}$ Method

This method is due to Coppage and London [4] in 1953. The dimensionless groups in this method are formulated in such a way that when the influence of additional groups is negligible, the remaining groups parallel to those of the recuperators. We will derive these dimensionless groups now by first making the governing differential equations in a dimensionless form.

To simplify the differential equations for the hot gas flow period, Eqs. (12) and (14), the independent variables $x$ and $\tau$ are made dimensionless as

$$X^* = x/L \quad (22)$$

$$\tau_h^* = \frac{1}{P_h}\left(\tau_h - \frac{x}{L}\tau_{d,h}\right) \quad (23)$$

---

[†] Note that for regenerators there are no counterparts of other flow arrangements of recuperators such as crossflow, multipass cross-counterflow, etc.

# THERMAL DESIGN THEORY FOR REGENERATORS

With this specific choice, the implicit functional relationships with the old and new sets of variables are

$$t_h = \phi(x, \tau_h) \longrightarrow t_h = \phi(X^*, \tau_h^*) \tag{24}$$

$$t_w = \phi(x, \tau_h) \longrightarrow t_w = \phi(\tau_h^*) \tag{25}$$

Thus using the following partial derivative relationships

$$\frac{\partial t_h}{\partial x} = \frac{\partial t_h}{\partial X^*}\frac{\partial X^*}{\partial x} + \frac{\partial t_h}{\partial \tau_h^*}\frac{\partial \tau_h^*}{\partial x} \tag{26}$$

$$\frac{\partial t_h}{\partial \tau_h} = \frac{\partial t_h}{\partial \tau_h^*}\frac{\partial \tau_h^*}{\partial \tau_h} \tag{27}$$

$$\frac{\partial t_w}{\partial \tau_h} = \frac{\partial t_w}{\partial \tau_h^*}\frac{\partial \tau_h^*}{\partial \tau_h} \tag{28}$$

Eqs. (12) and (14) reduce to

$$\frac{\partial t_h}{\partial X^*} = N_{tu,h}(t_w - t_h) \tag{29}$$

$$\frac{\partial t_w}{\partial \tau_h^*} = \frac{N_{tu,h}}{C_{r,h}^*}(t_h - t_w) \tag{30}$$

where

$$N_{tu,h} = (hA)_h / C_h \tag{31}$$

$$C_{r,h}^* = C_{r,h}/C_h \tag{32}$$

Similarly, the differential equations for the cold gas flow period, Eqs. (15) and (16), reduce to

$$\frac{\partial t_c}{\partial X^*} = N_{tu,c}(t_c - t_w) \tag{33}$$

$$\frac{\partial t_w}{\partial \tau_c^*} = \frac{N_{tu,c}}{C_{r,c}^*}(t_w - t_c) \tag{34}$$

where

$$N_{tu,c} = (hA)_c / C_c \tag{35}$$

$$C_{r,c}^* = C_{r,c}/C_c \tag{36}$$

$$\tau_c^* = \frac{1}{P_c}\left(\tau_c - \frac{x}{L}\tau_{d,c}\right) \tag{37}$$

Generally, the dwell times $\tau_{d,h}$ and $\tau_{d,c}$ are much smaller than the respective periods of operation $P_h$ and $P_c$. And since $x/L$ varies from 0 to 1, the last terms for $\tau_h^*$ and $\tau_c^*$ in Eqs. (23) and (37) are negligible. In this case,

$$\tau_h^* = \tau_h/P_h \qquad \tau_c^* = \tau_c/P_c \qquad (38)$$

In case of cryogenics and Stirling engine regenerators, the rotational speed or the valve switching frequency is so high that $\tau_{d,h}$ and $\tau_{d,c}$ may not be so small as to be negligible compared to $P_h$ and $P_c$. This effect is further discussed in Section 7.

Introducing the following definitions of dimensionless temperatures,

$$t_h^* = \frac{t_h - t_{c,i}}{t_{h,i} - t_{c,i}}, \qquad t_c^* = \frac{t_c - t_{c,i}}{t_{h,i} - t_{c,i}}, \qquad t_w^* = \frac{t_w - t_{c,i}}{t_{h,i} - t_{c,i}} \qquad (39)$$

Eqs. (29), (30), (33) and (34) reduce to

$$\frac{\partial t_h^*}{\partial X^*} = N_{tu,h}(t_w^* - t_h^*) \qquad (40)$$

$$\frac{\partial t_w^*}{\partial \tau_h^*} = \frac{N_{tu,h}}{C_{r,h}^*}(t_h^* - t_w^*) \qquad (41)$$

$$\frac{\partial t_c^*}{\partial X^*} = N_{tu,c}(t_c^* - t_w^*) \qquad (42)$$

$$\frac{\partial t_w^*}{\partial \tau_c^*} = \frac{N_{tu,c}}{C_{r,c}^*}(t_w^* - t_c^*) \qquad (43)$$

The boundary conditions and periodic equilibrium conditions of Eqs. (17)-(20) reduce to

$$t_h^*(0, \tau_h^*) = 1 \qquad \text{for } 0 \le \tau_h^* \le 1 \qquad (44)$$

$$t_c^*(1, \tau_c^*) = 0 \qquad \text{for } 0 \le \tau_c^* \le 1 \qquad (45)$$

$$t_w^*(X^*, \tau_h^*=1) = t_w^*(X^*, \tau_c^*=0) \qquad \text{for } 0 \le X^* \le 1 \qquad (46)$$

$$t_w^*(X^*, \tau_h^*=0) = t_w^*(X^*, \tau_c^*=1) \qquad \text{for } 0 \le X^* \le 1 \qquad (47)$$

It is clear from Eqs. (40)-(47) that the dependent temperatures are functions of

$$t_h^*, t_c^*, t_w^* = \phi\{X^*, \tau_h^*, \tau_c^*, N_{tu,h}, N_{tu,c}, C_{r,h}^*, C_{r,c}^*\} \qquad (48)$$

Thus we are able to reduce 15 independent variables and parameters to 7 with an additional idealization that

$$\tau_{d,h} \ll P_h \qquad \tau_{d,c} \ll P_c \qquad (49)$$

built into Eq. (38).

For overall regenerator performance, we are interested in determining the average fluid outlet temperatures. In a rotary regenerator, the outlet temperatures vary as a function of the angular coordinate $\theta$. If $\theta_h$ and $\theta_c$ represent the angles for the sectors through which hot and cold gases flow respectively, the space average outlet temperatures are

$$\bar{t}_{h,o} = \frac{1}{\theta_h} \int_0^{\theta_h} t_{h,o}(\theta) \, d\theta, \qquad \bar{t}_{c,o} = \frac{1}{\theta_c} \int_0^{\theta_c} t_{c,o}(\theta) \, d\theta \qquad (50)$$

where $t_{h,o}(\theta)$ and $t_{c,o}(\theta)$ represent the angular coordinate dependent fluid temperatures at the regenerator outlet. However, for an observer riding on the rotary regenerator matrix, the fluid outlet temperatures are functions of the time $\tau$. In this case, the time average outlet temperatures are

$$\bar{t}_{h,o} = \frac{1}{P_h} \int_0^{P_h} t_{h,o}(\tau)\, d\tau, \qquad \bar{t}_{c,o} = \frac{1}{P_c} \int_0^{P_c} t_{c,o}(\tau)\, d\tau \qquad (51)$$

Here $t_{h,o}(\tau)$ and $t_{c,o}(\tau)$ are the time dependent fluid temperatures at the regenerator outlet. $\bar{t}_{h,o}$ and $\bar{t}_{c,o}$ in Eqs. (50) and (51) respectively represent the space average and time average fluid temperatures at the regenerator outlet. Equation (51) is valid for both the fixed-matrix and rotary regenerators, provided that an observer rides on the matrix for the rotary regenerator.

Thus the functional relationship of Eq. (48) for the dependent regenerator outlet temperature is

$$t^*_{h,o},\ t^*_{c,o} = \phi\{N_{tu,h},\ N_{tu,c},\ C^*_{r,h},\ C^*_{r,c}\} \qquad (52)$$

A bar on these dimensionless outlet temperatures is eliminated for convenience.

These outlet temperatures are conveniently expressed by the regenerator effectiveness $\varepsilon = q/q_{max}$. Using the outlet temperatures defined by Eq. (51), the actual heat transfer rate in the regenerator is

$$q = C_h(t_{h,i} - \bar{t}_{h,o}) = C_c(\bar{t}_{c,o} - t_{c,i}) \qquad (53)$$

To determine $q_{max}$, we have to define a "perfect" heat exchanger. In this section, we have considered either a rotary regenerator or all matrices of a fixed-matrix regenerator as a system. The hot and cold fluids flow <u>continuously</u> in and out of such a system. A perfect exchanger having the same fluid flow rates and fluid inlet temperatures is a <u>counterflow</u> recuperator of infinite surface area. $q_{max}$ for this perfect heat exchanger is

$$q_{max} = C_{min}(t_{h,i} - t_{c,i}) \qquad (54)$$

where $C_{min}$ is the minimum of $C_h$ and $C_c$. The regenerator effectiveness is thus

$$\varepsilon = \frac{C_h(t_{h,i} - \bar{t}_{h,o})}{C_{min}(t_{h,i} - t_{c,i})} = \frac{C_c(\bar{t}_{c,o} - t_{c,i})}{C_{min}(t_{h,i} - t_{c,i})} \qquad (55)^\dagger$$

Then for $C_c = C_{min}$

$$t^*_{h,o} = 1 - \varepsilon C^* \qquad t^*_{c,o} = \varepsilon \qquad (56)$$

And Eq. (52) presented in terms of $\varepsilon$ is

$$\varepsilon = \phi\{N_{tu,h},\ N_{tu,c},\ C^*_{r,h},\ C^*_{r,c}\} \qquad (57)$$

---

$^\dagger$In the $\Lambda$-$\Pi$ method, we will consider only one matrix of a fixed-matrix regenerator as a system at a time. In that case, a different definition of the regenerator effectiveness will be introduced.

Since these independent dimensionless groups, defined in Eqs. (31), (32), (35) and (36) do not parallel to those of a recuperator (a direct transfer type exchanger), let us define a related set as follows.

$$N_{tu,o} = \frac{1}{C_{min}} \left[ \frac{1}{(1/(hA)_h + 1/(hA)_c)} \right] \quad (58)$$

$$C^* = C_{min}/C_{max} \quad (59)$$

$$C_r^* = C_r/C_{min} \quad (60)$$

$$(hA)^* = \frac{(hA) \text{ on the } C_{min} \text{ side}}{(hA) \text{ on the } C_{max} \text{ side}} \quad (61)$$

It is then valid to express

$$\varepsilon = \phi\{N_{tu,o},\ C^*,\ C_r^*,\ (hA)^*\} \quad (62)$$

based on Eq. (57). Here $N_{tu,o}$ is the modified number of transfer units. Since there is no direct heat transfer from the hot to the cold fluid in a regenerator (similar to that in a recuperator), UA does not come into picture directly for the regenerator. However, if the bracketed term of Eq. (58) is designated as $U_o A$, with $U_o$ as a modified overall heat transfer coefficient, then

$$\frac{1}{U_o A} = \frac{1}{(hA)_h} + \frac{1}{(hA)_c} \quad (63)$$

A comparison of this expression with that of a recuperator [5] reveals $U_o A$ is the same as UA when the wall thermal resistance and scale resistances are zero. Note that all the surface in a regenerator is primary surface (no fins) and hence the total surface temperature effectiveness $\eta_o = 1$. The definition of $N_{tu,o}$ parallels that of $N_{tu}$, and in the limiting case of $C_r^* = \infty$, the numerical solutions demonstrate that the regenerator has the same performance as a recuperator with its $N_{tu}$ identical to $N_{tu,o}$. The newly defined dimensionless groups are related to those of Eq. (57) as follows for $C_c = C_{min}$.

$$N_{tu,o} = \frac{1}{(C_{r,h}^*/C_{r,c}^*)/N_{tu,h} + 1/N_{tu,c}} \quad (64)$$

$$C^* = C_{r,h}^*/C_{r,c}^* \quad (65)$$

$$C_r^* = C_{r,c}^* \quad (66)$$

$$(hA)^* = \frac{N_{tu,c}}{N_{tu,h}} \frac{C_{r,h}^*}{C_{r,c}^*} \quad (67)$$

Since $\varepsilon = \phi(N_{tu},\ C^*)$ for a specified flow arrangement of a recuperator [5], comparing it with Eq. (62) reveals that the effectiveness for a regenerator is dependent upon two additional parameters, $C_r^*$ and $(hA)^*$. Since the thermal energy is stored and delivered by the matrix wall periodically, the wall temperature is going to be dependent upon (1) the storage heat capacity rate, and (2) the thermal conductances between the matrix wall and the hot fluid, $(hA)_h$, and between the matrix wall and the cold fluid $(hA)_c$. And as a

# THERMAL DESIGN THEORY FOR REGENERATORS

result, two additional dimensionless groups $C_r^*$ and $(hA)^*$ come into picture.

Solutions for a counterflow regenerator. No closed-form solutions are available to Eqs. (40)-(47). A number of approximate solutions have been obtained by researchers by employing different mathematical techniques. One of the more complete solutions has been obtained by Lambertson [6] by employing a finite difference method. He considered two coupled crossflow heat exchangers with fluids unmixed[†] as shown in Fig. 3. In this figure, for the heat exchanger on the left, one fluid stream is the hot gas and the other fluid stream is the matrix material "stream." For the heat exchanger on the right, the cold gas and the matrix material are the two "streams".

For the analysis, the temperature distribution $t_{r,i}$ of the matrix material stream is assumed initially; the hot and cold fluid inlet temperatures are known. A row-by-row marching procedure is applied starting with the upper left-hand corner element. An application of the energy balance and rate equations to this element yields the outlet temperatures. For the second element of the first row, now both inlet temperatures are known. The outlet temperatures are then obtained by applying the energy balance and rate equations. An application of this procedure to the remaining elements down the first row yield inlet temperatures for the elements of the second row as well as the outlet temperature of the hot fluid from the first row. This procedure is continued starting with the first element of the second row down to the second row, and so on, until $t_{h,o}$, $t_{c,o}$ and $t_{r,o}$ distributions are obtained. This finite difference procedure is iterated until the temperature distributions $t_{r,i}$ and $t_{r,o}$ are matched satisfying the "steady-state" periodic-flow conditions. Lambertson obtained the solution by dividing each crossflow unit first into 16x16 division, then into 32x32 division, then extrapolated the results to $\infty \times \infty$ divisions. He presented the solution in terms of $\varepsilon$ as a function of four

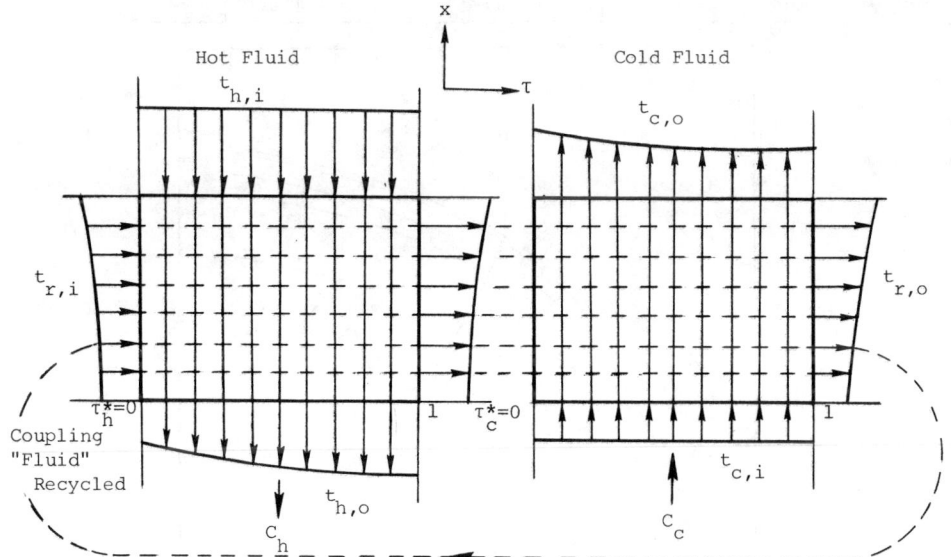

Fig. 3 Representation of a regenerator as two coupled crossflow exchangers.

---

[†] The temperature of either fluid is a function of position and time. $t = t(x, \tau)$.

Table 1. Counterflow regenerator $\varepsilon$ as a function of $N_{tu,o}$ and $C_r^*$ for $C^* = 1.0$ [5]

| $N_{tu,o}$ | $\varepsilon$ for the indicated matrix capacity-rate ratios, $C_r/C_{min}$ | | | | | | | |
|---|---|---|---|---|---|---|---|---|
| | 1.0 | 1.25 | 1.5 | 2.0 | 3.0 | 5.0 | 10.0 | $\infty$ |
| 0 | 0 | 0 | 0 | 0 | 0 | 0 | 0 | 0 |
| 0.5 | 0.322 | 0.326 | 0.328 | 0.330 | 0.332 | 0.333 | 0.333 | 0.333 |
| 1.0 | 0.467 | 0.478 | 0.485 | 0.491 | 0.496 | 0.499 | 0.500 | 0.500 |
| 1.5 | 0.548 | 0.566 | 0.576 | 0.586 | 0.594 | 0.598 | 0.599 | 0.600 |
| 2.0 | 0.601 | 0.623 | 0.636 | 0.649 | 0.659 | 0.664 | 0.666 | 0.667 |
| 2.5 | 0.639 | 0.665 | 0.679 | 0.694 | 0.705 | 0.711 | 0.713 | 0.714 |
| 3.0 | 0.667 | 0.696 | 0.712 | 0.728 | 0.740 | 0.746 | 0.749 | 0.750 |
| 3.5 | 0.690 | 0.721 | 0.738 | 0.755 | 0.767 | 0.774 | 0.777 | 0.778 |
| 4.0 | 0.709 | 0.741 | 0.759 | 0.776 | 0.789 | 0.796 | 0.799 | 0.800 |
| 4.5 | 0.724 | 0.758 | 0.776 | 0.794 | 0.807 | 0.814 | 0.817 | 0.818 |
| 5.0 | 0.738 | 0.772 | 0.791 | 0.809 | 0.822 | 0.829 | 0.832 | 0.833 |
| 5.5 | 0.749 | 0.785 | 0.803 | 0.821 | 0.834 | 0.842 | 0.845 | 0.846 |
| 6.0 | 0.759 | 0.796 | 0.814 | 0.832 | 0.845 | 0.853 | 0.856 | 0.857 |
| 6.5 | 0.768 | 0.805 | 0.824 | 0.842 | 0.855 | 0.862 | 0.865 | 0.867 |
| 7.0 | 0.776 | 0.814 | 0.833 | 0.850 | 0.863 | 0.870 | 0.874 | 0.875 |
| 7.5 | 0.784 | 0.822 | 0.840 | 0.858 | 0.871 | 0.878 | 0.881 | 0.882 |
| 8.0 | 0.790 | 0.829 | 0.847 | 0.865 | 0.877 | 0.884 | 0.888 | 0.889 |
| 8.5 | 0.796 | 0.835 | 0.854 | 0.871 | 0.883 | 0.890 | 0.894 | 0.895 |
| 9.0 | 0.802 | 0.841 | 0.859 | 0.876 | 0.888 | 0.895 | 0.899 | 0.900 |
| 9.5 | 0.807 | 0.846 | 0.864 | 0.881 | 0.893 | 0.900 | 0.904 | 0.905 |
| 10.0 | 0.811 | 0.851 | 0.869 | 0.886 | 0.898 | 0.904 | 0.908 | 0.909 |
| 20.0 | 0.865 | 0.906 | 0.922 | 0.935 | 0.943 | 0.948 | 0.951 | 0.952 |
| 50.0 | 0.914 | 0.951 | 0.962 | 0.970 | 0.975 | 0.978 | 0.980 | 0.980 |
| 90.0 | 0.935 | 0.969 | 0.977 | 0.982 | 0.986 | 0.987 | 0.988 | 0.989 |
| 100.0 | 0.939 | | 0.979 | 0.984 | | 0.989 | 0.989 | 0.990 |
| 500.0 | 0.974 | | 0.995 | 0.996 | | 0.998 | 0.998 | 0.998 |

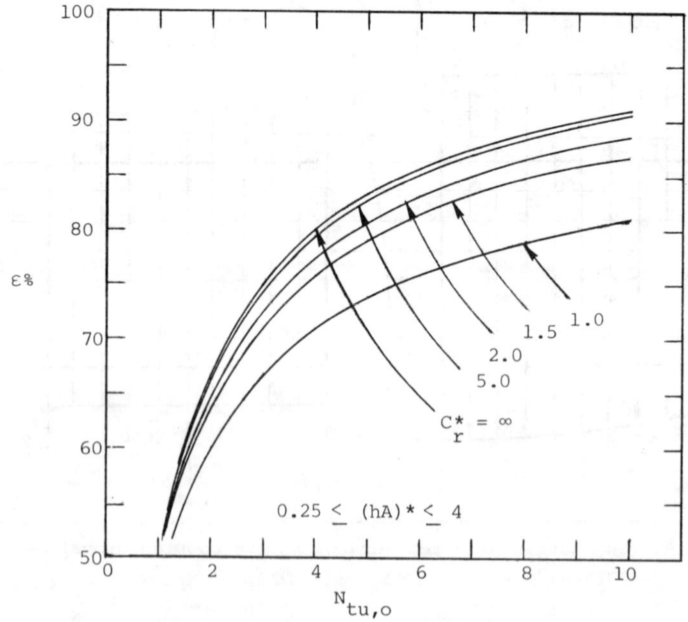

Fig. 4 Counterflow regenerator $\varepsilon$ as a function of $N_{tu,o}$ and $C_r^*$ for $C^* = 1$, from Table 1.

dimensionless groups of Eq. (62). He covered the following range of the parameters: $1 \leq N_{tu,o} \leq 10$, $0.1 \leq C^* \leq 1.0$, $1 \leq C_r^* \leq \infty$, and $0.25 \leq (hA)^* \leq 1$. His partial results for $C^* = 1$ are presented in Table 1 and Fig. 4. The results for other $C^*$ values are presented by Kays and London [5].

The following observations may be made by reviewing Fig. 4 and the results of [5] for $C^* < 1$: (1) For a specified $C_r^*$ and $C^*$, the heat exchanger effectiveness increases with increasing $N_{tu,o}$. For all $C_r^*$ and $C^*$, $\varepsilon \to 1$ as $N_{tu,o} \to \infty$. (2) For a specified $N_{tu,o}$ and $C^*$, $\varepsilon$ increases with increasing values of $C_r^*$. (3) For a specified $N_{tu,o}$ and $C^*$, $\varepsilon$ increases with decreasing values of $C^*$. (4) For $\varepsilon < 40\%$ and $C_r^* > 0.6$, $C^*$ and $C_r^*$ do not have a significant influence on the exchanger effectiveness.

Now let us further discuss the reasons why we chose the set of dimensionless independent groups of Eq. (62) instead of those of Eq. (57): (1) The separate influence of $(hA)^*$ on $\varepsilon$ is negligibly small for $0.25 \leq (hA)^* \leq 4$ as shown by Lambertson among others. The maximum error occurs at $C^* = 1$ and is shown in Table 2 for a range of $N_{tu,o}$ and $C_r^*$. In the normal operating range of a regenerator, $C_r^* \geq 0.7$, the maximum error at most is 2 points on %$\varepsilon$. Thus we can effectively eliminate $(hA)^*$ from Eq. (62).

$$\varepsilon = \phi(N_{tu,o}, C^*, C_r^*) \qquad (68)$$

Table 2. The error in $\varepsilon (= \varepsilon_1 - \varepsilon_2)$ for $C^* = 1$; $\varepsilon_1$ is for $(hA)^* = 1$ and $\varepsilon_2$ is for $(hA)^* = 0.25$.

| $N_{tu,o}$ | $C_r^*$ | | | | | |
|---|---|---|---|---|---|---|
| | 1.0 | 0.95 | 0.90 | 0.80 | 0.70 | 0.50 |
| 3 | -0.0012 | 0.0009 | 0.0033 | 0.0088 | 0.0154 | 0.0303 |
| 6 | -0.0016 | 0.0013 | 0.0044 | 0.0119 | 0.0196 | 0.0331 |
| 9 | -0.0015 | 0.0018 | 0.0054 | 0.0131 | 0.0208 | 0.0312 |

(2) When $C_r^* \to \infty$, the effectiveness $\varepsilon$ of a regenerator approaches that of a recuperator. The difference in $\varepsilon$ for $C_r^* > 5$ and that for $C_r^* = \infty$ is negligibly small and may be ignored for the design purpose. Thus by the selection of the Eq. (62) set, we have demonstrated the similarities and differences of the regenerator and recuperator.

The influence of $C_r^*$ on $\varepsilon$ can be presented by an empirical correlation for $\varepsilon \leq 90\%$. Lambertson presented a correlation which agreed within 1% of his numerical results for $3 \leq N_{tu,o} \leq 9$, $0.90 \leq C^* \leq 1$, and $1.25 \leq C_r^* \leq 5$. This correlation, after the exponent modified from 1.87 to 1.93 by Kays and London [5], is

$$\varepsilon = \varepsilon_{cf}\left[1 - \frac{1}{9(C_r^*)^{1.93}}\right] \qquad (69)$$

where $\varepsilon_{cf}$ is the counterflow recuperator effectiveness as follows.

$$\varepsilon_{cf} = \frac{1 - \exp[-N_{tu,o}(1 - C^*)]}{1 - C^* \exp[-N_{tu,o}(1 - C^*)]} \qquad (70)$$

Equation (69) agrees within 1% with the tabular data [5] for $C^* = 1$ for the following ranges: $2 \leq N_{tu,o} \leq 14$ for $C_r^* \leq 1.5$, $N_{tu,o} \leq 20$ for $C_r^* = 2$, and a complete range of $N_{tu,o}$ for $C_r^* \geq 5$. For decreasing values of $C^*$, the error due to approximation increases with lower values of $C_r^*$. For example, to obtain accuracy within 1%, $C_r^* \geq 1.5$ for $C^* = 0.9$, and $C_r^* \geq 2.0$ for $C^* = 0.7$.

Razelos [7] proposed the following approximate procedure to calculate the regenerator effectiveness $\varepsilon$ for the case of $C^* < 1$. For the known values of $N_{tu,o}$, $C^*$ and $C_r^*$, calculate "equivalent" values of $N_{tu,o}$ and $C_r^*$ for a balanced regenerator ($C^* = 1$), designated with a subscript m, as follows.

$$N_{tu,o,m} = 2N_{tu,o} C^*/(1 + C^*) \qquad (71)$$

$$C_{r,m}^* = 2C_r^* C^*/(1 + C^*) \qquad (72)$$

With these values of $N_{tu,o,m}$ and $C_{r,m}^*$, obtain the value of $\varepsilon_r$ from Table 1 or from the following approximate equation.

$$\varepsilon_r = \frac{N_{tu,o,m}}{1 + N_{tu,o,m}} \left[ 1 - \frac{1}{9(C_{r,m}^*)^{1.93}} \right] \qquad (73)$$

Subsequently, calculate $\varepsilon$ from

$$\varepsilon = \frac{1 - \exp\{\varepsilon_r (C^{*2} - 1)/[2C^*(1 - \varepsilon_r)]\}}{1 - C^* \exp\{\varepsilon_r (C^{*2} - 1)/[2C^*(1 - \varepsilon_r)]\}} \qquad (74)$$

A comparison of $\varepsilon$ from this procedure with tabular values of $\varepsilon$ of Kays and London [5] shows that the Razelos approximation yields more accurate values of $\varepsilon$ compared to that from Eq. (69) for $C^* < 1$. It can be shown that $\varepsilon$'s of Eqs. (69) and (74) are identical for $C^* = \infty$. Note that either by employing the foregoing approximate method or for a direct use of Eq. (69), we at most need Table 1 or Fig. 4; thus the tabular data for $C_r^* < 1$ are not needed.

<u>Solutions for a parallel flow regenerator</u>. The differential equations and boundary conditions for the parallel flow regenerator are the same as those of Eqs. (40)-(47) except for Eq. (45). The boundary condition of Eq. (45) for this case is

$$t_c^*(0, \tau_c^*) = 0 \qquad \text{for} \quad 0 \leq \tau_c^* \leq 1 \qquad (75)$$

The solution may be presented in terms of $\varepsilon$ as a function of the same four dimensionless groups as for a counterflow regenerator, Eq. (62). Theoclitus and Eckrich [8] obtained the solution numerically by a finite difference method. They covered the following ranges of the parameters: $1 \leq N_{tu,o} \leq 10$, $0.5 \leq C^* \leq 1.0$, $0.2 \leq C_r^* \leq \infty$, and $0.25 \leq (hA)^* \leq 1$. Their results for $C^* = 1$ are presented in Table 3 and Fig. 5.

From a review of Fig. 5, it is interesting to note that the effectiveness of a parallel flow regenerator oscillates above and below that for a parallel flow recuperator ($\varepsilon = 50\%$). The oscillations decrease in amplitude as $C_r^*$ increases. The maximum effectiveness for the parallel flow regenerator is reached at $C_r^* \simeq 1$, and it exceeds the effectiveness of a parallel flow recuperator. These results contrast with those for the counterflow regenerator where the limiting effectiveness, represented by the counterflow recuperator, is never exceeded, but is approached asymptotically as $C_r^*$ increases.

Table 3. Parallel flow regenerator $\varepsilon$ as a function of $N_{tu,o}$ and $C_r^*$ for $C^* = 1$ and $(hA)^* = 1$ [8]

| $N_{tu,o}$ | \multicolumn{9}{c}{$\varepsilon$ for the indicated matrix capacity rate ratios $C_r^*$} |
|---|---|---|---|---|---|---|---|---|---|
|  | 0.4 | 0.6 | 1.0 | 1.25 | 1.5 | 2.0 | 3.0 | 5.0 | $\infty$ |
| 1.0 | 0.3442 | 0.4025 | 0.4277 | 0.4311 | 0.4322 | 0.4329 | 0.4332 | 0.4332 | 0.4323 |
| 2.0 | 0.3854 | 0.4951 | 0.5315 | 0.5299 | 0.5145 | 0.5046 | 0.4967 | 0.4930 | 0.4908 |
| 3.0 | 0.3953 | 0.5371 | 0.5879 | 0.5625 | 0.5402 | 0.5175 | 0.5050 | 0.5007 | 0.4988 |
| 4.0 | 0.3983 | 0.5595 | 0.6277 | 0.5861 | 0.5485 | 0.5142 | 0.5026 | 0.5007 | 0.4999 |
| 5.0 | 0.3993 | 0.5726 | 0.6586 | 0.6031 | 0.5498 | 0.5049 | 0.4989 | 0.5000 | 0.5000 |
| 6.0 | 0.3997 | 0.5809 | 0.6837 | 0.6167 | 0.5480 | 0.4932 | 0.4962 | 0.4988 | 0.5000 |
| 7.0 | 0.3999 | 0.5863 | 0.7044 | 0.6282 | 0.5449 | 0.4807 | 0.4949 | 0.4999 | 0.5000 |
| 8.0 | 0.4000 | 0.5900 | 0.7218 | 0.6382 | 0.5415 | 0.4679 | 0.4950 | 0.5000 | 0.5000 |
| 9.0 | 0.4000 | 0.5926 | 0.7367 | 0.6471 | 0.5381 | 0.4553 | 0.4962 | 0.5001 | 0.5000 |
| 10.0 | 0.4000 | 0.5945 | 0.7496 | 0.6550 | 0.5348 | 0.4430 | 0.4986 | 0.5002 | 0.5000 |

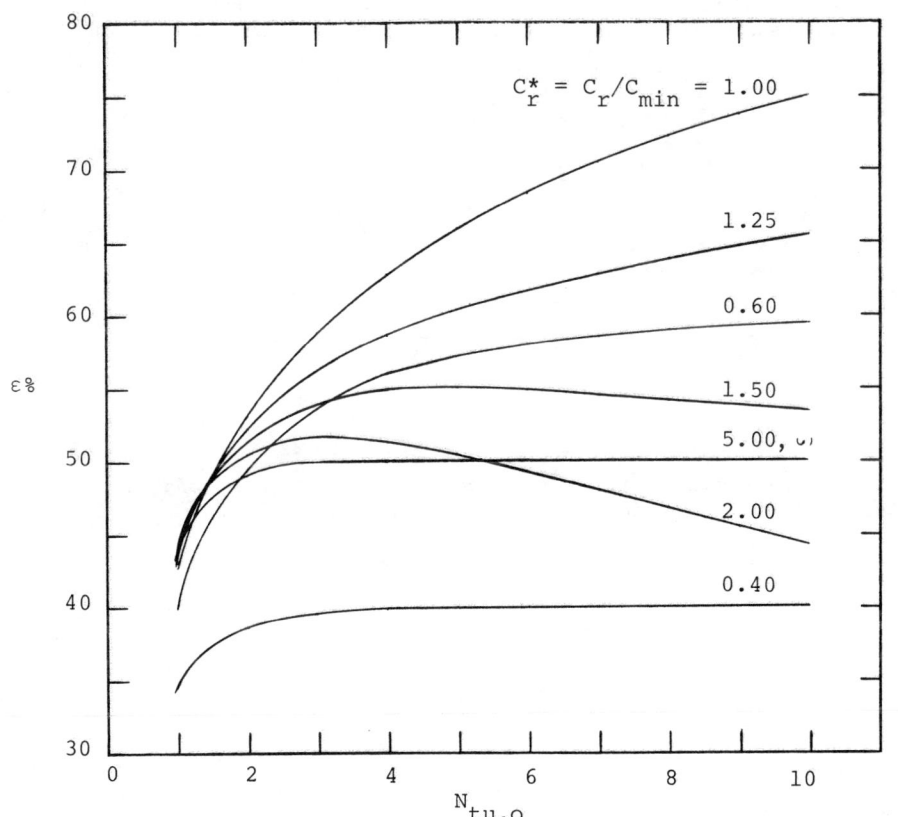

Fig. 5 Parallel flow regenerator $\varepsilon$ as a function of $N_{tu,o}$ and $C_r^*$ for $C^* = 1$ and $(hA)^* = 1$, from Table 3.

## 4.2 The Λ-Π Method

This method is due to Hausen [9] in 1929. He analyzed <u>one matrix</u> of a fixed-matrix regenerator starting with Eqs. (12), (14)-(16). He introduced the following Schumann dimensionless independent variables to make these equations dimensionless.

$$\xi_h = \frac{h_h A_h}{C_h} \frac{x}{L} \qquad \qquad \xi_c = \frac{h_c A_c}{C_c} \frac{x}{L} \tag{76}$$

$$\eta_h = \frac{h_h A_h}{\bar{C}_{r,h}}\left(\tau_h - \frac{x}{L}\tau_{d,h}\right) \approx \frac{h_h A_h}{\bar{C}_{r,h}}\tau_h \qquad \eta_c = \frac{h_c A_c}{\bar{C}_{r,c}}\left(\tau_c - \frac{x}{L}\tau_{d,c}\right) \approx \frac{h_c A_c}{\bar{C}_{r,c}}\tau_c \tag{77}$$

and $\qquad \bar{C}_{r,h} = (M_w)_h C_w = M_w c_w \qquad \qquad \bar{C}_{r,c} = (M_w)_c c_w = M_w c_w \tag{78}$

Here $A_h = A_c = A$ represents the total heat transfer area of a single matrix, and $(M_w)_h = (M_w)_c = M_w$ represents the total mass of a single matrix. Substituting these variables into Eqs. (12), (14)-(16) yield the following. For hot gas flow period,

$$\frac{\partial t_h}{\partial \xi_h} = (t_w - t_h) \tag{79}$$

$$\frac{\partial t_w}{\partial \eta_h} = (t_h - t_w) \tag{80}$$

For the cold gas flow period,

$$\frac{\partial t_c}{\partial \xi_c} = (t_c - t_w) \tag{81}$$

$$\frac{\partial t_w}{\partial \eta_c} = (t_w - t_c) \tag{82}$$

Note that these equations are now parameter-free. The boundary and periodic-flow conditions for these equations are still those of Eqs. (17)-(20).

$\xi$ and $\eta$ of Eqs. (76) and (77) are represented as

$$\xi = \left(\frac{hA}{CL}\right)x = bx \propto x \tag{83}$$

$$\eta = \left(\frac{hA}{\bar{C}_r}\right)\tau = c\tau \propto \tau \tag{84}$$

Here b and c are the constants since $h$, $A$, $C$, $L$, $\bar{C}_r$ all are constants due to the idealizations listed in Section 2. For this reason, the variable $\xi$ and $\eta$ are designated as the reduced length and reduced period variables, respectively.

If the temperatures $t_h$, $t_c$, and $t_w$ are made dimensionless as before, Eq. (39), it is evident from Eqs. (79)-(82) that

$$t_h^*, t_c^*, t_w^* = \phi\{\xi_h, \xi_c, \eta_h, \eta_c\} \tag{85}$$

# THERMAL DESIGN THEORY FOR REGENERATORS

For overall regenerator performance, we are interested in determining the time average fluid outlet temperatures. These temperatures again are obtained from the definitions of Eq. (51) with $\tau$ replaced by $\eta_h$ and $\eta_c$ for the hot and cold gas flow periods, respectively. These temperatures are expressed by the regenerator effectiveness. Since we have considered only one matrix in the foregoing analysis, the hot and cold fluid flows are intermittent. In an ideal "steady-state" periodic condition, the actual heat transfer will be

$$Q = C_h P_h (t_{h,i} - \bar{t}_{h,o}) = C_c P_c (\bar{t}_{c,o} - t_{c,i}) \tag{86}$$

The maximum possible heat transfer will be in a counterflow regenerator of infinite surface area and having the same fluid flow rates and fluid inlet temperatures. Thus this maximum possible heat transfer rates during hot and cold gas flow periods respectively are

$$Q_{max,h} = C_h P_h (t_{h,i} - t_{c,i}) \qquad Q_{max,c} = C_c P_c (t_{h,i} - t_{c,i}) \tag{87}$$

Thus the regenerator effectiveness during hot and cold gas flow periods are

$$\varepsilon_h = \frac{Q_h}{Q_{max,h}} = \frac{C_h P_h (t_{h,i} - \bar{t}_{h,o})}{C_h P_h (t_{h,i} - t_{c,i})} = \frac{t_{h,i} - \bar{t}_{h,o}}{t_{h,i} - t_{c,i}} \tag{88}$$

$$\varepsilon_c = \frac{Q_c}{Q_{max,c}} = \frac{C_c P_c (\bar{t}_{c,o} - t_{c,i})}{C_c P_c (t_{h,i} - t_{c,i})} = \frac{\bar{t}_{c,o} - t_{c,i}}{t_{h,i} - t_{c,i}} \tag{89}$$

The overall effectiveness of a single matrix may be defined as [10]

$$\varepsilon_r = \frac{Q_h + Q_c}{Q_{max,h} + Q_{max,c}} = \frac{2Q}{Q_{max,h} + Q_{max,c}} \tag{90}$$

Using Eqs. (86)-(90), it can be shown that

$$\frac{1}{\varepsilon_r} = \frac{1}{2}\left(\frac{1}{\varepsilon_h} + \frac{1}{\varepsilon_c}\right) \tag{91}$$

Now returning back to Eq. (85), the dependent fluid outlet temperatures averaged over respective periods are presented as functions of $\varepsilon_r$, $\varepsilon_h$ or $\varepsilon_c$. The four independent variables of Eq. (85) in this case are evaluated at $x = L$ and $\tau_h = P_h$ and $\tau_c = P_c$. These independent variables are

$$\Lambda_h = \xi_h(L), \quad \Lambda_c = \xi_c(L), \quad \Pi_h = \eta_h(P_h), \quad \Pi_c = \eta_c(P_c) \tag{92}$$

Thus

$$\varepsilon_r, \varepsilon_h, \varepsilon_c = \phi\{\Lambda_h, \Lambda_c, \Pi_h, \Pi_c\} \tag{93}$$

again a function of four dimensionless groups. Since

$$\Lambda = bL \quad \text{and} \quad \Pi = cP_h \quad \text{or} \quad cP_c \tag{94}$$

from Eqs. (83) and (84), and since b and c are constants, $\Lambda$ and $\Pi$ are designated as <u>reduced length</u> and <u>reduced period</u> respectively for the regenerator. The reduced length $\Lambda$ also designates the dimensionless heat transfer size of the regenerator. And this method is referred to as <u>the $\Lambda$-$\Pi$ method</u>. It has

Table 4. Designation of various types of regenerators.

| Terminology | $\Lambda-\Pi$ Method | $\varepsilon-N_{tu,o}$ Method |
|---|---|---|
| Balanced Regenerators | $\Lambda_h/\Pi_h = \Lambda_c/\Pi_c$ | $C^* = 1$ |
| Unbalanced Regenerators | $\Lambda_h/\Pi_h \neq \Lambda_c/\Pi_c$ | $C^* \neq 1$ |
| Symmetric and Balanced Regenerators | $\Lambda_h = \Lambda_c$, $\Pi_h = \Pi_c$ | $(hA)^*=1, C^*=1$ |
| Unsymmetric but Balanced Regenerators | $\Lambda_h/\Pi_h = \Lambda_c/\Pi_c$ | $(hA)^* \neq 1, C^*=1$ |
| Long Regenerators | $\Lambda/\Pi > 3$ | $C_r^* > 5$ |

been primarily used for the design of fixed-matrix regenerators. The terminology used for describing various types of fixed-matrix regenerators is described in Table 4.

For the effectiveness of the most general unbalanced and unsymmetric regenerator, Razelos [10] proposed an alternate set of four dimensionless groups, instead of those of Eq. (93), as

$$\varepsilon_r, \varepsilon_h, \varepsilon_c = \phi(\Lambda_m, \Pi_m, \gamma, R^*) \tag{95}$$

where $\Lambda_m$ and $\Pi_m$ are the mean reduced length and mean reduced period respectively. They have been proposed by Hausen [11] as the "harmonic means" in the following sense.

$$\frac{1}{\Pi_m} = \frac{1}{2}\left[\frac{1}{\Pi_h} + \frac{1}{\Pi_c}\right] \tag{96}$$

$$\frac{1}{\Lambda_m} = \frac{1}{2\Pi_m}\left[\frac{\Pi_h}{\Lambda_h} + \frac{\Pi_c}{\Lambda_c}\right] \tag{97}$$

and $\gamma$ and $R^*$ are

$$\gamma = \frac{\Pi_c/\Lambda_c}{\Pi_h/\Lambda_h} = \frac{C_c P_c}{(M_w)_c c_w} \frac{(M_w)_h c_w}{C_h P_h} = \frac{C_c}{C_h} \tag{98}$$

$$R^* = \frac{\Pi_h}{\Pi_c} = \frac{(hA)_h}{(hA)_c} \frac{P_h}{(M_w)_h c_w} \frac{(M_w)_c c_w}{P_c} = \frac{(hA)_h}{(hA)_c} \tag{99}$$

In Eqs. (98) and (99), note that $(M_w)_h c_w/P_h = C_{r,h}$, $(M_w)_c c_w/P_c = C_{r,c}$, and $C_{r,h} = C_{r,c} = C_r$ from Eq. (8). Razelos [10] also showed that the influence of $R^*$ on $\varepsilon_r$ is negligible for $1 \leq R^* \leq 5$. Thus $\varepsilon_r$

$$\varepsilon_r = \phi(\Lambda_m, \Pi_m, \gamma) \tag{100}$$

He also pointed out that

$$\varepsilon_r(\gamma) = \varepsilon_r(1/\gamma) \tag{101}$$

---

†Since $R^* = 1/(hA)^*$ for $C_c = C_{min}$, it has been already shown by Lambertson [6] that the influence of $(hA)^*$ is negligible on $\varepsilon$ for $0.25 \leq (hA)^* \leq 4$.

# THERMAL DESIGN THEORY FOR REGENERATORS

and therefore the tabulation of $\varepsilon_r$ is needed only for $\gamma \leq 1$. Note that since

$$\varepsilon_h = \gamma \varepsilon_c \tag{102}$$

the regenerator effectiveness $\varepsilon_r$ of Eq. (91) may also be presented as

$$\varepsilon_r = \frac{2\gamma}{\gamma + 1} \varepsilon_c \tag{103}$$

Before presenting the specific results, let us compare the dimensionless groups of the $\Lambda$-$\Pi$ method versus those of the $\varepsilon$-$N_{tu,o}$ method.

A Comparison of the $\varepsilon$-$N_{tu,o}$ and $\Lambda$-$\Pi$ methods. The functional relationships for these methods are given by Eqs. (62) and (93) or (95). For comparison purpose, we will consider $C_c = C_{min}$. The regenerator effectiveness $\varepsilon$ is related to $\varepsilon_r$, $\varepsilon_h$ and $\varepsilon_c$ as

$$\varepsilon = \begin{cases} (\gamma + 1)\varepsilon_r/2\gamma \\ \varepsilon_h/\gamma \\ \varepsilon_c \end{cases} \tag{104}$$

The independent variables of Eqs. (95) and (62) are related as follows for $C_c = C_{min}$.

$$\frac{\Lambda_m(1 + \gamma)}{4\gamma} = \left[\frac{C^*}{N_{tu,h}} + \frac{1}{N_{tu,c}}\right]^{-1} = N_{tu,o} \tag{105}$$

$$\gamma = C^* \tag{106}$$

$$\frac{\Lambda_m(1 + \gamma)}{2\gamma \Pi_m} = C_r^* \tag{107}$$

$$\frac{1}{R^*} = (hA)^* \tag{108}$$

Thus there is one-to-one correspondence between the dimensionless groups of the $\Lambda$-$\Pi$ and $\varepsilon$-$N_{tu,o}$ methods.

The independent variables of Eqs. (93) and (62) are related as follows for $C_c = C_{min}$.

$$\Lambda_h = \frac{(hA)_h}{C_h} = N_{tu,h} = C^*\left[1 + \frac{1}{(hA)^*}\right] N_{tu,o} \tag{109}$$

$$\Lambda_c = \frac{(hA)_c}{C_c} = N_{tu,c} = \left[1 + (hA)^*\right] N_{tu,o} \tag{110}$$

$$\Pi_h = \frac{(hA)_h}{\bar{C}_{r,h}} P_h = \frac{N_{tu,h}}{C^*_{r,h}} = \frac{1}{C^*_r}\left[1 + \frac{1}{(hA)^*}\right] N_{tu,o} \tag{111}†$$

---

†If the hot and cold gas dwell times are not neglected in the definition of $\Pi_h$ and $\Pi_c$, the right-hand term of Eqs. (111) and (112) should be multiplied by $(1-\tau_{d,h}/P_h)$ and $(1-\tau_{d,c}/P_c)$, respectively.

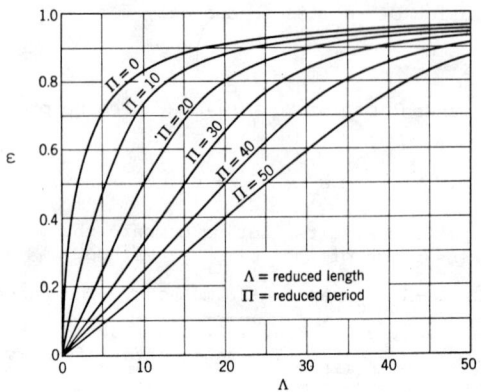

Fig. 6  Hausen's effectiveness chart for a balanced counterflow regenerator [12].

$$\Pi_c = \frac{(hA)_c}{\overline{C}_{r,c}} \qquad P_c = \frac{N_{tu,c}}{C^*_{r,c}} = \frac{1}{C^*_r}\left[1 + (hA)^*\right]N_{tu,o} \qquad (112)^\dagger$$

Noting the relationships of Eqs. (105)-(112), it is clear that there is one-to-one correspondence between the dimensionless groups of $\varepsilon$-$N_{tu,o}$ and $\Lambda$-$\Pi$ methods.

Solutions for a counterflow regenerator. Hausen [12] in 1930 obtained the approximate solution for a balanced and symmetric counterflow regenerator by the finite difference and heat pole methods. In this case,

$$\varepsilon = \phi(\Lambda, \Pi) \qquad (113)$$

where $\Lambda = \Lambda_h = \Lambda_c$ and $\Pi = \Pi_h = \Pi_c$. His results are shown in Fig. 6. From the review of Table 4 and relationships of Eqs. (109)-(112), we have

$$\Lambda = 2N_{tu,o} \qquad \Pi = 2N_{tu,o}/C^*_r \qquad (114)$$

Using the results of Table 1 and Eq. (114), the effectiveness of Fig. 6 can be determined for the balanced and symmetric regenerators.

Solution for a parallel flow regenerator. Hausen [12] also obtained the solution for a balanced parallel flow regenerator as shown in Fig. 7. The oscillations in $\varepsilon$ above and below $\varepsilon = 0.5$ are clearly observed in this figure.

## 5. INFLUENCE OF LONGITUDINAL WALL HEAT CONDUCTION

Longitudinal heat conduction in the wall may not be negligible, particularly for a high effectiveness regenerator having a short flow length L. Longitudinal wall heat conduction reduces the exchanger effectiveness, hence it is important that its influence on $\varepsilon$ is determined quantitatively. In this case, the

---
[†] See the footnote on the previous page.

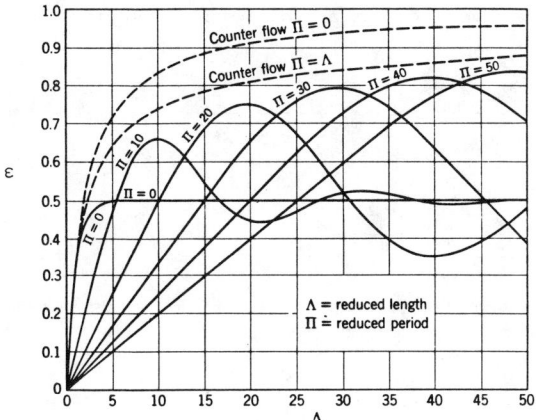

Fig. 7  Hausen's effectiveness chart for a balanced parallel flow regenerator [12].

differential equations for the hot and cold sides do not change, but those for the wall change. For finite axial heat conduction in the wall during the hot gas flow period, Fig. 2b is modified by adding the corresponding conduction terms as shown in Fig. 8.

Applying the first law of thermodynamics to the differential element of the wall in Fig. 8, we get

$$-k_w A_{k,h} \left( \frac{\partial t_w}{\partial x} + \frac{\partial^2 t_w}{\partial x^2} dx \right) + k_w A_{k,h} \frac{\partial t_w}{\partial x} - h_h \left( \frac{A_h dx}{L} \right) (t_h - t_w) + \left( \bar{C}_{r,h} \frac{dx}{L} \right) \frac{\partial t_w}{\partial \tau_h} = 0 \quad (115)$$

Upon simplification,

$$\frac{\partial t_w}{\partial \tau_h} = \frac{(hA)_h}{\bar{C}_{r,h}} (t_h - t_w) + \frac{k_w A_{k,h} L}{\bar{C}_r} \frac{\partial^2 t_w}{\partial x^2} \quad (116)$$

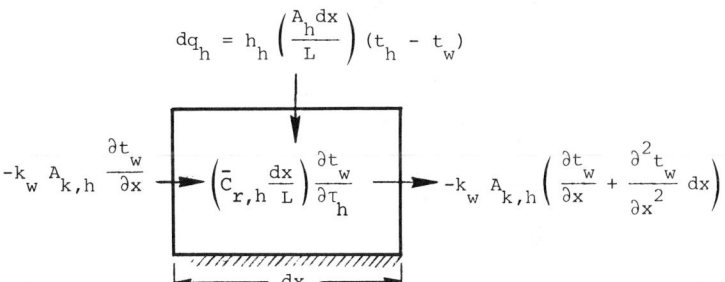

Fig. 8  Energy rate terms associated with the element dx of matrix during hot gas flow period; longitudinal heat conduction has been included.

This equation is made dimensionless as follows by employing the previous definitions of $t_w^*$, $\tau_h^*$, $X^*$, $N_{tu,h}$, $C_{r,h}^*$.

$$\frac{\partial t_w^*}{\partial \tau_h^*} = \frac{N_{tu,h}}{C_{r,h}^*}(t_h^* - t_w^*) + \frac{\lambda_h}{C_{r,h}^*}\frac{\partial^2 t_w^*}{\partial X^{*2}} \tag{117}$$

where

$$\lambda_h = \frac{k_w A_{k,h}}{L C_h} \tag{118}$$

Similarly the governing differential equation for the matrix wall temperature during the cold gas flow period is

$$\frac{\partial t_w^*}{\partial \tau_c^*} = \frac{N_{tu,c}}{C_{r,c}^*}(t_w^* - t_c^*) + \frac{\lambda_c}{C_{r,c}^*}\frac{\partial^2 t_w^*}{\partial X^{*2}} \tag{119}$$

where

$$\lambda_c = \frac{k_w A_{k,c}}{L C_c} \tag{120}$$

Since Eqs. (117) and (119) are second order partial differential equations with respect to $X^*$, in order to get a complete solution, we need to add four boundary conditions for the matrix wall temperatures: two during the hot gas flow period and two during the cold gas flow period. The simplest and fairly reasonable boundary conditions are the adiabatic boundary conditions for each period at $x = 0$ and $L$.

$$\frac{\partial t_w^*(0, \tau_h^*)}{\partial X^*} = \frac{\partial t_w^*(1, \tau_h^*)}{\partial X^*} = 0 \quad \text{for} \quad 0 \leq \tau_h^* \leq 1 \tag{121}$$

$$\frac{\partial t_w^*(0, \tau_c^*)}{\partial X^*} = \frac{\partial t_w^*(1, \tau_c^*)}{\partial X^*} = 0 \quad \text{for} \quad 0 \leq \tau_c^* \leq 1 \tag{122}$$

Thus the inclusion of the effect of longitudinal heat conduction adds two dimensionless groups $\lambda_h$ and $\lambda_c$ on which the exchanger effectiveness $\varepsilon$ would depend. Bahnke and Howard [13] suggested an alternate set of two dimensionless groups.

$$\lambda = \frac{k_w A_{k,t}}{L C_{min}} \tag{123}$$

$$A_k^* = \frac{A_k \text{ on the } C_{min} \text{ side}}{A_k \text{ on the } C_{max} \text{ side}} \tag{124}$$

where $A_{k,t}$ is the total solid area for longitudinal conduction,

$$A_{k,t} = A_{k,h} + A_{k,c} = A_{fr} - A_o = A_{fr}(1-\sigma) \tag{125}$$

Note that $\lambda$ and $A_k^*$ are related to $\lambda_h$ and $\lambda_c$ as follows for $C_c = C_{min}$.

# THERMAL DESIGN THEORY FOR REGENERATORS

$$\lambda = \lambda_c + \lambda_h/C^* \tag{126}$$

$$A_k^* = C^*(\lambda_c/\lambda_h) \tag{127}$$

This choice of dimensionless groups offers an advantage that the resulting $\varepsilon$ is not significantly affected by $A_k^*$ for $0.25 \leq A_k^* \leq 1$ [13]. Thus the effect of longitudinal heat conduction in the wall is accounted for by $\lambda$ and is added to the functional relationship for $\varepsilon$ of Eq. (68).

$$\varepsilon = \phi \left\{ N_{tu,o}, C^*, C_r^*, \lambda \right\} \tag{128}$$

In order to obtain a rigorous solution for this problem, Eqs. (40), (42), (117) and (119) need to be solved using the periodic and boundary conditions of Eqs. (44)-(47), (121) and (122). There is no closed-form solution to these equations. Bahnke and Howard [13] obtained numerical solutions by a finite difference method. They determined the exchanger effectiveness and conduction effect over the following ranges of dimensionless parameters: $1 \leq N_{tu,o} \leq 100$, $0.9 \leq C^* \leq 1$, $1 \leq C_r^* \leq \infty$, $0.01 \leq \lambda \leq 0.32$, $0.25 \leq (hA)^* \leq 1$, $0.25 \leq A_k^* \leq 1$. The ineffectivenesses $(1-\varepsilon)$ as a function of $N_{tu,o}$ and $\lambda$ are shown in Figs. 9 and 10 for $C_r^* > 5$ and $C^* = 1$ and 0.95 respectively.

Bahnke and Howard's results are correlated by Shah [14] as

$$\varepsilon = \varepsilon_{cf}\left[1 - \frac{1}{9C_r^{*1.93}}\right]\left[1 - \frac{1}{2-C^*}\left(\frac{1}{1 + N_{tu,o}(1 + \lambda\Phi)/(1 + \lambda N_{tu,o})} - \frac{1}{1 + N_{tu,o}}\right)\right] \tag{129}$$

where

$$\Phi = \left[\frac{\lambda N_{tu,o}}{1 + \lambda N_{tu,o}}\right]^{1/2} \tanh\left[\frac{N_{tu,o}}{\{\lambda N_{tu,o}/(1 + \lambda N_{tu,o})\}^{1/2}}\right] \tag{130a}$$

$$\approx \left[\frac{\lambda N_{tu,o}}{1 + \lambda N_{tu,o}}\right]^{1/2} \quad \text{for } N_{tu,o} \geq 3 \tag{130b}$$

The regenerator effectiveness $\varepsilon$ of Eq. (129) agrees well within ±0.5% with the results of Bahnke and Howard for the following range of parameters: $3 \leq N_{tu,o} \leq 12$, $0.9 \leq C^* \leq 1$, $2 \leq C_r^* \leq \infty$, $0.5 \leq (hA)^* \leq 1$, and $0 \leq \lambda \leq 0.04$. It agrees within ±1% for the following range of parameters: $1 \leq N_{tu,o} \leq 20$, $0.9 \leq C^* \leq 1$, $2 \leq C_r^* \leq \infty$, $0.25 \leq (hA)^* \leq 1$, and $0 \leq \lambda \leq 0.08$.

No detailed temperature distributions were obtained by either Lambertson [6] or Bahnke and Howard [13]. Mondt [15] obtained these temperature distributions by solving the differential equations numerically for some values of the associated dimensionless groups. Illustrative results are shown in Figs. 11, 12, and 13.

In Fig. 11, the matrix wall temperatures $t_w^*$ at x=0, x=L/2 and x=L are shown as functions of a dimensionless time for the $\lambda=0$ case. Also imposed are the hot and cold gas inlet and outlet temperatures. Experimental points shown for $t_{c,o}^*$ are in good agreement with the theoretical predictions. The wall temperatures are linear with time except for the sections of hot and cold fluid inlets.

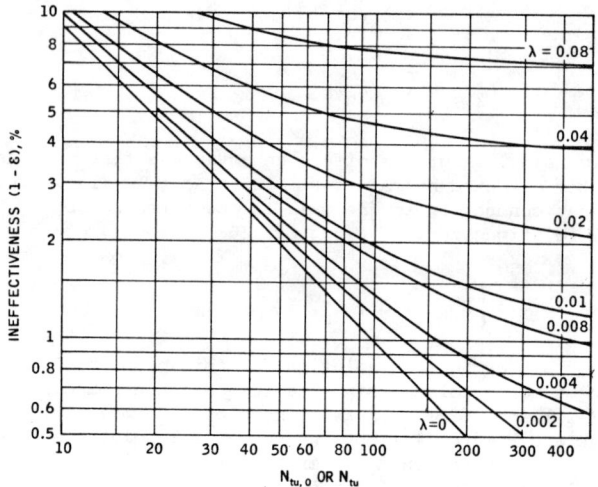

Fig. 9  Influence of longitudinal wall heat conduction on the performance of storage-type and direct transfer type counterflow exchangers; $C^* = 1$, $C_r^* > 5$ [5].

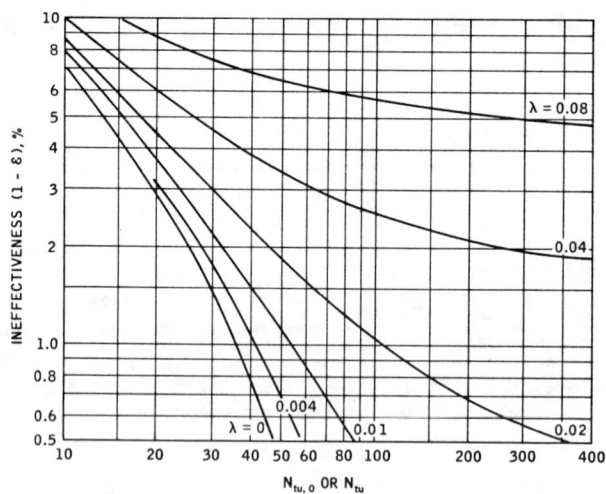

Fig. 10  Influence of longitudinal wall heat conduction on the performance of storage-type and direct transfer type counterflow exchangers; $C^* = 0.95$, $C_r^* > 5$ [5].

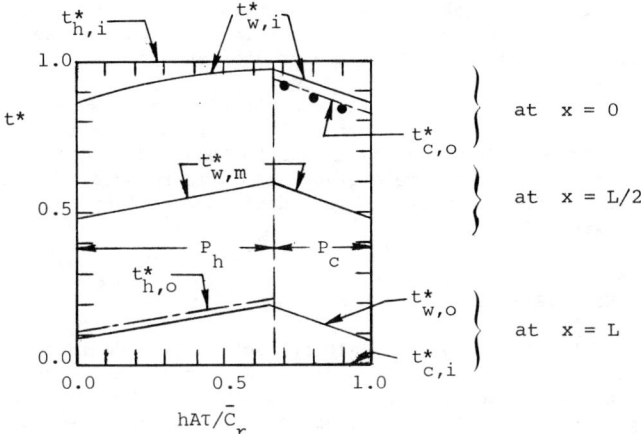

Fig. 11  Cyclic temperature fluctuations in the matrix at the entrance, midway and exit of a regenerator [15].

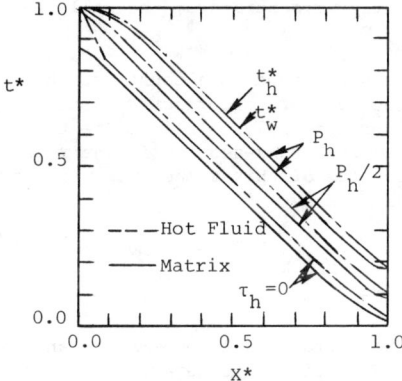

Fig. 12  Fluid and matrix wall temperature excursion during hot gas flow period [15].

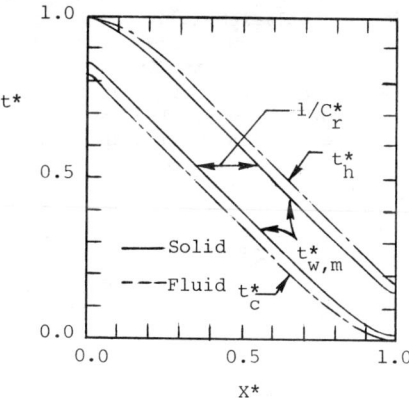

Fig. 13  Balanced regenerator temperature distributions at switching instants [15].

In Fig. 12, hot gas and matrix temperatures are shown as a function of the flow length $X^*$ for $\tau_h=0$, $P_h/2$ and $P_h$. Here again these temperature distributions are linear except for the regenerator ends.

In Fig. 13, the matrix wall temperatures are shown at switching times $\tau=P_h$ and $P_c$. The reduction in matrix wall temperature gradients at $X^*=0$ and increase in these gradients at $X^*=1$ due to longitudinal heat conduction is evident. Note that the time average $t^*_{c,\varrho}$ is reduced which in turn indicates the exchanger effectiveness $\varepsilon$ is reduced due to longitudinal wall heat conduction, as expected.

## 6. INFLUENCE OF WALL THERMAL RESISTANCE

One of the idealizations (#10 in Section 2) made in the foregoing regenerator design theory is that the wall thermal resistance is zero. This idealization is invoked in deriving Eqs. (14) and (16). The temperature gradient in the wall thickness direction in Fig. 2b is zero. It was also shown that the wall thickness resistance in $U_oA$ of Eq. (63) is zero. The zero wall thermal resistance represents a good approximation for metal matrices having high thermal conductivity. For those matrices having low thermal conductivity, such as ceramic matrices for fixed-matrix regenerators, the wall resistance may not be negligible.

A simplified method is now outlined to account for the influence of the wall thermal resistance on the regenerator effectiveness. An additional but essential idealization made for the analysis[†] is: The temperatures of hot and cold gases and the wall at any cross-section in the regenerator are linear with time, and the numerical value of this temperature-time gradient at any point in the wall is the same. This means

$$\frac{\partial t_w}{\partial \tau} = \frac{\partial t_{w,o}}{\partial \tau} = \frac{\partial t_{w,m}}{\partial \tau} = \text{constant} \qquad (131)$$

Here $t_{w,o}$ and $t_{w,m}$ are the surface wall temperature and mean wall temperature at a given instant of time. This linear temperature-time relationship represents a good approximation to the actual temperature profile in the greater part of either period along the most of the regenerator length as shown in Fig. 11.

Now let us consider a differential element of the regenerator matrix wall as shown in Fig. 14 during the hot gas flow period. Here $A_w$ represents the conduction area for the wall for heat conduction in the y direction. For continuous flow passages, $A_w = A$ (the convective heat transfer surface area), and $A_w \simeq A$ for non-continuous flow passages. The energy balance on the element of Fig. 14b yields the well-known one-dimensional transient conduction equation valid at each x coordinate.

$$\alpha_w \frac{\partial^2 t_w}{\partial y^2} = \frac{\partial t_w}{\partial \tau} \qquad (132)$$

where $\alpha_w = k_w/\rho_w c_w$ is the thermal diffusivity of the matrix material. The appropriate boundary conditions are

---

[†]In this section, the analysis is made for finite thermal resistance in the wall thickness direction and infinite thermal resistance in the longitudinal direction (the zero longitudinal heat conduction case).

# THERMAL DESIGN THEORY FOR REGENERATORS

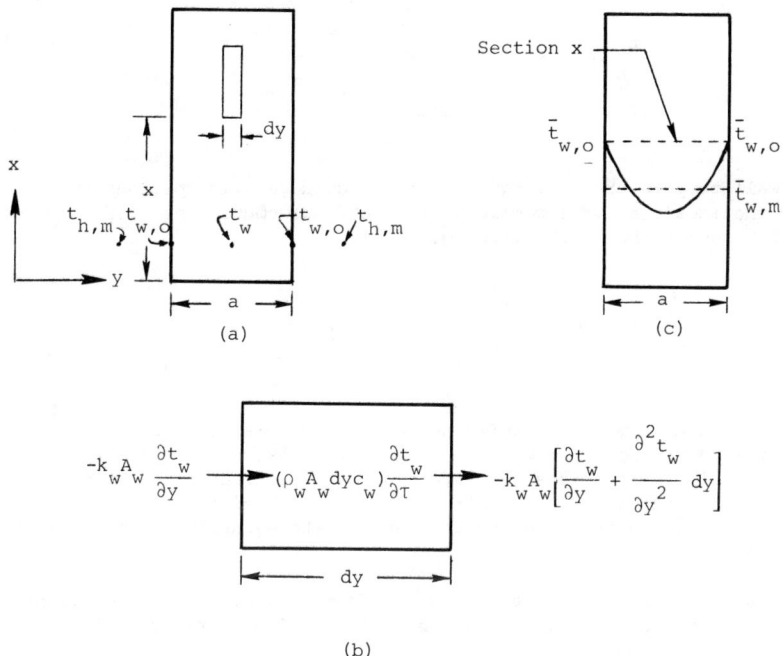

Fig. 14  (a) Matrix wall of thickness a, (b) energy transfer terms across the differential wall element, (c) a parabolic temperature distribution in the wall at a given x.

$$t_w = t_{w,o} \qquad \text{at } y = 0 \text{ and } a \qquad (133)$$

Now let us first obtain the temperature distribution in the wall by a double integration of Eq. (132) using the boundary conditions of Eq. (133). We get

$$\bar{t}_w - \bar{t}_{w,o} = -\frac{1}{2\alpha_w}\left(\frac{\partial t_w}{\partial \tau}\right)(a - y)y \qquad (134)$$

Note that $\partial t_w/\partial \tau$ has been treated as constant during the integration due to the idealization of Eq. (131). Since $t_w$ and $t_{w,o}$ then do not depend upon $\tau$, these are designated by a bar in Eq. (134) to denote them as time average quantities. The temperature profile of Eq. (134) is parabolic at each x as shown in Fig. 14c. The mean wall temperature $\bar{t}_{w,m}$ is obtained by integrating Eq. (134) with respect to y across the wall thickness as

$$\bar{t}_{w,m} - \bar{t}_{w,o} = -\frac{a^2}{12\alpha_w}\left(\frac{\partial t_w}{\partial \tau}\right) \qquad (135)$$

Thus the difference between the mean wall temperature and the surface wall temperature during hot and cold gas flow periods are

$$(\bar{t}_{w,m} - \bar{t}_{w,o})_h = -\frac{a^2}{12\alpha_w}\left(\frac{\partial t_w}{\partial \tau}\right)_h \tag{136}$$

$$(\bar{t}_{w,m} - \bar{t}_{w,o})_c = -\frac{a^2}{12\alpha_w}\left(\frac{\partial t_w}{\partial \tau}\right)_c \tag{137}$$

The mean wall temperatures during the hot and cold gas flow periods must be equal for the idealized true cyclic (periodic flow) conditions, i.e., $\bar{t}_{w,m,h} = \bar{t}_{w,m,c}$. Subtracting Eq. (137) from Eq. (136),

$$(\bar{t}_{w,o})_h - (\bar{t}_{w,o})_c = -\frac{a^2}{12\alpha_w}\left[\left(\frac{\partial t_w}{\partial \tau}\right)_c - \left(\frac{\partial t_w}{\partial \tau}\right)_h\right] \tag{138}$$

In order to determine the influence of wall thermal resistance on the regenerator effectiveness, we want to arrive at an expression for $R_w$ for a flat (plain) wall. This means we want to derive an expression for an equivalent UA for the regenerator for the finite wall thermal resistance case. This will be achieved by writing an energy balance and a rate equation for an element $dx$ during hot and cold gas periods.

The heat transfer by conduction to and from a matrix element of length $dx$ and wall thickness $a$ during the hot and cold gas flow periods is

$$dQ_h = \rho_w c_w a dA \left(\frac{\partial t_w}{\partial \tau}\right)_h P_h \tag{139}$$

$$dQ_c = -\rho_w c_w a dA \left(\frac{\partial t_w}{\partial \tau}\right)_c P_c \tag{140}^\dagger$$

Substituting values of the temperature gradients from these equations into Eq. (138) and noting $dQ_h = dQ_c = dQ$ for the idealized cyclic conditions,

$$(\bar{t}_{w,o})_h - (\bar{t}_{w,o})_c = \frac{a}{12k_w}\left[\frac{1}{P_h} + \frac{1}{P_c}\right]\frac{dQ}{dA} \tag{141}$$

Now the rate equations during hot and cold gas flow periods are

$$dQ_h = h_h\, dA\,[\bar{t}_h - (\bar{t}_{w,o})_h]P_h \tag{142}$$

$$dQ_c = h_c\, dA\,[\bar{t}_c - (\bar{t}_{w,o})_c]P_c \tag{143}$$

where $\bar{t}_h$ and $\bar{t}_c$ are the hot and cold fluid time average temperatures at a section $x$ during the hot and cold gas flow periods, respectively. Substituting $(\bar{t}_{w,o})_h$ and $(\bar{t}_{w,o})_c$ from these equations into Eq. (141) and noting $dQ_h = dQ_c = dQ$, a rearrangement yields

---

$^\dagger$The equations are valid for a fixed-matrix regenerator for which the surface area $dA$ at a given $x$ remains the same during hot and cold gas periods. For a rotary regenerator, these equations are valid for the observer riding on the matrix.

# THERMAL DESIGN THEORY FOR REGENERATORS

$$dQ = \left[\frac{1}{h_h P_h} + \frac{1}{h_c P_c} + \frac{a}{12k_w}\left(\frac{1}{P_h} + \frac{1}{P_c}\right)\right] dA(\bar{t}_h - \bar{t}_c) \tag{144}$$

Since $Q = Q_h = Q_c$ represents total heat transfer per cycle in time $(P_h + P_c)$, the average heat transfer rate during one cycle is

$$q = Q/(P_h + P_c) \tag{145}$$

and hence

$$dq = dQ/(P_h + P_c) \tag{146}$$

Substituting $dQ$ from Eq. (144) into Eq. (146), we get

$$dq = U_o \, dA(\bar{t}_h - \bar{t}_c) \tag{147}$$

where

$$\frac{1}{U_o} = \left[\frac{1}{h_h P_h} + \frac{1}{h_c P_c} + \frac{a}{12k_w}\left(\frac{1}{P_h} + \frac{1}{P_c}\right)\right](P_h + P_c) \tag{148}^\dagger$$

An integration of Eq. (147) along the length of a counterflow regenerator yields

$$q = U_o A \, \Delta t_{\ell m} \tag{149}$$

where

$$\Delta t_{\ell m} = \frac{(\bar{t}_{h,i} - \bar{t}_{c,o}) - (\bar{t}_{h,o} - \bar{t}_{c,i})}{\ln[(\bar{t}_{h,i} - \bar{t}_{c,o})/(\bar{t}_{h,o} - \bar{t}_{c,i})]} \tag{150}$$

The bar on $t$ represents the corresponding (hot or cold) period time average temperature. $A$ in Eq. (149) represents the total surface area $(A_h + A_c)$ in contrast to either $A_h$ or $A_c$ for a recuperator.

Now dividing both sides of Eq. (148) by the total surface area $A$ and introducing the definitions of $A_h$ and $A_c$ from Eqs. (9) and (10), we get

$$\frac{1}{U_o A} = \left[\frac{1}{h_h A_h} + \frac{1}{h_c A_c} + \frac{a}{12k_w}\left(\frac{1}{A_{w,h}} + \frac{1}{A_{w,c}}\right)\right] \tag{151}$$

where the relationships $A_{w,h} = A_h$ and $A_{w,c} = A_c$, as noted just before Eq. (132), are used in the term of parenthesis. Now this equation is valid for both the rotary and fixed-matrix regenerators as long as proper values of $A_h$ and $A_c$ from Eqs. (9) and (10) are used. From a comparison of this equation with that of a recuperator, the equivalent wall thermal resistance for the regenerator is

$$R_w = \frac{a}{12k_w}\left(\frac{1}{A_{w,h}} + \frac{1}{A_{w,c}}\right) \tag{152}$$

---

†It should be emphasized that $U_o$ is explicitly defined by Eq. (148). In contrast, the UA product is defined for a recuperator and $A$ must be specified to calculate $U$ for a recuperator.

For a special case of 50/50 split of flow areas in a rotary regenerator or $P_h = P_c$ in a fixed-matrix regenerator, $A_{w,h} = A_{w,c} = A_w/2$. Thus for this case

$$R_w = \frac{a}{3k_w A_w} \qquad (153)$$

Now for a recuperator having plain walls of thickness a, conduction area $A_w$ and thermal conductivity $k_w$, the wall thermal resistance is

$$R_w = \frac{a}{k_w A_w} \qquad (154)$$

From a comparison of Eqs. (153) and (154), it is evident that the wall thermal resistance of a regenerator is 1/3 that of an equivalent recuperator. Alternatively, the regenerator is equivalent to a recuperator of 1/3 wall thickness.

One of the basic idealizations made in the foregoing analysis is that the temperatures $t_h$, $t_c$ and $t_w$ are all linear with time. Since this is not true at the switching moment and near regenerator inlet and outlet, Hausen [11] suggested to modify $R_w$ of Eq. (152) by a factor $\Phi^*$ so that

$$\frac{1}{U_o A} = \left[ \frac{1}{h_h A_h} + \frac{1}{h_c A_c} + R_w \Phi^* \right] \qquad (155)$$

where for a plain wall,

$$\Phi^* = \begin{cases} 1 - \dfrac{a^2}{60\alpha_w}\left(\dfrac{1}{P_h} + \dfrac{1}{P_c}\right) & \text{for } \dfrac{a^2}{2\alpha_w}\left(\dfrac{1}{P_h} + \dfrac{1}{P_c}\right) \leq 10 \\[2ex] 2.142\left[0.3 + \dfrac{a^2}{2\alpha_w}\left(\dfrac{1}{P_h} + \dfrac{1}{P_c}\right)\right]^{-1/2} & \text{for } \dfrac{a^2}{2\alpha_w}\left(\dfrac{1}{P_h} + \dfrac{1}{P_c}\right) \geq 10 \end{cases} \qquad (156)$$

Razelos and Lazaridis [16] presented the values of $\Phi^*$ for hollow cylinders. Generally, the wall thermal resistance is much smaller than the hot or cold gas film resistance. Hence the wall thermal resistance formula of Eq. (152) is adequate for rating and sizing problems of most applications. This correction factor $\Phi^*$ is not adequate to accurately determine the temperature distribution in the wall immediately after the changeover and near the regenerator ends. In those cases, a numerical method of Heggs and Carpenter [17] is suggested to account for the wall thermal resistance effect. Heggs and Carpenter designate this effect as the intraconduction effect.

## 7. INFLUENCE OF PRESSURE AND CARRYOVER LEAKAGES

In both rotary and fixed-matrix regenerators, flow leakages from the cold to hot fluid streams and vice-versa occur due to carryover and pressure differences. The pressure leakage is defined as any leakage due to the pressure difference between the cold and hot sides of the regenerator. Generally, the cold gas is at a pressure higher than that for the hot gas. This pressure leakage can occur at valves in a fixed-matrix regenerator. In a rotary regenerator, it can occur at face seals, through the pores in the matrix, or through the circumferential gap between the disk and the housing. The carryover leakage is defined as any intermixing of the hot and cold side fluids as a result of the fluid trapped in the void volume of the matrix being pushed out from one side to the other side of the regenerator. The fluid carryover occurs either just after the valve switching in a fixed-matrix regenerator or just downstream of the face seals in a rotary regenerator. In the following, a simplified theory due to Klopfer [18] is presented to account for the influence of these leakages separately. Harper [19] presented a theory to account for the combined effect of carryover and pressure leakages. Banks and Ellul [20] presented a more general theory based on the similar idealizations. However, the leakage terms more detailed than considered here are difficult to measure experimentally.

7.1 Pressure Leakages

This leakage occurs at valves in a fixed-matrix regenerator. In a rotary regenerator, it occurs at face seals separating the hot and cold side, through the pores in the matrix itself, or through the circumferential gap between the disk and the housing. While the latter circumferential leakage is sometimes referred to as the side bypass leakage, the formers are referred to as the cross bypass leakage. The influence of the total pressure leakage on the regenerator effectiveness may be small and is evaluated quantitatively next. However, this total leakage represents a loss in the cold gas supply to the process or the thermodynamic system. It may have a substantial influence on the process or cycle efficiency. For example, in a gas turbine power plant, a 6% cold high pressure air leakage to the exhaust gas stream in a regenerator means a 6% reduction in the net power output, a significant penalty!

A simplified analysis is made now to analyze the influence of pressure leakage on the regenerator effectiveness. Let $\Delta W_p$ be the total mass flow leakage rate from the high pressure cold side to the low pressure hot side. In addition to all the idealizations made in Section 2, we will idealize that the one-half of this leakage occurs at one end of the regenerator and the other one-half at the other end of the regenerator as shown in Fig. 15. Also idealize that $C_c = C_{min}$. Since it is generally not possible to measure separately the side and cross bypass leakages, both are lumped into $\Delta W_p$. Define a pressure leakage factor $\chi_p$ as

$$\chi_p = \Delta W_p / W_c \tag{157}$$

In order to derive a decrease in the regenerator effectiveness due to the pressure leakages, we need to calculate the actual and internal regenerator effectivenesses. Since the cold gas stream is the minimum heat capacity rate stream, it will also be the $C_{min}$ stream in the presence of the leakage $\chi_p$. The regenerator ideal or internal effectiveness $c_i$ is derived based on the fluid enthalpies at the inlet and outlet of the regenerator boundary shown as a rectangle in Fig. 15. Since $C_c = C_{min}$, the cold side fluid having the flow rate of $(1 - \chi_p/2)C_c$ is the $C_{min}$ fluid. The heat transfer rate within the regenerator is

$$q_{act,i} = (1 - \chi_p/2)C_c(T_{c,o} - T_{c,i}) \tag{158}$$

The thermodynamically limited maximum possible heat transfer rate is

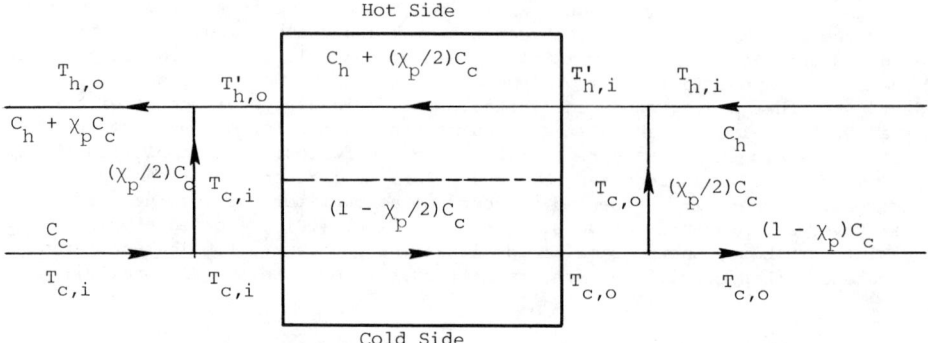

Fig. 15 Flow diagram and associated temperatures for a regenerator with pressure leakages.

$$q_{max,i} = (1 - \chi_p/2)C_c(T'_{h,i} - T_{c,i}) \tag{159}$$

where $T'_{h,i}$ is obtained as follows from an energy balance between the incoming streams on the hot side of Fig. 15.

$$T'_{h,i} = \frac{C_h T_{h,i} + (\chi_p/2)C_c T_{c,o}}{C_h + (\chi_p/2)C_c} \tag{160}$$

Thus the regenerator ideal or internal effectiveness $\varepsilon_i$ is

$$\varepsilon_i = \frac{q_{act,i}}{q_{max,i}} = \frac{T_{c,o} - T_{c,i}}{T'_{h,i} - T_{c,i}} \tag{161}$$

For actual regenerator effectiveness $\varepsilon$, the actual and maximum possible heat transfer rates are

$$q_{act} = (1 - \chi_p)C_c T_{c,o} - C_c T_{c,i} \tag{162}$$

$$q_{max} = (1 - \chi_p)C_c T_{h,i} - C_c T_{c,i} \tag{163}$$

and hence the actual regenerator effectiveness is

$$\varepsilon = \frac{q_{act}}{q_{max}} = \frac{(1 - \chi_p)C_c T_{c,o} - C_c T_{c,i}}{(1 - \chi_p)C_c T_{h,i} - C_c T_{c,i}} \tag{164}$$

Substituting $T'_{h,i}$ from Eq. (160) into Eq. (161) and eliminating $T_{c,o}$ from the resultant equation and Eq. (164) yields

$$\varepsilon_i = \frac{[(\chi_p/2) + 1/C^*][\varepsilon(T^*-1) + \chi_p(1 - \varepsilon T^*)]}{(1 - \chi_p)[(T^* - 1)/C^* + (\chi_p/2)\varepsilon T^*] - (\chi_p/2)(\varepsilon - \chi_p)} \tag{165}$$

Here again $C^* = C_c/C_h$ and $T^* = T_{h,i}/T_{c,i}$. This equation in terms of $\varepsilon$ as a function $\varepsilon_i$ is

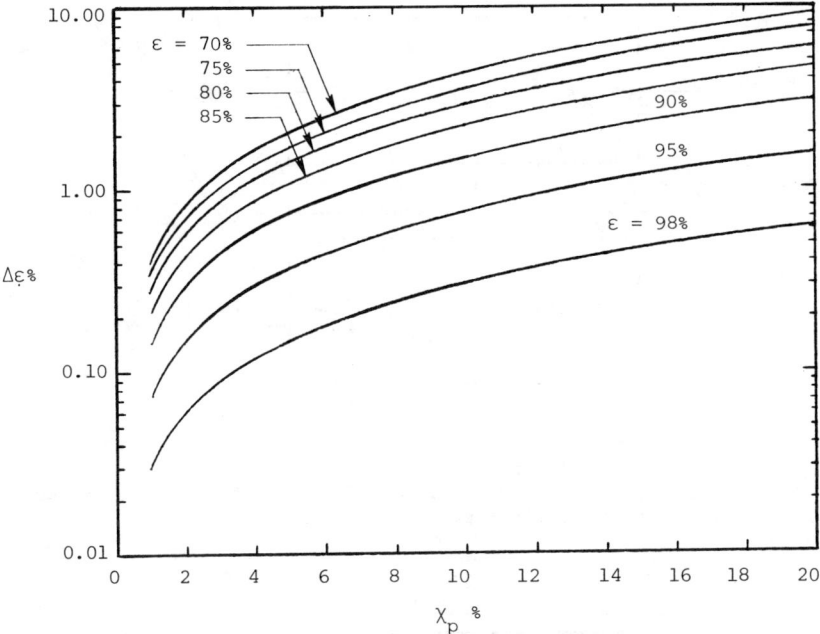

Fig. 16  Reduction in the regenerator effectiveness $\Delta\varepsilon$(%) as a function of $\chi_p$ (percentage pressure leakage rate) and $\varepsilon$ for $T^* = 2$ and $C^* = 1$.

$$\varepsilon = \frac{[(1 - \chi_p)(T^* - 1)/C^* - \chi_p^2/2]\varepsilon_i - (\chi_p/2 + 1/C^*)\chi_p}{(\chi_p/2)[(1 - \chi_p)T^* - 1](1 - \varepsilon_i) + [(1 - \chi_p)T^* - 1]/C^*} \quad (166)$$

The decrease in the regenerator effectiveness due to the pressure leakage rate, using Eqs. (165) and (166), can be presented in two alternative forms as follows.

$$\Delta\varepsilon = \varepsilon_i - \varepsilon = \begin{cases} \dfrac{(\chi_p/2)(1 - \varepsilon)[\varepsilon(T^* - 1) + \chi_p(1 - \varepsilon T^*) + 2/C^*]}{(1 - \chi_p)[(T^* - 1)/C^* + (\chi_p/2)\varepsilon T^*] - (\chi_p/2)(\varepsilon - \chi_p)} & (167) \\[2ex] \dfrac{(\chi_p/2)[(1-\chi_p)T^*-1](1-\varepsilon_i)\varepsilon_i + (\chi_p^2/2)(1+\varepsilon_i) + \chi_p(1-\varepsilon_i)/C^*}{(\chi_p/2)[1-\chi_p)T^*-1](1-\varepsilon_i) + [(1-\chi_p)T^*-1]/C^*} & (168) \end{cases}$$

Figure 16 shows $\Delta\varepsilon$ as a function of $\chi_p$ and $\varepsilon$ for typical values of $C^* = 1$ and $T^* = 2$. $\Delta\varepsilon$ is generally small and is significant for low values of $\varepsilon$ and high values of $\chi_p$.

## 7.2  Carryover Leakage

In both the $\varepsilon\text{-}N_{tu,o}$ and $\Lambda\text{-}\Pi$ theories, it is idealized that the gas particle dwell times $\tau_{d,h}^o$ and $\tau_{d,c}^o$ are negligible compared to the hot and cold gas flow periods $P_h$ and $P_c$, respectively, and are hence treated as zero.

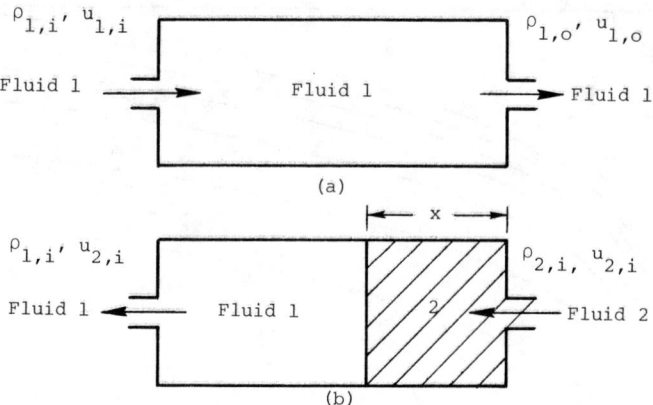

Fig. 17 Counterflow regenerator changeover: (a) The end of period 1; (b) Fluid 2 has been flowing for $x/u_{2,i}$ seconds, and $x/u_{2,i} < \tau_{d,2}$. Note that it is idealized that $u_{2,i}$ does not change in the matrix due to heat transfer for time $\tau \leq \tau_{d,2}$.

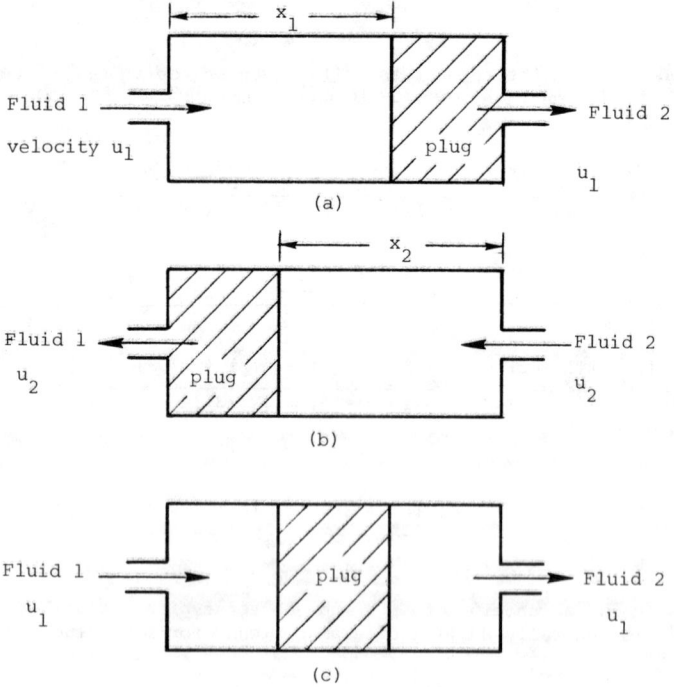

Fig. 18 Idealized plug formation: (a) The end of Fluid 1 period $(x_1/u_1)$, (b) The end of Fluid 2 period $(x_2/u_2)$, and (c) Fluid 1 halfway through the period.

This idealization was invoked in Eqs. (49) and (77). In reality, the dwell time is finite. Hence immediately after switching Fluid 2 to the matrix, it flows through the matrix passages, and pushes out Fluid 1 trapped in the void volume of the matrix as shown in Fig. 17. This trapped Fluid 1 is subsequently carried over in the Fluid 2 stream. The amount of Fluid 1 carried away with Fluid 2 is referred to as carryover leakage of Fluid 1 to Fluid 2 stream. Similarly, carryover leakage of Fluid 2 to Fluid 1 occurs immediately after switching Fluid 1 to the matrix.

In an extreme case, if the gas dwell time is greater than the gas flow period, an idealized "plug" of the gas is formed and trapped in a counterflow regenerator as shown in Fig. 18. In this case, carryover leakage is 100%, i.e., only Fluid 1 flows in the outlet pipe of Fluid 2, and only Fluid 2 flows in the outlet pipe of Fluid 1. The fluid streams are completely "contaminated". This situation should not arise in any actual regenerators.

Generally, the carryover leakage is very small and its influence on the regenerator effectiveness is also negligibly small for most regenerator applications, except possibly for the cryogenics regenerator and the Stirling engine regenerator.

Klopfer [18] has presented a simplified theory to determine the influence of the carryover leakage on the regenerator effectiveness. His idealized model is shown in Fig. 19. The derivations for $\varepsilon$ and $\varepsilon_i$ parallel those for the pressure leakage effects. For $\tau_{d,h} < P_h$ and $\tau_{d,c} < P_c$, the following are the results.

$$\varepsilon = \frac{q_{act}}{q_{max}} = \frac{[(1-\chi_c)C_c + \chi_h C_h]T_{c,o} - C_c T_{c,i}}{[(1-\chi_c)C_c + \chi_h C_h]T_{h,i} - C_c T_{c,i}} \quad (169)$$

$$\varepsilon_i = \frac{T'_{c,o} - T_{c,i}}{T_{h,i} - T_{c,i}} \quad (170)$$

where $T'_{c,o}$ from an energy balance between the fluid streams at the exit face of the cold side of the regenerator,

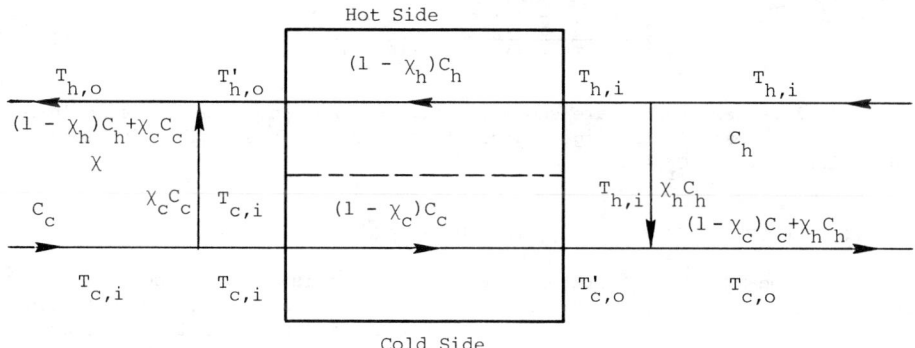

Fig. 19 Flow diagram and associated temperatures for a regenerator with carryover leakages.

$$T'_{c,o} = \frac{[(1 - \chi_c)C_c + \chi_h C_h]T_{c,o} - \chi_h C_h T_{h,i}}{(1 - \chi_c)C_c} \qquad (171)$$

and the fractional carryover leakages from the cold side to the hot side and from the hot side to the cold side respectively are

$$\chi_c = \frac{\Delta W_c}{W_c} = \frac{\Delta W_c \, c_{p,c}}{C_c} = \left(\frac{\rho_c \, c_{p,c}}{\rho_w \, c_w}\right)\left(\frac{\sigma + V^*}{1 - \sigma}\right) C_r^* = C_r^*/K \qquad (172)$$

$$\chi_h = \frac{\Delta W_h}{W_h} = \frac{\Delta W_h \, c_{p,h}}{C_h} = \left(\frac{\rho_h \, c_{p,h}}{\rho_w \, c_w}\right)\left(\frac{\sigma + V^*}{1 - \sigma}\right) C_r^* C^* \qquad (173)$$

$$K = \left(\frac{\rho_w \, c_w}{(\rho c_p)_c}\right)\left(\frac{1 - \sigma}{\sigma + V^*}\right) \qquad (174)$$

The regenerator internal effectivness $\varepsilon_i$ and the actual effectiveness $\varepsilon$ are expressed in terms of each other as

$$\varepsilon_i = \varepsilon + (1 - \varepsilon)\frac{(1 - p^*)}{(T^* - 1)} \frac{\chi_c}{(1 - \chi_c)} \qquad (175)$$

$$\varepsilon = \varepsilon_i - \frac{(1 - \varepsilon_i)(p^* - 1)}{p^*(T^* - 1)/\chi_c - p^* T^* + 1} \qquad (176)$$

where $p^* = p_{c,i}/p_{h,i}$ and $T^* = T_{h,i}/T_{c,i}$.

The decrease in the regenerator effectiveness due to carryover leakage, using Eqs. (175) and (176), can be presented in two alternative forms as follows.

$$\Delta\varepsilon = \varepsilon_i - \varepsilon = \begin{cases} (1 - \varepsilon)\dfrac{(1 - 1/p^*)}{(T^* - 1)} \dfrac{C_r^*}{K - C_r^*} & (177) \\ \\ \dfrac{(1 - \varepsilon_i)(p^* - 1)C_r^*}{Kp^*(T^* - 1) - C_r^*(p^* T^* - 1)} & (178) \end{cases}$$

Some typical values of $\Delta\varepsilon$ are shown in Fig. 20. As can be seen, the influence of carryover leakage on the regenerator effectiveness is negligible for most applications having $C_r^* < 10$.

## 8. INFLUENCE OF MATRIX MATERIAL, SIZE AND ARRANGEMENT

In a fixed-matrix regenerator, the outlet temperature of the cold gas (air) decreases as a function of time during the cold gas flow period. The difference in the outlet temperature from $\tau = 0$ to that at $\tau = P_c$ is referred to as the <u>temperature swing</u>. This temperature swing should be minimized so that the heated air from the regenerator is at a relatively constant temperature for the process downstream. The design of three and four stove systems [1] has been developed to minimize this temperature swing. Since the temperature swing is dependent upon the heat capacity of the matrix material, it can be minimized by

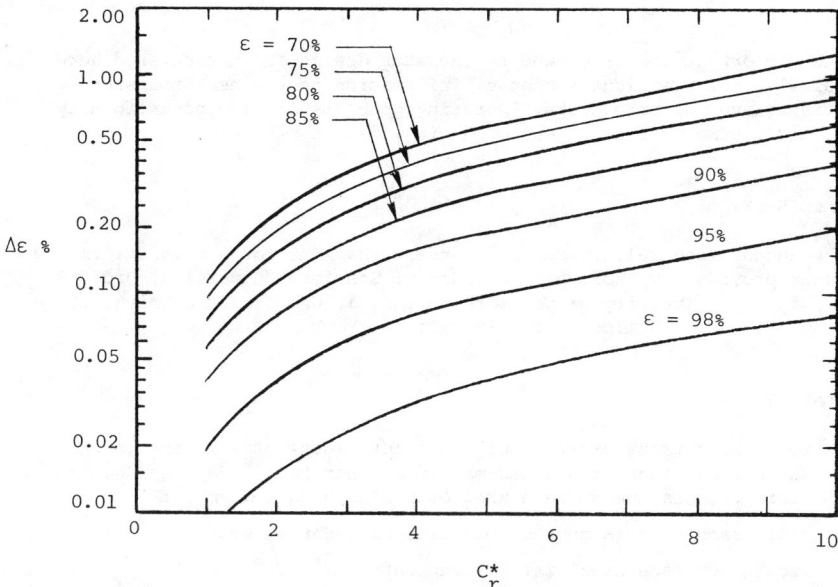

Fig. 20 Reduction in the regenerator effectiveness $\Delta\varepsilon(\%)$ as a function of $C_r^*$ and $\varepsilon$ for $K = 200$, $p^* = 4$, $T^* = 2$.

employing a high heat capacity material in the matrix such as the silicon carbide or corundum instead of the fireclay. However, the high heat capacity material is much more expensive. Use of proper matrix material size and arrangement along with different materials in a two or three zoned regenerator results in an optimum regenerator.

Heggs and Carpenter [21] among others proposed the use of high heat capacity material near the outlet end of the cold gas flow path and low heat capacity material near the inlet end of the cold gas flow path. Only 10% of the heat transfer surface area is required to have higher heat capacity material, the rest 90% area with the lower heat capacity material. Heggs and Carpenter also found that thick bricks be used near the outlet end and thin bricks near the inlet end of the cold gas flow path in a two-zoned regenerator. Either of the alternatives will reduce the temperature swing. The design of a two-zoned regenerator is carried out by considering two zones in series, the outlet fluid temperature from the first zone is then the inlet fluid temperature to the second zone.

## 9. SUMMARY

An in-depth analysis is presented for the understanding of dimensional and dimensionless variables and parameters associated with the thermal design of both rotary and fixed-matrix regenerators. Both the $\varepsilon$-$N_{tu,o}$ and $\Lambda$-$\Pi$ methods are introduced for the design of "steady-state" periodic-flow performance of regenerators. Similarity and subtle differences between these methods are outlined. It is clearly shown that both methods are equivalent. In the literature, it is frequently erroneously mentioned that the $\varepsilon$-$N_{tu,o}$ method is valid only for $P_h = P_c$. However, the solutions are indeed valid for $P_h \neq P_c$. The rotary regenerator generally has $\theta_h/\theta_c = P_h/P_c = 1.2$. Analyses are also presented to account for the influence of longitudinal heat conduction, wall thermal resistance, pressure leakages and carryover leakages on the regenerator effectiveness.

Many important topics of the regenerator design are not covered here, particularly, the transient response, information on the heat transfer coefficients and pressure drops, operating problems of flow nonuniformity, fouling and thermal stresses, and mechanical design problems.

ACKNOWLEDGEMENTS

The author sincerely appreciates and is thankful for a critical review on this paper provided by Prof. A. L. London of Stanford University, Prof. P. Razelos of City University of New York, Dr. P. J. Heggs of the University of Leads, and Dr. A. J. Willmott of University of York.

NOMENCLATURE

| | |
|---|---|
| $A$ | total heat transfer area (cold and hot sides) of a rotary regenerator and all matrices of a fixed-matrix regenerator, except in Section 4.2 it represents the surface area of a single matrix, $m^2$, $ft^2$ |
| $A_c$ | heat transfer area on the cold side of the regenerator, $m^2$, $ft^2$ |
| $A_{fr}$ | frontal or face area of the regenerator, $m^2$, $ft^2$ |
| $A_h$ | heat transfer area on the hot side of the regenerator, $m^2$, $ft^2$ |
| $A_k$ | total wall cross-sectional area for longitudinal conduction, subscripts c, h, and t denote cold side, hot side, and total regenerator, $m^2$, $ft^2$ |
| $A_k^*$ | ratio of $A_{k,c}$ to $A_{k,h}$, dimensionless |
| $A_o$ | minimum free flow (or open) area, $m^2$, $ft^2$ |
| $A_w$ | total wall area for transverse heat conduction (in the matrix wall thickness direction), $m^2$, $ft^2$ |
| $a$ | wall thickness, m, ft |
| $C$ | flow stream heat capacity rate with a subscript c or h, $Wc_p$, W/°C, Btu/hr°F |
| $C^*$ | heat capacity rate ratio, $C_{min}/C_{max}$, dimensionless |
| $\overline{C}$ | flow stream heat capacitance, $Mc_p$, $C\tau_d$, W·s/°C, Btu/°F |
| $C_{max}$ | maximum of $C_c$ or $C_h$, W/°C, Btu/hr °F |
| $C_{min}$ | minimum of $C_c$ or $C_h$, W/°C, Btu/hr °F |
| $C_r$ | total heat capacity rate of a regenerator, defined by Eq. (3); see Eq. (7) for the hot and cold side matrix heat capacity rates $C_{r,h}$ and $C_{r,c}$, W/°C, Btu/hr °F |
| $C_r^*$ | total matrix heat capacity rate ratio, $C_r/C_{min}$, $C_{r,h}^* = C_{r,h}/C_h$, $C_{r,c}^* = C_{r,c}/C_c$, dimensionless |
| $\overline{C}_r$ | total matrix wall heat capacitance, $M_w c_w$, see Eq. (5) for the hot and cold side matrix heat capacitance $\overline{C}_{r,h}$ and $\overline{C}_{r,c}$, W·s/°C, Btu/°F |
| $c_p$ | specific heat of fluid at constant pressure, J/kg °C, Btu/lbm °F |
| $h$ | heat transfer coefficient, W/m$^2$ °C, Btu/hr ft$^2$ °F |
| $(hA)^*$ | convection conductance ratio, defined by Eq. (61), dimensionless |
| $k_w$ | thermal conductivity of the matrix wall material, W/m °C, Btu/hr ft °F |

# THERMAL DESIGN THEORY FOR REGENERATORS

| | |
|---|---|
| $L$ | regenerator matrix length for gas flows, m, ft |
| $M$ | mass of the fluid in the regenerator at any instant of time, kg, lbm |
| $M_w$ | mass of all matrices of a regenerator, kg, lbm |
| $N$ | rotational speed for a rotary regenerator, rev/s, rev/sec |
| $N_{tu,c}$ | number of heat transfer units on the cold side, $(hA)_c/C_c$, dimensionless |
| $N_{tu,h}$ | number of heat transfer units on the hot side, $(hA)_h/C_h$, dimensionless |
| $N_{tu,o}$ | modified number of heat transfer units, defined by Eq. (58), dimensionless |
| $P_c$ | cold gas flow period, duration of the cold gas stream in the matrix or duration of matrix in the cold gas stream, s, sec |
| $P_h$ | hot gas flow period, duration of the hot gas stream in the matrix or duration of matrix in the hot gas stream, s, sec |
| $P_r$ | reversal period for switching from hot to cold gas stream or vice versa in a fixed-matrix regenerator, $P_r$ is generally negligible, s, sec |
| $P_t$ | total period, $P_t = P_c + P_h + P_r$, $\approx P_c + P_h$, s, sec |
| $p$ | fluid static pressure, Pa, psf |
| $p^*$ | ratio of cold fluid inlet pressure to hot fluid inlet pressure, $p_{c,i}/p_{h,i}$, dimensionless |
| $Q$ | heat transfer in a specified period or time, W·s, Btu |
| $q$ | actual heat transfer rate in the regenerator, W, Btu/hr |
| $R_w$ | wall thermal resistance, °C/W, hr °F/Btu |
| $R^*$ | a ratio of the hot to cold reduced periods, $\Pi_h/\Pi_c$, dimensionless |
| $T$ | temperature on the absolute scale, °K, °R |
| $T^*$ | ratio of hot fluid inlet temperature to cold fluid inlet temperature, $T_{h,i}/T_{c,i}$, dimensionless |
| $t$ | fluid static temperature to a specified arbitrary datum, °C, °F |
| $t_w$ | wall temperature, °C, °F |
| $t^*$ | temperature defined by Eq. (39), dimensionless |
| $U$ | overall heat transfer coefficient, W/m$^2$ °C, Btu/hr ft$^2$ °F |
| $u_m$ | fluid mean axial velocity, m/s, ft/sec |
| $V$ | heat exchanger total volume, m$^3$, ft$^3$ |
| $V^*$ | ratio of the header volume to the matrix total volume, dimensionless |
| $W$ | fluid mass flow rate, kg/s, lbm/hr |
| $X^*$ | axial distance, x/L, dimensionless |
| $x$ | Cartesian coordinate along the flow direction, m, ft |
| $y$ | Cartesian coordinate along the matrix wall thickness direction, m, ft |
| $\alpha_w$ | thermal diffusivity of the matrix material, $k_w/\rho_w c_w$, m$^2$/s, ft$^2$/sec |
| $\beta$ | heat transfer surface area density, a ratio of total heat transfer area to total volume of a regenerator, m$^2$/m$^3$, ft$^2$/ft$^3$ |
| $\gamma$ | ratio of heat capacity rates, defined by Eq. (98), dimensionless |
| $\varepsilon$ | regenerator effectiveness, defined by Eq. (55), dimensionless |

$\varepsilon_c$ temperature effectiveness on the cold fluid side, defined by Eq. (89), dimensionless

$\varepsilon_h$ temperature effectiveness on the hot fluid side, defined by Eq. (88), dimensionless

$\varepsilon_r$ the regenerator effectiveness of a single matrix, defined by Eq. (90), dimensionless

$\eta$ reduced time variable for a regenerator, defined by Eq. (77), dimensionless

$\theta$ angular coordinate in the cylindrical coordinate system, rad, deg

$\Lambda$ reduced length for a regenerator, defined by Eqs. (92), (109) and (110), dimensionless

$\lambda$ longitudinal wall heat conduction parameter based on the total conduction area, $k_w A_{k,t}/LC_{min}$, $\lambda_c = k_w A_{k,c}/LC_c$, $\lambda_h = k_w A_{k,h}/LC_h$, dimensionless

$\xi$ reduced length variable for a regenerator, defined by Eq. (76), dimensionless

$\Pi$ reduced period for a regenerator, defined by Eqs. (92), (111), (112), dimensionless

$\rho$ density, $kg/m^3$, $lbm/ft^3$

$\sigma$ ratio of free flow area to frontal area, $A_o/A_{fr}$, or the volumetric porosity for regenerators, dimensionless

$\tau$ time variable, s, sec

$\tau_d$ dwell time, residence time or transit time of a fluid particle, s, sec

$\tau_c^*$ time variable for the cold fluid, defined by Eq. (37), dimensionless

$\tau_h^*$ time variable for the hot fluid, defined by Eq. (23), dimensionless

$\chi_c$ cold fluid carryover leakage factor in a regenerator, defined by Eq. (172), dimensionless

$\chi_h$ hot fluid carryover leakage factor in a regenerator, defined by Eq. (173), dimensionless

$\chi_p$ pressure leakage factor in a regenerator, defined by Eq. (157), dimensionless

Subscripts

c    cold fluid side
h    hot fluid side
i    inlet to the regenerator
m    mean
max  maximum
o    outlet to the regenerator
p    pressure leakage
w    wall

REFERENCES

1. R.K. Shah, Classification of heat exchangers, in <u>Heat Exchangers-Thermohydraulic Fundamentals and Design</u>, edited by S. Kakac, A.E. Bergles, and F. Mayinger, Hemisphere Publishing Corp., New York (1981).

2. O.A. Saunders and S. Smoleniec, Heat transfer in regenerators, IME-ASME

General Discussion on Heat Transfer, pp. 443-445, London, England (Sept. 1951).
3. R.K. Shah and A.L. London, Laminar Flow Forced Convection in Ducts, Supplement 1 to Advances in Heat Transfer, Academic Press, New York (1978).
4. J.E. Coppage and A.L. London, The periodic-flow regenerator - A summary of design theory, Trans. ASME, Vol. 75, 779-787 (1953).
5. W.M. Kays and A.L. London, Compact Heat Exchangers, Second Edition, McGraw-Hill, New York (1964).
6. T.J. Lambertson, Performance factors of a periodic-flow heat exchanger, Trans. ASME, Vol. 80, 586-592 (1958).
7. P. Razelos, Personal communication, Dept. Appl. Sci., City Univ. of New York, St. George Campus, 130 Stuyvesant Place, Staten Island, NY (1980).
8. G. Theoclitus and T.L. Eckrich, Parallel flow through the rotary heat exchanger, Proceedings of the Third International Heat Transfer Conference, Vol. I, 130-138 (1966).
9. H. Hausen, Über die theorie von Wärmeaustauches in Regeneratoren, Zeitschrift für Angewandte Mathematik und Mechanik, Vol. 9, 173-200 (1929).
10. P. Razelos, An analytic solution to the electric analog simulation of the regenerative heat exchanger with time-varying fluid inlet temperatures, Wärme-und Stoffübertragung, Vol. 12, 59-71 (1979).
11. H. Hausen, Vervollständigte Berechnung des Wärmeaustausches in Regeneratoren (accomplished calculations of heat exchange in regenerators), Z. VDI Beiheft Verfahrenstechnik, No. 2, 31-43 (1942).
12. H. Hausen, Wärmeübertragung im Gegenstrom, Gleichstrom und Kreuzstrom, (Heat Transfer in Counterflow, Parallel flow and Crossflow), Second Edition, Springer Verlag, Berlin (1976).
13. G.D. Bahnke and C.P. Howard, The effect of longitudinal heat conduction on periodic-flow heat exchanger performance, Trans. ASME, Journal of Engineering for Power, Vol. 86, Series A, 105-120 (1964).
14. R.K. Shah, A correlation for longitudinal heat conduction effects in periodic-flow heat exchangers, Trans. ASME, Journal of Engineering for Power, Vol. 97, Series A, 453-454 (1975).
15. J.R. Mondt, Vehicular gas turbine periodic-flow heat exchanger solid and fluid temperature distributions, Trans. ASME, Journal of Engineering for Power, Vol. 86, Series A, 121-126 (1964).
16. P. Razelos and A. Lazaridis, A lumped heat-transfer coefficient for periodically heated hollow cylinders, Int. J. Heat Mass Transfer, Vol. 10, 1373-1387 (1967).
17. P.J. Heggs and K.J. Carpenter, A modification of the thermal regenerator infinite conduction model to predict the effects of intraconduction, Trans. I. Chem. E., Vol. 57, 228-236 (1979).
18. G.H. Klopfer, The design of periodic-flow heat exchangers for gas turbine engines, Technical Report HE-1, Dept. of Mechanical Engineering, Stanford University, Stanford, CA (Aug. 1969).
19. D.B. Harper, Seal leakage in the rotary regenerator and its effect on rotary regenerator design for gas turbines, Trans. ASME, Vol. 79, 233-245 (1957).
20. P.J. Banks and W.M.J. Ellul, Predicted effects of by-pass flows on regenerator performance, The Institution of Engineers, Australia, Mechanical and Chemical Engineering Transactions, MC9, 10-14 (1973).
21. P.J. Heggs and K.J. Carpenter, The effects of packing material, size and arrangement of the performance of thermal regenerators, Heat Transfer 1978, Vol. 4, 321-326, Hemisphere Publishing Co. (1978).
22. R.K. Shah, Compact Heat Exchangers, in Heat Exchangers-Thermohydraulic Fundamentals and Design, Edited by S. Kakac, A.E. Bergles, and F. Mayinger, Hemisphere Publishing Corp., New York (1981).

# Analytical and Experimental Thermal-Hydraulic Optimization of Finned Tube Bundle Heat Exchangers for Dry Cooling Towers

**O. FISCHER, K. H. BUCHER**
Swiss Federal Institute for Reactor Research (EIR)
CH-5303 Würenlingen, Switzerland

ABSTRACT

A computer code has been developed which evaluates the total investment costs for a dry cooling tower as a function of geometric parameters of a Forgo type heat exchanger. In order to minimize the total costs an optimized geometry results from parameter variation.

A test facility with open air cycle and electric heating supports the physical formulation by performance tests of water-air heat exchangers. Measurement method and results are presented.

1.  INTRODUCTION

Dry cooling systems are so far neither thermodynamically nor economically competitive with wet or once-through cooling, because they work with an additional thermic barrier. An increased temperature difference and much more cooling surface are required. As a result tower dimensions and hydraulic systems become large and expensive. Nevertheless there are good reasons to investigate dry cooling systems.

1.1 Task

Most of the cooling tower heat exchangers in operation today were developed shortly after the second world war and only slightly improved since then. It is common practice for heat exchangers to be designed for ease of manufacturing. Although used for all kinds of applications, the geometry remains fixed and the desired duty is achieved by adjusting the heat transfer area. However the above procedure is not satisfactory for instance on thermodynamic grounds.

Advanced high temperature reactor systems with gas turbine cycles reject waste heat at high temperature levels, thus heat transfer requirements allow reasonable dry cooling tower dimensions. Independence from water resources increase possible site locations and decrease environmental pollution, thus favouring dry cooling. Optimized head exchanger elements reduce total investment cost considerably for the case of large modern nuclear power plants.

A project to redesign and optimize water-air heat exchanger elements was

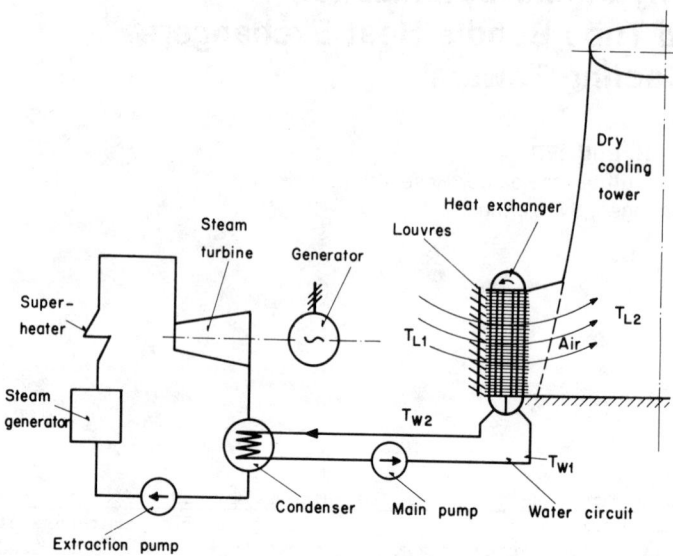

Fig. 1: Steam cycle power plant with a dry cooling tower. Circuit of indirect system with surface condenser.

initiated, that includes analytical optimization of the heat exchanger geometry and performance tests on existing and optimized heat exchangers [1].

1.2 Heat Exchanger Systems

The direct system uses an air heat exchanger to which the steam from the turbine is exhausted through a large duct and condensed. Its application is restricted to steam cycle power plants up to 200 $MW_e$ [2].

The indirect system needs spray- or surface-condensers in steam cycle plants and gas-liquid heat exchangers in closed gas turbine cycles. The heated water is pumped to the tower, where it is cooled down in dry cooling elements (Fig. 1). Up to now three different types of heat exchangers have been used in such systems [2]:

- Birwelco: aluminium band wound up on a steel tube
- GEA:     elliptic tube with punched fins (steel)
- Forgò:   all aluminium finned tube bundle.

Very extensive experience with Forò type heat exchangers (Fig. 2, Fig. 5) has been acquired in coal-, gas- and oil-fired power stations (approx. 2000 $MW_e$ in operation). Due to the relatively high heat exchange rate and economical aspects the present study focused on Forgò type heat exchangers. A standard Forgò element has a height of 5 m, a width of 2.4 m and a thickness of 0.15 m. It has 240 cooling tubes arranged for two counter cross flow water passes. Up to four elements can be stacked into a column. Two columns are combined to a

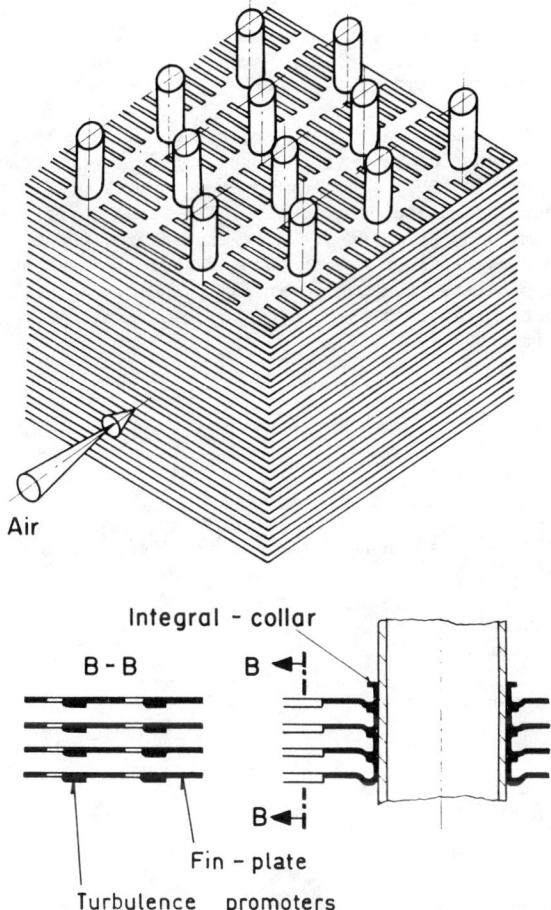

Fig. 2: Forgó type heat exchanger. Structure and details of fin design.

"delta" with an angle of $60°$. These units are arranged around the base of the cooling tower.

### 1.3 Optimization Criterion

Some purely thermohydraulic optimization criteria are known, e.g. the correlation between air pressure drop and the air-side heat transfer coefficient. But an optimization of such correlations would certainly not produce an economic solution. Cost distribution for a cooling system is approximately [3]:

- 60 % heat exchangers

- 33 % tower structure
- 7 % hydraulics.

manufacturing costs are about 75 % of the total.

Any improvement of the heat exchanger performance will be reflected in less costs of the cooling elements as well as in more compact tower structures and a cheaper hydraulic system. In short, cooling tower economics profit from the multiple effects of a thermohydraulic optimization. The total investment costs of the main cooling system, consisting of the heat exchangers, the tower structure, hydraulics, ducts, valves, fittings and so on, were selected as the final optimization criterion. Capitalized pumping power is included. As the heat exchangers dominate the tower cost, minimizing this leads to an optimal heat exchanger design for given conditions.

## 2. ANALYTICAL OPTIMIZATION

### 2.1 Overview

Forgó type heat exchangers can be described by a large number of geometric parameters (Fig. 2, Fig. 5) [3]:

- longitudinal tube pitch
- transversal tube pitch
- tube diameter
- number of tubes per tube row
- fin gap
- fin thickness
- number of slots
- distance between slots
- width of slots
- number of water flow passes
- water chamber design.

Important operating conditions are [4].

- inlet and outlet water temperatures
- air inlet temperature and density
- velocity of water
- velocity of air
- effect of fouling layer.

A comprehensive computer code has been developed for thermohydraulic and tower cost optimization. A simplified flowchart with the main features is given in Fig. 3. The necessary cost information was obtained by a very close cooperation with the industry.

The optimization is a step by step analysis. The influence of each parameter is investigated first. The gradient of the resulting cost function indicates the main parameters, i.e. those having the strongest influence on performance and costs. Not all of them show a significant minimum and some have to be treated as constants or constrained. These limitations result from thermodynamic, technical or interfacing requirements. The investigation of the individual para-

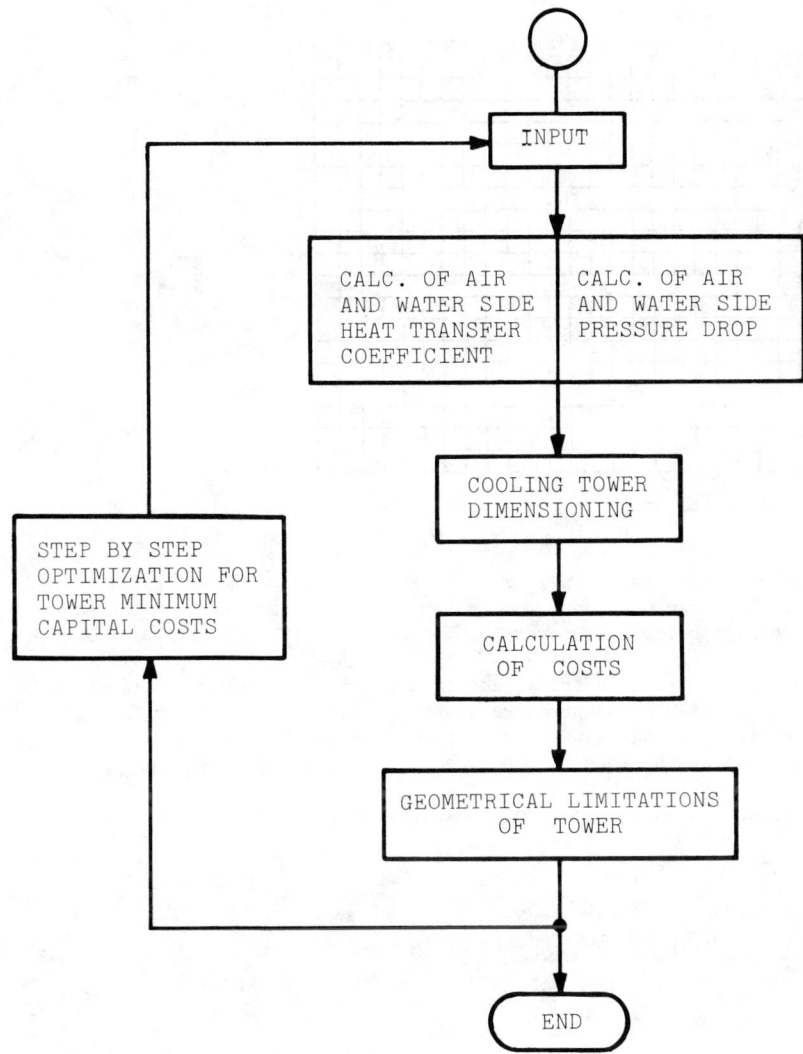

Fig. 3: Evaluation procedure for analytic optimization.

meters allows a better choice of the subsequent improvement step.

2.2 Air-Side Heat Transfer Coefficient

The air-side heat transfer coefficient is evaluated as follows [5]. To model the complex fin geometry and its influence on heat transfer, the finite element method is applied. The symmetrical design of the fin plates allows con-

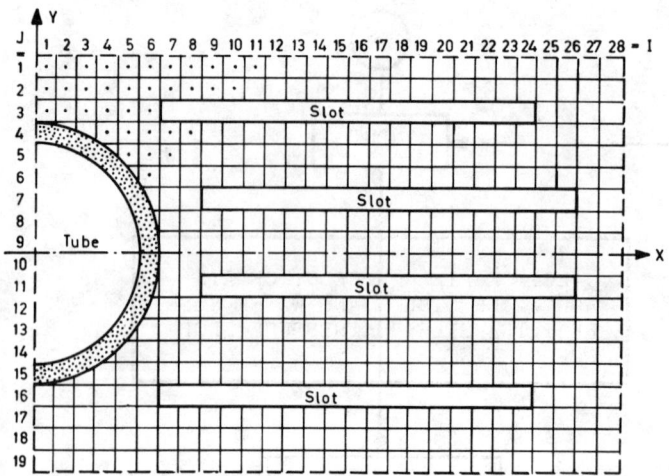

Fig. 4: Calculational section of a fin with finite element mesh.

sideration of a section as shown in Fig. 4. To get representative results, medium water temperature in the tubes $T_W$, medium air temperature $T_L$ and the air velocity $v_o$ resulting from the tower lay-out are assumed.

The equations for an electrical analog are solved by relaxed difference equations to calculate the temperature distribution $T_F(i,j)$ on the fin plate. Boundary conditions are the water temperature $T_W$ along the tube and the temperature equilibrium at the interface to the surrounding sections.

The local heat transfer coefficient is given by Colburn

$$\text{laminar} \quad \alpha(i,j) = 0.332 \frac{\lambda_F}{x} \text{Re}_x^{1/2} \text{Pr}^{1/3} \tag{1}$$

$$\text{turbulent} \quad \alpha(i,j) = 0.0292 \frac{\lambda_F}{x} \text{Re}_x^{4/5} \text{Pr}^{1/3} \tag{2}$$

$$\text{with} \quad \text{Re}_x = \frac{\rho v_o x}{\mu} \tag{3}$$

where x is the distance between the node considered and a preceding slot in the flow direction. The heat balance (4) is given in a simplified form here, omitting an additional term describing the effect of the turbulence promoters (Fig. 2). $\alpha_L$ ensues from (4).

$$\dot{Q} = 2\alpha_L A_F (\overline{T}_F - \overline{T}_L) = \sum_{i,j} 2\alpha(i,j) A(k,j) \left[ T_F(i,j) - \overline{T}_L \right] \tag{4}$$

### 2.3 Result for Gas Turbine Cycle Application

Investigations for high temperature reactor plants with gas turbine cycle

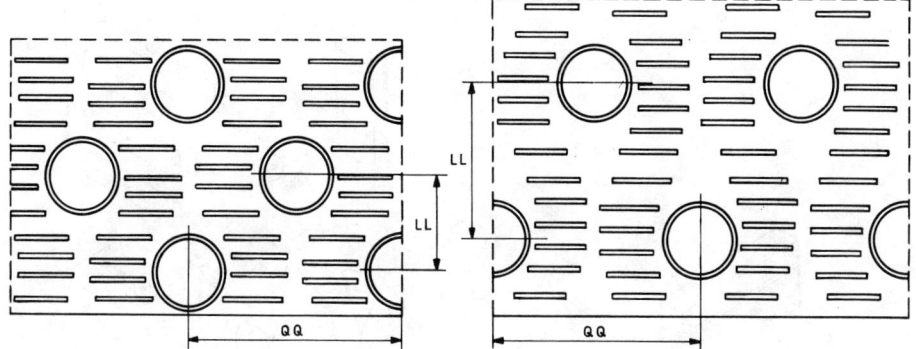

Fig. 5: Comparison between standard and optimized fin geometry.

lead to an optimized geometry as presented in Fig. 5. The large increase of the longitudinal tube pitch LL is striking. This was expected after the first trend calculations. Performance tests will be carried out in the near future.

|  | Standard | Optimized |
|---|---|---|
| Transversal tube pitch | 60 mm | 57 mm |
| Longitudinal tube pitch | 25 mm | 45 mm |
| Tube inner diameter | 16.5 mm | 17.0 mm |
| Fin gap | 2.6 mm | 2.3 mm |

## 3. TEST FACILITY

### 3.1 Disposition

Although several heat exchanger test facilities were available, a proper one had to be designed and erected for some reasons related to this special task [6]. Most of the facilities are too small to be relevant to the cooling tower elements under consideration and true cooling tower operating conditions can be achieved only in an air-side open loop.

Fig. 6 shows a schematic lay-out of the water loop. Measured mass flows and temperatures are indicated. Air is taken in through the heat exchangers to be tested. Those are mounted at the inlet of a wind tunnel with controlled air flow perpendicular or oblique to the air flow, thus simulating the delta arrangement (Fig. 8). The air mass flow is measured with 1 to 6 Venturi tubes according to the quantity, which is controlled by outlet-throttling. Water mass flow is given by an induction flow meter and an orifice. Temperatures are measured with thermocouple elements and thermistors. A special integral thermometer has been developed to measure the average air temperature $T_{L2}$: a titan wire is put on a frame to indicate electric resistance as a function of the temperature.

Performance range of the test facility is:

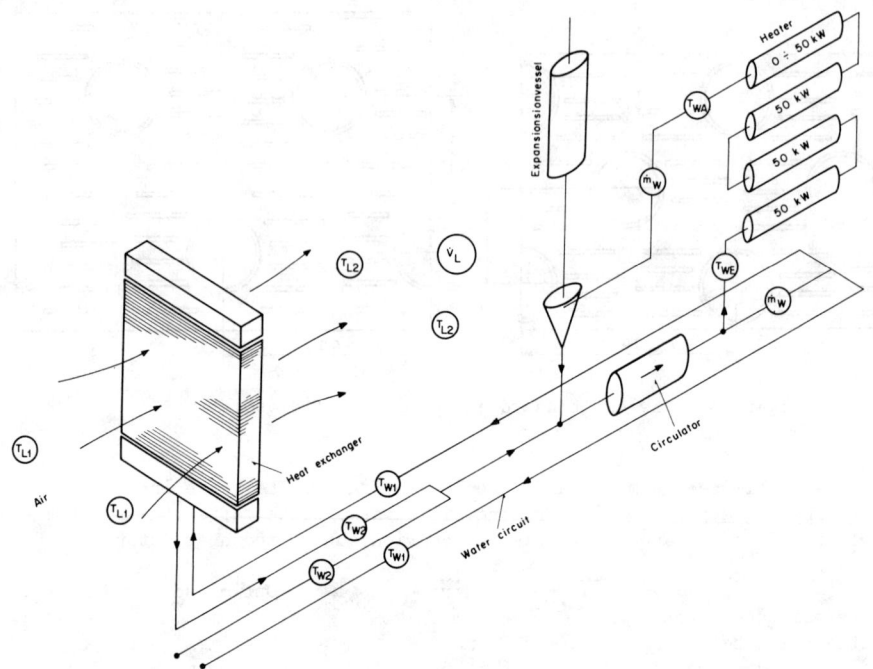

Fig. 6: Test facility. Water loop circuit and measured variables.

- air mass flow            0 ÷ 15 kg/s
- water mass flow       0 ÷ 12 kg/s
- heat load                0 ÷ 200 kW

The measured data are scanned by a data logger system with punch tape output.

## 3.2 Evaluation

The evaluation is carried out by a special computer code, which checks plausibility and supplies the results.

To get the air-side heat transfer coefficient the heat rejection load $\dot{Q}$, a representative temperature difference between water and air $\Delta T$, the water-side heat transfer coefficient $\alpha_W$, the conductivity of aluminium $\lambda$ and the heat exchanging areas $A_L$, $A_W$ should be known.

$$\dot{Q} = K_A A_A \Delta T \tag{5}$$

$$\frac{1}{K_A} = \frac{A_A}{A_W} \frac{1}{\alpha_W} + \frac{A_A}{A_W} \frac{d}{\lambda} + \frac{A_A}{A_L} \frac{1}{\alpha_L} \tag{6}$$

The parameters $A_L$, $A_W$, d and $\lambda$ are easy to determine. The water-side heat transfer coefficient $\alpha_W$ is calculated from the equation for turbulent tube flow given by Hausen [7].

To evaluate the heat load, four different heat balances are available:

- electric heating power
- air heating in heat exchanger
- water cooling in heat exchanger
- water warming in heater.

Due to the restricted size of the test elements temperature differences are small. The water warming balance is more accurate, as the reduced bypass flow $\dot{m}_W$ leads to an increased temperature difference.

A representative temperature difference cannot be measured within the limitations of reasonable expenses. However the overall heat transfer coefficient K can be evaluated with the given measured data, due to the performance characteristic following Bosnjakovic [8].

4. MEASUREMENTS

4.1 Benchmark Measurement

The first heat exchanger to be tested was a counter cross flow with three

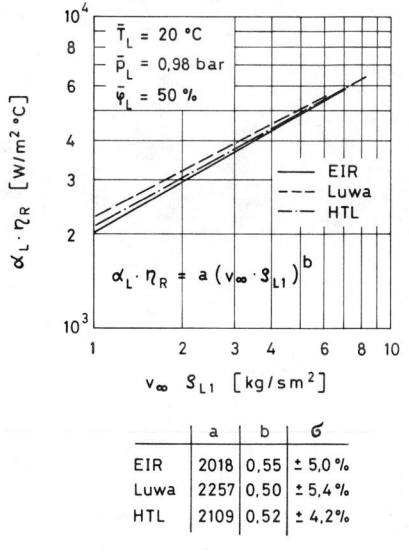
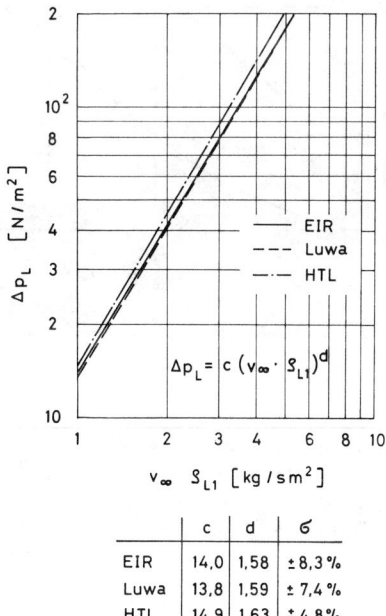

Fig. 7: Results of benchmark measurements.

water passes. It had been tested already in similar test sections, thus giving us the opportunity to make a benchmark measurement [6]. The results obtained from the different test facilities are in excellent agreement with one another (Fig. 7).

4.2 Standard Forgó Element

Extensive tests have been carried out on standard cooling tower elements of the Forgó type. The effect of the delta angle variation has been investigated (Fig. 8). Air-side heat transfer coefficient and the pressure drop are nearly identical for an angular variation between $40°$ and $70°$ and free air velocities up to 3 m/s. Consequently more heat exchanger elements could be placed at the lower periphery of a cooling tower [9].

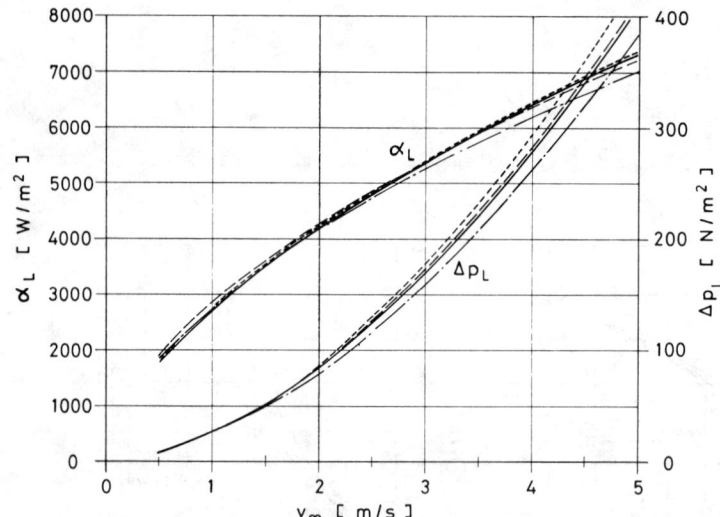

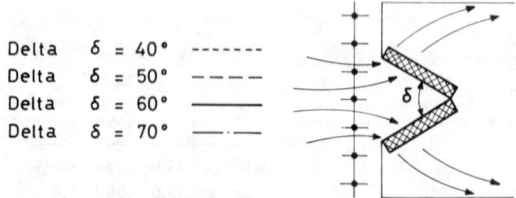

Fig. 8: Forgó standard element. Angle variation of delta arrangement.

# FINNED TUBE BUNDLE HEAT EXCHANGERS FOR DRY COOLING TOWERS

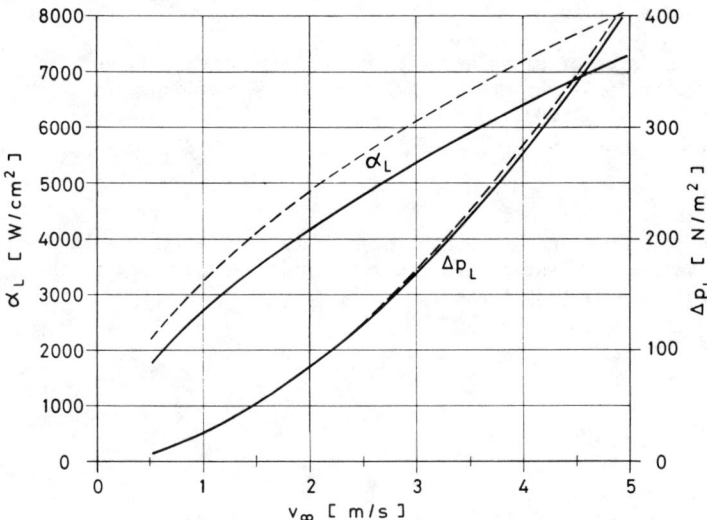

Fig. 9: Comparison between Forgó standard (———) and an element with electrophoretic corrosion protection coating (---), $\delta = 60°$.

## 4.3 Corrosion Protection Coating

Electrophoretic coating [10] has an enormous effect on the air-side heat transfer coefficient. It is increased by approx. 15 %, whereas the pressure drop remains nearly constant (Fig. 9).

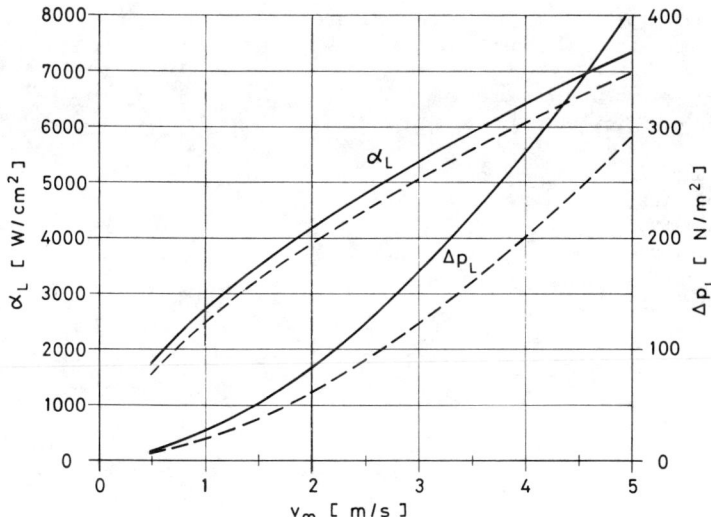

Fig. 10: Comparison between Forgó standard (———) and an element without slots and ribs (---), $\delta = 60°$.

## 4.4 Influence of Slots and Ribs

As expected the omission of slots and ribs on the fin plate reduces the air-side heat transfer by 10 %. But at the same time the air-side pressure drop was diminished about 25 % (Fig. 10).

## 4.5 Further Work

The described effects will need further investigations. It will be interesting to see if the behaviour of heat exchangers with optimized geometry is the same in comparison to those already tested.

## SYMBOLS

| | |
|---|---|
| A | area |
| d | thickness |
| (i,j) | node number |
| K | overall heat transfer coefficient |
| LL | longitudinal tube pitch |
| $\dot{m}$ | mass flow |
| p | pressure |
| $\Delta p$ | pressure drop |
| Pr | Prandtl-number |
| $\dot{Q}$ | heat load |
| QQ | transversal tube pitch |
| Re | Reynolds-number |
| T | temperature |
| $\Delta T$ | temperature difference |
| $\dot{V}$ | flow rate |
| $v_o$ | medium air velocity in heat exchanger |
| $v_\infty$ | free air velocity perpendicular to frontal area |
| x | distance between slot and node |
| X | coordinate |
| Y | coordinate |
| $\alpha$ | heat transfer coefficient |
| $\delta$ | delta angle |
| $\phi$ | humidity of the air |
| $\eta_R$ | efficiency of ribs |
| $\lambda$ | heat conductivity |
| $\mu$ | dynamic viscosity |
| $\rho$ | density |
| $\theta$ | statistic strewing |

## Indices

| | |
|---|---|
| A | frontal area |
| F | fin |
| L | air-side |
| L1 | air inlet |
| L2 | air outlet |
| W | water-side |
| W1 | water (heat exchanger inlet) |

W2  water (heat exchanger outlet)
WE  water (heater inlet)
WA  water (heater outlet)
x   distance between slot and node

REFERENCES

1. Markòczy, G.: Private communication, also IEA Project Group Heat Transfer and Heat Exchangers. Subtask 3: Analytical and Experimental Investigation of Finned Tube Bundle Gas-Liquid Heat Exchanger. Optimization of the Heat Exchanger Design. Short Description of the Programme. EIR, Würenlingen, 12.7.1977.

2. Heeren, H.; Holly, L.: Trockenkühler entlasten Gewässer. Energie, 23: 298 (1971), 23: 385 (1971).

3. Ospina, C.: Private communication also Analytical Study and Optimization of Heat Exchangers and Dry Cooling Towers. TM-HHT-216, confidential. EIR, Würenlingen, 12.7.1978.

4. Ospina, C.: Private communication, also ELTOP - A Computer Programme for Heat Exchanger and Dry Cooling Tower Optimization. TM-HHT-224, confidential. EIR, Würenlingen, 16.1.1979.

5. Ospina, C.: Private communication, also Warmetechnische Analyse des Modifizierten Forgo-Kühlelementes für Trockenkühltürme. TM-HHT-208, confidential. EIR, Würenlingen, 25.9.1978.

6. Stämpfli, E.: Candreia, P.: Private communication, also Prüfstand für thermohydraulische Messungen an Kühlelementen. TM-HHT-253, confidential. EIR, Würenlingen, 10.8.1979.

7. Hausen, H.: Neue Gleichungen für die Wärmeübertragung. Allgemeine Wärmetechnik, 9: 75 (1959).

8. Bosnjakovic, F. et al.: Einheitliche Berechnung von Rekuperatoren. VDI-Forschungsheft 432, 17 (1951).

9. Bucher, K.H. et al.: Private communication, also IEA Project Group Heat Transfer and Heat Exchangers. Subtask 3: Analytical and Experimental Investigation of Finned Tube Bundle Gas-Liquid Heat Exchangers. Report 1977-1979. EIR, Würenlingen, 22.4.1980.

10. Höfling, E.; Maly, Z.: Umfassendes Korrosionsschutzsystem für Lamellenkühler. Schweizerische Aluminium Rundschau, 6: 245 (1979).

# Offshore Heat Exchanger Design Practice throughout a Project

**M. A. TAYLOR**
British National Oil Company
London, England

ABSTRACT

As a project progresses through its various stages the design of equipment improves from rule of thumb through to computer optimised and guaranteed. This paper highlights some of the factors considered by an offshore oil operator during this gestation period.

1. MODELLING THE PROJECT

At the start of most projects the Process Engineer will have a number of alternative process parameters to study. He achieves this in the oil industry by computer modelling techniques. Each stage of the process can be represented as a heat or mass transfer step in which the various streams change in temperature, pressure, composition or state.

The modelling progress adopted is typically -

1. Draw a rough flowsheet.

2. Use experience to select key operating temperatures and pressures.

3. Do trial computer run with guessed worst case parameters for overall equipment design.

4. Re-draw flowsheet and adjust parameters on basis of earlier computer run.

5. Re-run computer model until optimal yield versus power and equipment size balance is achieved.

6. Re-run with all alternative cases.

7. Select worst case/s for each equipment item.

8. Call in equipment specialist to design item.

Using a computer model requires that temperature approaches and pressure drops be set by the process engineer at an early stage. He usually sets these figures based on his own experience, although in tricky or unusual cases he might refer to a specialist. Because computer modelling is an

Fig.1. Offshore Production Platform

Fig.2. A Module for an Offshore Production Modules.

Fig.3. A Module for an Offshore Production Modules

expensive procedure, there is little chance of changing the equipment selection later, so it had better be right to start with.

## 2. SELECTION CRITERIA FOR OFFSHORE HEAT EXCHANGERS

Knowledge of the capabilities of different heat exchanger types helps the process engineer to make the right selection at the start. The Process Engineer has to match these capabilities against his desired criteria which, for an offshore platform, can be summarised as follows:

### 2.1 Weights and Spaces

Figures 1, 2, and 3 are photographs of a module for an offshore production platform and models of two of its production modules, from which it can be seen how cramped is the available plant space. A typical offshore platform topside plan area is only 80m. x 60m., and might weigh 20,000 tonnes above the supporting structure. The cost of such a facility in the North Sea is around $1.5 billion in 1980, so both weight and space are immensely expensive.

### 2.2 Operability and Reliability

The cost of lost production is enormous, so the equipment must be reliable and it must be easy to operate in difficult surroundings. Bear in mind that it is a major job to remove any piece of equipment and return it to land for repair or replacement. Offshore maintenance also has to be kept fairly basic to avoid increasing offshore weight and space. So unsophisticated, robust equipment is usually preferred to more complex, possibly more efficient equipment.

### 2.3 Cost and Delivery

To the first cost on land of an exchanger has to be added the cost of getting it offshore, hooking it up, commissioning and running it; there is also a penalty applied for its weight and the space it occupies. These factors are not proportional to the first cost, and it is generally better to select on the basis of weight, space, operability and reliability first and only use on-land cost as a "tie breaker" in equipment evaluations.

Delivery from European fabricators is usually within the time span for offshore projects unless the exchanger is very special, e.g., a compressed gas cooler working at 4,000 psia would require special forgings to be made.

## 3. CAPABILITIES OF EXCHANGER TYPES USED OFFSHORE

The process engineer usually has a list of good and bad points for heat exchanger types, some of which are listed here with specific reference to offshore North Sea.

### 3.1 Air Cooled Heat Exchangers

#### Advantages

Reduces need for cooling water.

Disadvantages

> Need to go on outside surface of platform.
> Large plan area.
> Require power.
> Fin corrosion (although claims have been made for resin coated fins).
> Ambient air temperature varies much more than sea water temperature.
> Maintenance required on fans.
> Noisy.

Few installed on North Sea platforms as main coolers although small "air blast" coolers are usually used for diesel engine coolers.

Saving in cooling water duty is not valid if spent cooling water is to be used for injection into the reservoir.

3.2 Shell and Tube Heat Exchangers

Advantages

> Well proven technology.
> OK for high pressures and high temperatures.
> Wide range of materials available.
> Relatively cheap.
> Fairly robust.
> Can handle all fluids in all states.

Disadvantages

> Heavy.
> Difficult to clean, especially on shellside.
> Fairly spacious.

Despite disadvantages, they are used extensively chiefly because of temperature, pressure and proven technology advantages.

3.3 Plate Heat Exchangers

Advantages

> Good surface to volume and plan area ratios.
> Good surface to weight ratio.
> Small temperature approach possible.
> Very easy to clean.
> Easy to add surface later.
> Good heat transfer and low fouling tendency.
> Well proven technology.

### Disadvantages

Limited to 300 psi. and 300°F. approx.

Gasket material not always compatible with fluid.

Vaporising fluids difficult to handle.

Single phase gas cooling difficult.

Have found favour in offshore use because they are compact and perform well. Particularly useful for coolant/sea water heat interchange where their low temperature approach is of great advantage. Also very frequently used for cooling potentially fouling fluids such as crude oil.

## 3.4 Plate Fin Heat Exchangers

### Advantages

Very compact.

Very high surface to weight ratio.

Very small temperature approach possible.

Can exchange more than two streams.

Can operate at very low temperatures.

Fairly high operating pressures (approx. 800 psi.)

Relatively cheap per unit surface.

### Disadvantages

Impossible to clean.

Materials of construction limited - usually aluminium.

Unproven offshore as yet.

Fouling tendency on unclean fluids.

Relatively limited application offshore so far. One has been ordered for a partial condenser.

Chief advantage is close temperature approach in refrigeration circuits which reduces refrigeration horsepower.

If aluminium is used, it should be protected by steel sheathing to prevent thermic sparks if struck by ferrous materials.

Must only be used on really clean fluids.

## 4. SPECIAL FEATURES SOMETIMES CONSIDERED

### 4.1 Enhanced Surfaces

Occasionally Lofin tubing is used if it can be shown to save significantly on weight and space. It can sometimes be of use in gas coolers where glycol is injected to prevent hydrates forming. These exchangers tend to operate best within relatively narrow tubeside velocity bands which can lead to inefficient designs with plain tubes.

Sintered metal extended surfaces would not often be selected for a new plant because a small degree of fouling can block their pores. However, they might be used for very clean services in a de-bottlenecking exercise.

4.2 Coatings and Linings

Non-metallic coatings and linings are not favoured by us because we have found that modifications often have to be made offshore which sometimes mean cutting and welding lines. If these are coated or lined, then they are damaged and are not easily repaired. Damage to coated or lined heat exchangers is less likely, but they can suffer from fabrication "holidays" which go undetected until in use; also, some linings are vulnerable to physical shock which can often occur during hook-up. Metallic linings are acceptable if these have been applied by weld overlay or explosion cladding.

4.3 Special Baffles

Baffling such as the "no tube in window" type or ROD baffles (the Phillips Petroleum design) are acceptable when used to avoid vibration. "Support-only" baffles used in conjunction with the "no tube in window" type can produce a good design for low shellside pressure loss with little chance of vibration, and these are also acceptable. Otherwise, conventional baffles are preferred.

## 5. STAGES IN THE DESIGN OF HEAT EXCHANGERS

During the progress of a project from its conception to its completion, there are a number of self-contained stages. The following paragraphs describe the type of heat exchanger design work which is appropriate to each of these stages:

5.1 Feasibility Studies

The project has yet to be proved viable, so only minimal design work is done. The Process Engineer will calculate MTDs and heat duties, but guess film coefficients and fouling factors to calculate the surfaces of tubular exchangers. He will probably refer to Kern or in-house curves or nomographs for an idea of length and diameter, and he will "guestimate" the weight based on other projects. For plate heat exchangers he will probably consult the manufacturers' catalogues or contact a sales engineer for a quick budget size. He would not be concerned with the size of a plate fin heat exchanger at this stage.

5.2 Front End Design Studies

The project is now "on". But the Process Engineer has much to optimise and heat exchangers do not usually play a controlling role in his studies; so he won't fully design them for every alternative. Note that the modelling program he uses doesn't design heat exchangers. He will need a better idea of size and weight at this stage for overall comparisons. He will probably do a hand design using published or in-house short cut methods, or he may have a quick sizing program for use on a desk top computer; if he is lucky, he will have a colleague who will run a few cases through on the appropriate HTFS or HTRI program.

For non-tubular exchangers he will probably rely on manufacturers to supply him with enough examples to be able to pro-rate to cover the remaining cases. He will also pro-rate the tubular designs.

5.3 Worst Case Equipment Design

When the process has been optimised, there will still remain many different parameter sets for a given exchanger - start-up, re-cycle, high reservoir temperatures, with and without water saturation, etc., for example. The Process Engineer will write a data sheet for each case he considers may control some aspect of exchanger design; for example, a low flow case may govern because of the metal wall temperature or a high flow case could govern because of erosion velocity. He will issue these data sheets to a specialist either in-house or from a Consultant, or to a vendor. He now needs to know as much as possible about the exchanger because the next stage will be to complete the Piping and Instrument Drawings and start on Piping General Arrangement and Isometric Drawings. Sometimes the Client will have the exchanger completely designed and guaranteed at this stage, but this usually commits him to a fee to a Consultant or a promise to purchase to a vendor.

It is important to the project timetable not to have to repeat work, but it is also essential that the design actually works in practice. The Client will probably insist therefore on an acceptable thermal rating program being used, such as those of HTRI and HTFS. He may succeed in getting the mechanical design roughed out as well.

For non-tubular heat exchangers, the Client will get firm guaranteed designs with a full quotation from the manufacturers at this stage.

5.4 Procurement

The Client now needs firm guarantees that the equipment will work. Usually fabricators are not happy to accept the Client's own design without having it checked either by themselves or a Consultant. But if the Client has already had his exchanger designed and guaranteed by a reputable Consultant, this design will be good enough. It is quite common practice for Clients to ask vendors merely to guarantee the Client's own design, making no improvements to it. From the Client's point of view, so long as the equipment will work, this can be the best method since no dimensions need to be changed.

6. SPECIFICATIONS AND STANDARDS FOR TUBULARS

The TEMA standards are usually mandatory. Class "R" is usually called for, but some relaxation can be permitted. Tube diameters down to $\frac{1}{2}$" are often acceptable, and pitch/diameter ratios of 1.2 are sometimes allowed for refrigerant condensers. Standard tube lengths are not required offshore since tubes will be cut to length for tube replacement, or a complete new bundle will be installed. The tube gauges of TEMA "C" are sometimes acceptable.

Vibration has occurred in some offshore tubulars with near to TEMA maximum baffle pitches. Replacement of a bundle offshore is a major undertaking, so my Company insists on a guarantee being given by the vendor against vibration of any form, despite what TEMA may state. It is felt that sufficient is now known about the likely causes of vibration for this to be an acceptable guarantee for vendors to give.

Experience with tubulars in service offshore indicates that designers are now designing very tightly, both for heat transfer and pressure drop. Whilst we could not reasonably complain, this situation leaves no margin for our errors in predicting flows, temperatures and pressures. We will probably in the future add a small margin into the flows and heat duties on the data sheets, which will still allow for competitive designs, but will give us greater design tolerance.

The Process Engineer tends to be mean with his allowed pressure drops, often specifying only 10 or 15 psi. allowable drop in gas streams at 300 psia, for example. The thermal designer can, of course, always produce a design to match any pressure drop, but it may be huge. Very often, due to tubular geometry it is not possible to use up all the available pressure drop in one area but, unless the Client has permitted it, he will be unable to use the balance elsewhere on the same stream where it is more useful. Process Engineers would do well to specify a total permitted pressure drop for a stream passing through several exchangers, where possible. Strictly, the Process Engineer should do a balance between compressor or pump horsepower and tubular exchanger size, but this is seldom if ever carried out.

The type of TEMA unit is usually a matter of Client standards. We do not permit expansion bellows offshore for example. Generally, despite their inefficiency and weight penalties, many offshore operators prefer conservative designs like pull-through types for ease of maintenance.

The most space-efficient design offshore is often short and fat, as opposed to the more usual rule of "long and thin is best". Thermal engineers may need some guidance from the Client in this respect.

More specific requirements, based on current offshore experience, are dealt with in the next paper.

REFERENCES

1. D.Q. Kern, Process Heat Transfer, McGraw-Hill, New York, 1950.

# Problems Facing the Designer of Offshore Heat Exchangers

**M. A. TAYLOR**
British National Oil Company
London, England

ABSTRACT

Oil production in the hostile North Sea imposes a difficult set of restraints on both the Process and Thermal Engineer. Equipment must be light, small, robust, reliable and easy to operate and maintain. To achieve all these aims is usually impossible. This paper highlights some of the problems encountered during a project, and indicates how they have been overcome.

1. A TYPICAL OFFSHORE PROCESS

Even the apparently simple process of extracting oil and gas from a reservoir and piping the products ashore can prove quite a complex process. A typical offshore oil and gas production process is shown in Figure 1.

In this example, reservoir fluid is choked at the wellhead to feed the first of three stages of oil and gas separation. As the oil stream drops in pressure, it becomes less volatile and the pressures of the separation stages are optimised to maximise the product whilst keeping to pipeline specification and not using too much horsepower to recompress the gas. Reservoir fluids are sometimes very hot, so it may be necessary to cool the oil product before shipment, E1.

The gas released from the lowest pressure separator, V3, is first cooled by E3 and condensate is separated in V4. The gas is compressed by C1 to the second stage pressure where it combines with gas from separator V2. The combined stream is cooled by E4, separated in V5 and the gas compressed by C2 to the high pressure stage where it is combined with gas from separator V1. The combined stream is then cooled by E5 to about $80°F$ and water and condensate is separated in V6. (Reservoir fluid usually contains increasing quantities of water as the field ages, and so gas is assumed water saturated). If we wish to sell the gas, it will be necessary to drop its hydrocarbon and water dewpoints well below the lowest possible pipeline temperature, which for the North Sea is $40°F$, otherwise two phase flow may occur and hydrates may form. Hydrates are gum-like solids which are a clathrate combination of the lower hydrocarbons with water; they tend to form between 50 and $70°F$. The gas in this example is chilled in E6 by a refrigerant on the shellside, but to avoid hydrates forming as the temperature falls, ethylene glycol solution is sprayed into the exchanger tubes to mop up water as it condenses. The cold products are then separated in V7 into a gas and two liquid phases. The water rich glycol phase is sent for regeneration by hot oil or an electric heater, and the hydrocarbon is sent to the stabiliser, T1. The gas is now

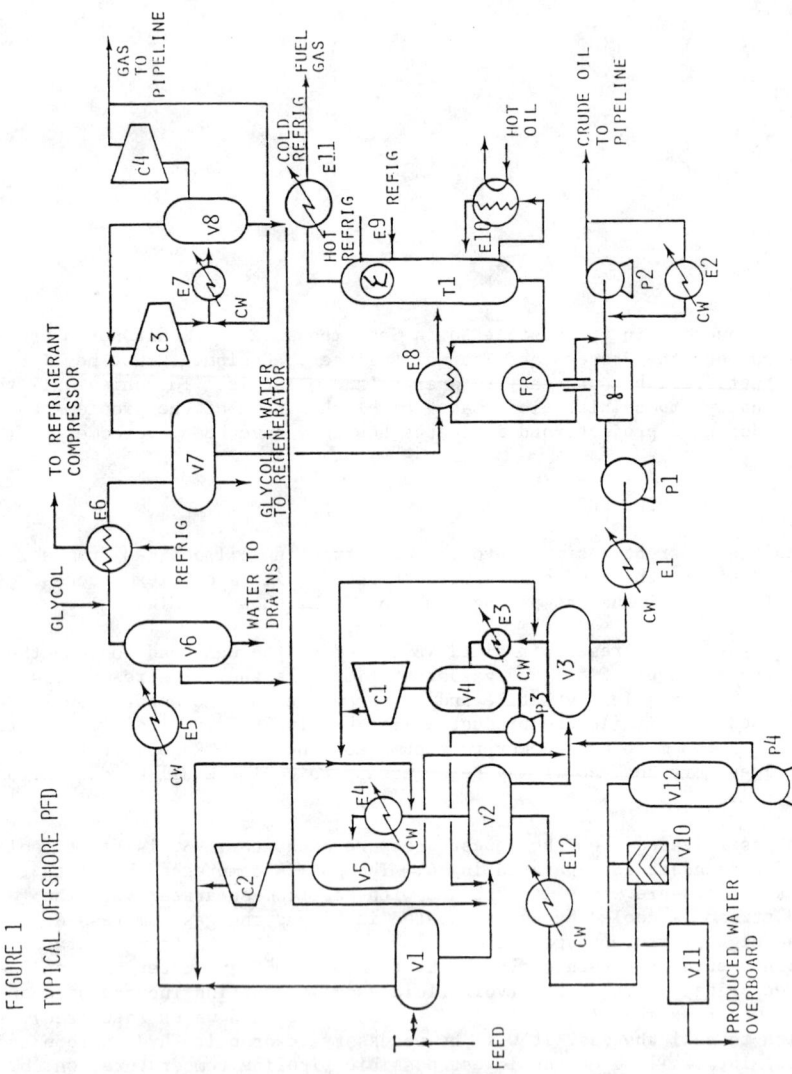

FIGURE 1
TYPICAL OFFSHORE PFD

Fig.1. Schematic Flow Diagramme of a Typical Offshore Oil and Gas Production Process.

ready for final compression before export via the pipeline. However, it is often impossible or uneconomic to compress, typically from 300 to 2,500 psia., in a single stage; it is necessary to do the compression in two stages, C3 and C4, sometimes within a single casing, but always with an intercooler, E7. There is a knock-out pot V8 downstream of the intercooler, but it is really for calamities only, since the gas is usually beyond its critical pressure at this stage.

The recovered condensate is stabilised to recover the $C_4+$ fraction for addition to the crude. It is sometimes possible to ship this condensate with the gas stream, but it usually earns more revenue in the crude stream. The condensate exchanges with the stabiliser column bottoms in E8 before entering the stabiliser column T1 where it is contacted with reflux from the refrigerated partial condenser, E9. The light ends are used for fuel gas, and any balance is recycled to the second stage gas compressor. The stabiliser bottoms products is controlled so that when it is mixed with the crude oil product the combined fluid meets the pipeline specification. Control is achieved by regulating the heat input to the stabiliser from the reboiler E10, and by the adjusting the reflux rate, temperature and pressure.

The water which flows in with the oil is known as "produced" water. It can be separated from any stage of gas/oil separation as a third phase. It is often collected from the second stage separator as shown here. It is then cooled by E12, passed through a gravity separation vessel V10 and a flotation cell V11 before being discharged overboard, clean of oil. Skimmed oil is returned to the process stream via sump V12 and pump P4. Produced water is cooled to meet both process and govermental requirements.

Apart form the heat exchangers shown on the flowscheme, there are many others used offshore. Some of these include -

- Coolant/seawater interchanger;
- Refrigerant condenser.
- Glycol regeneration interchanger.
- Fuel gas superheater.
- Lube oil coolers.
- Compressor seal oil coolers.
- Compressed air coolers.
- Air conditioning coolers and heaters.

2. VARIATIONS IN PROCESS PARAMETERS

Although oil production is a "continuous" process, it is not steady state. Initially, there will be a build-up from very little oil containing little water to a maximum plateau oil rate as more wells are drilled and come on line. Sooner or later with most North Sea fields water starts to enter with the oil, and will usually reach its peak rate at the end of the field's economic life when oil flow is at a minimum. The gas released during the field life generally follows the same profile as the oil; exceptions are when oil from a different reservoir with differing properties is produced, or if gas is recycled back down the wells to give "gas lift" to the oil to overcome the increasing static head when water enters the well with the oil.

The process engineer, therefore, has to design his plant to meet a very large turn down, 10 to 1 is typical, and to cover significantly varying product rates and properties. For each piece of equipment he has to consider the most arduous duty it must perform throughout its lifetime. Sometimes the most arduous duty occurs when the plant as a whole is producing very little product. For example, compressors used offshore are usually centrifugal, and whether they are driven by variable speed gas turbines or fixed speed electric motors, they will invariably need to be run on recycle at some stage in the life of the plant, and also at each start-up. Otherwise, at low flow-rates, surging will occur which can severely damage the compressors. The suction cooler E3 may have more duty imposed on it during re-cycle than at any other time.

At start-up and under low flow conditions the large horsepower main oil line pumps P2 may have to recycle some oil through cooler E2 to operate efficiently without overheating.

Variation in reservoir fluid temperature and pressure have to be considered. Temperature for example will usually increase with increasing water cut and will affect the performance of both crude and produced water exchangers E1 and E12 as well as all the gas coolers E3, E4 and E5.

Condenser duties will increase significantly when the hydrocarbon gas phase is water saturated, so even though in theory the stream entering E5 is not fully saturated, it is generally assumed to be so.

Well fluid composition will often change with time as new parts of the reservoir are brought on stream. Usually these are slight changes which do not affect the process but they can be serious; if possible the process design is checked at the conceptual stage for the effects of a composition change.

Alternative specifications sometimes have to be met by the same process plant. For example, on the Thistle platform we can export oil with a high vapor pressure through a pipeline, or with a low vapor pressure to a tanker. The tanker route has been used extensively whilst the plant was being commissioned and afterwards until the pipeline was brought into service; it was again used recently when the pipeline was out of service due to damage. It is important to keep the oil flowing - each day the platform exports oil worth about $4 million.

3. OPTIMISATIONS AFFECTING HEAT EXCHANGERS

3.1 Equipment Position in Flowscheme

The process engineer will seek to place equipment within his flowscheme so that weight and space are minimised. For example, the position of compressor cooler E7 is chosen so it can act as a recycle cooler for both compressors C3 and C4, but as a result it will have to be designed for the combined flows and heat loads of both compressors, since both will have to recycle simultaneously when starting up or when the gas throughput is less than the surge capacity of the compressors.

Another example is whether it is better to put crude oil cooler E1 upstream or downstream of the booster pump P1. If it is upstream there will be a pressure loss through the exchanger before the crude enters the pump impeller. Normally the sub-cooling of the saturated crude oil will more than compensate for the head loss through E1 and the NPSH into the pump will improve. But what happens when E1 is being serviced? Then the NPSH may fall to zero and P1 will cavitate. Putting E1 on the discharge side of P1 might overcome this problem, provided pump P1 is significantly below separator V3 outlet nozzle, but the pressure rating of E1 will then increase and will probably prevent a plate heat exchanger being used. Usually this dilemma is overcome by adding a standby unit to E1 so that the exchanger can be cleaned or serviced whilst production continues unabated.

3.2 Energy Conservation

Reasonable attempts are made to optimise the use of energy offshore but not at the expense of safety, reliability or operability. Usually each system is designed as an entity with no cross connections with other systems, so that should one system fail it may still be possible to produce oil. For example, if the refrigeration system failed, oil could still be produced, provided gas were flared. The cold end approach temperature for gas coolers is sometimes optimised against savings in compressor horsepower; the smaller the approach the larger the exchanger but the less horsepower required for gas compression. Usually however the minimum gas temperature is governed more by the wish to avoid hydrate formation than by any horsepower savings which could be made.

3.3 Metallurgical Considerations

Well fluids often turn "sour" after a time. This can happen if sulphate reducing bacteria are allowed to contaminate the reservoir; they break down the naturally occurring sulphates in the reservoir water to form $H_2S$. Carbon dioxide is often present as well and the combination of $H_2S$ and $CO_2$ in the presence of water is highly corrosive. In such cases an early decision is usually made to use "exotic" metal surfaces in all contacts with wet gas or hot, wet oil. Suitable materials for this case would be Titanium or Incalloy. There is therefore a case for using plate heat exchangers wherever possible, and the stage separation pressures may even be lowered to permit a plate cooler to be used on the crude oil.

Non-metallic linings can be used to cut costs but our experience has not been good with such materials. One problem with lined pipe and equipment is how to repair damage or make good the lining after offshore modifications have been made.

Where a tubular has to be used, the fluid will pass through the tubes, for otherwise the shell, outside of tubes and floating head, baffles and shellside of tubesheet/s would all have to be in exotic metal. If this tubular is a cooler what then would be the coolant? If sea water were used, then corrosion is a problem in pipework and on the shellside of the tubular unless non-ferrous materials are used, so a closed circuit coolant might be preferred, in which case carbon steel is acceptable. But coolant will be warmer than sea water by the amount of the temperature approach in the coolant/sea water interchanger.

3.4 Cooling Medium Optimisation

The cooling requirements for an offshore platform can be met by the use of sea water in an open circuit, or by interchange of heat between sea water and a closed circuit coolant, or a hybrid of both systems. The only alternative to the sea as a heat sink is air, but cooling by air is seldom used offshore because of the large plan area required for the air cooled heat exchangers; also, ambient air temperature extremes are wider apart than those for sea water, which makes the thermal design of the system more complex.

Sea water, drawn from 80 metres below sea level, always lies between 40 and 50°F.

The advantages of direct over indirect cooling include the possibility of obtaining lower process temperatures and less pumping horse-powers, but the disadvantages include the obligation to use non-ferrous contact materials, the greater fouling and scaling potential and, as a result, less flexibility in selecting process sides in a shell and tube exchanger.

3.5 Open circuit cooling

This system uses sea water for cooling throughout. Because the quantities are large, it is impractical to treat the water prior to circulation. There is, therefore, a risk of corrosion, erosion, scaling and freezing. All sea water pipework has, therefore, to be fabricated from one of the cupro-nickel alloys, and should be lagged against freezing on shut-down; equipment has also to be non-ferrous, and titanium is often used for heat exchangers of both plate and tubular type. The use of non-ferrous materials offshore makes mo ifications difficult, apart from its high first cost.

Open circuit coolant is 10 or 15°F. cooler than a closed curcuit coolant, since there is no sea water/coolant interchanger. Furthermore, there is some saving in equipment since no interchanger, coolant pumps or head tank is required. However, there is often a corresponding weight penalty in the coolers. Sea water will foul and scale, even on non-ferrous surfaces, so it must always be run through straight tubes, or else, in the case of high-pressure process fluids when sea water must flow on the shellside, the tubes must be on a 1.25 square or square rotated pitch/diameter ratio and the bundle must be removable for cleaning. Because it is impractical to pull bundles of greater weight than 10 tonnes offshore, the total surface per shell is restricted. It is also not practical offshore to use single pass removable bundles, so the number of shells may need to be increased to prevent temperature crosses. Furthermore, the combination of the open pitch requirement, the wider bundle to shell clearances required for removable bundles and the lower velocities required to avoid erosion make for inefficient thermal design.

If sea water is used directly to cool hot fluids, such as crude oil or produced water, scaling may occur if the surface temperature exceeds about 95°F. because inverse solubility salts such as sulphates can precipitate. Scaling within heat exchanger tubes is at a maximum between 0.5 and 1.0 m/s., above which it decreases with velocity, but above 3.5 m/s. erosion/corrosion can occur. The heat exchanger designer, therefore, tends to restrict his design to around the 2 m/s. mark. Often produced water will also scale as it is cooled; we have experienced scale formation on both sides of a produced water plate cooler.

All coolers using sea water will require de-fouling and possibly descaling at regular intervals. If plate heat exchangers can be used for these services, then the cleaning duty is much easier than for shell and tube exchangers, particularly if their bundles have to be pulled. A fouling/scaling factor of .0004 $m^2$ °C/W is required for sea water flowing through tubulars.

Thus, although there is advantage in the lower temperature of sea water over closed circuit coolant, especially for low mean temperature differences, in compressor suction coolers, for example, there are considerable hidden penalties in using direct sea water cooling as well as the more obvious ones.

## 3.6 Closed circuit cooling

Treated water coolant is circulated through a plate heat interchanger to lower its temperature to about $10°F$. above ambient sea water; a plate heat interchanger is used because it contains a large surface area per volume occupied, can be readily taken apart for cleaning and is thermally very effective at getting close temperature approaches.

Closed circuit pipework can be in carbon steel, and need not be lagged since treated water neither fouls nor freezes at the lowest ambient temperatures. The coolant is less likely to erode, so velocities can be higher. Coolers can be made of carbon steel on the coolant side, and it is immaterial whether coolant flows on shell or tube side since it does not foul or scale, so theoretically does not require bundles to be removable. However, our commissioning experience has shown the exchanger shellside to be a most efficient filter for millscale, rust, rubbish and dirt of all kinds, and for a year after start-up these bundles had to be pulled regularly to remove the rubbish. We have also experienced thermal breakdown of the redox agent which has caused some fouling. These problems could probably be overcome by the use of large and robust duplex line filters and, if so, then bundles could be non-removable; in which case where high pressure, corrosive or fouling process fluids have to flow through the tubes, the coolant can flow in direct counter current on the shell side, where tight pitches on triangular layout and close bundle to shell clearances ensure good heat transfer. The need to pull bundles is avoided, and generally only chemical cleaning of the coolant side is required. The fouling factor used for closed curcuit coolant is typically $.0002\ m^2. °C/W$ which leads to smaller surfaces than the equivalent open circuit cooler.

Treated water coolant can be subjected to higher surface temperatures than sea water, since there are no inverse solubility salts present to scale, but take care to avoid inhibitor degradation.

The disadvantages of closed circuit cooling lie in the additional equipment and pumping power required, and its higher temperature. Advantages are better controllability, greater design flexibility and less maintenance. The first cost of both closed and open circuit systems must be calculated, but the extra cost of the closed circuit equipment is offset by the cheaper materials which can be used and the smaller coolers required in most cases.

## 3.7 Hybrid cooling

In this case selected coolers, such as the crude product cooler E1 and possibly the produced water cooler E12 (but consider the likelihood of scaling) are directly sea water cooled, whilst the remaining coolers are on closed circuit coolant. The rationale is that the crude and produced water coolers are heavy duties which can be carried out in plate heat exchangers which are relatively easy to clean; a considerable load is thereby taken off the interchanger, head tank and collant circulating pump. However, all the material penalties apply to these services - lines have to be non-ferrous and lagged, etc. The injection of anti-scalant and anti-fouling chemicals into the direct cooling water will mitigate fouling of exchanger surfaces, but is expensive.

## 3.8 Waste Heat Recovery

Often there is need for process heating duties. On our example heat would be required for the column reboiler E10, and (not shown) for the regeneration of the glycol and superheating of the fuel gas. Space heating is also required for personnel, particularly in the accommodation block; modules in general, however, tend to need cooling even in winter because of the heat given off by process fluids, electric motors and turbines.

Electric heating can be used, but is inefficient and expensive to install. Waste heat recovery from the turbine exhaust can be an attractive alternative, and usually can produce more heat than is required. Heating oil is used as the transfer medium, and can be pumped anywhere on the platform.

The waste heat recovery exchangers exert a back pressure on the turbines which therefore produce slightly less power than otherwise; usually there is adequate power available, so this is not a big disadvantage. The exhaust ducting, already large, has to be much bigger and heavier to accept the exchangers, and this is a more serious disadvantage. Many turbine manufacturers are now offering "standard" waste heat recovery packages on their turbine exhausts.

3.9 Standby and Spares

If a unit is vital to oil production and if it is likely that it may have to come out of service at any time, then general practice is to add a standby unit. As only one standby is required, an optimisation is necessary to decide which combination is best: 2 x 100%, 3 x 50%, or 4 x 33 1/3%. The crude oil cooler E1 is often 3 x 50%, for example.

For items which are less vital, a bypass may be installed around them, so production may continue, albeit less efficiently.

Spare bundles are not stored offshore because of weight and space limitations. However, for bundles which have to be regularly pulled, a spare might be stored onshore to cover the risk of damage. Spare gaskets would always be stored offshore.

Standardisation of bundles is seldom possible without paying heavy penalties in efficiency.

Standard tube lengths are relatively unimportant because the thermal designer needs every freedom to optimise on the use of space; besides which the stores will stock the longest tube length and cut it to size if one or two tubes need replacing. Major re-tubing of a bundle would not take place offshore.

4. PROBLEMS ENCOUNTERED OFFSHORE

4.1 Surface Area and Pressure Drop

In many cases we have found exchangers to be the bottleneck of a process. Between the time of specification and start-up, changes in flowrate or specifications may have occurred which may affect the heat exchangers. In former times one generally had a good margin on heat exchanger designs because the thermal engineer had not the time to screw the design down tightly. With modern design programs it is well to remember that the resultant design will be based on best fit curves passing through the middle of the research data. If the Process Engineer has any qualms about the exchangers becoming a bottleneck, he should increase the flows and duties on his data sheet to reflect the safety margin he thinks safe - the Thermal Engineer won't do it in today's world.

## 4.2 Vibration

Exchangers with lengths approaching the TEMA maximum have vibrated. One platform is reported as having to replace six exchangers as a result. The difficulty of getting the old exchangers out and the new ones in are enormous; often other equipment has to be removed before it is possible. My Company now requires an anti-vibration guarantee at the thermal design stage.

## 4.3 Wax Deposits

Control of the crude oil temperature out of E1 can be achieved by regulating the coolant flow or by by-passing some of the crude oil. Whilst using the latter method, we have experienced wax precipitation on the overly cooled plates.

## 4.4 Scale Deposits

Although treated coolant itself is non-scaling, non-fouling and non-corrosive, some of the smaller coolant exchangers have been regularly plugged with scale and dirt from upstream equipment and pipelines. Unfortunately, hook-up work in the North Sea is difficult and the weather can rust equipment, millscale cannot always be removed prior to commissioning and dirt and rubbish build-up has often sheared off the strainer itself to add to the downstream debris. Large, robust, quick-opening, duplex cartridge filters have been used successfully to overcome the problem, but until these were installed bundles had to be pulled regularly, even on "clean" services.

## 4.5 Inhibitor

We have experienced some coolant loss through weeping thermal relief valves. On adding make up fresh water and a new dose of inhibitor we found the new slug of inhibitor acted as quite an effective de-scalant! The scale then, of course, finished up in some of the downstream coolers. If the inhibitor is over-heated (above $150^\circ$F., say), it can break down and act as a fouling agent.

## 4.6 Chemical Cleaning

We have found 10% HCl to be effective. All exchangers are required to have provision for chemical cleaning, even plate heat exchangers, because chemical cleaning is faster than dismantling and physically cleaning them. Nitrile gaskets, often used in plate heat exchangers, do not stand up to the acid, however.

## 4.7 Other Cleaning Methods

High pressure water jets are the most effective at cleaning exchangers, but can't be used where there is a risk of water entering electric or instrument controls or switches.

Other methods used successfully are power brushes and rods.

## 4.8 Davits

Where cleaning is often required, it is worthwhile installing davits on the channel to swing off the cover plate.

### 4.9 Bundle Pulling

Overhead runway beams are necessary to take the bundle weight as it is pulled. Bundles above 10 tonnes and over 12 foot length are difficult to handle offshore, particularly, if bundles have to be removed from the exchanger area. Bundles are pulled using block and tackle, and pulling eyes installed in the tube sheet make this job easier.

Annual planned maintenance shutdowns need to be short to keep lost revenue to a minimum; therefore it is good if as many bundles as possible are pulled simultaneously. This requirement calls for no stacked shells and plenty of space in between, which contradicts the requirement to squeeze the most into available space.

### 4.10 Nozzles

It is necessary to insert a spade into the process line before opening an exchanger for safety reasons. To do this requires "springing" the pipework which is not easy if the exchanger nozzle is vertical, goes immediately through the floor and into the top of the next vessel! For maintenance purposes, a side or hooked nozzle followed by a flanged pipe bend is preferred, despite its again occupying more space.

### 4.11 Re-assembly of Girth Flanges

We prefer girth flanges to use stud bolts. Then we can use "Pilgrim" nuts to re-assemble girth flanges quickly and accurately.

### 4.12 Water Leakage

If water leaks into tube or seal oil via the coolers, the consequences may be serious. Expanded tube to tubesheet joints have been known to weep, so we require all tube to tubesheet joints to be welded.

### 4.13 Vents and Drains

On long exchangers if only one vent and drain connection is supplied it may not be possible to thoroughly empty and purge the vessel. We require vents and drains _both_ ends.

## 5. CONCLUSION

I hope in this brief discussion I have been able to explain some of the conflicting requirements that the Process and Thermal Engineers must resolve in order to provide offshore plant which performs well yet is not too costly in terms of weight, space and money.

I am indebted to my company, the British National Oil Corporation (Development), Ltd., for permission to publish these papers.

## REFERENCES

1. D.A. Kern, <u>Process Heat Transfer</u>, McGraw-Hill, New York, 1950.
2. The British National Oil Corporation (Development), Ltd., Publications (Classified).

# A Survey of Dry Cooling Tower Technology for Power Generation Application

**ALI MONTAKHAB**
Tehran University of Technology
Department of Mechanical Engineering
P.O. Box 3406, Tehran, Iran

ABSTRACT

Conventional waste heat disposal systems for thermal power generating stations require significant amounts of cooling water and are subject to stringent environmental regulations which affect the economy and siting of the power station. Siting restrictions and water requirements are eliminated by using air cooling, which, in spite of the present technical and economic drawbacks, may be needed at some U.S. locations before 1990.

This paper is the result of a literature survey; it summarizes major design considerations and discusses the economic and technical aspects of using dry and wet/dry cooling systems for thermal electric power stations.

1. INTRODUCTION

The three most frequently used waste heat disposal systems for conventional power generating stations are (1) once-through cooling, (2) cooling ponds, and (3) evaporative cooling towers [1,2][1]. There are siting limitations associated with the first method [3-6], especially because of the trend toward increasingly higher electrical output and stringent environmental regulations limiting the use of river, lake, and ocean water [7]. The second and third methods require a significant amount of water, which may affect the economy and siting of the power station.

Siting restrictions and water requirements are minimized by using air cooling. Air cooling, or dry cooling as it is often termed, may be used alone or it may be combined with wet cooling towers to reduce the cooling water makeup requirements of a wet-cooled power station. In the wet/dry system, the dry cooling tower operates all year round, whereas the wet peaking tower is used only above certain ambient dry bulb temperatures. Dry and wet/dry systems have attracted a great deal of attention during the last few years [8-13] and may well be needed at some U.S. locations before 1990 [3,10].

About two dozen electrical generating plants in Europe, Asia, and South Africa use air-cooled steam condensers [11,12]. These plants have electrical outputs ranging from 4 to 200 MW(e). A small unit in Germany began operating

---

[1] Numbers in brackets denote references listed at the end of the paper.

in 1939 [13], and a 13-MW(e) unit has been operating in Luxembourg since 1956. The 120-MW(e) Rugeley station in England, which was the first large-scale natural-draft dry cooling installation, went into commercial service in 1961 [14]. The Neil Simpson (Wyodak) plant at Gillette, Wyoming, is the only current U.S. dry-cooled central power station. It is a coal-fired mine mouth station located in a region with a severely limited water supply [15]. There are two air-cooled units at Wyodak, a 3-MW(e) plant operated since 1962 and a 20-MW(e) plant operated since 1969. A 330-MW(e) plant under construction at Wyodak will be the world's largest, single dry-cooled unit, with operation planned for 1978 [16]. All power plants currently using dry-cooled condensers are fossil-fired, although application of the dry cooling concept to nuclear steam and gas turbine cycles appears feasible [17-19].

2. GENERAL DESIGN CONSIDERATIONS

Dry cooling requires a higher investment expenditure than once-through cooling, cooling ponds, and wet cooling cowers [20,21] and would not be employed if cooling water were available and evaporative cooling were environmentally acceptable. There are three general approaches to air cooling of large steam power plants: (1) condensation of the steam turbine exhaust in an air-cooled condenser; (2) direct contact condensation of the turbine exhaust steam by liquid water followed by air cooling of the recirculated water; and (3) condensation in a conventional liquid water-to-steam surface heat exhanger followed by air cooling of the recirculated water. The first approach has the advantage of simplicity. However, a provision must be made for delivering a very large vapor volume flow rate from the turbine exhaust to the air-cooled heat exchanger. The second approach has the advantage of using a fairly compact condenser and a liquid-to-air heat exchanger operating at near atmospheric pressure on the fluid side. However, for the same steam turbine back pressure condition, the temperature potential is less than that for the air-cooled condenser, and consequently a larger heat transfer surface area is required. Nuclear plants may require the third approach to provide the necessary isolation of the primary coolant, which is recycled through the reactor and therefore must be contained within a reliable closed loop.

2.1. Cost and Performance

In recent years, a large number of economic studies have been conducted on dry-cooled power plants [19-22]. These studies cited the high plant cost and poor thermal efficiency of dry-cooled plants. Because of air cooling, the thermal efficiency of a dry-cooled plant is poorest under high ambient temperature conditions, reducing the plant peak load capability during the summer. Most studies have shown that if dry cooling is used, the cost of electric power generation will be 10% to 20% higher than that for wet cooling towers. This range is a result of the different climatic conditions at the plant sites considered. The cost of dry cooling is higher because of the higher plant capital costs and poorer thermal efficiency of the dry-cooled power station. The high plant capital costs derive from the requirement for an extensive finned-tube heat transfer surface area associated with the dry cooling tower, and the poor thermal efficiency is the result of higher turbine exhaust pressure. Because turbine exhaust pressure is dependent on condenser coolant temperature, which in turn is dependent on ambient dry-bulb temperature, high turbine exhaust pressures (8.0 to 12.0 in. Hg abs) can be expected during summer operation [23-25]. Consequently, dry cooling tower plants are substantially penalized in terms of kilowatt output capability and thermal efficiency compared with plants employing once-through cooling, cooling ponds, or evaporative cooling towers.

To achieve dry cooling tower plant economic optimization, the trade-offs between capital and fuel costs have to be investigated. Higher fuel prices would justify a larger and more expensive dry cooling tower facility, which would result in a lower turbine exhaust pressure and improved plant performance. However, there is a rapid diminishing investment return condition for a larger dry cooling tower because the cost of the dry cooling tower versus the gain in lower turbine exhaust pressure tends to increase asymptotically.

## 2.2 Turbine Selection

Another major factor affecting dry cooling tower power plant economics is turbine selection [24,25]. Steam turbine characteristics suitable for plants with dry cooling towers are illustrated in Fig. 1. Curve 1 represents a turbine with a short last-stage blade length (LSBL) and curve 2 a turbine with a long LSBL. The exhaust annulus area of the curve 1 turbine is approximately 60% of that of the curve 2 turbine. The turbine with the smaller LSBL is more effecient at exhaust pressures higher than about 7.0 in. Hg abs because of its lower exhaust losses at high exhaust pressures, and it is less efficient at low exhaust pressures. Studies have favored a turbine with a short LSBL because the power capability between low and high ambient temperatures during summer operation is increased. Its loss in heat rate during low ambient temperature operation would be compensated for by its gain during summer operation.

## 2.3 Economic Optimization

In spite of the above drawbacks, there may be times when the installation of a dry cooling tower would not only be a necessity but might also be economically justifiable. The task confronting power engineers would be to minimize the power generation cost of a plant with a dry cooling tower. To do this, a comprehensive economic optimization would be necessary. This optimization would have to consider plant capacity factors, system load characteristics, site and ambient temperature constraints, and present and projected fuel prices on a case-by-case basis.

Several papers have addressed the problem of economic optimization of dry-cooled plants. Leung [26] proposes a method based on economic loading during operation. In this method, the dry tower plant relegates a portion of its generation capability to other plants during the summer months and limits its back pressure to approximately 6.0 in. Hg abs at a reduced load. During winter, it is assigned a base load role, operating at a lower turbine back pressure and a maximum load.

Another option for economic optimization of dry-cooled power plants is to use a dry cooling tower with wet tower peaking [27,28]. This results in a cooling system which can use conventional low-exhaust-pressure turbines and can meet any reduced makeup water situation a utility might face.

Englesson, Hu, and Savage [29] have made extensive economic studies on dry and wet/dry cooling towers (Table 1). Table 1 shows the number of cooling tower cells for several different designs ranging from all-dry using high-back-pressure turbines to all-wet using conventional turbines, with various combinations of wet/dry towers spanning the middle ground. All the designs are based upon a plant sized for a base output of 1094 MW(e).

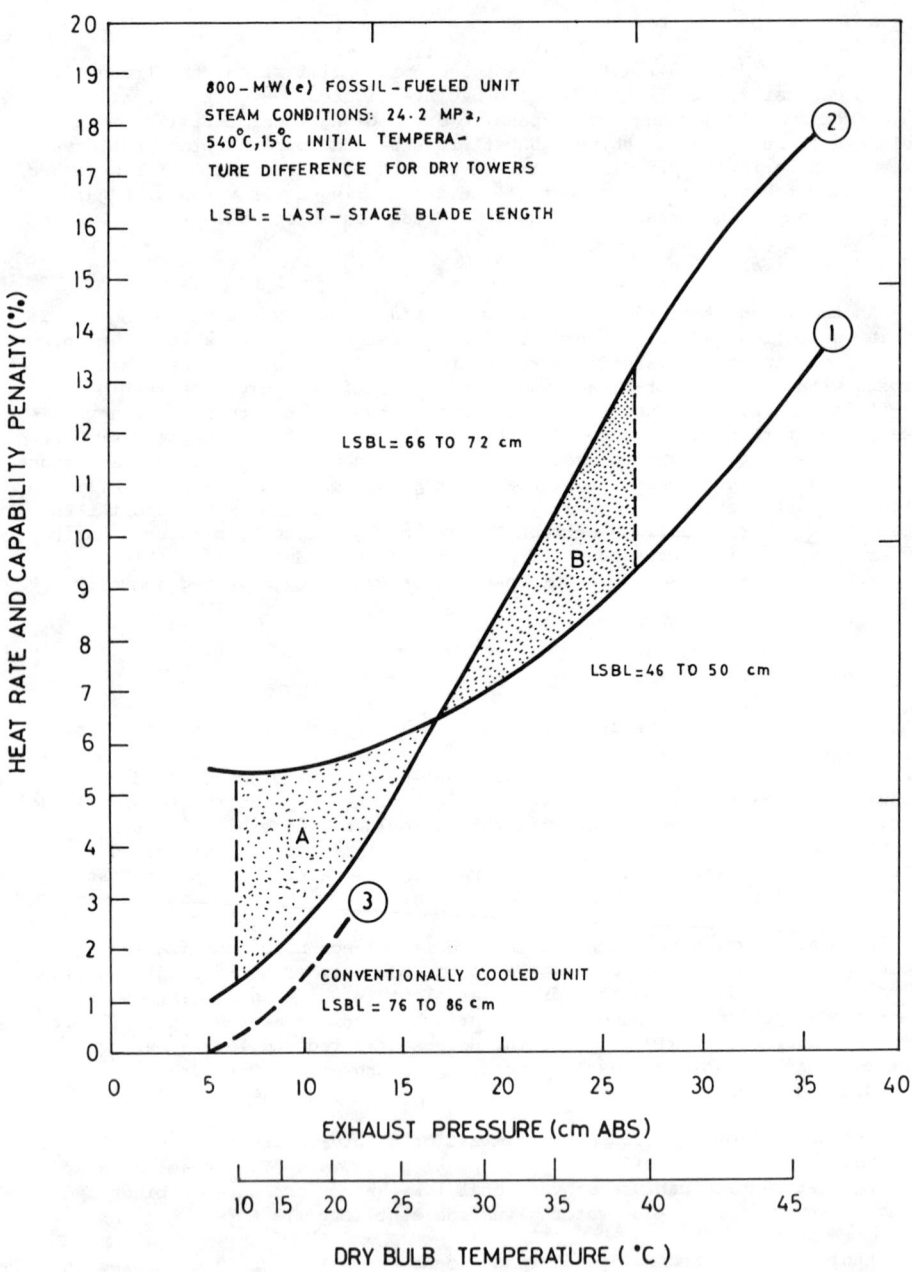

Fig.1. Dry Tower Plant Turbine Selection for Base Load with Summer Peak Demand [25]

It is apparent from Table 1 that the all-dry system, which uses a conventional low-back-pressure turbine, requires more than twice as much cooling surface as the dry system with a high-back-pressure turbine. This leads to a greater capital cost, as shown in Table 2. This higher capital cost could be offset by the greater efficiency of the low-back-pressure turbine and resultant lower replacement energy cost. However, although the cost of decreased capacity and replacement energy is less, the overall cost penalty is greater for the all-dry system with the low-back-pressure turbine. As shown in Table 2, the total cost for a dry system using high-back-pressure turbines is less than that for a dry system with low-back-pressure turbines. Thus, it appears that the optimum all-dry system should be designed to use high-back-pressure turbines.

In the wet/dry approach, the dry tower is used during most of the year to conserve water, and the wet system is used only as a peaking unit to carry the dry unit over periods of high ambient temperature. The effects of wet peaking on the overall capacity of a power plant are illustrated in Fig. 2, which shows plant capacity as a function of ambient temperature for a plant with a base output of 1094 MW(e).

Wet peaking has the advantages of (1) reducing turbine back pressure, making it possible to use conventional low-back-pressure turbines; (2) reducing initial capital cost; and (3) increasing overall plant capacity. The major disadvantage of the wet/dry system is the complexity caused by the addition of the wet tower to the system. However, this is offset by increased efficiency and reduced cost (the cost saving is shown in Table 2). Table 2 shows that a wet/dry system designed to achieve a 60% decrease in water consumption will result in a 70% increase in the total evaluated cost over the all-wet cooling system (compared with over a 200% increase in cost for the least costly all-dry cooling system with a high-back-pressure turbine).

Even if water were extremely scarce, it appears that a wet/dry system would be more economical than an all-dry system. This can best be seen by comparing the cost of the all-dry system using high-back-pressure turbines with that of a wet/dry system which would reduce water consumption by 99%. The all-dry cooling system would cost $297 million, more than 200% over the cost of the all-wet cooling system. The wet/dry system, with only 1% water consumption, would cost $233 million, only 140% over the all-wet system cost. Thus, the wet/dry system has advantages in the areas of water conservation, total cost, and overall plant capacity. Plant efficiency and capacity will grow in importance as the cost of fuel rises in the future.

The wet and dry sections of a wet/dry cooling system can be integrated in several ways. There may be separate towers for each type of cooling, such as those in Table 2, with the water flowing in series or in parallel. The system may consist of a single tower in which wet and dry sections are integrated within a common structure with parallel or series flow of the air in each section and with either parallel or series flow of water through the sections. Another proposed arrangement uses a portion of the dry heat transfer surface as the wet section packing. This approach, called deluge cooling, leads to even greater integration of the wet and dry sections, which could reduce tower size and cost.

Of the various alternative arrangements for wet/dry cooling, the scheme using separate towers has gained the most attention [27-29]. This is partly due to the possibility of air-side corrosion in the highly humid atmosphere of the integrated system and the need for expensive water chemistry control

TABLE 1

COMPARISON OF DRY, WET/DRY, AND WET COOLING SYSTEMS(a)

| Item | Mechanical Dry ($H$)(b) | Mechanical Dry ($L$)(c) | Percentage Makeup Requirement Mechanical Series Wet/Dry | | | | | Mechanical Wet |
|---|---|---|---|---|---|---|---|---|
| | | | 1% | 10% | 20% | 30% | 40% | |
| Number of tower cells, wet tower/dry tower(d) | 0/156 | 0/338 | 13/192 | 19/136 | 26/114 | 27/90 | 30/79 | 33/0 |
| Maximum operating back pressure $P_{max}$ (cm Hg abs) | 31.8 | 12.8 | 12.7 | 11.4 | 10.2 | 10.2 | 10.2 | 9.9 |
| Cross plant output at $P_{max}$ [MW(e)] | 946.7 | 1046.8 | 1048.4 | 1059.5 | 1069.9 | 1069.9 | 1069.9 | 1071.9 |
| Heat load at $P_{max}$ [MW(t)] | 2226.7 | 2127.2 | 2124.2 | 2115.4 | 2103.7 | 2103.7 | 2103.7 | 2100.7 |
| Heat load distribution at $P_{max}$ wet tower/dry tower(%) | 0.0/100.0 | 0.0/100.0 | 42.7/57.3 | 63.7/36.3 | 73.8/26.2 | 78.2/21.8 | 80.5/19.5 | 100.0/0.0 |
| Annual makeup water for wet towers ($10^6$ m³) | 0.0 | 0.0 | 0.165 | 1.669 | 3.205 | 5.040 | 6.200 | 16.057 |

(a) Based on Ref. [29]. Site: Middletown, U.S.; base output: 1094 MW(e); wet/dry type: mechanical series.
(b) High-back-pressure turbine.
(c) Conventional low-back-pressure turbine.
(d) The number of tower cells represents the amount of cooling surface area built into the towers.

TABLE 2
COST OF DRY, WET/DRY, AND WET COOLING SYSTEMS[a]

| Item | Mechanical Dry (H)[b] | Mechanical Dry (L)[c] | Percentage Makeup Requirement Mechanical Series Wet/Dry | | | | | Mechanical Wet |
|---|---|---|---|---|---|---|---|---|
| | | | 1% | 10% | 20% | 30% | 40% | |
| Capital cost ($10$^6$) | | | | | | | | |
| Cooling tower | 54.42 | 116.90 | 74.56 | 58.59 | 55.07 | 47.27 | 45.22 | 19.48 |
| Condenser | 15.20 | 20.88 | 15.98 | 14.11 | 13.64 | 13.62 | 13.25 | 13.61 |
| Circulating water system | 10.02 | 17.92 | 14.74 | 13.01 | 12.23 | 12.35 | 11.77 | 8.22 |
| Electrical Equipment | 7.32 | 15.72 | 12.18 | 9.46 | 8.52 | 7.37 | 6.91 | 2.25 |
| Indirect coat | 21.86 | 44.12 | 29.37 | 23.79 | 22.36 | 20.15 | 19.29 | 10.88 |
| Total capacity cost | 108.82 | 215.54 | 146.83 | 118.96 | 111.82 | 100.76 | 96.44 | 54.44 |
| Penalty cost ($10$^6$) | | | | | | | | |
| Capacity | 88.97 | 28.33 | 27.36 | 20.72 | 14.40 | 14.46 | 14.44 | 13.27 |
| Auxiliary power | 19.35 | 38.02 | 24.49 | 20.23 | 19.44 | 18.49 | 17.73 | 9.38 |
| Replacement energy | 55.56 | 0.29 | 5.48 | 11.39 | 11.34 | 13.47 | 14.25 | 3.07 |
| Auxiliary energy | 19.23 | 33.03 | 22.33 | 17.71 | 17.02 | 16.89 | 16.34 | 9.02 |
| Makeup water | 0 | 0 | 0.06 | 0.65 | 1.25 | 1.97 | 2.42 | 6.28 |
| Cooling system maintenance | 5.37 | 10.57 | 7.28 | 6.24 | 5.84 | 5.39 | 5.22 | 2.64 |
| Total penalty cost | 188.48 | 110.24 | 87.01 | 76.94 | 69.35 | 70.94 | 70.40 | 43.66 |
| Total evaluated cost ($10$^6$) (sum of capital and penalty costs) | 297.30 | 325.78 | 233.84 | 195.90 | 181.17 | 171.70 | 166.84 | 98.10 |

(a) Based on Ref. [29]. Site: Middletown. U.S.; base output: 1094 MW(e); wet/dry type: mechanical series.
(b) High-back-pressure turbine.
(c) Conventional low-back-pressure turbine.

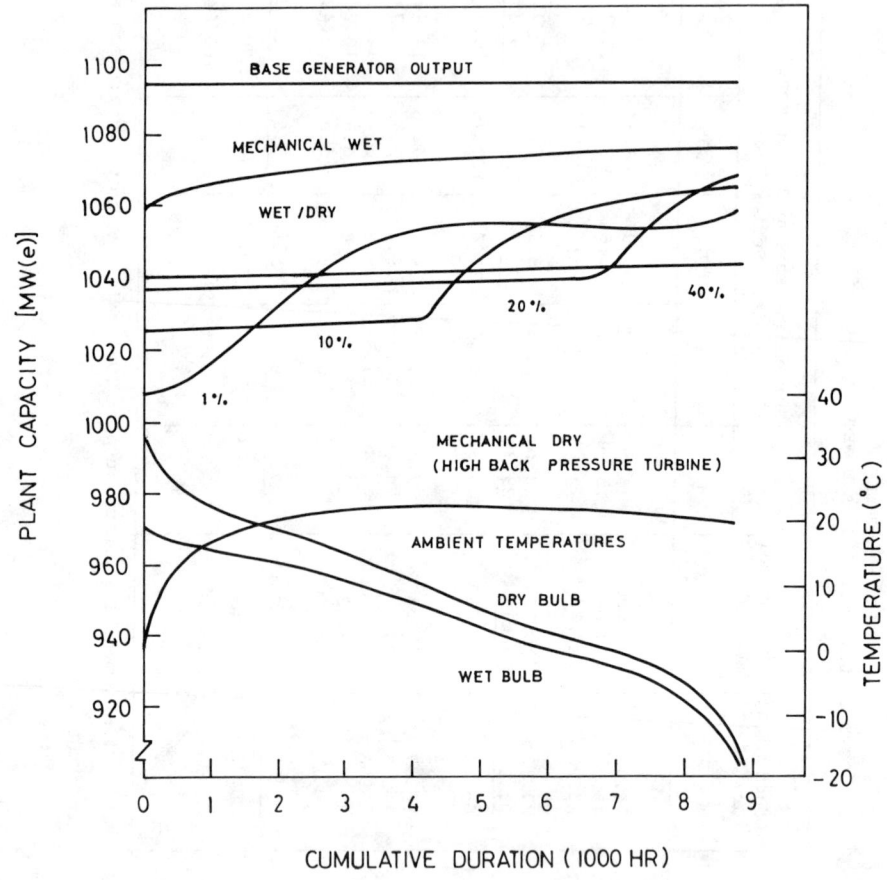

Fig.2. Net Generator Output as a Function of Temperature [29]

facilities to minimize the strong likelihood of air-side fouling and corrosion in the deluge system.

## 2.4 Weather Data

Economic optimization is dependent upon weather data which enable adequate predictions to be made of the conditions to be encountered during the opeating life of the plant. Several professional societies and industrial organizations have published excellent weather data for cooling system design and performance evaluation (e.g., [30]). Because of the large height of natural-draft dry cooling towers and their susceptibility to wind and atmospheric effects, attention has recently been given to providing diurnal and seasonal climatic data at selected sites in the U.S. [31].

## 3. DESIGN CONSIDERATIONS FOR LONG SERVICE LIFE

Proper selection of material and manufacturing technique is an important step toward reducing costs and avoiding corrosion in dry cooling towers. In order to find the optimum heat exchanger system, detailed investigations must be carried out on the requirements of a particular application, and a careful evaluation must be made of the criteria involved. Site conditions such as climate, air pollution by neighboring industrial plants, and suspended matter in the cooling air, expected life of the total plant, plant operating characteristics, and medium to be cooled are some of the basic items which require detailed investigation.

The expected life of the total plant is significant in establishing the criteria for selection of the heat exchanger material. For power stations with a service life of 30 to 40 yr, the high investment costs for the dry cooling tower are justified only if its life expectancy is the same. The life of a dry cooling tower depends largely on the ability of the fin tube material to resist corrosion as well as performance deterioration due to fouling and impairment of fin-tube bondage, and on how the surface area is designed and manufactured.

### 3.1 Corrosion

Johnson, Pratt, and Zima [16] provide an excellent summary of atmospheric corrosion literature, including extensive information relevant to dry cooling tower candidate materials. In general, however, the literature on atmospheric corrosion applies to only finned tubes under shutdown conditions, when the metal is at ambient temperature. When the plant is on line, the finned tube temperatures are well above the dew point, and the effects of corrosion are minimal except when there is fog, rain, or snow. In addition, the corrosive effects of some contaminants are not well defined in the dry cooling tower regime.

It is well known that corrosion may occur on surface crevices and cracks and can subsequently spread over the entire surface. The use of homogeneous surfaces is one way to avoid corrosion attacks. Crevices between the core tube and fins are difficult to avoid with fin tubes made of several components and lacking a homogeneous coating, and some extruded tube materials are subject to considerable plastic deformation during fabrication. Thus, cracks produced during fabrication, especially at the fin tips, contribute to corrosion.

If the composition of the atmosphere is known and corrosion occurs uniformly over the entire surface, it is possible to predict the service life of a cooling system. If the corrosion attacks are local, there is no clear correlation between material thickness and service life. Local corrosion attacks occur only if the protective coating is fully or partly destroyed and the remainder of the coating is not able to provide a protective cathodic effect.

Aluminum surfaces are compatible with certain atmospheric conditions, but may suffer greatly when exposed to a corrosive environment. With few exceptions, aluminum has performed satisfactorily in a large number of industrial air coolers and a few dry cooling towers [11,16]. Serious air-side corrosion at the Rugeley plant in England [11,14] was caused by a combination of a contaminated environment, crevices on the plate-type aluminum fin assembly, and horizontal fin orientation. This fin orientation maximized the accumulation of moisture and contaminants in the very humid atmosphere. Severe fouling by coal dust occurred on plate-type aluminum fins at the Ibbenburen plant in Germany [11]; this was mostly caused by location of the dry towers alongside a wet tower and a coal storage pile.

When an uncoated, pure aluminum fin is exposed to clean air, it produces a protective coating. Although this coating is extremely thin, it prevents the fin from rapid destruction by corrosion. However, if halogen gases or chlorine ions are in the air, formation of the protective coating is impeded, and the existing coating is destroyed. Degradation of the protective coating must be considered during the life of long-service plants, because destructive elements will always be present in the atmosphere. Dust and harmful deposits which cover the fins and are not likely to be removed when the plant is cleaned will be exposed to rain or high humidity during winter. Corrosive elements in the form of acids and bases may be highly concentrated in the dust. Therefore, pure aluminum is not recommended for fins which require a long service life and are exposed to destructive environmental conditions, although it may be used if the ambient air does not contain substances which could destroy the protective coating.[1] Corrosion-resistant aluminum alloys are attractive choices for heat exchanger material because of their light weight, high thermal conductivity, corrosion resistance, and low cost. Examples of aluminum alloys employed in tubular heat exchangers are given in Table 14 of Ref.[16].

Corrosion of dry-cooled finned surfaces can be minimized with an organic coating, which must be very thin to prevent degredation of heat transfer. The coating process is expensive, however, and thin coatings have microscopic pores and crevices which might retain air humidity and promote corrosion. Furthermore, there has not been enough experience with these coatings to assure that they will last over a 30-yr plant life without considerable maintenance. Most of the European dry-cooled installations are equipped with galvanized steel fin tubes. Although there has been some air-side corrosion on these installations, shutdown has not been required. Galvanized steel corrosion has not been a problem in clean environments.

3.2  Dirt Deposits

Dirt deposits increase air-side pressure drop, promote corrosion, and

---

[1]Reports from aluminum fin tube manufacturers confirm that aluminum corrosion is promoted by moisture coming in contact with dirt deposits [32]. Since neither moisture nor dirt deposits can be avoided in outdoor installations, the use of proper cleaning procedures is very important.

reduce heat transfer performance. For this reason, the environmental conditions and capability of the air-side surface area to be cleaned must be considered when designing the plant. The most suitable site must be selected and the appropriate fin and tube pitch determined, especially if fin tube bundles with large depths are involved. Dust and rainwater, for example, produce a firmly adhering dirt deposit which quite often withstands cleaning by compressed air or low-velocity jets. Thus, high-velocity jets are required for cleaning. Jet water cleaning is confined to special finned tube designs which can withstand the mechanical stress. Furthermore, satisfactory cleaning depends on how the tubes are arranged within the bundles because only the part of the fin surface hit by the water jet will be cleaned; the portion shielded by the core tube will not be cleaned.[1] It is recommended that the strength of the fin tube bundle be considered at an early stage of the design, since the bundle must withstand considerable loads during manufacture, transport, erection, cleaning, and additional mechanical stresses such as hailstorms.

3.3   Fin-Tube Bonding

The heat transfer performance of a finned tube surface depends not only on the fin design, but also on the heat transfer between the fin and the prime surface. Therefore, the bond between the core tube and the fins should not be impaired by thermal loads, corrosion, or mechanical stress or vibration (especially in view of the requirement for a long service life). The thermal expansion for the fin and the prime surface may differ considerably for different types of fin tube surfaces. This means that repeated temperature changes and involuntary temperature peaks may reduce the thermal bond between the tube and the fin. In addition, flow-induced vibration of the fin tubes and the corrosion setup in the crevices between the prime tube and the embedded or L-shaped would fin may considerably reduce the heat transfer rate over a long period of time. Galvanizing the core tube prior to winding the fins will not prevent crevice corrosion, since in the case of grooved fins, the protective layer may be destroyed; the manufacturing process used for L-shaped fins produces crevices on the protective layer, promoting future corrosion. It is, therefore, important that the finned tube design and manufacturing process should ensure a good thermal and corrosion-proof bond between the tube and the fin.

3.4   Atmospheric Effects

Atmospheric conditions such as wind and temperature inversions affect the performance of dry cooling towers [14,33,34]. However, the physical mechanisms whereby these performance changes occur are not understood. Moore [35] studied the effects of atmospheric conditions on three operating natural-draft dry cooling towers of various designs and concluded that daily wind and temperature inversions might cause a generation loss of about 1% to 4%. This study led to the conclusion that the choice of tower design was more crucial than the metaorological features of the plant location. It also suggested that the wind effect might be reversed and made favorable by incorporating design features which would use the dynamic head of the wind to augment the draft. However, in certain cases, the observed wind effects were too large to be the result of uniform dynamic wind pressure. Other factors such as ingestion of cold air at

---

[1] De steese and Simhan [11] report that at the Ibbenburen dry-cooled plant the fouling material which remained after all cleaning attempts caused a permanent reduction in heat transfer capability.

the tower top and heat exchanger losses due to flow obliquity may be the underlying cause of the wind effects.

Wind tunnel and field tests have played an important role in the development and performance assessment of large, natural-draft evaporative cooling towers. Gardner [36] describes a scale-model tower test of a large evaporative cooling tower aimed at studying the effect of wind on overall tower performance; in this study, the packing was simulated by electrically heated wires interspaced with perforated plates. Christopher and Forster []4] describe model tests designed to evaluate the effect of wind on the Rugeley dry cooling tower; this experiment examined the internal aerodynamic losses through the tower and assessed performance losses as a function of wind speed. Khalil and Saber Fahmy [37] describe a series of wind tunnel experiments performed on differently shaped cooling tower modles to determine the influence of tower shell geometry on performance. Scale-model and field tests have also been used to investigate downwind temperature distribution, plume rise, and recirculation in dry cooling towers [38,39].

## 4. DESIGN THEORY

The literature on dry and wet/dry cooling tower design is related to the highly developed heat exchanger design theory (Kays and London [40] and Bowman, Mueller, and Nagle [41]) and the original work of Merkel [42] on evaporative cooling towers. The latter theory is based on a formulation of the heat and mass transfer phenomena in evaporative cooling towers. Merkel [42] lumps sesible and latent heat transfer together and considers a single enthalpy difference driving force for the total energy transfer. This theory has been developed and experimentally confirmed by several investigators in the U.S. [43,44] and Europe [45]. Chilton [46], Wood and Betts [47,48], Rish [49], and Furzer [50]offer typical calculational procedures, and computeri ed methods are presented by Park and Vance [51] and by Winiarski and Tichenor [52]. Yadigaroglu and Pastor [53] present the exact differential equations governing heat and mass transfer and air flow in an evaporative cooling tower, and the Merkel equation is derived starting from this formulation and showing all the approximations involved. The effect of the approximations inherent in the Merkel equation is then analyzed under various ambient and load conditions.[1]

The use of dry and wet/dry cooling towers has been stimulated by a growing demand for electrical energy [54], a shortage of water at potential sites of large power stations [5,6], and stringent environmental legislation [7]. Symposia concerned with the problems associated with developing dry and wet/dry cooling towers [55,56] have been held to accelerate advancement of the technology.

Moore [57-59] has investigated the possibility of minimizing the size and height of natural-draft dry cooling towers; a similar study on mechanical-

---

[1]This study concludes that use of the Merkel equation causes an error of less than 10% compared with an "exact" formulation of the heat and mass transfer phenomena. The Merkel formulation consistently overpredicts the cooling range. This positive error increases with increasing air inlet temperature and humidity.

draft towers was done by Johnson and Dickinson [60]. Dry cooling tower studies were stimulated by a nuclear gas turbine program in the U.S. [61,62]; these resulted in the development of analytical design methods [17,63] for the economic evaluation of conceptual designs [18].

## 5. SUMMARY AND CONCLUSIONS

Conventional cooling systems for thermal power generating stations require significant amounts of cooling water, and are subject to stringent environmental regulations that affect the economy and siting of the power station. Siting restriction and water requirements are minimized by using air cooling, which, in spite of the present technical and economic drawbacks, may be needed at some locations before 1990. Air cooling, often called dry cooling, may be used alone or in combination with wet cooling towers to reduce the cooling water makeup requirements of a wet-cooled power station.

Dry cooling requires a higher investment expenditure than once-through cooling ponds, and wet cooling towers, and would not be employed if cooling water available and evaporative cooling were environmentally acceptable.

To achieve dry cooling tower plant economic optimization, the trade-offs between capital and fuel costs have to be investigated. Higher fuel prices would justify a larger and more expensive dry cooling tower facility, which would result in a lower turbine exhaust pressure and improved plant performance.

Turbine selection is another factor affecting the design and economics of dry-cooled power stations; turbine selection, however, is not a problem for power stations using wet/dry cooling towers.

For power stations with a service life of 30 to 40 yr, the high investment costs for the dry cooling tower are justified only if its life expectancy is the same. The life of a dry cooling tower depends largely on the ability of the fin tube material to resist corrosion; other influences are performance deterioration due to fouling and impairment of fin-tube bondage, and how the surface area is designed and manufactured.

Proper slection of material and manufacturing techniques is an important step toward reducing costs and avoiding corrosion in dry cooling towers. Aluminum surfaces are compatible with certain atmospheric conditions, but may suffer when exposed to a corrosive environment. Corrosion-resistant aluminum alloys are attractive choices for heat exchanger material because of their light weight, high thermal conductivity, corrosion resistance, and low cost. Most of the European dry-cooled installations are equipped with galvanized steel fin tubes. Galvanized steel corrosion has not been a problem in clean environments.

Dirt deposits increase air-side pressure drop, promote corrosion, and reduce heat transfer performance. For this reason, the environmental conditions and the capability of the air-side surface area to be cleaned must be considered when designing the plant.

The heat transfer performnce of a finned-tube surface depends on the bond between the fin and the prime surface. Therefore, the bond between the core tube and the fins should not be impaired by thermal loads, corrosion, or mechanical stress and vibration.

Atmospheric conditions such as wind and temperature inversions affect the performance of dry cooling towers. In view of the lack of engineering data, the effect of these atmospheric conditions on the performance of a dry cooling tower is difficult to evaluate analytically. Wind tunnel and field tests are being used instead.

ACKNOWLEDGEMENT

The encouragement and technical guidance of Professor A.L. London of Stanford University is gratefully acknowledged.

REFERENCES

1. Glicksman, L.R., "Thermal Discharge From Power Plants," ASME Paper 72-WA/Ener -2.

2. Moy, H. C., "Waste Heat Management of Steam Electric Power Plants," ASME Paper 79-ENAs-44.

3. Peterson, D. E., et al., "Thermal Effects of Projected Power Growth - The National Outlook," Hanford Engineering Development Laboratory Report HEDL-TME-73-45, July 1973.

4. Rossie, J.P., and E.A. Cacil, "Research on Dry Type Cooling Towers for Thermal Electric Generation," Parts I and II, Water Pollution Control Research Series 16130 EES 11/70, Environmental Protection Agency, November 1970.

5. U.S. Office of Science and Technology, "Considerations Affecting Steam Power Plant Site Selection," 1968.

6. Hauser, L.G., "Growing Water Requirements for the Growing Thermal Generating Additions of the Electric Utility Industry," in Proceedings of the American Power Conference, Vol. 31, 1969.

7. U.S. Federal Water Pollution Control Administration, "Waste Heat From Steam Electric Generating Plants Using Fossil Fuels and its Control," 1968.

8. Sneck, H.J., and D.H. Brown, "Plume Rise from Large Thermal Sources Such as Dry Cooling Towers," Transactions of the ASME, Journal of Heat Transfer, May 1974.

9. Heeren, H., and H. Holly, "Air Cooling for Condensation and Exhaust Heat Rejection in Large Generating Stations," in Proceedings of the American Power Conference, Vol. 32, 1970.

10. "U.S. ERDA Dry Cooling Tower Development Program Plan," Rev. 1, ERDA Report BNWL-B-393, Battelle Northwest Laboratory, July 1975.

11. De Steese, J.G., and K. Simhan, "European Dry Cooling Tower Operation Experience," ERDA Report BNWL-1995, Battelle Northwest Laboratory, March 1976.

12. Forgo, L., "Past and Future of Dry Cooling for Power Stations," Paper Presented at the Winter Annual Meeting of the ASME, Detroit, November 11-15 1974.

13. Miliaras, E.S., *Power Plants With Air-Cooled Condensing Systems*, The MIT Press, Cambridge, 1974.

14. Christopher, P.J., and V.T. Forster, "Rugeley Dry Cooling Tower System," *Proceedings of the Institute of Mechanical Engineers*, Vol. 184, Part 1, No. 11, 1969-70, pp. 197-222.

15. Westre, W.J., "Economics and Operating Experience with Air-Cooled Condensers - U.S. Experience," Paper Presented at the American Power Conference, Chicago, April 1971.

16. Johnson, A.B., D.R. Pratt, and G.E. Zima, "A Survey of Materials and Corrosion Performance in Dry Cooling Applications", ERDA Report BNWL-1958, Battelle Northwest Laboratory, March 1976.

17. Foster, A.R., S.A. Lamkin, and P. Kwok, "Design of Dry Cooling Towers for Use with a Direct Cycle HTGR," ASME Paper 73-Pwr-7.

18. Schoene, T.W., E.O. Winkler, and P. Fortescue, "The Gas Turbine HTGR Plant Economical Dry Cooling or Wet-Cooled High Efficiency Binary Configurations," Paper Presented at the American Power Conference, Chicago, April 21-23, 1975.

19. Rossie, J.P., R.O. Mitchell, and R.D. Young, "Economics of the Use of Surface Condensers with Dry-Type Cooling Systems for Fossil-Fueled and Nuclear Generating Plants," R.W. Beck Associates, December 1973.

20. Rossie, J.P., and W.A. Williams, Jr., "The Economics of Using Conventional Nuclear Steam Turbine-Generators with Dry Cooling Systems," Paper Presented at the Winter Annual Meeting of the ASME, Detroit, November 11-15, 1973.

21. Rossie, J.P., E.A. Cecil, and R.D. Young, "Cost Comparison of Dry-Type and Conventional Cooling Systems for Representative Nuclear Generating Plants," USAEC Report TID-26007, March 1972.

22. Rossie, J.P., and W.A. Williams, Jr., "The Cost of Energy from Nuclear Power Plants Equipped with Dry Cooling Systems," ASME Paper 72-Pwr-4.

23. Smith, E.C., and M.W. Larinoff, "Power Plant Siting, Performance and Economy with Dry Cooling Tower Systems," in *Proceedings of the American Power Conference*, Vol. 32, 1970.

24. Leung, P., G.R. Reti, and J.R. Schilling, "Dry Cooling Tower Plant Thermodynamics and Economic Optimization," ASME Paper 72-Pwr-5.

25. Silvestri, G.J., and J. Davids, "Effects of High Condenser Pressure on Steam Turbine Design," Paper Presented at the American Power Conference, 33rd Annual Meeting, April 1971.

26. Leung, P., "Dry Cooling Tower Plant Operation: An Economic Loading Approach," Paper Presented at the Winter Annual Meeting of the ASME, Detroit, November 11-15, 1973.

27. Larinoff, M.W., and L.L. Forster, "Dry and Wet-Peaking Tower Cooling Systems for Power Plant Application", *Transactions of the ASME, Journal of Engineering for Power*, July 1976.

28. Li, K.W., "Analytical Studies of Dry/Wet Cooling Systems for Power Plants," Paper Presented at the Witner Annual Meeting of the ASME, Detroit, November 11-15, 1973.

29. Englesson, G.A., M.C. Hu, and W.C. Savage, "Wet/Dry Cooling for Water Conservation," Paper Presented at the Waste Heat Management and Utilization Conference, Miami Beach, May 9-11, 1977.

30. "Evaluated Weather Data for Cooling Equipment Design," Fluor Products.

31. Torrance, K.E., S.H. Black, and L.G. Lykins, "Five-Year Averages of Diurnal and Seasonal Climate in the First 150 Meters of the Atmosphere at Selected Sites in the United States," Cornell University Report EPR-76-6, November 15, 1976.

32. "Report on Current Standards for Air-Cooled Heat Exchangers," R.W. Beck and Associates, May 1973.

33. Goecke, E., et al., "The Condenser System of the 150 MW Unit in the Power Station Ibbenburen der Preussag AG," Ver. Ind. Kraftwirtschaft Ber., May 969.

34. Van der Walt, et al., "The Design and Operation of a Dry Cooling Tower System for a 200 MW Turbo-Generator at Grootvlei Power Station, South Africa," Paper Presented at the Ninth World Energy Conference, Detroit, September 22-27, 1974.

35. Moore, F.K., "Equations for Estimating Performance Losses of Natural-Draft Dry Cooling Towers," Cornell University Report EPR-76-4, September 1976.

36. Gardner, B.R., "The Development of the Assisted-Draft Cooling Towers," Central Electricity Research Laboratories Report, December 1975.

37. Khalil, K.H., and Saber Fahmy, Discussion of Christopher and Forster's Paper on Rugeley Dry Cooling Tower System, Proceedings of the Institute of Mechanical Engineers, Vol. 184, Part 1, No. 11, 1969-1970.

38. Sneck, H.J., and D.H. Brown, "Plume Rise from Large Thermal Sources such as Dry Cooling Towers," Transactions of the ASME, Journal of Heat Transfer, May 1974.

39. Hansen, E.P., "Dry Towers and Wet/Dry Towers for the Indirect Power Plant Cycle," Paper Presented at the Winter Annual Meeting of the ASME, Detroit, November 11-15, 1973.

40. Kays, W.M., and A.L. London, Compact Heat Exchangers, 2nd ed., McGraw-Hill, New York, 1964.

41. Bowman, R.A., D.C. Mueller, and W.M. Nagle, "Mean Temperature Difference in Heat Exchanger Design," Transactions of the ASME, Vol. 62, 1940, p. 283

42. Merkel, F., "Verdunstungskuhlung," Zeitschrift des Vereines deutscher Ingenieure, Vol. 70, 1926, pp. 123-128.

43. Nottage, H.B., "Merkel's Cooling Diagram as a Performance Correlation for

Air-Water Evaporative Cooling Systems," *Transactions of the American Society of Heating and Ventilating Engineers*, Vol. 47, 1941, p. 429.

44. Lichtenstein, J., "Performance and Selection of Mechanical-Draft Cooling Towers," *Transactions of the ASME*, Vol. 65, 1943, p. 779.

45. Jackson, J., "The Testing of Mechanical Draught Water Cooling Towers," *Engineer*, Vol. 189, 1950, p. 140.

46. Chilton, H., "Performance of Natural-Draft Water-Cooling Towers," *Proceedings of the Institute of Mechanical Engineers*, Vol. 99, Part 2, No. 71, 1952, pp. 440-456.

47. Wood, B., and P. Betts, "A Temperature - Total Heat Diagram for Cooling Tower Calculations," *Engineer*, Vol. 189, 1950, pp. 337 and 349.

48. Wood, B., and P. Betts, "A Contribution to the Theory of Natural Draught Cooling Towers," *Proceedings of the Institute of Mechanical Engineers*, Vol. 163, 1950, p. 54.

49. Rish, R.F., "The Design of a Natural Draught Cooling Tower," in *International Heat Transfer Conference, August 28-September 1, 1961, Boulder, Colorado*, ASME, New York, 1961, pp. 951-958.

50. Furzer, I.A., "Natural Draft Cooling Tower: An Approximate Solution," *Ind. Eng. Chem. Process Design and Development*, Vol. 7, 1968, pp. 555-560.

51. Park, J.E., and J.M. Vance, "Computer Model of Crossflow Towers," in *Cooling Towers*, American Institute of Chemical Engineers Technical Manual, 1972, pp. 122-124.

52. Winiarski, L.D., and B.A. Tichenor, "Model of Natural Draft Cooling Tower Performance," *Proc. Am. Soc. Civil Engrs. J. Sanit. Eng. Div.*, Vol. 96, 1970, pp. 927-943.

53. Yadigaroglu, G., and E.J. Pastor, "An Investigation of the Accuracy of the Merkel Equation for Evaporative Cooling Tower Calculations," ASME Paper No. 74-HT-59, 1974.

54. "A Review and Comparison of Selected United States Energy Forecasts," Battelle Memorial Institute, December 1969.

55. Webb, R.L., and R.E. Barry (eds.), "Dry and Wet/Dry Cooling Towers for Power Plants," Papers Presented at the Winter Annual Meeting of the ASME, Detroit, November 11-15, 1973.

56. Rubin, A.M., (ed.), *Workshop on Dry Cooling Systems*, Sponsored by National Science Foundation Division of Engineering, the Franklin Institute Research Laboratories, Drexel University, and Oklahoma State University, Philadelphia, July 15-16, 1975.

57. Moore, F.K., "Scaling Law for Dry Cooling Towers with Combined Mechanical and Natural Draft," Cornell University Report 72-19, August 1972.

58. Moore, F.K., "On the Minimum Size of Natural-Draft Dry Cooling Towers for Large Power Plants," ASME Paper 72-WA/HT-60.

59. Moore, F.K., "On the Minimum Size of Large Dry Cooling Towers with

Combined Mechanical and Natural Draft," *Transactions of the ASME, Journal of Heat Transfer*, August 1973.

60. Johnson, B.M., and D.R. Dickinson, "On the Minimum Size of Forced Draft Dry Cooling Towers for Power Generating Plants," Paper Presented at the Winter Annual Meeting of the ASME, Detroit, November 11-15, 1973.

61. Goodjohn, A.J., and R.D. Kenyon, "The High Temperature Gas-Cooled Reactor - An Advanced Nuclear Power System for the 1980s," ASME Paper 73-Pwr-8.

62. Adams, R.G., *et al.*, "HTGR Gas Turbine Power Plant Configuration Studies," ASME Paper 73-WA/Pwr-7.

63. Montakhab, A., "Analysis of Dry Cooling Tower for Use with the Gas Turbine HTGR," Paper Presented at the Energy Conference, Shiraz, Iran, April 1975.

# ADVANCED SURFACE SELECTION AND PERFORMANCE

# Principles of Heat Transfer Augmentation. I: Single-Phase Heat Transfer

**ARTHUR E. BERGLES**
Department of Mechanical Engineering
Iowa State University
Ames, Iowa 50011 USA

ABSTRACT

This chapter provides a general introduction to the augmentation of single-phase heat transfer as well as to subsequent chapters on augmentation of two-phase heat transfer and on applications of qugmentation in industrial heat exchangers. A detailed discussion is given of the many techniques which are available for augmentation of single-phase heat transfer in both free and forced convection.

1. INTRODUCTION

1.1 General Background

Energy and materials saving considerations, as well as economic incentives, have led to recent expansion of efforts to produce more efficient heat-exchange equipment. Common thermal-hydraulic goals are to reduce the size of a heat exchanger required for a specified heat duty, to upgrade the capacity of an existing heat exchanger, to reduce the approach temperature difference for the process streams, or to reduce the pumping power.

The study of improved heat transfer performance is referred to as heat-transfer augmentation, enhancement, or intensification. In general, this means an increase in heat-transfer coefficient. Attempts to increase "normal" heat-transfer coefficients have been recorded for over a century and there is a large store of information. A recent report [1] cites 1967 technical publications, excluding patents and manufacturers' literature. The recent growth of activity in this area is clearly evident from the yearly distribution of the publications presented in Fig. 1.

1.2 Classification of Heat Transfer Augmentation Techniques

Augmentation techniques can be classified as <u>passive</u> methods, which require no direct application of external power and as <u>active</u> schemes, which require external power. The effectiveness of both types of techniques is strongly dependent on the mode of heat transfer, which could range from single-phase free convection to dispersed flow film boiling. Brief descriptions of passive techniques follow.

<u>Treated surfaces</u> involve fine-scale alternation of the surface finish or

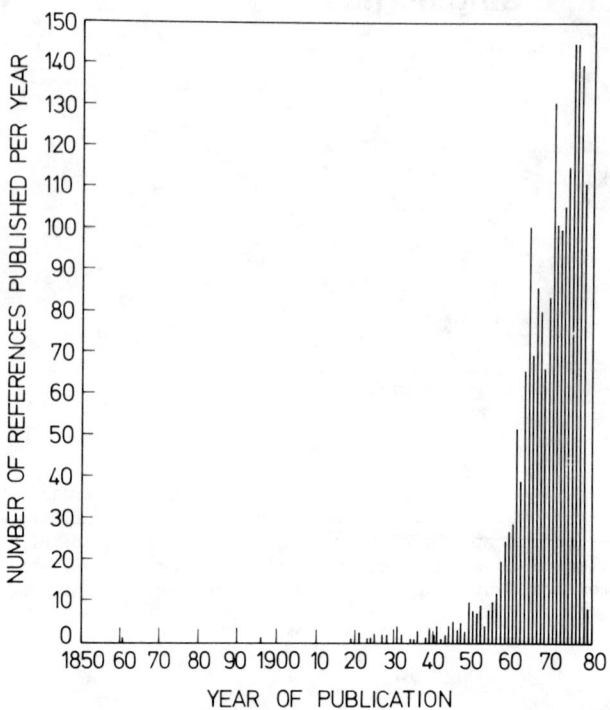

Fig. 1. References on Heat Transfer Augmentation Versus Year of Publication, to Early 1979 [1].

coating (continuous or discontinuous). They are used for boiling and condensing; the roughness height is below that which affects single-phase heat transfer.

Rough surfaces are produced in many configurations ranging from random sand-grain-type roughness to discrete protuberances. The configuration is generally chosen to promote turbulence rather than to increase the heat-transfer surface area. Application of rough surfaces is directed primarily toward single-phase flow.

Extended surfaces are routinely employed in many heat exchangers. Of particular interest is the development of new types of extended surfaces, such as integral inner-fin tubing, and improvement of heat-transfer coefficients on extended surfaces by shaping or interrupting the surfaces.

Displaced enhancement devices are inserted into the flow channel so as to indirectly improve energy transport at the heated surface. They are used with forced flow.

Swirl flow devices include a number of geometrical arrangements or tube inserts for forced flow that create rotating and/or secondary flow: coiled tubes, inlet vortex generators, twisted-tape inserts, and axial-core inserts with a screw-type winding.

Surface tension devices consist of wicking or grooved surfaces to direct the flow of liquid in boiling and condensing.

Additives for liquids include solid particles and gas bubbles in single-phase flows and liquid trace additives for boiling systems.

Additives for gases are liquid droplets or solid particles, either dilute phase (gas-solid suspensions) or dense phase (fluidized beds).

Mechanical aids involve stirring the fluid by mechanical means or by rotating the surface. Surface "scraping", widely used for viscous liquids in the chemical process industry, can also be applied to duct flow of gases. Equipment with rotating heat exchanger ducts is found in commercial practice.

Surface vibration at either low or high frequency has been used primarily to improve single-phase heat transfer.

Fluid vibration is the more practical type of vibration enhancement due to the mass of most heat exchangers. The vibrations range from pulsations of about 1 Hz to ultrasound. Single-phase fluids are of primary concern.

Electrostatic fields (d.c. or a.c.) are applied in many different ways to dielectric fluids. Generally speaking, electrostatic fields can be directed to cause greater bulk mixing of fluid in the vicinity of the heat transfer surface.

Injection is utilized by supplying gas to a flowing liquid through a porous heat-transfer surface or by injecting similar fluid upstream of the heat-transfer section. Surface degassing of liquids can produce augmentation similar to gas injection. Only single-phase flow is of interest.

Suction involves vapor removal, in nucleate or film boiling, or fluid withdrawal, in single-phase flow, through a porous heated surface.

Two or more of the above techniques may be utilized simultaneously to produce an enhancement which is larger than either of the techniques operating separately. This is termed compound augmentation.

It should be emphasized that one of the motivations for studying enhanced heat transfer is to assess the effect of an inherent condition on heat transfer. Some practical examples include roughness produced by standard manufacturing, degassing of liquids with high gas content, surface vibration resulting from rotating machinery or flow oscillations, fluid vibration resulting from pumping pulsation, and electrical fields present in electrical equipment.

## 1.3 Literature

In the area of single-phase augmentation, one of the first formal studies was published by Joule in 1861 [3]. He reported significant improvement in the "conductivity" or overall heat transfer coefficient for in-tube condensation of steam when a wire, spiralled around the condenser tube, was inserted in the cooling water jacket. In 1896, Whitham [4] reported increases in fire-tube boiler efficiency up to 18% when "retarders" or twisted-tape inserts were inserted in the tubes. It was suggested that the inserts should be used only when "the boiler plant is pushed and the draft is strong". Royds [5] presented in his 1924 textbook more detailed heat transfer and pressure drop data for cooling of hot gas flowing in a tube with and without twisted-tape inserts.

The study of enhancement by sound fields and vibrations originates over a half century ago. Tucker and Paris [6] noted in 1921 that hot wires used in microphones decreased in temperature as the intensity level of the sound field increased. In 1923, Richards [7] found that the thermal resistance of an oscillating hot wire decreased as the amplitude and frequency were increased.

For convenience, the world technical literature represented in Fig. 1 has been organized according to augmentation technique and mode of heat transfer. The distribution of the citations is shown in Table 1. Citations not included in this chapter can be obtained from the report by Bergles et al. [1] or the series by Bergles and Webb [2]. Updating of this bibliography, which is stored in a computerized information retrieval system, continues in the Iowa State University Heat Transfer Laboratory.

Let us turn now to the various techniques which have been found effective in increasing heat transfer coefficients in free convection.

## 2. FREE CONVECTION

### 2.1 Passive Techniques

With the exception of the familiar technique of providing extended surfaces, the passive techniques have little to offer in the way of augmented heat transfer. This is because the velocities are usually too low to cause flow separation or secondary flow. A recent review [8] of the limited data for free convection from machined or formed rough surfaces, with air, water, and oil, concludes that increases in heat transfer coefficient up to 100% can be obtained with air, but that the increases with liquids are very small.

Design procedures for single fins and fin arrays are well established; however, little attention has been given to interrupted extended surfaces. The restarting of thermal boundary layers is expected to increase coefficients so as to more than compensate for the lost area. The effectiveness of this concept has been demonstrated in the wire-loop fins which are used for baseboard hot water heaters or "convectors" [9]. This problem is also of interest in electronic cooling, where the "heat sinks" are often discontinuous fins, and in natural draft cooling of finned tube banks neccessitated by loss of fan power. Of considerable interest recently is the heat transfer performance of medium integral fin tubes for use in hot water heaters and solar thermal storage. The established equations for large fins in air, as found in baseboard convectors, are not applicable. The single-tube correlations of Gorenflo [10] are recommended as a starting point in design:

$$Nu_o = 0.6(Ra_o)^{0.25} \text{ for } Ra_o < 2 \times 10^7 \tag{1a}$$

$$Nu_o = 0.155(Ra_o)^{1/3} \text{ for } Ra_o > 2 \times 10^7 \tag{1b}$$

Caution should be exercised when using these equations for coils and other vertical arrays found in storage tanks. Manufacturer's data should be used in these situations.

The only clear-cut example of a beneficial fluid additive appears to be injection of vapor bubbles at the bottom of a heated surface. Tomari and

Table 1. Classification of Augmentation Techniques [1]

| | Single-Phase Natural Convection | Single-Phase Forced Convection | Pool Boiling | Forced-Convection Boiling | Condensation | Mass Transfer |
|---|---|---|---|---|---|---|
| **Passive Techniques (No External Power Requirements)** | | | | | | |
| Treated Surfaces | -- | -- | 73 | 12 | 32 | -- |
| Rough Surfaces | 4 | 282 | 21 | 43 | 37 | 22 |
| Extended Surfaces | * | 269 | 34 | 33 | 87 | 15 |
| Displaced Enhancement Devices | -- | 36 | 1 | 9 | 3 | 5 |
| Swirl Flow Devices | -- | 89 | -- | 67 | 9 | 5 |
| Coiled Tubes | -- | 118 | -- | 44 | 3 | 5 |
| Surface Tension Devices | -- | -- | 11 | 1 | -- | 1 |
| Additives for Liquids | 3 | 10 | 43 | 23 | -- | 3 |
| Additives for Gases | -- | 155 | -- | -- | 5 | 2 |
| **Active Techniques (External Power Required)** | | | | | | |
| Mechanical Aids | 4 | 34 | 19 | 1 | 17 | 10 |
| Surface Vibration | 44 | 21 | 9 | 2 | 6 | 9 |
| Fluid Vibration | 40 | 89 | 15 | 4 | 1 | 32 |
| Electric or Magnetic Fields | 37 | 30 | 23 | 7 | 13 | 8 |
| Injection or Suction | 3 | 17 | 7 | + | 6 | 2 |
| **Compound Enhancement (Two or More Techniques)** | + | 32 | 2 | 4 | 4 | 2 |

-- Not applicable
\* Not considered in this survey
+ No citations located.

Nishikoma [11] observed increases in average heat transfer coefficients when air was injected into water or ethylene glycol. The improvement is due primarily to the circulation induced by the rising bubbles.

## 2.2 Active Techniques

Mechanically aided heat transfer is a standard technique in the chemical and food industries when viscous liquids are involved. The effects of surface scrapes, paddles, etc. are discussed by Uhl [12].

As an example of a rotating heat exchange surface, consider the heated cylinder rotating about its own axis in a bulk stagnant fluid. At large rotational Reynolds numbers, where rotation dominates, Tang and McDonald [13] correlated data for several liquids by the following equation:

$$Nu = 0.12 \left[ \frac{d^2 \omega}{2 \nu} \right]^{2/3} Pr^{1/3} \qquad (2)$$

Surface vibration has been extensively studied in the laboratory. The predominant geometry has been the horizontal cylinder vibrated either horizontally or vertically. As shown in Fig. 2, heat transfer coefficients can be increased tenfold for low-amplitude/high-frequency situations. Similar results were obtained with high-amplitude/low-frequency experiments. While the improvements can be dramatic, it must be recognized that natural convection is inherently an inefficient mode of heat transfer. Since at maximum enhancement, the average velocity of the surface over a cycle is less than 1 m/s, it is often more practical to provide steady forced flow.

Several studies examined the effects of vibrating an entire enclosure containing a heated section immersed in liquid. Smith and Forbes [15] observed improvements in free convection heat transfer coefficients of up to

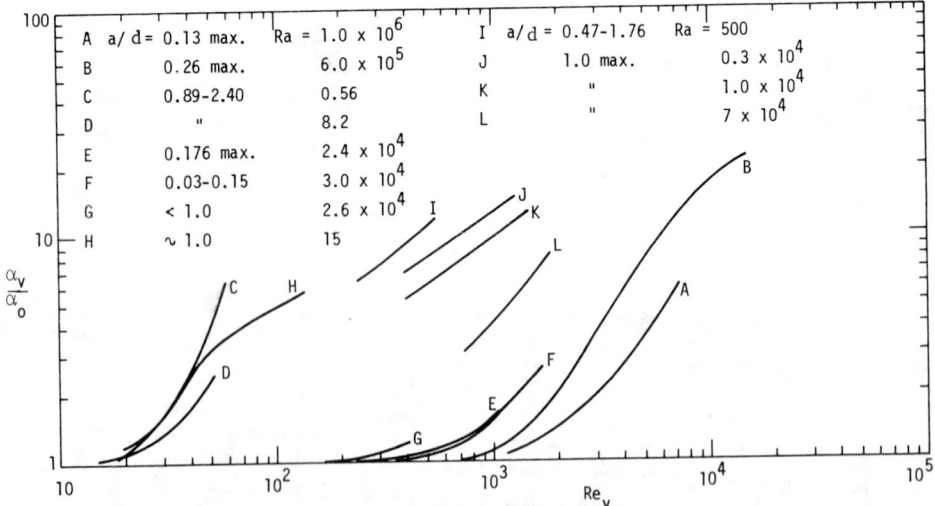

Fig. 2. Influence of Mechanical Vibration on Heat Transfer from Horizontal Cylinders, a/d <1 [14].

33% with a heated sidewall, and Pak et al. [16] recorded improvements of up to 200% with a small diameter wire.

Since it is usually difficult to apply surface vibrations to practical equipment, the alternative technique is often utilized whereby vibrations are applied to the fluid and focused toward the heated surface. Generators employed range from flow interrupters to piezo-electric transducers, thus covering the range from pulsation of 1 Hz to ultrasound of $10^6$ Hz. Much research has been reported on the effects of acoustic vibrations on heat transfer to gases from horizontal cylinders. Increases in average coefficients are observed only above an intensity of about 140 db, which is far in excess of human ear tolerance. The maximum increases reported are usually 100-200%. With proper ultrasonic transducer design, it is also possible to improve heat transfer to simple heaters immersed in liquids by several hundred percent. Cavitation is generally the dominant enhancement mechanism. As an example, a vibration study of the effect of ultrasonic vibration on heat transfer to water was presented by Yukawa et al. [17]. The maximum improvement reported was a 500% increase in coefficient; however, little improvement was noted with degassed water. In general, with liquids there is considerable difficulty in designing systems to transmit vibrational energy to large surfaces. The extensive literature on both surface and fluid vibrations is discussed in Ref. 14. Few studies have been reported in recent years, probably in recognition of practical limitations.

Electric fields can be utilized to increase heat transfer coefficients in free convection. The configuration may be a heated wire in a concentric tube maintained at a high voltage relative to the wire, or a fine wire electrode may be utilized with a horizontal plate. Dielectrophoretic or electrophoretic (especially with ionization of gases) forces cause greater bulk mixing in the vicinity of the heat transfer surface. Reference 14 cites studies which report heat transfer coefficients improved by as much as a factor of 40. The more recent paper by Yabe et al. [18] provides a good example of the large local improvements which can be expected in heat transfer coefficient. Recent activity has centered on the application of corona discharge cooling to practical problems. Cooling of cutting tools by point electrodes was proposed by Blomgren and Blomgren [19], while Reynolds and Holmes [20] have used parallel wire electrodes to improve the heat dissipation of a standard, horizontal, finned tube. Heat transfer coefficients can be increased by several hundred percent when sufficient voltage is supplied. Depending on the fluid and geometry, 1000 to 100,000 volts may be required.

Gas injection into a liquid through a porous heated plate, which has been used to similate nucleate boiling, can be regarded as an augmentative technique. Heat transfer coefficients can be increased several hundred percent (Sims et al. [21]); however, the difficulty of supplying and removing the gas must be overcome.

It is thus concluded that many augmentation techniques produce impressive improvements in free convection heat transfer coefficients. The discussion now turns to single-phase forced convection since this is the dominant mode of heat transfer in commercial heat exchange equipment.

## 3. FORCED CONVECTION

### 3.1 Rough Surfaces

Surface roughness has been used extensively to augment forced convection heat transfer. Integral roughness may be produced by the traditional manufacturing processes of machining, forming, casting, or welding. Various inserts can also provide surface protuberances. In view of the infinite number of possible geometric variations, it is not surprising that, even after more than 200 studies [1], no unified treatment is available.

Laminar flow data for tubes are provided by studies which considered convoluted tubes (Blumenkrantz and Taborek [22], Rozalowski and Gater [23]) and transverse fins in an annular passage (Zappa and Geiger [24]). Improvements of as much as 100% are observed. Augumentation is widely utilized in plate heat exchangers. Pescod [25] reported on a study of the improvements obtained by using spikes and ripples to enhance nominally laminar flow of air in parallel plate channels of large aspect ratios. Most plate heat exchangers utilize corrugated surfaces, for structural reasons as well as augmentation. It is generally agreed that the heat transfer and pressure drop characteristics of commercial corrugated surfaces used in plate exchangers are quite similar for both laminar and turbulent flow.

The diversity of results obtained for turbulent flow heat transfer to water in all types of roughened tubes is indicated in Figs. 3 and 4. Here the simplest coordinates are chosen for illustrative purposes. All calculations are based on the base area of the tube with no allowance for increases in surface area. While the heat transfer coefficients are increased at the most about 4 times, friction factors are increased as much as 58 times. Within this matrix of data lie surfaces which are "efficient" as far as both heat transfer and pressure drop are concerned. An evaluation of these data in terms of one of the standard performance evaluation criteria will be presented in the third chapter on this subject. Attention is directed here to several of the more recent semi-empirical correlations of rough surface behavior.

It has been demonstrated that the analogy between heat transfer and fric-

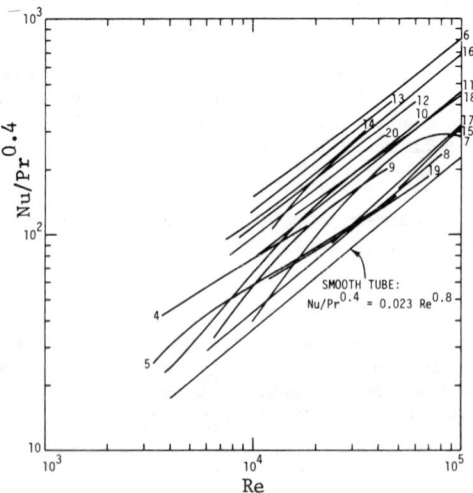

Fig. 3. Representative Heat Transfer Data for Tubes with Various Types of Internal Roughness [26].

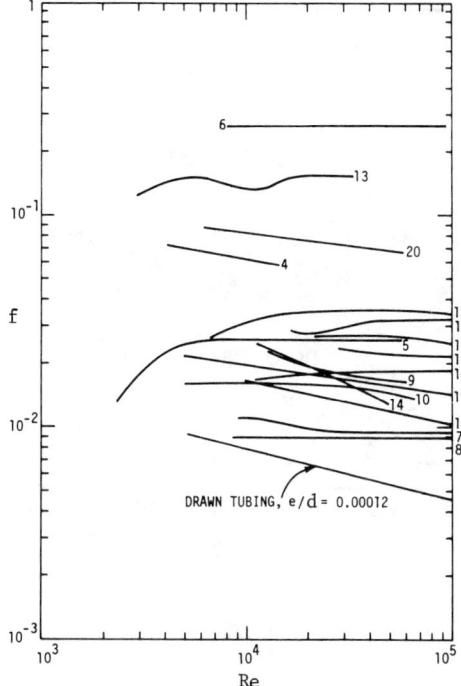

Fig. 4. Friction Data for Tubes with Internal Roughness (see Fig. 3) [26].

tion for rough surfaces in turbulent flow is dependent on the type of roughness. An analogy solution for a "sand-grain-type roughness" was developed by Dipprey and Sabersky [27]. Recent work has considered surfaces which can be produced commercially. Webb et al. [28] have correlated heat transfer coefficients for tubes with square repeated-rib roughness by first correlating the friction factor using Nikuradse's similarity function, $U^+$; the roughness Reynolds number, $e^+$; and $p/e$. By application of the heat transfer - momentum transfer analogy, the Stanton number is given by

$$St = \frac{f/2}{1 + (f/2)^{0.5} [\bar{g}(e^+)F(Pr) - U_e^+(e^+, p/e)]} \qquad (3)$$

Excellent correlation of their data was achieved, as shown in Fig. 5. Equation 3 also successfully predicted air data of other investigations, even for other rib profiles [29]. This similarity correlation method should be valid for any roughness type. It must be recognized, however, that extensive experimental data are required to get the various functional relations.

A different type of correlation technique has been proposed by Lewis [30]. Basically, the detailed behavior of roughness elements is required: form drag coefficients, heat transfer coefficient distribution, and separation length

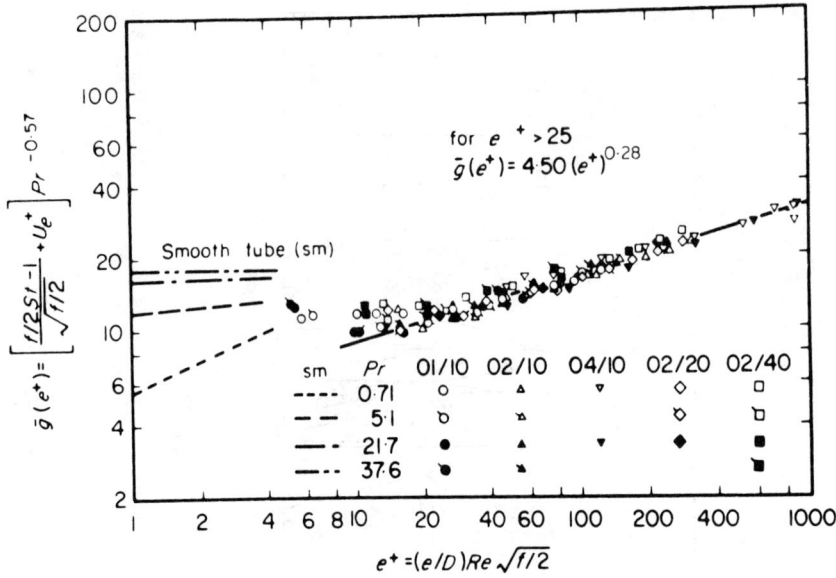

Fig. 5. Correlation of Heat Transfer Data for Tubes with Internal Repeated-Rib Roughness [28].

behind an element. When this information is available, a prediction can be formulated without recourse to data for the actual rough channel. The agreement with experimental data such as that in [28] is surprisingly good, considering the "separate effects" character of the model.

Considerable progress in the commercial production of tubes with internal roughness has been made in recent years. The configurations represented in Figs. 3 and 4 include tubes with rolled-in ridges, convoluted tubes, spirally-indented or rope tubes, tubes with spiral coil inserts, and tubes with inner spiral ribs. These tubes have been produced in commercial quantities, and several are available as standard catalog items.

3.2 Extended Surfaces

Extended surfaces can be considered "old technology" as far as most applications are concerned. Tube-and-plate fins, plate fins, finned tube banks, etc., are widely employed, and design data are available, at least for uninterrupted fins [31]. Attention will be directed here only to two topics of current interest: use of interrupted or perforated fins and integral inner fin tubes.

Compact heat exchangers of the plate-fin or tube-and-center variety use several augmentation techniques: offset strip fins, louvered fins, perforated fins, or corrugated fins. A comprehensive review of perforated plate heat exchanger performance was presented by Shah [32]. He concluded that, except for surfaces with many small perforations (plate open area greater than 20% and $d \leq 0.08$ cm), no improvement in heat transfer was observed in the laminar region. Perforated plates typically exhibit transition at Re = 500-1000. Many industrial plate fin heat exchangers operate in the Reynolds number range of 1000 to 5000. In this transition and early turbulent region, both j and f for the perforated surfaces are usually much higher than the values for unperforated surfaces. The heat transfer improvements shown in Fig. 6 are partially attributed to surface and fluid vibration. Lee and Yang [33] recommend operation in a "second laminar flow regime" to obtain enhancement without vibration and noise problems.

In general, heat transfer, flow friction, and acoustic characteristics cannot be predicted for perforated surfaces. A promising start in analysis has been made, however, by Sparrow et al. [34], who obtained a numerical solution for laminar flow of air in a plate heat exchanger with interrupted walls. The model approximates a tube-and-plate fin exchanger with wide slots in the fins. (Test conditions can be closer to the model if tubes are eliminated and transient testing is utilized.) These predictions are in reasonable agreement with data for appropriate geometries. It was suggested that perforation burrs and bent edges could explain the discrepancies. A good status report of airside (external) heat transfer in finned tube heat exchangers was recently presented by Webb [35].

Internally finned circular tubes are available in aluminum and copper (or copper alloys). The following correlations of Watkinson et al. [36] are available for laminar flow:

Spiral fin tubes:

$$Nu_e/Pr^{1/3}(d_e/L)^{1/3}(\eta_b/\eta_w)^{0.14} \phi_e = 19.2(b/p)^{0.5}Re_e^{0.26} \quad (4a)$$

where

$$\phi_e = 2.25(1 + 0.01Gr_e^{1/3})/\log Re_e \quad (4b)$$

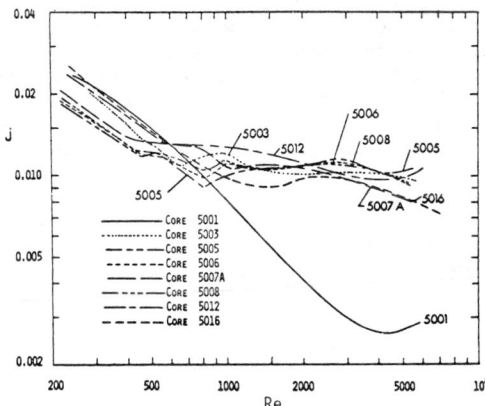

Fig. 6. Representative Heat Transfer Data for Perforated Plate-Fin Heat Exchangers [32].

Straight fin tubes:

$$Nu_e/Pr^{1/3}(d_e/L)^{1/3}(\eta_b/\eta_w)^{0.14} \phi_e = 2.43(1/n)^{0.5} Re_e^{0.46} \tag{5}$$

Isothermal friction factors for all tubes:

$$f_e = 16.4(d_e/d)^{1.4}/Re_e \tag{6}$$

These correlations are based on data for oil in horizontal tubes having approximately uniform temperature (steam heating). Other data for both water and ethylene glycol in both steam heated and electrically heated tubes, are in approximate agreement with the correlations [37]. As noted in Reference 38, the analytical results for uniformly heated tubes are not in good agreement with data.

Several studies have assessed the thermal-hydraulic performance of internally finned tubes (all from one manufacturer) in turbulent flow. In the latest report of this work, Carnavos [39] recommends the following equations:

$$Nu_e = 0.023 Pr^{0.4} Re_e^{0.8} (A_{fa}/A_{fc})^{0.1} (A_n/A_a)^{0.5} (\sec \psi)^3 \tag{7}$$

$$f = 0.046 Re_e^{-0.2} (A_{fa}/A_{fn})^{0.5} (\sec \psi)^{0.75} \tag{8}$$

It is noted that fin inefficiency corrections must be incorporated when applying these equations. These correlations provide a reasonable description of the available data for air, water, and ethylene/glycol water; however, there are significant differences among data from different investigations for essentially the same tube [38].

The first analytical study to predict the performance of tubes with straight inner fins for turbulent air flow was conducted by Patanker et al. [40]. The mixing length in the turbulence model was set up so that just one constant was required from the experimental data [42]. Expansion of analytical efforts to fluids of higher Prandtl number, tubes with practical contours, and tubes with spiraling fins would be desirable. It would be particularly significant if the analysis could predict with reasonable expenditure of computer time the optimum fin parameters for a specified fluid, flow rate, etc.

Internally finned tubes can be "stacked" to provide multiple internal passages of small hydraulic diameter. Carnavos [41] demonstrated the large increases in heat transfer coefficient (based on outer tube nominal area) which can be obtained in these tubes with air flow. Of course, pressure drop is also increased with these tubes.

Mass transfer techniques are being extensively utilized to study heat transfer in complex geometries where temperature sensors would be very combersome and probably would disturb the flow. Goldstein and Sparrow [43], for example, used naphthalene sublimation techniques to study corrugated-wall channels. The nethod can be applied to many of the configurations of interest for practical heat exchangers.

## 3.3 "Enhanced" Tubes

Many proprietary surface configurations have been produced by deforming the tube wall. The "convoluted", "corrugated", "spiral", or "spirally fluted" tubes have multiple-start spiral corrugations along the tube length. Both area increase and roughness effects are present. A systematic survey of the performance of eleven commercially available tubes in surface steam condenser service (condensing steam outside, water inside) is given by Marto et al. [44]. Up to 400% increase in the nominal inside heat transfer coefficient (based on diameter of smooth tube of same maximum inside diameter) was noted; however, pressure drops were about 20 times higher.

## 3.4 Displaced Enhancement Devices

Displaced enhancement devices are typically in the form of inserts, with elements arranged to promote transverse mixing ("Static Mixers"). They are used primarily for viscous liquids, to promote either heat transfer or mass transfer. Figure 7 illustrates typical improvements which can be gained from the use of static mixers. There are no broad-based correlations available due to the many geometrical arrangements and the strong influence of fluid properties and heating conditions. In general, the higher the heat transfer coefficient, the higher the pressure drop; the percentage increase in heat transfer is generally greater than the heat transfer increase. For example, the SMX Standard mixer exhibits a 7-fold increase in heat transfer and a nearly 70-fold increase in pressure drop. Additional data for the Kenics mixer are given in Reference 37. Similar inserts or packings have been used for turbulent flow; however, this application is recommended only for short sections with high heat fluxes since the pressure drop is so high [46].

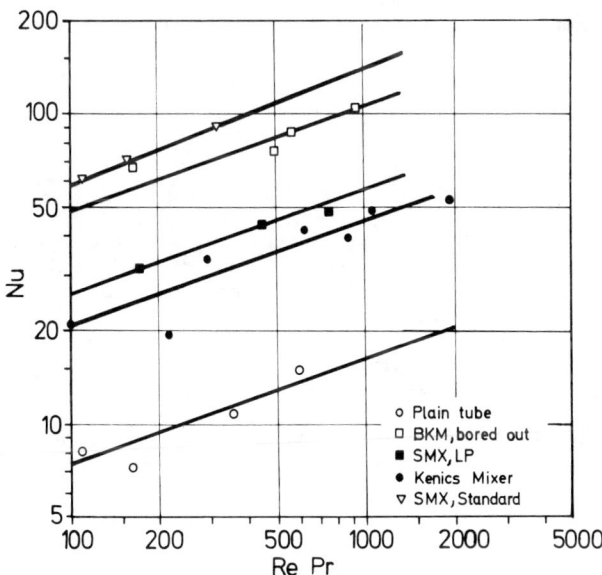

Fig. 7. Heat Transfer Data for Typical Static Mixer Inserts in Laminar Flow [45].

## 3.5 Swirl Flow Devices

Twisted-tape inserts have been widely used to improve heat transfer in both laminar and turbulent flow. The following correlation is recommended for fully developed laminar flow in a uniformly heated tube [47]:

$$Nu = 5.172(1 + 5.484 \times 10^{-3} (Pr)^{0.7} (Re)(y)^{-1.25})^{0.5} \tag{9}$$

Note that the correlation was established for a tape with no heat transfer. The correlation does not seem to be applicable to heating or cooling with a constant wall temperature [37]. At Re <100, the isothermal friction factors can be approximated by the expression for a semi-circular tube

$$f = 42.2(Re)^{-1} \tag{10}$$

Turbulent flow heat transfer in tubes with twisted-tape inserts has been correlated by [48]

$$Nu_e = F(0.023(1 + (\pi/2y)^2)^{0.4} (Re_e)^{0.8} (Pr)^{0.4}$$
$$+ 0.193((Re_e)^2 (y)^{-1} (d_e/d)(\Delta\rho/\rho)(Pr))^{1/3}) \tag{11}$$

The fin factor, F, which represents the ratio of total heat transfer to the heat transferred by the walls alone, can be estimated from conduction calculations. Isothermal friction factors are given by the following expression [48]:

$$f_e = 0.046(y)^{-0.046} (Re_e)^{-0.2} \tag{12}$$

## 3.6 Coiled Tubes

Heat transfer coefficients can be substantially higher in coiled tubes than in straight tubes due to the secondary flow promoted by the curvature. The following sources are recommended for the correlations:

>   Laminar flow in tubes with uniform wall temperature [49]
>   Laminar flow in tubes with uniform heat flux [50]
>   Transition and turbulent flow [49].

These correlations will not be repeated here due to their complexity and the fact that coils in themselves represent a basic heat exchanger configuration of rather specialized nature. Reference 1 can be consulted for the extensive literature on the subject.

## 3.7 Fluid Additives

Modest improvements in heat transfer are observed when gas bubbles or solid particles are added to liquids. Kenning and Kao [51] found heat transfer increases up to 50% when nitrogen bubbles were injected into turbulent water flow. Warkins et al. [52] studied suspensions of polystyrene spheres in oil to determine the mechanism of enhancement for addition of solid particles to laminar flows. Maximum improvements of 40% were observed.

Gas-side heat transfer can be enhanced by adding a small volumetric fraction of solid particles. The particles are carried along with the stream and separated for reuse, in the case of a once-through system, or circulated, in the case of a closed system. The enhancements of up to four times the pure gas heat transfer coefficients, shown in Fig. 8, are attributed to thinning of the viscous sublayer and higher thermal conductivity in that layer. Design information and a guide to the extensive literature are given in Reference 54.

Fluidized beds are being considered for many industrial applications. Heat transfer coefficients to tubes within a bed can be enhanced by a factor of 20, compared to pure gas flow at the same flow rate. Chen [55] reviews available correlations for plain tubes in fluidized beds and suggests that most correlations have an uncertainty band of 100%. This subject is treated in more detail in another lecture at this Institute.

When liquid droplets are added to a flowing gas stream, heat transfer is enhanced by sensible heating of the two-phase mixture, evaporation of the liquid, and disturbance of the boundary layer. Thomas and Sunderland [56] demonstrated that heat transfer coefficients for airflow can be increased as much as a factor of 30 if a continuous liquid film is formed on the heated surface.

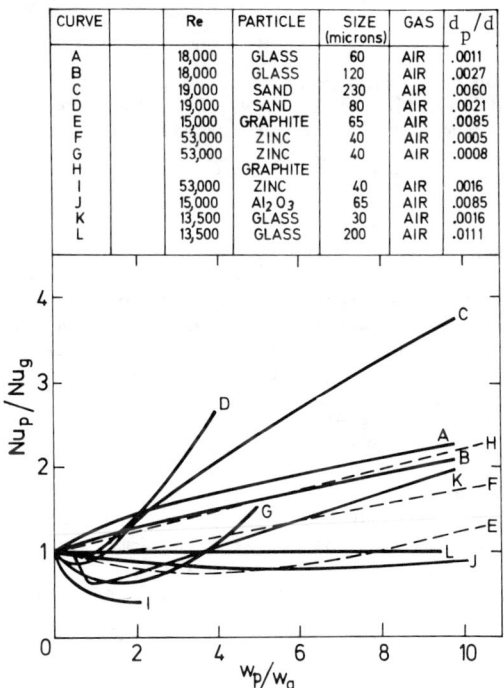

Fig. 8. Composite Heat Transfer Data for Gas-Solid Suspensions [53].

A more realistic indication of practical enhancement was proveded by Yang and Clark [57] who applied spray cooling to a compact heat exchanger core. The maximum improvement of 40% was attributed to formation of a partial liquid film and sensible heating of that film.

## 3.8 Mechanical Aids

Mechanically-aided heat transfer in the form of surface scraping can increase forced convection heat transfer. Unfortunately, the necessary hardware is not particularly compatible with most heat exchangers. A recent application of scraped-surface heat transfer to air flows is reported by Hagge and Junkhan [58]; a tenfold improvement in heat transfer coefficient was reported for laminar flow over a flat plate.

The other aspect of this technique is rotating surfaces. Moderate increases in heat transfer coefficients have been reported for laminar flow in a) a straight tube rotating about its own axis, McElhiney and Preckshot [59], b) a straight tube rotating around a parallel axis, Mori and Nakayama [60], c) a rotating circular tube, Mori et al. [61], and d) the rotating, curved, circular tube, Miyazaki [62]. Increases in turbulent flow are less; for example, in [61] maximum improvements of 350% were recorded for laminar flow, but for turbulent flow the maximum increase was 25%. In general, these are examples of naturally occurring phenomena which result in enhancement: cooling windings of rotating electrical machinery, cooling of gas turbine rotor blades, etc.

## 3.9 Surface Vibration

Surface vibration has been demonstrated to improve heat transfer to both laminar and turbulent duct flow of liquids [14]. The largest improvements (up to 200%) are observed with laminar or transitional flow utilizing a concentric tube heat exchanger with the inner tube vibrated transversely or a rectangular channel with a flexible, vibrating side.

## 3.10 Fluid Vibration

Fluid vibration has been extensively studied for both air (loudspeakers and sirens) and liquids (flow interruptors, pulsators, and ultrasonic transducers) [14]. The gas results are not encouraging, as intensities above 120 db are required and the effect is largely one of triggering fully turbulent flow at transitional Reynolds numbers. Moissis and Maroti [63] subjected a compact heat exchanger core to high-intensity acoustic vibrations and observed a maximum improvement in the gas-side heat transfer coefficient of 30%.

Pulsations are relatively simple to apply to low velocity liquid flows, and improvements of several hundred percent can be realized as noted in Fig. 9. Turbulence triggering and cavitation appear to be important enhancement mechanisms. Application of high frequency vibrations is difficult, and only modest improvements in heat transfer coefficients are recorded [14].

## 3.11 Electrostatic Fields

Some very impressive enhancements have been recorded with electric fields, particularly in the laminar flow region [14]. The recent studies of Porter

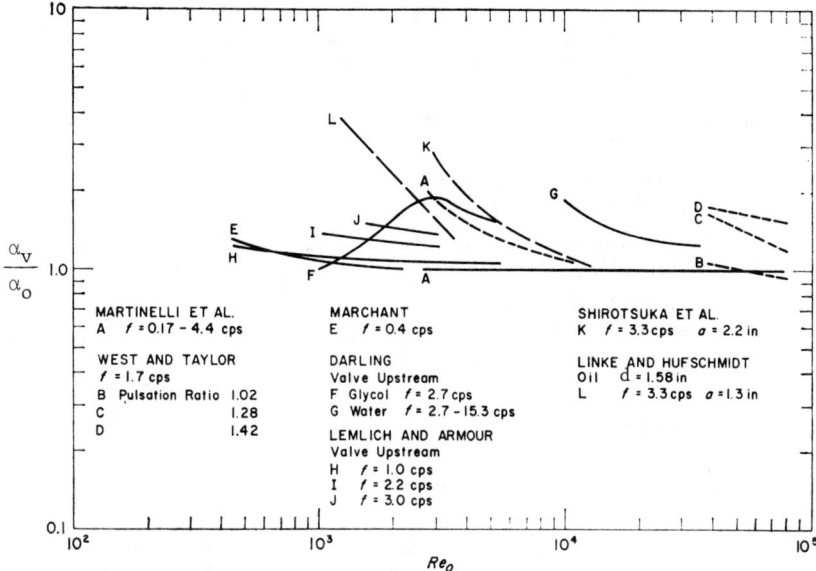

Fig. 9. Effect of Upstream Pulsations on Heat Transfer to Liquids Flowing in Pipes [14].

and Poulter [64], Savkar [65], and Newton and Allen [66] demonstrated improvements of at least 100% when voltages in the 10 kv range were applied to transformer oil. These conditions occur naturally in power transformers.

Mizushina et al. [67] found that even with intense fields, the enhancement disappears as turbulent flow is approached in a circular tube with a concentric inner electrode. Reynolds and Holmes [20] found little effect of corona wind even at low air velocities, with the exception of tests with three electrodes under a finned tube. In this case, a 60% increase in heat transfer coefficient was noted.

3.12 Injection or Suction

Single-phase heat transfer can be enhanced by injecting gas into a liquid through a porous heated surface [14]. Tauscher et al. [68] have demonstrated up to fivefold increases in local heat transfer coefficients by injecting similar fluid into a turbulent tube flow. The effects are similar to that produced by an orifice plate; in both cases the effect has died out after about 10 L/d.

Large increases in heat transfer coefficient are predicted for laminar (Kinney [69]) and turbulent flow (Kinney and Sparrow [70]) with surface suction. The general characteristics of the latter predictions were confirmed by experiments conducted by Aggarwal and Hollingsworth [71].

3.13 Compound Techniques

Compound techniques are a slowly emerging area of enhancement which holds promise for practical applications since heat transfer coefficients can usually be increased above each of the several techniques acting alone. Some examples are

 Rough tube wall with twisted-tape insert,
 Bergles et al. [72]

 Rough cylinder with acoustic vibrations,
 Dryukov and Boykov [73]

 Internally-finned tube with twisted-tape insert,
 Van Rooyen and Kroeger [74]

 Finned tubes in fluidized beds,
 Bartel and Genetti [75]

 Externally finned tubes subjected to vibrations,
 Zozulya and Khorunzhii [76]

 Gas-solid suspension with an electric field,
 Min and Chao [77]

 Fluidized bed with pulsations of air,
 Bhattacharya and Harrison [78].

It is interesting to note that some compound attempts are unsuccessful. Masliyah and Nandakumar [79], for example, found analytically that average Nusselt numbers for internally finned, coiled tubes were lower than for plain coiled tubes.

## 4. CONCLUSIONS

This review provides ample evidence that many techniques are available for augmentation of single-phase heat transfer in both free and forced convection. There is extensive documentation of laboratory studies which have established that a particular embodiment of an augmentation technique is effective in augmenting heat transfer. The rapidly increasing pool of worldwide literature includes data for laminar and turbulent flow of various fluids with various heat exchange configurations. Most studies of forced convection include pressure drop as well as heat transfer data. In a number of areas promising starts have been made in analyzing the complex flow and temperature fields produced by augmentation. The third chapter on this subject will consider a number of the techniques which have made the transition from the laboratory to full-scale industrial equipment.

NOMENCLATURE

$A_a$  = total heat transfer surface area

$A_{fa}$  = actual free flow area in duct

$A_{fc}$  = open core free flow area in duct

$A_{fn}$  = nominal free flow area based on inside diameter

$A_n$  = nominal surface area based on inside diameter

$a$  = amplitude of vibration

$b$  = average distance between fins

$c_p$  = specific heat at constant pressure

$d$  = inside diameter of tube or outside diameter of cylinder, unless otherwise noted

| | | |
|---|---|---|
| e | = | rib or roughness height |
| $e^+$ | = | roughness Reynolds number [28], see Fig. 5 |
| F | = | Prandtl number function [28], $F = Pr^{0.57}$ in Fig. 5; fin factor for twisted tape [48] |
| f | = | Fanning friction factor; vibrational frequency |
| Gr | = | Grashof number, $g_n d^3 \Delta\rho/\rho \nu^2$ |
| $\bar{g}$ | = | heat transfer roughness function [28], see Fig. 5 |
| $g_n$ | = | gravitational acceleration |
| j | = | Colburn factor, $StPr^{2/3}$ |
| Nu | = | Nusselt number, $\alpha d/\lambda$ |
| n | = | number of fins in a tube |
| Pr | = | Prandtl number, $\eta c_p/\lambda$ |
| p | = | rib spacing or fin pitch |
| Ra | = | Rayleigh number, GrPr |
| Re | = | Reynolds number, $\rho u d/\eta$ |
| St | = | Stanton number, $\alpha/\rho u c_p$ |
| $U_e^+$ | = | Nikuradse similarity function [28] |
| u | = | average flow velocity |
| w | = | mass flow rate |
| y | = | tube diameters per 180° turn of twisted tape |

Greek

| | | |
|---|---|---|
| $\alpha$ | = | heat transfer coefficient, based on actual surface area |
| $\eta$ | = | dynamic viscosity |
| $\lambda$ | = | thermal conductivity |
| $\nu$ | = | kinematic viscosity |
| $\rho$ | = | density |
| $\Delta\rho$ | = | density difference between liquid and vapor |
| $\phi_e$ | = | free convection factor, Eq. (4b) |
| $\psi$ | = | helix angle of internal spiral fins |
| $\omega$ | = | rotational velocity |

Subscripts

| | | |
|---|---|---|
| b | = | evaluated at bulk fluid temperature |
| e | = | based on equivalent or hydraulic diameter |
| g | = | refers to gas flow in gas-solid suspension |
| o | = | based on outside diameter of finned tube; refers to no augmentation |
| p | = | refers to particles in gas-solid suspension |
| v | = | vibrational condition |
| w | = | evaluated at wall temperature |

REFERENCES

1. Bergles, A. E., Webb, R. L., Junkhan, G. H., and Jensen, M. K., "Bibliography on Augmentation of Convective Heat and Mass Transfer", HTL-19, ISU-ERI-Ames-79206, Iowa State University, 1979.

2. Bergles, A. E. and Webb, R. L, "Bibliography on Augmentation of Convective Heat and Mass Transfer - Part 1 to Part 6, Previews of Heat and Mass Transfer, Vol. 4, No. 2, pp. 61-73, 1978 to Vol. 6, No. 3, pp. 242-313, 1980.

3. Joule, J. P., "On the Surface-Condensation of Steam", Phil. Trans. of the Royal Society of London, Vol. 151, pp. 133-160, 1861.

4. Whitham, J. M., Street Railway Journal, Vol. 12, p. 374, 1896.

5. Royds, R., Heat Transmission by Radiation, Conduction, and Convection, First Edition, Constable and Company, London, pp. 191-201, 1921.

6. Tucker, W. S. and Paris, E. T., "A Selective Hotwire Microphone", Transactions of the Royal Society, pp. 389-430, 1921.

7. Richards, R. C., "The Resistance of a Hot Wire in an Alternating Air Current", Philosophical Magazine and Journal of Science", Vol. 45, pp. 926-934, 1923.

8. Bergles, A. E., Junkhan, G. H., and Webb, R. L.,"Energy Conservation Via Heat Transfer Enhancement",COO-4649-5, Iowa State University, 1979.

9. Benforado, D. M., and Palmer, J., "Wire Loop Finned Surface - A New Application (Heat Sink for Silicon Rectifiers)",Chemical Engineering Progress Symposium Series, Vol. 61, No. 57, pp. 315-321, 1965.

10. Gorenflo, D., "Zum Waermeuebergang bei Blasenverdampfung an Rippenrohren", Dissertation, Technische Hochschule, Karlsruhe, 1966.

11. Tamari, M., and Nishikawa, K., "The Stirring Effect of Bubbles Upon the Heat Transfer of Liquids", Heat Transfer-Japanese Research, Vol. 5, No. 2, pp. 31-44, 1976.

12. Uhl, V. W., "Mechanically Aided Heat Transfer to Viscous Materials", in Augmentation of Convective Heat and Mass Transfer, ASME, New York, pp. 109-117, 1970.

13. Tang, I., and McDonald, T. W., "A Study of Boiling Heat Transfer from a Rotating Horizontal Cylinder", Int. J. Heat Mass Transfer, Vol. 14, pp. 1643-1658, 1971.

14. Bergles, A. E., "Survey and Evaluation of Techniques to Augment Convective Heat and Mass Transfer", in Progress in Heat and Mass Transfer, ed. U. Grigull and E. Hahne, Vol. 1, pp. 331-424, Pergamon, Oxford, 1969.

15. Smith, G. V., and Forbes, R. E., "The Effect of Random Vibration on Natural Convective Heat Transfer in Rectangular Enclosures", in Augmentation of Convective Heat Transfer, ASME, New York, pp. 158-162, 1970.

16. Pak, H. Y., Winter, E. R. F., and Schoenhals, R. J., "Convection Heat Transfer in a Contained Fluid Subjected to Vibration", in Augmentation of Convective Heat Transfer, ASME, New York, pp. 148-157, 1970.

17. Yukawa, H., Hoshino, T., and Saito, H., "Effect of Ultrasonic Vibration on Free Convective Heat Transfer from an Inclined Plate in Water", Heat Transfer-Japanese Research, Vol. 5, No. 4, pp. 1-16, 1976.

18. Yabe, A., Mori, Y., and Hijikata, K., "Heat Transfer Augmentation Around a Downward-Facing Flat Plate by Non-Uniform Electric Fields", Heat Transfer 1978, Vol. 3, Hemisphere, Washington, D. C., pp. 171-176, 1978.

19. Blomgren, O. C., Sr., and Blomgren, O. C., Jr., "Method and Apparatus for Cooling the Workpiece and/or the Cutting Tools of a Machining Apparatus", U. S. Patent No. 3,670,606, 1972.

20. Reynolds, B. L., and Holmes, R. E., "Heat Transfer in a Corona Discharge", Mechanical Engineering, pp. 44-49, October 1976.

21. Sims, G. E., Akturk, U., and Evans-Lutterodt, K. O., "Simulation of Pool Boiling Heat Transfer by Gas Injection at the Interface", Int. J. Heat Mass Transfer, Vol. 6, pp. 531-535, 1963.

22. Blumenkrantz, A. R., and Taborek, J., "Heat Transfer and Pressure Drop Characteristics of Turbotec Spirally Grooved Tubes in the Turbulent Regime", Heat Transfer Research, Inc. Report 2439-300-7, 1970.

23. Rozalowski, G. R., and Gater, R. A., "Pressure Loss and Heat Transfer Characteristics for High Viscous Flow in Convoluted Tubing", ASME Paper No. 75-HT-40, 1975.

24. Zappa, R. F., and Geiger, G. E., " Effect of Artificial Surface Roughness on Heat Transfer and Pressure Drop for a High Prandtl Number Fluid in Laminar Flow", ASME Paper No. 71-HT-36, 1971.

25. Pescod, D., "The Effects of Turbulence Promoters on the Performance of Plate Heat Exchangers", in Heat Exchangers: Design and Theory Sourcebook, Scripta, Washington, pp. 601-616, 1974.

26. Bergles, A. E., and Jensen, M. K., "Enhanced Single-Phase Heat Transfer for OTEC Systems", Proceedings Fourth Annual Conference on Ocean Thermal Energy Conversion, New Orleans, Louisiana, pp. VI-41 - VI-54, March 1977.

27. Dipprey, D. F., and Sabersky, R. H., "Heat and Momentum Transfer in Smooth and Rough Tubes at Various Prandtl Numbers", Int. J. Heat Mass Transfer, Vol. 6, pp. 329-353, 1963.

28. Webb, R. L., Eckert, E. R. G., and Goldstein, R. J., "Heat Transfer and Friction in Tubes with Repeated Rib Roughness", Int. J. Heat Mass Transfer, Vol. 14, pp. 601-618, 1971.

29. Webb, R. L., Eckert, E. R. G., and Goldstein, R. J., "Generalized Heat Transfer and Friction Correlations for Tubes with Repeated-Rib Roughness", Int. J. Heat Mass Transfer, Vol. 15, pp. 180-184, 1972.

30. Lewis, M. J., "An Elementary Analysis for Predicting the Momentum-and Heat-transfer Characteristics of a Hydraulically Rough Surface", J. Heat Transfer, Vol. 97, pp. 249-254, 1975.

31. Kays, W. M., and London, A. L., Compact Heat Exchangers, Second Edition, Mcgraw-Hill, New York, 1964.

32. Shah, R. K., "Perforated Heat Exchanger Surfaces. Part 2 - Heat Transfer

and Fluid Flow Characteristics", ASME Paper No. 75-WA/HT-9, 1975.

33. Lee, C. P. and Yang, W. J., "Augmentation of Convective Heat Transfer from High-Porosity Perforated Surfaces", *Heat Transfer 1978*, Vol. 2, Hemisphere, Washington, pp. 589-594, 1978.

34. Sparrow, E. M., Baliga, B. R., and Patankar, S. V., "Heat Transfer and Fluid Flow Analysis of Internal-Wall Channels, with Applications to Heat Exchangers," *J. Heat Transfer*, Vol. 99, pp. 4-11, 1977.

35. Webb, R. L., "Air-side Heat Transfer in Finned Tube Heat Exchangers", *Heat Transfer Engineering*, Vol. 1, No. 3, pp. 33-49, 1980.

36. Watkinson, A. P., Miletti, D. C., and Kubanek, G. R., "Heat Transfer and Pressure Drop of Internally Finned Tubes in Laminar Oil Flow", ASME Paper No. 75-HT-41, 1975.

37. Marner, W. J., and Bergles, A. E., "Augmentation of Tubeside Laminar Flow Heat Transfer by Means of Twisted-Tape Inserts, Static-Mixer Inserts, and Internally Finned Tubes", *Heat Transfer 1978*, Vol. 2, Hemisphere, Washington, pp. 583-588, 1978.

38. Bergles, A. E., "Enhancement of Heat Transfer", *Heat Transfer 1978*, Vol. 6, Hemisphere, Washington, pp. 89-108, 1978.

39. Carnavos, T. C., "Heat Transfer Performance of Internally Finned Tubes in Turbulent Flow", in *Advances in Enhanced Heat Transfer*, ASME, pp. 61-67, 1979.

40. Patankar, S. V., Ivanovic, M., and Sparrow, E. M., "Analysis of Turbulent Flow and Heat Transfer in Internally Finned Tubes and Annuli", *J. Heat Transfer*, Vol. 101, pp. 29-37, 1979.

41. Carnavos, T. C., "Cooling Air in Turbulent Flow with Multipassage Internally Finned Tubes", ASME Paper No. 78-WA/HT-52, 1978.

42. Carnavos, T. C., "Cooling Air in Turbulent Flow with Internally Finned Tubes", *Heat Transfer Engineering*, Vol. 1, No. 2, pp. 41-46, 1979.

43. Goldstein, Jr., L. and Sparrow, E. M., "Experiments on the Transfer Characteristics of a Corrugated Fin and Tube Heat Exchanger Configuration", J. Heat Transfer, Vol. 98, pp. 26-34, 1976.

44. Marto, P. J., Reilly, D. J., and Fenner, J. H., "An Experimental Comparison of Enhanced Heat Transfer Condenser Tubing", in *Advances in Enhanced Heat Transfer*, ASME, New York, pp. 1-9, 1979.

45. Pahl, M. H., and Muschelknautz, E., "Einsatz and Auslegung Statischer Mischer", *Chem.-Ing.-Tech.*, Vol. 51, pp. 347-364, 1979.

46. Megerlin, F. E., Murphy, R. W., and Bergles, A. E., "Augmentation of Heat Transfer in Tubes by Means of Mesh and Brush Inserts", *J. Heat Transfer*, Vol. 96, pp. 145-151, 1974.

47. Hong, S. W., and Bergles, A. E., "Augmentation of Laminar Flow Heat Transfer by Means of Twisted-Tape Inserts", J. Heat Transfer, Vol. 98, pp. 251-256, 1976.

48. Lopina, R. F., and Bergles, A. E., "Heat Transfer and Pressure Drop in Tape Generated Swirl Flow of Single-Phase Water", *J. Heat Transfer*, Vol.

91, pp. 434-442, 1969.

49. Schmidt, E. F., "Waermeuebergang and Druckverlust in Rohrschlangen", Chem.-Ing.-Tech., Vol. 39, pp. 781-789, 1967.

50. Abul-Hamayel, M. A., and Bell, K. J., "Heat Transfer in Helically-Coiled Tubes with Laminar Flow", ASME Paper No. 79-WA/HT-11, 1979.

51. Kenning, D. B. R., and Kao, Y. S., "Convective Heat Transfer to Water Containing Bubbles: Enhancement Not Dependent on Capillarity", Int. J. Heat Mass Transfer, Vol. 15, pp. 1709-1718, 1972.

52. Watkins, R. W., Robertson, C. R., and Acrivos, A., "Entrance Region Heat Transfer in Flowing Suspensions", Int. J. Heat Mass Transfer, Vol. 19, pp. 693-695, 1976.

53. Bergles, A. E., Junkhan, G. H., and Hagge, J. K., "Advanced Cooling Systems for Agricultural and Industrial Machines", SAE Paper No. 751183, 1976.

54. Depew, C. A. and Kramer, T. J., "Heat Transfer to Flowing Gas-Solid Mixtures", Advances in Heat Transfer, Vol. 9, pp. 113-180, 1973.

55. Chen, J.C., "Heat Transfer to Tubes in Fluidized Beds", ASME Paper No. 76-HT-75, 1976.

56. Thomas, W. C., and Sunderland, J. E., "Heat Transfer Between a Plane Surface and Air Containing Water Droplets", Ind. and Eng. Chemistry Fundamentals, Vol. 9, pp. 368-374, 1970.

57. Yang, W.-J., and Clark, D. W., "Spray Cooling of Air-Cooled Compact Heat Exchangers", Int. J. Heat Mass Transfer, Vol. 18, pp. 311-317, 1975.

58. Hagge, J. K., and Junkhan, G. H., "Mechanical Augmentation of Convective Heat Transfer in Air", J. Heat Transfer, Vol. 97, pp. 516-520, 1975.

59. McElhiney, J. E., and Preckshot, G. W., "Heat Transfer in the Entrance Length of a Horizontal Rotating Tube", Int. J. Heat Mass Transfer, Vol. 20, pp. 847-854, 1977.

60. Mori, Y., and Nakayama, W., "Forced Convection Heat Transfer in a Straight Pipe Rotating Around a Parallel Axis", Int. J. Heat Mass Transfer, Vol. 10, pp. 1179-1194, 1967.

61. Mori, Y., Fukada, T., and Nakayama, W., "Convective Heat Transfer in a Rotating Radial Circular Pipe (2nd Report)", Int. J. Heat Mass Transfer, Vol. 14, pp. 1807-1824, 1971.

62. Miyazaki, H., "Combined Free and Forced Convective Heat Transfer and Fluid Flow in a Rotating Curved Circular Tube", Int. J. Heat Mass Transfer, Vol. 14, pp. 1295-1309, 1971.

63. Moissis, R., and Maroti, L. A., "The Effect of Sonic Vibrations on Convective Heat Transfer in an Automotive Type Radiator Section", Dynatech Corporation Report No. 322, July 1962.

64. Porter, J. E., and Poulter, R., "Electro-Thermal Convection Effects with Laminar Flow Heat Transfer in an Annulus", Heat Transfer 1970, Vol. 2, Paper FC3.7, Elsivier Publishing Company, Amsterdam, 1970.

65. Savkar, S. D., "Dielectrophoretic Effects in Laminar Forced Convection Between Two Parallel Plates", *The Physics of Fluids*, Vol. 14, pp. 2670-2679, 1971.

66. Newton, D. C., and Allen, P. H. G., "Senftleben Effect in Insulating Oil Under Uniform Electric Stress", *Letters in Heat and Mass Transfer*, Vol. 4, No. 1, pp. 9-16, 1977.

67. Mizushina, T., Ueda, H., and Matsumoto, T., "Effect of Electrically Induced Convection on Heat Transfer of Air Flow in an Annulus", *J. Chemical Engineering Japan*, Vol. 9, No. 2, pp. 97-102, 1976.

68. Tauscher, W. A., Sparrow, E. M., and Lloyd, J. R., "Amplification of Heat Transfer by Local Injection of Fluid Into a Turbulent Tube Flow", *Int. J. Heat Mass Transfer*, Vol. 13, pp. 681-688, 1970.

69. Kinney, R. B., "Fully Developed Frictional and Heat Transfer Characteristics of Laminar Flow in Porous Tubes", *Int. J. Heat Mass Transfer*, Vol. 11, pp. 1393-1401, 1968.

70. Kinney, R. B., and Sparrow, E. M., "Turbulent Flow, Heat Transfer, and Mass Transfer in a Tube with Surface Suction", *J. Heat Transfer*, Vol. 92, pp. 117-125, 1970.

71. Aggarwal, J. K., and Hollingsworth, M. A., "Heat Transfer for Turbulent Flow with Suction in a Porous Tube", *Int. J. Heat Mass Transfer*, Vol. 16, pp. 591-609, 1973.

72. Bergles, A. E., Lee, R. A., and Mikic, B. B., "Heat Transfer in Rough Tubes with Tape-Generated Swirl Flow", *J. Heat Transfer*, Vol. 91, pp. 443-445, 1969.

73. Kryukov, Y. V., and Boykov, G. P., "Augmentation of Heat Transfer in an Acoustic Field", *Heat Trans.-Soviet Res.*, Vol. 5, No. 1, pp. 26-28, 1973.

74. Van Rooyen, R. S., and Kroeger, D. G., "Laminar Flow Heat Transfer in Internally Finned Tubes with Twisted-Tape Inserts", *Heat Transfer 1978*, Vol. 2, Hemisphere, Washington, pp. 577-581, 1978.

75. Bartel, W. J., and Genetti, W. E., "Heat Transfer from a Horizontal Bundle of Bare and Finned Tubes in an Air Fluidized Bed", *AIChE Symposium Series No. 128*, Vol. 69, pp. 85-93, 1973.

76. Zozulya, N. V., and Khorunzhii, Y., "Heat Transfer from Finned Tubes Moving Back and Forth in Liquid", *Chemical and Petroleum Engineering*, No. 9-10, pp. 830-832, 1968.

77. Min, K., and Chao, B. T., "Particle Transport and Heat Transfer in Gas-Solid Suspension Flow Under the Influence of an Electric Field", *Nuclear Science and Engineering*, Vol. 26, pp. 534-546, 1966.

78. Bhattacharya, S. C., and Harrison, D., "Heat Transfer in a Pulsed Fluidized Bed", *Trans. Institution of Chemical Engineers*, Vol. 54, pp. 281-286, 1976.

79. Masliyah, J. H., and Nandakumar, K., "Fluid Flow and Heat Transfer in Internally Finned Helical Coils", *Canadian J. Chemical Engineering*, Vol. 55, pp. 27-36, 1977.

# The Development of a High Performance Heat Transfer Surface

**C. M. B. RUSSELL**
Heat Transfer Division
CE—Lummus
London, England

ABSTRACT

A novel form of plate-fin, using vortex generators to enhance heat transfer has been developed using a four-stage program. Flow visualization tests, transient heat transfer tests, and tests with a small and with a large matrix are described and the results presented.

1. INTRODUCTION

A requirement arose in Heat Transfer Division of CE-Lummus for a plate-fin surface. This type of surface is widely used for automobile radiators and in the air conditioning industry, as well as for general use in heat transfer equipment. Fig. 1 shows the principle of such surface, which is used where there is a large difference in the heat transfer coefficient of the hot and cold streams, typically air and water. The hot fluid is in flattened tubes, to which are fixed plate fins, in the plane at right angles to the plane of the flattened tube.

The better conducting fluid (water, steam, refrigerant) passes inside the tube, and the other fluid - usually air - in cross-flow over the fins. The optimum dimensions expected for the CE-Lummus project indicated that the tube would be 25mm wide x 150mm long; the fins 25mm high; and the air gap between fins 2mm.

The airside heat transfer matrix is an obvious target for heat transfer enhancement. Such enhancement is described in the classic work of Kays and London (1), but most commercial surfaces have been developed by individual companies, and their performance is confidential. Three types of surface are frequently used, with turbulators, with louvres, and with plates curved in the direction of the airflow.

Turbulators increase the heat transfer by creating turbulence in the wake of projections on the fin surface. The heat transfer is increased, but at the expense of airside pressure drop. Because the heat transfer coefficient has increased, the weight of metal used in the matrix is decreased. However, the increased pressure drop leads to a lower frontal air velocity, and hence to an exchanger with greater frontal area. The extra cost of the headers and side frames hardly compensates for the reduced cost of the matrix, and the additional space requirement has contributed to the present unpopularity of turbulators.

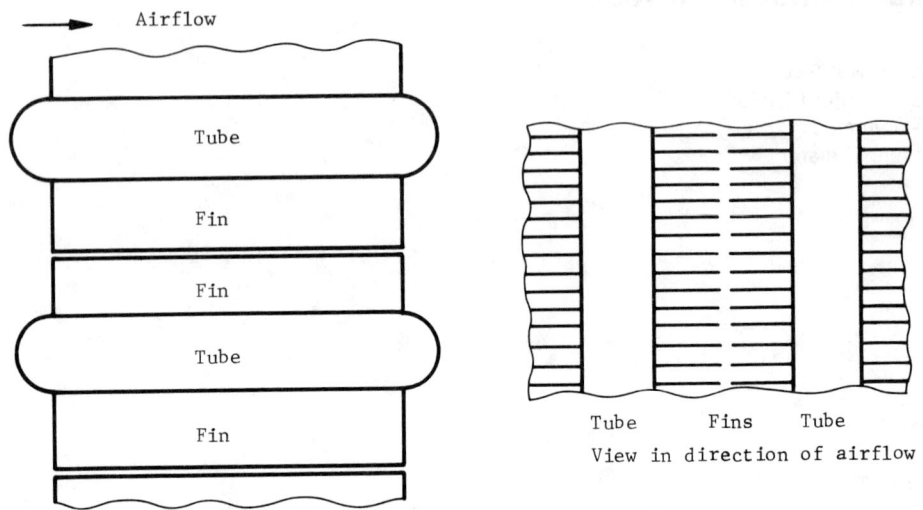

Fig.1. Diagram of Plate-Fin Extended Surface

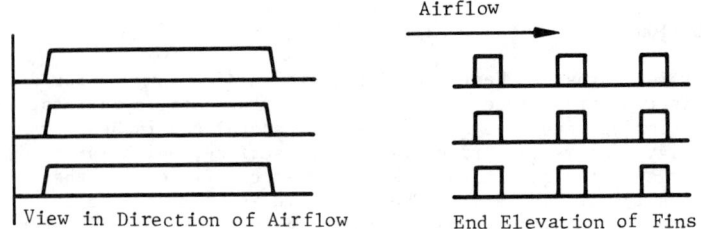

Fig.2. Louvres Punched up from the Fins are Often Used for Heat Transfer Enhancement

Louvres consist of punched-up strips in the fin material (Fig.2). These probably are effective because they break up the boundary layer over the fin. Kays and London report that typically louvres will double the heat transfer at the expense of a tripling of the pressure drop.

If the fins are curved - usually "wavy" - in the direction of the airflow, increased heat transfer occurs. The results reported by Kays and London are not good, but further work along these lines has produced commercial heat exchangers with excellent performance.

The use of louvred, and to a lesser extent, curved surface limits the closeness of the fin pitch. In general, there is an advantage in having as many fins as possible in the matrix, the fin density being limited by manufacturing possibilities or by the necessity for cleaning the matrix. Louvres will close up the minimum gap in the matrix, and curved fins decrease the

the possibilities of cleaning. It is often the case that plain fins can be used on a closer pitch than the enhanced fin, and this factor accounts for the continuing popularity of this type of fin.

None of the accepted fin types seemed suitable for the CE-Lummus project, so it was decided to attempt the development of a new surface. An account of this development follows.

2.  CHOICE OF ENHANCEMENT TYPE

A severe constraint was placed on the design of the new matrix by the necessity to keep the strength of the fin unimpaired. For this reason any type of louvred fin was not considered; both strength and cleaning considerations caused the rejection of curved fins.

Attention therefore focussed on the use of turbulators, but with a difference. If turbulators generating vortices - vortex generators or "VG"'s were used - it can be hoped that the increased mixing and velocity of the airstream will lead to better heat transfer, and local temperature difference; but a vortex demands much less energy than does random turbulence; so a modest increase in drag may be hoped for.

In particular, it was decided to try to generate vortices ot the type shown in Fig. 3, which is a cross-section of the air passage. The vortex rotation is about an axis parallel to the airflow, and a series of vortices exist across the passage. These are arranged so that adjacent vortices have opposite rotation, which gives a very persistant vortex. The wing-tip vortices of an aircraft are of this type, and may persist 50 km astern of a "747".

In view of the potential advantages to be drawn from vortices in heat transfer, a literature survey produced disappointing results.

The extensive review of enhanced heat transfer by Bergles et al (2) and Bergles' comprehensive review of the subject (3), did not yield any directly applicable reference. A computer search of the Heat Transfer and Fluid Flow Service's library of over 50,000 papers yielded the one reference discussed below. Searches under keywords such as "Flat Plates", or for "Vort" in the title, produced several hundred titles, clearly showing the interest of the heat transfer community in heat transfer improvement by vorticity for intube laminar flow. Small and Young (4) have pointed out the importance of vortices produced by a cylinder in cross-flow in a rod baffled exchanger. Edwards and Alker (5) reported work on the heat transfer in the wake of turbulators (cubes) on the floor of a tunnel, and of triangular vortex generators to give counter-rotation as Fig. 3, and co-rotation. Heat transfer was measured by applying a uniform heat flux to the tunnel floor and measuring the floor temperature by a luminescent phosphor. A good improvement in heat transfer was observed with all three systems, the turbulators (cubes) giving the greatest improvement, but over a shorter distance. The most persistant improvement was found with the counter-rotating vortex generators. No pressure drop measurements were possible in the wind tunnel used.

The general principles of the fin type having been decided, a test program in four stages was envisaged. These stages were :-

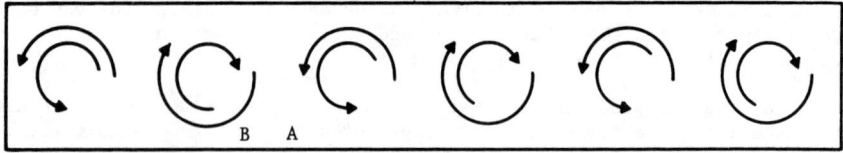

Fig.3. View of the Flow Passage Between Fins in the Direction of Airflow, Showing the Vortex Pattern Desired

1. Flow visualization tests to determine the pattern of VG's.
2. Heat transfer tests using transient techniques to determine the optimum design of VG's.
3. Heat transfer tests on a small matrix to check 1 and 2, and,
4. Heat transfer tests on a large matrix to determine "j" and "f" factors.

A description of these four stages follows.

FLOW VISUALISATION TESTS

A tunnel representing fourteen fins with six spaces was built, as shown in Fig. 4. The model is at six times full size, so the fins are 900mm long, and the passages are 15mm wide. The central gap between fins was not represented, so a pair of fins each 25mm high was represented by a plate 300mm high. One side of the tunnel was in clear plastic, and also two plates near the clear side. The space beyond the three clear plates was used for flow visualization tests with smoke.

Flow was induced in the tunnel with a small axial fan. Air entry was smoothed with a bell-mouth and honeycomb.

A yaw meter was used to determine the effectiveness of the VG's. Referring to Fig. 3, it will be realized that the flow will be axial at point "A", but that there will be an angled flow at point "B". Thus the vorticity of the flow can be measured by a yaw meter.

The tunnel was generally run with an airspeed of around 0.3 m/sec, to give clear appreciation of the smoke used, at a Reynolds number comparable to that anticipated at full scale.

The tunnel proved very effective for the rapid examination of various patterns of VG's. Triangular VG's which will produce vortices like a schoolboy's dart, or the "Concorde", proved disappointing. The best results were gained with barriers about half the width of the flow passages, and set at an angle to the flow. These act as very low aspect ratio wings, and spill vortex downstream like the wing-tip vortex of an aircraft. Results at low angles of incidence were again disappointing, and VG's at an angle of incidence of 20-30° seemed the most promising. Actual VG's could be either punched out or embossed on the fin surface. It was found that a punched out VG with the hole on the downstream side was much more effective than are with the hole upstream.

The best disposition of VG's found is shown on Fig. 5. The vortices were tending to become rather weak towards the end of the fin, and two sets of VG's

# THE DEVELOPMENT OF A HIGH PERFORMANCE HEAT TRANSFER SURFACE

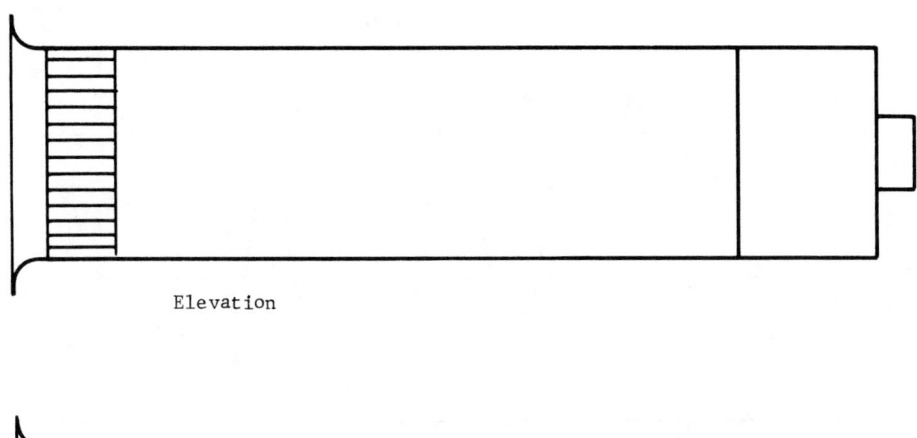

Fig.4. Tunnel Used for Flow Visualization Tests

in series was considered advisable with the disposition of Fig. 5. A plot of yaw angle across the channel is shown in Fig. 6. As can be seen, the flow pattern is very much as predicted. The yaw vane was only used with models of a single fin passage at 12 x full scale.

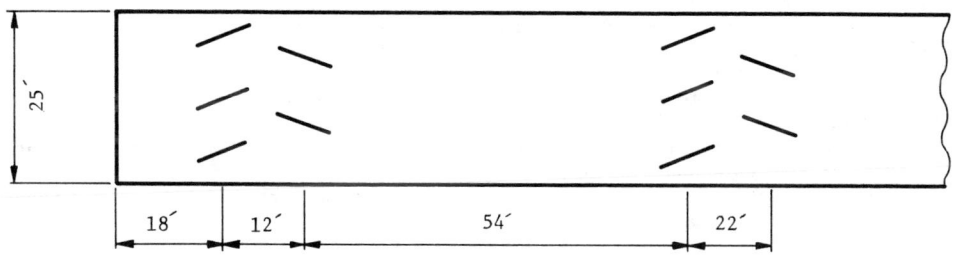

Fig.5. Disposition of Vortex Generators. Angle of Incidence $20^{o}$

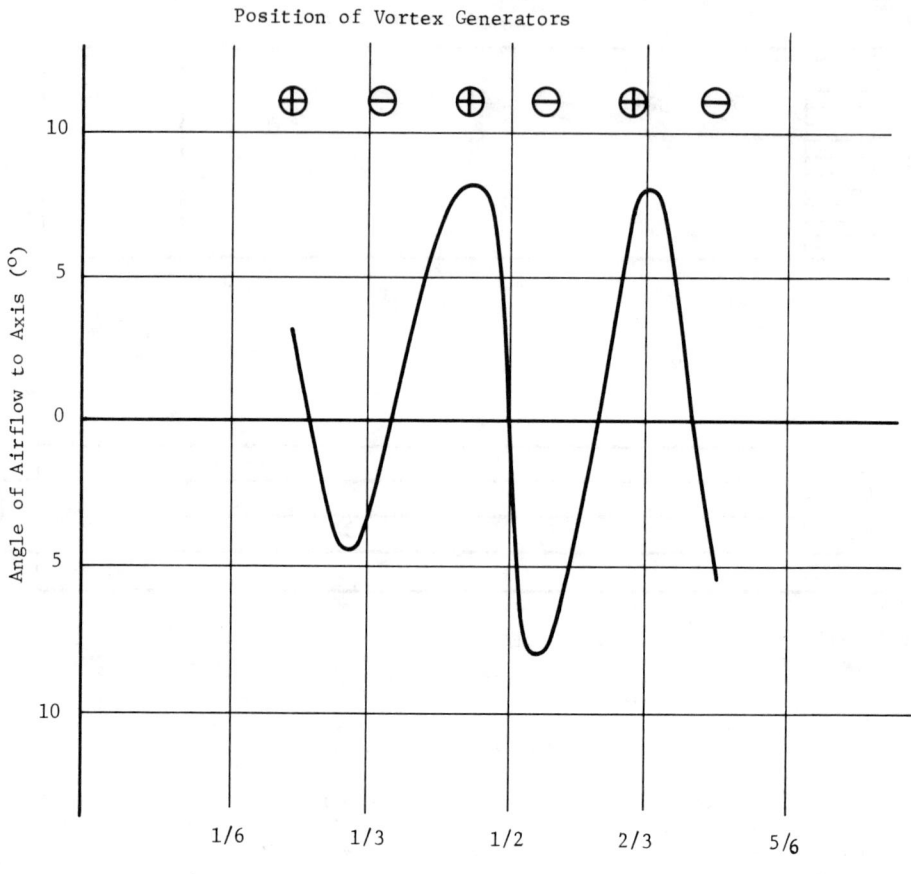

Fig.6. Yaw Gauge Readings Across Flow Channel

TRANSIENT HEAT TRANSFER TESTS

If a thin piece of material is introduced into an airstream, the rate of rise of temperature of the material will be proportional to the heat transfer rate to it, to the temperature difference between the material and the airstream, and inversely proportional to its heat capacity. Were the material painted with temperature sensitive paint, the time necessary to melt the paint at any point will be inversely proportional to the heat transfer rate at that point (assuming constant temperature and constant material thickness). Throckmorton and Stone (6) described the method, and Jones and Russell (7) have reported some applications in the field of fintubes.

Fig. 7 shows the tunnel used for the experiments. A model in plastic at 3.6 times full size was made of the fins. A hot airstream was generated by the heater in the tunnel, and the cool model was inserted into the stream. The melt line of temperature sensitive point was observed through a television camera with a VTR.

By these means, the heat transfer from a series of VG types, varying angles

# THE DEVELOPMENT OF A HIGH PERFORMANCE HEAT TRANSFER SURFACE

and dimensions, was measured. The tunnel is also capable of measuring pressure drop, and the actual performance of all the VG's can be measured.

TESTS ON SMALL MATRIX

The tunnel shown in Fig. 7 was also used for a series of tests to check the performance of the surface. A matrix was made, with 25mm high fins, in a 50mm wide passage, thus using two sets of fins. The passage was 114mm high, and the fins were spaced at 320/metre. Fins were of aluminium, 0.4mm thick. The walls of the tunnel, which held the fins, were heated by atmospheric steam. Airflow was varied, but the heater was not used.

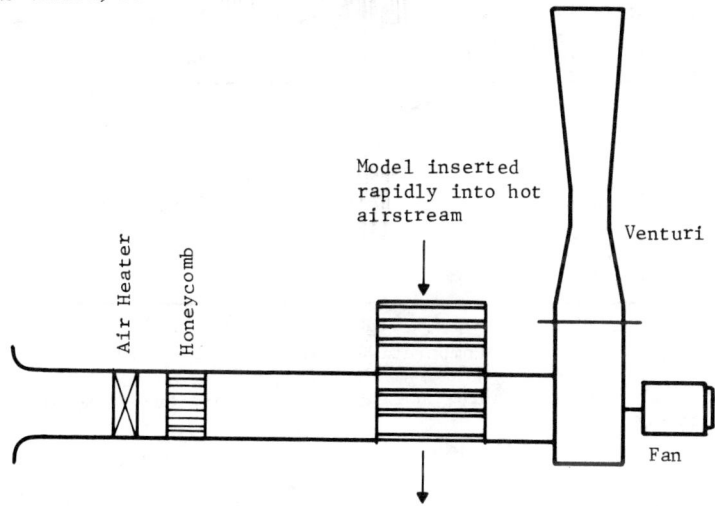

Fig.7. Tunnel Used for Transient Tests

This tunnel has been in use for several years, and consistently produces accurate heat balances. However, the matrix is considered too small to give absolute results. It is only used to compare heat transfer rates and pressure drops.

Fins with and without VG's were tested, and the heat transfer results are shown in Fig. 8. The heat transfer coefficient used in calculating the Stanton number is the film coefficient from the air to the fin, due allowance being made for the steam-side film coefficient, metal wall resistance, and fin efficiency. The fin efficiency is only some 75%, so errors in its calculation will lead to appreciable errors in the calculation of the heat transfer coefficient.

Fig. 8 only shows the relative performance of plain fins and fins with VG's to emphasise the comparative, not absolute, nature of the results.

There is some 50% increase of heat transfer, while the pressure drop increased by 43%. These results are very close to those predicted from the transient tests, and justified the testing of a larger matrix.

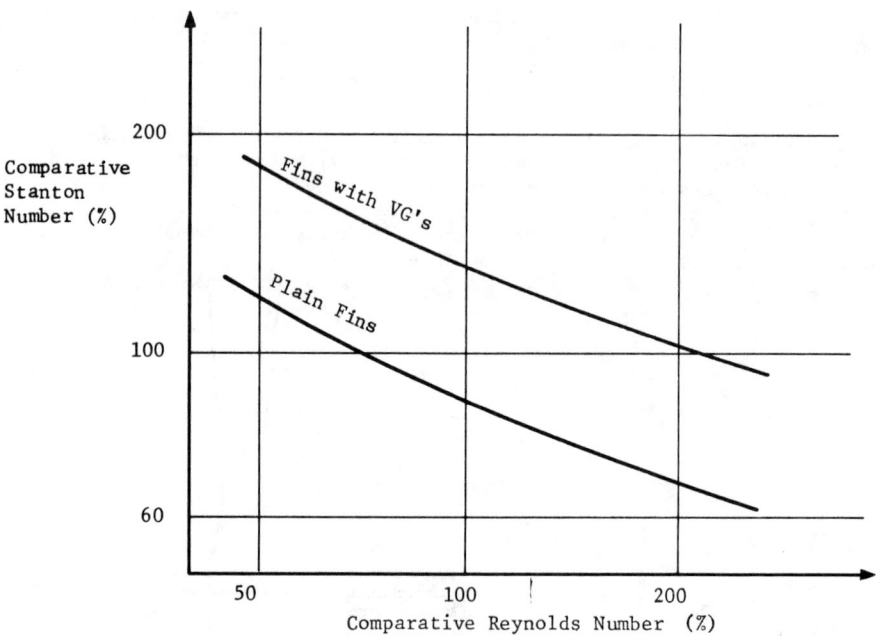

Fig.8. Stanton Number Vs Reynolds Number for Plain Fins and Fins with Vortex Generators

LARGE MATRIX TESTS

None of the tests described above will give heat transfer and pressure drop information of the accuracy required for heat exchanger design. A matrix of the fintube, of 450mm square frontal area, was constructed. The tubes were of brass, 178mm long and 25mm wide. Sweated onto the brass tubes were fins 150mm long x 25mm wide. The fins had a flange on each end, 2mm wide, to form a "Z" section. This gave good contact of the base of the fin, and stiffened the fin tip. The fins were 0.5mm thick, and spaced at 400/metre .

Figs 9 and 10 show a tube and the vortex generators on the fin.

The apparatus used for the tests is shown in Fig. 11. Air is drawn through the test matrix from a suction chamber, its quantity being measured in a venturi. Steam is generated in a boiler below the matrix, and heats the tubes.

The heat transferred during tests is calculated from the air temperature rise and the air quantity. It is necessary to provide an excess of steam blowing from the tubes to ensure that all the tube inside surface is swept by live steam; so no heat balance between electrical heat input to the boiler, and heat input to the air, is possible; however, heat balance runs, without using excess steam, were carried out before the heat transfer tests, and balances of $\pm 1\frac{1}{2}\%$ or better were obtained.

Heat transfer and pressure drop measurements were taken at varying airflow.

Fig.9. Photograph of Tube Tested

Fig.10. Photograph of Vortex Generators

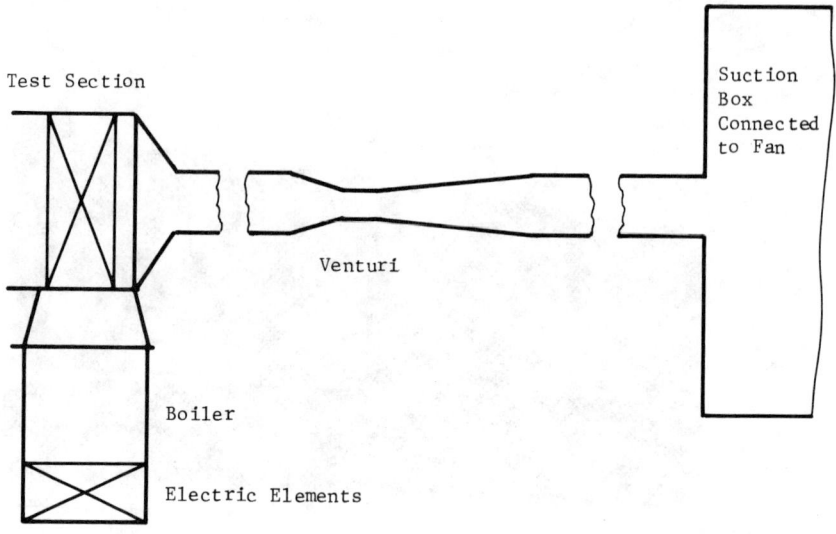

Fig.11. Rig for Heat Transfer Tests of Large Matrix

The results are shown on Figs. 12 and 13.

Reynolds number was calculated using a hydraulic diameter of 3.76mm, and the mean bulk velocity of the air. The mass flow used for Reynolds and Stanton numbers was that between the fins. Air properties at mean bulk air temperature were used.

The heat transfer rate was calculated by computing the overall heat transfer resistance from the measured heat load and logarithmic mean temperature difference; a value of $0.000176 \, m^2\text{-K/W}$ was subtracted from this to account for the steam condensing film coefficient, fouling, and the tube wall; and a correction was made for fin efficiency by the methods of Gardner (9). The thicker copper fin will give higher fin efficiencies than were obtained with the earlier tests - of the order of 90%. Thus errors in computing this will not be very important.

The heat transfer coefficient is related to the finned surface of $17.5 m^2$.

The ordinates in Fig. 11 are the "j" used by Kays and London (1), where

$$j = \text{Stanton Number} \times \text{Prandtl Number}^{2/3}$$

The ordinates of Fig. 12 are a dimensional friction factor giving the pressure drop :-

$$dp = f \, G^2 \, v$$

# THE DEVELOPMENT OF A HIGH PERFORMANCE HEAT TRANSFER SURFACE

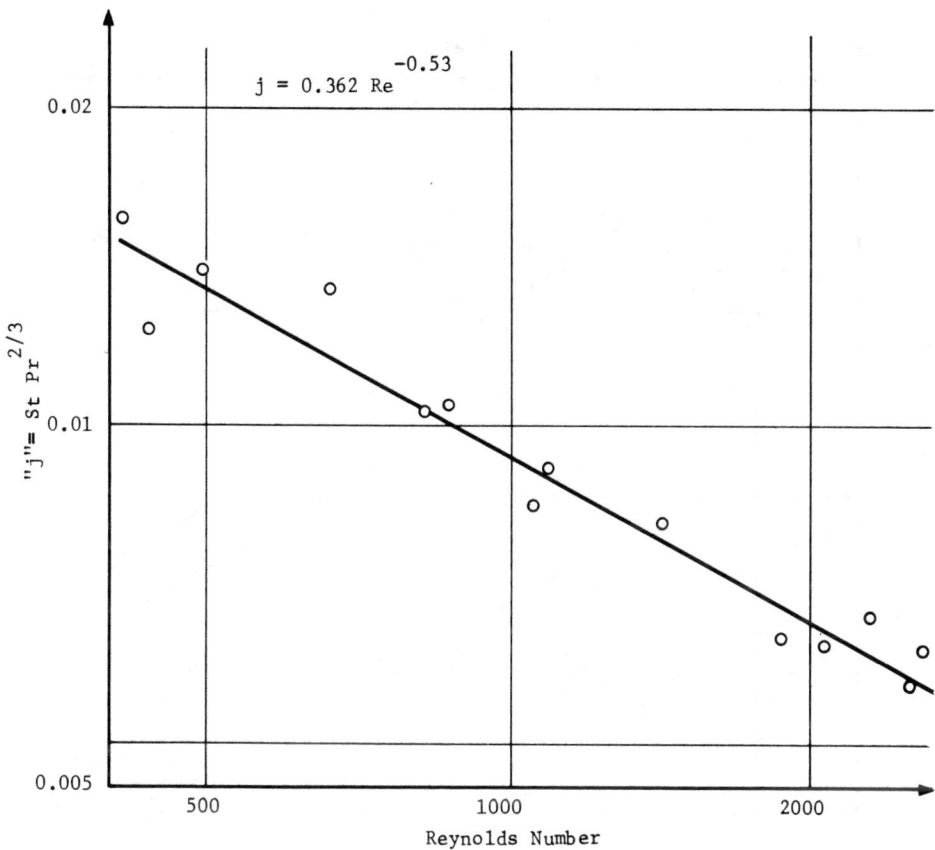

Fig.12. Heat Transfer Vs Reynolds Number

where

dp = Airside pressure drop $(N/m^2)$

f = Dimensional friction factor

G = Air mass flow $(kg/s-m^2)$

v = Mean air specific volume $(m^3/kg)$

## CONCLUSION

The methods outlined in this paper have led to a surface whose overall performance seems excellent. Should another attempt be made to develop a surface, the author would have sufficient confidence in the heat transfer measurements by injection into the hot airstream, to omit the third stage, measurements on a small matrix, described in the paper.

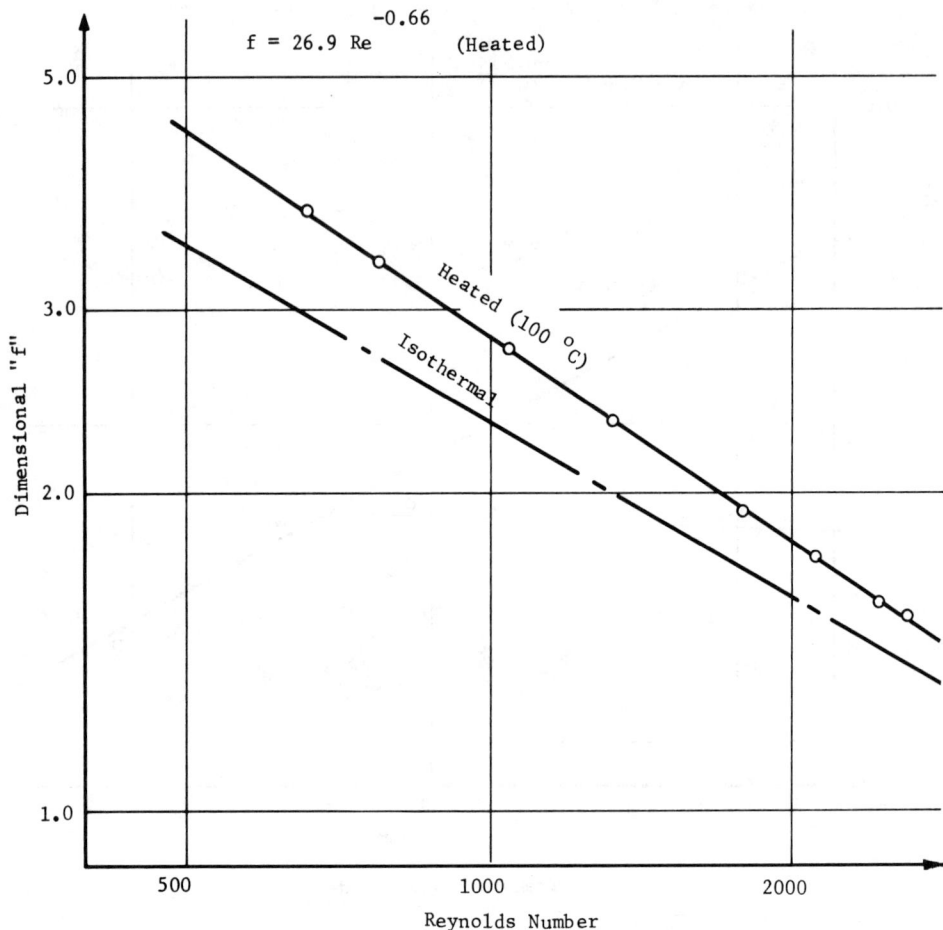

Fig.13. Dimensional Friction Factor Vs Reynolds Number

ACKNOWLEDGEMENTS

The tests were carrried out in the Osney Laboratories of the Department of Engineering Science of Oxford University. The help of the staff of the Department, and particularly of Dr T V Jones, is gratefully acknowledged.

Mr G H Lee did most of the experimental work, and reduced much of the data.

REFERENCES

1. Kays, W.M. and London, A.L. 1955. Compact Heat Exchangers. National Press.

2. Bergles, A.E. et al, 1979. Bibiography on Augmentation of Convective Heat and Mass Transfer. D. of E. HTL-19

3. Bergles, A.E. 1978. Enhancement of Heat Transfer. VIth Int. Heat Transfer Conference, Toronto.

4. Small, W.M. and Young, R.K. 1979. The Rod Baffle Heat Exchanger. Heat Transfer Engineering, Vol.1, No. 2, pp 21-27.

5. Edwards, F.J. and Alker, C.J.R. 1974. The Improvement of Forced Convection Heat Transfer using Surface Protrusions in the Form of of (a) Cubes and (b) Vortex Generators, Vth Int. Heat Transfer Conf.

6. Throckmorton, D.A. and Stone, D.R. 1974. Model Wall and Recovery Temperature Effects on Experimental Heat Transfer Data Analysis. AIAA Journal, Vol 12, No. 9, pp 1169-1170.

7. Jones, T.V. and Russell, C.M.B. 1978. Heat Transfer Distribution on Annular Fins. AIAA-ASME Thermophysics and Heat Transfer Conference, Palo Alto.

8. Gardner, K.A. 1945. Efficiency of Extended Surfaces, Trans. ASME Vol. 67, pp 621-631.

# Principles of Heat Transfer Augmentation. II: Two-Phase Heat Transfer

**ARTHUR E. BERGLES**
Department of Mechanical Engineering
Iowa State University
Ames, Iowa 50011 USA

ABSTRACT

This chapter discusses the many techniques which are available for augmentation of boiling and condensing, under both free and forced convection conditions. Both passive and active techniques have been shown to be effective in increasing phase change heat transfer coefficients for a wide range of conditions.

1. INTRODUCTION

Enhanced surfaces for boiling originate with the pioneering experiments of Jakob and Fritz [1] who demonstrated that large reductions in the wall superheat for pool nucleate boiling result from roughening the surface. They showed, for example, that the surface roughened by machining ("roughness screen") had an increase in the boiling heat transfer coefficient of as much as 250%.

A spirally corrugated tube was used to enhance shellside condensation in commercial shell-and-tube heat exchangers at least as early as 1921. The Arlberger catalog [2] describes steam-heated hot water heaters which had 50% more heat duty than similar units with smooth tubes.

All of the basic techniques described in the previous chapter can be used for phase change. However, there are important differences in the mechanism of augmentation which relate to the fundamental characteristics of phase-change heat transfer. For example, in the case of nucleate boiling, roughening the surface produces a more favorable size distribution of active nucleation sites rather than creating conditions which will disrupt the laminar sublayer.

The literature in this area is extensive; Table 1 of the previous chapter indicates that over one-third of the studies cited in the Iowa State University bibliography [3] have application to boiling or condensing. There is considerable worldwide research and development activity as the heat transfer improvements can be very large and there are many examples of successful commercialization.

2. POOL BOILING

2.1 Treated and Rough Surfaces

Treated surfaces involve surface condition adjustments other than those normally encountered. This excludes the well-known (but unpredictable) effects of surface material, finish, fouling, oxidation, etc., on nucleate boiling and critical heat flux. A novel technique for promoting nucleate boiling was proposed by Young and Hummel [4]. Spots of Teflon or other non-wetting material, either on the heated surface or in pits, were found to promote nucleation at less than 2° F. wall superheat. As shown in Fig. 1, there was a general reduction in $\Delta T_s$ at constant $\dot{q}$ by a factor of 3-4.* This technique is not effective for refrigerants as there are no "Freonphobic" materials (Bergles et al. [5]).

Thin insulating coatings are effective in increasing rates of heat transfer in pool boiling where heater temperature is the controlling parameter [6]. When surface temperatures are originally in the film boiling range, a thin coating of polytetrafluoroethylene, for example, reduces the fluid-surface interface temperature to the level where transition or nucleate boiling occurs. Bergles and Thompson [7] found large reductions in quench times due to scale or oxide coatings which promote destabilization of film boiling.

With well-wetting fluids (refrigerants, cryogens, organics, alkalai liquid metals), doubly re-entrant cavities are required to insure vapor trapping. This is illustrated in Fig. 2. The probability of having such active nucleation sites present is increased by machining, forming, or coating the surface. Furthermore, large cavities can be created which result in steady-state boiling at low $\Delta T_s$. The surfaces may appear either rough or smooth (as if treated), depending on the manufacturing procedure. Hence, the usual Treated and Rough classifications are lumped together for purposes of this discussion. Examples of these "structured" boiling surfaces are given in Table 1.

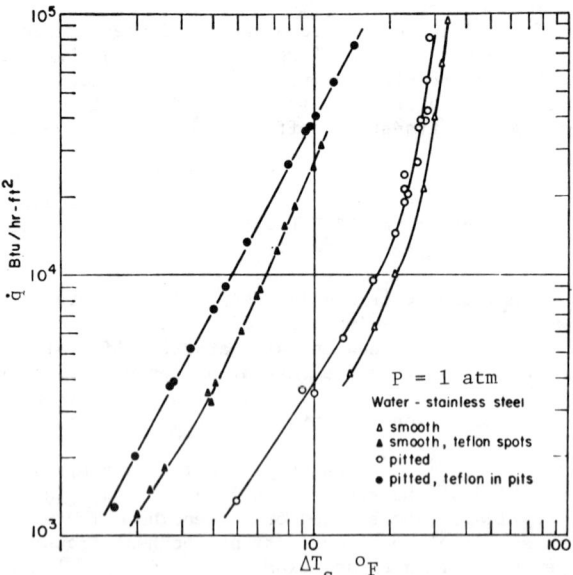

Fig. 1. Influence of Surface Treatment on Saturated Pool Boiling [4].

---

* This comparison, as well as others in this chapter, is based on the area of the plain heating surface.

Table 1. Examples of Structured Boiling Surfaces

| Category | Report | Procedure | Result |
|---|---|---|---|
| Machined | Kun and Czikk [8] | Cross grooved and flattened | Irregular matrix of re-entrant cavities |
| Formed | Webb [9] | Standard low-fin tubing with fins bent to reduce gap | Helical circumferential re-entrant cavities |
| | Zatell [10] | Above with additional variations | Helical circumferential re-entrant cavities |
| | Nakayama et al. [11] | Rolled, upset, and brushed | Helical circumferential or groove-type re-entrant cavities with periodic openings |
| | Wieland-Werke [12] | Standard low-fin tubing rolled to form a "T" | Helical circumferential re-entrant cavities |
| Multilayer | Ragi [13] | Stamped sheet with pyramids, open at the top, attached to surface | Regular matrix of re-entrant cavities |
| Coated | Marto and Rohsenow [14] | Poor weld | Irregular matrix of surface and re-entrant cavities |
| | O'Neill et al. [15] | Sintering or brazing | Irregular matrix of surface and re-entrant cavities |
| | Danilova et al. [16] | Sintering or brazing | Irregular matrix of surface and re-entrant cavities |
| | Oktay and Schmeckenbecher [17] | Electrolytic deposition | Irregular matrix of surface and re-entrant cavities |
| | Dahl and Erb [18] | Flame spraying | Irregular matrix of surface and re-entrant cavities |
| | Fujii et al. [19] | Particles bonded by plating | Irregular matrix of surface and re-entrant cavities |
| | Janowski et al. [20] | Metallic coating of a foam substrate | Irregular matrix of surface and re-entrant cavities |

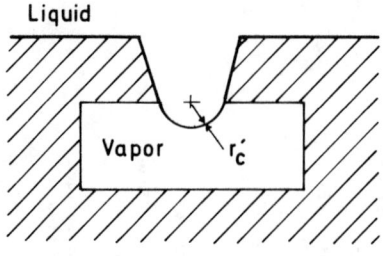
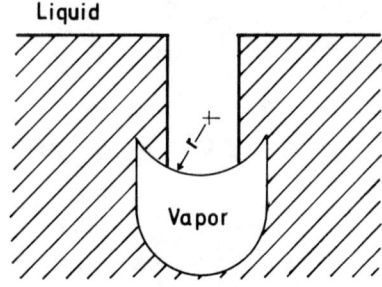

a. Square Re-entrant Cavity with $\beta = 90°$.

b. Doubly Re-entrant cavity with $\beta = 0°$.

Fig. 2. Re-entrant Cavities Used to Improve Boiling Stability (Radii shown correspond to maximum subcooling which can be sustained without flooding cavity).

Superheat reductions of up to a factor of 10 have been reported with these surfaces. It should be noted that the mechanism of vaporization is different for these surfaces than for normal cavity boiling. Here, the liquid flows to the interior where thin film evaporation occurs over a large surface area; the vapor is then ejected by "bubbling" (Czikk and O'Neill [21], Nakayama et al., [22]). It should be emphasized that the performance of these special surfaces is very sensitive to surface geometry and fluid condition. Additionally, very low temperature differences are involved; hence, it is necessary to be especially careful, or at least consistent, in measuring wall temperatures and saturation temperatures (pressures). The first comprehensive comparison of the nucleate boiling performances of various structured surfaces was reported recently by Taborek [23]. As shown in Fig. 3, all 3 surfaces have boiling curves well to the left of that for a single tube. It is evident that if only low temperature differences are available, high heat fluxes can be realized only with structured surfaces. Note that the heat flux is based on the area of the equivalent smooth tube for a particular O.D.

Limited evidence indicates that the critical or burnout heat flux, $\dot{q}_c$, with structured surfaces is usually as high or higher than that with plain surfaces (O'Neill et al. [15]).

The use of structured surfaces to enhance thin film evaporation has also been considered recently. Here, in contrast to the flooded pool experiments noted above, the liquid to be vaporized is sprayed or drips on heated horizontal tubes so as to form a thin film. The heat transfer coefficients can be quite high for pure convection without boiling if the film flow rate is either low or high enough to cause turbulence. When temperature differences are large enough to cause boiling, the boiling curves are similar to those for flooded pool boiling. If the available temperature difference is modest, structured surfaces can be used to promote boiling so as to improve the overall heat transfer coefficient. Few studies have been conducted in this area; however, it appears that with structured surfaces boiling curves for thin film evaporation are similar to those for flooded pool boiling. (See, for example, Chyu's results for water evaporating on heated tubes prepared with High Flux and Gewa-T surfaces [24]).

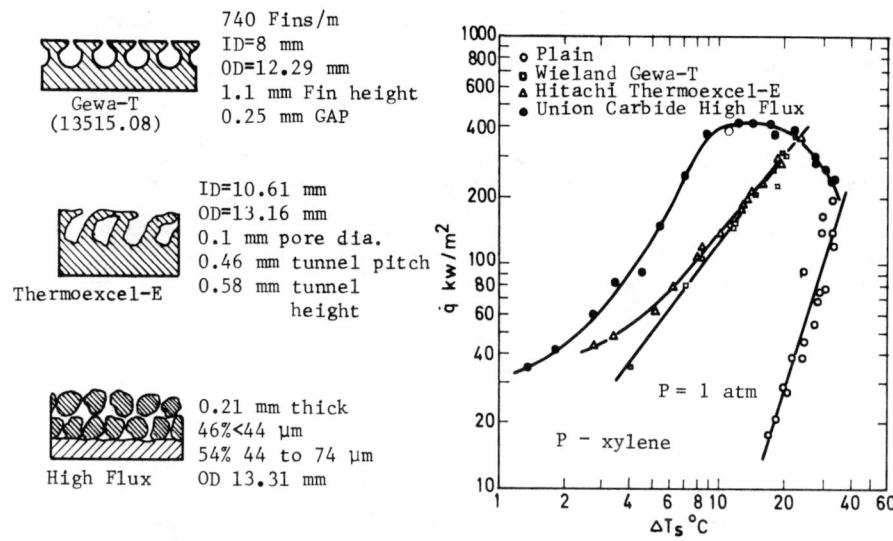

a) Sketch of Cross-Sections of Three Enhanced Heat Transfer Surfaces Tested

b) Boiling Curves for Three Enhanced Tubes and Smooth Tube

Fig. 3. Pool Boiling from Smooth and Structured Surfaces on the Same Apparatus [23].

2.2 Extended Surfaces

Low and medium fin tubes with external, circumferential fins are produced by many manufacturers for boiling of refrigerants and organics. Gorenflo [25] reported single-tube boiling of R-11 which indicated that the heat transfer coefficients based on total area were higher for all six tubes tested than for the reference plain tube. Hesse [26] found that $\alpha$ for pool boiling of R-114 was higher for finned tubes than for a plain tube; however, $\dot{q}_c$ was lower. Both comparisons were on a total area basis. When referenced to the projected area at the maximum outside diameter, the critical heat fluxes were about equal to those of the plain tube. The reduction in local critical heat flux is apparently due to bubble interference between fins. Westwater [27] suggests that the fins may be spaced as close as the departure diameter of a nucleate boiling bubble ( ∿1.5 mm for Freon 113).

Data for a three-dimensional finned (square, straight-sided fins in a square array) tubular test section were reported by Corman and McLaughlin [28]. Significant reductions in temperature differences were observed in low flux nucleate boiling; however, temperature differences were larger at high fluxes. The critical heat flux was increased.

A thorough survey of large-scale finned surfaces utilized for pool boiling is given by Westwater [27]. Properly shaped or insulated fins promote very large heat transfer rates if the base temperature is in the film boiling range.

The use of finned surfaces for enhancement of thin film evaporation is discussed in recent reports, for example, Sideman and Moalem-Maron [29]. These results are presented in detail in the lecture by Prof. Sideman.

2.3 Displaced Enhancement Devices

Lienhard et al. [30] suggested that the critical heat flux on a large horizontal plate can be augmented by placing an "egg crate" structure on vertical walls directly above the plate. This is an exploitation of hydrodynamic instability theory which predicts a large increase in critical flux as heater width is reduced below one wavelength. Experiments carried out for saturated boiling of various fluids indicate that the critical flux can be doubled by this technique.

2.4 Surface Tension Devices

Excluding heat pipes, the primary technique which relies on capillary forces involves the bonding of wicking materials to heated surfaces. Wicking may be utilized when there is an unreliable supply of liquid to the heated surface, e.g., electronic cooling systems in aircraft undergoing violent maneuvers or spacecraft operating in near-zero-gravity. Allingham and McEntire [31] found that the heat-transfer coefficient for saturated pool boiling was improved at low heat fluxes (probably due to the large nucleation sites provided by the wicking) but was reduced at moderate fluxes. Costello and Redeker [32] found that the critical heat flux was raised by as much as 200% when the wicking was not too dense and a narrow channel was maintained at the top for easy escape of vapor. Corman and McLaughlin [28] recently presented more extensive data for wick-augmented surfaces which confirm these trends.

2.5 Liquid Additives

Under certain circumstances, addition of agents produce substantial improvements in heat transfer [33]. Small additive concentrations ($<5$ wt-percent) are used so that the desirable properties of the working fluid are maintained. Increases of about 40% in the heat transfer coefficient for saturated pool boiling of water with volatile additives have been recorded; however the thermodynamics of binary mixtures seems to indicate that boiling performance should be decreased. The critical heat flux for saturated pool boiling from small diameter ($\sim 2$ mm) wires can be increased by over 100%, with an increase of over 200% having been reported for water/1-pentanol. Van Stralen's extensive program of testing and modeling of mixtures has been extended to film boiling [34]. In this study, it was found that a 4.1 wt-percent mixture of 2-butanone in water improved coefficients by up to 80%.

2.6 Mechanical Aids

Tang and McDonald [35] found that when heated cylinders are rotated at high speeds in saturated pools, convective coefficients are so high that boiling can be suppressed. This constitutes an augmentation of pool boiling. Marto and Gray [36] found that critical heat fluxes were elevated in a rotating-drum boiler where the vaporization occurred at the inside of the centrifuged liquid annulus. With proper liquid feed conditions to the heated surface, exit qualities in excess of 99% were obtained.

Tleimat [37] domonstrated that high heat transfer coefficients can be achieved with thin film evaporation produced by wiper blades.

## 2.7 Surface Vibration

Only several of the many studies of surface vibration have involved boiling. The consensus is that low flux boiling is improved due to the imposed velocity, but there is negligible change in the boiling curve once boiling is fully established [33].

Fuls and Geiger [38] studied the effect of enclosure vibration on pool boiling. A slight increase in the nucleate-boiling heat transfer coefficient was observed.

## 2.8 Fluid Vibration

Vibrations have little effect on fully-developed pool boiling; however, the critical heat flux can be increased substantially [33]. Acoustic disturbances are effective in disrupting film boiling so that quenching times can be reduced (Loosle and Holdredge [39]).

## 2.9 Electrostatic Fields

Impressive improvements in boiling heat transfer due to electrostatic fields have been noted in small-scale experiments. The typical arrangement is a pool of dielectric liquid with a small-diameter heated wire surrounded by a large-diameter cylindrical conductor. The available data indicate that fully developed nucleate boiling is unaffected; however, the critical heat flux can be increased by factors of 1.6 to 6 [33]. Film boiling can also be improved by a factor of two. In more recent work, Basu [40] found that electric fields are effective in eliminating boiling curve hysteresis for organic liquids.

## 2.10 Suction

Heat transfer coefficients in film boiling can be improved by continuously removing vapor through a porous heated-surface if a porous block is placed on the surface to stabilize the flow of liquid toward the surface. Wall superheats for water can be kept low for high heat fluxes [33]. A more recent study, which includes data for methanol and water, was reported by Raiff and Wayner [41]. The requirements of a porous heated-surface, a flow control element, and a vapor removal or recirculation system appear to limit the applications of suction boiling.

## 2.11 Compound Techniques

Compound augmentation, which involves two or more techniques applied simultaneously, has also been studied to a limited extent. For instance, the addition of surface roughness to the evaporator side of a rotating evaporator-condenser increased the overall coefficient by 10% (Bromley et al. [42]).

## 3. FORCED CONVECTION BOILING

### 3.1 Treated and Rough Surfaces

The effect of surface finish on flow boiling is much less pronounced than on pool boiling. Non-wetting spots produce no appreciable change in the boiling curve for forced flow of water; this behavior has been attributed to the initial activation and subsequent dominance of natural nucleation sites [43].

While it is more difficult to modify the inside surface of a tube, several of the surface preparations noted in Section 2.1 have been studied with internal flow. The High Flux surface does not improve high-flux, subcooled flow boiling; however, the annoying boiling curve hysteresis with fluorocarbons is eliminated (Murphy and Bergles [44]). On the other hand, substantial increases in heat transfer coefficients for forced convection vaporization have been reported with the same type of surface [45].

Burck et al. [46] found that surface roughness increased subcooled critical heat flux by only about 10%.

A variety of roughness configurations have been developed for evaporator tubes. Dual-diameter tubes, slotted helical inserts, helical ribs, and machined protuberances improve vaporization, critical heat flux, and post dryout [33]. More recently, Withers and Habdas [47] found that corrugation of the tube improved heat transfer coefficients for vaporization of Freon-12 by as much as 200%. This type of tubing, optimized according to a geometrical "severity parameter", is in commercial service.

Coiled wire inserts were considered for mercury once-through boilers for space power plants [33]. These inserts increase the turbulence level, thereby promoting diffusion of droplets to the heated surface. Ackerman [48] demonstrated that internally-ribbed tubes suppress pseudo-film boiling with supercritical water and permit operation at higher heat fluxes than is possible with smooth tubes.

3.2 Extended Surfaces

Tubes with integral or inserted internal fins increase heat transfer rates for refrigerant evaporation by several hundred percent over the smooth tube values (Schluender and Chwala, [49], and Kubanek and Miletti, [50]). Data from the latter study are shown in Figure 4; the heat transfer coefficients are based on the surface area of the smooth tube of the same diameter. "Enhanced" heat transfer tubes are available for vertical and horizontal evaporators. While the distorted tubes (e.g. doubly fluted and spirally corrugated) have been developed primarily for augmentation of condensation on the outside wall, it has been demonstrated that heat transfer coefficients for the evaporating fluid on the inside of the tube are also increased (Johnson et al. [51]). In falling film evaporation, Thomas and Young [52] found that $\alpha$ can be increased by a factor of over ten by use of internally finned tubes.

3.3 Displaced Enhancement Devices

Several investigators have modified flow channels to disturb the flow near the heated surface so as to increase bulk-boiling critical heat flux [33]. Tong et al. [53] demonstrated that power for a simulated boiling water reactor rod bundle can be increased by roughening the shroud so as to keep more liquid on the heated surface. Ryabov et al. [54] summarized a major study of increasing critical power in rod bundles by use of special spacers, inserts, etc. Megerlin et al. [55] reported subcooled boiling data with mesh and brush inserts. Critical heat fluxes were increased by about 100%; however, wall temperatures were very high due to onset of partial film boiling.

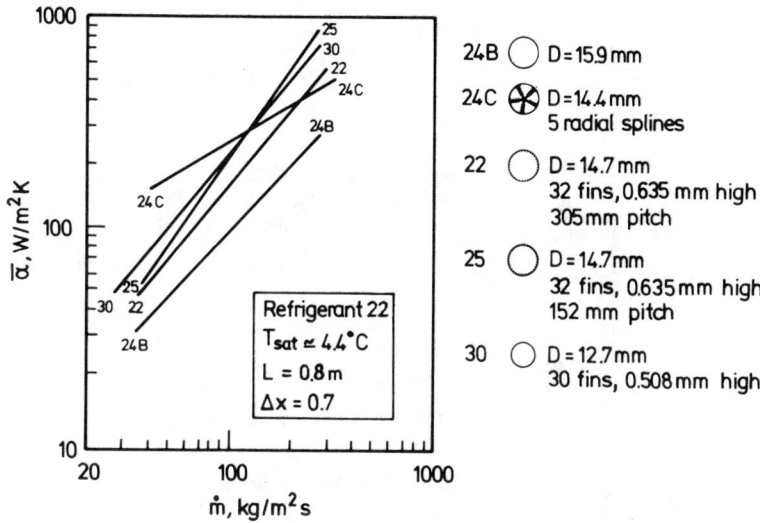

Fig. 4. Heat Transfer Coefficients for Evaporation in Internally Finned Tubes [50].

### 3.4 Swirl Flow Devices

A variety of devices have been proposed to augment flow boiling by imparting a swirling of secondary motion to the flow. Inlet vortex generators of the spiral-ramp or tangential-slot variety have been used to accommodate very large heat fluxes for subcooled flow boiling of water. One of the highest fluxes on record, $\dot{q}_c$ = 1.73 x $10^8$ W/m$^2$, has been obtained with this technique by Gambill and Greene [56]. Inlet swirl is effective in increasing $\dot{q}_c$ for subcooled boiling of water in a tube (Mayinger et al. [57]) or in an annulus with the inner tube heated (Ornatskiy et al. [58]).

Twisted tapes are quite popular due to their simplicity and adaptability to existing heat exchange equipment. They are ideal for hotspot applications since a short tape can cure the thermal problem while having little effect on the overall pressure drop. Boiling curves for subcooled boiling with twisted tapes are similar to those with empty tubes (Lopina and Bergles [59]); however, $\dot{q}_c$ can be increased by up to 100% (Gambill et al. [60]), as shown in Fig. 5. Loose-fitting tape inserts have been used by Sephton [61] with tubes intended for vertical tube evaporators for seawater desalination. These inserts are also effective for once-through vaporization of cryogenic fluids (Bergles et al. [62]) or steam (Cumo et al. [63], Hunsbedt and Roberts [64]) since all two-phase regimes are beneficially affected.

### 3.5 Coiled Tubes

Coiled tube vapor generators have advantages in terms of packing and generally higher heat transfer performance. As indicated in the recent survey of Jensen [65], the augmentation of boiling is very sensitive to geometrical and flow conditions. Modest improvements in $\alpha$ (circumferential average) for forced convection vaporization are obtained, with the improvement increasing as coil

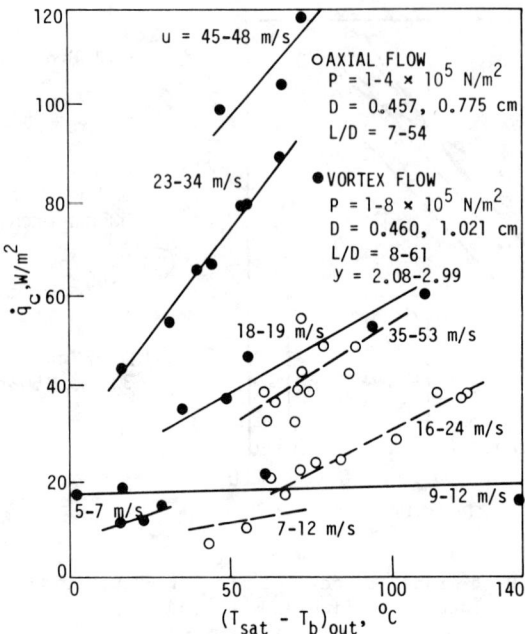

Fig. 5. Influence of Twisted-tape Inserts on Critical Heat Flux for Subcooled Boiling of Water (Gambill et al. [60]).

diameter is decreased. In the subcooled region, $\dot{q}_c$ is lower than for a comparable straight tube; however, $\dot{q}_c$ or $x_c$ is usually substantially higher than the straight tube value at outlet qualities of 0.2 and higher. This was recently demonstrated by Jensen [65] as shown in Fig. 6. The post dryout $\alpha$ is also increased with helical coils.

### 3.6 Additives for Liquids

Trace additives are not particularly effective in subcooled nucleate boiling; $\Delta T_s$ is reduced with some additives, but increased for others [33]. On the other hand, the overall $\alpha$ was doubled when a surfactant was added to seawater evaporating in a vertical (upflow) tube (Sephton [66]).

### 3.7 Surface Vibration

Few studies have been reported on the effect of surface vibration on flow-boiling heat transfer. Nucleate boiling coefficients for low velocity flow of water were improved somewhat at low heat flux but not at high heat flux [33]. Pearce [67] found insignificant changes in critical heat flux when a boiler tube was vibrated transversely. These results are expected, as fully-developed boiling is generally not affected by variations in flow velocity or other convective disturbances.

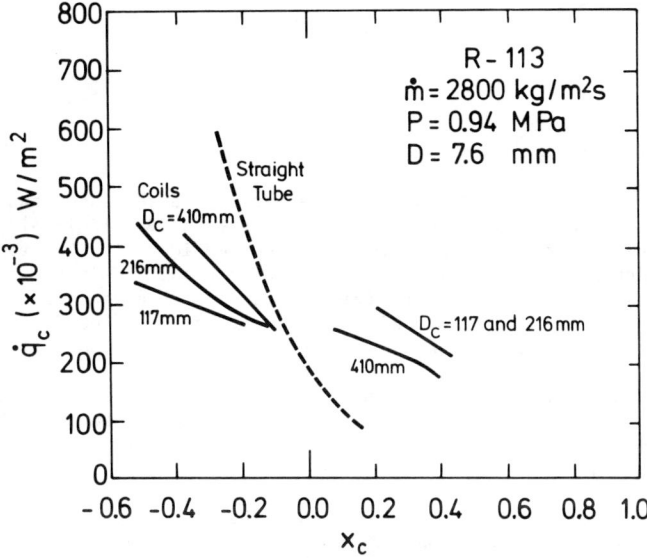

Fig. 6. Critical Heat Flux for Helical Coils of Various Diameter Compared with Straight Tube [65].

3.8 Fluid Vibration

In channel flow, it is usually necessary to locate the transducer some distance upstream or downstream of the test channel; as a consequence, the sound field is greatly attenuated. The several studies conducted in this area conclude that fluid vibrations, sonic or ultrasonic, do not appreciably affect flow boiling [33].

3.9 Electrostatic Fields

Data are available for flow in an annulus (inner wall heated) with high voltages imposed across the annular gap. Durfee et al. [68] reported that nucleate boiling, critical heat flux, and film boiling were all improved with isopropanol and deionized water in forced flow. The electrical field (3000 vdc) increased the critical heat flux by over 100% and provided proportional increases in the transition boiling region. An operating cost comparison, which included pumping power, showed that the electro-hydrodynamic system was somewhat superior to a conventional system.

3.10 Compound Techniques

Several studies of combined enhancement techniques have been reported. Sephton [61, 66] found that overall coefficients could be doubled by the addition of a surfactant to seawater evaporating in spirally corrugated or doubly fluted tubes (vertical upflow). Van der Most et al. [69] found only slight improvements with a surfactant additive for falling film evaporation in spirally corrugated tubes.

## 4. VAPOR SPACE CONDENSATION

### 4.1 Treated Surfaces

This first section on condensation augmentation considers condensation under conditions of low velocity so that shear forces are negligible. Surface treatment has been extensively investigated in connection with the promotion of dropwise condensation. Dropwise condensation is highly desirable since heat transfer coefficients for steam, for example, are as much as an order of magnitude higher than those for film condensation. It has been found that promoters are washed off condenser tubes in a relatively short period of time under normal industrial conditions. As a consequence, it is necessary periodically to inject the promoter into the steam passing to the condenser. The promoter then deposits on the tubes, forming the desired hydrophobic coating. "Permanent" coatings of noble metals or Teflon overcome this difficulty but at a substantial increase in cost.

Dropwise condensation will not be discussed further here since efforts to promote the phenomenon can be thought of as efforts to bring about the process rather than to augment it. A recent survey by Tanasawa [70] can be consulted for general background. It should be noted that the only real application is for steam condensers, as nonwetting substances are not available for most other working fluids. For example, no dropwise condensation promoters ("Freon-phobic") have been found for refrigerants (Iltscheff [71]). The augmentation of dropwise condensation, beyond inducing the process by selection of an effective, durable promoter, is fruitless since the heat transfer coefficients are already so high.

Glicksman et al. [72] showed that average coefficients for film condensation of steam on horizontal tubes can be improved up to 20% by strategically placing strips of Teflon or other nonwetting material around the tube circumference.

### 4.2 Rough Surfaces

Rough surfaces augment condensation primarily by introducing turbulence in the film. Increased surface area is a factor; however, the surface extensions usually do not greatly increase the area. An example of work in this area is the study of Nicol and Medwell [73]. At low film rates and at high flow rates (or film thickness), the roughness elements had no effect as they were buried within the laminar sublayer. The best performing roughness doubled the heat transfer coefficient for steam condensing outside a vertical tube.

### 4.3 Extended Surfaces

Surface extensions can be divided into the following types: loosely attached wires, contoured surfaces, and integral fins. Surface-tension-driven flows and condensate drainage are important considerations for all these types. Thomas [74] demonstrated that vertical surfaces equipped with vertical wires improve coefficients by as much as 800%. The condensate is drawn to the wires so that effective thin-film condensation prevails over much of the surface. The augmentation is limited by flooding.

Condenser tubes may be shaped, as shown in Fig. 7, to exploit the Gregorig effect, whereby condensation occurs primarily at the tops of convex ridges. Surface tension forces then pull the condensate into the concave grooves where it runs off. The resulting average heat transfer coefficient is substantially greater than that for a uniform film thickness. Several analyses have been proposed; a recent analysis for the optimum two-dimensional Gregorig surface is presented by Webb [76].

A unique double-grooved surface was developed by Markowitz et al. [77] for horizontal plate (facing down) condensers. Condensing coefficients (based on nominal surface area) for Freon 113 were improved by almost 100%.

Horizontal tubes with eliptical shape are superior to circular tubes, as shown by the analysis of Moalem and Sideman [78]. Horizontal tubes may also be contoured, and "roped" or "corrugated" tubes are available for desalination or steam condenser applications. More commonly, low integral circular fin tubes, made by many manufacturers, are used in shell-and-tube heat exchangers. The paper by Katz et al. [79] is frequently referred to for design. Average condensing coefficients (based on total area) for finned tubes are higher than for plain tubes of similar diameter.

Three-dimensional extended surfaces have been proposed for horizontal tube condensers, e.g., pin-finned tubes (Chandran and Watson, [80]). The surface described by Nakayama et al. [71] is shown in Fig. 8. The performance is compared to plain and finned horizontal tubes in Fig. 9. Webb and Gee [81] recently presented an analytical basis for the superior performance of condensing surfaces with three-dimensional fins.

4.4 Enhanced Surfaces

A variety of enhanced heat transfer surfaces have been developed for desalination systems. In the case of vertical tube evaporators (VTE), the outer wall is shaped to augment the condensing side according to the Gregorig effect noted previously; the inner surface is also distorted so that the evaporation coefficient is also enhanced. In the case of horizontal tube evaporators, condensing occurs on the inside of the tube and evaporation on the outside. Horizontal tube condensers (external) are also required for multistage flash

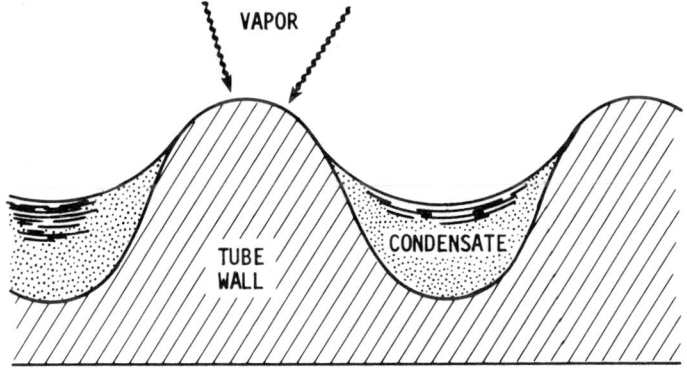

Fig. 7. Profile of the Condensing Surface Developed by Gregorig [75].

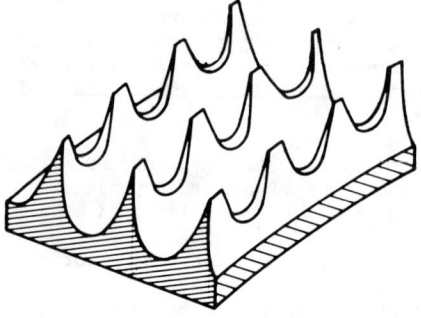

Fig. 8. Formed Condensing Surface "Thermoexcel-C" (Nakayama et al. [11]).

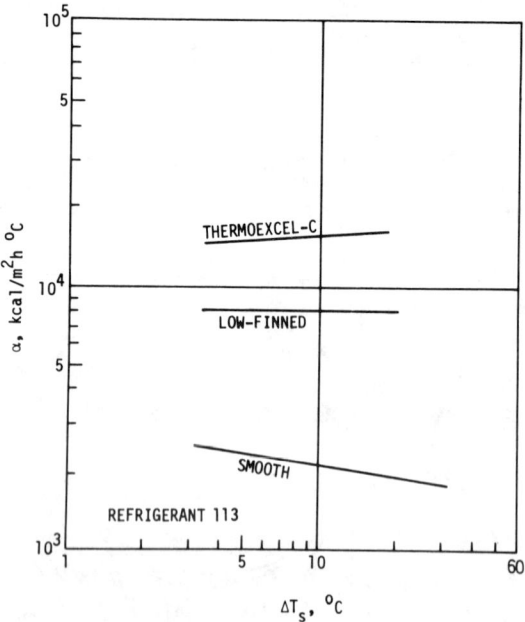

Fig. 9. Relative Performance of Condenser Tubes According to Nakayama et al. [11].

units. One of the major VTE studies has been reported by Alexander and Hoffman [82]. As shown in Fig. 10, the best commercial surfaces exhibit increases in overall coefficients of 200%.

## 4.5 Mechanical Aids

A number of studies have considered evaporation on rotating cylinders, disks, and square tubes, with the heat transfer coefficient usually being several hundred percent above the stationary case for water and organic fluids, e.g., Nicol and Gacesa [83] and Chandra et al. [84].

## 4.6 Surface Vibration

Recent studies in this area include those of Dent [85] and Brodov et al. [86] who obtained maximum increases of 10-15% by vibrating a horizontal condenser tube.

## 4.7 Electrostatic Fields

Electrical fields have been used to substantially improve condensation of Freon (Seth and Lee [87]). The fluid agitates and even destabilizes the condensate film.

## 4.8 Suction

The typical studies of laminar film condensation by Antonir and Tamir [88] and Lienhard and Dhir [89] indicate that heat transfer coefficients can be improved by as much as several hundred percent when the film thickness is reduced by suction. This is expected as the thickness of the condensate layer is the main parameter affecting the heat transfer rate in film condensation.

## 4.9 Compound Techniques

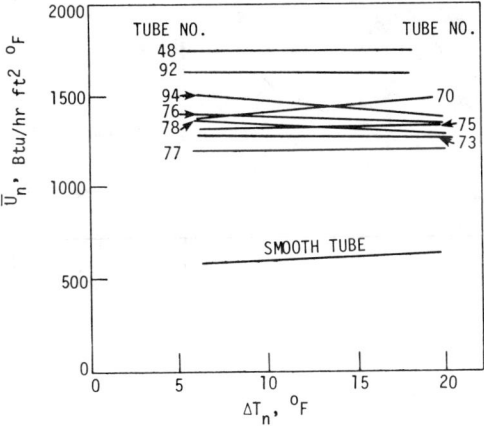

Fig. 10. Heat Transfer Enhancement in Vertical Tube Evaporators by Use of Shaped Tubes [82].

Several examples of combined techniques as used with vapor space condensation are rotating finned tubes (Chandran and Watson [80]), rotating rough disc, (Bromley et al. [42]), and rotating disc with suction (Chary and Sarma [90]). Moderate increases in condensing coefficient are reported.

## 5. FORCED-CONVECTION CONDENSATION

### 5.1 Rough Surfaces

Tubes having rough interior surfaces have been installed in horizontal tube evaporators where the brine evaporates on the outside of the condensing-steam-heated tubes. Internally grooved and knurled tubes were tested by Cox et al. [91]. Substantial increases in condensing performance were observed for several of the configurations. Optimization of the roughness appears to be necessary. Luu and Bergles [92] have found that tubes with repeated-rib roughness increase condensing coefficients for R-113 by over 100%.

### 5.2 Extended Surfaces

Vrable et al. [93] studied horizontal in-tube condensation of R-12 in internally-finned tubes. The heat transfer coefficient based on nominal surface area was increased by about 200%. Reisbig [94] also condensed R-12 (with some oil present) in internally finned tubes having an increased area of up to 175%. The nominal heat transfer coefficient was increased by up to 300%. Royal and Bergles [95,96] presented heat transfer and pressure drop data, with correlations, for condensation of steam inside tubes with straight or spiralled fins. Up to 150% increases in average coefficients were observed for complete condensation (see Fig. 11: Tubes D, E, F, and G). Similar improvements were obtained by Luu and Bergles [92] with the same tubes for R-113.

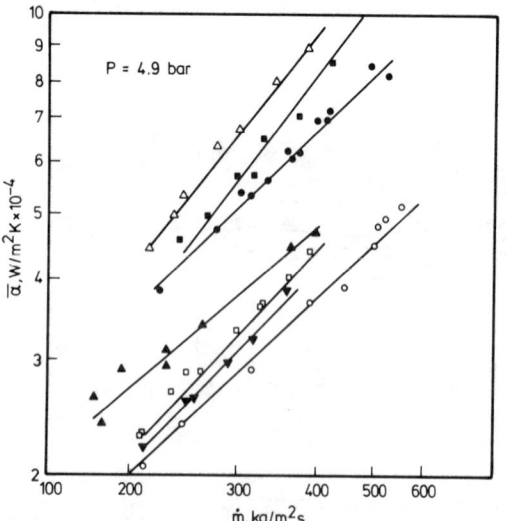

Fig. 11. Enhancement of In-Tube Condensation by Use of Internally Finned Tubes and Tubes with Twisted-Tape Inserts (Royal and Bergles [95]).

## 5.3 Displaced Enhancement Devices

Azer et al. [97] reported data for condensation in tubes having standard static mixer inserts. These inserts qualify as a displaced enhancement device due to bulk mixing of the flow by the elements which have right-and left-hand helices. Substantial improvements in heat transfer coefficients were reported; however, the increases in pressure drop were very large.

## 5.4 Swirl Flow Devices

Royal and Bergles [95,96] found that twisted-tape inserts improved in-tube condensation of water by 30% (Fig. 10: Tubes B and C); however, the pressure drop was also high. Luu and Bergles [92] report similar results for R-113.

Shklover and Gerasimov [98] used an interesting baffling technique to create a spiralling motion in the vapor condensing outside a tube bundle. With vapor velocities approaching sonic velocities, the condensing coefficients were high; however, no base data were included for comparison.

## 5.5 Coiled Tubes

Condensation in coiled tubes was studied by Miropolskii and Kurbanmukhamedov [99], and condensation in tube bends was studied by Traviss and Rohsenow [100]. In both cases, modest increases in condensation rates were observed relative to straight tubes.

## 5.6 Mechanical Aids

Weiler et al. [101] subjected a tube bundle, with nitrogen condensing inside the tubes, to high accelerations (essentially normal to the tube axis). At 325 g's, the overall heat transfer coefficient was increased by a factor of over 4.

## 5.7 Fluid Vibration

Mathewson and Smith [102] investigated the effects of acoustic vibrations on condensation of isopropanol vapor flowing downward in a vertical tube. A siren was used to generate an acoustic field of up to 186 db at frequencies ranging from 50-330 Hz. The maximum improvement in condensing coefficient was found to be about 60% at low vapor flow rates.

## 5.8 Electrostatic Fields

Choi [103] reported data for condensation of Freon-113 on the outside wall of an annulus in the presence of a radial electric field. With the maximum applied voltage of 30 kv, the average heat transfer coefficient was increased by 100%.

## 5.9 Compound Techniques

Weiler et al. [101] condensed nitrogen inside rotating tubes which were treated with a porous coating which increased coefficients above those for a rotating smooth tube.

## 6. CONCLUSIONS

Many techniques are available for augmentation of boiling and condensing, in both free convection and forced convection. Many laboratory studies have established conditions under which heat transfer can be improved. Few analyses are available owing to the difficulty of predicting the basic phase-change processes themselves. Current activity focuses on passive techniques such as structured surfaces for enhancement of nucleate pool boiling and thin film evaporation. There is much interest in commercialization, as will be documented in the following chapter.

## NOMENCLATURE

$D$ = inside diameter of tube
$D_c$ = coil diameter
$L$ = heat transfer length
$\dot{m}$ = mass velocity
$P$ = pressure
$\dot{q}$ = heat flux
$\dot{q}_c$ = critical heat flux
$r$ = radius of vapor-liquid interface
$T_b$ = mixed-mean fluid temperature
$T_{sat}$ = saturation temperature
$\Delta T_n$ = shell-side to tube-side exit temperature difference
$\Delta T_s$ = wall-minus-saturation temperature difference
$\bar{U}_n$ = average overall heat transfer coefficient based on nominal tube outside diameter
$u$ = velocity
$x_c$ = quality at critical heat flux location
$\Delta x$ = quality change during vaporization
$y$ = tube diameters per $180°$ tape twist

### Greek

$\alpha$ = heat transfer coefficient
$\bar{\alpha}$ = average heat transfer coefficient
$\beta$ = contact angle

## REFERENCES

1. Jakob, M. and Fritz, W., "Versuche ueber den Verdampfungsvorgang", Forschung auf dem Gebiete des Ingenieurwesens, Vol. 2, pp. 435-447, 1931.

2. Alberger Heater Company, Buffalo, New York, Catalog No. 3, 1921.

3. Bergles, A. E. et al., "Bibliography on Augmentation of Convective Heat and Mass Transfer", HTL-19, ISU-ERI-Ames-79206, Iowa State University, 1979.

4. Young, R. K., and Hummel, R. L., "Improved Nucleate Boiling Heat Transfer", Chem. Engng. Prog., Vol. 60, No. 7, pp. 53-58, 1964.

5. Bergles, A. E., Bakhru, N., and Shires, J. W., "Cooling of High-Power-Density Computer Components", EPL Report 70712-60, Massachusetts Institute of Technology, 1968.

6. Zhukov, V. M. et al., "Heat Transfer in Boiling of Liquids on Surfaces Coated with Low Thermal Conductivity Films", Heat Transfer-Soviet Research, Vol. 7, 1975.

7. Bergles, A. E. and Thompson, W. G., Jr., "The Relationship of Quench Data to Steady-State Pool Boiling Data", Int. J. Heat Mass Transfer, Vol. 13, pp. 55-68, 1970.

8. Kun, L. C. and Czikk, A. M., "Surface for Boiling Liquids", U.S. Patent No. 3,454,081, July 8, 1969.

9. Webb, R. L., "Heat Transfer Surface Having a High Boiling Heat Transfer Coefficient", U.S. Patent 3,696,861, October 10, 1972.

10. Zatell, V. A., "Method of Modifying a Finned Tube for Boiling Enhancement", U.S. Patent No. 3,768,290, October 30, 1973.

11. Nakayama, W. et al., "High-flux Heat Transfer Surface 'Thermoexcel'", Hitachi Review, Vol. 24, No. 8, pp. 329-333, 1975.

12. Wieland-Werke AG, P. O. Box 4420, Ulm, West Germany, "High-Performance Tubes for Flooded Evaporators", Wieland Product Information SAW-15e-06.78.

13. Ragi, E., "Composite Structure for Boiling Liquids and Its Formation", U.S. Patent No. 3,684,007, August 15, 1972.

14. Marto, P. J., and Rohsenow, W. M., "Effects of Surface Conditions on Nucleate Pool Boiling of Sodium", J. Heat Transfer, Vol. 88, pp. 196-204, 1966.

15. O'Neill, P. S., Gottzmann, C. F. and Terbot, C. F., "Novel Heat Exchanger Increases Cascade Cycle Efficiency for Natural Gas Liquefaction", Advances in Cryogenic Engineering, Vol. 17, pp. 421-437, 1972.

16. Danilova, G. N. et al., "Enhancement of Heat Transfer During Boiling of Liquid Refrigerants at Low Heat Fluxes", Heat Transfer-Sov. Res., Vol. 8, No. 4, pp. 1-8, 1976.

17. Oktay, S, and Schmeckenbecher, A. F., "Preparation and Performance of Dendritic Heat Sinks", J. Electrochemical Soc., Vol. 21, pp 912-918, 1974.

18. Dahl, P. J., and Erb, L. D., "Liquid Heat Exchanger Interface Method", U. S. Patent 3,990,862, November 9, 1976.

19. Fujii, M., Nishiyama, E., and Yamanaka, G., "Nucleate Pool Boiling Heat Transfer from Micro-Porous Heating Surface", in *Advances in Enhanced Heat Transfer*, ASME, pp. 45-51, 1979.

20. Janowski, K. R., Shum, M. S., and Bradley, S. A., "Heat Transfer Surface", U. S. Patent 4,129,181, December 12, 1978.

21. Czikk, A. M., and O'Neill, P. S., "Correlation of Nucleate Boiling from Porous Metal Films", in *Advances in Enhanced Heat Transfer*, ASME, pp. 53-60, 1979.

22. Nakayama, W., et al., "Dynamic Model of Enhanced Boiling Heat Transfer on Porous Surfaces", in *Advances in Enhanced Heat Transfer*, ASME, pp. 31-43, 1979.

23. Yilmaz, S., Hwalek, J.J., and Westwater, J.W., "Pool Boiling Heat Transfer Performance for Commercial Enhanced Tube Surfaces", ASME Paper No. 80-HT-41, presented at 19th National Heat Transfer Conference, July 1980.

24. Chyu, M. C., Institut fuer Verfahrenstechnik, Universitaet Hannover, Hannover, West Germany, personal communication, June 1980.

25. Gorenflo, D., "Zum Waermeuebergang bei Blasenverdampfung an Rippenrohren", Dissertation, Technische Hochschule, Karlsruhe, 1966.

26. Hesse, G., "Heat Transfer in Nucleate Boiling, Maximum Heat Flux and Transition Boiling", *Int. J. Heat Mass Transfer*, Vol. 16, pp. 1611-1627, 1973.

27. Westwater, J. W., "Development of Extended Surfaces for Use in Boiling Liquids", *AIChE Symp. Ser.*, Vol. 69, No. 131, pp. 1-9, 1973.

28. Corman, J. C. and McLaughlin, M. H., "Boiling Heat Transfer with Structured Surfaces", *ASHRAE Transactions*, Vol. 82, Part 1, pp. 906-918, 1976.

29. Sideman, S. and Moalem-Maron, D., "Transport Characteristics of Thin Films: Evaporation, Condensation and Mass Transfer on Smooth and Grooved Horizontal Conduits", in *Two-Phase Momentum, Heat and Mass Transfer in Chemical, Process and Energy Engineering Systems*, Vol. 2, Hemisphere, Washington, pp. 877-896, 1979.

30. Lienhard, J. H., Dhir, V. K., and Riherd, D. M., "Peak Pool Boiling Heat-Flux Measurements on Finite Horizontal Flat Plates", *J. Heat Transfer*, Vol. 95, pp. 477-482, 1973.

31. Allingham, W. D. and McEntire, J. A., "Determination of Boiling Film Coefficient for a Heated Horizontal Tube in Water-Saturated Wick Material", *J. Heat Transfer*, Vol. 83, pp. 71-76, 1961.

32. Costello, C. P. and Redeker, E. R., "Boiling Heat Transfer and Maximum Heat Flux for a Surface with Coolant Supplied by Capillary Wicking", *Chemical Engineering Progress Symposium Series*, Vol. 59, No. 41, pp. 104-113, 1963.

33. Bergles, A. E., "Survey and Evaluation of Techniques to Augment Convective Heat and Mass Transfer", in *Progress in Heat And Mass Transfer*, U. Grigull

and E. Hahne (Editors), Vol. 1, pp. 331-334, Pergamon, Oxford, 1969.

34. Van Stralen, S. J. D., "Nucleate Boiling in Binary Systems", in Augmentation of Convective Heat and Mass Transfer, ASME, New York, pp. 133-147, 1970.

35. Tang, S. I. and McDonald, T. W., "A Study of Heat Transfer from a Rotating Horizontal Cylinder", Int. J. Heat Mass Transfer, Vol. 14, pp. 1643-1658, 1971.

36. Marto, P. J. and Gray, V. H., "Effects of High Accelerations and Heat Fluxes on Nucleate Boiling of Water in an Axisymmetric Rotating Boiler", NASA TN D-6307, 1971.

37. Tleimat, B. W., "Performance of a Rotating Flat-Disk Wipid-Film Evaporator", ASME Paper No. 71-HT-37, 1971.

38. Fuls, G. M. and Geiger, G. E., "Effect of Bubble Stabilization on Pool Boiling Heat Transfer", J. Heat Transfer, Vol. 97, pp. 635-640, 1970.

39. Loosle, D. G. and Holdredge, R. M., "The Effects of Acoustic Vibrations on the Cooldown Time of Bodies in Cryogenic Liquids", Heat Transfer 1970, Vol. IV, Paper NC4.6, Elsevier, Amsterdam, 1970.

40. Basu, D. K, "Effects of Electric Field on Boiling Hysteresis in Carbon Tetracloride", Int. J. Heat Mass Transfer, Vol. 16, pp. 1322-1324, 1973.

41. Raiff, R. J. and Wayner, P. C., Jr., "Evaporation from a Porous Flow Control Element on a Porous Heat Source", Int. J. Heat Mass Transfer, Vol. 16, pp. 1919-1930, 1973.

42. Bromley, C. A, Humphreys, R. F., and Murray, W., "Condensation on and Evaporation from Radially Grooved Rotating Disks", J. Heat Transfer, Vol. 88, pp. 80-93, 1966.

43. Bergles, A. E., Bakkru, N., and Shires, J. W., Jr., "Cooling of High-Power-Density Computer Components", EPL Report 70712-60, Massachusetts Institute of Technology, 1968.

44. Murphy, R. W. and Bergles, A. E., "Subcooled Flow Boiling of Fluorocarbons - Hysteresis and Dissolved Gas Effects on Heat Transfer", Proceedings of the 1972 Heat Transfer and Fluid Mechanics Institute, Stanford University Press, pp. 400-416, 1972.

45. O'Neill, P. S., Linde Division of Union Carbide Corp., Personal Communication, June 9, 1976.

46. Burck, E., Hufschmidt, W., and DeClercq, E., "Einfluss Kuenstlicher Rauigkeiten auf die Kritische Waermestromdichte bei Erzwunger Konvektion", Atomkernenergie, Vol 14, pp. 305-308, 1969.

47. Withers, J. G. and Habdas, E. P., "Heat Transfer Characteristics of Helical Corrugated Tubes for Intube Boiling of Refrigerant R-12", AIChE Symposium Series, Vol. 70, No. 138, pp. 98-106, 1974.

48. Ackerman, J. W., "Pseudoboiling Heat Transfer to Supercritical Pressure Water in Smooth and Ribbed Tubes", J. Heat Transfer, Vol. 92, pp. 490-498, 1970.

49. Schluender, E. U., and Chwala, M., "Oertlicher Waermeuebergang and Druck-abfall bei der Stroemung Verdampfender Kaeltemittel in Innenberippten, Waggerechten Rohren", Kaeltetechnik-Klimatisierung, Vol. 21, No. 5, 136-139, 1969.

50. Kubanek, G. R., and Miletti, D. L., "Evaporative Heat Transfer and Pressure Drop Performance of Internally-Finned Tubes with Refrigerant 22", J. Heat Transfer, Vol. 101, 447-452, 1979.

51. Johnson, B. M., Jansen, G., and Otwzarski, P. C., "Enhanced Evaporating Film Heat Transfer from Corrugated Surfaces", ASME Paper No. 71-HT-33, 1971.

52. Thomas, D. G., and Young, G., "Thin Film Evaporation Enhancement by Finned Surfaces", Ind. Eng. Chem. Proc. Des. Dev., Vol. 9, pp. 317-323, 1970.

53. Tong, L. S. et al., "Critical Heat Flux of a Heater Rod in the Center of Smooth and Rough Square Sleeves, and in Line-Contact with an Unheated Wall", ASME Paper No. 67-WA/HT-29, 1967.

54. Ryabov, A. N. et al., "Boiling Crisis and Pressure Drop in Rod Bundles with Heat Transfer Enhancement Devices", Heat Transfer - Soviet Research, Vol. 9, No. 1, pp. 112-122, 1977.

55. Megerlin, F. E., Murphy, R. W. and Bergles, A. E., "Augmentation of Heat Transfer in Tubes by Means of Mesh and Brush Inserts", J. Heat Transfer, Vol. 96, pp. 145-151, 1974.

56. Gambill, W. R., and Greene, N. D., "A Preliminary Study of Boiling Burnout Heat Fluxes for Water in Vortex Flow", Chem. Eng. Prog., Vol. 54, No. 10, pp. 68-76, 1958.

57. Mayinger, F., Schad, O., and Weiss, E., "Investigations into the Critical Heat Flux in Boiling", MAN Report No. 09.03.01, 1966.

58. Ornatskiy, A. P., et al. "A Study of the Heat Transfer Crisis with Swirled Flows Entering an Annular Passage", Heat Transfer - Sov. Res., Vol. 5, No. 4, pp. 7-10, 1973.

59. Lopina, R. F., and Bergles, A. E., "Subcooled Boiling of Water in Tape-Generated Swirl Flow", J. Heat Transfer, Vol. 95, pp. 281-283, 1973.

60. Gambill, W. R., Bundy, R. D., and Wansbrough, R. W., "Heat Transfer, Burnout, and Pressure Drop for Water in Swirl Flow Tubes with Internal Twisted Tapes", Chem. Eng. Prog. Symp. Ser., Vol. 57, No. 32, pp. 127-137, 1961.

61. Sephton, H. H., "Interface Enhancement for Vertical Tube Evaporator: A Novel Way of Substantially Augmenting Heat and Mass Transfer", ASME Paper No. 71-HT-38 , 1971.

62. Bergles, A. E., Fuller, W. D., and Hynek, S. J., "Dispersed Film Boiling of Nitrogen with Swirl Flow", Int. J. Heat Mass Transfer, Vol 14, pp. 1343-1354, 1971.

63. Cumo, M., et al., "The Influence of Twisted Tapes in Subcritical, Once-Through Vapor Generator in Counter Flow", J. Heat Transfer, Vol. 96, pp. 365-370, 1974.

64. Hunsbedt, A., and Roberts, J. M., "Thermal-Hydraulic Performance of a 2MWT Sodium Heated, Forced Recirculation Steam Generator Model", J. Eng. Power, Vol. 96, pp. 66-76, 1974.

65. Jensen, M. K., and Bergles, A.E., "Boiling Heat Transfer and Critical Heat Flux in Helical Coils", ASME Paper No. 80-ht-49, 1980.

66. Sephton, H. H., "Upflow Vertical Tube Evaporation of Sea Water with Interface Enhancement; Process Development by Pilot Plant Testing", Desalination, Vol. 16, pp. 1-13, 1975.

67. Pearce, H. R., "The Effect of Vibration on Burnout in Vertical, Two-Phase Flow", AERE-R 6375, 1970.

68. Durfee, R. L. et al., "Boiling Heat Transfer of Electric Field (EHD)", Atomic Energy Commission Report NYO-24-04-76, 1966.

69. Van der Mast, V. C., Read, S. M., and Bromley, L. A., "Boiling of Natural Sea Water in Falling Film Evaporators", Desalination, Vol. 18, pp. 71-94, 1976.

70. Tanasawa, I., "Dropwise Condensation - the Way to Practical Applications", Heat Transfer 1978, Proceedings of the Sixth International Heat Transfer Conference, Vol. 6, Hemisphere, pp. 393-405, 1978.

71. Iltscheff, S., "Ueber Einige Versuche zur Erzielung von Tropfkondensation mit Fluorierten Kaéltemitteln, Kaeltetechnik-Klimatisierung, Vol. 23, pp. 237-241, 1971.

72. Glicksman, L. R., Mikic, B. B., and Snow, D. F., "Augmentation of Film Condensation on the Outside of Horizontal Tubes", American Institute of Chemical Engineers Journal, Vol. 19, pp. 636-637, 1973.

73. Nicol, A. A., and Medwell, J. O., "The Effect of Surface Roughness on Condensing Steam", Canadian Journal of Chemical Engineering, pp. 170-173, 1966.

74. Thomas, D. G., "Enhancement of Film Condensation Rate on Vertical Tubes by Longitudinal Fins", American Institute of Chemical Engineers Journal, Vol. 14, pp. 644-649, 1968.

75. Gregorig, R., "Hautkondensation an Feingewellten Oberflaechen bei Beruecksichtigung der Oberflaechenspannungen", Zeitschrift fuer Angewandte Mathematik und Physik, Vol. 5, pp. 36-49, 1954.

76. Webb, R. L., "A Generalized Procedure for the Design and Optimization of Fluted Gregoric Condensing Surfaces", Proceedings of the Fifth Ocean Thermal Energy Conversion Conference, Vol. 3, VI-123 - VI-145, 1978.

77. Markowitz, A., Mikic, B. B., and Bergles, A. E., "Condensation on a Downward Facing, Horizontal Rippled Surface", J. Heat Transfer, Vol. 94, pp. 315-320, 1972.

78. Moalem, D., Sideman, S., "Theoretical Analysis of a Horizontal Condenser-Evaporator Elliptical Tube", Journal of Heat Transfer, Vol. 19, pp. 259-270, 1976.

79. Katz, D. L., Young, E. H., and Balekjian, G., "Condensing Vapors on Finned

tubes", *Petroleum Refiner*, Vol. 33, No. 11, pp. 175-178, 1954.

80. Chandran, R., and Watson, F. A., "Condensation on Static and Rotating Pinned Tubes", *Transactions of the Institution of Chemical Engineers*, Vol. 54, pp. 65-72, 1976.

81. Webb, R. L., and Gee, D. L., "Analytical Predictions for a New Concept Spine-Fin Surface Geometry", *ASHRAE Trans.* Vol. 85, Pt. 2, pp. 274-283, 1979.

82. Alexander, L. G. and Hoffman, H. W., "Performance Characteristics of Corrugated Tubes for Vertical Tube Evaporators", ASME Paper No. 71-HT-30, 1971.

83. Nicol, A. A., and Gacesa, M., "Condensation of Steam on a Rotating Vertical Cylinder", *J. Heat Transfer*, Vol. 97, pp. 144-152, 1970.

84. Chandra, S., Houghton, A. V., and Castonguay, T. T., "The Condensation of Vapor on a Horizontal Rotating Square Tube", *Proceedings of the 1974 Heat Transfer and Fluid Mechanics Institute*, Stanford University Press, Stanford, pp. 21-37, 1974.

85. Dent, J. C., "On the Calculation of Heat Transfer for Condensation of Steam on a Vibrating Vertical Tube", *Int. J. Heat Mass Transfer*, Vol. 12, pp. 991-996, 1969.

86. Brodov, Y. M. et al., "The Effect of Vibration on Heat Transfer and Flow of Condensing Steam on a Single Tube", *Heat Transfer - Soviet Research*, Vol. 9, No. 1, pp. 152-156, 1977.

87. Seth, A. K. and Lee, L., "The Effect of an Electric Field on the Presence of Noncondensible Gas on Film Condensation Heat Transfer", *J. Heat Transfer*, Vol. 96, pp. 257-258, 1974.

88. Antonir, I., and Tamir, A., "The Effect of Surface Suction on Condensation in the Presence of a Noncondensible Gas", *J. Heat Transfer*, Vol. 99, pp. 496-499, 1977.

89. Lienhard, J., and Dhir, V., "A Simple Analysis of Laminar Film Condensation with Suction", *J. Heat Transfer*, Vol 94, pp. 334-336, 1972.

90. Chary, S. P. and Sarma, P. K., "Condensation on a Rotating Disk with Constant Axial Suction", *J. Heat Transfer*, Vol. 98, pp. 682-684, 1976.

91. Cox, R. B., Matta, G. A., Pascale, A. S., and Stromberg, K. G., "Second Report on Horizontal-Tubes Multiple-Effect Process Filot Plant Tests and Design", Office of Saline Water Research and Development Progress Report 592, 1970.

92. Luu, M., and Bergles, A. E., "Augmentation of In-Tube Condensation of R-113", HTL-23, ISU-ERI- Ames-80175, Iowa State University, 1980.

93. Vrable, D. L., Yang, W. J., and Clark, J. A., "Condensation of Refrigerant -12 Inside Horizontal Tubes with Internal Axial Fins", *Heat Transfer 1974*, Vol. III, The Japan Society of Mechanical Engineers, pp. 250-254, 1974.

94. Reisbig, R. L., "Condensing Heat Transfer Augmentation Inside Splined Tubes", ASME Paper No. 74-HT-7 presented at AIAA/ASME Thermophysics and

Heat Transfer Conference, Boston, July, 1974.

95. Royal, J. H., and Bergles, A. E., "Augmentation of Horizontal In-Tube Condensation by Means of Twisted-Tape Inserts and Internally-Finned Tubes", Journal of Heat Transfer, Vol. 100, pp. 17-24, 1978.

96. Royal, J. H., and Bergles A. E., "Pressure Drop and Performance Evaluation of Augmented In-Tube Condensation", Heat Transfer 1978, Vol. 2, Hemisphere, Washington, pp. 459-464, 1978.

97. Azer, N. Z., Fan, L. T., and Lin, S. T., "Augmentation of Condensation Heat Transfer with In-Line Static Mixers", Proceedings of the 1976 Heat Transfer and Fluid Mechanics Institute, Stanford University Press, Stanford, pp. 512-526, 1976.

98. Shklover, G. G., and Gerasimov, A. V., "Heat Transfer of Moving Steam in Coil-Type Heat Exchangers", Teploenergetika, Vol. 10, No. 5, pp. 62-65, 1963.

99. Miropolskii, Z. L., and Kurbanmukhamedov, A., "Heat Transfer with Condensation of Steam within Coils", Thermal Engineering, No. 5, pp. 111-114, 1975.

100. Traviss, D. P., and Rohsenow, W. M., "The Influence of Return Bends on the Downstream Pressure Drop and Condensation Heat Transfer in Tubes", ASHRAE Transactions, Vol. 79, Part 1, pp. 129-137, 1973.

101. Wieler, D. K., Czikk, A. M., and Paul, R. S., "Condensation in Smooth and Porous Coated Tubes Under Multi-g Accelerations", Chemical Engineering Progress Symposium Series, Vol. 62, No. 64, pp. 143-149, 1966.

102. Mathewson, W. F., and Smith, J. C., "Effect of Sonic Pulsation on Forced Convective Heat Transfer to Air and on Film Condensation of Isopropanol", Chemical Engineering Progress Symposium Series, Vol. 59, No. 41, pp. 173-179, 1963.

103. Choe, H. Y., "Electrohydrodynamic Condensation Heat Transfer", J. Heat Transfer, Vol. 90, pp. 98-102, 1968.

# Applications of Heat Transfer Augmentation

**ARTHUR E. BERGLES**
Department of Mechanical Engineering
Iowa State University
Ames, Iowa 50011 USA

ABSTRACT

   This lecture considers the factors which enter into the decision to use a heat transfer augmentation technique, either in the original equipment or as a retrofit. Additional resource material in the form of patents and commercial literature is mentioned. The suitability of available data is analyzed and thermal-hydraulic performance evaluations are outlined. Manufacturing and operational considerations are discussed in detail. Finally, examples of successful applications of augmentation technology are presented.


1. INTRODUCTION

   The two previous lectures on this subject discuss the many techniques which are available for augmenting single-phase or two-phase heat transfer. Many of these techniques have made the transition from the laboratory to full-scale industrial equipment. It is appropriate to view augmentation techniques as "second generation" heat transfer technology. Such new technology normally requires several phases of development for successful commercialization i.e., marketing and selling the product. These phases include:

   (1) Basic performance data for heat transfer and pressure drop, if applicable, must be obtained. General correlations must be developed to predict heat transfer and pressure drop as functions of the geometrical characteristics and other parameters.

   (2) Design methods must be developed to facilitate selection of the optimum parameters, e.g. surface geometry, for the various techniques and particular applications.

   (3) Manufacturing technology and cost of manufacture must be available for the desired technique and material.

   (4) Pilot plant tests of the proposed technique are required to confirm the design, establish long-term fouling and corrosion characteristics, and to permit a complete economic evaluation.

This lecture will discuss a number of considerations which enter into each of these phases and affect the decision to use an augmentation technique, either in the original equipment or as a retrofit.

2. SOURCES OF INFORMATION

In addition to the several thousand technical papers and reports on this subject [1], as noted in the first lecture, there exists a substantial body of patent literature. Even though this literature is generally less publicized, it is important because granted patents are generally indicative of industrial interest in commercialization of a technique. Patents can be used to identify variations of techniques now in commercial use or available for commercial use, to obtain descriptions of major manufacturing methods for current applications and to discover possible applications of techniques not now commercially applied. It must be recognized, however, that most patents stress legal rather than technical aspects. Of the approximately 500 patents which have been reviewed at Iowa State University and Pennsylvania State University only about 25% disclose actual heat transfer data [2]. It is interesting to note that over 60% of the patents listed are directed toward applications in single-phase forced convection, while about 85% of the patents involve the passive techniques.

Product information supplied by manufacturers is direct evidence of commercialization of augmentation techniques. The situation is very dynamic, as new products are continually being introduced while others are being discontinued. The major international marketing activity is observed in Germany, Japan, Sweden, United Kingdom, and the United States. Several hundred manufacturers are involved with products ranging from tube inserts to complete heat exchange systems which include augmented heat exchangers [3].

## 3. SUITABILITY OF BASIC PERFORMANCE DATA

Caution must always be exercised when extracting heat transfer and pressure drop data from the published literature and applying these data to a different set of conditions. For instance, the performance of special boiling surfaces for a given fluid and pressure level, can be quite sensitive to geometry, e.g., groove pitch and pore diameter in the case of Thermoexcel-E [4] and groove width in the case of GEWA-T [5]. Normal manufacturing tolerances are such that the actual dimensions may be quite far from the intended optimum dimension. The performance and optimum geometry are expected to be different as the fluid and operating pressure change. The situation can be even more serious than this, as illustrated by an example in the category of additives.

As shown in Fig. 1, large increases in critical heat flux (CHF) can be realized when a small percentage of 1-pentanol is added to water. However, when the data for the two studies with heaters of different diameter are compared, it is evident that the increase in CHF is much less with the larger heater. Furthermore, even with small-diameter heaters, a substantial <u>decrease</u> in CHF is observed in subcooled pool boiling and in CHF for forced convection subcooled boiling, except at very high subcooling [6].

The previous example basically concerns the problem of "scale-up" and "separate effects": How valid are small-scale, isolated laboratory test data when it comes to the design of large-scale, integrated heat exchange equipment? Usually, some adjustments to the available correlations will be required. Consider the typical finned coil used in a solar-assisted domestic water heater. Standard finned tubes (Fig. 2a) are wound to different shapes (Fig. 2b) and inserted in a storage tank. Warm water from the solar collector flows inside the coil heating the domestic water in the tank.

Fig. 3 presents a comparison of the heat transfer correlation for single horizontal finned tubes and heat transfer coefficients "backed out" from overall performance data for a horizontal coil. The fact that the actual coef-

APPLICATIONS OF HEAT TRANSFER AUGMENTATION 885

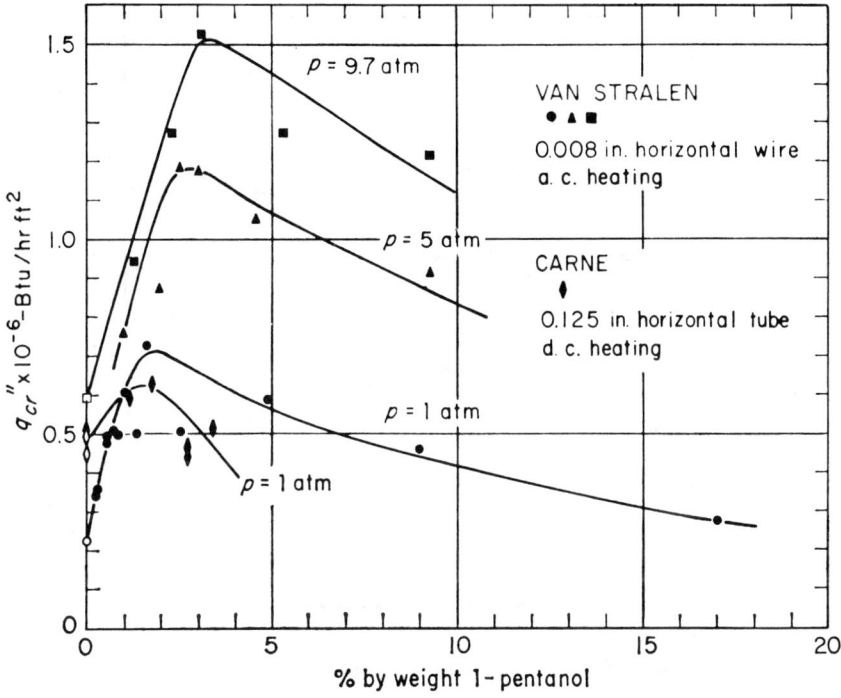

Fig. 1. Effect of Addition of 1-Pentanol on CHF for Saturated Pool Boiling of Water [6].

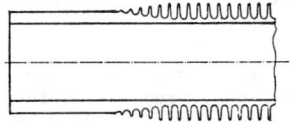

a. Tubing with Medium Height Integral Outer Fins.

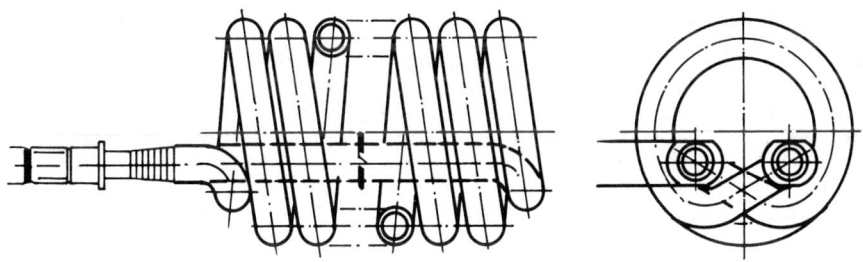

b. Heat Exchanger Coil for Solar Heating of Domestic Water (Wieland SW10 [8]).

Fig. 2. Finned Tubing Used for Water Heaters.

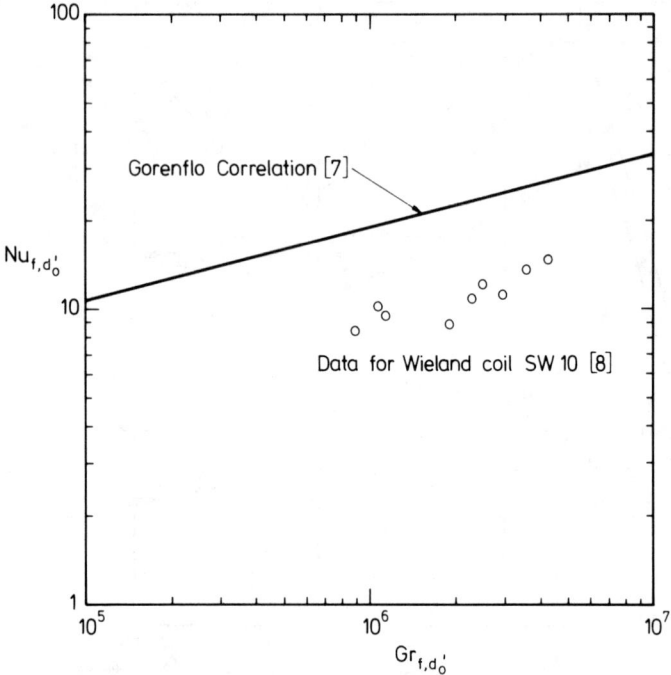

Fig. 3. Comparison of "Separate Effects" Correlation with Data Derived from Coil Performance.

ficients are lower is not surprising; but it is unfortunate that each coil configuration must be tested to get the necessary performance information.

In the area of boiling and condensing, data are often obtained for single tubes of the correct diameter with the intent of applying the data to a large bundle of tubes. Katz et al. [9] and Nakajima and Shiozawa [10], for example, found that coefficients on the upper finned tubes in a bundle are higher than coefficients on the lower tubes, due to bubble-enhanced circulation. Similar results are reported by Arai et al. [4] for a bundle of Thermoexcel tubes. It is quite probable that with certain types of augmented tubes it is sufficient to use the special tubes only in the lower rows since the augmented circulation in the upper rows is so high that augmentation is not effective there. The problem of vapor crossflow in large-scale falling film evaporators is discussed by Yung et al. [11]. The high vapor velocities usually present result in redistribution of the working fluid and incomplete wetting of the tubes. The redistribution is expected to be different for plain and augmented tubes. Several other lectures at this Institute consider this problem. For condensing in tube bundles, Withers and Young [12] found that the vertical row effect for corrugated tubes was different from that for bare tubes; in particular, the augmented tubes were less sensitive to the number of rows.

Scale-up problems also occur when applying electric fields, vibrations, and sound fields or pulsations. It is usually not too difficult to influence boiling or condensing from small test sections; however, industrial size heat exchangers with large surface areas and flow path lengths are less tractable.

For example, even though a patent has been granted on the use of acoustic vibrations in a shell-and-tube heat exchanger, by placing the transducer in a header [13], it is doubtful whether the vibrations would be felt in the tubes.

Finally, one should not be hesitant to critically examine the accuracy of the experimental data. Accuracy is particularly important with phase change processes since they are low thermal resistance precesses in general. Wall temperatures and fluid (or saturation) temperatures must be determined with precision if the temperature difference or the heat transfer coefficient is to be accurately established. The boiling curves for most special boiling surfaces have wall superheats of the order of 1 K. Such low $\Delta T_s$'s can only be accurately established when very accurate wall temperature and saturation temperature (pressure) measurements are made.

## 4. THERMAL-HYDRAULIC PERFORMANCE EVALUATIONS

### 4.1 General Considerations and Single-Phase Applications

For two-fluid heat exchangers, the ratio of thermal resistances between the two fluid streams is of primary importance in determining whether augmentation will be of benefit. Augmentation should be considered for the stream which has the larger thermal resistance. If the thermal resistances of both streams are approximately equal, augmentation of both streams may be considered.

Methodologies have been suggested for assessing the thermal-hydraulic performance of augmented surfaces in single-phase flow (Webb and Eckert, [14], Bergles et al., [15, 16]). The proposals contained in these papers are contained in Table 1. Here, consideration is given to the basic single-pass shell-and-tube heat exchanger configuration: N tubes of length L (or total area A) with a total flow rate w, and fixed inlet temperatures of both streams. The heat duty is q and pressure drop and pumping power are $\Delta p$ and P, respectively. The rather large number of performance criteria is a consequence of the different parameters which may be considered fixed and the different thermal/hydraulic goals which may be of interest. It should be noted that the thermal and hydraulic characteristics of both sides of the heat exchanger are important; however, the main emphasis here is on augmentation of the in-tube heat transfer coefficient. The thermal resistance of the tube wall and external flow can be included in the algebraic formulations, but the external pumping power must be considered separately.

As an example of the application of these criteria consider the problem of inside augmentation of a proposed shell-and-tube heat exchanger for an ocean thermal energy conversion (OTEC) power plant. The baseline plant is described in Ref. 17. Heat transfer and friction factor data for water flow in tubes with a wide variety of roughness configurations are summarized in Figs. 4 and 5. The many studies involved are described in Ref. 18.

Appropriate constraints for OTEC boilers or evaporators are constant heat duty and constant pumping power; thus, criterion 5 from Table 1 is applicable. In order to screen the tubes, it is assumed that the characteristic temperature difference does not change (low NTU) for augmented and baseline heat exchangers and that the internal surface provides all of the thermal resistance. The calculation proceeds by simply combining expressions for rate of heat transfer

$$q = hA \ \Delta T_{\ell m} \qquad (1)$$

Table 1. Performance Evaluation Criteria

| Criterion | Fixed Parameters ||||| Thermal/Hydraulic Goal |||| Parameter of Interest | Consequences |||| Note |
|---|---|---|---|---|---|---|---|---|---|---|---|---|---|---|---|
| | Geom | w | $\Delta p$ | P | q | Geom↓ | $\Delta p$↓ | P↓ | q↑ | | Geom | w | $\Delta p$ | P | |
| 1 [15] | N,L | X | | | | | | | X | $q_a/q_o$ | | | | | f |
| 2 [15] | N,L | | X | | | | | | X | $q_a/q_o$ | | → | ← | | a,f |
| 3 [15] | N,L | | | X | | | | | X | $q_a/q_o$ | | → | | → | f |
| 4 [15] | N,L | | | | X | | | X | | $P_a/P_o$ | | → | ← | | b |
| 5 [15] | | | X | X | X | X | | | | $A_a/A_o$ or $(NL)_a/(NL)_o$ | L(N) →↑ | → | ← | | c |
| 6 [15] | | | X | X | X | X | | | | $A_a/A_o$ or $(NL)_a/(NL)_o$ | L(N) →↑ | → | → | → | |
| 7 [15] | | X | | | X | X | | | | $A_a/A_o$ or $(NL)_a/(NL)_o$ | L(N) →↑ | | | | |
| A [14], 8[15] | A | X | (X) | X | X | X | | | | $A_a/A_o$ or $(NL)_a/(NL)_o$ | L(w),N(w) →↑ | | | | d |
| B [14] | A | X | (X) | X | X | | | | X | $q_a/q_o$ | L(w),N(w) →↑ | | | | |
| C [14] | A | X | | X | X | | | X | | $P_a/P_o$ | L(w),N(w) →↑ | | → | → | e,f |

a) $\Delta p$↓ is alternate objective, $q$↑ a consequence.
b) $\Delta p$↓ is alternate objective, P↓ a consequence.
c) For low NTU, $A_a/A_o$ comes out directly.
d) $(A_a/A_o)_A = (A_a/A_o)5$ at low NTU
e) $(q_a/q_o)_B = (q_a/q_o)3$.
f) $\Delta T_i$ ↓ is alternate objective, for fixed q.

Type of Roughness

Wire coil insert: 4,20
Knurling: 5
High transverse ribs: 6
Threads: 7,8
Roughness screen: 9
Low transverse ribs: 10,11,12,19
Spirally fluted: 13,15,17
Transverse grooves: 14,18
Sand grain: 16

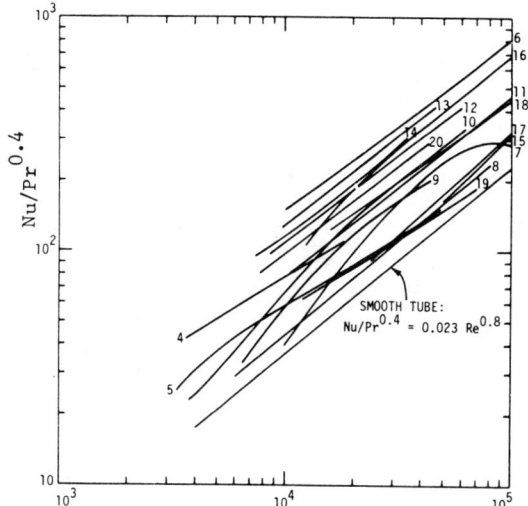

Fig. 4. Summary of Heat Transfer Data for Water Flowing in Internally Roughened Tubes [18].

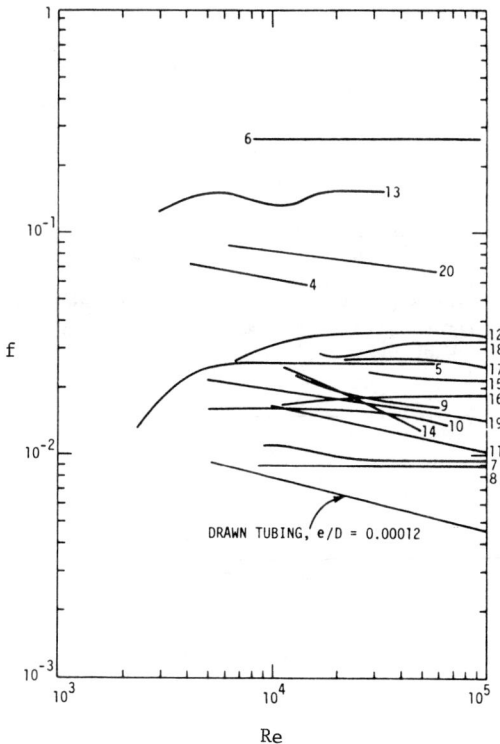

Fig. 5. Summary of Friction Factor Data for Water Flowing in Internally Roughened Tubes [18].

and pumping power

$$P = NV \frac{\pi D^2}{4} 4f \frac{L}{D} \frac{\rho V^2}{2g_o} \qquad (2)$$

for both augmented and reference exchangers to give the following relation:

$$R_5 = \frac{A_a}{A_o} = \frac{h_o}{h_a} = \frac{V_o^3 f_o}{V_a^3 f_a} \qquad (3)$$

In terms of dimensionless groups the desired area ratio is given by

$$\frac{A_a}{A_o} = \frac{Nu_o/Pr^{0.4}}{Nu_a/Pr^{0.4}} = \frac{Re_o^3 f_o}{Re_a^3 f_a} \qquad (4)$$

Since Nusselt numbers and friction factors are available in terms of Reynolds numbers, $R_5$ can be readily obtained as a function of one of the Reynolds numbers. The usual method of solution is to use smooth tube heat transfer and friction factor correlations, e.g., $Nu_o/Pr^{0.4} = 0.023 Re_o^{0.8}$ and $f_o = 0.046/Re_o^{0.2}$. By choosing $Re_a$, $f_a$ and $Nu_a/Pr^{0.4}$ can be picked off plots such as Figs. 4 and 5. Equation (4) is then solved for $Re_o$, and $A_a/A_o$ can be calculated.

The results shown in Fig. 6 suggest that among the more manufacturable tubes, the low transverse ribs have the best performance. Pursuing this configuration further, the full analysis with variable temperature difference and external thermal resistances is applied. As shown in Fig. 7, without enhancement of the outside of the tube (evaporating ammonia) the area can be decreased by about 30%. With a reasonable ammonia enhancement, area reductions of more than 50% are possible. The area reduction depends on the number of tubes (or frontal area); the length decreases faster than the frontal area increases. Since heat exchanger surface areas in projected OTEC systems are

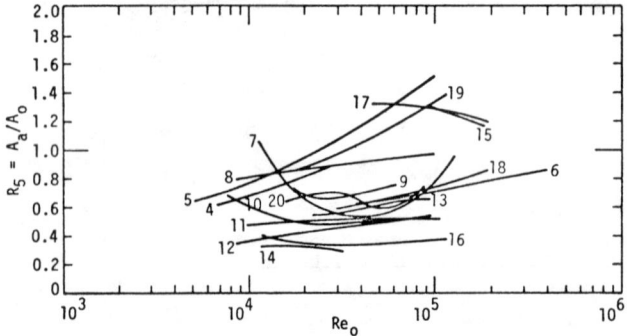

Fig. 6. Constant Pumping Power Performance of Rough Tubes (Derived from Data Given in Figs. 4 and 5) [18].

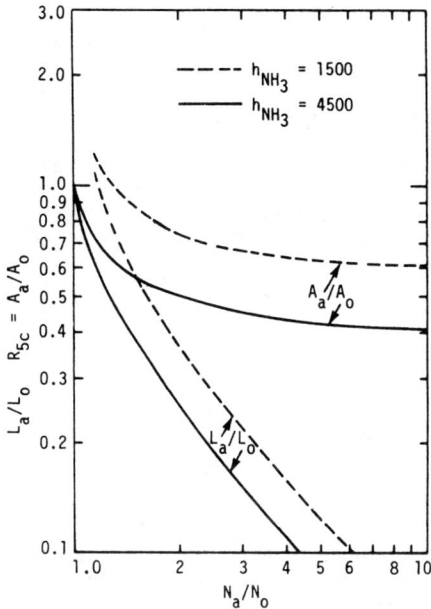

Fig. 7. Constant Pumping Power and Heat Duty Performance of an OTEC Evaporator [18].

extremely large ([0]10,000m$^2$/MW(e) net), augmentation is being seriously considered for these systems.

If generalized correlations for heat transfer and flow friction are available, it is possible to optimize the roughness configuration to provide the munimum area augmented heat exchanger. Webb and Eckert [14] describe such a procedure for Criteria A, B, and C in Table 1.

The subject of performance evaluation criteria for single-phase flows is of considerable current interest. Discussions of the many thermal/hydraulic "figures of merit" which have been proposed are presented by Bergles et al. [19] and Shah [20]. Recent applications include those by Webb et al. [21,22], who stress optimization of augmented tubes for material reduction. Bejan and Pfister [23] describe a formal procedure to use entropy generation as a figure or merit for evaluation of augmented heat transfer surfaces.

The extension of these performance evaluation criteria to two-phase heat transfer is complicated by the dependence of the heat transfer coefficient on the local temperature difference and or the quality. Royal and Bergles [24] and Luu and Bergles [25] have evaluated in-tube condensation by two parameters

$$R_h = \left[\frac{A_a}{A_o}\right]_{q,w} = \left[\frac{\bar{h}_o}{\bar{h}_a}\right]_{q,w} \tag{5}$$

$$R_{\Delta p} = \left[\frac{\Delta P_a}{\Delta P_o}\right]_{q,w} \tag{6}$$

The first parameter represents the area (length) of a condenser relative to the smooth tube (same nominal diameter) unit for equal heat duty, flow rate, and similar temperature or pressure level. The ratio is given simply by the ratio of the heat transfer coefficients, assuming, of course, that the condensing side resistance controls heat transfer. The second parameter represents the pressure drop of the augmented condenser relative to the smooth tube unit, subject to the same constraints. In calculating this ratio, the various components of pressure drop must be considered. The results from this evaluation of data for the tubes with twisted-tape inserts and internally finned tubes (Fig. 8) are shown in Figs. 9 and 10. This analysis suggests that Tube 5 has a clear advantage over the other tubes.

Several measures of performance have been proposed for subcooled boiling CHF. Gambill et al. [26] reported that the CHF's for twisted-tape inserts were approximately twice those for straight flow at the same test-section pumping power. Megerlin et al. [27] suggested, in view of the fact that the cooling passage volume is important when cooling high power electronic devices, that the maximum volumetric heat transfer rate be compared with the pumping power per unit volume.

The preceding lecture on augmentation of two-phase heat transfer described some of the heat transfer data of Kubanek and Miletti [28] for vaporization of refrigerants in internally finned tubes. Since the evaporation pressure drop is a major consideration in direct expansion water chillers, they suggested an "enhanced performance ratio"

$$\left[ \frac{h_a/h_o}{\Delta P_a / \Delta P_o} \right]_{x, \bar{p}, G, L} \tag{7}$$

This ratio is different from the ones suggested in Eqs. 5 and 6; however, the performance ranking of tubes should be the same regardless of which method is used. As shown in Table 2, Tube 25 is clearly the best performer.

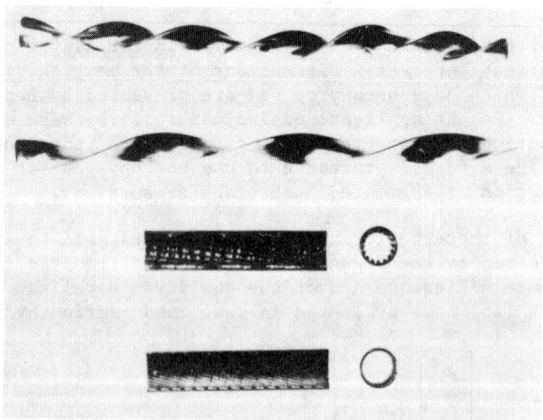

Fig. 8. Twisted Tape Inserts and Inner-Fin Tubes Used to Augment In-Tube Condensation. Top to Bottom: Tubes 2,3,4,5. For Reference, Bottom Tube is 15.9mm o.d. [25].

# APPLICATIONS OF HEAT TRANSFER AUGMENTATION

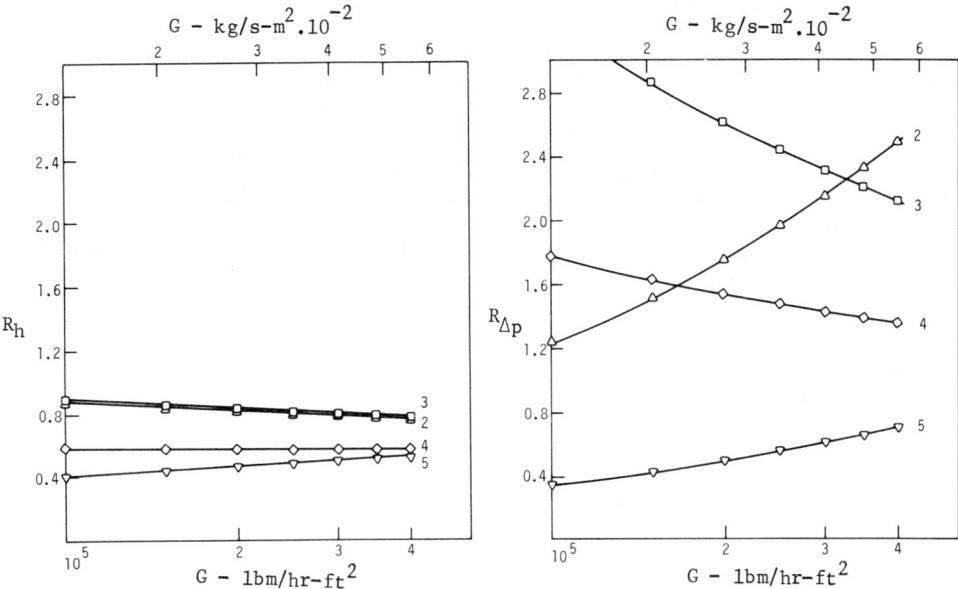

Fig. 9. Condenser Size Reduction Index vs. Mass Velocity, $P_{in}$ = 6.55 bar [25].

Fig. 10. Condenser Pressure Drop Index vs. Mass Velocity, $P_{in}$ = 6.55 bar [25].

Table 2. Comparison of Heat Transfer Enhancement and Pressure Drop Increase [28].

| Test Section length, m | Refrigerant Quality | | | Enhanced Performance Ratio for Tube | | | |
|---|---|---|---|---|---|---|---|
| | Inlet $x_1$ | Outlet $x_2$ | Change $\Delta x$ | 24C | 22 | 25 | 30 |
| 0.8 | 0.2 | 0.4 | 0.2 | 0.33 | 1.00 | 2.29 | 1.06 |
| | 0.4 | 0.6 | 0.2 | 0.16 | 0.76 | 1.89 | 0.80 |
| | 0.6 | 0.8 | 0.2 | 0.14 | 0.83 | 1.64 | 0.80 |
| | 0.2 | 0.9 | 0.7 | 0.48 | 1.53 | 2.11 | 1.54 |
| 2.44 | 0.2 | 0.4 | 0.2 | 0.48 | 0.95 | – | – |
| | 0.4 | 0.6 | 0.2 | 0.25 | 0.76 | – | – |
| | 0.6 | 0.8 | 0.2 | 0.23 | 0.92 | – | – |
| | 0.2 | 0.9 | 0.7 | 0.21 | 0.78 | 1.23 | – |

Basis: $G = 100 \times 10^3$ lb/(hr) (ft²)

It must be pointed out that these thermal-hydraulic comparisons consider only the basic heat exchanger. A full analysis of the effect of the augmentation may require consideration of the entire system if flow rates or temperature levels change as a result of the augmentation.

## 5. MANUFACTURING CONSIDERATIONS AND CODES

There are several important questions which must be raised when the thermal-hydraulic analysis is favorable and the decision is made to use augmentation in a heat exchanger. In most cases, the heat exchanger materials are specified for the application; however, the augmentation technique (tube configuration, insert, etc.) may be limited to the use of a particular material in its manufacture (Gilbert, [29]).

Codes and standards must be checked to see if use of the augmentation technique is allowed. For instance, certain fabrication procedures which deform the tube wall reduce the strength below the minimum specified by pressure vessel codes. The bending strength may also be affected, as in the case of some spirally fluted tubes; this might necessitate extra support plates to prevent vibration and fretting corrosion. Differences in national standards also complicate production. It appears to be cheaper, for example, to purchase tubing in Germany, ship it to the U.S. where a special boiling surface is applied, and return the tubes to Germany rather than fabricate the tubing in the U.S. to German standards.

By modifying the rod-spacer configuration or the shroud, the CHF can be increased for nuclear reactor fuel element rod clusters (Tong et al. [30], Ryabov et al. [31]). However, strict nuclear codes usually prohibit use of such devices in actual reactors.

## 6. OPERATIONAL CONSIDERATIONS

### 6.1 Operating History

Boiling performance may be quite sensitive to operating history. Large temperature overshoots which lead to boiling curve "hysteresis" are observed with highly wetting fluids. This problem may be reduced with special boiling surfaces, due to increased probability of doubly re-entrant cavities; however, reports of hysteresis with special surfaces are beginning to appear. Lewis and Sather [32] reported erratic heat transfer behavior in a 279-tube, "High Flux" flooded evaporator intended for OTEC applications. This was attributed to deactivation of nucleation sites by the flooding of the sites with liquid ammonia. The lowest value of the overall heat transfer coefficient, $U_o$, was estimated at 600 Btu/hr ft$^2$ F (3407 W/m$^2$K). The excellent heat transfer behavior of the High Flux surface was restored by drying out the bundle prior to filling and initiation of heating. After "soaking" in liquid ammonia, a five-day run was initiated to determine the reactivation period. As shown in Fig. 11, $U_o$ increased slowly with time and reached a steady value of 785 Btu/hr ft$^2$ F only after about 100 hours of operation.

Subsequent tests at Argonne National Laboratory were run with a sprayed bundle, Fig. 12, (Hillis et al. [33]) and disclosed similar deactivation/activation behavior. However it was also found that the heat transfer coefficient can be decreased by dryout if done when the surface is already fully activated, as shown in Fig. 13. This was attributed to difficulty in rewetting the dry heated surface by the thin falling film of ammonia. Brief mention of hysteresis effects with the Hitachi Thermoexcel tubing during flooded evaporation of ammonia is given by Torii et al. [34].

The above-mentioned studies of the Linde surface are for multiple tubes where the boiling heat transfer coefficients must be derived from the overall coefficients. A fundamental study of single-tube behavior, with the flooded-tube boiling coefficients obtained directly, was carried out by Chyu [35]. As shown in Figs. 14 and 15, the temperature overshoots with the High Flux surface are quite dramatic. Explanations are offered as to the effect of surface ageing (preboiling or drying), pool subcooling at start of test run, and rate of heat flux change on the boiling curves. These data confirm that special operating procedures are necessary to insure stable, low $\Delta T$ boiling of special

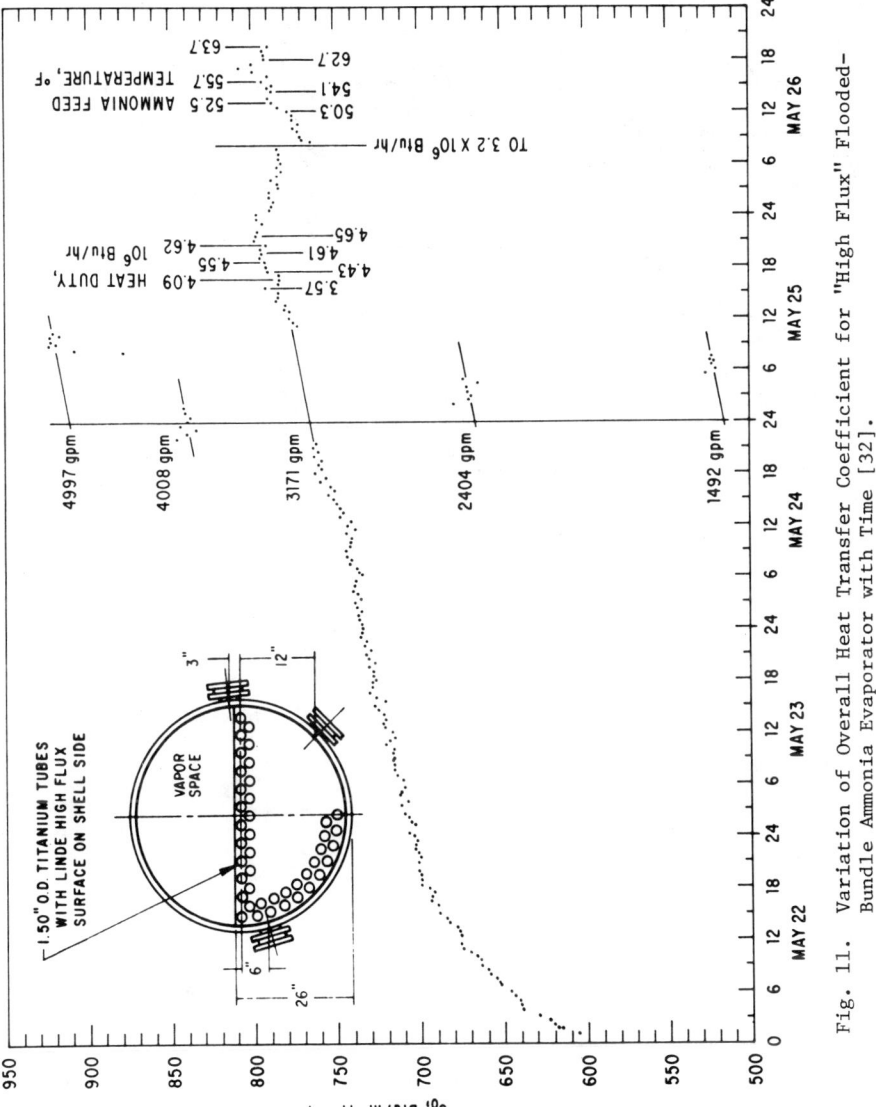

Fig. 11. Variation of Overall Heat Transfer Coefficient for "High Flux" Flooded-Bundle Ammonia Evaporator with Time [32].

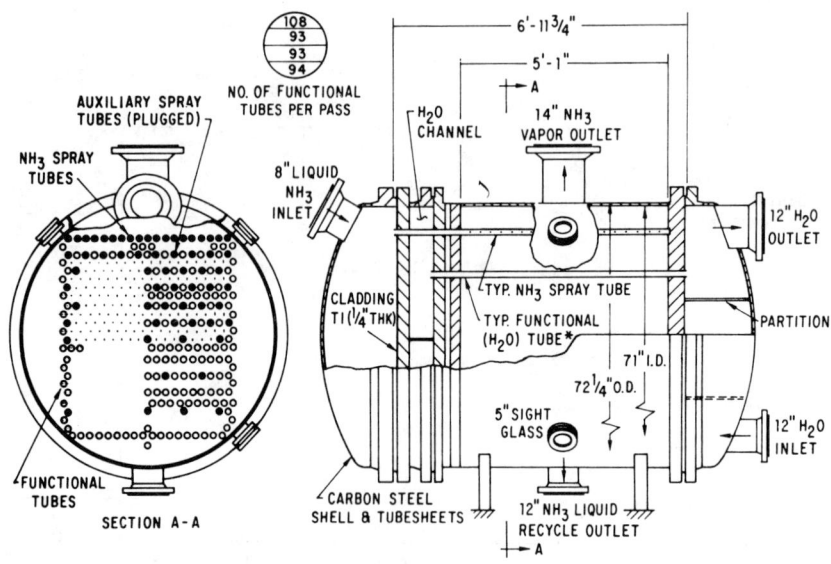

Fig. 12. Diagram of Union Carbide Sprayed-Bundle Evaporator [33].

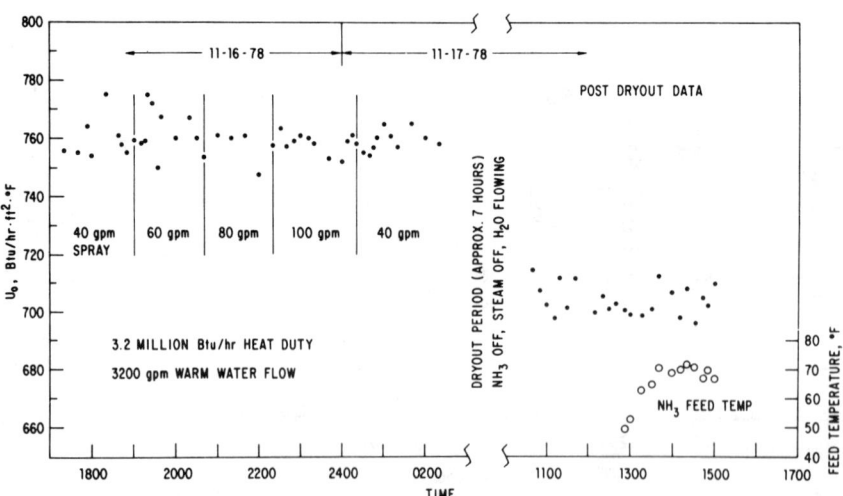

Fig. 13. Variation of Overall Heat Transfer Coefficients for High Flux Sprayed Bundle Evaporator, with Time and Procedure [33].

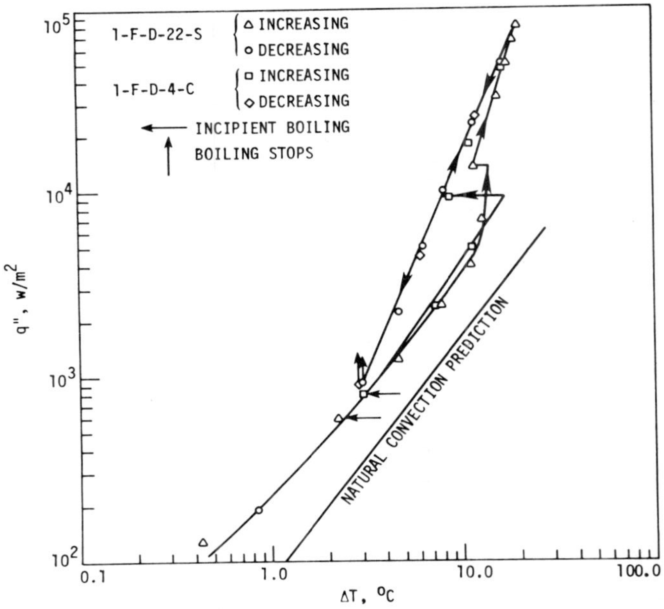

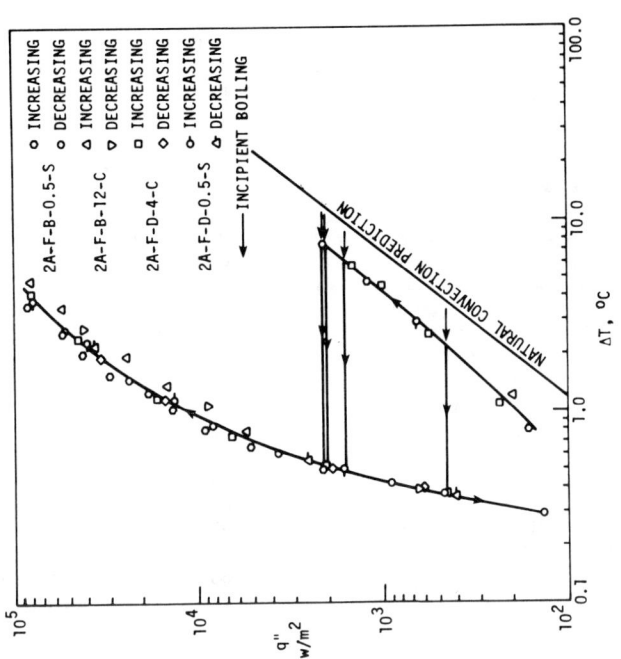

Fig. 15. Saturated Pool Boiling of R-113 on a High Flux Surface, Chyu [35].

Fig. 14. Saturated Pool Boiling of R-113 on a Smooth Copper Surface, Chyu [35].

boiling surfaces in flooded evaporators. Preliminary results of studies at Institut für Verfahrenstechnik, Universität Hannover, suggest that temperature overshoot and resulting hysteresis effects are less pronounced for falling film evaporation on horizontal tubes.

6.2 Fouling and Corrosion

While fouling is still perhaps "The Major Unresolved Problem in Heat Transfer" [36], relatively little work has been done on fouling of augmented surfaces. Some examples, however, can be cited.

With the increasing emphasis on waste heat recovery, much effort is directed roward minimizing the problems of fouling, abraision, and corrosion caused by dirty gases. According to Csathy [37], fin density is more important than fin height and segmented fins are effective since they cut larger particles and also allow better sootblower steam jet penetration.

Turning to internal turbulent flow of water, Watkinson and co-workers [38,39] have conducted studies of the accelerated fouling (calcium carbonate scaling) of spirally fluted and inner-fin tubes. Their conclusions were that at asymptotic fouling conditions, fouled augmented tubes exhibit heat transfer coefficients (nominal area basis) 35 - 100% above those for the fouled plain tubes. In fact, the "self-cleaning" action of spirally fluted tubes has been the basis for a patent [40].

Recent tests by Heat Transfer Research Inc. (HTRI) have established that at typical operating conditions the fouling resistances for corrugated-type plate heat exchangers are much lower than those used for tubular equipment [41]. This is shown in Fig. 16 where the Tubular Exchanger Manufacturers Association (TEMA) Recommended Value is also shown.

In general, special boiling surfaces should be used with clean fluids, as concentrations of impurities or corrosion products could plug up the small pores. Few data are available; however, laboratory experiments, e.g., Chyu [35], suggest that there is no difficulty in boiling distilled water for ex-

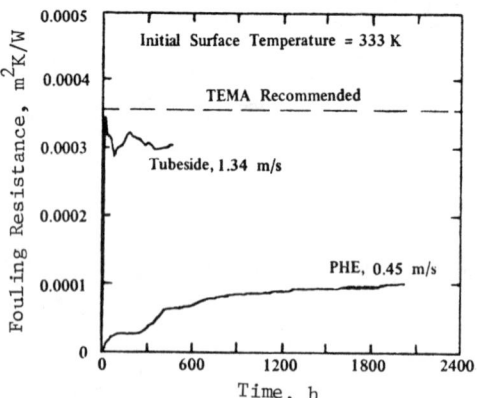

Fig. 16. Comparison of Tubeside and Plate Heat Exchanger Fouling from Typical Cooling Tower Water [41].

# APPLICATIONS OF HEAT TRANSFER AUGMENTATION

tended periods with High Flux surfaces. On the other hand, the rather grim descriptions of boiler "crud" presented by Macbeth [42] suggest that many enhancement techniques will not be effective for long in industrial boiler environments. It is interesting to note, though, that some investigators have reported enhanced heat transfer with porous magnitite deposits, apparently due to increased nucleation sites and "wick boiling".

Special boiling surfaces have potential wide application in refrigeration. Here, the fluid is inherently very clean; however, oil is frequently present. Arai et al. [4] report a rather substantial reduction in the boiling coefficient with addition of lubricating oil (up to 3.4% by weight) to R-12 if there is no foaming. However, if the tubes are immersed in the foaming region, the nearly-oil-free coefficient is restored.

Care must be taken that the augmentation device itself does not lead to accelerated fouling or corrosion. For instance, twisted-tape inserts could promote deposits at the tube-tape junction. To avoid such deposits, Pai and Pasint [43] advocated a gap between the wall and the tape used to improve a once-through steam generator.

Certain augmentation techniques may be considered under fouling conditions if cleaning is possible. This requirement may restrict the type of enhancement device. For example, tube inserts or large surface extensions would preclude mechanical cleaning of evaporator tubes.

## 6.3 Reliability

If an augmented exchanger is to be successful, the augmentation must perform reliably. The active techniques are particularly suspect from this viewpoint as loss of auxiliary power or failure of a piece of equipment could wipe out the augmentation. For instance, the loss of power, failure of the field generator, or breakdown of the fluid would obviate the beneficial effects of electrohydrodynamic augmentation.

The possibility of accelerated fouling or corrosion is a major concern in industrial plants, and thus often inhibits use of augmentation techniques. Although this concern may be unfounded, there is reluctance to take the chance that heat exchanger failure could shut down the entire plant, at a financial loss which would far exceed the original savings in the heat exchanger.

Tube bundle vibrations of substantial amplitude are often observed in commercial heat exchangers, as described in another lecture at this Institute. Some heat transfer augmentation is expected; however, it is dangerous to accentuate the vibrations for this purpose as prolonged vibrations could lead to serious damage in equipment. In the objectionable but not harmful category are whistling sounds eminated from some compact heat exchangers (particularly those with perforated plates) operated in the laminar-turbulent transition range [44]. There is significant heat transfer augmentation; however, the fluctuating fluid pressure can set up plate vibrations which lead to material failure.

## 6.4 Safety

It is apparent that there are objections to use of certain augmentation techniques because of the potential hazards. Electric fields are an example; the heat transfer coefficient for film condensation of R-113 can be increased several times by the application of a transverse electric field (Velkoff and Miller [45]); however, it is not apparent how 30,000 - 40,000 volts can be safely provided in a practical condenser.

In the area of acoustic augmentation, sonic fields result in some enhancement, e.g., a 60% increase in the heat transfer coefficient for condensation of isopropanol vapor flowing downward in a vertical tube when subjected to a siren providing sound at 50-330 Hz (Mathewson and Smith, [46]). The intensity required, however, was 186 db! The precautions which would be required to shield personnel would probably cost more than the original equipment savings.

When both safety and reliability considerations are taken into account, it is apparent that few active augmentation techniques are of great practical interest. The primary use of the available information on these techniques would appear to be in situations where the external influence is naturally present, e.g., rotating or vibrating equipment.

## 7. ECONOMICS

Basic cost goals must be established early. Augmentation usually will not be attractive unless the augmented heat exchanger offers a cost advantage relative to a conventional heat exchanger. A frequent complication is that reliable manufacturing cost data are not available for the desired surface geometry and material. For instance, manufacturers of enhanced tubing expect a lower unit price if the market expands, but generally have very little experience to back up the expectation. In general, the costing should be applied to the entire system and done on a life cycle basis. One such study indicates that seawater enhancement of OTEC heat exchangers may not be desirable when the total system is optimized on the basis of lowest cost [47]. This is contrary to what is expected from the analysis of the heat exchanger alone (Section 4).

Cost can be included as an additional parameter in the matrix of performance evaluations given in Table 1. An example of the heat transfer performance of inner-fin tubes relative to that for plain tubes, subject to fixed total (capital and operating) costs, is given in Ref. 15.

As an example of partial cost analysis, consider the use of augmented tubing in an aromatics reboiler. Table 3 lists the manufacturer's analysis of this application. The point is that the High Flux boiling surface on the inside of the tube together with a fluted exterior condensing surface increases the overall coefficient so much that the driving temperature difference is reduced, with the result that a large annual saving in fuel costs for the heating steam is expected. The increased tubing cost is not cited, but a reasonable payback period is expected.

Ziegler and Bergles [49] recently presented an analysis of the economics of using augmented surfaces in plate-fin gas-to-gas waste heat recuperators. The rate-of-return of a typical recuperator is improved over a wide range of operating conditions through use of augmented surfaces (strip fins, louvered fins, and wavy fins).

## 8. EXAMPLES OF APPLICATIONS OF AUGMENTATION

After sounding many warnings about possible complications in applying augmentation techniques to commercial heat exchangers, it is appropriate to

Table 3. Example of Energy Saving with High Flux Tubing in Aromatics Reboiler Application [48].

|  | BARE TUBE DESIGN | HIGH FLUX DESIGN |
|---|---|---|
| Heat Duty, Btu/hr | $10 \times 10^6$ | $10 \times 10^6$ |
| Boiling Fluid | BTX Aromatic | BTX Aromatic |
| Condensing Fluid | Steam | Steam |
| Steam Pressure, psig | 400 | 125 |
| Steam Condensing Temperature, °F | 448 | 352 |
| Boiling Temperature, °F | 325 | 325 |
| Mean Temperature Difference, °F | 123 | 27 |
| Exchanger Area, ft² | 6000 | 6000 |
| Exchanger Size (diameter x length) | 52in.x240in. | 52in.x240in. |
| Overall Heat Transfer Coefficient, Btu/hrft²F | 95 | 435 |
| Steam Cost, $/1000 lb | 2.50 | 1.50 |
| Energy Cost, $/yr | 1,924,320 | 1,030,600 |
| Energy Savings, $/yr |  | 893,000 |

cite some further examples of actual or proposed commercial useage. Only representative applications will be given, in rather sketchy form - in keeping with the way the performance and cost are usually reported. Such reporting does, however, go back many years. For example, in 1896, Whitham [50] reported increases of up to 18% in fire-tube boiler efficiency when "retarders" (twisted-tape inserts) were installed in the tubes.

## 8.1 Chemical Process Industry

Wolf et al. [51] report the use of Linde High Flux surfaces in distillation, and note how rising energy costs can be mitigated by better cycle arrangement and use of augmented heat exchanges. Many other examples of the use of this particular surface in distillation and gas liquifaction have been cited by the manufacturer. (See also Section 6.5.)

## 8.2 Power Industry

Augmented heat transfer surfaces are widely used to improve gas-side heat transfer (furnaces, economizers, and air-cooled condensers). Relative performance and cost analyses are not generally available. There is strong interest in developing more effective surfaces for dry cooling towers in view of the large surface areas involved. Renz and Becker [52] report some economic comparisons of finned tubes and "Kühlrohre" (air flows inside tubes with roughness elements) for this service.

In the area of waste heat recovery, reference is made to a "$35,000 annual saving in gas costs that plant officials attribute primarily to a $10,000 expenditure for turbulators in each of those boilers".

Graham and Aerni [53] discuss the potential for dropwise condensation in surface condensers, particularly for marine applications. A surface area reduction of 30% and a total weight reduction of 15% is typically predicted. The more recent review by Tanasawa [54] should be consulted for an assessment of the practical difficulties, including removal of non-condensables and maintenance of the promoter, which must be surmounted before the high heat transfer coefficients of dropwise condensation can be realized in practical equipment.

Recent news from the power industry indicates that Wolverine Tube Division of Universal Oil Products is presently refitting a Tennessee Valley Authority power plant condenser with "Korodense" tubes.

## 8.3 Refrigeration Industry

The air conditioning and refrigeration industry has been a leader in the use of augmentation for single-phase forced convection, e.g., air side fins and water side inserts. Internally finned tubes of various types are quite common for refrigerant evaporators of either the shell-and-tube or coaxial type. The bulk of the design information as well as performance and cost evaluations are treated as proprietary. An early estimate of refrigerant evaporator size reduction through use of the Dunham-Bush "Inner Fin" tube is given by Boling et al. [55].

Low finned tubes are widely used for enhancing pool boiling in flooded chillers; however, much current effort is directed toward use of the new surfaces described in a previous lecture. As an example, Fig. 17 depicts the improvement in evaporator performace when the standard low fin tubes (GEWA) are replaced by the new GEWA-T tubes. An approximate estimate of this situation is that a 25% increase in heat duty is possible with a 15% increase in tubing cost.

Package chillers can be reduced in size and cost by augmentation of both evaporators and condensers. Arai et al. [4] reported that use of enhanced boiling (Thermoexcel-E) and condensing surfaces (Thermoexcel-C), along with repeated ribs on the interior surfaces of the tubing in both the evaporator and condenser led to a 28% reduction in the heat exchanger length. The tubing cost was stated as being about 20% above that of conventional low-fin tubes.

## 8.4 Desalination

Much development work in enhanced heat transfer tubing has been motivated by desalting applications. There is a large potential market due to the large surface areas required. A rather detailed economic analysis by Cox et al. [57] was performed for Long Tube Vertical (LTV) and Horizontal Tube Multiple Effect (HTME) plants. Internally grooved tubes (with some effects using these tubes with knurled exterior as well) were shown to reduce the capital cost by about 10%. An important screening study of enhanced tubes for LTV systems was

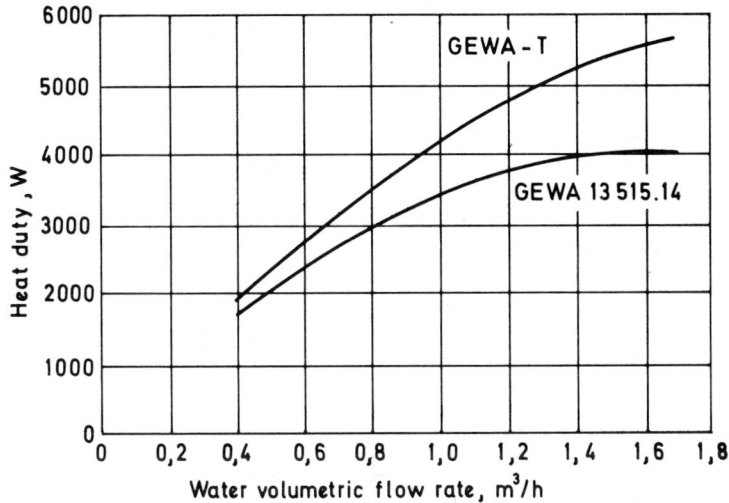

Fig. 17. Illustration of Overall Increase in Performance of Refrigerant Evaporator through Use of GEWA-T Surface [56].

reported by Alexander and Hoffman [58]. Increases of over 200% in the overall heat transfer coefficient were observed with several tubes.

In Multistage Flash (MSF) plants, only condensing surface is required. Rather than comparing smooth and enhanced tubes in a series of full-scale plant designs, Newson [59] elected to use the more common single-tube analysis. A commercial "roped" tube was chosen with tubeside (single-phase brine coolant) and shell-side (condensing pure water) enhancements of 1.63 and 1.67, respectively. The heat duty was fixed. The smooth tubes were 18.1 mm i.d., while the enhanced tubes were 24.5 mm i.d. As shown in Fig. 18, in the region of 1 m/s water velocity in the enhanced tube, the smooth tube cost is 40-50% more than the enhanced tube. Process and economic data are given in the figure. It is noted that here the enhanced tubing premium is only 7 1/2%. The most important factor becomes the operating cost or pumping power; thus, the comparison is very sensitive to water velocity.

Withers and Young [12] analyzed MSF steam condenser fabricated from corrugated ("Korodense") tubes. The combination of increased condensing coefficient and increased internal water coefficient led to a surface area reduction of about 30%, subject to constraints of constant heat duty, fixed pressure drop, and fixed coolant flow rate.

The comprehensive survey of Thomas [60] gives interesting speculations as to maximum heat transfer coefficients which might be realized with enhanced surfaces for desalination. The upper limit is suggested as about 6000 Btu/hr $ft^2$ F (34000 W/m$^2$ K) which was at that time, and still is, about 3 times the state-of-the-art coefficient. Fouling is a major problem, as even a small fouling factor can drastically reduce the coefficients predicted on the basis of clean surface data.

8.6 Electronics Industry

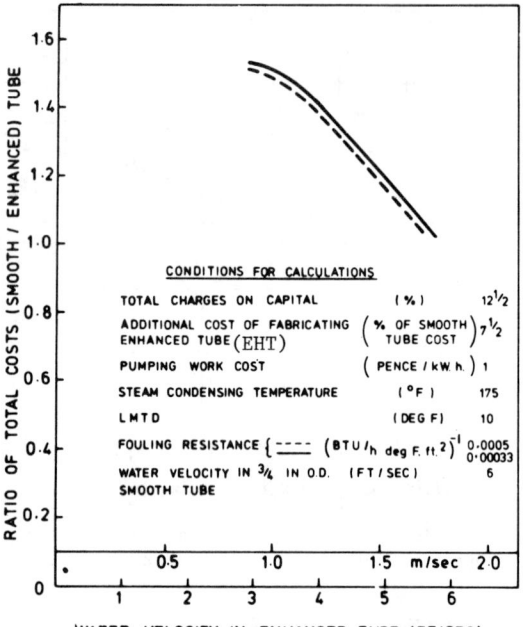

Fig. 18. Cost Ratio for Tubing in Typical MSF Condenser. Note: Also fixed are Ts and $N_{EHT}$. As $V_{ENT}$ varies, $L_{EHT}$ and $N_{smooth}$ vary. [59].

Extended surfaces of the interrupted type are widely used as "heat sinks" for electronic devices or electrical components. Liquid cooling with boiling or evaporation is occasionally required to dissipate very large heat generation rates. An established application of extended surfaces is found in the "Vapotron" series of cooling arrangements for high power communication tubes. As described by Beurtheret [61], the fins are shaped so as to maximize the average boiling heat transfer coefficient for either pool boiling or forced convection boiling.

## 8.5 Status of Application

Table 4 summarizes the present and potential use of special geometries and other passive augmentation techniques according to mode of heat transfer, as suggested in a recent evaluation [62]. The first part of the table considers augmenting the inner and outer surfaces of circular tubes while the second part considers plate-type and plate-fin heat exchangers. A coding system is employed to define the degree to which an enhancement technique has found application. A dash entry indicates not applicable or not intended to augment that heat transfer mode. The third column lists typical materials used for commercial manufacture. The last column indicates the authors' assessment of the performance potential, independent of whether a commercial manufacturing technique exists. It is evident that only a limited number of the available enhancement techniques are commercially utilized. In only a few cases are techniques considered to be in common use. Notably, some manufacturers of a given type of enhanced surface have not been able to market their product in significant quantities.

Table 4. Applications of Heat Transfer Augmentation Techniques [62].

| | Commercial Availability | Forced Convection | Boiling | Condensation | Typical Material | Performance Potential |
|---|---|---|---|---|---|---|
| | | | Mode | | | |
| *Inside Tubes* | | | | | | |
| Metal Coatings | Yes | — | 2 | — | Al, Cu, St | Hi |
| Integral Fins | Yes | 2 | 3 | 4 | Al, Cu | Hi |
| Flutes | Yes | 4 | 4 | 4 | Al, Cu | Mod |
| Integral Roughness | Yes | 2 | 3 | 4 | Cu, St | Hi |
| Wire Coil Inserts | Yes | 3 | 4 | 4 | Any | Mod |
| Displaced Promoters | Yes | 2 | 4 | 4 | Any | Mod (Lam) |
| Twisted Tape Inserts | Yes | 2 | 3 | 4 | Any | Mod |
| *Outside Circular Tubes* | | | | | | |
| Coatings | | | | | | |
|   Metal | Yes | — | 2 | 4 | Al, Cu, St | Hi (Boil) |
|   Non-Metal | No | — | 4 | 4 | "Teflon" | Mod |
| Roughness (Integral) | Yes | 3 | 2 | 4 | Al, Cu | Hi (Boil) |
| Roughness (Attached) | Yes | 3 | 4 | — | Any | Mod (For Conv) |
| Axial Fins | Yes | 1 | 4 | 4 | Al, St | Hi (For Conv) |
| Transverse Fins | | | | | | |
|   Gases | Yes | 1 | — | — | Al, Cu, St | Hi |
|   Liquids/Two-Phase | Yes | 1 | 1 | 1 | Any | Hi |
| Flutes | | | | | | |
|   Integral | Yes | — | — | 2 | Al, Cu | Hi |
|   Non-Integral | Yes | — | — | 4 | Any | Hi |
| *Plate-Fin Heat Exchanger* | | | | | | |
| Metal Coatings | Yes | — | 3 | — | Al | Hi |
| Surface Roughness | Yes | 4 | 3 | 4 | Al | Hi (Boil) |
| Configured or Interrupted Fins | Yes | 1 | 2 | 2 | Al, St | Hi |
| Flutes | No | — | — | 4 | Al | Mod |
| *Plate Type Heat Exchanger* | | | | | | |
| Metal Coatings | No | — | 4 | — | St | Lo |
| Surface Roughness | No | 4 | 4 | 4 | St | Lo |
| Configured Channel | Yes | 1 | 3 | 3 | St | Hi (For Conv) |

Use Code
1. Common Use
2. Limited Use
3. Some Special Cases
4. Essentially No Use

## 9. CONCLUDING REMARKS

It is evident that many factors enter into the decision to use a heat transfer augmentation technique. Many of the available techniques have gone through all of the steps required for commercialization, as listed in Section 1. Table 5 attempts to summarize the status of technology development for several of the more important enhancement techniques applied to circular tubes. Commercialization represents the ultimate stage of development; however, even commercial products usually require additional development work.

The further development of any technique will depend on two factors: 1) its performance potential, and 2) manufacturing cost. Presently, cost-effective manufacturing technology is probably the most significant barrier to commercial application of high performance enhancement techniques. Researchers appear intent on identifying surface geometries which offer the highest performance. It is suggested that more communication between researchers and manufacturing engineers is needed to identify the most cost-effective concepts.

Table 5. Status of Augmentation Technology Development [62].

| Inside Tubes | Single Phase Forced Convection | Boiling | Condensation |
|---|---|---|---|
| *Inside Tubes* | | | |
| Coatings | — | 3, 5 | — |
| Roughness | 2, 3, 5 | 1 | 1 |
| Internal fins | 1, 2, 3, 5 | 1, 5 | 1 |
| Flutes | 5 | 5 | — |
| Insert devices | 5 | 1, 5 | 1, 5 |
| *Outside Tubes* | | | |
| Metal coatings | — | 3, 5 | — |
| Non-metal coatings | — | 1 | 4 |
| Roughness | 1, 2 | 1, 2, 5 | 1 |
| Extended surface | 5 | 5 | 5 |
| Flutes | — | — | 5 |

Code:  1. Basic performance data
       2. Design methods
       3. Manufacturing technology
       4. Heat exchanger application
       5. Commercialization

An example of a recent research workshop intended to promote this type of communication is described by Junkhan et al. [63].

ACKNOWLEDGMENTS

These lectures on heat transfer augmentation were prepared while at the Institut für Verfahrenstechnik der Universität Hannover with the assistance of an award from the Alexander von Humboldt Foundation.

NOMENCLATURE

A = heat transfer area (nominal area basis)
D = tube inside diameter
e = protuberance height
f = Fanning friction factor
G = mass velocity
Gr = Grashof number based on root diameter of finned tube, coil average
h = heat transfer coefficient (nominal area basis unless otherwise specified)
L = length of tube bundle
N = number of tubes in tube bundle
NTU = number of transfer units
Nu = Nusselt number
$Nu_{d_o'}$ = Nusselt number based on local h and root diameter of finned tube
P = Pumping power
Pr = Prandtl number

p = pressure
$\Delta$p = pressure drop
q" = heat flux
$q''_{cr}$ = critical heat flux
Re = Reynolds number (no allowance for change in cross-sectional area due to roughness)
$R_h$ = performance ratio defined in Eq. (5)
$R_{\Delta p}$ = performance ratio defined in Eq. (6)
$R_5$ = performance ratio from Table 1, no external resistance
$R_{5c}$ = performance ratio from Table 1, external resistance included along with variable temperature difference
$\Delta T$ = wall-minus-saturation temperature difference
$\Delta T_i$ = approach temperature difference for two-fluid heat exchanger
$\Delta T_{\ell m}$ = log-mean temperature difference
$U_o$ = overall heat transfer coefficient, based on outside area
V = mean flow velocity
x = distance along heated length

Greek

$\rho$ = density

Subscripts

a = refers to augmented tubes
f = properties evaluated at film temperature
in = condition at inlet of tube
o = refers to normal smooth tubes
P = evaluated at constant pumping power
q = evaluated at constant heat duty
w = evaluated at constant mass flow rate

REFERENCES

1. Bergles, A. E., Webb, R. L., Junkhan, G. H., and Jensen, M. K., "Bibliography on Augmentation of Convective Heat and Mass Transfer", HTL-19, ISU-ERI-Ames-79206, Iowa State University, May 1979.

2. Webb, R. L., Junkhan, G. H., and Bergles, A. E., "Bibliography of U.S. Patents on Augmentation of Convective Heat and Mass Transfer", Heat Transfer Laboratory Report, Iowa State University, 1980.

3. Junkhan, G. H., Bergles, A. E., and Webb, R. L., "Energy Conservation Via Heat Transfer Enhancement", Quarterly Progress Report, COO-4649-7, Iowa State University, August 1979.

4. Arai, N. et al., "Heat Transfer Tubes Enhancing Boiling and Condensation in Heat Exchanger of a Refrigerating Machine", ASHRAE Trans., Vol. 83, Part 2, pp. 58-70, 1977.

5. Saier, M., Kästner, H.-W., Klöckler, R., "Y and T-Finned Tubes and Methods and Apparatus for their Making", U.S. Patent No. 4,179,911, December 25, 1979.

6. Bergles, A. E., "Survey and Evaluation of Techniques to Augment Convective Heat and Mass Transfer", in Progress in Heat and Mass Transfer, U. Grigull and E. Hahne (Editors), Vol.1,Pergamon, Oxford, pp. 331-334, 1969.

7. Gorenflo, D., "Zum Wärmeübergang bei Blasenverdampfung an Rippenrohren", Dissertation, Technische Hochschule, Karlsruhe, 1966.

8. Gschaider, J., "Berechnungsunterlagen für Wendelförmig Gewickelte Solarwärmeübertrager", Diplom-Arbeit, Fachhochschule für Technik Ulm, May 1980.

9. Katz,D. L. et al., "Boiling outside Finned Tubes", Petroleum Refiner, Vol. 34, pp. 113-116, 1955.

10. Nakajima, K. and Shiozawa, A., "An experimental Study on the Performance of a Flooded Type Evaporator", Heat Transfer - Japanese Research, Vol. 4, No. 4, pp. 49-66, 1975.

11. Yung, D., Lorenz, J. J., and Ganic, E., "Vapor/Liquid Interaction and Entrainment in Falling Film Evaporators", J. Heat Transfer, Vol. 102, pp. 20-25, 1980.

12. Withers, J. G. and Young, E. H., "Steam Condensing on Vertical Rows of Horizontal Corrugated and Plain Tubes", Industrial and Engineering Chemistry, Process Design and Development, Vol. 10, pp. 19-30, 1971.

13. Worn, G. A. and Rubin, F. L., "Method and Apparatus Employing Sonic Waves in Heat Exchangers", U.S. Patent No. 2,664,274, December 29, 1953.

14. Webb, R. L. and Eckert, E. R. G., "Application of Rough Surfaces to Heat Exchanger Design", Int. J. Heat Mass Transfer, Vol. 15, 1647-1658, 1972.

15. Bergles, A. E., Blumenkrantz, A. R. and Taborek, J., "Performance Evaluation Criteria for Enhanced Heat Transfer Surfaces", AIChE Preprint 9 for 13th National Heat Transfer Conference, Denver, Colorado, August, 1972.

16. Bergles, A. E., Bunn, R. L. and Junkhan, G. H., "Extended Performance Evaluation Criteria for Enhanced Heat Transfer Surfaces", Letters in Heat and Mass Transfer, Vol. 1, pp. 113-120, 1974.

17. Trimble, L. C. et al., "Ocean Thermal Energy Conversion System Study Report", Proceedings, Third Workshop on Ocean Thermal Energy Conversion (OTEC), APL/JHU SR 75-2, pp. 3-21, August 1975.

18. Bergles, A. E. and Jensen, M. K., "Enhanced Single-Phase Heat Transfer for OTEC Systems", Proceedings, Fourth Annual Conference on Ocean Thermal Energy Conversion, University of New Orleans, pp. VI-41 - VI-54, March 1977.

19. Bergles, A. E., Junkhan, G. H., and Bunn, R. L., "Performance Criteria for Cooling Systems on Agricultural and Industrial Machines", SAE Transactions, Vol. 85, pp. 38-48, 1976.

20. Shah, R. K., "Compact Heat Exchanger Surface Selection Methods", Heat Transfer 1978, Proceedings of the Sixth International Heat Transfer Conference, Vol. 4, Hemisphere, Washington, pp. 193-201, 1978.

21. Webb, R. L. and Scott, M. J., "A Parametric Analysis of the Performance of Internally Finned Tubes for Heat Exchanger Applications", J. Heat Transfer, Vol. 102, pp. 38-43, 1980.

22. Webb, R. L. and Hong, J. T., "Water-Side Enhancement for OTEC Shell-and Tube Evaporators", Paper III 13/4 presented at 7th Ocean Energy Conference, June 1980.

23. Bejan, A. and Pfister, P. A., Jr., "Evaluation of Heat Transfer Augmentation Techniques Based on Their Impact on Entropy Generation", Letters in Heat and Mass Transfer, Vol. 7, pp. 97-106, 1980.

24. Royal, J. H. and Bergles, A. E., "Pressure Drop and Performance Evaluation of Augmented In-Tube Condensation", Heat Transfer 1978, Proceedings of the Sixth International Heat Transfer Conference, Vol. 2, pp. 459-464, 1978.

25. Luu, M and Bergles, A. E., "Experimental Study of the Augmentation of In-Tube Condensation of R-113", ASHRAE Trans., Vol.85, Part 2, pp. 132-145, 1979.

26. Gambill, W. R., Bundy, R. D., and Wansbrough, R. W., "Heat Transfer, Burnout, and Pressure Drop for Water in Swirl Flow Tubes with Internal Twisted Tapes", Chem. Eng. Prop. Symp. Ser., Vol.57, No. 32, pp. 127-137, 1961.

27. Megerlin, F. E., Murphy, R. W. and Bergles, A. E., "Augmentation of Heat Transfer in Tubes by Means of Mesh and Brush Inserts", J. Heat Transfer, Vol 96, pp. 145-151, 1974.

28. Kubanek, G. R., and Miletti, D. L., "Evaporative Heat Transfer and Pressure Drop Performance of Internally-Finned Tubes with Refrigerant 22", J. Heat Transfer, Vol. 101, 447-452, 1979.

29. Gilbert, P. T., "Selection of Materials for Heat Exchangers", Paper presented at Sixth International Congress on Metallic Corrosion, Sydney, Australia, December 1975.

30. Tong, L. S. et al. "Critical Heat Flux of a Heater Rod in the Center of Smooth and Rough Sleeves, and in Line-Contact with an Unheated Wall", ASME Paper No. 67-WA/HT-29, 1967.

31. Ryabov, A. N. et al., "Boiling Crisis and Pressure Drop in Rod Bundles with Heat Transfer Enhancement Devices", Heat Transfer-Sov. Res., Vol. 9, No. 1, pp. 112-121, 1977.

32. Lewis, L. G. and Sather, N. F., "OTEC Performance Tests of the Union Carbide Flooded-Bundle Evaporator", ANL-OTEC-PS-1, Argonne National Laboratory, December 1978.

33. Hillis, D. L. et al., "OTEC Performance Tests of the Union Carbide Sprayed-Bundle Evaporator", ANL/OTEC-PS-3, Argonne National Laboratory, May 1979.

34. Torii, T. et al., "The Use of Heat Exchangers with Thermoexcels' Tubing in Ocean Thermal Energy Power Plants", ASME Paper No. 78-WA/HT-65, 1978.

35. Chyu, M-C., "Boiling Heat Transfer from a Structured Surface", MS Thesis in Mechanical Engineering, Iowa State University, 1979.

36. Taborek, J. et al., "Fouling, the Major Unresolved Problem in Heat Transfer, Part I", Chem. Eng. Prog., Vol. 68, No. 2, pp. 59-67, 1972.

37. Csathy, D., "Heat Recovery from Dirty Gas", Paper presented at Sixth Conference on Energy and the Environment, Pittsburgh, May 1979.

38. Watkinson, A. P. and Martinez, O., "Scaling of Indented Heat Exchanger Tubes", J. Heat Transfer, Vol. 97, pp. 490-492, 1975.

39. Watkinson, A. P., Louis, L., and Brent, R., "Scaling of Enhanced Heat Exchanger Tubes", The Canadian Journal of Chemical Engineering, Vol. 52, 1974, pp. 558-562.

40. Withers, J. G., Jr., "Heat Exchanger Apparatus and Method of Controlling Fouling Therein", U.S. Patent No. 4,007,774, February 15, 1977.

41. Cooper, A., Suitor, J. W., and Usher, J. D., "Cooling Water Fouling in Plate Heat Exchangers", Heat Transfer Engineering, Vol. 1, No. 3, pp. 50-56, 1980.

42. Macbeth, R. V., "Fouling in Boiling Water Systems", in Two-Phase Flow and Heat Transfer, D. Butterworth and G. F. Hewitt (Editors), Oxford University Press, pp. 323-342, 1977.

43. Pai, R. H. and Pasint, D., "Research at Foster Wheeler Advances Once-Through Boiler Design", Electric Light and Power, pp. 66-70, January, 1965.

44. Shah, R. K., "Perforated Heat Exchanger Surfaces. Part 1 - Flow Phenomena, Noise and Vibration Characteristics", ASME Paper No. 75-WA/HT-8, 1975.

45. Velkoff, H. R. and Miller, J. H., "Condensation of Vapor on a Vertical Plate with a Transverse Electric Field", J. Heat Transfer, Vol. 87, pp. 197-201, 1965.

46. Mathewson, W. F. and Smith, J. C., "Effect of Sonic Pulsation on Forced Convective Heat Transfer to Air and on Film Condensation of Isopropanol", Chemical Engineering Progress Symposium Series, Vol. 59, No. 41, pp. 173-179, 1963.

47. Westinghouse Electric Corporation, Power Generation Divisions, Lester, Pa, "Ocean Thermal Energy Conversion Power System Development, Phase I Conceptual Design Briefing Document",1978.

48. Linde Division, Union Carbide Corp., Tonawanda, New York. "Technical Information - High Flux Tubing", September 1977.

49. Ziegler, R. P. and Bergles, A. E, "Economic Evaluation of Augmented Surfaces in Waste Heat Recuperator Design", HTL-20, ISU-ERI-Ames-80024, Iowa State University, May 1979.

50. Whitham, J. M., Street Railway Journal, Vol. 12, p. 374, 1896.

51. Wolf, C. W., Weiler, D. W. and Ragi, E. G., "Energy Costs Prompt Improved Distillation", Oil and Gas Journal, September 1, 1974.

52. Renz, U. and Becker, N., "Vergleich von Kühlsystemen für Trockenkühltürme", Brennst.-Wärme-Kraft, Vol. 30, pp. 368-370, 1979.

53. Graham, C. and Aerni, E. F., "Dropwise Condensation - a Heat Transfer Process for the 70's", Naval Ship Systems Command 7th Annual Technical Symposium, 1970.

54. Tanasawa, I, "Dropwise Condensation - the Way to Practical Applications", Heat Transfer 1978, Proceedings of the Sixth International Heat Transfer Conference, Vol. 6, Hemisphere Publishing Corp., Washington, D.C., pp. 393-405, 1978.

55. Boling, C., Donovan, W. J., and Decker, A. S., "Heat Transfer of Evaporating Freon with Inner Fin Tubing", Refrigerating Engineering, Vol. 61, pp. 1338-1340, 1384, 1953.

56. Wieland-Werke AG, P.O. Box 4420, Ulm, West Germany, "High Performance Tubes for Flooded Evaporators", Wieland Product Information SAW-15e-06.78.

57. Cox. R. B. et al., "Second Report on Horizontal Tube Multiple Effect Process Pilot Plant Test and Designs", U.S. Office of Saline Water Research and Development Progress Report No. 592, 1970.

58. Alexander, L. G., and Hoffman, H. W., "Performance Characteristics of Corrugated Tubes for Vertical Tube Evaporators", ASME Paper No. 71-HT-30, 1971.

59. Newson, I. H., "Enhanced Heat Transfer Condenser Tubing for Advanced Multistage Flash Distillation Plants", Proceedings of the Fifth International Symposium on Fresh Water from the Sea, Vol. 2, pp. 107-115, 1976.

60. Thomas, D. G., "Prospects for Further Improvements in Enhanced Heat Transfer Surfaces", Desalination, Vol. 12, pp. 189-215, 1973.

61. Beurtheret, C. A., "Transfer de Flux Supêrieur a 1 kw/cm$^2$ par Double Changement de Phase Entre une Paroi non Isotherme et un Liquid en Convection Forcêe", Heat Transfer 1970, Proceedings of the Fourth International Heat Transfer Conference, Vol. V, Paper B4.2, Elsevier, Amsterdam, 1970.

62. Bergles, A. E., Webb, R. L. and Junkhan, G. H., "Energy Conservation Via Heat Transfer Enhancement", Energy, Vol. 4, pp. 193-200, 1979.

63. Junkhan, G. H., Bergles, A. E. and Webb, R. L., "Research Workshop on Energy Conservation through Enhanced Heat Transfer", HTL-21, ISU-ERI-Ames-80063, Iowa State University, October, 1979.

# OPERATIONAL CONSIDERATIONS

# The Transient Response of Heat Exchangers

**RAMESH K. SHAH**
Harrison Radiator Division
General Motors Corporation
Lockport, New York 14094 USA

ABSTRACT

The transient performance of heat exchangers, when used in a system, should be known so that the initial design is made properly for the maintenance free desired life of the exchanger; also the control engineer could properly control the system or process involved. The transient response analysis is presented for direct transfer type (recuperative) and regenerative exchangers. This includes the problem formulation, exchanger variables (dimensional and dimensionless) and specific solutions for counterflow ($C_{min}/C_{max}$ = 0 and 1), crossflow, and periodic-flow exchangers. All the solutions presented in the paper are for the responses to step changes in inlet temperatures and/or flow rates. Frequency response and impulse response are not considered, however, they can be determined from the step responses.

1. INTRODUCTION

A heat exchanger is generally a part of the system that may be exposed to a number of planned or unplanned transients or start-ups and shutdowns in a certain time (either a day, a month, or a year); or the system may be designed to work at varying loads. The transients may also arise due to the change in operating conditions such as a change in inlet temperatures or flow rates. The transients may also be due to the inherent instability in the system such as flow oscillations in a two-phase system. The transients could produce such undesirable effects as reduced heat transfer performance, severe thermal stresses and eventual mechanical failure. Thus it is important to know and be able to predict the transient response or dynamic characteristics of a heat exchanger, in addition to its steady-state performance. When the heat exchanger is massive and/or is a part of a complex system such as a process plant, a power plant or an air-conditioning system, it is also essential to know its transient response in order to provide an effective control system. This is because the outlet-fluid temperature response of a heat exchanger to a change in one of the inlet-fluid temperatures is not instantaneous, and that affects the system performance. The lag of the response is influenced by the thermal capacitances of the solid wall ($M_w c_w$) and the fluids ($Wc_p \tau_d$), and also the thermal resistances between the fluids and wall ($1/\eta_o hA$). Thus this lag is due to thermal capacitance-resistance effects and not due to "thermal inertia" as frequently implied in the literature. There is no natural analog to inertia (or electrical inductance) in any thermal (heat transfer) circuit.

There are three types of transient responses: (1) step response, (2) frequency response, and (3) impulse response. The step response characterizes the behavior of the heat exchanger subjected to a sudden change in operating conditions (inlet temperatures or flow rates). The step response asymptotically approaches the steady state value corresponding to new operating conditions. The frequency response describes the behavior of the heat exchanger subjected to a change varying periodically with time. The response also varies periodically, but its maximum amplitude and phase angle relative to the input change depend upon the input frequency. An impulse response characterizes the behavior of the device subjected to a disturbance having infinite amplitude but infinitesimal duration. The frequency response method of analysis is most useful when the heat exchanger is a part of the system and an overall dynamic behavior of several pieces of equipment needs to be synthesized. However, the disturbances that occur in most heat exchangers are approximated closely by step functions than by periodic or impulse functions. Hence, in this paper, we will describe the transient behavior of heat exchangers subjected to a step function at the inlet. The frequency or impulse response can be determined from the step response as described by Raven [1] among others.

In this paper, we will restrict the transient response analysis and discussion to single-phase two-fluid heat exchangers. Even then, the mathematical solutions for the transient performance of the exchangers are difficult to obtain due to a large number of dimensionless groups (to be described next) associated with the problem. Only some simplified solutions, available in the open literature, are presented in this paper. Although these solutions provide a feel and understanding of the transient problem, they may be too approximate in reality. In order to determine accurate transient response for a specific application, it is necessary to solve the problem numerically without invoking idealizations that are common in the simplified analyses. The current trend is to use sophisticated computer programs for the transient analysis of heat exchangers.

In the following sections, first, we will formulate the transient problem by specifying idealizations, deriving governing differential equations, outlining dimensional and dimensionless variables, and providing an implicit solution. Next, some solutions will be summarized. The transient performance problem will be outlined separately for the direct transfer type and storage-type heat exchangers.

## 2. DIRECT TRANSFER TYPE EXCHANGERS

In a direct transfer type heat exchanger, two fluids are separated by a thin wall (parting plates or tube walls) through which heat flows. There are no moving parts in the exchanger. Although simultaneous flow of both fluids is required in the exchanger, there is no mixing of two fluids. This type of exchanger is also referred to as a <u>recuperator</u>. Plate-fin and tube-fin extended surface exchangers and shell-and-tube exchangers are the common examples.

First the transient response problem is formulated for a step change in inlet temperature of the hot fluid as an example. The problem formulation includes a list of idealizations, dimensional and dimensionless variables for the problem, and a generalized implicit form of solution. Next some specific solutions are presented for a step change in inlet temperature of either fluid for the following exchangers: $C^* = 0$ exchanger, counterflow exchanger with $C^* = 1$, crossflow exchanger, and shell-and-tube exchanger.

### 2.1 Problem Formulation

<u>Idealizations</u>. The following idealizations are built into the governing differential equations presented in the next subsection.

# THE TRANSIENT RESPONSE OF HEAT EXCHANGERS

1. The temperatures of both fluids and the wall are functions of time $\tau$ and the position $x$ (for counterflow or parallel flow exchanger) or $x$ and $z$ (for crossflow exchanger only).

2. Heat transfer between the exchanger and the surroundings is negligible. There are no thermal energy sources within the exchanger. No phase change occurs in the exchanger.

3. The mass flow rates of both fluids, although may be different, do not vary with time. The fluids are uniformly distributed.

4. The velocity and temperature of each fluid at the inlet are uniform over the flow cross section and are constant with time except for the imposed step change.

5. The convective heat transfer coefficient on each side, and the thermal properties of both fluids and the wall are constant, independent of temperature, time and position.

6. Longitudinal heat conduction within the fluids and wall as well as the transverse conduction within the fluid is neglected.

7. The heat transfer surface area on each fluid side is uniformly distributed.

8. Either the wall thermal resistance and fouling resistances are negligible or they are lumped with convective thermal resistances on hot and cold sides.

9. The thermal capacitance of the shell of the exchanger is considered negligible relative to that of the heat transfer surface.

The second through seventh idealizations parallel to those usually made for the steady-state behavior of exchangers. The eighth and ninth idealizations are necessary for the simplified transient analysis to follow.

Governing differential equations and dimensional variables. On the basis of the foregoing idealizations, let us derive the governing differential equations and boundary conditions. Consider a counterflow recuperator of Fig. 1.[†] An elemental flow passage of each fluid and the associated wall (heat transfer surface) are presented in Fig. 1a. Control volumes of length $dx$ each passing through a hot fluid passage, wall, and a cold fluid passage internally in a heat exchanger are shown in Fig. 1b. Various energy transfer terms associated with three control volumes are also shown in Fig. 1b.

Before setting up the differential equations, let us define the heat capacitance terms $\bar{C}_w$, $\bar{C}_h$ and $\bar{C}_c$ for the fluids and the wall. The heat capacitance of the wall is defined as

$$\bar{C}_w = M_w c_w \qquad (1)$$

Here $M_w$ is the mass of wall (mass of all heat transfer surface available on hot and cold fluid sides), and $c_w$ is the specific heat of the wall material. The hot fluid heat capacitance $\bar{C}_h$ and heat capacity rate $C_h$ are defined and

---

[†] Although the differential equations are derived for a counterflow recuperator, the results are presented for counterflow, crossflow and all flow arrangements with $C^* = 0$.

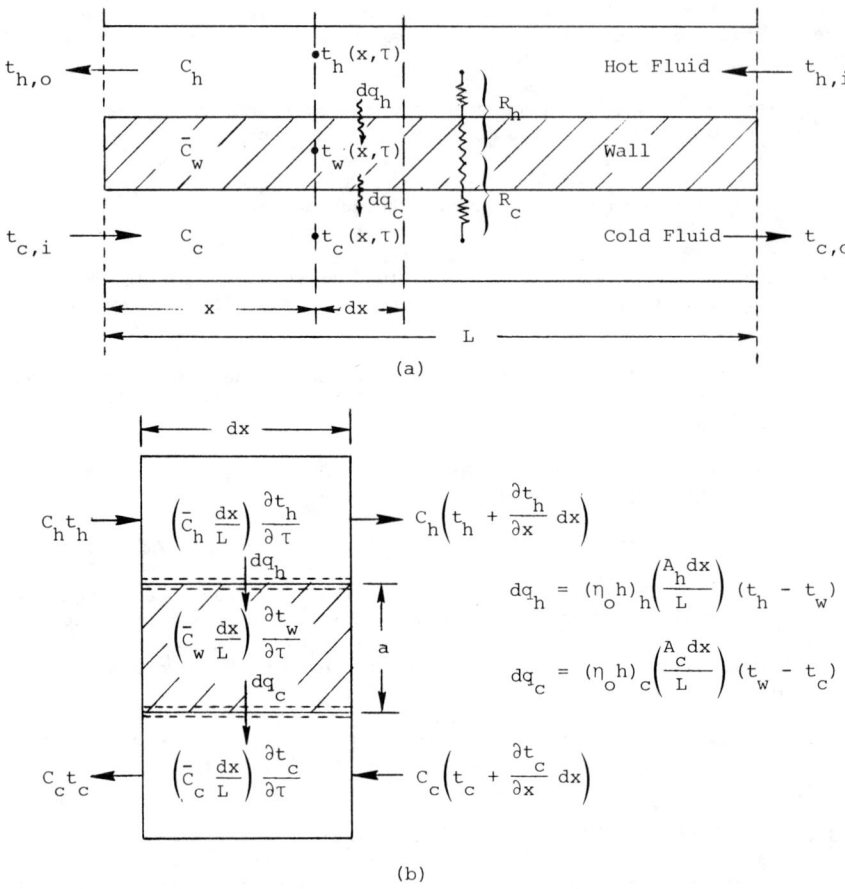

Fig. 1 (a) Elemental fluid flow passages and associated wall of a counter-flow exchanger showing sections x and dx, (b) energy transfer terms associated with control volumes for hot fluid, wall, and cold fluid dx elements.

related as follows.

$$\bar{C}_h = M_h c_{p,h} \qquad C_h = W_h c_{p,h} = \frac{u_{m,h} \bar{C}_h}{L} = \frac{\bar{C}_h}{\tau_{d,h}} \qquad (2)$$

Here $M_h$ is the mass of the hot fluid contained in the exchanger at any instant of time, $c_{p,h}$ is the specific heat of the hot fluid, $u_{m,h}$ is the mean axial velocity of the hot fluid, L is the exchanger length, and $\tau_{d,h}$ is the hot fluid dwell time. Similarly, the cold fluid heat capacitance and capacity rate are

$$\bar{C}_c = M_c c_{p,c} \qquad C_c = W_c c_{p,c} = \frac{u_{m,c} \bar{C}_c}{L} = \frac{\bar{C}_c}{\tau_{d,c}} \qquad (3)$$

# THE TRANSIENT RESPONSE OF HEAT EXCHANGERS

Now let us review Fig. 1b. During its flow through the control volume, the hot fluid transfers heat to the wall by convection resulting in reduction in its outlet enthalpy and internal stored thermal energy. Applying the energy balance, the first law of thermodynamics, to the hot fluid control volume, we get

$$C_h \left( t_h + \frac{\partial t_h}{\partial x} dx \right) + (\eta_o h)_h \left( \frac{A_h dx}{L} \right)(t_h - t_w) - C_h t_h + \left( \bar{C}_h \frac{dx}{L} \right)\frac{\partial t_h}{\partial \tau} = 0 \qquad (4)$$

Simplification and rearrangement yields

$$\bar{C}_h \frac{\partial t_h}{\partial \tau} + C_h \frac{\partial t_h}{\partial (x/L)} + (\eta_o hA)_h (t_h - t_w) = 0 \qquad (5)^\dagger$$

Similar energy balances on the cold fluid and wall result

$$\bar{C}_c \frac{\partial t_c}{\partial \tau} - C_c \frac{\partial t_c}{\partial (x/L)} - (\eta_o hA)_c (t_w - t_c) = 0 \qquad (6)$$

$$\bar{C}_w \frac{\partial t_w}{\partial \tau} - (\eta_o hA)_h (t_h - t_w) + (\eta_o hA)_c (t_w - t_c) = 0 \qquad (7)$$

The initial conditions for the fluids and wall temperature are the corresponding steady-state temperatures before the step input on the hot fluid inlet temperature. These steady-state temperatures are designated with a single quotation mark.

$$t_h(x,0) = t_h'(x,0) \qquad (8)$$

$$t_c(x,0) = t_c'(x,0) \qquad (9)$$

$$t_w(x,0) = t_w'(x,0) \qquad (10)$$

The boundary conditions are

$$t_h(0,\tau) = \begin{cases} t_{h,i} & \text{for } \tau \geq 0 \\ t_{h,i}' & \text{for } \tau < 0 \end{cases} \qquad (11)$$

$$t_c(L,\tau) = t_{c,i} \qquad (12)$$

Based on the foregoing differential equations, initial conditions and boundary conditions, the dependent fluid and wall temperatures are functions of the following variables and parameters.

---

$^\dagger \eta_o$ is the temperature effectiveness of total heat transfer area on the one side of an extended surface heat exchanger, $\eta_o = 1 - (1-\eta_f)A_f/A$ where $\eta_f$ is the fin efficiency and $A_f$ is the fin surface area.

$$\underbrace{t_h, t_c, t_w}_{\substack{\text{Dependent} \\ \text{variables}}} = \phi \Big\{ \underbrace{x, t_{h,i}, t_{c,i}, C_h, C_c, (\eta_o hA)_h, (\eta_o hA)_c, \text{flow arrangement,}}_{\text{Steady-state variable and parameters}}$$

$$\underbrace{\tau, \bar{C}_w, \bar{C}_h, \bar{C}_c}_{\substack{\text{Transient variable} \\ \text{and parameters}}} \Big\} \qquad (13)$$

Notice that in the steady-state analysis, $(\eta_o hA)_h$ and $(\eta_o hA)_c$ appear only if we are interested in determining the wall temperature distribution. Otherwise, these parameters are lumped into only one parameter, UA, the overall conductance.

*Dimensionless groups and implicit solutions.* Eleven independent variables and parameters exist for the dependent fluid and wall temperatures for a given flow arrangement. The independent variables and parameters will be reduced in number by formulating appropriate dimensionless groups. The specific form of these groups is somewhat optional. We will formulate them in such a way that we will retain all the steady-state dimensionless groups and come up with additional groups that are due to the transient problem. Since we are interested in the fluid temperature responses for a step change in the inlet temperature of one of the fluids, let us first define them in a dimensionless form. $t_h^*$ is defined as the hot fluid temperature change at any $x$ and at any $\tau$ from its initial value normalized with respect to the ultimate change at time equal to infinity. The initial temperature and the ultimate change in temperature are specified by the boundary and initial conditions and are constant with time.

$$t_h^*(x,\tau) = \frac{\Delta t_h(x,\tau)}{\Delta t_h(x,\infty)} = \frac{t_h(x,\tau) - t_h(x,0)}{t_h(x,\infty) - t_h(x,0)} \qquad (14)$$

where the various temperatures are defined in Fig. 2. Similarly the cold fluid dimensionless temperature response $t_c^*$ is defined as

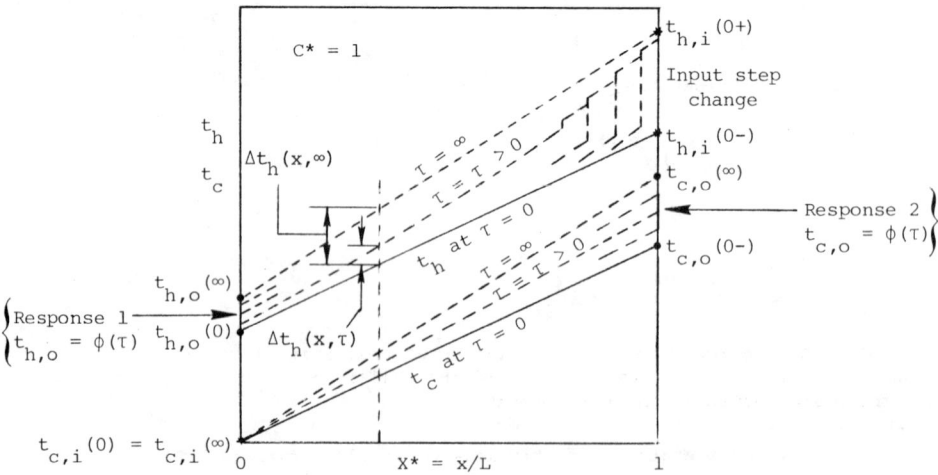

Fig. 2 Identification of the temperature responses for counterflow exchanger.

# THE TRANSIENT RESPONSE OF HEAT EXCHANGERS

$$t_c^*(x,\tau) = \frac{t_c(x,\tau) - t_c(x,0)}{t_c(x,\infty) - t_c(x,0)} \tag{15}$$

Note that

$$t_h^* = 0 \text{ at } \tau = 0, \qquad t_h^* \to 1 \text{ at } \tau \to \infty \tag{16}$$

$$t_c^* = 0 \text{ at } \tau = 0, \qquad t_c^* \to 1 \text{ at } \tau \to \infty \tag{17}$$

Now let us formulate the independent dimensionless variables and parameters for the transient problem. First the steady-state independent dimensionless groups are

$$X^* = x/L = \text{dimensionless flow length variable} \tag{18}$$

$$N_{tu} = UA/C_{min} = \text{number of transfer units} \tag{19}$$

$$C^* = C_{min}/C_{max} = \text{capacity rate ratio of fluids} \tag{20}$$

$$R^* = \frac{(\eta_o hA) \text{ on the } C_{max} \text{ side}}{(\eta_o hA) \text{ on the } C_{min} \text{ side}} = \text{thermal resistance ratio} \tag{21}$$

Here $C_{min}$ is the minimum of $C_h$ and $C_c$, and $C_{max}$ is the maximum of $C_h$ and $C_c$. $UA$ is defined as

$$\frac{1}{UA} = \frac{1}{(\eta_o hA)_h} + \frac{1}{(\eta_o hA)_c} \tag{22}$$

The wall resistance and fouling resistances, if important, are included in the hot and cold side thermal resistances, as mentioned in the idealization 8. Note that even if we are not interested in the wall temperature response, $R^*$ will appear as an independent parameter for the transient problem as long as $\bar{C}_w^*$ is not zero. $\bar{C}_w^*$ is defined in Eq. (24).

The additional dimensionless groups for the transient problem will now be defined. The generalized time and wall heat capacitance groups are defined straightforward.

$$\tau^* = \tau/\tau_{d,min} = \text{dimensionless time variable} \tag{23}$$

$$\bar{C}_w^* = \bar{C}_w/\bar{C}_{min} = \text{wall capacitance ratio} \tag{24}$$

The fluid capacitance ratio can also be defined as $\bar{C}_{min}/\bar{C}_{max}$. However, since

$$\frac{\bar{C}_{min}}{\bar{C}_{max}} = \frac{C_{min} \tau_{d,min}}{C_{max} \tau_{d,max}} = C^* \tau_d^* \tag{25}$$

and $C^*$ is already considered as a dimensionless parameter, the new dimensionless parameter is then the dwell time ratio $\tau_d^*$.

$$\tau_d^* = \frac{\tau_{d,min}}{\tau_{d,max}} = \frac{\tau_d \text{ on the } C_{min} \text{ side}}{\tau_d \text{ on the } C_{max} \text{ side}} = \text{dwell time ratio} \qquad (26)$$

Note that the values of $\tau_d^*$ and $R^*$ may be greater than, less than, or equal to unity. The subscript "min" refers to the quantity associated with the minimum of $C_c$ and $C_h$; and $\tau_{d,min}$ is not necessarily less than $\tau_{d,max}$.

Thus the dependent fluid temperature responses (to a step change in the inlet temperature of one of the fluids) are functions of three steady-state groups and four additional groups due to the transient problem.

$$t_h^*, t_c^* = \phi\{\underbrace{X^*, N_{tu}, C^*}_{\text{Steady-state}}, \underbrace{\tau^*, R^*, \bar{C}_w^*, \tau_d^*}_{\text{Transient}}\} \qquad (27)$$

Since we are primarily interested in the fluid output temperature responses, we can eliminate the variable $X^*$ by evaluating $t_h^*$ at $X^* = 0$ and $t_c^*$ at $X^* = 1$ for the example of Fig. 2. Since the step change in the inlet temperature can be imposed on either of the two fluids, we will have the responses $t_{h,o}^*$ and $t_{c,o}^*$ for each case of the imposed step change. An alternative designation for these responses is proposed by Cima and London [2] in terms of the time dependent exchanger effectiveness as follows.

For steady-state heat exchanger analysis, the dimensionless temperature is the exchanger effectiveness defined as the actual heat transfer rate divided by thermodynamically limited maximum possible heat transfer rate [3].

$$\varepsilon = \frac{C_h(t_{h,i}-t_{h,o})}{C_{min}(t_{h,i}-t_{c,i})} = \frac{C_c(t_{c,o}-t_{c,i})}{C_{min}(t_{h,i}-t_{c,i})} \qquad (28)$$

The concept of this effectiveness may be extended to the time-dependent effectiveness for each fluid as

$$\varepsilon_h(\tau) = \frac{C_h[(t_{h,i}(\tau) - t_{h,o}(\tau)]}{C_{min}[t_{h,i}(\tau) - t_{c,i}(\tau)]} \qquad (29)$$

$$\varepsilon_c(\tau) = \frac{C_c[t_{c,o}(\tau) - t_{c,i}(\tau)]}{C_{min}[t_{h,i}(\tau) - t_{c,i}(\tau)]} \qquad (30)$$

In order to represent the transient response, a generalized effectiveness for each fluid may now be defined as follows.

$$\varepsilon_{f,h}^*(\tau) = \frac{\varepsilon_h(\tau) - \varepsilon_h(\tau=0+)}{\varepsilon_h(\tau=\infty) - \varepsilon_h(\tau=0+)} \qquad (31)$$

$$\varepsilon_{f,c}^*(\tau) = \frac{\varepsilon_c(\tau) - \varepsilon_c(\tau=0+)}{\varepsilon_c(\tau=\infty) - \varepsilon_c(\tau=0+)} \qquad (32)$$

Consider the example of Fig. 2 to further define these effectivenesses explicitly. A step change in temperature is imposed on the hot fluid at inlet. In this case, we will designate $\varepsilon^*_{f,h}(\tau)$ of Eq. (31) as $\varepsilon^*_{f,1}(\tau)$. Now evaluate $\varepsilon_h(\tau=0+)$, $\varepsilon_h(\tau)$ and $\varepsilon_h(\tau=\infty)$ from the definition of Eq. (29) and the following relationships.

$$t_{h,i}(0+) = t_{h,i}(\tau) = t_{h,i}(\infty) \tag{33}$$

and
$$t_{c,i}(0+) = t_{c,i}(\tau) = t_{c,i}(\infty) \tag{34}$$

Substituting the resulting expressions into Eq. (31) yields

$$\varepsilon^*_{f,1} = \frac{t_{h,o}(\tau) - t_{h,o}(0)}{t_{h,o}(\infty) - t_{h,o}(0)} \tag{35}^\dagger$$

Similarly, we get the following expression for the dimensionless transient response in the outlet temperature of the cold fluid $\varepsilon^*_{f,c}(\tau)$, designated as $\varepsilon^*_{f,2}$, for the example of Fig. 2.

$$\varepsilon^*_{f,2} = \frac{t_{c,o}(\tau) - t_{c,o}(0)}{t_{c,o}(\infty) - t_{c,o}(0)} \tag{36}^\dagger$$

Note that $\varepsilon^*_{f,1}$ and $\varepsilon^*_{f,2}$ represent the ratio of the corresponding fluid temperature change from the initial value to its ultimate change at a time is equal to infinity.

$$\varepsilon^*_{f,1} = 0 \text{ at } \tau = 0, \qquad \varepsilon^*_{f,1} \to 1 \text{ at } \tau \to \infty \tag{37}$$

$$\varepsilon^*_{f,2} = 0 \text{ at } \tau = 0, \qquad \varepsilon^*_{f,2} \to 1 \text{ at } \tau \to \infty \tag{38}$$

Although $\varepsilon^*_{f,1}$ and $\varepsilon^*_{f,2}$ are defined above for the specific example of Fig. 2, the subscripts 1 and 2 have a more general meaning. The subscript 1 denotes the response on leaving the exchanger of that fluid which had the step input imposed on it at its inflow section. The subscript 2 denotes the response of the other fluid at its outlet section. $\varepsilon^*_{f,1}$ and $\varepsilon^*_{f,2}$ are referred to as the self response and cross response, respectively. Thus the magnitude of $\varepsilon^*_{f,1}$ or $\varepsilon^*_{f,2}$ may be considered as a measure of degree of approach of the outlet temperature in question to the new equilibrium condition brought on by the step function input. $\varepsilon^*_{f,1}$ and $\varepsilon^*_{f,2}$ are related to previously defined $t^*_{h,o}$ and $t^*_{c,o}$ as follows. For the step change in the inlet temperature of the hot fluid

---

$^\dagger$For constant properties, a step change in inlet temperature of one fluid does not affect $N_{tu}$ and $C^*$. Hence the steady-state exchanger effectiveness $\varepsilon$ after the step input remains the same as that before the step input. The steady-state fluid outlet temperatures after the step input [needed for Eqs. (35) and (36)] are calculated from the known $\varepsilon$, $C^*$, inlet temperatures and the definition of $\varepsilon$.

$$\varepsilon^*_{f,1} = t^*_{h,o} \qquad \varepsilon^*_{f,2} = t^*_{c,o} \qquad (39)$$

For the step change in the inlet temperature of the cold fluid

$$\varepsilon^*_{f,1} = t^*_{c,o} \qquad \varepsilon^*_{f,2} = t^*_{h,o} \qquad (40)$$

Thus in general, the transient fluid temperature responses (to a step change in one of the fluid inlet temperatures) are dependent upon six dimensionless groups for each different flow arrangement of the two-fluid exchanger.

$$\varepsilon^*_{f,1}, \varepsilon^*_{f,2} = \phi\{\tau^*, N_{tu}, C^*, \bar{C}^*_w, R^*, \tau^*_d\} \qquad (41)$$

The general transient heat exchanger problem is quite complex since it involves a solution of three simultaneous partial differential equations for temperatures as functions of time and position. A variety of initial and boundary conditions are possible. Also the solution depends upon seven independent groups [see Eq. (27)]. Hence, no general solution has been obtained for a transient two-fluid exchanger problem. Solutions are available only for a few special cases of technical interest. Kays and London [3] provide a comprehensive summary of eighteen solutions obtained up to 1964. In these solutions, $C^*$ is specialized to be either zero or unity. Thus Eq. (41) reduces to

$$\varepsilon^*_{f,1}, \varepsilon^*_{f,2} = \phi\{\tau^*, N_{tu}, \bar{C}^*_w, R^*, \tau^*_d\} \qquad (42)$$

An unwiedly group of five independent parameters still remain. Therefore, some further constaints are imposed such as $\bar{C}^*_w$ being very large (for gas side of the exchanger) or zero (for condensing and/or evaporating fluids on both sides), or $\tau^*_d = 1$ (fluids having the same velocity and flow lengths) for special solutions provided by Kays and London [3].

The solutions for the transient response are presented now separately for heat exchangers with $C^* = 0$, counterflow exchangers with $C^* = 1$, counterflow exchangers with $C^* < 1$, single-pass crossflow exchangers, and shell-and-tube exchangers. The following restrictions are implied for these solutions: (1) The initial conditions, Eqs. (8)-(10), existing at time $\tau=0-$ always correspond to the steady-state conditions. (2) Most of the solutions presented have the boundary condition of Eq. (11), i.e., either a step change in $t_{h,i}$ or $t_{c,i}$. Only some solutions are for a step change in the flow rate. (3) Solutions are presented only for the fluid temperature response at the exchanger outlet section. The fluid temperature responses within the exchanger or the wall temperature responses are not presented.

## 2.2 Heat Exchangers with $C^* = 0$

In many heat exchangers, such as condensers, evaporators, intercoolers, pre-coolers, and liquid-to-gas heat exchangers, the heat capacity rate of one fluid is much larger than that of the other fluid. For these exchangers, the temperature of the $C_{max}$ fluid can be approximated as constant throughout the exchanger, and hence $\bar{C}^*_{max} \to \infty$, and $\tau_{d,max} \to 0$. We will consider the temperature responses to a step change in inlet temperature of (a) constant temperature

# THE TRANSIENT RESPONSE OF HEAT EXCHANGERS

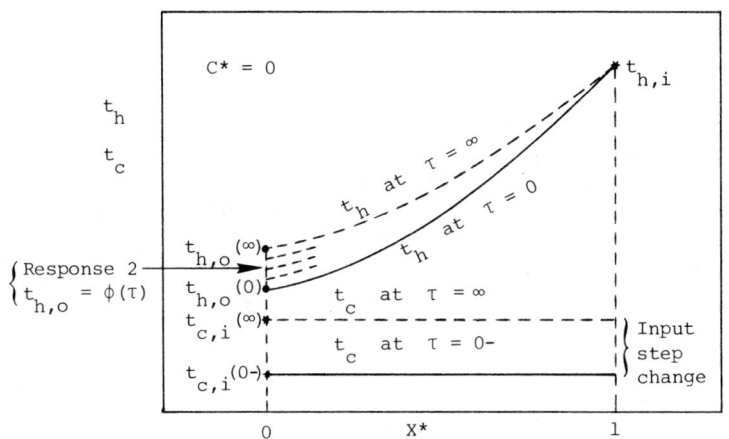

Fig. 3  Temperature response for the $C_{min}$ fluid for a step change in temperature of the $C_{max}$ fluid.

($C_{max}$) fluid, Fig. 3, and (b) variable temperature ($C_{min}$) fluid, as shown later in Fig. 8. The results are applicable to the counterflow, parallel flow, crossflow, or any other arrangement heat exchanger.

Step change in inlet temperature of the $C_{max}$ fluid. In this case, the sudden changes in the inlet temperature of the condensing or evaporating fluid occur due to sudden changes in the system pressure. Or a step change is imposed on the inlet temperature of liquid coolant. A typical temperature response for the $C_{min}$ fluid in an evaporator is shown in Fig. 3. In this case, since $C^* = 0$ and $\tau_d^* = \infty$, Eq. (41) for $\varepsilon_{f,2}^*$ reduces to

$$\varepsilon_{f,2}^* = \phi\{\tau^*, N_{tu}, \bar{C}_w^*, R^*\} \tag{43}$$

and by the problem specification,

$$\varepsilon_{f,1}^* = 1 \tag{44}$$

The solution $\varepsilon_{f,2}^*$ has been obtained as a function of $N_{tu}$ by Rizika [4], London et al. [5], and Myers et al. [6] for different ranges of $\tau^*$, $\bar{C}_w^*$ and $R^*$.

Rizika [4] obtained the following exact solution for $0 \leq \tau^* \leq 1$ valid for all $\bar{C}_w^*$ and $R^*$.

$$\varepsilon_{f,2}^* = \frac{1 - e^{-X}[Y \sinh(X/Y) + \cosh(X/Y)]}{1 - \exp(-N_{tu})} \tag{45}$$

where
$$X = \frac{N_{tu}(1 + R^*)(1 + R^* + \bar{C}_w^*)\tau^*}{2R^*\bar{C}_w^*} \tag{46}$$

$$Y = \left[1 - \frac{4R^*\bar{C}^*_w}{(1 + R^* + \bar{C}^*_w)^2}\right]^{-1/2} \tag{47}$$

Kays and London [3] present the solution of Eq. (45) graphically by their Fig. 3-6. For specific values of $N_{tu}$ and $R^*$, this solution is presented later in Fig. 5 for $\tau^* \leq 1$.[†] It is observed from this figure that Rizika's solution is useful only for $\bar{C}^*_w < 1$ for $N_{tu} = 1$ and $\bar{C}^*_w \leq 3$ for which $\varepsilon^*_{f,2} \approx 0.9$ at $\tau^* = 1$. For a condenser/evaporator problem with a phase changing fluid on one side and a liquid on the other side, $\bar{C}^*_w$ may be an order of magnitude 1, and Rizika's solution would be useful. However, if a gas instead of a liquid is on the other side (the $C_{min}$ side) of the exchanger, $\bar{C}^*_w \approx 1,000$ as in an air-cooled condenser, an intercooler or a precooler. Then a limit of $\tau^* \leq 1$ would limit the calculable $\varepsilon^*_{f,2}$ from Rizika's solution to less than 0.01. Note that $\varepsilon^*_{f,2}$ at $\tau^* = 1$ represents the outlet temperature response for the $C_{min}$ fluid after one dwell time ($=\tau_d$), that is the fluid particle which experienced the step change at inlet has just reached the outlet. The temperature of this fluid particle will change substantially only if $\bar{C}^*_w(=\bar{C}_w/\bar{C}_{min})$ is low. For this reason Eq. (45) is used for $\bar{C}^*_w \leq 1$.

London et al. [5] also present analytical solutions for two limiting cases, $R^* = \infty$ and 0 as follows.

For $R^* = \infty$ and any $\bar{C}^*_w$

$$\varepsilon^*_{f,2} = \begin{cases} \dfrac{1 - \exp(-N_{tu}\tau^*)}{1 - \exp(-N_{tu})} & \text{for } \tau^* \leq 1 \\[6pt] 1 & \text{for } \tau^* \geq 1 \end{cases} \tag{48}$$

For $R^* = 0$ and any $\bar{C}^*_w$

$$\varepsilon^*_{f,2} = \begin{cases} \dfrac{1 - \exp\{-N_{tu}[\tau^*/(1 + \bar{C}^*_w)]\}}{1 - \exp(-N_{tu})} & \text{for } \dfrac{\tau^*}{(1+\bar{C}^*_w)} \leq 1 \\[6pt] 1 & \text{for } \dfrac{\tau^*}{(1+\bar{C}^*_w)} \geq 1 \end{cases} \tag{49}$$

The temperature responses of Eqs. (48) and (49) are shown in Fig. 4. It can be shown that these temperature responses are also valid for the cases of any value of $R^*$ and $\tau^*_w = 0$ [5].

In addition to the foregoing cases, Kays and London also present additional solutions, obtained by electromechanical analog tests, for some specific values of $N_{tu}$, $\bar{C}^*_w$ and $R^*$. However, improvements in these solutions have been made by Myers et al. [6] who obtained rigorous finite difference solutions for intermediate and large values of $\bar{C}^*_w$. Figures 5, 6, and 7 are prepared using the results obtained from the computer program of Myers [7]. The following observations may be made from these figures: (1) Two independent groups $\tau^*$ and $\bar{C}^*_w$ have been combined into one $(\tau^*-1)/\bar{C}^*_w$ as plotted in abscissa. As found from Fig. 5, the transient is essentially complete by $(\tau^*-1)/\bar{C}^*_w = 1$. This observation has been substantiated by 27 combinations of $N_{tu}$, $\bar{C}^*_w$, and $R^*$ that Myers et al. [6] considered. This means $\tau^* = 1 + \bar{C}^*_w$ is the dimensionless time required for the heat exchanger to respond. Thus the larger the wall capacitance, the longer is the response time. (2) An increase in the value of $N_{tu}$ reduces the response

---

[†]The results for $\tau^* > 1$ in Fig. 5 are from the finite difference solution of Myers et al. [6] and are discussed in the last paragraph of this page.

# THE TRANSIENT RESPONSE OF HEAT EXCHANGERS

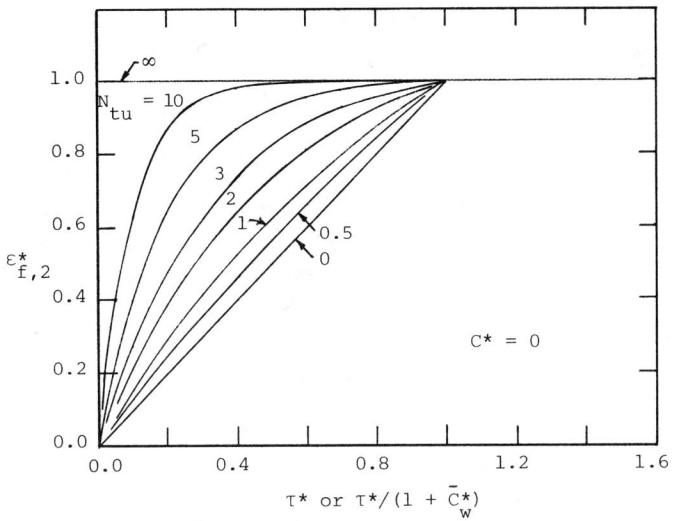

Fig. 4 The transient response solutions of Eqs. (48) and (49).

time as found from Fig. 6. (3) An increase in the value of $R^*$ also reduces the response time as found from Fig. 7. For $R^* = \infty$, a 100% response is realized in one dwell time. In contrast, for $R^* = 0$, from Eq. (49) or Fig. 7, a 100% response is realized in $\tau^*/(1+\bar{C}_w^*) = 1$ or $\tau^* \simeq \bar{C}_w^*$ for a large $\bar{C}_w^*$. That means $\tau = \bar{C}_w^* \tau_{d,min}$ or $\bar{C}_w^*$ dwell times are required for the 100% response for $R^* = 0$. The design range for $R^*$ is 1-3 by incorporating extended surfaces on the gas side.

Myers et al. [6] also proposed the following approximate equation valid for $1 < \bar{C}_w^* \leq 2{,}000$ and $\tau^* > 1$.

$$\varepsilon_{f,2}^* = 1 - \tilde{A}\exp[-\tilde{B}(\tau^*-1)/\bar{C}_w^*] \tag{50}$$

where
$$\tilde{A} = 1 - \frac{1 - e^{-Z}[Y\sinh(Z/Y) + \cosh(Z/Y)]}{1 - \exp(-N_{tu})} \tag{51}$$

$$\tilde{B} = \frac{1}{\tilde{A}}\left[\frac{2N_{tu}(1+R^*)\bar{C}_w^*}{(1 + R^* + \bar{C}_w^*)}\right]\left[\frac{Y e^{-Z}\sinh(Z/Y)}{1 - \exp(-N_{tu})}\right] \tag{52}$$

$$Z = \frac{N_{tu}(1 + R^*)(1 + R^* + \bar{C}_w^*)}{2R^*\bar{C}_w^*} \tag{53}$$

and $Y$ is defined by Eq. (47). A detailed comparison of this approximate solution with the finite difference solution for all 27 cases indicated an excellent agreement between the two solutions for $R^* \geq 1$. In heat exchangers with $C^* \simeq 0$, the thermal resistance on the $C_{min}$ side is always much greater than that on the $C_{max}$ side, and hence $R^* > 1$. Thus the closed-form expression of Eq. (50) represents an excellent approximation to determine the transient response within the specified ranges of $\bar{C}_w^*$ and $\tau^*$.

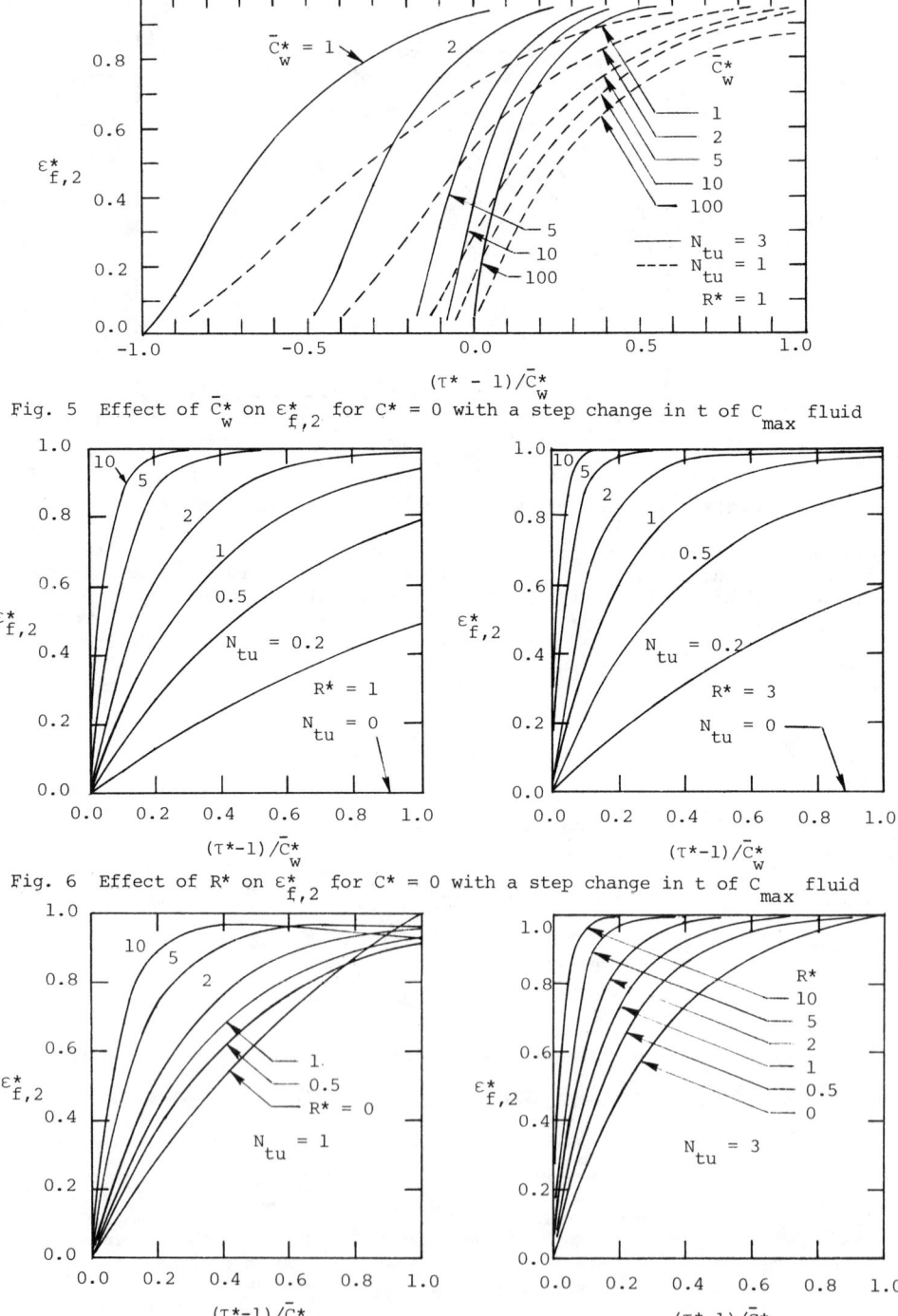

Fig. 5 Effect of $\bar{C}_w^*$ on $\varepsilon_{f,2}^*$ for $C^* = 0$ with a step change in t of $C_{max}$ fluid

Fig. 6 Effect of $R^*$ on $\varepsilon_{f,2}^*$ for $C^* = 0$ with a step change in t of $C_{max}$ fluid

Fig. 7 Effect of $N_{tu}$ on $\varepsilon_{f,2}^*$ for $C^* = 0$ with a step change in t of $C_{max}$ fluid

For real large values of $\bar{C}_w^*$, Myers et al. [6] presented the following asymptotic solution.

$$\varepsilon_{f,2}^* = \frac{1}{1 - \exp(-N_{tu})} \left\{ 1 - e^{-N_{tu}\zeta_1} - e^{-N_{tu}} \phi_o \left[ N_{tu}(1+R^*)\zeta_1/R^*, N_{tu}/R^* \right] \right.$$

$$\left. + e^{-N_{tu}\zeta_1} \phi_o \left[ N_{tu}\zeta_1/R^*, N_{tu}(1+R^*)/R^* \right] \right\} \tag{54}$$

where
$$\zeta_1 = (1+R^*)(\tau^*-1)/\bar{C}_w^* \tag{55}$$

and
$$\phi_o(x,y) = e^{-y} \int_{\tau=0}^{x} e^{-\tau} I_o(2\sqrt{y}\sqrt{\tau}) d\tau \tag{56}$$

The function $\phi_o(x,y)$ has been tabulated by Binkley et al. [8].

Step change in the inlet temperature of the $C_{min}$ fluid. In this case, the response of the outlet temperature of the $C_{min}$ fluid is obtained for a step change in its inlet temperature. A typical temperature response of the $C_{min}$ fluid is shown in Fig. 8. This type of transient problem is more common in practice compared to the one in the preceding section that has a step change in the inlet temperature of the $C_{max}$ fluid.

For the present case, the independent dimensionless groups for the solution $\varepsilon_{f,1}^*$ are the same as those of the preceding section, Eq. (43).

$$\varepsilon_{f,1}^* = \phi\{\tau^*, N_{tu}, \bar{C}_w^*, R^*\} \tag{57}$$

The $C_{max}$ fluid has infinite heat capacitance and hence its temperature does not change at all in this case, i.e., $\varepsilon_{f,2}^* = 1$. Alternatively, if the $C_{max}$ fluid is the cold fluid as shown in Fig. 8, then $t_{c,o}(\infty) = t_{c,o}(0)$.

Kays and London [3] outline the explicit solutions of Eq. (57) in their Figs. 3-11, 3-12, and 3-13 for specific values of $N_{tu}$, $\bar{C}_w^*$, and $R^*$. These approximate solutions were obtained by the electromechanical analog results. These results have been found incorrect by Myers et al. [9] who analyzed the problem analytically and showed that four independent variables of Eq. (57) can be reduced to two.

$$\varepsilon_{f,1}^* = \phi(\zeta_2, N_{tu}/R^*) \tag{58}†$$

where
$$\zeta_2 = (1+R^*)^2(\tau^*-1)/\bar{C}_w^* \tag{59}$$

Myers et al. [6] provided the following exact solution for $\tau^* > 1$.‡

---

†The variable $R^*$ of Myers et al. [6,9] in the present terminology is $1/R^*$. Throughout this chapter, $R^*$ is consistently defined as Eq. (21) unless specified otherwise.

‡Note that for $\tau^* < 1$, $\varepsilon_{f,1}^* = 0$, and $\tau^* = 0$ represents a point of singularity.

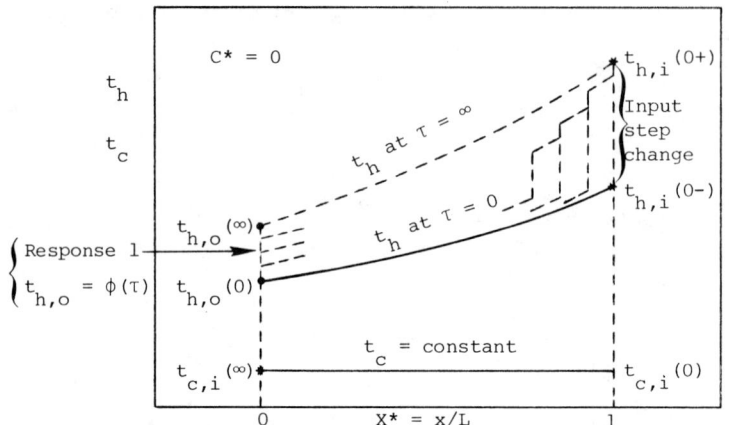

Fig. 8 Outlet temperature response for the $C_{min}$ fluid.

$$\varepsilon^*_{f,1} = \phi_o[\zeta_2 N_{tu}/R^*, N_{tu}/R^*] + \exp(-N_{tu}/R^*)\exp(-\zeta_2 N_{tu}/R^*) I_o[(2N_{tu}/R^*)\zeta_2^{1/2}] \qquad (60)$$

Here $I_o(\ )$ represents the modified Bessel function of the first kind and zero order and the function $\phi_o(x,y)$ is defined in Eq. (56). The temperature response expressed by Eq. (60) is shown in Fig. 9. To illustrate the order of magnitude of the time required for approximately 90% temperature response, $\zeta_2 \simeq 2$ from Fig. 9. Thus from Eq. (59),

$$\tau = \left[1 + \frac{2\bar{C}^*_W}{(1+R^*)^2}\right]\tau_{d,min} \qquad (61)$$

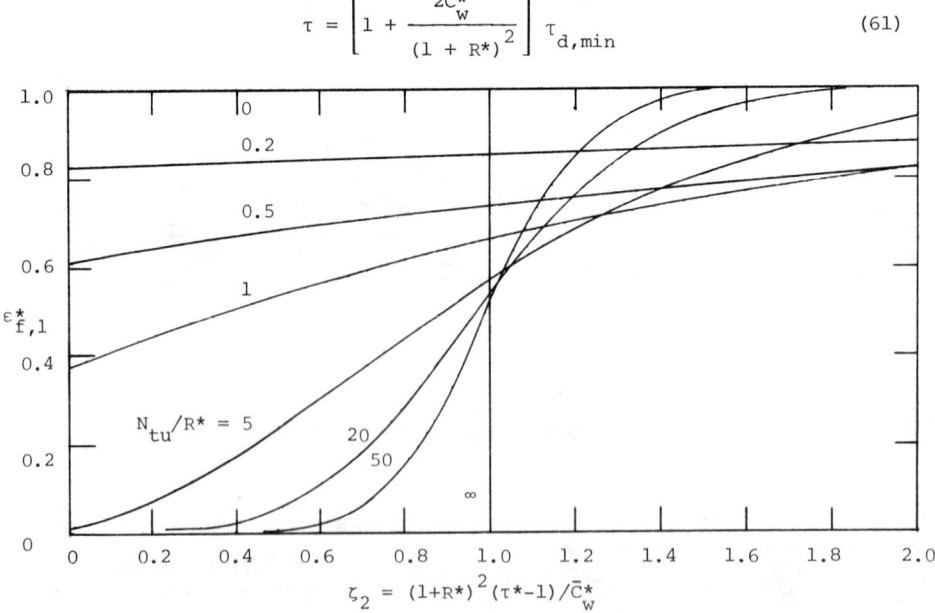

Fig. 9 Outlet temperature response for a step change in inlet temperature of the $C_{min}$ fluid for the $C^* = 0$ case [6].

THE TRANSIENT RESPONSE OF HEAT EXCHANGERS                                              931

Thus the response time will be from one dwell time for $R^* = \infty$ to $(1 + 2\bar{C}^*_w)$ dwell times for $R^* = 0$. For a well-designed exchanger, $R^* \simeq 1$, and hence the response time will be approximately $(1 + \bar{C}^*_w/2)$ dwell times.

Step change in the flow rate of the $C_{min}$ fluid for $C^* = 0$. London et al. [5] conducted analog tests and found that a 100% temperature response to a flow rate change in the $C_{min}$ fluid is virtually attained in one dwell time. As discussed in the preceding subsections, there is a large variation in the dwell time required for 90-100% temperature response for a step change in the inlet temperature. This variation is dependent upon the values of $N_{tu}$, $\bar{C}^*_w$ and $R^*$.

A step change in the flow rate of the $C_{min}$ fluid will change $N_{tu}$. The new steady state effectiveness is then calculated from [3]

$$\varepsilon = 1 - \exp(-N_{tu}) \tag{62}$$

It is seen from this relationship that a sizable flow rate change has only a small influence on the steady-state effectiveness, and as a result, only a small direct influence on the outlet fluid temperature.

2.3 Counterflow Heat Exchangers with $C^* = 1$

In gas turbine regenerators and some heat exchangers in process industries, the heat capacity rate of both fluids is approximately the same, and $C^* \simeq 1$ is a good approximation. For these exchangers, the temperature distribution for each fluid in the steady state at time $\tau = 0-$ is shown by solid lines in Fig. 2. By imposing a step change in the inlet temperature of the hot fluid at $\tau = 0+$, the temperatures of hot and cold fluids vary as functions of time and position as shown by dotted incomplete lines in Fig. 2. Finally, after a "long" time ($\tau \to \infty$), the temperature distribution will reach another asymptotic steady-state condition shown by the dotted complete lines in Fig. 2. Even though a step change is imposed on the hot fluid inlet temperature in Fig. 2, the results presented below are valid regardless of whether the step change is imposed in the inlet temperature of the hot fluid or that of the cold fluid.

The temperature responses are dependent upon five dimensionless groups as shown in Eq. (42). To present the solutions in compact form that could be easily understood, we will reduce the dimensionless groups for specific cases of technical interest.

Step change in inlet temperature: The $\bar{C}^*_w > 100$ case. In a gas-to-gas heat exchanger, $\bar{C}_w \gg \bar{C}_h$ or $\bar{C}_c$ on the order of 1,000 fold. The dwell-time ratio $\tau^*_d$ is generally between 1/4 and 4. Based on the order of magnitude analysis for each term of the differential equations (5)-(7), London et al. [5] showed that $\tau^*_d$ is not a significant parameter for large $\bar{C}^*_w$ and falls out of the differential equations. Based on the analog tests, they also demonstrated that $\tau^*$ and $\bar{C}^*_w$ can be combined conveniently into a single parameter $\tau^*/(1.5 + \bar{C}^*_w)$ at least for $R^* = 1$. Based on further electromechanical analog experiments, they found that $R^*$ in the range $1/4 \leq R^* \leq 4$ has no measurable influence on $\varepsilon^*_{f,2}$; however, there is an $R^*$ influence on $\varepsilon^*_{f,1}$. Hence for the $\bar{C}^*_w > 100$ case, Eq. (42) reduces to

$$\varepsilon^*_{f,1} = \phi \left\{ N_{tu}, \frac{\tau^*}{1.5 + \bar{C}^*_w}, R^* \right\} \tag{63}$$

$$\varepsilon^*_{f,2} = \phi \left\{ N_{tu}, \frac{\tau^*}{1.5 + \bar{C}^*_w} \right\} \tag{64}$$

London et al. [5] obtained $\varepsilon^*_{f,1}$ and $\varepsilon^*_{f,2}$ for $1.5 \leq N_{tu} \leq 8$ by a seven-lump electromechanical analog and a numerical method. For large $\bar{C}^*_w$, $1.5 << \bar{C}^*_w$, and the second independent parameter of Eq. (63) is approximately $\tau^*/\bar{C}^*_w$. Based on the analog data, they then combined $\tau^*/\bar{C}^*_w$ and $R^*$ successfully in the following independent group

$$\left\{ \frac{\tau^*}{\bar{C}^*_w} - 0.4 \left[ \frac{R^*-1}{R^*+1} \right] \right\}$$

so that $\varepsilon^*_{f,1}$ is now dependent upon this group and $N_{tu}$. This group is applicable within $0.25 \leq R^* \leq 4$ and $\bar{C}^*_w > 100$. Thus the temperature responses $\varepsilon^*_{f,1}$ and $\varepsilon^*_{f,2}$ for the $\bar{C}^*_w > 100$ case are each dependent upon two independent parameters. These temperature responses are shown in Fig. 10.

One interesting observation from the results of Fig. 10 is that the magnitudes of $\varepsilon^*_{f,1}$ and $\varepsilon^*_{f,2}$ are functions of $N_{tu}$ for a given value of abscissa. For a high effectiveness exchanger, $N_{tu}$ is high. At high $N_{tu}$, $\varepsilon^*_{f,1}$ represents a much smaller temperature change than $\varepsilon^*_{f,2}$ at a specified time for an input step of a given size and everything else remaining the same. For a counterflow exchanger with $C^* = 1$, it can be shown that the temperature change associated with $\varepsilon^*_{f,1}$ is proportional to the steady-state ineffectiveness $(1-\varepsilon)$ and the temperature change associated with $\varepsilon^*_{f,2}$ is proportional to the steady-state exchanger effectiveness $\varepsilon$. Thus for $\varepsilon \stackrel{>}{=} 90\%$, the $\varepsilon^*_{f,1}$ response is 1/9 of the $\varepsilon^*_{f,2}$ response for a given step input. Alternatively, the time required for the same response ($\varepsilon^*_{f,1} = \varepsilon^*_{f,2}$) will be much longer for the fluid

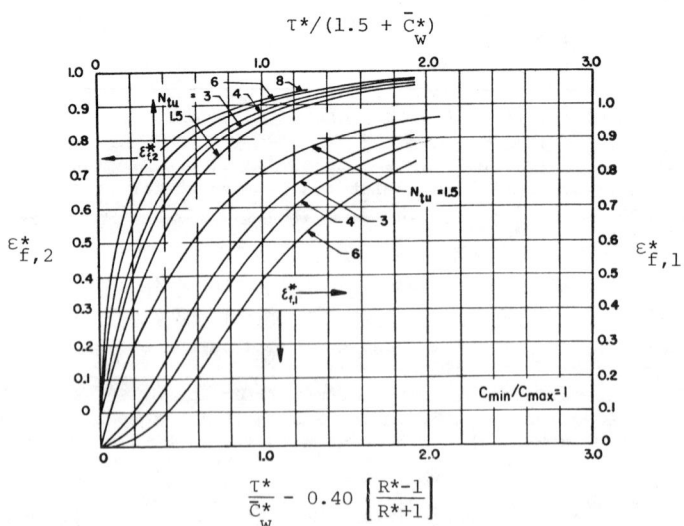

Fig. 10  Temperature responses for a step change in temperature of one fluid for a counterflow exchanger with $C^* = 1$ and $\bar{C}^*_w > 100$. Also $0.25 \leq R^* \leq 1$ for $\varepsilon^*_{f,1}$ [3].

on which the step input is imposed at inlet compared to that for the fluid on the other side.

Step change in inlet temperature: The $\bar{C}_w^* = 0$ case. For a water-cooled steam condenser, $\bar{C}_w^* \simeq 0.3$. Hence the $\bar{C}_w^* = 0$ case is useful as a limiting case for some applications. London et al. [5] showed that $\varepsilon_{f,2}^*$ of Fig. 10 is also applicable for this case with the restriction as shown on the following parameters: $C^* = 1$, $\tau_d^* = 1$, and no limitation on $R^*$. No solution is available for $\varepsilon_{f,1}^*$.

Step change in inlet temperature: The $0 \leq \bar{C}_w^* \leq 100$ case. The temperature response $\varepsilon_{f,2}^*$ of Fig. 10 is also applicable for this case provided that $R^* = 1$, $\tau_d^* = 1$. No solution is available for $\varepsilon_{f,1}^*$.

An alternative approach for the transient response solution has been considered by Hansen [10] wherein he developed formulas for the characteristic times for self ($\varepsilon_{f,1}^*$) and cross ($\varepsilon_{f,2}^*$) responses for parallel flow and counterflow heat exchangers. These characteristic response times are difficult to define precisely because of the mathematical approximations involved for obtaining closed form solutions. However, as a first approximation, these characteristic response times may be considered as the time required to obtain a 63% response after the step input. Thus Hansen's results are in terms of the characteristic response time and are independent of the time variable $\tau$, although his results are valid for the complete range of $C^*$ from zero to unity. Hansen's results being very complicated are not presented here.

Step change in the flow rate. When a step change in the flow rate is imposed at inlet, its response at outlet is instantaneous for incompressible fluids and very fast for compressible fluids so that it is also considered instantaneous. Of course, the changes in the flow rate on one side do not have any influence on the flow rate on the other side. Hence of prime interest is to determine the temperature responses $\varepsilon_{f,1}^*$ and $\varepsilon_{f,2}^*$ due to the change in the flow rate on one side.

Fux [11] investigated the response of the outlet fluid temperature due to the flow rate change of either fluid for a counterflow exchanger having $C^* = 1$. He also included the effect of variations in the heat transfer coefficient due to the change in the flow rate. He obtained expressions for the mean delay time of the temperature responses. Since the results are very complicated, they are not presented here.

Cima and London [2] obtained analog solutions for two step function changes in $N_{tu}$ from 1.5 to 1 and from 1 to 1.5. A step change in $N_{tu}$ is achieved by a step change in the flow rate of the $C_{min}$ fluid. The temperature response is presented in Fig. 11. The results were obtained for constant $\bar{C}_w^*$ magnitudes. Since $\bar{C}_w^* = \bar{C}_w/C_{min}$ and $\bar{C}_w$ is constant, $\bar{C}_w^*$ constant means $\bar{C}_{min} = C_{min_{d,min}}$, $\tau_{d,min}$ is constant. For a constant fluid density, if $C_{min}$ increases, then $\tau_{d,min}$ will decrease accordingly such that $C_{min}$ will remain constant. If the fluid density is not a constant, $\bar{C}_w^* = $ constant is an approximation. From the results of the preceding subsection, as long as $\bar{C}_w^*$ is large, it has only a minor influence on the temperature response. Based on the results of Fig. 11 and other tests, Cima and London [2] concluded

$$\varepsilon_{f,1}^* = \phi(\tau^*) \tag{65}$$

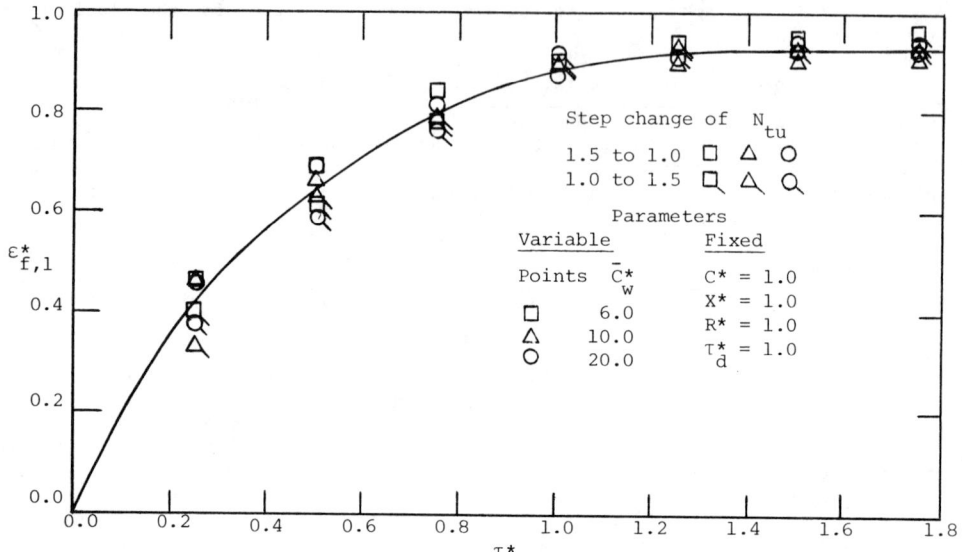

Fig. 11 $\varepsilon_{f,1}^*$ response to a step change in $N_{tu}$ from 1.5 to 1 and 1 to 1.5 [2].

It is interesting to note from Fig. 11 that a 90% response is achieved within one dwell time ($\tau^* \simeq 1$). For a gas turbine regenerator application, one dwell time is about 0.5-1 second.

## 2.4 Crossflow Heat Exchanger

The analysis for the transient response of a crossflow heat exchanger is much more difficult than that of a counterflow heat exchanger. This is because the temperature is a function of time and two position variables (x and z). All of the dimensionless parameters for a counterflow exchanger are still applicable for the crossflow exchanger.

Myers et al. [9] analyzed by an integral method the transient response of a crossflow heat exchanger with one fluid mixed and the other unmixed. They considered the fluid with a "stepped" inlet temperature as mixed, and the "unstepped" fluid as unmixed. For the large wall capacitance ($\bar{C}_w^*$ large), they presented the following approximate solution.

$$\varepsilon_{f,1}^* = \phi_o[B(\tau^*-1), A'] + e^{-A'}e^{-B(\tau^*-1)} I_o\{2[B(\tau^*-1)A']^{1/2}\} \quad \text{for } \tau^* > 1 \quad (66)$$

$$\varepsilon_{f,2}^* = \frac{P'}{1-e^{-P'}} \int_{X^*=0}^{\tau^*} e^{-P'X^*} \phi_o[B(\tau^*-X^*), A'X^*] dX^* \quad \text{for } \tau^* \leq 1 \quad (67)$$

$$\varepsilon_{f,2}^* = \frac{P'}{1-e^{-P'}} \int_{X^*=0}^{1} e^{-P'X^*} \phi_o[B(\tau^*-X^*), A'X^*] dX^* \quad \text{for } \tau^* \geq 1 \quad (68)$$

where
$$A' = \frac{N_{tu,1}(a_1+R^*)}{1+R^*} \quad (69)$$

$$B = \frac{N_{tu,1}(1+R^*)}{\bar{C}^*_w(a_1+R^*)} \tag{70}$$

$$P' = \frac{N_{tu,1}(1-a_1)}{1+R^*} \tag{71}$$

$$a_1 = 1 - \frac{1-e^{-N_{tu}}}{N_{tu}} \tag{72}$$

$$R^* = R_2/R_1 \tag{73}$$

$$\bar{C}^*_w = \bar{C}_w/\bar{C}_1 \tag{74}$$

Here the subscript 1 denotes values for the stepped fluid and the subscript 2 denotes values for the unstepped fluid. The function $\phi_o(\ )$ is defined in Eq. (56). Myers et al. [9] presented graphically the criterion for the magnitude of how large $\bar{C}^*_w$ should be. This criterion is presented in terms of functions of $\bar{C}^*_w$, $N_{tu}$, $R^{*w}$ and $C^*$.

Since Eqs. (66)-(68) incorporate a large number of parameters, a compact graphical presentation of the results is not feasible. A summary of the results are presented in terms of the time required to attain 90% response for $\tau^* > 10$ and large $\bar{C}^*_w$ as shown in Fig. 12. The dimensionless groups A', B and P' of this figure are defined by Eqs. (69)-(71).

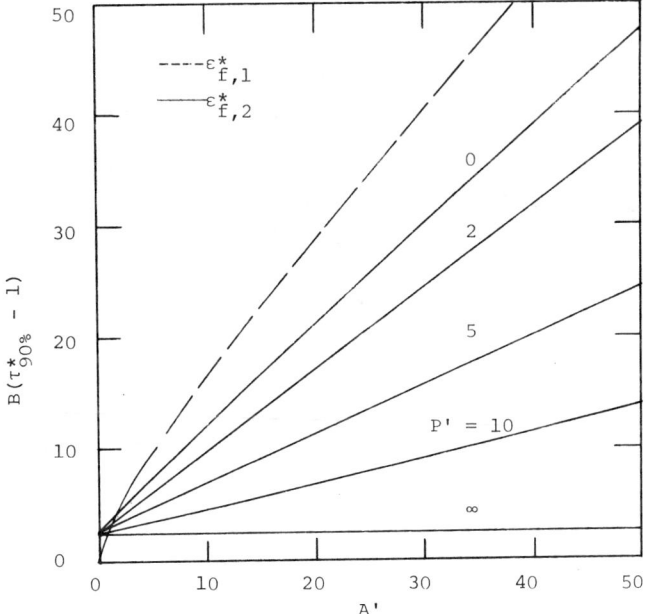

Fig. 12  90% response times for $\tau^* > 10$ for a crossflow heat exchanger with "stepped" fluid mixed, "unstepped" fluid unmixed [9].

Yamashita et al. [12] analyzed the transient response of a crossflow exchanger with both fluids unmixed. They imposed a step change in the inlet temperature of the hot fluid. Using finite difference methods, they determined and graphically presented the dimensionless outlet temperatures $t_h^*$ and $t_c^*$ as functions of $\tau^*$ or $\tau^*/\bar{C}_w^*$. They considered one of five groups ($\tau_d^*$, $1/\bar{C}_w^*$, $R^*$, $N_{tu}$, and $C^*$) as a parameter, the value of the other four groups being unity in their graphical presentation.

## 2.5 Shell-and-Tube Exchangers

One of the most comprehensive solutions for the transient response of shell-and-tube heat exchangers has been obtained by Tan and Spinner [13] by employing a Laplace transformation technique. They considered the shellside fluid having infinite heat capacity rate so that its temperature does not change. Thus they effectively considered the exchanger having $C^* = 0$ instead of the commonly used one shell pass two tube pass shell-and-tube exchanger. However, they considered finite shellside thermal resistance and finite tube-wall heat capacity. They presented closed form solutions for the transient response of the tubeside outlet fluid temperature for the following cases: (1) a step change in the inlet temperature of the tube fluid, (2) a step change in the inlet temperature of the shell fluid, and (3) a step change in the tube fluid velocity (flow rate) with the tubeside heat transfer coefficient either constant or variable dependent upon the velocity. Some typical results of Tan and Spinner are presented in Figs. 13-16. In these figures, the subscript t stands for the tube side and s stands for the shell side. The argument (0) for $U_m$, $N_{tu,t}$ and $R^*$ in these figures represents the corresponding values before the step change (at time $\tau=0$).

In a shell-and-tube exchanger, fluid flows and temperature distributions are three-dimensional. And they must be accounted for in a rigorous transient performance analysis. As indicated earlier, even with the one-dimensional analysis, available are only a limited number of solutions. Hence the only recourse left is a numerical analysis of the problem for specific applications. An

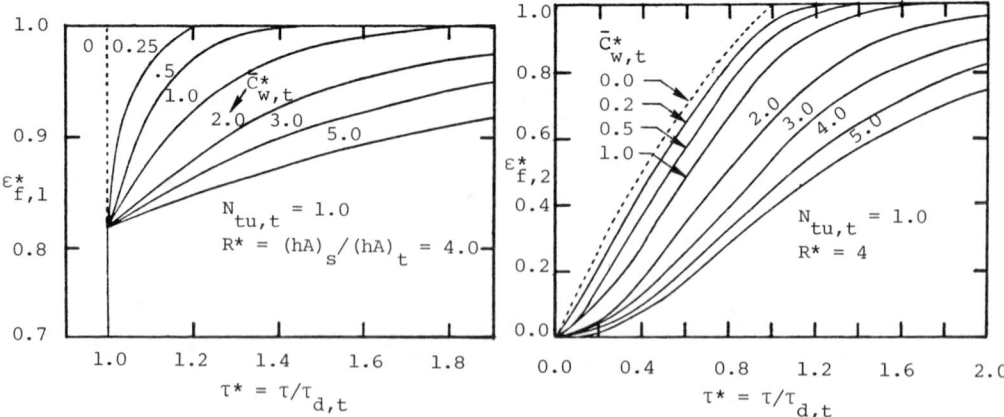

Fig. 13 Temperature response at outlet of the tube-fluid for a step change in tube-fluid inlet temperature.

Fig. 14 Temperature response at outlet of the tube-fluid for a step change in shell-fluid temperature.

# THE TRANSIENT RESPONSE OF HEAT EXCHANGERS

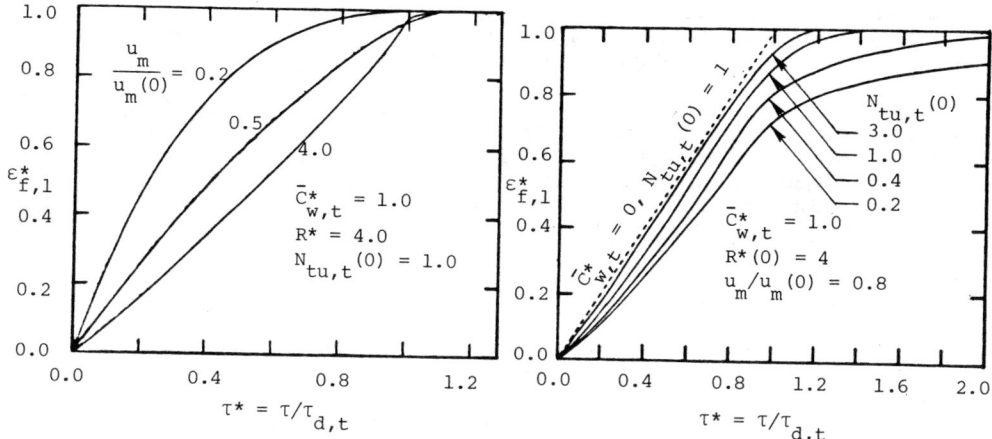

Fig. 15  Temperature response to step changes in tube-fluid velocity with constant h.

Fig. 16  Temperature response to a step change in tube-fluid velocity for $h \propto u_m^{0.8}$.

attempt at a three-dimensional numerical analysis has been made by Patankar and Spalding [14]. They analyzed the steady-state and transient behavior of a shell-and-tube exchanger. For the transient problem, they considered $UA/C_t = 1$, $C^* = 1$, $\tau_d^* = 1$, and $\bar{C}_w^* = 0$. The transient responses $\varepsilon_{f,1}^*$ and $\varepsilon_{f,2}^*$ were obtained for two cases: a step change in the shellfluid inlet temperature, and a step change in the tubefluid inlet temperature. They then extended the analysis for $\bar{C}_w^* \neq 0$ and $R^* \neq 1$. Patankar and Spalding did not attempt to analyze in detail the transient response problem. However, they tried to show how powerful the numerical method can be and also the results obtained can be very interesting and sometimes unexpected.

## 3. STORAGE-TYPE EXCHANGERS

In a storage-type exchanger, the thermal energy is alternatively stored and rejected by the matrix wall. When hot gas flows through the flow passages, heat is transferred to the matrix wall. In a chronological time interval, when the same flow passages are occupied by the cold gas, heat is transferred from the matrix wall to the cold fluid. This storage-type exchanger, simply referred to as a regenerator, may be categorized as either a rotary regenerator or a fixed-matrix regenerator [15]. In order to have a continuous operation in a regenerator, either the matrix must be moved periodically in and out of the fixed streams of gases as in a rotary regenerator or the gas flows must be divered to and from the fixed matrices as in a fixed-matrix regenerator. Thus for a continuous operation, a fixed-matrix regenerator has at least two matrices operated in parallel, but usually three or four [15].

The transient response for a rotary regenerator for gas turbine plant is of particular interest to the designer of a fuel control system because the lag introduced by the exchanger influences the shaft output transient behavior. The transient performance of fixed-matrix regenerators is also of particular interest because some of them (such as the Cowper stove) operate under varying thermal loads and may not attain the periodic equilibrium conditions between shutdowns. Air conditioning regenerators are required to operate in varying climatic conditions. Thus it is important to know the transient performance (response) of the storage-type exchangers.

First the transient response problem is formulated for a step change in inlet temperature of either fluid. Next some specific solutions are presented for this problem. Finally methods and literature sources are presented for any arbitrary temperature change at inlet, a step change in the fluid flow rate, and a change in both the inlet temprature of one fluid and the fluid flow rate of either fluid.

## 3.1 Problem Formulation

There are subtle differences in the transient performance of rotary and fixed-matrix regenerators. Once these differences are clearly understood, the results to be presented are applicable to both types of regenerators. For a rotary regenerator, the fluid outlet temperatures are dependent upon time and position. These temperatures, after averaged over the exposed flow area at every instant of time, are <u>continuous</u> functions of time. For a fixed-matrix regenerator, the fluid outlet temperatures are functions of time only in a given period. Hot and cold fluids flow alternatively in this regenrator for the specified periods $P_h$ and $P_c$. Hence if the outlet temperatures are averaged over time in a given period, these average temperatures will be different for the successive periods and will thus be dependent upon time. Thus "one period" time averaged outlet temperatures will be <u>discrete</u> (discontinuous) functions of time (see Fig. 18). A smooth curve drawn through these "period" average outlet temperatures for a fixed-matrix regenerator is identical to the continuous curve for the spatial average outlet temperatures for a rotary regenerator, if all other parameters have the same magnitudes. With this understanding, the results to be presented are applicable to both types of regenerators. Willmott and Burns [16] further expound on the regenerator-recuperator analogy.

To formulate the problem, starting with the idealizations for the analysis, the basic differential equations are derived next. Dimensional variables and and dimensionless groups for the problem are then outlined with a general functional relationship for the solution.

<u>Idealizations</u>. All the idealizations of Section 2.1, made for the direct transfer type exchangers, are also invoked here. Additional idealizations made are as follows.

1. The temperature of the fluid and the matrix wall are only functions of time $\tau$ and position $x$ in the flow direction. Hence, when the step change is imposed, the temperature distribution in the matrix wall in the angular direction on each fluid side of a rotary regenerator is neglected.

2. The carryover and pressure leakages are negligible.

<u>Differential equations</u>. The basic differential equations, boundary conditions and periodic conditions are identical to those for a regenerator having periodic equilibrium conditions. These have been derived in detail in [15]. The only difference is a step change is imposed on one or more of the operating conditions at time $\tau = 0$. For completeness, these differential equations, and boundary and periodic conditions are provided next for a specific case of a step change imposed on the inlet temperature of the hot fluid at time $\tau = 0$. It is idealized that all operating conditions remain unchanged until periodic equilibrium is reestablished.

For the flow tube in the hot gas flow region,

Fluid:   $\bar{C}_h \dfrac{\partial t_h}{\partial \tau} + C_h \dfrac{\partial t_h}{\partial (x/L)} + (hA)_h (t_h - t_w) = 0$ (75)

Matrix:  $\bar{C}_r \dfrac{\partial t_w}{\partial \tau} - (hA)_h (t_h - t_w) = 0$ (76)

For the flow tube in the cold gas flow region,

Fluid:   $\bar{C}_c \dfrac{\partial t_c}{\partial \tau} + C_c \dfrac{\partial t_c}{\partial (x/L)} + (hA)_c (t_w - t_c) = 0$ (77)

Matrix:  $\bar{C}_r \dfrac{\partial t_w}{\partial \tau} - (hA)_c (t_w - t_c) = 0$ (78)

where $\bar{C}_r = M_w c_w = \bar{C}_w$. In contrast to two time variables $\tau_h$ and $\tau_c$ considered for the regenerator in [15], here we have considered only one continuous time variable $\tau$ for the rotary regenerator application. The discrete time variable for the fixed-matrix regenerator will be defined later.

The boundary conditions of Eqs. (11) and (12) are also considered for the regenerator. To be specific, inlet fluid temperatures of both hot and cold flow periods remain constant during any cycle. A step change in the inlet temperature of one fluid is made at $\tau=0+$ (at the beginning of a period for a fixed-matrix regenerator); operating conditions then remained unchanged until cyclic equilibrium is reestablished. For the periodic conditions, the matrix wall temperature distribution at the beginning of any period is related to the distribution at the end of the preceeding period.

$t_w(x, \tau=0+) = t_w(x, \tau=P_t-)$ (79)

$t_w(x, \tau=\tau_h-) = t_w(x, \tau=\tau_h+)$ (80)

The terms on the left-hand side correspond to the hot gas flow period, and the terms on the right-hand side correspond to the cold gas flow period.

Based on the foregoing differential equations, boundary conditions, and periodic conditions, the dependent fluid and wall temperatures are functions of the following variables and parameters.

$\underbrace{t_h, t_c, t_w}_{\text{Dependent variables}} = \phi \{ x, \tau, \underbrace{t_{h,i}, t_{c,i}}_{\text{Indepen. variables}}, \underbrace{C_h, C_c, C_r, \bar{C}_h, \bar{C}_c,}_{\text{Operating conditions}}$

$\underbrace{\bar{C}_r, (hA)_h, (hA)_c, L}_{\text{Parameters under designer's control}} \}$ (81)

$P_t$ is not included in the foregoing list since $P_t = \bar{C}_r / C_r$.

**Dimensionless groups and implicit solutions.** Thirteen independent variables and parameters exist for the dependent fluid and wall temperatures. We will reduce them in number by formulating appropriate dimensionless groups as those already used for the periodic equilibrium performance of the regenerator plus some new. Two methods of analysis of the regenerator thermal performance are the $\varepsilon\text{-}N_{tu,o}$ method and the $\Lambda\text{-}\Pi$ method [15]. Customarily, the $\varepsilon\text{-}N_{tu,o}$ method is used for the rotary regenerators and the $\Lambda\text{-}\Pi$ method is used for the fixed-matrix regenerators, although both methods are equivalent as shown in [15]. We will describe the dimensionless groups for the transient problem first for the $\varepsilon\text{-}N_{tu,o}$ method and then for the $\Lambda\text{-}\Pi$ method.

In the $\varepsilon\text{-}N_{tu,o}$ method, the regenerator effectiveness for the periodic equilibrium condition is expressed as [3,15]

$$\varepsilon = \phi\{N_{tu,o}, C^*, C_r^*, (hA)^*\} \tag{82}$$

When deriving the independent groups on the right-hand side, the idealization made was that the fluid particle residence times were small compared to the total period of cyclic operation. Hence $\bar{C}_h$, $\bar{C}_c$, and $\bar{C}_r$ of Eq. (81) were effectively eliminated before arriving at the dimensionless groups of Eq. (82). For the determination of transient performance, we may not neglect them for some applications. Two obvious dimensionless groups that can be formulated from $\bar{C}_h$, $\bar{C}_c$ and $\bar{C}_r$ are

$$\bar{C}^* = \frac{\bar{C}_{min}}{\bar{C}_{max}} = \frac{C_{min}\tau_{d,min}}{C_{max}\tau_{d,max}} = C^*\tau_d^* \tag{83}$$

and

$$\bar{C}_r^* = \frac{\bar{C}_r}{\bar{C}_{min}} = \frac{C_r P_t}{C_{min}\tau_{d,min}} = C_r^* \frac{P_t}{\tau_{d,min}} \tag{84}$$

Since $C^*$ and $C_r^*$ are already considered, the two new dimensionless parameters are $\tau_d^*$ and $P_t/\tau_{d,min}$. Two independent dimensionless variables are defined as

$$X^* = \frac{x}{L} \tag{85}$$

$$\tau_r^* = \frac{\tau}{P_t} \tag{86}$$

The dependent variables $t_h$ and $t_c$ are normalized in a manner similar to those for the direct transfer type exchanger, and are designated in terms of temperature responses $\varepsilon_{f,1}^*$ and $\varepsilon_{f,2}^*$. The definitions of $\varepsilon_{f,1}^*$ and $\varepsilon_{f,2}^*$ are first given next separately for rotary and fixed-matrix regenerators.

For a rotary regenerator, the fluid outlet temperatures are dependent upon time and position. These temperatures, after averaged over the exposed flow area at any instant of time, are functions of time only, and are used in the definitions of $\varepsilon_{f,1}^*$ and $\varepsilon_{f,2}^*$ that are identical to those of a direct transfer type exchanger, Eqs. (35) and (36). For example, for a step input in the hot fluid inlet temperature

# THE TRANSIENT RESPONSE OF HEAT EXCHANGERS

$$\varepsilon^*_{f,1} = \frac{\bar{t}_{h,o}(\tau) - t_{h,o}(0)}{\bar{t}_{h,o}(\infty) - t_{h,o}(0)} \tag{87}$$

$$\varepsilon^*_{f,2} = \frac{\bar{t}_{c,o}(\tau) - t_{c,o}(0)}{\bar{t}_{c,o}(\infty) - t_{c,o}(0)} \tag{88}$$

Here all $t_{h,o}$ are averaged over the hot fluid flow area and all $t_{c,o}$ are averaged over the cold fluid flow area. $\varepsilon^*_{f,1}$ and $\varepsilon^*_{f,2}$ are then expressed as a continuous function of time.

For a fixed-matrix regenerator, the fluid outlet temperatures are dependent upon the time only in a given period. The outlet temperature responses are not a continuous function of time because of hot and cold fluids flowing alternatively for the periods $P_h$ and $P_c$ respectively, through the same fixed matrix. It is idealized that the step changes are made at the beginning of the period under consideration. The temperature responses are measured in terms of chronological time average outlet temperatures for each cycle ($\bar{t}_{h,o}, \bar{t}_{c,o}$) following the step change. For example, for a step change in the hot fluid inlet temperature at the beginning of the hot period, the temperature responses for the n-th cycle are

$$\varepsilon^{*(n)}_{f,1} = \frac{\bar{t}_{h,o}(n) - t_{h,o}(0)}{\bar{t}_{h,o}(\infty) - t_{h,o}(0)} \tag{89}$$

$$\varepsilon^{*(n)}_{f,2} = \frac{\bar{t}_{c,o}(n) - t_{c,o}(0)}{\bar{t}_{c,o}(\infty) - t_{c,o}(0)} \tag{90}$$

A plot is then made of $\varepsilon^{*(n)}_{f,1}$ or $\varepsilon^{*(n)}_{f,2}$ versus a dimensionless time variable that incorporates a dimensional time, $\tau = nP_t$. Such a plot will have discrete points through which a smooth curve could be plotted. For example, see Fig. 18. The discrete points correspond to the responses after integer number of cycles for a fixed-matrix regenerator. A smooth curve passing through these points would represent the response for a rotary regenerator as mentioned earlier. With this difference between the temperature responses of fixed-matrix and rotary regenerators, the superscript (n) in Eqs. (89) and (90) will now be dropped for convenience, also, a bar on the average outlet temperatures will be dropped. The temperature responses will be designated simply as $\varepsilon^*_{f,1}$ and $\varepsilon^*_{f,2}$.

Note that the regenerator is in periodic equilibrium conditions just before a step change is imposed in the inlet temperature of one of the fluids so that $\varepsilon^*_{f,1} = \varepsilon^*_{f,2} = 0$ at $\tau=0-$. Once periodic equilibrium is attained, $\varepsilon^*_{f,1} = \varepsilon^*_{f,2} = 1$ at $\tau \to \infty$. The new regenerator effectiveness $\varepsilon(\infty)$ is identical to $\varepsilon(0)$ before the step change in the inlet temperature of either fluid. This is because we have idealized fluid properties and thermal resistances independent of temperatures, and a review of Eq. (82) or (95) indicates no change in the independent parameters for $\varepsilon(\infty)$ due to a step change in the inlet temperature. In contrast, if a step change is imposed on the flow rate of either fluid, one or more of the independent groups of Eq. (82) or (95) will change, resulting in $\varepsilon(\infty)$ different from $\varepsilon(0)$. It will be necessary to first calculate $\varepsilon(\infty)$ and then $t_{h,o}(\infty)$ and $t_{c,o}(\infty)$ from $\varepsilon(\infty)$ for $\varepsilon^*_{f,1}$ and $\varepsilon^*_{f,2}$. Here

$$\varepsilon(\infty) = \frac{C_h[t_{h,i}(\infty) - t_{h,o}(\infty)]}{C_{min}[t_{h,i}(\infty) - t_{c,i}(\infty)]} = \frac{C_c[t_{c,o}(\infty) - t_{c,i}(\infty)]}{C_{min}[t_{h,i}(\infty) - t_{c,i}(\infty)]} \tag{91}$$

where $t_{h,i}(\infty) = t_{h,i}(0)$ and $t_{c,i}(\infty) = t_{c,i}(0)$ and are known since they are kept constant.

From the foregoing list of dimensionless groups for the transient response problem, $\varepsilon^*_{f,1}$ and $\varepsilon^*_{f,2}$ are functions of the following groups.

$$\varepsilon^*_{f,1}, \varepsilon^*_{f,2} = \phi \left\{ \tau^*_r, N_{tu,o}, C^*, C^*_r, R^*, \tau^*_d, \frac{P_t}{\tau_{d,min}} \right\} \qquad (92)$$

Note that $R^* = 1/(hA)^*$. Since $X^* = 1$ or $0$ for $\varepsilon^*_{f,1}$ and $\varepsilon^*_{f,2}$ respectively, it is eliminated from the list of right-hand side groups.

It may be noted that the wall capacitance parameter $\bar{C}^*_w$ of Eq. (41) for the direct transfer type exchanger appears to have no direct counterpart in Eq. (92). However, $\bar{C}^*_r$ defined as in Eq. (84) would be analogous to $\bar{C}^*_w$, a wall-to-fluid capacitance ratio. Further, noting that

$$\tau^* = \frac{\tau}{\tau_{d,min}} = \tau^*_r \frac{P_t}{\tau_{d,min}} \qquad (93)$$

Eq. (92) is alternatively presented in a form paralleling Eq. (41) as

$$\varepsilon^*_{f,1}, \varepsilon^*_{f,2} = \phi \left\{ \tau^*, N_{tu,o}, C^*, \bar{C}^*_r, R^*, \tau^*_d, \frac{P_t}{\tau_{d,min}} \right\} \qquad (94)$$

where only $P_t/\tau_{d,min}$ parameter appears as an extra parameter for the storage-type exchanger.

In the $\Lambda$-$\Pi$ method, the regenerator effectiveness for the periodic equilibrium condition is expressed as [15]

$$\varepsilon = \phi(\Lambda_h, \Lambda_c, \Pi_h, \Pi_c) \qquad (95)$$

where

$$\Lambda_h = \frac{(hA)_h}{C_h} = N_{tu,h} = C^* \left[ 1 + \frac{1}{(hA)^*} \right] N_{tu,o} \qquad (96)$$

$$\Lambda_c = \frac{(hA)_c}{C_c} = N_{tu,c} = \left[ 1 + (hA)^* \right] N_{tu,o} \qquad (97)$$

$$\Pi_h = \frac{(hA)_h}{\bar{C}_{r,h}} P_h = \frac{N_{tu,h}}{C^*_{r,h}} = \frac{1}{C^*_r}\left[ 1 + \frac{1}{(hA)^*} \right] N_{tu,o} \qquad (98)†$$

$$\Pi_c = \frac{(hA)_c}{\bar{C}_{r,c}} P_c = \frac{N_{tu,c}}{C^*_{r,c}} = \frac{1}{C^*_r}\left[ 1 + (hA)^* \right] N_{tu,o} \qquad (99)†$$

Depending upon the values of $\Lambda$'s and $\Pi$'s, the regenerators are

---

†If the hot and cold gas dwell times are not neglected in the definition of $\Pi_h$ and $\Pi_c$, the right-hand term of Eqs. (98) and (99) should be multiplied by $(1-\tau_{d,h}/P_h)$ and $(1-\tau_{d,c}/P_c)$, respectively.

# THE TRANSIENT RESPONSE OF HEAT EXCHANGERS

designated as follows: Symmetric and balanced regenerators have $\Lambda_h = \Lambda_c$ and $\Pi_h = \Pi_c$; unsymmetric but balanced regenerators (sometimes simply referred to as balanced regenerators) have $\Lambda_h/\Pi_h = \Lambda_c/\Pi_c$. The unbalanced regenerators have $\Lambda_h/\Pi_h \neq \Lambda_c/\Pi_c$.

For a <u>balanced</u> regenerator, Willmott and Burns [17] showed that the transient responses are dependent upon

$$\varepsilon^*_{f,1}, \varepsilon^*_{f,2} = \phi(\eta, \Lambda, \Pi) \qquad (100)$$

where

$$\eta = n(\Pi_h + \Pi_c) \qquad (101)$$

$$\Lambda = \Lambda_h = \Lambda_c \qquad (102)$$

$$\Pi = \Pi_h = \Pi_c \qquad (103)$$

For unsymmetric but balanced regenerators, Willmott and Burns [17] showed that the transient response $\varepsilon^*_{f,1}$ is a function of

$$\varepsilon^*_{f,1} = \phi(\eta_m, \Lambda_m, \Pi_m, R^*) \qquad (104)$$

where $\Lambda_m$ and $\Pi_m$ are the harmonic means defined as follows.

$$\frac{1}{\Lambda_m} = \frac{1}{2}\left[\frac{1}{\Lambda_h} + \frac{1}{\Lambda_c}\right] \qquad (105)\dagger$$

$$\frac{1}{\Pi_m} = \frac{1}{2}\left[\frac{1}{\Pi_h} + \frac{1}{\Pi_c}\right] \qquad (106)$$

and the unsymmetry factor $R^*$ is defined as

$$R^* = \frac{\Lambda_h}{\Lambda_c} = \frac{\Pi_h}{\Pi_c} \qquad (107)$$

also

$$\eta_m = n(2\Pi_m) \qquad (108)$$

For the unbalanced regenerators, Willmott and Burns [17] showed that the transient response $\varepsilon^*_{f,1}$ is a function of

$$\varepsilon^*_{f,1} = \phi(\eta_m, \Lambda_m, \Pi_m, \gamma) \qquad (109)$$

where the unbalance factor $\gamma$ is defined as

---

$\dagger$The harmonic mean $\Lambda_m$ is correctly defined by Hausen [18] as $1/\Lambda_m = [\Pi_h/\Lambda_h + \Pi_c/\Lambda_c]/2\Pi_m$. Equation (105) is correct only for the <u>balanced</u> and symmetric or unsymmetric regenerators.

$$\gamma = \frac{\Lambda_h/\Pi_h}{\Lambda_c/\Pi_c} \tag{110}$$

## 3.2 Specific Solutions for a Step Input in Inlet Temperature†

Equations (75)-(78) or their counterpart in dimensionless form have seven independent variables and parameters as shown in Eq. (92). Because of the seven groups, general solutions to these equations with appropriate initial and boundary conditions cannot be presented graphically in a compact form. To obtain the solutions and present the results in a manageable form, the number of independent dimensionless groups must be reduced. Some specific solutions of technical interest are presented next.

$C^* = 1$, $\bar{C}_r^* > 100$. London et al. [19] obtained the transient responses of a counterflow regenerator having $C^* = 1$ and partial constraints on other variables as

$$\bar{C}_r^* = C_r^* \frac{P_t}{\tau_{d,min}} > 100, \qquad 2 \leq C_r^* \leq 20, \qquad \frac{1}{4} \leq R^* \leq 4 \tag{111}$$

Similar to the $\bar{C}_w^* > 100$ case for direct transfer type exchangers, $\tau_r^*$ is not an important parameter for the $\bar{C}_r^* > 100$ case. London et al. [19] also showed analytically that $P_t/\tau_{d,min}$ is not an important parameter for $\bar{C}_r^* > 100$. Hence the number of independent groups in Eq. (92) reduces to four.

$$\varepsilon_{f,1}^*, \varepsilon_{f,2}^* = \phi\left\{\tau_r^*, N_{tu,o}, C_r^*, R^*\right\} \tag{112}$$

Based on the electromechanical analog results and arguments paralleling those for the counterflow direct transfer type exchanger, London et al. [19] showed that

$$\varepsilon_{f,1}^* = \phi\left\{\left[\frac{\tau_r^*}{C_r^* - 0.6} - 0.53\left(\frac{R^*-1}{R^*+1}\right)\right], N_{tu,o}\right\} \tag{113}$$

$$\varepsilon_{f,2}^* = \phi\left\{\frac{\tau_r^*}{C_r^*}, N_{tu,o}\right\} \tag{114}$$

Note that $R^*$ has no influence on $\varepsilon_{f,2}^*$ for $1/4 \leq R^* \leq 4$. The temperature responses for this case are shown in Fig. 17.

It is interesting to note that all responses approach asymptotically to unity (equilibrium), and in the asymptotic region, $\varepsilon_{f,1}^*$ and $\varepsilon_{f,2}^*$ are dependent upon $N_{tu,o}$ only. Also notice that $\varepsilon_{f,1}^*$ converges more slowly to equilibrium compared to $\varepsilon_{f,2}^*$ at a given $N_{tu,o}$. A physical explanation for this significant difference in $\varepsilon_{f,1}^*$ and $\varepsilon_{f,2}^*$ responses may be given as follows. Step change in inlet temperature of hot gas will have to travel through the regenerator flow length. Since it is idealized that $\bar{C}_r^* > 100$, the matrix wall capacitance is much higher than that of the hot gas. So it will take a significant time before the effect of step change is realized at the outlet. In contrast, the cold gas flows in the counterflow direction, and the cold gas near the outlet is immediately affected due to the step change in the hot gas inlet temperature because the

---

†Since the differential equations (75)-(78) are linear, $\varepsilon_{f,1}^*$ and $\varepsilon_{f,2}^*$ are independent of the magnitude of the step change in the inlet temperature.

# THE TRANSIENT RESPONSE OF HEAT EXCHANGERS

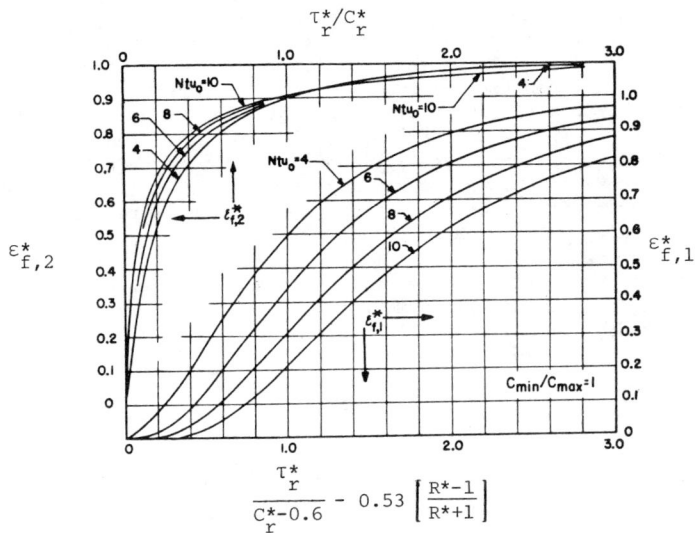

Fig. 17  Temperature responses for a step change in temperatue of one fluid for a counterflow regenerator with $C^* = 1$, $\bar{C}_r^* > 100$, $2 \leq C_r^* \leq 20$, and $1/4 \leq R^* \leq 4$ [3].

hot gas inlet and cold gas outlet are on the same side. Hence $\varepsilon_{f,2}^*$ response is much faster.

Willmott and Burns [17] also analyzed the transient response of a symmetric balanced† fixed-matrix regenerator by a finite difference method. Their results are in agreement with the results in Fig. 17 of London et al. For example, their typical results are shown in Fig. 18 for the symmetric balanced regenerators. Clearly, $\varepsilon_{f,1}^*$ and $\varepsilon_{f,2}^*$ are independent of $\Pi$ in this figure. Since $\Lambda/\Pi \geq 100$ in this figure and $\Lambda/\Pi \simeq C_r$, the results are independent of $C_r^*$ for $C_r^* > 100$.

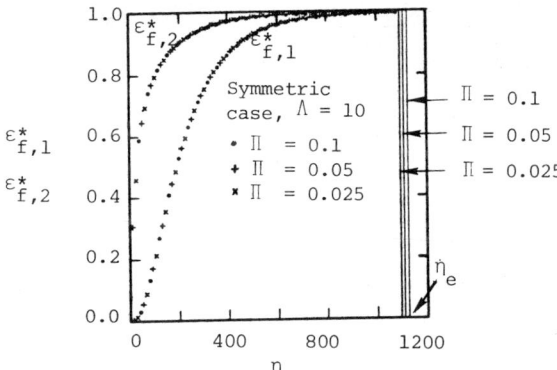

Fig. 18  $\varepsilon_{f,1}^*$ and $\varepsilon_{f,2}^*$ for a step change in hot gas inlet temperature for $\Lambda/\Pi > 100$ [17].

---

†$\Lambda_h = \Lambda_c = \Lambda$, $\Pi_h = \Pi_c = \Pi$, or $C^* = 1$, $R^* = 1$, $\tau_{d,h}/P_h = \tau_{d,c}/P_c$.

Since the equilibrium periodic conditions are reached asymptotically, a criterion is needed to define when the equilibrium conditions are reached. London et al. [5,19] define that the equilibrium conditions are approximately reached on each side when the magnitudes of $\varepsilon^*_{f,1}$ and $\varepsilon^*_{f,2}$ are 0.9. Willmott and Burns [17] set up a different criterion for equilibrium conditions. Instead of considering the response in terms of $\varepsilon^*_{f,1}$ and $\varepsilon^*_{f,2}$, they considered the equilibrium conditions approximately reached when the regenerator effectiveness $\varepsilon$ during transient conditions has the magnitude $0.9999\,\varepsilon(\infty)$. Here $\varepsilon(\infty)$ is the new regenerator effectiveness when the equilibrium is reached at time approaching infinity.†

Willmott and Burns [17] showed that the dimensionless transient response time to reach equilibrium (defined in the sense described above) for both hot and cold fluids depends only upon $\Lambda$ for the symmetric balanced regenerator by

$$\eta_e = 0.662\,\Lambda^2 + 4.144\,\Lambda + 6.464 \tag{115}‡$$

where the dimensionless time $\eta_e$ to attain equilibrium condition is defined as

$$\eta_e = \frac{(hA)_h}{\bar{C}_r}(\tau_e - \tau_{d,h}) = n[\Pi_h + \Pi_c] = n(2\Pi) \tag{116}$$

The term after the first equality from the left is for the rotary regenerator and the term after the second equality is for the fixed-matrix regenerator. $\tau_e$ is the dimensional time to reach equilibrium.

Although as mentioned earlier and shown in Fig. 18, the response $\varepsilon^*_{f,2}$ is faster than $\varepsilon^*_{f,1}$, the dimensionless time $\eta_e$ required to attain $\varepsilon^*_{f,1}$ and $\varepsilon^*_{f,2}$ to converge to $0.9999\,\varepsilon_h(\infty)$ and $0.9999\,\varepsilon_c(\infty)$ is the same. Here $\varepsilon_h(\infty)$ and $\varepsilon_c(\infty)$ are the new temperature effectivenesses on the hot and cold sides of the regenerator at time approaching infinity. For this reason, Willmott and Burns [17] then considered only $\varepsilon^*_{f,1}$ to obtain correlations for $\eta_e$ for unsymmetric and unbalanced regenerators to be discussed now.

<u>Unbalanced/unsymmetric regenerators.</u> Since the transient responses to a single step change in the inlet temprature are of exponential nature as shown in Fig. 18, Willmott and Burns [20] approximated as

$$\varepsilon^*_{f,1} = 1 - \exp[-(\eta_m - C_1)/\eta_\tau] \tag{117}$$

$$\varepsilon^*_{f,2} = 1 - \exp[-(\eta_m + C_2)/\eta_\tau] \tag{118}$$

Here $C_1$, $C_2$ and $\eta_\tau$ are positive real constant, and $\eta_m$ is the dimensionless harmonic mean time. $\eta_m = n(2\Pi_m)$ where n is the total number of cycles from the time when a step change in inlet temperature is imposed, $\eta_\tau$ is the time

---

†As noted earlier, $\varepsilon(\infty)$ is identical to $\varepsilon(0)$ before the step change in the inlet temperature of either fluid.

‡The time $\eta_e$ required to reestablish equilibrium is independent of the total period $P_t$ and is made up of many short cycles or a few long cycles depending upon the magnitude of $P_t$.

constant of the transient response. Willmott and Burns [20] showed that $C_1$, $C_2$ and $\eta_\tau$ are functions of $\Lambda_m$ and $\gamma$ and are independent of $\Pi_m$ for $\Lambda_m^2/\Pi_m > 3$. Other observations that they made based on their study are as follows: (1) The response $\varepsilon_{f,1}^*$ always lags the response $\varepsilon_{f,2}^*$. (2) The time lag $(C_1+C_2)$ between the responses increases with the reduced length $\Lambda_m$. (3) For any regenerator, the two responses $\varepsilon_{f,1}^*$ and $\varepsilon_{f,2}^*$ always have identical time constants. (4) The time constant is determined by the final operating parameters of the regenerator, particularly $\Lambda_m$ and $\gamma$. The time constant is independent of the magnitude of the step change in either inlet gas temperature or gas flow rate.

Based on finite difference solutions, Willmott and Burns [17] correlated the transient response time $\eta_e$ [see Eq. (116)] to reach the equilibrium condition for $\varepsilon_{f,1}^*$. For unsymmetric but balanced regenerators $(\Lambda_h/\Pi_h = \Lambda_c/\Pi_c$ or $R^* \neq 1$, $C^* = 1$ and $\tau_{d,h}/P_h = \tau_{d,c}/P_c)$, they presented

$$\eta_e = \frac{1}{4}\left[0.622 \Lambda_m^2 + 4.144 \Lambda_m + 6.464\right](1+R^*)(1 + \frac{1}{R^*}) \tag{119}$$

where

$$\eta_e = n(2\Pi_m) \tag{120}$$

For unbalanced regenerators, they presented the time $\eta_e$ as a function of $\Lambda_m$ and $\gamma$ as shown in Fig. 19.

It is clear from this figure that the transient response time required to attain a periodic equilibrium (for a step change in the inlet temperature) is largest for the symmetric-balanced regenerator ($\gamma = 1$ or $C^* = 1$, $R^* = 1$). This time is significantly reduced for a departure from the $\gamma = 1$ or $C^* = 1$ condition.

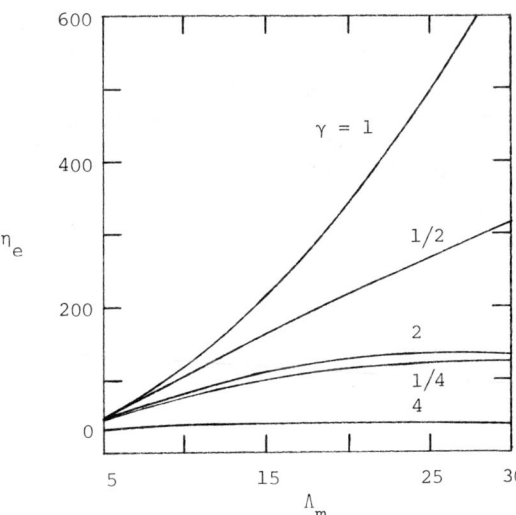

Fig. 19  Time $\eta_e$ to reestablish equilibrium for an unbalanced regenerator [17].

## 3.3 Solutions for Arbitrary Change in Inlet Temperature

The solutions described so far have a step change in the inlet temperature, and they are valid regardless of the step size. Since the differential equations are linear and homogeneous, a sum of solutions is again a solution to that set of equations. Thus it is possible to construct a temperature response solution by superposition for the following cases: (1) a simultaneous step change in inlet temperatures of the both fluids, (2) a multiple step change in the inlet temperature of one fluid, and (3) any arbitrary specified variation in the inlet temperature. For the last case, the solution is obtained by merely breaking up the inlet temperature change into a number of constant temperature steps and summing or superposing the corresponding temperature response solutions for each step by Duhamel's theorem as described by Cima and London [2]. Duhamel's theorem is also used to obtain the solution for the first two cases.

## 3.4 Solutions for a Step Change in the Fluid Flow Rate

A step change in the flow rate of one of the fluids may influence the mean heat transfer coefficient on that side, particularly if the flow is turbulent. Thus, a step change in the flow rate will affect a number of independent groups of Eq. (92) or (109). In the $\varepsilon$-$N_{tu,o}$ method, it may influence $N_{tu,o}$, $C^*$, $R^*$, and $\tau_d^*$. If the step change is on the $C_{min}$ side, it will also affect $C^*$ and $P_t/\tau_{d,min}$. In the $\Lambda$-$\Pi$ method, it may influence $\Lambda$'s and $\Pi$'s. If it is idealized that the mean heat transfer coefficient $h$ is approximately linearly proportional to $W$,[†] then $\Lambda$ is independent of the flow rate. In this case, a step change in hot gas flow rate can be investigated simply by considering a step change in $\Pi_h$ alone.

Willmott and Burns [17] investigated the transient response for the case of a step change in the hot fluid flow rate with the idealization that $h$ is linearly proportional to $W$. Thus they considered a step change in $\Pi$ alone. As seen in Fig. 20, they found that the time taken to reestablish the equilibrium after a step change in hot gas flow rate increases with the size of the step change up to a certain threshold. Beyond this threshold value in the percentage change in $\Pi_h$, the transient response is dependent only upon the final operating conditions. The threshold value of $\Pi_h$ increases with the reduced length $\Lambda_h$. For example, it represents a 10% change in $\Pi_h$ for $\Lambda_h = 10$ and a 35% change in $\Pi_h$ for $\Lambda_h = 20$ in Fig. 20. For changes in $\Pi_h$ below the threshold value, the transient responses $\varepsilon^*_{f,1}$ and $\varepsilon^*_{f,2}$ are dependent upon the magnitude of the step change in the flow rate.[‡] Willmott and Burns showed that this parameter has only a small influence on the balanced regenerators. Beyond the threshold value for the step change in flow rate (25-40%), $\varepsilon^*_{f,1}$ and $\varepsilon^*_{f,2}$ become independent of the step change size, and are very approximately the same as for the step changes in the inlet gas temperature.

For the general case of unbalanced regenerators, Willmott and Burns [17] show that the transient response is dependent upon $\Lambda_m$ and $\gamma$ only for the step change in $\Pi_h$ greater than the threshold value.

---

[†] In fully developed turbulent flow, $h \propto W^{0.8}$. In fully developed laminar flow, $h$ is independent of the flow rate $W$.

[‡] Note that $\varepsilon^*_{f,1}$ and $\varepsilon^*_{f,2}$ are independent of the magnitude of the step change in inlet temperature of one of the fluids.

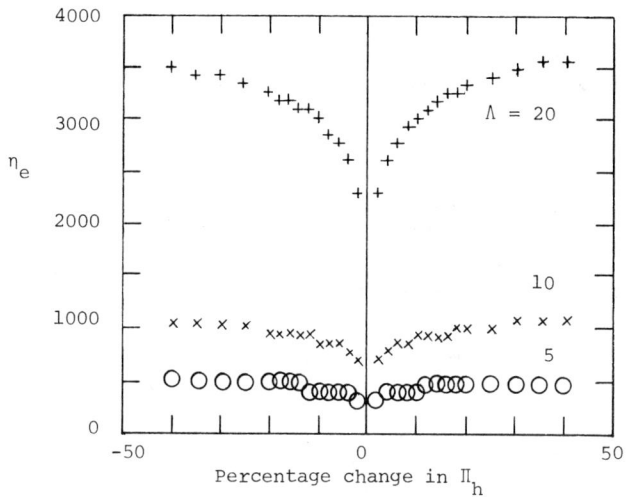

Fig. 20 $\eta_e$ for a step change in $\Pi_h$; the final state of the regenerator is symmetric with reduced period $\Pi_h = \Pi_c = 4$ [17].

### 3.5 Simultaneous Change in Fluid Flow Rate and Inlet Temperatures

A step change in fluid flow rate results into nonlinearity in the differential equations (75)-(78). As a result, a transient response solution cannot be obtained for a simultaneous step change in both fluid flow rates by superimposing the individual solutions. However, the superposition approach can be extended if a step change in one fluid flow rate is followed by any arbitrary changes in inlet temperatures of one or both fluids. The details of this procedure are outlined by Burns and Willmott [21].

Burns and Willmott [21] presented three specific transient response solutions involving simultaneous changes in the flow rate and inlet temperatures: (1) A step increase in $\Pi_h$ along with a step increase in $t_{h,i}$; (2) a step decrease in $\Pi_h$ along with a step increase in $t_{h,i}$; and (3) a step increase in $\Pi_c$ along with a step increase in $t_{h,i}$.

## 4. SUMMARY

In the text, first the transient response problem is formulated by providing the idealizations, governing differential equations and boundary conditions, and a list of dimensional and dimensionless variables and parameters associated with the problem. A large number of dimensionless groups associated with the problem clearly indicates how complicated the transient response problem is. Next, all available transient response solutions are presented for the case of a step change in the inlet temperature or the flow rate of one or both fluids. The solutions are presented for the exchanger with $C^* = 0$, counterflow exchanger with $C^* = 1$, crossflow exchangers with one or both fluids unmixed, shell-and-tube exchanger with variable shellside resistance and tubeside heat transfer coefficient, and the regenerators (periodic-flow exchangers). Since only a limited number of solutions are available in the literature, and the transient response problem has become more important in the recent energy conscientious era, it is recommended to investigate this problem thoroughly.

## ACKNOWLEDGEMENTS

The author is grateful to Mr. R. S. Johnson for obtaining computer results of Figs. 5-7 and to Mr. E. H. Sanderson for preparing some of the figures.

## NOMENCLATURE

| | |
|---|---|
| $A$ | heat transfer surface area (primary plus secondary, if any) on one side of a heat exchanger, m$^2$, ft$^2$ |
| $A_c$ | total heat transfer surface area (both primary and secondary, if any) on the cold side of the exchanger, m$^2$, ft$^2$ |
| $A_h$ | total heat transfer surface area (both primary and secondary, if any) on the hot side of the exchanger, m$^2$, ft$^2$ |
| $C$ | flow stream heat capacity rate with a subscript c or h, $Wc_p$, W/°C, Btu/hr °F |
| $C^*$ | heat capacity rate ratio, $C_{min}/C_{max}$, dimensionless |
| $\bar{C}$ | flow stream heat capacitance, $Mc_p$, $C\tau_d$, W·s/°C, Btu/°F |
| $C_{max}$ | maximum of $C_c$ or $C_h$, W/°C, Btu/hr °F |
| $C_{min}$ | minimum of $C_c$ or $C_h$, W/°C, Btu/hr °F |
| $C_r$ | heat capacity rate of a regenerator, $M_w c_w N$ or $M_w c_w/P_t$, W/°C, Btu/hr °F |
| $C_r^*$ | matrix heat capacity rate ratio, $C_r/C_{min}$, dimensionless |
| $\bar{C}_r$ | matrix wall heat capacitance, $M_w c_w$, W·s/°C, Btu/°F |
| $\bar{C}_r^*$ | a ratio of $\bar{C}_r$ to $\bar{C}_{min}$, dimensionless |
| $\bar{C}_w$ | matrix wall heat capacitance, $M_w c_w$, W·s/°C, Btu/°F |
| $\bar{C}_w^*$ | a ratio of $\bar{C}_w$ to $\bar{C}_{min}$, dimensionless |
| $c_p$ | specific heat of fluid at constant pressure, J/kg °C, Btu/lbm °F |
| $c_w$ | specific heat of wall material, J/kg °C, Btu/lbm °F |
| $h$ | heat transfer coefficient, W/m$^2$ °C, Btu/hr ft$^2$ °F |
| $(hA)^*$ | convection conductance ratio, $(hA)$ on the $C_{min}$ side divided by $(hA)$ on the $C_{max}$ side, dimensionless |
| $L$ | heat exchanger length for fluid flows, m, ft |
| $M$ | mass of the fluid in the heat exchanger at any instant of time, kg, lbm |
| $M_w$ | mass of the heat exchanger core, kg, lbm |
| $N$ | rotational speed of a rotary regenerator, rev/s, rev/sec |
| $N_{tu}$ | number of heat transfer units, $UA/C_{min}$, dimensionless |
| $N_{tu,c}$ | number of heat transfer units based on the cold side, $(\eta_o hA)_c/C_c$, dimensionless |
| $N_{tu,h}$ | number of heat transfer units based on the hot side, $(\eta_o hA)_h/C_h$, dimensionless |
| $N_{tu,o}$ | modified number of transfer units for a regenerator, see [3,15] for the definition, dimensionless |
| $n$ | cycle number after a step change in the inlet temperature is imposed |
| $P_c$ | cold gas flow period, duration of the cold gas stream in the matrix or duration of matrix in the cold gas stream, s, sec |

| Symbol | Description |

$P_h$     hot gas flow period, duration of the hot gas stream in the matrix or duration of matrix in the hot gas stream, s, sec

$P_t$     total period between the start of two successive heating (or cooling) periods in a regenerator, $P_t \simeq P_h + P_c$, s, sec

$R^*$     ratio of thermal resistances on the $C_{min}$ to $C_{max}$ side, $1/(\eta_o hA)^*$ or $\Lambda_h / \Lambda_c$ [see Eq. (107)], dimensionless

$t$     fluid static temperature to a specified arbitrary datum, °C, °F

$t_c^*$     cold fluid temperature response, defined by Eq. (15), dimensionless

$t_h^*$     hot fluid temperature response, defined by Eq. (14), dimensionless

$U$     overall heat transfer coefficient, W/m² °C, Btu/hr ft² °F

$u_m$     fluid mean axial velocity, m/s, ft/sec

$W$     fluid mass flow rate, kg/s, lbm/hr

$X^*$     axial distance, $x/L$, dimensionless

$x$     Cartesian coordinate along the flow direction, m, ft

$z$     Cartesian coordinate along the width direction, m, ft

$\gamma$     the unbalance factor, $\Lambda_h \Pi_c / \Lambda_c \Pi_h$ or $C^*$, dimensionless

$\varepsilon$     heat exchanger effectiveness, defined by Eq. (28), dimensionless

$\varepsilon_{f,1}^*$     temperature self response, defined by Eq. (35), dimensionless

$\varepsilon_{f,2}^*$     temperature cross response, defined by Eq. (36), dimensionless

$\eta$     reduced time variable for a regenerator, $hA\tau/M_w c_w$, dimensionless

$\eta_e$     time required to attain periodic equilibrium to $0.9999\varepsilon(\infty)$, defined by Eqs. (116) and (120), dimensionless

$\eta_m$     "harmonic mean" time for equilibrium, $n(2\Pi_m)$, dimensionless

$\Lambda$     reduced length for a regenerator, defined by Eqs. (96) and (97), dimensionless

$\Lambda_m$     harmonic mean reduced length, defined by Eq. (105), dimensionless

$\Pi$     reduced period for a regenerator, defined by Eqs. (98) and (99), dimensionless

$\Pi_m$     harmonic mean reduced period, defined by Eq. (106), dimensionless

$\tau$     time variable, s, sec

$\tau_d$     dwell time, residence time or transit time of a fluid particle, s, sec

$\tau_{d,min}$     dwell time of the $C_{min}$ fluid, s, sec

$\tau_{d,max}$     dwell time of the $C_{max}$ fluid, s, sec

$\tau^*$     time variable, $\tau/\tau_{d,min}$, dimensionless

$\tau_d^*$     dwell time ratio, $\tau_{d,min}/\tau_{d,max}$, dimensionless

$\tau_r^*$     time variable, $\tau/P_t$, dimensionless

Subscripts

c     cold fluid side

h     hot fluid side

i     inlet to the exchanger

| | |
|---|---|
| m | bulk mean or harmonic mean |
| o | outlet to the exchanger |
| s | shellside |
| t | tubeside |

REFERENCES

1. F.H. Raven, <u>Automatic Control Engineering</u>, McGraw-Hill, New York, pp. 345-347 (1961).

2. R.M. Cima and A.L. London, The transient response of a two-fluid counterflow heat exchanger — The gas-turbine regenerator, <u>Trans. ASME,</u> Vol. 80, 1169-1179 (1958).

3. W.M. Kays and A.L. London, <u>Compact Heat Exchangers</u>, Second Edition, McGraw-Hill, New York (1964).

4. J.W. Rizika, Thermal lags in flowing incompressible fluid systems containing heat capacitors, <u>Trans. ASME</u>, Vol. 78, 1407-1413 (1956).

5. A.L. London, F.R. Biancardi, and J.W. Mitchell, The transient response of gas-turbine plant heat exchangers - regenerators, intercoolers, precoolers, and ducting, <u>Trans. ASME, J. of Engng. for Power</u>, Vol. 81, Series A, 433-448 (1959).

6. G.E. Myers, J.W. Mitchell, and C.F. Lindeman, Jr., The transient response of heat exchangers having an infinite capacity rate fluid, <u>Trans. ASME, J. of Heat Trans.</u>, Vol. 92, Series C, 269-275 (1970).

7. G.E. Myers, Personel communication, Mech. Eng. Dept., Univ. of Wisconsin, Madison, WI (1979).

8. S.R. Binkley, H.E. Edwards, and R.W. Smith, Table of the temperature distribution function for heat exchange between a fluid and a porous solid, U.S. Bureau of Mines, Pittsburgh, PA (1952).

9. G.E. Myers, J.W. Mitchell, and R.F. Norman, The transient response of crossflow heat exchangers, evaporators and condensers, <u>Trans. ASME, J. of Heat Trans.</u>, Vol. 89, Series C, 75-80 (1967).

10. P.D. Hansen, The dynamics of heat exchange processes, Sc.D. Thesis, Mech. Eng. Dept., Massachusetts Institute of Tech., Cambridge, MA (1960).

11. A. Fux, Dynamic models for practical heat exchangers, M.S. Thesis, Mech. Eng. Dept., Massachusetts Institute of Tech., Cambridge, MA (1959).

12. H. Yamashita, R. Izumi and S. Yamaguchi, Analysis of the dynamic characteristic of cross-flow heat exchangers with both fluids unmixed, <u>Bulletin of the JSME</u>, Vol. 21, No. 153, 479-485 (1978).

13. K.S. Tan and I.H. Spinner, Dynamics of a shell-and-tube heat exchanger with finite tube-wall heat capacity and finite shell-side resistance, <u>Industrial and Engineering Chemistry Fundamentals</u>, Vol. 17, 353-358 (1978).

14. S.V. Patankar and D.B. Spalding, A calculation procedure for the transient and steady-state behavior of shell-and-tube heat exchangers, in <u>Heat</u>

Exchangers: Design and Theory Sourcebook, edited by N. Afgan and E.U. Schlünder, McGraw-Hill, New York, pp. 155-176 (1974).

15. R.K. Shah, Thermal design theory for regenerators, in Heat Exchangers-Thermohydraulic Fundamentals and Design, edited by S. Kakac, A.E. Bergles, and F. Mayinger, Hemisphere Publishing Corp., New York (1981).

16. A.J. Willmott and A. Burns, The recuperator analogy for the transient performance of thermal regenerators, International Journal of Heat and Mass Trans., Vol. 22, 1107-1115 (1979).

17. A.J. Willmott and A. Burns, Transient response of periodic-flow regenerators, International Journal of Heat and Mass Transfer, Vol. 20, 753-761 (1977).

18. H. Hausen, Vervollständigte Berechung des Wärmeaustausches in Regeneratoren (accomplished calculations of heat exchange in regenrators), Z. VDI Beiheft Verfahrenstechnik, No. 2, 31-43 (1942).

19. A.L. London, D.F. Sampsell, and J.G. McGowan, The transient response of gas turbine plant heat exchangers - additional solutions for regenerators of the periodic-flow and direct-transfer types, Trans. ASME, Journal of Engineering for Power, Vol. 86, Series A, 127-135 (1964).

20. A.J. Willmott and A. Burns, Periodic-flow regenerators: parameter identification for transient performance, Heat Transfer 1978, Vol. 4, 297-302, Hemisphere Publishing Corp., New York (1978).

21. A. Burns and A.J. Willmott, Transient performance of periodic flow regenerators, International Journal of Heat and Mass Transfer, Vol. 21, 623-627 (1978).

# Dynamic Behaviour of Double-Phase-Change Heat-Exchangers

**FRANZ MAYINGER and MANFRED SCHULT**
Institut für Verfahrenstechnik
der Universität Hannover
Hannover, FRG

ABSTRACT

To predict the dynamic behaviour of a double phase change heat exchanger the thermo- and fluiddynamic effects - especially pressure drop and heat transfer with boiling and condensation - have to be exactly known, which can be achieved by a careful literature study. The system of continuity and constitutive differential equations is not linear and cannot be solved in an analytical way for most cases of applications. Therefore a numerical solution in a computer code is necessary.

For simple geometrical conditions - boiling refrigerant in a tube and condensing water vapour in an annulus around it - measured and theoretically predicted transfer functions are compared and it is shown, that good agreement can be achieved as long as no two-phase flow instabilities occur, the perturbation frequency is not to high and the system pressure is not changed to much. Flow pattern have a strong influence onto the dynamic behaviour. There is a strong coupling between the boiling and condensation side.

Although the dynamic behaviour of the system is not linear, it was possible to evaluate the experimental results by aid of a Fourier analysis and to plot the data in form of transfer functions, demonstrating the amplification coefficient and the phase angle between the perturbated and the corresponding parameter. This analysis can be only performed if the deviation from the linearity is not to large.

1. INTRODUCTION

In power and in chemical engineering frequently heat-exchangers are used, where by condensing saturated vapour on one side liquid of the same or another substance is evaporated. Such an apparatus for heat transport working on both sides with liquid-vapour-mixtures is called a heat-exchanger with double-phase-change. The energy transport in a two-phase flow system has the following advantages.

Even with small temperature differences between the primary and the secondary side of the heat exchanger, an economic heat transport is possible.

The mass flow rate for the energy transport is small due to the large latent heat of evaporization.

Operation is possible also with free convection only because of the large density differences between the phases.

With increasing problems in the world wide energy supply the application of double-phase heat-exchangers may become more and more important especially with safing low temperature energy. In chemical engineering, where aqueous solutions have to be concentrated by partial evaporation using saturated steam as energy supply phase-change has also some benefits if heat has to be transported over long distances. Last not least, low temperature heat can be used in refrigerant processes and in heat pump systems. Also here the primary heat is often available in form of latent heat of evaporization of saturated vapour.

## 2. FLUID DYNAMIC FUNDAMENTALS

The transient and dynamic behaviour of heat exchangers with single phase flow on both sides was treated in detail in the paper by Shah /1/ and by Schöne /2/. All theoretical models describing the dynamic behaviour of heat exchangers start from the wellknown laws of conservation for mass, energy and momentum.

$$\partial \rho / \partial t + \operatorname{div}(\rho \vec{w}) = 0 \tag{1}$$

$$\frac{\partial(\rho h)}{\partial t} + \operatorname{div}(\rho h \vec{w}) = \dot{q} U_b / A_c \tag{2}$$

$$\frac{\partial(\rho \vec{w})}{\partial t} + \operatorname{div}(\rho \vec{w} \vec{w}) = \operatorname{grad} p + \eta \Delta \vec{w} \tag{3}$$

As shown in the literature the methods describing the dynamic behaviour of single phase heat exchangers are mainly different in the procedure evaluating approximativ solutions for these non linear partial differential equations. With phase change on one side or on both sides additional difficulties in describing the dynamic behaviour arise from a physically correct modelling of the typical two-phase flow phenomena, like slip ratio, pressure drop and heat transfer with boiling and condensation.

Studies of the dynamic behaviour of steam generators - with phase change on one side only - were mainly concerned with controlling and with stability problems. With phase change both phases have to be taken in account in the balance equations

$$\frac{\partial}{\partial z}\left[\rho_F w_F(1-\varepsilon) + \rho_D w_D \varepsilon\right] + \frac{\partial}{\partial t}\left[\rho_F(1-\varepsilon) + \rho_D \varepsilon\right] = 0 \tag{4}$$

$$\frac{\partial}{\partial z}\left[\rho_F w_F h_F(1-\varepsilon)+\rho_D w_D h_D \varepsilon\right]+\frac{\partial}{\partial t}\left[\rho_F h_F(1-\varepsilon)+\rho_D h_D \varepsilon\right]=\dot{q} U_b/A_c \quad (5)$$

$$\frac{\partial}{\partial z}\left[\rho_F w_F^2(1-\varepsilon)+\rho_D w_D^2 \varepsilon\right]+\frac{\partial}{\partial t}\left[\rho_F w_F(1-\varepsilon)+\rho_D w_D \varepsilon\right]=$$
$$=\frac{\partial p}{\partial z}-R_{diss}-g\left[\rho_F(1-\varepsilon)+\rho_D \varepsilon\right]; \quad (6)$$

which involves a number of additional unknown variables. Profos /3/ developed a theoretical model for calculating the dynamic behaviour of the outlet conditions in a steam generator assuming that the heat flux is independed from the pipe-wall temperature. In continuing this work Isermann /4/ incorporated the coupling between pipe-wall temperature and heat transfer to the pipe. Hasenkopf /5/ additionally regarded the coupling of the fluid temperature and the gas temperature on the combustion side of a boiler. Numerical integration of the differential equations instead of looking for analytical solutions by using linearization methods has the advantage that local alterations of the thermodynamic conditions can be taken better in account /6, 7/. The geometrical extension of the heat exchanger or of the steam generator, however, has then to be nodalized in a proper way and the difference between the numerical solution and the real condition becomes smaller with a more detailed subdivision of the flow channel.

With double phase change a number of special two phase flow peculiarities arise which do not exist in single phase flow. With phase change various flow pattern can occur which have consequences with respect to heat transfer and pressure drop. Changes in mass flow rate, pressure and temperature may cause flow instabilities in a boiling channel. Bergles /8/ gave a detailed classification of all possible appearances of these instabilities. He distinguishes mainly between a static and a dynamic instability. Static instability results in a new steady state condition, whereas dynamic instabilities produce a periodical change in the flow around the old or a new mean value.

For dynamic instabilities usually more than one mechanism is responsible. The two main origins for dynamic instabilities are density waves and pressure drop changes. The later ones mainly occur in fluid dynamic systems, filled with compressible volumes /9/.

A theoretical model was developed and experiments were performed by Schult /1o/ researching the dynamic behaviour of double phase change heat exchangers. The theoretical model starts from the conservation laws (equations 4 - 6) and is coupling both sides of the heat exchanger - the condensing and the evaporating one - via the heat fluxes transported and the heat stored. In the model coaxial flow of both substances in a tube and in an annulus is assumed with vapour condensing in the annulus, as illustrated in fig. 1.

In order, not to overstress the mathematical treatment and to stay within economical computer times, the model contains the following simplifications:

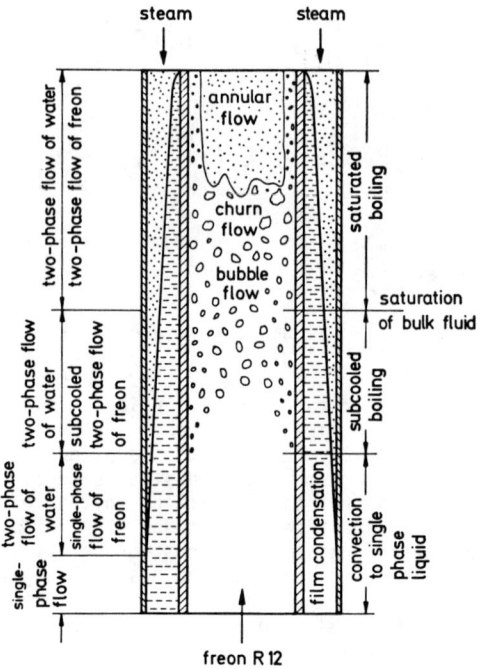

Fig. 1: Flow regions of the vapour transformer

The flow on each side of the heat exchanging wall is one-dimensional

Vapour and liquid are in thermodynamic equilibrium

The temperature of the solid wall is only a function of the length and of the time, i.e. it is assumed that the wall is very thin or has very high heat conductivity.

On the boiling side the fluid enters the channel sub-cooled and

on the condensing side the vapour is saturated.

There is no special assumption for the flow patterns on the boiling side, however, on the condensing side falling film-flow of the liquid is regarded to be present. The coupling conditions, between the condensing and the boiling side, can be easily derived with the help of fig. 2, illustrating a volumetric unit of this single tube annular heat exchanger. The energy balance for the inner tube - boiling side - reads

$$\dot{Q}_{sp} = \dot{Q}_{ax} + \dot{Q}_w - \dot{Q}_f \qquad (7)$$

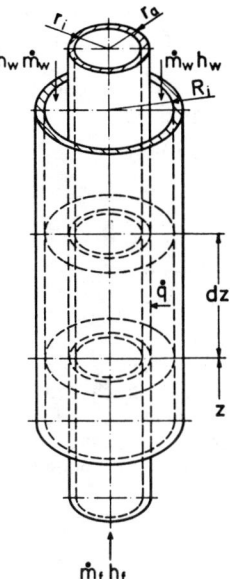

Fig. 2: Volume element of the test section

where the various heat fluxes are defined as follows:

$$\dot{Q}_{sp} = m_R \cdot c_p \cdot \partial \vartheta_R / \partial t , \qquad (8)$$

$$\dot{Q}_w = \alpha_w U_{wb} (\vartheta_w - \vartheta_R) dz , \qquad (9)$$

$$\dot{Q}_f = \alpha_f U_{fb} (\vartheta_R - \vartheta_f) dz , \qquad (10)$$

$$\dot{Q}_{ax} = \lambda A_R (\partial^2 \vartheta / \partial z^2) dz . \qquad (11)$$

The axial heat flux $Q_{ax}$ is small compared with the heat fluxes $Q_w$ and $Q_f$ in radial direction, due to the fact that the heat transfer coefficients on the side of condensation and of evaporation are very high. With negligible axial temperature gradient $d\vartheta/dz$ in the wall, the wall temperature at the elevation z and at the time $t_1$ reads

$$\vartheta_{R_1} = \vartheta_{R_o} \left[ 1 + e^{-\left(\frac{1}{\tau_w} + \frac{1}{\tau_f}\right) \Delta t} \right] \qquad (12)$$

In equation 12, the ratio of the heat capacity versus the heat added or subtracted are expressed in the form of the time constants $\tau_w$ and $\tau_f$

$$\tau_w = \frac{m_R \cdot c_R}{\alpha_w A_{wb}} \qquad (13)$$

$$\tau_f = \frac{m_R \cdot c_R}{\alpha_f \cdot A_{fb}} \qquad (14)$$

The heat transfer coefficients $\alpha_w$ and $\alpha_f$ are functions of the fluid-dynamic and thermodynamic conditions on both sides of the heat exchanging wall. For solving the system of differential equations (equation 4 - 6) together with the radial energy balance, constitutive equations have to be developed describing the following physical phenomena

subcooled boiling on the evaporating side

two-phase flow pressure drop on the evaporating and on the condensing side

slip ratio on the evaporating and on the condensing side

heat transfer with boiling and with condensation.

Most informations for describing these phenomena can be taken from the literature in form of empirical equations. However, one has to be very careful to select these equations in such a way that they meet the real conditions as good as possible.

## 2.1 Subcooled Boiling

With high heat flux densities vapour bubbles are formed at the wall though the mean temperature of the fluid is below the saturation temperature. This phenomenon, called subcooled boiling, improves the heat transfer coef-

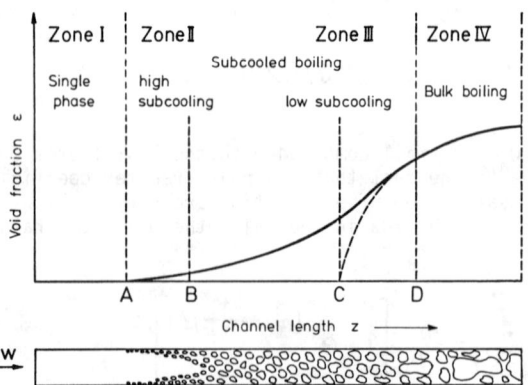

Fig. 3: Vapour void fraction along the length of a heated channel

ficient considerably. 4 axial zones are distinguished with subcooled boiling as demonstrated in fig. 3. In zone I the heat is transported from the wall to the fluid by single phase forced convection. By heat addition along the channel the temperature of the wall and of the fluid is increased. In zone II the first vapour nuclei are formed and the evaporation is still increasing in zone III. Finally in zone IV, the mean temperature of the liquid reaches the saturation temperature corresponding to the pressure being present at this elevation. There is a number of papers in the literature, describing the phenomena with subcooled boiling. Reference is only made to a few of them /11 - 14/.

In the model of Schult /10/ it is assumed that up to the position B in fig. 3, where the vapour void rises considerably, the heat transfer is governed by single phase flow and down stream from this point, heat transfer correlations for fully developed boiling were used. The position for this sudden change in the heat transfer coefficient was correlated using the theory of Saha and Zuber /12/ and of Ünal /15/. Both theories are valid for a variety of liquid substances.

## 2.2 Friction Pressure Drop

In the literature there is a large number of papers dealing with two-phase flow friction pressure drop. For the boiling side, the Baroczy-Chisholm equation /16/ gave best results compared to calibration tests performed in an experimental facility, discussed later. The Baroczy-Chisholm equation gives the ratio between the two-phase flow friction pressure drop and the friction pressure drop if the liquid part would flow only in the channel.

$$\frac{\Delta p_{2Ph}}{\Delta p_{Fo}} = 1 + (\Gamma^2 - 1)\left[B\dot{x}^{0,9}(1-\dot{x})^{0,9} + \dot{x}^{1,8}\right] \qquad (15)$$

This equation contains a term for the density and the viscosity ratio

$$\Gamma = (\rho_F/\rho_D)^{0,5}(\eta_D/\eta_F)^{0,125} \qquad (16)$$

and the constant B can be taken from tab. 1. Equation 15 contains the local quality $\dot{x}$ which continueously changes along the boiling channel.

Tab. 1: Values of B for smooth tubes

| Γ | ṁ (kg/m²s) | B |
|---|---|---|
| ≤ 9,5 | ≤ 500 | 4,8 |
|  | 500 < ṁ < 1900 | 2400/ṁ |
|  | ≥ 1900 | 55/ṁ^(1/2) |
| 9,5 < Γ 28 | ≤ 600 | 520/( ṁ^(1/2)) |
|  | > 600 | 21/Γ |
| ≥ 28 |  | 15000/(Γ²ṁ^(1/2)) |

In the literature the friction pressure drop in a condensing vapour flow usually is treated as a single phase gas flow, however, taking in account that the volumetric mass flow rate is decreasing due to the condensation effect. This method is certainly correct as long as the condensed liquid film is thin compared to the cross flow area of the vapour. In double phase change heat exchangers usually all vapour is condensed and then the friction pressure losses have to be calculated in a similar way as usual for two-phase flow, i.e. with a two-phase friction multiplier. For detailed information reference is made to the thesis by Schult /10/.

## 2.3 Slip Ratio

For boiling flow as a good prediction of the slip ratio s between the phases

$$S = \frac{W_D}{W_F} \tag{17}$$

the correlation by Nabizadeh /17/ can be used, which is based on the theory by Zuber and Findlay /18/. Nabizadeh is not directly giving a correlation for the slip ratio s but for the volumetric void fraction $\varepsilon$ is correlated with the slip ratio s via

$$S = \frac{\dot{x}}{1-\dot{x}} \cdot \frac{1-\varepsilon}{\varepsilon} \cdot \frac{\rho_F}{\rho_D} \tag{18}$$

Using Zuber/Findlay's theory, the mean void fraction $\bar{\varepsilon}$ reads

$$\bar{\varepsilon} = \frac{\dot{x}}{\rho_D} \left[ C_o \left( \frac{\dot{x}}{\rho_D} + \frac{1-\dot{x}}{\rho_F} \right) + \frac{1{,}18}{\dot{m}} \left( \frac{\sigma_g (\rho_F - \rho_D)}{\rho_F^2} \right)^{0{,}25} \right]^{-1} \tag{19}$$

where the factor $C_o$ is given by Nabizadeh as:

$$C_o = \bar{\varepsilon} \left[ 1 + \frac{1}{n} + Fr_o^{-0{,}1} \left( \frac{\rho_D}{\rho_F} \right)^n \left( \frac{1-\dot{x}}{\dot{x}} \right)^{1{,}22n} \right] \tag{20}$$

with

$$n = \sqrt{0{,}6 \frac{\rho_F - \rho_D}{\rho_F}} \tag{21}$$

and

$$Fr_o = \dot{m}^2 / (\rho_F^2 g d)$$

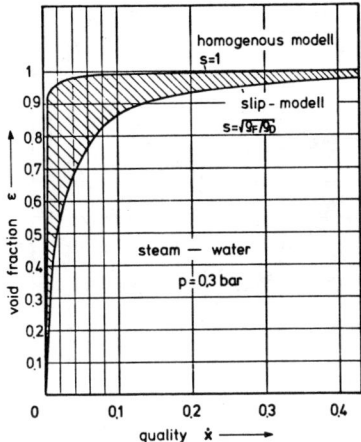

Fig. 4: Void pre-slip model, dicted by slip model and with homogeneous flow

Principally the theory by Zuber and Findlay modified by Nabizadeh, can also be used for the condensing vapour flow. For these parts of the condensing channels where the quality is larger than 0,3, there is almost no or a negligible difference in the prediction of the void fraction by using a slip model or a homogeneous model with identical velocities of vapour and liquid, as demonstrated in fig. 4. Therefore with a good approximation homogeneous flow can be assumed on the condensing side.

2.4 Heat Transfer

The local heat transfer with condensing is dependent whether the falling film flow is laminar or turbulent. Correlations in the literature differ sometimes widely even for identical fluid dynamic and thermodynamic conditions as illustrated in fig. 5. Best agreement with measured values was found for laminar film flow using Hewitt's /19/ equation and for turbulent condition with Kosky /20/ equation.

With boiling two regions of heat transfer mechanism have to be distinguished as wellknown in the literature. In the first one, nucleate boiling with bubbles formed at the heated wall prevails and in the second one an only thin liquid film is present at the wall evaporating at its surface. For the film evaporation an equation given by Ahrens /21/

$$\frac{\alpha_{2Ph}}{\alpha_{Fo}} = C_3 \left[ B_o \cdot 10^{-4} + C_4 (1/X_{tt})^r \right] (1 + d_i/L)^s$$

$$C_3 = 0{,}85 \qquad C_4 = 4{,}5 \qquad r = 0{,}35$$

$$s = (1/X_{tt})^t \qquad t = 0{,}41$$

(22)

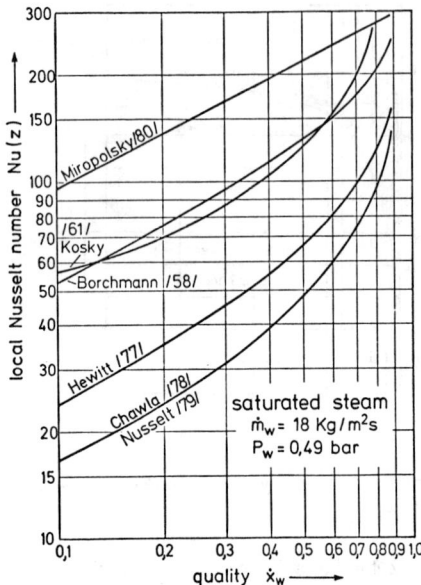

Fig. 5: Local Nusselt numbers for condensing steam as predicted by different authors

$$X_{tt} = \left(\frac{\rho_D}{\rho_F}\right)^{0,5} \left(\frac{\eta_F}{\eta_D}\right)^{0,1} \cdot \left(\frac{1-\dot{x}}{\dot{x}}\right)^{0,9} \quad (22)$$

can be used. Equation (22) in its principal form is wellknown from the literature.

From the numerous correlations predicting heat transfer with bubble boiling under free and forced convective conditions, the equation by Lavin /22/

$$\dot{q} = 0,25 \frac{\lambda_F}{d_a}\left[\frac{d_a^2}{\eta_F d_i}\right]^{0,69} \cdot Pr^{0,69} \left[\frac{\rho_F}{\rho_D} - 1\right]^{0,31} \left[\frac{p\, d_a}{\sigma}\right]^{0,31} \cdot \dot{q}^{0,69} \quad (23)$$

gave best agreement with the measurements.

Under certain conditions especially if the temperature difference between the condensing vapour and the boiling liquid is large, critical heat flux conditions may arise with a consequent sudden change in the heat transfer coefficient. There are several equations in the literature /10/ predicting the boiling crisis.

## 3. EXPERIMENTAL SET UP

The above mentioned balance equations and constitutive correlations were combined in a computer program to predict the dynamic behaviour of a heat exchanger with phase change on both sides. A pure theoretical treatment without experimental assessment could give results far away from the real behaviour of the nature. Already the selection of equations describing single phenomena and separate effects, like heat transfer with boiling and condensation or two-phase pressure drop needs some calibration tests to find a correlation best fitted to the geometrical and fluid dynamic conditions in the heat exchanger. The computer code therefore was tested in an experimental program, which was performed in a test rig with a very simple double phase change heat exchanger, consisting only of two vertical concentric tubes, where in the inner one, under up-flow conditions, the refrigerant R 12 was evaporated and in the annulus, between the inner and the outer tube, under downflow conditions, water vapour was condensed.

The design of the annular test section with the main measuring devices, is shown in fig. 6. On both sides of this simple heat exchanger, the mass flow rate, the pressure and the temperature were measured at the inlet as

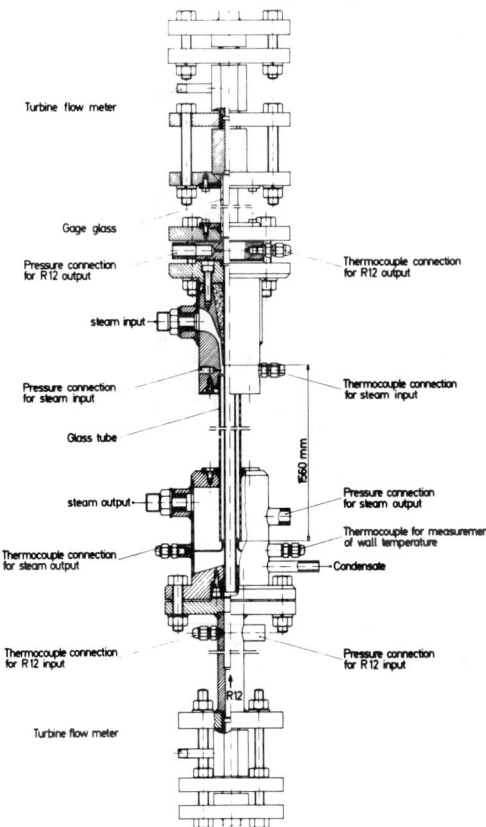

Fig. 6: The double-pipe evaporator - test object

well as at the outlet. The pressure was picked up by electrical pressure gauges and the temperature was determined by thermocouples. Much care was taken to get correct readings of the mass flow rate changes. It pointed out, that turbine mass flow meters were the best sensors for single phase flow as well as for two phase flow. For the later conditions a special calibration procedure combined with a theoretical treatment is necessary to get correct results /23/. In single phase flow the turbine impeller is mainly driven by the momentum of the flow, while the shear stress and consequently the friction on the turbine vanes may be neglected. For a two-phase flow the momenta of vapour and liquid have to be added to a total momentum including the rotation of the turbine wheel. Liquid and vapour in a two-phase flow have different velocities and usually also a large difference in density. Therefore the quality and the void fraction have to be known. The void fraction was measured by the $\gamma$ ray attenuation method. For the dynamic measurements the intensity of the $\gamma$ source had to be high enough to get correct readings also for oscillations with high frequency. A 2 Curie $\gamma$ source therefore was used.

The double phase change heat exchanger was incorporated in a test rig consisting of 2 loops as demonstrated in fig. 7. One of the loops provides the saturated water vapour condensing in the test section, and the other one, the liquid refrigerant evaporating in the heat exchanger.

The refrigerant is only partially evaporated in the heat exchanger, which consists on both sides of stainless steel. The pump 1 delivers liquid refrigerant to the test section 6. The two-phase mixture produced there, is condensed and subcooled in the condenser 9, and from there, the refrigerant flows back to the pump 1. On the condensing side of the test section - the annular heat exchanger - saturated steam of 25 bar, produced in a steam generator 14, is reduced to the test conditions by means of a diaphragm valve 15. By this reduction in pressure the steam becomes superheated and it has therefore to be cooled down to saturation temperature. This is performed in the injection cooler 16.

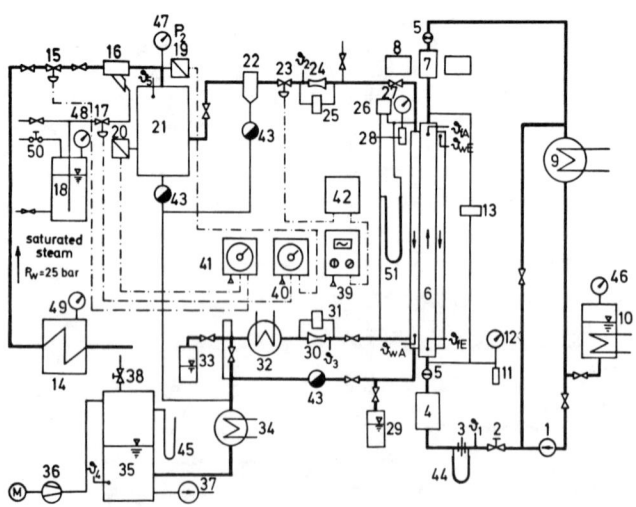

Fig. 7: Schematic of the coupled water-loop and freon R 12-loop

Sinusoidal perturbations can be imposed onto the mass flow rate of the steam and of the refrigerant at the inlet of each side to the test section. A PI-controller 24, is used for mass flow rate control. It receives its nominal value from a pneumatic sinus transmitter, and its actual value from a differential pressure sensor, which in this case can be either a venturi tube or a turbine flow meter. The signal of the command variable is transmitted to a diaphragm valve. Due to the fact that not all of the steam entering the test sections is always condensed a two-phase mixture may leave the test section and therefore liquid and vapour is separated to measure each flow rate separately in addition.

## 4. THEORETICAL AND EXPERIMENTAL RESULTS

The influence of the various flow pattern and of the different heat transfer mechanisms onto the energy transport between condensing steam and evaporating refrigerant, were tested under steadystate conditions at the beginning. This was not only necessary for calibrating the measuring technique and for selecting the equations describing the physical phenomena test, but also for getting a better understanding of the dynamic effects observed later. In fig. 8 the courses of temperature, heat transfer coefficient and heat flux density along the test section, are demonstrated. In this test the water vapour was condensed only partially, so that with neglecting the small pressure drop, the temperature staid constant anlong the whole test section.

The refrigerant R 12 enters the heat exchanger subcooled and undergoes after a certain distance from the inlet position subcooled boiling. In the example, shown in fig. 8, the refrigerant reaches saturation temperature after approximately flowing half the way of the test section. From this position on the temperature of the refrigerant slightly decreases due to the pressure drop in the tube. The heat flux density $\dot{q}$ is governed by the temperature difference between the condensing steam $\vartheta_w$ and the evaporating refrigerant $\vartheta_f$. This temperature difference decreases continuously in the first half of the test section and from thereon it slightly increases.

The heat transfer coefficient $\alpha_w$ on the steam side, is mainly depending from the thickness of the water film on the wall and from the flow con-

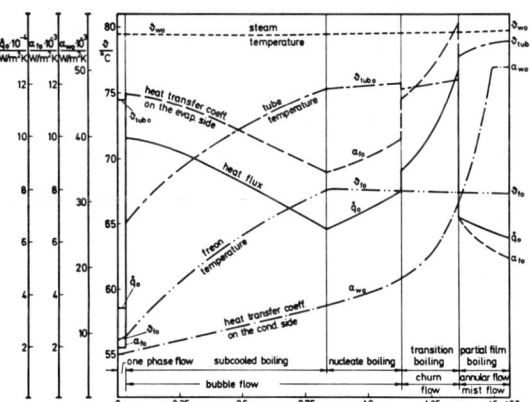

Fig. 8: Steady-state results of a caldulated vapour transformer

ditions- laminar or turbulent -. In the upperpart of the heat exchanger - right side of fig. 8 - the liquid film is thin and the heat transfer coefficient therefore high.

The heat flux density $\dot{q}$ is slightly diminuated in the region of subcooled boiling, because the temperature difference between steam and refrigerant becomes smaller. The heat transfer coefficient $\alpha_f$ on the evaporation side and the heat flux density $\dot{q}$ are mainly coupled in the region of bubble boiling, because $\alpha_f$ is a function of the heat flux density there.

When saturation temperature is reached in the refrigerant, the heat flux density increases, because the temperature difference between the tube wall and the fluid, as well as the heat transfer coefficient on the steam side are rising. This effect is supported by the pressure drop on the refrigerant side. The bubble boiling is followed by two-phase flow heat transfer - surface evaporation - and the heat flux is still increased by a slight improvement of the heat transfer coefficient on the refrigerant side. This is due to the fact that the velocity of the two-phase flow is increased by vapour production.

In the example, shown in fig. 8, boiling crisis occured in the upper part of the test section. This is demonstrated by the sudden deterioration of the heat transfer coefficient $\alpha_f$ on the boiling side. Consequently also the heat flux density $\dot{q}$ is decreasing.

The data measured in the dynamic tests were evaluated by using the Fourier analysis. The perturbations were always sinusoidal, however, the responding functions did not have exactly sinusoidal character, but were somewhat distorted, as demonstrated in fig. 9. This is due to the nonlinear behaviour of the heat exchanger. So the question rises, how the dynamic behaviour can be presented in a simple graphical way. Evaluating the experiments, it pointed out, that the measured data could be represented by super imposing the basic oscillation of the frequency $\omega$ with the first harmonic of the frequency $2\omega$. So the basic oscillation can be presented as a transfer function. The absolute value of the transfer function can be

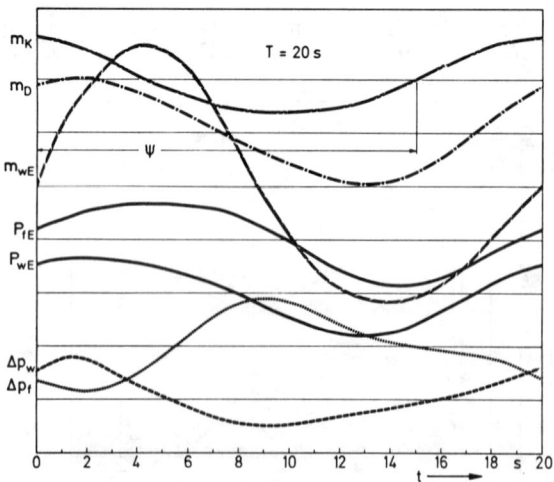

Fig. 9: Temporal course of fluiddynamic parameters corresponding sinusoidal perturbances

written in dimensionless form




and is called the amplification coefficient. $Y_0$ and $X_0$ are steadystate values. Y is the amplitude of the response function and X is the amplitude of the disturbance function. In addition a phase angle $\psi$ can be defined. It is determined from the difference between the two zero passages of the disturbing oscillation and the response oscillation. When the oscillation of the response lags behind the disturbance oscillation, the phase angle is defined as positiv.

There are three main classes of perturbations possible in a double phase change heat exchanger, namely perturbations of

> mass flow rate
>
> temperature and
>
> pressure.

Principally these perturbations can occur on both sides of the heat exchanger, however, on the condensing side temperature changes are always strictly linked with pressure variations, because of the saturation conditions in the steam atmosphere. Pressure changes usually are not regarded as normal operation transients and their modelling needs additional correlations for physical phenomena, like propagation effects of pressure waves, not described here, and not incorporated in the above mentioned computer code. Therefore the discussion will be concentrated to oscillations and perturbations of mass flow rate and temperature.

Discussing the consequences of perturbations one has strictly to distinguish whether the transient behaviour is stable or whether instabilities - for example due to boiling crisis or pressure drop effects - occur. In stable behaviour the system is adjusted to a new quasi steadystate around which the thermo- and fluiddynamic status is oscillating. Instabilities show a very rapid and strong change in the fluiddynamic conditions and the oscillations are not quasi steady.

## 4.1 Perturbations On The Primary Side

The condensing steam on the primary side is saturated and therefore temperature fluctuations would be necessarily linked with pressure changes. The deliberations on the dynamic behaviour under normal operation transients therefore can be concentrated to mass flow rate perturbations on the primary side. The correlations discussed in chapter 2, do not fully take in account instability effects, especially if they are consequences of sudden changes in the flow pattern. Therefore deviations between measured and theoretically predicted values were mainly observed in connection with flow pattern changes.

Best agreement between theory and experiment was found with smaller frequences of the perturbations, which easily can be understood remembering the assumptions in the theoretical model which neglected the heat stored in the tube walls. In fig. 10 the transfer function and the shift of the phase angle is demonstrated for the mass flow rate $\dot{m}_D$ of the vapour produced by

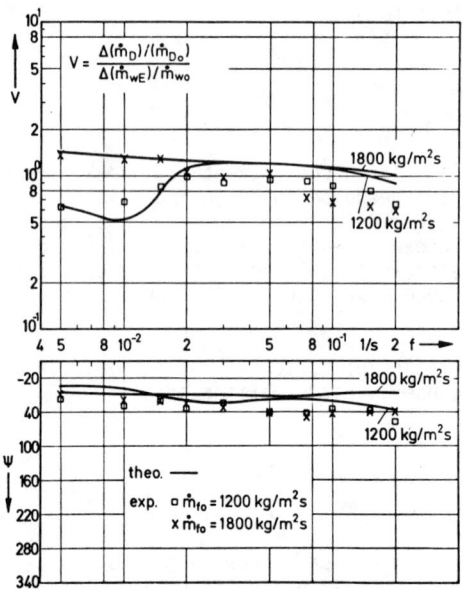

Fig. 10: Transfer functions of the vapour mass flow at the exit of the evaporating side.
Comparison between the calculated and measured data.

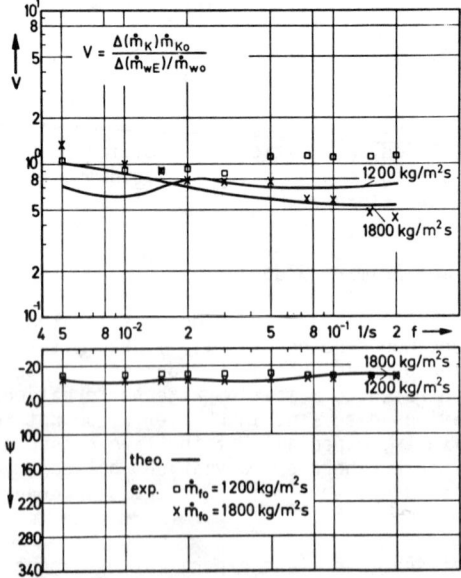

Fig. 11: Transfer functions of the condensate mass flow at the exit of the condensing side. Comparison between the calculated and measured data.

evaporation of the refrigerant. The mass flow rate $\dot{m}_{wo}$ of the water vapour, before the oscillations started was 18 kg/m²s and the perturbations in this mass flow rate $\Delta(\dot{m}_{wo})$ at the inlet of the test section, imposed later on, were 6,5 kg/m²s. The steadystate mass flow rates on the refrigerant side $\dot{m}_{fo}$ were 1.800 kg/m²s and 1.200 kg/m²s respectively. In the figure the amplification coefficient and the phase angle is plotted. Deviation between measurement and experiment are mainly in regions where instabilities were observed.

In fig. 11 the transfer functions for the mass flow rate of the condensed water vapour $\dot{m}_K$ is illustrated.

In the correlations discussed in chapter 2 it is assumed that on both sides of the heat exchanger vapour and liquid are in thermodynamic equilibrium. This is not fully the case in reality because the liquid boundary layer on the boiling side is superheated and the liquid film of the condensed water on the other side is slightly subcooled. With periodic changes of the temperature in these boundary layers effects of heat storage occur influencing the amplification coefficient and the phase angle on the refrigerant side as well as on the water vapour side.

From the fig. 10 and 11 it can be seen that at low perturbation frequences the agreement between theory and experiment is better than at the higher ones. With increasing frequency the damping effects of the fluid boundary layers on both sides of the heat exchanger gain more influence. These damping effects are reducing the maxima and the minima of the amplification coefficient, which are therefore overpredicted in the theory compared to the experiment.

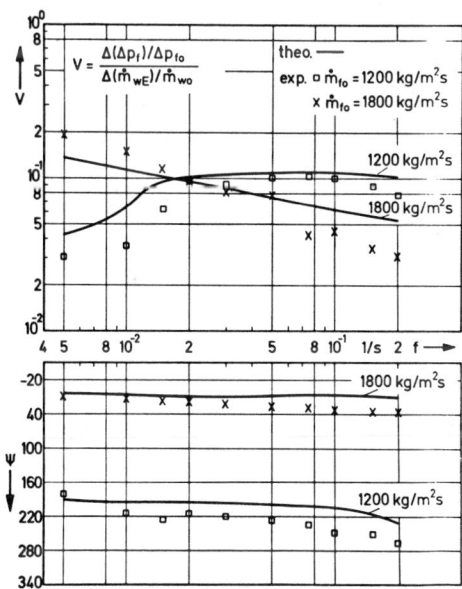

Fig. 12: Transfer functions of the pressure drop on the evaporating side. Comparison between the calculated and measured data.

The pressure drop is more sensitiv to perturbations than the mass flow rate. On the evaporation side as demonstrated in fig. 12, the agreement between calculated and measured values is still not to bad. However, on the condensing side there are large quantitative discrepances although qualitatively there is agreement in the tendency of the amplification coefficients versus frequency, as shown in fig. 13. These discrepances can be explained by the different components contributing to the pressure drop on the condensing side. The total pressure drop on this side is mainly influenced by friction and momentum change. Both have opposit signs because the decreasing momentum due to condensing vapour results in a pressure increase and friction is always reducing the pressure. Under certain conditions both terms may be of the same order and then the pressure change is almost zero. As long as there is no stagnant liquid water in the annulus of the condensing side, the potential term has no influence because the gravity force is in equilibrium with the wall friction of the liquid falling film flow. The amplification coefficient is related to the steadystate value and therefore with almost zero steadystate pressure drop a very small mistake in predicting the steadystate conditions gives large errors in the transfer functions. As shown in fig. 13, the amplification factor for the pressure drop in the condensing flow can be in the order of ten.

In the fig. 10 - 13 the transfer functions were always related to a steadystate mass flow rate.

The theoretical model, however, is not restricted to certain fluiddynamic or thermodynamic states. It can be used in a wide range of mass flow rates, pressures and temperatures, as shown in the fig. 14 and 15. In both fig. as the previous ones, the mass flow rate of the saturated vapour entering the annulus and condensing there, was perturbated. Instead of constant mass flow rates of the refrigerant, now constant pressure (fig. 14) and constant subcooling (fig. 15) on the evaporating side - i.e. at the entrance of the re-

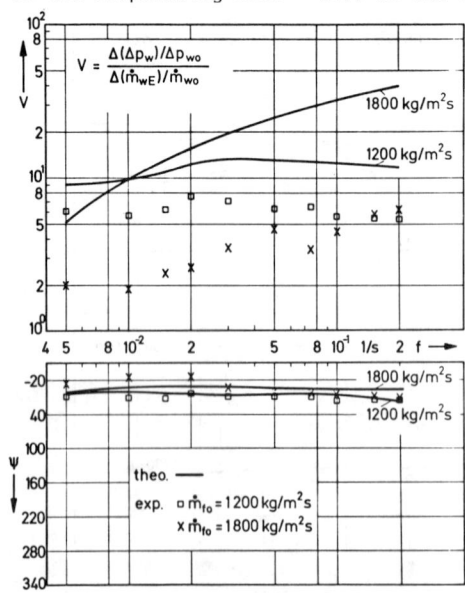

Fig. 13: Transfer functions of the pressure drop on the condensing side. Comparison between the calculated and measured data.

# DYNAMIC BEHAVIOUR OF DOUBLE-PHASE-CHANGE HEAT-EXCHANGERS

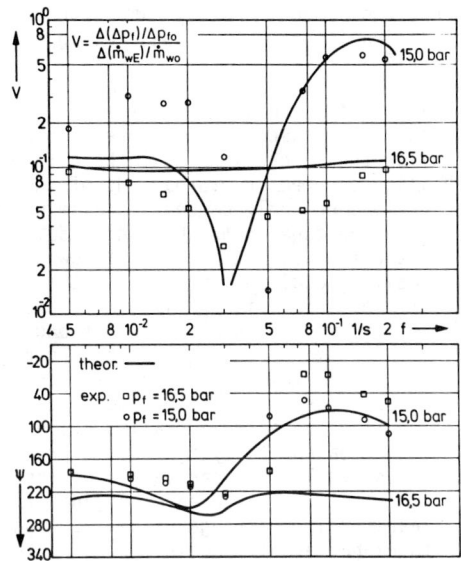

Fig. 14: Amplification factor and phase shift measured and calculated for pressure drop on boiling side. Mass flow rate of condensing vapour perturbated.

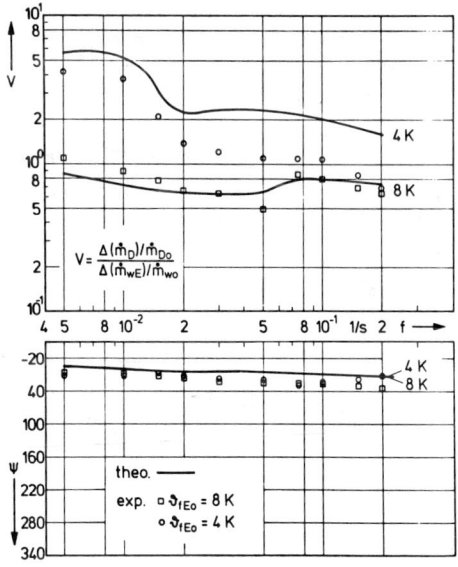

Fig. 15: Amplification factor and phase shift of evaporated refrigerant mass flow rate (measured and calculated). Mass flow rate of condensing vapour perturbated. Constant subcooling

frigerant flow - are regarded. Theory and experiment show a sudden and
large change in the amplification factor, as well as in the phase angle for
the boiling pressure drop.

To understand physically this behaviour, the wall temperature along
the heat exchanger has to be discussed. Temperature readings in the upper
part of the tube wall allow the conclusion, that there temporarily and
locally partial film boiling occured. Perturbations of the condensing
vapour flow on the water side, not only caused and eliminated this partial
film boiling up and down in the tube on the boiling side. These instabilities strongly influenced the transfer functions. At low perturbation
frequences partial film boiling could be observed in the upper part of the
heat exchanger over the whole oscillation period. With increasing frequency,
the response behaviours became more pronounced and strong pulsations in
the wall temperature and in the mass flow rate of the refrigerant could be
observed. This pulsations can be explained by the fact that high condensing
vapour flow rate strongly produces film boiling and in the period of low
vapour flow rate bubble boiling can be re-installed again. With very high
frequences the secondary side cannot fully follow the primary perturbations
due to storage effects in the wall. The mean values of the heat flux density were below critical heat flux and nucleate boiling staid stable in the
refrigerant flow which resulted in a stronger amplification of the pressure
drop oscillations, due to higher evaporation rates.

In the fig. 10 - 15 the perturbation amplitude in the vapour mass flow
rate was the same. In fig. 16 the validity of the theoretical model is
illustrated for different perturbation amplitudes. The transfer functions
for the mass flow rate of the vapour produced by boiling are in good agreement for high perturbation amplitudes, as well as for low ones.

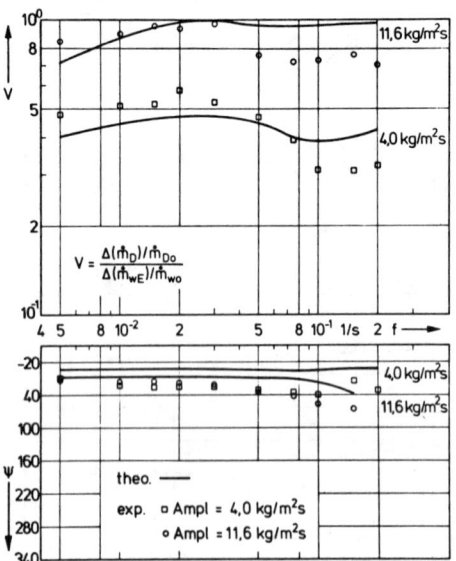

Fig. 16: Amplification factor and phase shift of evaporated refrigerant
mass flow rate (measured and calculated). Mass flow rate of
condensing vapour perturbated. Constant mixture flow rate.

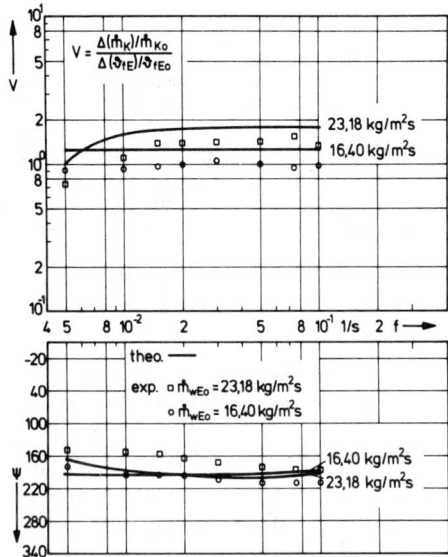

Fig. 17: Amplification factor and phase shift of condensed steam flow rate (measured and calculated) perturbation of inlet temperature on boiling side.

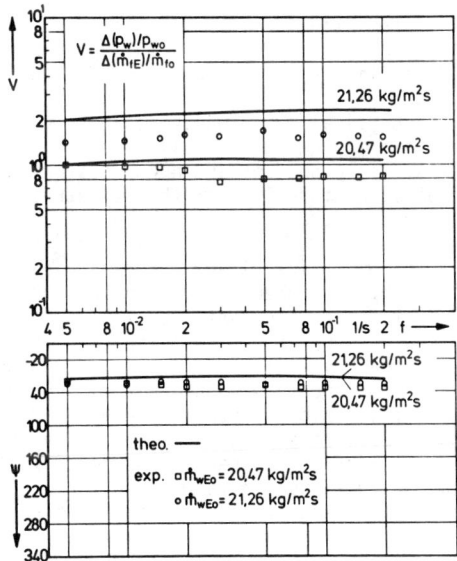

Fig. 18: Amplification factor and phase shift of pressure drop (calculated and measured), perturbated mass flow on boiling side.

4.2 Perturbations On The Secondary Side

With perturbations on the secondary side, the numerical procedure by solving the differential equations in the code becomes a little more complicated, however, the physical models are the same as with perturbations in the condensing steam flow. As demonstrated in the fig. 17 and 18, the system stays more stable with perturbations on the secondary side than on the primary one. The reason for this is very easily to be understood. The heat source on the primary side is a system of constant temperature. If the mass flow rate of the boiling refrigerant is changed - see fig. 17 - the heat transfer coefficient in the boiling region, inclucing the subcooled boiling area, is not to much affected. Only if single phase heat transfer prevails which is the case at higher mass flow rates, the amplification coefficient becomes larger and deviates remarkably from one.

In fig. 18 is shown how fluctuations in the mass flow rate of the refrigerant influence the pressure on the condensing side. With decreasing mass flow rate of the refrigerant, the heat transport is slightly more deteriorated than with increasing mass flow rate. This effect, however, is partially balanced by a prolongation of the subcooled boiling area. To transport the same amount of heat with a somewhat lower heat transfer coefficient a higher temperature difference between primary and secondary side is needed. This is automatically achieved by rising the temperature and via the saturation conditions also the pressure of the condensing steam. So the amplification coefficient for this transfer function is near one.

## 5. CONCLUSIONS

The computer code predicting the dynamic behaviour of a double phase change heat exchanger is in good agreement with experimental results as long as no fluiddynamic instabilities occur and as the frequency of the fluctuations is not to high. The later restriction is a consequence of the simplified assumptions made in the physical model which neglect the heat storage in the structure material and thermodynamic disequilibrium in the fluids. Instabilities are strongly system dependant and a correct quantitative prediction needs a very detailed modelling of the geometric data of the whole loop.

In the code a number of empirical and semi-empirical correlations is incorporated describing physical phenomena, like pressure drop or heat transfer, which may be strongly dependant from the geometrical conditions too. Therefore by using the code it has to be checked whether these constitutive equations fit the physical effects influencing the dynamic behaviour well enough also for the design under study. For heat exchangers strongly deviating from the design, discussed here a careful literature survey is needed. For an apparatus which is already operating, the easiest way to adjust the correlations and to verify the code, is to do calibration tests under steadystate conditions and to measure the pressure drop and if possible also the heat transfer coefficients in the range of interest.

REFERENCES

1. Shah, R.K., "Transient behaviour of heat exchangers", Proceedings of the Advanced Study Institute on Heat Exchangers, Istanbul 1980

2. Schöne, A., "Das dynamische Verhalten von Wärmetauschern und seine Beschreibung durch Näherungen", Verlag R. Oldenburg, Minden - Wien, 1966

3. Profos, P., "Die Regelung von Dampfanlagen", Springer-Verlag, Berlin/Göttingen/Heidelberg, 1962

4. Isermann, R., "Das regeldynamische Verhalten von Überhitzern", Fortschritt-Ber. VDI-Z., Reihe 6, Nr. 4, 1965

5. Hasenkopf, O., "Übertragungsverhalten eines Dampferzeugers in Zusammenwirken mit einem gasgekühlten Kernreaktor", Fortschritt-Ber. VDI-Z. Reihe 6, Nr. 36

6. Varcop, L, "Die Dynamik zwangsdurchströmter Verdampfersysteme unter Berücksichtigung von Druckänderungen des Strömungsmediums", Regelungstechnik 15, 1967, S. 4o4/412

7. Proska, F., "Verfahren zur Berechnung der Frequenzgänge von Gleichstrom- und Gegenstromwärmetauschern", Regelungstechnik 10, 1962, S. 206/210 und S. 256/260

8. Bergles, A.E., "Review of Instabilities in Two-Phase Systems", Proceedings of the NATO Advanced Study Institute on Two-Phase Flows and Heat Transfer, Istanbul, Aug. 1976

9. Veziroglu, T.N., Lee, S.S. and Kakac, S., "Fundamentals of Two-Phase Flow Oscillations and Experiments in Single channel Systems", Proceedings of the NATO Advanced Study Institute on Two-Phase Flows and Heat Transfer, Istanbul, Aug. 1976

10. Schult, M., "Untersuchungen zum dynamischen Verhalten von Doppelrohrwärmetauschern mit doppeltem Phasenwechsel", Dissertation der Universität Hannover, 1979, Institut für Verfahrenstechn.

11. Bucher, B., "Beitrag zum Siedebeginn beim unterkühlten Sieden mit Zwangskonvektion", Dissertation am Institut für Verfahrenstechnik der Universität Hannover, 1979

12. Saha, P., Zuber, N., "Point of Net Vapour Generation and Vapour Void Fraction in Subcooled Boiling", Proceedings of the Heat Transfer Conference, B 4.7, Tokyo 1970

13. Rouhani, S.Z., "Calculation of Void, Volume Fraction in the subcooled and Quality Boiling Regions", Int. J. Heat Mass Transfer, Vol. 13, 1970, pp. 383/393

14. Dix, G.E., "Vapour Void Fractions for Forced Convection with Subcooled Boiling at Low Flow Rates", General Electric Report No. NEDO-10491

15. Ünal, H.C., "Determination of the Initial Point of Net Vapour Generation in Flow Boiling Systems", Int.J.Heat Mass Transfer, Vol. 18, 1975, pp. 1095/1099

16. Chisholm, D., "Pressure Gradients Due to Friction during the Flow of Evaporating Two-Phase Mixtures in Smooth Tubes and Channels", Int. J. Heat Mass Transfer, Vol. 16, 1973, pp. 347/358

17. Nabizadeh, H., "Modellgesetze und Parameteruntersuchungen für den volumetrischen Dampfgehalt in einer Zweiphasenströmung", Dissertation am Institut für Verfahrenstechnik der Universität Hann., 1977

18. Zuber, N., Findlay, J.A., "Average Volumetric Concentration in Two-Phase Flow System", J. Heat Transfer, No. 9, 1960, pp. 453/468

19. Hewitt, G.F., Hall-Taylor, N.S., "Annular Two-Phase Flow, Pergamon Press, 1970

20. Kosky, P.G., Staub, F.W., "Local Condensing Heat Transfer Coefficient in the Annular Flow Regime", AIChE Journal, Vol. 17, No. 5, 1971, pp. 1037/1043

21. Ahrens, K.-H., Mayinger, F., "Boiling Heat Transfer in the Transition Region from Bubble Flow to Annular Flow", Proc. Int. Seminar Momentum, Heat & Mass Transfer in Two-Phase Energy and Chemical Systems, I.C.H.M.T. Dubrovnik, Yugoslavia, 1978

22. Lavin, J.G., Young, E.H., "Heat Transfer to Evaporating Refrigerants in Two-Phase Flow", AIChE Journal, Vol. 11, 1965, pp. 1124

23. Belda, W., "Dryoutverzug bei Kühlmittelverlust in Kernreaktoren", Dissertation am Institut für Verfahrenstechnik der TU Hannover, 1975

Table 2

Additional information on fluddynamic parameters, used in tests and calculations, presented in figures 10 - 18:

Fig. 10: $P_{fEo}$ = 18 bar $\quad\quad$ $P_{wEo}$ = 0,46 bar
$\Delta\vartheta_{fEo}$ = 12° C $\quad\quad$ $\dot{m}_{wo}$ = 18 kg/m²s
$\quad\quad\quad\quad\quad\quad\quad$ $\Delta(\dot{m}_{wE})$ = 6,5 kg/m²s

Fig. 11: $P_{fEo}$ = 18 bar $\quad\quad$ $P_{wEo}$ = 0,46 bar
$\Delta\vartheta_{fEo}$ = 12° C $\quad\quad$ $\dot{m}_{wo}$ = 18 kg/m²s
$\quad\quad\quad\quad\quad\quad\quad$ $\Delta(\dot{m}_{wE})$ = 6,5 kg/m²s

Fig. 12: $P_{fEo}$ = 18 bar $\quad\quad$ $P_{wEo}$ = 0,46 bar
$\Delta\vartheta_{fEo}$ = 12° C $\quad\quad$ $\dot{m}_{wo}$ = 18 kg/m²s
$\quad\quad\quad\quad\quad\quad\quad$ $\Delta(\dot{m}_{wE})$ = 6,5 kg/m²s

Fig. 13: $P_{fEo}$ = 18 bar $\quad\quad$ $P_{wEo}$ = 0,46 bar
$\Delta\vartheta_{fEo}$ = 12° C $\quad\quad$ $\dot{m}_{wo}$ = 18 kg/m²s
$\quad\quad\quad\quad\quad\quad\quad$ $\Delta(\dot{m}_{wE})$ = 6,5 kg/m²s

Fig. 14: $\dot{m}_{fo}$ = 1200 kg/m²s $\quad$ $\dot{m}_{wo}$ = 18,4 kg/m²s
$\Delta\vartheta_{fEo}$ = 12 K $\quad\quad$ $P_{wEo}$ = 0,46 bar
$\quad\quad\quad\quad\quad\quad\quad$ $\Delta(\dot{m}_{wE})$ = 6,5 kg/m²s

Fig. 15: $\dot{m}_{fo}$ = 1500 kg/m²s $\quad$ $\dot{m}_{wo}$ = 18 kg/m²s
$P_{vEo}$ = 18 bar $\quad\quad$ $P_{wEo}$ = 0,48 bar
$\quad\quad\quad\quad\quad\quad\quad$ $\Delta(\dot{m}_{wE})$ = 6,5 kg/m²s

Fig. 16: $\dot{m}_{fo}$ = 1200 kg/m²s $\quad$ $\dot{m}_{wo}$ = 18 kg/m²s
$P_{fEo}$ = 18 bar $\quad\quad$ $P_{wo}$ = 0,45 bar
$\Delta\vartheta_{fEo}$ = 12 K

Continue Table 2:

Fig. 17: $\dot{m}_{fo}$ = 1200 kg/m²s  $\dot{m}_{wo}$ = 18 kg/m²s
$P_{fEo}$ = 18 bar  $P_{wo}$ = 0,45 bar
$\Delta\vartheta_{fEo}$ = 12 K

Fig. 18: $\dot{m}_{fo}$ = 1200 kg/m²s  $P_{wEo}$ = 0,48 bar
$P_{fEo}$ = 18 bar  $\Delta(\vartheta_{fE})$ = 4,0 K
$\Delta\vartheta_{fEo}$ = 12 K

# Vibration in Heat Exchangers

**FRANZ MAYINGER and HANS GÜNTER GROSS**
Institute für Verfahrenstechnik
der Universität Hannover
Hannover, FRG

## 1. INTRODUCTION

To improve the thermal efficiency heat exchangers are commonly equiped with baffles. These devices produce a cross flow around the tube bundles which is favourable for the heat transport, however, which also may induce vibrations. If the amplitudes of the vibrations become to high, fretting corrosion and and erosion of the tubes at the position of the baffles may occur.

Since mor than 30 years research activities are underway to study the vibration phenomena in cross flow bundles of heat exchangers. Mostly it is assumed that oscillations are excited by vortices departing from the tubes, then the strongest vibrations should be observed if the departure frequency of the vortices and the resonance frequency of the tubes are identical. A safe layout of the tube banks would then be not too difficult, one just has to avoid the coincidence of these both frequencies.

## 2. MECHANISMS EXCITING VIBRATIONS

If the vibrations are excited by the vortices, which again are produced by the flow around a tube, the exciting frequency can be easily predicted by the Strouhal-number /1/

$$S = \frac{f_w \cdot d}{w} \qquad (1)$$

which gives the relationship between the vortex frequency $f_w$, the diameter of the tube d and the vlocity w of the flow, in front of the tube or the cylinder. In the literature it is clearly stated /2 - 4/ that there is a linear connection between the vortex frequency and the flow velocity, which means that the Strouhal-number is constant. After a careful literature survey Chen /5/ found that the value of the Strouhal-number should be between 0,17 and 0,21, for Reynolds-numbers from 300 up to more than $2 \cdot 10^5$.

Buffeting

Owen /15/ studied the turbulence behind the tubes and found that this may cause vibration. He called this aeroelastic exciting phenomenon buffeting. These turbulent and stochastic velocity fluctuations are mainly due to the perturbation of the boundary layer on the rear side of the tube. These fluctuations have a very wide frequency spectrum.

A special case is the resonant buffeting, which also has a statistical energy distribution, however, in addition a periodical velocity fluctuation is superimposed. If the frequency of this periodical fluid dynamic exciting force coincides with the resonance frequency of the tubes, vibrations of large amplitudes may be the consequence.

Galloping - Wake Galloping

In civil engineering another vibration exciting phenomenon is wellknown, which is called galloping - for a single obstacle - or wake galloping - for a group of obstacles like cylinders -. Galloping was observed with chimneys or with the cables of high voltage transportation lines /16, 17/. As shown in fig. 1 for the example of a tube or rod bundle with 3 rows galloping can produce a lifting force due to the partial deflection of the flow. This phenomenon mainly in the second and in the third row may perform a vibration rectangular to the direction of the inlet flow.

In civil engineering galloping was mainly observed with non-cylindrical obstacles, like prismatic rods, where one has to distinguish between a lift force

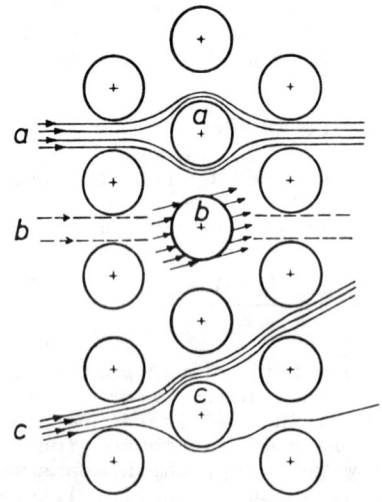

Fig. 1: Flow paths in rod bundle

$$L = \tfrac{1}{2} \cdot c_L \cdot H \cdot \rho \cdot w_{rel}^2 \qquad (2)$$

and a drag force

$$D = \tfrac{1}{2} \cdot c_D \cdot H \cdot \rho \cdot w_{rel}^2 \qquad (3)$$

In tube banks of staggered arrangement the lift force may be due to the mentioned deflection of the flow. The critical velocity can be formulated by solving the differential equation, describing the exciting and damping effects as well as the inertia forces

$$w_{krit} = - \frac{2 \cdot d^*}{\rho \cdot H \cdot K} \qquad (4)$$

In equation (4) $d^*$ represents a damping factor, H a characteristic length of the profile and K a factor describing the combination of lift and drag forces. Compared to the before mentioned phenomena - exciting by vortices and by turbulent buffeting - where only at a single frequency high amplitudes of vibration occur, galloping or wake galloping can produce strong vibrations above a certain flow velocity over a wide spectrum of frequencies.

Aeroelastic Coupling

Another mode exciting vibrations in cylindrical tube banks may be the aeroelastic coupling, first mentioned by Livesey /18/ and later by Connors /13/. In contrast to the vortex or galloping induced vibration, the tubes not only move rectangular to the flow but also in the flow direction.

The aeroelastic coupling is a consequence of the movement of the rods. If the rod leaves its original or stationary position, the fluiddynamic forces around the rod, which are influenced by the relative position to the neighbouring rods, change. So a new exciting force for vibration may be created. The frequency of the vibration, however, is then not only depending on the flow velocity, but also on the resonant frequency of the surrounding rods. The main influencing factor with aeroelastic coupling is the movement of the neighbouring tubes, which means that a small vibration of a few rods in the tube bank may excite other rods by fluiddynamic forces and therefore this phenomenon is called aeroelastic coupling. Contrary to the exciting modes discussed before - like buffeting or galloping - with aeroelastic coupling no favoured vibration direction can be observed. The vibration movement of each rod is depending on and influenced by the movement of its neighbouring rods.

Spivack /6/, Dumpleton /7/ and Gregorig /8 - 10/ found that the vortices cannot be the only and main reason for inducing vibrations in tube banks. Most of the experiments in the literature studiing vibrations in heat exchangers used tube banks where only one tube could freely move and all others were fixed /11/. Owen /12/ was one of the first who doubted about the influence of the vortices on exciting vibrations and found that freely moving and interfering tubes may have other mechanisms to produce vibration. Also Connors /13/ observed vibration phenomena, which could not be explained by the vortex theory. He introduced the idea of aeroelastic coupling between the different tubes in the bank.

The vibration induced by vortices was described by Magnus /14/, starting from the equation of movement

$$m\ddot{y} + d^*\dot{y} + ky = L \sin\Omega t \qquad (5)$$

where

L is a lifting force

$\Omega$ the departing frequency of the vortices

$d^*$ a damping factor

k a constant, representing the reacting force.

Introducing the damping factor of Lehr $D^*$ and the resonance frequency $\omega_0$ of the tube one can predict the maximum vibration amplitude

$$|y_{max}| = \frac{L}{k} \cdot \frac{1}{\sqrt{\left(1 - \frac{\Omega^2}{\omega_0^2}\right)^2 + 4D^* \frac{\Omega^2}{\omega_0^2}}} \qquad (6)$$

by solving equation (5). From equation (5) and (6) one can calculate the velocity $w_{res}$ of the flow at which resonance of the tube is to be expected

$$w_{res} = \frac{\omega_0 d}{2\pi S} = \frac{f_0 d}{S} \qquad (7)$$

Practical experience, however, shows that vibrations in heat exchangers occur sometimes far before this velocity is reached and sometimes also much later.

## 3. FLOW PULSATIONS AND ENERGY DENSITY DISTRIBUTIONS

In a turbulent flow fluctuations are superimposed to the mainflow direction which may be of periodic or stochastic nature. Using the root-mean-square value

$$\sqrt{\overline{f(t)^2}} = \sqrt{\frac{1}{2T} \cdot \int_{-T}^{+T} f^2(t)\, dt} \qquad (8)$$

of these fluctuations the grade of turbulence

$$Tu = \frac{\sqrt{1/3\,(\overline{u_x^2} + \overline{u_y^2} + \overline{u_z^2})}}{w} \qquad (9)$$

can be defined as wellknown in the literature.

Like the kincetic energy of the flow in the main direction, also the turbulence represents a certain fluctuation energy. This energy can be diversificated according to the frequency spectrum of the turbulent flow pulsations. If in addition one refers the energy of a special pulsation frequency to the mean or integral value of the energy of the whole spectrum, one can define a reduced energy density function

$$F_{red}(f) = \frac{\overline{E^2(f_m)}}{b\,\overline{E^2}} \qquad (10)$$

for a given frequency f, where E is the voltage measured at a hot wire anemometer. In equation (10) b is the range of the frequency under consideration for this function.

In the flow there are turbulent and periodic components of velocities

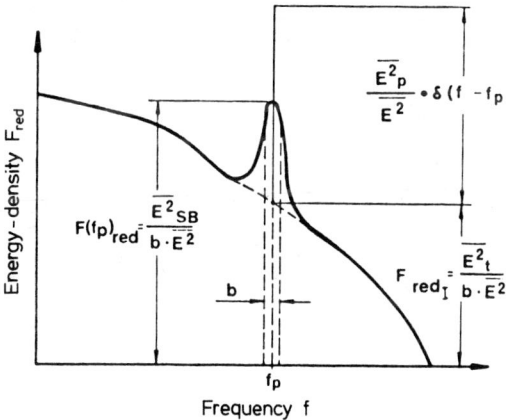

Fig. 2: Energy - density - spectrum (schematic)

$$\overline{u^2} = \overline{u_t^2} + \overline{u_p^2} \qquad (11)$$

and the total energy exciting vibrations is the sum of both components.

$$\overline{u^2} \cdot \int_0^\infty F_{red}\, df = \overline{u_t^2} \cdot \int_0^\infty F_{red_t}\, df + \overline{u_p^2} \cdot \int_0^\infty F_{red_p}\, df \qquad (12)$$

In equation (12) $\overline{u_t^2}$ represents the kinetic energy of the turbulent fluctuation and $\overline{u_p^2}$, that of the periodic fluctuations of the frequency $f_p$. A schematic distribution of the energy density of a flow with turbulent and periodic fluctuation is illustrated in fig. 2. The energy distribution shows a pronounced peak at the frequency of the periodic fluctuations. From this energy peak the vibration of the rods may obtain its exciting forces.

4. EXPERIMENTAL OBSERVATIONS

Detailed experiments studiing the vibration phenomena in tube banks and also in tube or rod arrays of one row only were performed by Gross /19/. The main aim of these research activities was to get information which of the above mentioned exciting phenomena may have the main influence on the development of vibrations. In the experimental setup, staggered and inline arrangements of tube banks in airflow were tested. Contrary to the usual habit in the literature all tubes in the bank were movably suspended. The reacting forces in the suspension device were made rather weak and large vibration amplitudes were allowed to be able to study the influence of even small exciting forces and to get a well readable signal of the rod movement. For a single row of rods the course of the movement - i.e. the vibration pattern - is illustrated in fig. 3. There is a clear alternation in the direction of the vibration course. From this observation one can conclude that the vibration is mainly excited by aeroelastic coupling. This assumption is supported by comparing the critical velocity, when the vibration started, with the Strouhal-criterion in equation (1). The measured critical velocity was by a factor of 2 higher than that

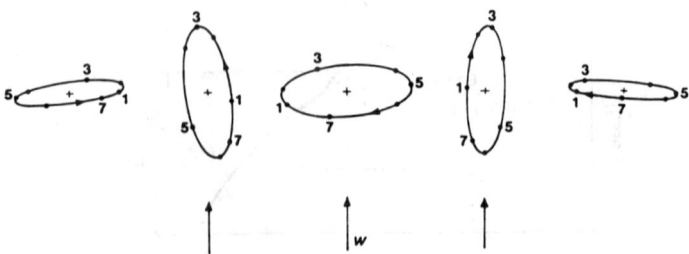

Fig. 3: Vibration pattern in a single row

one predicted by Chen /5/ with the Strouhal-criterion. Therefore vortices leaving periodically the lee-side of the tubes could not be the main reason for inducing these vibrations.

Similar observations could be made with a two-row arrangement. Measurements with this configuration were made under two different conditions, namely with all rods freely suspended and with the first row kept firmly in position. Fig. 4 gives information about the vibration amplitude of the rods in the first and in the second row as a function of the flow velocity. With all rods movably suspended, the second row starts earlier - i.e. at a lower flow velocity - to vibrate then with the first row fixed. Also the vibration amplitudes are larger then in the partially fixed case. The fact that with fixed rods in the first row, the onset of the vibration in the second row could be largely delayed to higher flow velocities, seems to be a clear hint that aeroelastic coupling plays an important role, which, however, refers only to the second row. For the first row there must be an additional exciting mechanism, for example galloping or wake-galloping.

Staggered Arrangement of Tube Banks

Besides the flow velocity and the suspension of the rods also geometrical parameters may have an influence on the beginning and the amplitudes of vibrations. A significant geometrical parameter is the ratio τ between the distances of the rods and their diameter d. This ratio was varied in the tests of Gross /19/ in the range 1,1 to 1,5. Also the arrangement of the rods - staggered or in line - certainly plays an important role. The velocities in the narrowest gap between the rods were varied between 10 and 60 m/s corresponding to Reynolds-numbers from 20.000 to 120.000. The diameter of the rods to which this Reynolds-number is refered to was 30 mm. The suspension of the rods was made in such a way that the elasticity was approximately the same as that of heat exchanger tubes between two baffles. With up to 10 mm the absolute amplitude A of the possible rod movement was rather large.

In the fig. 5, 6, 7 the observed vibration amplitudes for s over d ratios of 1,1, 1,3 and 1,5 are illustrated.

Only with the narrowest spacing the rods in the first row begin to vibrate before a movement of the other rods can be observed. With larger spacing - 1,3 and 1,5 - the second and the third row are mostly effected by vibration. Rows downstream and especially the last row in the bundle, which in these tests was the 6th one, have much lower amplitudes. The velocities where the vibration starts are by a factor 5 higher than the Strouhal-criterion would predict. Although calculations based on resonant buffeting, i.e. exciting by turbulent fluctuations, would result in a much lower critical velocity. Galloping may have influence in the lower velocity range on the vibration as can be concluded from stroboscopic observations giving information about the course of the rod movement. When the first vibration starts the rods in the second row show a oscillation movement almost exclusively rectangular to the main flow direction. With

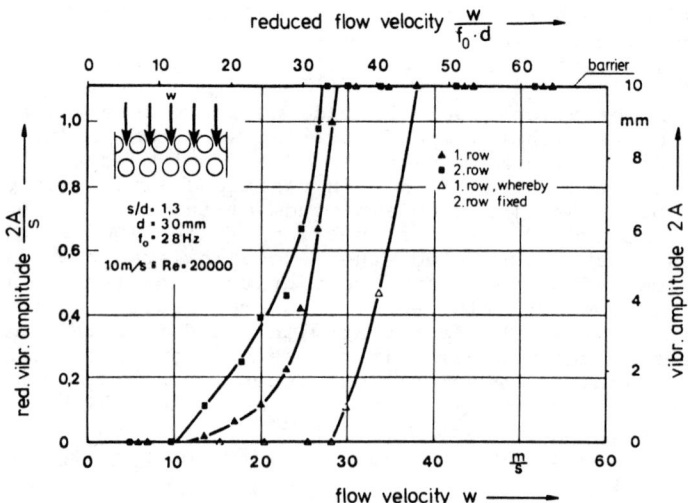

Fig. 4: Vibration amplitude of rods in a 2 rows arrangement

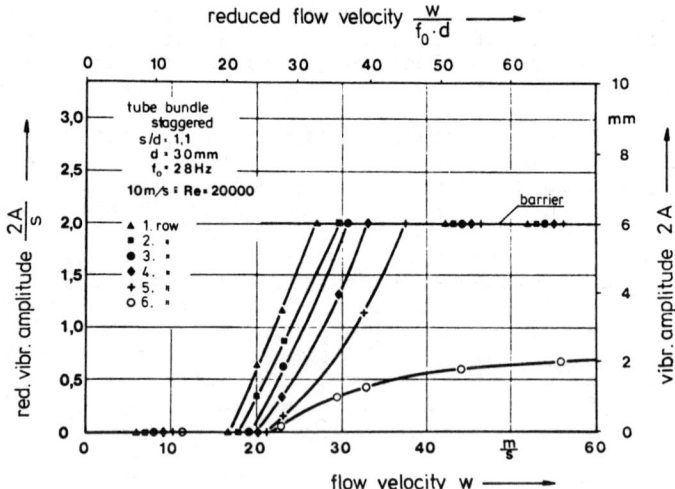

Fig. 5: Vibration amplitude of a staggered cross flow bundle (s/d = 1,1)

increasing velocity the movement of the rows has a more stochastic character, as illustrated in fig. 8, changing slowly with time. From fig. 8, it can be clearly seen, that the second row undergoes the strongest vibrations and that the first one is only weakly influenced by the vibration inducing effects.

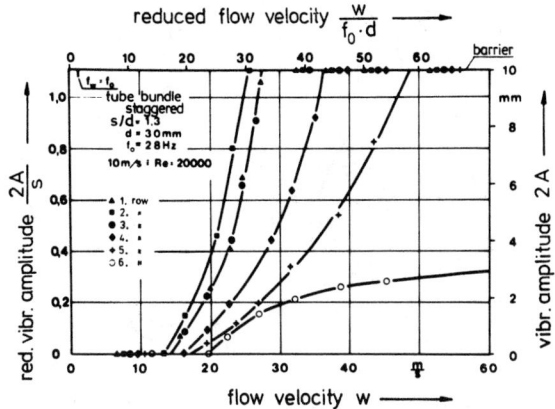

Fig. 6: Vibration amplitude of a staggered cross flow bundle (s/d = 1,3)

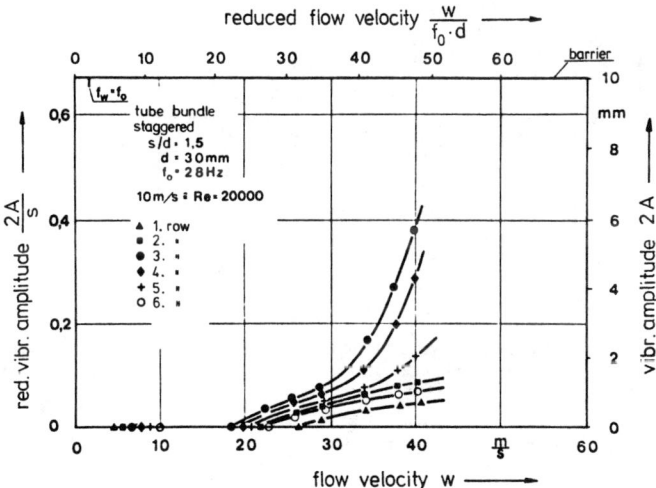

Fig. 7: Vibration amplitude of a staggered cross flow bundle (s/d = 1,5)

The s/d ratio τ has a pronounced influence not only onto the beginning of vibration but also onto the amplitudes with τ = 1,3 vibrations were observed at the lowest flow velocities as demonstrated in fig. 9. This is valid not only for the critical row in the bundle - usually the second row - but also for the bundle as a whole.

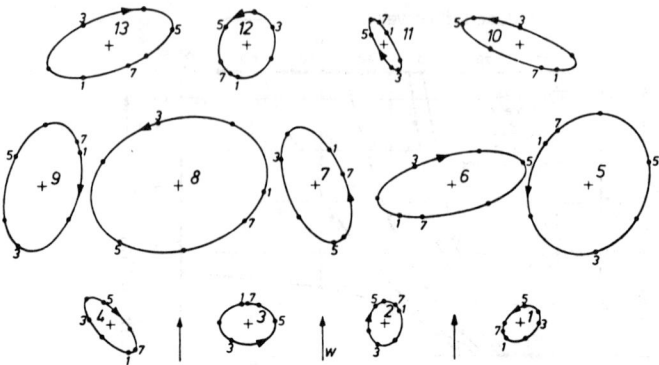

Fig. 8: Vibration pattern in a cross flow bundle (first 3 rows)

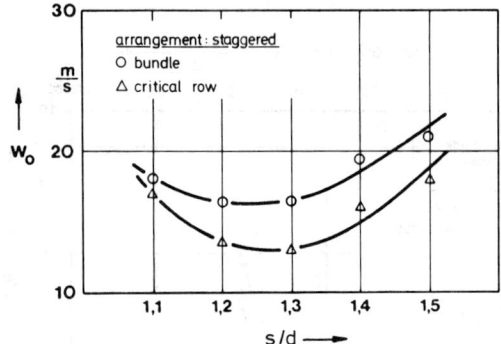

Fig. 9: Onset of vibrations as function of distance - diameter ratio

Rod Bundles with Inline Arrangement

Similar to the staggered conditions also for inline arrangement the amplitudes are increasing contineously with the flow velocity, and the vibrations were also mostly pronounced in the first 3 rows. From the fig. 10, 11 and 12 one can conclude that the position of critical row moves downstream with increasing s/d ratio. For the narrowest spacing the first row is undergoing the largest amplitudes and starts first with vibrations as illustrated in fig. 10. At a s/d ratio of 1,3 the vibration amplitudes of the first two rows are almost identical. This exponential sensitivity to vibrations is restricted to the second row only at the s/d ratio of 1,5. The first and the second row - as illustrated in fig. 12 - show first symptoms of damping effects which are clearly pronounced for the 4th and the following rows.

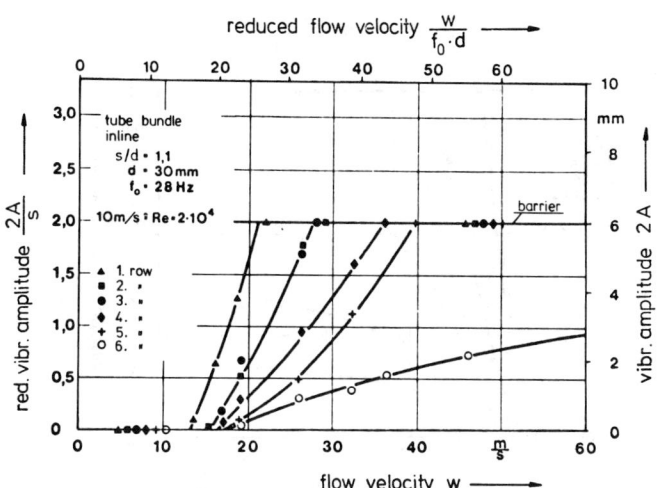

Fig. 10: Vibration amplitude of in line cross flow bundle (s/d = 1,1)

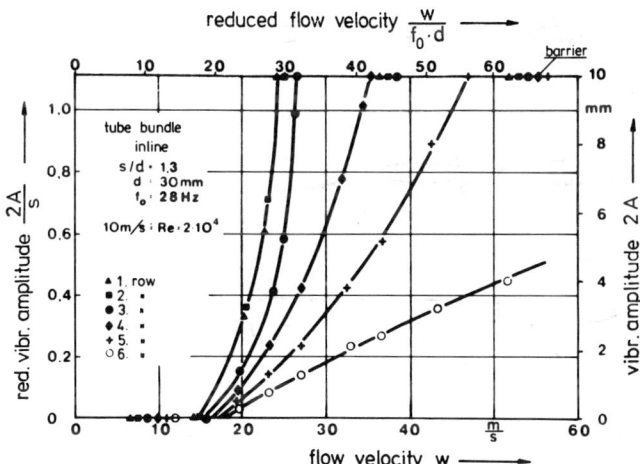

Fig. 11: Vibration amplitude of in line cross flow bundle (s/d = 1,3)

The course of the vibration movement is in the inline arrangement strongly different from that under staggered conditions. As illustrated in fig. 13 which shows the vibration course in the first three rows of an arrangement with a s/d ratio of 1,3,

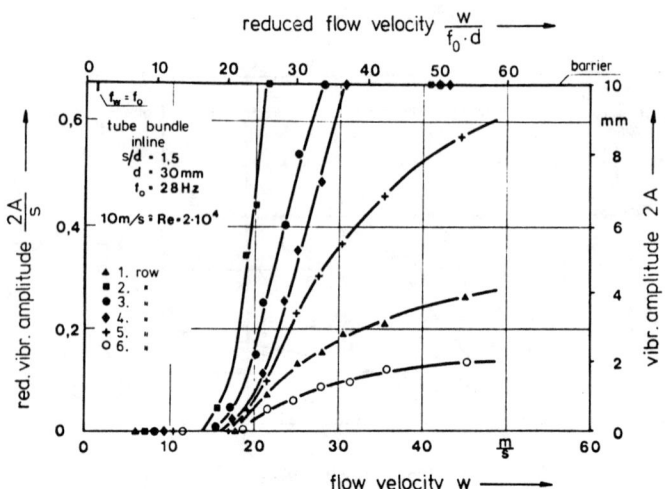

Fig. 12: Vibration amplitude of inline cross flow bundle (s/d = 1,5)

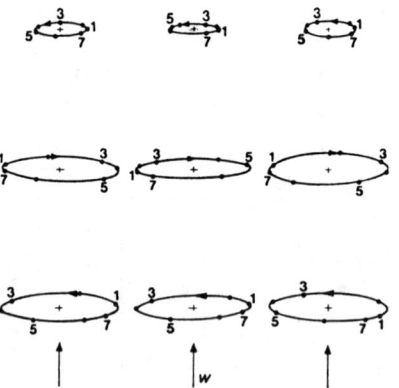

Fig. 13: Vibration pattern, in line arrangement (first 3 rows)

the rods are oscillating perpendiculary to the flow. Within one row all rods are oscillating in phase. With respect to the next row, however, a phase shift of 180° was observed. The vibration phenomenon of a purely perpendicular movement as well as the range of velocities at which the first vibrations occured give the hint that galloping is the main exciting effect. Fig. 13 also gives a good visuell impression that the first two rows are undergoing the largest amplitude which decrease tremendously already for the third row. Aeroelastic coupling seems to have no or negligible

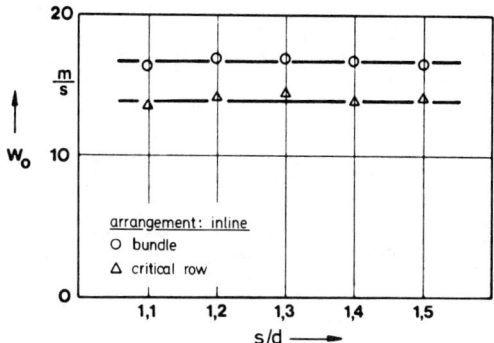

Fig. 14: Onset of vibrations as function of distance-diameter ratio

influence and for exciting by vortices the critical velocity is to high.

The critical velocity, i.e. the flow velocity at which vibrations start, is independent from the s/d ratio as shown in fig. 14. This applies for the first row as well as for the whole bundle. There is a simple explanation for this observation. As we learned from fig. 8 with a staggered arrangement the rods moved in all directions while vibrating within the bundle. This caused a contineous change of the gaps between the rods which again was the reason for aeroelastic coupling. The relative displacement and its consequences of the rods within the bundle as well as the sensitivity of the aeroelastic coupling to this displacement undergoes a maximum and therefore at a certain s/d ratio the bundle has the lowest stability against vibrations. In the inline arrangement aeroelastic coupling plays a subordinated role and therefore back fitting effects by the displacement of the rods are not present and so the s/d ratio does not influence the sensitivity against vibrations. In other words due to the in phase movement of the rods the gaps between the rods remain constant and no additional fluctuation energy exciting the vibrations is produced.

The influence of the aeroelastic coupling can be obviously demonstrated by measuring the drag forces acting on the rod along its vibration course. The product of the drag force times the way of displacement is the energy absorbed by the rod due to the flow fluctuations. One can do this test twice the first time with movable suspended neighbouring rods and then with fixed ones. Under the first conditions - movable rods - one measures an absorbed energy which is by the factor 8 larger than in the fixed surrounding.

Cinematographic observations show that the vibrating tube takes another way during its upflow movement than at the downflow one. Due to this it can be easily understood that the absorbed energy especially from periodical fluctuations excited by the neighbouring rods is different. The consequence is an aeroelastic coupling. Measurements with a hot wire anemometer showed that periodical vortices are only existing behind the rods in the first two rows whilst in the following rows only stochastic turbulent fluctuations could be observed.

## 5. CONCLUSIONS AND MEASURES TO REDUCE VIBRATIONS

There seem to be two effects mainly influencing the vibration of tube- or rod bundles, namely the wake galloping and the aeroelastic coupling. From this information the conclusion can be drawn that two measures could be taken in account to reduce the sensibility of a bundle against vibrations namely

   increasing of the inlet turbulence

   putting out of tune the resonance frequencies
   of neighbouring rods.

The inlet turbulence can be easily increased by placing a grid upstream of the first row of the bundle. Already Vickery /20/ found a reduction of the oscillating pressure onto a prismatic rod in the order of 100 % by increasing the inlet turbulence. Using a punched plate with wholes of 10 mm diameter and placed 20 cm upstream of the first row a remarkable improvement of the stability against exciting vibrations could be observed /19/. By this measure the grade of turbulence which in the tests before was in the order of 0,7 % could be rised up to 50 %.

The improvement was much more pronounced for the staggered arrangement than for the inline one and the onset of vibrations could be shifted to velocities which were almost twice of that under low turbulence conditions, as illustrated in fig. 15. The vibration behaviour shown there, has to be compared with the results presented in fig. 5, where the same geometrical of the large difference in turbulence. The dotted line in fig. 15 represents the amplitude of the critical row at low turbulence corresponding the conditions in fig. 5. Comparing both figures, one can see that the grade of turbulence mainly influences the vibration behaviour of the critical rows, which are usually the three inlet rows. Further downstream there is no effect of the turbulence promotor which can be easily explained by the fact that then the grade of turbulence is anyhow high enough due to the perturbation of the flow in the first rows. It can be assumed that the increased inlet turbulence affects the drag coefficient rectangular to the flow direction and reduces by this the onset of aeroelastic coupling.

Whilst the increasing of the inlet turbulence is certainly a measure of practical use, the misstuning of the resonance frequency of neighboured rods seems to be more of academic interest. Never the less it should be briefly pointed out here that this can reduce the vibration amplitudes remarkably. It, however, does not change the critical velocity as shown in fig. 16 for the most sensitive s/d ratio of 1,3 and a staggered arrangement. The measured results in this figure again have to

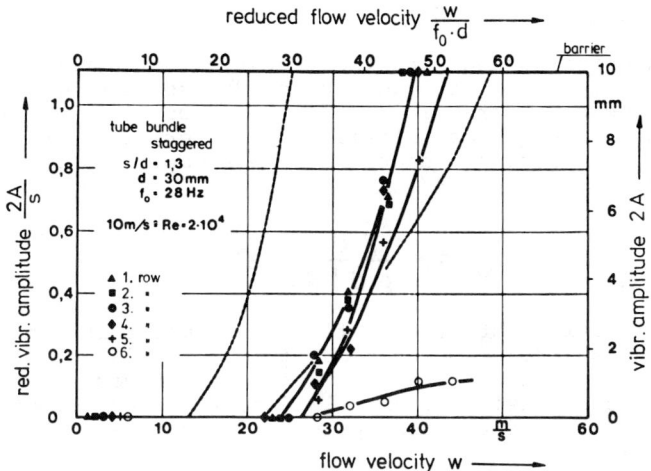

Fig. 15: Vibration amplitude with increased inlet turbulence

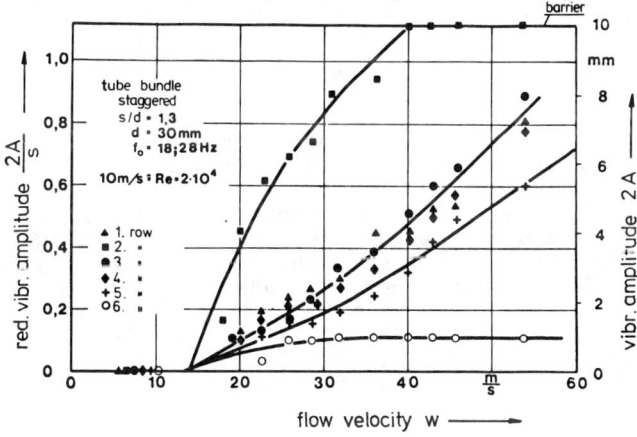

Fig. 16: Vibration amplitude with misstuned resonance frequency

be compared with that of fig. 5. The resonance frequency of the rods in fig. 5 was 28 Hz. The frequency of the bundle was then misstuned in that way that each second rod in a row was adjusted to a resonance frequency of 18 Hz simply by using a higher mass, i.e. a thicker wall of the tube. A remarkably reduced amplitude mainly in the 1st and 3rd row was the consequence. The situation can be still improved if instead of two different resonance frequencies three are applied.

In a staggered arrangement the vibration is mainly induced by aeroelastic coupling - as we can conclude from the experimental results discussed before. This aeroelastic coupling is introduced by the beginning of oscillations in the critical row. From the experiments one also can deduce a simple empirical correlation for the critical velocity /19/.

$$W_{krit} = -\frac{2 \cdot d^*}{\rho \cdot d \cdot l \cdot K} \qquad (13)$$

at which the vibrations start in a flow of normal turbulence. This velocity - as already pointed out in chapter 2 - is mainly influenced by the damping factor $d^*$, a constant K representing the drag force, the density of the fluid $\rho$, the diameter of the tube d and the length of the tube between two baffles.

Exciting by periodic vortices or by resonant buffeting can only expected in the first row at the very beginning of the vibration. The amplitudes in this region, however, are low and no rod damage should be expected.

There is certainly a wide and interesting field of further research work for better understanding vibration phenomena and improving the design of heat exchanger tube banks.

REFERENCES

1. Strouhal, V. 1878. Ueber eine besondere Art der Tonerregung. Wien Ann. Physik und Chemie, Neue Folge, Bd. 5, S. 216 - 251

2. von Karman, Th. 1911. Über den Mechanismus des Flüssigkeitswiderstandes, den ein bewegter Körper in einer Flüssigkeit erfährt. Nachrichten von der Königlichen Gesellschaft der Wissenschaften in Göttingen, S. 509-517

3. Shair, F.H. et.al. 1963. The Effect of Confining Walls on the Stability of the Steady Wake Behind a Circular Cylinder. Journal of Fluid Mechanics, Vol. 17, pp. 546-550.

4. Tritten, D.J. 1959. Experiments on the Flow Past a Circular at Low Reynolds Number. Journal of Fluid Mechanics, Vol. 6, pp. 547 - 567.

5. Chen, Y.N. 1968. Flow-Induced Vibration and Noise in Tube-Bank Heat Exchangers Due to von Karman Streets. Journal of Engineering for Industry, Trans. ASME, Series B, Vol.90, No. 1, Feb. 1968, pp. 134 - 146.

6. Spivak, H.M., 1946. Vortex Frequency and Flow Pattern in the Wake of Two Parallel Cylinders at Varied Spacing Normal to an Air Stream. Journal of the Aeronautical Sciences, pp. 289 - 301

7.  Dumpleton, 1959. Gas Excited Vibrations of Boiler Tubes. ED Report 4 EC/1, Babcock Wilcox, Ltd.

8.  Gregorig, R., Clasen, P., 1971. Ein Schwingungskriterium eines querangeströmten Rohres. Teil 4: Schwingversuche im Rohrbündel Chem.-Ing.-Technik, 43, 17, S. 982-985.

9.  Funakawa, M., 1973. Vibration of Tube Banks by Wake Forces. Proc. Int. Symp. Vibration Problems in Industry, Keswick.

10. Heinecke, E., 1976. Model Tests on Fluid-Elastic Vibrations in Heat Exchangers with Tubes in Cross Flow. Proc. Gas Cooled Reactors with Emphasis on Advanced Systems. Vol II, Intern. Atomic Energy Agency, Vienna.

11. Chen, S.S., 1977. Dynamics of Heat Exchanger Tube Banks. J. Fluids Eng., Trans. ASME, 99, No. 3, pp 462-469.

12. Owen, P.R., 1965. Buffeting Excitation of Boiler Tube Vibration. Jnl. of Mechanical Engineering Science, Vol.7, No. 4, pp. 431 - 439.

13. Connors, J.J., 1970. Fluidelastic Vibration of Tube Arrays Excited by Cross Flow. Winter Annual Meeting of ASME, New York.

14. Magnus, K., 1969. Schwingungen. B.G. Teubner-Verlag

15. Owen, P.R., 1965. Buffeting Excitation of Boiler Tube Vibration. J. Mech. Eng. Scie., 7, 4, pp. 431 - 439.

16. Den Hartog, J.P., 1954. Proc.Nat. Acad. Sci. of USA, 40, 155 - 157

17. Parkinson, G.V. and Brooks, N.P.H., 1961. On the Aeroelastic Instability of Bluff Cylinders. Transactions of the ASME Journal of Applied Mechanics, pp. 252 - 259.

18. Livesey, J.L. and Dye, R.C.F., 1962. Vortex Excited Vibration of a Heat Exchanger Tube Row. Journal Mechanical Engineering Science, Vol. 4, No. 4.

19. Groß, H.G., 1975. Untersuchung aeroelastischer Schwingungs-Mechanismen und deren Berücksichtigung bei der Auslegung von Rohrbündelwärmetauschern, Dissertation der Universität Hannover.

20. Vickery, B.J., 1966. Fluctuating lift and drag on a long Cylinder of Square cross-section in a Smooth and in a turbulent stream. J. Fluid Mech., Vol. 25, part 3, 481 - 494.

# Heat Exchanger Fouling and Corrosion

J. G. COLLIER
United Kingdom Atomic Energy Authority
Harwell, UK

ABSTRACT

Fouling of heat transfer surfaces introduces perhaps the major uncertainty into the design and operation of heat exchange equipment. After a brief description of the various types of fouling the chapter goes on to review the current theories of fouling including the "turbulent burst" theory of Cleaver and Yates. Fouling in equipment involving boiling and evaporation is often more severe than in single phase heat exchangers and moreover, in aqueous systems, is frequently associated with corrosion. The reasons for this are identified and illustrated by reference to corrosion in nuclear power plant steam generators. Finally the modification of heat transfer and pressure drop characteristics by fouling layers is briefly reviewed.

1. INTRODUCTION

Epstein (1978) in his Keynote paper to the 6th International Heat Transfer Conference in Toronto, defined fouling as the *undesired accumulation of solid material at phase boundaries*. Although fouling is by no means confined to heat transfer equipment, it is in this particular field that its unwanted presence is perhaps most acutely felt. As research work on the various aspects of single-phase and two-phase heat transfer have progressed so the uncertainties in heat transfer rates from clean surfaces have been markedly reduced. However, in practice industrial heat exchangers rarely operate with non-fouling fluids. Low temperature cryogenic heat exchangers are perhaps the only exception. The probability that fouling will occur in a heat exchanger is therefore normally taken into account at the design stage by the use of an assumed fouling resistance or *fouling factor*. However, few systematic investigations of fouling have been carried out and the uncertainty in the fouling factor now greatly exceeds the uncertainty in the other terms of the overall heat transfer equation

$$\frac{1}{U} = \frac{1}{h_1} + R_{f1} + \frac{x}{k_W} + \frac{A_2}{A_1}\frac{1}{h_2} + R_{f2} \qquad (1)$$

where U is the overall heat transfer coefficient, $h_1$ and $h_2$ are the individual film coefficients for sides 1 and 2 respectively, $R_{f1}$ and $R_{f2}$ are the fouling resistances for sides 1 and 2 respectively and $x/k_w$ is the tube wall thermal resistance.

After discussing sources of information on fouling, this chapter will describe the different types of fouling which can occur and will review the current theories of heat exchanger fouling. Fouling in equipment involving boiling or evaporation is often more severe than in single-phase heat exchangers and, moreover, in aqueous systems, is frequently the site of rapid corrosion, for example, in nuclear or waste heat boilers. Finally, fouling also influences both the local heat transfer and pressure drop characteristics as well as introducing an insulating layer between the heat transfer fluid and the surface.

## 2. SOURCES OF INFORMATION

Apart from the excellent review by Epstein (1978) already referred to, there are other valuable sources of information on heat exchanger fouling. In 1972, Taborek et al (1972) published an important two-part paper on predictive methods for fouling in heat exchangers. More recently, the first international conference on the fouling of heat transfer equipment was held at Rensselaer Polytechnic Institute, New York 13-17 August 1979. The proceedings of this conference are due to be published in the second half of 1980. Finally, a quarterly current awareness digest entitled "Fouling Prevention Research Digest" is now being published by Harwell and is available on the basis of an annual subscription.

## 3. TYPES OF FOULING

Epstein has delineated six classes or types of fouling depending upon the immediate cause of the fouling, viz

(a) *Scaling* involves the crystallisation of inverse solubility salts (such as $CaCO_3$, $CaSO_4$, $Na_2SO_4$ in water) onto a superheated heat transfer surface. This process can occur under both evaporating or non-evaporating conditions.

(b) *Particulate Fouling* involves the deposition of particulates suspended in the fluid stream onto the heat transfer surface. This process includes *sedimentation*, i.e. settling under gravitational forces as well as other deposition mechanisms.

(c) *Chemical Reaction Fouling* involves deposits caused by some form of chemical reaction within the fluid stream itself (but not with the heat transfer surface). Polymerisation, cracking and coking of hydrocarbon liquids at high temperatures are prime examples.

(d) *Corrosion Fouling* involves a chemical reaction between the heat transfer surface and the fluid stream to produce corrosion products which, in turn, foul the surface. Examples of this would be the on-load aqueous corrosion processes often experienced within nuclear and waste heat boilers.

(e) *Biofouling* involves the accumulation of biological organisms at the heat transfer surface.

(f) *Freezing Fouling* occurs as a result of the crystallisation of a pure liquid or one component from a liquid phase on to a subcooled heat transfer surface.

Not all these mechanisms are mutually exclusive; often more than one mechanism will be occurring simultaneously.

## 4. FOULING CURVES

The amount of material deposited per unit area (m) is related to the *fouling resistance* ($R_f$) and the density ($\rho_f$), thermal conductivity ($k_f$) and thickness of the deposit (x) by the equation

$$m = \rho_f x = \rho_f k_f R_f \qquad (1)$$

If either the amount of material per unit area (m) or the fouling resistance ($R_f$) is determined by experiment as a function of time (t) then the curves so obtained are referred to as *fouling curves*. Three types of behaviour are commonly observed (Figure 1) viz

(i) a linear increase of m (or $R_f$) with time (t)

(ii) a rate of deposition which falls off with increasing time (t)

and (iii) an asymptotic behaviour where the value of m (or $R_f$) finally becomes constant independent of time (t).

Often, there will be a period, the *initiation* or *delay* period before deposition or fouling starts. Whether or not this incubation period is present depends on the type of fouling which is occurring.

## 5. MODELS OF FOULING

Fouling is usually considered to be the difference between two simultaneous processes - a deposition process and a re-entrainment process. The net fouling rate can then be depicted as the difference between a deposition mass flux ($\dot{m}_D$) and re-entrainment mass flux ($\dot{m}_E$). Thus

$$\frac{dm}{dt} = \dot{m}_D - \dot{m}_E \qquad (2)$$

One of the earliest models of fouling was that by Kern and Seaton (1959). In this model it was assumed that $\dot{m}_D$ remained constant with t but that $\dot{m}_E$ was proportional to m and therefore increased with time to approach $\dot{m}_D$ asymptotically. Thus if $\dot{m}_D$ = bm then integration of equation (2) from the initial condition m = 0 @ t = 0 gives

$$m = m^* (1 - e^{-t/t_c}) \qquad (3)$$

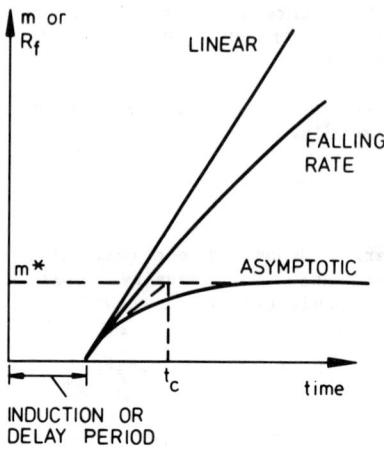

Figure 1  Fouling Curves

where $m^*$ is the asymptotic value of m, $t_c = 1/b$. The time constant, $t_c$ represents the average residence time for an element of fouling material at the heating surface.

Epstein has reviewed the large number of alternative models available for the calculation of $\dot{m}_D$ and $\dot{m}_E$. In most models the rate of deposition is given as the product of the concentration of particulate material (c) times a mass transfer coefficient (k)

$$\dot{m}_D = k\,c \tag{4}$$

The mass transfer coefficient (k) is a sensitive function of the size of the particulate material being deposited.  Four distinct regions can be identified (Claxton and Collier (1973)) with increasing size of particle

(a) deposition of molecular dissolved species

(b) small particles that still transport like dissolved species but with smaller diffusion coefficients

(c) particles sufficiently large their inertia becomes important and they will only deposit if they enter the laminar flow region near the walls with sufficient radial momentum imposed upon them by eddies in the turbulent flow regions;

(d) as particles become larger still their stopping distance becomes of the same order as the pipe diameter. The response of such particles to turbulent fluctuations becomes limited and the mass transfer coefficient levels out and, as the particle size increases still further, starts to decline.

There are fewer models of the re-entrainment process. The model proposed by Taborek et al (1972) indicates that

$$\dot{m}_E = \text{const} \left( \frac{\tau_i m^i}{\sigma} \right) \qquad (5)$$

This equation indicates that the rate of deposit spalling is proportional to the force trying to remove the deposit, the shear stress, $(\tau_i)$, and inversely proportional to the strength of the bonding of the deposit $(\sigma)$.

One difficulty with these types of model is that the processes of deposition and re-entrainment are assumed to occur simultaneously over the whole surface. Such a picture does not appear physically realistic.

Recently, our understanding of near-wall turbulent flows has indicated that fluid is ejected out of the wall layer region in a series of "turbulent bursts". This phenomenon has been examined by Cleaver and Yates (1973, 1975, 1976) who have provided a convincing explanation of the re-entrainment process. Their view is that the re-entrainment process occurs as a result of these spatially and temporally randomly distributed "turbulent bursts". The bursts originate at the surface but at any instant of time cover less than ½% of the surface area. Each burst acts like a miniature tornado, lifting deposited material from the surrounding surface and dispersing it in the fluid stream.

If each turbulent burst was successful in re-entraining all the material adjacent to it then $t_c$, the residence time for material at the surface, would be given by

$$t_c = \frac{\text{Average time between bursts}}{\text{Surface fraction covered by bursts at any instant}} \qquad (6)$$

Measurements have been made of the residence time of deposits on the fuel surfaces of nuclear reactors using the radioactive properties of the corrosion products. These measurements indicate 60-90 day residence times for $^{60}Co$ for example; much longer than would be derived from a straightforward application of the Cleaver-Yates model. The probable explanation is that the re-entrainment process is considerably less than 100% efficient and many bursts succeed in removing no deposit and others only some of the deposit beneath them.

## 6. FOULING IN ONCE-THROUGH BOILERS

Deposition and corrosion is particularly prevalent on steam generating surfaces. With a once-through evaporator all dissolved salts and suspended matter entering with the liquid feed to the evaporator must either:

(a) remain in the evaporator to form a deposit,

(b) be mechanically entrained in the outflow of steam,

(c) be actually dissolved in the steam outflow.

A comprehensive review of solubility of various salts in steam was published by Styrikovich and Martynova (1963). For subcritical systems the partitioning of the salt between the water and steam phases is required. The partition (or distribution) coefficient denoted by $K_D$ is:

$$K_D = \frac{c_G}{c_L} = \text{fn}\left(\frac{\rho_G}{\rho_L}\right) = \left(\frac{\rho_G}{\rho_L}\right)^m \tag{7}$$

where $c_G$ and $c_L$ are the concentrations of salt in the steam and water phases respectively. The partition coefficient is found to be a simple power law of the ratio of the phase densities, the value of m varying from salt to salt. (Figure 2)

If it is not possible to establish that all or most of the salts present in the feed are soluble in the effluent steam to a greater extent than the feed concentration then a continuous build-up of salt deposits will result. It is, of course, not possible to state the converse, i.e. that evaporators in which the saturation levels are not exceeded will be free from deposits. Collier and

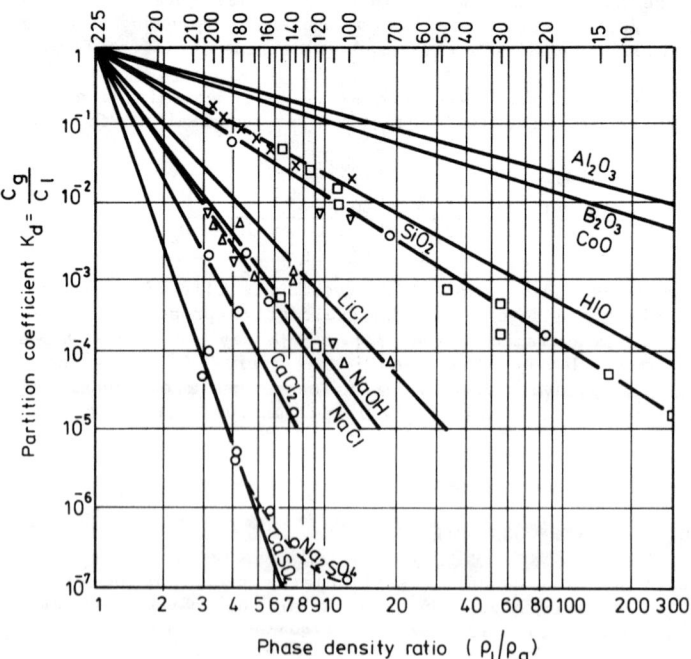

Figure 2  The Variation of Partition Coefficients with the Ratio of Densities of the Two-Phases

Pulling (1963) studied the deposition of silica from unsaturated solutions in a low-pressure steam-heated vertical tube evaporator. The pattern of deposition appeared to be closely related to the heat transfer regions encountered. In the saturated nucleate boiling region, deposition appeared in the form of annular rings around a large number of discrete sites on the surface. An explanation for the form of these deposits can be given in terms of an examination of the microstructure of the boiling process. Vapour bubbles grow at active nucleation sites leaving, as they grow, a very thin "microlayer" of liquid beneath the bubble. This thin layer is totally evaporated leaving the solid content of the "microlayer" deposited on the surface. The bubble detaches and unsaturated liquid contacts the surface again. A new bubble forms and the process is repeated. Hospeti and Mesler (1965) employing $^{35}S$ as a tracer in a saturated solution of calcium sulphate have made use of this phenomenon to measure the thickness of the microlayer. Since the number of active sites is proportional to the square of the heat flux, the results of Palen and Westwater (1966) which indicate a sharp increase in the rate of fouling of surfaces with calcium sulphate as the heat flux is increased under pool boiling conditions, are readily explained.

In the two-phase forced convection region where nucleation had been suppressed, Collier and Pulling (1963) found the surface of the evaporator free from deposits. Further up the tube a region of continuous deposit was found and it was suggested that this corresponded to the "liquid deficient region". However, because of the use of steam heating, the tube wall temperature did not rise appreciably at dryout and although there was no liquid film on the surface, water droplets were able to impinge on the tube wall and subsequently to evaporate to dryness. In a constant heat flux situation, corresponding to a sharp rise in temperature at the dryout point, droplets would be unable to "wet" the wall and the formation of a deposit by this means is unlikely. Deposition, however, may occur as a result of decreased solubility in the steam phase as its temperature is raised.

Careful studies have been carried out on the deposition of magnetite in once-through boilers. The magnetite exists in boiler feedwater as a colloidal suspension of particles and agglomerates in the size range from below 1 μm to 50 μm. For the turbulent flow of water in unheated pipes, the rate of deposition increases with the Reynolds number of the flow.

In the core region of the flow, away from the wall, turbulent eddies carry the magnetite particles in a radial direction. As the wall is approached the "turbulent bursts" are compensated for by gentler fluid "sweep backs" due to the much larger surface area not covered by the "bursts". Deposition occurs via the back sweeps. In practice, deposition of magnetite occurs initially in the natural valleys formed in commercially rough tubes.

The rate of deposition of magnetite is relatively insensitive to the presence of a heat flux through the tube wall as long as boiling does not occur. Once subcooled boiling and subsequently saturated boiling occur the rate of deposition is dramatically increased.

For boiling systems we have

$$\frac{d R_f}{dt} = K \phi^n \qquad (8)$$

Under intense nucleate boiling conditions the rate of deposition increases sharply with heat flux and n = 2. Other workers have concluded that the value of n should be nearer unity and have shown a strong dependence of the deposition rate on the local steam quality, the rate of deposition increasing as the quality is increased. It is generally observed that mass velocity has little effect on deposition under boiling conditions.

At the dryout point, under conditions of constant heat flux, the rate of deposition drops sharply.

Corrosion at boiling surfaces

"On-load corrosion" is a particularly rapid and severe attack of boiler tubes in zones where steam is raised. As a result of intensive research into this problem of the past few years a considerable amount is known about the reasons for this type of corrosion. In particular, work carried out by Masterson, Castle and Mann (1969) has elucidated three mechanisms by which corrosive salts may be concentrated in a boiler. The first mechanism is that of dryout, either complete as in a once-through boiler, or partial as may occur due to stratification in horizontal tubes of waste-heat boilers. A second concentration mechanism is that occurring in crevices in the heating surface, particularly where the heating is applied asymmetrically; for example at the tube support plate in a water reactor steam generator. The third mechanism is closely associated with the build-up of porous deposits on the heat transfer surfaces. Liquid is drawn into the deposit between the micron-size magnetite particles by a "wicking" effect whilst steam is liberated into tunnels in the deposit. Very high concentration factors in excess of $10^4$ have been measured for the liquid held up within such a deposit.

Porous magnetite is not itself a serious barrier to heat transfer but if the pores should become in-filled with copper salts from the feed train, the thermal conductivity of the deposit is drastically reduced. Cracking of the deposit may occur under the thermal stress and it may locally separate from the heat transfer surface resulting in a low conductivity steam gap between the deposit and the heat transfer surface. Copper bearing alloy feed heater tube bundles should therefore be avoided in nuclear and waste heat boiler circuits.

Mann (1975) has developed a model for the behaviour of aggressive agents such as sodium chloride or sodium hydroxide in a once-through boiler. His results show that the equilibrium indicated by equation (7) is not achieved at dryout and that concentrated solutions form as rivulets which penetrate a short distance beyond the dryout point. Mann postulated that the level of solute concentration in the liquid film is determined primarily by the temperature of the tube wall and the pressure in the system. Finally, a point is reached where sodium chloride will exceed its solubility limit in the liquid phase and be precipitated. However, the concentration of sodium hydroxide can continue to very high levels.

7. THE INFLUENCE OF DEPOSITION UPON HEAT TRANSFER AND PRESSURE DROP

Deposits can and do have an influence on the hydrodynamic performance of heat exchangers and steam generators. As previously mentioned, magnetite particles reaching the wall initially fill in the micro-roughnesses of the internal tube surface causing a small decrease in the pressure drop. This may be accompanied by a corresponding decrease in the heat transfer rate. Reductions in the dryout heat flux and the liquid deficient heat transfer rates have been observed to result from deposition. Further deposits will lead to an increase in pressure drop and this increase can be many times that anticipated from just the change in flow cross-sectional area since the deposit is often rippled and acts as a surface roughness.

Cohen (1969) discusses the heat transfer behaviour of fouled surfaces. Under non-boiling conditions the heat transfer through the deposit is by conduction. The effective conductivity of the deposit may be computed from Maxwell's formula

$$k_{eff} = k_f \left[ \frac{1 - (1-a\, k_p/k_f)\, b}{1 + (a-1)\, b} \right] \qquad (9)$$

where $k_f$ is the conductivity of the liquid (or vapour) phase
$k_p$ is the conductivity of the particulate phase
$a = 3k_f/(2k_f + k_p)$
$b = V_p/(V_f + V_p)$
$V_p$ is the total volume of the particulate phase
$V_f$ is the total value of the fluid phase.

The equation is valid up to values of $b \approx 0.5$. For magnetite deposits $k_p = 1.7$ W/m$^2$ °C/m. For a deposit of 65% porosity in water at 300°C $k_{eff}$ is calculated to be 0.78 W/m$^2$ °C/m. If the continuous phase is steam then the value drops to 0.17 W/m$^2$ °C/m. The temperature rise for a constant heat flux surface when dryout occurs within a magnetite deposit is therefore accentuated by the reduction in its effective conductivity.

Nucleate boiling from surfaces covered by porous magnetite deposits has been studied by Cohen and Taylor (1967). The number of nucleation sites in the case of the fouled surface was considerably greater than for a clean surface. Figure 3 shows the experimental boiling curves for both the fouled and clean surfaces. In the single-phase forced convective region the fouled surface temperature exceeded that of the clean surface by 16°C. Nucleation occurred at a lower heat flux in the case of the fouled surface and for heat fluxes above about 1.2 MW/m$^2$ the temperature of the fouled surface fell below that of the clean surface. The remarkable improvement of the performance of the deposit covered surface may be understood when the behaviour of tubing with porous metal coatings under boiling conditions is considered. Clearly the deposit acts to provide large stable nucleation sites.

The presence of a deposit upon a steam generating surface appears to reduce the dryout or critical heat flux by a small amount; of the order of 5-10% (Macbeth (1977)). In special cases the presence of an additional thermal resistance in a system heated by a primary fluid may significantly improve the overall thermal performance of the system. This is illustrated in Figure 4. The heat input characteristic for the initial system is shown as curve (1). This intersects the heat removal characteristic corresponding to the liquid deficient heat transfer region at point A. The inclusion of an additional resistance either on the primary fluid side or on the boiling surface side will produce a new characteristic (curve 2) which will now intersect the heat removal characteristic corresponding to the nucleate boiling heat transfer region at point B. This results in an appreciably higher heat flux than in the case without the resistance. The optimum value of the resistance is one which allows B to be just below the dryout heat flux corresponding to the local boiling side condition.

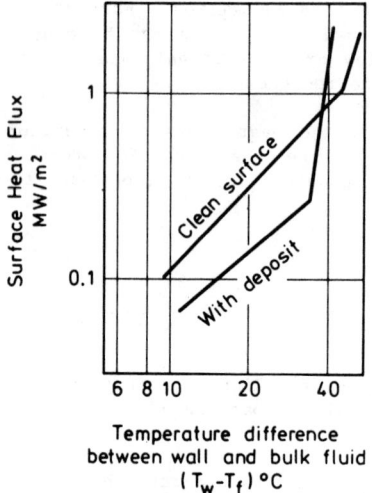

Figure 3  Nucleate Boiling on Clean and Fouled Surfaces
(Cohen and Taylor)

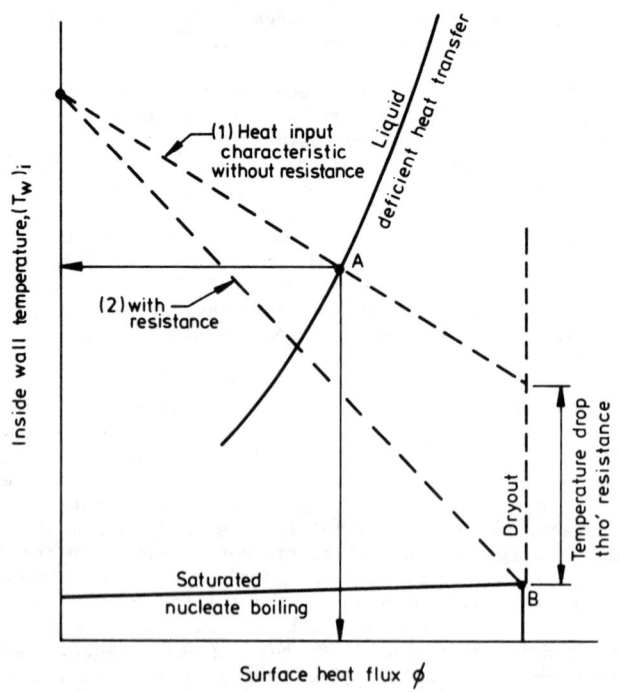

Figure 4  The influence of a resistance in increasing the overall heat flux

# HEAT EXCHANGER FOULING AND CORROSION

One of the most important influences of fouling layers in heat exchanger design is that on the pressure drop of the system. The fouling layer is usually rough and increases the pressure drop. However, the effect of surface roughness on pressure drop is often much less in two-phase flow than it is in single-phase flow. Thus, the two-phase pressure drop multiplier is often smaller for rough channels than for smooth channels. This phenomenon was discussed by Shires (1972) and Figure 5 illustrates the sort of effect that can occur. The reason for the smaller influence of roughness in two-phase flow is that the pressure drop is dominated by the roughness at the liquid film - vapour core interface and, thus, the effect of the surface roughness is not so important.

## 8. CONCLUDING REMARKS

Fouling and corrosion within heat exchangers represent a significant economic loss. It has been suggested (Pritchard (1979)) that the extra capital cost incurred in the UK alone in 1977 due to fouling was of the order of £100 million and that additional fuel costs were a further £50-60 million. With the costs of energy rising fouling can be seen to be a subject of increasing importance and one which merits much greater attention than it has received hitherto. One of the difficulties to date has been the diverse character of the particular industrial situations encountered. Pritchard (1979) has clearly stated what industry's needs are

- better methods of removing foulants, preferably on-stream.

- improved designs of heat exchanger which take greater account of the need to minimise deposition on the heat transfer surfaces.

- the need to develop better antifoulants and to provide more reliable techniques to determine whether a process stream has a high or low propensity to foul a heat transfer surface.

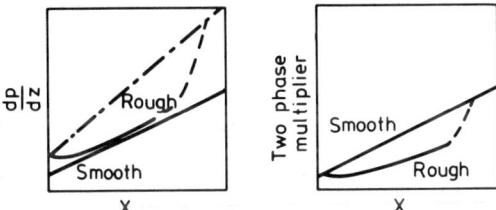

Figure 5    Effect of surface roughness on two-phase pressure drop and two-phase friction multiplier.

REFERENCES

Claxton, K.T. and Collier, J.G. (1973). "Effects of corrosion product transport in fast reactors", Proc. of Int. Conf. on Liquid Alkali Metals, Nottingham University 4-6th April 1973, pp.290-300 (1973).

Cleaver, J.W. and Yates, B. (1973). "Mechanism of detachment of colloidal particles from a plant substrate in turbulent flow", J. Coll and Interf. Sci. Vol.44 p.464, Sept. 1973.

Cleaver, J.W. and Yates, B. (1975). "A sub-layer model for deposition of particles from a turbulent flow", Chem. Eng. Sci. Vol.30, p.983 (1975).

Cleaver, J.W. and Yates, B. (1976). "The effect of re-entrainment on particle deposition", Chem. Eng. Sci. Vol.31, p.147 (1976).

Cohen, P. (1969). "Water coolant technology of power reactors", published by Gordon & Breach (1969).

Cohen, P. and Taylor, G.R. (1967). "Discussion of Boiling Heat transfer at Low Heat Flux", by W.C. Elrod et al, Trans. ASME(C) J. of Heat Transfer $\underline{89}$, 242, (1967).

Collier, J.G. and Pulling, D.J. (1963). "The deposition of solids in a vertical tube evaporator", Industrial Chemist, Vol.39, pp.129-133, 200-203 (1963).

Epstein, N. "Fouling in Heat Exchangers". Keynote paper KS-18 6th International Heat Transfer Conference Toronto, Canada, August 7-11 (1978), Volume 6 of Proceedings.

Hospeti, N.B. and Mesler, R.B. (1965). "Deposits formed beneath bubbles during nucleate boiling of radioactive calcium sulphate solutions", A.I.Ch.E. J. $\underline{11}$, 662, (1965).

Kern, D.Q. and Seaton, R.E. (1959). "A theoretical analysis of thermal surface fouling", Brit. Chemical Engng., Vol.4, p.258 May 1959.

Macbeth, R.V. (1977). "Fouling in boiling water systems", Chapter of book "Two-phase flow and Heat transfer", edited by Butterworth & Hewitt and published by OUP (1977).

Macbeth, R.V., (1971). "An investigation into the effect of 'crud' deposits on surface temperature, dryout and pressure drop, with forced convection boiling of water at 69 bar in an annular test section", AEEW-R705.

Mann, G.M.W., (1975). "Distribution of sodium chloride and sodium hydroxide between steam and water dryout in an experimental once-through boiler", Chem. Engng. Sci. 30 (2), 249-260, [27].

Masterson, H.G., Castle, J.E. and Mann, G.M.W. (1969). "Waterside corrosion of power station boiler tubes", Chemistry and Industry, pp.1261-1266 (September 1969).

Palen, J. and Westwater, J.W. (1966). Chem. Engng. Prog. Symp. Series Vol.62, No.64, pp.77-86 (1966).

Pritchard, A.M. (1979). "Heat Exchanger Fouling in British Industry", Fouling Prevention Research Digest, Vol.1, No.1-4, pp.iv-vi (1979).

Shires, G.L. (1972). "The influence of Surface Roughness on Two-Phase Pressure Drop", Winfrith RDD Note No.196 (1972).

Somerscales, E.F.C. and Knudsen, J.G. (1980). "Fouling of Heat Transfer Equipment", International Conference 1979, Published by Hemisphere Publishing Corp.

Styrikovich, M.A. and Martynova, O.I. (1963). "Contamination of the steam in boiling reactors from solution of water impurities", Sov. At. Energy 15 (3) 917 (1963).

Taborek, J., Aoki, T., Ritter, R.B., Palen, J.W. and Knudsen, J.G. (1972). "Fouling: the Major Unresolved Problem in Heat Transfer", Chem. Engng. Progress Vol.68, No.2, Feb. pp.59-67, and CEP Vol.68, No.7 July pp.69-78 (1972).

# Fouling of Heat Transfer Surfaces

**JOÃO DE DEUS R. S. PINHEIRO**
Universidade do Minho—Grupo de Engenharia
Braga, Portugal

ABSTRACT

Over the last decade increasing efforts have been directed towards a better understanding of fouling, particularly in what concerns the mechanisms involved and the effective role played by the operating parameters likeky to affect the fouling processes.

At the same time, several fouling models were proposed in an attempt to rationalize the prediction of fouling allowances in design and optimize the operation of heat transfer equipment.

The majority of fouling models make use of a material balance in which the net fouling rate (dR/dt) is expressed as the difference between a deposition rate and a removal function and arrive, ultimately, at an equation for the time dependence of the fouling resistance of the same form $|R = R_\infty (1 - e^{-\beta t})|$. Yet, despite these similarities, the functional dependences of $R_\infty$ and $\beta$ on the operating variables predicted by the various models are substantially different, due mainly to different assumptions regarding the deposition and removal functions. As a consequence it is virtually impossible to make full use of the information accumulated for a given type of fouling when a distinct mechanism is involved.

Bearing in mind the above considerations, a critical review of the problem of fouling is undertaken, with special emphasis being placed on the mechanisms and models more closely associated with -solubility, particulate and reaction fouling.

An attempt is made to encompass the referred models into a more general approach, as a first step towards a sistematic treatment of existing data.

1. INTRODUCTION

The term "fouling", originally a descriptive expression used in the oil industry, became established in the literature to mean any undesirable deposit on heat transfer surfaces which increases the resistance to heat transmission. Examples of fouling may be found in almost every heat transfer process, ranging from the usual deposits in home-cooking apparatus to the highly complex dirt that builds up in industrial exchangers.

For the common case of a cylindrical tube with flow and fouling at both the inside and outside surfaces represented in Fig. 1, the total resistance to heat

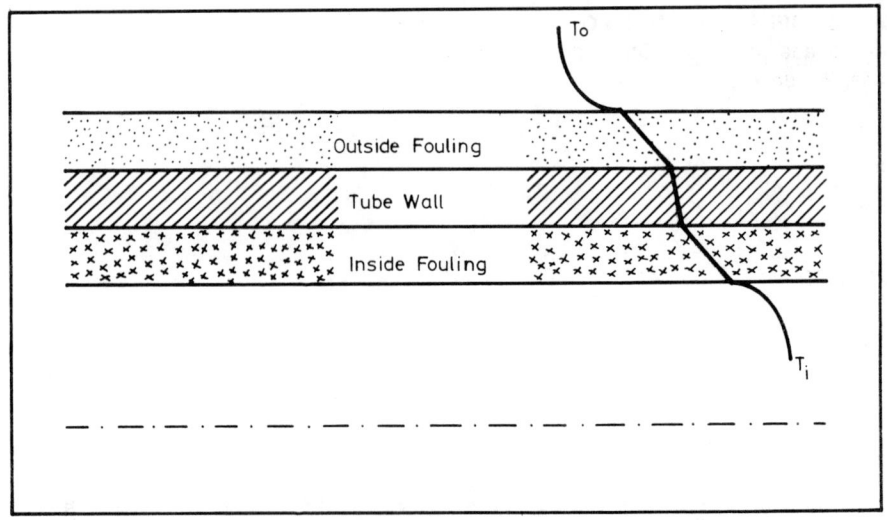

Fig.1. Scheme of a fouled tube

transfer between the two streams, $R_T$, is

$$R_T = \frac{1}{h_i A_i} + \frac{R_i}{A_i} + \frac{\Delta r}{k_t A_{m\ell}} + \frac{R_o}{A_o} + \frac{1}{h_o A_o} \tag{1}$$

where the symbols have the usual meaning, except for $R_i$ and $R_o$, known as "fouling factors" which represent unit fouling resistances.

The overall heat transfer coefficient in fouled conditions may thus be expressed in terms of the overall heat transfer coefficient in clean condition and on overall fouling resistance envolving the terms related to fouling:

$$(\frac{1}{U_o})_f = \underbrace{\frac{1}{h_i}\frac{A_o}{A_i} + \frac{\Delta r}{k_t}\frac{A_o}{A_{m\ell}} + \frac{1}{h_o}}_{(1/U_o)_c} + \underbrace{R_i \frac{A_o}{A_i} + R_o}_{R_f} \tag{2}$$

or,

$$R_f = (\frac{1}{U_o})_f - (\frac{1}{U_o})_c \tag{3}$$

which clarifies the physical meaning of the fouling resistance, $R_f$.

In its simplicity, the definition of fouling encompasses a great variety of fouling problems. In fact, the vast range of industrial stream and conditions appear to make most fouling situations virtually unique, rendering comprehensive understanding and treatment rather difficult.

In an attempt to sistematize the subject, some fouling classifications have been suggested, one of which, by Bott [1] that appears to be well accepted, considers five broad types of fouling

# FOULING OF HEAT TRANSFER SURFACES

- Particulate fouling – related to the deposition of particles suspended in fluid streams, e.g., sand, muds, corrosion debris, dust in gas streams.
- Reaction fouling – comprising the fouling originated by a chemical reaction at the heat transfer surface, namely cracking and polymerization of hydrocarbons.
- Solubility fouling – which induces the precipitation and deposition of dissolved material due to solubility changes with temperature.
  Wax from kerosene, calcium carbonate and calcium sulphate from cooling waters are examples.
- Corrosion fouling – a special case in which "in situ" deposits are formed with the assistance of surface material.
- Biological fouling – involving the growth of microorganisms in heat transfer surfaces. Slime formation in cooling towers is a typical case [2].

Regardless of its specific nature, the fouling of heat transfer surfaces represents always an economic penalty estimated at £ 400 M/year in the U.K. alone [3], a figure likely to increase for higher fuel prices.

Broadly speaking it is possible to identify as major components of the fouling costs

- The increase in capital expenditure due to the oversizing of the exchangers
- The energy losses associated with poorer performances of the equipment
- The maintenance costs involved in cleaning
- The loss of production during the plant shut-down for cleaning operations

In general, at the design or selection stage, these factors are tradded-off one against each other, anti-fouling treatment costs are estimated, and a final compromise is reached.

In assessing heat exchanger requirements, the engineer usually evaluates the overall heat transfer coefficient for clean conditions and then adds up a "fouling allowance" ($R_f$), to obtain the design value $(U_o)_f$ referred in eqs (2) and (3).

The value of $R_f$ to be allowed may be based on previous experience or, as it is often the case, use is made of the recommendations of the Tubular Exchanger Manufacturers' Association – TEMA – an extract of which is shown in Table 1. With regard to this some comments are worthwhile

- While fouling is obviously a time-function starting with zero and proceeding, usually, along some pseudo-assymptotic or linear relationship (see Fig. 2), TEMA factors are nearly constant. One would, then, expect TEMA factors to represent the assymptotic values, $R_f$. Yet, Bott and Walker [4] have shown that this is not the case.
- TEMA factors depend on some cases on velocity and temperature in a somewhat clear way. This may be, however, rather misleading since the dependence of $R_f$ on operating variables is still an open question as discussed latter.

TABLE 1

FOULING FACTOR (h.ft. °F/BTU) FOR HEAT TRANSFER EQUIPMENT
(from TEMA Tables)

| Types of Water | u < 3 ft/sec | u > 3 ft/sec |
|---|---|---|
| Seawater | 0.0005 | 0.0005 |
| Distilled | 0.0005 | 0.0005 |
| Cooling tower (untreated) | 0.003 | 0.003 |
| River water | 0.003 | 0.002 |
| Muddy | 0.003 | 0.002 |
| Hard | 0.003 | 0.003 |

| Type of Fluid | (any condition) |
|---|---|
| Machinery and transformer oil | 0.001 |
| Fuel-oil | 0.005 |
| Organics vapours | 0.0005 |
| Steam, exhaust | 0.001 |
| Refrigerating vapours | 0.002 |
| Air | 0.002 |
| Orgains liquids | 0.001 |
| Brine (cooling) | 0.001 |

Recent papers on fouling [4,5,6] have stressed this and other difficulties derived from the wide variety of situations likely to occur. Yet, many authors agree with Taborek et al. [7,8] in claiming for a better understanding of the fouling mechanisms in order to overcome "the major unresolved problem in heat transfer". Under these circumstances the present work is an attempt to bring together some "apparently distinct" contributions that have been published on the matter.

2. FOULING MECHANISMS AND MODELS

2.1 Basic Model

Most models that have been proposed are highly simplified in the sense that they are based upon several simplifying assumptions, such as

- Only one type of fouling is present
- The fouling layer is homogeneous throughout its thickness
- Fouling roughness can be neglected
- Changes in physical properties of the streams are omitted
- The initial condition of the surface is not considered

Furthermore, no model attempts to encompass all the variable likely to affect fouling, but concentrated analysis on the so-called "transfer variables", i.e.

- Time
- Velocity
- Temperature
- Concentration

leaving behind other factors like

- Nature and condition of surface
- Properties of foulant stream
- Nature of the process
- Design of the equipment
- Flutuactions in operation

and, specially

- Simultaneous action of different mechanisms

It is antecipated that the dependence of $R_f$ on temperature and concentration predicted by some models has agreed reasonably well with the experiment. But, with regard to velocity the conflict is marked.

Not surprisingly, many authors focus their attention on the influence of velocity on fouling build-up and, most often, model comparison and evaluation is made through the predicted velocity dependences.

This tendency is also reflected in this work.

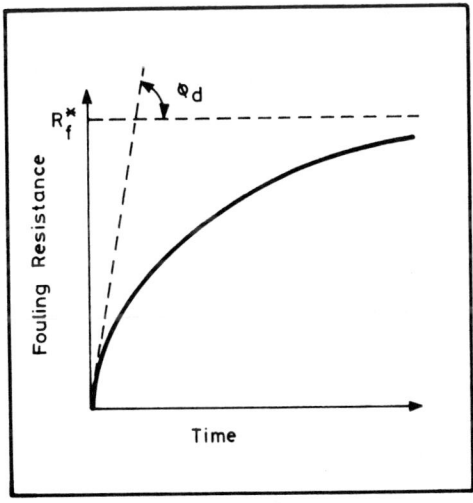

Fig.2. Typical fouling curve.

The first real attempt to derive a general fouling model was by Kern and Seaton [9] who noted that experimentally observed fouling curves followed a typical pattern in which after an initial period of fast fouling build-up, the fouling resistance (or the fouling thickness) tended to remain nearly constant (sse Fig. 2). This behaviour could be described by the following equation

$$R_f = R_f^* \left[ 1 - \exp(-\beta t) \right] \tag{4}$$

where $R_f^*$ is the assymptotic fouling resistance for $t = \infty$, t is the time and $\beta$ is a coefficient representing the inverse of a relaxation time.

In trying, by intuitive reasoning, to predict the dependence of the parameters $R_f^*$ and $\beta$ on the operating variables, Kern and Seaton made the important suggestion that the net fouling rate, $dR_f/dt$, was the result of two opposing rates: — a deposition rate, $\phi_d$, and a removal rate, $\phi_r$:

$$\frac{dR_f}{dt} = \phi_d - \phi_r \tag{5}$$

Kern and Seaton consider further that $\phi_d$ is independent from time (or from $R_f$, which is the same) whereas $\phi_r$ depends directly on the actual value of $R_f$:

$$\phi_d = \left(\frac{dR_f}{dt}\right)_{t=0} \tag{6}$$

$$\phi_r = \beta R_f \tag{7}$$

Under these conditions, the integration of eq. (5) yields

$$R_f = \frac{\phi_d}{\beta} \left[ 1 - \exp(-\beta t) \right] \tag{8}$$

which is similar to eq. (4) with $R_f^* = \phi_d/\beta$.

This basic model of Kern and Seaton constitutes the fundamentals behind most models that have appeared in the literature, which differ essentialy on the functional dependence of $\phi_d$ and $\phi_r$.

## 2.2 Particulate Fouling

Particulate fouling is widely experienced in cooling water systems specially those of the type "once through" using sea and river water, or when some process species or contaminator acts as promoter for the agglomeration of microscopic dust particles [10].

Kern and Seaton [9] considered that the deposition function was linearly dependent on foulant concentration in the bulk, $c_b$, and on fluid velocity, u:

$$\phi_d = k_1 c_b u \tag{9}$$

where $k_1$ is a coefficient that encompasses the dependence on all the variables besides the transfer variables. (Other k coefficients with a numerical subscript will hereafter bear the same physical meaning).

As to the removal rate function, $\phi_r$, it was considered to be depend on the shear stress at the wall, $\tau$, and on the actual fouling thickness, $x_f$:

# FOULING OF HEAT TRANSFER SURFACES

$$\phi_r = k_2 \tau x_f = k_3 u^2 R_f \tag{10}$$

Substituting into the general rate equation - eq. (5) - and integrating, one obtains

$$R_f = \frac{k_1}{k_3} \frac{c_b}{u} \left[1 - \exp(-k_3 u^2 t)\right] \tag{11}$$

The functional dependences on $\underline{u}$ predicted by this model of Kern and Seaton, namely

$$\phi_d \alpha u \; ; \; R_f^* \alpha 1/u \; ; \; \beta \alpha u^2$$

were experimentally vindicated by the work of Watkinson |11| with sand-water slurries.

Yet, when studying the particulate fouling in a gas-oil stream Watkinson |12| obtained different dependences of those parameters on $\underline{u}$.

Furthermore, while the sand-water slurries studies showed no significant influence of temperature on the range 79 - 91 ºC, the opposite was observed with the gas-oil fouling.

Under these circumstances Watkinson suggested a new deposition-removal model comprising threee basic steps

- Transfer of foulant to the wall.
- Adhesion of foulant to the solid surface.
- Removal (by a mechanism identical to the one proposed by Kern and Seaton).

Watkinson related the first two steps to the deposition rate by considering that this could be described as the product between the particles mass flux to the wall, N, and a sticking probability, S, i.e., the probability that any particle reaching the wall would stick to it:

$$\phi_d = N S \tag{12}$$

N is given by the usual product of a mass transfer coefficient, $K_d$, to the driving force for mass transfer, $c_b - c_s$, i.e., the difference between the foulant concentrations in the bulk and at the wall:

$$N = k_d (c_b - c_s) \tag{13}$$

For values of the Schmidt Number much greater than unity, $k_d$ may be evaluated by the Metzner-Friend equation [13]

$$k_d = C_1 \frac{\sqrt{f}}{Sc^{2/3}} u \simeq k_4 u \tag{14}$$

where $c_1$ is a numerical coefficient and f is the Fanning friction factor.

Combining eqs (13) and (14) one gets

$$N = k_4 u (c_b - c_s) \tag{15}$$

To determine S, Watkinson suggests that a particle will adhere if the physico chemical adhesive forces between the particle and the surface overcome the drag forces exerted near the wall:

$$S = S_o F_{adh}/F_{drag} \tag{16}$$

where $S_o$ is some appropriate constant.

Assuming that the bonding is governed by an Arrhenius-type expression, the adhesion force is written

$$F_{adh} = F_a \exp(-E/RT_s) \tag{17}$$

were $F_a$ is a constant, E is the activation energy for bonding, R is the universal gas constant and $T_s$ is the surface temperature. As to the drag force, Watkinson considered the usual dependence on the drag coefficient, $C_D$, cross-sectional area of the particle in the flow direction, Ap, fluid density, $\rho$, and velocity, u:

$$F_{drag} = F_d C_D A_p \rho u^2 \tag{18}$$

Therefore, S may be written as

$$S = k_5 \frac{\exp(-E/RT_s)}{u^2} \tag{19}$$

and consequently

$$\phi_d = k_4 k_5 \frac{(c_b - c_s) \exp(-E/RT_s)}{u} \tag{20}$$

Using the Kern-Seaton expression for $\phi_r$ - eq.(10) - Watkinson's model yields

$$R_f = \frac{k_4 k_5}{k_3} \frac{(c_b - c_s) \exp(-E/RT_s)}{u^3} \left[1 - \exp(-k_3 u^2 t)\right] \tag{21}$$

Although the theoretical dependence of $\phi_d$ - eq. (20) - showed very good agreement with the experimental results with gas-oil, the same was not true in what concerns the dependence on $\underline{u}$ predicted by eq. (21). Moreover, the model was at variance with the experimental results obtained with sand-water slurries.

In view of this, the model is presented another way, by considering that depending on the actual temperature and velocity ranges, two limiting situations can be visualized

- Mass Transfer Control when $S \rightarrow 1$ in which case all particles reaching the wall stick to it (due, for instance, to a very high temperature). In this case, also, $c_s = 0$, and therefore $\phi_d = N$. As a result the model for mass transfer control becomes identical to the model of Kern and Seaton, as shown in Table 2.
- Surface Adhesion Control for low temperatures or high velocities, in which case Watkinson postulates the independence of $\phi_d$ from velocity, an assumtion that does not agree either with the original formulation - eq. (20) - or with the experiment.

TABLE 2

PARTICULATE FOULING

| Mass Transfer Control<br>($S \to 1$, $c_s \to 0$) | $u \rightleftarrows T$ | Surface Adhesion Control | |
|---|---|---|---|
| $\phi_d = k_D u = k_4 u c_b$ (22) | | $\phi_d = k_6 \exp(-E/RT_s)$ | (25) |
| (Kern-Seaton: $\phi_d = k_1 u c_b$) | | $\phi_r = k_3 u^2 R_f$ | |
| $\phi_r = k_3 u^2 R_f$ (23) | | $R_f = \dfrac{k_6}{k_3} \dfrac{\exp(-E/RT_s)}{u^2} \left|1 - \exp(-k_3 u^2 t)\right|$ | (26) |
| $R_f = \dfrac{k_4}{k_3} \dfrac{c_b}{u} \left|1 - \exp(-k_3 u^2 t)\right|$ (24) | | Gudmundson: $\phi_r = k_7 u R_f$ | (27) |
| | | $R_f = \dfrac{k_6}{k_7} \dfrac{\exp(-E/RT_s)}{u} \left|1 - \exp(-k_7 u\, t)\right|$ | (28) |
| (Independence on T) | | (Independence on $c_b$) | |

A modified Watkinson - model for Adhesion Control was used by Gudmundson [14] by considering that, in this case, the removal rate function was a function of $\underline{u}$ rather than $u^2$. This suggestion, based to some extent on the experimental results obtained by Gudmundson with magnetite particulate deposition in water, correlates also the gas-oil results when used together with eq. (20).

A summary of the relevant equations is presented in Table 2. In Table 3 a comparison is made between the various models for particulate fouling in what respects velocity dependence. The suggestion, according to the limited data available is that for the particulate fouling dependence of the fundamental parameters - $\phi_d$ and $\phi_r$ - on $\underline{u}$ and $\underline{T}$ will depend on the prevailing operating conditions.

TABLE 3

SUMMARY OF DEPENDENCE ON FLUID VELOCITY (u)

| | $\phi_d$ | $R_f$ | $\beta$ | |
|---|---|---|---|---|
| Kern-Seaton, Watkinson (MT) | u | 1/u | $u^2$ | (Sand-Water) |
| Watkinson (MT + Adhesion) | 1/u | $1/u^3$ | $u^2$ | |
| Gudmundson (MT + Adhesion) | 1/u | $1/u^2$ | u | (Gas-Oil) |
| Watkinson (Adhesion) | - | $1/u^2$ | $u^2$ | |
| Gudmundson (Adhesion) | - | 1/u | u | |

This means, for instance, that the initial fouling rate $(dR_f/dt)_{t=0} = \phi_d$ may increase or decrease with velocity depending on the controlling factor.

2.3 Reaction Fouling

Reaction fouling occurs frequently in hydrocarbon streams, associated with a large variety of situations ranging from the thermal cracking responsible for the cocking to the formation of polymers and others long-chain compounds [15,16].

In most cases, it is not easy to establish the actual reaction that produces the foulant due to the complex kinetics that are usually involved [17].

Among the factors likely to affect reaction fouling one may refer

- The temperature.
- The presence of catalysts, which may be in some instances, the pipe wall itself or metallic impurities.
- The composition of the process stream, including contaminants and, specially, dissolved oxygen.

This last case is of great importance since in many cases the oxygen may react with the hydrocarbon and produce unstable peroxides that decompose to extremely reactive free radicals that induce polymerization by self-renewal mechanisms of the type [18]

$$RH + O_2 \longrightarrow ROOH \longrightarrow RO\cdot + \cdot OH$$
$$\downarrow +R'$$
$$ROR'\cdot \xrightarrow{+R'} ROR'R'\cdot \quad \text{etc.}$$

Considerable work has been done on the kinetics underneath reaction fouling, mainly in the field of catalysis. Here, though, only the simple (but widely experienced) case in which fouling reasults from a first reaction is considered:

$$\text{Foulant Precursor (P)} \longrightarrow \text{Foulant (D) + Light Products}$$

Furthermore and following the basic treatment of Crittenden and Kolackowski [19] it is also assumed that

- The precursor concentration in the bulk is constant ($C_{Pb}$ = const.)
- Foulant may diffuse away from the reaction zone to the bulk
- Mass transfer of light products is not limiting

Under those conditions the deposition rate is

$$\phi_d = k_8 (N_P - N_D) \tag{29}$$

where $N_P$ is the flux of P to the reaction zone and $N_D$ the flux of D to the bulk.

Using, again, the stagnant film model one may write

$$N_P = k_P (c_{Pb} - c_{Pi}) \tag{30}$$
$$N_D = k_D (c_{Di} - c_{Db}) \tag{31}$$

with

$c_{Pi}$, $c_{Di}$ — precursor and foulant concentrations at deposit/fluid interface

$c_{Pb}$ — precursor concentration in the bulk, $c_{Pb} \cong$ const.

$c_{Db}$ — foulant concentration in the bulk, $c_{Db} \approx 0$

The mass flux of the precursor must be balanced by the rate of reaction under consideration, i.e.

$$N_P = k_R \, c_{Pi} \tag{32}$$

where $k_R = A \exp(-E/RT_s)$ is the velocity constant of the reaction.

Now, eliminating $c_{Pi}$ between eqs. (30) and (32) and rearranging, eq. (29) becomes

$$\phi_d = k_8 \left| \frac{c_{Pb}}{\frac{1}{k_P} + \frac{1}{k_R}} - K_D \, c_{Di} \right| \tag{33}$$

where

$$k_P = \frac{c_1 \sqrt{f}}{Sc_P^{2/3}} \, u \quad \text{and} \quad k_D = \frac{c_1 \sqrt{f}}{Sc_D^{2/3}} \, u$$

according to the Metzner-Friend equation [13].

For low solubilities and/or diffusivities of species D (as it is often the case) the backdiffusion term in eq. (33) may be neglected and the deposition function written simply

$$\phi_d = k_8 \frac{c_{Pb}}{\frac{1}{k_P} + \frac{1}{k_R}} \tag{34}$$

The removal funcion used by Grittenden and Kolackowski is basically the same as the one suggested by Taborek et al. [7] and Gudmundson [14], i.e., making use of a parameter $\psi$ - "fouling layer bond resistance" - wich appears to be dependent on velocity. Thus,

$$\phi_r = k_3 \, u^2 \, R_f / \psi \tag{35}$$

and assuming that

$$\psi = k_9 \, u^\alpha \tag{36}$$

then,

$$\phi_r = k_{10} \, u^{2-\alpha} \, R_f \tag{37}$$

The general equation ($dR_f/dt = \phi_d - \phi_r$) may now be integrated with the help of eqs. (34) and (37) to yield

$$R_f = \frac{k_8}{k_{10}} \frac{1}{u^{1-\alpha}} \left| \frac{c_{Pb}}{\frac{1}{k_4} + \frac{u}{A\exp(-E/RT_s)}} \right| \left| 1 - \exp(-k_{10} u^{2-\alpha} t) \right| \qquad (38)$$

Besides this general case it is worthwhile to consider two limiting cases:

- $k_R \gg k_p \longrightarrow$ mass transfer control
- $k_R \ll k_p \longrightarrow$ surface reaction control

For mass transfer control

$$\phi_d = (k_8 k_4) u \, c_{Pb} \qquad (39)$$

$$R_f = \left(\frac{k_8 k_4}{k_{10}}\right) \frac{c_{Pb}}{u^{1-\alpha}} \left| 1 - \exp(-k_{10} u^{2-\alpha} t) \right| \qquad (40)$$

For surface reaction control

$$\phi_d = (k_8 A 0 \, \exp(-E/RT_s)) \qquad (41)$$

$$R_f = \frac{k_8 A}{k_{10}} \frac{\exp(-E/RT_s)}{u^{2-\alpha}} \left| 1 - \exp(-k_{10} u^{2-\alpha} t) \right| \qquad (42)$$

It is interesting to note that the expression for $\phi_d$ - eq. (39) and (41) - are in good agreement with the latest model of Watkinson, in what regards eqs. (22) and (25).

As to the dependence of $R_f^*$, they are determined by the actual value of the exponential $\alpha$. For $\alpha = 0$ it becomes identical to the dependences predicted by Watkinson; for $\alpha = 1$ it is similar to the formulation of Gundmundson; for $\alpha = 2$ it yields a new type of model.

## 2.4 Solubility Fouling

As mentioned earlier, solubility results from the precipitation of dissolved matter due to solubility changes with temperature beyond he saturation point [20].

Two distinct situations may occur

- Deposition of normal (or direct) solubility compounds onto cool surfaces as in the case of silica deposition from geothermal waters [21] or wax from kerosene [22].
- Deposition of inverse solubility salts near the hor surface after reaching the supersaturation. This is also referred to as "scaling" and is more frequent in cooling water systems where calcium and magnesium carbonates and sulphates are the predominant foulants [23,24].

Whilst supersaturation is a necessary condition for crystallization fouling to occur, it is not a sufficient condition. In fact, for a crystal to grow it is required that its Gibbs' function decreases with increasing crystal size. If the crystal is smaller than a certain size, its Gibbs' function increases with rising crystal size and thus the crystal will tend to redissolve in the liquid. Therefore, most often the nucleation sites are provided by solid boundaries or by the suspended particles (and this shows the importance of simultaneous particles/solubility fouling as well as the influence of surface nature and condition).

Once sufficient nucleation sites are formed the process of growth takes place, involving three basic steps as observed by McCabe [25]:

- Diffusion of material from bulk to the wall
- Incorporation of material into the crystal lattice
- Removal of heat generated in the process

Most authors that worked on scaling in cooling waters considered step 2 as the controlling step due either to the lattice arrangement required or to the influence of equilibriums likely to affect the surface process. For $CaCO_3$ scaling, for instance, the following equilibriums explain to some extent the influence of temperature and pressure (through the concentration of $CO_2$) or the effect of pH [26]:

$$2 H_2O + CO_2 \rightleftharpoons H_3O^+ + HCO_3^-$$

$$HCO_3^- + H_2O \rightleftharpoons H_3O^+ + CO_3^{2-}$$

Other factors do affect the scaling processes [24,26]. Of these, mention must be made to the characteristics of the process stream as studied by Taborek and co-workers [7,8] who noted that

- If a solution contains predominantly a single salt, the crystalline deposits are strong and show great adherence forces to the surface. In this case, not only the "plateau" of the fouling resistance vs. time curve is comparatively high (when observed), but there exists also an induction period during which no significant fouling is observed - see Fig. 3 a) -. Clearly, the induction period is related to the initial nucleation, i.e., to the first "layer" of fouling, and depends strongly on the nature and condition of the surface. On the other hand, no sensible effect of velocity upon fouling thickness was noticed.
- If various salts are presented, the scaling is of the normal type observed by Kern and Seaton - see Fig. 3 b) -, the crystalline clusters are highly irregular and weakness planes were detected.

An important work on scalling was reported by Taborek et al. [7,8] in which an overall review is undertaken and an extensive research program developed. The model suggested by Taborek and co-workers is focused on the surface process, i.e., mass transfer from the bulk to the wall is not considered as the controlling step. This is reasonable for cooling water systems in which water velocities are comparatively high.

The model by Taborek et al. is also of the deposition-release type previously discussed, in which the deposition rate is assimilated to a nth order reaction such that

$$\phi_d = C_o (c_b)^n \exp(-E/RT_s) \qquad (43)$$

where E is the activation energy for bonding and $C_o$ is a coefficient dependent

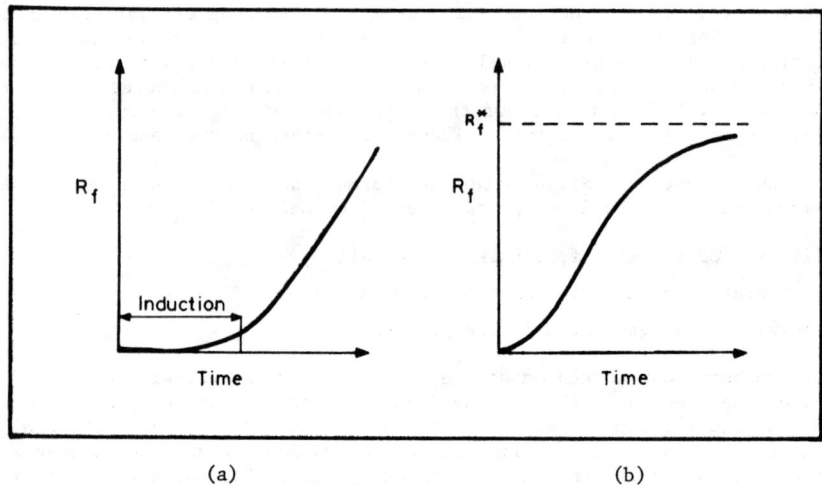

Fig.3. Solubility fouling. (a) Single salt. (b) Mixed salts.

upon several parameters, namely, on the residence-time of foulant material. The coefficient is, to some extent a "sticking probability", though of a diferent nature from Watkinson's.

It is convenient to express the dependences on $\underline{u}$ in a explicit form and therefore $C_o$ is written as

$$C_o = k_o / u^\gamma \qquad (44)$$

and eq. (43) becomes

$$\phi_d = k_o \frac{c_b^n \exp(-E/RT_s)}{u^\gamma} \qquad (45)$$

As to the removal function, Taborek et al. suggest an equation already mentioned (eq. (37)):

$$\phi_r = k_{10} u^{2-\alpha} R_f \qquad (37)$$

Thus, the integration of the general rate equation yields for

$$R_f = \frac{k_o}{k_{10}} c_b^n \frac{\exp(-E/RT_s)}{u^{2+\gamma-\alpha}} \left| 1 - \exp(-k_{10} u^{2-\alpha} t) \right| \qquad (46)$$

Although no numerical data is given by Taborek et al., a good correlation of experimental results is claimed by these authors.

This, in fact, is not surprising, since equations (45),(37) and (46) contain the dependences predicted by other models in what respects velocity, temperature

# FOULING OF HEAT TRANSFER SURFACES

and concentration, providing that appropriate values are given to the exponents $\alpha$, $\gamma$ and n.

This suggests the concentration of a generalizes model comprising the fundamental dependences on the transfer variables

## 3. GENERALIZED MODEL

The importance of having a general fouling model is apparent by noting that in most situations there are simultaneous mechanisms at work, making somehow questionable the straight utilization of particular fouling models.

Bearing in mind the derivations presented regarding some of the more recent fouling models, it is possible to visualize a significant area of common approach, despite the differences in presentation and context of the original models.

It is, thus, noticed that

1) In most fouling situations, the dependence of the fouling resistance on time may be described by an equation of the type

$$R_f^* = R_f |1 - \exp(-\beta t)| \qquad (47)$$

2) The net fouling rate, $dR_f/dt$, can be taken as the difference between a deposition and a removal rate

$$\frac{dR_f}{dt} = \phi_d - \phi_r \qquad (48)$$

3) The fouling deposition comprises two basic steps

   · Transport of foulant to the solid/fluid surface
   · Incorporation of foulant into the fouling layer

   each of which may control deposition

4) The more general equation for the deposition rate, $\phi_d$, in what regards the dependence on the transfer variables is for a first order surface process

$$\phi_d = k_8 \left| \frac{c_{Pb}}{\frac{1}{k_P} + \frac{1}{k_R}} - k_D c_{Di} \right| \qquad (49)$$

where $k_P$ and $k_D$ are given by the Metzner-Friend equation mentioned earlier, and $K_R$ is

$$k_R = k_n \frac{\exp(-E/RT_s)}{u^\gamma} \qquad (50)$$

5) The removal rate, $\phi_r$, results from the shearing action of the fluid on the weaker planes of the fouling layer. These "weak points" are, in principle, related to the actual conditions in which the fouling was formed (namely, velocity, temperature, concentration, impurities) and increase in number as the layer gets thicker

$$\phi_r = k_2 \tau x_f/\psi$$

$$= k_{10} u^{2-\alpha} R_f \qquad (51)$$

6) The more general equation is, therefore

$$R_f = \frac{k_8 \left| \dfrac{c_{Pb}}{\dfrac{1}{k_p} + \dfrac{1}{k_R}} - k_D c_{Di} \right|}{k_{10} u^{2-\alpha}} \left| 1 - \exp(-k_{10} u^{2-\alpha} t) \right| \qquad (52)$$

or, for the cases in which backdiffusion is unimportant

$$R_f = \frac{k_8}{k_{10}} \frac{c_b}{u^{2-\alpha}} \left( \frac{1}{\dfrac{1}{k_p} + \dfrac{1}{k_R}} \right) \left| 1 - \exp(-k_{10} u^{2-\alpha} t) \right| \qquad (53)$$

For the two limiting areas of mass tranfer or surface process control the appropriate equations are shown in Table 3, together with the comparison with some experimental results.

It may be noted that for surface process control the parameter $\gamma$ appears to be close to unity. Since, on the other hand, $\gamma = 0$ for mass transfer control, one may conclude that the range of variation of $\gamma$ is, in principle, $0 < \gamma < 1$.

With regard to the values of $\alpha$, Table 3 suggests a range of variation between 0 and 2 for both types of control, which requires a deeper analysis of the factors behind this parameter, namely the fouling structure.

It is noted first that contrary to the assumptions introduced in deriving the models, the fouling layer has not an homogeneous structure. This has been referred by some authors and was clearly visualized by Taborek et al. [8] using time-lapse-movies techniques. Three zones were identified (see Fig. 4)

- Upper layer - A - loosely packed like a sand dune
- Middle layer - B - probably a crystallization front initiated by a progressive increase in temperature under constant heat flux conditions
- Lower layer - C - definite crystalline formation

A similar observation was made by Atkins [28] who noted that fouling in fired process heater tubes manifests itself in two main layers: Hard coke on the tube surface and porous, tarry material between the hard coke and the fluid.

This heterogeneous structure explains well the sinusoidal-type curve of fouling or, in ither words, the fouling fluctuations with time (see Fig. 5)

# FOULING OF HEAT TRANSFER SURFACES

## TABLE 3
### GENERALIZED MODEL

| Mass Transfer Control<br>($k_R \gg k_P$) | | Surface Process Control<br>($k_R \ll k_P$) | |
|---|---|---|---|
| $\phi_d = k_P c_{Pb} = k_4 c_b u$ (54) | | $\phi_d = k_8 k_R c_b = k_8 k_{11} \dfrac{\exp(-E/RT_s)}{u^\gamma}$ | (56) |

$$\phi_r = k_{10} u^{2-\alpha} R_f$$

| | | | |
|---|---|---|---|
| $R_f = \dfrac{k_4}{k_{10}} \dfrac{c_b}{u^{1-\alpha}} \left\|1-\exp(-k_{10} u^{2-\alpha} t)\right\|$ (55) | | $R_f = \left\|\dfrac{k_8 k_{11}}{k_{10}} \dfrac{\exp(-E/RT_s)}{u^{2+\gamma-\alpha}}\right\| \left\|1-\exp(-k_{10} u^{2-\alpha} t)\right\|$ | (57) |

### Dependences on velocity

| | |
|---|---|
| − $\phi_d \; \alpha \; u$ | − $\phi_d \; \alpha \; 1/u^\gamma$ |
| − $R_f^* \; \alpha \; 1/u^{1-\alpha}$ | − $R_f^* \; \alpha \; 1/u^{2+\gamma-\alpha}$ |
| − $\beta \; \alpha \; u^{2-\alpha}$ | − $\beta \; \alpha \; u^{2-\alpha}$ |

### Comparison with experiment

| | |
|---|---|
| | Watkinson (gas-Oil) : $\alpha = 1$, $\gamma = 1$ |
| Watkinson (Sand-Water) : $\alpha = 0$ | Newson [27] (Magnet.-Water): $(\gamma-\alpha) = 0.5$ |
| Thomas [27] (Magnet.-Water): $\alpha = 1.5$ | Gudmundson (Magnet.-Water): $\alpha = 1$, $\gamma = 1$ |
| Burril [27] (Magnet.-Water): $\alpha \to 0$ | Taborek (Ca $CO_3$ scale) : $\alpha > 0$, $\gamma > 0$ |
| | Bott − Gudm. (Wax-Kerosene) : $\alpha = 1$, $\gamma = 1$ |

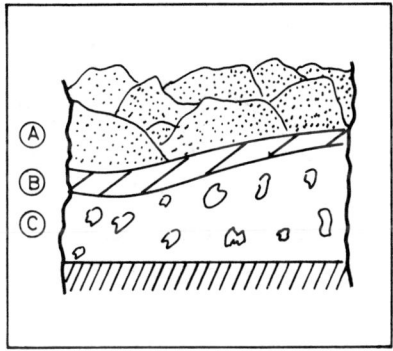

Fig.4. Fouling layers [7,8]

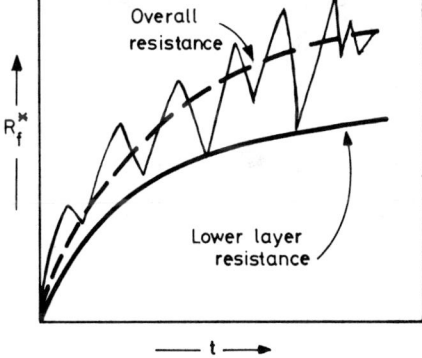

Fig. 5. Fluctuations in $R_f$ (or $x_f$)

since layer A may suffer shearing-off much more easily than the lower layer.

The thickness and characteristics of the upper layer is naturally a function of the variables affecting the process. For high velocities the upper layer is probably not allowed to form which means that the fluctuations in $R_f$ with time will be less significant. This is supported by the experimental curves shown in Fig. 6 in which it may be seen that for a lower Reynolds Number the fluctuations are much greater.

Another factor that may affect the structure of the fouling layer is the actual pressure on the walls, High pressures will promote compact layers of greater density and will most probably decrease the fluctuations in $R_f$.

As to the temperature, it is difficult to predict on fouling structure since it favors, on one hand, the adhesion of foulant, but, on the other hand it may favor reactions near the interface that may produce deposits not immediatetly incorporated in the hard zone which may suffer erosion.

Another interesting aspect worth mentioning is the potential effect of the upper layer with regard to the development of the lower layer. Taborek et al. |8| and Newson [27] verified that a sudden increase in fluid velocity promoting the shearing-off of the upper layer tends to induce higher assymptotic fouling resistances, as illustrated in Fig. 7. This suggests that the upper layer was acting as an inhibitor for the progress of the lower layer, probably by trapping some reaction product formed in layer B. Thus, the upper layer may have two conflicting effects

i) The promotion of the development of the hard zone by inducing temperature an increase inside the fouling layer

ii) The inhibition of reaction inside the fouling layer by trapping some reaction product liberated in the hard-up process.

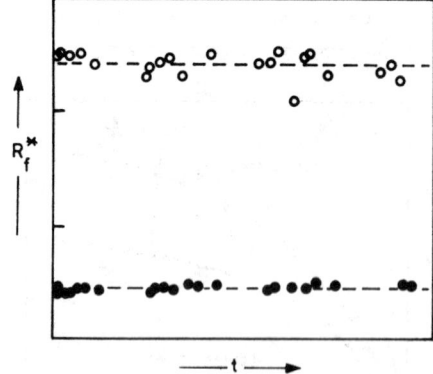

Fig.6. Effect of velocity upon the fluctuations in $R_f$ |8|.

○ $R_e \sim$ 9 000
● $R_e \sim$ 25 000

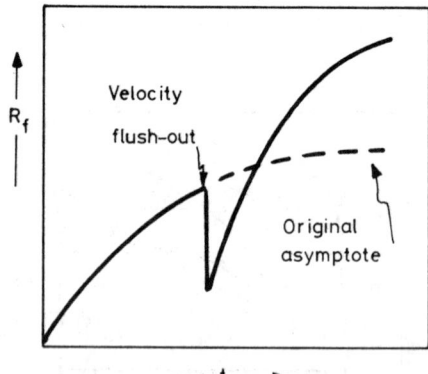

Fig.7. Effect of high velocity flush-out.

It is not possible at the present stage to determine the factors that will affect the bond strength of the fouling layer and consequently it is not possible, either, to predict the actual value of the exponent α in a given process. So, the only conclusion in this regard is that α may, most probably, vary between 0 and 2 either for mass transfer or surface process control, depending on the operating conditions.

Different combinations between α and γ are, therefore, possible, yielding different dependences on u as shown in Table 4. This means, also, that in trying by experimental observation to determine the influence of u or T ( or other variables)on fouling resistance, one may start by having a certain pair of values ($α_1$, $γ_1$) for a given set of conditions and a different pair ($α_2$, $γ_2$) for different conditions. If it is so the experimental curves obtained would be qualitatively similar to these exemplified in Fig. 8.

In fact, the somewhat peculiar curves obtained by Watkinson, Louis and Brent [29] or by Bott and Walker [30] may serve as good examples of changes in α or γ during the experiments. (See Figs. 9 and 10).

It is therefore necessary to explore further the factors that will determine the actual value of α and γ since, in fact, their values will affect strongly the fouling response to the operating variables.

In any case it appears that the similarities between different fouling mechanims are sugnificant enough to warrant the effortof translading into a common language the scarce data available.

TABLE 4

GENERALISED MODEL - DEPENDENCES ON u

| Control | α | $\phi_d$ | $R_f^*$ | β | γ |
|---|---|---|---|---|---|
| Mass Transfer | 0 | u | 1/u | $u^2$ | |
| | 1 | u | - | u | |
| | 2 | u | u | - | γ = 0 |
| Surface Process | 0 | - | $1/u^2$ | $u^2$ | |
| | 1 | - | 1/u | u | |
| | 2 | - | - | - | |
| | 0 | 1/u | $1/u^3$ | $u^2$ | |
| | 1 | 1/u | $1/u^2$ | u | γ = 1 |
| | 2 | 1/u | 1/u | - | |

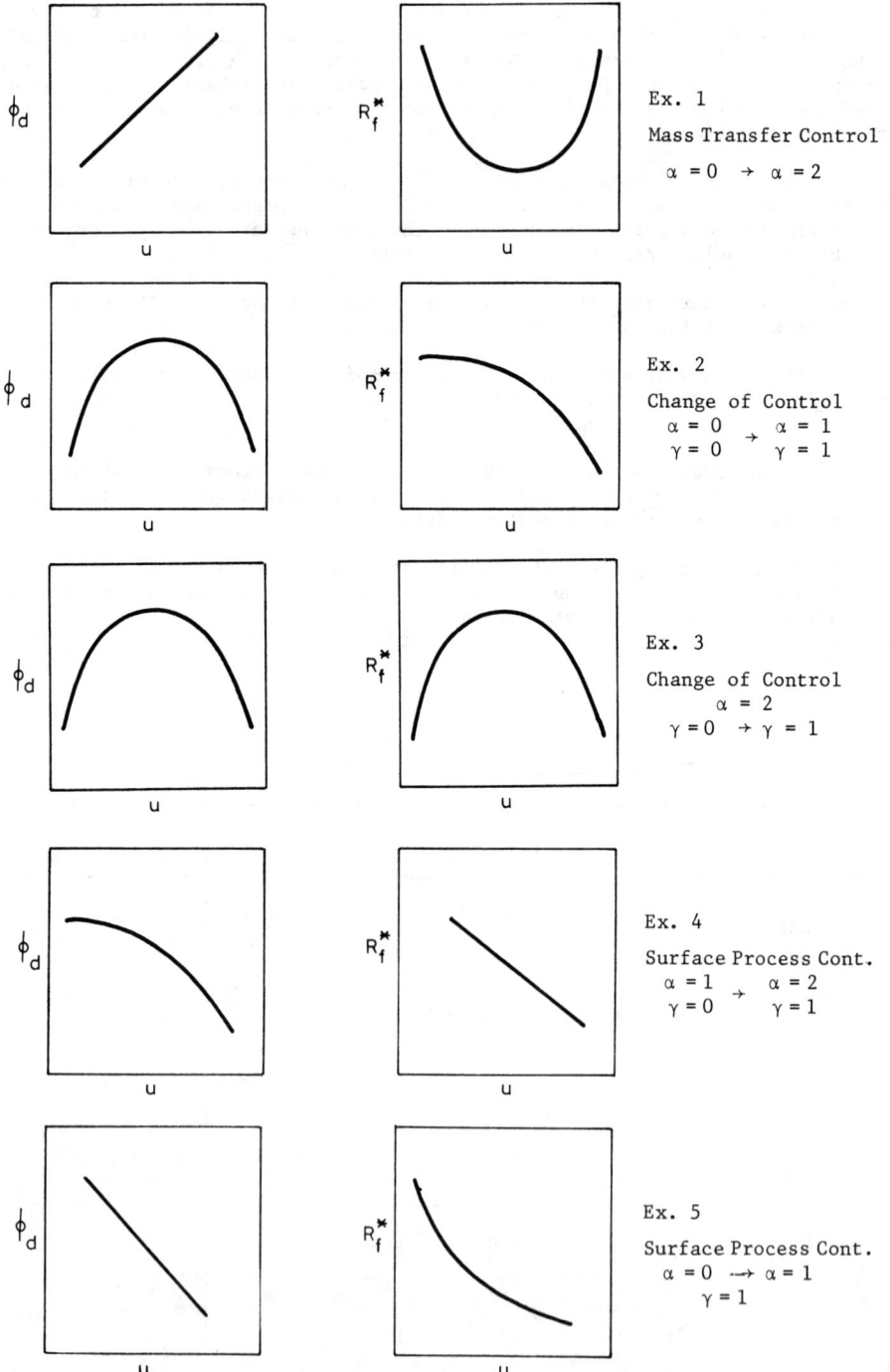

Fig.8. Examples of possible curves for $R_f$ vs. u and $\phi_d$ vs. u

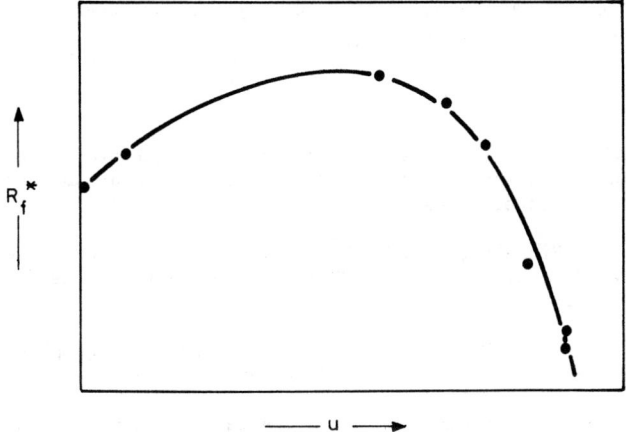

Fig.9. Experimental curve for $CaCO_3$ scale [29]

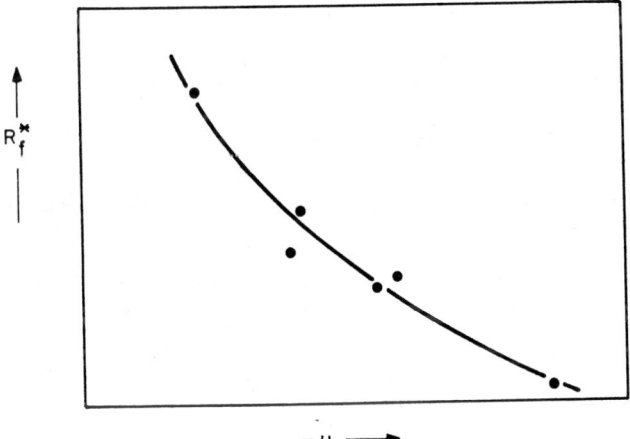

Fig.10. Experimental curve for wax deposition from kerosene [14].

ACKNOWLEDGEMENTS

The author wishes to express his gratitude to Dr. T. R. Bott (University of Birmingham) for the encouragment and helpfull discussions regarding the research program on fouling of the author.

REFERENCES

1.  Bott, T.R. 1973. Fouling in shell-and-tube heat exchangers. Paper n° 5, Advances in Thermal and Mechanical Design of Shell-and-Tube Heat Exchangers, Birnichill Institute, Glasgow.

2.  Bott, T.R. and Pinheiro, M.M.V.P.S. 1976. Biological fouling-velocity and temperature effects. Paper 76, CSME/SCChE, 16th National Heat Transfer Conf., St. Louis, U.S.A.

3.  Thackery, P.A. 1979. The cost of fouling in heat exchange plant. Proceedings Conf. Fouling-Art or Science, p.1, Univ. Surrey, U.K.

4.  Bott, T.R. and Walker, R.A. 1971. Fouling in heat transfer equipment. The Chem. Engn., 391.

5.  Suitor, J. Marner, W. and Ritter, R. 1977. The history and status of research in fouling of heat exchangers in water service. Can. J. Chem. Engn. $\underline{35}$ p. 347.

6.  Braun, R. 1977. The nature of petroleum process fouling. Mater Perf. ($\underline{35}$), Nov.

7.  Taborek, J. and Aoki, T. Ritter. R. and Pallen, J. Knudsen, J. 1972. Fouling: the major unresolved problem in heat transfer. Chem. Eng. Prog. $\underline{68}$(2), p. 59.

8.  Taborek, J. and Aoki, T. Ritter, R. and Pallen. J. Knudsen, J. 1972. Predictive methods for fouling behaviour. Chem. Eng. Prog., $\underline{68}$(7), p. 69.

9.  Kern, D. and Seaton, R. 1959. Surface fouling – how to calculate limits. Chem. Eng. Prog. $\underline{55}$(6), p. 71.

10. Birchall, G.A. 1979. The control of fouling within cooling water systems. 1979. Proc. Conf. Souling – Art or Science. p. 92, Univ. Surrey, U.K.

11. Watkinson, A. and Epstein, N. 1970. Particulate fouling of sensible heat exchangers. 4th Int. Heat Transfer Conf., Versailles, France.

12. Watkinson, A. and Epstein, N. 1969. Gas oil fouling in a sensible heat exchanger. Chem. Eng. Prog. Symp. Ser. $\underline{65}$(95), p. 84.

13. Metzner, A. and Friend, W. 1958. Theoretical analogies between heat, mass and momentum transfer. Can. J. Chem. Engn., p. 235, Dec.

14. Gudmundson, J. 1977. Heat transfer fouling. Ph.D. Thesis. Univ. Birmingham, U.K.

15. Taylor, W. and Wallace, T. 1968. Kinetics of deposit formation from hydrocarbons. I.P.E.C. prod. Res. Dev. $\underline{7}$(3), p. 199.

16. Lawler, D. 1979. Fouling by crude oil in refinery heat transfer exchangers. Proc. Conf. Fouling – Art or Science, p. 155, Univ. Surrey, U.K.

17. Hauster, R. and Thalmayer, C. 1975. Fouling and corrosion in feed effluent exchangers. A.I.P., Midyear Meeting, Chicago.

18. Gillies, W. 1979. Fouling and its control by chemical additives in hydrocarbon streams. Proc. Conf. Fouling - Art or Science, p. 188, Univ. Surrey, U.K.

19. Crittenden, B. and Kolaczkowski, S. 1979. Mass transfer and chemical kinetics in hydrocarbon fouling. Proc. Conf. Fouling - Art or Science, p. 169, Univ. Surrey, U.K.

20. Watkinson, A. and Martinez, O. 1975. Trans. ASME, Series C, J. Heat Transfer, (97), p. 504.

21. Gudmundson, J. and Bott, T.R. 1977. Deposition - The geothermal constraint. I. Chem. Eng. Symp. Ser. 48, 27.

22. Bott, T.R. and Gudmundson, J. 1977. Deposition of paraffin wax from kerosene in cooled heat exchanger tubes. Can. J. Chem. Engn., $\underline{55}$, p. 381.

23. Hasson, D. et al. 1968. Mechanism of calcium carbonate scale deposition on heat transfer surfaces. I of E.C. Fundam. $\underline{7}(1)$, p, 59.

24. Mansfield, G. 1979. Some aspects of cooling water treatment. Proc. Conf. Fouling - Art or Science, p. 10, Univ. Surrey, U.K.

25. Bridgwater, J. 1979. Crystallization fouling - a review of fundamentals. Proc. Conf. Fouling - Art or Science. p. 82, Univ. Surrey, U.K.

26. Basic principles of water treatment for cooling water systems. 1976. Ed. Dearborn Chemicals Ltd., Widnes, U.K.

27. Newson, I. 1979. Studies of particulate deposition from flowing suspensions. Proc. Conf. Fouling - Art or Science. p. 35, Univ. Surrey, U.K.

28. Atkins, G. 1962. Petro/Chem Eng. $\underline{34}(4)$ p. 20,.

29. Watkinson, A. Louis, L. and Brent, R. 1974. Scaling of enhanced heat exchanger tubes. Can. J. Chem. Engn. $\underline{52}$ p. 558.

30. Bott, T.R. and Walker, R. 1974. Fouling in heat exchanger tubes - some observations. DSIR-SAIChE-SAIMech.E. Symp. Heat Transfer Design and Operation of Heat Exchangers, Johannesburg, South Africa.

# Fouling of Heat Transfer Equipment: Summary Review

**MICHAEL G. O'CALLAGHAN**
Department of Mechanical Engineering
Massachusetts Institute of Technology
Cambridge, Massachusetts 02139 USA

ABSTRACT

   The fouling of heat exchangers is reviewed with special emphasis on basic fouling mechanisms and heat transfer equipment design aspects.

1. INTRODUCTION

   The fouling of heat transfer equipment may be defined as the deposition of unwanted material on heat exchange surfaces causing a degradation in performance. It is one of the most important issues facing the heat exchanger designer and yet is one of the most poorly understood. It has, in fact, been called "the major unresolved problem in heat transfer" [1].

   Fouling can typically degrade the performance of heat exchangers by as much as 80%, and can sometimes cause complete failure. In the past, heat exchanger designers have assumed that this degradation is constant in time (which will be shown is not true) and have increased heat transfer surface areas to overcome the loss in performance. Often this "oversurfacing" accounts for more than half of the required clean area, thereby substantially increasing the cost of the equipment. Also, the fact that the equipment is initially clean and operates at a much higher performance level than will eventually be needed, is rarely examined by designers. An over-sized heat exchanger, which may have been designed as a sensible heat exchanger, may initially operate as a boiler, with potentially dangerous and costly results.

   This paper is meant to be a <u>summary review</u> of the fouling of heat exchangers with particular emphasis on the basic fouling models. It is intended to supplement rather than replace earlier reviews of the field [1-4]. The present review also limits itself to fluid-soild interfaces which are important in all "separated flow" heat exchangers such as those of the shell and tube and plate-fin types. Phenomena which occur at liquid-liquid or liquid-vapor interfaces are not addressed here.

2. TYPES OF FOULING

   Fouling is commonly classified by the immediate cause or mechanisms of the process involved. Four categories are needed to describe the fouling of solid-fluid interfaces:

   (1) Precipitation Fouling: The deposition of material on a heat transfer

surface by crystallization from a liquid solution. This mechanism is most severe in applications involving boiling but also occurs in single-phase heat transfer equipment.

(2) Particulate Fouling: The deposition of <u>suspended</u> solids on heat transfer surfaces. The foulant in this case is not structurally attached to the surface, but may have substantial attachment forces nonetheless.

(3) Chemical Reaction Fouling: Deposition that occurs by virtue of chemical recombination that involves the heat transfer medium and possibly, the heat transfer surface. This fouling category is primarily of importance in the petroleum industry. Deposition that involves chemical reaction with the heat transfer surface is also called corrosion.

(4) Biological Fouling: The attachment and subsequent growth of both macro-organisms (barnacles, mussels, etc.) and micro-organisms (bacteria, algae, etc.). A by-product of the growth of these organisms may be a slime-like substance which can also adhere to and insulate the heat transfer surface. Current interest in Ocean Thermal Energy Conversion (OTEC) has greatly stimulated research in this area.

Categories (1) through (3) exhibit the same basic characteristics experimentally and may be modelled in a similar fashion. In the remainder of this paper, a non-specific reference to a fouling type may be assumed to apply equally well to categories (1) through (3). Further, we will largely limit our discussion to exclude the highly specialized aspects of bio-fouling. For a review of progress on bio-fouling the reader is referred to Epstein [2] and Jensen [5].

## 3. EXPERIMENTAL CHARACTERISTICS OF FOULING

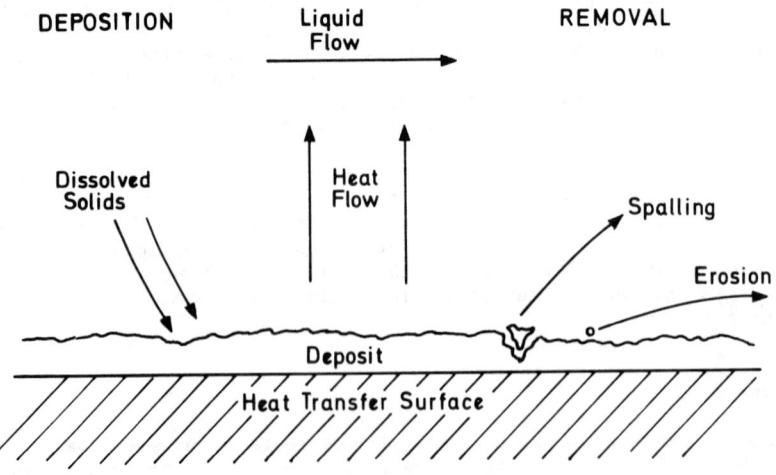

Figure 1. Schematic Representation of Foulant Deposition and Removal

# FOULING OF HEAT TRANSFER EQUIPMENT: SUMMARY REVIEW

## 3.1 Measurements of the Effects of Fouling.

A schematic representation of the deposition of fouling is shown in Fig. 1. The functional effect of fouling on a heat transfer surface may be expressed by the thermal fouling resistance [6] given by

$$R_f = \frac{X_f}{K_f} \tag{1}$$

where $X_f$ is the thickness of the deposit and $K_f$ is the thermal conductivity of the deposit.

Based on Equation (1), the fouling rate may be given by

$$\frac{dR_f}{dt} = \frac{1}{K_f}\frac{dX_f}{dt} \tag{2}$$

Equation (2) relates the <u>functional</u> effect of fouling ($R_f$) to the directly measurable thickness of the fouling layer.* However, this method of determining the effect of fouling is seldom practical since the heat transfer equipment must be disassembled before the fouling thickness can be measured. In addition, the thermal conductivity of the deposit may not be known, and may vary with the thickness.

From a functional point of view, a thermal method of measuring fouling resistance has substantial advantage over the thickness measurement discussed above. Such a method would allow the resistance to be measured <u>in situ</u>, while the heat transfer equipment is actually in operation. The thermal measurement technique is most convenient when the equipment is operated under a condition of constant heat flux. The local value of $R_f$ may then be obtained by measuring the heat transfer surface temperature, the adjacent fluid temperature and using the known value of heat flux. Under these conditions, the overall thermal conductance between the wall surface and the heat transfer fluid may be expressed as

$$U = \left[ R_f + \frac{1}{h} \right]^{-1} \tag{3}$$

where h = the convective heat transfer coefficient between the surface of of the deposit and the heat transfer fluid, at time t.

Similarly, under the initially clean condition, the conductance is given by

$$U_o = h_o \tag{4}$$

where $h_o$ = the convective heat transfer coefficient between the clean heat transfer wall and the adjacent fluid.

In terms of the wall and fluid temperature, Equations (3) and (4) may be ex-

---

* In obtaining Equation (2) it was assumed that the thermal conductivity is invariant with thickness.

pressed as

$$U = \frac{\dot{q}}{(T_w - T_f)} \tag{5}$$

and

$$U_o = \frac{\dot{q}}{(T_{wo} - T_f)} \tag{6}$$

where $T_w$ and $T_{wo}$ are the wall temperatures measured at time t and at time zero, respectively, and $T_f$ is the bulk fluid temperature.

Combining Equations (3) through (6), we obtain

$$R_f = \frac{(T_w - T_{wo})}{\dot{q}} + \left[\frac{1}{h_o} - \frac{1}{h}\right] \tag{7}$$

This equation expresses the fouling resistance in terms of the change in wall temperature at constant heat flux, and the change in the convective heat transfer resistance (1/h). If we assume that the bracketed term in Equation (7) is negligible, we tacitly assume that there is no change in h due to flow blockage caused by presence of the foulant or due to surface roughness effects. The increase in h due to the presence of a rough layer of deposit can actually cause $R_f$ + 1/h to be less than 1/ho! In this case $R_f$ would take on slightly <u>negative</u> values for a brief period of time, followed by a normal increase.

The three types of fouling curves normally found experimentally are shown in Fig. 2. The three behavior modes are called linear, falling rate and asymptotic, respectively. The asymptotic mode is of greatest practical importance since it raises the possibility of indefinite operation of heat transfer equipment without additional fouling or required maintenance. The

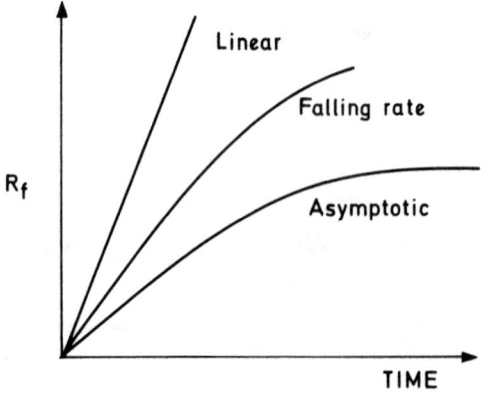

Figure 2. Linear, Falling Rate, and Asymptotic Fouling Curves.

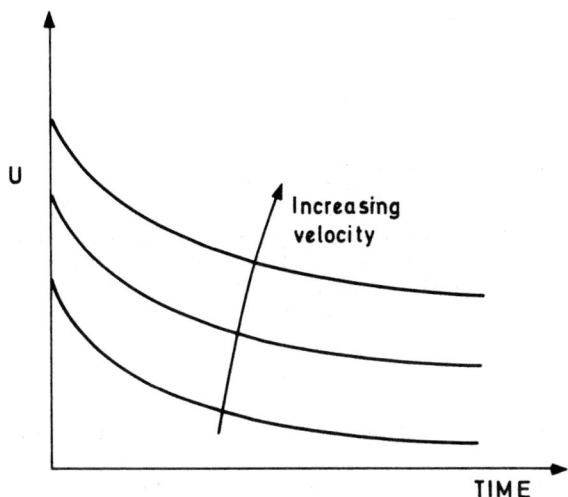

Figure 3. Effect of Asymptotic Fouling on the Overall Heat Transfer Coefficient.

reader should note that the linear and falling rate modes may, in fact, be the early stages of asymptotic behavior. For this reason, a fouling experiment must be carried out for very long periods of time before it can be classified in the linear or falling rate mode. The effect of asymptotic fouling on the overall coefficient U is shown in Figure 3.

The asymptotic nature of fouling resistance has been verified experimentally by Hasson et al [7], Watkinson and Martinez [8] and Bott and Gudmundsson [9] on a variety of systems exhibiting fouling of various types. It was also found that the asymptotic fouling resistance decreases with increasing wall shear stress. This presumably demonstrates that the fluid flow actually ablates deposit from heat transfer surfaces. The higher the shear stress, the higher the rate at which material is removed.

In addition to wall shear stress, the strength of the scale is also an influence on the level of fouling. Morse and Knudson [10] have reported that the characteristic time to an asymptotic value is a function of the deposits' mechanical strength: the longer the time to reach an asymptote, the higher the strength of the deposit. Indirect support for this theory is exhibited in data from several investigations of calcium carbonate scaling which indicate an absence of asymptotic fouling until a threshold shear stress of about $0.8 \text{ N/m}^2$ is reached. This value of shear stress is presumably the value necessary to exceed the yield strength of the deposit. A similar experiment involving river water corrosion [11], which is much stronger than calcium cabonate, yields a threshold shear stress of over $25 \text{ N/m}^2$.

4. ANALYTICAL BASIS FOR FOULING BEHAVIOR

The most widely accepted basis for fouling behavior is due to Kern and Seaton [12] who expressed the net fouling rate as the difference between a deposition and a removal function:

$$\frac{dR_f}{dt} = \theta_d - \theta_r \tag{8}$$

where  $\theta_d$ = the rate of fouling deposition, and
       $\theta_r$ = the rate of "Mechanical" removal.

The functions $\theta_d$ and $\theta_r$ have been given many different forms by various investigators (see below). The classical assumption if that $\theta_d$ is constant with time and $\theta_r$ is directly proportional to the fouling resistance. Thus, if $\theta_r = bR_f$, Equation (8) may be integrated to give

$$R_f = \frac{\theta_d}{b}(1 - e^{-bt}) \tag{9}$$

Equation (9) describes the fouling behavior shown in Figure (2) at least qualitatively, and in fact may be fit very accurately to a variety of data [7,8,10,12]. The quantity 1/b is defined as the time constant of the fouling curve, i.e. the time required for 63% of the asymptotic fouling level to be reached. It may also be interpreted as the mean residence time of an element of fouling deposit [13]. The time constant may be used as a tool to distinguish falling rate behavior from true asymptotic behavor. After $3\theta_c$ has elapsed, the fouling resistance should be at about 95% of the asymptotic value and should be nearly constant with time. If instead, $R_f$ is still increasing, then falling rate but not asymptotic behavior may be assumed.

It should be noted that Equation (9) does not apply <u>in general</u> to biological fouling as it does to other types. Thus, bio-fouling must be considered on a specifific case-by-case basis to obtain information on $R_f$ with time. The analysis in this section should be considered to apply to other types of fouling.

Modeling efforts in the past 25 years have centered on evaluating the functions $\theta_d$ and $\theta_r$ for specific fouling situations. We will now discuss some of the basic aspects of recent developments in this area.

4.1 Deposition Models

Precipitation, particulate and chemical reaction fouling may be described by serial 4 step deposition model as follows:

(1) Molecular diffusion of the deposition components or reactants from the bulk solution to the deposition-liquid interface.

(2) Chemical reaction (if any) at the deposition-liquid interface possibly involving the heat transfer surface material (corrosion).

(3) Crystallization of the reaction products (if any) on the deposition surface. This is the process by which freshly formed products from step (2) order themselves into the solid structure.

(4) Diffusion of the reaction product (if any) into the bulk solution.

In most situations, one of the above steps will be rate limiting. For example, if the reaction rate (step 2) or the crystallization rate (step 3) controls the overall deposition rate, then

$$\theta_d = C_1 \exp \frac{-E}{RT_d} (C_w)^r \tag{10}$$

where

- $C_w$ = the solute concentration under reaction conditions at the deposit surface
- $r$ = the reaction order, which is defined as zero for crystallization
- $E$ = the kinetic activation energy
- $R$ = the universal gas constant
- $T_d$ = the temperature at the deposit surface

and $C_1$ = an empirically determined constant.

Equation (10) exhibits "falling rate" behavior due to the decrease in $T_d$ as the deposit thickness increases with time.

Diffusion becomes the rate-limiting mechanism for low flow velocities and for chemical species with low diffusion coefficients but high reaction rates. The general equation for this case is:

$$\theta_d = C_2 K_m (C_f - C_w) \tag{11}$$

where $C_w$ = concentrations in the bulk and at the surface, respectively

$K_m$ = the turbulent mass transfer coefficient which is a function of the Reyonolds and Schmidt numbers

and $C_2$ = an empirically determined constant.

The rate of deposition in this case increases with bulk concentration and with flow velocity. However, a substantial increase in flow velocity will increase the diffusion to such a high level that chemical reaction is again the rate-limiting process. For cases where both diffusion and reaction rates are important, the reader is referred to the analysis by Hasson [14].

### 4.2 Re-entrainment Models

In contrast with the deposition factor, very little theoretical development has been done on the mechanism of deposit removal. The fouling removal term, $\theta_R$, characterizes the mechanical removal of deposit by the liquid flowing in contact with the deposit. Kern and Seaton [12] postulated that this term would have the form

$$\theta_R = C_3 \tau X_f \tag{12}$$

where $\tau$ = the shear stress at the deposit surface, and
$X_f$ = the thickness of the deposit.

The rate of removal of deposit given by Equation (12) is proportional to the mechanical stress imposed on the deposit surface (shear stress) and is also proportional to the scale thickness since particles are more likely to be ablated at larger thicknesses. Other fouling removal factors have been forwarded by Taborek et al [1] and Beal [15]. Taborek assumed a function of the form of Equation (12) but included a deposit strength factor to account for the specific resistance to re-entrainment by a particular foulant. Beal's latest model distinguishes between loose deposit, which is assumed to be removable by the circulating liquid and the more adherent interior deposit which is not subject to such removal. Beal also assumes that loose deposit is gradually sintered into adherent deposit at the interface between the two. Quantitative details of this process were not given.

The most important contribution to the modeling of hydrodynamic removal in many years is that of Cleaver and Yates [16,17]. Their work provides information on the "turbulent burst" phenomenon in which local high vorticity flows impinge on the heat transfer surface. These vortices, which are not unlike miniature tornadoes, would provide a much higher particle removal rate than would the classical picture of turbulent flow. Epstein [13] reviewed the work of Cleaver and Yates and derived an expression for the time constant of the asymptotic fouling curve based on the "turbulent burst" flow picture. Based on that expression the re-entrainment fuction will have the form

$$\theta_r = C_4 \frac{\tau_s \beta}{\mu \psi} \qquad (13)$$

where  $\tau_b$ = the shear stress at the deposit-liquid interface
$\beta$ = the fraction of the heat transfer surface covered by bursts at any instant
$\mu$ = the absolute viscosity of the liquid
$\psi$ = a deposit strength factor
and $C_4$ = an empirically determined constant

Attempts to experimentally verify Equation (13) have not yet been published.

## 5. HEAT TRANSFER EQUIPMENT DESIGN FOR MINIMUM FOULING

### 5.1 Fluid Flow Velocity

The use of higher fluid flow velocities has two conflicting effects that tend to partially offset each other. For diffusion-rate controlled deposition, Equation (11) indicates that the rate of <u>deposition</u> will increase with fluid velocity. In contrast, Equations (12) and (13) indicate that re-entrainment will also be augmented by increased flow velocity. The dominating effect must be determined experimentally for each specific case, but for cooling water purposes, it is always advantageous to use the highest practicable velocity [2].

### 5.2 Design of Heat Transfer Equipment for the Asymptotic Fouling Condition

An acceptable practice is to over-size the required heat transfer surface area by an amount sufficient to offset the reduction in the overall heat transfer coefficient. However, there is unfortunately no general model or calculation method available that will yield values for $R_f$ for any situation. It is therefore necessary to obtain $R_f$ from experimental sources such as reference [18].

The heat transfer equipment designer should realize that the equipment will be over-sized when it is clean and provide a higher performance level.

# FOULING OF HEAT TRANSFER EQUIPMENT: SUMMARY REVIEW

Precautions should be taken to insure that the additional heat transfer will not create an undesirable situation.

## 5.3 Adjustments of Surface Temperature Distribution due to Fouling Deposits

In most cases, fouling deposits will preferentially occur where the surface to fluid temperature difference is the largest. As a result, the heat flux (proportional to the <u>deposit</u> surface to fluid temperature difference) will become more uniform as fouling proceeds. If the heat transfer in the clean heat exchanger is dominated by local areas of high heat flux, then the <u>overall</u> performance degradation of the device may by underestimated by an average value of $R_f$. Experimental verification of this temperature redistribution is given in [19].

## NOMEMCLATURE

| | | |
|---|---|---|
| $b$ | = | deposit constant $\sec^{-1}$ |
| $C_1$ | = | empirical constant for Eq. (10) $\left[\frac{mK}{N}\left(\frac{m^3}{mol}\right)^r\right]$ |
| $C_2$ | = | empirical constant for Eq. (11) $\left[\frac{m^3\text{-K-sec}}{N\text{-mol}}\right]$ |
| $C_3$ | = | empirical constant for Eq. (12) $\left[\frac{m^2 K}{N^2}\right]$ |
| $C_4$ | = | empirical constant for Eq. (13) $\sec^{-1}$ |
| $C$ | = | solute concentration at the wall or in the bulk fluid, respectively $\left[\frac{mol}{m^3}\right]$ |
| $E$ | = | activation energy $\left[\frac{N\text{-m}}{mol}\right]$ |
| $h$ | = | convective heat transfer coefficient $\left[\frac{N}{m\text{-sec-K}}\right]$ |
| $K_f$ | = | thermal conductivity of foulant $\left[\frac{N}{\sec\text{-K}}\right]$ |
| $\dot{q}$ | = | heat flux $\left[\frac{N}{\sec\text{-m}}\right]$ |
| $r$ | = | reaction order in Eq. (10) [dimensionless] |
| $R$ | = | universal gas constant $\left[\frac{N\text{-m}}{\text{-mol}}\right]$ |
| $R_f$ | = | fouling resistance $\left[\frac{\sec\text{-m-K}}{N}\right]$ |
| $t$ | = | time [sec] |
| $T$ | = | temperature [K] |
| $U$ | = | overall heat transfer coefficient $\left[\frac{N}{m\text{-sec-K}}\right]$ |

Greek Nomenclature

β = fraction of the heat transfer surface covered by bursts at any instant [dimensionless]

θ = rate of deposit removal of fouling $\left[\dfrac{n-K}{N}\right]$

μ = dynamic viscosity $\left[\dfrac{N\text{-}s}{m^2}\right]$

τ = fluid shear stress [$N/m^2$]

ψ = scale strength factor [dimensionless]

subscripts

o       clean condition

w      wall

f       bulk fluid

r       removal

         deposit

REFERENCES

1. Taborek, J., Knudson, J., Aoki, T., Ritter, R.B., and Palen, J.B. 1972. "Fouling: The Major Unresolved Problem in Heat Transfer", Parts I and II, Chem. Eng. Prog., 68, 59-67 and 69-78.

2. Epstein, N. 1978. "Fouling in Heat Exchangers", Proc. Sixth Int. Heat Transf. Conf., 6, 235-53.

3. Bott, R.R. and Walker, R.A. 1971. "Fouling in Heat Transfer Equipment", The Chemical Engineer, 255, 391-93.

4. Suitor, J.W., Marner, W.J. and Ritter, R.B. 1977. "The History and Status of Research in Fouling of Heat Exchangers in Cooling Water Service", Can. Journ. Chem. 55, 374-80.

5. Jensen, L.D. (editor). 1977. "Biofouling Control Precedures", Marcel Dekker Publishers, New York.

6. Rohsenow, W.M. and Choi, H.Y. 1961. "Heat, Mass, and Momentum Transfer", Prentice-Hall, New York, Chapt. 6.

7. Hasson, D. and Zahavi, J. 1970. "Mechanism of Calcium Sulfate Scale Deposition on Heat Transfer Surfaces", IEC Fundamentals, 9, 1-10.

8. Watkinson, A.P. and Martinez, O. 1976. "Scaling of heat Exchanger Tubes by Calcium Carbonate", J. of Heat Transfer, Trans. ASME, 76-HT-8.

9. Bott, T.R. and Gudmundsson, J.S. 1976. "Deposition of Paraffin Wax from Kerosene in Cooled Heat Transfer Tubes", Presented at the 16th Nat. Heat Transfer Conference, St. Louis, Missouri.

10. Morse, R.W. and Knudsen, J.G. 1977. Can. J. Chem. Eng., 55, 272-9.

11. McAllister, R.A., Eastham, D.H., Dougharty, N.A., and Hollier, M. 1961. Corrosion, 17, 579-88.

12. Kern, D.Q. and Seatton, R.E. 1959. "A Theoretical Analysis of Thermal Surface Fouling", Brit. Chem. Eng., 4, 258-62.

13. Epstein, N. 1979. "Fouling: Technical Aspects", presented at the 1st Int. Conference on the Fouling of Heat Exchangers, Troy, New York, (Aug.)

14. Hasson, D., Avriel, R., Resnick, W., Rozenman, T., and Windreich, S. 1968. "Mechanism of Calcium Carbonate Scale Deposition on Heat Transfer Surfaces, IEC Fundamentals, 7, 59-65.

15. Beal, S.K. 1973. Trans., Amer. Nuc. Soc., 17, 163-72.

16. Cleaver, J.W. and Yates, B. 1973. "Mechanism of Detachment of Colloidal Particles from a Flat Substrate in a Turbulent Flow", J. Coll. Interf. Sci., 44, 464-74.

17. Cleaver, J.W. and Yates B. 1976. "The Effect of Re-entrainment on Particle Deposition", Chem. Eng. Sci., 31, 147-51.

18. Rohsenow, W.M. and Hartnett, J.P. 1973. Handbook of Heat Transfer, McGraw-Hill, New York, Chapter 18.

19. Hauster, R.H. and Thalmayer, C.E., 1975. Presented at the 40th mid-year meeting of the American Petroleum Institute, Chacago, (May).

# Dynamic Response of Tubes in Cross Flow Subjected to Axial Forces near the Buckling Load

**ERNST P. HEINECKE**
Institut für Reaktorbauelemente
Kernforschungsanlage Jülich GmbH
517o Jülich, Postfach 1913, Germany

ABSTRACT

In heat exchangers with tubes in cross flow unsteady flow fluctuations are the reason for tube vibrations. Due to inevitable temperature differences in the primary circuit some of the tubes may be subjected to such high axial forces, that the buckling load can be reached. The influence of dynamic pressure and axial forces on the response of the tubes is described for staggered and in-line tube banks.

1. INTRODUCTION

Many authors have reported on experimental and theoretical studies on acoustical and mechanical vibrations in heat exchangers. Mainly three kinds of excitation mechanisms for tube vibrations are described. The tubes are forced to vibrations either by vortices which develop in the wake of the tubes, by turbulent flow fluctuations, or both vortices and turbulence, and by fluidelastic forces the latter being generated by a coupling of the vibrating tubes and the flow field araound them. A survey on literature is given in /1/, /2/, and /3/.

For the safe lay out of a heat exchanger with respect to vibrations two facts should be known: (1) the aerodynamic loads acting on the tubes, and (2) the mechanical properties of the tubes (mass, damping, end conditions, elasticity etc.). Due to experimental difficulties there exists no exact knowledge on the aerodynamic forces, whereas, within a scatterband, the mechanical behaviour of the tubes in a full-size heat exchanger is known. Because the unsteady loads are unknown modell-tests are performed in which the dynamic response of the tubes is measured. The results of the experiments are extrapolated to the full-size heat exchanger.

The usefulness of such experiments depends upon the skill of the experimentator who has to decide which operating conditions of the full-size apparatus must be simulated in a model test. In the following section the influence of the main parameters is outlined.

2. MODELLING RULES

The natural frequencies of elastical beams are described by the formula (see /4/)

$$f = C\left(\frac{EI}{m_L L^4}\right)^{1/2} \tag{1}$$

where E = Young's module, I = momentum of inertia, $m_L$ = mass/unit length, L = length of beam between 2 supports, C = constant depending on end conditions.

If D and d stand for the outer and inner diameter of a tube equation (1) can be written as follows:

$$f = \left(\frac{E}{\rho}\right)^{1/2} \frac{C}{4L^2} D\left[1 + \left(\frac{d}{D}\right)^2\right]^{1/2} \tag{2}$$

where $\rho$ = density of tube material.

In case a tube is simply supported at both ends C has the value of 1.57. If a compressive force, e. g. due to temperature differences, acts on the tube the natural frequency decreases according to equation (3), (see /5/).

$$\frac{f^*}{f} = \left(1 - \frac{P}{P^*}\right)^{1/2} \tag{3}$$

where P = compressive force, P* = buckling load.

The buckling load is calculated after equation (4)

$$P^* = \frac{EI\pi^2}{L_o^2} \tag{4}$$

where $L_o$ is proportional to L and depends on end conditions.

Equation (3) shows that with increasing compressive forces P the frequency f decreases to zero. Combining equations (2) and (3) on gets the resulting frequency of a tube subjected to compressive forces:

$$f^* = \left(\frac{E}{\rho}\right)^{1/2} \frac{D}{4L^2} \left[1 + \left(\frac{d}{D}\right)^2\right]^{1/2} C \left(1 - \frac{P}{P^*}\right)^{1/2} \tag{5}$$

The frequency of the exciting aeriodynamic forces is described by equation (6)

$$f = \frac{SU}{D} \tag{6}$$

where S = Strouhal number, U = characteristic velocity, D = diameter of tube.

The response of a tube to unsteady forces depends, apart from the absolute height of the forces and internal damping, on the ratio of the Eigenfrequency f* of the tube and the frequency f of the exciting forces, see equation (7).

$$\frac{f^*}{f_s} = \underbrace{\frac{1}{SU}}_{1} \underbrace{\left(\frac{E}{\rho}\right)^{1/2}}_{2} \cdot \underbrace{\frac{1}{4}\left(\frac{L}{D}\right)^2 \left[1 + \left(\frac{d}{D}\right)^2\right]^{1/2}}_{3} \underbrace{C\left(1 - \frac{P}{P^*}\right)^{1/2}}_{4} \tag{7}$$

The first three items in equation (7) correspond to the unsteady flow conditions, material properties of the tube, and the geometrical conditions. The last item describes the influence of end conditions and tensile forces. With respect to the influence of compressive forces the fourth item deserves special attention. The frequency of a tube decreases with increasing axial load, and consequently the energy which is needed to excite the tube to vibrations decreases proportional to 1/f*, assuming that damping forces are indenpendent on frequency. This means, that high amplitude vibrations must be expected in tube bundles if some of the tubes are subjected to compressive forces. The experimental set-up and the results of experiments are described in the following paragraphs.

## 3. EXPERIMENTAL SET-UP

In gas-cooled reaktors gas-gas heat exchangers are used. In Fig. 1 a schematical description of a heat exchanger is given. The tubes in the inlet and outhlet region regions of the heat exchanger are exposed to a flow normal to the tube axis. In this case, as is well known, the tubes may be excited to mechanical vibrations. Because of the lenght of the heat-exchanger and inevitable temperature differencies ( $\leq$ 40 °C) axial loads near the buckling load must be exspected. (Fig. 1)

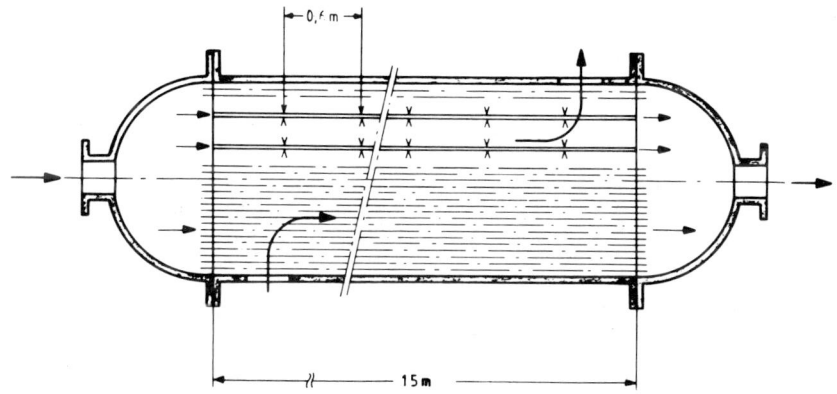

Fig. 1 Gas - Gas heat exchanger

The aim of the experiments was to find out the influence of axial load and tube arrangement on the mechanical response of tubes. As a first approximation a section modell was built, by which the flow conditions and the mechanical response of tubes in this region could be simulated. In a solid steel frame six rows of tubes with a free length L of 0,6 m and a dimension of 12 x 1 mm were installed. One of the Tubes could be exposed to an axial load (Fig. 2) of defined height. The other tubes were fixed to the side walls of the modell by bolts. The response of the tubes was measured by strain gauges.

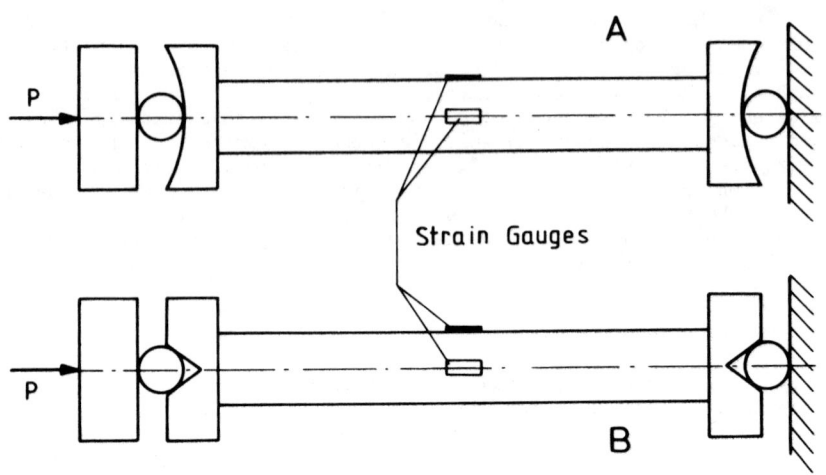

Fig. 2. Test tubes, different end conditions

Flow fluctuations were measured by condensor microphones. All experiments were performed in an open circuit wind tunnel.

## 4. RESULTS

In preliminary tests without flow the mechanical behaviour of a test tube was in vestigated. In Fig. 3 the theoretical influence of a compressive axial load on the Eigenfrequency of a tube is described (full line). The measuring points show the results of the test tube with the end conditions denoted by "A" in Fig. 3. The results are in good agreement to the theoretical prediction. In Fig. 4 the reference length for different end conditions is given.

In Figures 5, 6 und 7 the response of tubes under compressive axial loads for different tube arrays is shown. The first example in Fig. 5 describes the behaviour of a tube in a staggered arrangement ($S_t/D$ = 2.6, $S_L/D$ = 0.75). Three main regimes can be discerned. In the first one the mechanical response (q < 460 N/m$^2$) is moderate, the relative amplitude x/D is small. If the dynamic pressure q is exceeding a given value (460 < q < 1200 Nm$^2$) the amplitude of tube vibrations suddenly increases by a factor 15 in a certain range of axial load. A further increase of the axial load leads to the result, that high amplitude vibrations no longer occure. The mechanical frequency of the test tube was 78 Hz (with no axial load). Additionaly to the tube vibrations the frequency of flow fluctuations was measured. For the tube array mentioned above a Strouhal S number of 0.71 was found. It is surprising, and this must be kept in mind for the results of the other arrays, that the frequency of flow fluctuations and that of the tubes are very wide apart (factor 25). Acoustical resonance effects in the heat exchanger duct are in excellent agreement with the measurement of flow fluctuations.

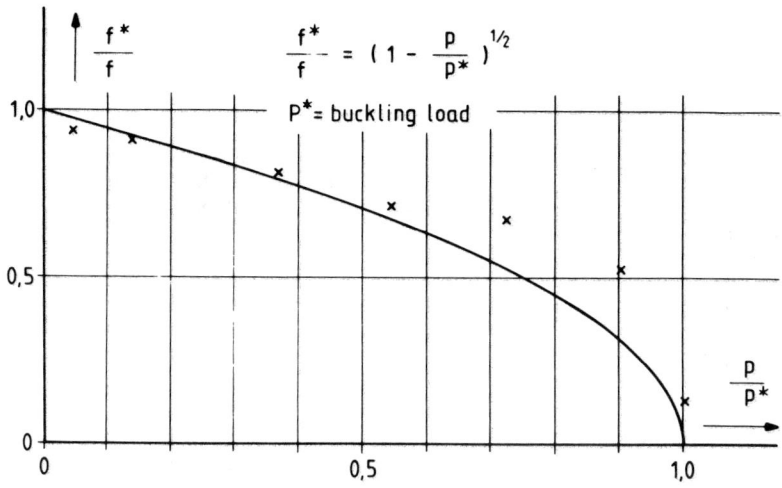

Fig. 3. Eigenfrequency as function of axial load

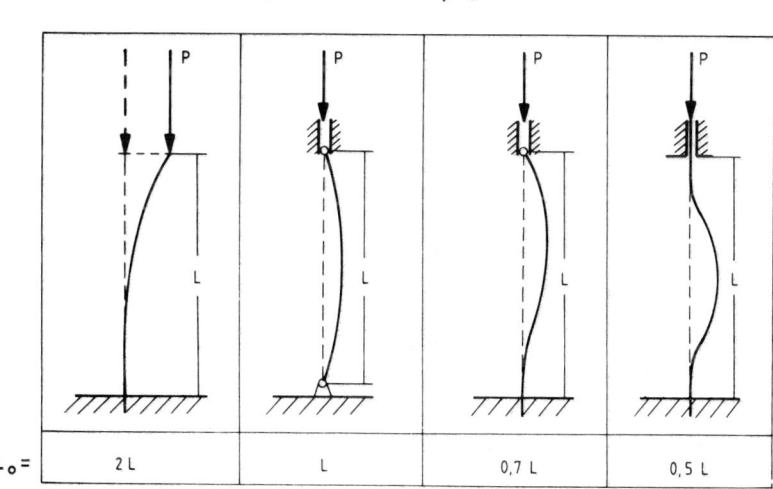

Fig. 4. Buckling load, reference length (Euler)

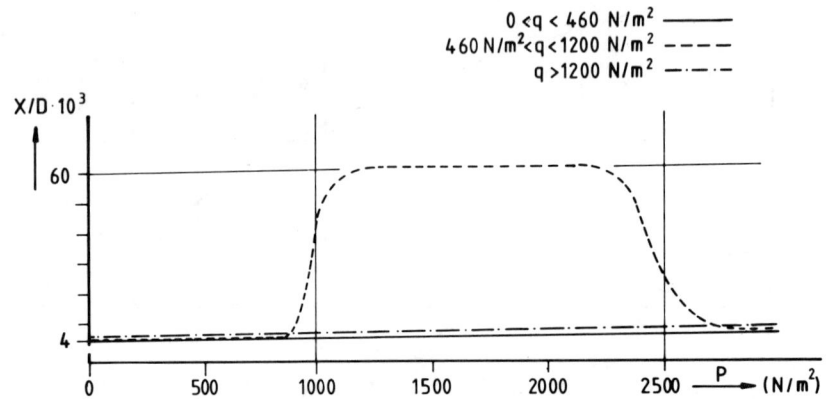

Fig. 5. Relative amplitude depending on p and q, staggered ($S_t/D = 2,6$; $S_l/D = 0,75$; $1,2 \cdot 10^4 < Re < 4 \cdot 10^4$)

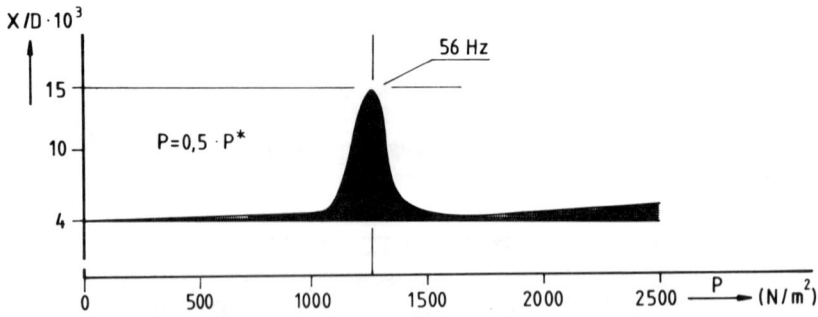

Fig. 6. Relative amplitude depending on p and q, staggered ($S_t/D = 1,5$; $S_l/D = 1,3$; $1,2 \cdot 10^4 < Re < 4 \cdot 10^4$)

In Fig. 6 the results of another staggered arrangement is shown. ($S_{t/D}$ = 1.5, $S_{L/D}$ = 1.3). The response of the test tube is very much different from that in the first array. With increasing axial load the amplitude increases only within a very short range of P. After a certain load is exceeded the vibrations fall down to a very low value. The maximum of the vibrations is near half the theoretical value of the buckling load. The Strouhal number S had a value of 0.61.

At last experiments with an in-line arrangement were performed. In this case the onset of high amplitude vibrations happens af a very much lower value of P than was observed for the staggered tube arrays (see Fig. 7). Additionally a much higher amplitude was observed. From earlier experiments it is known that in in-line arrays fluid-elastic vibrations are most probably the cause for excessive tube vibrations (see [2] and [3] ). In all experiments, as well with the staggered as well with the in-line arrays the test tube was positioned in different rows. No influence of the position on the mechanical behaviour of the tube could be measured.

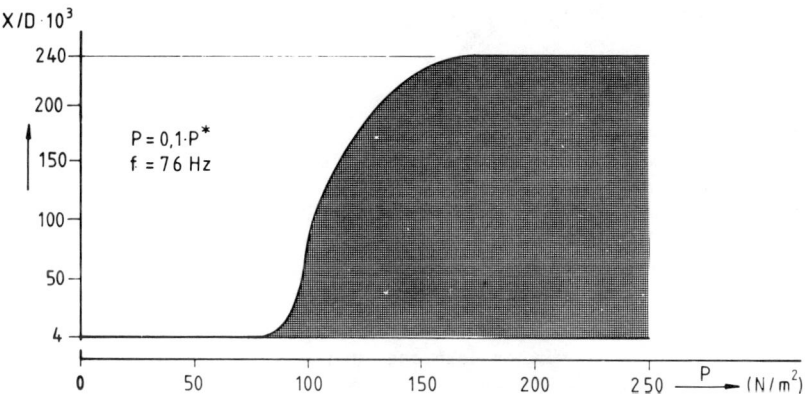

Fig. 7. Relative amplitude depending on p and q, in-line
($S_t/D$ = 2,2 ; $S_l/D$ = 1,44 ; $10^4$ < Re < 4,5·$10^4$)

## 5. SUMMARY

Tube vibrations in two staggered array and one in-line tube bank have been measured in a range of $10^4 \leq Re \leq 4 \cdot 10^4$. One of the tubes in the test array could be subjected to compressive axial forces. The response of the test tube was different in all three arrangements. The highest amplitudes were observed in the in-line tube bank. It is surmised that in this case fluid-elastic coupling between the vibrating tube and the flow field around the tube occurs. In case of the staggered arrays an explanation of the excitation mechanism cannot be given at the moment. Usually the assumption is made that vortex shedding is the reason for tube vibrations. In the experiments described above however the frequency of the vortex shedding, which was well pronounced, was much higher (factor 20 ÷ 25) than the natural frequency of the test tube. The only reason for these effects might be that the tube had a non-linear spring constant. In that case a subharmonic excitation may occur (see [5] ). Further experiments on this topic are planned.

## 6. ACKNOWLEDGEMENT

The author wants to thank Mrs. I. Esser who did the type writing and Mr. G. Schumacher for making the drawings. He is expecially indebted to Mr. K. Hansen, who performed the experiments with great skill and endurance.

## NOMENCLATURE

| | |
|---|---|
| C | constant depending on end conditions |
| D | outer tube diameter, m |
| d | inner tube diameter, m |
| E | Young's module, $N/m^2$ |
| f | natural frequency of a tube, 1/s |
| f* | frequency of a tube under axial load, 1/s |
| $f_s$ | frequency of flow fluctuations, 1/s |
| I | momentum of inertia, $m^4$ |
| L | length of tube, m |
| $L_o$ | effective tube length, m |
| P | axial load, N |
| P* | buckling load, N |
| q | dynamic pressure, $N/m^2$ |
| $\rho$ | density of tube material, $kg/m^3$ |
| Re | Reynolds number |
| S | Strouhal number |
| $S_L$ | longitudinal pitch, m |
| $S_t$ | traverse pitch, m |
| U | characteristic velocity, m/s |

## 7. REFERENCES

1. Heinecke, E., 1970, Strömungstechnische und aeroakustische Erscheinungen in Zylindergittern, Jül-815-RB

2. Heinecke, E., 1975, Modell tests on fluid-elastic vibrations in heat exchangers with tubes in cross flow, IAEA-SM-2oo/31, 13-17.1o.1975

3. Heinecke, E., 1978, Fluid-elastic vibrations in heat exchangers with tubes in cross-flow. B.N.E.S-Vibration in nuclear plant, Keswick, U.K. May 1978

4. Freberg, C.R., and Kemmler, E.S., Aircraft vibration and flutter, Wiley a Sons, Inc. 1944

5. Timoshenko, S., Young, D.H. and Weaver, W. Jr., Vibration problems in engineering. John Wiley a Sons, fourth edition.

# Why Laminar Flow Heat Exchangers Can Perform Poorly

**WARREN M. ROHSENOW**
Department of Mechanical Engineering
Massachusetts Institute of Technology
Cambridge, Massachusetts 02139 USA

ABSTRACT

Heat exchangers with non-interconnecting passages (parallel tubes or plates) are usually sized assuming uniform flow in each of the passages. Particularly in laminar flow their performance can depart from these predictions due to non-uniform flow distribution resulting from

(a) superimposed gross natural convection in horizontal orientation,

(b) the effect of the viscosity temperature relation permitting two different flow rates for the same pressure drop when liquids are cooled,

(c) having non-uniform size passages resulting from large manufacturing tolerances or errors.

Also the effect of conduction along the passages can reduce performance.

III.5.1 INTRODUCTION

We shall focuss on laminar flow inside parallel tubes assuming this heat transfer resistance is controlling and the heat transfer coefficient on the other shell side is very much higher with either uniform temperature evaprating or condensing fluids so that the wall temperature is essentially uniform.

To illustrate the nature of non-uniformities we also neglect entrance and exit pressure drops and hydrodynamic and thermal entrance effects in the tubes. Then the heat transfer coefficient for uniform wall temperature is given by a constant Nusselt number

$$\left. \begin{array}{ll} \text{Tubes:} & \text{Nu} \equiv \dfrac{hD}{k} = 3.66 \\[6pt] \text{Parallel Plates:} & \text{Nu} \equiv \dfrac{hD_h}{k} = 7.60 \end{array} \right\} \qquad (1)$$

and pressure drop is calculated as:

$$\frac{dp}{dx} = \frac{4f}{D} \frac{G^2}{\rho\, 2 g_o} \qquad (2)$$

For tubes

$$f = 16/(GD/\mu)$$

for parallel plates  (3)

$$f = 24 \,(GD/\mu)$$

Then since $G = w_1/(\pi D^2/4)$ then in length L for tubes:

$$\frac{dp}{dx} = \frac{128}{\pi g_o} \frac{L\mu}{\rho} \frac{w_1}{D^4} \quad (4a)$$

and in length L

$$\Delta P = \frac{128}{\pi g_o} \frac{L\mu}{\rho} \frac{w_1}{D^4} \quad (4b)$$

where $w_1$ is the flow rate in a single tube passage, $W/n$.

In a parallel passage exchanger with uniform flow in all passsages the temperature distribution, Fig. III.5.1 in each passage with uniform wall temperature would be given by:

$$\frac{T_w - T_2}{T_w - T_1} = \exp\left(-\frac{h\pi DL}{w_1 c}\right) \quad (5)$$

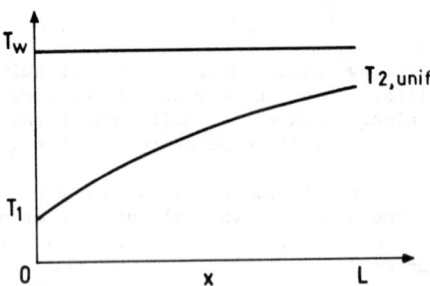

FIG. III.5.1 Temperature Distribution For Uniform Flow Distribution

Here h and $\Delta P$ would be uniform for all passages and given by Equations (1) and (4).

## III.5.2 NATURAL CONVECTION IN HORIZONTAL EXCHANGERS

A horizontal exchanger of height $H$ consisting of parallel tubes or plates is shown in Fig. III.5.1. In each header there is a hydrostatic pressure gradient which is different in each header because $\rho_1 > \rho_2$.

$$\left.\begin{array}{ll} P_{1z} - P_{1t} = \rho_1 z & P_{1b} - P_{1t} = \rho_1 H \\ \underline{P_{2z} - P_{2t} = \rho_2 z} & \underline{P_{2b} - P_{2t} = \rho_2 H} \\ \Delta P_z - \Delta P_{2t} = (\rho_1 - \rho_2)z & \Delta P_b - \Delta P_t = (\rho_1 - \rho_2)H \end{array}\right\} \quad (6)$$

In laminar flow exchangers this difference in hydrostatic $\Delta P$ can be of the order of magnitude of the $\Delta P$ between inlet and outlet. As flow decreases, $T_2$, $(\rho_1 - \rho_2)$ and $(\Delta P_b - \Delta P_t)$ all increase while $P_1 - P_2$ decreases. Hence, as flow decreases this effect becomes more significant. This problem was dicussed by Mueller [1].

From Eq (6) $\Delta P_b > \Delta P_t$ therefore from Eq. (4) $w_{1t} < w_{1b}$ ; hence from Eq. (5) $T_{2t} > T_{2b}$ .

The flow rate variation is linear in $z$ if $(\rho_{2b} - \rho_{2t}) \gg (\rho_1 - \rho_2)_{avg}$
From Eqs. (4) and (6)

$$w_z - w_t = \frac{\pi g_o \rho D^4}{128 L \mu} (\rho_1 - \rho_2) z = \Gamma z \qquad (7)$$

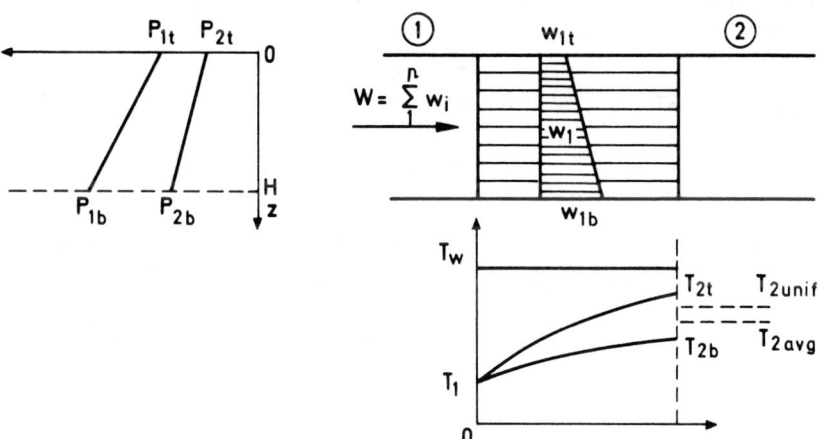

FIG. III.5.2 Non-Uniform Flow and Temperature Distribution due to Hydrostatic Head Difference in Headers in Horizontal Exchangers

Then with $w_1 = w_z$, Eq. (5) becomes:

$$\frac{T_w - T_{2z}}{T_w - T_1} = \exp\left(-\frac{h\pi DL}{c(w_t + \Gamma z)}\right) \tag{8}$$

For parallel plates of the same width Eqs. (7) and (8) may be integrated to obtain average exit temperature in the header.

Since $w_z$ is linear in $z$

$$w_{avg} = \frac{1}{H}\int_0^H (w_t + \Gamma z)dz = w_t + \frac{\Gamma H}{2}$$

$$= w_t + \frac{w_b - w_t}{2} = \frac{w_t + w_b}{2} = \frac{W}{n} \tag{9}$$

The enthalpy average exit temperature, $T_{2\,avg}$, is the flow weigthed mean of the exit temperature with Eq. (5):

$$\frac{T_w - T_{2\,avg}}{T_w - T_1} = \frac{1}{\frac{W}{n}H}\int_0^H w_z \exp\left(-\frac{h\pi DL}{cw_z}\right)dz$$

$$= \frac{1}{\frac{W}{n}H}\int_0^H (w_t + \Gamma z)\exp\left(-\frac{h\pi DL}{c(w_t + \Gamma z)}\right)dz \tag{10}$$

Here $T_{2z}$ does not vary linearly with $z$. The result is that this average exit header temperature $T_{2avg}$ is less than the exit temperature. $T_{2unif}$, from Eq. (5) for uniform flow, $W/n$, in each passage. This is due to Second Law of Thermodynamics entropy increase when mixing streams of different temperatures.

While each one of the passages is experiencing a uniform $h$ given by $Nu = 3.66$, the effective $h$ or $Nu$ calculated from the everage exit temperature $T_{2avg}$

$$h_{eff} = \frac{q}{A\,\Delta T_{\ell m}} = \frac{Wc(T_{2avg} - T_1)}{n\pi DL(T_{2avg} - T_1)}\ell n\frac{T_w - T_1}{T_w - T_{2avg}} \tag{11}$$

is less than this magnitude, $Nu_{eff} < 3.66$ for tubes or 7.60 for parallel plates. This difference increases as Re decreases, Fig. III.5.3.

As Reynolds number or flow rate W decreases the skew in the flow rate distribution becomes worse and at some lower flow rate the flow in the upper passages can reverse while the throughput flow W is still positive. In this situation some of the heated flow in the exit header flows back to the inlet header mixing with the cold inlet fluid, resulting in greatly reduced performance of the exchanger.

The effect illustrated here for cooling in Fig. III.5.2 also occurs in heating. In that case $w_t > w_b$, but effect on performance is the same.

Obviously the reduced exchanger performance due to the effect discussed here can be eliminated by changing the exchanger orientation to vertical with the colder header at the bottom to eliminate this superimposed natural

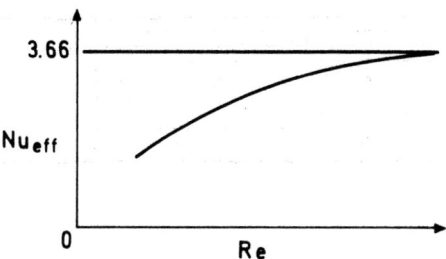

FIG. III.5.3  Effective Nusselt Number For Overall Exchanger Performance Due to Hydrostatic Head Effects in Headers

convection.

III.5.3  VISCOSITY INDUCED NON-UNIFORM FLOW

A typical viscosity-temperature curve for liquids and gases is shown in Fig. III.5.4.

When cooling a liquid (or heating a gas) in parallel passages, this viscosity -temperature relation can result in two different flow rates for the same passage pressure drop. This problem was also discussed by Mueller [1].

Consider the case of an oil being cooled in a parallel plate (or tube) exchanger with uniform wall temperature, Fig. III.5.5. Here the exchanger is vertical with the cold header down to eliminate the superimposed natural convection effect discussed in Sect. III.5.2.

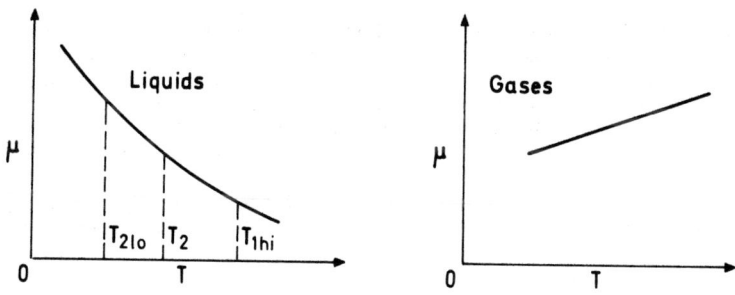

FIG. III.5.4  Typical Viscosity Curves

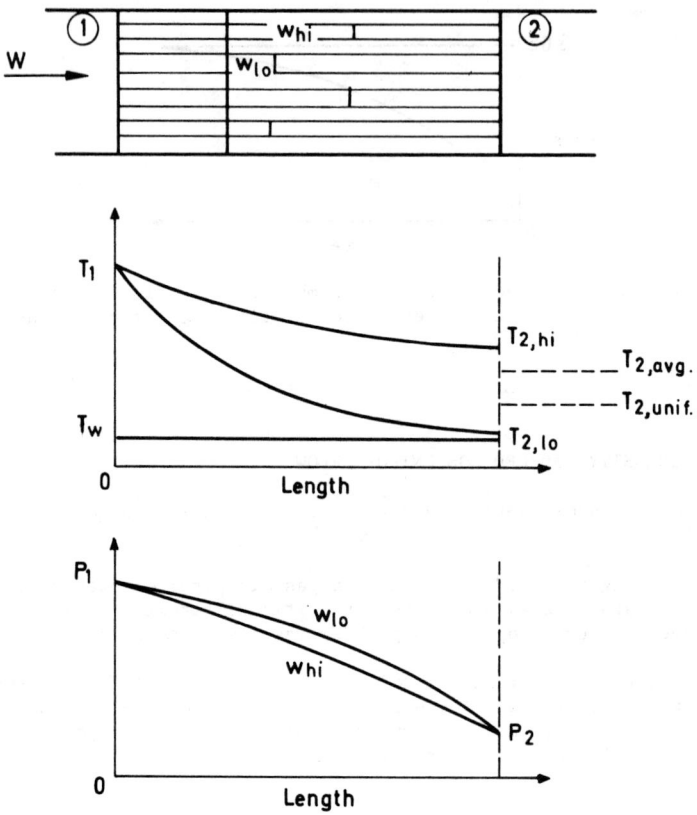

FIG. III.5.5  Non-Uniform Flow, Temperature and Pressure Due to Viscosity-Temperature Effects

Postulate for the moment that two different flow rates, $w_{hi} > w_{lo}$, can exist for the same $\Delta P$ between headers. Then the temperature distribution, Eq. (5), shown in Fig. III.5.5 could exist. From Fig. III.5.4 the average viscosity for the $w_{hi}$ stream is less than for the $w_{lo}$ stream. For many cases the product of $\mu w$ could be the same for the two streams. From Eq. (4) $\Delta P \sim \mu w$; therefore both $w_{hi}$ and $w_{lo}$ can exist for the same $\Delta P$ between headers.

This is a kind of metastable situation. If there are $n_{hi}$ and $n_{lo}$ streams the total flow rate is $W = n_{hi} w_{hi} + n_{\ell} w_{\ell o}$ and the average mixed exit temperature is:

$$T_{2\,avg} = \frac{(n w T_2)_{hi} + (n w T_2)_{\ell o}}{W} \qquad (12)$$

# WHY LAMINAR FLOW HEAT EXCHANGERS CAN PERFORM POORLY

Here $T_{2avg}$ is greater than $T_{2unif}$ which would result if uniform flow, $W/n$, existed in each tube. Each tube is operating with $Nu = 3.66$ but the $h_{eff}$ calculated with $T_{2avg}$ in Eq. (11) gives $Nu_{eff} < 3.66$. This difference is greater as Re decreases resulting in an overall performance curve as shown in Fig. III.5.5.

In Eq. (4a), $(dP/dx) \sim \mu$. As $\mu$ increases down the tube the pressure gradient increases resulting in pressure distributions in the high and low flow streams as shown in Fig. III.5.5. If small tubes interconnecting the through flow passages were placed at many positions down the length to permit cross-flows the two pressure curves would approach each other and the flows would become more uniform, and $Nu_{eff}$ would approach 3.66. Obviously this is a practical manufacturing impossibility.

This suggests that the only solution to the non-uniform flow problem described here is to have interconnected passages. The simplest way to accomplish this is to put the oil to be cooled on the shell side of a shell-and-tube exchanger with either flow parallel to the tubes or cross-flow with or without baffles. Of course, cleaning the shell side is more difficult than cleaning the inside of tubes.

This flow maldistribution can be reduced by multipassing the tube side flow taking smaller temperature drops in each pass. The effect on performance is reduced but not eliminated.

If oil is heated and we postulate that two different flow rates can exit with temperature distributions as shown in Fig.III.5.6 it is seen from the viscosity-temperature curve that the average $\mu$ for the high flow is greater than for the low flow; therefore $(\mu w)_{hi} > (\mu w)_{\ell o}$, and with equal pressure drops in the two passages (between headers) the flow rates are the same. In this case uniform flow and uniform exit temperatures exist resulting in $Nu_{eff} = 3.66$.

For gases, viscosity increases with temperature, Fig. III.5.4 therefore this non-uniform flow distribution can occur when heating gases but not when cooling gases, the reverse of what happens with liquids. For modest temperature rise this effect is small, but for large temperature rises the effect can be quite large.

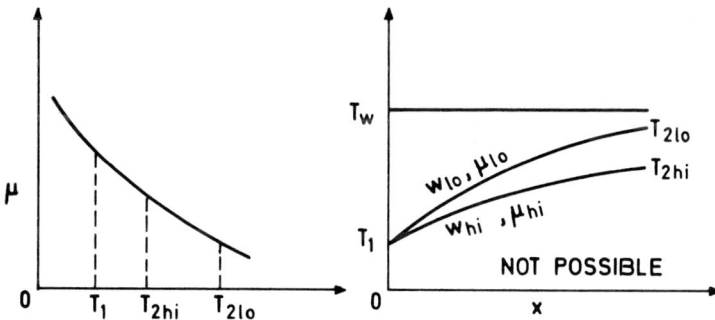

FIG. III.5.6 Impossible Non-Uniform Flows For Heating of Oil

## III.5.4 NON-UNIFORM PASSAGE SIZE

Non-uniform flow obviously can result from having non-uniform size passages as shown by Shah and London [2]. In Eq. (4b) for the same $\Delta P$, $L$ and fluid properties $w_1 \sim D^4$; hence $\Delta w_1 = \Delta D^4$.

To illustrate this effect consider a tubular exchanger with uniform wall temperature with half the tubes of $D_a$ and half of $D_b$. Then from Eq. (4b)

$$w_a = \Gamma D_a^4 \quad ; \quad w_b = \Gamma D_b^4 \tag{13}$$

$$\Gamma \equiv \frac{\pi g_o \rho \Delta P}{128 \mu L} \tag{14}$$

and Eq. (5)

$$\frac{T_w - T_{2a}}{T_w - T_1} = \exp\left(-\frac{M}{D_a^3}\right) \tag{15}$$

$$M \equiv \frac{h \pi L}{\Gamma c} \tag{16}$$

with a similar equation for $T_{2b}$ vs. $D_b$.

The average exit temperature is:

$$T_w - T_{2avg} = \frac{w_a}{w_a + w_b}(T_w - T_{2a}) + \frac{w_b}{w_a + w_b}(T_2 - T_{2b}) \tag{17}$$

Define a hydraulic diameter as:

$$D_h = \frac{4(\text{area})}{\text{perimeter}} = \frac{D_a^2 + D_b^2}{D_a + D_b} \approx \frac{D_a + D_b}{2} \tag{18}$$

With Eqs. (13), (14), (16) and (18)

$$\frac{M}{D_a^3} = \frac{M}{D_h^3} \frac{D_h^3}{D_a^3} = 4(3.66) \frac{L/D_h}{Re_{D_h} Pr} \frac{(1+R^4)(1+R^2)}{(1+R)^2} \tag{19}$$

$$\frac{M}{D_b^3} = 4(3.66) \frac{L/D_h}{Re_{D_h} Pr} \frac{(1+R^4)(1+R^2)}{R^4(1+R)^2} \tag{20}$$

where $3.66 = Nu_{Da} = Nu_{Db}$ and $R \equiv D_b/D_a$

From Eq. (11)

$$h_{eff} = \frac{W c}{n \pi D_h L} \ln \frac{T_w - T_1}{T_w - T_{2avg}} \tag{21}$$

Substitute Eq. (13) in (17) then in Eq. (21)

$$h_{eff} = - \frac{W\,c}{n\,\pi\,D_h\,L} \ln\left[\frac{1}{1+R^4} \exp\left(-\frac{M}{D_a^3}\right) + \frac{R^4}{1+R^4} \exp\left(-\frac{M}{D_b^3}\right)\right] \quad (22)$$

For $n$ uniform diameter tubes of the same total flow area and $W$, $D_u = [(D_a^2 + D_b^2)/2]^{1/2}$. Then the ratio of heat transfer areas is $(A_u/A) = \sqrt{2}\,(D_a^2 + D_b^2)^{1/2}/(D_a + D_b)$ which is only 1.008 for $D_b/D_a = 1.3$; so we neglect this effect.

With $D_a = D_b = D_u$ in Eq. (22), it becomes

$$h_{unif} = \frac{W\,c}{n\,\pi\,D_u\,L}\,\frac{M}{D_u^3} \quad (23)$$

also

$$\frac{L}{D_h}\,Re_{D_h} = \frac{L}{D_u}\,Re_{D_u}$$

and

$$\frac{4\,(3.66)\,L/D}{Re_D\,Pr} = \frac{A\,h}{W\,c} \equiv NTU \quad (24)$$

where $h = 3.66\,k\,D_u$

Divide Eq. (22) by Eq. (23) and substitute Eqs. (19), (20), (18) and (24),

$$\frac{h_{eff}}{h_{unif}} = -\frac{(1+R)}{(1+R^2)^{1/2}}\,\frac{1}{NTU}\,\ln\left[\frac{1}{1+R^4}\exp\left(-NTU\,\frac{(1+R^4)(1+R^2)}{(1+R)^2}\right) + \frac{R^4}{1+R^4}\exp\left(-NTU\,\frac{(1+R^4)(1+R^2)}{R^4(1+R)^2}\right)\right] \quad (25)$$

This equation is shown plotted in Fig. 3.4.7.

Each tube performs individually with $Nu = 3.66$, but because of the mixing of the non-uniform exit temperature streams the $h_{eff}$ calculated from this average exit temperature is less than this.

Similar calculations can be made for other distributions of tube diameters. The results are similar but, of course, numerically different. Shah and London [2] present results for various distributions of rectangular passage sizes for uniform heat flux and uniform wall temperature. Also London [3] and Mondt [4] show the effect of non-uniform shaped passages.

III.5.5. AXIAL CONDUCTION IN TUBE WALL

When an axial temperature gradient exists in the tube wall heat flows along the tube wall and results in a decrease in the total heat transfer in the

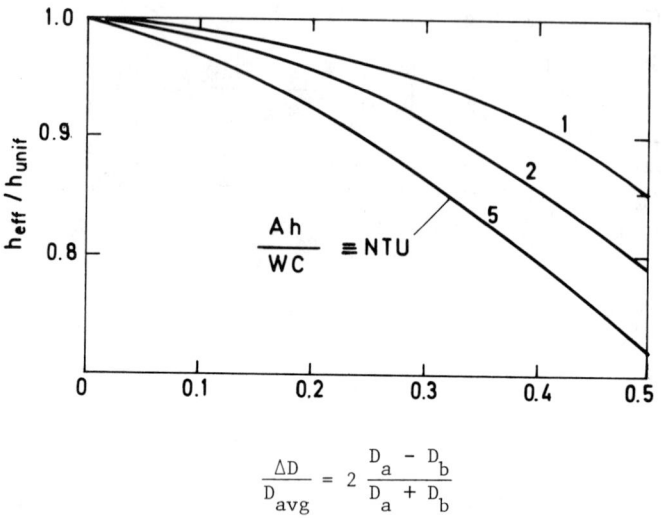

$$\frac{\Delta D}{D_{avg}} = 2\frac{D_a - D_b}{D_a + D_b}$$

FIG. III.5.7  Reduction In Performance Due to Non-Uniform Diameter Tubes

exchanger compared with the heat transferred if the axial metal conductivity were zero. Fig.III.5.8 shows the heat flow paths. The dotted curve represents the wall temperature distribution with zero axial conductivity and the solid curve is the wall temperature with finite wall axial conductivity. Shown also are heat flow paths from hot fluid to wall, then along the wall, and on to the cold fluid. An energy balance on a wall element, Fig. III.5.8 is:

$$h_h(T_h - T_w)P - h_c(T_w - T_c)P + k_w A_w \frac{d^2 T_w}{dx^2} = 0$$

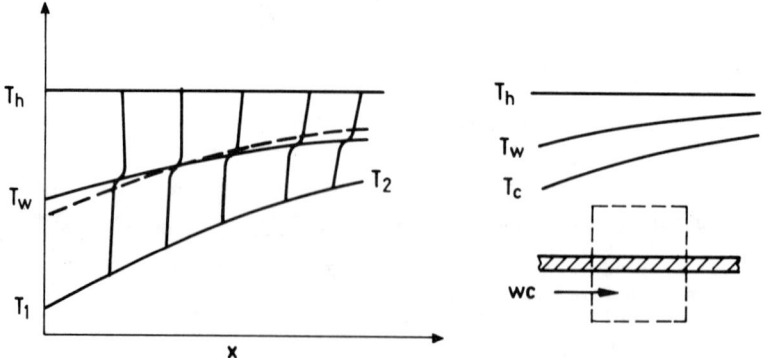

FIG. III.5.8  Effect of Axial Wall Conductivity of Heat Flow Paths

and for the cold fluid

$$w c \frac{dT_c}{dx} = h_c (T_w - T_c) P$$

where P is the perimeter and $A_w$ the wall cross-sectional area.

These equations can be non-dimensionalized with:

$$z \equiv x/L$$

$$Mo \equiv \frac{k_w A_w}{L w c}$$

$$NTU_c \equiv \frac{P L h_c}{w c} \qquad (26)$$

$$NTU_h \equiv \frac{P L h_h}{w c}$$

where M is the Mondt number [5].

The above energy equations then become

$$\left. \begin{array}{l} NTU_h (T_h - T_w) - NTU_c (T_w - T_c) + Mo \dfrac{d^2 T_w}{dz^2} = 0 \\[6pt] \dfrac{dT_c}{dz} = NTU_c (T_w - T_c) \end{array} \right\} \qquad (27)$$

These equations may be solved simultaneously on a computer for $T_w$ vs. z and $T_c$ vs. z for given magnitudes of $NTU_c$, $NTU_h$ and M, taking $dT_w/dz = 0$ at $z = 0$ and $z = 1$, insulated ends. The effectiveness $\varepsilon \equiv (T_2 - T_1)/(T_h - T_1)$ may be compared with the case where $M = 0$, $\varepsilon_o$. Then $(\varepsilon_o - \varepsilon)/\varepsilon_o$ is a function of Mo, $NTU_c$, and $NTU_h$.

The extreme cases, when $Mo = \infty$ are readily solved. Here the wall temperature would be uniform. Consider the case of $Mo = \infty$, $h_c = h_h$ or $NTU_c = NTU_h$ with $T_h$ uniform, Fig. III.5.9.

For $Mo = 0$, from Eq. (5) where $U = h/2$, since $h_c = h_h$,

$$\varepsilon_o = \frac{T_2 - T_1}{T_h - T_1} = 1 - \exp\left(-\frac{NTU}{2}\right) \qquad (28)$$

For $Mo = \infty$, where $T_w$ is uniform,

$$\frac{T_2 - T_1}{T_w - T_1} = 1 - \exp(-NTU) \qquad (29)$$

also

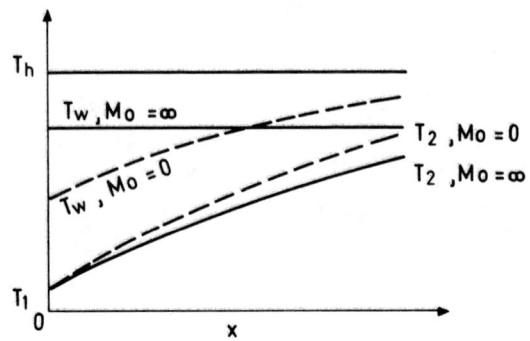

FIG. III.5.9  Temperature Distributions With $Mo = 0$ and $Mo = \infty$, $w_c c_c / w_h c_h = 0$

$$q = w c (T_2 - T_1) = h P L (T_h - T_w) \qquad (30)$$

Combine Eqs. (29) and (30) to solve for $T_w$ and evaluate $\varepsilon_\infty$,

$$\varepsilon_\infty = \frac{T_2 - T_1}{T_h - T_1} = \frac{1}{\frac{1}{NTU} + \frac{1}{1 - \exp(-NTU)}} \qquad (31)$$

Then from Eqs. (28) and (31)

| $NTU_c$ | $\varepsilon_o$ | $\varepsilon_\infty$ | $\Delta\varepsilon/\varepsilon_o$ |
|---|---|---|---|
| 0 | 0 | 0 | 0 |
| 1 | 0.390 | 0.387 | 0.00769 |
| 2 | 0.632 | 0.6036 | 0.0448 |
| 5 | 0.9179 | 0.8287 | 0.097 |
| 10 | 0.9933 | 0.9091 | 0.0848 |
| $\infty$ | 1.0000 | 1.0000 | 0 |

As $NTU_c$ increases the decrease in performance goes through a maximum.

The case of $w_c c_c = w_h c_h$ in counterflow has a greater deterioration of performance due to axial wall conduction. The temperature distributions for $Mo = 0$ and $\infty$ are shown for $h_c = h_h$ in Fig. III.5.10. Obviously the uniform wall temperature prevents the effectiveness, $\varepsilon$, from becoming greater than 0.5.

For $Mo = 0$ with $h_c = h_h$, $U = h/2$ and $NTU_u = NTU_h/2$

$$\varepsilon_o = \frac{T_{c_2} - T_{c_1}}{T_{h_1} - T_{c_1}} = \frac{NTU/2}{1 + NTU/2} \qquad (32)$$

For $Mo = \infty$ and $h_c = h_h$, $(T_w - T_{c_1}) = (T_{h_1} - T_w)$ and

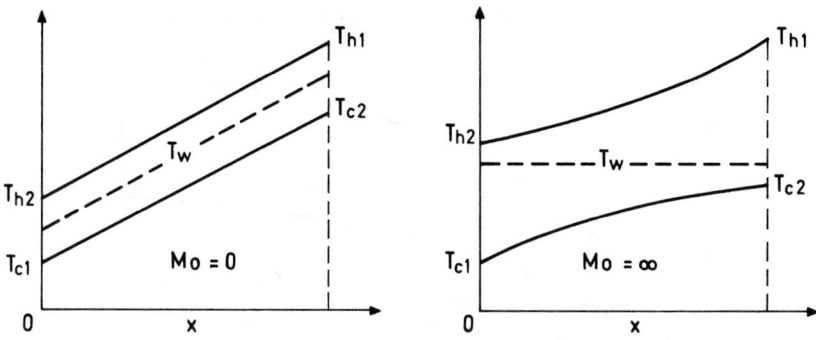

FIG. III.5.10 Temperature Distributions For $w_c c_c = w_h w_c$ Counterflow With $h_c = h_h$

and

$$\frac{T_{c_2} - T_{c_1}}{T_w - T_{c_1}} = \frac{T_{h_1} - T_{h_2}}{T_{h_1} - T_w} = 1 - \exp(-NTU) \tag{33}$$

Then

$$\varepsilon_\infty = \frac{T_{c_2} - T_{c_1}}{T_{h_1} - T_{c_1}} = \frac{1}{2}\left(1 - e^{-NTU}\right) \tag{34}$$

Numerical results are as follows:

| NTU | $\varepsilon_o$ | $\varepsilon_\infty$ | $\Delta\varepsilon/\varepsilon_o$ |
|---|---|---|---|
| 0 | 0 | 0 | 0 |
| 1 | 0.3333 | 0.3161 | 0.0517 |
| 2 | 0.5000 | 0.4323 | 0.1354 |
| 5 | 0.7143 | 0.4966 | 0.3048 |
| 10 | 0.8333 | 0.4999 | 0.4001 |
| ∞ | 1.000 | 0.5000 | 0.5000 |

For finite magnitudes of Mo the actual performance lies between $\varepsilon_o$ and $\varepsilon_\infty$.

In cases in which the actual wall temperatures are uniform, obviously there is no effect of wall conduction. Two examples are uniform $T_h$ and $T_c$ and equal $wc$ in parallel flow with $h_c = h_h$.

For cases of finite Mo a family of equations such as Eq. (27) must be solved simultaneously to determine wall and fluid temperatures. When $T_h$ is not uniform, an energy balance equation for the hot fluid is added to Eq. (27). Many cases have been solved — Chiou [6], Landau and Hlinka [7] and Barron and Yeh [8] are examples.

To illustrate the nature of the results, Fig. III.5.11 shows $\Delta\varepsilon/\varepsilon_o$ for a single pass, unmixed crossflow exchanger with $w_c c_c = w_h c_h$.

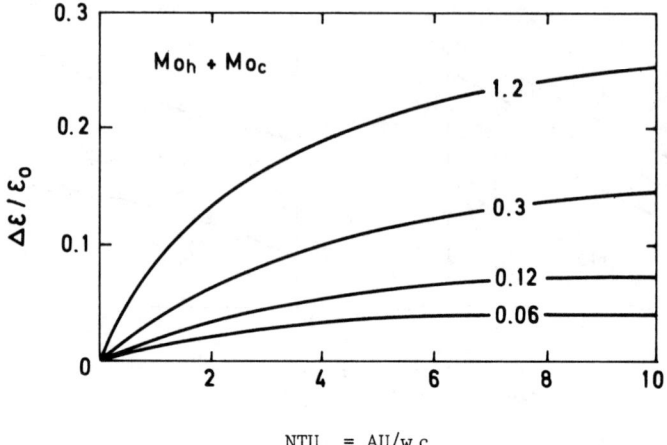

FIG. III.5.11 Single Pass Unmixed Cross Flow, $w_c c_c = w_h c_h$ (Chiou 6)

Here $Mo_h$ and $Mo_c$ are for $wc$ and $A_w/L$ in the direction of flow for each side. In general the major effects on $\Delta\varepsilon/\varepsilon$ are due to $Mo_h + Mo_c$ and $NTU_{min}$. Other quantities $(hA)_{min}/(hA)_{max}$ and $Mo_h/Mo_c$ also affect $\Delta\varepsilon/\varepsilon$, but only in a minor way.

## III.5.6  CONCLUSIONS

The following suggestions for designing laminar flow heat exchangers follow from the results of all of the effects discussed.

1. Operate in a vertical orientation with the cold header down.

2. When cooling liquids, particularly higher viscosity oils, place the oil in interconnecting passages such as the shell side of shell-and tube exchangers.

3. Minimize or eliminate non-uniformity in passage size.

4. Use the minimum wall thickness, $t$, allowable from stress considerations, and if $(t/k_w) \ll (1/h)_{min}$, use lower thermal conductivity wall material if corrosion and cost permit it.

## NOMENCLATURE

| | |
|---|---|
| A | heat transfer area |
| c | specific heat at constant pressure |
| $D_h$ | hydraulic diameter |
| f | friction factor |
| G | mass flow rate |
| $g_o$ | constant, 32.17 lbm ft/lbf sec$^2$, 4.17x10$^8$ lbm ft/lbf hr$^2$ |
| h | heat transfer coefficient |

| | |
|---|---|
| H | height |
| k | thermal conductivity |
| L | length |
| M | defined by Eq. (16) |
| NTU | $Ah/wc$ |
| n | number of passages |
| P | perimeter |
| P,p | pressure |
| R | $D_b/D_a$ |
| T | temperature |
| W | total flow rate |
| w | flow rate per passage |
| x | distance along passage |
| z | $x/L$ |
| $\Gamma$ | defined in Eq. (7) |
| $\epsilon$ | exchanger heat transfer effectiveness |
| $\mu$ | viscosity |
| $\rho$ | mass density |

## REFERENCES

1. Mueller, A.C. 1956. Modern developments in heat transfer. Lectures at two-week summer course, M.I.T. (June).

2. Shah, R.K. and London A.L. 1979. Effects of non-uniform passages on compact heat exchanger performance. ASME paper 79 Wa/GT-9 (Dec.)

3. London, A.L. 1970. Laminar flow regenerators - Influence of manufacturing tolerences., ASME J. of Power, 92A, pp. 46-56.

4. Mondt, J.R. 1976. Effect of non-uniform passages on deepfold heat exchangers. ASME paper 76-Wa/HT-32 (Dec.).

5. Mondt, J.R.. 1961. Effect of longitudial wall conduction on apparent convection behavior of plate-fin surfaces. Int. J. Ht. Trans. Conf., Part III, 73, pp. 607-613.

6. Chiou, J.P. 1976. Performance deterioration in crossflow heat exchangers due to longitudinal conduction in the wall. ASME paper 76-Wa/HT-8.

7. Landau, H.G. and Hlinker, J.W. 1960. Steady state temperature distribtion in counter flow exchanger with longitudinal wall conduction. ASME paper 60-Wa/HT- 236 (Dec.)

8. Barron, R.F. and Yeh, S.L. 1976. Longitudinal conduction in a three-fluid heat exchanger. ASME paper 76-Wa/HT-9 (Dec.)

# Operational Problems Encountered in Heat Exchangers Used for Geothermal Energy Utilization

**ORHAN YEŞIN and ZEKAI KUTAN**
Middle East Technical University
Ankara, Turkey

ABSTRACT

The heat exchangers which are used for geothermal energy utilization have some peculiar operational problems and current efforts in geothermal applications have been concentrated to minimize or eliminate these problems. This paper briefly surveys the types of heat exchangers used in geothermal energy applications, basic problems encountered during operation and the methods employed to minimize them. The opportunity is also taken to report the current geotermal heat exchanger tests carried out in Turkey.

1. INTRODUCTION

Geothermal energy is an alternative source of energy which is utilized mainly for electricity generation, space heating, for industrial and agricultural applications.

Type of utilization is fundamentally based on the thermodynamical state of geothermal fluid:

a) Superheated or dry steam is preferably used in steam turbines for the production of electricity. After a simple filtration, steam is fed directly into turbines located at the geothermal field site.

b) Pressurized hot water of relatively high enthalpy and with temperature above 100 °C is also prefered for electricity production. Part of pressurized water flashes into steam while the rest remains as liquid just below the boiling point. Wet mixture is sent into cyclone separators at the well head. The steam obtained from this type of geothermal fluid with temperatures down to 130 °C is considered feasible to be used for power generation. However, the efficiencies for converting geothermal resources occuring at temperatures below 250 °C to electricity are less than those of fossil-fuel-fired or nuclear power plants [1]. The separated geothermal hot water may further be utilized for space heating purposes or it may be pumped into specially designed heat exchangers to heat up some liquid hydrocarbons such as isobutane or freon which boil below 100 °C. Hydrocarbon vapour then drives another series of turbines to generate more electricity.

c) Hot water with relatively low enthalpy and with temperatures below 100 °C is not suitable for power production by conventional methods. This type of fluid is rather utilized for residential heating or for industrial and agricultural applications.

Because of the high temperatures and the presence of some impurities, geothermal hot water is seldom used as it is extracted from the borehole. Heat transfer is usually provided between geothermal fluid and a working fluid in a heat exchanger which is either down-hole type or surface-mounted type.

1.1 Down-Hole Type Heat Exchangers

Most of the down-hole heat exchangers are merely standard black iron pipe of sufficient length. A conventional U-bend pipe (hairpin) is placed in a geothermal borehole and it allows very effective operation with gravitational circulation (thermo-syphon). Fig.1 illustrates a good example for a down-hole heat exchange system which is used for geothermal residential heating purposes in Klamath Falls, Oregon, U.S.A. [2].

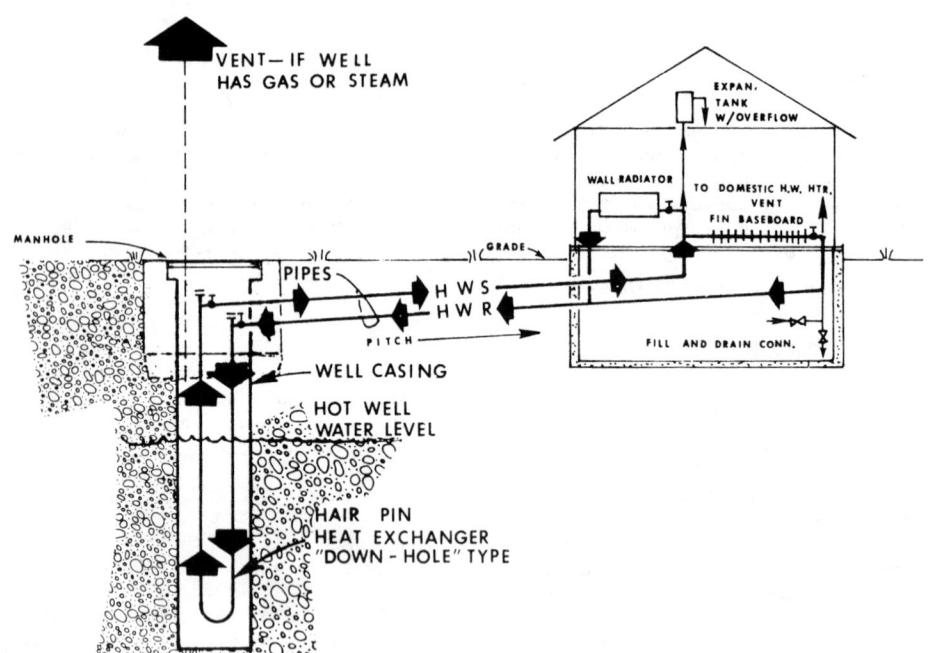

Fig.1. Down-Hole Heat Exchanger (From reference 2)

1.2 Surface-Mounted Heat Exchangers

Surface-mounted heat exchangers are usually located near the borehole but may also be installed at or near the site to be heated or other applications are to be carried out.

Heat exchangers are either of the single-tube and shell type used for domestic water heating and for small central heating systems or the multiple-tube type. They are cheap and easily constructed for small scale applications

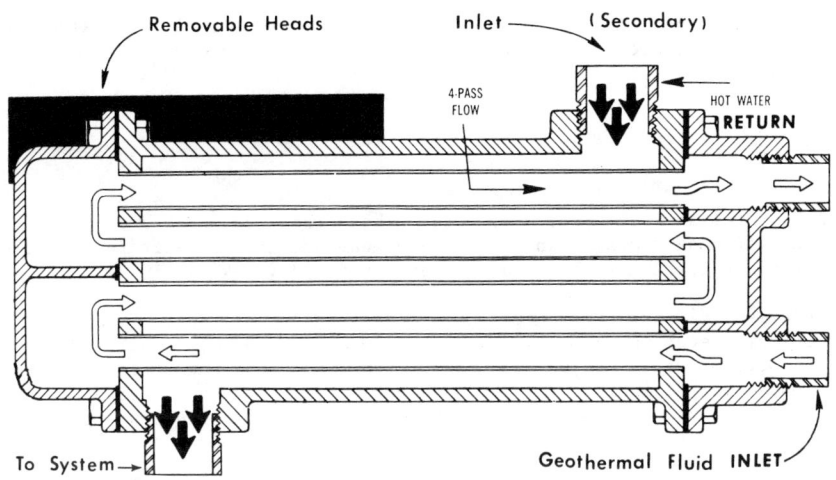

Fig.2. Surface-Mounted Heat Exchanger (From reference 2)

as long as aggressive geothermal fluids are not involved.

A typical surface-mounted heat exchanger is shown in Fig.2 which has straight tubes usually made from stainless steel to control corrosion. The geothermal fluid flows inside the tubes and the secondary fluid (clean water or liquid hydrocarbon) flows through the shell. This arrangement permits easier cleaning and inspection for minimum maintenance since the heads are fully removable [2].

2. HEAT EXCHANGER PERFORMANCE

Heat exchanger tubes may be coated with tantalum or plastics as a protection against corrosion. This causes an additional thermal resistance over the heat transfer surfaces. The "fouling" of the surfaces by deposition of chemicals in the geothermal fluid increases the thermal resistance forthermore.

$$R_s = \frac{1}{A h_s} \quad (1)$$

where $R_s$ is the thermal resistance due to scale formation, $h_s$ is the scale coefficient of heat transfer, $1/h_s$ is the fouling factor and A is the heat transfer area of the surface before scale precipitation.

Fouling has other adverse effects than loss of heat transfer capabilities, especially when it occurs on the coolant side of high-temperature heat exchanger which may cause tube overheat and failure. Furthermore, porous or spongy deposits may act as concentration cells for corrosive agents which, in combination with a higher corrosion rate at elevated temperature may also lead to tube failure. Scale deposits also increase the pressure drop through tubes [3].

To estimate the surface area of the heat exchangers, realistic values for heat transfer coefficients for both fluids, wall resistance and fouling factors should be known.

$$A = \frac{q}{U \Delta T_m} \quad (2)$$

where A is total surface area, q is rate of heat transfer, U is over-all coefficient of heat transfer and $\Delta T_m$ is the mean temperature difference between primary fluid (geothermal fluid) and secondary fluid (working fluid).

Since many different working fluids operating in both subcritical and supercritical states may be under consideration in geothermal applications, no uniquely optimum design can be proposed. Furthermore, the uncertain character of the geothermal fluid itself is such that the type of heat exchanger, flow configuration, tube size and baffling arrangements are difficult to specify in advance [1].

Since the geothermal fluid side of the heat exchanger should be liable to fouling and corrosion, geothermal fluid would be circulated through the tubes to ease cleaning and reduce material costs. For liquid-dominated systems, the tube side coefficient of heat transfer in turbulent flow can be predicted using a Dittus-Boelter type correlation:

$$N_{Nu} = 0.023 \, N_{Re}^{0.8} \, N_{Pr}^{0.33} \quad (3)$$

for $N_{Re} > 10,000$ and $0.7 < N_{Pr} < 700$. For nonsupercritical flow on the shell side, similar correlations can be used for crossflow conditions on the outside of the tube bundle

$$N_{Nu} = 0.330 \, N_{Re}^{0.6} \, N_{Pr}^{0.33} \quad (4)$$

For the same Reynolds number, higher coefficients will appear on the shell side because of the induced turbulence in crossflow. Eqs. 3 and 4 are not suitable for fluids near the critical state [1].

Because of the chemical compositions and temperatures of geothermal fluids vary widely from field to field, it is extremely difficult to estimate the extent of a potential scaling problem for the heat exchanger unless field tests are performed. Calcium Carbonate ($CaCO_3$) and Silica ($SiO_2$) scales are commonly observed in a number of natural geothermal systems [4].

Fouling on the geothermal fluid side may be assumed to be equivalent to a heat transfer coefficient of 2837 W/m$^2$K. If values much lower than this are present, direct contact heat transfer might be an alternative method. Since the working fluid must be relatively free of contamination an equivalent fouling coefficient of 11349 W/m$^2$K may be used. For a heat exchanger with supercritical flow on the shell side and pressurized geothermal fluid in the tubes, an overall coefficient of heat transfer in the range 738 to 908 W/m$^2$K is typical [1].

For subcritical cases, the heat exchanger may be divided into preheat, boiling and superheat sections and coefficients applicable for each section can be calculated. The typical design conditions for a 100 MW(e) power plant utilizing 150 °C - 2000 kg/s geothermal resource and operating on R-32[*] binary-fluid cycle with working fluid flow rate 2140 kg/s indicate that over-all heat transfer coefficient at the preheat section is 823 W/m$^2$K, at the boiling section is 965 W/m$^2$K and at the superheat section is 681 W/m$^2$K. Minimum heat removal temperature is 26.7 °C, working fluid outlet temperature is 135 °C and geothermal fluid outlet temperature is 57.8 °C. The primary heat exchange area is estimated as 58,238 m$^2$ [1].

3. BASIC OPERATIONAL PROBLEMS OF GEOTHERMAL HEAT EXCHANGERS

At present the extensive use of geothermal resources is limited due to some basic operational problems encountered in the heat exchangers of the conventional tube and shell design. These problems are as follows:
 a) The fouling of the heat transfer surfaces by deposition of some chemicals contained in the geothermal fluid.
 b) Presence of non-condensable gases in the geothermal fluid.
 c) Corrosion caused by geothermal fluid as it flows through the tubes.
 d) Environmental pollution caused by the disposal of geothermal fluid.

In most geothermal sources the fluid contains some non-condensable gases such as nitrogen ($N_2$), ammonia ($NH_3$), hydrogen sulfide ($H_2S$) and carbondioxide ($CO_2$). These non-condensable gases can be separated and removed by piping the geothermal fluid into a deaerator in which the fluid is allowed to boil slightly by reduction in the pressure and the gases are thus removed. This procedure is also used to adjust the acidity of water to minimize subsequent corrosion problems. The non-condensable gases may also cause noise in the heat exchanger tubes and block the circulation.

Geothermal fluids are well known for their corrosive character. The materials which can be used in geothermal applications should be corrosion-resistant. The 13% chrome ferritic stainless steels at Rockwell B-60 to C-28 hardness, Monel metal, Inconel, Hastalloy, Zirconium and Ceramics are suitable materials. Coating with tantalum or with some high temperature plastics, such as Teflon also gives satisfactory results against corrosion [5].

In most of the geothermal fields over the world $CO_2$ constitutes a high percentage (about 90-95 %) of all non-condensable gases. The release of $CO_2$ from the geothermal fluid is due to the following chemical reactions

$$2\ HCO_3^- \longrightarrow CO_3^- + CO_2 \uparrow + H_2O$$
$$Ca^{++} + CO_3^- \longrightarrow CaCO_3 \downarrow$$
(5)

in which calcium carbonate ($CaCO_3$) precipitates in the tubes at a rate proportional to the partial pressure variation of $CO_2$ in the fluid. The depositions of calcium carbonate may be in the form of hard calcite or soft aragonite which do not actually depend on the absolute concentration of calcium but is related to carbonate equilibrium. In order to maintain this equilibrium

---

[*] Critical temperature and pressure of R-32 ($CH_2F_2$-difluoromethane) are 78.6 °C and 583 bar respectively.

it is necessary to provide a certain pressure on geothermal fluid so that $CO_2$ remains in dilute solution. The nucleation and growth of scale deposits, particularly silica ($SiO_2$) are controlled by many factors such as supersaturation, foreign ion effects ($Mg^{+2}$, $Al^{+3}$, $Fe^{+2}$, $Fe^{+3}$), hydrodynamic conditions and heat transfer rates.

Disposal of geothermal fluids after being used in the heat exchangers imposes some restrictions and problems. Particularly in space heating applications, geothermal hot water with low calorific value is used. Thus the quantity of fluid requiring disposal is in a great order of magnitude than that in electric power generation.

There are two major disposal methods involved in geothermal systems:
  a) Surface disposal to rivers, lakes, sea etc. with the following limitations from environmental pollution considerations. Normal stream temperature plus a few degrees Celsius is the limit in tightly regulated environmental situation. Generally the temperature of the cooled geothermal fluid must be below 40 °C. Salinity exceeding 300 ppm is undesirable if stream water is to be used for agricultural irrigation. Boron represents a possible source of chemical pollution for the stream. Boron concentration should be kept below 1 ppm.
  b) Reinjection of the geothermal fluid into inactive wells. This method obviously increases the operational cost but eliminates the disadvantages of surface disposal.

## 4. DEVELOPMENT OF ADVANCED GEOTHERMAL HEAT EXCHANGERS

The cost of heat exchangers is a large fraction of the total cost of a geothermal energy conversion system. In a geothermal electricity production plant which operates on binary cycle, the cost of heat exchangers is about one-half of the capital cost of the plant [6].

Extensive research and development programmes for advanced heat exchangers to be used in geothermal energy utilization are carried out by Energy Research and Development Administration of U.S.A. Three main heat exchange methods being currently investigated and developed are as follows [6,7]:

### 4.1 Sand Scouring Method

Sand is injected into geothermal fluid (5-10 % by weight) before it enters the heat exchanger tubes and sand is removed after the heat exchange. Numerous tests have shown that the heat transfer surfaces are maintained in a polished and clean condition when sand is used, otherwise fouling takes place in a few hours. The fouling factor achieved and measured with laboratory simulated fluids is about 1/10 what is expected under most geothermal use conditions. Highly effective centrifugal type separation has been developed for the end of the heat exchange cycle, removing over 99% of the sand for re-use. The amount of sand and flow rates should be adjusted in order to keep the metalic erosion at minimum. The application of the sand scouring method in a geothermal power plant is diagramatically shown in Fig.3.

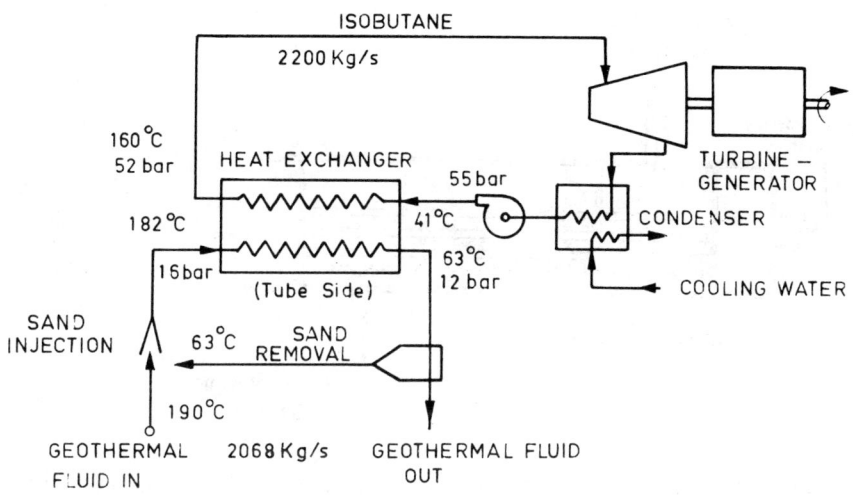

Fig.3. Application of the Sand Scouring Method (from reference 6)

4.2 Fluidized Bed Method

Application of fluidized bed concept to geothermal heat exchangers has two additional advantages besides the enhanced heat transfer of the conventional fluidized bed. The geothermal fluid passes through a bed of fine sand, and the sand acts as nucleation centers for scale deposition. Furthermore, the gentle scouring action on the tubes and fins keep their surfaces clean.

Various tests carried out on fluidized bed method have shown good success. No scale deposition has occured on the tubes within the bed, while those just out of the bed quickly accumulate scale and later develop spot corrosion under the deposited material.

Heat transfer coefficients are excellent, with significant improvement for the total heat transfer per tube surface area compared to conventional liquid-liquid heat exchange across clean tubes in standard heat exchangers. The reason is that the shell side usually has lower heat transfer coefficient compared to the tube side in standard heat exchangers, but introducing fluidized bed on the shell side would enhance the heat transfer rate. Furthermore, the organic fluid is on the tube side in the fluidized bed method, giving better film coefficients for the organic fluid compared to the shell side in conventional heat exchangers.

The cost of fluidized bed heat exchanger is relatively higher than a standard shell and tube type because of larger size requirement and the staged construction necessary. Fig.4 shows a typical geothermal heat exchanger which is composed of a series of isothermal beds with the geothermal fluid directed from stage to stage by a series of baffles.

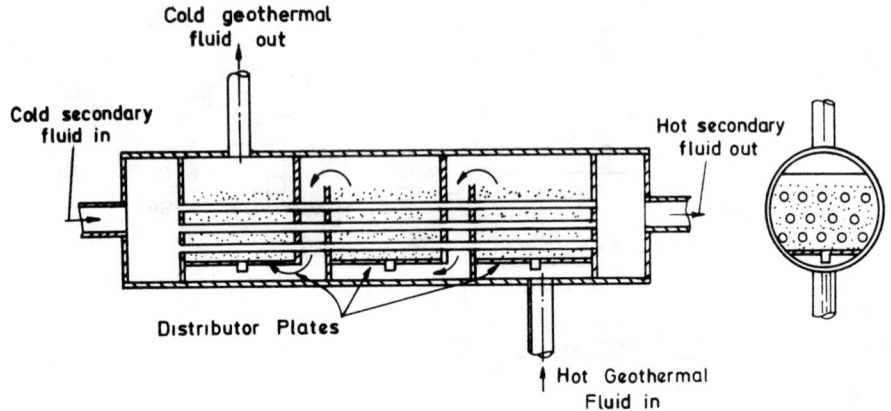

Fig.4. Fluidized Bed Type Geothermal Heat Exchanger (from reference 6)

4.3 Direct Contact Heat Exchange Method

Another possible mode of heat exchange is the direct mixing of the primary and secondary fluids and then the separation of them, if the two fluids are essentially immiscible. The construction cost is reduced since the tubes are not required.

The direct contact is achieved by either spraying the organic fluid onto the geothermal fluid surface or into the volume occupied by geothermal fluid. Separation of the fluids afterwards can be accomplished by applying standard chemical industry separation methods, namely, either using density differences or fractional distillation, even to sub-atmospheric conditions. Direct contact heat exchange method is illustrated in Fig.5 where organic fluid is sprayed onto the surface of geothermal fluid.

## 5. HEAT EXCHANGER TESTS PERFORMED IN THE GEOTHERMAL FIELDS OF TURKEY

In 1962 the Minerals Research and Exploration Institute of Turkey started to collect the geothermal field exploration data. Western region of Turkey was found to have a geothermal potential of commercially exploitable size. The geothermal fluid of Kızıldere field is a two-phase mixture of 15% quality and 200 °C maximum temperature and is suitable for electricity production. Afyon field is more promising for district heating since the local geothermal fluid is hot water at 95 °C temperature.

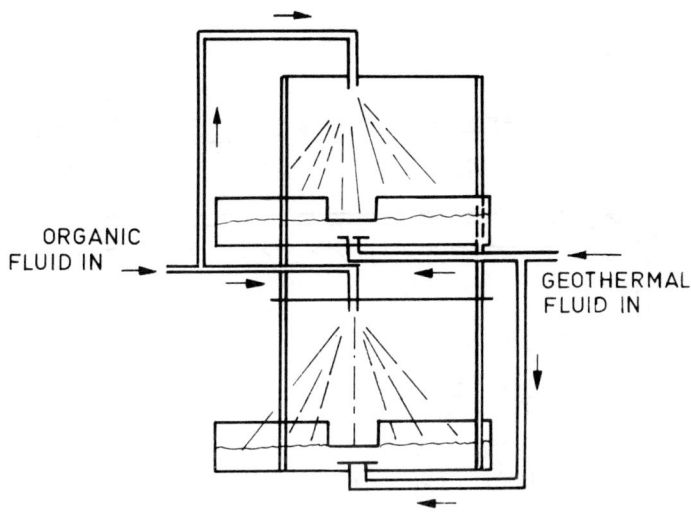

Fig.5. Direct Contact Heat Exchange Method (from reference 6)

A series of heat exchanger tests were performed in the geothermal fields of Turkey to determine if there were any scale deposition on the tube walls and if so the rate and composition of the deposits [8,9]. The heat exchanger used in the tests is a counter-flow type and is made from two concentric and straight tubes, 30.5 m long each, 2" diameter galvanized inner tube for the hot geothermal water and 4" diameter outer tube for the cold geothermal water. The hotter fluid was supplied from a borehole directly, whereas the secondary (colder) fluid was taken from another borehole and passed through a number of pools in series to cool the water and to slowdown the flow in order to encourage the precipitation of silica from solution before being forced through the outer tube by a centrifugal pump. The pressure inside the inner pipe was controlled by two valves placed at the entrance and at the exit of the heat exchanger. Fig.6 is an illustration of a deepwell pump installed at one of the boreholes and coupled to a surface mounted heat exchanger similar to the one tested.

The first series of tests were run at maximum discharge pressure of a borehole which was 20.6 bar (gage). The hot geothermal fluid (single phase) was cooled down to temperatures as low as 60 $^{\circ}$C from 195 $^{\circ}$C. At the same time cold water was heated from 30 $^{\circ}$C to 50 $^{\circ}$C. At the end of first seven months no significant scale deposition was observed due to cooling of geothermal fluid inside the inner tube. Total thickness of the deposit was of the order of 0.1 mm which was identified as zinc carbonate, arsenic sulphide and ferrous sulphide.

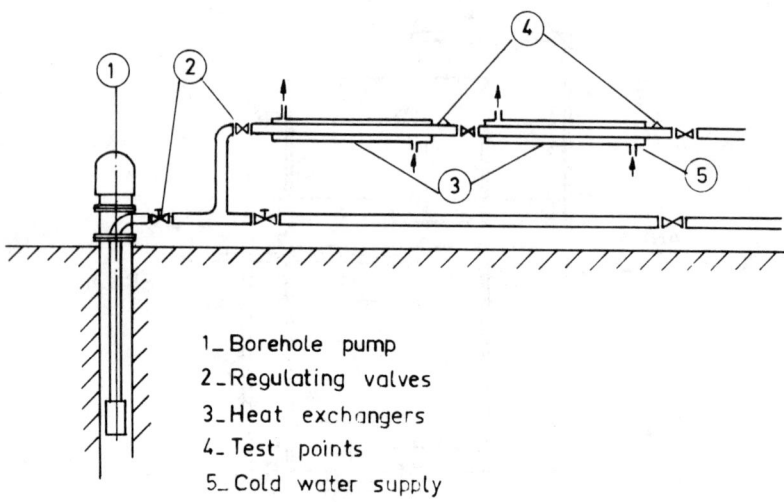

Fig.6. Heat Exchanger Test in Geothermal Field

A second series of tests were performed at approximately the same temperature conditions but the pressure of hot geothermal fluid was reduced from a wellhead pressure of about 15.7-18.6 bar (gage) in steps to 4.9 bar (gage). Under these conditions a two-phase steam-water mixture entered the heat exchanger. Every test at a particular pressure level continued about one week. At the end of each test the heat exchanger was disassembled and examined. Scale deposition in the inner tube wall had a thickness of the order of 0.1 mm and it had a chemical composition similar to that of previous tests.

Calcite deposition was observed between the upstream valve and the upper end of the 4" tube where the cold water left the heat exchanger. This means that calcite deposition did not take place in a 5 cm length when the fluid was subjected to external cooling. As the operation pressure was reduced, the thickness of the calcite deposition increased from about 1 mm at 13.7 bar (gage) between the upstream valve and cold water outlet.

A possible explanation of this fact might be that the cold water in the outer tube lowers the temperature of the laminar sublayer of hot geothermal fluid adjacent to the inner tube wall and hence increases the solubility of calcium carbonate within that layer [10]. It should be noted that calcium carbonate has a retrograde solubility which means that its solubility increases with a fall in its temperature. This increase in solubility successfully inhibits scale deposition on the metal wall adjacent to the laminar sublayer. If this explanation is the correct one, then the reverse would be true so that heating of geothermal fluid is likely to result in decreasing solubility of the water in the laminar sublayer followed by deposition. This has been observed during the tests to take place in the geothermal fluid flowing through the secondary circuit.

The heat exchanger tests performed in geothermal fields have shown that calcium carbonate precipitation inside the transmission pipes could be prevented by cooling the pipes from outside independent of operating pressure.

NOMENCLATURE

| | |
|---|---|
| A | heat transfer area ($m^2$) |
| $h_s$ | scale coefficient of heat transfer ($W/m^2 K$) |
| $N_{Nu}$ | Nusselt number |
| $N_{Pr}$ | Prandtl number |
| $N_{Re}$ | Reynolds number |
| q | rate of heat transfer (W) |
| $R_s$ | thermal resistance due to scale formation (K/W) |
| $\Delta T_m$ | mean temperature difference ($^\circ C, K$) |
| U | over-all coefficient of heat transfer ($W/m^2 K$) |

REFERENCES

1. Milora, S.L. and Tester, J.W. 1976. Geothermal energy as a source of electric power. The MIT Press, U.S.A.

2. Wehlage, E.F. 1976. The basics of applied geothermal engineering. Published by Geothermal Information Services, California, U.S.A.

3. Gardner, K.A. 1974. Anticipation of operating problems in the design of heat transfer equipment. Heat Exchangers: Design and Theory Source Book (Chp. 2). Mc-Graw-Hill, U.S.A.

4. Kruger, P. and Otte, C. Edit. 1973. Geothermal Energy: resources, production, stimulation. Stanford University Press. California, U.S.A.

5. Berman, E.R. 1975. Geothermal Energy. Noyes Data Corporation, New Jersey, U.S.A.

6. Kunze, J.F. 1976. Advanced heat exchanger developments for geothermal application. Idaho National Engineering Laboratory. Report to USERDA.

7. Blake, F.G. 1977. General survey of geothermal energy. Proceed. of CENTO Symposium on Geothermal Energy. Ankara. pp. 17-23.

8. Tan, E. 1972. Heat exchanger tests on Kızıldere geothermal field. Technical report published by Mineral Research and Exploration Institute of Turkey.

9. Kutan, Z. 1975. Heat exchanger test report of Afyon geothermal field. Technical report published by Mineral Research and Exploration Institute of Turkey.

10. James, R. 1972. Further tests on the Kızıldere field for power production and a pre-feasibility study. Technical report published by Mineral Research and Exploration Institute of Turkey.

# PROBLEMS AND PROSPECTS FOR THE FUTURE

# Unresolved Problems in Heat Exchanger Design*

**D. BUTTERWORTH**
Heat Transfer and Fluid Flow Service (HTFS)
AERE Harwell, Oxon OX11 ORA, UK

ABSTRACT

Five practically-important but unresolved problems in heat exchanger design are discussed: viz. flow induced tube vibration, fouling, mixture boiling, flow distribution in two-phase flow and detailed turbulence flow modelling. For each of these topics, the current state of knowledge is reviewed and the outstanding problems highlighted. The question is then posed whether economic solutions to these outstanding problems are possible, or whether alternative methods of avoiding them are preferable. The preferred solution is shown to be different in each case.

1. INTRODUCTION

There are an enormous number of uncertainties in heat exchanger design but this paper concentrates on five problems where the uncertainties are particularly acute. An important criterion in choosing these problems is that they are of practical importance.

For each of the problems chosen, an attempt has been made to suggest ways out of the difficulties. One way is to concentrate research efforts on the problem in question and to continue this research until satisfactory solutions are obtained. The opposite approach is to allow for the uncertainties in design by, say, giving large margins for error. Of course, instead of carrying out an exhaustive test programme, a much smaller programme could be conducted in order to reduce the error margins rather than remove them altogether. A third way of solving the problem is to change the design in some fundamental way, such that the problem no longer exists. This paper indicates which of these ways is likely to be most profitable in the near future, for solving the particular problem under discussion.

2. FLOW INDUCED TUBE VIBRATION

2.1 Present Knowledge

Many types of heat exchanger involve bundles of tubes, across which fluid flows. The most important class of these is shell-and-tube exchangers.

---

* This paper was first presented at the Interflow '80 Conference, Harrogate, 6-8 February 1980 and has been published in I Chem E Symposium Series No 60. It is reproduced here in a slightly modified form with the kind permission of the Institution of Chemical Engineers, UK.

The typical sizes of these exchangers have been growing over the years, and the velocities in them increasing. The larger size of the exchanger has tended to involve larger unsupported tube spans which, when combined with the higher fluid velocities, has led to an increased likelihood of flow induced tube vibration. This problem comes to the attention of the designer when the vibration is severe enough to lead to damage.

The causes of vibration can be classified [1] as resonance and fluid-elastic instability. Resonance occurs when some excitation frequency produced by the flow corresponds to the natural frequency of the tube. The excitation frequency, which is flow dependant, may be higher or lower than the tube natural frequency in order to avoid vibration. This means that the fluid velocities may be higher or lower than some critical velocity. Fluidelastic instability occurs because of the extra forces which the fluid flow imparts to the tube as the tube moves through some vibration cycle. If, as the tube moves through this cycle, the forces give more energy to the movement of the tube than is dissipated by damping, the tube will vibrate. The damping can be that due to the fluid and to the support system. There is known to be some critical velocity below which fluidelastic instability will not occur for a given tube bundle and fluid. Beyond the critical velocity, however, the amplitude of vibration increases very rapidly indeed. So much so that exceeding the critical velocity will almost certainly lead to damaging vibration. This is in contrast to resonance induced vibration, where a resonance may occur without this leading to damage. A further difference is that a resonance may disappear if the velocity is increased beyond the critical velocity but instabilities of the fluidelastic type can only get worse with increasing velocity.

Whichever source of vibration is being checked for in design, it is necessary to know the natural frequency of the tube. The simplest, but least accurate, method of doing this is to assume that each span of tube between adjacent supports behaves independently of any other span. Since this is not physically reasonable, modern practice is to treat the tube as a whole, assuming usually clamped supports at the tube plates and pinned supports at the baffles. Simple analytical or graphical solution techniques are possible for uniform (and slightly non-uniform) baffle spacing, as is discussed by Soper [1]. For more complicated arrangements, it is necessary to resort to computer programs which solve the detailed mechanics of tube movement, usually using finite-element techniques. Of course, tubes not only have a fundamental natural frequency but they have higher natural frequencies corresponding to higher modes of vibration. These higher frequencies do present a problem with multi-supported tubes because the frequencies of each mode can be very close together thus making it almost impossible to find an operating region where there is a good separation between the excitation frequency and any tube natural frequency. This problem may be avoided by keeping the excitation frequency below the tube fundamental frequency but this would normally lead to uneconomic design if handling gas on the shell side.

The excitation frequencies for resonance can arise because of vortex shedding or because of turbulence buffeting. The vortex shedding frequency is usually characterised by means of a dimensionless group known as the Strouhal number:

$$S\ell = f_e D_o / u_m \qquad (1)$$

Maps which give the Strouhal number as a function of tube layout have been produced by Fitz-Hugh [2]. Deep inside a tube bundle, vortex shedding is not usually well correlated over the tube length. This means that no strong net forces are present to move the tubes. One phenomenon which may, however, cause the vortices to shed in phase, is that of acoustic resonance, which can occur with gas flows. This occurs when the vortex shedding frequency corresponds to the frequency for an acoustic standing wave in the shell. There is very little information on excitation frequencies due to turbulence buffeting. Some information has, however, been produced by Owen [3], but it is of doubtful general application.

Fluidelastic instability can be predicted by the Connors [4] equation:

$$u_c = Kf_n \left(\frac{m_e \delta}{\rho_o}\right)^{\frac{1}{2}}. \qquad (2)$$

The proportionality constant, K, in this equation depends upon the tube layout and pitch-to-diameter ratio. Detailed information on this is currently proprietary but the trends in the HTFS data are shown in Figure 1.

2.2 Major Uncertainties

The knowledge presented in the previous section has been gained mainly from laboratory experiments on idealised equipment. Data have been obtained, however, on normal heat exchangers which have been operated up to failure. By comparing exchangers which have operated without problem with similar exchangers which have failed, it should be possible to see how the

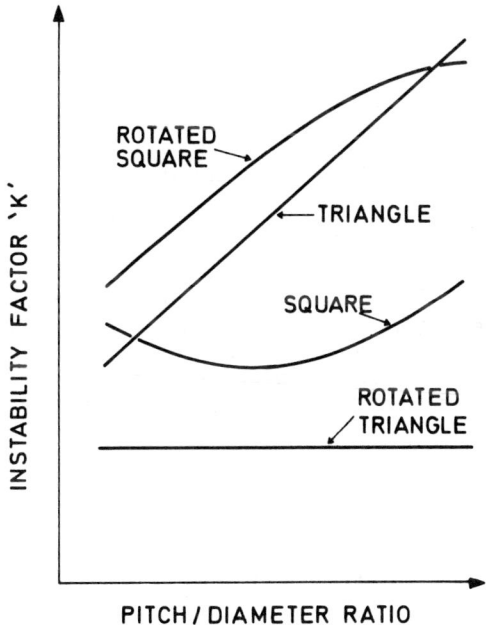

Fig.1. Trend in data for Connors constant

above calculation methods work out in practice. The first difficulty which arises in doing this is that it is never clear which of the various mechanisms for vibration have led to failure. It should be possible to decide which of the mechanisms is causing the problem by applying each criterion in turn and then seeing which of the failure criteria apply to a particular case. The difficulty is that there are usually such large uncertainties in the parameters needed for the calculation methods that any failure mechanism may be confirmed by a suitable choice of these parameters. This problem is complicated by the fact that the various conditions are not constant throughout the heat exchanger. For example, some tubes are differently supported from others, and local velocities vary considerably throughout the unit. Because of these difficulties, there is very little agreement between the different teams of experts as to which vibration mechanisms are of importance in particular circumstances in normal operating heat exchangers. The recommendation for design is therefore usually to assume that any of the mechanisms could occur, and so produce designs which avoid all of them. Hence, shellside velocities must be less than the critical velocity for fluidelastic instability, and it must also be well away from a velocity which could induce resonance.

When applying the various techniques, it is still necessary, of course, to have good values of the various parameters: eg. Strouhal number, Connors K-value, tube natural frequency, tube damping etc. Of these various parameters, that which gives the greatest problem at the moment is the log decrement, $\delta$, in equation (2). This is a measure of the damping on the tube. Field data can be used to suggest the global values of $\delta$ in practical exchangers, but this approach is fraught with dangers for the reasons already given and also because it assumes that $\delta$ is a universal constant. An alternative approach is to measure damping directly, either in practical exchangers or in experimental rigs which simulate tubes with realistic supports.

The log decrement is defined as follows:

$$\frac{a_n}{a_{n+1}} = e^{\delta} \qquad (3)$$

where $a_n$ and $a_{n+1}$ are successive amplitudes as the vibration decays after some initial disturbance. Equation (3) and hence equation (2) assume a linear system, whereas heat exchanger tubes may not behave in a linear manner. In fact, experiments conducted in HTFS on a multi-supported tube, have indicated severe non-linear effects in that the effective value of $\delta$ can vary by a factor of 10 dependant upon the amplitude of the vibration. Such results throw grave doubts on the applicability of the Connors equation but unfortunately do not suggest an alternative method. Indeed, alternative approaches which take full account of non-linearities are expected to be very complicated.

2.3 Avoidance Techniques

Considerable research is being undertaken throughout the world to try to improve our understanding of tube vibration and hence reduce some of the uncertainties noted above. To date, this research has proved reasonably profitable in that good progress is being made for the money being spent on this research. There seems no reason to doubt that future research will also be worthwhile. Pending results of such research, however, designs have

# UNRESOLVED PROBLEMS IN HEAT EXCHANGER DESIGN

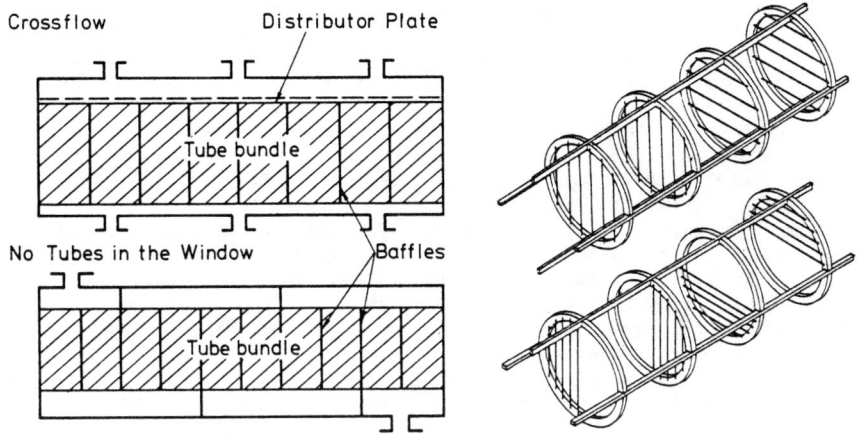

Fig.2. Crossflow and no-tubes-in-the-window designs

Fig.3. Rod baffles

been evolved which very much reduce the likelihood of vibration.  Two such are the crossflow and no-tubes-in-the-window designs, as illustrated in Figure 2.  Both these designs allow for the introduction of additional tube support baffles without causing any other consequential design problems. The number and location of such support baffles can be chosen to satisfy any criteria for avoiding vibration with whatever margin of safety the designer requires.  A somewhat different method of avoiding crossflow induced vibration is to eliminate crossflow itself.  This can be done, at least over most of the exchanger, by changing the form of tube support so that axial flow can occur.  This approach has been used by Phillips Petroleum [5] in developing their so-called "rod-baffle designs" in which the tubes are supported by a grid of rods, as illustrated in Figure 3.

3. FOULING

3.1 Present Knowledge

Support for including fouling in this paper is given by the title of an important review of the subject by Taborek et al [6]: "Fouling: the major unresolved problem in heat transfer".  Further indication of the uncertainties involved in this subject is suggested by the title chosen for a recent conference at the University of Surrey (28th March 1979): "Fouling - Science or Art?".

There are a number of processes which can give rise to fouling: viz. crystallization, particulate deposition, chemical reaction, polymerisation, coking, growth of biological materials and, finally, corrosion on the surface.  Experience has been built up by some organisations in dealing with some of these types of fouling for particular fluid streams.  Crystallization fouling can, for example, be avoided by avoiding supersaturation, and when this is impossible, using additives which change the crystal structure in order to weaken the deposit. Additives may also be used to inhibit corrosion fouling and biocides used to prevent the growth of biological deposits. For some of the fouling mechanicms, notably crystallization and particulate deposition, equations have been devised for predicting the build-up of fouling resistance with time. This approach is reviewed elsewhere in this conference. Such methods are heavily dependant upon empirical data and are limited to particular situations. They certainly do not give general methods for predicting fouling.

At present, the normal way of accounting for the fouling resistance in design is for a value to be chosen which is based upon experience. Such experience has usually been built up by the heat exchanger user on the basis that a previous exchanger designed with such-and-such a fouling resistance worked satisfactorily (or didn't work satisfactorily, as the case may be). It is very rare for overall performance data to have been obtained on an exchanger in order to determine the actual fouling resistances in operation. In some instances fouling resistances have been measured by by-passing part of the flow through special rigs which have an instrumented test section for measuring the fouling resistance. Usually, however, the choice of a suitable fouling resistance to use in a given design is in the nature of an inspired guess.

3.2 Future Research Possibilities

The cost of fouling in the UK has been estimated by Thackery [7] to be in the range of £300M - £500M per year. In view of this, and of the apparent lack of good prediction methods, it would be expected that fouling is the subject in heat transfer on which most research is being conducted at present. This is not, however, the case and the reason for this is that fouling is not a single problem but a whole set of different problems which just happen to produce the same symptoms. Considerable research would be necessary on just one of these problems in order to understand it and it alone. Hence, in defining a research programme, one is faced with a choice of either attacking a very specific problem which is only of narrow interest, or attempting to attack everything at once, which would involve an overall research effort many times greater than the whole of the research effort currently devoted to heat transfer. Despite these difficulties, it would seem sensible to choose some fairly specific problems and attack these in detail. A research programme along these lines has been recently set up jointly by Harwell and the National Engineering Laboratory funded by the Department of Industry via the Chemicals and Minerals Requirements Board. It is hoped that by approaching the problems in a fundamental way, and by using a multi-disciplinary team, a general understanding of fouling will be obtained over and above what is learant about the particular problems chosen.

3.3 Avoidance Techniques

Little is known about avoidance techniques for fouling, mainly because little is known about the causes of fouling. For example, it is unclear why certain particles will stick to a given surface while others will not. Also, there is little idea what causes deposits to break away from a surface. For example, crystalline deposits may break off due to an excessive build-up of thermal stresses or, alternatively, they may break away due to fluid forces.

Despite the lack of understanding of the mechanisms involved in fouling, designers tend to use basic rules of good design in order to try to avoid fouling. Such a rule would be to put the most fouling fluid on the tubeside.

This has two benefits. The first is that there is less danger of low velocity regions or stagnant regions than there would be if the flow were on the shellside where the flow paths are much more complicated. Such low velocity, or stagnant regions, are felt to be those in which the worse fouling occurs. The second reason for putting the most fouling fluid on the tubeside is the rather practical one that it is usually easier to clean inside the tubes than outside them after any fouling layer has built up. When fouling fluids are present on the shellside, in addition to designing in such a way that the shell can be cleaned, it is normal to design with sensible combinations of baffle cut to baffle pitch, which hopefully will avoid regions of low flow or re-circulating flow.

Some types of exchangers are regarded as less prone to fouling than others. One method of avoiding fouling is therefore to choose an exchanger which is less prone to fouling. For example, plate exchangers are supposed to foul less than shell and tube exchangers, assuming that both are well designed [8]. The explanation given for this is that the corrugations on the plates give rise to high turbulence which tends to scour the surface and keep it clean. Similar claims have been made recently about rod baffle exchangers.

4. MIXTURE BOILING

4.1 Present Knowledge

The present status of knowledge on boiling mixtures has been reviewed by Stephan [9], whose paper should be consulted for greater detail than can be given here. It has been known for many years that, when boiling liquid in a pool, the heat transfer coefficients for a binary mixture may be much lower than would be expected from the data for boiling the individual components in that mixture. This is illustrated in Figure 4. The reason for this is quite well understood and is due to the fact that as the bubble at the surface grows, the liquid near the vapour-liquid interface becomes denuded in the more

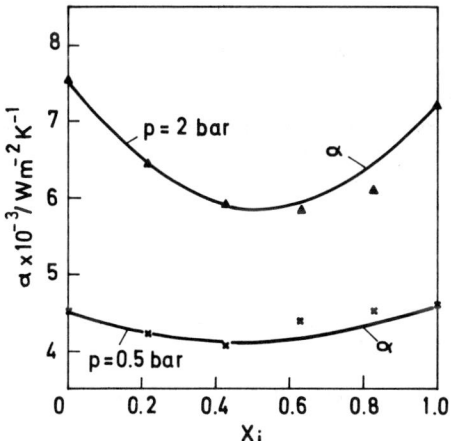

Fig.4. Heat transfer coefficients for boiling binary mixtures

volatile component, thus setting up a concentration gradient between the interface and the bulk of the liquid. The more volatile component then has to diffuse through the less volatile component in order to reach the interface, thus introducing a mass transfer resistance which is not present when boiling single components.

Although these processes are understood qualitatively, they are difficult to quantify, largely because nucleate boiling of single components is difficult to predict without introducing the extra complication of mixtures. The most successful prediction methods currently available involve empirical corrections to data for boiling the pure components. The method of Stephan and Körner [10] appears to be one of the best of these methods. Instead of correlating the heat transfer coefficient, they correlate the wall superheat, $\Delta T$, obtained for a given wall heat flux. This is in effect the reciprocal of the heat transfer coefficient. They then define an ideal wall superheat for the binary mixture as follows:

$$\Delta T_I = x_i \Delta T_i + x_j \Delta T_j \tag{4}$$

where the subscripts i and j refer to the two components in the mixture. $\Delta T_i$ is the temperature difference for boiling pure component i at the specified heat flux. For a binary mixture,

$$x_i + x_j = 1. \tag{5}$$

The actual temperature difference is given by correcting the ideal temperature difference as follows:

$$\Delta T = \Delta T_I (1 + |\theta|) \tag{6}$$

where $\theta$ is given by

$$\theta = A_{ij} (y_i^* - x_i). \tag{7}$$

The parameter $A_{ij}$ is a binary interaction parameter which must be obtained empirically for the mixture and pressure in question. Stephan and Körner were able to remove the pressure dependance effects, as follows:

$$A_{ij} = A_{ij}^o (0.88 + 0.12 \times 10^{-5} p) \tag{8}$$

where $A_{ij}^o$ now depends only on the binary mixture. Stephan and Körner list values of this constant for a limited number of binary pairs.

Stephan and Preuser [11] have extended the Stephan and Körner method to multicomponent mixtures, as follows:

$$\Delta T_I = \sum_{i=1}^{n} x_i \Delta T_i \tag{9}$$

$$\theta = \sum_{j=1}^{n-1} A_{ij} (y_j^* - x_j) \tag{10}$$

where $A_{ij}$ are the binary interaction coefficients. In using these equations, Stephan and Preuser listed the components in ascending order of their boiling point. If this is not done, different answers may be obtained for the same system.

Normally, in heat exchanger design, we are interested in the convective boiling of mixtures rather than in their pool boiling. Chen[12] has shown that for boiling pure components, the boiling-side coefficient is made up of nucleate and convective components as follows:

$$\alpha = \alpha_{nuc} + \alpha_{con} \tag{11}$$

The correction factors produced by Stephan and Körner would be expected to apply only to $\alpha_{nuc}$ and not $\alpha_{con}$. Corrections for the latter have been devised but they appear to be small [13]. An effect not accounted for by equation (11) is that arising from the fact that the temperature of the mixture being boiled rises with passage through the exchanger because of the rise in bubble point of the unvapourised liquid. This means that some of the heat must go into heating up the liquid and the vapour as well as into boiling the liquid, whereas equation (11) only deals with the latter problem. A general equation for handling this has not been quoted in the literature, but it is reasonable to suppose that the method devised for condensation will apply. The methods of Silver [13] and Bell and Ghaly [15] can therefore be applied to boiling as follows:

$$\frac{1}{\alpha_{eff}} = \frac{1}{\alpha_{nuc} + \alpha_{con}} + \frac{Z}{\alpha_g} \tag{12}$$

where $\alpha_{eff}$ is the effective boiling side coefficient based on the difference between the wall temperature and the temperature of the vapour-liquid mixture if brought to equilibrium at that point in the unit. $\alpha_g$ is the gas-phase heat transfer coefficient and Z is given by

$$Z = d\dot{Q}_g/d\dot{Q} \tag{13}$$

which is the ratio (locally) of the heat added for sensible heating of the gas phase to the total heat added to the system.

4.2 Future Research Possibilities

It is now clear that considerable recent progress has been made on the understanding and correlation of pool boiling (see the review by Stephan [9]). The major region on uncertainty is now that of forced convective boiling which is, of course, the situation of major practical interest. Some ideas on the extension of the pool boiling methods to convective boiling are given above and a new correlation by Bennett and Chen [16] has just been produced. However, further work in this area is required to confirm and extend the available methods.

Unlike the previous questions discussed in this paper, mixture boiling is not a problem which can be designed around.

## 5. FLOW DISTRIBUTION IN TWO-PHASE FLOW

### 5.1 Nature of the Problem

It often happens that two-phase flow must enter a number of parallel tubes in a heat exchanger. An example of this is a tubeside condenser with multiple passes where two-phase flow, generated in an earlier pass, discharges into a header and subsequently must enter the tubes in the next pass. The question then arises as to what fractions of liquid and vapour enter each tube in the next pass. At present, it is almost impossible to predict where the liquid and vapour will go when two-phase flow is split, and this is illustrated with two examples given in the next two paragraphs.

The first example concerns the draining of water from a horizontal steam pipe which was known to have some water flowing along the bottom. A pipe section was designed, as shown in Figure 5(A), which contained a drainage line in the bottom of the pipe, together with a weir just beyond the drainage hole, which it was supposed would stop the liquid continuing along the pipe and thus allow it to run down the drain. Surprisingly little liquid did, however, run down the drain and a new method of draining the liquid was therefore required. Before designing a new pipe section with a special separator, it was decided to turn the pipe section around, as shown in Figure 5(B), so that the weir was now upstream of the drain. Much to everyone's amazement, this second arrangement worked perfectly.

The above example was told to the author by the person who did the investigation, but the second example was obtained by the author himself. It is a well known fact that when two-phase flow enters a side tube from the main tube, there is a marked tendency for gas to enter the side tube and for the liquid to continue straight on. In an attempt to redress this imbalance, for a horizontal tube with a smaller tube running off horizontally at right-angles, the author proposed the arrangement shown in Figure 6(A). The

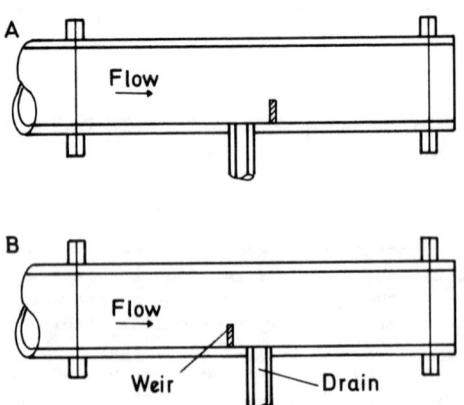

Fig.5. Example of liquid drainage

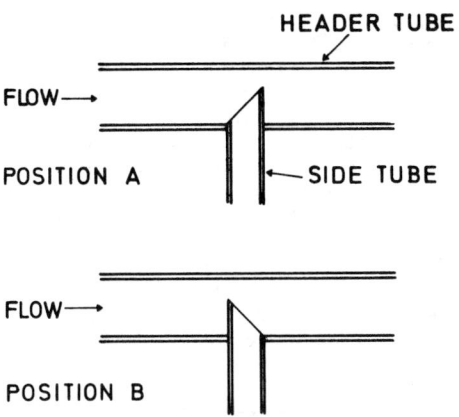

Fig.6. Devices for modifying flow split

results from this arrangement were compared with the case when the T section was the other way around, as shown in Figure 6(B), as well as a case where the side tube was flush with the walls of the main tube. Figure 7 shows the performance of these three arrangements for a given flow of air and water along the main tube. The broken line in this Figure shows an ideal arrangement

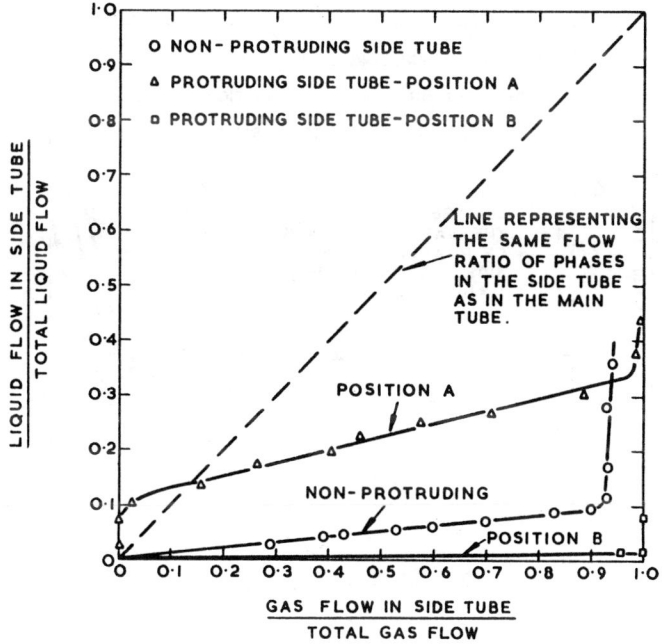

Fig.7. Flow split results for one inlet air-water flow condition

in which the liquid and vapour are split in the same proportion. As can be seen, the side tube in arrangement (A) tended to be much closer to this ideal arrangement than for the non-protruding side tube. Furthermore, the tube in position (B) was much worse in its capacity for equally splitting the flow. So far, the results were according to expectations. However, it was decided to carry out a further test with a different flow of air and water in the main tube and the results of this test are shown in Figure 8. These results were entirely unexpected and show the results for the non-protruding side tube to be closer to the idealised line than when the tube was in either position (A) or (B). Furthermore, these latter two cases give rather similar results with tube (B) sometimes behaving better than tube (A).

These two examples serve to show how difficult it can be to split two-phase flow evenly in simple equipment. Although these two examples are curiosities, they are typical of the sorts of problems which do arise. In fact, after considerable research by a number of organisations for the last 10 years, we are no closer than we were to understanding how two-phase flows split. Also, there is no evidence at present that a continuation of such research would produce such an understanding in the near future, except for one or two simple cases.

5.2 A Possible Solution

Despite the difficulties just mentioned, designers may occasionally have to allow for two-phase flow splitting in their designs. Clearly, a precise solution is out of the question. It may be, though, that simple

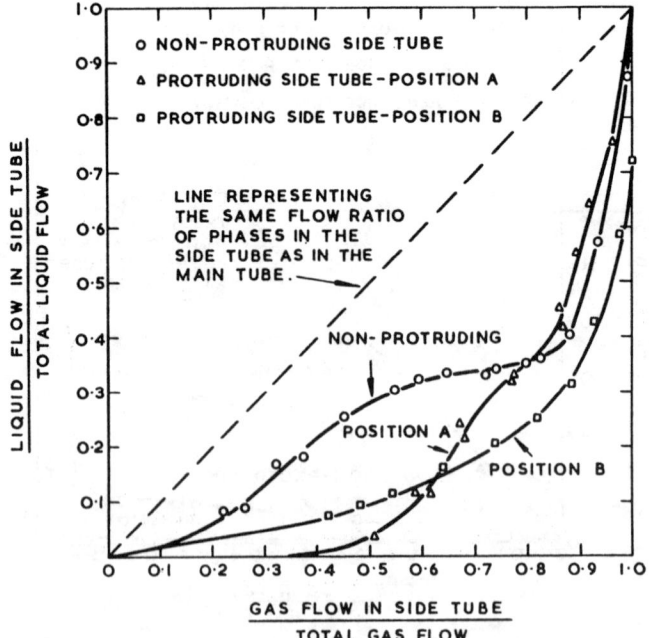

Fig.8. Unexpected flow split results for another set of flow conditions

# UNRESOLVED PROBLEMS IN HEAT EXCHANGER DESIGN

tests on similar equipment may allow the designer to get some qualitative picture about how the two-phase flow will behave. From this, he may be able to design for extreme cases.

Extreme cases can also be assessed analytically. For example, it could be supposed that, for one extreme, the total flows and flow fractions at entry to each tube are identical. For this situation, the heat transfer and pressure drop for flow in tubes may be calculated from standard two-phase flow correlations. The opposite extreme may also be considered, of all liquid entering one set of tubes and all the gas entering the other set. Assuming that no change in quality occurs in the tubes, this problem may be readily analysed to give both the fraction of tubes full of liquid and the pressure drop along the tubes. The results obtained are [17]:

$$\varepsilon_\ell = 1/(1 + 1/x^{1.11}) \qquad (14)$$

$$\phi_g^2 = (1 + x^{1.11})^{0.9}. \qquad (15)$$

The above equations are written in terms of the Lockhart and Martinelli [18] parameters defined by:

$$x^2 = \Delta p_\ell/\Delta p_g \qquad (16)$$

$$\phi_g^2 = \Delta p/\Delta p_g. \qquad (17)$$

Figure 9 shows the predictions of equation (17) compared with those of the Lockhart and Martinelli correlation which would apply for low velocity,

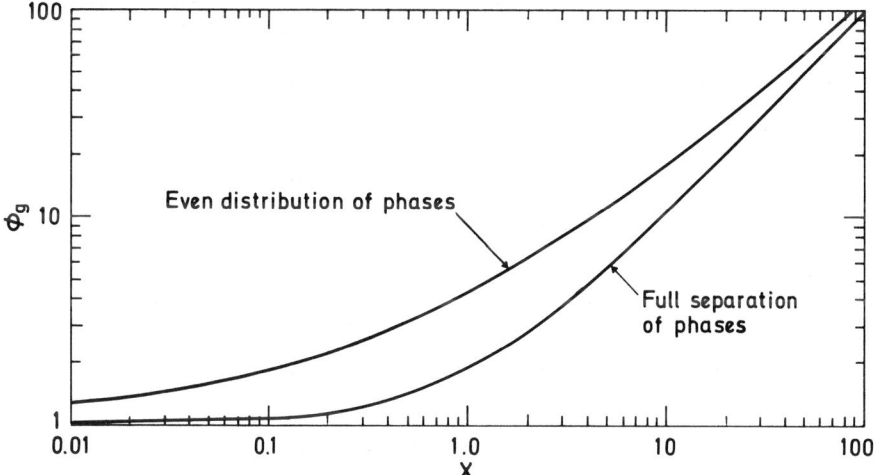

Fig. 9. Theoretical extremes for flow split into large number of parallel tubes

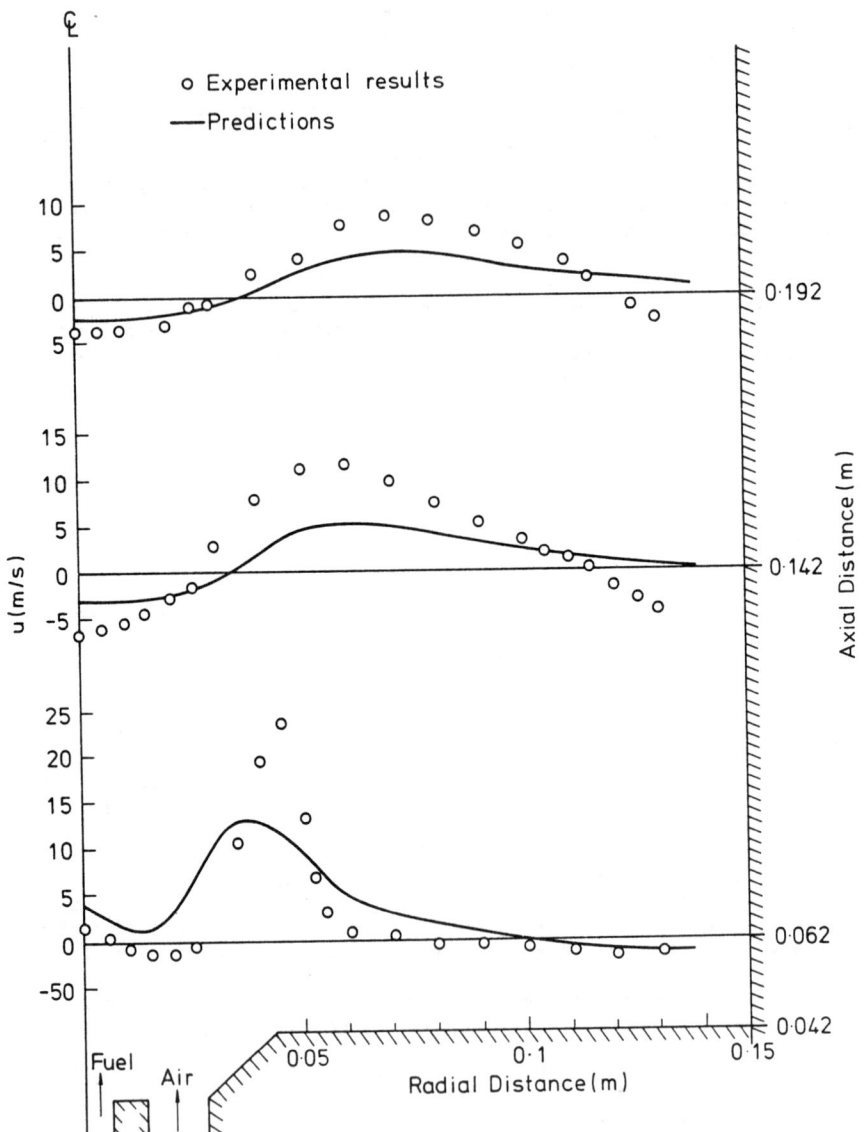

Fig.10. Comparison of furnace performance data with predictions from TUFC

needed on this topic but it is unlikely that this will lead to general prediction methods in the foreseeable future. Mixture boiling is something which has to be faced up to when it occurs; ie. there is not usually any practical way of avoiding it. However, progress is being made on our understanding of the subject and some simple prediction methods already exist. A relatively small investment in research at this stage would be expected to advance the subject considerably. If a two-phase flow stream needs to be split in a heat exchanger, we have no way at present of knowing where the liquid and where the gas will go. This is unlikely to be a problem which can be solved by a reasonable research investment. The designer should therefore attempt to avoid splitting two-phase flow but when this is not possible he should design for the worst case. Detailed turbulence flow modelling is a complex and difficult problem but one which, when solved, can often give extremely valuable detailed information in equipment behaviour. Despite fundamental difficulties in the solution of this sort of problem, rapid strides have been made in predicting practical situations and future research would be expected to progress things even further.

## NOMENCLATURE

$A_{ij}$ = Interaction constant in equation (7)

$A_{ij}^o$ = Interaction constant for boiling binary mixtures

$a_n$ = Amplitude for n'th oscillation (m)

$D_o$ = Tube outside diameter (m)

$f_e$ = Excitation frequency ($s^{-1}$)

$f_n$ = Tube natural frequency ($s^{-1}$)

$K$ = Parameter in equation (2)

$k$ = Turbulence kinetic energy ($m^2/s^2$)

$\dot{Q}$ = Heat load from entry of exchanger up to the point in question (W)

$\dot{Q}_g$ = Heat load for sensible heating of the gas phase from entry of exchanger up to the point in question (W)

$p$ = Pressure (Pa)

$S\ell$ = Strouhal number defined by equation (1)

$T$ = Temperature (K)

$u_c$ = Critical velocity for fluidelastic instability - calculated at minimum flow area (m/s)

$u_m$ = Crossflow velocity at minimum flow area (m/s)

$X$ = Lockhart-Martinelli parameter defined by equation (7)

| | | |
|---|---|---|
| $x_i$ | = | Mole fraction of component i in the liquid phase |
| $y_i^*$ | = | Mole fraction of component i in the gas phase which is in equilibrium with $x_i$ |
| Z | = | Parameter defined by equation (13) |

Greek

| | | |
|---|---|---|
| $\alpha$ | = | Heat transfer coefficient (W/m$^2$K) |
| $\alpha_{nuc}$ | = | Nucleate boiling component of heat transfer coefficient (W/m$^2$K) |
| $\alpha_{con}$ | = | Convective component of heat transfer coefficient (W/m$^2$K) |
| $\alpha_g$ | = | Gas phase heat transfer coefficient (W/m$^2$K) |
| $\alpha_{eff}$ | = | Effective boiling side heat transfer coefficient (W/m$^2$K) |
| $\delta$ | = | Logarithmic decrement defined by equation (3) |
| $\Delta$ | = | Difference operator |
| $\varepsilon$ | = | Dissipation rate (m$^2$/s$^3$) |
| $\varepsilon_\ell$ | = | Fraction of tubes running full of liquid |
| $\theta$ | = | Parameter defined by equation (6) |
| $\nu_{turb}$ | = | Turbulent kinematic viscosity (m$^2$/s) |
| $\rho_o$ | = | Density of shell side fluid (kg/m$^3$) |
| $\phi_g$ | = | Lockhart-Martinelli parameter defined by equation (18) |

REFERENCES

1. Soper, B.M.H. 1979, "Flow induced vibration in shell and tube heat exchangers", Int. Meeting on Industrial Heat Exchange and Heat Recovery, Liege, Belgium, November.

2. Fitz-Hugh, J.S. 1973, "Flow induced vibration in heat exchangers", Paper 427, Int. Symposium on Vibration Problems in Industry, Keswick, England.

3. Owen, P.R. "Buffeting excitation of boiler tube vibration", J.Mech.Eng. Science, Vol 74, No 4, pp 431-439, 1965.

4. Connors, H.J. 1970, "Fluidelastic vibration of tube arrays excited by cross flow", Symposium on Flow Induced Vibration in Heat Exchangers, ASME Winter Annual Meeting.

5. Small, W.M. and Young, R.K, "Exchanger design cuts tube vibration failure", Oil and Gas J, Vol 75, No 37, pp 77-80, 1977.

6. Taborek, J., Aoki, T., Ritter, R.B. and Palen, J.W, "Fouling - the major unresolved problem in heat transfer", Chem. Engng. Progress Vol 68, No 2, pp 59-67, 1972.

7. Thackery, P.A, 1979 "The cost of fouling in heat exchanger plant", Guildford Conference, "Fouling - Science or Art", March.

8. Cooper, A., Suitor, J.W, and Usher, J.D, "Cooling Water Fouling in Plate Heat Exchangers", 6th Int. Heat Transfer Conference, Toronto, August, Vol 4, pp 403-406, 1979.

9. Stephan, K., 1980 "Heat transfer with natural convection boiling in multicomponent mixtures", Advanced Study Institute on Heat Exchangers, Istanbul, Turkey, August 4-15.

10. Stephan, K., and Körner, M., "Calculation of heat transfer of evaporating binary mixtures", Chem.Ing.Tech. Vol 41, No 7, pp 409-417, 1969.

11. Stephan, K., and Preusser, P., "Heat transfer in natural convection boiling of polynary mixtures", 6th Int. Heat Transfer Conf. Toronto, Vol 1, pp 187-192, 1978.

12. Chen, J.C, "Correlation for boiling heat transfer to saturated fluids in convective flow", Ind. Engng. Chem. Process Design Develop, Vol 58, No 3, pp 322-329, 1966.

13. Shock, R.A.W, 1976, "Evaporation of binary mixtures in upward annular flow", Int.J.Multiphase Flow, Vol 2, No 4, pp 411-433.

14. Silver, L., "Gas cooling with aqueous condensation", Trans. Instn. Chem. Engrs. Vol 25, pp 30-42, 1947.

15. Bell, K.J, and Ghaly, M.A, "An approximate generalised method for multicomponent partial condensers", Vol 69, No 131, pp 72-79, 1973.

16. Bennett, D.L, and Chen, J.C, 1980, "Forced convective boiling in vertical tubes for saturated pure components and binary mixtures, AIChE J. Vol 26, No 3, pp 454-461.

17. Butterworth, D., and Grant, I D R, "Two phase flow and boiling in shell and tube heat exchangers", Lecture Notes for Two Phase Flow and Heat Transfer Course, Harwell/Winfrith, UK, 1975.

18. Lockhart, R.W, and Martinelli, R.C, "Proposed correlation of data for isothermal two-phase, two-component flow in pipes", Chem.Engng.Prog. Vol 45, No 1, pp 39-48, 1949.

19. Deardorff, J.W, 1974, "Three dimensional numerical study of turbulence in an entraining mixed layer", Boundary Layer Meteorology, Vol 7, p 199.

20. Jones, W.P, and Launder, B.E, 1972, "The prediction of laminarization with a two-equation model of turbulence", Vol 15, pp 301-314, 1972.

21. Launder, B.E, Morse, A., Spalding, D.B, and Rodi, W., 1973, "The prediction of free shear stress flows - a comparison of the performance of six turbulence models". Proc. Langley Free Shear Flows Conf., NASA SP320.

# Suggestions for Further Research and Development on Heat Exchangers

E. ACHENBACH, K. J. BELL, A. E. BERGLES, D. BUTTERWORTH,
J. G. COLLIER, E. N. GANIĆ, F. MAYINGER, M. D. MIKHAILOV,
M. N. ÖZIŞIK, C. M. B. RUSSELL, R. K. SHAH, K. STEPHAN,
G. YADIGAROGLU, and A. A. ŽUKAUSKAS

E. ACHENBACH: Though the knowledge on heat exchangers has been brought to a high standard, there are some open questions which must be answered for being able to design heat exchangers exactly. The list of problems given in this suggestion is far from being complete, but exhibits those problems to which the author is envisaged during his work on heat exchangers in cross flow:
(1) The main point is to increase the efficiency by increasing the heat transfer and decreasing the pressure drop. This may be a problem of skillful combining the geometrical arrangement of the tubes with appropriate surface roughness.
(2) One of the most serious engineering problems is the onset of vibrations induced by the non-steady seperation of the flow past the tubes. The vibration may be accoustical vibration due to accoustical resonance occurring between the shell walls or vibrations of the tubes themselves, combined with fluid elastic phenomena. The knowledge about this subject is not as advanced as should be desireable.
(3) A further challenge to the engineer comes from the oblique flow past tube banks, particularly, the equations describing the pressure drop are far from representing the experimental evidence.
(4) The occurrence of hot streaks in heat exchangers leads to the problem of thermal stress. Therefore the decay of hot peaks must be predictable. For such computations, the engineer is missing mixing coefficients.
(5) Finally, a problem associated with the preceeding one is the occurrence of by pass flows particularly at the walls, which leads to non-uniform temperature profiles across the bundle. Here the question arises about pressure drop coefficients of the flow in the vicinity of the walls.

K. J. BELL: I fully endorse the selection of unresolved topics given by Butterworth in his section though I think I might differ in the strategy I would employ to resolve them.
    Additionally, however, I would like to suggest the importance of coming to grips with the problem and consequences of uncertainty in the design of heat exchangers and particularly heat exchange systems. While future research work will reduce the uncertainty in the values of the physical properties we use and will improve the range and accuracy of the heat transfer and pressure drop correlations and design methods, at the same time the need to design to closer temperature approaches is making the effects of the remaining uncertainties more serious. Once we develop ways to analyze those effects and identify critical components (in their effect upon the outcome), we can devise system modifications that will compensate for, or at least minimize, the adverse effects of uncertainty.

A. E. BERGLES: <u>Correlations</u> are needed to predict heat transfer and pressure drop, if applicable, for enhanced surfaces and other augmentation techniques as functions of the geometrical characteristics and other parameters. Both single-phase and two-phase heat exchangers are of interest.
<u>Performance Evaluation Criteria</u> should be developed further so that they can be readily applied to give standard evaluations of new data. For example, how does a new surface rate according to various objective functions: increased heat transfer rate, reduced heat exchanger size, reduced pressure drop, or reduced antropy generation? An alternate use of the criteria is to utilize the correlations and select optimum geometrical parameters for a particular application.
<u>Cost Effective Manufacturing</u> is required for those augmentation techniques which have been demonstrated to be effective in laboratory tests.
<u>Fouling and Corrosion Characteristics</u> must be established for augmentation techniques. This should be a feedback process so that "anti-fouling" surfaces are developed.

D. BUTTERWORTH: Before making any suggestions, I must restate a point made strongly in my paper "Unresolved problems in heat exchanger design." This point is that there is more than one route of solving a heat exchanger problem and the designer and researcher should together decide which is the best route to an acceptable solution. These routes can be summarised in the following diagram:

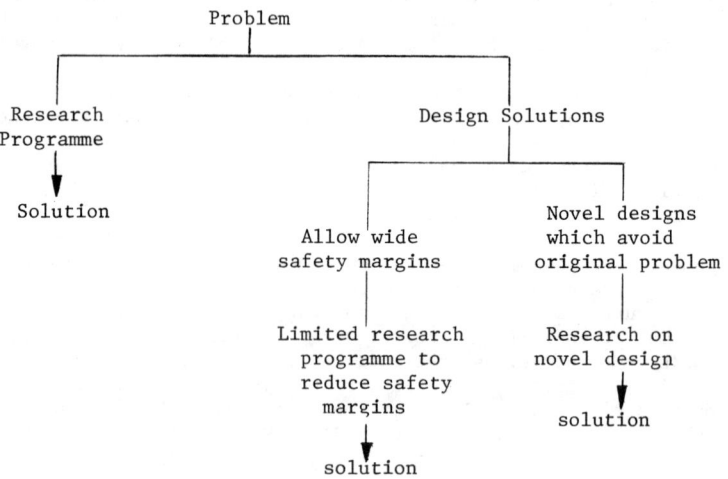

Five practically-important problems were described in the paper and possible solution routes suggested. These were:
(1) Flow induced vibration: Direct research has yielded some valuable information but some serious barriers to progress are now appearing. However novel designs to avoid vibration are now being devised and perhaps more effort could be put in here.
(2) Fouling: This is a huge problem which is difficult to grasp but we should attempt the direct research approach, perhaps using multidisciplinary research teams. Some thought could however be given to designs which give lower fouling.
(3) Mixture boiling: There is no way of avoiding mixture boiling with alternative designs. We must therefore study it by direct research. Dr. Stephan has shown that we now have some understanding of multicomponent

pool boiling but we still need to study convective boiling.
(4) Flow distribution in two phase flow: When you try to split two phase flow in a header or manifold, anything may happen! The process is so unpredictable that research is unlikely to give useful information. We should, therefore, always try to avoid splitting two phase flow when we design heat exchangers. If this is not possible we must allow wide safety margins.
(5) Detailed turbulence flow modelling: Recent rapid progress is being made into research on this topic and further research is likely to be profitable. We now need to apply this information in design.

J. G. COLLIER: As this is an Institute about heat exchangers and I spoke on the topic of boiling and evaporation. I will confine myself to just two or three topics of special relevance in this area.

Basically, for too long researchers in the field of boiling have either concentrated on pool boiling from single tubes or plates using various fluids or alternately have looked at forced convection boiling inside tubes but using only one fluid, water. Unfortunately most industrial process heat exchangers are not concerned with evaporation of water, usually hydrocarbons. People do not like doing research on hydrocarbons because of the safety problems. Furthermore very few fluids in industrial process heat transfer are single component pure fluids. More work must therefore be done to help the process heat exchanger designer. He is concerned with both in-tube evaporation within kettle type re-boilers.

Specifically I would recommend more research be done on:
(1) Forced convection boiling within a tube to measure particularly the CHF for:
   (a) Pure organic liquids mainly hydrocarbons.
   (b) Binary mixtures of organic liquids where one of the fluid is water and also where it is not.
   (c) Multicomponent mixtures.
(2) Similar work be carried out for plate fin type channels as used in LNG liquefacation and evaporation.
(3) Forced convection in cross flow over a single rod of diameters in range 10 - 40 mm using
   (a) Water to look at effect of diameter of CHF,
   (b) Binary mixtures,
   (c) Multicomponent mixtures,
(4) Extend (3) to look at heat transfer rates for pure liquids, binary mixtures + multicomponent mixtures in tube bundles of various sizes.

E. N. GANIĆ: One of the main difficulties encountered in designing falling film heat exchangers is maintaining the complete wettability of the tubes. When the flow rate of a falling film is very low and/or the surface is very hot, the liquid film breaks down and dry patches form on the surface. This results in an abrupt decrease in the heat transfer coefficient. More experimental data on the minimum wetting flow rate (i.e., the minimum liquid flow rate required to wet the tube surface at a given local heat flux) is needed. The experiments should be performed with various fluids to find the effects of surface tension and contact angle on film breakdown and to develop a correlation for design application.

The vapor/liquid interaction and entrainment on the shell side of horizontal tube falling film evaporators should be also studied in more detail. As a result of vapor penetration, a vapor crossflow exists on the shell side. The droplets and liquid columns falling from a given tube may be partially or totally deflected away from the next tube by vapor crossflow, thereby causing liquid redistribution and incomplete wetting of lower tubes in the bank.

The above mentioned problems are especially important in the case of heat exchangers like those used for solar energy application. Such heat exchangers require very large heat transfer areas to compensate for the low $\Delta T$'s, and the pumping power needed to recycle the liquid affects the efficiency and feasibility of these systems.

F. MAYINGER: Heat exchangers are the cheapest apparatus in chemical plants and in power plants. Therefore the users are not often willing to spend large amounts of money for applied research and the manufacturers do not have the budget to do it. Nevertheless there is a large reservoir of detailed knowledge from <u>fundamental</u> research, mainly concerned with heat transfer and pressure drop in simple configurations. Difficulties however still exist to apply this knowledge to technical apparatus. There is a gap between fundamental research and applied techniques. Applied research - not restricted to special users or interest - should be encouraged.

In detail the following research activities may be of interest:
<u>Heat Transfer (single-phase)</u>: <u>High viscous</u> fluids especially also for food industry.
<u>Low Pressure</u> gases (0.1 Ton ÷ 100 Tons). Here, in spite of high flow velocities, Low Reynolds numbers are present. The region between Knudsen flow (molecular flow) and Laminar flow still needs research.
<u>Pressure Drop</u>: At low absolute pressures the pressure drop has to be much more exactly known, than at ambient and high pressures.
<u>Heat Transfer (Two-phase-flow)</u>: Heat and mass transfer in multicomponent fluids with boiling is still of great interest. The same is valid for condensation of gaseous mixtures especially with non-condensable components.
<u>Interface Heat Transfer at Phase Boundaries</u>: In absorption and extraction processes, in chemical reactions as well as in subcooled boiling, in spray condensers or in injection evaporators interface heat transfer between the phase boundaries occurs. Here usually not only the heat transfer coefficient but also the transfer area are unknown. Experiments in this field are very difficult and need very sophisticated measuring techniques. Experimental and theoretical activities should be encouraged in this field.
<u>Transient Conditions</u>: Up to what frequencies or transient time gradients can we treat heat transfer processes quasi-steady? Mainly for safety deliberations in chemical plants and in power stations, heat transfer under transient conditions is of great interest. There is a great gap of knowledge in this field.
<u>Better Understanding of Heat Transfer Mechanism</u>: For a more precise and more reliable layout of technical apparatus, a better understanding of the physical phenomena is important. Research activites for better understanding of the fluid dynamics and thermodynamic conditions should therefore be encouraged. One experimental possibility for a better success may be the application of optical measuring methods.

In steam generators especially for pressurized water reactors corrosion is still a problem. The main influence certainly plays here a good water treatment and a correct selection of the material the tube are made from. But also proper fluid dynamics conditions can reduce corrosion, avoiding an increase of mineral concentration in certain parts. Research should also be done in this field.

M. D. MIKHAILOV: For the mathematical modelling, analytical and numerical solutions to the heat exchanger problems, I can make the following suggestions:
(1) Heat exchangers should be classified into groups for the purpose of developing sufficiently general mathematical models and computer programs to solve the fluid flow and heat transfer problems associated with each class.

(2) The steady-state and transient heat transfer problems of heat exchangers can be transformed, in the most general case, into the solution of a complex eigen value problem, as well as the problem of inversion of Laplace transform. Therefore, efforts should be directed towards the analysis of complex eigen value problems as well as the numerical inversion of the Laplace transform.
(3) General computer programs should be developed for the analysis of heat transfer and fluid dynamics problems of heat exchangers.

M. N. OZISIK: With increased use of heat exchangers for high temperature applications, there is pressing need to account for, in heat transfer analysis, the effects of radiation more rigorously. In most analysis of heat exchangers, radiation effects are introduced very approximately, by treating the radiation part of the problem as a surface phenomena only, but not as a bulk phenomena. In fact, in many applications, the fluid in heat exchangers at high temperatures acts as a participating medium. Therefore, research should be directed to study heat transfer for combined convection and radiation, by treating the fluid as an absorbing, emitting and scattering medium, and compile the resulting heat transfer coefficients for use in heat transfer applications for heat exchangers.

C. M. B. RUSSELL: During a recent investigation into the use of plastic tubes for dry cooling towers, I came to the conclusion that they were not suitable for this application; however, they are light, cheap and easily worked and very resistant to corrosion. Heat conduction is poor. They are good thermal insulators.
  Research into the possibilities of plastic exchangers and into the possibilities of modifying the thermal properties of plastics could lead to valuable results.
    The reaction between the fire elements in a process plant
      - Energy input - Energy (Export)
      - Heat interchange - Heat rejection.
and exo-or endothermic heat is not well understood, particulars as it affects design of fired heaters, air (or water) coolers and heat exchangers. Research by network analysis of such systems should be most worthwhile.

R. K. SHAH: The following research topics will be very valuable contributions to the fundamentals and to the design of heat exchangers:
(1) Obtain single-phase design date (j and f vs. Re) experimentally for compact heat exchanger surfaces by systematically varying the geometrical parameters associated with each type of surface.
(2) Develop the numerical methods to determine analytically the single-phase design data of compact heat exchanger surfaces. The constants for the numerical modelling may be based on the sophisticated model experiments and flow visualization. The final results should be approximated by simplified equations and correlations so that the designer could maintain the feel and understanding.
(3) Arrive at correlations for the influence of temperature-dependent fluid properties on j (or Nu) and f data for noncircular flow passages, for various thermal boundary conditions, and for developing flows.
(4) Perform experiments and numerical analysis for proper header design and for determining the influence of gross malflow distribution on the exchanger performance.
(5) Obtain the transient response solutions for at least the most commonly used flow arrangements for better understnading of the heat exchanger and the system dynamic behavior.
(6) Conduct an extensive testing to arrive at two-phase (boiling and condensation correlations for compact heat exchanger surfaces. This is one

of the most neglected areas of the research that has a very substantial potential for immediate industrial applications.

K. STEPHAN: As a result of the energy crisis, there is an urgent need to do research and development work in the enhancement of heat transfer in all fields, i.e., free and forced convection, boiling, condensation and also radiation.

Heat transfer to multicomponent fluids has been too much neglected in the recent years. Special emphasis should be laid on multicomponent condensation, which has not yet been well explored. Little has also been done on forced and natural convection flow of multicomponent liquids and gases. Because, Prandtl and Grashof number in some cases depend roughly on the composition, very peculiar boundary layer effects may be observed.

Another field that has been too much neglected are heat transfer phenomena in chemically reacting systems.

G. YADIGAROĞLU: Most of the heat exchanger design methods have been essentially one-dimensional in a number of applications, however, for which precise performance predictions are essential, the flows in the heat exchange systems are clearly three dimensional. In several of these applications the situation is further complicated by the simultaneous pressence of two phases. Therefore there is a clear need to develop efficient analytical techniques and adequate models for analyzing three dimensional two-phase flows with heat and mass exchange between phases and to test these against experiments.

Examples are:
(1) Flows through the tube matrix in the shells of steam generator or condensers.
- boiling in the shell of a PWR steam generator.
- condensation in shell-and-tube condensers.
- flow in the shell side of LMFBR steam generators (boiling inside the tubes).
(2) Waste heat rejection in spray ponds: In this case one has to calculate the complicated three dimensional flow field of the air and the trajectories of the water droplets.

Dynamics of heat exchangers and steam generators.
Some high-performance heat exchangers (intermediate heat exchanger of LMFBR; LMFBR or PWR steam generator; power-plant steam generator, etc.) are expected to undergo severe transients during their operation. It is therefore necessary to have adequate analytical tools for predicting transient behavior to assure that thermal stresses remain at acceptable levels. Numerical methods are generally applied; with long channel lengths and transient times, however, the computation times often become prohibitive. Development of efficient numerical techniques in conjunction with state-of-the -art analytical models of two-phase flow are needed for dealing with such problems.

A. A. ŽUKAUSKAS: I suggest the following areas on which research is likely to be most fruitful:
(1) Looking for a new method of the heat transfer augmentation.
(2) Application of new materials, metals and alloys together with the development of new technologies to manufacture compact heat exchangers.
(3) Computer aided design of optimal compact heat exchangers.
(4) Determination of the local values of different parameters in heat exchangers.
(5) Vibration problems in heat exchangers.
(6) Analytical and experimental thermal-hydraulic optimization of tube bundle heat exchangers.

# Index

Aceotropic mixtures, 318-320, 324
Acetone in boiling mixtures, 320, *322*, 326, *328-330*, 332
Achenbach, E., 85-96, 1107
Ackermann film theory, 337, 343
Acoustic vibrations, 825, 1107
Additives, fluid:
    in heat transfer augmentation, 821, *823*, 832-833, 862, 866, 884
Aeroelastic coupling in vibrating tube banks, 983-984, 987, 992-994, 996
Aerosol in deposition, 387
Agitated beds and submerged surfaces, 177, 193-198, 203
Air coolant advantages, 51-53
Air-cooled exchangers, 3, 49-81
    applications of, 52-53, 647, 654, 783-784, 799
    characteristics of, 3, 29, 783-784
    crossflow, 3, 49-81
    (*See also* Dry cooling)
Air cooled steam condensers, 799-800
Air flow:
    draft, 10
    in regenerators, *33*, 35-37
Air preheaters, 9, *11*, 12, *33*-35
Agarwal, J. K., deposition data of, 387-389
Algorithm:
    for concurrent flow exchangers, 4, 153-155, 159-160
    for Delaware method, 581
Alternate energy applications, 537
    (*See also* Ocean Thermal Energy Conversion; Solar energy applications)
Aluminum dry-cooled exchanger fins, 808, 811
American Society of Mechanical Engineers (ASME) codes, 170, 172, 581
Ammonia, 173, 619
    dimensional equations for, *245*
    in evaporators, 708, *710*, 894, *895*
Annular (vapor belt) distributor, 556
Applications:
    of heat transfer augmentation, 883-906 passim
    of plate exchangers, 173-174
    of radiative heat exchange, 401, 404-405

    for various systems, 6, 7, 10, 14, 20, 22, 27, 29, 34, 35, 37, 39, 45, 213-214, 235, 253, 537
    (*See also* Designs)
Argonne National Laboratory (Idaho Falls, Idaho), 894
Arithmetic mean temperature difference (AMTD), 567-568
Arrays, finite element method of, 461-462, 474-478
    (*See also* Tube banks; Tube bundles)
Arrhenius expression, 1020
Asbestos gaskets, 23
ASME (American Society of Mechanical Engineers), 170, 172, 581
Assemblies, complex, finite element method of, 461-462, 468-474, 478
Atmospheric effects, 809-810
Augmentation (*see* Heat transfer augmentation)
Axial flow in yawed tubes, 98-100, 109
Axial load (force) on tubes, 7, 1049-1056

Babcock and Wilcox steam generators, 623-624
Bačlić, Branislav S., 481-494
Baffles, 42$n$, 306, 663
    classified, 16, *18*, 19, 44
    described, 542-544, 648, *649*, 786
    in shell and tube exchangers, 15-19, 542-543, *549*, 551, 553, 555, 563-564, 582-585, 590-592, 598, 600-602, 609-611
    spacing of, 563-564, 582, 584-586, 598, 607, 614
    vibrations from, 981
Bar and plate exchangers, 23, *24*
Baskakov's correlation, 193, 201$n$
Beds:
    gas-fluidized, 211, 213, 218-219, 223
    packed, 4, 177, 181-186, 203, 221
    perfectly mixed, 178, 182-193
Bell, Kenneth J., 165-175, 537-558, 559-579, 581-617, 1107
Bellows, internal, 546, *547*
Benzene in boiling mixtures, 320, *323*, 324, 350
Bergles, Arthur, 819-842, 857-881, 883-911, 1108
Bernoulli equation, 446, 447

---
[1] Page numbers in *italics* denote figures and tables.

1113

# INDEX

Bingham plastic fluid flow in double pipes, 153, 159-163
Biofouling, 1000, 1015, 1038, 1042, 1091, 1092
Biot number in fluidized beds, 189
Blasius relationship in pressure drop, 305
Boiler design, 6, 619-646
Boilers:
   fire-tube, 620
   water-tube, 620
Boiling, 4, 6, 1109, 1111, 1112
   basic processes of, 235-256, 261-288
   defined, 235
   in double-phase-change exchangers, 955-956, 958, 961, 964, 965, 968, 969, 971, 973-976
   in falling liquid films, 366, 708, 716
   forced convection, 4, 5, 1109
      heat augmentation for, *823*, 857, 863-868, 894-898, *905*, 906
      of single component liquids, 4, 261-285
      in vertical tubes, 261-285, 1109
   nucleate: "fully developed," 270, 271, 285
   of soluble multicomponent mixtures, 315-334
   subcooled, 960-961, 1005
   (*See also* Pool boiling)
"Boiling bed," 211
Boiling mixtures, problems unresolved about, 8, 1087, 1093-1096, 1103, 1108-1109
Boiling surfaces, 4, 857-*859*, 874, 884, 887, 898-899
   (*See also* Surfaces)
Boiling water reactor (BWR) steam generation, 621-622
Boltzmann constant, *352*
Boundary conditions, radiation, 5, 399, 406, 411-418, 422
Box volume in compact surfaces, 12, 14, 23, 111, 113
Brazing roughness, 136, 142, 146
Brine in desalination, *431*, 683-684, 686, 687-689, 691-693, 696-698, 872
British Electrical and Allied Manufacturers' Association standards, 653
Brownian motion, 377
Bubble growth, 4, 364-367
   in boiling/evaporation, 235, 239-242, 255
   in condensing films, 364, 366-367
   detachment in, 4, 213, 216, 222-227, 235, 240
   in fluidized bed exchangers, 209, 211-213, 215-229, 231
   frequency of, 4, 216, 221-226, 235, 240
   in multicomponent mixtures, 316-320, 324
   in subcooled boiling, 960-961, 963
Bubbling bed fluidization, 209-231
   obstacle test in, 215-217, 221-229
Bucher, K. H., 765-777
Buckling load:
   on fins, 142
   on tubes, 7, 1049-1055
Bulk mean temperature difference and fluid properties, 498-501, 503, 511
Burnout, 247, 370
Butanol in boiling mixtures, 320, *322*
Butterworth, D., 289-313, 647-679, 1087-1106, 1107-1109
Bypass (leakage), 114, 582-586, 591, 753, 1107

$C^* = 0$ exchanger, 916, 924-934
   (*See also* Heat capacity rate ratio)
Calcium carbonate in geothermal tubes, 1076-1077, 1081-1083
Carryover leakages, 6, 723, 752, 753, 755-759
CE-Lummus (Heat Transfer Division), plate-fin development by, 843
Channel curves, 541
Checkers, 37
Chemical reaction:
   in fluidized beds, 213-214
   fouling by, 1000, 1015, 1038, 1042, 1091
Chen correlation in horizontal tube flow, 636
Chun-Seban relation in falling liquid film, 708, *710*
Circulation limiting heat flux in boiling, *254*, 255
Clausius-Clapeyron equation, 237
Cleaning, 20, 21, 22, 170, *171*, 559, 591, 648-650, 797, 809
Close tube exchanger, 168
Coatings, non-metallic, 786
Co-counter-current flow in coupled passes, *488*, 491
Co-current condensers, 660-667
Co-current flow, 660-667, 669
   in coupled passes, 483, *485*, 487, 492
   (*See also* Parallel flow)
Coke in fouling, 1028
Colburn-Drew method, 303
Colburn-Hougen method (analogy) in gas phase resistances, 300-303
Colburn number in compact surfaces, 116, 125
Collier, J. G., 235-259, 261-288, 619-646, 999-1011, 1109

# INDEX

Compact exchangers, 3, *11*, 111–146, 1112
  construction types of, 113–114
  defined, 12, 14–15, 111–113
  design procedures for, 15, 455–458, 495–536
  surface geometric properties of, 495–498, 509–511
Compact exchanger surfaces, 111–146
Compact heat transfer surface, distinguished, 113*n*
Compactness defined, 111–113
Computer code:
  for double-phase-change behavior, 955, 965, 969, 976
  for dry cooling tower costs, 765, 768, 772–773
  for shell and tube design logic, 559–561, 565–566
Computer modelling techniques, 779–783
Computer programs, 160, 560–563, 591–592, 926, 929
  for crossflow effectiveness, 481–493
  finite element method for, 462–463, 465
  verification of, 565–566
Computer time, 591–592, 1100, 1101
Concurrent flow in double pipes, 4, 153–163
Condensate film resistances, 291–299
Condensation, 4, 45–46, 289–307, 337–352, 1111–1112
  in desalination, 6, 357–376
  design basis of, 337, 343–344
  in double-phase-change exchangers, 955–960, 962–967, 969–970, 972–976
  in film evaporation, 357–371
  on a flat plate, 344–350
Condensation, forced convection, 857, 872–874, *905, 906*
Condensation, vapor space, 857, 860–872, *905, 906*
Condensers:
  crossflow, 667–668
  direct-contact: described, 656–657
    thermal evaluation for, 671–674
  fluid flow: basics, 6, 289–310
  heat transfer: basics, 6, 52, 289–310
    augmentation of, 868–874
  introduction to, 9, *11*, 14, 16, 20, 647–648
  of multicomponent mixtures, 4–5, 337–352
    with homogenous liquid phase, 337
    with immiscible liquid phases, 337–343, 350–352
  operational failure, 674–675
  thermohydraulic design of, 6, 647–675

Connors equation, 1090
Construction features, 3, 9, *11*, 15–37, 46
Convection:
  in fluidized beds, 218
  forced, 5, 45
    to absorbing/emitting gas with radiation, 401, 402, 416–418
    through circular tubes, 412–415
    heat transfer augmentation for, 819, *823*, 825–836, 857, 905
    in pool boiling, 235, 251
    to transparent gas, 401, 402, 411–416
  forced and free: in condensation, 349–350
  free, 45, 819, 822–825, 857
  natural, 4
    boiling multicomponent mixtures with, 315–334
    in horizontal parallel tubes, 1059–1061
    of multicomponent liquid/gas, 1112
    in pool boiling, 235, 241, 242, 247, 250–252
  single phase, 3, *11*, 45
  two-phase, *11*, 45–46
Cooling, dry (*see* Air-cooled exchangers; Dry cooling design; Dry cooling towers)
Core mass velocities in compact design, 519–521, 525–526
Corrosion in various systems, 20, 21, *171*, 173, 358, 430, 539–541, 557, 559, 627, 682, 793–795, 803, 807, 809, 811, 883, 894, 898–899, 999, 1000, 1003, 1006, 1009, 1015, 1075–1076, 1108
Costing:
  for dry cooling tower exchangers, 765, 767–*769*
  for optimal design, 458, 565, 801–807
Couette flow, inter-plate, 418–420
Counter-co-current flow in coupled passes, 484, *488, 490*
Counter crossflow in Fòrgo exchanger, 773–774
Countercurrent condensers, 660–667
Countercurrent flow, 660–667, 669
  in coupled passes, 483, *486*, 552–553
  in plate exchangers, 172–173
  (*See also* Counterflow)
Counterflow:
  in complex assemblies: finite element model, 472, *473*
  heat distribution and, 434–436, 438–441
Counterflow exchanger, *11*, 16, 21–23, 30, 42, 43
  described, 38–39, 115

Counterflow exchanger (*Cont*):
   temperature distribution in, 122, *135*-138, 499-501
   transient response of, 7, 915, 917, 918, 924, 925, 931-934, 949
Counterflow header, 134-*135*
COUPFLOW program, 160
Cowper stove, 35, 37
Critical heat flux (CHF):
   in heat transfer augmentation, 884, *885*, 892, 894, 1109
   improvement of, 858, 860, 862, 864, 866-867
   of natural convection boiling mixtures, 317, 328
   of nucleate boiling, 241, 247-248, 251, *252*-255, 265, 273-278
Cross-counterflow exchanger, *11, 38*, 41-42, 115, *441*
Cross-parallel flow, overall, *11*, 42, 115
Crossed yaw tube bundles, 105
Crossflow:
   boiling outside tubes in, 251-252, 1076
   exact explicit equations of, 481-493
   of falling film vapor, 715-716
   in ideal tube bank, 581-587, 590, 598-606
   research: heat transfer rates, 1109
   temperature distribution in, 3, 436-*437*, 439-*440*
   in yawed tubes, 98, 100-102, 105-106, 108-109
Crossflow effectiveness equations, 481-493
Crossflow exchanger, coupled, 733
Crossflow exchangers, 5, 11, 16, 38-43, 115
   flow distribution in, 118, 138-140
   gas-to-air single pass: rating for, 509
   for nuclear reactors, 3
   temperature distribution in, 499-501
   transient response analysis of, 915, 917, 924, 925, 927, 934-936, 949
   vibrations in, 7, 1049, 1091
Cryogenics, 3, 10, 13, 27$n$, 27, 35, 239, 331, 647, 655, 730, 757, 999
Crystallization fouling, 1091-1092
Cylindrical flows, 15

Dammköhler's formula, 193
Darcy's law, 229
Delaware method:
   rating problem of, 583-*584*
   for shell-side design, 6, 581-613
Deposition:
   in geothermal tubes, 1075, 1077, 1079, 1081, 1082
   hydrodynamic influence of, 1000-1009, 1013, 1015, 1019, 1024-1052, *1038*, 1091, 1092
   models of, 1042-1043
   rates/removal of, *1038*, 1039, 1042-1044
Deposition motion of droplets, 377-393
   data compared, 386-393
Deposition velocity, defined, 381
Desalination, thermal, 5, 6, 357-376, 695-701, 902-*904*
   film evaporation in, 357-376, 695-701, 869
   schematic presentation of, *683*
   thermodynamic considerations of, 684-689
Desalination plant manufacturers, 358, 365, 694, 696-698
Design modification program, 563-565
Design optimization, 458, 530-533, 565
Design process, 560-562
Designer problems, 561-563, 1103
Designs, 5-9, 31, 38, 43, 45, 429-816, 1087-1096, 1107
   compact exchanger procedures in, 495-533
   for compact surfaces, 142-146
   major uncertainties of, 1089-1090
   for offshore oil exchangers, 779-798
   overview of, 455-459
   for regenerators, 721-760
   shell and tube options of, 537-557
Dimensionless groups, 3
   in bed applications, 203
   in compact surfaces, 116-117
   in regenerator flows, 722, 723, 728-742
   for regenerator transient response, 915, 920-924, 940-946, 949
Direct contact exchangers, 9-11, 647, *648*, 656-657
Direct spectrum analysis, 5
Direct transfer (recuperative) exchangers (regenerators): transient performance of, 7, 10, *11*, 766, 915-937
Disk rotary regenerators, *11*, 31-*33*, 146, 721, 723, *724*, 726
Displaced enhancement devices, 820, *823*, 831, 862, 864, 873
Dittus-Boelter equation, 125, 267, 1076
Divided flow, shell fluid mixed, *11*, 43, *44*
Domestic water heating:
   geothermal energy for, 1074-1075
   solar energy for, 884-885
Double-phase-change exchangers, 7
   dynamic behavior of, 955-980
Doublepipe exchanger, *11*, 20, 155-157
   heat transfer in, 153-163

INDEX                                                                                                          **1117**

Droplets:
    deposition motion of, 5, 298, 301, *306*, 377-395
    in falling film evaporators, 714-716
    in immiscible liquid phases, 350-*351*
    in spray condensation, 671-674, 1112
Drum rotary regenerators, *11*, 31-32
Dry cooling design, 800-812
Dry cooling towers, 9, 14, 52, 807-812
    finned tube bundle optimization for, 765-769
    optimization of, 800-807, 811, 901-902
    technology survey of, 6, 799-816, 1111
Drying in agitated particles, 177, 183, 187, 188, 197-198
Dryout:
    in boiler tubes, 631-635
    of boiling tubes, 265, *266*, 273, 276, 278, 281, 283-285
    in falling liquid films, 709-712
    of fouled surfaces, 1005-1007
Dukler prediction in falling liquid films, 708

E shell, 16, *17, 545*, 582, *586, 587*, 665, 666
    condensers with, *650*-651, 663-667
    TEMA standards for, *545*, 550-551, 553, 554, 660
Economics, 801-807, 900
    (*See also* Costing)
Eigenfrequency of a tube, 1050, 1052, *1053*
Eigenfunction in plug flow models, 154, 159-160
Electrostatic fields in heat transfer augmentation, 821, *823*, 825, 834-835, 862, 867, 871, 873
Enclosure radiation, 402-404
Energy density distributions, 985-986
Enthalpy, 6
    of bed particulates, 186
Entrainment, 6, 650
    in droplet deposition, 380, 383, 386, 388, 391
    of falling film evaporators, 714-716
    limited in tube bundles, *254*, 255
Equations:
    of conservation: for dynamic behavior, 956
    for cross-flow effectiveness, 481-493
Equations, differential:
    for evaporators, 955, 956
    for regenerator effectiveness, 722, 725-759
    for regenerators: transient response in, 938-940, 948, 949
Ethanol in boiling mixtures, *326*, 332

Eulerian integral scales, 383-385
Evaluation procedure for optimal design, 455, 458
Evaporation, 4, 7, 1109
    basic processes of, 235-236, 261-288, *682*, 684, 689-690
    in desalination, 6, 681-701
    in double-phase-change exchangers, 955, 957-960, 963, 965, 967-975
    in film condensers, 357-371
    inside tubes, 261-285
Evaporators, 9, *11,* 14, 46, 235, 252-253, 261, 887
    falling film, *683*-684, 696-701
    flash, *683*-684, 689-693
    pool-boiling, *683*-684
    in reactors: thermal-hydraulics, 623, 628, 634, 635
    thin film, in water desalination, 681-701
    vapor-compression falling film, *683*-684, 688
    vapor-compression submerged, *683*-684, 688
"Excess-temperature" ratio, 325
Extended surfaces:
    analytical solution for, 128-132
    radiation of, 401-424 passim
Extended surface (indirect) exchanger, 5, 9-*11*, 15, 23-29, 41, 113, 114
    surface density of, 12, 113, 114

$F_N$ method, 421
F shell, *545*, 551-553, *650*
    Delaware method applied to, 590-591
Falling liquid films, 705-719
    breakdown of, 6, 709, 711-713
    condensation in, 343-344, 350-352
Falling liquid film exchangers, 1109
    shell and tube evaporators for, 705-717
Fanning friction factor, 116, 121, 122, 124-127, 507, 1019
Feasibility studies for offshore oil use, 786-787
Film, falling:
    in evaporator/condenser, 362-367
Film boiling:
    saturated, 266, *280*
    subcooled, 266, 278, *280*
    in vertical tubes, 241, 249-250, 252-253, 266, 278-280
Film evaporator/condenser:
    in desalination, 5, 357-376, 695-700

Film evaporator/condenser (*Cont*):
  horizontal tube, 5, 6, 46, 357, 358, *360–361*, 364–368
  vertical tube, 358–*359*, 362–366, 368
Fin effectiveness in compact exchangers, 505, 513–514, 522, 526–527
Fin efficiency, 430–431
Fin geometry in Forgò tubes, *767*–771, 776
Fin surface to base radiation, 401–402, 409, 411
Fin-tube bonding, 809, 811
Finite element analysis, 461–480
Finned tube bundle:
  in dry cooling tower, 6, 765, 768–771, 776, 807–809
  (*See also* Tube bundles)
Finned tube exchanger, 27, 29
Finned tubes, 16, *29*, 113
  in air-cooled crossflow, 73–81
  fluid dynamics of, 75–77
  heat transfer augmentation for, 829–831, 835, 861–862, 869, 884–886, *885*, 892
  heat transfer surfaces of, 15, *31*, 49, 73–81, 589, 600, 605
Fins, 9, *25*, 114
  in compact exchangers, 15, 113, 117, 128–131, 141, 142
  in extended surface, 23–*30*
  in plate-fin exchangers, 15, 23–27, 32
  in shell and tube exchangers, 538–539
  on tube-fin exchangers, 27–29
Fins, radiating:
  base and surface interaction, 404, 409–411
  heat transfer from, 404–409
Fischer, O., 765–777
Flashing evaporation, 684, 689–693
Floating head shell and tube exchangers, *545*, 547–550, 577, 593
Flooding limited in tube bundles, *254*, 255
Flow (*see* Axial flow in yawed tubes; Bingham plastic fluid flow in double pipes; Fluid flow; Knudsen flow; Laminar flow; Malflow distribution; Mixed flow; Oblique flow; Parallel flow; Plug flow; Slug flow; Split flow in shell fluid; Swirl flow; Turbulence; Vortical flow)
Flow arrangements of various systems, 10, 38–44, 115, 172–173, 439
Flow distribution:
  in compact exchangers, 113, 132–146 passim
  in double pipes, 20
  two-phase: problems unresolved about, 1087, 1096–1100, 1103, 1109

Flow friction of compact surfaces, 11, 114, 115, *117*, 121
Flow oscillations and transient response, 915
Flow pattern:
  for condenser shell-side pressure drop, 306–*307*
  in convection boiling, *263*
  in plain tube banks, 54–60
Flow perturbations in double-phase-change, 969–976
Flow pulsations, equations for, 985–986
Flow regimes, 3
  in air coolers, 49–80 passim
Flow regimes, critical, 49, 54–56, 65–67
Flow regimes, subcritical:
  in air coolers, 49, 60
Flow regimes, supercritical, in air coolers, 49, 54, 62–65, 67
Flow reversal, 34, 42, 45, 146
Flow velocity, 12
  in equipment fouling, 1044–1045
  and fluidized beds, 12
  nonuniform: correction for, 444–447
Flow visualization of plate-fin exchanger, 843, 846–847
Fluid dynamics, 956–964, *979–980*
Fluid flow:
  arrangements of, 38–*45*
  baffles and, 19–20, 542–543, 553
  in boiler design, 632–643
  of condensers: basic, 284–310
  in falling film evaporators, 705–717
  -mixing in passes, 481–492
  in regenerator transient response, 948–949
  in shell and tube exchangers, 538, 542–543
  temperature distribution of, 433–438, 538
Fluid flow friction, 3, 440–443
Fluid pressure:
  in shell and tube channels, 556–557
  (*See also* Pressure drop)
Fluid properties in rating and sizing compact exchangers, 498–501, 511–512, 524–525
Fluid temperature in compact exchanger design, 495, 498–505, 511–517, 524–525
Fluid temperatures, mean mixed:
  in multipass crossflow, 482–493
Fluidelastic instability, 1088, 1107
Fluidization, 210–230
Fluidized beds, 4, 10, *11*, 12
  in geothermal exchangers, 1079–*1080*
Fluidized bed combustion plants, 621
Fluidized bed evaporator (FBE), 692–693

INDEX

Heat transfer coefficient (*Cont.*):
  in beds/submerged surfaces, 178–186, 188–*196*, 198–203
  in boiling vertical tubes, 262, 264–267, 270–273, 275–278, 283, 284
  for clean condition, 1014–1015
  compact fin effectiveness and, 505–506, 513–514, 526–527
  in compact surfaces, 113, 114, 130, 137, 142
  in evaporation/condensation, *360*, 363–364, 366–371
  to film boiling, 252, 253, 277–*279*, 283
  to fluidization, 213, 216, 218, 220, 221, 230
  in fluidized beds, 621
  fouling factors and, 429–432
  geometric parameters of, 883–884, 886, 887, 891, 893, 894
  in horizontal boiler tubes, 636–638
  increase in, 819, 821, 824–827, 830–836, 857, 860–874 passim, 891, 894, 903
  to liquid falling film, 705–707, 712
  to multicomponent condensation, 337, 342–346, 349, 351–352
  to natural convection boiling mixtures, 315–334
  in plate exchangers, 170, 175
  to pool boiling, 236, 242–243, 254
  in shell and tube design, 562–566, 570–574, 576
  for shell-side, 583–585, 588–589, 593, 598, 601, 610
  to tube condensation, 289, 291–300, 303
  of various systems, 4–7, 12, 14, 20, 22, 23, 26, 27, 538, 543, 557, 1107, 1109–1111
  in yawed tubes, 102–103, 106, 108, 109
Heat transfer mechanisms, 3, 9–46 passim
Heat transfer regimes, 4
  in boiling surface convection, 261, *263*
Heat Transfer Research, Inc., 898
Heat transfer resistance for mixed beds, 183–186
Heat transfer surface:
  in fluidized bed wall, 214, 218, 221, 222
  fouling of, 1000, 1003, 1005–1007, 1009, 1013–1014
  in gas-to-liquid compacts, 14–15, 111, 113
  high performance of, 843–854
  improvement of, 819–874
  of plate exchangers, 165, 168–170, 172, 174–175

  of regenerators, 722–723, 725, 731–732, 739, 745–750, 759
  in various systems, 9, 10, 20, 21, 23, 27, 800, 803
  of vertical tubes, 262, 266
Heavy water reactor (HWR), steam generator in, 621–623
Heinecke, Ernst P., 1049–1056
Helical exchangers, 98, 105
Helical fin tubes in air-cooled crossflow, 73–75
Helmholtz instability, 247
n-Heptane-methylcyclohexane in boiling mixtures, 320–*321*
Herringbone matrix, 32
High Flux flooded evaporator (Union Carbide), 365, 894, *895*
High-temperature gas-cooled reactor (HTGR), 97, 98
Hitachi Thermoexcel tubing, 894
Hitachi Zosen (Innoshima, Japan), desalination unit by, 365, 697
Horizontal tube multiple effect plant, 902
Hydraulic design methodology, 455–457
Hydraulic diameter (*see* Surface compactness/density)
Hydraulic resistance coefficient in yawed tubes, 100, 102–*107*, 109
Hydrocarbons, 46, 648
  dimensional equations for, *245*, 249, 250
  in geothermal systems, 1073, 1075
  in multicomponent boiling mixtures, 320–324, 328–332
  problems with, 1109
  shell and tube design coefficient of, *571*–573
Hydrogen as fluidization gas, 191, *192*, 198, 199

Immiscible liquids in condensation, 337–343, 350–352
Inconel, 623
Indirect transfer exchangers, 766
Information:
  on fouling, 1000
  on heat transfer augmentation, 822, 857, 884
Injection, augmentation by, 821, *823*, 825, 835
Inlet temperatures for regenerators, 938–941, 944–949
Inspections, 170, *171*
Institute of Physical and Technical Problems of Energetics (Lithuania), 54

Instrumentation, measuring:
  for bubbling beds, 222
  for cooling tower elements, 771-772
  for deposition motion of droplets, 389-390
  for double-phase-change evaporator, 965-967
  for yawed tube heat transfer, 101
Interface, immiscible liquid phase, 337-352
Interfaces, planar:
  evaporation at, 235-237
Interfacial resistance, 291, 302, 345, 352
Interrupted fins, 26
Iowa State University Heat Transfer laboratory, heat transfer augmentation information from, 822, 857, 884

J Shell, 533, *545*, 551, *650*-651, 660, 667
Jet condenser, 10

K-$\epsilon$ model for turbulent flow, 1101
K shell, *545*, 551, 553-554, *650*, 651
Kakaç, S., 3-8
Kármań-Nikuradse equation, 124
Kettle reboiler, 22
Kilkiş, Birol İ., 209-231
Kirchhoff law in radiative fin surface, 406, 412
Knudsen flow, 1110
Korodense tubing, 902, 903
Kutan, Zekai, 1073-1083
Kutateladze correlation, 248

Lagrangian integral (functions), 383-384
Lamella exchanger, *11*, 165, 167, 171
Lamella type surfaces, *11*, 23, 697
Laminar films, 357, 362, 364, 365, 367
Laminar flow, 6, 8
  in air-cooled crossflow, 54-57, 60, 62, 69, 80, 86, 88-95
  in closed conduits, 440-444, 449-450
  in compact surfaces, 116-*118*, 122-124, 126-127, 140-143
  condensate films and, 291-293, 303
  in double pipes, 154, 155, 159-163
  in falling film evaporators, 705, 707
  performance, 6, 8, 115
Laminar flow exchanger can perform poorly, 1057-1070
Laminar flow surfaces, 115
Laplace transform in crossflow, 482, 1111
Laser-Doppler diagnostics of two-phase flow, 5

Laser-Doppler velocimeters, 5
Leakage:
  in baffles, 582-585, 586, 591, 600-602, 606
  in various systems, 6, 15, 20-22, 30-32, 114, 166, *171*, 539, 541, 543, 544, 798
  (*See also* Carryover leakages)
Legendre polynomials, 554
Light water reactor (LWR) (*see* Boiling water reactor steam generation; Pressurized water reactor)
Linde High Flux surfaces, 901
Liquid:
  in fluidized beds, 211
  shell and tube design coefficients of, *571*-573
  (*See also* Vapor-liquid interaction; vapor-liquid interface)
Liquid-deficient region in boiling tubes, 266, 281-284
Liquid metal fast breeder reactor (LMFBR), 46, 621, 628, *629*, 1112
Liquid natural gas (LNG), 1109
Ljungstrom rotary air preheater, 12, *33*, 34
Lockman Column multi-flash evaporator, 696
Lofin tubing, 785-786
Logarithmic Mean Temperature Difference (LMTD): in design equations, 552, 567-568, 574, 578, 590, 611, 661, 666
  and $N_{tu}$, 10, 15, 16, 496n, 498-499
Long tube vertical plant (LTV), 902-903
Longitudinal heat conduction in regenerator walls, 723, 727, 742-748, 917
Lorenz-Yung correlation, *708*, 716
Loschmidt number, 179
Louver fins, 23, 32, 128-130, 132, 141, 828
  in plate-fin exchangers, 26, 27n, 114, 843-845

Magnetite in fouling, 1005-1007
Malflow distribution, gross, 136-140, 142-143, 146
Manifolds, 9
Manufacturing codes, 894-895
Manufacturing non-uniformities of non-interconnecting passages, 1057, 1064-1065, 1070
Marangoni-effect shear stresses, *319*, 320
Mass in compact surfaces, 14, 23, 111, 113
Mass flux density in multicomponent condensation, 339-342

INDEX                                                                                                                               **1123**

Mass transfer:
  in heat transfer augmentation, *823*
  in multicomponent mixtures, 4
  in water cooling condensers, 10
Mass transfer coefficient:
  of incondensable gas vapor, 300-301
  of multicomponent condensation, 342-344, 346, 349
  of particulate deposition, 1002, 1019
Mass transfer control in fouling, 1020-1021
Mastanaish, K., 377-395
Matrix:
  in plate-fin tests, 843, 845, 846, 849-850
  for rotary regenerator, 29, 31, 32, 34, 721, 722, 724, 731, 758-759
  of various systems, 9, 10, 12-15, 29, 31, 32, 35
Matrix heat exchangers, 23, 29-31, 120, 721
  (*See also* Regenerator)
Matrix method for finite element models, 462-478
Matrix wall, regenerator, 10, 12, 721-728, 732, 747-749, 753, 757-759, 938-944
  material for, 733, 758-759, 938
Maxwell's formula, 1007
Mayinger, Franz, 955-997, 1110
MEA (monoethanolamine) gas absorption system, 173
Mean temperature difference (MTD), 542, 567, 578, 589, 593, 786
Mechanical design methodology, 455-458
Merkel equation, 810
Metals:
  for grooved tubes, 369-370
  in plate exchangers, 170-173
Methane, 619
  dimensional equations for, *245*
Methanol in boiling mixtures, 320, *323, 326,* 329, *330,* 332, 340, *341,* 344, 348
Methylene chloride, *245,* 351
Metzner-Friend equation, 1023, 1027
Mikhailov, M. D., 153-163, 461-479, 1110-1111
Minerals Research and Exploration Institute (Turkey), geothermal fields and, 1080
Minimum fluidization condition, 211-212
Mixed flow, finite element model of, *473*-474
Mixture boiling (*see* Boiling mixtures)
Modelling, 1110-1112
Mole fraction in multicomponent liquid mixtures, 315-319, *321*-326, 329, 330, 339, 344
Momentum in falling film, 362
Mondt number, 1067

Montakhab, Ali, 799-816
MSF distillation scheme, 358, *360-361,* 365
Multicomponent mixtures, 4-5, 315-352
  heat transfer with condensation in, 337-352
  natural convection boiling in, 315-334, 1108-1109, 1112
Multidimensional shell-side flows in condensers, 670-671
Multipass exchanger, *11,* 31, 41-*45*
  described, 38, 115
Multiple shells in series, 552-553
Multistage Flash plant (MSF), 903
National Engineering Laboratory (East Kilbride, Scotland), data from, 174-175, 1092
Navier-Stokes equations, 1100
Newtonian flow, laminar, 160, 163
Newtonian fluid, 153-157, 159-163
Nikuradse's function, 827
No-tubes-in-the-window exchanger, 19, 555, *556,* 1093
  Delaware method applied to, 590, 593, 598, 606, 786
Noncircular cross sections for circular pipes, 443-444
Non-uniform passages, laminar flow in, 1064-1065, 1070
Nozzles, 9, 16, 21, 44, 540-*542,* 550, 553, 554, 556, 671, 798
NTU ($N_{tu}$) (*see* Number of Transfer Units)
Nuclear power plants, 3, 9, 12, 27, 53, 85, 213, 218, 1073, 1112
  corrosion in, 999, 1000, 1006
  dry cooling towers of, 765, 770, 800
  gas-to-gas exchangers in, 3, 1051
  steam generation for, 619, 621-*629*
Nucleate boiling, 6, 39, 241-248, 251, 253, 364, 860, 1094
  dimensional equations for, *245*
  of multicomponent mixtures, 331-334
  onset of (ONB), 236, 241-*245,* 258, 267-276, 278, 860, 863
  saturated, 270-271
  subcooled, 267-270, 866
Nucleation, 4, 235, 255, 261-285, 352
  heterogeneous, 235-238
  homogeneous, 235-237
  siting of, 4, 235, 237-242
  on surface deposition, 1005-1008
Number of (heat) Transfer Units (NTU, $N_{tu}$), 10, 12, 15, 887-*888*
  in compact exchanger effectiveness, 496, 498-499, 514-515, 520-525
  thermal design, 111, 115, 119, 120, 143

Number of (heat) Transfer Units (NTU, $N_{tu}$), (*Cont.*):
  for counterflow transients, 922, 923*n*, 940, 942, 944, *945*, 948
  design method of, 437–*441*
  for flow arrangements, 506
  and heat capacity rate ($C^*$), 506
  and log-mean temperature difference (LMTD), 496*n*, 498–499
  multipass crossflow formulas, 481–493
  in non-uniform tubes, *1066*–1070
  rotary regenerator analysis, 721–722, 728–732, 734–737, 740–742, 744–745, 755, 759
  in sizing problems, 520–525
Nusselt equation, 291, 365
Nusselt liquid film theory:
  in condensate film resistance, 293–295
  in multicomponent mixtures, 337, 339, 347, 351, 352, 363–365, 705, 708
Nusselt number:
  in compact surface, 116, 122–123, 125–127
  for condensing steam, 964
  constant: for parallel flow, 1057, *1061*, 1064
  in fin effectiveness, 505, 836
  for laminar flow, *449*
  local: in radiation-convection, 422–*423*
  for pseudoplastic flows, 153–163
  for tube bundles, 86


Oblique flow, 133
  around yawed cyclinders, 98–101, 109
Oblique flow header, 133–134, *136*, 146
O'Callaghan, Michael G., 1037–1047
Ocean Thermal Energy Conversion (OTEC), 887, 890, 891, 1038
Offset fins, 26
Offshore exchangers on oil platform, 6, *778*–788, 789–798
Oil production, 789–792
One-dimensional flow in closed conduits, 444–447
1-n exchanger, 43
1-2 exchanger, 42–43
Organic liquids:
  design coefficients of: classified, *571*–573
  in multicomponent condensation, 340, *341*, 344, 350–352
  (*See also* Hydrocarbons; Refrigerants)
Oversurfacing, 1037
Özişik, M. Necati, 399–425, 461–479, 1111

P-1 approximation, 421
Parallel counterflow, shell fluid mixed, *11*, 42–43
Parallel flow, *11*, 39–42, 142, 438, 439, 925
  finite element model of, 469–*473*
Parallel flow headers, 133–*135*
Parallel plates multipass, *11*, 44, *45*, 429
Parallel yawed tube bundles, 105
Particle convection in fluidized beds, 188–189, 198
Particles:
  in deposition motion, 377–387
  in fluidization, 211, 220–222, 226–228
  motion of, 177–178, 181–184, 190–198, 377–387
  radiation scattered by, 401, 402
  in surface-bed exchanger, 12, 177–204
Particulates in fouling, 7, 1000, 1002–1007, 1015, 1018–1022, 1025, 1038, 1091, 1092
Parting plates, 10, 23
Pass divider, 541–542
Passage-to-passage non-uniformity, 142–146
Patents for heat transfer augmentation, 884
Péclet number, 723
Penetration model in submerged surfaces/beds, 177, 183–193, 198, 203
Perforated fins, 26, 114, 129, 141
Perforated plates, 828–829
Periodic-flow exchangers, 7
Periodic-flow regenerator, 29, 35
PI-controller, 24, 967
Piezo-electric transducer, 825
Pin fins, 114, 128–130, 869
Pinheiro, João de Deus R. S., 1013–1035
Pipe surfaces (*see* Surfaces)
Piping, 9, 21, 39
  fluidization near, 215–232
  in reactor steam generators, 623, 627, 632
Planck function, 403
Plastic tubing, thermal properties of, 1111
Plate and frame exchanger, 4
  details and applications of, 165–*166*, 170–172
Plate baffle exchanger, *11*
Plate exchangers, 4, 10, *11*, 15, 20–23, 41, 165–175, 647, 784
  advantages/disadvantages of, 784–785
  flow arrangements of, 172–173
  surface density of, 12–14
Plate-fin and tube exchanger, 11, 27*n*
  in air-cooled crossflow, 73–74

# INDEX

Plate-fin exchanger, 1, *11*, 13–15, 26, 32, 39, *40*–41, 655–656, 1109
   surfaces of: characteristics, 111, 114, 128–130, 142, 785
      heat transfer enhancement for, 843–854
Plate fins, logitudinal:
   radiation of, 401, 404–409
Plesset and Zwick asymptotic solution, 240
Plug flow:
   in double pipes, 154, 159–163
   in fluidization, 219
Pohlhausen solution, 348
Polytetrafluorethylene coating, 858
Pool boiling, 4, 16, 1095
   heat transfer: augmentation for, *823*, 857–863, 884, *885*, 894, 897, 1109
   regimes in, 235, 241–247, 265, 325, 331
Power generation plants, 7, 799–800
Prandtl number, 830, 1112
   bulk mean temperature and, 498
   in compact surface, 116, 117, 126
   condensate, 296
   for gases, 60, 71
   for liquid film, 705–707
   in multicomponent condensation, 339
Prandtl tube, 101
Precipitation fouling, 1037–1038
Pressure drop:
   in air-cooled exchangers, 49–81 passim
   in compact exchangers, 113–146 passim
      design procedures, 495, 506–509, 515–517, 528–530
      with Delaware method: shell-side fluid, 581, 582, 585–588, 593, 601, 604–614
   in double-phase-change, 956, 957, 960–976
   across evaporating channels, 638–*640*, 643
   in finned tube banks: crossflow, 77–81
   in fluidized beds, 12
   in Forgò exchanger, 767, *769*, 774
   geometric parameters of, 883–887, *893*, 903
   in plate exchangers, 168–170, 173–175
   research suggestions on, 1107–1110
   in shell and tube exchangers, 16, 559, 562–566
   in tube condensation, 303–*307*
   of various systems, 3, 4, 7, 8, 19, 20, 23, 26, 27, 32, 34, 1107
   from wall deposition, 1006–1009
   in yawed tubes, 99, 101, 103, 105–108
Pressure drop coefficient of crossflow tube bundles, 85–86, 88, 91, 95
Pressure fluctuation in bubbling bed, 212, 222–225, 229, 230

Pressure leakage in regulators, 752–755
Pressure in plate exchangers, 167, 168, 170
Pressurized water reactor (PWR), 622–*627*, 1112
   shell and tube exchangers in, 537
   Soviet 440 MW(e) steam generator, 623, *625*
Process engineer problems, 561, 783, 786–789, 792–797
Prototype Fast Reactor (U.K.), evaporator in, *628*
Pseudoplastic fluid flows, 153, 159–163

Radiation, 5, 45, 400–423, 1111
   in fluidized beds, 188
   in gas convection, 416–420
   from space vehicles, 401, 404–*405*
Radiative heat exchange between fin surfaces, 401–404, 409–411
Radiative heat flux, 401, 403, 407, 410, 413, 416–423
   in parallel plane medium, 420–423
Radiative heat transfer, 45, 46, 241, 399–425
Radiator, vehicular, *13*, 14, 45, 52, 132, *133*, 134
Rating problems for compact exchangers, 495–518, 533
Rayleigh instability, 715
RC-318 ($C_4F_8$), dimensional equations for, *245*
Reboiler, aromatic, high flux tubing in, 901
Recuperator:
   air coolers, 53
   described, 10, 31, 721, 723, 915
   regenerator flow compared to, 731, 735, 736, 752
   steady state test in, 117–118
Reduced length-reduced period method ($\Lambda$-$\Pi$):
   for fixed-matrix regenerators, 721, 722, 728, 739–743, 755, 759
   for regenerator transient response, 942–949
Re-entrainment in fouling models, 1001, 1003, 1043–1044
Refrigerants:
   in boiling mixtures, 320–*321*, 331
   dimensional equation for, 244, *245*, 254
   in double-phase-change, 955, 956, 966–968, 971–973
   fouling factor of, *431*
   in multicomponent condensation, 350–*352*
   R-11, *244*, *245*, 861
   R-12, 244, 245, 872, 965–968, 971
   R-32, 1077
   R-112, 350, *352*

Refrigerants (*Cont.*):
  R-113, 244, *245*, 350–*357*, 872, 873, 899–900
  R-114, 861
  R-115, 244, *245*
Refrigeration industry application, 902
Regenerator (regenerative exchanger), 6, 10–12, 15, 29–31, 37, 45, 53, 113, 114, 721–760
  described, 29, 31–38, 114, 721
  periodic flow theory for, 722–723
  surface density of, 14, 31, 114
  surface geometry of, 111, 132–136, 142
  thermal design theory for, 6, 721–760
  transient response analysis, 7, 915, 938–948
  unbalanced/unsymmetrical, 946–947
Regenerator, counterflow, 721, 723, 728, 733–736, 742, *746*, 751, 756
Regenerator, fixed-matrix, 6, *11*, 16, 29, 31, 32, 35–38, 46, 114, 721–723, 725–728, 731–738, 740, 748, 750*n*–752
  transient performance of, 937–941, 945
Regenerator, parallel flow, 722, 736–*743*
Regenerator, rotary, 6, 9, *11, 13*, 29, 31–34, 114, 142, 146, 721, 723–728, 730–731, 752–753
  disk type, *11*, 31–*33*, 146, 721, 723, *724*, 726
  drum type, *11*, 31–32
  transient performance of, 937–939
Reliability, 899
Rensselaer Polytechnic Institute (New York City), 1979 international conference on fouling, 1000
Repair, *171*
Research and development, 1107–1112
Research possibilities, 1092, 1095
Resonance frequency in tube vibration, 981
Resonance vibration, 1088, 1107
Retrofit, heat transfer augmentation in, 883
Reversible heat accumulator, 29
Reynolds "flow," 344
Reynolds number:
  for compact surfaces, 116–119, 122–129, 131–132, 141–143
    design characteristics, 496, 502, 512–513
  critical, 3, 85, 86, 88, *90*, 91, 93–95
  for falling film liquids, 709, *710*
  in non-circular cross sections, 444
  in plate exchangers, 168, 364–365, 370
  for shell-side fluid, 589, 604–609
  in single phase vapor, 284, 1110
  for surface roughness, 827
  for vertical flow deposition, 378
  for vibration flows, 981, 987
Reynolds number, film:
  for condensate film resistance, 292, 294, 362–367, 370
  for falling film liquids, 705–712
  for laminar flow, 292–294
Reynolds number, liquid, 293–294, 296–297
Reynolds number, subcritical:
  in yawed tubes, 97, 99–100, 105, 109
Reynolds number range:
  for air-coolers, 3, 32, 49, 54–81
  for fluidized bed: plug flow, 219
  for perforated plates, 829
  subcritical: yawed tubes, 97, 99–109 passim
  for surface characteristics, 120, 128
  for tube bundles, 85–95
ROD baffles, *11*, 786
Rod baffle shell and tube exchanger, 554–555
Rod bundles, in-line vibrations of, 990–994
Rohsenow, Warren M., 429–454, 1057–1071
Rohsenow correlation, 708
Rosenblad (ROSCO) vertical wall, 697
Rotating drum apparatus, penetration model of, 186–188
Rotation, surface, 6
  heat transfer augmentation by, 821, *823*, 834, 862, 871
Runge-Kutta method, 159
Russell, C. M. B., 843–855, 1111

Safety, 899–900
  margins of, 1109
Salts in corrosion, 698, 1000, 1003–1006, 1015, 1025
  (*See also* Desalination; Water, salt; Water, sea)
Sand scouring in geothermal tubes, 1078–1079
Scale formation, 7, 52, 358, 682, 698, 794, 795, 797, 1000, 1015, 1024–1025, 1041, 1075, 1076, 1079, 1081, 1082
Scaling up for heat transfer augmentation, 884, 886–887
Schlünder, E. U., 177–208
Schmidt number:
  and Brownian diffusion, 377
  in multicomponent condensation, 339
  in surface fouling, 1019, 1043
Schult, Manfred, 955–980, 981
SCON program, 670
Seals, 9, 31–34, 582, 586, 591, 592, 606, *649*, 650, 753
Sedimentation, fouling by, 1000
Segmented fins, 26

# INDEX

Sensible heat exchangers, 9
Serrated fins, 26, *28*
Shah, Ramesh K., 9-46, 111-151, 455-459, 495-536, 721-763, 915-953, 1111-1112
Shah correlation in stratified flow, 636-638
Shear stress:
    in compact surfaces, 117
    intube condensation and, 296, 298, 308, 367-368
    in surface fouling, 1018, 1030, 1041, 1043, 1044
Shell and tube condensers, thermal evaluation of, 657-671, 1112
Shell and tube exchangers, 5-22, 41, 43, 113, 170, 171, 348-350, 429, 436, 537-613, 705-717
    advantages/disadvantages of, 784
    construction features of, 15, 113, 348-354 passim, 537-557
    Delaware method for, 581-615
    design coefficients for, 570-573
    falling film breakdown in, 705-717
    geometric parameters of, 562-*563*, 574-579
    head types for, 15-17
    inspection/cleaning of, *171*, 648, 650, 784
    performance evaluation of, 887-*889*
    preliminary design of, 6, 15, 457, 559-579
    rating program for, 562-563
    shell-side design of, 6, 562-563, 581-613
    shell types for, 15-*17*, 20, *545*-554, *650*-651, 663-667
    sizing for, 450-453
    stream allocation in, 556-557
    temperature distribution in, 462, 465-468, 478
    transient response of, 20, 936-937, 949
Shell-side design, 563, 581-613
Shell-side rating, 562-563, 581-613
Shell surface (*see* Surfaces)
Shells:
    with expansion joint (roll), 546
    in series, 568-570
    for shell and tube exchangers, 15-*17*, 540-541
Shippingport PWR (Shippingport, Pennsylvania), original steam generator at, 623-*624*
Shishedjiev, B. K., 153-163
Shot, metal, 37
Sideman, Samuel, 357-376, 681-703
Simplified zone analysis, 402, 404
Single-blow transient test, 120
Singlepass exchangers, *11*, 31, 38-43

Single-phase liquid, 267, 272
Sizing problems:
    for compact exchangers, 495, 496, 518-530, 533
    for counterflow shell and tube, 450-453
Skin friction, 126, 130, 132
Slip ratio in double-phase-change, 956, 960, 962-963
Sludge deposition, 627
Slug flow, radiation in, 421-423
SMX standard mixer, 831
Sodium in breeder reactors, 619, 628
Solar domestic water collectors, 884, *885*
Solar energy applications, 709, 884-*885*, 1110
Solubility in fouling models, 7
Space heating, geothermal hot water in, 1073, 1074
Spectroscopy, high resolution interference, 5
Spiral plate exchangers, 4, *11*, 20, 22, 165-167, *171*
Spiral tube exchanger, *11*, 20
Split flow in shell fluid, 44, 1096-1100, 1103, 1109
Spray condensers, 6, 647, 656-657, 671-674, 766
ST-4 computer program, 591
Stanton number, 849, 852
    in compact surface, 116
    in surface roughness friction, 827
Stationary regenerator, 35
Steam:
    in condensers, 654, *673*, 868, 872
    in desalination, *683*-688, 693, 694, 696, 698
    in geothermal exchangers, 1073
    shell and tube design coefficients of, 566, 567, *571*
Steam boilers, 54, 619-646
Steam generators, 20, 214, 619-*629*, 956-957
    corrosion in, 999, 1000, 1003, 1006, 1007
    boiler design for, 619
    in Soviet PWR, 623, *625*
Steam turbine, 766
    for dry cooling tower, 800-804, 811
    selection of, 801, *802*
Steam-water in reactor generation, 622-625, 630-636, 638-643
Steel alloys, 52
Stefan-Boltzman constant, 250, 403
    of radiative surface, 403
Stefan-Boltzman fourth-power law, 401, 411, 416
Stein's solutions, 154-156, 159-160, 163

Stephan, K., 315-336, 337-355, 1112
Stirling engine regenerator, 37, 730, 757
Stokes law of resistance, 377, 378, 380, 383-385, 387
Storage-type exchangers, 10, 12, 29, 721, 937-949
   (See also Regenerator)
Stresses, thermal, 20, 38, 39, 211, 430
   between shell and tube-bundle, 537-550 passim, 559
Strip fins, 23, 26, 32
   surfaces of, 114, 129-132, 141, 509, 513, 828
Strouhal number, 981, 986, 987, 1050, 1052, 1055, 1088-1090
Sturm-Liouville equations, 154
Submerged surfaces, 4, 177-204
Suction, vapor removal by, 821, *823*, 835, 863, 871
Suction effect, wall, 345, 347-349
Superheat in heterogeneous nucleation, 237-238, 241, 242
Surface adhesion control in fouling, 1020-1021
Surface compactness/density, 12, *13*, 111, *112*
   classification by, *11*-15, 113
   of regenerators, 14, 31, 34, 113
   of shell and tubes, 12-14
Surface condition in boiling, 256
Surface exchangers, 10
Surface tension:
   in binary mixtures, 317-320
   in bubble growth, 238, 240, 317-320
   in falling liquid film: breakdown, 706-707, 712
   flow in, 1109
Surface tension devices for heat transfer augmentation, 821, *823*, 862, 868-869
Surfaces:
   in rating and sizing: compact exchangers, 496-498, 502, 509-511, 518, 527-528
   selection and performance of, 7, 817-911, 1111-1112
Surfaces, extended, 5, 872
   finite element method for, 461-462, 474-478
   heat transfer augmentation by, 820, *823*, 828-830, 864, 868-869
   in forced-convection boiling, 863-864
Surfaces, primary (direct), 9
Surfaces, rough:
   boiling heat transfer on, 331, 858
   in compact exchanger, 117
   developed-flow along, 440-443

fouling of, 239, 1006-1009, 1040
heat transfer augmentation by, 820, 823, 825-828, 857-860, 864, 868, 887, *889*-891
   in forced convection, 863-864, 872
   in pool boiling, 857-860
   for vapor space condensation, 686
Surfaces, smooth, 3, 5, 857, 858, *861*, *889*
Surfaces, treated:
   heat transfer augmentation by, 819-820, *823*, 868
   in pool boiling, 857-860
   for vapor space condensation, 868
Swirl flow, 820, *823*, 832, 864, 873

T tube (pull-through floating head), *545, 548*, 577
Tank (distributor), 9
Taylor, M.A., 779-788, 789-798
Taylor instability, 247, 249
Teflon, 23, 170, 858, 868, 1077
TEMA (Tubular Exchanger Manufacturers Association) standard for tubes, 15, *17*, 44, 543-*545*, 550, *577*, 581, 590-593, 601, 648-651, 787-788, *897*, 898, 1015, *1016*
Temperature:
   fouling adjustments of, 1045
   in plate exchanger, 167-168
   (See also Critical heat flux; Heat transfer; Wall temperatures)
Temperature-dependent properties in rating, 502-505
Temperature distribution:
   of fluids, 433-437
   in various systems, 5, 7, 499-501
Tennessee Valley Authority, 902
Test facility for special cooling tower elements, 765, 771-776
Test techniques for compact exchangers, 117-121
THA (horizontal tube evaporator), 358, *360-361*
Thermal capacitances and transient response, 915, 917-920
Thermal design methodology, 455-458
Thermal expansion, 7, 43
Thermal-hydraulic optimization of finned tube bundle, 6, 765-777
Thermal-hydraulic performance evaluation, 883, 887-893

INDEX

**1129**

Thermal-hydraulic phases:
  single, 47-230
    heat transfer augmentation for, 819-842
  two, 235-395
    heat transfer augmentation for, 857-881
Thermal resistance:
  equation for, 119
  in multicomponent mixtures, 337-338, 351-352
  order of magnitude of, 432-433
  of regenerator wall, 6
Thermodynamic equilibrium, 262, 285, 325
Thermodynamic theory of mixture bubble formation, 317-320
Thermoexcel surfaces, *870*, 884, 886, 902
Thermosiphon reboilers, horizontal and vertical, 16, 22, 561-562
Thin film evaporators/condensers, 5, 6, 357-376, 681-701
Three Mile Island (Middletown, Pennsylvania), 626
Toluene, 324
Transient boiling region, 266, 241, 248, 277-278
Transient response problem, 7, 915-950
Tray condensers, 674
Tube and shell exchangers (*see* Shell and tube exchangers)
Tube arrangements (arrays), 14, 27, *30*, 101, 114
  pitch of, 3, 7
Tube banks:
  in air crossflow, 49-81
  arrangements of: vibration and, 981, 983-984, 987-989, 991, 994, 996
  of finned tubes, 73-81
  ideal, 581-585
  of plain tubes, 53-69
  of rough surface tubes, 69-73
Tube bundles, 3, 12, 15, 16, 23, 29
  boiling processes in, 235, 252-254
    limiting heat flux in, 253, 255
  crossflow vibrations in, 981-983
Tube bundles, horizontal, condensation in, 358, *360*, 366
  immersed, 12, 222
  in-line, 3, 54, 56-57, *59*, 60, 81, 85-87, 92-95
  pulling of: offshore rigs, 794, 796, 798
  staggered, 3, 54, 57, *59*, 60, 81, 85-95
  surface density of, 14
  with yawed tubes, 97-109

Tube-fin exchangers, *11*, 27, 29, *40*, 114
  surface characteristics of, 111, 114, 115, 128, 129
Tube geometry, 15, 904, *905*
Tube passes in multipass exchangers, 41-*44*
Tube pitch in shell-side exchangers, 593-598
Tube sheets, 539-540, *548*, 574, *575*
Tube surfaces (*see* Surfaces)
Tube surface temperatures, 412, 415
  (*See also* Shell and tube condensers; Shell and tube exchangers, temperature distribution in)
Tube walls, 10, 23
  axial conduction in, 1065-1070
Tubes (cylinders), 3-7
  axial forces on, 1049-1056
  forced convection in, 235, 251-252, 255
  heat transfer enhancement, 904-*906*
  shapes of, 6
  in shell and tube exchangers, 538-539
  in shell-side design: geometry parameters for, 538, 592-600, 602, *604*
  pitch of, 593-598
  pressure drop of, 585-590, 604-606, 610-612
  simplified mechanisms, 582-589
Tubes, coiled, heat transfer augmentation by, 832, 865-866, 873, *905*
Tubes, elliptical, 369
Tubes, fluted, 365, 368-370, *905*
Tubes, grooved, 368-370
Tubes, low-fin, 538-539, 557, 589, 600
  Delaware method applied to, 589, 600
Tubes, parallel, 7-8, 1057-1070
Tubes, roped, 903
Tubes, rough, in air-cooled crossflow, 49, 69-73, 81
Tubes, smooth (plain):
  in air-cooled crossflow, 49, 53-69, *72, 73*, 81
  in desalination, 368-370
Tubes, vertical, boiling in, 261, *263*-285 passim
Tubular exchanger, 10, *11*, 15-20, 23, 113
  surface density of, *13*
Tubular Exchanger Manufacturers Association (*see* TEMA)
TUFC computer program, 1101, *1102*
Turbine, steam, 766, 800-803, 811
Turbulators, 128, 843, 845
Turbulence:
  in air-cooled crossflow, 54-57, 60-61, 69, 76, 80, 86, 88-95
  baffles and, 19
  in closed conduits, 440, 443, 446

Turbulence (*Cont.*):
   in compact surfaces, 116-*118*, 124-128, 141, 142, 146
   in falling film evaporators, 705, 707
   in plates, 21
   tube vibration and, 19, 982-983, 985-987, 994, 1049
   in vertical tube: core flow, 377-380, 383-387, 393
   vorticity and, 7, 54, 55, 982-983, 1049
Turbulent flow modelling problems, 8, 1087, 1100-1101, 1103, 1109
Turbulent bursts (upsweep), 379, 999, 1003, 1005, 1044
Turbulent films, 357, 362, 364, 366-367, 369
Two-phase exchangers, 4-7
Two-phase flow:
   deposition motion in, 377-393
   forced-convection region of, *263*, 271-272
   pattern difficulties in, 261-262
   research and development for, 5, 8, 1109, 1110, 1112
   split in tubes, 1096-1100, 1103, 1109

U tube, 546-*548*, 553, 593, 623, 626, 631, 648, 655, 668
U tube, inverted, 623, 626, 631
U tube exchanger, example estimate for, *577*-579

Valves in rotary regenerators, 35, 721, 730, 753
Vapor:
   compression-condensation of, 682, 684-685, 687-688, 693-695
   in condenser types, 651-652, 654, 656, 672-674, 886
   in systems, 4, 5, 10, 16, 22, 119, 166, 172, 337
Vapor compression distillation, 693-695
Vapor compression-evaporation (VCE), 682-685, 687-688, 693-695
Vapor formation:
   in boiling/evaporation, 235-242, 247, 248, 251, 255, 261-266, 278, *282*-285
   in- and out-tube condensation, 289-290, 293, 298, 301, 303, *304*, 306
Vapor-liquid interaction, 4-6
   in bubble growth, 239
   in falling film evaporator shells, 714-716
Vapor-liquid interface:
   in boiling/evaporation, 239, 302-303
   in condensation, 337, 341, 345-347, 349

Vapor phase, multicomponent mixture at, 337, 339-344, 349
Vapor recompression, 358, *359*, 362
Vapotron communication tubes, 904
Venting of condensers, 650, 653
Vertical tube evaporator, 696-697, 869, 871
Vibration, 981-997, 1087-1091, 1107, 1108, 1112
   buckling load from, 1049, 1051-1052, 1055
   heat transfer augmentation by, 821, *823*, 824-825, 834, 862, 866, 871, 873, 886-887, 894
   problems unresolved about, 211, 588, 1087-1091, 1101
   sources of, 787, 1088-1089
   ultrasonic, 825
   in various systems, 7, 8, 16, 19, 541, 543, 555, 559, 588, 590, 643, 797
Viscosity, 8, 9, 22, 334
Viscosity-temperature, 8
   -induced non-uniform flow, 1057, 1061-1062, 1070
Void fraction in fluidized beds, 191, *192*, 198, *199*, 201, 221
Void fraction correlations in tubes, *641*-643
Vortex generators, 7, 820, 843-*851*, 865
Vortex shedding and tube vibration, 1055, 1088-1089
Vortex theory, 981, 983-984, 1044, 1049
Vortical flow, 54-55

Wall heat conduction, longitudinal, 6, 155
Wall temperatures:
   in compact surface, 119, 122
   in condensers, 289, 363, 367, 370
   in convection boiling tubes, 262, *263*, 265, 283
   in multicomponent condensation, 339, 341, 343, 345, 351
   in pool boiling, 241, 242
   for regenerator matrix, 721, 728, 744-752 passim, 938-942
   in various systems, 4, 32, 39, 887
Wall thermal resistance, 184
   in regenerators, 748-752, 921
Wall-to-particle heat transfer coefficient, 178, 180-181, 189-191
Wallis correlation in vapor flow, 290
Waste heat boilers, design of, 619, 628, 630-632
Waste heat disposal systems, 799

# INDEX

Water:
- design coefficients to, *571*
- dimensional equations for, *245*, 250
- in falling films, *710*
- fouling factors of, 173, *1016*, 1018-1019, 1029
- in research, 1109
- salt (saline), 696-697
  - fouling from, 429, *431*, 683, 1078
- sea (brine), *710*
  - in desalination, 358, *360*, 365, 367-368, 370, *683*-692, 695
  - fouling factor, *1016*
  - as heat sink, 173
  - in open circuit cooling, 794, 795
  - shell and tube design coefficient for, *571*-573

Water cooling tower, 10
Water spray, 10
Wavy fins, 129, 130, 132, 141
Weber number, 707
Wet-cooling, 799, 800, *804-805*, 810, 811, 898
Wet/dry cooling, 799-810 passim

Wilke's relation in falling liquid films, 706, 716
Wind tunnel, atmospheric, 101, 810, 812
Wolverine Tube Division, Universal Oil Products, 902
  ST Truflin low-fin tube, 570
Wood pulping process evaporators, 681, 682

X shell (crossflow) exchanger, 551, *545*, 554, *650*, 651

Yadigaroglu, G., 1112
Yaw angle in pressure distribution, 98, 100-109
Yawed tube bundles, 97-109
Yeşin, Orhan, 1073-1083

Zivi equation, 295
Zuber ideal flow, 247-249
Žukauskas, A. A., 49-83, 1112